THE NATURE OF COMPUTATION

The Nature of Computation

Cristopher Moore

Santa Fe Institute

Stephan Mertens

Otto-von-Guericke University, Magdeburg
and
Santa Fe Institute

OXFORD
UNIVERSITY PRESS

Great Clarendon Street, Oxford OX2 6DP

Oxford University Press is a department of the University of Oxford.
It furthers the University's objective of excellence in research, scholarship,
and education by publishing worldwide in

Oxford New York

Auckland Cape Town Dar es Salaam Hong Kong Karachi
Kuala Lumpur Madrid Melbourne Mexico City Nairobi
New Delhi Shanghai Taipei Toronto

With offices in

Argentina Austria Brazil Chile Czech Republic France Greece
Guatemala Hungary Italy Japan Poland Portugal Singapore
South Korea Switzerland Thailand Turkey Ukraine Vietnam

Oxford is a registered trade mark of Oxford University Press
in the UK and in certain other countries

Published in the United States
by Oxford University Press Inc., New York

First published 2011

British Library Cataloguing in Publication Data

Data available

Library of Congress Cataloging in Publication Data

Data available

Printed in Great Britain
on acid-free paper by
CPI Group (UK) Ltd, Croydon, CR0 4YY

ISBN 978–0–19–923321–2

For Tracy and Doro

Contents

Figure Credits

Preface

> The familiar essayist didn't speak to the millions; he spoke to *one* reader, as if the two of them were sitting side by side in front of a crackling fire with their cravats loosened, their favorite stimulants at hand, and a long evening of conversation stretching before them. His viewpoint was subjective, his frame of reference concrete, his style digressive, his eccentricities conspicuous, and his laughter usually at his own expense. And though he wrote about himself, he also wrote about a *subject*, something with which he was so familiar, and about which he was often so enthusiastic, that his words were suffused with a lover's intimacy.
>
> Anne Fadiman, *At Large and At Small*

> It is not incumbent upon you to finish the work,
> yet neither are you free to desist from it.
>
> Rabbi Tarfon

The sciences that awe and inspire us deal with fundamentals. Biology tries to understand the nature of life, from its cellular machinery to the gorgeous variety of organisms. Physics seeks the laws of nature on every scale from the subatomic to the cosmic. These questions are among the things that make life worth living. Pursuing them is one of the best things humans do.

The theory of computation is no less fundamental. It tries to understand why, and how, some problems are easy while others are hard. This isn't a question of how fast our computers are, any more than astronomy is the study of telescopes. It is a question about the *mathematical structures* of problems, and how these structures help us solve problems or frustrate our attempts to do so. This leads us, in turn, to questions about the nature of mathematical proof, and even of intelligence and creativity.

Computer science can trace its roots back to Euclid. It emerged through the struggle to build a foundation for mathematics in the early 20th century, and flowered with the advent of electronic computers, driven partly by the cryptographic efforts of World War II. Since then, it has grown into a rich field, full of deep ideas and compelling questions. Today it stands beside other sciences as one of the lenses we use to look at the world. Anyone who truly wants to understand how the world works can no more ignore computation than they can ignore relativity or evolution.

Computer science is also one of the most flexible and dynamic sciences. New subfields like quantum computation and phase transitions have produced exciting collaborations between computer scientists, physicists, and mathematicians. When physicists ask what rules govern a quantum system, computer scientists ask what it can compute. When physicists describe the phase transition that turns water to ice, computer scientists ask whether a similar transition turns problems from easy to hard.

This book was born in 2005 when one of us was approached by a publisher to write a book explaining computational complexity to physicists. The tale grew in the telling, until we decided—with some hubris—to explain it to everyone, including computer scientists. A large part of our motivation was to write the book we would have liked to read. We fell in love with the theory of computation because of the beauty and power of its ideas, but many textbooks bury these ideas under a mountain of formalism. We have not hesitated to present material that is technically difficult when it's appropriate. But at every turn we have tried to draw a clear distinction between deep ideas on the one hand and technical details on the other—just as you would when talking to a friend.

Overall, we have endeavored to write our book with the accessibility of Martin Gardner, the playfulness of Douglas Hofstadter, and the lyricism of Vladimir Nabokov. We have almost certainly failed on all three counts. Nevertheless, we hope that the reader will share with us some of the joy and passion we feel for our adopted field. If we have reflected, however dimly, some of the radiance that drew us to this subject, we are content.

We are grateful to many people for their feedback and guidance: Scott Aaronson, Heiko Bauke, Paul Chapman, Andrew Childs, Aaron Clauset, Varsha Dani, Josep Díaz, Owen Densmore, Irit Dinur, Ehud Friedgut, Tom Hayes, Robert Hearn, Stefan Helmreich, Reuben Hersh, Shiva Kasiviswanathan, Brian Karrer, David Kempe, Greg Kuperberg, Cormac McCarthy, Sarah Miracle, John F. Moore, Michel Morvan, Larry Nazareth, Sebastian Oberhoff, Ryan O'Donnell, Mark Olah, Jim Propp, Dana Randall, Sasha Razborov, Omer Reingold, Paul Rendell, Sara Robinson, Jean-Baptiste Rouquier, Amin Saberi, Jared Saia, Nicolas Schabanel, Cosma Shalizi, Thérèse Smith, Darko Stefanović, John Tromp, Vijay Vazirani, Robin Whitty, Lance Williams, Damien Woods, Jon Yard, Danny Yee, Lenka Zdeborová, Yaojia Zhu, and Katharina Zweig.

We are also grateful to Lee Altenberg, László Babai, Nick Baxter, Nirdosh Bhatnagar, Marcus Calhoun-Lopez, Timothy Chow, Nathan Collins, Alex Conley, Will Courtney, Zheng Cui, Wim van Dam, Tom Dangniam, Aaron Denney, Hang Dinh, David Doty, Taylor Dupuy, Bryan Eastin, Charles Efferson, Veit Elser, Leigh Fanning, Steve Flammia, Matthew Fricke, Michel Goemans, Benjamin Gordon, Stephen Guerin, Samuel Gutierrez, Russell Hanson, Jacob Hobbs, Neal Holtschulte, Peter Høyer, Luan Jun, Valentine Kabanets, Richard Kenyon, Jeffrey Knockel, Leonid Kontorovich, Maurizio Leo, MatjažLeonardis, Phil Lewis, Scott Levy, Chien-Chi Lo, Jun Luan, Shuang Luan, Sebastian Luther, Jon Machta, Jonathan Mandeville, Bodo Manthey, Pierre McKenzie, Pete Morcos, Brian Nelson, ThanhVu Nguyen, Katherine Nystrom, Olumuyiwa Oluwasanmi, Boleszek Osinski, John Patchett, Robin Pemantle, Yuval Peres, Carlos Riofrio, Tyler Rush, Navin Rustagi, George Saad, Gary Sandine, Samantha Schwartz, Oleg Semenov, David Sherrington, Jon Sorenson, George Stelle, Satomi Sugaya, Bert Tanner, Amitabh Trehan, Yamel Torres, Michael Velbaum, Lutz Warnke, Chris Willmore, David Wilson, Chris Wood, Ben Yackley, Yiming Yang, Rich Younger, Sheng-Yang Wang, Zhan Zhang, and Evgeni Zlatanov. We apologize for any omissions.

We express our heartfelt thanks to the Santa Fe Institute, without whose hospitality we would have been unable to complete this work. In particular, the SFI library staff—Margaret Alexander, Tim Taylor, and Joy LeCuyer—fulfilled literally hundreds of requests for articles and books.

We are grateful to our editor Sönke Adlung for his patience, and to Alison Lees for her careful copy-editing. Mark Newman gave us invaluable help with the LaTeX Memoir class, in which this book is typeset, along with insights and moral support from his own book-writing experiences. And throughout the process, Alex Russell shaped our sense of the field, separating the wheat from the chaff and helping us to decide which topics and results to present to the reader. Some fabulous monsters didn't make it onto the ark, but many of those that did are here because he urged us to take them on board.

Finally, we dedicate this book to Tracy Conrad and Doro Frederking. Our partners, our loves, they have made everything possible.

Cristopher Moore and Stephan Mertens
Santa Fe and Magdeburg, 2019

How to read this book

Outside a dog a book is a man's best friend.
Inside a dog it's too dark to read.

Groucho Marx

We recommend reading Chapters 1–7 in linear order, and then picking and choosing from later chapters and sections as you like. Even the advanced chapters have sections that are accessible to nearly everyone.

For the most part, the only mathematics we assume is linear algebra and some occasional calculus. We use Fourier analysis and complex numbers in several places, especially Chapter 11 for the PCP Theorem and Chapter 15 on quantum computing. Mathematical techniques that we use throughout the book, such as asymptotic notation and discrete probability, are discussed in the Appendix. We assume some minimal familiarity with programming, such as the meaning of **for** and **while** loops.

Scattered throughout the text you will find Exercises. These are meant to be easy, and to help you check whether you are following the discussion at that point. The Problems at the end of each chapter delve more deeply into the subject, providing examples and fleshing out arguments that we sketch in the main text. We have been generous with hints and guideposts in order to make even the more demanding problems doable.

Every once in a while, you will see a quill symbol in the margin—yes, like that one there. This refers to a note in the Notes section at the end of the chapter, where you can find details, historical discussion, and references to the literature.

Since theoretical computer science is a rapidly evolving field, we invite the reader to check periodically for updates and addenda at www.nature-of-computation.org. There we will include new problems and exercises, and notes and references for new results.

A note to the instructor

We have found that Chapters 1–8, with selections from Chapters 9–11, form a good text for an introductory graduate course on computational complexity. We and others have successfully used later chapters as texts or supplementary material for more specialized courses, such as Chapters 12 and 13 for a course on Markov chains, Chapter 14 for phase transitions, and Chapter 15 for quantum computing. Some old-fashioned topics, like formal languages and automata, are missing from our book, and this is by design.

The Turing machine has a special place in the history of computation, and we discuss it along with λ-calculus and partial recursive functions in Chapter 7. But we decided early on to write about computation as if the Church-Turing thesis were true—in other words, that we are free to use whatever model of computation makes it easiest to convey the key ideas. Accordingly, we describe algorithms at a "software" level, as programs written in the reader's favorite programming language. This lets us draw on the reader's experience and intuition that programs need time and memory to run. Where necessary, such as in our discussion of LOGSPACE in Chapter 8, we drill down into the hardware and discuss details such as our model of memory access.

Please share with us your experiences with the book, as well as any mistakes or deficiencies you find. We maintain errata at www.nature-of-computation.org. We can also provide a solution manual on request, which currently contains solutions for over half of the problems.

Note on the 2018 printing

This printing corrects a large number of typos throughout the book, fixes a number of aesthetic typesetting issues, and updates many of the Notes to take recent advances into account. With the help of our dedicated readers, we have also improved the discussion and readability of the book and corrected some mathematical errors, for instance in the proof that quadratic Diophantine equations are NP-complete (Section 5.4.4 and Problems 5.24 and 5.25). There are a number of new problems, added at the end of the Problems sections so as not to disturb the numbering of previous ones. In particular, readers interested in pseudorandomness and derandomization will be pleased to learn that Nisan's generator for fooling space-bounded computation now appears, with many hints to guide the reader, as Problem 11.18.

Chapter 1

Prologue

Some citizens of Königsberg
Were walking on the strand
Beside the river Pregel
With its seven bridges spanned.

"O Euler, come and walk with us,"
Those burghers did beseech.
"We'll roam the seven bridges o'er,
And pass but once by each."

"It can't be done," thus Euler cried.
"Here comes the Q.E.D.
Your islands are but vertices
And four have odd degree."

William T. Tutte

1.1 Crossing Bridges

We begin our journey into the nature of computation with a walk through 18th-century Königsberg (now Kaliningrad). As you can see from Figure 1.1, the town of Königsberg straddles the river Pregel with seven bridges, which connect the two banks of the river with two islands. A popular puzzle of the time asked if one could walk through the city in a way that crosses each bridge exactly once. We do not know how hard the burghers of Königsberg tried to solve this puzzle on their Sunday afternoon walks, but we do know that they never succeeded.

It was Leonhard Euler who solved this puzzle in 1736. As a mathematician, Euler preferred to address the problem by pure thought rather than by experiment. He recognized that the problem depends only on the set of connections between the riverbanks and islands—a *graph* in modern terms. The graph

corresponding to Königsberg has four vertices representing the two riverbanks and the two islands, and 1.1
seven edges for the bridges, as shown in Figure 1.2. Today, we say that a walk through a graph that crosses

FIGURE 1.1: Königsberg in the 17th century.

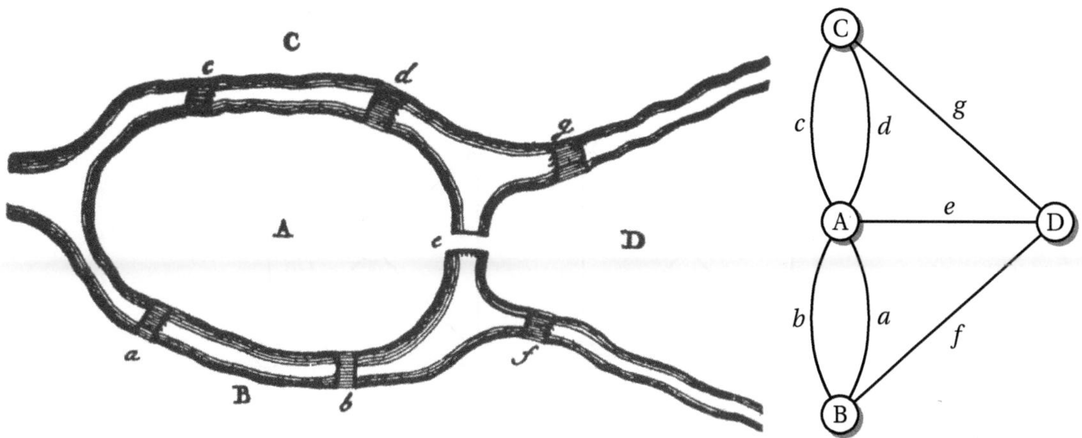

FIGURE 1.2: The seven bridges of Königsberg. Left, as drawn in Euler's 1736 paper, and right, as represented as a graph in which each riverbank or island is a vertex and each bridge an edge.

each edge once is an *Eulerian path*, or an *Eulerian cycle* if it returns to its starting point. We say that a graph is Eulerian if it possesses an Eulerian cycle.

Now that we have reduced the problem to a graph that we can doodle on a sheet of paper, it is easy to explore various walks by trial and error. Euler realized, though, that trying all possible walks this way would take some time. As he noted in his paper (translated from the Latin):

> As far as the problem of the seven bridges of Königsberg is concerned, it can be solved by making an exhaustive list of possible routes, and then finding whether or not any route satisfies the conditions of the problem. Because of the number of possibilities, this method of solutions would be too difficult and laborious, and in other problems with more bridges, it would be impossible.

Let's be more quantitative about this. Assume for simplicity that each time we arrive on an island or a riverbank there are two different ways we could leave. Then if there are n bridges to cross, a rough estimate for the number of possible walks would be 2^n. In the Königsberg puzzle we have $n = 7$, and while 2^7 or 128 routes would take quite a while to generate and check by hand, a modern computer could do so in the blink of an eye.

But Euler's remark is not just about the bridges of Königsberg. It is about the entire *family of problems of this kind*, and how their difficulty grows, or *scales*, as a function of the number of bridges. If we consider the bridges of Venice instead, where $n = 420$—or Pittsburgh, where $n = 446$—even the fastest computer imaginable would take longer than the age of the universe to do an exhaustive search. Thus, if searching through the space of all possible solutions were the only way to solve these problems, even moderately large cities would be beyond our computational powers.

Euler had a clever insight which allows us to avoid this search completely. He noticed that in order to cross each edge once, any time we arrive at a vertex along one edge, we have to depart on a different edge. Thus the edges of each vertex must come in pairs, with a "departure" edge for each "arrival" edge. It follows that the *degree* of each vertex—that is, the number of edges that touch it—must be even. This

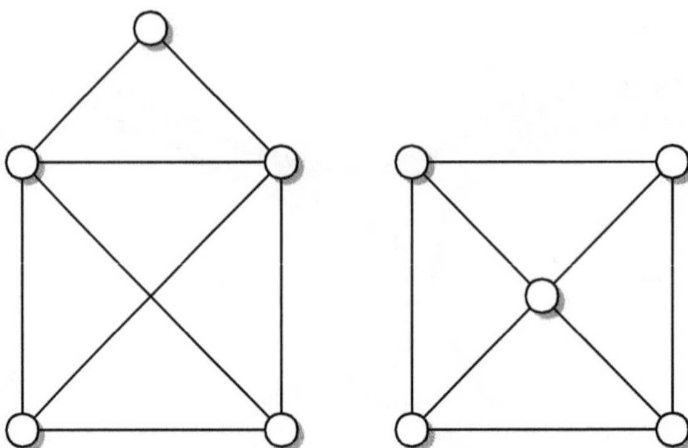

FIGURE 1.3: A classic children's puzzle. Can you draw these graphs without lifting the pen from the paper, or drawing the same edge twice? Equivalently, do they have Eulerian paths?

holds for all vertices except the vertices where the path starts and ends, which must have odd degree unless they coincide and the path is a cycle.

This argument shows that a necessary condition for a graph to be Eulerian is for all its vertices to have even degree. Euler claimed that this condition is also sufficient, and stated the following theorem:

Theorem 1.1 *A connected graph contains an Eulerian cycle if and only if every vertex has even degree. If exactly two vertices have odd degree, it contains an Eulerian path but not an Eulerian cycle.*

This theorem allows us to solve the bridges of Königsberg very quickly. As the poem at the head of this chapter points out, all four vertices have odd degree, so there is no Eulerian path through the old town of Königsberg.

Beyond solving this one puzzle, Euler's insight makes an enormous difference in how the complexity of this problem scales. An exhaustive search in a city with n bridges takes an amount of time that grows *exponentially* with n. But we can check that every vertex has even degree in an amount of time proportional to the number of vertices, assuming that we are given the map of the city in some convenient format. Thus Euler's method lets us solve this problem in *linear* time, rather than the exponential time of a brute-force search. Now the bridges of Venice, and even larger cities, are easily within our reach.

Exercise 1.1 *Which of the graphs in Figure 1.3 have Eulerian paths?*

In addition to the tremendous speedup from exponential to linear time, Euler's insight transforms this problem in another way. What does it take to *prove* the existence, or nonexistence, of an Eulerian path? If one exists, we can easily prove this fact simply by exhibiting it. But if no path exists, how can we prove that? How can we convince the people of Königsberg that their efforts are futile?

Imagine what would have happened if Euler had used the brute-force approach, presenting the burghers with a long list of all possible paths and pointing out that none of them work. Angry at Euler for spoiling their favorite Sunday afternoon pastime, they would have been understandably skeptical. Many would have refused to go through the tedious process of checking the entire list, and would have held out

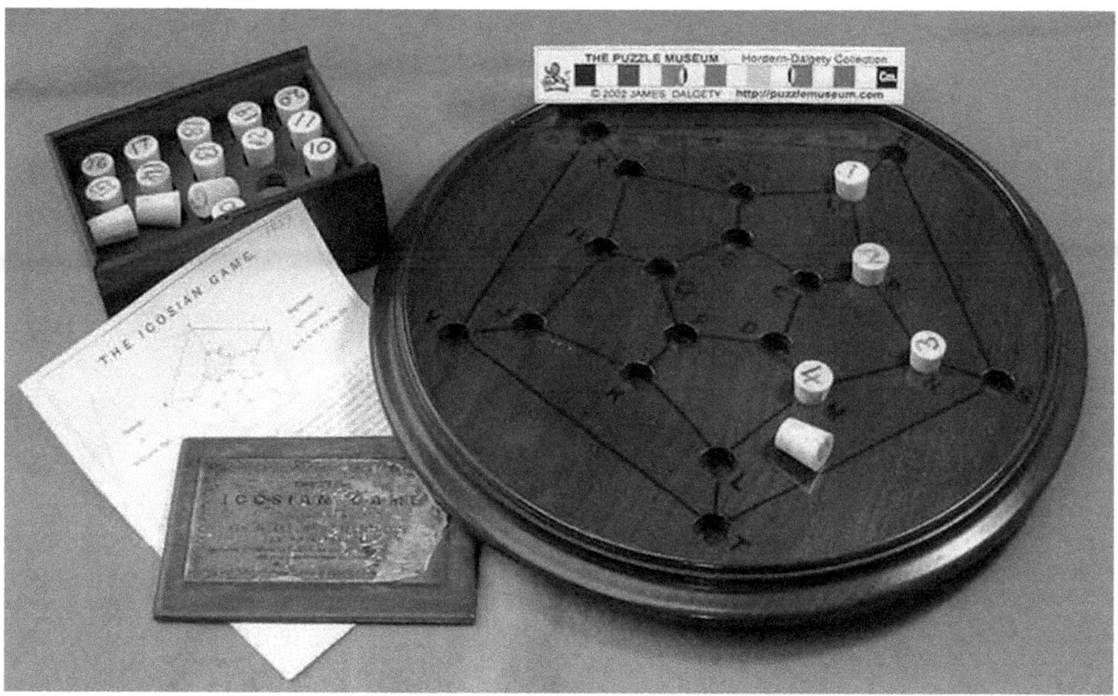

FIGURE 1.4: Hamilton's Icosian game.

the hope that Euler had missed some possibility. Moreover, such an ungainly proof is rather unsatisfying, even if it is logically airtight. It offers no sense of *why* no path exists.

1.2 In contrast, even the most determined skeptic can follow the argument of Euler's theorem. This allows Euler to present a proof that is simple, compact, and irresistible: he simply needs to exhibit three vertices with odd degree. Thus, by showing that the existence of a path is equivalent to a much simpler property, Euler radically changed the *logical structure* of the problem, and the type of proof or disproof it requires.

1.2 Intractable Itineraries

The next step in our journey brings us to 19th-century Ireland and the Astronomer Royal, Sir William Rowan Hamilton, known to every physicist through his contributions to classical mechanics. In 1859, Hamilton put a new puzzle on the market, called the "Icosian game," shown in Figure 1.4. The game was a commercial failure, but it led to one of the iconic problems in computer science today.

The object of the game is to walk around the edges of a dodecahedron while visiting each vertex once and only once. Actually, it was a two-player game in which one player chooses the first five vertices, and the other tries to complete the path—but for now, let's just think about the solitaire version. While such walks had been considered in other contexts before, we call them *Hamiltonian* paths or cycles today, and 1.3

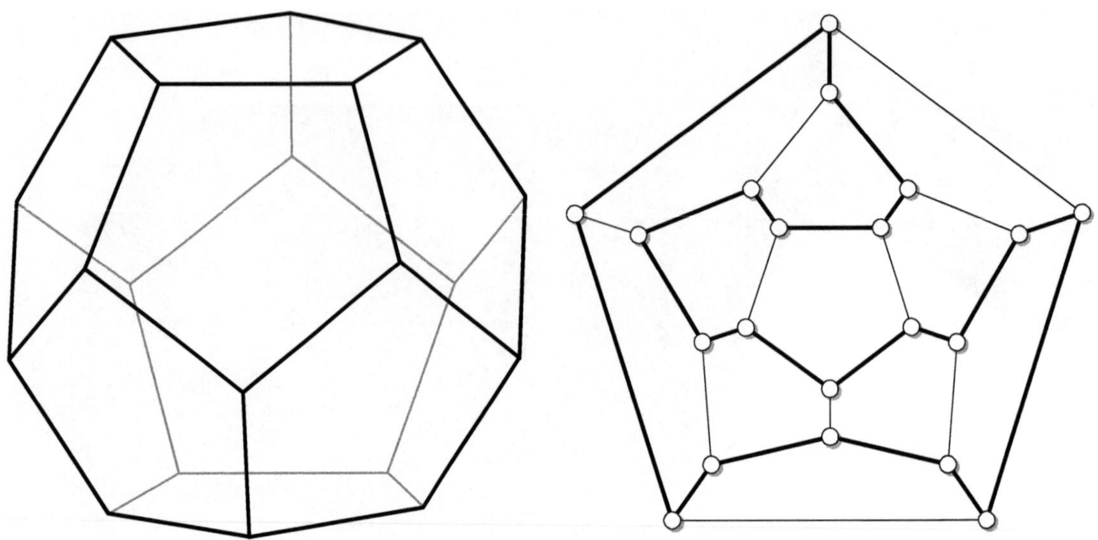

FIGURE 1.5: Left, the dodecahedron; right, a flattened version of the graph formed by its edges. One Hamiltonian cycle, which visits each vertex once and returns to its starting point, is shown in bold.

we say that a graph is Hamiltonian if it possesses a Hamiltonian cycle. One such cycle for the dodecahedron is shown in Figure 1.5.

At first Hamilton's puzzle seems very similar to the bridges of Königsberg. Eulerian paths cross each edge once, and Hamiltonian paths visit each vertex once. Surely these problems are not very different? However, while Euler's theorem allows us to avoid a laborious search for Eulerian paths or cycles, we have no such insight into Hamiltonian ones. As far as we know, there is no simple property—analogous to having vertices of even degree—to which *Hamiltonianness* is equivalent.

As a consequence, we know of no way of avoiding, essentially, an exhaustive search for Hamiltonian paths. We can visualize this search as a tree as shown in Figure 1.6. Each node of the tree corresponds to a partial path, and branches into child nodes corresponding to the various ways we can extend the path. In general, the number of nodes in this search tree grows exponentially with the number of vertices of the underlying graph, so traversing the entire tree—either finding a leaf with a complete path, or learning that every possible path gets stuck—takes exponential time.

To phrase this computationally, we believe that there is no program, or *algorithm*, that tells whether a graph with n vertices is Hamiltonian or not in an amount of time proportional to n, or n^2, or any polynomial function of n. We believe, instead, that the best possible algorithm takes exponential time, 2^{cn} for some constant $c > 0$. Note that this is not a belief about how fast we can make our computers. Rather, it is a belief that finding Hamiltonian paths is *fundamentally harder* than finding Eulerian ones. It says that these two problems differ in a deep and qualitative way.

While finding a Hamiltonian path seems to be hard, *checking* whether a given path is Hamiltonian is easy. Simply follow the path vertex by vertex, and check that it visits each vertex once. So if a computationally powerful friend claims that a graph has a Hamiltonian path, you can challenge him or her to

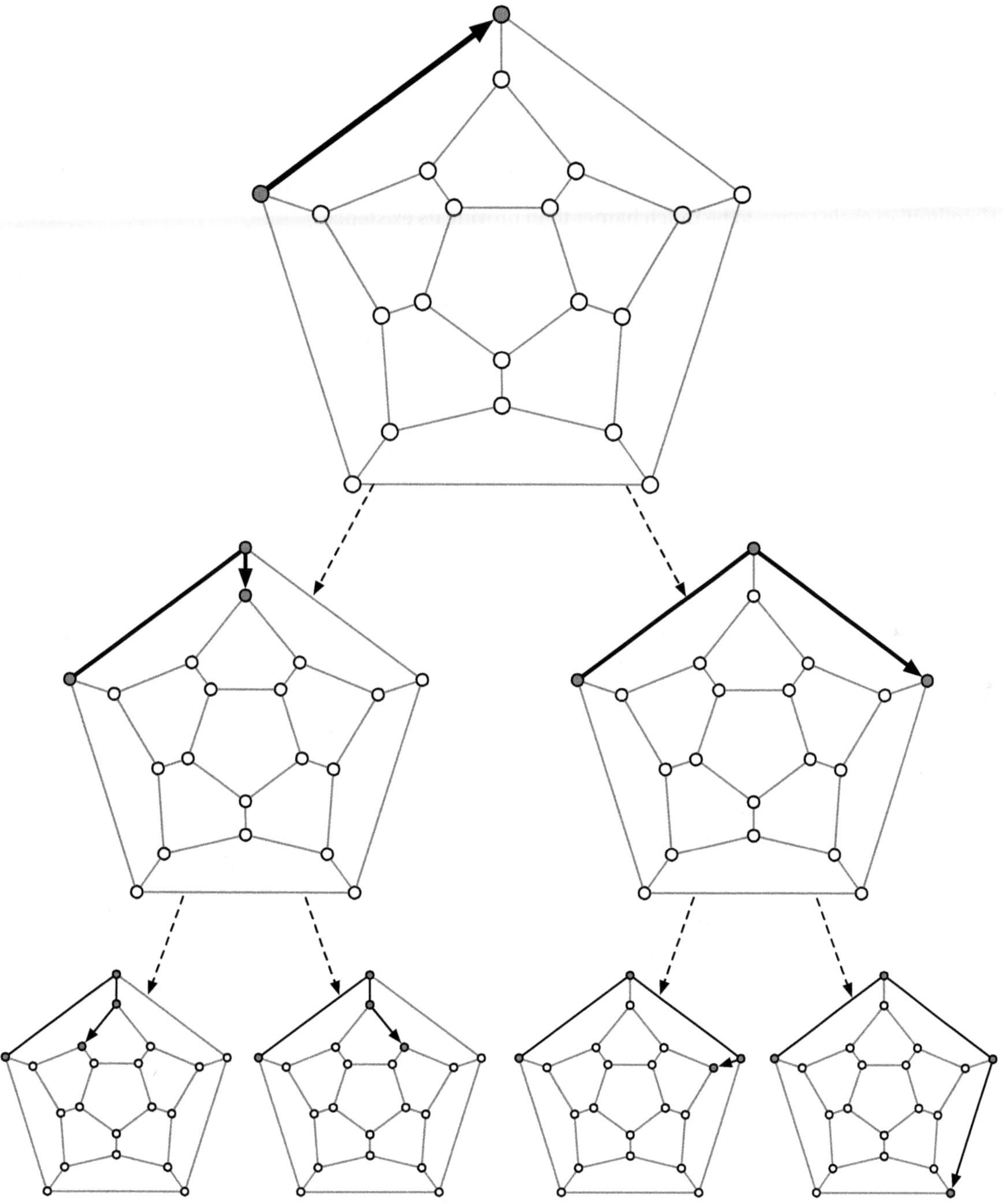

FIGURE 1.6: The first two levels of the search tree for a Hamiltonian path.

prove that fact by showing it to you, and you can then quickly confirm or refute their claim. Problems like these are rather like finding a needle in a haystack: if I show you a needle you can confirm that it is one, but it's hard to find a needle—at least without a magnetic insight like Euler's.

On the other hand, if I claim that a haystack has no needles in it, the only way to prove this is to sift carefully through all the hay. Similarly, we know of no way to prove that no Hamiltonian cycle exists without a massive search. Unlike Eulerian cycles, where there is a simple proof in either case, the logical structure of Hamiltonianness seems to be fundamentally asymmetric—proving the nonexistence of a Hamiltonian cycle seems to be much harder than proving its existence.

Of course one might think that there is a more efficient method for determining whether a graph is Hamiltonian, and that we have simply not been clever enough to find it. But as we will see in this book, there are very good reasons to believe that *no such method exists.* Even more amazingly, if we are wrong about this—if Hamilton's problem can be solved in time that only grows polynomially—then so can thousands of other problems, all of which are currently believed to be exponentially hard. These problems range from such classic search and optimization problems as the Traveling Salesman problem, to the problem of finding short proofs of the grand unsolved questions in mathematics. In a very real sense, the hardness of Hamilton's problem is related to our deepest beliefs about mathematical and scientific creativity. Actually *proving* that it is hard remains one of the holy grails of theoretical computer science.

1.3 Playing Chess With God

> It's a sort of Chess that has nothing to do with Chess, a Chess that we could never have imagined without computers. The Stiller moves are awesome, almost scary, because you know they are the truth, God's Algorithm—it's like being revealed the Meaning of Life, but you don't understand one word.
>
> Tim Krabbé

As we saw in the previous section, the problem of telling whether a Hamiltonian cycle exists has the property that finding solutions is hard—or so we believe—but checking them is easy. Another example of this phenomenon is factoring integers. As far as we know, there is no efficient way to factor a large integer N into its divisors—at least without a quantum computer—and we base the modern cryptosystems used by intelligence agents and Internet merchants on this belief. On the other hand, given two numbers p and q it is easy to multiply them, and check whether $pq = N$.

1.4

This fact was illustrated beautifully at a meeting of the American Mathematical Society in 1903. The mathematician Frank Nelson Cole gave a "lecture without words," silently performing the multiplication

$$193\,707\,721 \times 761\,838\,257\,287 = 147\,573\,952\,588\,676\,412\,927$$

on the blackboard. The number on the right-hand side is $2^{67} - 1$, which the 17th-century French mathematician Marin Mersenne conjectured is prime. In 1876, Édouard Lucas managed to prove that it is composite, but gave no indication of what numbers would divide it. The audience, knowing full well how hard it is to factor 21-digit numbers, greeted Cole's presentation with a standing ovation. Cole later admitted that it had taken him "three years of Sundays" to find the factors.

On the other hand, there are problems for which even *checking* a solution is extremely hard. Consider the Chess problems shown in Figure 1.7. Each claims that White has a winning strategy, which will lead

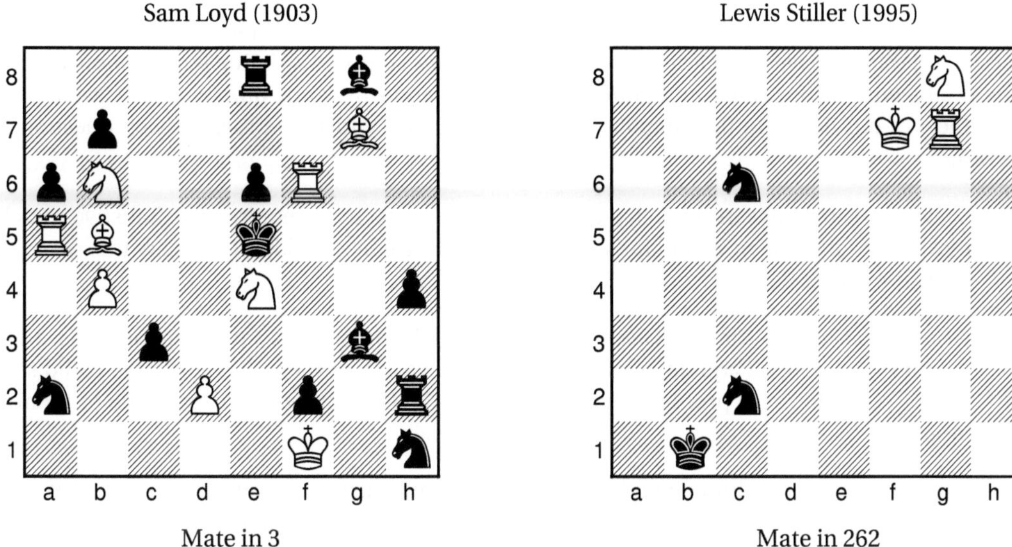

FIGURE 1.7: Chess problems are hard to solve—and hard to check.

inexorably to checkmate after n moves. On the left, $n = 3$, and seeing how to corner Black—after a very surprising first move—is the work of a pleasant afternoon. On the right, we have a somewhat larger value of n: we claim that White can force Black into checkmate after 262 moves.

But how, dear reader, can we prove this claim to you? Unlike the problems of the previous two sections, we are no longer playing solitaire: we have an opponent who will do their best to win. This means that it's not enough to prove the existence of a simple object like a Hamiltonian path. We have to show that there exists a move for White, such that no matter how Black replies, there exists a move for White, such that no matter how Black replies, and so on… until, at most 262 moves later, every possible game ends in checkmate for Black. As we go forward in time, our opponent's moves cause the game to branch into a exponential tree of possibilities, and we have to show that a checkmate awaits at every leaf. Thus a strategy is a much larger object, with a much deeper logical structure, than a path.

There is indeed a proof, consisting of a massive database of endgames generated by a computer search, that White can mate Black in 262 moves in the position of Figure 1.7. But verifying this proof far exceeds the capabilities of human beings, since it requires us to check every possible line of play. The best we can do is look at the program that performed the search, and convince ourselves that it will run correctly. As of 2007, an even larger search has confirmed the long-standing opinion of human players that Checkers is a draw under perfect play. For humans with our finite abilities, however, Chess and Checkers will always keep their secrets. 1.5

1.4 What Lies Ahead

As we have seen with our three examples, different problems require fundamentally different kinds of search, and different types of proof, to find or verify their solutions. Understanding how to solve problems as efficiently as possible—and understanding how, and why, some problems are extremely hard—is the subject of our book. In the chapters ahead, we will ask, and sometimes answer, questions like the following:

- Some problems have insights like Euler's, and others seem to require an exhaustive search. What makes the difference? What kinds of strategies can we use to skip or simplify this search, and for which problems do they work?
- A host of problems—finding Hamiltonian paths, coloring graphs, satisfying formulas, and balancing numbers—are all equally hard. If we could solve any of them efficiently, we could solve all of them. What do these problems have in common? How can we transform one of them into the other?
- If we could find Hamiltonian paths efficiently, we could also easily find short proofs—if they exist—of the great unsolved problems in mathematics, such as the Riemann Hypothesis. We believe that doing mathematics is harder than this, and that it requires all our creativity and intuition. But does it really? Can we prove that finding proofs is hard?
- Can one programming language, or kind of computer, solve problems that another can't? Or are all sufficiently powerful computers equivalent? Are even simple systems, made of counters, tiles, and billiard balls, capable of universal computation?
- Are there problems that no computer can solve, no matter how much time we give them? Are there mathematical truths that no axiomatic system can prove?
- If exact solutions are hard to find, can we find approximate ones? Are there problems where even approximate solutions are hard to find? Are there others that are hard to solve perfectly, but where we can find solutions that are as close as we like to the best possible one?
- What happens if we focus on the amount of memory a computation needs, rather than the time it takes? How much memory do we need to find our way through a maze, or find a winning strategy in a two-player game?
- If we commit ourselves to one problem-solving strategy, a clever adversary can come up with the hardest possible example. Can we defeat the adversary by acting unpredictably, and flipping coins to decide what to do?
- Suppose that Merlin has computational power beyond our wildest dreams, but that Arthur is a mere mortal. If Merlin knows that White has a winning strategy in Chess, can he convince Arthur of that fact, without playing a single game? How much can Arthur learn by asking Merlin random questions? What happens if Merlin tries to deceive Arthur, or if Arthur tries to "cheat" and learn more than he was supposed to?

- If flipping random coins helps us solve problems, do we need truly random coins? Are there strong pseudorandom generators—fast algorithms that produce strings of coin flips deterministically, but with no pattern that other fast algorithms can discover?
- Long proofs can have small mistakes hidden in them. But are there "holographic" proofs, which we can confirm are almost certainly correct by checking them in just a few places?
- Finding a solution to a problem is one thing. What happens if we want to generate a random solution, or count the number of solutions? If we take a random walk in the space of all possible solutions, how long will it take to reach an equilibrium where all solutions are equally likely?
- How rare are the truly hard examples of hard problems? If we make up random examples of a hard problem, are they hard or easy? When we add more and more constraints to a problem in a random way, do they make a sudden jump from solvable to unsolvable?
- Finally, how will quantum computers change the landscape of complexity? What problems can they solve faster than classical computers?

Problems

A great discovery solves a great problem, but there is a grain of discovery in the solution of any problem. Your problem may be modest, but if it challenges your curiosity and brings into play your inventive faculties, and if you solve it by your own means, you may experience the tension and enjoy the triumph of discovery.

George Pólya, *How To Solve It*

1.1 Handshakes. Prove that in any finite graph, the number of vertices with odd degree is even.

1.2 Pigeons and holes. Properly speaking, we should call our representation of Königsberg a *multigraph*, since some pairs of vertices are connected to each other by more than one edge. A *simple* graph is one in which there are no multiple edges, and no self-loops.

Show that in any finite simple graph with more than one vertex, there is at least one pair of vertices that have the same degree. Hint: if n pigeons try to nest in $n-1$ holes, at least one hole will contain more than one pigeon. This simple but important observation is called the *pigeonhole principle*.

1.3 Proving Euler's claim. Euler didn't actually prove that having vertices with even degree is sufficient for a connected graph to be Eulerian—he simply stated that it is obvious. This lack of rigor was common among 18th century mathematicians. The first real proof was given by Carl Hierholzer more than 100 years later. To reconstruct it, first show that if every vertex has even degree, we can cover the graph with a set of cycles such that every edge appears exactly once. Then consider combining cycles with moves like those in Figure 1.8.

1.4 Finding an Eulerian path. Let's turn the proof of the previous problem into a simple algorithm that constructs an Eulerian path. If removing an edge will cause a connected graph to fall apart into two pieces, we call that edge a *bridge*. Now consider the following simple rule, known as Fleury's algorithm: at each step, consider the graph G' formed by the edges you have not yet crossed, and only cross a bridge of G' if you have to. Show that if a connected graph has two vertices of odd degree and we start at one of them, this algorithm will produce an Eulerian path, and that if all vertices have even degree, it will produce an Eulerian cycle no matter where we start.

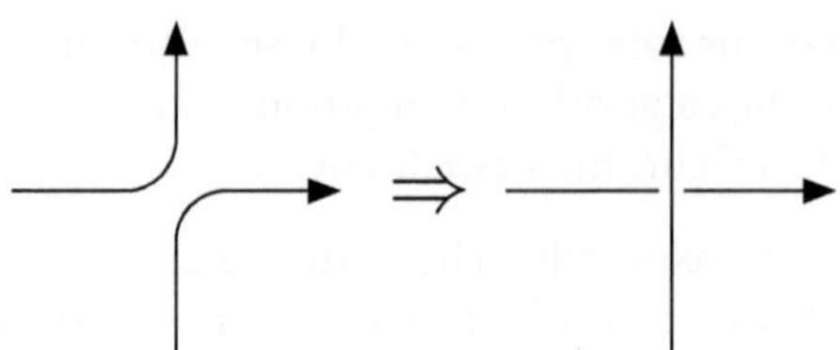

FIGURE 1.8: Combining cycles at a crossing.

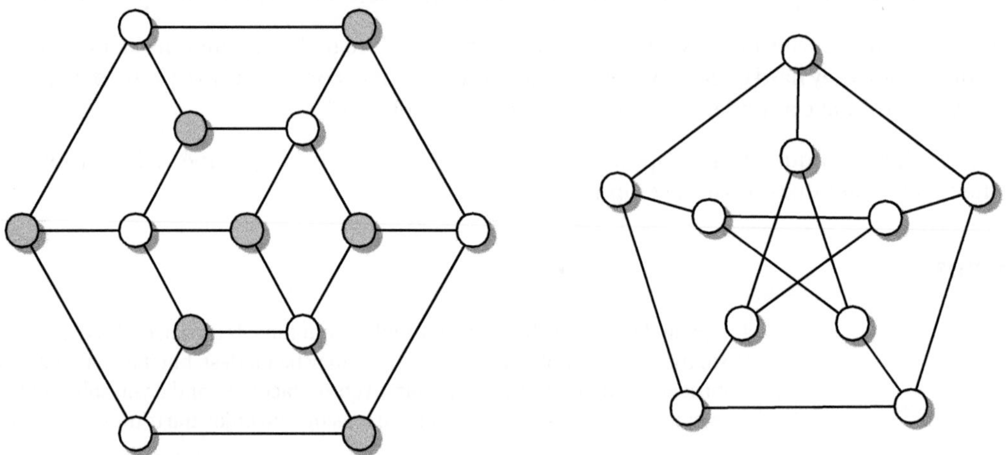

FIGURE 1.9: Two graphs that have Hamiltonian paths, but not Hamiltonian cycles.

1.5 One-way bridges. A *directed graph* is one where each edge has an arrow on it. Thus there can be an edge from u to v without one from v to u, and a given vertex can have both incoming and outgoing edges. An Eulerian path would be one that crosses each edge once, moving in the direction allowed by the arrow. Generalize Euler's theorem by stating under what circumstances a directed graph is Eulerian.

1.6 Plato and Hamilton. Inspired by Hamilton's choice of the dodecahedron, consider the other four Platonic solids, and the graphs consisting of their corners and edges: the tetrahedron, cube, octahedron, and icosahedron. Which ones are Hamiltonian? Which are Eulerian?

1.7 A rook's tour. Let G be an $m \times n$ grid—that is, a graph with mn vertices arranged in an $m \times n$ rectangle, with each vertex connected to its nearest neighbors. Assume that $m, n > 1$. Prove that G is Hamiltonian if either m or n is even, but not if both m and n are odd.

1.8 Symmetry and parity. Show that each graph in Figure 1.9 has a Hamiltonian path, but no Hamiltonian cycle. Hint: use the colors for the one on the left. For the one on the right, called the Petersen graph, exploit its symmetry.

1.9 The Chinese postman. The *Chinese postman problem*, named in honor of the Chinese mathematician Mei-Ko Kwan, asks for the shortest cyclic tour of a graph that crosses every edge *at least* once. If the graph is not Eulerian, the postman has to repeat some edges. Show that the shortest postman's tour crosses each edge at most twice, and that this worst case only occurs if the graph is a tree. Show, moreover, that the repeated edges form a set of paths that connect pairs of vertices of odd degree, creating a *matching* where each odd-degree vertex has another one as a partner. It turns out (see Note 5.6) that there is a polynomial-time algorithm that finds the matching that minimizes the total length of these paths, and hence the shortest postman's tour.

1.10 Existence, search, and the oracle. We can consider two rather different problems regarding Hamiltonian cycles. One is the *decision problem*, the yes-or-no question of whether such a cycle exists. The other is the *search problem* or *function problem*, in which we want to actually find the cycle.

Suppose that there is an oracle in a nearby cave. She will tell us, for the price of one drachma per question, whether a graph has a Hamiltonian cycle. If it does, show that by asking her a series of questions, perhaps involving modified versions of the original graph, we can find the Hamiltonian cycle after spending a number of drachmas that grows polynomially as a function of the number of vertices. Thus if we can solve the decision problem in polynomial time, we can solve the search problem as well.

Notes

1.1 Graph theory. Euler's paper on the Königsberg bridges [272] can be regarded as the birth of graph theory, the mathematics of connectivity. The translation we use here appears in [111], which contains many of the key early papers in this field. Today graph theory is a very lively branch of discrete mathematics with many applications in chemistry, engineering, physics, and computer science.

We will introduce concepts from graph theory "on the fly," as we need them. For an intuitive introduction, we recommend Trudeau's little treatise [796]. For a more standard textbooks, see Bollobás [119] or the *Handbook on Graph Theory* [354]. Hierholzer's proof, which we ask you to reconstruct in Problem 1.3, appeared in [400]. The Chinese Postman of Problem 1.9 was studied by Mei-Ko Kwan in 1962 [512].

The river Pregel still crosses the city of Kaliningrad, but the number of bridges and their topology changed considerably during World War II. See also [614].

1.2 Persuasive proofs. According to the great Hungarian mathematician Paul Erdős, God has a book that contains short and insightful proofs of every theorem, and sometimes humans can catch a glimpse of a few of its pages. One of the highest forms of praise that one mathematician can give another is to say "Ah, you have found the proof from the book." [26].

In a sense, this is the difference between Eulerian and Hamiltonian paths. If a large graph G lacks an Eulerian path, Euler's argument gives a "book proof" of that fact. In contrast, as far as we know, the only way to prove that G lacks a Hamiltonian path is an ugly exhaustive search.

This dichotomy exists in other areas of mathematics as well. A famous example is Thomas Hales' proof of the Kepler conjecture [362], a 400-year-old claim about the densest packings of spheres in three-dimensional space. After four years of work, a group of twelve referees concluded that they were "99% certain" of the correctness of the proof. Hales' proof relies on exhaustive case checking by a computer, a technique first used in the proof of the Four Color Theorem. Such a proof may confirm that a fact is true, but it offers very little illumination about *why* it is true. See [220] for a thought-provoking discussion of the complexity of mathematical proofs, and how computer-assisted proofs will change the nature of mathematics.

1.3 A knight's tour. One instance of the Hamiltonian path problem is far older than the Icosian game. Around the year 840 A.D., a Chess player named al-Adli ar-Rumi constructed a *knight's tour* of the chessboard, shown in Figure 1.10. We can think of this as a Hamiltonian path on the graph whose vertices are squares, and where two vertices are adjacent if a knight could move from one to the other. Another early devotee of knight's tours was the Kashmiri poet Rudrata (circa 900 A.D.). In their lovely book *Algorithms* [215], Dasgupta, Papadimitriou and Vazirani suggest that Hamiltonian paths be called Rudrata paths in his honor.

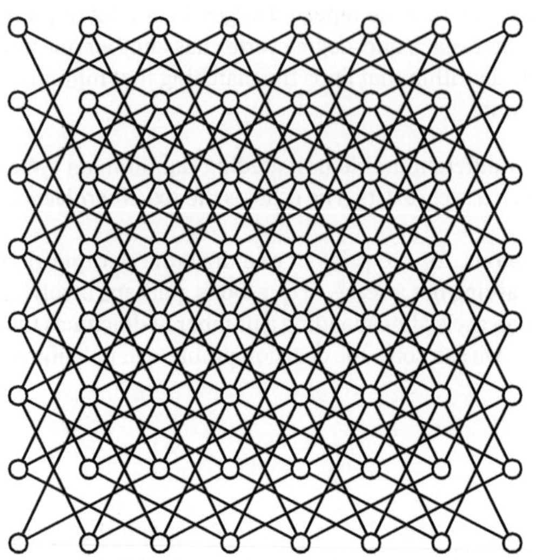

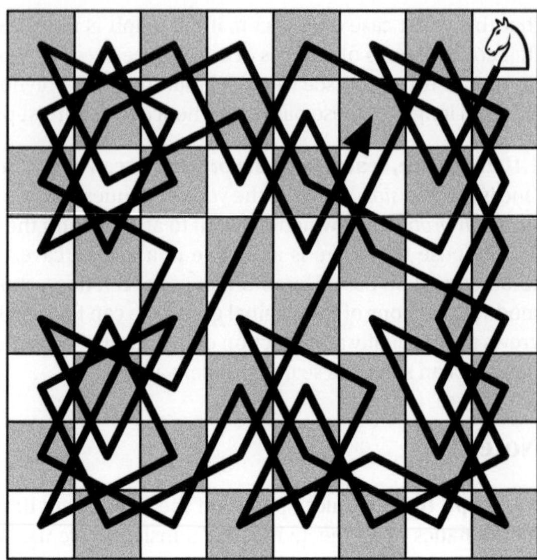

FIGURE 1.10: On the left, the graph corresponding to the squares of the chessboard with knight's moves between them. On the right, Al-Adli's Hamiltonian tour, or knight's tour, of this graph. Note that with one more move it becomes a Hamiltonian cycle.

1.4 Cryptic factors. As we will see in Chapter 15, it is not quite right to say that cryptosystems like RSA public-key encryption are based on the hardness of factoring: they are based on other number-theoretic problems, which might be easier than factoring. However, if we can factor large integers efficiently, we can solve these problems too.

1.5 Endgame. The *endgame tablebase* is a computerized database of all endgame positions in Chess with a given number of pieces, that reveals the value (win, loss or draw) of each position and and how many moves it will take to achieve that result with perfect play [789]. The 262-move problem in Figure 1.7 was found by Stiller [774], who extended the tablebase to six-piece endgames. Of all positions with the given material that are a win, it has the longest "distance to mate."

The proof that Checkers is a draw [729] uses a similar endgame database that contains the value of all Checkers positions with ten pieces or fewer. There are 3.9×10^{13} of these positions, compressed into 237 gigabytes of diskspace. The proof traces all relevant lines of play from the starting position to one of the positions in this database. This proof constitutes what game theorists call a *weak* solution of a game. For a *strong* solution, we would have to compute the optimal move for every legal position on the board, not just the opening position. We will discuss the enormous difficulty of this problem in Chapter 8.

Can we hope for a weak solution of Chess? The six-piece endgame database for Chess is more than five times bigger than the ten-piece database for Checkers. This is no surprise since Checkers uses only half of the squares of the 8×8 board and has only two types of pieces (man and king). In addition, the rules of Checkers, in which pieces can only move forwards and captures are forced, keep the tree from branching too quickly. For Chess, even a weak solution seems out of reach for the foreseeable future.

Chapter 2

The Basics

> An algorithm is a finite answer to an infinite number of questions.
>
> Stephen Kleene

As we saw in the Prologue, there seems to be some mysterious difference between Eulerian and Hamiltonian paths that makes one of them much harder to find than the other. To put this differently, Hamiltonianness seems to be a qualitatively more subtle property than Eulerianness. Why is one of these problems easy, while the other is like searching for a needle in a haystack?

If we want to understand the nature of these problems, we need to go beyond particular puzzles like the bridges of Königsberg or the edges of the dodecahedron. We need to ask how hard these problems are in general—for cities with any number of bridges, or graphs of any size—and ask how their complexity grows with the size of the city or the graph. To a computer scientist, the "complexity" of a problem is characterized by the amount of computational resources required to solve it, such as how much time or memory we need, and how these requirements grow as the problem gets larger.

In order to measure the complexity of a problem, we will think about the best possible *algorithm*, or computer program, that solves it. However, as we will see, computational complexity theory is not about how to write better programs, any more than physics is about building better spaceships. It is about understanding the underlying structure of different problems, and asking fundamental questions about them—for instance, whether they can be broken into smaller pieces that can be solved independently of each other.

2.1 Problems and Solutions

Let's start this chapter by saying precisely what we mean by a "problem," and what constitutes a "solution." If you have never thought about computational complexity before, our definitions may seem slightly counterintuitive. But as we will see, they give us a framework in which we can clearly state, and begin to answer, the questions posed in the Prologue.

2.1.1 *What's the Problem?*

Any particular instance of a problem, such as the Königsberg bridges, is just a finite puzzle. Once we have solved it, there is no more computation to do. On the other hand, Euler's generalization of this puzzle,

> EULERIAN PATH
> Input: A graph G
> Question: Does there exist an Eulerian path on G?

is a worthy object of study in computational complexity theory—and we honor it as such by writing its name in elegant small capitals. We can think of this as an infinite family of problems, one for each graph G. Alternately, we can think of it as a function that takes a graph as its input, and returns the output "yes" or "no."

To drive this point home, let's consider a somewhat comical example. How computationally complex is Chess? Well, if you mean the standard game played on an 8×8 board, hardly at all. There are only a finite number of possible positions, so we can write a book describing the best possible move in every situation. This book will be somewhat ungainly—it has about 10^{50} pages, making it difficult to fit on the
2.1 shelf—but once it is written, there is nothing left to do.

Now that we have disposed of Chess, let's consider a more interesting problem:

> GENERALIZED CHESS
> Input: A position on an $n \times n$ board, with an arbitrary number of pieces
> Question: Does White have a winning strategy?

Now you're talking! By generalizing to boards of any size, and generalizing the rules appropriately, we have made it impossible for any finite book, no matter how large, to contain a complete solution. To solve this problem, we have to be able to solve Chess problems, not just look things up in books. Moreover, generalizing the problem in this way allows us to consider how quickly the game tree grows, and how much time it takes to explore it, as a function of the board size n.

Another important fact to note is that, when we define a problem, we need to be precise about what input we are given, and what question we are being asked. From this point of view, cities, graphs, games, and so on are neither complex nor simple—specific questions about them are.

These questions are often closely related. For instance, yes-or-no questions like whether or not a Hamiltonian cycle exists are called *decision problems*, while we call the problem of actually finding such a cycle a *search problem* or *function problem*. Problem 1.10 showed that if we can solve the decision version of HAMILTONIAN CYCLE, then we can also solve the search version. But there are also cases where it is easy to show that something exists, but hard to actually find it.

2.1.2 *Solutions and Algorithms*

Now that we've defined what we mean by a problem, what do we mean by a solution? Since there are an infinite number of possible graphs G, a solution to EULERIAN PATH can't consist of a finite list of answers we can just look up. We need a general method, or *algorithm*, which takes a graph as input and returns
2.2 the correct answer as output.

While the notion of an algorithm can be defined precisely, for the time being we will settle for an intuitive definition: namely, a series of elementary computation steps which, if carried out, will produce the desired output. For all intents and purposes, you can think of an algorithm as a computer program written in your favorite programming language: C++, JAVA, HASKELL, or even (ugh) FORTRAN. However, in order to talk about algorithms at a high level, we will express them in "pseudocode." This is a sort of informal programming language, which makes the flow of steps clear without worrying too much about the syntax.

As our first example, let us consider one of the oldest algorithms known. Given two integers a and b, we would like to know their greatest common divisor $\gcd(a,b)$. In particular, we would like to know if a and b are *mutually prime*, meaning that $\gcd(a,b)=1$.

We can solve this problem using Euclid's algorithm, which appears in his *Elements* and dates at least to 300 B.C. It relies on the following fact: d is a common divisor of a and b if and only if it is a common divisor of b and $a \bmod b$. Therefore,

$$\gcd(a,b)=\gcd(b,a \bmod b). \tag{2.1}$$

This gives us an algorithm in which we repeatedly replace the pair (a,b) with the pair $(b,a \bmod b)$. Since the numbers get smaller each time we do this, after a finite number of steps the second number of the pair will be zero. At that point, the gcd is equal to the current value of the first number, since 0 is divisible by anything.

We can express this as a *recursive* algorithm. When called upon to solve the problem $\gcd(a,b)$, it calls 2.3
itself to solve the simpler subproblem $\gcd(b,a \bmod b)$, until it reaches the *base case* $b=0$ which is trivial to solve. Note that if $a<b$ in the original input, the first application of this function will switch their order and call $\gcd(b,a)$. Here is its pseudocode:

```
Euclid(a,b)
begin
    if b = 0 then return a;
    return Euclid(b, a mod b);
end
```

Calling this algorithm on the pair (120, 33), for instance, gives

$$\begin{aligned}
&\texttt{Euclid}(120,33)\\
=&\texttt{Euclid}(33,21)\\
=&\texttt{Euclid}(21,12)\\
=&\texttt{Euclid}(12,9)\\
=&\texttt{Euclid}(9,3)\\
=&\texttt{Euclid}(3,0)\\
=&3.
\end{aligned}$$

Is this a good algorithm? Is it fast or slow? Before we discuss this question, we bring an important character to the stage.

2.1.3 *Meet the Adversary*

> The Creator determines and conceals the aim of the game, and it is never clear whether the purpose of the Adversary is to defeat or assist him in his unfathomable project... But he is concerned, it would seem, in preventing the development of any reasoned scheme in the game.
>
> H. G. Wells, *The Undying Fire*

Computer scientists live in a cruel world, in which a malicious adversary (see Figure 2.1) constructs instances that are as hard as possible, for whatever algorithm we are trying to use. You may have a lovely algorithm that works well in many cases, but beware! If there is any instance that causes it to fail, or to run for a very long time, you can rely on the adversary to find it.

The adversary is there to keep us honest, and force us to make ironclad promises about our algorithms' performance. If we want to promise, for instance, that an algorithm always succeeds within a certain amount of time, this promise holds in *every* case if and only if it holds in the *worst* case—no matter what instance the adversary throws at us.

This is not the only way to think about a problem's hardness. As we will see in Chapter 14, for some problems we can ask how well algorithms work on average, when the instance is chosen randomly rather than by an adversary. But for the most part, computational complexity theorists think of problems as represented by their worst cases. A problem is hard if there *exist* hard instances—it is easy only if *all* its instances are easy.

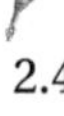

2.4 So, is finding the gcd more like EULERIAN PATH or HAMILTONIAN PATH? Euclid's algorithm stops after a finite number of steps, but how long does it take? We answer this question in the next section. But first, we describe how to ask it in the most meaningful possible way.

2.2 Time, Space, and Scaling

> It is convenient to have a measure of the amount of work involved in a computing process, even though it be a very crude one... We might, for instance, count the number of additions, subtractions, multiplications, divisions, recordings of numbers, and extractions of figures from tables.
>
> Alan M. Turing, 1947

Time and space—the running time of an algorithm and the amount of memory it uses—are two basic computational resources. Others include the number of times we evaluate a complicated function, the number of bits that two people need to send each other, or the number of coins we need to flip. For each of these resources, we can ask how the amount we need *scales* with the size of our problem.

2.2.1 *From Physics to Computer Science*

The notion of scaling is becoming increasingly important in many sciences. Let's look at a classic example from physics. In 1619, the German astronomer and mathematician Johannes Kepler formulated his "harmonic law" of planetary motion, known today as Kepler's third law. Each planet has a "year" or orbital

FIGURE 2.1: The adversary bringing us a really stinky instance.

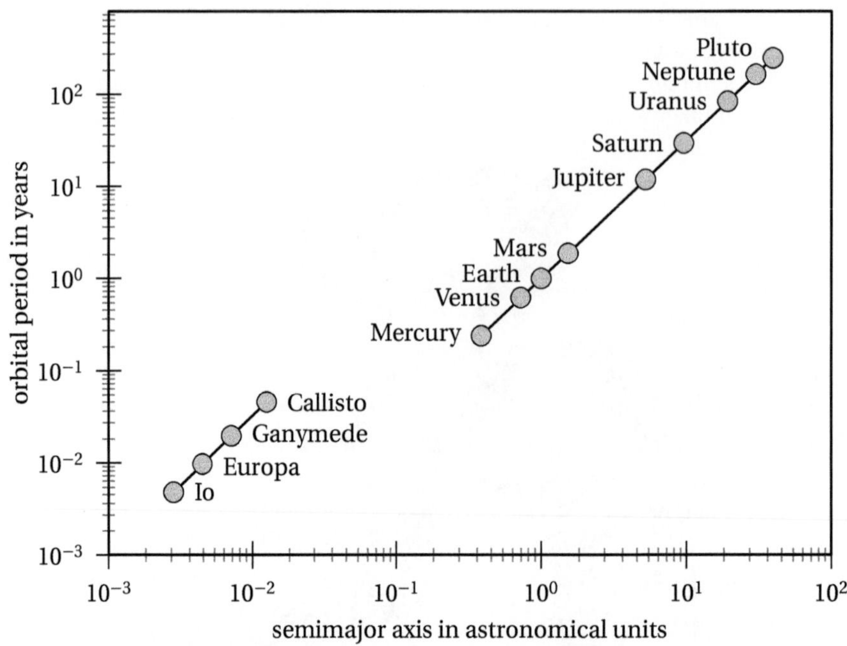

FIGURE 2.2: Scaling in physics: Kepler's harmonic law.

period T, and a distance R from the sun, defined as its semimajor axis since orbits are ellipses. Based on extensive observations, Kepler found that the ratio between T^2 and R^3 is the same for all planets in the solar system. We can write this as

$$T = C \cdot R^{3/2}, \tag{2.2}$$

for some constant C. If we plot T vs. R on a log–log plot as in Figure 2.2, this becomes

$$\log T = C' + \frac{3}{2} \log R,$$

where $C' = \log C$ is another constant. Thus all the planets—including Pluto, for tradition's sake—fall nicely on a straight line with slope $3/2$.

What happens if we plot T vs. R for other systems, such as the four Galilean moons of Jupiter? The constant C in (2.2) depends on the central body's mass, so it varies from system to system. But the exponent $3/2$, and therefore the slope on the log–log plot, remains the same. The scaling relationship (2.2) holds for any planetary system in the universe. It represents a fundamental property of celestial dynamics, which turns out to be Isaac Newton's celebrated law of gravitation. The constants don't matter—what matters is the way that T scales as a function of R.

In the same way, when we ask how the running time of an algorithm scales with the problem size n, we are not interested in whether your computer is twice as fast as mine. This changes the constant in front of the running time, but what we are interested in is how your running time, or mine, changes when n changes.

What do we mean by the "size" n of a problem? For problems like EULERIAN PATH, we can think of n as the number of vertices or edges in the input graph—for instance, the number of bridges or riverbanks in the city. For problems that involve integers, such as computing the greatest common divisor, n is the number of bits or digits it takes to express these integers. In general, the size of a problem instance is the amount of information I need to give you—say, the length of the email I would need to send—in order to describe it to you.

2.2.2 *Euclidean Scaling*

Let's look at a concrete example. How does the running time of Euclid's algorithm scale with n? Here n is the total number of digits of the inputs a and b. Ignoring the factor of 2, let's assume that both a and b are n-digit numbers.

The time it takes to calculate $a \bmod b$ depends on what method we use to divide a by b. Let's avoid this detail for now, and simply ask how many of these divisions we need, i.e., how many times Euclid's algorithm will call itself recursively before reaching the base case $b = 0$ and returning the result. The following exercise shows that we need at most a linear number of divisions.

Exercise 2.1 *Show that if $a \geq b$, then $a \bmod b < a/2$. Conclude from this that the number of divisions that Euclid's algorithm performs is at most $2\log_2 a$.*

If a has n digits, then $2\log_2 a \leq Cn$ for some constant C. Measuring n in bits instead of digits, or defining n as the total number of digits of a and b, just changes C to a different constant. But the number of divisions is always linear in n.

Following Kepler, we can check this linear scaling by making observations. In Figure 2.3 we plot the worst-case and average number of divisions as a function of n, and the relationship is clearly linear. If you are mathematically inclined, you may enjoy Problems 2.6, 2.7, and 2.8, where we calculate the slopes of these lines analytically. In particular, it turns out that the worst case occurs when a and b are successive Fibonacci numbers, which as Problem 2.4 discusses are well-known in the field of rabbit breeding.

Now that we know that the number of divisions in Euclid's algorithm is linear in n, what is its total running time? In a single step, a computer can perform arithmetic operations on integers of some fixed size, such as 64 bits. But if n is larger than this, the time it takes to divide one n-digit number by another grows with n. As Problem 2.11 shows, if we use the classic technique of long division, the total running time of Euclid's algorithm scales as n^2, growing polynomially as a function of n.

Let's dwell for a moment on the fact that the size n of a problem involving an integer a is the number of bits or digits of a, which is roughly $\log a$, rather than a itself. For instance, as the following exercise shows, the problem FACTORING would be easy if the size of the input were the number itself:

Exercise 2.2 *Write down an algorithm that finds the prime factorization of an integer a, whose running time is polynomial as a function of a (as opposed to the number of digits in a).*

However, we believe that FACTORING cannot be solved in an amount of time that is polynomial as a function of $n = \log a$, unless you have a quantum computer.

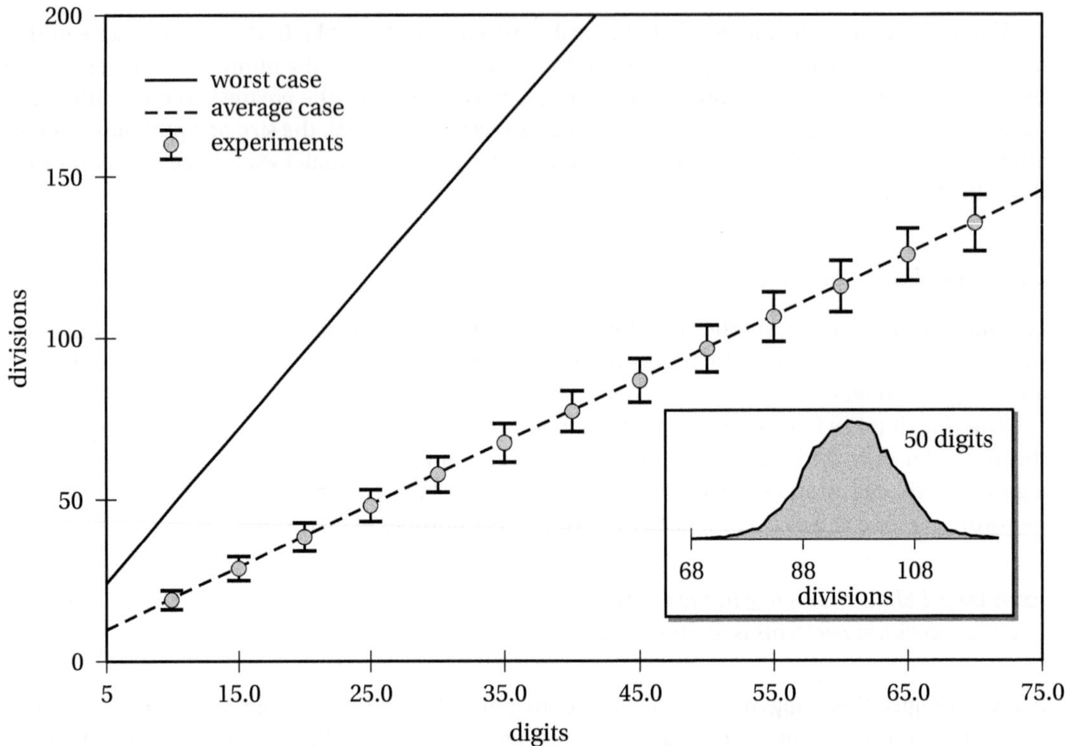

FIGURE 2.3: Scaling in computer science: the number of divisions done by Euclid's algorithm grows linearly with n when given n-digit inputs. In the inset, the typical distribution of a set of random inputs.

2.2.3 *Asymptotics*

In order to speak clearly about scaling, we will use asymptotic notation. This notation lets us focus on the qualitative growth of a function of n, and ignores the constant in front of it.

For instance, if the running time $f(n)$ of an algorithm grows quadratically, we say that $f(n) = \Theta(n^2)$. If it grows at most quadratically, but it might grow more slowly, we say that $f(n) = O(n^2)$. If it definitely grows less than quadratically, such as n^c for some $c < 2$, we say that $f(n) = o(n^2)$. Buying a faster computer reduces the constant hidden in Θ or O, but it doesn't change how $f(n)$ grows when n increases.

We define each of these symbols in terms of the limit as $n \to \infty$ of the ratio $f(n)/g(n)$ where $g(n)$ is whatever function we are comparing $f(n)$ with. Thus $f(n) = o(n^2)$ means that $\lim_{n\to\infty} f(n)/n^2 = 0$, while $f(n) = \Theta(n^2)$ means that $f(n)/n^2$ tends neither to zero nor to infinity—typically, that it converges to some constant C. We summarize all this in Table 2.1. In Appendix A.1 you can find formal definitions and exercises involving these symbols.

Exercise 2.3 *When we say that $f(n) = O(\log n)$, why don't we have to specify the base of the logarithm?*

These definitions focus on the behavior of $f(n)$ in the limit $n \to \infty$. This is for good reason. We don't really care how hard HAMILTONIAN PATH, say, is for small graphs. What matters to us is whether there is an

symbol	$C = \lim_{n\to\infty} \frac{f(n)}{g(n)}$	roughly speaking...
$f(n) = O(g(n))$	$C < \infty$	"$f \le g$"
$f(n) = \Omega(g(n))$	$C > 0$	"$f \ge g$"
$f(n) = \Theta(g(n))$	$0 < C < \infty$	"$f = g$"
$f(n) = o(g(n))$	$C = 0$	"$f < g$"
$f(n) = \omega(g(n))$	$C = \infty$	"$f > g$"

TABLE 2.1: A summary of asymptotic notation.

efficient algorithm that works for all n, and the difference between such an algorithm and an exhaustive search shows up when n is large. Even Kepler's law, which we can write as $T = \Theta(R^{3/2})$, only holds when R is large enough, since at small distances there are corrections due to the curvature of spacetime.

Armed with this notation, we can say that Euclid's algorithm performs $\Theta(n)$ divisions, and that its total running time is $O(n^2)$. Thus we can calculate the greatest common divisor of two n-digit numbers in polynomial time, i.e., in time $O(n^c)$ for a constant c. But what if Euclid's algorithm isn't the fastest method? What can we say about a problem's *intrinsic* complexity, as opposed to the running time of a particular algorithm? The next section will address this question with a problem, and an algorithm, that we all learned in grade school.

2.3 Intrinsic Complexity

For whichever resource we are interested in bounding—time, memory, and so on—we define the *intrinsic complexity* of a problem as the complexity of the *most efficient* algorithm that solves it. If we have an algorithm in hand, its existence provides an upper bound on the problem's complexity—but it is usually very hard to know whether we have the most efficient algorithm. One of the reasons computer science is such an exciting field is that, every once in a while, someone achieves an algorithmic breakthrough, and the intrinsic complexity of a problem turns out to be much less than we thought it was.

To illustrate this point, let's discuss the complexity of multiplying integers. Let $T(n)$ denote the time required to multiply two integers x and y which have n digits each. As Figure 2.4 shows, the algorithm we learned in grade school takes time $T(n) = \Theta(n^2)$, growing quadratically as a function of n.

This algorithm is so natural that it is hard to believe that one can do better, but in fact one can. One of the most classic ideas in algorithms is to *divide and conquer*—to break a problem into pieces, solve each piece recursively, and combine their answers to get the answer to the entire problem. In this case, we can break the n-digit integers x and y into pairs of $n/2$-digit integers as follows:

$$x = 10^{n/2} a + b \quad \text{and} \quad y = 10^{n/2} c + d\,.$$

Here a and b are the *high-order* and *low-order* parts of x, i.e., the first and second halves of x's digit sequence, and c and d are similarly the parts of y. Then

$$xy = 10^n ac + 10^{n/2}(ad + bc) + bd\,. \tag{2.3}$$

Of course, on a digital computer we would operate in binary instead of decimal, writing $x = 2^{n/2} a + b$ and so on, but the principle remains the same.

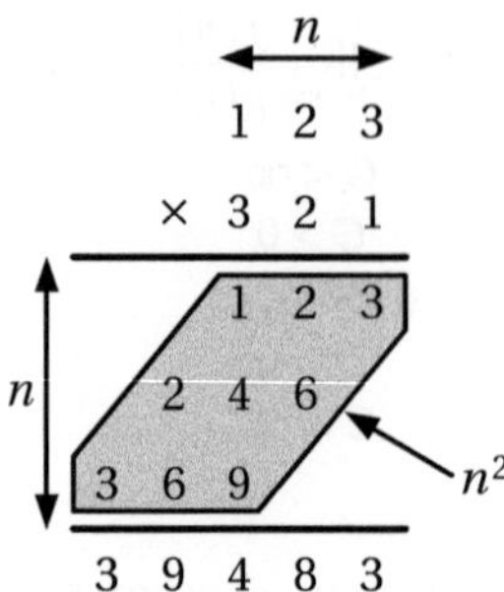

FIGURE 2.4: The grade-school algorithm for multiplication, which takes $\Theta(n^2)$ time to multiply n-digit integers.

This approach lets us reduce the problem of multiplying n-digit integers to that of multiplying several pairs of $n/2$-digit integers. We then reduce this problem to that of multiplying $n/4$-digit integers, and so on, until we get to integers so small that we can look up their products in a table. We assume for simplicity that n is even at each stage, but rounding $n/2$ up or down makes no difference when n is large.

What running time does this approach give us? If we use (2.3) as our strategy, we calculate four products, namely ac, ad, bc, and bd. Adding these products together is much easier, since the grade-school method of adding two n-digit integers takes just $O(n)$ time. Multiplying an integer by 10^n or $10^{n/2}$ is also easy, since all we have to do is shift its digits to the left and add n or $n/2$ zeros. The running time $T(n)$ then obeys the equation

$$T(n) = 4T(n/2) + O(n). \tag{2.4}$$

If $T(n)$ scales faster than linearly, then for large n we can ignore the $O(n)$ term. Then the running time is dominated by the four multiplications, and it essentially quadruples whenever n doubles. But as Problem 2.13 shows, this means that it grows quadratically, $T(n) = \Theta(n^2)$, just like the grade-school method. So we need another idea.

The key observation is that we don't actually need to do four multiplications. Specifically, we don't need ad and bc separately—we only need their sum. Now note that

$$(a+b)(c+d) - ac - bd = ad + bc. \tag{2.5}$$

Therefore, if we calculate $(a+b)(c+d)$ along with ac and bd, which we need anyway, we can obtain $ad + bc$ by subtraction, which like addition takes just $\Theta(n)$ time. Using this trick changes (2.4) to

$$T(n) = 3T(n/2) + O(n). \tag{2.6}$$

Now the running time only triples when n doubles, and using Problem 2.13 gives

$$T(n) = \Theta(n^\alpha) \text{ where } \alpha = \log_2 3 \approx 1.585.$$

So, we have tightened our upper bound on the complexity of multiplication from $O(n^2)$ to $O(n^{1.585})$.

Is this the best we can do? To be more precise, what is the smallest α for which we can multiply n-digit integers in $O(n^\alpha)$ time? It turns out that α can be arbitrarily close to 1. In other words, there are algorithms whose running time is less than $O(n^{1+\varepsilon})$ for any constant $\varepsilon > 0$. On the other hand, we have a lower bound of $T(n) = \Omega(n)$ for the trivial reason that it takes that long just to read the inputs x and y. These upper and lower bounds almost match, showing that the intrinsic complexity of multiplication is essentially linear in n. Thus multiplication turns out to be much less complex than the grade-school algorithm would suggest.

2.5

2.4 The Importance of Being Polynomial

> For practical purposes the difference between polynomial and exponential order is often more crucial than the difference between finite and non-finite.
>
> Jack Edmonds, 1965

Finding an algorithm that multiplies n-digit integers in $O(n^{1.585})$ time, instead of $O(n^2)$, reveals something about the complexity of multiplication. It is also of practical interest. If $n = 10^6$, for instance, this improves the running time by a factor of about 300 if the constants in the Os are the same.

However, the most basic distinction we will draw in computational complexity is between polynomial functions of n—that is, n raised to some constant—and exponential ones. In this section, we will discuss why this distinction is so important, and why it is so robust with respect to changes in our definition of computation.

2.4.1 *Until the End of the World*

One way to illustrate the difference between polynomials and exponentials is to think about how the size of the problems we can handle increases as our computing technology improves. Moore's Law (no relation) is the empirical observation that basic measures of computing technology, such as the density of transistors on a chip, are improving exponentially with the passage of time.

A common form of this law—although not Moore's original claim—is that processing speed doubles 2.7
every two years. If the running time of my algorithm is $\Theta(n)$, doubling my speed also doubles the size n of problems I can solve in, say, one week. But if the running time grows as $\Theta(2^n)$, the doubling the speed just increases n by 1.

Thus whether the running time is polynomial or exponential makes an enormous difference in the size of problems we can solve, now and for the foreseeable future. We illustrate this in Figure 2.5, where we compare various polynomials with 2^n and $n!$. In the latter two cases, solving instances of size $n = 100$ or even $n = 20$ would take longer than the age of the universe. A running time that scales exponentially implies a harsh bound on the problems we can ever solve—even if our project deadline is as far away in the future as the Big Bang is in the past.

Exercise 2.4 *Suppose we can currently solve a problem of size n in a week. If the speed of our computer doubles every two years, what size problem will we be able to solve in a week four years from now, if the running time of our algorithm scales as* $\log_2 n$, $\sqrt{n}$, n, n^2, 2^n, *or* 4^n?

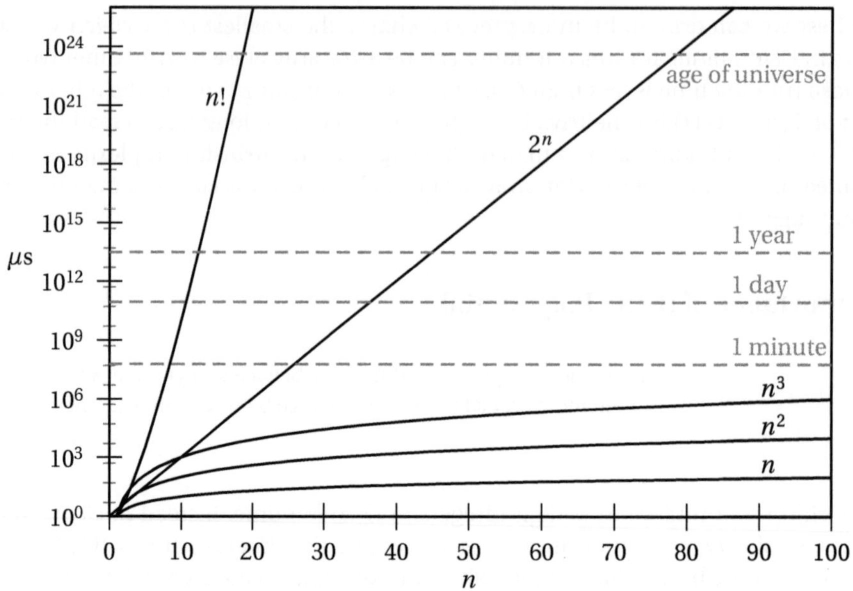

FIGURE 2.5: Running times of algorithms as a function of the size n. We assume that each one can solve an instance of size $n = 1$ in one microsecond. Note that the time axis is logarithmic.

```
Euler
input: a graph G = (V, E)
output: "yes" if G is Eulerian, and "no" otherwise
begin
    y := 0 ;
    for all v ∈ V do
        if deg(v) is odd then y := y + 1;
        if y > 2 then return "no";
    end
    return "yes"
end
```

FIGURE 2.6: Euler's algorithm for EULERIAN PATH. The variable y counts the number of odd-degree vertices.

2.4.2 Details, and Why they don't Matter

In the Prologue we saw that Euler's approach to EULERIAN PATH is much more efficient than exhaustive search. But how does the running time of the resulting algorithm scale with the size of the graph? It turns out that a precise answer to this question depends on many details. We will discuss just enough of these details to convince you that we can and should ignore them in our quest for a fundamental understanding of computational complexity.

In Figure 2.6 we translate Euler's Theorem into an algorithm, and express it in pseudocode. Quantifying the running time of this algorithm is not quite as trivial as it seems. To get us started, how many times will the **for** loop run? In the worst case, all vertices—or all but the last two—have even degree. Thus the **for** loop will, in general, run $O(|V|)$ times.

Next we need to specify how we measure the running time. The physical time, measured in seconds, will vary wildly from computer to computer. So instead, we measure time as the number of *elementary steps* that our algorithm performs. There is some ambiguity in what we consider "elementary," but let us assume for now that assignments like $y := 0$, arithmetical operations like $y + 1$, and evaluations of inequalities like $y > 2$ are all elementary and take constant time.

The next question is how long it takes to determine the degree of a vertex, which we denote $\deg(v)$. Clearly this is not very hard—we just have to count the number of neighbors v has. But analyzing it precisely depends on the format in which we are given the input graph G.

One common format is the *adjacency matrix*. This the $|V| \times |V|$ matrix A such that

$$A_{ij} = \begin{cases} 1 & \text{if } (i,j) \in E \text{, i.e., there is an edge from vertex } i \text{ to vertex } j \\ 0 & \text{otherwise}. \end{cases}$$

We could give this matrix directly as a $|V| \times |V|$ array of 0s and 1s. However, if G is a *sparse* graph that only has a few edges, then there are just a few pairs i, j such that $A_{ij} = 1$. As Problem 2.18 shows, we can then describe G more efficiently by giving a list of these pairs, or equivalently a list of edges.

Determining $\deg(v)$ takes different numbers of steps depending on which of these formats we use. Given the entire adjacency matrix, we would use a **for** loop with i ranging from 1 to $|V|$, and increment a counter each time $A_{vi} = 1$. Given a list of edges (i, j), we could scan the entire list and increment a counter each time either $i = v$ or $j = v$, and so on.

However, it is not even obvious that we can carry out instructions like "check if $A_{ij} = 1$" in a single elementary step. If the input is stored on a magnetic tape—an ancient memory technology which our reader is surely too young to remember—it might take a long time to roll the tape to the location of the data we wish to read. Among theoretical models of computation, a Turing machine, which we will discuss in Chapter 7, takes t steps to move to the tth location on its tape, while a machine with random access memory (RAM) can access any location in its memory in a single step. Thus moving from one of these models to another could change our running time considerably.

Finally, we need to agree how we specify the *size* of an instance. In general, this is the number n of bits it takes to describe it—if you like, the length of the email I would have to send you to tell you about it. This depends on our choice of input format, and n can be smaller or larger depending on whether this format is efficient or inefficient.

All these considerations make it difficult to quantify the running time precisely, and how it scales with the input size, without going into a great deal of detail about our input format, the particular implementation of our algorithm, and the type of machine on which we run our program. These are worthy engineering questions, but the goal of computational complexity theory is to take a larger view, and draw deep qualitative distinctions between problems. So, rather than studying the art of grinding lenses and mirrors, let us turn our attention to the stars.

2.4.3 Complexity Classes

As we saw in the previous section, one of the most basic questions we can ask about a problem is whether it can be solved in polynomial time as a function of its size. Let's consider the set of all problems with this property:

> P is the class of problems for which an algorithm exists that solves instances of size n in time $O(n^c)$ for some constant c.

Conversely, a problem is outside P if *no algorithm exists* that solves it in polynomial time—for instance, if the most efficient algorithm takes exponential time $2^{\varepsilon n}$ for some $\varepsilon > 0$.

P is our first example of a *complexity class*—a class of problems for which a certain kind of algorithm exists. We have defined it here so that it includes both decision problems, such as "does there exist an Eulerian path," and function problems, such as "construct an Eulerian path." Later on, many of our complexity classes will consist just of decision problems, which demand a yes-or-no answer.

More generally, for any function $f(n)$ we can define TIME($f(n)$) as follows:

> TIME($f(n)$) is the class of problems for which an algorithm exists that solves instances of size n in time $O(f(n))$.

In particular, P contains TIME(n), TIME(n^2), and so on, as well as noninteger exponents like TIME($n^{\log_2 3}$) which we met in Section 2.3. Formally,

$$\mathsf{P} = \bigcup_{c>0} \mathsf{TIME}(n^c).$$

The essential point is that we allow any exponent that is *constant* with respect to n. Exponents that grow as n grows, like $n^{\log n}$, are excluded from P. Throughout the book, we will use poly(n) as a shorthand for "$O(n^c)$ for some constant c," or equivalently for $n^{O(1)}$. In that case, we can write

$$\mathsf{P} = \mathsf{TIME}(\mathrm{poly}(n)).$$

If we wish to entertain running times that are exponentially large or even greater, we can define EXP = TIME($2^{\mathrm{poly}(n)}$), EXPEXP = TIME($2^{2^{\mathrm{poly}(n)}}$), and so on. This gives us a hierarchy of complexity classes, in which
2.8 the amount of computation we can do becomes increasingly astronomical:

$$\mathsf{P} \subseteq \mathsf{EXP} \subseteq \mathsf{EXPEXP} \subseteq \cdots$$

But back down to earth. Why is the question of whether a problem can be solved in polynomial time or not so fundamental? The beauty of the definition of P is that it is extremely robust to changes in how we measure running time, and what model of computation we use. For instance, suppose we change our definition of "elementary step" so that we think of multiplying two integers as elementary. As long as they have only a polynomial number of digits, each multiplication takes polynomial time anyway, so this at most changes our running time from a polynomial to a smaller polynomial.

2.9 Similarly, going from a Turing machine to a RAM, or even a massively parallel computer—as long as it has only a polynomial number of processors—saves at most polynomial time. The one model of computation that seems to break this rule is a quantum computer, which we discuss in Chapter 15. So, to be clear, we define P as the class of problems that *classical* computers, like the ones we have on our desks and in our laps today, can solve in polynomial time.

The class P is also robust with respect to most input formats. Any reasonable format for a graph, for example, has size n which is polynomial in the number of vertices, as long as it isn't a multigraph where some pairs of vertices have many edges between them. Therefore, we can say that a graph problem is in P if the running time is polynomial in $|V|$, and we will often simply identify n with $|V|$.

However, if we change our input format so drastically that n becomes exponentially larger or smaller, the computational complexity of a problem can change quite a bit. For instance, we will occasionally represent an integer a in *unary* instead of binary—that is, as a string of a ones. In that case, the size of the input is $n = a$ instead of $n = \log_2 a$. Exercise 2.2 shows that if we encode the input in this way, FACTORING can be solved in polynomial time.

Of course, this is just a consequence of the fact that we measure complexity as a function of the input size. If we make the input larger by encoding it inefficiently, the problem becomes "easier" in an artificial way. We will occasionally define problems with unary notation when we want some input parameter to be polynomial in n. But for the most part, if we want to understand the true complexity of a problem, it makes sense to provide the input as efficiently as possible.

Finally, P is robust with respect to most details of how we implement an algorithm. Using clever data structures, such as storing an ordered list in a binary tree instead of in an array, typically reduces the running time by a factor of n or n^2. This is an enormous practical improvement, but it still just changes one polynomial to another with a smaller power of n.

The fact that P remains unchanged even if we alter the details of our computer, our input format, or how we implement our algorithm, suggests that being in P is a *fundamental property of a problem*, rather than a property of how we humans go about solving it. In other words, the question of whether HAMILTONIAN PATH is in P or not is a mathematical question about the nature of Hamiltonian paths, not a subjective question about our own abilities to compute. There is no reason why computational complexity theory couldn't have been invented and studied thousands of years ago, and indeed there are glimmers of it here and there throughout history. 2.10

2.5 Tractability and Mathematical Insight

It is often said that P is the set of *tractable* problems, namely those which can be solved in a reasonable amount of time. While a running time of, say, $O(n^{10})$ is impractical for any interesting value of n, we encounter such large powers of n very rarely. The first theoretical results proving that a problem is in P sometimes give an algorithm of this sort, but within a few years these algorithms are usually improved to $O(n^3)$ or $O(n^4)$ at most.

Of course, even a running time of $O(n^3)$ is impractical if $n = 10^6$—for instance, if we are trying to analyze an online social network with 10^6 nodes. Now that fields like genomics and astrophysics collect vast amounts of data, stored on stacks of optical disks, containing far more information than your computer can hold at one time, some argue that even linear-time algorithms are too slow. This has given rise to a new field of *sublinear* algorithms, which examine only a small fraction of their input.

But for us, and for computational complexity theorists, P is not so much about tractability as it is about mathematical insight into a problem's structure. Both EULERIAN PATH and HAMILTONIAN PATH can be solved in exponential time by exhaustive search, but there is something different about EULERIAN PATH that yields a polynomial-time algorithm. Similarly, when we learned in 2004 that the problem of telling whether an n-digit number is prime is in P, we gained a fundamental insight into the nature of PRIMALITY, even though the resulting algorithm (which we describe in Chapter 10) is not very practical.

The difference between polynomial and exponential time is one of kind, not of degree. When we ask whether a problem is in P or not, we are no longer just computer users who want to know whether we can finish a calculation in time to meet a deadline. We are theorists who seek a deep understanding of why some problems are qualitatively easier, or harder, than others.

2.11

Problems

> If there is a problem you can't solve, then there is an easier problem you can solve: find it.
>
> George Pólya, *How to Solve It*

2.1 Upgrades. The research lab of Prof. Flush is well-funded, and they regularly upgrade their equipment. Brilliant Pebble, a graduate student, has to run a rather large simulation. Given that the speed of her computer doubles every two years, if the running time of this simulation exceeds a certain T, she will actually graduate earlier if she waits for the next upgrade to start her program. What is T?

2.2 Euclid extended. Euclid's algorithm finds the greatest common divisor $\gcd(a,b)$ of integers a and b. Show that with a little extra bookkeeping it can also find (possibly negative) integers x and y such that

$$ax+by=\gcd(a,b). \tag{2.7}$$

Now assume that $b<a$ and that they are mutually prime. Show how to calculate the multiplicative inverse of b modulo a, i.e., the y such that $1\le y<b$ and $by\equiv 1 \bmod a$.

Hint: the standard algorithm computes the remainder $r=a \bmod b$. The extended version also computes the quotient q in $a=qb+r$. Keep track of the quotients at the various levels of recursion.

2.3 Geometrical subtraction. Euclid's original algorithm calculated $a \bmod b$ by repeatedly subtracting b from a (by marking a line of length b off a line of length a) until the remainder is less than b. If a and b have n or fewer digits, show that this method can take exponential time as a function of n.

2.4 Fibonacci's rabbits. Suppose that I start my first year as a rabbit farmer with one baby rabbit. It takes a year for a baby rabbit to mature, and mature rabbits produce one baby per year. (Note that rabbits are immortal and reproduce asexually; elsewhere on the farm there are spherical cows.) If F_ℓ is the rabbit population in the ℓth year, show that the first few values of F_ℓ are $1,1,2,3,5,8,\ldots$ and that in general, these obey the equation

$$F_\ell=F_{\ell-1}+F_{\ell-2}, \tag{2.8}$$

with the initial values $F_1=F_2=1$. These are called the *Fibonacci numbers*. Show that they grow exponentially, as rabbits are known to do. Specifically, show that

$$F_\ell=\Theta(\varphi^\ell)$$

where φ is the "golden ratio,"

$$\varphi=\frac{1+\sqrt{5}}{2}=1.618...$$

In other words, find constants A and B for which you can prove by induction on ℓ that for all $\ell \geq 1$,

$$A\varphi^\ell \leq F_\ell \leq B\varphi^\ell .$$

Hint: φ is the largest root of the quadratic equation $\varphi^2 - \varphi - 1 = 0$. Equivalently, it is the unique positive number such that

$$\frac{\varphi}{1} = \frac{1+\varphi}{\varphi} .$$

2.5 Exponential growth, polynomial time. Using Problem 2.4, show that the problem of telling whether an n-digit number x is a Fibonacci number is in P. Hint: how many Fibonacci numbers are there between 1 and 10^n?

2.6 Euclid at his worst. Let's derive the worst-case running time of Euclid's algorithm. First, prove that the number of divisions is maximized when a and b are two adjacent Fibonacci numbers. Hint: using the fact that the smallest a such that $a \bmod b = c$ is $b + c$, work backwards from the base case and show that, if $b \leq a$ and finding $\gcd(a, b)$ takes ℓ divisions, then $a \geq F_{\ell+1}$ and $b \geq F_\ell$.

Now suppose that a and b each have n digits. Use Problem 2.4 to show that the number of divisions that Euclid's algorithm performs is at most $\log_\varphi a = C_{\text{worst}} n + O(1)$ where

$$C_{\text{worst}} = \frac{1}{\log_{10} \varphi} \approx 4.785 .$$

This is the slope of the upper line in Figure 2.3.

2.7 Euclid and Gauss. Now let's derive the average-case running time of Euclid's algorithm. First, let x denote the ratio b/a. Show that each step of the algorithm updates x as follows,

$$x = g(x) \text{ where } g(x) = \frac{1}{x} \bmod 1 . \tag{2.9}$$

Here by a number mod 1, we mean its fractional part. For instance, $\pi \bmod 1 = 0.14159...$ This function $g(x)$ is called the *Gauss map*, and its graph is shown in Figure 2.7. It also plays an important role in the theory of continued fractions: as we iterate (2.9), the integers $\lfloor 1/x \rfloor = (1/x) - (1/x \bmod 1)$ label the different "branches" of $g(x)$ we land on, and give the continued fraction series of x.

If we start with a random value of x and apply the Gauss map many times, we would expect x to be distributed according to some probability distribution $P(x)$. This distribution must be *stationary*: that is, it must remain the same when x is updated according to (2.9). This means that $P(x)$ must equal a sum of probabilities from all the values that x could have had on the previous step, adjusted to take the derivative of g into account:

$$P(x) = \sum_{y:g(y)=x} \frac{P(y)}{|g'(y)|} .$$

Show that one such distribution—which turns out to be essentially unique—is

$$P(x) = \frac{1}{\ln 2} \left(\frac{1}{x+1} \right) .$$

Of course, we haven't shown that the probability distribution of x converges to this stationary distribution. Proving this requires much more sophisticated arguments.

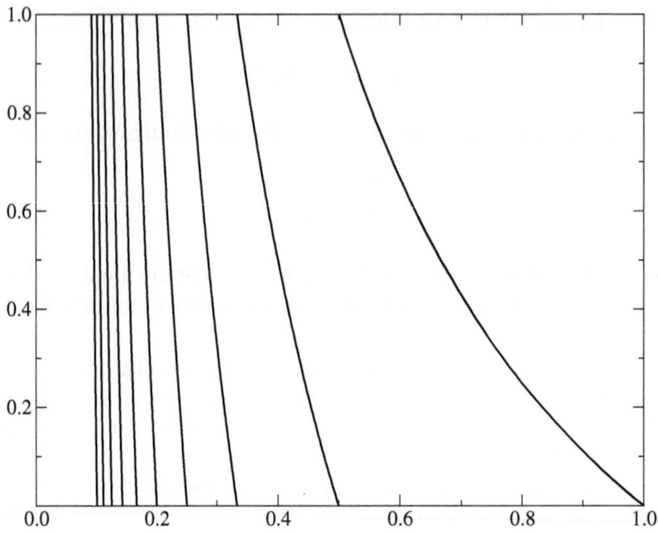

FIGURE 2.7: The Gauss map $g(x) = (1/x) \bmod 1$.

2.8 Euclid on average. Continuing from the previous problem, argue that the average number of divisions Euclid's algorithm does when given n-digit numbers grows as $C_{\text{avg}} n$ where

$$C_{\text{avg}} = -\frac{1}{\mathbb{E}[\log_{10} x]}.$$

Here $\mathbb{E}[\cdot]$ denotes the expectation given the distribution $P(x)$,

$$\mathbb{E}[\log_{10} x] = \int_0^1 P(x) \log_{10} x \, dx.$$

Evaluate this integral to obtain

$$C_{\text{avg}} = \frac{12 \cdot \ln 2 \cdot \ln 10}{\pi^2} \approx 1.941.$$

This is the slope of the lower line in Figure 2.3. Clearly it fits the data very well.

2.9 The golden ratio again. To connect Problems 2.6 and 2.7, show that $1/\varphi = 0.618...$ is the largest fixed point of the Gauss map. In other words, it is the largest x such that $g(x) = x$. This corresponds to the fact that if a and b are successive Fibonacci numbers, the ratio $x = b/a$ stays roughly constant as Euclid's algorithm proceeds. Then show that, since $\lfloor \varphi \rfloor = 1$, the golden ratio's continued fraction expansion is

$$\varphi = 1 + \cfrac{1}{1 + \cfrac{1}{1 + \cfrac{1}{\ddots}}}$$

```
        3 1 4
1 1 3 ) 3 5 5 0 0
        3 3 9
          1 6 0
          1 1 3
            4 7 0
            4 5 2
              1 8
```

FIGURE 2.8: Using long division to calculate $a \bmod b$. In this case $a = 35500$, $b = 113$, $\lfloor a/b \rfloor = 314$, and $a \bmod b = 18$.

Finally, show that cutting off this expansion after ℓ steps gives an approximation for φ as the ratio between two successive Fibonacci numbers,

$$\varphi \approx \frac{F_{\ell+1}}{F_\ell}.$$

2.10 Euclid and Fibonacci. Use Euclid's algorithm to show that any two successive Fibonacci numbers are mutually prime. Then, generalize this to the following beautiful formula:

$$\gcd(F_a, F_b) = F_{\gcd(a,b)}.$$

Note that we use the convention that $F_1 = F_2 = 1$, so $F_0 = 0$. Hint: you might want to look ahead at Problem 3.19.

2.11 Long division. Show that if a has n digits, $b \le a$, and the integer part $\lfloor a/b \rfloor$ of their ratio has m digits, then we can obtain $a \bmod b$ in $O(nm)$ time. Hint: consider the example of long division shown in Figure 2.8, where $n = 5$ and $m = 3$. If you are too young to have been taught long division in school, humble yourself and ask your elders to teach you this ancient and beautiful art.

Now consider the series of divisions that Euclid's algorithm does, and show that the total time taken by these divisions is $O(n^2)$.

2.12 Divide and conquer. Show that for any constants $a, b > 0$, the recursive equation

$$T(n) = a\,T(n/b) \tag{2.10}$$

has the exact solution

$$T(n) = C n^{\log_b a},$$

where $C = T(1)$ is given by the base case.

2.13 Divide and a little more conquering. Now show that if $a > b > 1$ and we add a linear term to the right-hand side, giving

$$T(n) = a\,T(n/b) + Cn,$$

then $T(n)$ is still $T(n) = \Theta(n^{\log_b a})$. In other words, prove by induction on n that there are constants A and B, depending on C and the initial conditions $T(1)$, such that

$$A n^{\log_b a} \le T(n) \le B n^{\log_b a}$$

for all $n \ge 1$. Feel free to assume that n is a power of b.

2.14 Toom's algorithm, part 1. After reading the divide-and-conquer algorithm for multiplying n-digit integers in Section 2.3, the reader might well ask whether dividing these integers into more than two pieces might yield an even better algorithm. Indeed it does!

To design a more general divide-and-conquer algorithm, let's begin by thinking of integers as polynomials. If x is an n-digit integer written in base b, and we wish to divide it into r pieces, we will think of it as a polynomial $P(z)$ of degree $r-1$, where $z = b^{n/r}$. For instance, if $x = 314\,159\,265$ and $r = 3$, then $x = P(10^3)$ where

$$P(z) = 314z^2 + 159z + 265\,.$$

Now, if $x = P(z)$ and $y = Q(z)$, their product xy is $R(z)$, where $R(z) = P(z)Q(z)$ is a polynomial of degree $2r-2$. In order to find R's coefficients, it suffices to sample it at $2r-1$ values of z, say at integers ranging from $-r+1$ to $r-1$. The coefficients of R are then linear combinations of these samples.

As a concrete example, suppose that $r = 2$, and that $P(z) = az + b$ and $Q(z) = cz + d$. Then $R(z) = Az^2 + Bz + C$ is quadratic, and we can find A, B, and C from three samples, $R(-1)$, $R(0)$ and $R(1)$. Write the 3×3 matrix that turns $(R(-1), R(0), R(+1))$ into (A, B, C), and show that the resulting algorithm is essentially identical to the one described in the text.

2.15 Toom's algorithm, part 2. Now let's generalize the approach of the previous problem to larger values of r. Each sample of $R(z)$ requires us to multiply two numbers, $P(z)$ and $Q(z)$, which have essentially n/r digits each. If we ignore the time it takes to multiply by M, the running time of this algorithm is

$$T(n) = (2r+1)\,T(n/r)$$

which by Problem 2.12 has the solution

$$T(n) = \Theta\left(n^{\alpha}\right) \text{ where } \alpha = \frac{\log 2r-1}{\log r}\,. \tag{2.11}$$

Show that α tends to 1 as r tends to infinity, so by taking r large enough we can achieve a running time of $O(n^{1+\varepsilon})$ for arbitrarily small ε.

2.16 Toom's algorithm, part 3. Let's continue from the previous problem. As r grows, the constant hidden in the Θ of (2.11) grows too, since we have to multiply the vector of samples by a matrix M of size $O(r^2)$. This suggests that the optimal value of r is some function of n. Show that, in the absence of any information about M's structure, a nearly optimal choice of r is

$$r = 2^{\sqrt{\log n}}\,.$$

Then show that the running time of the resulting algorithm is

$$T(n) = n\,2^{O(\sqrt{\log n})}$$

which is less than $n^{1+\varepsilon}$ for any $\varepsilon > 0$.

2.17 Fast matrix multiplication. Suppose we need to compute the product C of two $n \times n$ matrices A, B. Show that the naive algorithm for this takes $\theta(n^3)$ individual multiplications. However, we can do better, by again using a divide-and-conquer approach. Write

$$A = \begin{pmatrix} A_{1,1} & A_{1,2} \\ A_{2,1} & A_{2,2} \end{pmatrix},\; B = \begin{pmatrix} B_{1,1} & B_{1,2} \\ B_{2,1} & B_{2,2} \end{pmatrix},\; \text{and } C = \begin{pmatrix} C_{1,1} & C_{1,2} \\ C_{2,1} & C_{2,2} \end{pmatrix},$$

where $A_{1,1}$ and so on are $n/2 \times n/2$ matrices. Now define the following seven $n/2 \times n/2$ matrices,

$$\begin{aligned}
M_1 &= (A_{1,1} + A_{2,2})(B_{1,1} + B_{2,2}) \\
M_2 &= (A_{2,1} + A_{2,2})B_{1,1} \\
M_3 &= A_{1,1}(B_{1,2} - B_{2,2}) \\
M_4 &= A_{2,2}(B_{2,1} - B_{1,1}) \\
M_5 &= (A_{1,1} + A_{1,2})B_{2,2} \\
M_6 &= (A_{2,1} - A_{1,1})(B_{1,1} + B_{1,2}) \\
M_7 &= (A_{1,2} - A_{2,2})(B_{2,1} + B_{2,2}) .
\end{aligned}$$

Then show that C is given by

$$\begin{aligned}
C_{1,1} &= M_1 + M_4 - M_5 + M_7 \\
C_{1,2} &= M_3 + M_5 \\
C_{2,1} &= M_2 + M_4 \\
C_{2,2} &= M_1 - M_2 + M_3 + M_6 .
\end{aligned}$$

Again using the fact that the cost of addition is negligible, show that this gives an algorithm whose running time is $\Theta(n^\alpha)$ where $\alpha = \log_2 7 \approx 2.807$. The optimal value of the exponent α is still not known.

2.6

2.18 How to mail a matrix. Given a graph $G = (V, E)$ with $|V| = n$ vertices and $|E| = m$ edges, how many bits do we need to specify the adjacency matrix, and how many do we need to specify a list of edges? Keep in mind that it takes $\log n$ bits to specify an integer between 1 and n. When are each of these two formats preferable?

In particular, compare *sparse* graphs where $m = O(n)$ with *dense* graphs where $m = \Theta(n^2)$. How do things change if we consider a *multigraph*, like the graph of the Königsberg bridges, where there can be more than one edge between a pair of points?

2.19 We all live in a yellow subroutine. Another sense in which P is robust is that if one polynomial-time program uses another as a subroutine, then the total running time is still polynomial—at least if we're careful.

Suppose an algorithm B runs in polynomial time, and computes a function, e.g., a string or integer. Now suppose an algorithm A performs a polynomial number of steps, one of which calls B as a subroutine. Show that A also runs in polynomial time. Note that the input A sends to B might not be A's original input, but rather some other input that A is interested in. Hint: show that the set of polynomial functions is closed under composition. In other words, if $f(n)$ and $g(n)$ are both poly(n), so is their composition $f(g(n))$.

On the other hand, give an example where calling a polynomial-time algorithm B as a subroutine a polynomial number of times can give a total running time which is exponential. Hint: B's output might be bigger than its input. Can you think of other types of subroutine use where we can, or can't, guarantee polynomial time?

2.20 A little bit more than polynomial time. A *quasipolynomial* is a function of the form $f(n) = 2^{\Theta(\log^k n)}$ for some constant $k > 0$, where $\log^k n$ denotes $(\log n)^k$. Let us define QuasiP as the class of problems that can be solved in quasipolynomial time. First show that any quasipolynomial $f(n)$ with $k > 1$ is $\omega(g(n))$ for any polynomial function $g(n)$, and that $f(n) = o(h(n))$ for any exponential function $h(n) = 2^{\Theta(n^c)}$ for any $c > 0$. Thus $\mathsf{P} \subseteq \mathsf{QuasiP} \subseteq \mathsf{EXPTIME}$.

Then show that the set of quasipolynomial functions is closed under composition. Therefore, QuasiP programs can use each other as subroutines in the same sense that P programs can (see Problem 2.19) even if the main program gives an instance to the subroutine whose size is a quasipolynomial function of the original input size.

Notes

2.1 The Book of Chess. Schaeffer et al. [730] estimated that the number of legal positions in Checkers is 10^{18}. For Chess, the number of possible positions in a game 40 moves long was estimated at 10^{43} by Claude Shannon [746] in 1950. In 1994, Victor Allis [36] proved an upper bound of 5×10^{52} for the number of Chess positions and estimated the true number to be 10^{50}.

2.2 Dixit Algorizmi. The word *algorithm* goes back to the Persian Astronomer Muhammad ibn Musa al-Khwarizmi, born about 780 A.D. in Khwarezm (now Khiva in Uzbekistan). He worked in Baghdad, serving the caliph Abd-Allah al Mamun, son of the caliph Harun al-Rashid of *1001 Arabian Nights* fame. Al-Khwarizmi brought the Hindu number system to the Arab world, from where it spread to Europe, allowing us to write $31 \times 27 = 837$ instead of $\mathrm{XXXI} \times \mathrm{XXVII} = \mathrm{DCCCXXXVII}$.

In medieval times, arithmetic was identified with al-Khwarizmi's name, and the formula *dixit Algorizmi* (thus spake al-Khwarizmi) was a hallmark of clarity and authority. Al-Khwarizmi's legacy is also found in the Spanish word *guarismo* (digit) and in the word *algebra*, which can be traced back to al-Khwarizmi's book on the solution of equations, the *Kitab al-muhtasar fi hisab al-gabr w'al-muqabalah*. A good reference on al-Khwarizmi, and the role of algorithms in Mathematics and Computer Science, is Knuth [493].

2.3 Euclid's algorithm. Euclid's algorithm appeared in his *Elements* in the 3rd century B.C. in Proposition 2 of Book VII. However, there is some evidence that it was known to Aristarchus and Archimedes [149]. The first proof that it finds the gcd of two n-digit numbers in $O(n)$ steps was given by Pierre-Joseph-Étienne Finck in 1841, who used the argument of Exercise 2.1 to get an upper bound of $2\log_2 a = (2\log_2 10)n$ on the number of divisions. This is probably the first nontrivial mathematical analysis of the running time of an algorithm. Three years later, Gabriel Lamé gave the bound of $5\log_{10} a = 5n$ using Fibonacci numbers, and the first part of Problem 2.6 is called Lamé's theorem. For a history of these results, see Shallit [741]. For a history of the average running time, discussed in Problems 2.7 and 2.8, see Knuth [495].

2.4 The adversary. Why should we focus on worst cases? Certainly other sciences, like physics, assume that problems are posed, not by a malicious adversary, but by Nature. Norbert Wiener draws a distinction between two kinds of devil, one that works cleverly against us, and another that simply represents our own ignorance [823, pp. 34–36]. Nature, we hope, is in the second category:

> The scientist is always working to discover the order and organization of the universe, and is thus playing a game against the arch-enemy, disorganization. Is this devil Manichaean or Augustinian? Is it a contrary force opposed to order or is it the very absence of order itself?
>
> ...This distinction between the passive resistance of nature and the active resistance of an opponent suggests a distinction between the research scientist and the warrior or the game player. The research physicist has all the time in the world to carry out his experiments, and he need not fear that nature will in time discover his tricks and method and change her policy. Therefore, his work is governed by his best moments, whereas a chess player cannot make one mistake without finding an alert adversary ready to take advantage of it and to defeat him. Thus the chess player is governed more by his worst moments than by his best moments.

There are actually three kinds of instance we might be interested in: worst-case ones, random ones, and real ones. In Problem 2.8 we derived the performance of Euclid's algorithm on random numbers, and in Chapter 14 we will consider problems based on random graphs and random Boolean formulas. But for many problems, there doesn't seem to be any natural way to define a random instance. Real-world problems are the most interesting to an engineer, but they typically have complicated structural properties that are difficult to capture mathematically. Perhaps the best way to study them is empirically, by going out and measuring them instead of trying to prove theorems.

Finally, one important reason why computer science focuses on the adversary is historical. Modern computer science got its start in the codebreaking efforts of World War II, when Alan Turing and his collaborators at Bletchley Park broke the Nazis' Enigma code. In cryptography, there really is an adversary, doing his or her best to break your codes and evade your algorithms.

2.5 Fast multiplication: Babbage, Gauss, and Fourier. The idea of multiplying two numbers recursively by dividing them into high- and low-order parts, and the fact that its running time is quadratic, was known to Charles Babbage—the 19th-century inventor of the Differential and Analytical Engines, whom we will meet in Chapter 7. He wanted to make sure that his Analytical Engine could handle numbers with any number of digits. He wrote [69, p. 125]:

> Thus if $a \cdot 10^{50} + b$ and $a' \cdot 10^{50} + b'$ are two numbers each of less than a hundred places of figures, then each can be expressed upon two columns of fifty figures, and a, b, a', b' are each less than fifty places of figures... The product of two such numbers is
>
> $$aa'10^{100} + (ab' + a'b)10^{50} + bb'.$$
>
> This expression contains four pairs of factors, $aa', ab', a'b, bb'$, each factor of which has less than fifty places of figures. Each multiplication can therefore be executed in the Engine. The time, however, of multiplying two numbers, each consisting of any number of digits between fifty and one hundred, will be nearly four times as long as that of two such numbers of less than fifty places of figures...
>
> Thus it appears that whatever may be the number of digits the Analytical Engine is capable of holding, if it is required to make all the computations with k times that number of digits, then it can be executed by the same Engine, but in an amount of time equal to k^2 times the former.

The trick of reducing the number of multiplications from four to three, and the resulting improvement in how the running time scales with the number of digits, is the sort of thing that Babbage would have loved. We will spend more time with Mr. Babbage in Chapter 7.

The first $O(n^{\log_2 3})$ algorithm for multiplying n-digit integers was found in 1962 by Karatsuba and Ofman [455]. However, the fact that we can reduce the number of multiplications from four to three goes back to Gauss! He noticed that in order to calculate the product of two complex numbers (where $\imath = \sqrt{-1}$)

$$(a + b\imath)(c + d\imath) = (ac - bd) + (ad + bc)\imath$$

we only need three real multiplications, such as ac, bd, and $(a + c)(b + d)$, since we can get the real and imaginary parts by adding and subtracting these products. The idea of [455] is then to replace $\imath$ with $10^{n/2}$, and to apply this trick recursively.

Toom [794] recognized that we can think of multiplication as interpolating a product of polynomials as described in Problems 2.14–2.16, and thus achieved a running time of $O(n^{1+\varepsilon})$ for arbitrarily small ε. This is generally called the Toom–Cook algorithm, since Stephen Cook also studied it in his Ph.D. thesis.

In 1971, Schönhage and Strassen [733] gave an $O(n \cdot \log n \cdot \log\log n)$ algorithm. The idea is to think of an integer x as a function, where $x(i)$ is its ith digit. Then, except for carrying, the product xy is the convolution of the corresponding functions, and the Fourier transform of their convolution is the product of their Fourier transforms. They then use the Fast Fourier Transform algorithm, which as we will discuss in Section 3.2.3 takes $O(n \log n)$ time. We outline this algorithm in Problems 3.15 and 3.16; we can also think of it as a special case of the Toom–Cook algorithm, where we sample the product of the two polynomials at the $2r$th roots of unity in the complex plane. An excellent description of this algorithm can be found in Dasgupta, Papadimitriou and Vazirani [215].

In 2007, Fürer [306] improved this algorithm still further, obtaining a running time of $n \cdot \log n \cdot 2^{O(\log^* n)}$. Here $\log^* n$ is the number of times we need to iterate the logarithm to bring n below 2; for instance, $\log^* 65536 = 4$ and $\log^* 10^{10000} < 5$. Since $\log^* n$ is "nearly constant," it seems likely that the true complexity of multiplication is $\Theta(n \log n)$. And in fact, this has been announced by Havey and Van Der Hoeven at the time we write this [374].

2.6 Matrix multiplication. The problem of calculating *matrix* products also has a long and interesting history. Multiplying two $n \times n$ matrices requires n^3 multiplications if we use the textbook method, but algorithms that work in time $O(n^{\alpha})$ have been achieved for various $\alpha < 3$. In 1969, Strassen obtained the algorithm of Problem 2.17, for which $\alpha = \log_2 7 \approx 2.807$. Coppersmith and Winograd [202] presented an algorithm with $\alpha \approx 2.376$. Their algorithm has recently been improved [826, 219, 522] and as of 2017, the fastest variant has $\alpha \approx 2.3728639$.

While clearly $\alpha \geq 2$ since we need $\Omega(n^2)$ time just to read the input, it is not known what the optimal value of α is. However, there is some very promising recent work on algebraic approaches by Cohn and Umans [190] and Cohn, Kleinberg, Szegedy and Umans [188]. These include reasonable conjectures which would imply that $\alpha = 2$, or more precisely, that we can multiply matrices in time $O(n^{2+\varepsilon})$ for any $\varepsilon > 0$.

2.7 Moore's Law. Gordon Moore, a co-founder of Intel, originally claimed in 1965 that the number of transistors in an integrated circuit roughly doubled each year. He later changed the doubling time to two years, and "Moore's Law" came to mean a similar claim about speed, memory per dollar, and so on. While clock speeds have recently leveled off, the real speed of computation measured in instructions per second continues to rise due to improvements in our computers' architecture, such has having multiple processors on a single chip, speeding up memory access by cleverly predicting what data the program will need next, and so on.

It could be argued that, at this point, Moore's Law has become a self-fulfilling prophecy driven by consumer expectations—now that we are used to seeing multiplicative improvements in our computers every few years, this is what we demand when we buy new ones.

Some technologists also believe that improvements in computing technology are even better described by Wright's Law. This states that as manufacturers and engineers gain more experience, technology improves polynomially as a function of the number of units produced. In this case, the exponential improvements we see are due to the exponential growth in the number of computers produced so far.

However, these improvements cannot continue forever without running up against fundamental physical constraints. If the current exponential growth in chip density continues, by around 2015 or 2020 our computers will use one elementary particle for each bit of memory. At these scales, we cannot avoid dealing with quantum effects, such as electron "tunneling" through potential barriers and jumping from one location to another. These effects will either be a source of noise and inconvenience—or, as we discuss in Chapter 15, of new computational power.

2.8 Exponentials. Some readers may find it jarring that functions of the form 2^{n^c} for any constant c are called simply "exponential." However, allowing the exponent to be polynomial in n, rather than simply linear, gives the class EXP the same robustness to the input format that P possesses.

2.9 Elementary steps. We need to be somewhat careful about how liberally we define the notion of an "elementary step." For instance, Schönhage [732] showed that machines that can multiply and divide integers of arbitrary size in a single step can solve PSPACE-complete problems in polynomial time (we will meet the class PSPACE in Chapter 8). Hartmanis and Simon [371] found that the same is true of machines that can perform bitwise operations on strings of arbitrary length in a single step. Finally, Bertoni, Mauri, and Sabadini [108] showed that such machines can solve #P-complete problems, which we will meet in Chapter 13. A review can be found in van Emde Boas [806].

Both PSPACE and #P are far above P in the complexity hierarchy. Thus if we assume that arithmetic operations can be performed in constant time regardless of the size of the numbers involved, we lose our ability to draw distinctions between hard problems and easy ones. We get a more meaningful picture of complexity if we assume that the cost of arithmetic operations is logarithmic in the size and accuracy of the numbers, i.e., linear in the number of digits or bits that are being manipulated. And this assumption is certainly more realistic, at least where digital computers are concerned.

2.10 The history of polynomial time. Computational complexity theory as we know it began with the 1965 paper of Juris Hartmanis and Richard Stearns [372], for which they received the Turing Award in 1993. Their paper defines classes of functions by how much time it takes to compute them, proves by diagonalization (as we will discuss in Chapter 6) that increasing the computation time yields an infinite hierarchy of more and more powerful classes, and notes that changing the type of machine can alter the running time from, say, $\Theta(n)$ to $\Theta(n^2)$. They also make what has turned out to be a rather impressive understatement:

> It is our conviction that numbers and functions have an intrinsic computational nature according to which they can be classified... and that there is a good opportunity here for further research.

At around the same time, the idea that polynomial time represents a good definition of tractable computation appeared in the work of Cobham [181] and Edmonds [260]. Cobham says:

> The subject of my talk is perhaps most directly indicated by simply asking two questions: first, is it harder to multiply than to add? and second, why? ... There seems to be no substantial problem in showing that using the standard algorithm it is in general harder—in the sense that it takes more time or more scratch paper—to multiply two decimal numbers than to add them. But this does not answer the question, which is concerned with ... properties intrinsic to the functions themselves and not with properties of particular related algorithms.

He goes on to define a class $\mathscr{L}$ of functions that can be computed in polynomial time as a function of the number of digits of their input, and recognizes that changing from one type of machine to another typically changes the power of the polynomial but preserves $\mathscr{L}$ overall.

Edmonds studied a polynomial-time algorithm for MAX MATCHING, an optimization problem that asks how to form as many partnerships as possible between neighboring vertices in a graph. He presents his result by calling it a *good* algorithm, and says:

> There is an obvious finite algorithm, but that algorithm increases in difficulty exponentially with the size of the graph. It is by no means obvious whether or not there exists an algorithm whose difficulty increases only algebraically [i.e., polynomially] with the size of the graph. The mathematical significance of this paper rests largely on the assumption that the two preceding sentences have mathematical meaning...

He then proposes that any algorithm can be broken down into a series of elementary steps, and that once we agree on what types of steps are allowed, the question of whether an algorithm exists with a given running time becomes mathematically well-defined.

For an even earlier discussion of whether mathematical proofs can be found in polynomial time, see the 1956 letter of Gödel to von Neumann discussed in Section 6.1.

2.11 The Robertson–Seymour Theorem. There are strange circumstances in which we can know that a problem is in P, while knowing essentially nothing about how to solve it. To see how this could be the case, let's start with a simple graph property.

A graph is *planar* if it can be drawn in the plane without any edges crossing each other. Kuratowski [511] and Wagner [816] showed that G is planar if and only if it does not contain either of the graphs K_5 or $K_{3,3}$ shown in Figure 2.9 as a *minor*, where a minor is a graph we can obtain from G by removing vertices or edges, or by shrinking edges and merging their endpoints. With some work, we can check for both these minors in polynomial time. While this is far from the most efficient algorithm, it shows that PLANARITY is in P.

Planarity is an example of a *minor-closed* property. That is, if G is planar then so are all its minors. Other examples of minor-closed properties include whether G can be drawn on a torus with no edge crossings, or whether it can be embedded in three-dimensional space in such a way that none of its cycles are knotted, or that no two cycles are linked. For any fixed k, the property that G has a vertex cover of size k or less (see Section 4.2.4) is also minor-closed.

Wagner conjectured that for every minor-closed property, there is a finite list $\{K_1, K_2, \ldots\}$ of excluded minors such that G has that property if and only if it does not contain any of them. After a series of 20 papers, Neil Robertson and

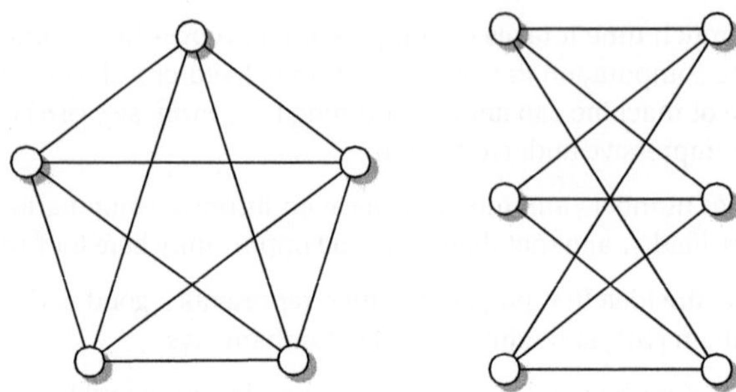

FIGURE 2.9: A graph is planar if and only if it does not contain either of these graphs as a minor.

Paul Seymour proved this conjecture in 2004 [705]. Along the way, they proved that for any fixed K, we can check in $O(n^3)$ time whether a graph with n vertices contains K as a minor.

As a result, we know that for any minor-closed property, the problem of telling whether a graph has it or not is in P. But Robertson and Seymour's proof is *nonconstructive*: it tells us nothing about these excluded minors, or how big they are. Moreover, while their algorithm runs in $O(n^3)$ time, the constant hidden in O depends in a truly horrendous way on the number of vertices in the excluded minors K_i (see [441] for a review). We are thus in the odd position of knowing that an entire family of problems is in P, without knowing polynomial-time algorithms for them, or how long they will take.

Chapter 3

Insights and Algorithms

> It has often been said that a person does not really understand something until he teaches it to someone else. Actually a person does not *really* understand something until he can teach it to a *computer*, i.e., express it as an algorithm... The attempt to formalize things as algorithms leads to a much deeper understanding than if we simply try to comprehend things in the traditional way.
>
> Donald E. Knuth

To the townspeople of Königsberg, the set of possible paths across the seven bridges seemed like a vast, formless mist. How could they find a path, or tell whether one exists, without an arduous search? Euler's insight parted this mist, and let them see straight to the heart of the problem.

Moreover, his insight is not just a sterile mathematical fact. It is a living, breathing algorithm, which solves the Bridges of Königsberg, or even of Venice, quickly and easily. As far as we know, whether a problem is in P or not depends on whether an analogous insight exists for it: some way we can guide our search, so that we are not doomed to wander in the space of all possible solutions.

But while mathematical insights come in many forms, we know of just a few major strategies for constructing polynomial-time algorithms. These include *divide and conquer*, where we break problems into easier subproblems; *dynamic programming*, where we save time by remembering subproblems we solved before; *greedy algorithms*, which start with a bad solution or none at all, and make small changes until it becomes the best possible one; *duality*, where two seemingly different optimization problems turn out to have the same solution, even though they approach it from opposite directions; and *reductions*, which transform one problem into another that we already know how to solve.

Why do these strategies work for some problems but not for others? How can we break a problem into subproblems that are small enough, and few enough, to solve quickly? If we start with a bad solution, can we easily feel our way towards a better one? When is one problem really just another one in disguise? This chapter explores these questions, and helps us understand why some problems are easier than they first appear. Along the way, we will see how to sort a pack of cards, hear the music of the spheres, typeset beautiful books, align genomes, find short paths, build efficient networks, route the flow of traffic, and run a dating service. 3.1

3.1 Recursion

We have already seen two examples of recursion in Chapter 2: Euclid's algorithm for the greatest common divisor, and the divide-and-conquer algorithm for multiplying n-digit integers. These algorithms work by creating "children"—new incarnations of themselves—and asking them to solve smaller versions of the same problem. These children create their own children in turn, asking them to solve even smaller problems, until we reach a base case where the problem is trivial.

We start this chapter with another classic example of recursion: the Towers of Hanoi, introduced by the mathematician Edouard Lucas under the pseudonym of "N. Claus de Siam." While this is really just a puzzle, and not a "problem" in the sense we defined in Chapter 2, it is still an instructive case of how a problem can be broken into subproblems. The story goes like this:

> In the great temple at Benares, beneath the dome that marks the centre of the world, rests a brass plate in which are fixed three diamond needles, each a cubit high and as thick as the body of a bee. On one of these needles, at the creation, God placed sixty-four disks of pure gold, the largest disk resting on the brass plate, and the others getting smaller and smaller up to the top one...
>
> Day and night, unceasingly, the priests transfer the disks from one diamond needle to another according to the fixed and immutable laws of Brahma, which require that the priest on duty must not move more than one disk at a time and that he must place this disk on a needle so that there is no smaller disk below it. When the sixty-four disks shall have been thus transferred from the needle which at creation God placed them to one of the other needles, tower, temple, and Brahmins alike will crumble into dust and with a thunderclap the world will vanish.

This appears to be a product of the French colonial imagination, with Hanoi and Benares chosen as suitably exotic locations. Presumably, if a Vietnamese mathematician had invented the puzzle, it would be called the Towers of Eiffel.

A little reflection reveals that one way to move all n disks from the first peg to the second is to first move $n-1$ disks to the third peg, then move the largest disk from the first to the second, and then move the $n-1$ disks from the third peg to the second. But how do we move these $n-1$ disks? Using exactly the same method. This gives the algorithm shown in Figure 3.1.

We can think of running a recursive algorithm as traversing a tree. The root corresponds to the original problem, each child node corresponds to a subproblem, and the individual moves correspond to the leaves. Figure 3.2 shows the tree for $n=3$, with the solution running from top to bottom.

If the number of disks is n, what is the total number of moves we need? Let's denote this $f(n)$. Since our algorithm solves this problem twice for $n-1$ disks and makes one additional move, $f(n)$ obeys the equation

$$f(n)=2f(n-1)+1. \tag{3.1}$$

The base case is $f(0)=0$, since it takes zero moves to move zero disks. The solution is given by the following exercise.

Exercise 3.1 *Prove by induction on n that the solution to* (3.1) *with the base case* $f(0)=0$ *is*

$$f(n)=2^n-1.$$

```
Hanoi(n,i,j) // move n disks from peg i to peg j
begin
    if n = 0 then return;
    Hanoi(n-1,i,k);
    move a disk from peg i to peg j ;
    Hanoi(n-1,k,j);
end
```

FIGURE 3.1: The recursive algorithm for solving the Towers of Hanoi. Here k denotes the third peg, other than i and j. Note that i, j, and k are "local variables," whose values change from one incarnation of the algorithm to the next. Note also that in the base case $n = 0$, the algorithm simply returns, since there is nothing to be done.

In fact, as Problem 3.1 asks you to show, this algorithm is the best possible, and $2^n - 1$ is the smallest possible number of moves. Thus the priests in the story need to perform $2^{64} - 1 \approx 1.8 \times 10^{19}$ moves, and it seems that our existence is secure for now. If the number of moves were only, say, 9 billion, we might be in trouble if the priests gain access to modern computing machinery.

3.2

3.2 Divide and Conquer

Like the solution to the Towers of Hanoi, many recursive algorithms work by breaking a problem into several pieces, finding the answer to each piece, and then combining these answers to obtain the answer to the entire problem. We saw this in Section 2.3, where we multiplied n-digit numbers by breaking them into pairs of $n/2$-digit numbers.

Perhaps the simplest example of this approach is *binary search.* If I want to look up a word w in a dictionary, I can compare w to the word in the middle, and then focus my search on the first or second half of the dictionary. Using the same approach lets me focus on one-quarter of the dictionary, and so on. Since each step divides the dictionary in half, I can find w in a dictionary of N words in just $\log_2 N$ steps.

In this section, we will see several important problems where a divide-and-conquer strategy works. These include sorting large lists, raising numbers to high powers, and finding the Fourier transform of an audio signal. In each case, we solve a problem recursively by breaking it into independent subproblems, solving each one, and then combining their results in some way.

3.2.1 *Set This House in Order: Sorting*

> We start with the smallest. Then what do we do?
> We line them all up. Back to back. Two by two.
> Taller and taller. And, when we are through,
> We finally will find one who's taller than who.
>
> Dr. Seuss, *Happy Birthday To You!*

Suppose I wish to sort a pack of cards using the divide-and-conquer strategy. I start by splitting the pack into two halves, and sorting each one separately. I then merge the two sorted halves together, so that the

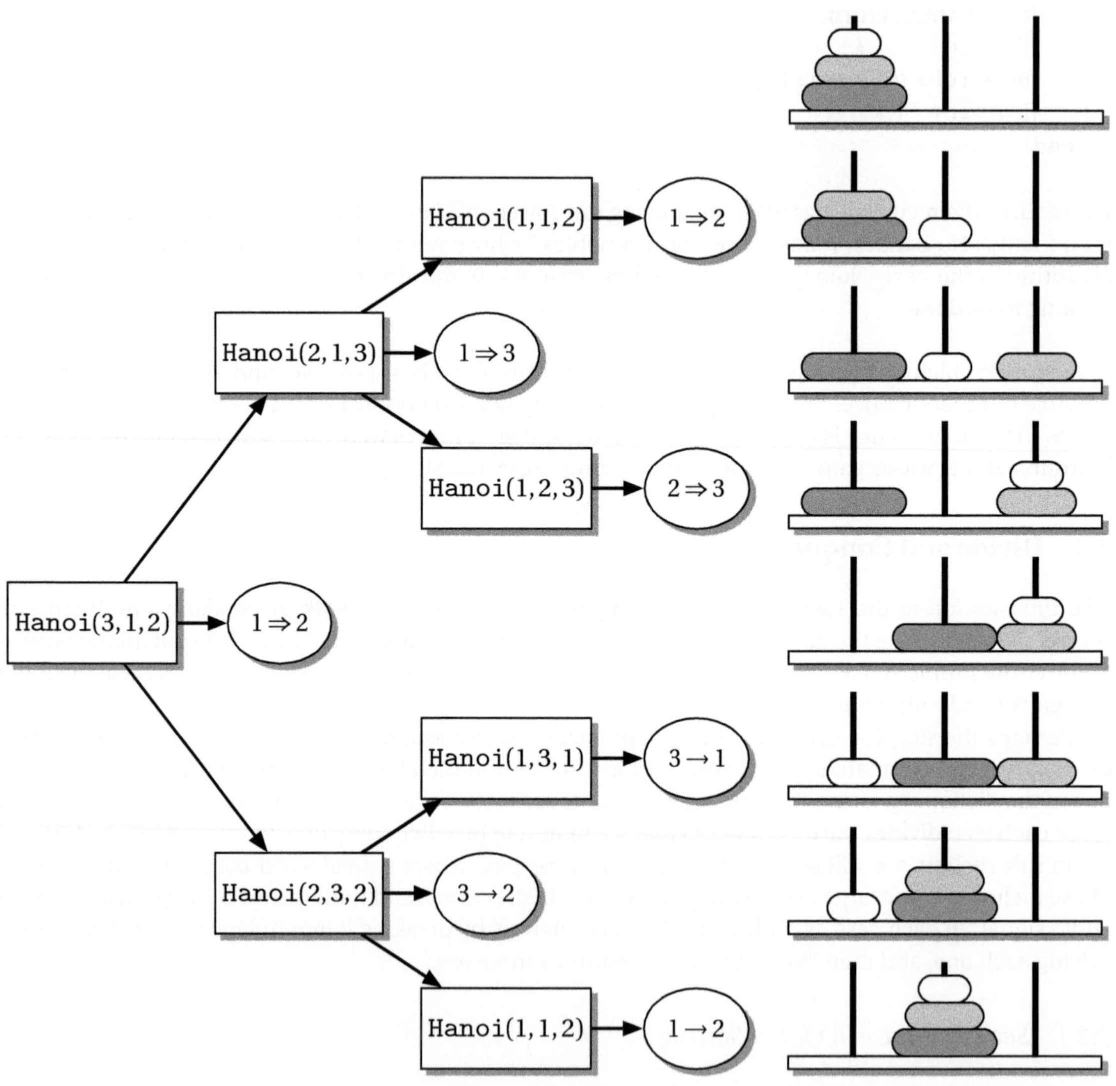

FIGURE 3.2: The tree corresponding to the recursive algorithm for the Towers of Hanoi with $n = 3$. Each node Hanoi(n, i, j) corresponds to the subproblem of moving n disks from peg i to peg j. The root node, corresponding to the original problem, is at the left. The actual moves appear on the leaf nodes (the ellipses), and the solution goes from top to bottom.

```
Mergesort(ℓ)
input: a list ℓ of n elements
output: a sorted version of ℓ
begin
    if |ℓ| ≤ 1 then return;
    ℓ1 := the first half of ℓ ;
    ℓ2 := the second half of ℓ ;
    ℓ1 :=Mergesort(ℓ1);
    ℓ2 :=Mergesort(ℓ2);
    return merge(ℓ1,ℓ2);
end
```

FIGURE 3.3: The `Mergesort` algorithm.

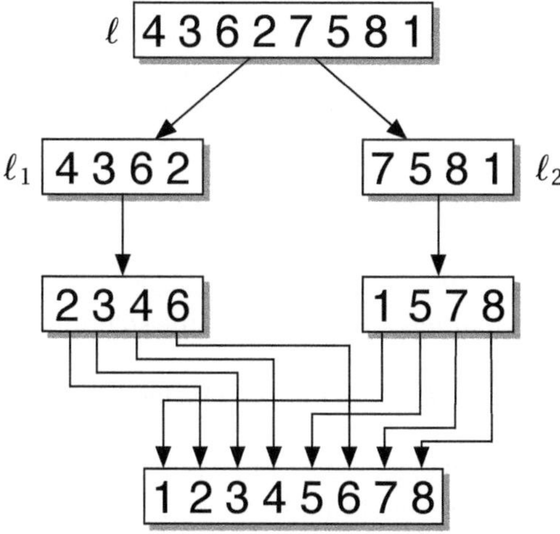

FIGURE 3.4: `Mergesort` splits the list ℓ into two halves, and sorts each one recursively. It then merges the two sorted halves, taking elements from ℓ_1 or ℓ_2, whichever is smaller.

entire pack is sorted: think of a careful riffle shuffle, where I let a card fall from either my left hand or my right, depending on which of the two cards should come first. This gives a recursive algorithm shown in Figure 3.3, which we illustrate in Figure 3.4.

To quantify the running time of this algorithm, let's count the number of times it compares one element to another. Let $T(n)$ be the number of comparisons it takes to sort an element of length n. Assuming for simplicity that n is even, sorting the two halves of the list recursively takes $2T(n/2)$ comparisons. How many comparisons does the `merge` operation take? We start by comparing the elements at the heads of ℓ_1 and ℓ_2, moving whichever one is smaller to the final sorted list, and continuing until ℓ_1 or ℓ_2 is empty. This takes at most $n-1$ comparisons, but for simplicity we'll assume that it takes n. Then the total number

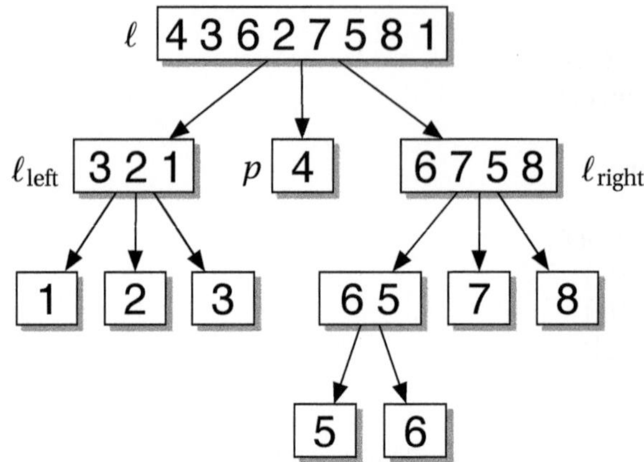

FIGURE 3.5: The recursive tree of Quicksort. At each step we choose a pivot p, and partition ℓ into sublists ℓ_{left} and ℓ_{right} depending on whether each element is smaller or larger than p. The leaves of the tree correspond to lists of size 1, for which no further sorting is necessary.

of comparisons is

$$T(n)=2T(n/2)+n\,. \tag{3.2}$$

Since it takes zero comparisons to sort a list of size 1, the base case is $T(1)=0$. As Problem 3.7 asks you to show, if we assume for simplicity that n is a power of two, the solution to (3.2) is

$$T(n)=n\log_2 n=\Theta(n\log n)\,.$$

Now let's look at `Quicksort`. Like `Mergesort`, it is a divide-and-conquer algorithm, but now we break the list up in a different way. Instead of simply breaking it into halves, we choose a *pivot* element p, compare all the other elements to it, and put them in the left or right sublist according to whether they are smaller or larger than p. We then recursively sort these sublists as in Figure 3.5, subdividing the original list until we are left with lists of size 0 or 1. This gives the pseudocode shown in Figure 3.6.

`Quicksort` is interesting to us mainly in that it offers a contrast between worst-case and average-case running time. The number $T(n)$ of comparisons it takes to sort a list of size n depends on where the pivot p falls in the list. Let's say that p is the rth smallest element. Then there are $r-1$ elements smaller than p, and $n-r$ elements larger than p, and these elements end up in the left and right sublists respectively. Counting the $n-1$ comparisons it takes to compare p to everyone else, this gives the equation

$$T(n)=T(r-1)+T(n-r)+n-1\,. \tag{3.3}$$

If we are very lucky and the pivot is always the median of the list, we have $r=n/2$. Ignoring the difference between n and $n-1$, this gives

$$T(n)=2T(n/2)+n$$

just as we had for `Mergesort`, and $T(n)=n\log_2 n$.

```
Quicksort(ℓ)
input: a list ℓ of n elements
output: a sorted version of ℓ
begin
    if n ≤ 1 then return;
    choose a pivot p ∈ ℓ ;
    forall the x ∈ ℓ do
        if x < p then put x in ℓ_left ;
        else put x in ℓ_right ;
    end
    return {Quicksort(ℓ_left), p, Quicksort(ℓ_right)} ;
end
```

FIGURE 3.6: The Quicksort algorithm.

On the other hand, if we are very unlucky, the pivot is always the smallest or largest in the list, with $r = 1$ or $r = n$. In this case, we have succeeded only in whittling down a list of size n to one of size $n-1$, and the running time is

$$T(n) = T(n-1) + n - 1 .$$

With the base case $T(1) = 0$, this gives the arithmetic series

$$T(n) = 1 + 2 + 3 + \cdots + n - 1 = \frac{n(n-1)}{2} = \Theta(n^2).$$

Since there is a large difference between the best case and the worst case, let's consider the *average* case, where r is uniformly random—that is, where it is equally likely to take any value between 1 and n. Averaging (3.3) over r gives us the following equation, where now $T(n)$ is the average number of comparisons:

$$\begin{aligned} T(n) &= n - 1 + \frac{1}{n} \sum_{r=1}^{n} \Big(T(r-1) + T(n-r) \Big) \\ &= n - 1 + \frac{2}{n} \sum_{r=1}^{n} T(r-1) . \end{aligned} \tag{3.4}$$

While the average case is presumably not as good as the best case, we might still hope that $T(n)$ scales as $\Theta(n \log n)$, since most of the time the pivot is neither the largest nor the smallest element. Indeed, as Problem 3.9 shows—and, using another method, Problem 10.3—when n is large the solution to (3.4) is

$$T(n) \approx 2n \ln n .$$

Since

$$\frac{2n \ln n}{n \log_2 n} = 2 \ln 2 \approx 1.386 ,$$

the average number of comparisons is only 39% greater than it would be if the pivot were always precisely the median.

Now, why might r be uniformly random? There are two reasons why this could be. One is if we choose the pivot deterministically, say by using the first element as in Figure 3.5, but if the input list is in random order, where all $n!$ permutations are equally likely. This is all well and good, but in the world of computer science assuming that the input is random is overly optimistic. If our friend the adversary knows how we choose the pivot, he can give us an instance where the pivot will always be the smallest or largest element—in this case, a list that is already sorted, or which is sorted in reverse order. Thus he can saddle us with the worst-case running time of $\Theta(n^2)$.

However, the other reason r might be random is if we choose the pivot randomly, rather than deterministically. If p is chosen uniformly from the list then r is uniformly random, no matter what order the input is in. Instead of averaging over inputs, we average over the algorithm's choices, and achieve an average running time of $2n \ln n$ no matter what instance the adversary gives us. We will return to this idea in Chapter 10, where we explore the power of randomized algorithms.

Having seen both these algorithms, the reader might wonder whether it is possible to sort qualitatively faster than `Mergesort` or `Quicksort`. We will see in Section 6.2 that if we use comparisons, sorting n elements requires at least $\log_2 n! \approx n \log_2 n$ steps. Therefore, the number of comparisons that `Mergesort` performs is essentially optimal. This is one of the few cases where we can determine the optimal algorithm.

3.3

3.2.2 *Higher Powers*

> That sweet joy may arise from such contemplations cannot be denied. Numbers and lines have many charms, unseen by vulgar eyes, and only discovered to the unwearied and respectful sons of Art. In features the serpentine line (who starts not at the name) produces beauty and love; and in numbers, high powers, and humble roots, give soft delight.
>
> E. De Joncourt, as quoted by Charles Babbage

Let's look at another example of the divide-and-conquer strategy. Given x and y as inputs, how hard is it to calculate x^y?

If x and y each have n digits, x^y could be as large as 10^{n10^n} and have $n10^n$ digits. It would take an exponential amount of time just to write this number down, regardless of how long it takes to calculate it. So, in order to keep the result to at most n digits, we define the following problem:

MODULAR EXPONENTIATION
Input: n-digit integers x, y, and p
Output: $x^y \bmod p$

As we will see later, this problem is important in cryptography and algorithms for PRIMALITY. Is it in P?

An obvious approach is to start with $x^0 = 1$ and do y multiplications, increasing the power of x by one and taking the result mod p each time. But since y is exponentially large, this would take exponential time. A much better approach is to start with x, square it, square its square, and so on. This gives the powers

$$x, x^2, x^4, x^8, \ldots$$

```
Power(x,y,p)
input: integers x,y,p
output: x^y mod p
begin
    if y = 0 then return 1;
    t := Power(x,⌊y/2⌋,p);
    if y is even then return t^2 mod p ;
    else return x t^2 mod p ;
end
```

FIGURE 3.7: The repeated-squaring algorithm for MODULAR EXPONENTIATION.

where we take the result mod p at each step. If y is a power of 2, we get x^y after just $\log_2 y = O(n)$ squarings. If y is not a power of 2, we can first derive x^{2^k} for all powers of 2 up to y, and then combine these according to y's binary digit sequence: for instance,

$$x^{999} = x \cdot x^2 \cdot x^4 \cdot x^{32} \cdot x^{64} \cdot x^{128} \cdot x^{256} \cdot x^{512}.$$

Since this product involves $O(n)$ powers, the total number of multiplications we need to do is still $O(n)$. Since we know how to multiply n-digit numbers in polynomial time, the total time we need is polynomial as a function of n. Therefore, MODULAR EXPONENTIATION is in P.

We can view this as a divide-and-conquer algorithm. Let $\lfloor y/2 \rfloor$ denote $y/2$ rounded down to the nearest integer. Then we calculate x^y recursively by calculating $x^{\lfloor y/2 \rfloor}$ and squaring it, with an extra factor of x thrown in if y is odd. This gives the algorithm shown in Figure 3.7.

The fact that we can get exponentially high powers by squaring repeatedly will come up several times in this book. For instance, by applying the same idea to matrix powers, we can find paths in graphs even when these graphs are exponentially large.

Modular exponentiation is also interesting because, as a function, it seems to be much harder to do backwards than forwards. Consider the following problem:

> DISCRETE LOG
>
> Input: n-digit integers x, z, and p
>
> Output: An integer y, if there is one, such that $z = x^y \bmod p$

We call this problem DISCRETE LOG since we can think of y as $\log_x z$ in the world of integers mod p.

Our current belief is that, unlike MODULAR EXPONENTIATION, DISCRETE LOG is outside P. In other words, if we fix x and p, we believe that $f(y) = x^y \bmod p$ is a *one-way function*: a function in P, whose inverse is not. We will discuss pseudorandom numbers and cryptosystems based on this problem, and quantum algorithms that would break them, in Chapters 10 and 15.

3.2.3 *The Fast Fourier Transform*

We end this section with one more divide-and-conquer algorithm—one which is used today throughout digital signal processing, from speech recognition and crystallography to medical imaging and audio

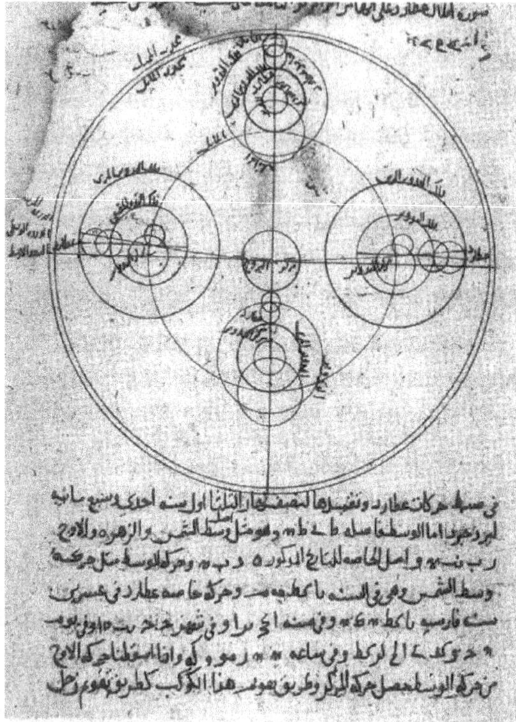

FIGURE 3.8: Fourier analysis through the ages. Left, Ibn al-Shāṭir's model of the motion of Mercury using six epicycles. Right, adjusting a coefficient in a tide-predicting machine.

compression. Readers who are unfamiliar with complex numbers should feel free to skip this section for now, but you'll need to understand it before we study quantum computing. First, some history.

Early Greek astronomers, much like modern physicists, were very fond of mathematical elegance. They regarded circles and spheres as the most perfect shapes, and postulated that the heavenly bodies move around the Earth in perfect circular orbits. Unfortunately, this theory doesn't fit the data very well. In particular, planets undergo *retrograde* motion, in which they move across the sky in the reverse of the usual direction.

To fix this problem while remaining faithful to the idea of circular motion, Ptolemy proposed that the planets move in *epicycles*, circles whose centers move around other circles. The position of each planet is thus a sum of two vectors, each of which rotates in time with a particular frequency. By adding more and more epicycles, we can fit the data better and better. By the 14th century, Islamic astronomers had produced Ptolemaic systems with as many as six epicycles (see Figure 3.8).

3.4 In a more terrestrial setting—but still with astronomical overtones—in 1876 Sir William Thomson, later Lord Kelvin, built a machine for predicting tides. It had a series of adjustable wheels, corresponding to combinations of the daily, lunar, and solar cycles. A system of pulleys, driven by turning a crank, summed these contributions and drew the resulting graph on a piece of paper. Tide-predicting machines like the one shown in Figure 3.8 were used as late as 1965.

The art of writing functions as a sum of oscillating terms is called Fourier analysis, in honor of Joseph Fourier, who studied it extensively in the early 19th century. We can do this using sums of sines and cosines, but a more elegant way is to use Euler's formula,

$$e^{\imath\theta} = \cos\theta + \imath\sin\theta\,.$$

Then we can write any smooth function $f(t)$ as

$$f(t) = \sum_{\alpha} \tilde{f}(\alpha)\, e^{\imath\alpha t}\,,$$

where the sum ranges over some set of frequencies α. The function $\tilde{f}$, which gives the coefficient for each α, is called the *Fourier transform* of f.

If rather than a continuous function, we have a discrete set of n samples $f(t)$ where $t = 0, 1, \ldots, n-1$, it suffices to consider frequencies that are multiples of $2\pi/n$. Let ω_n denote the nth root of 1 in the complex plane,

$$\omega_n = e^{2\imath\pi/n}\,.$$

Then this discrete set of frequencies $2\pi k/n$ gives us the *discrete Fourier transform*,

$$f(t) = \frac{1}{\sqrt{n}} \sum_{k=0}^{n-1} \tilde{f}(k)\, \omega_n^{kt}\,. \tag{3.5}$$

The reason for the normalization factor $1/\sqrt{n}$ will become clear in a moment.

Now suppose we have a set of samples, and we want to find the Fourier transform. For instance, we have a series of observations of the tides, and we want to set the parameters of our tide-predicting machine. How can we invert the sum (3.5), and calculate $\tilde{f}(k)$ from $f(t)$?

The thing to notice is that if we think of f and $\tilde{f}$ as n-dimensional vectors, then (3.5) is just a matrix multiplication:

$$\begin{pmatrix} f(0) \\ f(1) \\ f(2) \\ \vdots \end{pmatrix} = \frac{1}{\sqrt{n}} \begin{pmatrix} 1 & 1 & 1 & \\ 1 & \omega_n & \omega_n^2 & \cdots \\ 1 & \omega_n^2 & \omega_n^4 & \\ & \vdots & & \ddots \end{pmatrix} \cdot \begin{pmatrix} \tilde{f}(0) \\ \tilde{f}(1) \\ \tilde{f}(2) \\ \vdots \end{pmatrix},$$

or

$$f = Q \cdot \tilde{f} \text{ where } Q_{tk} = \frac{1}{\sqrt{n}}\, \omega_n^{kt}\,.$$

Thus we can calculate $\tilde{f}$ from f by multiplying by the inverse of Q,

$$\tilde{f} = Q^{-1} \cdot f\,.$$

As the following exercise shows, Q^{-1} is simply Q's complex conjugate Q^*, i.e., the matrix where we take the complex conjugate of each entry:

Exercise 3.2 *Prove that* $Q \cdot Q^* = \mathbb{1}$ *where* $\mathbb{1}$ *denotes the identity matrix.*

Since Q is symmetric, we can also say that Q^{-1} is the transpose of its complex conjugate. Such matrices are called *unitary*, and we will see in Chapter 15 that they play a crucial role in quantum computation.

We can now write

$$\tilde{f} = Q^* \cdot f$$

or, as an explicit sum,

$$\tilde{f}(k) = \frac{1}{\sqrt{n}} \sum_{t=0}^{n-1} f(t)\, \omega_n^{-kt}. \tag{3.6}$$

Turning now to algorithms, what is the most efficient way to evaluate the sum (3.6), or its twin (3.5)? We can multiply an n-dimensional vector by an $n \times n$ matrix with n^2 multiplications. Assuming that we do our arithmetic to some constant precision, each multiplication takes constant time. Thus we can obtain $\tilde{f}$ from f, or vice versa, in $O(n^2)$ time. Can we do better?

Indeed we can. After all, Q is not just any $n \times n$ matrix. It is highly structured, and we can break the product $Q^* \cdot f$ down in a recursive way. We again divide and conquer, by dividing the list of samples $f(t)$ into two sublists of size $n/2$: those where t is even, and those where it is odd. Assume n is even, and define

$$f_{\text{even}}(s) = f(2s) \text{ and } f_{\text{odd}}(s) = f(2s+1),$$

where s ranges from 0 to $n/2 - 1$. Also, write

$$k = (n/2)k_0 + k',$$

where $k_0 = 0$ or 1 and k' ranges from 0 to $n/2 - 1$. In particular, if n is a power of 2 and we write k in binary, then b is the most significant bit of k, and k' is k with this bit removed.

Now we can separate the sum (3.6) into two parts as follows:

$$\begin{aligned}
\tilde{f}(k) &= \frac{1}{\sqrt{n}} \left(\sum_{t \text{ even}} f(t)\, \omega_n^{-kt} + \sum_{t \text{ odd}} f(t)\, \omega_n^{-kt} \right) \\
&= \frac{1}{\sqrt{n}} \left(\sum_{s=0}^{n/2-1} f_{\text{even}}(s)\, \omega_n^{-2ks} + \omega_n^{-k} \sum_{s=0}^{n/2-1} f_{\text{odd}}(s)\, \omega_n^{-2ks} \right) \\
&= \frac{1}{\sqrt{2}} \frac{1}{\sqrt{n/2}} \left(\sum_{s=0}^{n/2-1} f_{\text{even}}(s)\, \omega_{n/2}^{-k's} + (-1)^{k_0}\, \omega_n^{-k'} \sum_{s=0}^{n/2-1} f_{\text{odd}}(s)\, \omega_{n/2}^{-k's} \right) \\
&= \frac{1}{\sqrt{2}} \left(\tilde{f}_{\text{even}}(k') + (-1)^{k_0}\, \omega_n^{-k'}\, \tilde{f}_{\text{odd}}(k') \right).
\end{aligned} \tag{3.7}$$

We used the following facts in the third line,

$$\begin{aligned}
\omega_n^2 &= \omega_{n/2} \\
\omega_{n/2}^{-k} &= \mathrm{e}^{-2\imath\pi k_0}\, \omega_{n/2}^{-k'} = \omega_{n/2}^{-k'} \\
\omega_n^{-k} &= \mathrm{e}^{-\imath\pi k_0}\, \omega_{n/2}^{-k'} = (-1)^{k_0}\, \omega_{n/2}^{-k'}.
\end{aligned}$$

Equation (3.7) gives us our divide-and-conquer algorithm. First, we recursively calculate the Fourier transforms $\tilde{f}_{\text{even}}$ and $\tilde{f}_{\text{odd}}$. For each of the n values of k, we multiply $\tilde{f}_{\text{odd}}(k')$ by the "twiddle factor" $\omega_n^{-k'}$. Finally, depending on whether k_0 is 0 or 1, we add or subtract the result from $\tilde{f}_{\text{even}}(k')$ to obtain $\tilde{f}(k)$.

Let $T(n)$ denote the time it takes to do all this. Assuming again that we do our arithmetic to fixed precision, $T(n)$ is the time $2T(n/2)$ it takes to calculate $\tilde{f}_{\text{even}}$ and $\tilde{f}_{\text{odd}}$, plus $O(1)$ for each application of (3.7). This gives

$$T(n)=2T(n/2)+\Theta(n).$$

If we assume that n is a power of 2 so that it is even throughout the recursion, then $T(n)$ has the same kind of scaling as `Mergesort`,

$$T(n)=\Theta(n\log n).$$

This is known as the Fast Fourier Transform, or FFT for short.

What if n is not a power of 2? If n is composite, the divide-and-conquer idea still works, since if n has a factor p, we can divide the list into p sublists of size n/p. As Problem 3.14 shows, this gives a running time of $\Theta(n\log n)$ whenever n is a so-called "smooth" number, one whose largest prime factor is bounded by some constant. Another type of FFT, described in Problem 3.17, works when n is prime. Thus we can achieve a running time of $\Theta(n\log n)$ for any n.

The FFT plays an important role in astronomy today, but in ways that Ptolemy could never have imagined—searching for extrasolar planets, finding irregularities in the cosmic microwave background, and listening for gravitational waves. And as we will see in Chapter 15, a quantum version of the FFT is at the heart of Shor's quantum algorithm for FACTORING. 3.5

3.3 Dynamic Programming

> Turning to the succor of modern computing machines, let us renounce all analytic tools.
>
> Richard Bellman, *Dynamic Programming*

Most problems are hard because their parts interact—each choice we make about one part of the problem has wide-ranging and unpredictable consequences in the other parts. Anyone who has tried to pack their luggage in the trunk of their car knows just what we mean.

This makes it hard to apply a divide-and-conquer approach, since there is no obvious way to break the problem into subproblems that can be solved independently. If we try to solve the problem a little at a time, making a sequence of choices, then each such sequence creates a different subproblem that we have to solve, forcing us to explore an exponentially branching tree of possible choices.

However, for some problems these interactions are limited in an interesting way. Rather than each part of the problem affecting every other part in a global fashion, there is a relatively narrow channel through which these interactions flow. For instance, if we solve part of the problem, the remaining subproblem might be a function, not of the entire sequence of choices we have made up to this point, but only of the most recent one. As a consequence, many sequences of choices lead to the same subproblem. Once we solve this subproblem, we can reuse its solution elsewhere in the search tree. This lets us "fold up" the search tree, collapsing it from exponential to polynomial size.

This may all seem a bit abstract at this point, so let's look at two examples: typesetting books and aligning genomes.

3.3.1 *Moveable Type*

Anyone who would letterspace lower case would steal sheep.
Frederic Goudy

The book you hold in your hands owes much of its beauty to the TEX typesetting system. One of the main tasks of such a system is to decide how to break a paragraph into lines. Once it chooses where the line breaks go, it "justifies" each line, stretching the spaces between words so that the left and right margins are straight. Another option is to stretch out the spaces between the letters of each word, a practice called *letterspacing*, but we agree with the great font designer Mr. Goudy that this is an abomination.

Our goal is to place the line breaks in a way that is as aesthetic as possible, causing a minimum of stretching. For simplicity, we will ignore hyphenation, so that each line break comes between two words. Thus given a sequence of words $w_1,\ldots,w_n$, we want to choose locations for the line breaks. If there are $\ell+1$ lines, we denote the line breaks $j_1,\ldots,j_\ell$, meaning that there is a line break after w_{j_1}, after w_{j_2}, and so on. Thus the first line consists of the words $w_1,\ldots,w_{j_1}$, the second line consists of the words $w_{j_1+1},\ldots,w_{j_2}$, and so on.

Justifying each of these lines has an aesthetic cost. Let $c(i,j)$ denote the cost of putting the words $w_i,\ldots,w_j$ together on a single line. Then for a given set of line breaks, the total cost is

$$c(1,j_1)+c(j_1+1,j_2)+\cdots+c(j_\ell+1,n). \tag{3.8}$$

How might $c(i,j)$ be defined? If it is impossible to fit $w_i,\ldots,w_j$ on a single line, because their total width plus the $j-i$ spaces between them exceeds the width of a line, we define $c(i,j)=\infty$. If, on the other hand, we can fit them one a line with some room to spare, we define $c(i,j)$ as some increasing function of the amount E of extra room. Let L be the width of a line, and suppose that a space has width 1. Then

$$\begin{aligned} E &= L-\left(|w_i|+1+|w_{i+1}|+1+\cdots+1+|w_j|\right) \\ &= L-(j-i)-\sum_{t=i}^{j}|w_t|, \end{aligned}$$

where $|w_i|$ denotes the width of w_i. A typical cost function, which is not too different from that used in TEX, is

$$c(i,j)=\left(\frac{E}{j-i}\right)^3 .$$

Here $E/(j-i)$ is the factor by which we have to stretch each of the $j-i$ spaces to fill the line, and the exponent 3 is just an arbitrary way of preferring smaller stretch factors. This formula is undefined if $i=j$, i.e., if the line contains just a single word, and it also ignores the fact that we don't justify the last line of the paragraph. However, it is good enough to illustrate our algorithm.

With this definition of $c(i,j)$, or any other reasonable definition, our goal is to choose the line breaks so that the total cost (3.8) is minimized:

> TYPESETTING
>
> Input: A sequence of words $w_1,\ldots,w_n$
>
> Output: A sequence of line break locations $j_1,\ldots,j_\ell$ that minimizes the total cost of the paragraph

How can we solve this problem in polynomial time?

We could try a *greedy* algorithm, in which we fill each line with as many words as possible before moving on to the next. However, this is not always the right thing to do. Consider the following paragraph, which we have typeset greedily:

Daddy, please buy me a
little baby
humuhumunukunukuāpua'a!

Arguably, this looks better if we stretch the first line more in order to reduce the space on the second line,

Daddy, please buy
me a little baby
humuhumunukunukuāpua'a!

This is precisely the kind of sacrifice that a greedy algorithm would refuse to make. The difficulty is that the different parts of the problem interact—the presence of a word on the last line can change how we should typeset the first line.

So, what kind of algorithm might work? The number of possible solutions is 2^n, since we could in theory put a line break, or not, after every word. Of course, this includes absurd solutions such as putting every word on its own line, or putting the entire paragraph on a single line. But even if we already know that there are ℓ line breaks and $\ell+1$ lines, finding the optimal locations $j_1,\ldots,j_\ell$ is an ℓ-dimensional optimization problem. If $\ell = n/10$, say, corresponding to an average of 10 words per line, there are $\binom{n}{n/10} = 2^{\Omega(n)}$ possible solutions. Thus, *a priori*, this problem seems exponentially hard.

The key insight is that each line break cuts the paragraph into two parts, which can then be optimized independently. In other words, each line break blocks the interaction between the words before it and those after it. This doesn't tell us where to put a line break, but it tells us that once we make this choice, it cuts the problem into separate subproblems.

So, where should the first line break go? The cost of the entire paragraph is the cost of the first line, plus the cost of everything that comes after it. We want to minimize this total cost by putting the first line break in the right place. Let $f(i)$ denote the minimum cost of typesetting the words $w_i,\ldots,w_n$, so that $f(1)$ is the minimum cost of the entire paragraph. Then

$$f(1) = \min_j \left(c(1,j) + f(j+1)\right).$$

The same argument applies to the optimal position of the *next* line break, and so on. In general, if we have decided to put a line break just before the ith word, the minimum cost of the rest of the paragraph is

$$f(i) = \min_{j:i\le j<n} \left(c(i,j) + f(j+1)\right), \tag{3.9}$$

and the optimal position of the next line break is whichever j minimizes this expression. The base case is $f(n+1)=0$, since at that point there's nothing left to typeset. This gives us a recursive algorithm for $f(i)$, which we show in Figure 3.9.

Exercise 3.3 *Modify this algorithm so that it returns the best location j of the next line break.*

```
f(i) // the minimum cost of the paragraph starting with the ith word
begin
    if i = n+1 then return 0 ;
    f_min := +∞ ;
    for j = i to n do
        f_min := min(f_min, c(i,j) + f(j+1)) ;
    return f_min ;
end
```

FIGURE 3.9: The recursive algorithm for computing the minimum cost $f(i)$ of typesetting the part of the paragraph starting with the ith word.

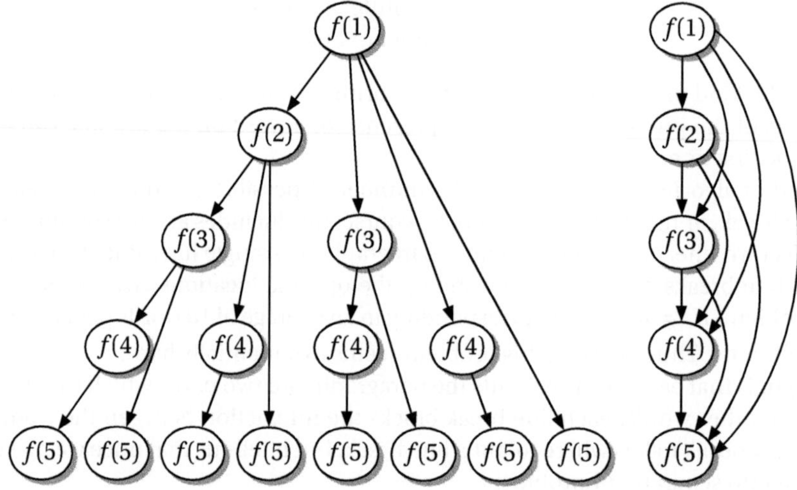

FIGURE 3.10: Calculating the optimal cost $f(1)$ using the recursive algorithm for TYPESETTING causes us to solve the same subproblems many times, taking exponential time. But if we memorize our previous results, we solve each subproblem only once. Here $n = 5$.

This is all well and good. However, if we're not careful, this algorithm will take exponential time. The reason is that it recalculates the same values of f many times. For instance, it calculates $f(4)$ whenever it calculates $f(1)$, $f(2)$, or $f(3)$; it calculates $f(3)$ whenever it calculates $f(1)$ or $f(2)$; and so on.

Exercise 3.4 *Show that the recursive algorithm for $f(1)$ will calculate $f(n)$ a total of 2^{n-2} times.*

To avoid this, we *memorize* values of the function we have already calculated, by placing each one in a table. Then, when we call the function for $f(i)$, we first check the table to see if we have already calculated it, and we only launch our recursive scheme if we haven't. By storing our previous results, we only have to calculate each $f(i)$ once (see Figure 3.10). Since there are n different values of i and the **for** loop runs through $O(n)$ values of j for each one, the total running time is $O(n^2)$.

This combination of recursion and memorization implements the algorithm from the top down. Another approach is to work from the bottom up—solving the simplest subproblems first, then the subproblems that depend on these, and so on. In this case, we would work backwards, first calculating $f(n)$, then $f(n-1)$, and so on until we reach the cost $f(1)$ of the entire paragraph. Programs based on these two implementations look rather different from each other, but the actual work done by the computer is essentially the same.

To summarize, the parts of the paragraph in TYPESETTING interact, but each line break blocks this interaction, and cuts the problem into independent subproblems. If we have chosen how to typeset part of the paragraph, then the subproblem we have left, namely typesetting the rest of the paragraph $w_i, \ldots, w_n$, depends only on our last choice of line break $j = i - 1$, and not on the entire sequence of choices before that. The total number of different subproblems we have to deal with is poly(n), one for each possible value of i, and it takes poly(n) time to combine these subproblems at each stage of the recursion. Thus if we save our previous results, solving each subproblem only once and reusing it when it appears again in the tree, the total running time is poly(n), and TYPESETTING is in P.

3.6

3.3.2 Genome Alignment

Let's use dynamic programming to solve another problem, which is important in genomics, spell checking, and catching term-paper plagiarism. Given two strings s and t, the *edit distance* $d(s,t)$ between them is the minimum number of insertions, deletions, or mutations needed to change s to t, where each mutation changes a single symbol. For instance, consider the following two important professions:

```
P A S T R Y C O   O K
  A S T R     O N O M E R
```

Such an arrangement of the two strings, showing which symbols correspond to each other and which are inserted or deleted, is called an *alignment*. Since this alignment involves deleting P, Y, and C, inserting N, E, and R, and mutating K to M, it shows that the edit distance between these two strings is at most 7. But how can we tell whether this is optimal?

Let's call EDIT DISTANCE and ALIGNMENT the problems of finding the edit distance and the optimal alignment respectively. Just as in TYPESETTING, the number of possible alignments grows exponentially in the length of the strings:

Exercise 3.5 *Suppose that s and t each have length n. Show that the number of possible alignments between them is at least 2^n. If you enjoy combinatorics, show that it is*

$$\sum_{j=0}^{n} \binom{n}{j}^2 = \binom{2n}{n}.$$

Hint: each alignment specifies a subset of the symbols of s, and a corresponding subset of the symbols of t.

So, *a priori*, it is not obvious that we can solve these problems in polynomial time.

However, we can again use dynamic programming for the following reason. Suppose we cut s into two parts, s_{left} and s_{right}. In the optimal alignment, this corresponds to cutting t somewhere, into t_{left} and t_{right}. Once we decide where the corresponding cut is, we can then find the optimal alignment of s_{left} with t_{left}, and, independently, the optimal alignment of s_{right} with t_{right}. Just as placing a line break separates a paragraph into two independent parts, making an initial choice about the alignment cuts it into two independent subproblems.

In particular, if we decide how to align the very beginning of the two strings, we can find the alignment between their remainders separately. So let s_1 denote the first symbol of s, let s' be the remainder of s with s_1 removed, and define t_1 and t' similarly. Now there are three possibilities. Either the optimal alignment consists of deleting s_1 and aligning s' with t; or it inserts t_1 and aligns s with t'; or it matches s_1 and t_1, mutating one into the other if they are different, and aligns s' with t'. The edit distance in the first two cases is $d(s',t)+1$ or $d(s,t')+1$ respectively. In the last case, it is $d(s',t')$ if s_1 and t_1 are the same, and $d(s',t')+1$ if they are different.

Since the edit distance is the minimum of these three possibilities, this gives us a recursive equation,

$$d(s,t)=\min\left(d(s',t)+1, d(s,t')+1, \begin{cases} d(s',t') & \text{if } s_1=t_1 \\ d(s',t')+1 & \text{if } s_1 \neq t_1 \end{cases}\right). \tag{3.10}$$

Evaluating (3.10) gives us a recursive algorithm, which calculates the edit distance $d(s,t)$ in terms of the edit distance of various substrings. Just as for our TYPESETTING algorithm, we need to memorize the results of subproblems we have already solved. As the following exercise shows, the number of different subproblems is again polynomial, so dynamic programming yields polynomial-time algorithms for EDIT DISTANCE and ALIGNMENT.

Exercise 3.6 *Show that if s and t are each of length n, there are only $O(n^2)$ different subproblems that we could encounter when recursively calculating $d(s,t)$.*

Exercise 3.7 *Write a recursive algorithm with memorization that outputs the optimal alignment of s and t, using an algorithm for the edit distance $d(s,t)$ as a subroutine.*

Both TYPESETTING and ALIGNMENT have a one-dimensional character, in which we solve a problem from left to right. The subproblem we have left to solve only "feels" our rightmost, or most recent, choices. Thus, while there are an exponential number of ways we could typeset the first half of a paragraph, or align the first halves of two strings, many of these lead to exactly the same remaining subproblem, and the number of different subproblems we ever need to consider is only polynomial.

To state this a little more abstractly, let's visualize the interactions between the parts of a problem as a graph. If this graph consists of a one-dimensional string, cutting it anywhere separates it into two pieces. Moreover, there are n places to cut a string of length n, and there are a polynomial number of substrings that can result from these cuts. For this reason, many problems involving strings or sequences can be solved by dynamic programming.

Strings are not the only graphs that can be cut efficiently in this way. As Problems 3.25, 3.26, and 3.28 show, dynamic programming can also be used to solve problems on trees and even on certain fractals. However, if the network of interactions between parts of a problem is too rich, it is too hard to cut into subproblems, and dynamic programming fails to give a polynomial-time algorithm.

3.7

3.4 Getting There From Here

> Go often to the house of thy friend, for weeds soon choke up the unused path.
>
> Scandinavian Proverb

Imagine that we are living on a graph. We start at one vertex, and we want to reach another. We ask a friendly farmer the classic question: can we get there from here?

REACHABILITY
Input: A (possibly directed) graph G and two vertices s, t
Question: Is there a path from s to t?

Now suppose the graph is *weighted*, so that for each pair of vertices i and j, the edge between them has a length w_{ij}. Then what is the shortest path from s to t?

SHORTEST PATH
Input: A weighted graph G and two vertices s, t
Question: How long is the shortest path from s to t?

Note that "shortest" here could also mean cheapest or fastest, if w_{ij} is measured in dollars or hours rather than in miles.

There are many ways to solve these two problems, including some of the algorithmic strategies we have already seen. However, so many other problems can be expressed in terms of REACHABILITY and SHORTEST PATH that they deserve to be considered algorithmic strategies in and of themselves. In this section we look at them from several points of view.

3.4.1 Exploration

Perhaps the simplest way to solve REACHABILITY is to start at the source s and explore outward, marking every vertex we can reach, until we have exhausted every possible path. This naive approach gives the algorithm `Explore` shown in Figure 3.11. At each step, we have a set Q of vertices waiting to be explored. We take a vertex u from Q, mark it, and add its unmarked neighbors to Q. When Q is empty, our exploration is complete, and we can check to see if t is marked.

The order in which `Explore` explores the graph depends on which vertex u we remove from Q, and this depends on what kind of data structure Q is. If Q is a *stack*, like a stack of plates, then it acts in a last-in, first-out way. When we ask it for a vertex u, it returns the one at the top, which is the one that was added most recently. In this case, `Explore` performs a *depth-first* search, pursuing each path as deeply as possible, following it until it runs out of unmarked vertices. It then backtracks to the last place where it had a choice, pursues the next path as far as possible, and so on.

Depth-first searches are easy to express recursively. As the program in Figure 3.12 calls itself, its children explore u's neighbors, its grandchildren explore u's neighbors' neighbors, and so on.

```
Explore(G,s,t)
input: a graph G and a vertex s
begin
    Q := {s} ;
    while Q is nonempty do
        remove a vertex u from Q ;
        mark u ;
        for all unmarked neighbors v of u do add v to Q
    end
end
```

FIGURE 3.11: This algorithm explores the graph and marks every vertex we can reach from s.

```
Explore(G,u)
begin
    mark u ;
    for all unmarked neighbors v of u do Explore(G,v)
end
```

FIGURE 3.12: A depth-first search exploration of the graph, written recursively.

On the other hand, Q could be a *queue*, like a line of people waiting to enter a theater. A queue operates in a first-in, first-out fashion, so the next u in line is the vertex that has been waiting in Q the longest. In that case, `Explore` performs a *breadth-first* search. It explores all of s's neighbors first, then s's neighbors' neighbors, and so on, expanding the set of marked vertices outward one layer at a time. If G is an unweighted graph, in which every edge has weight 1, the paths that `Explore` follows are among the shortest paths from s as shown in Figure 3.13.

3.4.2 *Middle-First Search*

Let's look at a rather different kind of algorithm for REACHABILITY. Recall from Section 2.4.2 that the *adjacency matrix* of a graph with n vertices is an $n \times n$ matrix A, where $A_{ij} = 1$ if there is an edge from i to j and 0 otherwise. Now consider the following useful fact, where $A^t = A \cdot A \cdots A$ (t times) denotes the tth matrix power of A:

$$(A^t)_{ij} \text{ is the number of paths of length } t \text{ from } i \text{ to } j\,.$$

For example,

$$(A^3)_{ij} = \sum_{k,\ell} A_{ik} A_{k\ell} A_{\ell j}$$

is the number of pairs k,ℓ such that there are edges from i to k, from k to ℓ, and from ℓ to j. Equivalently, it is the number of paths of the form $i \to k \to \ell \to j$. Consider the graph in Figure 3.14. The powers of its adjacency matrix are

$$A = \begin{pmatrix} 0 & 1 \\ 1 & 1 \end{pmatrix}, \quad A^2 = \begin{pmatrix} 1 & 1 \\ 1 & 2 \end{pmatrix}, \quad A^3 = \begin{pmatrix} 1 & 2 \\ 2 & 3 \end{pmatrix}.$$

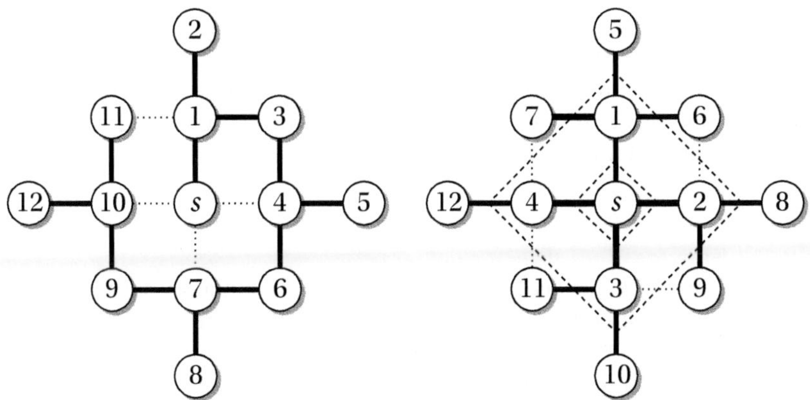

FIGURE 3.13: Left, a depth-first search which follows a path as deeply as possible, and then backtracks to its last choice. Right, a breadth-first search which builds outward from the source one layer at a time. The neighbors of each vertex are ordered north, east, south, west, and vertices are numbered in the order in which they are removed from the stack or queue.

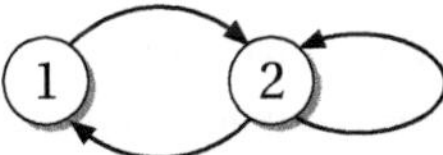

FIGURE 3.14: A little directed graph.

For instance, there are 3 paths of length 3 that begin and end at vertex 2.

If the graph does not already have *self-loops*—that is, edges from each vertex to itself, giving us the option of staying put on a given step—we can create them by adding the identity matrix $\mathbb{1}$ to A. Now consider the following exercise:

Exercise 3.8 *Given a graph G with n vertices and adjacency matrix A, show that there is a path from s to t if and only if $(\mathbb{1}+A)^{n-1}_{st}$ is nonzero.*

This offers us a nice way to solve REACHABILITY on all pairs of vertices s,t at once: simply raise the matrix $\mathbb{1}+A$ to the $n-1$st power. For simplicity, we might as well raise it to the nth power instead.

What is the most efficient way to compute $(\mathbb{1}+A)^n$? We could start with $B=\mathbb{1}$ and then repeat the following n times,

$$B \to B(\mathbb{1}+A).$$

This is a bit like breadth-first search, since each update extends the set of reachable vertices by another step. However, this will involve n matrix multiplications. A smarter approach is to start with $B=\mathbb{1}+A$ and square the matrix repeatedly:

$$B \to B^2,$$

or more explicitly,

$$B_{ij} \to \sum_k B_{ik} B_{kj}. \tag{3.11}$$

Just as in our divide-and-conquer algorithm for exponentiation in Section 3.2.2, we only need to repeat this update $\log_2 n$ times to reach the nth power, rounding up if n is not a power of 2.

What if we just want a Boolean matrix, where $B_{ij} = 1$ or 0 depending on whether there is a path from i to j or not, rather than counting paths of various lengths? We can do this in the style of (3.11), by replacing multiplication with AND and addition with OR. This gives

$$B_{ij} \to \bigvee_k (B_{ik} \wedge B_{kj}). \tag{3.12}$$

Here $\bigvee_k$ means the OR over all k, just as $\sum_k$ means the sum. This equation says, recursively, that we can get from i to j if there is a k such that we can get from i to k and from k to j. The base case of this recursion is the initial value $B = \mathbb{1} \vee A$—that is, j is reachable from i if $i = j$ or i and j are adjacent.

Since this strategy tries to find a vertex in between i and j, it is often called *middle-first search.* While it is not the most efficient in terms of time, the fact that it only needs $\log_2 n$ levels of recursion makes it very efficient in its use of memory. In Chapter 8, we will apply middle-first search to graphs of exponential size, whose vertices correspond to every possible state of a computer.

3.4.3 *Weighty Edges*

Now let's put a cost or length w_{ij} on each edge, and ask for a matrix B whose entries B_{ij} are the lengths of the shortest paths. This variant of the problem deserves a name:

ALL-PAIRS SHORTEST PATHS
Input: A weighted graph G with weights w_{ij} on each edge (i, j)
Output: A matrix where B_{ij} is the length of the shortest path from i to j

We start by defining w_{ij} for all pairs i, j, instead of just those connected by an edge of G. If $i \neq j$ and there is no edge from i to j, we set $w_{ij} = \infty$. Similarly, we set $w_{ii} = 0$ since it costs nothing to stay put. This gives us an $n \times n$ matrix W.

Now, in analogy to our matrix-product algorithm for REACHABILITY, we initially set $B_{ij} = w_{ij}$ for all i, j, and then "square" B repeatedly. But this time, we replace multiplication with addition, and replace addition with minimization:

$$B_{ij} \to \min_k (B_{ik} + B_{kj}). \tag{3.13}$$

In other words, the length B_{ij} of the shortest path from i to j is the minimum, over all k, of the sum of the lengths of the shortest paths from i to k and from k to j. We claim that, as before, we just need to update B according to (3.13) $\log_2 n$ times in order to get the correct value of B_{ij} for every pair.

This gives the algorithm shown in Figure 3.15 for ALL-PAIRS SHORTEST PATHS. For clarity, $B(m)$ denotes the value of B after m iterations of (3.13), so $B(0) = W$ and our final result is $B(\log_2 n)$. Assuming that min and + take $O(1)$ time, it is easy to see from these nested loops that this algorithm's total running time is $\Theta(n^3 \log n)$. Problem 3.34 shows how to improve this to $\Theta(n^3)$ by arranging these loops a little differently.

3.8 In a sense, this algorithm is an example of dynamic programming. Once we have chosen a midpoint k, finding the shortest path from i to j breaks into two independent subproblems—finding the shortest path from i to k and the shortest path from k to j. The pseudocode we give here is a "bottom-up" implementation, in which we calculate $B(m)$ from the previously-calculated values $B(m-1)$.

```
All-Pairs Shortest Paths(W)
input: a matrix W of weights w_ij
output: a matrix B where B_ij is the length of the shortest path from i to j
begin
    forall the i, j do B_ij(0) := w_ij ;
    for m = 1 to log_2 n do
        for i = 1 to n do
            for j = 1 to n do
                B_ij(m) := B_ij(m − 1) ;
                for k = 1 to n do
                    B_ij(m) := min(B_ij(m), B_ik(m − 1) + B_kj(m − 1));
    return B(log_2 n) ;
end
```

FIGURE 3.15: A middle-first algorithm for ALL-PAIRS SHORTEST PATHS. To distinguish each value of B from the previous one, we write $B(m)$ for the mth iteration of (3.13).

There is another reason that this algorithm deserves the name "dynamic." Namely, we can think of (3.13) as a *dynamical system,* which starts with the initial condition $B = W$ and then iterates until it reaches a fixed point.

Exercise 3.9 *Show how to modify these algorithms for* REACHABILITY *and* SHORTEST PATH *so that they provide the path in question, rather than just its existence or its length. Hint: consider an array that records, for each i and j, the vertex k that determined the value of B_{ij} in* (3.12) *or* (3.13).

Exercise 3.10 *Suppose some of the weights w_{ij} are negative. Can we still find the lengths of the shortest paths—some of which may be negative—by iterating* (3.13) *until we reach a fixed point? What happens if there is a* cycle *whose total length is negative?*

3.4.4 But How do we Know it Works?

Up to now, the *correctness* of our algorithms—that is, the fact that they do what they are supposed to do—has been fairly self-evident. As our algorithms get more complicated, it's important to discuss how to prove that they actually work.

Typically, these proofs work by induction on the number of layers of recursion, or the number of times a loop has run. We would like to establish that, after the algorithm has reached a certain stage, it has made some concrete type of progress—that it has solved some part of the problem, or reached a solution of a certain quality. Such a partial guarantee is often called a *loop invariant.* The next exercise asks you to use this approach to prove that our algorithm for ALL-PAIRS SHORTEST PATHS works.

Exercise 3.11 *Prove that, during our algorithm for* ALL-PAIRS SHORTEST PATHS, *$B_{ij}(m)$ is always an upper bound on the length of the shortest path from i to j. Then, show by induction on m that this algorithm satisfies the following loop invariant: after running the outermost loop m times, $B_{ij}(m)$ equals the length of the shortest path from i to j which takes 2^m or fewer steps in the graph. Conclude that as soon as $2^m \geq n$, the algorithm is complete.*

We conclude this section by urging the reader to solve Problems 3.20 and 3.21. These problems show that TYPESETTING and EDIT DISTANCE can both be recast in terms of SHORTEST PATH, demonstrating that SHORTEST PATH is capable of expressing a wide variety of problems.

3.5 When Greed is Good

We turn now to our next algorithmic strategy: greed. Greedy algorithms solve problems step-by-step by doing what seems best in the short term, and never backtracking or undoing their previous decisions. For many problems, this is a terrible idea, as current economic and environmental policies amply demonstrate, but sometimes it actually works. In this section, we look at a greedy algorithm for a classic problem in network design, and place it in a general family of problems for which greedy algorithms succeed.

3.5.1 Minimum Spanning Trees

The trees that are slow to grow bear the best fruit.

Molière

In the 1920s, Jindřich Saxel contacted his friend, the Czech mathematician Otakar Borůvka, and asked him how to design an efficient electrical network for South Moravia (an excellent wine-growing region). This led Borůvka to the following problem: we have a graph where each vertex is a city, and each edge e has a length or cost $w(e)$. We want to find a subgraph T that *spans* the graph, i.e., that connects all the vertices together, with the smallest total length $w(T)=\sum_{e\in T} w(e)$.

If the edge between two cities is part of a cycle, we can remove that edge and still get from one city to the other by going around the other way—so the minimum spanning subgraph has no cycles. Since a graph without cycles is a tree, what we are looking for is a *minimum spanning tree*. Thus Borůvka's problem is

MINIMUM SPANNING TREE
Input: A weighted connected graph $G=(V,E)$
Question: A spanning tree T with minimum total weight $w(T)$

We assume that G is connected, since otherwise asking for a spanning tree is a bit unfair.

How can we find the minimum spanning tree? Or, if there is more than one, one of the minimum ones? Once again, the number of possible solutions is exponentially large. If G is the *complete* graph on n vertices, in which all $\binom{n}{2}$ pairs of vertices have edges between them—as in Borůvka's original problem, since we could choose to lay an electrical cable between any pair of cities—then Problem 3.36 shows that the number of possible spanning trees is n^{n-2}. For $n=100$ this is larger than the number of atoms in the known universe, so we had better find some strategy other than exhaustive search.

Let's try a greedy approach. We grow the network step-by-step, adding edges one at a time until all the vertices are connected. We never add an edge between two vertices that are already connected to each other—equivalently, we never complete a cycle. Finally, we start by adding the lightest edges, using the heavier ones later if we have to. This gives *Kruskal's algorithm*, shown in Figure 3.17.

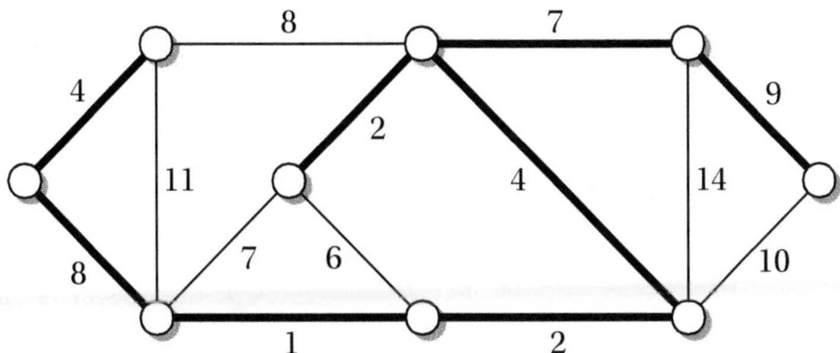

FIGURE 3.16: A weighted graph and a minimum spanning tree. Is it unique?

```
Kruskal(G)
input: a weighted graph G = (V, E)
output: a minimum spanning tree
begin
   F := ∅ ;
   sort E in order from lightest to heaviest ;
   for each edge e ∈ E do
      if adding e to F would not complete a cycle then add e to F ;
   return F;
end
```

FIGURE 3.17: Kruskal's algorithm for MINIMUM SPANNING TREE.

Let's prove that this algorithm works. At each step, the network F is a disjoint union of trees, which we call a *forest*. We start with no edges at all, i.e., with a forest of n trees, each of which is an isolated vertex. As F grows, we merge two trees whenever we add an edge between them, until F consists of one big tree. The following exercise shows that F spans the graph by the time we're done:

Exercise 3.12 *Show that if we complete the **for** loop and go through the entire list of edges, then F is a spanning tree.*

Alternatively, we can stop as soon as F contains $n-1$ edges:

Exercise 3.13 *Show that a forest with n vertices is a spanning tree if and only if it has $n-1$ edges.*

So, Kruskal's algorithm finds a spanning tree. But how do we know it finds one of the best ones? We would like to prove that adding the next-lightest edge is never a bad idea—that it's never a good idea to "sacrifice" by adding a heavier edge now in order to reduce the total weight later. We can prove this inductively using the following lemma.

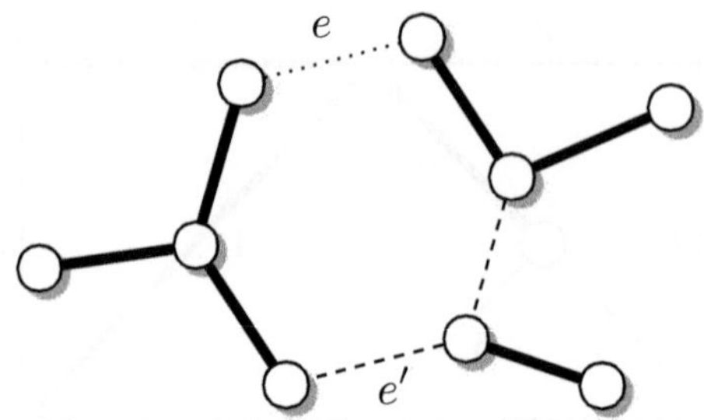

FIGURE 3.18: The proof of Lemma 3.1. Edges already in F are shown in bold, and additional edges in the spanning tree T are dashed. Adding e to T would complete a cycle, and we can obtain a new spanning tree T' by removing e'. If e is one of the lightest edges, T' is at least as light as T.

Lemma 3.1 *Suppose that F is a forest defined on the vertices of G, and that F is contained in a minimum spanning tree of G. Let e be one of the lightest edges outside F that does not complete a cycle, i.e., that connects one of F's trees to another. Then the forest $F \cup \{e\}$ is also contained in a minimum spanning tree.*

Proof Let T be a minimum spanning tree containing F, and assume that T does not contain e. Then T provides some other route from one end of e to the other, so adding e to T would complete a cycle. As shown in Figure 3.18, we could break this cycle by removing some other edge $e' \notin F$. This would give us a new spanning tree T' that also contains F, namely

$$T' = T \cup \{e\} - \{e'\}.$$

But we know that $w(e) \leq w(e')$ since by hypothesis e is one of the lightest edges we could have added to F. So T' is at least as light as T,

$$w(T') = w(T) + w(e) - w(e') \leq w(T).$$

Either we were wrong to assume that T is a minimum spanning tree, or both T and T' are minimum spanning trees. In either case the lemma is proved. □

Using Lemma 3.1 inductively, we see that at all times throughout Kruskal's algorithm, the forest F is a subgraph of some minimum spanning tree. In the terminology of Section 3.4.4, this is a loop invariant. This invariant holds all the way until F is itself is a spanning tree, so it must be a minimum one.

What is the running time of Kruskal's algorithm? Since adding e to F would complete a cycle if and only if there already a path from one of e's endpoints to the other, we can use one of our polynomial-time algorithms for REACHABILITY to check whether or not we should add e. This is far from the most efficient method, but it does show that MINIMUM SPANNING TREE is in P.

3.9

Exercise 3.14 *Run Kruskal's algorithm on the graph of Figure 3.16.*

Exercise 3.15 *Show that if the weights of the edges are distinct from each other, then the* MINIMUM SPANNING TREE *is unique.*

Exercise 3.16 *Find a polynomial-time algorithm that yields the* maximum*-weight spanning tree, and prove that it works.*

3.5.2 *Building a Basis*

Lemma 3.1 tells us that the greedy strategy for MINIMUM SPANNING TREE never steers us wrong—adding the next-lightest edge never takes us off the path leading to an optimal solution. However, this lemma and its proof seem rather specific to this one problem. Can we explain, in more abstract terms, what it is about the structure of MINIMUM SPANNING TREE that makes the greedy strategy work? Can we fit MINIMUM SPANNING TREE into a more general family of problems, all of which can be solved greedily?

One such family is inspired by finding a basis for a vector space. Suppose I have a list S of n-dimensional vectors. Suppose further that $|S| \geq n$ and that S has rank n, so that it spans the entire vector space. We wish to find a subset $F \subseteq S$ consisting of n linearly independent vectors that span the space as well. We can do this with the following greedy algorithm: start with $F = \emptyset$, go through all the vectors in S, and add each one to F as long as the resulting set is still linearly independent. There is never any need to go back and undo our previous decisions, and as soon as we have added n vectors to F, we're done.

This algorithm works because the property that a subset of S is linearly independent—or "independent" for short—obeys the following axioms. The first axiom allows us to start our algorithm, the second gives us a path through the family of independent sets, and the third ensures that we can always add one more vector to the set until it spans the space.

1. The empty set $\emptyset$ is independent.
2. If X is independent and $Y \subseteq X$, then Y is independent.
3. If X and Y are independent and $|X| < |Y|$, there is some element $v \in Y - X$ such that $X \cup \{v\}$ is independent.

Exercise 3.17 *Prove that these three axioms hold if S is a set of vectors and "independent" means linearly independent.*

A structure of this kind, where we have a set S and a family of "independent" subsets which obeys these three axioms, is called a *matroid*. In honor of the vector space example, an independent set to which nothing can be added without ceasing to be independent is called a *basis*.

What does this have to do with MINIMUM SPANNING TREE? Let S be the set E of edges of a graph, and say a subset $F \subseteq E$ is "independent" if it is a forest, i.e., if it has no cycles. Clearly the first two axioms hold, since the empty set is a forest and any subgraph of a forest is a forest. Proving the third axiom is trickier, but not too hard, and we ask you to do this in Problem 3.40. Thus, in any graph G the family of forests forms a matroid. Finally, a "basis"—a forest to which no edges can be added without completing a cycle—is a spanning tree.

Now suppose that each vector in S has some arbitrary weight, and that our goal is to find a basis whose total weight is as small as possible. We can generalize Kruskal's algorithm as follows: sort the elements of S from lightest to heaviest, start with $F = \emptyset$, and add each $v \in S$ to F if the resulting set is still independent. The following lemma, which generalizes Lemma 3.1 and which we ask you to prove in Problem 3.41, proves that this greedy algorithm works.

Lemma 3.2 *Let S be a set where the family of independent sets forms a matroid. Suppose an independent set F is contained in a minimum-weight basis. Let v be one of the lightest elements of S such that $F \cup \{v\}$ is also independent. Then $F \cup \{v\}$ is also contained in a minimum-weight basis.*

Therefore, for any matroid where we can check in polynomial time whether a given set is independent, the problem of finding the minimum-weight basis is in P.

3.5.3 Viewing the Landscape

A greedy algorithm is like a ball rolling down a hill, trying to find a point of minimum height. It rolls down as steeply as possible, until it comes to rest at a point where any small motion increases its altitude. The question is whether this is the *global* minimum, i.e., the lowest point in the entire world, or merely a *local* minimum where the ball has gotten stuck—whether jumping over a hill, or tunneling through one, could get us to an even lower point.

For some problems, such as MINIMUM SPANNING TREE, the landscape of solutions has one big valley, and a ball will roll straight to the bottom of it. Harder problems have a bumpy landscape, with an exponential number of valleys separated by forbidding mountain crags. In a landscape like this, greedy algorithms yield local optima—solutions such that any small change makes things worse. But in order to find the global optimum, we have to make large, complicated changes. In the absence of a good map of the landscape, or a good insight into its structure, our only recourse is an exhaustive search.

3.10 Of course, for a maximization problem the valleys in this metaphor become mountain peaks. Climbing uphill takes us to a local maximum, but the highest peak may be far off in the mist. Next, we will explore a problem where a kind of greedy algorithm finds the global maximum, but only if we define the landscape in the right way.

3.6 Finding a Better Flow

Wouldn't it be wonderful if you could tell whether your current solution to life's problems is the best possible? And, if it isn't, if you could tell how to improve it? If you could answer these questions efficiently, you could find the optimal solution with a kind of greedy algorithm: start with any solution, and repeatedly improve it until you reach the best possible one. In this section, we will apply this strategy to an important type of network flow problem.

My graduate students, my spouse, and I are looking forward to attending a glorious conference, where we will discuss the latest and most beautiful advances in complexity theory and gorge ourselves on fine food. Money is no object. But unfortunately, I have left the travel arrangements until rather late, and there are only a few seats available on each airplane flight that can get us from our university, through various intermediate cities, to the location of the conference. How many of my students can I get there by one route or another?

We can formalize this problem as follows. I have a network G, namely a directed graph where each edge $e=(u,v)$ has a nonnegative integer capacity $c(e)$. I am trying to arrange a *flow* from a source vertex s to a destination t. This flow consists of assigning an integer $f(e)$ to each edge—the number of students who will take that flight—such that $0 \le f(e) \le c(e)$. Just as for electric current, the total flow in and out of any vertex other than s and t must be zero. After all, I don't wish to leave any students or spouses, or pick up new ones, at any of the intervening airports.

My goal is to maximize the total flow out of s and into t, which we call the *value* of the flow. This gives the following problem:

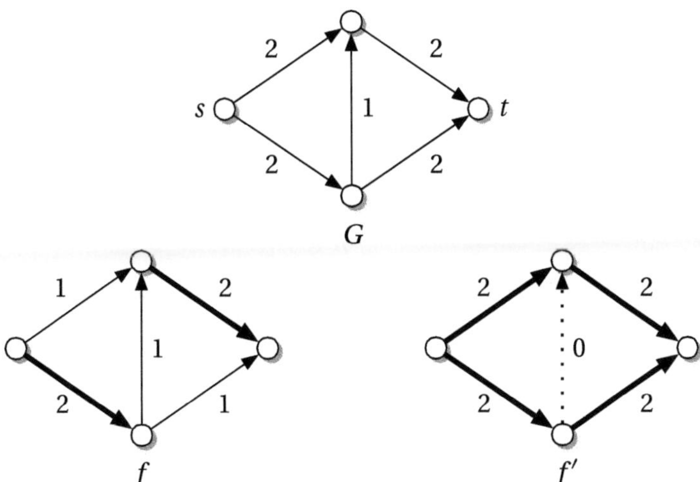

FIGURE 3.19: A network G and two possible flows on it, with edges shown dotted, solid or bold depending on whether the flow on that edge is 0, 1 or 2. On the lower left, a flow f whose value is 3. On the lower right, a better flow f' whose value is 4, which in this case is the maximum.

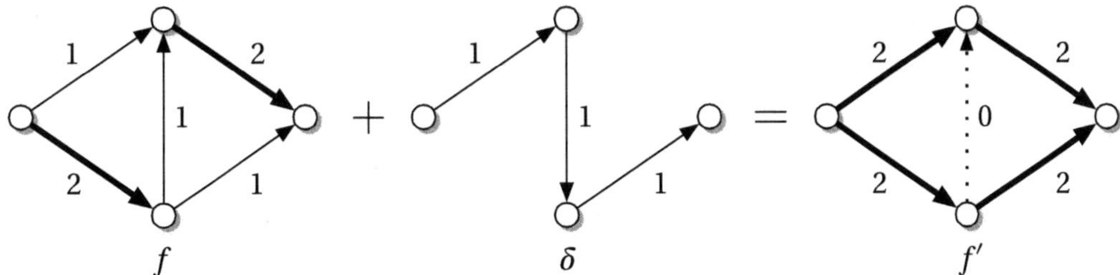

FIGURE 3.20: We can improve the flow, changing f to f', by adding flow along a path δ from s to t. In this case, one of the edges in this path is a reverse edge, and adding flow along it cancels f on the corresponding forward edge.

MAX FLOW
Input: A network G where each edge e has a nonnegative integer capacity $c(e)$, and two vertices s, t
Question: What is the maximum flow from s to t?

As an example, Figure 3.19 shows a simple network, and two flows on it. The flow f shown on the lower left has a value of 3, while the maximum flow f', shown on the lower right, has a value of 4.

Now suppose that our current flow is f. As Figure 3.20 shows, we can improve f by adding flow along a path δ from s to t. When does such a path exist? We can only increase the flow along an edge e if there is unused capacity there. So, given a flow f, let us define a *residual network* G_f where each edge has capacity $c_f(e) = c(e) - f(e)$. If there is a path from s to t along edges with nonzero capacity in G_f, we can increase the flow on those edges.

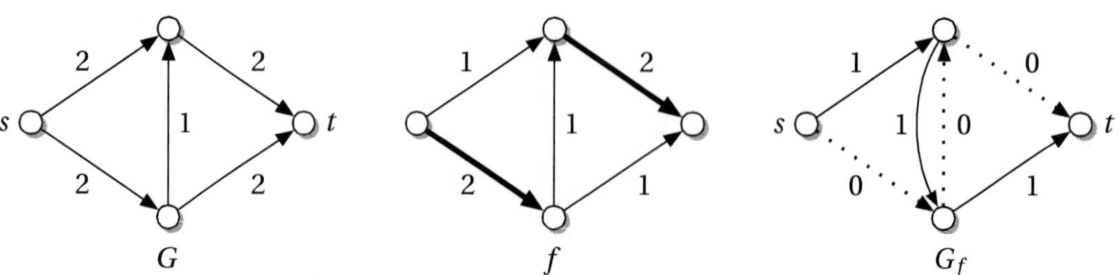

FIGURE 3.21: The residual network G_f for the flow f, showing the unused capacity $c_f(e) = c(e) - f(e)$ of each edge, and a reverse edge $\overline{e}$ with capacity $c_f(\overline{e}) = f(e)$. Other reverse edges are not shown.

However, in Figure 3.20, one of the edges of δ actually goes against the flow in f. By adding flow along this edge, we can cancel some—in this case, all—of the flow on the corresponding edge of G. To allow for this possibility in our residual graph G_f, for each "forward" edge $e = (u, v)$ of G, we also include a "reverse" edge $\overline{e} = (v, u)$. Since we can cancel up to $f(e)$ of the flow along e without making the flow negative, we give these reverse edges a capacity $c_f(\overline{e}) = f(e)$ as shown in Figure 3.21.

Given a flow f, we will call a path from s to t along edges with nonzero capacity in the residual network G_f an *augmenting path.* Now consider the following theorem:

Theorem 3.3 *A flow f is maximal if and only if there is no augmenting path. If there is an augmenting path, increasing the flow along it (and decreasing the flow wherever δ follows reverse edges) produces a new flow f' of greater value.*

Proof First suppose that an augmenting path δ exists. Each edge of δ is either a forward edge e where $f(e) < c(e)$ and $f(e)$ can be increased, or a reverse edge $\overline{e}$ where $f(e) > 0$ and $f(e)$ can be decreased. Thus adding a unit of flow along δ gives a new flow f' with $0 \le f'(e) \le c(e)$ along every edge, and whose value is greater than that of f.

Conversely, suppose that there is a flow f' whose value is greater than that of f. We can define a flow Δ on G_f as follows:

$$\Delta(e) = \max\left(0, f'(e) - f(e)\right), \ \Delta(\overline{e}) = \max\left(0, f(e) - f'(e)\right).$$

In other words, we put flow on e if $f'(e) > f(e)$, and on $\overline{e}$ if $f'(e) < f(e)$.

It is easy to check that the total flow Δ in and out of any vertex other than s or t is zero. Moreover, the flow Δ on each edge of G_f is nonnegative and less than or equal to the capacity c_f, since if $f'(e) > f(e)$ we have

$$0 < \Delta(e) = f'(e) - f(e) \le c(e) - f(e) \le c_f(e),$$

and if $f'(e) < f(e)$ we have

$$0 < \Delta(\overline{e}) = f(e) - f'(e) \le f(e) = c_f(\overline{e}).$$

Thus Δ is a legal flow on G_f. Moreover, Δ has positive value, equal to the value of f' minus that of f. No such flow can exist unless there is a path from s to t along edges with nonzero capacity in G_f, and we are done. □

Theorem 3.3 gives us a simple method for telling whether a flow f is maximal, and to produce an improved flow if it is not. All we have to do is construct the residual network G_f and ask REACHABILITY if there is a path δ from s to t. If there is such a path, we find one, and increase the flow along it as much as possible, i.e., by the minimum of $c_f(e)$ among all the edges $e \in \delta$. We then recalculate the residual capacities in G_f, and continue until no augmenting paths remain. At that point, the current flow is maximal. This is called the *Ford–Fulkerson algorithm.*

The astute reader will immediately ask how many times we need to perform this improvement. Since the value of the flow increases by at least one each time, the number of iterations is at most the sum of all the capacities. This gives us a polynomial-time algorithm if the capacities are only polynomially large. However, in a problem of size n, i.e., described by n bits, the capacities could be n-bit numbers and hence exponentially large as a function of n. As Problem 3.43 shows, in this case the Ford–Fulkerson algorithm could take an exponentially large number of steps to find the maximum flow, if we choose our augmenting paths badly.

Luckily, as Problems 3.44 and 3.45 show, there are several ways to ensure that the total number of iterations is polynomial in n, even if the capacities are exponentially large. One is to use the shortest path from s to t in G_f, and another is to use the "fattest" path, i.e., the one with the largest capacity. Either of these improvements proves that MAX FLOW is in P.

A priori, one could imagine that the maximum flow is fractional—that at some vertices it splits the 3.11
flow into noninteger amounts. But the Ford–Fulkerson algorithm gives us the following bonus:

Exercise 3.18 *Prove that if the capacities $c(e)$ are integers, there is a maximal flow f where $f(e)$ is an integer for all e. Hint: use the fact that the Ford–Fulkerson algorithm works.*

Exercise 3.19 *Is the maximal flow always unique? If not, what kind of flow is the difference $\Delta(e)$ between two maximal flows?*

It's interesting to note that if we don't allow reverse edges—if we only allow improvements that increase the flow everywhere along some path in G from s to t—then we can easily get stuck in local optima. In fact, the flow f of Figure 3.19 is a local optimum in this sense. In a sense, reverse edges let us "backtrack" a little bit, and pull flow back out of an edge where we shouldn't have put it.

To put this another way, recall the landscape analogy from Section 3.5.3, but with hills instead of valleys, since this is a maximization problem. Each step of our algorithm climbs from the current flow to a neighboring one, trying to get as high as possible. Without reverse edges, this landscape can be bumpy, with multiple hilltops. By adopting a somewhat larger set of moves, we reorganize the landscape, changing it to a single large mountain that we can climb straight up.

3.7 Flows, Cuts, and Duality

In our last episode, my students and I were trying to attend a conference. Now suppose that an evil competitor of mine wishes to prevent us from presenting our results. He intends to buy up all the empty seats on a variety of flights until there is no way at all to get to the conference, forcing us to seek letters of transit from a jaded nightclub owner. How many seats does he need to buy?

Let's define a *cut* in a weighted graph as a set C of edges which, if removed, make it impossible to get from s to t. The *weight* of the cut is the sum of its edges' weights. Alternately, we can say that a cut is a

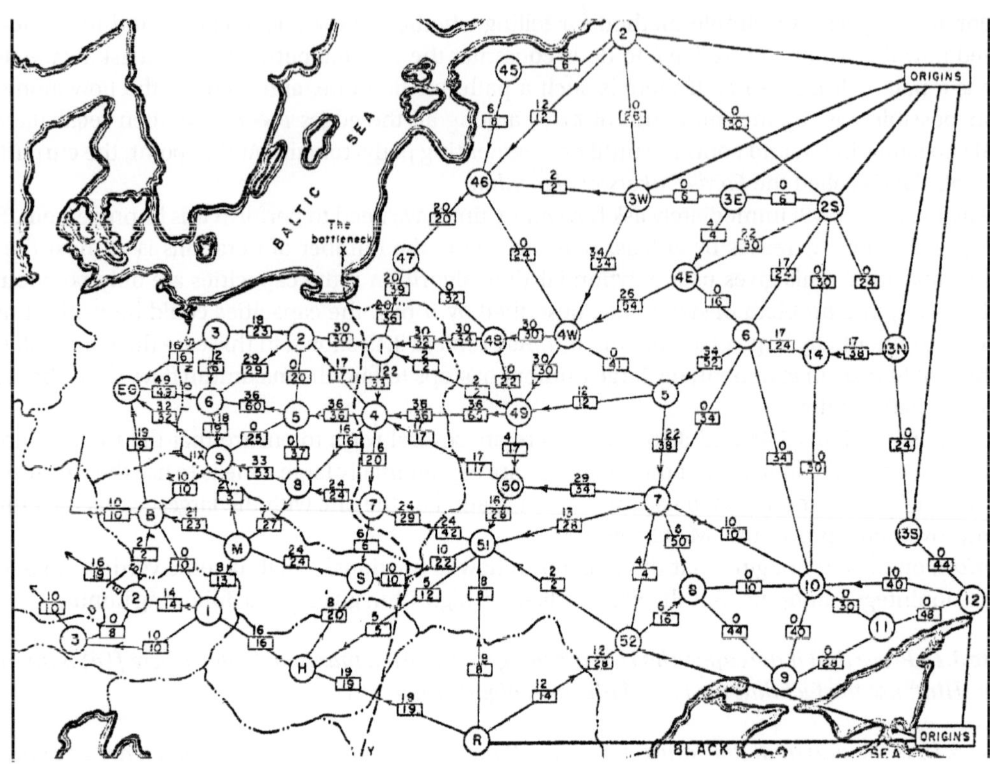

FIGURE 3.22: A MIN CUT problem from the Cold War. A 1955 technical report for the United States Air Force sought to find a "bottleneck" that would cut off rail transport from the Soviet Union to Europe.

partitioning of the vertices of G into two disjoint sets or "sides," S and T, such that $s \in S$ and $t \in T$. Then C consists of the edges that cross from S to T, and its weight is the sum of their capacities.

My competitor wishes to solve the following problem:

MIN CUT (s-t version)
Input: A weighted graph G and two vertices s, t
Question: What is the weight of the minimum cut that separates s from t?

In this section, we will see that MIN CUT and MAX FLOW have exactly the same answer—the weight of the minimum cut is exactly the value of the maximum flow. Thus MIN CUT is really just MAX FLOW in disguise.

To see this, we start by proving that

$$\text{value}(\text{MAX FLOW}) \leq \text{weight}(\text{MIN CUT}) . \tag{3.14}$$

Given a cut, the flow has to get from one side to the other, and the flow on any edge is at most its capacity. Therefore, the value of *any* flow is less than or equal to the weight of any cut. In particular, the value of the MAX FLOW is less than or equal to the weight of the MIN CUT.

The tighter statement

$$\text{value}(\text{MAX FLOW}) = \text{weight}(\text{MIN CUT}) \tag{3.15}$$

is less obvious. It certainly holds if G is simply a chain of edges from s to t, since then the MIN CUT consists of an edge with the smallest capacity, and the capacity of this "bottleneck" edge is also the value of the MAX FLOW. Does a similar argument work if there are many paths from s to t, and many vertices where the flow can branch and merge?

Indeed it does, and this follows from the same observations that led to the Ford–Fulkerson algorithm. Recall that Theorem 3.3 shows that if f is a maximal flow, there is no augmenting path. In other words, s is cut off from t by a set of edges whose residual capacity c_f is zero.

Now let S be the set of vertices reachable from s along edges with nonzero c_f, and let T be the rest of the vertices including t. Recall that $c_f(e) = c(e) - f(e)$. Since each edge that crosses from S to T has $c_f(e) = 0$, we have $f(e) = c(e)$. Thus each such edge is *saturated* by the flow f, i.e., used to its full capacity. The weight of the cut between S and T is the total weight of all these edges, and this equals the value of f.

Thus there exists a cut whose weight equals the maximum flow. But since any cut has weight at least this large, this cut is minimal, and (3.15) is proved. Since MAX FLOW is in P, this proves that MIN CUT is in P as well.

Exercise 3.20 *Solve* MIN CUT *in the graph G in Figure 3.19. Is the* MIN CUT *unique? If not, find them all.*

The inequality (3.14) creates a nice interplay between these problems, in which each one acts as a bound on the other. If I show you a cut of weight w, this is a proof, or "witness," that the MAX FLOW has a value at most w. For instance, the set of all edges radiating outward from s form a cut, and indeed no flow can exceed the total capacity of these edges. Conversely, if I show you a flow with value f, this is a proof that the MIN CUT has weight at least f.

From this point of view, (3.15) states that the MIN CUT is the *best possible upper bound* on the MAX FLOW, and the MAX FLOW is the *best possible lower bound* on the MIN CUT. What is surprising is that these bounds are tight, so that they meet in the middle. These two optimization problems have the same solution, even though they are trying to push in opposite directions.

This relationship between MAX FLOW and MIN CUT is an example of a much deeper phenomenon called *duality*. MAX FLOW is a constrained optimization problem, where we are trying to maximize something (the value of the flow) subject to a set of inequalities (the edge capacities). It turns out that a large class of such problems have "mirror images," which are minimization problems analogous to MIN CUT. We will discuss duality in a more general way in Section 9.5.

Finally, it is natural to consider the problem which tries to *maximize*, rather than minimize, the weight of the cut. Now that we know that MIN CUT is in P, the reader may enjoy pondering whether MAX CUT is as well. We will resolve this question, to some extent, in Chapter 5.

3.8 Transformations and Reductions

We close this chapter by introducing a fundamental idea in computational complexity—a transformation, or *reduction*, of one problem to another.

Suppose I am running a dating service. Some pairs of my clients are compatible with each other, and I wish to arrange as many relationships as possible between compatible people. Surprisingly, all my clients are monogamous, and have no interest in ménages à trois. So, my job is to find a set of compatible couples, such that no person participates in more than one couple.

I can represent my clients' compatibilities as a graph $G = (V, E)$, where each client is a vertex and each compatible couple is connected by an edge. I wish to find a *matching*, i.e., a subset $M \subseteq E$ consisting of disjoint edges, which covers as many vertices as possible. If I am especially lucky, there will be a *perfect matching*—an M that covers every vertex in G so that everyone has a partner.

We focus here on the unusual case where all my clients are heterosexual, in which case G is bipartite. This gives me the following problem,

> PERFECT BIPARTITE MATCHING
>
> Input: A bipartite graph G
>
> Question: Does G have a perfect matching?

More generally, I want to maximize the number of happy couples by finding the matching M with the largest number of edges:

> MAX BIPARTITE MATCHING
>
> Input: A bipartite graph G
>
> Question: What is the maximum matching?

There are an exponential number of possible matchings, and it is not obvious how to find the maximal one in polynomial time. The good news is that we already know how to do this. We just need to translate this problem into another one that we have solved before.

As Figure 3.23 shows, we can transform an instance of MAX BIPARTITE MATCHING into an instance of MAX FLOW. We turn each compatible couple (u, v) into a directed edge $u \to v$, pointing from left to right. We then add two vertices s, t with edges $s \to u$ for each u on the left, and $v \to t$ for each t on the right. Finally, we give every edge in this network a capacity 1.

We claim that the size of the maximum matching equals the value of the MAX FLOW on this network. We express this in the following exercise:

Exercise 3.21 *Show that there is a matching consisting of m edges if and only if this network has a flow of value m. Hint: use the fact proved in Exercise 3.18 that at least one of the maximal flows is integer-valued.*

Thus we can solve MAX BIPARTITE MATCHING by performing this transformation and then solving the resulting instance of MAX FLOW. Since MAX FLOW is in P, and since the transformation itself is easy to do in polynomial time, this proves that MAX BIPARTITE MATCHING and PERFECT BIPARTITE MATCHING are in P.

This type of transformation is called a *reduction* in computer science. On one level, it says that we can solve MAX BIPARTITE MATCHING by calling our algorithm for MAX FLOW as a subroutine. But on another

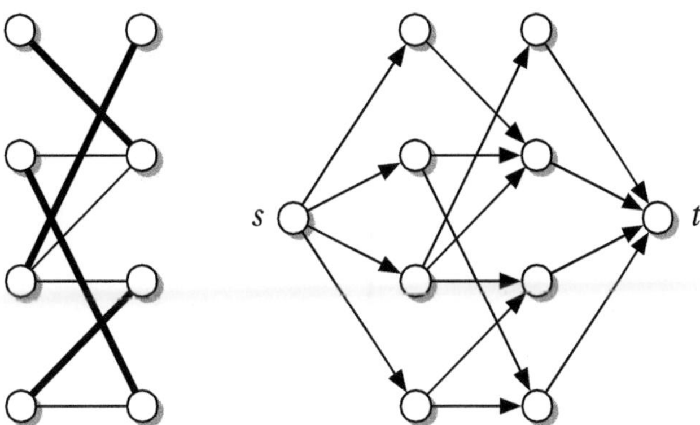

FIGURE 3.23: The reduction from MAX BIPARTITE MATCHING to MAX FLOW. The bipartite graph on the left has a perfect matching, and this corresponds to a flow of value 4 in the directed graph on the right. All edges have capacity 1.

level it says something much deeper—that MAX BIPARTITE MATCHING *is no harder than* MAX FLOW. We write this as an inequality,

$$\text{MAX BIPARTITE MATCHING} \leq \text{MAX FLOW}. \tag{3.16}$$

Consider the fact that, before we saw the polynomial-time algorithm for MAX FLOW, it was not at all obvious that either of these problems are in P. This reduction tells us that if MAX FLOW is in P, then so is MAX BIPARTITE MATCHING. In other words, as soon as we find a polynomial-time algorithm for MAX FLOW we gain one for MAX BIPARTITE MATCHING as well.

Reductions have another important application in computational complexity. Just as a reduction $A \leq B$ shows that A *is at most as hard as* B, it also shows that B *is at least as hard as* A. Thus we get a conditional lower bound on B's complexity as well as a conditional upper bound on A's. In particular, just as $B \in \mathsf{P}$ implies $A \in \mathsf{P}$, it is equally true that $A \notin \mathsf{P}$ implies $B \notin \mathsf{P}$.

If the reader finds the word "reduction" confusing, we sympathize. Saying that A can be reduced to B makes it sound as if B is smaller or simpler than A. In fact, it usually means the reverse: A can be viewed as a special case of B, but B is more general, and therefore harder, than A. Later on, we will put this another way—that B is *expressive enough* to describe the goals and constraints of A.

As we will discuss below, it is generally very hard to prove that a problem is outside P. However, for many problems B we are in the following curious situation: thousands of different problems can be reduced to B, and after decades of effort we have failed to find polynomial-time algorithms for any of them. This gives us very strong evidence that B is outside P. As we will see in the next two chapters, this is exactly the situation for many of the search and optimization problems we care about.

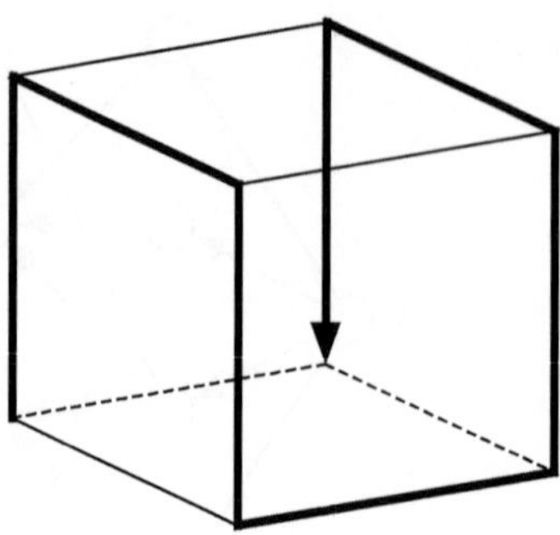

FIGURE 3.24: A Hamiltonian path on the 3-dimensional hypercube.

Problems

Problems worthy of attack
prove their worth by hitting back.

Piet Hein

3.1 Recursive priests are optimal. Prove that the recursive algorithm for the Towers of Hanoi produces the best possible solution, and that this solution is unique: i.e., that it is impossible to solve the puzzle in less than $2^n - 1$ moves, and that there is only one way to do it with this many. Hint: prove this by induction on n. A recursive algorithm deserves a recursive proof.

3.2

3.2 Hamilton visits Hanoi. Find a mapping between the recursive solution of the Towers of Hanoi and a Hamiltonian path on an n-dimensional hypercube (see Figure 3.24). They both have length $2^n - 1$, but what do the vertices and edges of the cube correspond to?

3.3 Hanoi, step by step. Find an *iterative*, rather than recursive, algorithm for the Towers of Hanoi. In other words, find an algorithm that can look at the current position and decide what move to make next. If you like, the algorithm can have a small amount of "memory," such as remembering the most recent move, but it should not maintain a stack of subproblems.

3.4 Our first taste of state space. The state space of the Towers of Hanoi puzzle with 2 disks can be viewed as as a graph with 9 vertices, as shown in Figure 3.25. Describe the structure of this graph for general n, and explain what path the optimal solution corresponds to.

3.5 Four towers. What if there are 4 pegs in the Towers of Hanoi, instead of 3? Here is one possible solution. We will assume that n is a *triangular number,*

$$n_k = 1+2+3+\cdots+k = k(k+1)/2\,.$$

Then to move n_k disks, recursively move the $n_{k-1} = n_k - k$ smallest disks, i.e., all but the k largest, to one of the four pegs; then, using the other 3 pegs, move the k largest disks using the solution to the 3-peg puzzle; then recursively

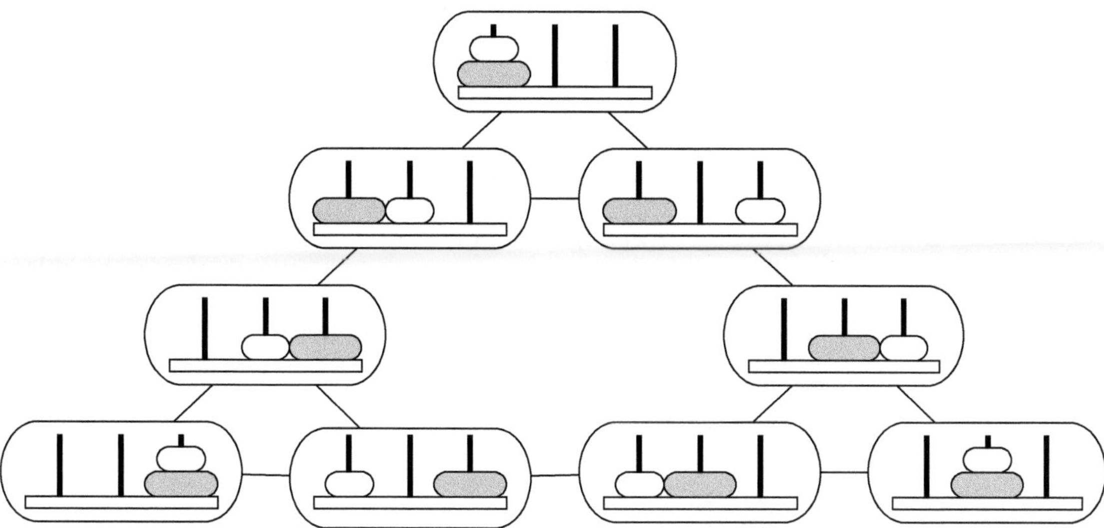

FIGURE 3.25: The state space of all possible positions of the Towers of Hanoi with $n = 2$. Each vertex is a position, and each edge is a legal move.

move the n_{k-1} smallest ones on top of the k largest, at which point we're done. Show that the total number of moves done by this algorithm obeys the equation

$$f(n_k) = 2f(n_{k-1}) + 2^k - 1$$

and that, with the base case $f(n_0) = 0$, the solution is

$$f(n_k) = (k-1)2^k + 1 = \Theta\left(\sqrt{n}\, 2^{\sqrt{2n}}\right).$$

Can you generalize this to 5 or more pegs?

3.6 Pick a card. You walk into a room, and see a row of n cards. Each one has a number x_i written on it, where i ranges from 1 to n. However, initially all the cards are face down. Your goal is to find a *local minimum*: that is, a card i whose number is less than or equal to those of its neighbors, $x_{i-1} \geq x_i \leq x_{i+1}$. The first and last cards can also be local minima, and they only have one neighbor to compare to. There can be many local minima, but you are only responsible for finding one of them.

Obviously you can solve this problem by turning over all n cards, and scanning through them. However, show that you can find such a minimum by turning over only $O(\log n)$ cards.

3.7 The running time of `Mergesort`. Show that the equation for the number of comparisons that `Mergesort` performs on a list of size n,

$$T(n) = 2T(n/2) + n,$$

with the base case $T(1) = 0$, has the exact solution

$$T(n) = n \log_2 n,$$

where we assume that n is a power of 2. Do this first simply by plugging this solution into the equation and checking that it works. Then come up with a more satisfying explanation, based on the number of levels of recursion and the number of comparisons at each level.

We can improve this slightly using the fact that the merge operation takes at most $n-1$ comparisons. Prove that the exact solution to the equation

$$T(n)=2T(n/2)+n-1$$

is

$$T(n)=n\log_2 n-n+1,$$

where as before we assume that n is a power of 2.

3.8 More pieces. Generalizing the previous problem, show that for any positive constants a, b, the equation

$$T(n)=a\,T(n/b)+n^{\log_b a},$$

with the base case $T(1)=0$, has the exact solution

$$T(n)=n^{\log_b a}\log_b n,$$

where we assume that n is a power of b.

3.9 Integrating quicksort. Let's solve the equation (3.4) for the average running time of `Quicksort`. First, show that if n is large, we can replace the sum by an integral and obtain

$$T(n)=n+\frac{2}{n}\int_0^n T(x)\,\mathrm{d}x. \tag{3.17}$$

Since we expect the running time to be $\Theta(n\log n)$, let's make a guess that

$$T(n)=An\ln n$$

for some constant A. Note that we use the natural logarithm since it is convenient for calculus. Substitute this guess into (3.17) and show that $A=2$, so that $T(n)\approx 2n\ln n$ as stated in the text.

3.10 Better pivots. In general, the running time $T(n)$ of `Quicksort` obeys the equation

$$T(n)=n+\int_0^n P(r)(T(r)+T(n-r))\,\mathrm{d}r,$$

where $P(r)$ is the probability that the pivot has rank r. For instance, (3.4) is the special case where the pivot is a random element, so $P(r)=1/n$ for all r.

One common improvement to `Quicksort` is to choose *three* random elements, and let the pivot be the median of these three. Calculate the probability distribution $P(r)$ of the pivot's rank in this case. Solve this equation in the limit of large n by converting it to an integral as in Problem 3.9, and again try a solution of the form $T(n)=An\ln n$. How much does this technique improve the constant A? What happens to A if the pivot is the median of 5 elements, 7 elements, and so on?

3.11 Partway to the median. Imagine that the pivot in `Quicksort` always divides the list into two sublists of size γn and $(1-\gamma)n$, so that the number of comparisons obeys the equation

$$T(n)=T(\gamma n)+T((1-\gamma)n)+n.$$

Show that $T(n)=A(\gamma)\,n\ln n$, and determine the constant $A(\gamma)$. How does $A(\gamma)$ behave when γ is close to 0 or to 1?

3.12 Factoring with factorials. Consider the following problem:

MODULAR FACTORIAL
Input: Two n-digit integers x and y
Output: $x! \bmod y$

This sounds a lot like MODULAR EXPONENTIATION, and it is tempting to think that MODULAR FACTORIAL is in P as well. However, show that

$$\text{FACTORING} \leq \text{MODULAR FACTORIAL},$$

in the sense that we can solve FACTORING by calling a subroutine for MODULAR FACTORIAL a polynomial number of times. Thus if MODULAR FACTORIAL is in P, then FACTORING is in P as well. Hint: first note that y is prime if and only if $\gcd(x!, y) = 1$ for all $x < y$.

3.13 Gamma function identities. (A follow-up to the previous problem.) There is a natural generalization of the factorial to noninteger values, called the *Gamma function*:

$$\Gamma(x) = \int_0^\infty t^{x-1} \mathrm{e}^{-t}\, \mathrm{d}t\,. \tag{3.18}$$

We have $\Gamma(x+1) = x\Gamma(x)$, and in particular $\Gamma(x) = (x-1)!$ if x is an integer. But $\Gamma(1/2)$, for instance, is $\sqrt{\pi}$.

Now consider the following formula, which *almost* gives a divide-and-conquer algorithm for $\Gamma(x)$:

$$\Gamma(x)\Gamma(x+1/2) = 2^{1-2x}\sqrt{\pi}\,\Gamma(2x)\,. \tag{3.19}$$

Suppose there were an identity of the form

$$\Gamma(2x) = f(x)\Gamma(x)^2$$

for some function f such that, when x is an integer, $f(x) \bmod y$ can be computed in polynomial time for n-digit integers x and y. Show that then MODULAR FACTORIAL, and therefore FACTORING, would be in P. 3.12

3.14 FFTs for smooth numbers. Let's generalize the running time of the Fast Fourier Transform to values of n other than powers of 2. If n has a factor p and we divide the list into p sublists of size n/p, argue that the running time obeys

$$T(n) = p\,T(n/p) + pn\,,$$

where for the purposes of elegance we omit the Θ on the right-hand-side. Let $\{p_i\}$ be the prime factorization of n, with each prime repeated the appropriate number of times: i.e., $n = \prod_i p_i$. Taking the base case $T(p) = p^2$ for prime p, show that the running time is

$$T(n) = n \sum_i p_i\,.$$

Now, a number is called *q-smooth* if all of its prime factors are smaller than q. Show that if n is q-smooth for some constant q, the running time is $\Theta(n \log n)$.

3.15 Convolutions and polynomials. Given two functions f, g, their *convolution* $f \star g$ is defined as

$$(f \star g)(t) = \sum_s f(s)\, g(t-s)\,.$$

Convolutions are useful in signal processing, random walks, and fast algorithms for multiplying integers and polynomials. For instance, suppose that we have two polynomials whose tth coefficients are given by $f(t)$ and $g(t)$ respectively:

$$P(z)=\sum_t f(t)z^t\,, \quad Q(z)=\sum_t g(t)z^t\,.$$

Then show that the coefficients of their product are given by the convolution of f and g,

$$P(z)Q(z)=\sum_t (f\star g)(t)z^t\,.$$

Now, as in Problem 2.14, let's represent integers as polynomials, whose coefficients are the integers' digits. For instance, we can write $729=P(10)$ where $P(z)=7z^2+2z+9$. Show that, except for some carrying, the product of two integers is given by the product of the corresponding polynomials, and so in turn by the convolution of the corresponding functions.

3.16 Convolving in Fourier space. Continuing from the previous problem, suppose we have two functions $f(t), g(t)$ defined on the integers mod n. Their convolution is

$$(f\star g)(t)=\sum_{s=0}^{n-1} f(s)\,g(t-s),$$

where $t-s$ is now evaluated mod n. How long does it take to calculate $f\star g$ from f and g? Evaluating the sum directly for each t takes $O(n^2)$ time, but we can do much better by using the Fourier transform.

Show that the Fourier transform of the convolution is the product of the Fourier transforms. More precisely, show that for each frequency k from 0 to $n-1$,

$$\widetilde{(f\star g)}(k)=\tilde{f}(k)\,\tilde{g}(k).$$

Thus we can find $f\star g$ by Fourier transforming f and g, multiplying them together, and then inverse Fourier transforming the result.

Conclude that if we use the Fast Fourier Transform, we can convolve two functions defined on the integers mod n in $O(n\log n)$ time. If necessary, we can "pad" the two functions out, repeating one and adding zeros to the other, so that n is a power of 2. This is the heart of the Schönhage–Strassen algorithm for integer multiplication discussed in Note 2.5. It is also at the heart of virtually all modern signal processing.

3.17 FFTs for primes. The Fast Fourier Transform described in the text works well if n has many small factors. But what do we do if n is prime?

First, note that when the frequency is zero, the Fourier transform is just proportional to the average, $\tilde{f}(0)=(1/\sqrt{n})\sum_t f(t)$, which we can easily calculate in $O(n)$ time. So, we focus our attention on the case where $k\neq 0$, and write (3.6) as follows:

$$\tilde{f}(k)=\frac{1}{\sqrt{n}}\sum_{t=0}^{n-1} f(t)\,\mathrm{e}^{-2\imath\pi kt/n}=\frac{1}{\sqrt{n}}\left(f(0)+S(k)\right)$$

where

$$S(k)=\sum_{t=1}^{n-1} f(t)\,\mathrm{e}^{-2\imath\pi kt/n}\,.$$

Now, for any prime n, there is an integer a whose powers $1, a, a^2, \ldots, a^{n-2}$, taken mod n, range over every possible t from 1 to $n-1$. Such an a is called a *primitive root*; algebraically, a is a generator of the multiplicative group $\mathbb{Z}_n^*$ (see Appendix A.7). By writing $k=a^\ell$ and $t=a^m$ and rearranging, show that S is the convolution of two functions, each of which is defined on the integers mod $n-1$. Since in the previous problem we showed how to convolve two functions in time $O(n\log n)$, this gives a Fast Fourier Transform even when n is prime.

3.18 Dynamic Fibonacci. Suppose I wish to calculate the nth Fibonacci number recursively, using the equation $F_\ell = F_{\ell-1} + F_{\ell-2}$ and the base case $F_1 = F_2 = 1$. Show that if I do this without remembering values of F_ℓ that I calculated before, then the number of recursive function calls is exponentially large. In fact, it is essentially F_ℓ itself. Then show that if I memorize previous values, calculating F_ℓ takes only $O(\ell)$ function calls. (Of course, this is exponential in the number of digits of ℓ.)

3.19 Divide-and-conquer Fibonacci. Let F_ℓ denote the ℓth Fibonacci number. Starting from the equation $F_\ell = F_{\ell-1} + F_{\ell-2}$ and the base case $F_1 = F_2 = 1$, prove that for any ℓ and any $m \geq 1$ we have

$$F_\ell = F_{m+1}\, F_{\ell-m} + F_m\, F_{\ell-m-1}\,. \tag{3.20}$$

In particular, prove the following:

$$\begin{aligned} F_{2\ell} &= F_\ell^2 + 2F_\ell\, F_{\ell-1} \\ F_{2\ell+1} &= F_{\ell+1}^2 + F_\ell^2\,. \end{aligned} \tag{3.21}$$

Then show that if ℓ and p are n-bit numbers, we can calculate $F_\ell \bmod p$ in poly(n) time. Hint: using (3.21) naively gives a running time that is polynomial in ℓ, but not in $n = \log_2 \ell$.

3.20 Getting through the paragraph. Show that the TYPESETTING problem can be described as SHORTEST PATH on a weighted graph with a polynomial number of vertices. What do the vertices of this graph represent, and what are the weights of the edges between them?

3.21 Alignments are paths. At first, calculating the minimum edit distance $d(s,t)$ seems like finding a shortest path in an exponentially large graph, namely the graph of all strings of length $n = \max(|s|,|t|)$ where two strings are neighbors if they differ by a single insertion, deletion, or mutation. However, it can be reduced to SHORTEST PATH on a graph of size $O(n^2)$ in the following way. For each $0 \leq i \leq |s|$ and $0 \leq j \leq |t|$, let v_{ij} be a vertex corresponding to a point in the alignment that has already accounted for the first i symbols of s and the first j symbols of t. Show how to assign weights to the edges between these vertices so that $d(s,t)$ is the length of the shortest path from $v_{0,0}$ to $v_{|s|,|t|}$.

3.22 Increasing subsequences. Given a sequence of integers $s_1, s_2, \ldots, s_n$, an *increasing subsequence* of length k is a series of indices $i_1 < i_2 < \cdots < i_k$ such that $s_{i_1} < s_{i_2} < \cdots < s_{i_k}$. For instance, the sequence

$$6,3,4,8,1,5,7,2,9$$

has an increasing subsequence of length 5, namely

$$3,4,5,7,9\,.$$

Even though there are an exponential number of possible subsequences, show that the problem of finding the longest one can be solved in polynomial time.

3.23 A pigeon ascending. Now prove that any sequence $s_1, \ldots, s_n$ of distinct integers has either an increasing subsequence, or a decreasing subsequence, of length at least $\lceil \sqrt{n} \rceil$. Hint: for each i, let a_i and b_i, respectively, be the length of the longest increasing and decreasing subsequence ending with s_i. Then show that for every $i \neq j$, either $a_i \neq a_j$ or $b_i \neq b_j$, and use the pigeonhole principle from Problem 1.2.

3.13

3.24 Multiplying matrices mellifluously. Suppose that I have a sequence of matrices, $M^{(1)}, \ldots, M^{(n)}$, where each $M^{(t)}$ is an $a_t \times b_t$ matrix and where $b_t = a_{t+1}$ for all $1 \leq t < n$. I wish to calculate their matrix product $M = \prod_{t=1}^n M^{(t)}$, i.e., the matrix such that

$$M_{ij} = \sum_{\ell_1=1}^{b_1} \sum_{\ell_2=1}^{b_2} \cdots \sum_{\ell_{n-1}}^{b_{n-1}} M^{(1)}_{i,\ell_1} M^{(2)}_{\ell_1,\ell_2} \cdots M^{(n)}_{\ell_{n-1},j}\,.$$

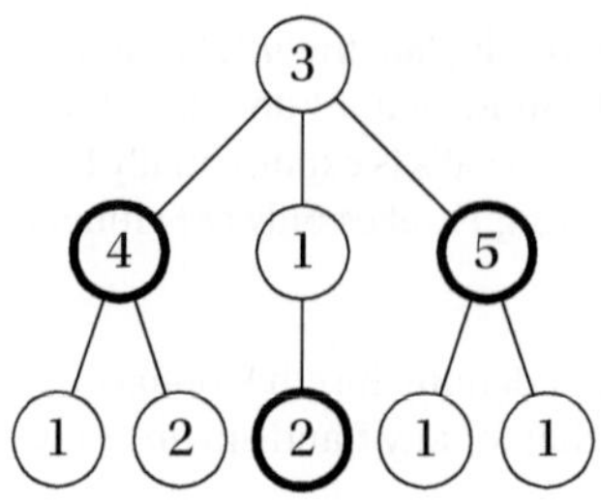

FIGURE 3.26: A tree with vertex weights and its maximum-weight independent set.

However, the number of multiplications I need to do this depends on how I parenthesize this product. For instance, if $M^{(1)}$ is a $1 \times k$ row vector, $M^{(2)}$ and $M^{(3)}$ are $k \times k$ matrices, and $M^{(4)}$ is a $k \times 1$ column vector, then

$$M = \left(M^{(1)} \cdot M^{(2)}\right) \cdot \left(M^{(3)} \cdot M^{(4)}\right)$$

takes $k^2 + k^2 + k$ multiplications, while

$$M = M^{(1)} \cdot \left(\left(M^{(2)} \cdot M^{(3)}\right) \cdot M^{(4)}\right)$$

takes $k^3 + k^2 + k$ multiplications. Show how to find the optimal parenthesization, i.e., the one with the smallest number of multiplications, in polynomial time.

3.25 Prune and conquer. Suppose I have a graph $G = (V, E)$ where each vertex v has a weight $w(v)$. An *independent set* is a subset S of V such that no two vertices in S have an edge between them. Consider the following problem:

MAX-WEIGHT INDEPENDENT SET
Input: A graph $G = (V, E)$ with vertex weights $w(v)$
Question: What is the independent set with the largest total weight?

Use dynamic programming to show that, in the special case where G is a tree, MAX-WEIGHT INDEPENDENT SET is in P.

Hint: once you decide whether or not to include the root of the tree in the set S, how does the remainder of the problem break up into pieces? Consider the example in Figure 3.26. Note that a greedy strategy doesn't work.

Many problems that are easy for trees seem to be very hard for general graphs, and we will see in Chapter 5 that MAX-WEIGHT INDEPENDENT SET is one of these. Why does dynamic programming not work for general graphs?

3.26 Prune and conquer again. Suppose I have a graph $G = (V, E)$ where each edge e has a weight $w(e)$. Recall that a *matching* is a subset $M \subseteq E$ such that no two edges in M share an endpoint. Consider the weighted version of MAX MATCHING, illustrated in Figure 3.27:

MAX-WEIGHT MATCHING
Input: A graph $G = (V, E)$ with edge weights $w(e)$
Question: What is the partial matching with the largest total weight?

Using dynamic programming, give a polynomial-time algorithm for this problem in the case where G is a tree, similar to that for Problem 3.25.

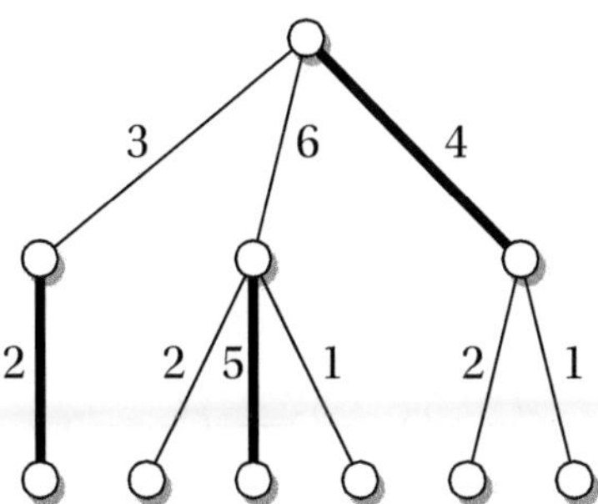

FIGURE 3.27: A tree with edge weights and its maximum-weight matching.

3.27 Perhaps donating it all to the public is a better idea. You own a piece of beachfront property, which we represent by the unit interval. There are n interested buyers, each of whom is willing to pay some price for a particular subinterval. Your goal, for better or worse, is to make as much money as possible: that is, to find the subset of buyers who will give you the largest total price, subject to the constraint that you can't sell overlapping subintervals.

There is no guarantee that the prices increase with the length of the subinterval, or even that the price of a subinterval is less than that of one that contains it, since different buyers value different sections of beach in arbitrary ways. Of course, the beach itself is public, and there is a large setback required in order to preserve the coral reef.

Formally, given the list of prices v_i and subintervals $[x_i, y_i]$ where $i = 1, \ldots, n$, your goal is to find the subset $S \subseteq \{1, \ldots, n\}$ that maximizes $\sum_{i \in S} v_i$ subject to the constraint that $[x_i, y_i] \cap [x_j, y_j] = \emptyset$ for all $i, j \in S$ with $i \neq j$. Show that we can think of this as a MAX-WEIGHT INDEPENDENT SET problem on a special kind of graph, called an *interval graph*, where each vertex represents a subinterval and two vertices are adjacent if the intervals overlap.

Consider various greedy algorithms for this problem, such as selling intervals in order starting with the highest price, or with the highest ratio of price to length. For each of these, construct examples for which they fail. Then show that we can solve this problem in polynomial time using dynamic programming.

3.28 Fractal algorithms. A *Sierpiński triangle* is a graph like that shown in Figure 3.28. It has a fractal structure, in which each triangle of side ℓ consists of three triangles of side $\ell/2$, which meet at three vertices. First, show that the number of vertices grows as $n = \Theta(\ell^{\log_2 3})$. Then show that if each vertex has a weight $w(v)$, which we are given as part of the input, then the corresponding MAX-WEIGHT INDEPENDENT SET problem (see Problem 3.25) can be solved in polynomial time.

3.29 Choices at the boundaries. Consider a graph consisting of n vertices arranged in a $\sqrt{n} \times \sqrt{n}$ square lattice, where each vertex is connected to its four nearest neighbors. Show that in this case, MAX-WEIGHT INDEPENDENT SET and MAX-WEIGHT MATCHING can be solved in $2^{O(\sqrt{n})}$ time. Hint: how can we divide this problem into subproblems, and how large is the boundary between them? What is the best running time we can achieve on a rectangle where one dimension is much larger than another, such as an $n^{1/3} \times n^{2/3}$ rectangle? What about an $n^{1/3} \times n^{1/3} \times n^{1/3}$ cube, or a d-dimensional cubic lattice?

3.30 Dynamic parsing. A *context-free grammar* is a model considered by the Chomsky school of formal linguistics.
The idea is that sentences are recursively generated from high-level mental symbols through a series of production
rules. For instance, if S, N, V, and A correspond to sentences, nouns, verbs, and adjectives, we might have rules such 3.14
as $S \to NVN$, $N \to AN$, and so on, and finally rules that replace these symbols with actual words. Each sentence

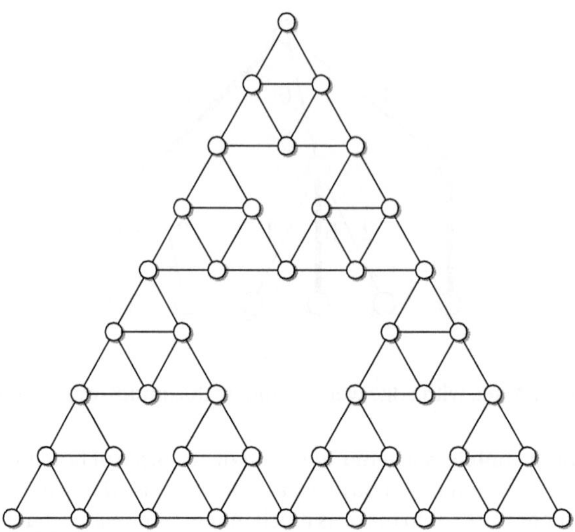

FIGURE 3.28: A Sierpiński triangle with $\ell = 8$ and $n = 42$.

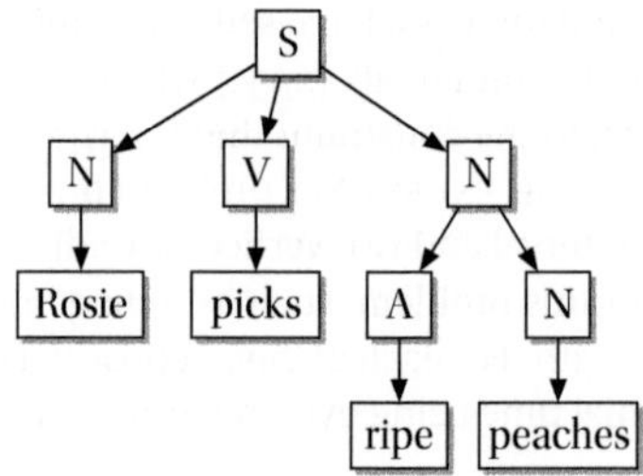

FIGURE 3.29: The parse tree for a small, but delightful, English sentence.

corresponds to a *parse tree* as shown in Figure 3.29, and *parsing* the sentence allows us to understand what the speaker has in mind.

Formally, a context-free grammar $G = (V, S, T, R)$ consists of a set V of *variables* which represent the speaker's internal symbols, an initial symbol $S \in V$ from which the entire string is generated, a set T of *terminal* symbols in which the final string must be written, and a set R of *production rules* of the form $A \to s$ where $A \in V$ and s is a string of symbols belonging to $V \cup T$. If $A \in V$ and w is a string of symbols in T, we write $A \rightsquigarrow w$ if there is some sequence of production rules that generates w from A. Finally, we say that w is *grammatical* if $S \rightsquigarrow w$, in which case each production corresponds to a node of the parse tree. For instance, if G is the following grammar,

$$V = \{S\}$$
$$T = \{x, \cdot, +, (,)\}$$
$$R = \{S \to (S \cdot S), (S + S), S \to x\},$$

then the set of grammatical strings consists of properly parenthesized expressions like x, $(x \cdot x)$, $(x \cdot (x + x))$, and so on. Now consider the following problem:

CONTEXT-FREE GRAMMAR MEMBERSHIP
Input: A grammar $G=(V,S,T,R)$ and a string w of symbols in T
Question: Is w grammatical?

Even though there are an exponential number of possible parse trees, show that this problem is in P, as is the problem of finding a parse tree if one exists. Feel free to assume that G is in *Chomsky normal form*: that is, every production is of the form $A \to BC$ where $A, B, C \in V$, or $A \to a$ where $A \in V$ and $a \in T$.

3.31 Faster Hamiltonian paths. Even when dynamic programming doesn't give a polynomial-time algorithm, it can sometimes give a better exponential-time one. A naive search algorithm for HAMILTONIAN PATH takes $n! \sim n^n e^{-n}$ time to try all possible orders in which we could visit the vertices. If the graph has maximum degree d, we can reduce this to $O(d^n)$—but d could be $\Theta(n)$, giving roughly n^n time again. Use dynamic programming to reduce this to a simple exponential, and solve HAMILTONIAN PATH in $2^n \operatorname{poly}(n)$ time. Hint: what do you have to know about a partial Hamiltonian path to determine whether it can be completed?

3.32 Leaner Hamiltonian paths. Given a graph G, focus on two vertices s,t. We will compute the number H of Hamiltonian paths that start at s and end at t. For any subset $S \subseteq V$, let $N(S)$ denote the number of paths from s to t of length $n-1$ that do *not* visit any vertex in S. However, these paths may visit other vertices any number of times. Show that

$$H=\sum_{S\subseteq V}(-1)^{|S|}N(S). \tag{3.22}$$

Then show that (3.22) can be evaluated in $2^n \operatorname{poly}(n)$ time and $\operatorname{poly}(n)$ memory.

Hint: use the inclusion–exclusion principle (see Appendix A.3.1).

3.33 Paths and eigenvectors. Consider again the graph shown in Figure 3.14 on page 61. Show that the number of paths of length ℓ that start at vertex 1 and end at vertex 2 is the Fibonacci number F_ℓ. Then, by diagonalizing the adjacency matrix A, show that F_ℓ can be written exactly as

$$F_\ell = a\varphi^\ell + b\left(-\frac{1}{\varphi}\right)^\ell,$$

where $\varphi = (1+\sqrt{5})/2$ is the golden ratio. Using the equations $F_0 = 0$ and $F_1 = 1$, find the constants a and b. In particular, this shows that $F_\ell = \Theta(\varphi^\ell)$ as in Problem 2.4.

3.34 Here, there, and everywhere. The Floyd–Warshall algorithm for ALL-PAIRS SHORTEST PATHS is shown in Figure 3.30. The only difference between it and the matrix-squaring algorithm we gave in Section 3.4.3 is that the loop over k is now outside the loops over i and j, rather than being the innermost. Surprisingly, this change means that we no longer need to repeat this process $\log_2 n$ times with an additional outer loop, so the running time is now $\Theta(n^3)$ instead of $\Theta(n^3 \log n)$.

Prove that this algorithm works. Hint: state a loop invariant that holds after running the outer loop k times, and use induction on k.

3.35 Squeezing through the bottleneck. Given a graph G where each edge has a capacity $c(e)$, the *bottleneck* $b(i,j)$ between two vertices i and j is the maximum, over all paths δ from i to j, of the minimum capacity of the edges along δ. For instance, in a MAX FLOW problem, $b(s,t)$ is the largest amount by which we can increase the flow along any one path, as in Problem 3.45 below.

Give a middle-first algorithm for calculating the matrix of bottlenecks, $B_{ij} = b(i,j)$. Hint: modify the definition of matrix multiplication in a way analogous to (3.12) for REACHABILITY or (3.13) for SHORTEST PATH, so that starting with the matrix C of capacities and "squaring" it repeatedly gives B.

```
Floyd-Warshall All-Pairs Shortest Paths(W)
input: a matrix W where W_ij is the weight of the path from i to j
output: a matrix B where B_ij is the length of the shortest path from i to j
begin
    B := W ;
    for k = 1 to n do
        for i = 1 to n do
            for j = 1 to n do
                B_ij = min(B_ij, B_ik + B_kj) ;
    return B ;
end
```

FIGURE 3.30: The Floyd–Warshall algorithm for ALL-PAIRS SHORTEST PATHS.

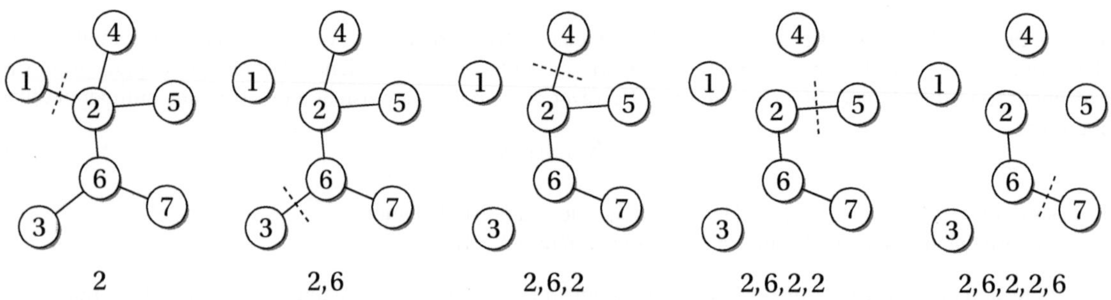

FIGURE 3.31: The Prüfer code of a labeled tree.

3.36 Counting trees. A labeled tree is a tree whose n vertices are labeled with the numbers $1,\ldots,n$. We consider two labeled trees distinct based not just on their topology, but on their labels. Cayley's formula states that the number T_n of labeled trees with n vertices is n^{n-2}. Here are the 3^1 labeled trees with $n=3$:

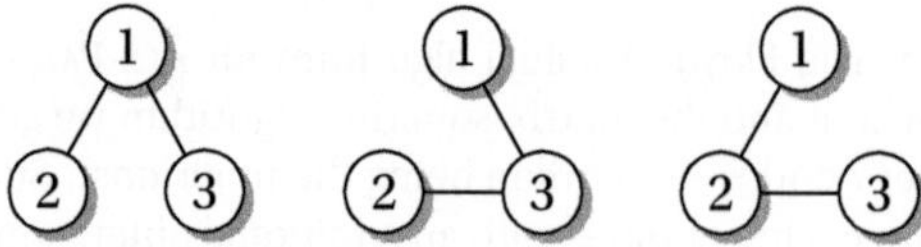

Draw the $4^2=16$ labeled trees with 4 vertices. How many consist of a star, i.e., a central vertex with 3 neighbors, and how many consist of a path?

We can prove Cayley's formula constructively by giving a one-to-one map from the set of labeled trees to the set of all sequences $(p_1,\ldots,p_{n-2})$ with $p_i \in \{1,\ldots,n\}$. The map works like this. Start with the empty sequence, and do the following until only two vertices remain: at each step, cut the leaf with the smallest label from the tree, and append the label of its neighbor to the sequence. This map is called the Prüfer code; see Figure 3.31 for an example.

Prove that the Prüfer code is one-to-one by showing how to reverse the map. That is show how to reconstruct the tree from a given sequence $(p_1,\ldots,p_{n-2})$. Hint: convince yourself of the fact that the degree of vertex i is one plus the number of occurences of i in the sequence. Then remember that the map cuts a vertex of degree one from the tree at each step; so, it might be helpful to maintain a list of all vertices that require exactly one more edge.

3.37 Borůvka's algorithm. Borůvka's original algorithm is a more parallel version of Kruskal's. Rather than adding a single edge to the forest, we add one edge for each tree as follows: for each tree t in F, find the minimum-weight edge with one end in t, and add all of these edges simultaneously. Prove, using Lemma 3.1, that this algorithm works.

3.38 Whittling the tree. Prove the following counterpart to Lemma 3.1: for any cycle C in a graph G, let e be the heaviest edge in C. Then there is a minimum spanning tree of G that does *not* include e. Use this to propose a greedy algorithm that starts with the entire graph, and whittles edges away until only a minimum spanning tree remains.

3.39 Spanning the plane. Consider the minimum spanning tree connecting n points located in the plane, where the distance between two points is the standard Euclidean distance. Show that no point has more than 6 neighbors in the tree, and that no point has more than 5 neighbors if the tree is unique. Hint: with how small an angle can two edges meet?

3.40 The forest matroid. Prove that for any graph G, the family of forests on the vertices of G satisfies the third matroid axiom. In other words, show that if F and F' are forests and F' has more edges than F, then there is some edge $e \in F'$ such that $F \cup \{e\}$ is still a forest.

3.41 Generalizing Kruskal to matroids. Prove Lemma 3.2. Hint: work in analogy to the proof of Lemma 3.1. Assume for the purposes of contradiction that F is contained in a minimum-weight basis B that does not contain v, and then show that there is some v' such that $B' = B \cup \{v\} - \{v'\}$ is a basis with $w(B') \le w(B)$.

3.42 Forests and subspaces. Both forests and linearly independent sets of vectors form a matroid, but we can make the analogy between these problems much tighter. Given a graph G with n vertices, show how to associate an n-dimensional vector with each edge such that a set of edges is linearly independent if and only if it is a forest. Hint: define these vectors so that the edges around a cycle sum to zero.

Then consider the $(n-1)$-dimensional subspace V consisting of n-dimensional vectors whose components sum to zero, and show that a set of edges is a basis for V if and only if it is a spanning tree. In other words, a forest spans G if and only if its vectors span V.

3.43 Small improvements. Give an example of a graph in which, if we do a poor job of choosing which augmenting paths to use, the Ford–Fulkerson algorithm can take exponential time. Hint: increase the capacities of the four outer edges in Figure 3.19.

3.44 Short paths, big improvements. In the Ford–Fulkerson algorithm, I am free to increase the flow along any augmenting path. Suppose I always use one of the *shortest* paths from s to t in G_f, where shortest means simply the number of edges, regardless of their capacity. Furthermore, I increase the flow as much as possible, i.e., by the minimum of $c_f(e)$ over all the edges e appearing in δ. Show that this improved algorithm finds the maximum flow in at most $|V||E| = O(n^3)$ iterations even if the capacities are exponentially large, and so that MAX FLOW is in P.

Hint: let d be the length of the shortest path from s to t in G_f, and let $F \subseteq E$ be the set of edges that occur in some path of length d. First show that if d stays the same, then the reverse edges created in the residual graph never get added to F. Then, show that each iteration either increases d or decreases $|F|$.

3.45 Fat paths. Here is another version of the Ford–Fulkerson algorithm which runs in polynomial time. Given an augmenting path, define its *bottleneck capacity* as the minimum of $c_f(e)$ over all edges e appearing in δ, i.e., the largest amount by which we can increase the flow along δ. Show that if we always use the path with the largest bottleneck capacity, we will find the maximum flow in at most

$$\frac{\log f_{\max}}{-\log(1-1/m)} \le m \ln f_{\max}$$

improvements. Here m is the number of edges in the graph, $f_{\max}$ is the value of the maximum flow, and we assume the capacities are integers.

Hint: let f be the current flow, and suppose that the best path has bottleneck capacity c. Consider the set S consisting of all the vertices reachable by edges e with $c_f(e) > c$. Show that $t \notin S$, so S defines a cut in G_f. Use this cut to show that $f_{\max} - f \leq mc$, and therefore that each improvement decreases $f_{\max} - f$ by a factor of $1 - 1/m$ or better.

3.46 Lightening the tree. Kruskal's algorithm for MINIMUM SPANNING TREE starts with no edges at all, and builds a tree one edge at a time. Design a new polynomial-time algorithm for MINIMUM SPANNING TREE which is analogous to the Ford–Fulkerson algorithm for MAX FLOW, in the sense that it starts with an arbitrary spanning tree and makes a series of improvements until it is as light as possible. What moves can you think of that turn a spanning tree into another one with smaller weight?

3.47 Happy couples. Hall's Theorem states that a bipartite graph with n vertices on each side has a perfect matching if and only if every subset S of the vertices on the left is connected to a set T of vertices on the right, where $|T| \geq |S|$. Prove Hall's theorem using the duality between MAX FLOW and MIN CUT and the reduction given in Figure 3.23. Note that this gives a polynomial-time algorithm for telling whether this is true of all subsets S, even though there are 2^n of them.

3.48 Many roads to Rome. Two paths in a graph are called *edge-disjoint* if they have no edges in common. However, they are allowed to share vertices. Consider the following generalization of REACHABILITY, which asks for multiple edge-disjoint paths from s to t:

DISJOINT PATHS
Input: A directed graph G, two vertices s, t, and an integer k
Question: Is there a set of k edge-disjoint paths from s to t?

Prove that this problem can be solved in polynomial time.

3.49 Euler helps out at the dating service. We say that a graph is *d-regular* if every vertex has d neighbors. Use Hall's Theorem to show that if G is bipartite and d-regular—that is, each vertex on the left is connected to d vertices on the right, and vice versa—then G has a perfect matching.

Now suppose that we actually want to find a perfect matching. We can do this using MAX FLOW. But, at least for some values of d, there is a more elegant approach. Suppose that d is even. Show that we can reduce the problem of finding a perfect matching on G to the same problem for a $(d/2)$-regular subgraph G', with the same number of vertices, but half as many edges.

Hint: cover G with one or more Eulerian paths. Assuming that we can find an Eulerian path in $O(m)$ time where $m = dn$ is the number of edges, show that this gives an $O(m)$ algorithm for finding perfect matchings in d-regular graphs whenever d is a power of 2.

3.50 Spin glasses. In physics, a *spin glass* consists of a graph, where each vertex i has a "spin" $s_i = \pm 1$ representing an atom whose magnetic field is pointing up or down. Each edge (i, j) has an *interaction strength* J_{ij}, describing to what extent s_i and s_j interact. Additionally, each vertex i can have an *external field* h_i. Given a state of the system, i.e., a set of values for the s_i, the *energy* is

$$E(\{s_i\}) = -\sum_{ij} J_{ij} s_i s_j - \sum_i h_i s_i .$$

As we will discuss in Section 12.1, the system tries to minimize its energy, at least in the limit of zero temperature. An edge (i, j) is called *ferromagnetic* if $J_{ij} > 0$, and *antiferromagnetic* if $J_{ij} < 0$. Then ferromagnetic edges want i and j to be the same, and antiferromagnetic edges want them to be different. In addition, an external field h_i that is positive or negative wants s_i to be $+1$ or -1 respectively.

Minimizing the energy—or as physicists say, finding the *ground state*—gives the following optimization problem:

> SPIN GLASS
>
> Input: A graph G with interaction strengths $\{J_{ij}\}$ and external fields $\{h_i\}$
>
> Output: The state $\{s_i\}$ that has the lowest energy.

Show that in the ferromagnetic case, where $J_{ij} \geq 0$ for all i, j, this problem is in P. Hint: reduce it to MIN CUT.

3.51 When a tree is not enough. In Section 3.5.1, we described an electrical network as a spanning tree. But such a network is rather fragile, since it falls apart into two pieces if we remove a single edge. A more robust network would be a spanning subgraph that is 2-*connected*, i.e., that remains connected if we remove any single edge. Equivalently, every pair of vertices is connected by at least 2 paths which are edge-disjoint as in Problem 3.48.

Consider the problem MINIMUM 2-CONNECTED SPANNING SUBGRAPH. Do you think a greedy strategy still works? If you think so, formulate and prove a lemma analogous to Lemma 3.1, write down a greedy algorithm, and prove that it works. If you don't think so, give an example of a graph where a greedy approach fails—for instance, where the lightest edge is not part of the solution. If you take the latter route, discuss whether you believe this problem is in P. Hint: what is the simplest 2-connected graph?

Notes

3.1 Books. In keeping with our large-scale point of view, our purpose in this chapter is not to explain how to implement algorithms in practice, or how to make them more efficient by using clever data structures. The art and science of efficient algorithms deserves an entire book, or even two. We recommend Kleinberg and Tardos [490] and Dasgupta, Papadimitriou, and Vazirani [215] for modern approaches to algorithm design, and Cormen, Leiserson, Rivest, and Stein [203] for an encyclopedic account of various data structures. Here we restrict ourselves to the major families of polynomial-time algorithms, and the structural properties that a problem needs to have for them to work.

3.2 Towers of Hanoi. Lucas introduced the Towers of Hanoi in [546]; N. Claus de Siam is an anagram of Lucas d'Amiens. The translation we quote here is by W. W. Rouse Ball [76]. The description of the solution as a Hamiltonian path on an n-dimensional hypercube, which we ask for in Problem 3.2, is closely related to the Gray code, in which integers are represented by binary digit sequences that change by just one bit when we go from one integer to the next:

$$000 \to 001 \to 011 \to 010 \to 110 \to 111 \to 101 \to 100 .$$

This is unlike standard binary notation, where carrying 1s can flip many bits. Schemes similar to the Gray code can be used to solve a number of other puzzles, including the "Baguenaudier" or "Chinese rings" [316].

Henry Dudeney introduced the four-post variant of Problem 3.5 in his book *The Canterbury Puzzles* [253]. For k pegs, it is known that $2^{\Theta(n^{1/(k-2)})}$ moves are necessary and sufficient, but the constant in Θ is not known [402, 778, 782]

3.3 Sorting. `Mergesort` was invented in 1945 by John von Neumann; see Knuth [496]. `Quicksort` was devised in 1961 by Tony Hoare [404], and remains one of the most popular algorithms in existence. Even though `Quicksort` performs more comparisons, it is often faster in practice because the partitioning operation can be carried out "in place," in the same section of memory that contained the original list.

3.4 Epicycles. Eudoxus of Cnidus, a contemporary of Plato and a celebrated mathematician who made important contributions to the theory of proportions, proposed a beautiful explanation for retrograde motion. He placed each planet on the equator of a sphere and then placed this sphere inside another with the same center, but rotating in the opposite direction along a slightly tilted axis. Finally, he placed these inside third and fourth spheres to give the planet its overall motion against the stars. While the resulting model doesn't fit very well with observations, it reproduces the qualitative aspects of retrograde motion, in which the planet performs a figure-8 relative to its overall motion.

Two centuries later, Hipparchus proposed epicyclic models of the sun and moon which fit reasonably well with observations. Three centuries after that, Ptolemy added several improvements to Hipparchus' model, including allowing the center of the circle to differ from the Earth's location—corresponding to adding a constant term in the Fourier series—and allowing the speed of rotation to be uniform around a point other than the center [225].

In the 14th century, Ibn al-Shāṭir used four epicycles for Venus and the outer planets and six for Mercury as shown in Figure 3.8. If one subtracts the motion of the Earth, al-Shāṭir's models are almost identical to those of Copernicus, which appeared 150 years later [474].

3.5 Fast Fourier Transforms. The Fast Fourier Transform is often called the Cooley–Tukey FFT after James Cooley and John Tukey, who described it in 1965 [200]. They subsequently learned [199] that (3.7) appeared in Danielson and Lanczos [211] in the form of sines and cosines, and before that in Runge [718]. They also noted the work of I. J. Good [346], who gave a fast algorithm in the case where n is the product of mutually prime factors.

Like so much else, it turns out that the Fast Fourier Transform was first discovered by Carl Friedrich Gauss in the early 1800s. In essence, he derived (3.7) and its generalization to factors other than 2, and used it to interpolate the orbits of the asteroids Juno and Pallas from $n = 12$ samples [320]. To be fair to Cooley and Tukey, however, they were the first to show that the FFT's running time scales as $\Theta(n \log n)$. Excellent reviews of all this history can be found in [712] and [393].

The algorithm of Problem 3.17, in which the Fourier transform for prime n becomes a convolution modulo $n-1$, is due to Rader [677]. Along with the Cooley–Tukey FFT, these algorithms allow us to perform the Fourier transform in $O(n \log n)$ time for any n.

3.6 Dynamic programming. The phrase "dynamic programming" was coined by Richard Bellman [95]. In addition to computer science, it has been applied to many problems in control theory, economics, and machine learning.

A classic example from economics is to find the investment strategy that maximizes our return. Abstractly, suppose that at each time-step t we can make a move x_t drawn from a finite set, and suppose that our gain at each step is some function $F(x_t, x_{t+1})$ of our moves on that step and the next one. If you believe that future value should be "discounted" at some rate $\gamma < 1$ (an assumption famously at odds with long-term sustainability, but which is mathematically convenient) the total value is

$$V = \sum_{t=0}^{\infty} \gamma^t F(x_t, x_{t+1}).$$

If our first move is x, the maximum possible value we can achieve is then

$$V_{\max}(x) = \max_y \left(F(x,y) + \gamma V_{\max}(y) \right),$$

and the y that maximizes this expression is the best move to make next. We can determine the optimal policy by a kind of infinite-time dynamic programming, iterating this system of equations until we find its fixed point.

If we can write a problem recursively in terms of $O(n^t)$ subproblems, and if each of these depends in a simple way on $O(n^e)$ smaller subproblems, the straightforward dynamic programming algorithm will take $O(n^{t+e})$ time. For instance, for TYPESETTING we have $t = e = 1$, and for ALIGNMENT we have $t = 2$ and $e = 0$, so both can be solved in $O(n^2)$ time. However, in many cases we can reduce the running time to a smaller power of n by using the convexity or concavity of the functions involved; see Galil and Park [312].

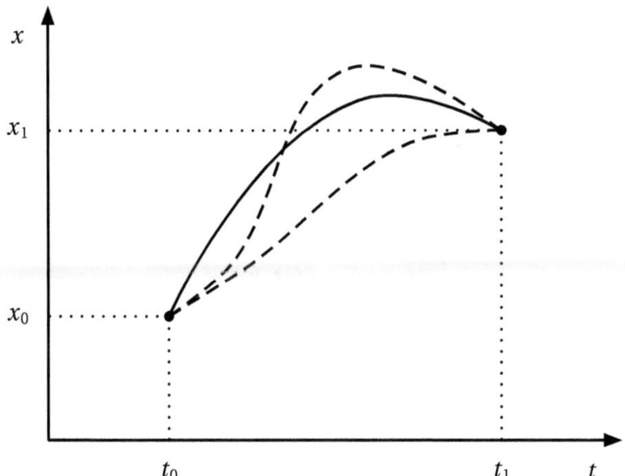

FIGURE 3.32: Among all possible ways (dashed) to go from x_0 to x_1 in time $t_1 - t_0$, a particle chooses the one (solid) that minimizes the action (3.23).

The dynamic programming algorithm for TYPESETTING was invented by Knuth and Plass, who also showed that it can be described as a SHORTEST PATH as in Problem 3.20. In [497] they describe many improvements and extensions to it, along with a marvelous history of typesetting by humans and computers.

Finally, we regret to inform the reader that, for whatever odd reason, computer scientists say *memoize* instead of *memorize*. We encourage the reader to go forth and expunge this bizarre neologism.

3.7 Dynamic programming in physics. Dynamic programming is also related to the *calculus of variations* [249]. We illustrate this relationship with an example from physics. Consider a particle of mass m that moves along the x axis subject to a force $F(x)$. The trajectory $x(t)$ of the particle is given by the solution of Newton's equation $ma = F(x)$, where $\mathrm{d}v/\mathrm{d}t$ is the acceleration and $\mathrm{d}x/\mathrm{d}t$ is the velocity. Let's assume that the force F can be written as $F = -\mathrm{d}V/\mathrm{d}x$ for some potential energy $V(x)$. The kinetic energy minus the potential energy is called the *Lagrangean*, $L(x,v) = (1/2)mv^2 - V(x)$, and its integral

$$\int_{t_0}^{t_1} L(x(t), v(t))\,\mathrm{d}t \tag{3.23}$$

is called the *action*. The *principle of least action* states that among all paths that go from x_0 to x_1 in time $t_1 - t_0$, the true trajectory $x(t)$ is the one that minimizes the action as shown in Figure 3.32.

This recasts Newtonian mechanics as an optimization problem,

$$S(t_0, x_0) = \min_{\eta} \int_{t_0}^{t_1} L\left(\eta, \frac{\mathrm{d}\eta}{\mathrm{d}t}\right) \mathrm{d}t\,, \tag{3.24}$$

where we minimize over all paths $\eta(t)$ such that $\eta(t) = x$ and $\eta(t_1) = x_1$. This is a somewhat unusual optimization problem, since the independent "variable" is a function instead of a real number. The calculus of variations was invented to deal precisely with this sort of problem.

In (3.24) we explicitly stated the dependence of S on the initial point (t_0, x_0). It also depends on the final point (t_1, x_1). But just as in our TYPESETTING algorithm, we will only vary the initial point and write (t, x) instead of (t_0, x_0). If the trajectory is smooth, the initial velocity v at at time t fixes the trajectory up to time $t + \mathrm{d}t$ for an infinitesimal time interval dt. Dynamic programming then tells us to optimize the initial velocity v:

$$\begin{aligned} S(t,x) &= \min_v \left\{ L(x,v)\,dt + S(t+dt, x+v\,dt) \right\} \\ &= \min_v \left\{ L(x,v)\,dt + S(t,x) + \frac{\partial S}{\partial t}\,dt + \frac{\partial S}{\partial x}\,v\,dt \right\}. \end{aligned} \tag{3.25}$$

At the optimum v the derivative of the right-hand side with respect to v must vanish,

$$\frac{\partial L(x,v)}{\partial v} + \frac{\partial S(t,x)}{\partial x} = 0. \tag{3.26}$$

We can't determine v yet since we don't know $\partial S/\partial x$. We can get another equation by subtracting $S(t,x)$ from both sides of (3.25) and dividing by dt, giving

$$L(x,v) + \frac{\partial S}{\partial t} + \frac{\partial S}{\partial x}\,v = 0.$$

Differentiating with respect to x gives

$$\frac{\partial L}{\partial x} + \frac{\partial^2 S}{\partial x\,\partial t} + \frac{\partial^2 S}{\partial x^2}\,v = \frac{\partial L}{\partial x} + \frac{d}{dt}\frac{\partial S}{\partial x} = 0, \tag{3.27}$$

where the time derivative of $\partial S/\partial x$ takes into account both the change in $S(t,x)$ with respect to t and the fact that x changes by $v\,dt$. If we take the time derivative of (3.26) and combine the result with (3.27), we finally arrive at the *Euler–Lagrange equation*:

$$\frac{d}{dt}\frac{\partial L(x,v)}{\partial v} - \frac{\partial L(x,v)}{\partial x} = 0. \tag{3.28}$$

This in turn can be transformed into Newton's equation by substituting the definition of L from (3.23) (exercise!)

Note that we didn't actually use the fact that the action is minimized—we simply set its derivative to zero. In fact, the trajectory is often a saddle point of the action, rather than a minimum [351].

3.8 Shortest paths. These matrix-squaring algorithms are far from the fastest algorithms for SHORTEST PATH if we are just interested in one pair of vertices, or even one source and all destinations. However, they are reasonably good for ALL-PAIRS SHORTEST PATHS. Moreover, unlike the `Explore` algorithm described at the beginning of Section 3.4, they only need a small amount of memory, as we will see in Chapter 8.

The algebra in which $a+b$ and ab are replaced by $\min(a,b)$ and $a+b$ respectively is called the *min-plus algebra* or, in honor of the Brazilian mathematician Imre Simon, the *tropical algebra.* The reason it fits so neatly into the matrix-powering algorithm is that it is a *semiring*, i.e., it obeys the associative and distributive laws: just as $a(b+c) = ab+ac$, $a+\min(b,c) = \min(a+b, a+c)$. This approach to ALL-PAIRS SHORTEST PATHS is called the Bellman–Ford algorithm, and was also proposed by Shimbel.

The Floyd–Warshall algorithm, discussed in Problem 3.34, was presented as an algorithm for REACHABILITY by Warshall and extended to the case of weighted graphs by Floyd. However, it was discovered a few years earlier by Bernard Roy. Another important case of dynamic programming, Dijkstra's algorithm [239], uses a weighted version of breadth-first search to find the shortest path from one source to all destinations. A good description of it can be found in [215]. A history of all these algorithms, and references, can be found in Schrijver [735].

3.9 Minimum spanning trees. Kruskal's algorithm appeared in [505]. A translation of Borůvka's original paper [128], along with a history of related algorithms for MINIMUM SPANNING TREE, can be found in [580]. It is interesting to note that Borůvka's original algorithm is the source of some of the fastest known modern algorithms.

Another common algorithm grows a single tree outward, always adding the lightest edge that connects this tree to the other vertices. This is usually called Prim's algorithm [670], although it was found earlier by the number theorist Jarník [427]. The proof that it works is essentially identical to our proof of Lemma 3.1. However, since not all subgraphs of a tree are trees, Prim's algorithm doesn't fit directly into the framework of matroids. Instead, the set of trees forms a more general structure called a *greedoid*. See [262, 396, 501, 642, 679] for discussions of matroids, greedoids and their generalizations. The Prüfer code of Problem 3.36 appeared in [674].

3.10 Greedy algorithms and making change. Greedy algorithms have an intuitive appeal that makes them very popular, but proving their correctness—or their quality as approximations—can be a major challenge. A classic example is change-making [742], the problem of representing a given value with as few coins as possible from a given set of denominations. Cashiers and waiters around the world use the greedy algorithm, which repeatedly takes the largest coin less than or equal to the amount remaining. But does this algorithm really give the smallest number of coins?

It can be shown that the greedy algorithm works for US and European coins, whose denominations in cents are $\{1,5,10,25,50,100\}$ and $\{1,2,5,10,20,50,100,200\}$ respectively. But it doesn't work in general. Take the denominations $\{1,6,10\}$ as an example: $12=6+6$, but the greedy algorithm would represent 12 with 3 coins, $12=10+1+1$. Another example where greedy change-making is not optimal is the old English system with its florins and half-crowns.

In fact, if we consider the set of denominations and the target value x as the input, the change-making problem turns out to be as hard as HAMILTONIAN PATH: that is, it is NP-complete (see Chapter 5). The problem of deciding whether the greedy solution is optimal for a given value of x is also very hard [503], whereas the problem of deciding whether the greedy approach works for all x is in P [654].

3.11 MAX FLOW and MIN CUT. The Ford–Fulkerson algorithm appeared in [286]. Edmonds and Karp [263] showed that using the shortest path as in Problem 3.44 or the path with the largest bottleneck capacity as in Problem 3.45 causes the Ford–Fulkerson algorithm to run in polynomial time. The shortest-path method was analyzed a few years earlier by Dinic [243], who gave a polynomial-time algorithm that works by adding a flow which saturates at least one edge along every path from s to t. The fact that MAX FLOW = MIN CUT was proved independently by Ford and Fulkerson [286] and Elias, Feinstein and Shannon [266]. We will see more examples of duality in Chapter 9.

The idea of reducing PERFECT BIPARTITE MATCHING on a d-regular graph to a $(d/2)$-regular subgraph using Eulerian paths as in Problem 3.49 comes from Gabow and Kariv [310]. The reduction from the ferromagnetic case of SPIN GLASS to MIN CUT in Problem 3.50 is due to Angles d'Auriac, Preissmann, and Rammal [217].

The Cold War MIN CUT example shown in Figure 3.22 is from a RAND Corporation technical report by Harris and Ross [368], which was declassified in 1999. We learned about it from Schrijver's review [735].

3.12 Factoring with factorials. We are indebted to Richard Lipton for Problems 3.12 and 3.13, who wrote about them on his wonderful blog rjlipton.wordpress.com. The duplication formula (3.19) is due to Legendre.

3.13 Increasing and decreasing sequences. The theorem of Problem 3.23 that any permutation of length n has an increasing or decreasing subsequence of length $\lceil\sqrt{n}\rceil$ is due to Erdős and Szekeres [270]. The pigeonhole principle proof we give here is from Seidenberg [737].

3.14 Formal grammars. Chomsky [164] proposed a hierarchy of increasingly complex kinds of grammars. Beyond context-free grammars, there are context-sensitive grammars that are as powerful as Turing machines (see Chapter 7) with $O(n)$ memory, and unrestricted grammars that are computationally universal. Many intermediate classes have also been defined, where variables have certain kinds of attributes that constrain the production rules. Whether or not these are good models of human languages, many programming languages are essentially context-free, and parsing the source code of a program is one of the first steps in compiling it.

Chapter 4

Needles in a Haystack: the Class NP

> If Edison had a needle to find in a haystack, he would proceed at once with the diligence of the bee to examine straw after straw until he found the object of his search... I was a sorry witness of such doings, knowing that a little theory and calculation would have saved him ninety per cent of his labor.
>
> Nikola Tesla

We turn now to the second problem from the Prologue. Given a graph, recall that a *Hamiltonian path* is a path that visits each vertex exactly once. As far as we know, the only way to determine whether our favorite city possesses a Hamiltonian path is to perform some kind of exhaustive, and exhausting, search.

However, the news is not all bad. If a well-travelled friend of ours claims to know of such a path, she can convince us of her claim by traversing it with us. We can verify that it is Hamiltonian by checking off each vertex as we visit it—stopping, of course, at a succession of delightful cafés along the way.

This may seem painfully obvious, but let's think about it a little more deeply. Consider the proverbial task of finding a needle in a haystack. In what way is this problem hard? It is easy to tell a needle from a blade of hay, but it is hard to find a single needle—or tell if there is one—if it is lost in a haystack of exponential size. Sitting on a needle provides immediate proof that one exists, but even if one has been 4.1
rolling in the hay for some time, one might remain ignorant about a needle's existence or location.

Similarly, while it seems hard to find a Hamiltonian needle, or tell whether there is one, buried in an exponential haystack of possible paths, it is easy to check whether a given path is Hamiltonian. What other problems have this character, and how are they related to each other? The class of such problems forms a complexity class of its own, called NP. In this chapter, we look at several definitions of NP, and think about proofs, witnesses, and lucky guesses. We will see that NP includes a wide variety of fundamental problems, including coloring maps, satisfying systems of constraints, and untying knots.

4.1 Needles and Haystacks

Consider the following problem:

> HAMILTONIAN PATH
> Input: A graph G
> Question: Does there exist a Hamiltonian path on G?

This is a decision problem, with a yes-or-no answer. Unlike EULERIAN PATH, we do not know of a polynomial-time algorithm for it. However, if the answer is "yes," there is an easily checkable proof of that fact—namely, the Hamiltonian path itself.

NP is the class of decision problems with this kind of logical structure, where yes-instances are easy to verify. Here's a first (informal) definition:

> A decision problem is in NP if, whenever the answer for a particular instance is "yes," there is a simple proof of this fact.

By "simple proof," of course, we mean a proof that we can check in polynomial time. For instance, we can check that a given path is Hamiltonian in essentially linear time as a function of the number of vertices.

Like HAMILTONIAN PATH, many problems in NP are of the form "Does there exist a thing such that...." If there is one, the proof of that fact is the thing itself. However, later in this chapter we will see some problems in NP that are not stated in this way, and where it is not so obvious that a simple proof exists.

It's important to realize that NP is a profoundly asymmetric notion. If I happen to know the path, it's easy to prove to you that a Hamiltonian graph is Hamiltonian—but how can I prove that a non-Hamiltonian graph has no such path? I somehow need to convince you that every possible path *fails* to be Hamiltonian. As we discussed in the Prologue, I can present you with a truckload of computer output, and claim that it shows an exhaustive search that eliminates every path. But this is hardly a simple proof, since it would take you ages to check, or even read, my work. Proving that a haystack is needle-free requires us to check all the hay.

If a problem is in P, it has a polynomial algorithm that is guaranteed to give the right answer. If the input is a yes-instance, the sequence of steps executed by this algorithm, which end with it returning "yes," constitutes a proof of that fact. Thus

$$\mathsf{P} \subseteq \mathsf{NP}, \tag{4.1}$$

where here by P we mean the set of decision problems solvable in polynomial time. However, we strongly believe that this inclusion is proper, i.e., that $\mathsf{P} \neq \mathsf{NP}$—that finding needles is harder than checking them.

Essentially, P is the subset of NP consisting of problems for which there is some mathematical insight that lets us avoid an exhaustive search. Such an insight might allow us to apply one of the algorithmic strategies that we explored in Chapter 3:

- break the problem into smaller parts (the Fast Fourier Transform),
- build the solution up from simpler subproblems (SHORTEST PATH),
- grow a solution greedily (MINIMUM SPANNING TREE),

- improve solutions until we reach the optimum (MAX FLOW), or
- reduce it to a problem we already know how to solve (MAX BIPARTITE MATCHING).

In the absence of such an insight, it seems hard to avoid searching through the entire haystack, and this takes exponential time if the number of possible solutions is exponentially large.

On the other hand, as we saw in Chapter 2, it is hard to know whether we have found the best possible algorithm, and it is very hard to prove that *no* polynomial-time algorithm works. So, while there are many problems in NP such as HAMILTONIAN PATH for which we strongly believe that no polynomial time algorithm exists, no one has been able to prove this.

The question of whether finding needles really is harder than distinguishing them from blades of hay—that is, whether or not $\mathsf{P} \neq \mathsf{NP}$—is considered one of the deepest and most profound questions, not just in computer science, but in all of mathematics. We will wrestle with this question in Chapter 6, after discussing the formal definition of NP, its historical background, and an amazing sense in which the hardest problems in NP are all equivalent to each other. But first, let us introduce some prominent members of this class that we will encounter later in our journey.

4.2 A Tour of NP

In this section, we meet some of the classic problems in NP. Each one asks whether a certain puzzle has a solution, or whether a certain type of object exists—whether we can color a given graph, or make a given formula true, or balance a given set of numbers, or fit a given number of people at a party without them getting into a fight. We will also see some reductions between these problems, transforming questions about graphs into questions about Boolean formulas or vice versa.

4.2.1 Colorful Graphs

At most scientific conferences, the vast majority of talks are boring. This makes it especially frustrating when two of the few talks you actually want to see take place at the same time in different rooms. Suppose, in a rare act of candor, each conference participant supplies the organizers with the list of talks they want to hear. Suppose further that there is a plentiful supply of lecture rooms, but only k time slots. Can we schedule the talks in such a way that makes everyone happy? That is, can we assign a time slot to each talk, so that no two talks take place simultaneously if someone wants to see them both?

We can formalize this as follows. Let G be a graph whose vertices are the talks, and let two talks be adjacent (i.e., joined by an edge) if there is someone who wants to attend both of them. We want to assign one of the k time slots to each vertex in such a way that adjacent vertices have different time slots.

Let's represent each time slot by a color. A k*-coloring* is an assignment of one of k different colors to each vertex, and a coloring is *proper* if no two adjacent vertices have the same color, as shown in Figure 4.1. Thus we wish to solve the following kind of problem:

> GRAPH k-COLORING
>
> Input: A graph G
>
> Question: Is there a proper k-coloring of G?

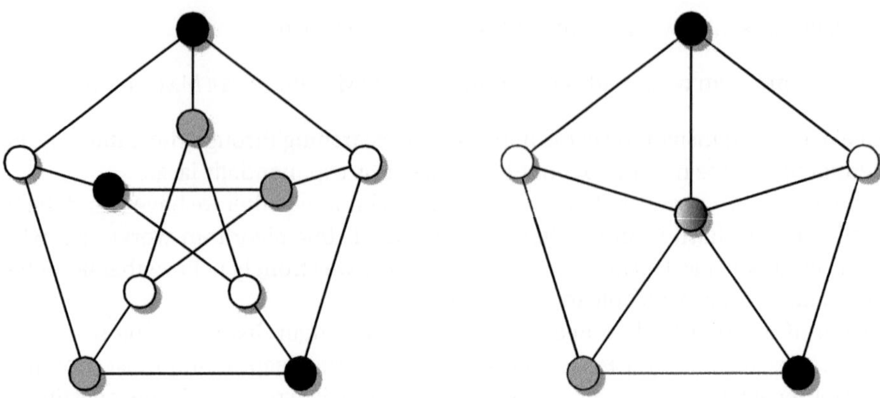

FIGURE 4.1: The Petersen graph (left) with a proper 3-coloring. The cart-wheel graph (right) cannot be colored with fewer than 4 colors.

Despite its colorful terminology, GRAPH k-COLORING is a serious problem with a wide range of applications. It arises naturally whenever one is trying to allocate resources in the presence of conflicts, as in our scheduling example. For instance, suppose we are trying to assign frequencies to wireless communication devices. If two devices are too close, they need to use different frequencies to avoid interference. This problem is equivalent to GRAPH k-COLORING where each vertex is a device, k is the number of frequencies, and there is an edge between two devices if they are close enough to interfere with each other.

GRAPH k-COLORING is in NP since it is easy to check that a coloring is proper. As for HAMILTONIAN PATH, however, this doesn't tell us how to find a proper coloring in the haystack of k^n possible colorings. How does the complexity of this problem depend on k?

A graph can be colored with one color if and only if it has no edges at all, so GRAPH 1-COLORING is trivially in P. The following exercise shows that GRAPH 2-COLORING is in P as well:

Exercise 4.1 *Prove that* GRAPH 2-COLORING *is in* P. *Hint: give a greedy strategy which finds a 2-coloring if one exists.*

For three or more colors, on the other hand, no polynomial-time algorithm is known. As we will see in Chapter 5, we have good reason to believe that no such algorithm exists, and that GRAPH 3-COLORING takes exponential time to solve.

What happens if we restrict the type of graph we want to color? Let's consider *planar* graphs, i.e., those that can be drawn in the plane without any edges crossing each other. These are famous because of their connection to the map coloring problem, in which we want to color territories so that no territories that share a border have the same color. This corresponds to coloring the *dual graph* whose vertices are territories and where adjacent territories are connected by an edge, as in Figure 4.2.

Such graphs are usually planar. However, we have to be careful at the Four Corners region where New Mexico, Arizona, Colorado and Utah meet, and in the figure we assume that two territories are not adjacent if they only touch at a corner. Moreover, the state of Michigan consists of two disconnected regions, and even old Königsberg is now in a part of the Russian Federation that is separated from the rest of Russia by Lithuania and Belarus. Requiring that all the lakes and oceans have the same color poses an additional wrinkle.

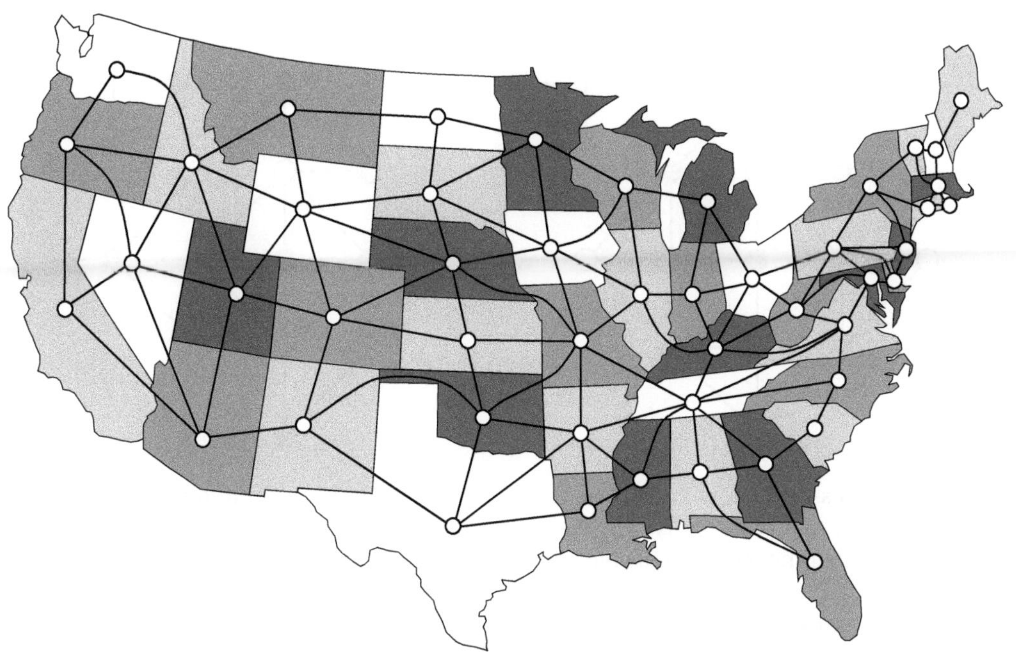

FIGURE 4.2: The dual graph of the continental United States, with a proper 4-coloring. Is it 3-colorable?

The famous Four Color Theorem tells us that every map, and therefore every planar graph, can be colored using no more than 4 colors. This gives us a very simple algorithm for PLANAR GRAPH k-COLORING whenever $k \geq 4$: return "yes." Of course, a lot of mathematical work went into proving that this algorithm works. But once this work is done, there is no computation left to do.  4.2

However, the Four Color Theorem doesn't tell us which planar graphs are 3-colorable. How hard is PLANAR GRAPH 3-COLORING, the special case of GRAPH 3-COLORING where the graph is planar?

Recall from Section 3.8 that if we have two problems A and B, we say that *A is reducible to B*, and write $A \leq B$, if there is a polynomial-time algorithm that translates instances of A to instances of B. In that case, B is at least as hard as A, since if we can solve B in polynomial time then we can solve A as well.

An instance of PLANAR GRAPH 3-COLORING is already an instance of GRAPH 3-COLORING, so we have the trivial reduction

$$\text{PLANAR GRAPH 3-COLORING} \leq \text{GRAPH 3-COLORING}.$$

Thus if we can solve GRAPH 3-COLORING in general, we can certainly solve the planar case. On the other hand, restricting a problem to a special case often makes it easier, and one might imagine that there are techniques for GRAPH 3-COLORING that only work for planar graphs. Surprisingly, it turns out that

$$\text{GRAPH 3-COLORING} \leq \text{PLANAR GRAPH 3-COLORING}.$$

In other words, we can reduce the general problem to the planar case—we can transform an arbitrary graph G to a planar graph G', such that G' is 3-colorable if and only if G is. Thus PLANAR GRAPH 3-COLORING is just as hard as GRAPH 3-COLORING in general, and the two problems have the same complexity.

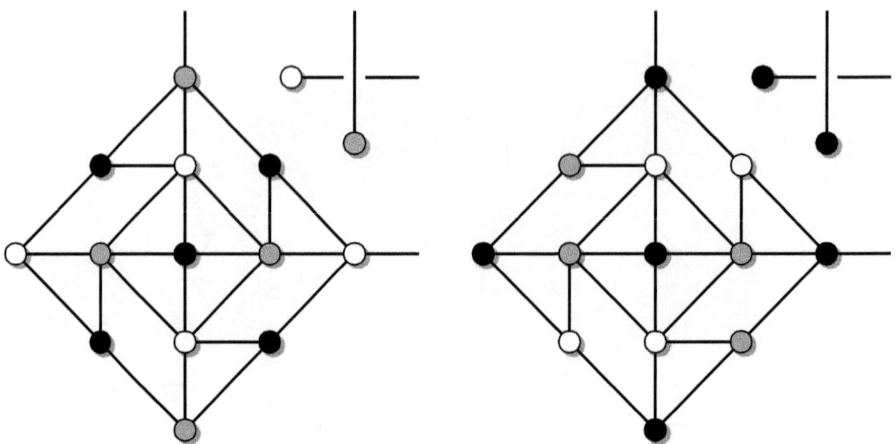

FIGURE 4.3: This crossover gadget allows us to convert an arbitrary graph into a planar one while preserving the property of 3-colorability. The gadget can be colored in two different ways up to symmetry, depending on whether the transmitted colors are different (left) or the same (right).

How does this transformation work? We start by drawing G on a sheet of paper, perhaps with some places where one edge crosses another. Then wherever such a crossing occurs, we replace it with the "crossover gadget" displayed in Figure 4.3. This gadget transmits the vertex colors from east to west, and from north to south, forcing the vertices at the other end of these edges to be a different color just as G's original edges do. With the help of the following exercise, we can carry out this transformation in polynomial time.

Exercise 4.2 *Suppose that G is a graph with n vertices. Show that in polynomial time, we can find a way to draw G in the plane so that it has $O(n^4)$ crossings.*

For planar graphs, increasing the number of colors to 4 makes the colorability problem trivial, although it was far from trivial to prove this. On the other hand, the next exercise shows that for general graphs, increasing the number of colors doesn't make the problem any easier:

Exercise 4.3 *Prove that* GRAPH k-COLORING $\leq$ GRAPH $(k+1)$-COLORING *for any k. Hint: add a vertex.*

Next we turn to problems with a logical flavor—where variables are true or false, and constraints on them are built of ANDs and ORs.

4.2.2 Can't Get No Satisfaction

Consider the following conundrum. I have in my larder a bottle of an excellent Pommard, a fresh Quail, and some delicious Roquefort. I wish to plan a dinner party for my favorite coauthor. However, there are some obvious constraints. Pommard pairs beautifully with Quail, so it would be crazy to serve the bird without the wine. However, such a light wine would be overwhelmed by the Roquefort, so I can't possibly serve it and the Roquefort together. On the other hand, my friend likes both Quail and Roquefort, so I would like to serve at least one of them—and I simply must have at least the Pommard or the Quail.

Let us represent our ingredients by three Boolean variables, p, q and r, indicating whether or not I will serve them. We use the symbols $\vee$ and $\wedge$ for OR and AND respectively, and we represent the negation of a variable x as $\overline{x}$. Then our culinary constraints can be expressed as a Boolean formula,

$$\phi(p,q,r)=(p\vee\overline{q})\wedge(\overline{p}\vee\overline{r})\wedge(q\vee r)\wedge(p\vee q). \tag{4.2}$$

If $\phi(p,q,r)=\texttt{true}$ for a particular assignment of truth values to the variables, we say that this assignment *satisfies* ϕ. We leave the problem of finding the satisfying assignment, which is unique and delicious, to the reader.

Unfortunately, my cat is very fond of Quail. If I serve it without some Roquefort to distract her, my guest will find his portion stolen away. This adds an additional constraint, giving a new formula 4.3

$$\phi'(p,q,r)=(p\vee\overline{q})\wedge(\overline{p}\vee\overline{r})\wedge(q\vee r)\wedge(p\vee q)\wedge(\overline{q}\vee r). \tag{4.3}$$

As the following exercise shows, there is now no way to satisfy all the constraints, and we are forced to order out for pizza.

Exercise 4.4 *Prove that the formula ϕ' of* (4.3) *has no satisfying assignments.*

Both ϕ and ϕ' are Boolean formulas with a special structure, called CNF for *conjunctive normal form.* This means that each one is the AND (conjunction) of a series of *clauses*, and each clause is the OR of a set of *literals*, where a literal is either a variable or its negation. Thus there are multiple ways to satisfy each clause, but we must satisfy every clause to satisfy the entire formula.

Let's say that ϕ is *satisfiable* if at least one satisfying assignment exists. Then we are wrestling with the following problem:

> SAT
>
> Input: A CNF Boolean formula $\phi(x_1,\ldots,x_n)$
>
> Question: Is ϕ satisfiable?

It's easy to check whether a given truth assignment is satisfying: we just have to go through all of ϕ's clauses, and make sure that at least one literal in each clause is true. This shows that SAT is in NP. On the other hand, with n variables we have a haystack of 2^n possible truth assignments to consider.

Just as each edge in GRAPH COLORING imposes the constraint that its two endpoints have different colors, each SAT clause demands that at least one of its variables take a particular value. Both of them are *constraint satisfaction* problems, where we have a set of constraints and we want to know if we can satisfy all of them simultaneously.

However, SAT is one of the most flexible such problems, since it is easy to express a wide variety of constraints in CNF form. Consider the following exercise:

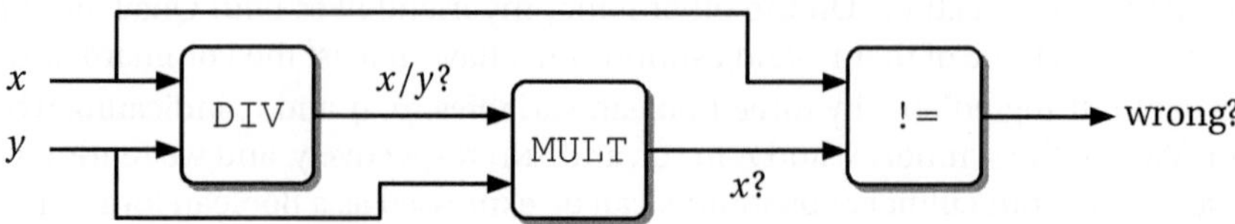

FIGURE 4.4: Verifying a chip: DIV works on every input if and only if there is no input x, y that makes this circuit output true. Equivalently, the formula ϕ corresponding to this circuit is unsatisfiable.

Exercise 4.5 *Show how to convert a graph G to a CNF formula ϕ such that G is 3-colorable if and only if ϕ is satisfiable, and thus provide a reduction*

$$\text{GRAPH 3-COLORING} \leq \text{SAT}.$$

Hint: for each vertex v, let v_r, v_g, and v_b be Boolean variables which are true if v is colored red, green, or blue respectively.

This ability to express other problems makes SAT a very useful tool. For instance, consider the problem of verifying computer hardware. Let's suppose that you have designed a chip for division that you want to verify. It purports to take two inputs x, y and output x/y. Let's also suppose that you already have a trustworthy chip for multiplication. Then you can use the multiplier to multiply y by the output x/y of the divider, and compare the result with x. Let the final output be true if these are different, and false if they are the same (up to whatever numerical accuracy you expect). The resulting circuit is shown in Figure 4.4.

This combination of chips comprises a circuit made of AND, OR and NOT gates. Suppose we have a set of Boolean variables that include both the input bits of x and y, and the truth values z carried internally by the wires of the chip. Then it is not too hard to write a CNF formula ϕ that expresses the following worrisome claim: each individual gate works correctly, but the divider gives the wrong answer. In order to prove that the divider always works correctly, we need to prove that for every combination of x, y and z, either one of the gates malfunctions, or the output is correct—in other words, that ϕ is unsatisfiable.

Our culinary examples above consisted of clauses with two variables each. Fixing the number of variables per clause gives important special cases of SAT:

k-SAT
Input: A CNF Boolean formula ϕ, where each clause contains k variables
Question: Is ϕ satisfiable?

We will not insult the reader by pointing out that 1-SAT is in P. On the other hand, 2-SAT is also in P, and the corresponding algorithm might entertain even our smartest readers.

The key idea is to represent a 2-SAT formula as a directed graph. First note that if one literal of a 2-SAT clause is false, the other one must be true. Therefore, a 2-SAT clause is equivalent to a pair of implications,

$$(\ell_1 \vee \ell_2) \Leftrightarrow (\overline{\ell_1} \rightarrow \ell_2) \text{ AND } (\overline{\ell_2} \rightarrow \ell_1). \tag{4.4}$$

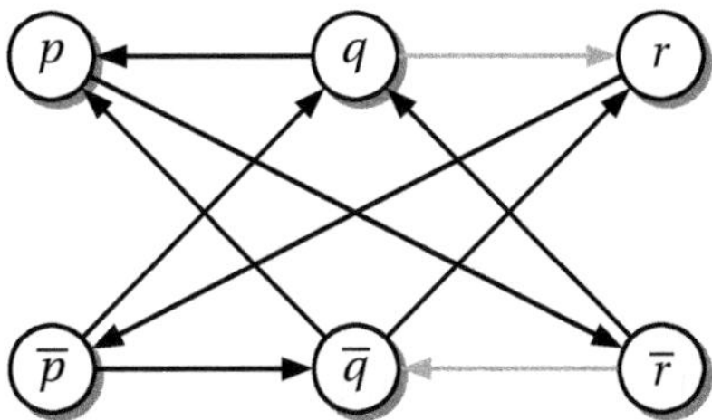

FIGURE 4.5: The directed graph of implications (4.4) corresponding to the 2-SAT problem ϕ from our dinner example (black edges) and the additional constraint imposed by my cat in ϕ' (gray edges).

Now suppose ϕ is a 2-SAT formula with n variables $x_1,\ldots,x_n$. We define a graph $G(\phi)$ with $2n$ vertices representing the $2n$ possible literals x_i and $\overline{x}_i$, and with a pair of directed edges between these vertices for each clause. This construction can clearly be carried out in polynomial time. For instance, the formula ϕ from our dinner party corresponds to the directed graph shown in Figure 4.5, and the gray edges correspond to the additional constraint imposed in ϕ'.

The following theorem shows that a 2-SAT formula is satisfiable if and only if there are no contradictory loops in this graph. In the process of proving it, we will see an important phenomenon in SAT: clauses get shorter, and more demanding, when variables are given values they don't like. We use the notation $x \rightsquigarrow y$ to mean that there is a path, i.e., a chain of implications, from the literal x to the literal y.

Theorem 4.1 *A* 2-SAT *formula ϕ is satisfiable if and only if there is no variable x with paths from $x \rightsquigarrow \overline{x}$ and $\overline{x} \rightsquigarrow x$ in $G(\phi)$.*

Proof If such a pair of paths exists then x must be `true` if it is `false`, and vice versa. Then ϕ is contradictory, and hence unsatisfiable.

Conversely, suppose that no such pair of paths exists. First note that a single one of these paths does not cause a contradiction. For instance, a path from x to $\overline{x}$ simply forces x to be `false`.

Now suppose we set the value of some variable. One way to view the implications in (4.4) is that setting x `true`, say, converts any clause of the form $(\overline{x} \vee y)$ into a one-variable clause (y), which demands that y be `true`. We call these one-variable clauses *unit* clauses. Satisfying a unit clause might turn some 2-SAT clauses into additional unit clauses, and so on. Each of these steps corresponds to following a directed edge in G, and following each one to its logical conclusion sets every literal reachable from x to `true`.

This process is called *unit clause propagation*. It can lead to two possible outcomes: either the cascade of unit clauses runs out, leaving us with a smaller 2-SAT formula on the remaining unset variables, or we are forced to set both y and $\overline{y}$ `true` for some y, giving a contradiction. Our claim is that the latter never occurs if we use the algorithm in Figure 4.6, unless there is a contradictory pair of paths.

To see this, suppose we start the unit clause propagation process by setting x `true`. Then a contradiction only occurs if there are paths $x \rightsquigarrow y$ and $x \rightsquigarrow \overline{y}$ for some variable y. But $G(\phi)$ has the following symmetry: for any literals ℓ_1,ℓ_2, there is a path $\ell_1 \rightsquigarrow \ell_2$ if and only if there is a path $\overline{\ell_2} \rightsquigarrow \overline{\ell_1}$. Thus there are also paths $y \rightsquigarrow \overline{x}$ and $\overline{y} \rightsquigarrow \overline{x}$, and hence a path $x \rightsquigarrow \overline{x}$. But in that case, we would have set x `false` instead. Similarly, if setting x `false` leads to a contradiction, there is a path $\overline{x} \rightsquigarrow x$, and we would have set x `true`.

This shows that, as long as there is no contradictory loop, the algorithm of Figure 4.6 succeeds in finding a satisfying assignment for ϕ—and therefore that ϕ is satisfiable. □

```
while there is an unset variable do
    choose an unset variable x ;
    if there is a path x ⇝ x̄ then set x = false ;
    else if there is a path x̄ ⇝ x then set x = true ;
    else set x to any value you like ;
    while there is a unit clause do unit clause propagation ;
end
```

FIGURE 4.6: A polynomial-time algorithm for 2-SAT.

Theorem 4.1 tells that that we can solve 2-SAT by solving $2n$ instances of REACHABILITY, checking to see if there are paths $x \rightsquigarrow \overline{x}$ and $\overline{x} \rightsquigarrow x$ for each variable x. We can think of this as a reduction

$$\text{2-SAT} \le \text{REACHABILITY},$$

with the understanding that each instance of 2-SAT corresponds to a polynomial number of instances of REACHABILITY, rather than a single one as in the reduction from MAX BIPARTITE MATCHING to MAX FLOW. Since REACHABILITY is in P, this shows that 2-SAT is in P as well.

Now that we know that 2-SAT is in P, what about 3-SAT? Unlike (4.4), there seems to be no way to treat a 3-SAT clause as an implication of one literal by another. We can certainly write $(x \vee y \vee z)$ as $\overline{x} \Rightarrow (y \vee z)$, but we would need some kind of branching search process to decide which of y or z is true.

It turns out that 3-SAT is just as hard as SAT in general. That is,

$$\text{SAT} \le \text{3-SAT}.$$

In particular, for any k we have

$$k\text{-SAT} \le \text{3-SAT}.$$

That is, any k-SAT formula ϕ can be converted to a 3-SAT formula ϕ' which is satisfiable if and only if ϕ is. To prove this, we will show how to convert a SAT clause c to a collection of 3-SAT clauses which are satisfiable if and only if the original clause is. If c has just one or two variables, we can add dummy variables, padding it out to a set of three-variable clauses:

$$(x) \Leftrightarrow (x \vee z_1 \vee z_2) \wedge (x \vee z_1 \vee \overline{z}_2) \wedge (x \vee \overline{z}_1 \vee z_2) \wedge (x \vee \overline{z}_1 \vee \overline{z}_2) \tag{4.5}$$

$$(x \vee y) \Leftrightarrow (x \vee y \vee z) \wedge (x \vee y \vee \overline{z}). \tag{4.6}$$

Exercise 4.6 *Show that the clauses* (x) *and* $(x \vee y)$ *are satisfied if and only if there exist values for* z_1, z_2, *and* z *such that the right-hand sides of* (4.5) *and* (4.6) *are satisfied.*

More interestingly, if $k > 3$ we can break a k-SAT clause into a chain of 3-SAT clauses, using dummy variables as the chain links. For instance, we can break a 5-SAT clause down like this:

$$(x_1 \vee x_2 \vee x_3 \vee x_4 \vee x_5) \Leftrightarrow (x_1 \vee x_2 \vee z_1) \wedge (\overline{z}_1 \vee x_3 \vee z_2) \wedge (\overline{z}_2 \vee x_4 \vee x_5). \tag{4.7}$$

Exercise 4.7 *Show that the* 5-SAT *clause on the left-hand side of* (4.7) *is satisfied if and only if there exist truth values for* z_1 *and* z_2 *such that the right-hand side is satisfied.*

Similarly, a k-SAT clause becomes $k-2$ 3-SAT clauses, linked together by $k-3$ dummy variables z_i.

This reduction sheds some light on the difference between 2-SAT and 3-SAT. Each clause in this chain except the first and last has to have three literals: the original literal it represents, and two to link it to the clauses on either side. If there were a way to break these clauses down even further, so that they have just two literals each, we would have 3-SAT $\le$ 2-SAT and 3-SAT would be in P. However, as Problem 4.9 shows, there is no simple way to do this. As far as we know, the computational complexity of k-SAT, like that of GRAPH k-COLORING, makes a huge jump upward when k goes from two to three.

Exercise 4.8 *Show how to reduce* GRAPH 2-COLORING *to* 2-SAT.

Exercise 4.9 *The size n of a problem is the total number of bits it takes to specify it. However, for k*-SAT *we will usually identify n with the number of variables. Explain why we can do this, as long as k is a constant, and as long as we only care about whether these problems are in* P *or not.*

4.2.3 Balancing Numbers

I have not altered the weights of the balance,
I have not tampered with the balance itself.

Egyptian *Book of the Dead*

A wealthy and eccentric uncle passes on from this world, and bequeaths to you an antique balance with a set of brass weights. You wish to display the balance on your mantelpiece. Can you place some of the weights on the left side of the scale, and the others on the right, so that the scale is balanced?

> INTEGER PARTITIONING
>
> Input: A list $S=\{x_1,\ldots,x_\ell\}$ of positive integers
>
> Question: Is there a *balanced partition*, i.e., a subset $A\subseteq\{1,\ldots,\ell\}$ s.t. $\sum_{i\in A} x_i=\sum_{i\notin A} x_i$?

Problems like this arise in packing and scheduling. Suppose we have a dual-processor machine, and we have a set of jobs we need to assign to the two processors. If S is the list of the running times of these jobs, then assigning them according to a balanced partition lets us carry them out in parallel, completing them all in time equal to half their total running time.

A close relative of INTEGER PARTITIONING asks whether I can balance a given weight with a subset of my uncle's weights—that is, whether a subset exists with a given total:

> SUBSET SUM
>
> Input: A set $S=\{x_1,\ldots,x_\ell\}$ of positive integers and an integer t
>
> Question: Does there exist a subset $A\subseteq\{1,\ldots,\ell\}$ such that $\sum_{i\in A} x_i=t$?

It is easy to reduce INTEGER PARTITIONING to SUBSET SUM, since INTEGER PARTITIONING is just the special case $t=W/2$ where $W=\sum_i x_i$ denotes the total of all the weights. Problem 4.13 asks you to give a reduction in the reverse direction, showing that these two problems are in fact equivalent. We also note that SUBSET SUM was used in an early proposal for public key cryptography.

INTEGER PARTITIONING and SUBSET SUM have a somewhat different flavor than the previous problems
4.4 in this chapter. We can think of them as constraint satisfaction problems on ℓ Boolean variables y_i, each of which determines whether or not $x_i \in A$. However, while GRAPH COLORING and SAT have many constraints, each of which involves just a few variables, INTEGER PARTITIONING and SUBSET SUM each have a single constraint, which involves all of them.

When one first encounters INTEGER PARTITIONING, it seems plausible that some kind of greedy strategy might succeed. For instance, we could sort S from smallest to largest, start at one end or the other, and add each number to whichever side of the balance has the smaller total. But unlike MINIMUM SPANNING TREE, this approach simply doesn't work. Consider the example $S = \{4,5,6,7,8\}$. The greedy algorithm that adds the smallest first gives $\{4,6,8 \mid 5,7\}$. Adding the largest first gets to $\{8,5 \mid 7,6\}$ after four steps, and since these are balanced it is doomed whichever side it puts 4 on. Neither of these strategies finds the perfect solution, $\{4,5,6 \mid 7,8\}$.

Another approach is dynamic programming. Consider the following function, which tells us whether or not we can balance a weight t using a subset of the weights x_j through x_ℓ:

$$f(j,t) = \begin{cases} \texttt{true} & \text{if there is an } A \subseteq \{j,\ldots,\ell\} \text{ such that } \sum_{i\in A} x_i = t \\ \texttt{false} & \text{otherwise.} \end{cases}$$

To solve an instance of SUBSET SUM, we need to calculate $f(1,t)$. Now, for each j we can either include x_j, adding it to the balance, or leave it out. This lets us write $f(j,t)$ recursively:

$$f(j,t) = f(j+1,t) \text{ OR } f(j+1,t-x_j). \tag{4.8}$$

Thus we go through the list from left to right, deciding which weights to include. The base case is given by $f(\ell,t) = \texttt{true}$ if $t = 0$ or $t = x_\ell$, and $\texttt{false}$ otherwise—or, if you prefer, $f(\ell+1,t) = \texttt{true}$ if $t = 0$ and $\texttt{false}$ otherwise, since if $j = \ell+1$ we have no weights left.

As in Section 3.3, if we calculate $f(1,t)$ recursively while remembering values of f we calculated before, the running time is proportional to the number of different subproblems we need to solve. Since j ranges from 1 to ℓ and t ranges from 0 to the total weight $W = \sum_i x_i$, the number of different subproblems is $O(W\ell)$. So, if W is only polynomial in n, we can solve INTEGER PARTITIONING in polynomial time.

Exercise 4.10 *Using dynamic programming, find a balanced partition of*

$$S = \{62,83,121,281,486,734,771,854,885,1003\}$$

if one exists.

This shows that INTEGER PARTITIONING is in P if the inputs x_i are given in unary, since then $W = O(n)$ where n is the total length of the input. However, if the x_i are given in binary then W is typically exponentially large, forcing the dynamic programming algorithm to take exponential time. To be more quantitative about this, let's assume that each x_i has b bits. As Figure 4.7 shows, the total size of the input is then $n = b\ell$. If $b = \ell = \sqrt{n}$, say, W could easily be $2^b\ell = 2^{\sqrt{n}}\sqrt{n}$.

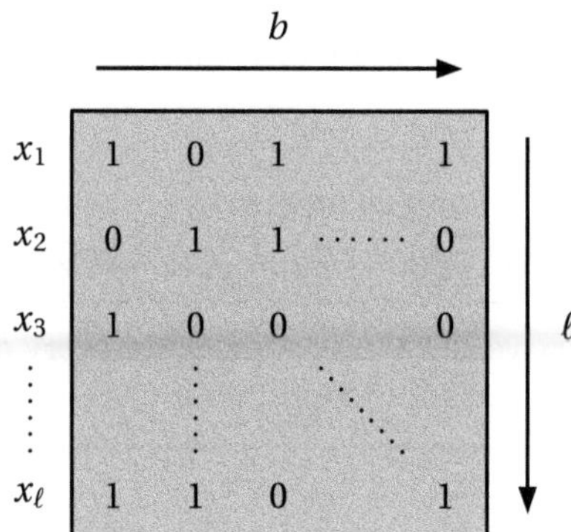

FIGURE 4.7: Instances of INTEGER PARTITIONING can grow in two dimensions: along the ℓ-axis (more numbers) or the b-axis (more bits per number).

In practical applications, the number of bits b is usually constant because we only need a partition that is balanced up to some fixed precision. Indeed, we can find an *approximately* balanced partition in polynomial time, as Problem 9.7 shows for a similar packing problem. But if the adversary hands us an instance with n-bit weights, we know of no polynomial-time algorithm that achieves perfect balance.

4.5

Exercise 4.11 *Show that* SUBSET SUM *can be solved in polynomial time if the target sum t is very small or very large—specifically, if* $\min(t, W-t) = \mathrm{poly}(n)$, *where* $W = \sum_i x_i$ *is the total weight.*

Exercise 4.12 *Show that* INTEGER PARTITIONING *and* SUBSET SUM *are also in* P *if we consider the other extreme, where S consists of a few large numbers rather than many small ones—that is, where b is unrestricted but* $\ell = O(\log n)$.

4.2.4 Rival Academics and Cliquish Parties

I am planning a cocktail party for my academic acquaintances. However, there are a number of rivalries among them—pairs of people who will glare at each other, insult each others' work, and generally render the conversation quite tiresome. I would like to arrange a gathering that is as large as possible, but I want to make sure that I don't invite both members of any of these pairs.

Formally, given a graph $G = (V, E)$, we say that a subset $S \subseteq V$ is *independent* if no two vertices in S are adjacent. If vertices are people and edges are rivalries, I want to find the largest possible independent set in my social graph.

Fundamentally, this is an optimization problem. The diligent reader has already met it, or rather its weighted version, in the Problems of Chapter 3. In order to formulate a decision problem, we will give a target size k as part of the input, and ask the yes-or-no question of whether there is a set at least this large.

> INDEPENDENT SET
>
> Input: A graph G and an integer k
>
> Question: Does G have an independent set of size k or more?

Now let's say that a subset $S \subseteq V$ is a *vertex cover* if for every edge $e \in E$, at least one of e's endpoints is in S. Asking for small vertex covers gives us another problem:

> VERTEX COVER
>
> Input: A graph G and an integer k
>
> Question: Does G have a vertex cover of size k or less?

Also, we say that S is a *clique* if every vertex in S is adjacent to every other, so that S forms a complete graph. Asking for large cliques gives us yet another problem:

> CLIQUE
>
> Input: A graph G and an integer k
>
> Question: Does G have a clique of size k or more?

The following exercise shows that these three problems are equivalent to each other. Indeed, the relationship between them is so simple that it hardly deserves to be called a reduction.

Exercise 4.13 *Given a graph $G = (V, E)$ and a subset $S \subseteq V$, show that the following three statements are equivalent:*

1. *S is an independent set,*
2. *$V - S$ is a vertex cover,*
3. *S is a clique in the complemented graph $\overline{G} = (V, \overline{E})$, where two vertices are adjacent if and only if they are not adjacent in G.*

Therefore, in a graph G with n vertices, there is an independent set of size k or more if and only if there is a vertex cover of size $n - k$ or less, and this is true if and only if there is a clique in $\overline{G}$ of size k or more.

In terms of my cocktail party, inviting an independent set means disinviting a vertex cover—in other words, making sure that for every rival pair, at least one of them is excluded. Edges in $\overline{G}$ describe compatible pairs, who can eat and drink with each other in peace, and in this case the invitees form a clique.

It is worth noting that all of these problems are in P if k is constant. In that case, we can solve them by brute force, going through all $\binom{n}{k} = O(n^k)$ possible sets of k vertices, and seeing if any of them are independent, or if they form a vertex cover, or a clique. The problem is that k is part of the input, so we cannot say that this algorithm runs in time $O(n^c)$ for a constant c. For that matter, if $k = \alpha n$ for some $0 < \alpha < 1$, then $\binom{n}{k}$ grows exponentially as a function of n.

Let's pause for a moment to consider the relationship between these decision problems and the corresponding optimization problems. These decision problems are in NP—for instance, I can prove that there is an independent set of size at least k by exhibiting one. On the other hand, consider the following slight variation:

> INDEPENDENT SET (exact version)
>
> Input: A graph $G = (V, E)$ and an integer k
>
> Question: Does G's largest independent set have size exactly k?

Is this problem in NP? If the answer is "yes," how can I prove that to you? Showing you an independent set of size k is not enough—I also need to prove that no larger one exists. As we will discuss in the next section, this version of the problem has a more complex logical structure, and places it in a complexity class somewhat higher than NP.

Yet another version—which is presumably the one we really want to solve—is the search for the largest possible independent set:

> MAX INDEPENDENT SET
> Input: A graph G
> Output: An independent set S of maximal size

We will discuss problems like these in Chapter 9. For now, we leave you with Problem 4.18, which shows that if any of these versions of INDEPENDENT SET can be solved in polynomial time, they all can be.

Exercise 4.14 *Show that if a graph with n vertices is k-colorable, it has an independent set of size at least n/k. Is the converse true?*

4.3 Search, Existence, and Nondeterminism

So far we have given an informal definition of NP, and some examples that illustrate the flavor of this complexity class. The time has come to discuss NP in more detail. In this section, we give precise definitions of NP from several points of view: checking a witness, verifying a logical statement, and running a miraculous kind of program that can try many solutions at once.

4.3.1 NP, NTIME, and Exhaustive Search

Here is a formal definition of NP:

> NP is the class of problems A of the following form:
>
> x is a yes-instance of A if and only if there exists a w such that (x, w) is a yes-instance of B,
>
> where B is a decision problem in P regarding pairs (x, w), and $|w| = \text{poly}(|x|)$.

We call w the *witness* of the fact that x is a yes-instance of A. It is also sometimes called a *certificate.*

This definition may seem a bit technical at first, but it is really just a careful way of saying that solving A is like telling whether there is a needle in a haystack. Given an input x, we want to know whether a needle w exists—a path, a coloring, or a satisfying assignment—and B is the problem of checking whether w is a bona fide needle for x. For instance, if A is HAMILTONIAN PATH then x is a graph, w is a path, and $B(x, w)$ is the problem of checking whether w is a valid Hamiltonian path for x.

Note that we require that the witness w can be described with a polynomial number of bits. This is usually obvious—for instance, if x is a graph with n vertices, the number of bits it takes to encode a Hamiltonian path w is $|w| = n \log_2 n = \text{poly}(|x|)$. Since B's input (x, w) has total size $|x| + |w|$ and B is in P, this requirement ensures that B's running time is $\text{poly}(|x| + |w|) = \text{poly}(|x|)$.

How long does it take to solve such a problem deterministically? Since $|w| = \text{poly}(n)$, there are $2^{\text{poly}(n)}$ possible witnesses. We can check them all by running B on each one of them, so

$$\mathsf{NP} \subseteq \mathsf{EXP}.$$

Here $\mathsf{EXP} = \mathsf{TIME}(2^{\text{poly}(n)})$, as defined in Section 2.4.3, is the class of problems that we can solve in exponential time. We believe that this is the best possible inclusion of NP in a deterministic time complexity class—in other words, that the only general method for solving problems in NP is exhaustive search.

To state this as strongly as possible, we believe there is a constant $\alpha > 0$ such that, for any $g(n) = o(2^{n^\alpha})$, we have

$$\mathsf{NP} \not\subseteq \mathsf{TIME}(g(n)).$$

Thus we believe that solving problems in NP generally takes exponential time, where "exponential" means exponential in n^α for some α. Note that this is much stronger than the belief that $\mathsf{P} \neq \mathsf{NP}$. For instance, it could be that $\mathsf{NP} \subseteq \mathsf{TIME}(g(n))$ for some $g(n)$ which is superpolynomial but subexponential, such as $n^{\log n}$. This would not imply that $\mathsf{P} \neq \mathsf{NP}$, but it would show that we can do qualitatively better than an exhaustive search.

More generally, we can define the complexity class of needle-in-a-haystack problems where needles can be checked in $O(f(n))$ time. This gives the following generalization of NP, analogous to the classes $\mathsf{TIME}(f(n))$ we defined in Chapter 2:

> $\mathsf{NTIME}(f(n))$ is the class of problems A of the following form:
>
> x is a yes-instance of A if and only if there exists a w such that (x, w) is a yes-instance of B,
>
> where B is a decision problem in $\mathsf{TIME}(f(n))$ regarding pairs (x, w), and where $|w| = O(f(n))$.

For instance, $\mathsf{NEXP} = \mathsf{NTIME}(2^{\text{poly}(n)})$ is the rather large class of problems where we are given an exponential amount of time to check a given needle, and where each needle might need an exponential number of bits to describe it. Since there are $2^{2^{\text{poly}(n)}}$ possible witnesses, solving such problems with exhaustive search takes doubly-exponential time, giving

$$\mathsf{NEXP} \subseteq \mathsf{EXPEXP}.$$

More generally, we have

$$\mathsf{NTIME}(f(n)) \subseteq \mathsf{TIME}(2^{O(f(n))}).$$

Once again, we believe this is optimal—that at all levels of the complexity hierarchy, there is an exponential gap between the classes NTIME and TIME.

4.3.2 *There Exists a Witness: The Logical Structure of* NP

> Hermione looked outraged. "I'm sorry, but that's ridiculous! How can I possibly prove it doesn't exist? Do you expect me to get hold of all the pebbles in the world and test them?"
>
> J. K. Rowling, *Harry Potter and the Deathly Hallows*

Another way to think about NP is as follows. We can associate a decision problem A with a property $A(x)$, where $A(x)$ is true if x is a yes-instance of A. We can say that a property B is in P if it can be checked in polynomial time. Then NP properties are simply P properties with a "there exists" in front of them: $A(x)$ if there exists a w such that $B(x,w)$, where $B(x,w)$ is the property that w is a valid witness for x.

This "there exists" is an *existential quantifier*, and we write it $\exists$. This gives us another definition of NP:

> NP is the class of properties A of the form
>
> $$A(x) = \exists w : B(x,w)$$
>
> where B is in P, and where $|w| = \text{poly}(|x|)$.

For instance, if x is a graph and $A(x)$ is the property that x is Hamiltonian, then $B(x,w)$ is the polynomial-time property that w is a Hamiltonian path for x.

Algorithmically, the quantifier $\exists$ represents the process of searching for the witness w. We can also think of it as a conversation, between a *Prover* who claims that x is a yes-instance, and a *Verifier* who is yet to be convinced. The Prover, who has enormous computational power, provides a proof of her claim in the form of the witness w. Then the Verifier—who, like us, is limited to humble polynomial-time computation—checks the witness, and makes sure that it works.

This logical definition makes it easy to see why NP treats yes-instances and no-instances so differently. The negation of a "there exists" statement is a "for all" statement: if there is no needle, everything in the haystack is hay. For instance, the claim that a graph is *not* Hamiltonian looks like this:

$$\overline{A(x)} = \overline{\exists w : B(x,w)} = \forall w : \overline{B(x,w)}.$$

Here $\forall$ is a *universal quantifier*, which we read "for all." In other words, for all sequences w of n vertices, w fails to be a Hamiltonian path.

What happens if we take a problem like HAMILTONIAN PATH, and switch yes-instances and no-instances? Consider the following problem:

> NO HAMILTONIAN PATH
>
> Input: A graph G
>
> Question: Is it true that G has no Hamiltonian Path?

This is a very different problem. Now there is a simple proof if the answer is "no," but it seems hard to prove that the answer is "yes" without an exhaustive search. Problems like this have a complexity class of their own, called coNP: the class of problems in which, if the input is a no-instance, there is a simple proof of that fact.

The complements of P properties are also in P, since we can just modify their polynomial-time algorithms to output "yes" instead of "no" and vice versa. So, coNP properties like "non-Hamiltonianness" take the form

$$A(x) = \forall w : B(x,w).$$

where B is in P. As we discussed in the Prologue, claims of this form do not seem to have simple proofs. But they have simple *dis*-proofs, since it takes just one counterexample to disprove a "for all." Just as we believe that $\mathsf{P} \neq \mathsf{NP}$, we believe that these two types of problems are fundamentally different, and that $\mathsf{NP} \neq \mathsf{coNP}$.

What about questions like the version of INDEPENDENT SET defined at the end of the last section, which asks whether the largest independent set of a graph has size exactly k? Here the answer is "yes" if there exists an independent set of size k, and if all sets of size $k+1$ or greater are not independent. Unless we can see a clever way to re-state this property in simpler terms, this requires us to use both $\exists$ and $\forall$.

To take another example, imagine that I am a chip designer, and that I have designed a circuit C which computes some function $f_C(x)$ of its input x. I claim that C is the most efficient circuit that does this—that there is no way to compute f_C with a smaller circuit C'. Logically, this means that for every smaller circuit C', there is some input x on which C' and C act differently. This gives a statement with two nested quantifiers, one $\forall$ and one $\exists$:

$$\forall C' < C : \exists x : f_{C'}(x) \neq f_C(x).$$

There seems to be no simple way to prove this claim, since there are an exponential number of alternate circuits C'. But if a rival claims that I'm wrong, I can challenge him to play a game with me. In each round, he names a smaller circuit C' that he claims also computes f_C. If I can reply with an x such that $f_{C'}(x) \neq f_C(x)$, then I win. If I have a winning strategy in this game—that is, if I always have a winning move x, no matter what C' he chooses—then my claim is true.

For an even deeper logical structure, recall our discussion of Chess in the Prologue. It is Black's move, and I claim that Black has a winning strategy from the current position. This is tantamount to saying that Black has a move, such that no matter how White replies, Black has a move, such that... and so on until Black puts White in checkmate. The logical structure of this statement has many nested quantifiers, alternating between $\exists$s and $\forall$s, and the number of quantifiers is the length of the longest possible game.

Just as NP consists of P with a single $\exists$, there is a series of complexity classes called the *polynomial hierarchy*, where we add more and more layers of quantifiers, and express properties of greater and greater logical depth. Games like Chess lead to even higher classes where the number of quantifiers can grow with the size of the problem. We will discuss these complexity classes in Chapters 6 and 8. For now, let's stick with problems that involve just one exponential haystack.

Exercise 4.15 *A friend of yours is addlepated and is confusing* coNP *with the complement of* NP. *Help him or her out.*

Exercise 4.16 *The class* $\mathsf{NP} \cap \mathsf{coNP}$ *is the class of problems where, for each* x, *there is a simple proof that* x *is a yes-instance or that it is a no-instance, whichever it is. Prove that* $\mathsf{P} \subseteq (\mathsf{NP} \cap \mathsf{coNP})$, *where here we restrict our definition of* P *to decision problems. Do you think that* $\mathsf{P} = (\mathsf{NP} \cap \mathsf{coNP})$?

Exercise 4.17 *Show that if* $\mathsf{P} = \mathsf{NP}$ *then* $\mathsf{NP} = \mathsf{coNP}$. *Is the converse necessarily true?*

4.3.3 What's the N For? The Peculiar Power of Nondeterministic Computers

I leave to the various futures (not to all) my garden of forking paths.

Jorge Luis Borges

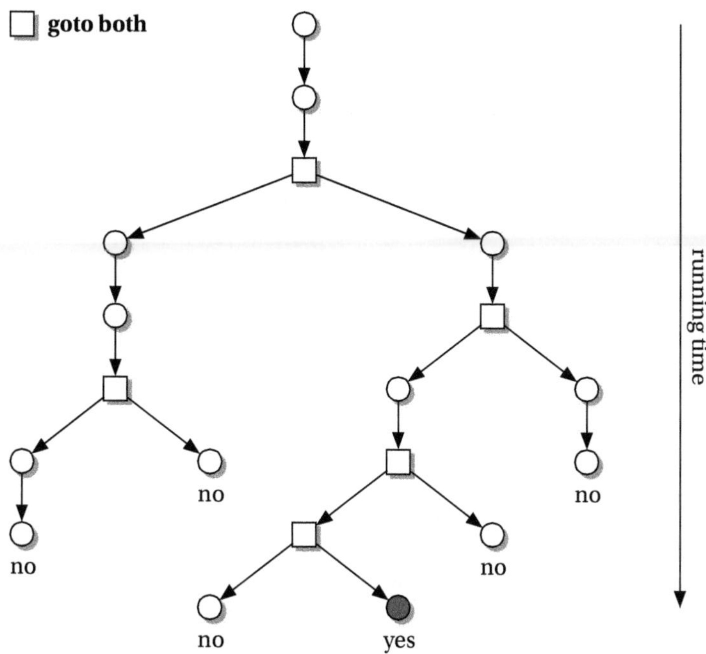

FIGURE 4.8: The branching computation tree of a nondeterministic program.

> If you come to a fork in the road, take it.
>
> Yogi Berra

In many textbooks, NP is defined not in terms of the existence of witnesses, but in terms of a curious kind of computer. The N in NP stands for "nondeterministic," and a nondeterministic computer is one that can make several possible choices at each step. Imagine a programming language in which we have the following magical instruction:

goto both line 1, line 2

This instruction splits the computation into two parallel processes, one continuing from line 1 and the other from line 2. As we encounter more and more such instructions, the program branches into a tree of computation paths as shown in Figure 4.8. We say that the running time is polynomial if every path ends within a polynomial number of steps. Finally, we define the output of a nondeterministic computer in the following way: it is "yes" if *any* of these paths leads to "yes," and "no" if *all* of them lead to "no."

Since we can branch on every step, the number of computation paths can grow exponentially as a function of the running time. This gives nondeterministic computers the amazing, and unrealistic, ability to search an exponentially large space of candidate solutions exhaustively in polynomial time. In the "needle in a haystack" metaphor, they can examine all the blades of hay in parallel. If a needle exists, they can find it in no time at all.

Exercise 4.18 *Write down a polynomial-time nondeterministic program for* 3-SAT *in your favorite programming language augmented with* ***goto both****.*

Exercise 4.19 *Show that if a nondeterministic program only executes* $O(\log n)$ ***goto both*** *instructions on each branch, the problem it solves is in* P.

Nondeterministic computers give us the following alternate definition of NP:

> NP is the class of problems for which a nondeterministic program exists that runs in time poly(n), such that the input is a yes-instance if and only if there exists a computation path that returns "yes."

Similarly, for any function $f(n)$ of the input size we can define the class NTIME($f(n)$) as follows:

> NTIME($f(n)$) is the class of problems for which a nondeterministic program exists that runs in time $O(f(n))$, such that the input is a yes-instance if and only if there exists a computation path that returns "yes."

Note that nondeterministic computers can also run *deterministic* programs, namely those which never use the **goto both** instruction. This gives another proof that P $\subseteq$ NP, and more generally that TIME($f(n)$) $\subseteq$ NTIME($f(n)$).

It is not hard to see that this definition of NP is equivalent to our previous one. We can simulate a nondeterministic program A with a deterministic program B by giving B an additional string of bits telling it which way to go at each **goto both**. This string is our witness w, and its length is at most the running time of A. Conversely, we can simulate both the Prover and a witness-checking program B with a nondeterministic program A that starts by guessing the witness—with $|w| = \text{poly}(n)$ **goto both** commands, writing a 0 or 1 on each bit of w—and then verifies it by running B.

There are many metaphors we could use to describe a nondeterministic computer. We could say that it spawns an exponential number of processes which work in parallel; we could say that it makes a lucky guess about which path to follow; we could say that it flips a coin at each step, and that it gets credit for finding the solution even if the probability of doing so is exponentially small. We have chosen to focus on the metaphor of a conversation between a Prover and a Verifier—a metaphor which will become far richer in Chapter 11.

Finally, it should be emphasized that "nondeterministic" computation has nothing to do with *probabilistic* or *randomized* computation. In the latter setting, the computer can make random choices, but must obtain the correct result with reasonably large probability. We will discuss this type of computation in Chapter 10.

4.4 Knots and Primes

All the NP problems we have seen so far are stated in the form "does there exist...," in which case the witness w is simply the thing we are looking for. However, not all problems in NP are of this form, at least not on the surface. Let us look at two problems that hide their $\exists$s rather than putting them on display, so that defining witnesses for them is not so obvious.

4.4.1 *Primality*

> There is no apparent reason why one number is prime and another not. To the contrary, upon looking at these numbers one has the feeling of being in the presence of one of the inexplicable secrets of creation.
>
> Don Zagier

One of the classic problems in mathematics, and one which many people have struggled to find efficient algorithms for, is the problem of whether a given number is prime:

PRIMALITY
Input: An n-bit integer p
Question: Is p prime?

In this section, we will show that PRIMALITY is in NP. We start with the following easy exercise. It consists of finding a simple witness for non-PRIMALITY, i.e., for COMPOSITENESS.

Exercise 4.20 *Show that* PRIMALITY *is in* coNP.

Finding a simple witness for PRIMALITY, on the other hand, is not quite so easy. The first thing that comes to mind is checking that $\gcd(p,t)=1$ for all t ranging from 2 to $p-1$. The number of such t is linear in p, but p is exponentially large as a function of the input size n, so there are exponentially many such t. We need a proof of primality that is far more compact—one whose size, and the time it takes to verify it, is polynomial in $n=O(\log p)$.

Such witnesses do exist, but defining them takes a little thought. Readers who aren't familiar with number theory may be tempted to skip this section, but the mathematics we learn here will be very useful when we look at randomized algorithms, cryptography, and the quantum algorithm for FACTORING.

First, a little modular arithmetic. We write $x \equiv_p y$ if x and y are equivalent mod p, i.e., if $x \bmod p = y \bmod p$. Now suppose p is a prime. If a is an integer greater than 1 and less than p, its *order* is the smallest $r>0$ such that $a^r \equiv_p 1$. Fermat's Little Theorem (see Appendix A.7) states that if p is prime, then

$$a^{p-1} \equiv_p 1$$

for any a. This does not mean that the order r of a is $p-1$. However, it does imply that r is a divisor of $p-1$, so that $a^{p-1}=a^{rk} \equiv_p 1^k = 1$ for some k.

This theorem also has a converse, also known as Lehmer's Theorem, which states that p is prime if and only if an a exists with order *exactly* $p-1$. This means that its powers $1, a, a^2, \ldots \bmod p$ cycle through every nonzero integer mod p. For instance, if $a=3$, its powers mod 7 are

$$1,3,2,6,4,5,1,\ldots$$

and its order is 6. Such an a is called a *primitive root mod* p. As we discuss in Appendix A.7, we also say that a generates the group $\mathbb{Z}_p^* = \{1,2,\ldots,p-1\}$ of integers with multiplication mod p. If such an a exists, it constitutes a proof, or witness, that p is prime—assuming that we can check that it is genuine.

So our Prover starts by giving us a primitive root a. To verify it, we start by checking that $a^{p-1} \equiv_p 1$. We can do this in polynomial time using the repeated-squaring algorithm for modular exponentiation described in Section 3.2.2.

However, this is not enough. Even if $a^{p-1} \equiv_p 1$, it is entirely possible that the order r of a is actually some smaller divisor of $p-1$ rather than $p-1$ itself. To rule out this possibility, we need to make sure that $a^t \not\equiv_p 1$ for all $0 < t < p-1$. It is easy enough to check this for each t, but there are p possible values of t in this range, and it would take exponential time to check all of them. Even if we restrict ourselves just to the divisors of $p-1$, we could still be in trouble, since Problem 4.23 shows that an n-bit number can have a superpolynomial number of divisors.

Happily, we can restrict ourselves further to the case where t is a special kind of divisor of $p-1$: namely, $p-1$ divided by a prime. Why is this? Suppose that a's order r is $(p-1)/s$ for some $s > 1$, and let q be one of the prime divisors of s. Then we have

$$a^{(p-1)/q} = a^{r(s/q)} \equiv_p 1^{s/q} = 1 .$$

Therefore, to confirm that a's order really is $p-1$ rather than a proper divisor of $p-1$, it suffices to check that $a^t \not\equiv_p 1$ where $t = (p-1)/q$ for each prime factor q of $p-1$. Moreover, the following exercise shows that $p-1$ has a polynomial number of distinct prime factors, so we only need to check a polynomial number of different values of t.

Exercise 4.21 *Show that any n-bit integer has at most n distinct prime factors.*

Unfortunately, in order to carry out this strategy we need to know the prime factors of $p-1$, and FACTORING seems to be a hard problem in itself. So, as part of the witness that p is prime, the Prover kindly gives us the prime factorization of $p-1$. In other words, in addition to a primitive root a, he gives us a list of primes $q_1, \ldots, q_m$ and integer powers $s_1, \ldots, s_m$ such that

$$q_1^{s_1} q_2^{s_2} \cdots q_m^{s_m} = p-1 , \tag{4.9}$$

a fact which we can easily verify. All that the Prover needs to prove is that the q_i are prime, which he does by giving us—yes, you guessed it—more witnesses.

Registered Certificate of Primality — Issued by the Primality Certification Board

N	Prime factors of N − 1	a	$a^{N-1} \bmod N = 1$	$a^{(N-1)/p} \bmod N \neq 1$, for prime factors p of N − 1
2444789759	2, 1222394879	11	✓	$11^{1222394879} \equiv 2444789758$, ✓ $11^2 \equiv 121$ ✓
1222394879	2, 611197439	19	✓	$19^{611197439} \equiv 1222394878$, ✓ $19^2 \equiv 361$ ✓
611197439	2, 305598719	13	✓	$13^{305598719} \equiv 611197438$, ✓ $13^2 \equiv 169$ ✓
305598719	2, 152799359	37	✓	$37^{152799359} \equiv 305598718$, ✓ $37^2 \equiv 1369$ ✓
152799359	2, 76399679	11	✓	$11^{76399679} \equiv 152799358$, ✓ $11^2 \equiv 121$ ✓
76399679	2, 38199839	11	✓	$11^{38199839} \equiv 76399678$, ✓ $11^2 \equiv 121$ ✓
38199839	2, 19099919	13	✓	$13^{19099919} \equiv 38199838$, ✓ $13^2 \equiv 169$ ✓
19099919	2, 37, 258107	11	✓	$11^{9549959} \equiv 19099918$, ✓ $11^{516214} \equiv 7921368$, ✓ $11^{74} \equiv 6206319$ ✓
258107	2, 23, 31, 181	2	✓	$2^{129053} \equiv 258106$, ✓ $2^{11222} \equiv 67746$, ✓ $2^{8326} \equiv 71301$, ✓ $2^{1426} \equiv 57204$ ✓

It is hereby confirmed that 2,444,789,759 has been certified prime.

Signed: [signature] Date: 1 September, 1975

★ PRIME ★ — Registered authentic

FIGURE 4.9: The Pratt certificate for the primality of 2444789759. We assume that no certificate is necessary for primes of 3 digits or fewer.

Thus, in its entirety, the witness for p consists of

1. A primitive root a mod p,
2. The prime factorization of $p-1$, i.e., a list of integers $q_1, \ldots, q_m$ and $s_1, \ldots, s_m$ such that (4.9) holds, and
3. A witness for the primality of each q_i.

This kind of witness, or "certificate," for PRIMALITY is called a *Pratt certificate* after its inventor. We show an example in Figure 4.9.

The recursive structure of Pratt certificates looks a bit worrying. In order to provide witnesses for the 4.6
primality of each q_i, we have to provide witnesses for the prime factors of each $q_i - 1$, and so on. At some point, we get down to primes that are small enough so that no witness is necessary—say, those of up to 3 digits. But how many bits long is the entire certificate?

Luckily, as Problem 4.24 shows, the total size of the certificate is only $O(n^2) = \text{poly}(n)$ bits. Moreover, the total time to verify it is also poly(n). Thus PRIMALITY is in NP, and along with Exercise 4.20 we have

$$\text{PRIMALITY is in NP} \cap \text{coNP}.$$

In other words, for any number p there is a simple proof of its primality if it is prime, and a simple proof of its compositeness if it is composite.

After a long and heroic series of steadily improving algorithms for PRIMALITY, including randomized algorithms that we will discuss in Section 10.8, a deterministic polynomial-time algorithm was finally found in 2004. Thus, we now know that PRIMALITY is in P. Nevertheless, Pratt certificates gave an important early insight into the hidden logical structure of PRIMALITY—the fact that the claim "p is prime" can be rephrased as the claim that a certain kind of witness exists.

Let's reflect on the fact that, even though PRIMALITY is in P, FACTORING still seems to be hard. In essence, this is because it is possible to know that p is composite without knowing anything about its factors. For instance, the so-called Fermat test seeks to prove that p is composite by finding an a such that $a^{p-1} \not\equiv_p 1$, violating the Little Theorem. Even when this test gives a witness a for the non-primality of p, there is no obvious way to extract from this witness any information about p's factors.

Thus, given our current state of knowledge, FACTORING seems to be significantly harder than PRIMALITY. As we discuss in Chapter 15, much of modern-day cryptography is based on this belief. On the other hand, we will see that FACTORING can be reduced to ORDER FINDING, the problem of finding the order of a given a mod p, and this reduction is at the heart of Shor's quantum algorithm for FACTORING.

4.4.2 Knots and Unknots

> "A knot!" said Alice, always ready to make herself useful, and looking anxiously about her. "Oh, do let me help to undo it!"
>
> Lewis Carroll, *Alice in Wonderland*

Knot theory is a rich and beautiful field of mathematics, with important connections to statistical physics, quantum field theory, molecular biology, and, of course, magic tricks. One basic decision problem is the following:

UNKNOT
Input: A closed non-intersecting curve C in $\mathbb{R}^3$
Question: Can C be untied without cutting it or passing it through itself?

To be a little more precise, by "untying" we mean a continuous deformation or *isotopy* of the knot—or rather of the three-dimensional space in which it is embedded—that maps C onto a simple loop, i.e., a circle. We say that a knot is an "unknot" if such an isotopy exists.

To make UNKNOT a computer science problem, we need to describe C's topology in some discrete way. One way to do this is with a *knot diagram* which projects C onto the plane and shows where and how it crosses itself. Formally, this is a planar graph where every vertex has degree four, and where edges are marked at their endpoints as underpasses or overpasses. We will think of the size n of this input as the number of crossings.

Telling if a knot is an unknot or not is not a simple matter. Consider the unknot shown in Figure 4.10. We can use it in a magic trick, where the loop is closed by a volunteer from the audience who holds the
4.7 ends of the rope. We raise a curtain, fiddle briefly with the knot, and voilá! The knot is gone.

The *Reidemeister moves,* shown in Figure 4.11, capture everything we can do to a knot without cutting or breaking it. Thus one possible witness for UNKNOT is a sequence of Reidemeister moves that convert C's knot diagram to a circle. However, it wasn't proved until very recently that we can always do this with a polynomial number of moves. As a cautionary note, there are unknots where the only way to untie them requires us to first *increase* the number of crossings.

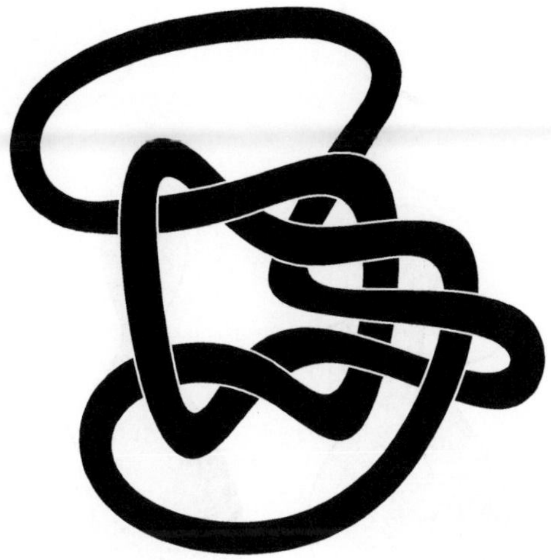

FIGURE 4.10: The Brazilian Unknot.

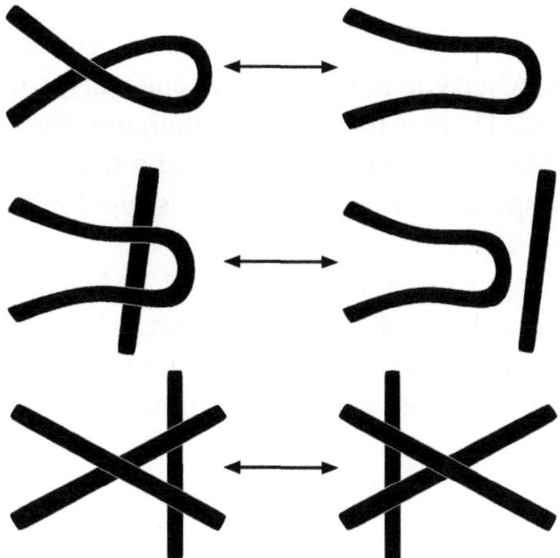

FIGURE 4.11: The Reidemeister moves.

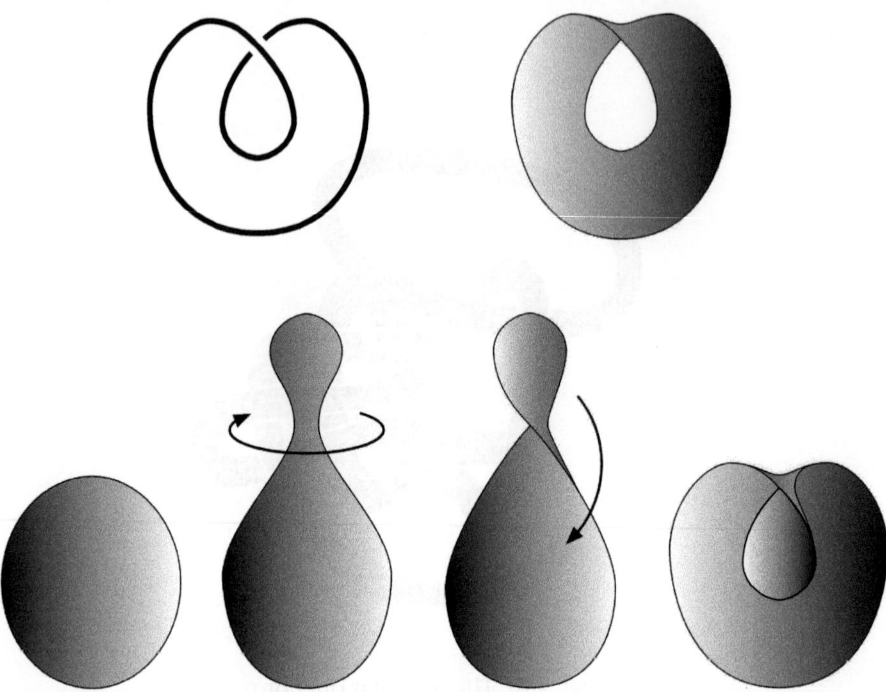

FIGURE 4.12: The curve C on the upper left is an unknot. There are two surfaces that have it as their boundary: the Möbius strip on the upper right, and a disk which has been twisted and folded. Such a disk exists if and only if C is an unknot, so a description of it can serve as a witness.

The first proof that UNKNOT is in NP relied on another type of witness. Imagine dipping a wire-frame model of C in soap-bubble liquid. When you pull it out, the soap will form a surface with C as its boundary. Let us say that this surface *spans* C. There is usually more than one such surface: for instance, the curve shown in Figure 4.12 is spanned both by a Möbius strip and a twisted disk. It can be shown that C is an unknot if and only if it can be spanned by a surface that is topologically a disk, so we can use such a *spanning disk* as a witness for UNKNOT.

The simplest spanning disk might be highly convoluted. After all, any sequence of Reidemeister moves that unties the knot also unfolds and flattens the disk. However, one can show using sophisticated ideas from topology that such a disk, if it exists, can be described with a polynomial number of bits. This established that UNKNOT is in NP—but as for PRIMALITY, finding this witness took some deep mathematical thought.

4.8

Problems

> If you don't make mistakes, you're not working on hard enough problems.
>
> Frank Wilczek

4.1 Ask the oracle. Suppose that we have access to an oracle, as in Problem 1.10, who can answer the decision problem SAT. Show that, by asking her a polynomial number of questions, we can find a satisfying assignment if one exists. Hint: first describe how setting a variable gives a smaller SAT formula on the remaining variables. Show that this is true of SUBSET SUM as well. This property is called *self-reducibility*, and we will meet it again in Chapters 9 and 13.

4.2 Coloring, with the oracle's help. Analogous to the previous problem, but a little trickier: suppose we have an oracle for the decision problem GRAPH k-COLORING. Show that by asking a polynomial number of questions, we can find a k-coloring if one exists. Hint: how can we modify the graph to indicate that two vertices are the same color, or that they are different?

4.3 Keeping it planar. Now suppose we have an oracle for PLANAR GRAPH 3-COLORING. Show that we can find a 3-coloring of a planar graph if one exists. Hint: the regions bounded by the edges of a planar graph are called *faces*. What questions can we ask the oracle that will allow us to break or collapse the faces until only triangles are left? And how can we 3-color a planar graph whose faces are all triangles? Finally, what happens if we want to find a 4-coloring of a planar graph?

4.4 Colors and degrees. Suppose that the maximum degree of any vertex in a connected graph G is D. An easy exercise is to show that G is $(D+1)$-colorable. A much harder (but fun) problem is to show that G is D-colorable, unless G is a complete graph or a cycle of odd length. This is known as Brooks' Theorem [140].

4.5 Redrawing the map. The Four Color Theorem only requires that countries have different colors if they share a border, rather than just a corner. Let us call a map *proper* if only 3 countries meet at each corner (unlike the Four Corners region of the United States). Prove that the Four Color Theorem holds if and only if it holds for proper maps.

4.6 Edges and bridges. A graph G is 3*-regular* if every vertex has degree exactly 3, and a *bridge* is an edge that will divide a graph into two pieces if removed. A k*-edge-coloring* is a coloring of the edges of G with k colors such that no two edges that share an endpoint have the same color, as shown in Figure 4.13. Now consider the following conjecture, put forward by the Scottish mathematician and physicist Peter Guthrie Tait [784]:

Conjecture 4.2 *Any planar,* 3*-regular graph without a bridge has a* 3*-edge-coloring.*

Show that this conjecture is equivalent to the Four Color Theorem.

4.7 Hamilton and Tait try to color the map. In 1886, Tait made another conjecture, where a *bridge* is defined as in the previous problem:

Conjecture 4.3 *Any planar,* 3*-regular graph without a bridge is Hamiltonian.*

Show that this conjecture would imply the Four Color Theorem. Unfortunately, it is false. In 1946, after the conjecture had been open for sixty years, the great mathematician and cryptographer William Thomas Tutte found a counterexample. This killed off one of the last hoped-for "easy proofs" of the Four Color Theorem.

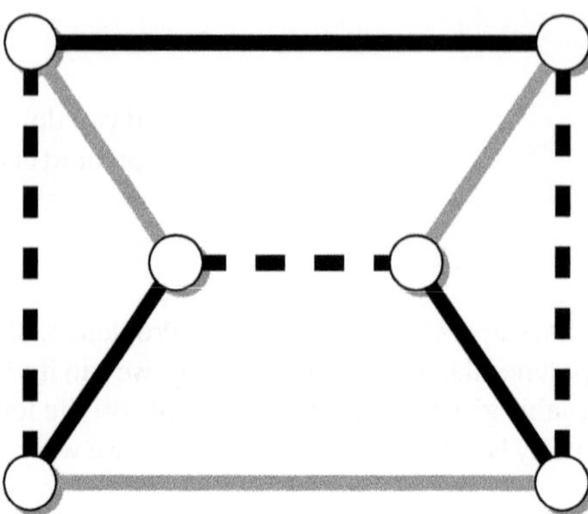

FIGURE 4.13: A 3-regular graph and a 3-edge-coloring of it.

4.8 Four colors can't cross over. Show that no crossover gadget analogous to the one in Figure 4.3 exists for k colors for any $k \geq 4$. More specifically, show that there is no planar graph with four outer vertices v_n, v_s, v_e, v_w where the pairs v_n, v_s and v_e, v_w are each forced to be the same color, but which are otherwise unconstrained. First prove this by using the Four Color Theorem, and then—rather harder—prove it on its own.

4.9 Can't break down from 3 to 2. Show that there is no way to build a 3-SAT clause $(x \vee y \vee z)$ out of 2-SAT parts. In other words, show that there is no 2-SAT formula ϕ involving x, y, z and any number of additional variables, such that ϕ is satisfiable if and only if at least one of x, y, or z is true. However, this does not prove that 3-SAT is not in P. Why not?

4.10 CNF vs. DNF. A Boolean formula is in *disjunctive normal form* (DNF) if it is an OR of clauses, each of which is the AND of a set of literals. Show that SAT for such formulas is in P.

Now show that a CNF formula can be converted into a DNF formula on the same variables. Since instances of 3-SAT consist of CNF formulas, why does this not prove that P = NP? Hint: consider the formula

$$(x_1 \vee y_1) \wedge (x_2 \vee y_2) \wedge \cdots \wedge (x_n \vee y_n).$$

What happens when we convert it to DNF form?

4.11 All for a good clause. In analogy with the degree of a vertex in a graph, say that the degree of a variable in a SAT formula is the number of clauses it appears in. Show that any k-SAT formula where no variable has degree greater than k is satisfiable, and that we can find a satisfying assignment in polynomial time. Hint: consider a bipartite graph with clauses on the left side and variables on the right, and use Hall's Theorem from Problem 3.47.

4.12 Same-sum subsets. Suppose that S is a list of 10 distinct integers, each ranging from 0 to 100. Show that there are two distinct, disjoint sublists $A, B \subset S$ with the same total. (Note that, unlike in INTEGER PARTITIONING, we don't demand that $A \cup B = S$.) That doesn't make it easy to find one!

4.13 Balancing the scales. Reduce SUBSET SUM to INTEGER PARTITIONING. Hint: add an element to the list.

4.14 When greed works. A sequence $a_1, \ldots, a_n$ of integers is called *superincreasing* if each element a_i is strictly greater than the sum of all previous elements. Show that SUBSET SUM can be solved in polynomial time if S is superincreasing. Hint: be greedy. What does your algorithm do in the case $S = \{1, 2, 4, 8, \ldots\}$?

4.15 Faster subsets. The naive search algorithm for SUBSET SUM takes 2^ℓ time, where $\ell = |S|$ is the number of weights. Describe an algorithm that is quadratically faster up to polynomial factors, i.e., whose running time is $2^{\ell/2}\,\text{poly}(\ell)$. Hint: divide the list S into two halves, and make a sorted list of all the totals you can make with each half. 4.5

4.16 Covers, cliques, and matchings. Prove that INDEPENDENT SET and VERTEX COVER are in P for bipartite graphs, and therefore that CLIQUE is in P for graphs whose complement is bipartite. Hint: reduce to MAX BIPARTITE MATCHING.

4.17 A cover of fixed size. Suppose we ask whether a graph G with n vertices has a vertex cover of size k or less. The naive exhaustive search algorithm would take about $\binom{n}{k}$ time to check all possible sets of k vertices. However, this problem can be solved in $O(f(k)n^c)$ time, where the exponent c does not depend on k. Thus VERTEX COVER is in P whenever $k = O(1)$, a phenomenon called *fixed-parameter tractability.*

Prove this in two ways. First, show that for any constant k, the property VC_k that a graph has a vertex cover of size at most k is *minor-closed.* That is, if it holds for G then it also holds for any graph G' formed by removing or contracting G's edges. The Robertson-Seymour Theorem (see Note 2.11) then shows that VC_k can be checked in $O(n^3)$ time, with a leading constant that depends on k.

Second, give an explicit algorithm that takes $2^k\,\text{poly}(n)$ time. Hint: for any edge (u, v), either u or v must be in the vertex cover.

4.18 Decision, optimization, and search. Of the three versions of INDEPENDENT SET defined in Section 4.2.4, only the first one is *a priori* in NP. Nevertheless, show that if any of the three versions is in P, they all are.

Hint: show that if we have an oracle for the first version, we can find the largest possible independent set—or, if it is not unique, one of the largest ones—by deciding, one at a time, which vertices to include in the set. This is yet another appearance of self-reducibility, as in Problem 4.1.

4.19 Small witnesses are easy to find. The logical definition of NP requires that the number of bits of the witness is polynomial in the number of bits of the input, i.e., $|w| = \text{poly}(n)$. Suppose we have a property for which witnesses of logarithmic size exist, $|w| = O(\log n)$. Show that any such property is in P.

4.20 Closure under concatenation. Let L be a set of finite strings. For example, L could be the set of strings of 0s and 1s that represent a prime in binary, or the set of grammatical sentences in your favorite language (assuming that your grammar is well-defined). Then let L^* be the set of finite strings that can be written as concatenations of words in L. In other words, each word $v \in L^*$ can be written $w_1 w_2 \cdots w_k$ for some k, where each of $w_1, w_2, \ldots, w_k$ is in L. Note that $k = 0$ is allowed, so the empty word is in L^*.

Now prove that if L is in P (i.e., if the problem of telling whether an input word is in L is in P) then so is L^*. Similarly, show that if L is in NP then so is L^*. Hint: this is easy for NP. For P it takes a little more work, since *a priori* there are an exponential number of ways to divide w up into words in L.

4.21 Small factors. Consider the following decision problem related to FACTORING:

> SMALL FACTOR
>
> Input: Two integers q and r
>
> Question: Does q have a factor smaller than r?

Show that this problem is in $\text{NP} \cap \text{coNP}$.

4.22 This many and no fewer. Given a graph G, its *chromatic number* $\chi(G)$ is the smallest number of colors we need to color it; that is, the smallest k such that it is k-colorable. Consider the following problem:

FIGURE 4.14: On the left, a classic peg solitaire game. In each move, a peg jumps over another, which is removed. Can you reduce this configuration of 8 pegs to a single peg in the center? On the right, a sliding-block puzzle. Can you move the big square block to the bottom center, where it can slide out of the box? The shortest solution has 81 moves.

CHROMATIC NUMBER
Input: A graph G and an integer k
Question: Is $\chi(G) = k$?

Do you believe this problem is in NP? Why or why not?

4.23 Many divisors. Let x be an n-bit integer. Show that the number of distinct divisors, including both prime and composite divisors, of x can be more than polynomial in n. In fact, show that it can be as large as $2^{\Omega(n/\log n)}$. Hint: let x be the product of the first t prime numbers, and use the fact that the tth prime number is $O(t \log t)$. This follows from the Prime Number Theorem, which we discuss in Section 10.6.1.

4.24 Pratt proves primality parsimoniously. Show that the Pratt certificate for an n-bit prime p has total size $O(n^2)$.

Hint: given a prime p, let's say its "tag" consists of a primitive root for $\mathbb{Z}_p^*$ and the prime factorization of $p-1$. Then the witness for p consists of p's tag, the tags of the prime factors q_i of $p-1$, and the witness for each q_i. Let $T(p)$ denote the total number of tags making up p's witness. Write an expression for $T(p)$ in terms of the $T(q_i)$, and prove by induction that $T(p) \leq \log_2 p$. Assume that no witness is necessary for $p = 2$, so we only need to provide witnesses for the odd factors q_i of $p-1$.

4.25 Puzzle classes. Consider the two puzzles shown in Figure 4.14. In peg solitaire, pegs can jump over each other, and the peg they jump over is removed. In a sliding-block puzzle, we slide blocks into empty spaces, until one or more blocks are moved to a desired position. Suppose we generalize these to problems called PEG SOLITAIRE and SLIDING BLOCKS, where the initial configuration and the final goal are given as part of the input. Which of these problems do you think are in NP, and why?

4.26 Products are easier than sums. Define a problem SUBSET PRODUCT analogous to SUBSET SUM. Show that if the inputs are integers, this problem is in P.

Notes

4.1 Agricultural terminology. What does one call the elements of a haystack? After a long debate we settled on a "blade of hay," following James Joyce [448].

4.2 The Four Color Theorem. The Four Color Theorem has a very colorful history. The conjecture that each map can be colored with at most four colors was made in 1852 by Francis Guthrie, who at that time was a student of Augustus de Morgan. After several early attempts at proving the theorem failed, Alfred Kempe published a proof in 1879 which was widely accepted. But 11 years later, Percy Heawood found a flaw in Kempe's proof, although Kempe's approach does give a proof of the *Five* Color Theorem for planar graphs. Another proof by Peter Guthrie Tait in 1880 enjoyed the same finite lifetime as Kempe's proof; it was shown to be incorrect in 1891 by Julius Petersen.

It is said that the eminent mathematician Hermann Minkowski once told his students that the Four Color Conjecture had not been settled because only third-rate mathematicians had dealt with it. But after a long period of futile attempts to prove it himself, he admitted, "Heaven is angered by my arrogance; my proof is also defective."

Progress was delayed until the 1960s, when the German mathematician Heinrich Heesch argued that the number of graphs that need to be checked to prove the Four Color Conjecture is finite. Heesch estimated that there are only about 8000 of such graphs, and he proposed to check these configurations by computer. But the German national science foundation cancelled his grant, preventing Heesch from pursuing this strategy.

The implementation of Heesch's idea had to wait until 1976, when Kenneth Appel and Wolfgang Haken boiled down the number of potential counterexamples down to less than 2000 and checked each one of them by computer [49, 48]. This was the first computer-assisted proof of a major theorem in mathematics. It began a debate on the nature of mathematical proof, which is still ongoing. Although some progress has been made in simplifying the original proof of Haken and Appel, even the shortest version still relies on exhaustive checking by computer [787].

This is an unsatisfying state of affairs. The best proofs do not simply certify that something is true—they illuminate it, and explain *why* it must be true. While the Four Color theorem may be proved, it is not well understood.

On the other hand, it is possible that there is no understanding to be had. Just as Erdős believed that God has a book containing all the most elegant proofs (see Note 1.2), some say that the Devil has a book that contains all the mathematical truths that have no elegant proofs—truths that can be stated easily, but for which the only proofs are long and ugly. At this time, it seems perfectly possible that the Four Color Theorem is in the Devil's book.

4.3 Satisfying banquets. This example of 2-SAT is a thinly veiled version of a puzzle by Brian Hayes [383], involving invitations of Peru, Qatar and Romania to an embassy banquet.

4.4 SUBSET SUM and cryptography. A public-key cryptosystem, as we will discuss in Section 15.5.1, provides an asymmetric pair of keys: a *public* and a *private* key. Anyone can encrypt a message using the public key, but only the owner of the private key can easily decrypt the message. In 1978, Merkle and Hellman [566] proposed a public-key cryptosystem based on SUBSET SUM. The private key consists of a superincreasing sequence $E=(e_1,\dots,e_n)$ of integers, a modulus m greater than the sum of the elements in S, and a multiplier w that is relatively prime to m. The public key is a sequence $H=(h_1,\dots,h_h)$ derived from the private key via

$$h_i = w e_i \bmod m \ .$$

The encryption of an n-bit message $X=(x_1,\dots,x_n)$ is the number

$$c = H\cdot X = \sum_{i=1}^{n} h_i x_i \, .$$

Decrypting the message amounts to solving $c = H \cdot X$ for $X = \{0,1\}^n$, which is of course equivalent to solving SUBSET SUM for H with target sum c. For all we know, this requires exponential time and makes it virtually impossible for any recipient to decrypt the message c. The owner of the private key, however, can simplify the decryption considerably by calculating

$$\begin{aligned} w^{-1} c &= w^{-1} H \cdot M \\ &\equiv_m w^{-1}\, w\, E \cdot M \\ &\equiv_m E \cdot X . \end{aligned}$$

Now the condition $m > \sum_i s_i$ allows us to replace '$\equiv_m$' by '$=$' in the last line, leaving us with an instance of SUBSET SUM for the superincreasing sequence E (and the target sum $w^{-1}c$). The latter, however, can be solved in polynomial time, as shown in Problem 4.14. Note that we had to choose m and w to be relatively prime in order to calculate w^{-1} such that $w^{-1} w \equiv_m 1$ (see Section A.7).

Note that in order to break the Merkle–Hellman cryptosystem, all we need is to find *some* pair (w', m') of relatively prime numbers such that $a_i = w' h_i \bmod m'$ is a superincreasing sequence, not necessarily identical to the original sequence E. Then we can apply the same decryption scheme as above and solve the resulting easy instance of SUBSET SUM. Suprisingly, such a pair (w', m') can be found in polynomial time, as shown by Adi Shamir [744], so the Merkle–Hellman cryptosystem is not secure. Nevertheless, it still serves as a simple example of public-key encryption.

4.5 Pseudopolynomial and exponential algorithms. Algorithms such as the dynamic programming algorithm for SUBSET SUM, which run in polynomial time as a function of the integers in the input, are sometimes called *pseudopolynomial.* Another example is the naive algorithm for FACTORING given in Exercise 2.2. As we commented in the text, such algorithms take exponential time as a function of the input size n if these integers are given in binary.

The $2^{n/2}\,\mathrm{poly}(n)$ time algorithm of Problem 4.15 is due to Horowitz and Sahni [412].

4.6 Prime witnesses. The Pratt certificates for PRIMALITY that we describe in Section 4.4.1 were found in 1976 by Vaughan Pratt [667].

4.7 Knots and unknots. The Brazilian Unknot shown in Figure 4.10 was discovered by Borelli and Kauffman during a seaside lunch on the island of Corsica [122]. It is a close relative of the Chefalo knot of parlor magic [58], which can be untied even if a volunteer from the audience holds both ends of the rope.

Figures 4.10 and 4.11 were made by Keith Wiley using the program Druid, developed by Wiley and Williams [824].

4.8 Knot witnesses. Haken [361] showed that C is an unknot if and only if it is spanned by some S that is topologically a disk. The fact that such a disk can be described with a witness of polynomial size, and thus that UNKNOT is in NP, was shown by Hass, Lagarias and Pippenger [376]. They start by triangulating the three-dimensional space around the knot, and showing that this can be done with only $O(n)$ tetrahedra where n is the number of crossings. They do not show that a spanning disk exists with only $\mathrm{poly}(n)$ triangles—if they had, this would imply a polynomial upper bound on the number of Reidemeister moves. Instead, they use a characterization given by Haken of the spanning disk in terms of a set of variables that describe how it intersects each tetrahedron in the triangulation. To form a spanning disk, these variables must satisfy a certain set of constraints. They show that to find a solution to these constraints, it suffices to consider values for these variables that are at most 2^{cn} for some constant c. Thus we have $O(n)$ variables, each of which has $O(n)$ bits, giving a witness with a total of $O(n^2)$ bits.

By analyzing how complicated it is to flatten the spanning disk described by [376], Hass and Lagarias [375] gave an exponential upper bound on the number of moves required to untie an unknot with n crossings: 2^{cn}, where $c = 10^{11}$! (Don't worry, the ! here is just for emphasis.) Finally, in 2014, Lackenby [514] showed that $O(n^{11})$ moves suffice, so we can use the sequence of Reidemeister moves directly as a witness of polynomial size.

In fact, there are also polynomial-size witnesses that a knot is nontrivial—i.e., that it is not an unknot. This was shown by Kuperberg [510] assuming the Generalized Riemann Hypothesis, and unconditionally by Lackenby [513]. Thus UNKNOT is in NP$\cap$coNP, and could conceivably be in P.

Chapter 5

Who is the Hardest One of All? NP-Completeness

A poet once said "The whole universe is in a glass of wine." We will probably never know in what sense he meant that, for poets do not write to be understood. But it is true that if we look at a glass closely enough we see the entire universe.

Richard Feynman

Do not ask God the way to heaven;
he will show you the hardest one.

Stanislaw J. Lec

In the previous two chapters, we have seen a variety of clever algorithms for problems like 2-SAT, MINIMUM SPANNING TREE, and MAX FLOW. On the other hand, we have also seen problems that seem to evade all our algorithmic strategies, like 3-SAT, GRAPH COLORING, and HAMILTONIAN PATH. Do these problems really require an exponential search through the haystack of solutions? Or have we simply not been clever enough to find efficient algorithms for them?

Our inability to solve these problems is no accident. There is a precise sense in which they are among the hardest problems in NP. Namely, each of these problems has the amazing ability to express all the others, or any other problem in NP. This remarkable property, which is called NP-*completeness*, implies that if any of these problems can be solved in polynomial time, they all can be, and P = NP. Conversely, if any of these problems requires exponential time, they all do. Thus unless finding needles in haystacks is much easier than we thought—that is, unless every kind of mathematical object that is easy to check is also easy to find—these problems really are as hard as they seem.

In this chapter we will define and explore the concept of NP-completeness. We will find NP-complete problems everywhere, from coloring maps to evaluating integrals. In the process, we will see how problems which appear to be vastly different can be transformed into each other, and how to build computers out of tiles, brass weights, and integer polynomials. 5.1

5.1 When One Problem Captures Them All

Consider the following audacious definition.

> A problem B in NP is NP-*complete* if, for any problem A in NP, there is a polynomial-time reduction from A to B.

Writing this as an inequality, we see that B is one of the hardest problems in NP:

> A problem $B \in \mathsf{NP}$ is NP-complete if $A \leq B$ for all $A \in \mathsf{NP}$.

How in the world could a problem be NP-complete? It must be possible to translate instances of any problem $A \in \mathsf{NP}$ into instances of B, so that if we could solve B, we could solve A as well. To put this differently, B must be so general, capable of expressing so many different kinds of variables and constraints, that it somehow contains the structure of every other problem in NP as well as its own.

Indeed, such problems do exist. At the risk of seeming tautological, consider the following problem:

> WITNESS EXISTENCE
>
> Input: A program $\Pi(x, w)$, an input x, and an integer t given in unary
>
> Question: Does there exist a w of size $|w| \leq t$ such that $\Pi(x, w)$ returns "yes" after t or fewer steps?

This problem is NP-complete because it repeats the very definition of NP. For instance, if we want to reduce HAMILTONIAN PATH to WITNESS EXISTENCE, we let Π be the witness-checking program that takes a graph x and a path w and returns "yes" if w is a Hamiltonian path in x.

Why do we express t in unary? This ensures that t is less than or equal to the total size n of WITNESS EXISTENCE's input, so both Π's running time and the size of the witness w are polynomial as functions of n. Thus we can check w in poly(n) time by running Π for t steps, and WITNESS EXISTENCE is in NP.

It is nice to know that at least one NP-complete problem exists. However, WITNESS EXISTENCE is rather dry and abstract, and it is hard to imagine how it could be connected to any natural problem we actually care about. The astonishing fact is that many of our favorite problems, including many of those introduced in Chapter 4, are NP-complete. Each of these problems is as hard as any problem in NP, and wrestling with any one of them means wrestling with the entire class. Before, when facing an individual problem like GRAPH COLORING or SAT, we might have imagined that there is a clever insight specific to that problem which would allow us to solve it in polynomial time. But NP-completeness raises the stakes. If any of these problems possess a polynomial-time algorithm, they all do:

> If any NP-complete problem is in P, then P = NP.

Most people would rejoice at this possibility, since there are many practical problems we wish we could solve more efficiently. But as we will see, to a theorist this would be a complete and devastating shift in our world view. Giants would walk the earth, lions would lie down with lambs, and humans would have no role left to play in mathematical discovery. The consequences of P = NP would be so great that we regard NP-completeness as essentially a proof that a problem cannot be solved in polynomial time—that there is no way to avoid an exponential search through its haystack. We will discuss the earth-shaking consequences that P = NP would have in Chapter 6.

We should emphasize that when discussing NP-completeness, we restrict ourselves to reductions that map single instances of A to single instances of B. This is different from the reduction from 2-SAT to REACHABILITY in Section 4.2.2, for example, in which we allowed ourselves to call a subroutine for REACHABILITY polynomially many times.

Moreover, we require our reductions to map yes-instances to yes-instances, and no-instances to no-instances. Then if x is a yes-instance of A, there is a witness for this fact: namely, the witness for the corresponding yes-instance $f(x)$ of B. Such reductions preserve the asymmetric nature of NP. In particular, if B is in NP and $A \leq B$, then A is in NP as well. Thus for the purposes of this chapter,

5.2

> A polynomial-time reduction from A to B is a function f, computable in polynomial time, such that if x is an instance of A, then $f(x)$ is an instance of B. Moreover, $f(x)$ is a yes-instance of B if and only if x is a yes-instance of A.

As we discussed in Section 3.8, the composition of two polynomial-time reductions is a polynomial-time reduction. Thus the relation $A \leq B$ that A can be reduced to B is transitive:

$$\text{If } A \leq B \text{ and } B \leq C, \text{ then } A \leq C.$$

Therefore, to prove that a given problem is NP-complete, it suffices to reduce to it from another problem which is already known to be NP-complete.

These reductions create a family tree of NP-complete problems. In the decades since NP-completeness was discovered, this tree has grown to include thousands of problems in graph theory, algebra, planning, optimization, physics, biology, recreational puzzles, and many other fields. Over the course of this chapter, we will watch this tree grow its first few branches, starting with its root WITNESS EXISTENCE.

Exercise 5.1 *Show that if* P = NP *then any problem in* P *is* NP*-complete, including the following trivial one:*

> ZERO OR ONE
>
> *Input: A bit b*
>
> *Question: Is $b = 1$?*

5.2 Circuits and Formulas

To get the ball rolling, in this section we will translate the witness-checking programs of WITNESS EXISTENCE into simpler mathematical objects: Boolean circuits and formulas. The claim that a witness exists will then become the claim that we can make these formulas true. By breaking these formulas down into small enough pieces, we will reduce WITNESS EXISTENCE to one of the most basic constraint satisfaction problems: 3-SAT, which we met in Chapter 4.

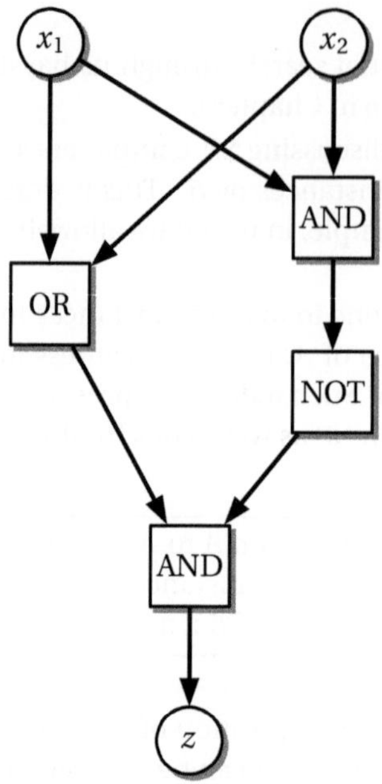

FIGURE 5.1: A Boolean circuit with inputs x_1, x_2 and output z.

5.2.1 From Programs to Circuits

A *Boolean circuit* is a directed graph with a set of source nodes called the inputs, and one or more sink nodes called the outputs. Each internal node, or "gate," is labeled AND, OR, or NOT, and produces the corresponding function of its inputs. This graph is *acyclic*, meaning that there are no loops—information flows in one direction from the inputs to the outputs.

We can divide a Boolean circuit into a series of layers, where the gates in each layer receive their inputs from the gates in the layers above them. Its *depth* is the length of the longest path from an input to an output, or equivalently the smallest number of layers we can divide it into. Its *width* is the largest number of gates in any one layer, and its *size* is the total number of gates.

Let's focus on circuits with one output. Just as for Boolean formulas, we call a circuit *satisfiable* if there is a truth assignment for its inputs that makes its output `true`. For instance, the circuit shown in Figure 5.1 is satisfiable, since the output is `true` if $x_1 = \texttt{true}$ and $x_2 = \texttt{false}$ or vice versa.

Now consider the following problem:

CIRCUIT SAT
Input: A Boolean circuit C
Question: Is C satisfiable?

We will argue that

$$\text{WITNESS EXISTENCE} \leq \text{CIRCUIT SAT},$$

and therefore that CIRCUIT SAT is NP-complete. To be more explicit, we will describe how to convert an instance (Π, x, t) of WITNESS EXISTENCE to a Boolean circuit C of polynomial size, such that C is satisfiable if and only if there is a witness w such that $\Pi(x, w)$ returns "yes" within t steps.

The reduction consists of "compiling" the program $\Pi(x, w)$ into a Boolean circuit C. We can do this since Boolean circuits are themselves a powerful enough model of computation to express any algorithm. If you doubt this claim, consider the fact that in the real world we compile our programs into sequences of instructions in machine language, which manipulate just a few bits at a time. We then carry out these instructions on a chip made of logical gates, each of which is made of a handful of transistors.

Each layer of C corresponds to a single step of Π. The wires leading from each layer to the next carry, in binary form, the values of all the variables that Π uses. If Π contains **for** or **while** loops, which execute the same instructions many times, we build C by "unrolling" these loops and repeating the corresponding layers of the circuit.

Since Π's running time is at most t steps, C has $O(t)$ layers. Moreover, Π only has time to access $O(t)$ different variables or locations in the computer's memory, so the number of bits that we need to carry from each layer to the next is only $O(t)$. Thus C's width and depth are both $O(t)$, and its total size is $O(t^2) = \text{poly}(t)$.

This gives us a circuit whose inputs correspond to the bits of x and w, and whose output is `true` if and only if w is a witness for x. (We assume here that t is large enough so that Π can check w in t steps or less.) If we restrict C by fixing the truth values of some of its inputs, we get a smaller circuit C' on its remaining inputs. So, if we fix the bits of x to those in our instance of WITNESS EXISTENCE, the remaining circuit C' is satisfiable if and only if a witness w exists for that x.

We can produce C' in poly(t) time, and since t is given in unary this is polynomial as a function of WITNESS EXISTENCE's input size. This completes the reduction from WITNESS EXISTENCE to CIRCUIT SAT, and proves that CIRCUIT SAT is NP-complete.

The details of this reduction may seem technical, but the idea is clear. Telling whether we can get a witness-checking program to return "yes" is NP-complete, since any NP-complete problem can be phrased that way. Boolean circuits can compute just as well as programs can, so telling whether we can get them to return `true` is just as hard. We will see this theme over and over—if an object is capable of computation, understanding its behavior is computationally complex.

Now that we have translated WITNESS EXISTENCE into a question about a concrete mathematical structure, we're off and running. Next stop: 3-SAT.

5.2.2 From Circuits to Formulas

We saw in Section 4.2.2 that 2-SAT is in P, while 3-SAT seems to be outside P. Here we will prove that 3-SAT is NP-complete, with the reduction

$$\text{CIRCUIT SAT} \leq \text{3-SAT}.$$

Thus, unless P = NP, going from $k = 2$ to $k = 3$ causes an enormous jump in k-SAT's computational complexity. The fact that we were able to reduce k-SAT to 3-SAT, but not all the way down to 2-SAT, was no accident.

To reduce CIRCUIT SAT to 3-SAT, we need to show how to transform a Boolean circuit C to a 3-SAT formula ϕ which is satisfiable if and only if C is. As in the chip design example of Section 4.2.2, we start by

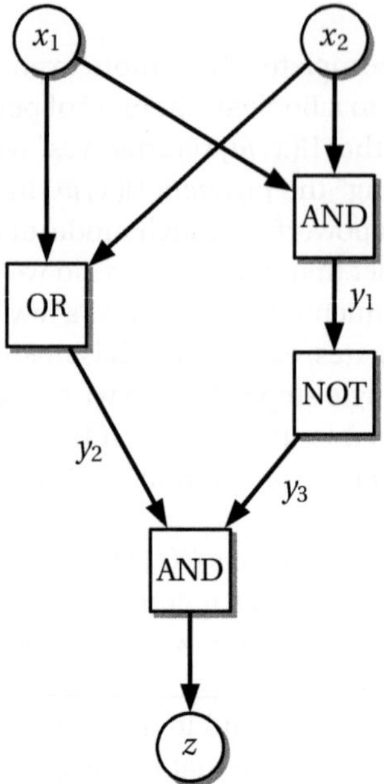

FIGURE 5.2: The Boolean circuit of Figure 5.1, with Boolean variables y_1, y_2, y_3 representing the truth values carried by each wire. We can transform the condition that the circuit works, and that z is `true`, into a SAT formula involving the variables x, y, and z.

introducing internal variables that represent the truth value carried by each wire as shown in Figure 5.2. We then ensure that the circuit is working properly by imposing, for each gate, the constraint that the output of that gate is correct.

For instance, suppose we have an AND gate with inputs x_1, x_2 and output y. The requirement that $y = x_1 \wedge x_2$ can be written in SAT form as

$$y = x_1 \wedge x_2 \Leftrightarrow (x_1 \vee \overline{y}) \wedge (x_2 \vee \overline{y}) \wedge (\overline{x}_1 \vee \overline{x}_2 \vee y).$$

The first two of these clauses force y to be `false` if x_1 or x_2 is `false`, and the last one forces y to be `true` if both x_1 and x_2 are. Using de Morgan's law,

$$y = x_1 \vee x_2 \Leftrightarrow \overline{y} = \overline{x}_1 \wedge \overline{x}_2,$$

the reader can easily find a similar SAT formula for an OR gate. For a NOT gate we have

$$y = \overline{x} \Leftrightarrow (x \vee y) \wedge (\overline{x} \vee \overline{y}).$$

Finally, the requirement that the output is `true` is just a one-variable clause. For instance, the Boolean circuit in Figure 5.2 becomes the formula

$$\begin{aligned}\phi = &(x_1 \vee \overline{y}_1) \wedge (x_2 \vee \overline{y}_1) \wedge (\overline{x}_1 \vee \overline{x}_2 \vee y_1) \\ &\wedge (\overline{x}_1 \vee y_2) \wedge (\overline{x}_2 \vee y_2) \wedge (x_1 \vee x_2 \vee \overline{y}_2) \\ &\wedge (y_1 \vee y_3) \wedge (\overline{y}_1 \vee \overline{y}_3) \\ &\wedge (y_2 \vee \overline{z}) \wedge (y_3 \vee \overline{z}) \wedge (\overline{y}_2 \vee \overline{y}_3 \vee z) \\ &\wedge z\,.\end{aligned}$$

This converts any circuit C with n gates to a SAT formula ϕ with $O(n)$ clauses, which is satisfiable if and only if C is. This formula has a mixture of 1-, 2-, and 3-variable clauses, but, as we discussed in Section 4.2.2, it is easy to convert it to a 3-SAT formula by adding dummy variables. This shows that CIRCUIT SAT can be reduced to 3-SAT, and therefore that 3-SAT is NP-complete.

3-SAT is one of the most basic NP-complete problems. In the next sections, we will use it to prove that a number of other problems are NP-complete as well, by showing that we can reduce 3-SAT to them. Seeing that these reductions work is usually easy, but designing them for the first time can be very hard.

5.3 Designing Reductions

CIRCUIT SAT is NP-complete because Boolean circuits are powerful enough to carry out any computation. 3-SAT is NP-complete because it is powerful enough to express the claim that a Boolean circuit works, and that its output is `true`. In this section, we will show that other problems are NP-complete because they are powerful enough to express the idea that Boolean variables are `true` or `false`, and to impose constraints on those variables just as 3-SAT does—and that other problems are powerful enough to express the structure of *these* problems.

As we design these reductions, it's important to remember that a reduction is a map from instances to instances. In order to prove that a reduction works, we will also think about maps from solutions to solutions. However, these maps are part of our analysis of the reduction, not of the reduction itself. Like an embassy translator, a reduction translates questions from one language to another. This translation has to be faithful—the answer, yes or no, has to remain the same—but the reduction's responsibility is to translate the question, not to answer it.

5.3.1 *Symmetry Breaking and* NAESAT

SAT has many variants, in which we demand that various kinds of constraints be satisfied. One of the most useful is NOT-ALL-EQUAL SAT or NAESAT for short. Just as a 3-SAT clause $(\ell_1 \vee \ell_2 \vee \ell_3)$ forbids truth assignments where all three literals are `false`, a NAE-3-SAT clause (ℓ_1, ℓ_2, ℓ_3) forbids assignments where all three are `false` or where all three are `true`. In other words, it requires at least one literal to be true, and at least one to be false.

Unlike SAT, NAESAT is symmetric with respect to switching the Boolean values `true` and `false`. As we will see, this makes it especially useful when we are trying to prove NP-completeness for problems with a similar symmetry.

Just as for k-SAT, we can define the special case NAE-k-SAT where there are k literals in each clause. The following exercise shows that, like 2-SAT, NAE-2-SAT is in P:

Exercise 5.2 *Show how to reduce* NAE-2-SAT *to* GRAPH 2-COLORING. *Hint: a* NAE-2-SAT *clause forces the truth values of its literals to be different. However, note that literals can be negated.*

On the other hand, we will prove that NAE-3-SAT is NP-complete, with the reduction

$$\text{3-SAT} \leq \text{NAE-3-SAT}.$$

Just like k-SAT, NAE-k-SAT goes from easy to hard when k goes from 2 to 3.

To construct this reduction, we need to show how to convert a SAT formula into a NAESAT one. The reader's first impulse might be to express a NAESAT clause as a pair of SAT clauses,

$$(x,y,z) = (x \lor y \lor z) \land (\overline{x} \lor \overline{y} \lor \overline{z}).$$

But this is going in the wrong direction. We need to show that any SAT clause can be expressed as a combination of NAESAT clauses, not the other way around.

The major difference between the two problems is the `true`–`false` symmetry of NAESAT. If a truth assignment satisfies a NAESAT formula then so does its complement, in which we flip the truth value of every variable. How can we build a SAT clause, which asymmetrically prefers that its literals are `true`, out of these symmetric building blocks?

The idea is to break this symmetry by introducing a "dummy" variable s. Suppose we have a 3-SAT formula

$$\phi = (x_1 \lor y_1 \lor z_1) \land \cdots \land (x_m \lor y_m \lor z_m).$$

We convert this to the following NAE-4-SAT formula,

$$\phi' = (x_1, y_1, z_1, s) \land \cdots \land (x_m, y_m, z_m, s).$$

Note that the same s appears in every clause.

To prove that this reduction works, we need to show that ϕ' is satisfiable if and only if ϕ is. In one direction, suppose that ϕ is satisfiable. Then there is a truth assignment such that every clause has at least one true literal. If we set $s =$ `false` then each clause of ϕ' has a false literal as well, satisfying the NAESAT formula ϕ'.

Conversely, suppose that ϕ' is satisfiable. Since satisfying assignments for NAESAT formula come in complementary pairs, ϕ' must have a satisfying assignment where $s =$ `false`. At least one of the other literals in each clause must be true, so the truth values of the other variables satisfy ϕ.

There is another way to put this. If ϕ' is satisfiable, consider a particular satisfying assignment for it. If $s =$ `false`, then ϕ' is equivalent to ϕ on the other variables. On the other hand, if s is `true`, it forces at least one of the other literals in each clause to be false, giving the 3-SAT formula

$$(\overline{x}_1 \lor \overline{y}_1 \lor \overline{z}_1) \land \cdots \land (\overline{x}_m \lor \overline{y}_m \lor \overline{z}_m).$$

But this is equivalent to ϕ if we switch the meanings of `true` and `false`. In other words, ϕ' is equivalent to ϕ, demanding that at least one literal in each clause is true, if by "true" we mean having a value *opposite to that of* s.

The lesson here is that when we design a reduction from A to B, we are free to encode A's variables in terms of B's variables in any way we like, as long as B's structure lets us express the appropriate constraints on them. There is nothing hallowed about the values `true` and `false`—they are simply two values that a variable in a 3-SAT problem can take, and we can translate them into NAESAT variables in whatever way is most convenient.

The only problem with this reduction is that it increases the number of variables per clause by one, giving

$$\text{3-SAT} \le \text{NAE-4-SAT}.$$

But no matter—we can reduce NAE-4-SAT to NAE-3-SAT, as the following exercise shows.

***Exercise 5.3** By using additional variables as we did for* SAT *in Section 4.2.2, show how to break a* NAE-k-SAT *clause, for any k, into a chain of* NAE-3-SAT *clauses.*

This completes the chain of reductions

$$\text{3-SAT} \le \text{NAE-4-SAT} \le \text{NAE-3-SAT}$$

and proves that NAE-3-SAT is NP-complete.

5.3.2 Choices, Constraints, and Colorings

Our next example of reduction design will prove that GRAPH 3-COLORING is NP-complete. Thus, just as we promised in Section 4.2.1, GRAPH k-COLORING is easy for $k = 2$ and hard for $k \ge 3$ unless $\mathsf{P} = \mathsf{NP}$. For aesthetic reasons we will use the following three colors: cyan, yellow, and magenta.

Designing a reduction is rather like building a computer. Computers are built out of physical components that carry variables from place to place, and that process these variables by imposing constraints on them where they meet. Similarly, a typical reduction consists of building two kinds of gadgets: "choice" gadgets which represent setting a variable to one of its possible values, and "constraint" gadgets which force two or more variables to obey a certain constraint.

Let's start by trying to reduce 3-SAT to GRAPH 3-COLORING. In this case, our choice gadgets need to represent setting a Boolean variable `true` or `false`, so there should be two different ways to color them. One possibility is shown in Figure 5.3. For each variable x, we have a gadget consisting of a pair of vertices, which are connected to each other and to a central vertex shared by all the gadgets. In any given coloring, the central vertex will have some color, and by symmetry we can assume that it is cyan. Then in each gadget, one vertex is yellow and the other is magenta. Just as we arbitrarily defined `true` as different from the dummy variable in s in our reduction from SAT to NAESAT, here we can arbitrarily label one vertex in the pair x and the other $\overline{x}$, and say that x's truth value is determined by whichever one is magenta.

Our next goal is to define a constraint gadget. This should be a small graph that we can connect to three choice gadgets, which can then be 3-colored if and only if at least one of the literals in the corresponding clause is `true`. If we connect the clause gadget to the vertex in the choice gadget corresponding to x or $\overline{x}$, depending on whether x is negated in the clause, we would like the gadget to be colorable if and only if at least one of those vertices is magenta.

But there's a problem. The constraints in GRAPH 3-COLORING are inherently symmetric with respect to permuting the colors. In particular, any gadget that is 3-colorable when connected to three magenta vertices is also 3-colorable when connected to three yellow ones. Since switching yellow and magenta corresponds to flipping the values of all the variables, this means that—unless we represent Boolean

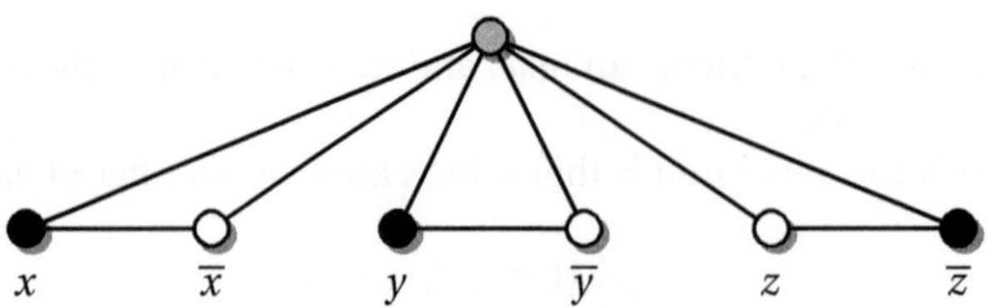

FIGURE 5.3: Choice gadgets for GRAPH 3-COLORING. Each variable has a gadget consisting of a pair of vertices, and all the gadgets are connected to a central vertex. Whatever color this central vertex has, each gadget has two possible colorings, which we associate with the truth values of that variable according to an encoding of our choice. Here we have three variables, and we show a coloring corresponding to the truth assignment $x = \texttt{true}$, $y = \texttt{true}$, and $z = \texttt{false}$.

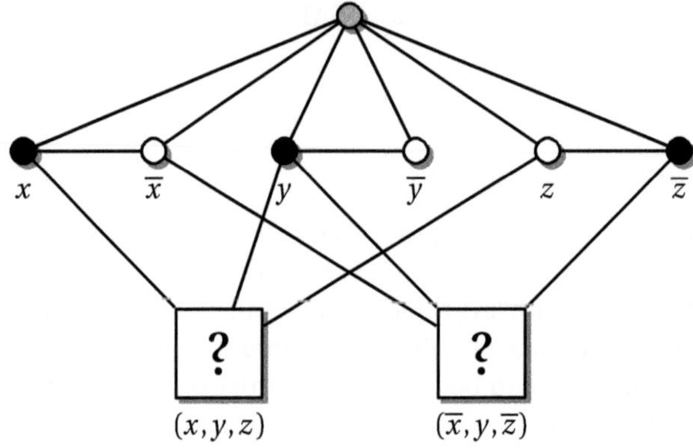

FIGURE 5.4: Connecting variables to constraints. In this case, we are trying to simulate the NAE-3-SAT formula $(x,y,z)\wedge(\overline{x},y,\overline{z})$. What gadget can we put in each box, so that it is 3-colorable if and only if the vertices it is connected to are not all the same color—that is, if the literals in that clause don't have the same truth value?

variables in some completely different way—we can only enforce constraints that are symmetric with respect to `true` and `false`.

Happily, we know an NP-complete problem that is symmetric in this way: NAESAT. So we will prove that GRAPH 3-COLORING is NP-complete by reducing to it from NAE-3-SAT instead of from 3-SAT. There's no shame in this. We have simply gained a better understanding of what kind of constraints are easy to represent in the world of GRAPH 3-COLORING. The transitivity of NP-completeness lets us prove that GRAPH 3-COLORING is NP-complete by reducing from whichever NP-complete problem is most convenient—whichever one is closest in spirit to GRAPH 3-COLORING.

We urge the reader to take a moment and try to fill in the black box in Figure 5.4—in other words, to design a gadget which is 3-colorable if and only if the three vertices attached to it are not all the same color. The answer, shown in Figure 5.5, is a triangle.

By generating a graph with a choice gadget for each variable and a constraint gadget for each clause, and wiring them up as shown in this example, we can convert any NAE-3-SAT formula ϕ with n variables and m clauses into a graph G with $2n+3m+1$ vertices, which is 3-colorable if and only if ϕ is satisfiable.

Clearly we can carry out this conversion in polynomial time. This completes the proof that

$$\text{NAE-3-SAT} \leq \text{GRAPH 3-COLORING},$$

and therefore that GRAPH 3-COLORING is NP-complete.

Note that the reduction does not color the graph, or satisfy the formula. It simply converts formulas to graphs, translating a question of satisfiability to one of colorability. We color the graph in Figure 5.5 to illustrate the fact that colorings of G correspond to satisfying assignments of ϕ.

Exercise 5.4 *In this reduction, is the mapping from satisfying assignments to colorings one-to-one? If not, how many colorings correspond to each satisfying assignment?*

What does this say about GRAPH k-COLORING for other values of k? In Exercise 4.3, you showed that

$$\text{GRAPH } k\text{-COLORING} \leq \text{GRAPH } (k+1)\text{-COLORING}$$

for any k. By induction, then, GRAPH k-COLORING is NP-complete for any $k \geq 3$.

Let's pause for a moment, and note that the vast majority of graphs don't look like the one in Figure 5.5. But what matters, for the worst-case hardness of GRAPH 3-COLORING, is that some of them do. It's entirely possible that GRAPH 3-COLORING is easy for most graphs in some sense, and in Chapter 14 we will explore what happens when we construct graphs randomly. But this reduction shows that if there exist hard instances of NAE-3-SAT, then there also exist hard instances of GRAPH 3-COLORING. Therefore, any polynomial-time algorithm that works for all instances of GRAPH 3-COLORING would yield a polynomial-time algorithm for NAE-3-SAT, and this would imply that $\mathsf{P} = \mathsf{NP}$.

In the case of planar graphs, we showed in Section 4.2.1 that

$$\text{GRAPH 3-COLORING} \leq \text{PLANAR GRAPH 3-COLORING}.$$

Therefore PLANAR GRAPH 3-COLORING is NP-complete as well. Exercise 4.1 shows that GRAPH k-COLORING is in P for $k \leq 2$ colors, and the Four Color Theorem implies that PLANAR GRAPH k-COLORING is in P for $k \geq 4$. Thus for planar graphs, the only hard case is for three colors—and this case is as hard as all of NP.

5.3.3 Setting a Threshold

In Section 4.2.4, we met INDEPENDENT SET, VERTEX COVER, and CLIQUE, and showed that they are equivalent to each other. Here we prove that all three are NP-complete, by proving that INDEPENDENT SET is.

The reductions we have seen so far in this chapter all transform one kind of constraint into another, and preserve the property that we can satisfy all of the constraints simultaneously. But INDEPENDENT SET is not a constraint satisfaction problem. Instead, it asks whether we can find an independent set of size k or more, where the threshold k is given to us as part of the input.

How can we design a reduction from a constraint satisfaction problem to a threshold problem like INDEPENDENT SET? The good news is that we can set the threshold k in any way we like. For instance, suppose we want to prove that INDEPENDENT SET is NP-complete by reducing GRAPH COLORING to it. Then we need to show how to take an arbitrary graph G and produce a pair (G', k) such that G' has an independent set of size k if and only if G is 3-colorable.

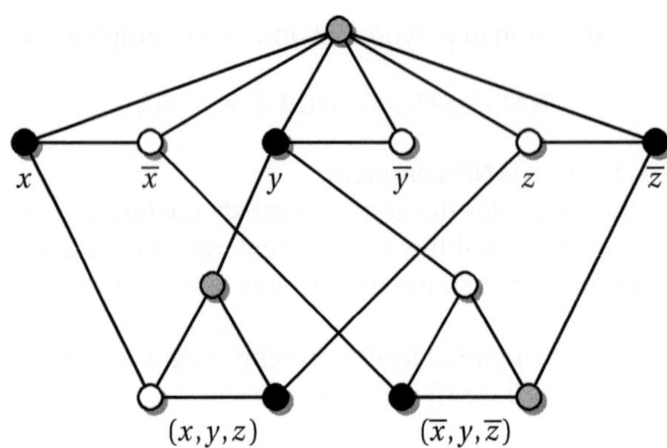

FIGURE 5.5: The constraint gadget is a triangle. A 3-coloring of a triangle requires all 3 colors, so we can color it if and only if the three vertices it is connected to are not all the same color. This enforces the NAE-3-SAT constraint that the three literals not have the same truth value. The two constraint gadgets shown correspond to the NAE-3-SAT formula $\phi = (x, y, z) \wedge (\overline{x}, y, \overline{z})$. To illustrate how the reduction works, we show the 3-coloring corresponding to the satisfying truth assignment $x =$ `true`, $y =$ `true`, $z =$ `false`.

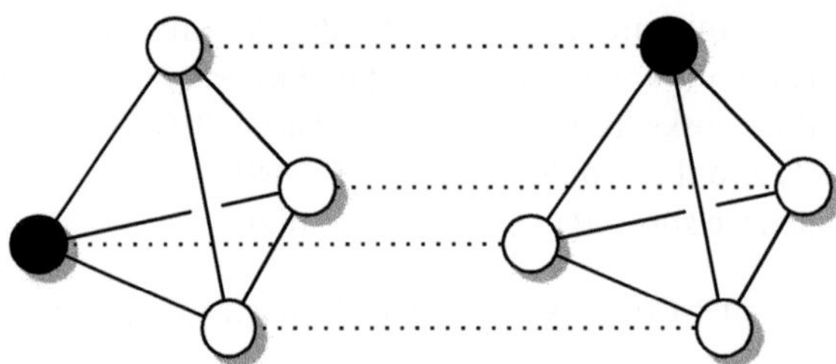

FIGURE 5.6: The choice gadget for our reduction from GRAPH 3-COLORING to INDEPENDENT SET: a complete graph with ℓ vertices has ℓ different independent sets, each consisting of a single vertex. By connecting corresponding vertices along the dotted edges, we can force neighboring gadgets to use different vertices, simulating the constraint that neighboring vertices of G have different colors.

To carry out this reduction, one gadget we could try is a complete graph with ℓ vertices. By setting the threshold k high enough, we can force the independent set to include at least one vertex in each such gadget—but no independent set can include more than one of these vertices. Thus it can act as a choice gadget where some variable can take ℓ different values. In particular, we can think of it as giving a vertex in G one of ℓ different colors.

Now we need to impose the constraint that two neighboring vertices of G have different colors. We can do this by wiring our choice gadgets to each other as shown in Figure 5.6, drawing an edge between corresponding pairs of vertices. The independent set can't include both vertices in any such pair, so neighboring choice gadgets have to take different values. This forces these gadgets to express the constraints of GRAPH ℓ-COLORING.

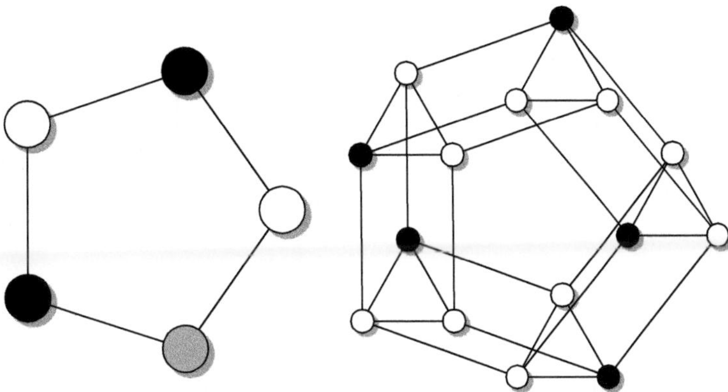

FIGURE 5.7: The reduction from GRAPH 3-COLORING to INDEPENDENT SET. On the left, a 3-coloring of a graph G with $n = 5$ vertices. On the right, the corresponding independent set of size $k = 5$ in the corresponding graph G'.

Since GRAPH 3-COLORING is NP-complete, we might as well set $\ell = 3$, in which case our choice gadgets are simply triangles. This gives the reduction shown in Figure 5.7. We change G to G' by replacing each vertex v of G with a triangle, and connect the vertices of each triangle to the corresponding vertices of its neighbors. To force each choice gadget to have a vertex in the independent set, we set k equal to the number of vertices in G. Thus if G has n vertices, G' has $3n$ vertices, and we set $k = n$.

This proves that

$$\text{GRAPH 3-COLORING} \le \text{INDEPENDENT SET},$$

and thus that INDEPENDENT SET is NP-complete.

5.3.4 *Fitting Things Together:* TILING

From the Alhambra to Tetris, tilings are an enduring source of beauty and puzzlement. Given a finite set of tiles and a finite region in the plane, I can ask you whether you can tile the region with these tiles, covering it without overlaps or gaps. I give you as many copies of each tile as you want, and I also allow you to rotate or reflect them.

TILING
Input: A set T of tile shapes and a region R
Question: Can R be tiled with copies of tiles in T?

For simplicity, let's assume that the tiles are *polyominoes*, which are made out of unit squares in the integer lattice, and that R is similarly a union of unit squares. We encode the tile shapes, and the region R, as 5.3
a pattern of 0s and 1s inside some rectangle. In that case, the total area of the region is at most the size of the input, so we can easily check a proposed tiling in polynomial time.

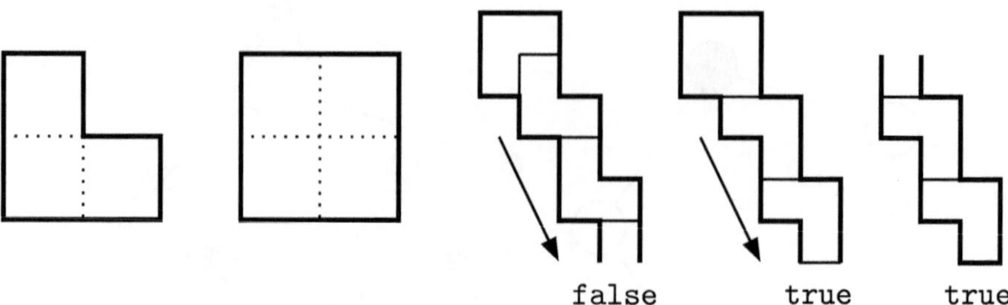

FIGURE 5.8: The right tromino and the 2×2 square. We also show choice gadgets corresponding to the circuit's inputs, which tile a wire in two different ways corresponding to `true` and `false`. Finally, we show how to end a wire so that the circuit's output is forced to be `true`.

Exercise 5.5 *Consider a version of* TILING *where the region* R *is an* $x \times y$ *rectangle, and where* x *and* y *are given as part of the input in binary. Do you think this version is in* NP*? Why or why not?*

We will show that TILING is NP-complete by reducing to it from CIRCUIT SAT. Our reduction is very physical: we will literally show how to build a circuit in the plane, creating a region consisting of "wires" and "gates" which is tileable if and only if the circuit is satisfiable. We will focus on the case where the set T consists of just the 2×2 square and the so-called *right tromino*, shown in Figure 5.8. If this special case is NP-complete, then so is the more general version where T is given as part of the input.

First we need wires. These should be able to carry truth values, so there should be two ways to tile them, and the choice of which tiling we use should propagate from one end of the wire to the other. We do this by extending wires with a series of knight's moves, so that they can be filled with trominoes in two different ways as in Figure 5.8. We give each wire a direction, and arbitrarily define one of these tilings as `true` and the other as `false`. Our choice gadgets are 2×2 square caps on the input wires, and we can set the inputs to `true` or `false` by filling each cap with a square or a tromino. Similarly, we can force the output of the circuit to be `true` by ending its output wire as shown in Figure 5.8.

Next, we need gates. Figure 5.9 shows a junction between three wires that acts as an AND gate. It and its surrounding wires can be tiled in four different ways, corresponding to the four possible values of its inputs, and in each case its output wire is forced to carry the correct value. We also need a NOT gate, a way to split a wire into two that carry the same truth value, and ways to bend a wire without breaking its truth value. These gadgets appear in the example shown in Figure 5.10, in which we convert a simple circuit to an instance of TILING.

The one thing we can't do, since all this takes place in two dimensions, is let two wires cross each other. Luckily, we don't need to. As Figure 5.11 shows, we can simulate a crossover with a planar gadget made of XOR gates, which return `true` if exactly one of their inputs is `true`. In Problem 5.18, we ask you to complete this gadget by building an XOR out of AND, OR, and NOT gates.

Just as the crossover gadget shown in Figure 4.3 lets us reduce GRAPH 3-COLORING to the planar case, this gadget gives the reduction

$$\text{CIRCUIT SAT} \leq \text{PLANAR CIRCUIT SAT},$$

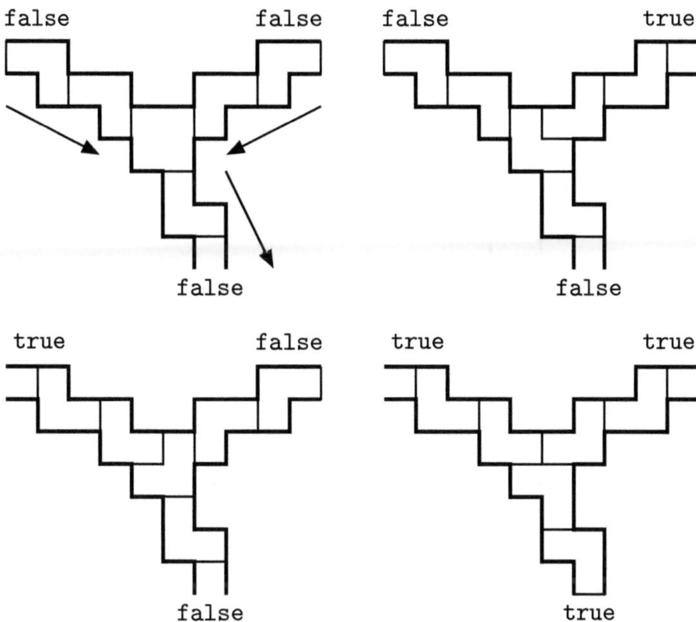

FIGURE 5.9: The four ways of tiling our AND gate, corresponding to the four possible truth values of its inputs. In each case, the truth value of its output (according to our encoding) is the AND of its inputs.

and shows that the planar case is just as hard as the general one. Along with the reduction

$$\text{PLANAR CIRCUIT SAT} \leq \text{TILING}$$

given by our other gadgets, this proves that TILING is NP-complete.

Strictly speaking, to complete this proof we need to argue that a planar circuit with n gates can be laid out on the lattice with these wires and gates, in such a way that the total length of the wires, and hence the area of R, are polynomial, and that we can find such a layout in polynomial time. We cheerfully omit this intuitive, but tedious, part of the proof.

While the reduction is much more complicated, it turns out that we can do without the square tile.  5.4
Even the following variant of TILING is NP-complete:

> TROMINO TILING
>
> Input: A region R
>
> Question: Can R be tiled with right trominoes?

On the other hand, if our tiles are just dominoes—that is, 1×2 rectangles—then the problem gets easier, and DOMINO TILING is in P. The reason is simple: if we define a graph where each vertex corresponds to a square in R and is connected to its neighboring squares, a domino tiling is just a perfect matching as shown in Figure 5.12. Moreover, this graph is bipartite, so we can use the polynomial-time algorithm for MAX BIPARTITE MATCHING we discussed in Section 3.8.

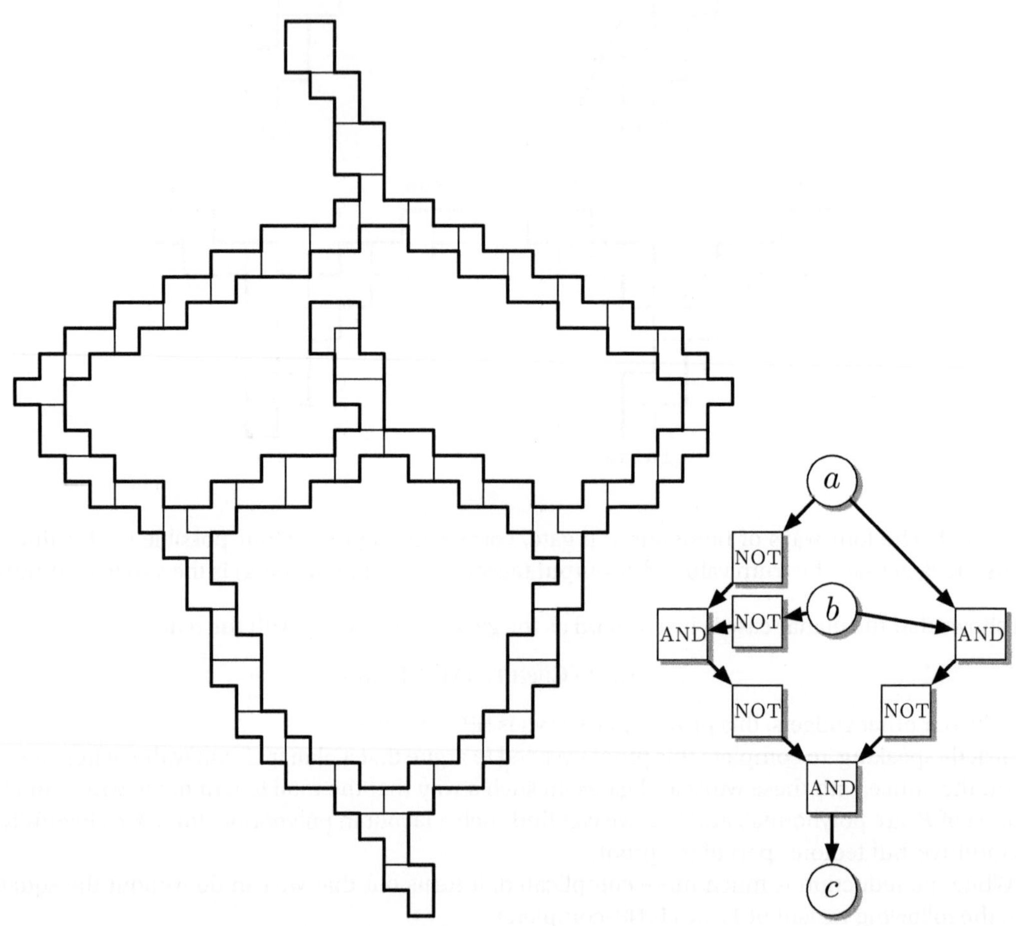

FIGURE 5.10: The instance of TILING corresponding to the CIRCUIT SAT problem where the inputs are a and b, the output is $c = \overline{\overline{a} \wedge \overline{b}} \wedge \overline{a \wedge b}$, and c must be `true`. We show the tiling corresponding to the satisfying assignment $a =$ `true` and $b =$ `false`.

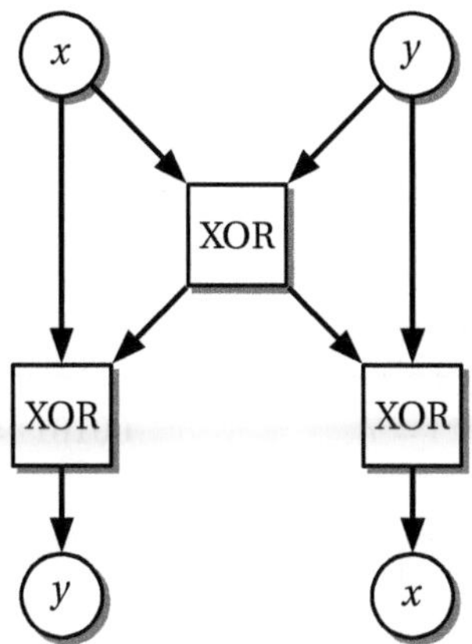

FIGURE 5.11: A crossover gadget that reduces CIRCUIT SAT to PLANAR CIRCUIT SAT, assuming you know how to build an XOR gate.

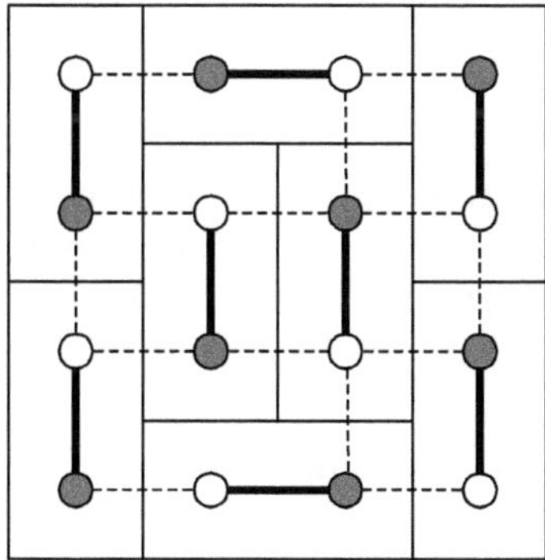

FIGURE 5.12: Domino tilings are perfect matchings of bipartite graphs, so DOMINO TILING is in P.

5.3.5 *Bit by Bit*

In Section 4.2.3 we met two problems involving sets of integers, SUBSET SUM and INTEGER PARTITIONING. Here we prove that SUBSET SUM is NP-complete—and therefore, given the reduction described in Problem 4.13, that INTEGER PARTITIONING is NP-complete as well.

These problems seem very different from the previous problems we've described. I have a set of weights, and I want to know if I can use a subset of them to balance a given object. How in the world could I use these weights, and the weight of the object, to encode a problem about formulas or graphs?

The trick is to think of these weights not as continuous quantities, but as strings of bits, and to use these bits to engineer our choices and constraints. If we express the weights in binary, we can think of each weight as a bundle of bits. We can take each bundle all at once, by including that weight in the subset, or not at all. For instance, if we have four bits to work with, then $8 = 1000$ consists of just the first bit, $5 = 0101$ consists of the second and fourth, and $2 = 0010$ consists of the third. The fact that $8+5+2 = 15 = 1111$ means that these three bundles, taken together, cover each bit exactly once.

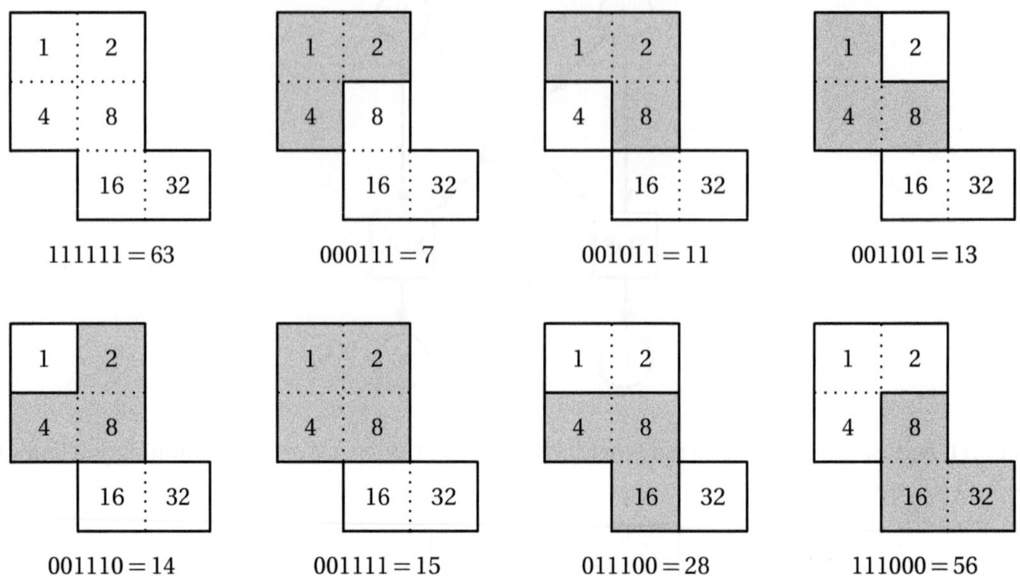

FIGURE 5.13: The reduction from TILING to SUBSET SUM. A subset of $S = \{7, 11, 13, 14, 15, 28, 56\}$ that sums to 63 corresponds to a tiling of the region. Here we use base 2 for simplicity. We can use a larger base to make sure that adding the weights doesn't cause any digits to be carried.

This sounds a lot like TILING. Given an instance of TILING with a region R, let's assign a bit to each of R's squares, giving the ith square a weight of 2^i. Then for each possible location, orientation, and shape of one of the tiles, we define a weight equal to the total weight of the squares it covers, as shown in Figure 5.13. A set of weights whose sum equals the total weight of R corresponds to a tiling, which covers all the squares without gaps or overlaps. Thus

$$\text{TILING} \le \text{SUBSET SUM}$$

and SUBSET SUM is NP-complete.

There is a slight problem with this reduction. We don't want to mistakenly allow the tiles to overlap, but still have the right total weight. This could happen if adding their weights causes some ones to be carried when we add their weights together. To avoid this, instead of using powers of 2, we use powers of k for some k which exceeds the largest number of possible tile positions that could cover any given square. If there are t types of tiles and each one consists of at most m squares then $k = 8tm + 1$ suffices, since each square could be covered by at most $8m$ rotations and reflections of a given type of tile. Then adding the weights in base k can't involve any carrying, so any solution is forced to cover each digit, and thus each square, exactly once.

Before the reader rushes to her machine shop to prepare a set of weights corresponding to her favorite tiling puzzle—say, the one in Figure 5.10—we should point out that the instances of SUBSET SUM this reduction leads to are somewhat unphysical. To encode an instance of TILING where the region and the tiles have n squares, we need to control the weights to n bits of accuracy, or even more if we use a larger base k. As the following exercise shows, it's hard to do this for, say, $n = 100$, even if we can machine our weights down to the atomic level:

Exercise 5.6 *Dust off your high school chemistry textbooks, and show that there are roughly* 2^{83} *atoms in a 1 kilogram brass weight.*

Constructing weights with 160 bits of accuracy would take more than the mass of the Earth. Thus "physical" instances of SUBSET SUM and INTEGER PARTITIONING only have a limited number of bits, and as we showed in Section 4.2.3, such instances can be solved in polynomial time. Nevertheless, when viewed as questions about large integers instead of about brass weights, these problems are NP-complete.

5.4 Completeness as a Surprise

In this section, we explore some additional problems whose NP-completeness is somewhat surprising. These surprises include optimization problems for which finding the minimum is easy but finding the maximum is hard (or vice versa), and others in which it is easy to tell if there is a *perfect* solution, but much harder to tell if there is a good one. We will see that NP-completeness even lurks in fields of mathematics that we know and love, like arithmetic and calculus.

5.4.1 When Maximizing is Harder than Minimizing

In Section 3.7, we saw that MIN CUT is in P, and we raised the question whether MAX CUT is in P as well. Recall that a *cut* of a weighted graph is a division of its vertices into two groups, and the weight of a cut is the total weight of the edges that cross from one group to the other.

MAX CUT
Input: A weighted graph G and an integer k
Question: Does there exist a cut in G of weight k or greater?

(Note that unlike MIN CUT, we don't demand that the cut separate a specific pair of vertices.) We will show that MAX CUT is NP-complete by proving

$$\text{NAE-3-SAT} \le \text{MAX CUT}.$$

Specifically, we will reduce NAE-3-SAT to unweighted MAX CUT, where every edge has weight 1 and the weight of the cut is simply the number of edges crossing from one side to the other. In that case, MAX CUT asks whether it is possible to color G's vertices black and white so that k or more edges have endpoints of different colors. Once we know that the unweighted version of MAX CUT is NP-complete, we know that the weighted version is NP-complete too, since it is at least as hard.

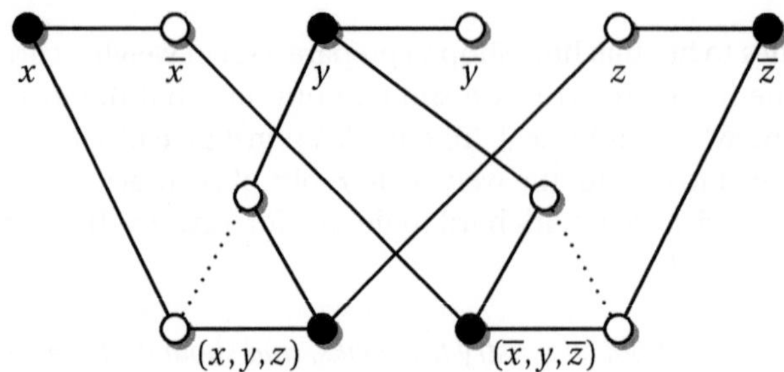

FIGURE 5.14: The reduction from NAE-3-SAT to MAX CUT. We use the same NAE-3-SAT formula $\phi = (x,y,z)\wedge(\overline{x},y,\overline{z})$, and illustrate it with the same satisfying truth assignment $x = \texttt{true}$, $y = \texttt{true}$, $z = \texttt{false}$ as in Figure 5.5. Every edge has weight 1. All but the dotted edges are in the cut, which has weight $n+5m = 13$.

As in our proof that INDEPENDENT SET is NP-complete, we get to choose both the graph G and the threshold k that MAX CUT has to reach. So, we will show how to convert an NAE-3-SAT formula ϕ to a pair (G,k) such that G has a cut of weight k if and only if ϕ is satisfiable.

The gadgets we use for the reduction are almost identical to those for GRAPH 3-COLORING. As shown in Figure 5.14, our choice gadgets again have a pair of vertices for each variable x with an edge between them. Our intention is to force MAX CUT to choose a truth value for x by putting these vertices on opposite sides, thus including all n of the edges between them in the cut.

As before, our constraint gadgets are triangles. Including the edges connecting them to the choice gadgets, each one has 6 edges. If its neighbors are not all on the same side of the cut—that is, if the literals in the corresponding NAESAT clause do not all have the same truth value—then we can choose sides for the gadget's vertices so that 5 of these edges are in the cut. But if its neighbors are all on the same side, no more than 4 of these edges can be in the cut (exercise!). Thus we can achieve a cut of weight $n+5m$ if and only if every clause can be satisfied, so we set our threshold at $k = n+5m$. Clearly we can carry out this reduction in polynomial time, so we're done.

The alert reader will have noticed that MAX CUT does not correspond exactly to the version of MIN CUT we defined in Section 3.7. That problem specified two vertices s and t, and demanded that the cut separates one from the other. The version of MAX CUT defined above asks for a "global" cut, i.e., which separates the graph into two pieces without regard to which vertices go in which piece. Of course, we can also define

> MAX CUT (s-t version)
>
> Input: A weighted graph G, two vertices s,t and an integer k
>
> Question: Does there exist a cut in G of weight at least k that separates s and t?

But as the following exercise asks you to show, this version of MAX CUT is just as hard:

Exercise 5.7 *Show that the s-t version of* MAX CUT *is* NP*-complete as well.*

5.4.2 *When Good is Harder than Perfect*

Constraint satisfaction problems like k-SAT ask whether we can satisfy every clause in a formula. But what if we just want to satisfy most of them? Consider the following problem:

> MAX-k-SAT
>
> Input: A k-SAT formula ϕ and an integer ℓ
>
> Question: Is there a truth assignment that satisfies ℓ or more clauses in ϕ?

For $k \geq 3$, MAX-k-SAT is trivially NP-complete, since we can solve k-SAT by setting ℓ equal to the total number of clauses. But in fact, MAX-k-SAT is NP-complete even for $k = 2$. We will prove that

$$\text{NAE-3-SAT} \leq \text{MAX-2-SAT}$$

with the following reduction. For each NAE-3-SAT clause (x, y, z), include in ϕ the following six 2-SAT clauses

$$(x \vee y), (y \vee z), (x \vee z)$$
$$(\overline{x} \vee \overline{y}), (\overline{y} \vee \overline{z}), (\overline{x} \vee \overline{z})$$

This may cause some clauses to appear in ϕ multiple times, but we count each appearance separately. A little thought shows that any truth assignment satisfies 3 or 5 of these 6 clauses, depending on whether or not it satisfies the corresponding NAE-3-SAT clause. Thus if the original NAE-3-SAT formula has m clauses, we set our threshold to $\ell = 5m$.

The fact that MAX-2-SAT is NP-complete shows that even if a constraint satisfaction problem is easy, the optimization problem associated with it can be very hard. Telling whether we can solve problems approximately can be harder than telling whether we can solve them perfectly.

To put this another way, let's visualize the set of truth assignments as a landscape, where the height is the number of satisfied clauses. This landscape is very bumpy, and is filled with an exponential number of local optima. We can tell in polynomial time whether some peak reaches all the way to perfection—but if none of them does, finding the highest peak is very hard.

Another example for which the quest for a good solution is harder than that for a perfect one is MAX CUT. A "perfect cut" would put the endpoints of every edge on opposite sides. But this is only possible if the graph is bipartite, or equivalently 2-colorable, and GRAPH 2-COLORING is in P.

Exercise 5.8 *Show that for any constant t, the case of* MAX-k-SAT *where we ask whether we can satisfy all but t of the clauses is in* P. *What happens when t is not constant?*

5.4.3 *Hitting the Bullseye*

Suppose we have a problem where both minimizing and maximizing are easy. Does this mean we can tell whether there is a solution in a given interval?

Recall the greedy algorithm for MINIMUM SPANNING TREE in Section 3.5.1. The same strategy works for MAXIMUM SPANNING TREE, so both "extremal" versions of the spanning tree problem are in P. The following problem asks for a spanning tree whose weight is in a specific interval between these extremes:

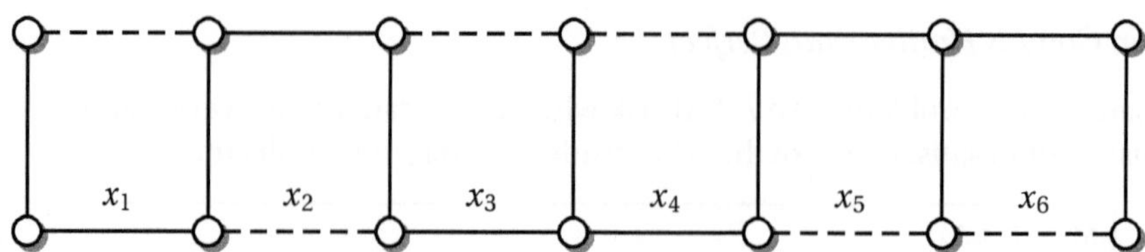

FIGURE 5.15: The reduction from SUBSET SUM to EXACT SPANNING TREE uses a ladder-shaped graph with weighted edges. The edges along one beam of the ladder have the weights $x_1,\ldots,x_n$ of SUBSET SUM, and all other edges have weight zero. The spanning tree in the example (solid edges) has weight $x_1+x_3+x_4$.

EXACT SPANNING TREE
Input: A weighted graph G and two integers ℓ and u
Question: Is there a spanning tree T with weight w such that $\ell \leq w \leq u$?

It's not clear how to generalize the greedy algorithm to EXACT SPANNING TREE. In fact, there is probably no efficient algorithm at all, because EXACT SPANNING TREE is NP-complete. We will prove that

$$\text{SUBSET SUM} \leq \text{EXACT SPANNING TREE}.$$

Our reduction works by constructing the ladder-shaped graph shown in Figure 5.15. Suppose that we have an instance of SUBSET SUM with weights $S=\{x_1,\ldots,x_n\}$ and target t. We construct G with $n+1$ rungs and $2(n+1)$ vertices. We place the weights $x_1,\ldots,x_n$ on the n edges along one beam of the ladder, and give the other edges weight zero. Then each spanning tree T of G uses a subset of edges from the weighted beam, and T's total weight is the weight of this subset. Setting $\ell=u=t$ completes the reduction.

Another problem for which aiming at a specified value is harder than finding the extreme values is SUBSET SUM itself. If we ask to maximize or minimize the target sum instead of specifying it, the problem becomes trivial—just take all of the weights, or none of them.

5.4.4 *Hard Arithmetic*

> The Mock Turtle went on. "We had the best of educations... Reeling and Writhing, of course, to begin with, and then the different branches of Arithmetic—Ambition, Distraction, Uglification, and Derision."
>
> Lewis Carroll, *Alice in Wonderland*

Most NP-complete problems are cast in terms of discrete structures like graphs or Boolean formulas. Among the problems we have seen so far, the arithmetic problems INTEGER PARTITIONING and SUBSET SUM are notable exceptions. But NP-completeness lurks even in ancient algebraic questions about polynomials with integer roots.

In a *Diophantine equation*, we ask for solutions which are natural numbers, i.e., nonnegative integers.
5.5 Let's start by asking for solutions to equations mod p:

MODULAR DIOPHANTINE EQUATION
Input: A polynomial $P(y_1,\ldots,y_n)$ and a prime p
Question: Is there an integer solution to the equation $P(y_1,\ldots,y_n) \equiv_p 1$?

We will prove that

$$\text{3-SAT} \leq \text{MODULAR DIOPHANTINE EQUATION}$$

and therefore that this problem is NP-complete.

First, we need some way to represent Boolean variables. We use Fermat's Little Theorem, which we met in Section 4.4.1. For any prime p, we have

$$y^{p-1} \equiv_p x \quad \text{where} \quad x = \begin{cases} 1 & \text{if } y \not\equiv_p 0 \\ 0 & \text{if } y \equiv_p 0. \end{cases}$$

This lets us represent a Boolean variable x as an integer y, where x is `false` if p divides y and `true` otherwise.

We can use the same trick to combine Boolean variables into clauses. For instance, we can represent the clause $c = (x_1 \vee x_2 \vee \overline{x}_3)$ as the expression

$$q_c = \left(y_1^{p-1} + y_2^{p-1} + 1 - y_3^{p-1}\right)^{p-1}.$$

For any $p > 3$, we have $q_c \equiv_p 1$ if c is satisfied, and $q_c \equiv_p 0$ if it is not.

To represent a 3-SAT formula ϕ, we let $P(y_1,\ldots,y_n)$ be the product of q_c over all the clauses c appearing in ϕ. In that case, $P \equiv_p 1$ if all the clauses are satisfied, and $P \equiv_p 0$ if any of them are violated. Thus the modular Diophantine equation $P \equiv_p 1$ has an integer solution if and only if ϕ is satisfiable.

This reduction is very pretty, but it has several drawbacks. First, we would like to understand the complexity of Diophantine equations over the integers, rather than the integers mod p. Secondly, the polynomial P involves many variables. Thirdly, its degree grows with the size of the 3-SAT formula.

So, how complex are Diophantine equations with just two variables, and whose degree is a small constant? Let's start by considering equations which are merely linear:

LINEAR DIOPHANTINE EQUATION
Input: Natural numbers a, b and c
Question: Are there natural numbers x and y such that $ax + by = c$?

Obviously the equation $ax + by = c$ has an infinite number of *rational* solutions, at least if a and b are both nonzero. The restriction to integers complicates the situation a little, but not much. In Problem 5.23, we ask you to prove that there are integer solutions if and only if $\gcd(a, b)$ divides c, and it is a simple matter to check whether any of these solutions has the property that x and y are both nonnegative. As we saw in Chapter 3, we can compute the gcd in polynomial time using Euclid's algorithm, so LINEAR DIOPHANTINE EQUATION is in P.

Now let's consider quadratic Diophantine equations, in just two variables. Surely these aren't much harder?

> QUADRATIC DIOPHANTINE EQUATION
>
> Input: Natural numbers a, b and c
>
> Question: Are there natural numbers X and Y such that $aX^2 + bY = c$?

Again the answer is easy if we allow X and Y to be real numbers. In contrast to the linear version, however, now the restriction to natural numbers is much more severe. In fact, QUADRATIC DIOPHANTINE EQUATION is NP-complete. First, let's establish that it's in NP:

Exercise 5.9 *Show that* QUADRATIC DIOPHANTINE EQUATION *is in* NP. *Hint: how large can X and Y be? Keep in mind that the size of the input, and the solution, is defined as the total number of bits of a, b, and c.*

We will prove that QUADRATIC DIOPHANTINE EQUATION is NP-complete using a clever reduction from SUBSET SUM. Recall that SUBSET SUM asks whether it is possible to select elements from $\{x_1, \ldots, x_n\}$ that sum up to a specified target t. If we define

$$S = 2t - \sum_i x_i ,$$

this is equivalent to asking whether the equation

$$S - \sum_{i=1}^{n} \sigma_i x_i = 0 \tag{5.1}$$

has a solution where each σ_i is $+1$ or -1. To see this, let $\sigma_i = +1$ if x_i is part of the subset that sums to t, and let $\sigma_i = -1$ if it is not.

Now, any reduction from SUBSET SUM to QUADRATIC DIOPHANTINE EQUATION must somehow encode $n+1$ numbers, namely the weights $x_1, \ldots, x_n$ and the target t, into just three numbers, the coefficients a, b and c. The main ingredient that allows us to do this is the Chinese Remainder Theorem (see Appendix A.7.3). First we choose m such that $2^m > |S| + \sum_i x_i$, so that the left-hand side of (5.1) is zero mod 2^m if and only if it is truly zero. This lets us rewrite (5.1) as

$$S - \sum_{i=1}^{n} \sigma_i x_i \equiv 0 \bmod 2^m \qquad (\sigma_i = \pm 1). \tag{5.2}$$

Next, choose n odd prime numbers $q_1, \ldots, q_n$. According to the Chinese Remainder Theorem, we can define $\theta_1, \ldots, \theta_n$ as the smallest positive integers such that

$$\begin{aligned} \theta_i &\equiv x_i \bmod 2^m \\ \theta_i &\equiv 0 \bmod q_j^m \quad \text{for all } j \neq i \\ \theta_i &\not\equiv 0 \bmod q_i . \end{aligned} \tag{5.3}$$

The first of these equations states that the last m bits of θ_i coincide with x_i for each i. Then we can rewrite our SUBSET SUM problem again, changing (5.2) to

$$S - \sum_{i=1}^{n} \sigma_i \theta_i \equiv 0 \bmod 2^m \qquad (\sigma_i = \pm 1). \tag{5.4}$$

At this point the second and third equations in (5.3) are somewhat mysterious, but they are key to the next step, in which we express (5.4) as a quadratic equation. Let

$$H=\sum_{i=1}^{n}\theta_i \qquad \text{and} \qquad K=\prod_{j=1}^{n}q_j^m\,. \tag{5.5}$$

Then any integer X of the form

$$X=\sum_{i=1}^{n}\sigma_i\theta_i \qquad (\sigma_i=\pm 1) \tag{5.6}$$

is a solution of the modular quadratic equation

$$H^2-X^2\equiv 0 \bmod K\,. \tag{5.7}$$

This is easily verified by squaring the sums involved in H and X and observing that

$$\theta_i\theta_j\equiv 0 \bmod K \text{ for all } i\neq j\,.$$

The following lemma shows that if we bound X and choose the q_i large enough, the converse is also true: any solution of (5.7) has the form (5.6). By adding one more equation, we can require $X\equiv\pm S$, so that either the θ_i in (5.6) or their negations correspond to a solution to the original SUBSET SUM problem.

Lemma 5.1 *Let θ_i, H and K be defined as in* (5.3) *and* (5.5). *It is possible to choose q_i of size* poly(n) *so that $2H<K$. In that case, any X such that $0\le X\le H$ and which fulfills the following equation is of the form* (5.6):

$$H^2-X^2\equiv 0 \bmod K\,. \tag{5.8}$$

The case of SUBSET SUM *where exactly one of the weights x_i is odd is* NP*-complete, so we can assume without loss of generality that S is odd. Then* (5.4) *has a solution if and only if* (5.8) *and the following hold,*

$$S^2-X^2\equiv 0 \bmod 2^{m+1}\,. \tag{5.9}$$

We leave the proof of this lemma to the reader in Problem 5.24.

Lemma 5.1 tells us that (5.4) is equivalent to (5.8) and (5.9). We will use the Chinese Remainder Theorem again to transform these into a single quadratic Diophantine equation. First observe that since K is a product of odd numbers, K and 2^{m+1} are relatively prime. Then consider this exercise, which gives two new parameters to play with:

Exercise 5.10 *Let a and b be relatively prime, and suppose that λ_1,λ_2 are integers such that* $\gcd(\lambda_1,b)=\gcd(\lambda_2,a)=1$. *Show that $\lambda_1 ax+\lambda_2 by\equiv 0 \bmod ab$ if and only if $x\equiv 0 \bmod b$ and $y\equiv 0 \bmod a$.*

Then for any λ_1 and λ_2 such that $\gcd(\lambda_1,K)=\gcd(\lambda_2,2^{m+1})=1$, we can write (5.8) and (5.9) as

$$\lambda_1 2^{m+1}(H^2-X^2)+\lambda_2 K(S^2-X^2)\equiv 0 \bmod 2^{m+1}K\,. \tag{5.10}$$

Equivalently, there are integers X and Y such that

$$\lambda_1 2^{m+1}(H^2-X^2)+\lambda_2 K(S^2-X^2)\equiv 0=2^{m+1}KY\,. \tag{5.11}$$

Rearranging gives an equation of the right form for QUADRATIC DIOPHANTINE EQUATION,

$$\underbrace{\left(\lambda_1 2^{m+1}+\lambda_2 K\right)}_{a} X^2+\underbrace{2^{m+1}K}_{b}\, Y=\underbrace{\lambda_1 2^{m+1}H^2+\lambda_2 K s^2}_{c}. \tag{5.12}$$

However, we still have to impose the constraint $X \le H$ from Lemma 5.1. Happily, Problem 5.25 shows that for a judicious choice of λ_1 and λ_2, this is equivalent to the constraint $Y \ge 0$, i.e., that Y is a natural number, which (along with $X \ge 0$) QUADRATIC DIOPHANTINE EQUATION gives us for free.

At this point we have successfully mapped an instance of SUBSET SUM to a quadratic Diophantine equation. However, we still need to show that this reduction can be carried out in polynomial time: specifically, that we can find suitable primes q_i and compute the θ_i. Since the q_i are only polynomial in n, we can find them easily. And as Problem 5.26 shows, we can make the Chinese Remainder Theorem algorithmic, so we can do both of these things in polynomial time.

5.4.5 *(Computationally) Complex Integrals*

Our previous problem showed that NP-completeness can show up in algebra. Next, we'll see that it appears even in calculus. Consider the following problem:

> COSINE INTEGRATION
>
> Input: A list of integers $x_1,\ldots,x_n$
>
> Question: Is $\int_{-\pi}^{\pi}(\cos x_1\theta)(\cos x_2\theta)\cdots(\cos x_n\theta)\,\mathrm{d}\theta \neq 0$?

This problem looks very different from all the problems we have seen so far in this book. For that matter, it is not even clear that it is in NP. In fact it is, but the witness is hidden behind some transformations.

Using the identity

$$\cos\theta=\frac{\mathrm{e}^{\imath\theta}+\mathrm{e}^{-\imath\theta}}{2},$$

we write the integrand as follows,

$$\prod_{j=1}^{n}\cos x_j\theta=\frac{1}{2^n}\prod_{j=1}^{n}\left(\mathrm{e}^{\imath x_j\theta}+\mathrm{e}^{-\imath x_j\theta}\right).$$

Expanding this product gives 2^n terms. Each one is the product of $\mathrm{e}^{\imath x_j\theta}$ for some j, and $\mathrm{e}^{-\imath x_j\theta}$ for the others. If A denotes the set of j where we take $\mathrm{e}^{\imath x_j\theta}$ instead of $\mathrm{e}^{-\imath x_j\theta}$, then we can write the integrand as a sum over all subsets $A\subseteq\{1,\ldots,n\}$:

$$\frac{1}{2^n}\sum_{A\subseteq\{1,\ldots,n\}}\left(\prod_{j\in A}\mathrm{e}^{\imath x_j\theta}\right)\left(\prod_{j\notin A}\mathrm{e}^{-\imath x_j\theta}\right)=\frac{1}{2^n}\sum_{A\subseteq\{1,\ldots,n\}}\mathrm{e}^{\imath\theta\left(\sum_{j\in A}x_j-\sum_{j\notin A}x_j\right)}.$$

For each A, the difference $\sum_{j\in A}x_j-\sum_{j\notin A}x_j$ is some integer y. Now, to compute the integral, we use the fact that $\int_{-\pi}^{\pi}\mathrm{e}^{\imath y\theta}$ for any integer frequency y depends only on whether or not y is zero:

$$\int_{-\pi}^{\pi}\mathrm{e}^{\imath y\theta}\,\mathrm{d}\theta=\begin{cases}2\pi & \text{if } y=0\\ 0 & \text{if } y\neq 0.\end{cases}$$

Thus the integral is just

$$\int_{-\pi}^{\pi} \left(\prod_{j=1}^{n} \cos x_j \theta \right) \mathrm{d}\theta = \frac{2\pi}{2^n} Z, \qquad (5.13)$$

where Z is the number of subsets A such that $y = 0$.

But such an A is a balanced partition in the sense of INTEGER PARTITIONING, i.e.,

$$\sum_{j \in A} x_j = \sum_{j \notin A} x_j .$$

Thus the integral is proportional to the number Z of balanced partitions, and it is nonzero if and only if there is at least one such partition. COSINE INTEGRATION is just INTEGER PARTITIONING in disguise.

Note that calculating this integral is equivalent to counting how many balanced partitions there are. This is, *a priori*, much harder than telling whether there is at least one. We will meet this kind of problem, where we want to count the solutions of an NP problem, in Chapter 13.

The NP-completeness of COSINE INTEGRATION is especially fascinating because we generally assume that we can estimate an integral like $\int_{-\pi}^{\pi} f(\theta)\,\mathrm{d}\theta$ by integrating it numerically, or by sampling $f(\theta)$ at a set of points. However, there are two problems with this. First, if there are only a small number Z of balanced partitions, then the integral $2\pi Z/2^n$ is exponentially small, so we would have to estimate it to exponential precision in order to tell whether or not a balanced partition exists. Secondly, any hard instance of INTEGER PARTITIONING involves exponentially large numbers x_j, so the function $f(\theta)$ oscillates with exponentially high frequency, as suggested by Figure 5.16.

These facts suggest that we have to use exponentially many sample points, or exponentially many steps in our numerical integration, in order to solve COSINE INTEGRATION. The fact that it is NP-complete tells us that, unless P = NP, there is no way around this.

5.5 The Boundary Between Easy and Hard

> Sad to say, but it will be many more years, if ever, before we really understand the Mystical Power of Twoness... 2-SAT is easy, 3-SAT is hard, 2-dimensional matching is easy, 3-dimensional matching is hard. Why? Oh, why?
>
> Eugene Lawler

We have seen many problems where one variant is in P and another is NP-complete. What makes the difference? Why are some problems easy while others, which are seemingly similar, are hard? In this section we look at a few of the most common reasons for this. These include how much choice we have when solving a problem, the "stringiness" or "webbiness" with which its variables interact, the barriers between one solution and another, and the dimensionality or loopiness of the graph on which it is defined.

We open this section with an important caveat, however. It is tempting to think that there is a finite number of explanations for why one problem is in P and another is NP-complete. However, if we could classify the structural properties that make a problem solvable in polynomial time, we would be able to prove that P ≠ NP. So, while arguments like those we are about to give certainly help build our intuition about which problems are which, we shouldn't expect a complete understanding of the boundary between these two complexity classes any time soon.

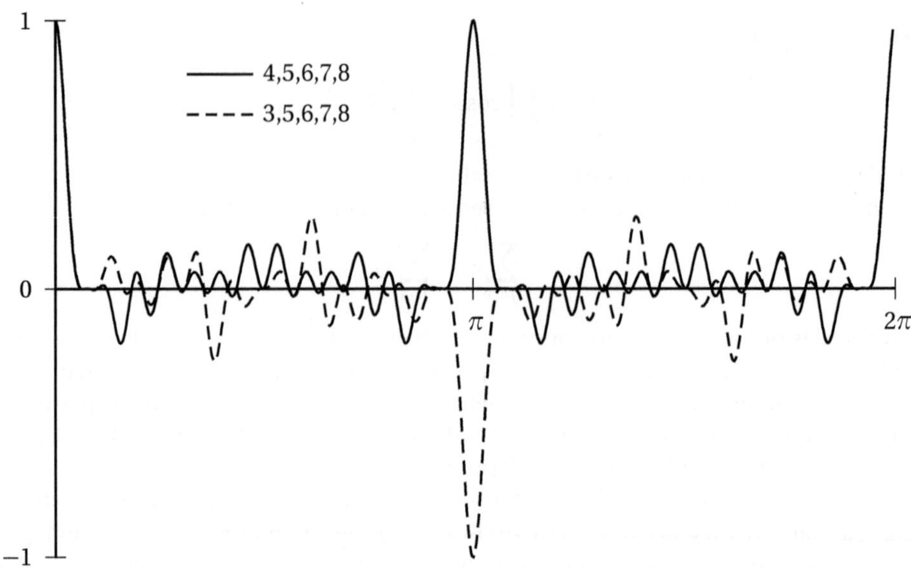

FIGURE 5.16: The functions $f(\theta)$ corresponding to two instances of INTEGER PARTITIONING. If we integrate them over this interval, one of them is zero and the other is nonzero, depending on whether or not a balanced partition exists. Telling which is which using numerical integration is difficult since these functions oscillate rapidly.

5.5.1 *Choice, or the Lack of it*

Both GRAPH k-COLORING and k-SAT are in P for $k = 2$, and are NP-complete for $k \geq 3$. Why do these problems cross the threshold from easy to hard when k goes from 2 to 3?

One explanation is the presence or absence of choice. In GRAPH 2-COLORING, once we have chosen the color of a vertex, this immediately forces its neighbors to be the opposite color. These alternating colors propagate automatically until we have colored the entire graph—or at least the connected component containing our initial vertex—or learned that it isn't colorable.

In GRAPH 3-COLORING, on the other hand, if we have colored a vertex cyan, we can choose whether to color each of its neighbors yellow or magenta. Each of their neighbors will also typically have two choices, and so on. In general, if we have a palette of k colors, a vertex with one colored neighbor has $k-1$ choices left. Thus a naive search algorithm takes roughly $(k-1)^n$ time, which is exponential for $k \geq 3$.

Similarly, suppose we set the truth value of a variable in a k-SAT formula. A clause that wanted that variable to take the opposite value becomes a $(k-1)$-SAT clause on its remaining variables. If $k = 2$ it becomes a unit clause, which immediately forces us to set another variable as described in Section 4.2.2. If $k \geq 3$, on the other hand, we have $k-1$ choices for how to satisfy it, giving an exponential search tree.

This also helps us understand why MAX-2-SAT is NP-complete, even for $k = 2$. If we set a variable in a way that disagrees with a 2-SAT clause, we still have two choices—we can satisfy it by setting its other variable, or we can ignore it and try to meet the threshold by satisfying other clauses instead.

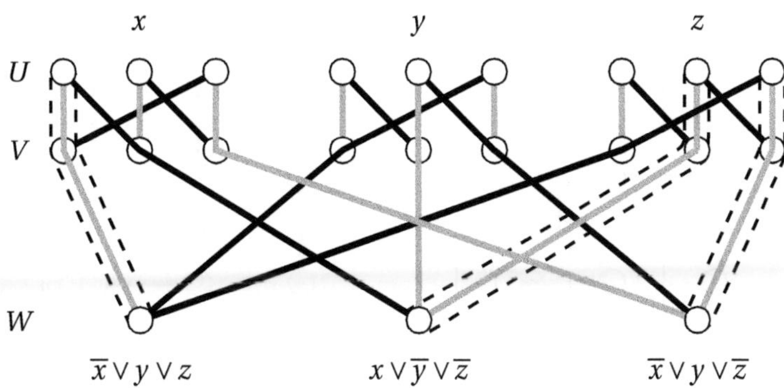

FIGURE 5.17: The reduction from 3-SAT to PERFECT 3-MATCHING. Compatible triplets are connected with black or gray lines, which correspond to setting a variable `true` or `false`. The zig-zag shape of the choice gadgets force these choices to be consistent. Each $U \times V$ pair is compatible with the clauses it could satisfy, and pairs not used to satisfy clauses are matched to additional dummy elements of W (not shown). The 3-matching here corresponds to setting $x = z =$ `false`.

5.5.2 *Paths, Webs, and Barriers*

The idea of choice doesn't always account for the "two vs. three" phenomenon. Sometimes something subtler is going on. Let's look at another example. Recall the problem BIPARTITE MATCHING, where we are running a dating service, and we want to pair up all our clients into compatible couples. We saw in Section 3.8 that this problem is in P, since it can be reduced to MAX FLOW.

Suppose we are now running a dating service for an alien species with *three* sexes, requiring one of each for a successful mating. Some triplets are compatible, others are not, and our goal is to choose a set of compatible triplets such that everyone is part of exactly one. Formally,

> PERFECT 3-MATCHING
>
> Input: Three sets U, V, W of size n and a set $S \subseteq U \times V \times W$
>
> Question: Is there a subset $T \subseteq S$ of size n that includes each element of $U \cup V \cup W$ exactly once?

We will prove that PERFECT 3-MATCHING is NP-complete by reducing 3-SAT to it. To describe the reduction, it's useful to think first of compatible pairs among elements of U and V, and then think of each such pair as compatible with some elements of W. For each variable x in the 3-SAT formula, we have a choice gadget consisting of t elements of U and t elements of V, where t is the number of clauses x appears in. As Figure 5.17 shows, we make pairs of these elements compatible in a zig-zag pattern, so that each element of U is compatible with exactly two elements of V. Choosing one of these two mates corresponds to setting x `true` or `false`, and the zig-zag ensures that all t matings in the gadget correspond to the same value of x.

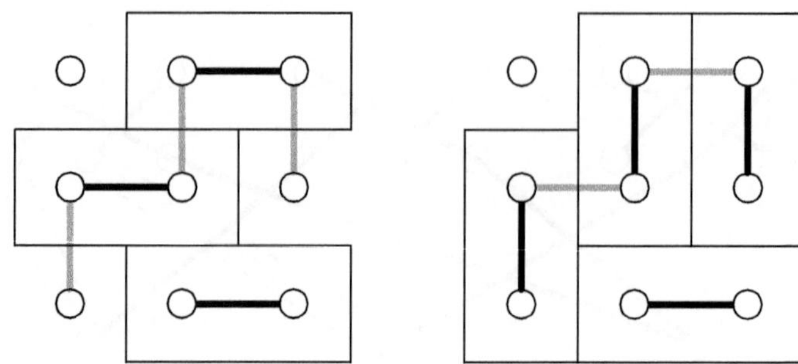

FIGURE 5.18: We can improve a partial domino tiling, or equivalently a partial matching, by flipping the edges along an alternating path.

Each constraint gadget consists of an element of W, which we make compatible with the pairs in $U \times V$ corresponding to variable settings that would satisfy that clause as shown in Figure 5.17. Finally, we add to W a set of dummy elements, which are compatible with every pair in $U \times V$, to mop up the unused elements of each choice gadget. Then a 3-matching exists if and only if the 3-SAT formula is satisfiable, and PERFECT 3-MATCHING is NP-complete.

Now pretend, for a moment, that we didn't already know about the reduction from BIPARTITE MATCHING to MAX FLOW. Then the mystery is not why PERFECT 3-MATCHING is NP-complete, but why BIPARTITE MATCHING isn't. After all, we have seen that many problems like TILING and INDEPENDENT SET, which try to fit things in a graph or region without letting them overlap, are NP-complete. At first blush, finding a matching looks very similar to these problems, since it consists of fitting in as many edges as we can without letting them overlap at their endpoints.

In some sense, the reason why BIPARTITE MATCHING is in P is the following. Suppose we have a graph $G=(V,E)$ and a partial matching $M \subseteq E$. We call a vertex *isolated* if it is not covered by an edge in M. An *alternating path* is a path that begins and ends at isolated vertices, and that alternates between edges in M and not in M. Then we have the following theorem, analogous to Theorem 3.3:

Theorem 5.2 *Suppose that $M \subseteq E$ is a partial matching of a bipartite graph $G=(V,E)$. Then M is maximal if and only if there is no alternating path.*

Proof If an alternating path exists, we can flip the edges along it, putting each one in M if it's out and taking it out of M if it's in. This gives us a new matching M' which has one more edge than M has, and which covers the two isolated vertices at the path's endpoints, as shown in Figure 5.18.

Conversely, as Figure 5.19 shows, the union of two partial matchings consists of several kinds of connected components: shared edges and paths and cycles whose edges alternate between the two matchings. If one matching has more edges than the other then at least one of these components gives an alternating path for the smaller matching, which will increase the number of edges if we flip it. □

Theorem 5.2 tells us that we can tell whether a matching is maximal in polynomial time. In fact, in the bipartite case, Problem 5.27 shows that these alternating paths correspond exactly to the augmenting paths that the Ford–Fulkerson algorithm uses to solve MAX FLOW. The situation for non-bipartite graphs

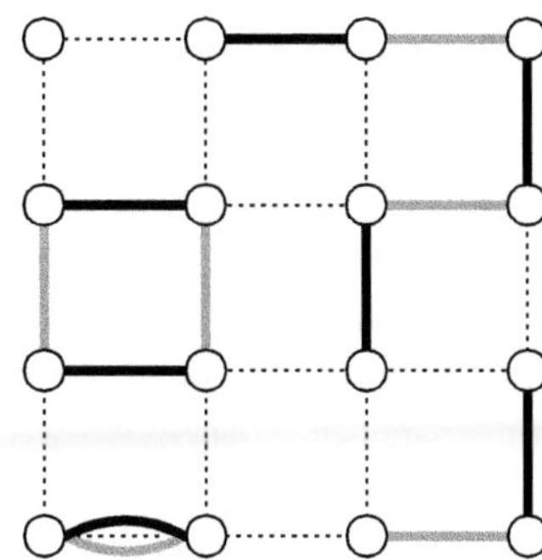

FIGURE 5.19: The union of two partial matchings consists of shared edges (lower left) and cycles and paths alternating between the two matchings.

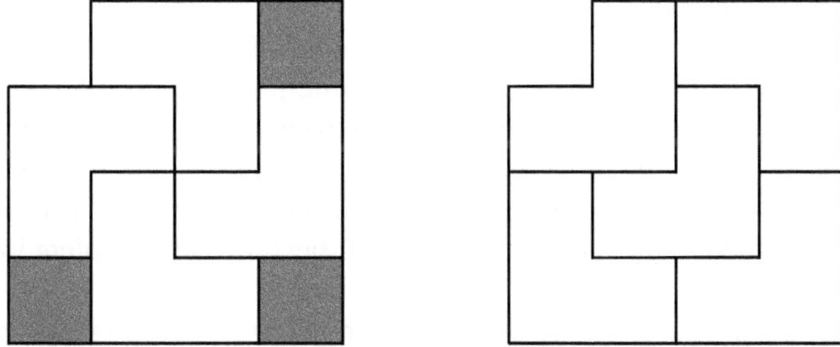

FIGURE 5.20: Improving a tromino tiling. We cannot fill the three uncovered squares, and tile the entire region, without a global rearrangement.

is slightly more complicated, since we need to exclude paths that contain loops of odd length, but we can still find them in polynomial time. Once we find an alternating path, we can use it to improve our solution greedily until we either get a perfect matching or a proof that there is none. 5.6

This is also why DOMINO TILING is in P, even though TROMINO TILING is NP-complete. Suppose we have a partial tiling of a region with dominoes. This corresponds to a partial matching as shown in Figure 5.18, where uncovered squares correspond to isolated vertices. An alternating path is a chain along which we can shift the dominoes, creating space for one more domino and covering the squares at both ends. We can find these paths easily, improve the tiling, and continue until we either cover all the squares or obtain a proof that we can't.

For TROMINO TILING, on the other hand, no result analogous to Theorem 5.2 is known. If we try to improve a partial tiling by rearranging the tiles to make space for new ones, this creates not a simple path from one uncovered square to another, but a branching, webby mess of new tiles and old ones. Finding such a rearrangement, or telling if one exists, is much harder than simply looking for a path from one place to another. Moreover, the improved tiling might have very little in common with the previous one, as shown in Figure 5.20. Using the landscape analogy from Section 3.5.3, there are many local optima with high barriers between them, and finding the global optimum requires some kind of global exploration.

This distinction between paths and webs shows up in k-SAT as well. If a 2-SAT formula is unsatisfiable, there is a simple witness of this fact—namely, a contradictory loop as described in Section 4.2.2. Because of the simple way that each clause links to the next, it is easy to prove that the subformula consisting of the clauses along this loop—and therefore, the entire formula—is unsatisfiable. In contrast, if a 3-SAT formula is unsatisfiable, this is not because of a simple chain of implications, but because of a complex, branching web of contradictions. As a result, finding the simplest unsatisfiable subformula, and proving that it is unsatisfiable, seems to require an exponential search.

5.5.3 One, Two, and Three Dimensions

We saw in Chapter 3 that many problems with a one-dimensional character, such as TYPESETTING and ALIGNMENT, can be solved in polynomial time. We can typically solve such problems by going from left to right and using dynamic programming, or by breaking them into two pieces and solving the pieces recursively. There is only one path along which information can flow, and by interrupting this path we can break the problem into independent pieces.

In two dimensions, on the other hand, information can flow from one part of the problem to another along many different paths, making it hard to use dynamic programming or recursion. For instance, breaking a grid into two pieces creates a long boundary between them, and the pieces can interact across it at every point. Moreover, as we saw in TILING, in two dimensions we have the freedom to draw a circuit in the plane, which can carry out an arbitrary witness-checking program. For these and other reasons, many problems whose one-dimensional versions are in P become NP-complete in two or more dimensions. Let's look at an example.

Cellular automata, or CAs for short, are popular discrete models of physical and biological systems. They take place on a finite-dimensional lattice, where the state at each location x and time t is given by a symbol $a_{x,t}$ belonging to some finite alphabet. At each step, we update all the sites simultaneously using a local transition function f, where the new state at each site depends only on its previous value and the previous values of its neighbors.

For example, suppose we have a one-dimensional lattice, where the neighborhood of each site consists of itself and its nearest neighbors. Then when we go from time t to time $t+1$, the new state of each site x is given by

$$a_{x,t+1} = f(a_{x-1,t}, a_{x,t}, a_{x+1,t}).$$

If the alphabet of possible states is $\{0,1\}$, a typical transition function might be

$$\begin{array}{cccccccc} 111 & 110 & 101 & 100 & 011 & 010 & 001 & 000 \\ 0 & 0 & 0 & 1 & 1 & 1 & 1 & 0 \end{array} \tag{5.14}$$

Since 00011110 encodes the integer 30 in binary, this is called rule 30. Its behavior is quite rich.

5.7 Given a string a of 0s and 1s, let $f(a)$ denote the new string resulting from applying f everywhere. For instance, if at $t=0$ we have the string $a = 10111001$, running the cellular automaton for two steps gives the following states:

$$\begin{array}{r@{\;}cccccccc} a: & 1 & 0 & 1 & 1 & 1 & 0 & 0 & 1 \\ f(a): & & 0 & 1 & 0 & 0 & 1 & 1 & \\ f(f(a)): & & & 1 & 1 & 1 & 1 & & \end{array}$$

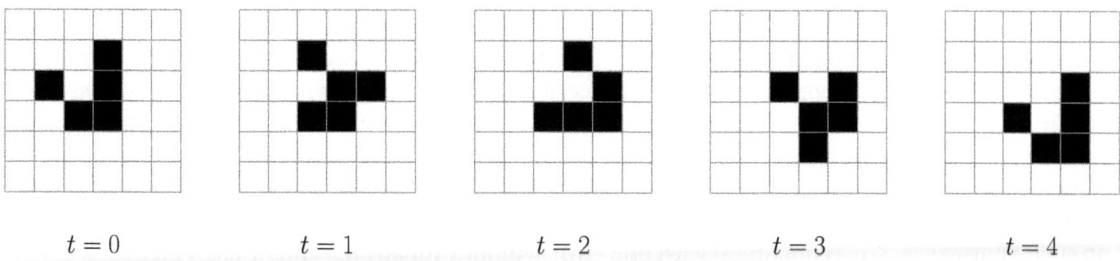

FIGURE 5.21: A glider in the Game of Life, which moves one cell down and to the right every four time steps. Black and white cells represent 1s and 0s respectively, and we assume that the surrounding cells are all 0s.

The string gets shorter at each step since we don't know the symbols outside it on either side. Alternately, we could use *cyclic boundary conditions*, and assume that the lattice wraps around at the ends.

The most famous CA rule is the "Game of Life," which was invented by the British mathematician John Horton Conway in 1970. It takes place on the square lattice, and each site has the 8 neighbors it could reach with a Chess king's move. We again have $S = \{0,1\}$, corresponding to "dead" and "live" cells. This species needs three parents to create a child, so a site that is currently 0 will become 1 if it has exactly three 1s in its neighborhood. A site that is currently 1 will survive, staying a 1, if it has two or three 1s in its neighborhood. If it has fewer than two or more than three, it dies—of loneliness or overcrowding—and becomes a 0.

The Game of Life is popular because it displays an amazing variety of behaviors. We give a taste of this in Figure 5.21, where a "glider" travels down and to the right while fluctuating between four shapes. We will see much more complicated objects in Section 7.6.4, including constructions of universal computers.

Running a cellular automaton forwards in time is simple enough—we just have to apply the transition function at every site. But what about running it backwards? Given the state of a finite lattice, can we tell whether it has a *predecessor*, which will lead to it in one step?

> CA PREDECESSOR
>
> Input: A cellular automaton f and a state a on a finite lattice
>
> Question: Is there a state a' such that $a = f(a')$?

Problem 5.29 shows that the one-dimensional version of this problem is in P. However, in two or more dimensions CA PREDECESSOR is NP-complete. We will reduce to it from TILING. The current state will correspond to the region we are trying to tile, and the predecessor will correspond to the tiling.

Here's how this works. We define the current state a by marking each cell of the region R with a symbol r, and the other cells with b (for "background"). To define the predecessor state a' we expand our alphabet further, with a unique symbol for each cell of each tile in each possible orientation. For instance, since the right tromino has four possible orientations and the square has just one, for the tile set we used in Section 5.3.4 we would have $3 \times 4 + 4 = 16$ symbols in addition to r and b. These symbols represent claims like "this cell is the west end of a tromino whose ends point south and west," or "this cell is the northeast corner of a square," and so on.

Given a state a' containing an array of these symbols, we can check that it corresponds to a valid tiling by examining the neighborhood of every site. So we define the transition function f as follows. If a site's neighborhood is inconsistent—for instance, if a tromino symbol is immediately to the west of the northeast corner of a square—we update that site to an "error" symbol x. Otherwise, if it corresponds locally to a consistent tiling, and we update it to the "region" symbol r. Finally, any site labeled b or x stays that way.

With this definition of f, we have $f(a')=a$ for any state a' corresponding to a valid tiling of R. But if no such tiling exists, $f(a')$ contains at least one error symbol x for any a'. Thus a has a predecessor if and only if the region R has a tiling. This proves that

$$\text{Tiling} \leq \text{CA Predecessor}\,,$$

and CA Predecessor is NP-complete.

CA Predecessor is a good illustration of the difference between one dimension and two. But what about the difference between two and three? Some NP-complete problems, like Graph 3-Coloring, are NP-complete even when restricted to planar graphs, either because a crossover gadget like the one in Figure 4.3 exists, or because we can reduce to them from some other planar problem. On the other hand, if $k \geq 4$ then the Four Color Theorem tells us that the planar case of Graph k-Coloring is in P, so for these problems even the two-dimensional version is easy.

One crucial difference between two and three dimensions is the Jordan Curve Theorem, which states that drawing a closed curve separates the plane into two parts. While no one knows how to prove the Four Color Theorem directly from this fact, it can be used to prove the (much easier) Five Color Theorem that every planar graph is 5-colorable. In some sense, for five or more colors, the Jordan Curve Theorem
4.2 lets us "divide and conquer" the graph by dividing it into the inside and the outside of a loop. In three or more dimensions, in contrast, closed curves no longer divide space into two parts. Edges can cross each other, there is no restriction on how vertices can be connected, and Graph k-Coloring is NP-complete, instead of trivial, even for $k=4$ or $k=5$.

Finally, many NP-complete problems are in P when restricted to trees, as we saw in Problem 3.25 for Independent Set. Like the one-dimensional lattice, trees have no loops, and removing a single vertex causes them to fall apart into separate pieces. For this reason, many graph problems are in P in the special case where the graph can be described as a combination of a small number of trees. We will see in Chapter 14 that it is the "loopiness" of a typical instance of 3-SAT that makes it hard.

5.6 Finally, Hamiltonian Path

We end this chapter by closing a circle, and returning to the problem of our Prologue. We will prove that

$$\text{3-SAT} \leq \text{Hamiltonian Path}\,,$$

and therefore that Hamiltonian Path is NP-complete. We focus on the case where the graph G is directed. In Problem 5.31, you can show Hamiltonian Path is NP-complete for undirected graphs as well.

The reduction is shown in Figure 5.22. For each variable x, we have a choice gadget consisting of a chain of t vertices, where t is at least as large as the number of clauses in which x appears. We can traverse this chain either left-to-right or right-to-left, and these choices correspond to setting x `true` or `false` respectively.

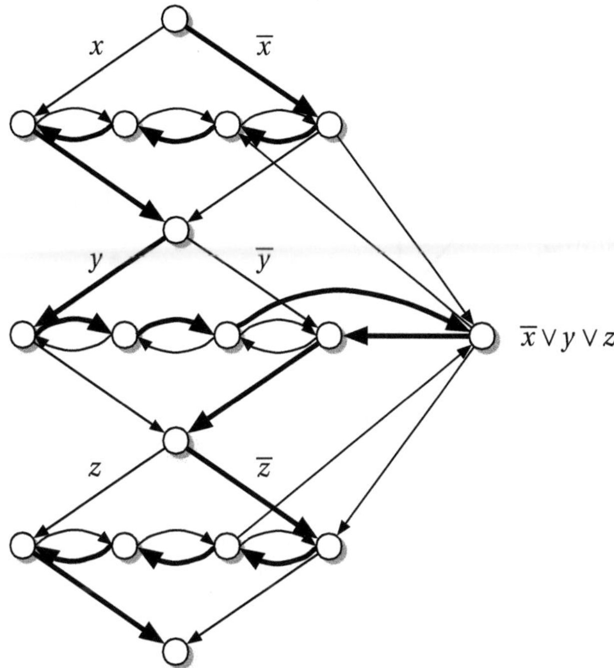

FIGURE 5.22: The reduction from 3-SAT to HAMILTONIAN PATH. Each choice gadget consists of a chain of vertices which can be covered left-to-right or right-to-left, corresponding to setting that variable `true` or `false`. Each variable has optional detours to the clauses it could satisfy. In this case the path, shown in bold, corresponds to setting $x = $ `false`, $y = $ `true`, and $z = $ `false`. The clause $\overline{x} \vee y \vee z$ is satisfied by both x and y. We could have visited it by taking a detour from x's gadget, but we chose to do so from y's gadget instead.

For each clause, our constraint gadget consists of a single vertex. To visit it, we add optional "detour" edges from each choice gadget to this vertex and back again, oriented so that we can only make the detour if we are going through the choice gadget in the direction that the clause agrees with.

Since every clause vertex has to be visited at some point, at least one of its variables must have the truth value it desires. Thus there is a Hamiltonian path which moves from top to bottom through the choice gadgets, making detours to visit each clause vertex on the way, if and only if the 3-SAT formula is satisfiable.

This shows that HAMILTONIAN PATH is NP-complete. Adding an edge from the bottom vertex in Figure 5.22 back to the top vertex changes the path to a cycle, proving that HAMILTONIAN CYCLE—the problem of whether we can visit each vertex once and return to our starting point—is NP-complete as well.

Exercise 5.11 *Show that* HAMILTONIAN CYCLE *is* NP*-complete by reducing* HAMILTONIAN PATH *directly to it.*

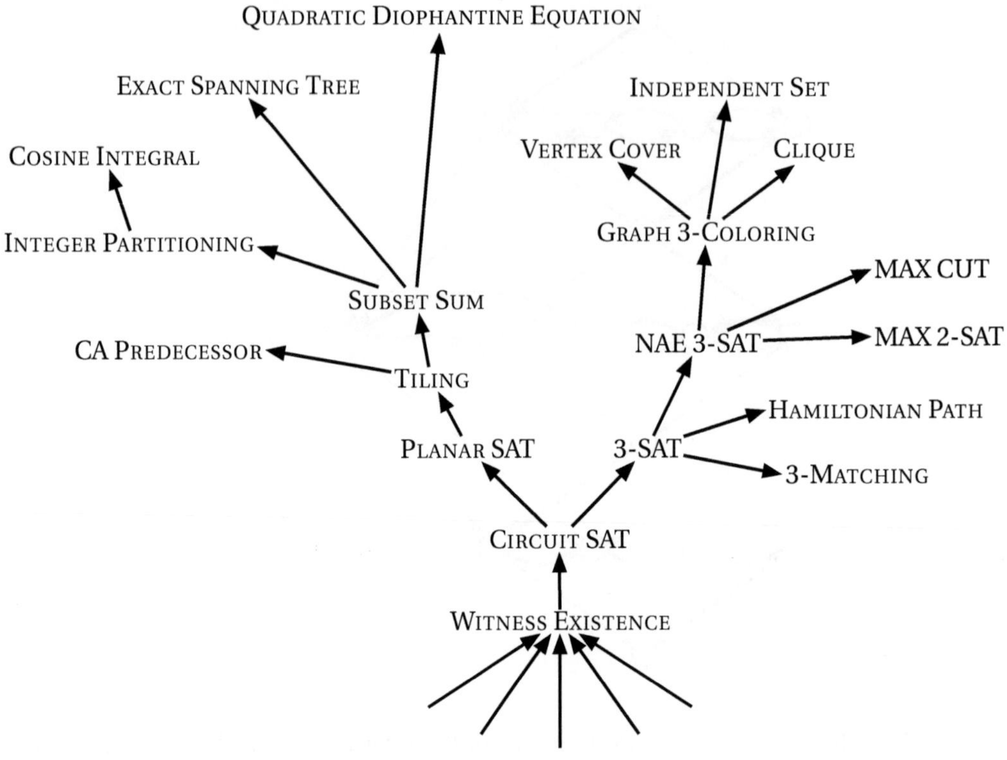

FIGURE 5.23: The beginnings of the family tree of NP-complete problems.

As we close this chapter, let us briefly gaze at a portrait of our family tree (Figure 5.23). The reader should visualize it as continuing on—in smaller and smaller print, or perhaps in hyperbolic space—to include thousands of problems. This tree shows the particular reductions that we used to prove these problems are NP-complete. But keep in mind that they can all be reduced to each other, if only by circling around and going back up through the root WITNESS EXISTENCE.

Problems

> All the problems of the world could be settled if people were only willing to think. The trouble is that people very often resort to all sorts of devices in order not to think, because thinking is such hard work.
>
> Thomas J. Watson

5.1 From circuits to NAESAT. Find a direct reduction from CIRCUIT SAT to NAE-3-SAT. Hint: consider the 3-SAT clause $(\overline{x}_1 \vee \overline{x}_2 \vee y)$ in our description of an AND gate. Does it ever happen that all three literals are true?

5.2 One for all. 1-IN-3 SAT is a variant of 3-SAT where each clause is satisfied if *exactly one* of its literals is true. Prove that 1-IN-3 SAT is NP-complete. In fact, prove that the special case POSITIVE 1-IN-3 SAT, where no variables are negated, is NP-complete. First, make a judicious choice of which NP-complete problem would be easiest to reduce from.

5.3 Easy majority? Consider MAJORITY-OF-3 SAT, in which each clause contains three literals and at least two of them must be true for it to be satisfied. Either show that this problem is in P by reducing it to 2-SAT, or show that it is NP-complete by reducing 3-SAT to it.

5.4 The power of positive thinking. Another variant of SAT is HORN-SAT, where each clause contains at most one positive (i.e., non-negated) variable. Show that HORN-SAT is in P. Hint: a clause such as $(\overline{x} \vee \overline{y} \vee z)$ is equivalent to the implication $(x \wedge y) \to z$. Design an algorithm that starts with the unit clauses and determines the smallest possible set of variables that have to be true. 5.8

5.5 Limited truth. Consider the following variant of k-SAT:

THRESHOLD k-SAT
Input: A k-SAT formula ϕ and an integer t
Question: Is there a satisfying assignment for ϕ where t or fewer variables are `true`?

Show that THRESHOLD k-SAT is NP-complete even when $k = 2$.

5.6 Forbidden colors. Consider the following variant of GRAPH 3-COLORING. In addition to the graph G, we are given a forbidden color f_v for each vertex v. For instance, vertex 1 may not be red, vertex 2 may not be green, and so on. We can then ask whether G has a proper 3-coloring in which no vertex is colored with its forbidden color:

CONSTRAINED GRAPH 3-COLORING
Input: A graph G with n vertices and a list $f_1, \ldots, f_n \in \{1,2,3\}$
Question: Does G have a proper 3-coloring $c_1, \ldots, c_n$ where $c_v \neq f_v$ for all v?

Either show that CONSTRAINED GRAPH 3-COLORING is NP-complete, or show that it is in P by reducing it to another problem in P. What about CONSTRAINED GRAPH 4-COLORING?

5.7 Yakety Yak. Consider the following model of cellphone conversations. We have an undirected graph $G = (V, E)$ where the vertices are people, and each edge indicates that two people are within range of each other. Whenever two people are talking, their neighbors must stay silent on that frequency to avoid interference. Thus a set of conversations consists of a set of edges $C \subset E$, where vertices in different edges in C cannot be neighbors of each other.

The *cellphone capacity* of G is the largest number of conversations that can take place simultaneously on one frequency, i.e., the size of the largest such set C. Now consider the following problem:

CELLPHONE CAPACITY
Input: A graph G and an integer k
Question: Is there a set C of conversations with $\lvert C \rvert \geq k$?

Prove that CELLPHONE CAPACITY is NP-complete.

5.8 MAX-2-SAT reloaded. Give a simpler proof that MAX-2-SAT is NP-complete by reducing to it from the unweighted version of MAX CUT. Hint: replace each edge with a pair of clauses.

5.9 Balanced colors. A k-coloring is *balanced* if exactly $1/k$ of the vertices have each color. Show that for any $k \geq 3$ the problem of whether a graph has a BALANCED k-COLORING is NP-complete. Then show that it is in P for $k = 2$.

5.10 Coloring with bounded degrees. Show that GRAPH 3-COLORING remains NP-complete even if we restrict ourselves to graphs whose maximum degree is 4. Hint: show how to replace a vertex of degree greater than 4 with a gadget where no "internal" vertex has degree more than 4, and where several "external" vertices are forced to be the same color. Generalize this to show that GRAPH k-COLORING is NP-complete for graphs of maximum degree $k+1$, providing a converse to Problem 4.4. For $k = 3$, can you make a *planar* gadget that does this?

5.11 Hypercoloring. A *hypergraph* $G = (V, E)$ is like a graph, except that each "edge" can be a set of more than 2 vertices. A hypergraph is k*-uniform* if every edge consists of k vertices. Finally, we say that a hypergraph is 2*-colorable* if we can color its vertices black and white in such a way that no edge is monochromatic; that is, every edge contains at least one black vertex and at least one white one.

Prove that HYPERGRAPH 2-COLORING is NP-complete, even for the special case of 3-uniform hypergraphs. (Don't reply that this is just NAE-3-SAT in disguise, since NAESAT allows variables to be negated.)

5.12 All the colors of the rainbow. For another hypergraph problem, say that a k-uniform hypergraph is *rainbow colorable* if its vertices can be colored with k different colors, so that every edge contains one vertex of each color. Show that for any $k \geq 3$, HYPERGRAPH RAINBOW k-COLORING is NP-complete.

5.13 The color wheel. Given a graph $G = (V, E)$ and an integer k, a *wheel* k*-coloring* assigns a color $c(v) \in \{0, 1, 2, \ldots, k-1\}$ to each $v \in V$, with the requirement that if $(u, v) \in E$ then either $c(u) \equiv_k c(v)+1$ or $c(u) \equiv_k c(v)-1$. Thus colors of neighboring vertices must not only be different—they must be adjacent on the "color wheel." If $k = 3$, this is the same as the usual GRAPH 3-COLORING, but for $k > 3$ it is more constrained. For instance, suppose $k = 5$. Then if $c(u) = 0$, $c(v)$ must be 1 or 4. If $c(u) = 1$, then $c(v)$ must be 0 or 2. And so on.

Prove that WHEEL 5-COLORING, the question of whether a given graph G has a valid wheel 5-coloring, is NP-complete. This is a case of a more general problem, called GRAPH HOMOMORPHISM or GRAPH H-COLORING: given a graph $G = (V, E)$ and a graph $H = (W, F)$, is there a mapping $\phi : V \to W$ such that if $(u, v) \in E$, then $(\phi(u), \phi(v)) \in F$? GRAPH k-COLORING corresponds to the case where H is the complete graph with k vertices, and WHEEL k-COLORING corresponds to the case where H is a cycle of size k.

5.14 INDEPENDENT SET direct from SAT. Using the same gadgets as shown in Figure 5.14 for the reduction NAE-3-SAT $\leq$ MAX CUT, it is possible to reduce 3-SAT directly to INDEPENDENT SET. Explain how.

5.15 Clique covers. We say that a graph G is covered by a set of cliques if each clique is a subgraph of G, and if every vertex in G is contained in one of the cliques, as shown in Figure 5.24. Now consider the following problem:

CLIQUE COVER
Input: A graph $G = (V, E)$ and an integer k
Question: Can G be covered with k or fewer cliques?

Show that this problem is NP-complete. Hint: note that if we can cover a graph with k cliques, we can also cover it with k disjoint cliques.

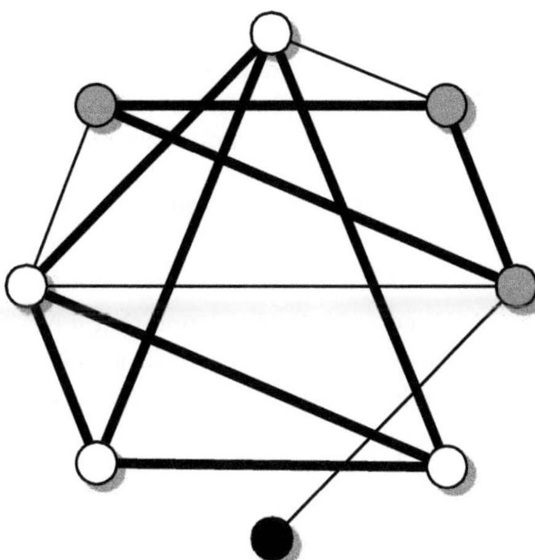

Figure 5.24: A graph covered by three cliques of sizes 4, 3, and 1. Their edges are shown in bold.

5.16 Cliques and colorings. In Problem 4.22, we defined the chromatic number $\chi(G)$ of a graph G as the smallest k such that G is k-colorable. Similarly, the *clique number* $\omega(G)$ is the size of the largest clique. Clearly $\chi(G) \geq \omega(G)$, since each vertex of a clique needs a different color. We will see in Problem 9.44 that for any graph G such that $\chi(G) = \omega(G)$, we can solve Graph Coloring and Clique in polynomial time. Prove, unfortunately, that telling if G has this property is NP-complete. First show that it is in NP.

5.17 Really independent sets. Given a graph $G = (V, E)$, say that a subset $S \subseteq V$ is *really independent* if there are no $u, v \in S$ that are two or fewer steps apart in the graph—that is, if for all $u, v \in S$, the length of the shortest path between u and v is at least 3. The problem Really Independent Set takes a graph G and an integer k as input, and asks if G has a really independent set of size k or more. Prove that this problem is NP-complete.

5.18 Crossover logic. Complete the crossover gadget of Figure 5.11, and hence the reduction from Circuit SAT to Planar Circuit SAT, by designing a planar circuit of and, or, and not gates than implements an xor gate. Note that the input and output wires need to be on the exterior of the gadget so that they can be wired up to other gates in a large circuit without a wire crossing. Hint: you must use at least one not gate.

5.19 Easy tilings. Suppose that our tile set consists of a single polyomino, and that rotations and reflections are not allowed. Show that in this case, Tiling is in P.

5.20 Tilings in a box. Let Rectangle Tiling be the problem of telling whether a region R can be covered with one tile each from a set of tiles T, where both R and each tile in T are rectangles. Show that Rectangle Tiling is NP-complete if the heights and widths of these rectangles are given in binary.

5.21 Straight trominoes. It turns out that Tiling is also NP-complete if our only tiles are *straight* trominoes, i.e., 1×3 and 3×1 rectangles. Assuming this, give an alternate proof that Perfect 3-Matching is NP-complete by showing  5.4

$$\text{Straight Tromino Tiling} \leq \text{Perfect 3-Matching}.$$

Hint: color the squares of the lattice with three colors so that each tile is forced to contain one square of each color.

5.22 Rep-tiles. Prove that if I remove any square from an 8×8 checkerboard, I can tile the remainder of the board with right trominoes. In fact, generalize this to $2^k \times 2^k$ checkerboards for any k. Is this tiling unique? This problem doesn't have that much to do with computational complexity, but the answer is recursive and quite beautiful. Hint: show that the tromino is a so-called "rep-tile," where a larger one can be tiled with smaller ones.

5.23 Linear Diophantine equations. Let a, b, and c be natural numbers. Show that the linear equation

$$ax+by=c \tag{5.15}$$

has integer solutions x and y if and only if $\gcd(a,b)$ divides c, and that it has either zero or infinitely many integer solutions. Then give a polynomial time algorithm that returns a solution (x,y) where $x,y\geq 0$ or reports that no such solution exists. Hint: if you haven't done Problem 2.2 on the extended Euclidean algorithm yet, do it now.

5.24 Quadratic Diophantine equations, step 1. Prove Lemma 5.1. First show that, if $m\geq 2$, we can ensure that $2H<K$ by choosing n consecutive primes q_i of size polynomial in n. Use the fact that, by the Prime Number Theorem, the tth prime is of size $O(t\log t)$.

To show that (5.8) implies (5.6), note that $H^2-X^2=(H+X)(H-X)$, and that each q_i divides either $H+X$ or $H-X$ but not both. Use these two possibilities to define σ_i.

To show that we can assume s is odd, reduce the general case of SUBSET SUM to the case where there is a single odd weight that we never use. Show that in that case, (5.9) implies that $X\equiv \pm s \bmod 2^m$ or $X=-s$, so that either the set given by X or its complement with the θ_is flipped is a solution.

5.25 Quadratic Diophantine equations, step 2. Show that the constraints $Y\geq 0$ and $X\leq H$ in (5.12) are equivalent if λ_1 and λ_2 are chosen properly. Hint: set $\lambda_2=-1$ and λ_1 large enough, and remember that $Y\geq 0$ means that the left-hand side of (5.10) is nonnegative. First show that $|S|\leq H$ (whether or not there is a solution!) and again use $2H<K$.

5.26 The Chinese Remainder algorithm. The Chinese Remainder Theorem (Appendix A.7.3) says that if $q_1,\ldots,q_\ell$ are mutually prime, the system of congruences $x\equiv r_i \bmod q_i$ for all $1\leq i\leq \ell$ has a unique solution $0\leq x<Q$ where $Q=q_1\cdots q_\ell$. Show that given the list of q_i and r_i, we can find x in polynomial time, as a function of the total number of bits of the q_i and r_i.

Hint: for each $1\leq i\leq \ell$, define $Q_i=Q/q_i$. Then use the extended Euclidean algorithm (Problem 2.2) to find the multiplicative inverse of Q_i modulo q_i. Finally, write x as a linear combination of the Q_i.

5.27 Augmented matchings. If we apply the reduction from MAX BIPARTITE MATCHING to MAX FLOW discussed in Section 3.8, a partial matching corresponds to a flow of one unit along each edge in the matching, as well as along the edges that attach these edges to s and t. The Ford–Fulkerson algorithm then looks for augmenting paths, i.e., paths from s to t along edges of nonzero capacity in the residual graph. Show that adding flow along an augmenting path is the same as flipping the edges along an alternating path, and increasing the size of the matching. Hint: consider Figure 5.25.

5.28 Steiner trees. In Section 3.5.1 we discussed the problem MINIMUM SPANNING TREE, in which we want to design the lightest network that reaches every vertex in a graph. But what if not everyone wants to be served by our network? Suppose that we again want to build the lightest possible tree, but that we are now only obligated to reach a subset S of the vertices. This gives us the following problem:

> STEINER TREE
>
> Input: A weighted graph $G=(V,E)$, a subset $S\subseteq V$, and an integer k
>
> Question: Is there a subtree of G of weight k or less which covers S?

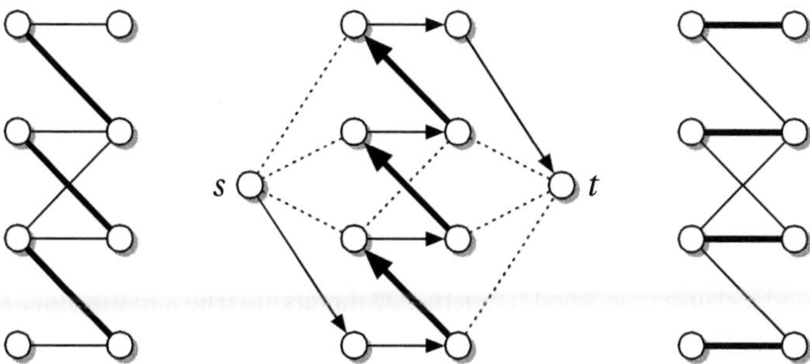

FIGURE 5.25: Increasing flow along an augmenting path corresponds to flipping edges along an alternating path, which increases the size of a partial matching.

Prove that STEINER TREE is NP-complete, even in the unweighted case where every edge has weight 1.

5.9 Hint: reduce from TROMINO TILING. Given a region R, define a graph with a central vertex, one vertex for each possible location and orientation of a tromino, and one vertex for each of R's squares. Decide how to connect these vertices together, and find a threshold k such that a Steiner tree of weight k exists if and only if the region can be tiled.

5.29 One-dimensional automata. Show that for one-dimensional cellular automata, CA PREDECESSOR can be solved in polynomial time. Hint: use dynamic programming, or reduce to REACHABILITY. Start by assuming that, like the example discussed in Section 5.5.3, the transition function f depends only on the nearest neighbors, but comment on how the complexity of this problem grows with the size of the neighborhood and the alphabet of local states.

5.30 Spin glasses. Recall the definition of an Ising spin glass from Problem 3.50. We have a graph G where each edge between vertices i, j has an interaction strength J_{ij}. Given a set of spins $s_i = \pm 1$ on the vertices, the energy is $E = -\sum_{ij} J_{ij} s_i s_j$. We wish to find the ground state, i.e., the values of the s_i that minimize the energy. Let's phrase this as a decision problem, in which we ask whether a state exists below a certain energy:

> SPIN GLASS
>
> Input: A graph G with interaction strengths J_{ij} and an energy E
>
> Question: Is there a state $\{s_i\}$ such that $-\sum_{ij} J_{ij} s_i s_j \le E$?

We showed in Problem 3.50 that in the ferromagnetic case $J_{ij} \ge 0$, this problem is in P even if we also have external fields h_i. However, things get much more complicated if some edges are antiferromagnetic. Show that this version of SPIN GLASS is NP-complete. Hint: show that this is true even if all the edges are antiferromagnetic, with $J_{ij} = -1$ for every edge (i, j). 5.10

5.31 No directions. Prove that UNDIRECTED HAMILTONIAN PATH is NP-complete by reducing DIRECTED HAMILTONIAN PATH to it. Hint: given a directed graph G, define an undirected graph G' where each vertex of G corresponds to a three-vertex gadget in G'. Figure out how to connect these gadgets so that Hamiltonian paths in G' are forced to simulate those in G.

5.32 The incremental path to enlightenment. If we have solved one instance of an NP-complete problem, how much does it help us solve another instance which is very similar? For instance, suppose we could figure out how adding a single edge (u, v) to a graph G changes the solution to INDEPENDENT SET. We can phrase this as the following problem:

INDEPENDENT SET CHANGE
Input: A graph G, a maximal independent set S of G, and a pair of vertices u, v
Output: A maximal independent set S' of the graph $G' = G \cup (u, v)$

Show that if this problem is in P then so is INDEPENDENT SET, in which case P = NP.

5.33 Half and half. Consider the following variant of SAT, called HALF-AND-HALF k-SAT. Each clause contains k literals where k is even, and a clause is satisfied if exactly $k/2$ literals are true and the other $k/2$ are false. Show that this problem is NP-complete whenever $k \geq 4$.

5.34 Quadratics and quartics. While quadratic equations over a single real variable are easy, systems of multiple quadratic equations over multiple variables $x_1, \ldots, x_n$ are NP-hard, i.e., any problem in NP can be reduced to them. Prove this by reducing from 3-SAT or SUBSET SUM. Then show that telling whether a single quartic equation, i.e., an equation of degree 4, has a solution over real variables $x_1, \ldots, x_n$ is NP-hard.

5.35 Matrix completion. Suppose we are given some of the entries M_{ij} of a matrix M, and we are asked to fill in the rest of the entries in such a way that M has rank 2. Show that it is NP-hard to tell if this is possible. Hint: M is of rank k if and only if there are k-dimensional vectors u_i and v_j such that $M_{ij} = u_i^T v_j$. Reduce from GRAPH 3-COLORING by replacing the colors with three unit-length vectors $2\pi/3$ apart.

Notes

5.1 History. The concept of NP-completeness was developed in the early 1970s by Stephen Cook [198] and Richard Karp [462], and independently by Leonid Levin [527]. Cook and Karp received the Turing Award in 1982 and 1985 respectively. Our NP-completeness proofs are taken from a variety of sources, including Sipser [759], Papadimitriou [646], and Moret [616]. A few of them are original, though obvious to the experienced practitioner.

5.2 Reductions. Reductions that map single instances of A to single instances of B, mapping yes-instances to yes-instances and no-instances to no-instances, are called *many–one reductions* or *Karp reductions.* If there is an oracle that can solve instances of B, such a reduction allows us to ask the oracle a single question, and we can immediately repeat the oracle's answer as our answer to A without any further processing.

Reductions in which we can use B as a subroutine, call it a polynomial number of times, and do anything we like with its answers, are called *Turing reductions.* If A is Turing-reducible to B and B is in P, then A is in P as well. For instance, Problems 1.10 and 4.1, where we find a Hamiltonian path or a satisfying assignment by asking an oracle a polynomial number of questions about whether or not one exists, are Turing reductions from these search problems to the decision problems HAMILTONIAN PATH and SAT.

Note that Turing reductions don't have to preserve the asymmetry of NP. For instance, by switching the oracle's answer from "yes" to "no," we can reduce a coNP-complete problem such as UNSAT to an NP-complete one. However, if A is Turing-reducible to B and $B \in$ P, then $A \in$ P as well.

Since Turing reductions are, *a priori,* more powerful than Karp reductions, it is conceivable that there are problems that are NP-complete under Turing reductions but not under Karp reductions. However, for nearly all known NP-complete problems, Karp reductions suffice. In any case, regardless of which kind of reduction we use in our definition of NP-completeness, we have P = NP if any NP-complete problem is in P.

5.3 Polyominoes. Polyominoes are a marvelous source of beautiful puzzles and of deep questions in combinatorics and statistical physics. The authoritative reference is the book by Solomon Golomb [344]. Up to rotation and reflection, there are one domino, two trominoes, five tetrominoes, and twelve pentominoes:

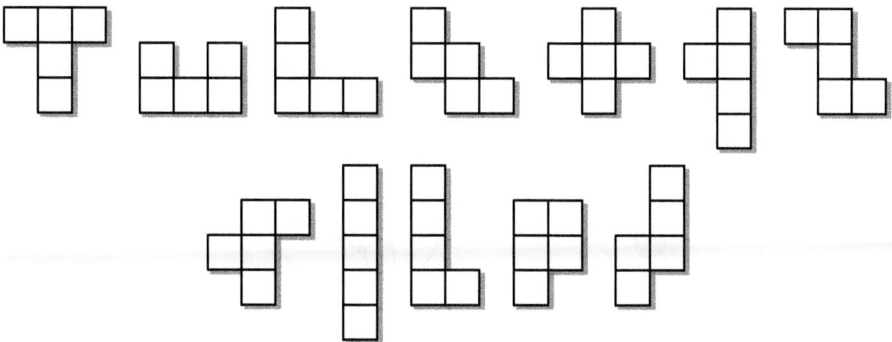

The asymptotic number of n-ominoes grows exponentially as K^n for some constant K, called *Klarner's constant*. The value of K is unknown, but is known to be between 3.98 [81] and 4.65 [485], and is estimated to be about 4.0626 [429].

5.4 Tilings. The proof we give here that TILING is NP-complete is from Moore and Robson [606]. Using a more complicated reduction, they showed that TILING is still NP-complete even if our only tile is the right tromino. In 3 or more dimensions, they showed NP-completeness for dominoes and straight trominoes, i.e., blocks consisting of two or three cubes in a row.

Beauquier, Nivat, Rémila, and Robson [92] showed that TILING is NP-complete if our tiles are straight trominoes, i.e., 1×3 rectangles, or even horizontal dominoes and vertical trominoes (where in this case we do not allow rotations). On the other hand, Kenyon and Kenyon [475] showed that TILING with these tile sets is in P in the case of simply-connected regions, i.e., regions without holes.

In Section 7.6.5, we will prove that tiling problems on the infinite plane can be *undecidable*, i.e., beyond the reach of any finite-time algorithm.

5.5 Diophantine equations and complex integrals. Diophantus of Alexandria was a Greek mathematician who lived in the 3rd century A.D. He worked on what we call today Diophantine equations: polynomial equations with integer coefficients, for which we seek integer solutions. The most famous Diophantine equation asks for integers x, y, and z such that

$$x^n + y^n = z^n \tag{5.16}$$

for a given integer exponent n. For $n = 2$, the solutions are known as Pythagorean triples, and Euclid proved that there are infinitely many of them, such as $(3,4,5)$, $(5,12,13)$, and $(8,15,17)$. The claim that no nontrivial solution exists for $n \geq 3$ entered the history of mathematics in 1637, when Pierre de Fermat wrote it on the margin of his copy of Diophantus' *Arithmetica*. He also wrote "I have a truly marvelous proof of this proposition which this margin is too narrow to contain." Fermat never published his marvelous proof, and it seems doubtful now that he had one. It took generations of mathematicians, and 357 years, before a proof was finally found by Andrew Wiles.

The proof that QUADRATIC DIOPHANTINE EQUATION is NP-complete is by Manders and Adleman [552]. We are grateful to Sebastian Oberhoff for a very careful reading of this section, vigilantly finding mistakes until we got it right. In general, telling whether a Diophantine equation has any solutions is undecidable, as we discuss briefly in Chapter 7. The proof that COSINE INTEGRATION is NP-complete is due to Plaisted [662].

5.6 Paths, trees, and flowers. In a non-bipartite graph, not all alternating paths can be used to improve a matching. Consider Figure 5.26, in which an alternating path goes around an odd loop. We can't actually swap the edges along it, and indeed the matching shown is maximal. Edmonds called such loops *flowers*, and devised a polynomial algorithm to detect and remove them. This leaves us just with genuinely *augmenting* paths, i.e., alternating paths that don't visit the same vertex twice, giving a polynomial-time algorithm for finding perfect matchings that maximize

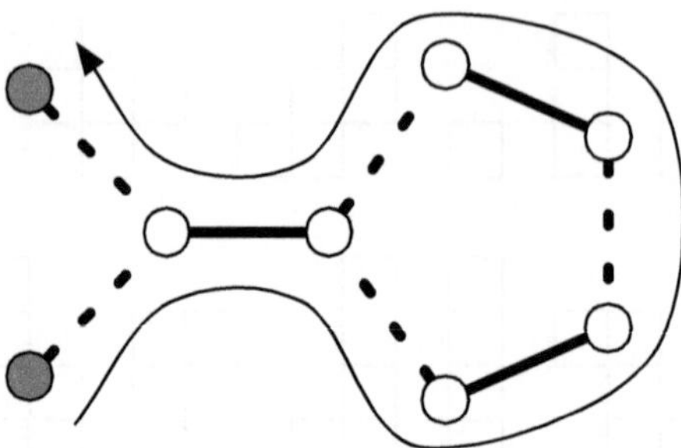

FIGURE 5.26: A partial matching of a non-bipartite graph. There is an alternating path between the two isolated vertices, but because it goes around an odd loop it doesn't help us improve the matching.

the total weight of their edges—or that minimize it, also giving a polynomial-time algorithm [261] for the Chinese postman problem discussed in Problem 1.9. In fact, Edmonds' "paths, trees, and flowers" algorithm is one of the first algorithms that was explicitly shown to run in polynomial time, and we quoted from his paper [260] in Note 2.10.

5.7 Cellular automata. Cellular automata go back to Stanislaw Ulam and John von Neumann, who studied them as models of crystal growth and self-replication. The Game of Life was invented by John Horton Conway. As we will see in Section 7.6.4, its dynamics are rich enough to allow for universal computation.

One-dimensional cellular automata were studied extensively by Stephen Wolfram in the 1980s (e.g. [831]), who classified the "elementary" rules with two states and nearest-neighbor interactions. There are 8 possible settings of the neighborhood, consisting of a site and its two neighbors, so there are $2^8 = 256$ possible rules. Even among these elementary rules there is at least one, rule 110, which is capable of universal computation. We discuss this in Section 7.6.4 as well.

5.8 Schaefer's Dichotomy Theorem. Having seen several variants of SAT, including those in Problems 5.2, 5.3, and 5.4, the reader might be curious to know which are NP-complete and which are not (assuming that $\mathsf{P} \neq \mathsf{NP}$). Suppose we have a set S of clause types, where each type specifies how many variables it has, and which truth values would satisfy it. For instance, 3-SAT has 4 clause types, in which 0, 1, 2, or 3 variables are negated. Then let's define S-SAT as the variant of SAT where each formula consists of clauses with types in S. Schaefer [727] showed that S-SAT is in P if and only if one of the following properties applies to the clause types in S:

1. they can all be satisfied by the all-true assignment,
2. they can all be satisfied by the all-false assignment,
3. they can all be expressed using 2-SAT clauses,
4. they can all be expressed using XORSAT clauses (page 366),
5. they can all be expressed using HORN-SAT clauses, or
6. they can all be expressed using anti-HORN-SAT clauses,

where an anti-HORN-SAT clause is one with at most one negated variable. Otherwise, S-SAT is NP-complete. Thus, in a sense, every variant of SAT with Boolean variables is either in P or is NP-complete.

Bulatov [145] proved a similar result for constraint satisfaction problems where variables can take 3 possible values, and it is conjectured that similar results hold no matter how many values the variables can take. However, one reason to question this conjecture is a result from Ladner [515], which we discuss in Section 6.6, showing that if $\mathsf{P} \neq \mathsf{NP}$ then there exist problems which float in between—which are outside P but which are not NP-complete either. For instance, FACTORING and GRAPH ISOMORPHISM are believed to lie in this intermediate zone.

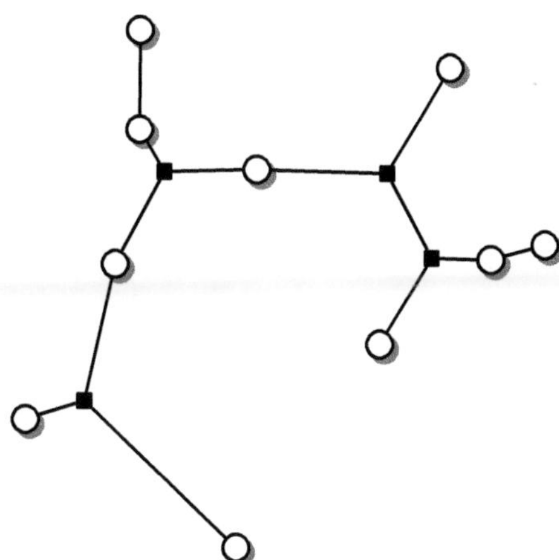

FIGURE 5.27: Euclidean minimum Steiner tree. In this example, 10 terminal vertices (circles) are optimally connected with the help of 4 Steiner points (squares).

5.9 Steiner trees. The problem STEINER TREE defined in Problem 5.28 is also NP-complete in the planar case. EUCLIDEAN STEINER TREE is a beautiful variant in which we are given points in the plane, we can use any point in the plane as an interior point of the tree, and the weight of each edge is its Euclidean length, see Figure 5.27. This version of the problem is NP-*hard*, meaning that any problem in NP can be reduced to it, but it is not known to be in NP. Do you see why it might not be? An excellent reference on STEINER TREE and all its variants is Prömel and Steger [671], the NP-hardness of EUCLIDEAN STEINER TREE is shown in [319].

5.10 Spin glasses. In physics, a spin glass is the canonical example of a system with a bumpy landscape. In a typical example, there are exponentially many local optima, where flipping the spin of any one vertex increases the energy. These optima are separated by high energy barriers, making it difficult for search algorithms to escape from them and find the global optimum. The ensemble of spin glasses where G is the complete graph and the J_{ij} are chosen randomly is known as the *Sherrington–Kirkpatrick model* [748].

Barahona [77] showed that SPIN GLASS is NP-complete on three-dimensional lattices, or more to the point, whenever G is not required to be planar. He also showed that it is NP-complete in the planar case if we include external fields as in Problem 3.50. However, as we will see in Section 13.7, the planar case without external fields is in P.

Chapter 6

The Deep Question: P vs. NP

> The evidence in favor of Cook's and Valiant's hypotheses is so overwhelming, and the consequences of their failure are so grotesque, that their status may perhaps be compared to that of physical laws rather rather than that of ordinary mathematical conjectures.
>
> Volker Strassen

The question of whether or not $\mathsf{P} \neq \mathsf{NP}$ is the central question of computational complexity theory. More broadly, it is widely recognized as one of the deepest unsolved problems in all of mathematics. It is one of seven Millennium Prize Problems which the Clay Mathematics Institute put forward as challenges for the 21st century. This ranks it along with the Riemann Hypothesis regarding the distribution of the prime numbers and the zeros of the zeta function; the Poincaré Conjecture (recently proved by Grigori Perelman) regarding high-dimensional spheres; the existence and uniqueness of solutions to the Navier–Stokes equations, which describe turbulent flows in hydrodynamics; and the solution to Yang–Mills theory, one of the basic building blocks of high-energy physics.

One could argue that the P vs. NP question is more fundamental than all of these. Primes, spheres, 6.1
turbulence, and elementary particles are important mathematical and physical objects, and they pose deep and difficult problems. But the P vs. NP question is about the nature of problem-solving itself. Since P is the class of problems where we can *find* solutions efficiently, and NP is the class of problems where we can *check* them efficiently, we can phrase it as follows:

Is it harder to find solutions than to check them?

Our intuition that this is so is very strong. It's harder to find a needle in a haystack, or tell if one exists, than it is to prick our finger with it. But proving this intuition—proving that finding Hamiltonian paths, say, or satisfying assignments to a 3-SAT formula, really is like finding a needle in a haystack—has turned out to be a very difficult problem in itself.

We start this chapter by discussing the consequences of the possibility that $\mathsf{P} = \mathsf{NP}$. We show that this would force us to revise how we think, not just about computer science, but about truth, creativity, and intelligence. We then explore why $\mathsf{P} \neq \mathsf{NP}$ is so hard to prove, including a set of "metatheorems" which show that most of the proof techniques people have proposed so far are doomed to fail.

We then take a look at the internal structure of NP. We show that it has subclasses that capture various types of mathematical proof, and that, unless P = NP, there are problems floating in limbo between P and NP-completeness. Finally, we discuss what loopholes might exist in the P vs. NP question, and why we believe, despite the enormous difficulties, that someday it will be resolved.

6.1 What if P=NP?

The most prosaic way to phrase the P vs. NP question is to ask whether SAT—or GRAPH COLORING, or CLIQUE, or TILING—can be solved in polynomial time. From the point of view of chip designers or suitcase packers, it is quite frustrating that these problems seem to be exponentially hard. Perhaps we have simply not been clever enough. Perhaps there is some mathematical insight that puts one of these problems, and therefore all of them, in P. Is P = NP really so unthinkable?

As we alluded to earlier, if P = NP the consequences would be far greater than better party planning and an easier time tiling our bathrooms. In this section we show that our picture of computational complexity would change drastically—not just for these two classes, but throughout the complexity hierarchy. More importantly, we would be forced to rethink the nature of intelligence, and rethink the role that creativity plays in mathematics and the sciences.

6.1.1 *The Great Collapse*

> There was a sound like that of the gentle closing of a portal as big as the sky, the great door of heaven being closed softly. It was a grand AH-WHOOM.
>
> Kurt Vonnegut, *Cat's Cradle*

In this section, we will show that if P = NP then many other complexity classes are equal to P as well. To start with, recall that coNP is the class of problems, such as NO HAMILTONIAN PATH, where there is a proof or witness we can check in polynomial time whenever the answer is "no." As Exercise 4.17 shows, if P = NP then P = coNP as well, since we can switch the outputs "yes" and "no" of our favorite polynomial-time algorithm.

Beyond NP and coNP, there is an infinite tower of complexity classes, of which P and NP are just the first two levels—or more precisely, the zeroth and the first. As we discussed in Section 4.3, problems in NP correspond to properties of the form $\exists w : B(x, w)$, where w is the witness and B is in P. Similarly, problems in coNP correspond to properties of the form $\forall w : B(x, w)$. Thus NP or coNP can be thought of as P with a single $\exists$ or $\forall$ in front of it.

We can define higher complexity classes with increasing numbers of quantifiers. Consider the following problem, which we briefly discussed in Section 4.3.2:

> SMALLEST BOOLEAN CIRCUIT
>
> Input: A Boolean circuit C that computes a function f_C of its input
>
> Question: Is C the smallest circuit that computes f_C?

Logically, we can express the claim that C is a yes-instance as follows:

$$\forall C' < C : \exists x : f_{C'}(x) \neq f_C(x).$$

We can think of this formula as a two-move game. First, the Skeptic chooses a circuit C' smaller than C, and claims that it computes the same function that C does. Then the Prover responds with an input x on which C' and C behave differently, i.e., for which $f_{C'}(x) \neq f_C(x)$. The statement is true if the Prover has a winning strategy—if she can demonstrate that every possible circuit C' fails on some input.

Since SMALLEST BOOLEAN CIRCUIT can be expressed with two quantifiers, a $\exists$ inside a $\forall$, it is in the following complexity class:

> Π_2P is the class of properties A of the form
>
> $$A(x) = \forall y : \exists z : B(x, y, z),$$
>
> where B is in P, and where $|y|$ and $|z|$ are polynomial in $|x|$.

When we say that $|y|$ and $|z|$ are polynomial in $|x|$, we mean that there is some polynomial function $f(n)$ such that the quantifiers $\forall y$ and $\exists z$ range over strings of length at most $f(|x|)$. In this case, y and z are strings that encode Boolean circuits and inputs respectively, and each of these is shorter than the description of the circuit C.

Continuing this way, we can define classes with k layers of quantifiers, alternating between $\exists$ and $\forall$:

> Σ_kP is the class of properties A of the form
>
> $$A(x) = \exists y_1 : \forall y_2 : \exists y_3 : \cdots : \mathsf{Q} y_k : B(x, y_1, \ldots, y_k), \quad (6.1)$$
>
> where B is in P, $|y_i| = \text{poly}(|x|)$ for all i, and $\mathsf{Q} = \exists$ or $\forall$ if k is odd or even respectively. Similarly, Π_kP is the class of properties A of the form
>
> $$A(x) = \forall y_1 : \exists y_2 : \forall y_3 : \cdots : \mathsf{Q} y_k : B(x, y_1, \ldots, y_k), \quad (6.2)$$
>
> where B is in P, $|y_i| = \text{poly}(|x|)$ for all i, and $\mathsf{Q} = \forall$ or $\exists$ if k is odd or even respectively.

The notation Σ_kP comes from the fact that the outer quantifier $\exists y_1$ is an OR, or a sum, over all possible y_1. Similarly, we write Π_kP since $\forall y_1$ is an AND, or a product, over all possible y_1.

Problems in these classes correspond to two-player games that last for k moves. For instance, if a Chess problem claims that White can mate in k moves, this means that there exists a move for White, such that for all of Black's replies, there exists a move for White, and so on until White has won. Given the initial position and the sequence of moves the players make, we can check in polynomial time whether White has checkmated her opponent. So if we are playing Generalized Chess on an $n \times n$ board, this type of claim is in Σ_kP. See also Problem 6.2.

Some simple remarks are in order. First, increasing the number of quantifiers can only make these classes larger—we can always add a quantifier on a dummy variable, inside or outside all the existing quantifiers. Therefore, for all $k \geq 0$ we have

$$\Sigma_k \subseteq \Sigma_{k+1}, \Sigma_k \subseteq \Pi_{k+1}, \Pi_k \subseteq \Sigma_{k+1}, \Pi_k \subseteq \Pi_{k+1}. \tag{6.3}$$

Secondly, as in Section 4.3, we can think of each $\exists$ as a layer of nondeterminism, that asks whether there is a witness that makes the statement inside that quantifier true. In analogy to the N in NP, we can write

$$\Sigma_k \mathsf{P} = \mathsf{N} \cdot \Pi_{k-1} \mathsf{P}.$$

In particular, since $\Sigma_0 \mathsf{P} = \Pi_0 \mathsf{P} = \mathsf{P}$ we have

$$\Sigma_1 \mathsf{P} = \mathsf{NP} \text{ and } \Pi_1 \mathsf{P} = \mathsf{coNP}.$$

More generally,

$$\Pi_k \mathsf{P} = \mathsf{co}{-}\Sigma_k \mathsf{P},$$

since the negation of a $\forall$ is a $\exists$ and vice versa. For instance, the statement that C is *not* the smallest possible circuit that computes f_C can be expressed as

$$\begin{aligned} &\overline{\forall C' < C : \exists x : f_{C'}(x) \neq f_C(x)} \\ &= \exists C' < C : \overline{\exists x : f_{C'}(x) \neq f_C(x)} \\ &= \exists C' < C : \forall x : f_{C'}(x) = f_C(x), \end{aligned}$$

so NOT THE SMALLEST BOOLEAN CIRCUIT is in $\Sigma_2 \mathsf{P}$. Similarly, the claim that White cannot mate in k moves is in $\Pi_k \mathsf{P}$.

These complexity classes are known, collectively, as the *polynomial hierarchy*. Taking their union over all k gives the class PH, which consists of problems that can be phrased with any constant number of quantifiers:

6.2

$$\mathsf{PH} = \bigcup_{k=0}^{\infty} \Sigma_k \mathsf{P} = \bigcup_{k=0}^{\infty} \Pi_k \mathsf{P}.$$

The fact that these two unions are identical follows from (6.3).

Just as we believe that $\mathsf{P} \neq \mathsf{NP}$, and that $\mathsf{NP} \neq \mathsf{coNP}$, we believe that the classes Σ_k and Π_k are all distinct—that they form an infinite hierarchy of increasingly powerful complexity classes as shown in Figure 6.1. That is, we believe that each time we add a quantifier, or a layer of nondeterminism, we get a fundamentally deeper kind of problem. While problems in NP require witnesses of polynomial size, problems in $\Sigma_k \mathsf{P}$ and $\Pi_k \mathsf{P}$ require winning strategies k moves deep, which are much larger objects. Even SMALLEST BOOLEAN CIRCUIT requires the Prover to have, for each smaller circuit C', an input x on which C' gets the wrong answer. Since there are an exponential number of circuits of a given size, such a strategy takes, *a priori*, an exponential number of bits to describe.

But what happens to all this if $\mathsf{P} = \mathsf{NP}$? This is tantamount to saying that

$$\text{if } B(x,y) \text{ is in } \mathsf{P}, \text{ then } A(x) = \exists y : B(x,y) \text{ is also in } \mathsf{P},$$

assuming, as always, that $|y| = \mathrm{poly}(|x|)$. In other words, if $\mathsf{P} = \mathsf{NP}$, then P can absorb existential quantifiers, as long as they range over witnesses of polynomial size. Since we also have $\mathsf{P} = \mathsf{coNP}$, we can absorb universal quantifiers too.

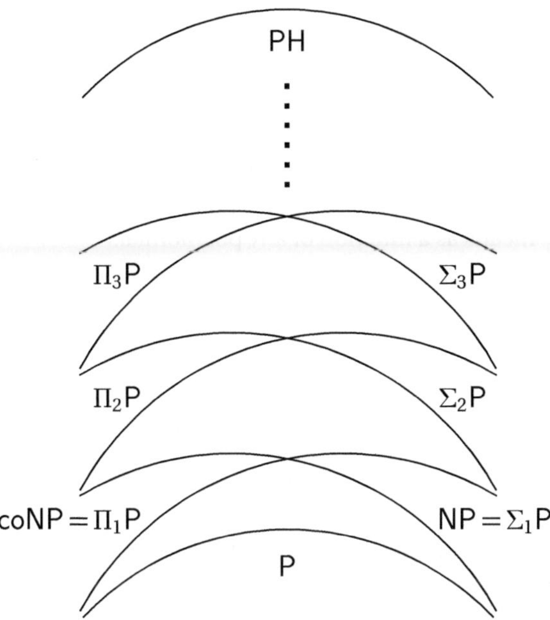

FIGURE 6.1: The polynomial hierarchy. P is the zeroth level, and NP and coNP are at the first level. We believe that all of these levels are distinct, so that they form an infinite tower of increasingly powerful complexity classes.

By absorbing the innermost quantifier in this way, we can reduce the number of quantifiers until there are none left, leaving us just with a polynomial-time property of the input x. Thus, if P = NP, the entire polynomial hierarchy collapses to P. To the extent that we find this possibility unbelievable, we can take it as additional evidence that P $\neq$ NP.

Similarly, the following exercise shows that if NP = coNP—in which case $\exists$s can always be replaced with $\forall$s—then the consequences for the polynomial hierarchy would be almost as great as if P = NP:

Exercise 6.1 *Suppose that* NP = coNP. *Show that in that case, the polynomial hierarchy collapses to its first level. That is,*

$$\mathsf{PH} = \mathsf{NP} = \mathsf{coNP}.$$

Hint: two quantifiers of the same type can be merged into one, since we can treat a pair of variables as a single variable. For instance, we can write $\exists x : \exists y$ *as* $\exists(x, y)$.

We take this as evidence that no problem in NP $\cap$ coNP can be NP-complete, since that would imply that NP = coNP. For instance, it seems very unlikely that FACTORING is NP-complete, since related decision problems like SMALL FACTOR (see Problem 4.21) are in NP $\cap$ coNP.

6.1.2 As Below, So Above

A common misconception is that the P vs. NP question is about polynomial time. However, it is really a question about the power of nondeterminism in general, and whether finding solutions is harder than checking them, regardless of how much time we have. In this section, we will show that if $\mathsf{P} = \mathsf{NP}$, then $\mathsf{TIME}(f(n)) = \mathsf{NTIME}(f(n))$ for essentially any class of functions $f(n)$ that grow at least polynomially.

Recall from Section 2.4.3 that EXP is the class of problems we can solve in an exponential amount of time, where "exponential" means as a function of any polynomial:

$$\mathsf{EXP} = \bigcup_k \mathsf{TIME}(2^{n^k}) = \mathsf{TIME}(2^{\mathrm{poly}(n)}),$$

and that $\mathsf{NEXP} = \mathsf{NTIME}(2^{\mathrm{poly}(n)})$ is the set of problems where we can check a solution or witness in exponential time. Then in analogy to $\mathsf{P} \neq \mathsf{NP}$, we can ask whether $\mathsf{EXP} \neq \mathsf{NEXP}$. Similarly, we can ask whether $\mathsf{EXPEXP} \neq \mathsf{NEXPEXP}$, and so on up the hierarchy.

However, these questions are not independent of each other. We will prove the following:

If $\mathsf{P} = \mathsf{NP}$, then $\mathsf{EXP} = \mathsf{NEXP}$, $\mathsf{EXPEXP} = \mathsf{NEXPEXP}$, and so on.

The proof is very simple. Recall from Section 2.4.3 that since we measure complexity as a function of the input size, hard problems become "easy" if we express the input in a very inefficient way. By "padding" the input to make it longer, we can translate a function in NEXP down to one in NP. We can then solve it deterministically in polynomial time as a function of its new, padded size—which is exponential as a function of its original size.

For instance, suppose that a problem A is in NEXP, so that witnesses can be checked in time $t(n) = 2^{O(n^c)}$ for some constant c. Now pad the input, making it $t(n)$ bits long, by adding $t(n) - n$ zeros. The resulting problem A' is in NP, since witnesses can be checked in time $t(n)$, linear in the size of its input. But if $\mathsf{P} = \mathsf{NP}$, then A' can be solved deterministically in time $\mathrm{poly}(t(n)) = 2^{O(n^c)}$. Thus A is in EXP, and $\mathsf{EXP} = \mathsf{NEXP}$.

We skipped one step in this proof—namely, the time it takes to pad the input. We can add $t(n) - n$ zeros in $O(t(n))$, but we have to compute $t(n)$ first. Here we use the fact that $t(n)$ is *time constructible*, meaning that it can be computed in $O(t(n))$ time. As Problem 6.3 shows, virtually any reasonable function is time-constructible, including functions like $t(n) = n$, n^c, and 2^{n^c} which we use to define complexity classes. In that case, padding takes $O(t(n))$ time, and the total computation time is still $\mathrm{poly}(t(n))$.

More generally, we can say the following. If $\mathsf{P} = \mathsf{NP}$, then for any time-constructible function $t(n) \geq n$, we have

$$\mathsf{NTIME}(t(n)) \subseteq \mathsf{TIME}\big(\mathrm{poly}(t(n))\big).$$

If C is a class of superpolynomial functions such that $t(n)^c \in C$ for any $t(n) \in C$ and any constant c, then

$$\mathsf{NTIME}(C) = \mathsf{TIME}(C).$$

In addition to the exponentials $2^{\mathrm{poly}(n)}$, such classes include the quasipolynomials $2^{O(\log^c n)}$, double exponentials $2^{2^{\mathrm{poly}(n)}}$, and so on. So, if $\mathsf{P} = \mathsf{NP}$, then we also have $\mathsf{QuasiP} = \mathsf{NQuasiP}$, $\mathsf{EXPEXP} = \mathsf{NEXPEXP}$, and so on up the complexity hierarchy.

A similar argument, given in Problem 6.4, shows that if EXP = NEXP, then EXPEXP = NEXPEXP, and so on. Thus equality between deterministic and nondeterministic classes propagates up the complexity hierarchy. This also means that inequality propagates down: if we could prove that EXP $\neq$ NEXP, or that EXPEXP $\neq$ NEXPEXP, this would establish that P $\neq$ NP.

6.1.3 *The Demise of Creativity*

We turn now to the most profound consequences of P = NP: that reasoning and the search for knowledge, with all its frustration and elation, would be much easier than we think.

Finding proofs of mathematical conjectures, or telling whether a proof exists, seems to be a very hard problem. On the other hand, *checking* a proof is not, as long as it is written in a careful formal language. Indeed, we invent formal systems like Euclidean geometry, or Zermelo–Fraenkel set theory, precisely to reduce proofs to a series of simple steps. Each step consists of applying a simple axiom such as *modus ponens*, "If A and $(A \Rightarrow B)$, then B." This makes it easy to check that each line of the proof follows from the previous lines, and we can check the entire proof in polynomial time as a function of its length. Thus the following problem is in P:

Proof Checking
Input: A statement S and a proof P
Question: Is P a valid proof of S?

This implies that the following problem is in NP:

Short Proof
Input: A statement S and an integer n given in unary
Question: Does S have a proof of length n or less?

Note that we give the length of the proof in unary, since the running time of Proof Checking is polynomial as a function of n. Note also that Short Proof is NP-complete, since S could be the statement that a given SAT formula is satisfiable.

Now suppose that P = NP. You can take your favorite unsolved mathematical problem—the Riemann Hypothesis, Goldbach's Conjecture, you name it—and use your polynomial-time algorithm for Short Proof to search for proofs of less than, say, a billion lines. For that matter, if P = NP you can quickly search for solutions to most of the other Millennium Prize Problems as well. The point is that *no proof constructed by a human will be longer than a billion lines anyway*, even when we go through the tedious process of writing it out axiomatically. So, if no such proof exists, we have no hope of finding one.

This point was raised long before the classes P and NP were defined in their modern forms. In 1956, the logician Kurt Gödel wrote a letter to John von Neumann. Turing had shown that the *Entscheidungsproblem*, the problem of whether a proof exists for a given mathematical statement, is undecidable—that is, as we will discuss in Chapter 7, it cannot be solved in finite time by any program. In his letter Gödel considered the bounded version of this problem, where we ask whether there is a proof of length n or less. He defined $\varphi(n)$ as the time it takes the best possible algorithm to decide this, and wrote: 6.3

> The question is, how fast does $\varphi(n)$ grow for an optimal machine. One can show that $\varphi(n) \geq Kn$. If there actually were a machine with $\varphi(n) \sim Kn$ (or even only $\varphi(n) \sim Kn^2$), this would have consequences of the greatest magnitude. That is to say, it would clearly indicate that, despite the unsolvability of the *Entscheidungsproblem*, the mental effort of the mathematician in the case of yes-or-no questions could be completely replaced by machines (footnote: apart from the postulation of axioms). One would simply have to select an n large enough that, if the machine yields no result, there would then be no reason to think further about the problem.

If our mathematical language has an alphabet of k symbols, then the number of possible proofs of length n is $N = k^n$. Even excluding those which are obviously nonsense leaves us with a set of exponential size. As Gödel says, we can solve SHORT PROOF in polynomial time—in our terms, P = NP—precisely if we can do much better than exhaustive search (in German, *dem blossen Probieren*, or "mere sampling") among these N possibilities:

> $\varphi \sim Kn$ (or $\sim Kn^2$) means, simply, that the number of steps vis-à-vis exhaustive search can be reduced from N to $\log N$ (or $(\log N)^2$).

Can SHORT PROOF really be this easy? As mathematicians, we like to believe that we need to use all the tools at our disposal—drawing analogies with previous problems, visualizing and doodling, designing examples and counterexamples, and making wild guesses—to find our way through this search space to the right proof. But if P = NP, finding proofs is not much harder than checking them, and there is a polynomial-time algorithm that makes all this hard work unnecessary. As Gödel says, in that case we can be replaced by a simple machine.

Nor would the consequences of P = NP be limited to mathematics. Scientists in myriad fields spend their lives struggling to solve the following problem:

> ELEGANT THEORY
>
> Input: A set E of experimental data and an integer n given in unary
>
> Question: Is there a theory T of length n or less that explains E?

For instance, E could be a set of astronomical observations, T could be a mathematical model of planetary motion, and T could explain E to a given accuracy. An elegant theory is one whose length n, defined as the number of symbols it takes to express it in some mathematical language, is fairly small—such as Kepler's laws or Newton's law of gravity, along with the planets' masses and initial positions.

Let's assume that we can compute, in polynomial time, what predictions a theory T makes about the data. Of course, this disqualifies theories such as "because God felt like it," and even for string theory this computation seems very difficult. Then again, if we can't tell what predictions a theory makes, we can't carry out the scientific method anyway.

With this assumption, and with a suitable formalization of this kind, ELEGANT THEORY is in NP. Therefore, if P = NP, the process of searching for patterns, postulating underlying mechanisms, and forming hypotheses can simply be automated. No longer do we need a Kepler to perceive elliptical orbits, or a Newton to guess the inverse-square law. Finding these theories becomes just as easy as checking that they fit the data.

We could go on, and point out that virtually any problem in design, engineering, pattern recognition, learning, or artificial intelligence could be solved in polynomial time if $\mathsf{P} = \mathsf{NP}$. But we hope our point is clear. At its root, our belief that $\mathsf{P} \neq \mathsf{NP}$ is about the nature of intelligence itself. We believe that the problems we face demand all our faculties as human beings—all the painstaking work, flashes of insight, and leaps of intuition that we can bring to bear. We believe that *understanding matters*—that each problem brings with it new challenges, and that meeting these challenges brings us to a deeper understanding of mathematical and physical truth. Learning that there is a comparatively simple algorithm that solves every problem—a mechanical process, with which we could simply turn the crank and get the answer—would be deeply disappointing.

So, if so much is riding on the P vs. NP question, why hasn't it been resolved? In the next sections, we will discuss why this question is so hard, and why it has defeated virtually every strategy that people have proposed to resolve it. 6.4

6.2 Upper Bounds Are Easy, Lower Bounds Are Hard

> If the Theory of making Telescopes could at length be fully brought into Practice, yet there would be certain Bounds beyond which Telescopes could not perform.
>
> Isaac Newton, *Opticks*

The most direct strategy to prove that $\mathsf{P} \neq \mathsf{NP}$ is to prove, for some problem A in NP, that it is not in P. How can we do that?

One of the overarching themes of computer science is that lower bounds on a problem's complexity are much harder to prove than upper bounds. To prove that a problem is in P, all we have to do is exhibit a polynomial-time algorithm that solves it. But to prove that it is not in P, we have to reason about *all possible* polynomial-time algorithms simultaneously, and prove that each one fails in some way.

This is a tall order. In Chapter 3, we described various strategies for polynomial-time algorithms, and discussed why these strategies succeed or fail for particular problems. We could conceivably prove that a given problem A cannot be solved in polynomial time by dynamic programming, say, or a greedy algorithm. Even this would be very challenging, since there are many ways that dynamic programming can be organized, and many types of moves that a greedy algorithm could make. But even if we could do this, it would not prove that A is outside P. After all, the list of strategies in Chapter 3 is far from exhaustive, and it seems prudent to assume that there are families of polynomial-time algorithms completely different from any we have seen before.

In the next few sections, we look at several situations in which we *can* prove lower bounds, showing that a problem is outside a certain complexity class, or that one class is strictly more powerful than another. However, we will also show that none of these techniques have any hope of proving that $\mathsf{P} \neq \mathsf{NP}$. Indeed, we will describe negative "metatheorems" that eliminate entire approaches to the P vs. NP question. At the time we write this, there are just a few ideas on the mathematical horizon—all of which seem extremely challenging to implement—that have any hope of resolving it.

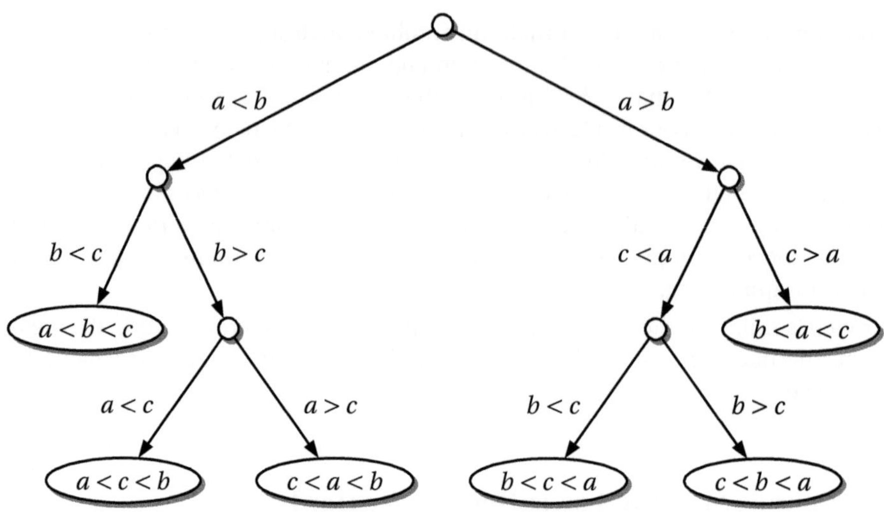

FIGURE 6.2: A decision tree for sorting a list of $n = 3$ items.

6.2.1 Sorting Black Boxes

> ...it is written that animals are divided into (a) those that belong to the emperor; (b) embalmed ones; (c) those that are trained; (d) suckling pigs; (e) mermaids; (f) fabulous ones; (g) stray dogs; (h) those that are included in this classification; (i) those that tremble as if they were mad; (j) innumerable ones; (k) those drawn with a very fine camel's-hair brush; (l) etcetera; (m) those that have just broken the flower vase; (n) those that at a distance resemble flies.
>
> Jorge Luis Borges, *The Analytical Language of John Wilkins*

An example for which it is comparatively easy to prove lower bounds is the rather mundane problem of sorting a list of n items. In Section 3.2.1, we looked at two algorithms, `Mergesort` and `Quicksort`. We measured their running time in terms of the number of comparisons they perform, i.e., the number of times that they ask whether one element is larger than another. We found that `Mergesort` performs about $n \log_2 n$ comparisons, and that `Quicksort` also performs $\Theta(n \log n)$ comparisons on average. Can we do better than this? Is there a sorting algorithm that makes $o(n \log n)$ comparisons, or even $An \log_2 n$ for some constant $A < 1$?

In the children's game of "Twenty Questions," where we ask a series of 20 yes-or-no questions, the largest number of different objects that we can hope to distinguish is 2^{20}—or 3×2^{20} if we start with the traditional question of "animal, vegetable, or mineral." If the set of possible objects is larger than this, then by the pigeonhole principle there is some pair of objects that yield the same sequence of answers, so we cannot tell them apart.

A sorting algorithm is faced with the same difficulty. As it compares pairs of elements, it travels down the branches of a *decision tree*, like that shown in Figure 6.2 for the case $n = 3$. In order to sort the list, it has to learn which of the $n!$ possible orders the input is in, so this tree has to have $n!$ leaves. A binary tree of depth t can have at most 2^t leaves, so the algorithm needs to make

$$t \geq \log_2 n!$$

comparisons. Using Stirling's approximation $n! \geq n^n \, \mathrm{e}^{-n}$, this gives

$$t \geq \log_2(n^n \, \mathrm{e}^{-n}) = n \log_2 n - O(n).$$

Thus, except for the $O(n)$ term, `Mergesort` performs the optimal number of comparisons. In fact, as Problem 3.7 shows, `Mergesort` even gets part of the $O(n)$ term right if we compute its running time more carefully.

Strictly speaking, this argument only showed that there exist input orders where the sorting algorithm needs to make $\log_2 n!$ comparisons—that is, that there exist leaves in the tree that are $\log_2 n!$ steps away from the root. But as Problem 10.2 shows, it is also true that if all $n!$ input orders are equally likely, then the average number of comparisons is at least $\log_2 n!$.

This "Twenty Questions" argument is about *information*, not computation. No matter how powerful an algorithm is computationally, it can't solve the problem if it doesn't have enough information to determine the answer. Since information is much easier to quantify and bound than computational power, this lets us prove a lower bound on the number of comparisons we need, without even thinking about what the algorithm will do with this information once it has it.

6.2.2 *But are the Boxes Really Black?*

Our $\Theta(n \log n)$ lower bound on sorting has a loophole. It assumes that making comparisons is the only way that the algorithm can access its input. If we can look directly at the bits of the input, we can do better. For instance, if the elements of the list are m-bit integers, then the *radix sort* algorithm can sort them in $O(mn)$ time. If m is a constant, as it is in many real applications, then its running time is $O(n)$ instead of $O(n \log n)$.

Radix sort can do this because its actions are not based on comparisons. Instead, it sorts the input one bit at a time, by slotting elements into an array. It takes advantage of the fact that when a and b are strings or integers, the relation $a < b$ is not a "black box"—it has internal structure that we can understand. This allows us to look inside the box and gain more information than we could if we were limited to asking it yes-or-no questions.

Consider the following analogy with the P vs. NP question. If we think of a 3-SAT formula as a black box to which we feed truth assignments, which simply responds "satisfied" or "unsatisfied" for no reason we can understand, then we have no choice but to try all 2^n assignments until we find one that works.

However, this point of view is far too pessimistic. We can see the formula and understand how the clauses interact, at least at a local level, and use this to guide our search. As Problem 6.6 shows, there are algorithms that solve 3-SAT in time $O(c^n)$ for a constant $c < 2$. These algorithms take exponential time, but they are exponentially faster than a naive exhaustive search. Thus while we think of NP-complete problems as searching for needles in a haystack, in reality we are not quite so helpless—they correspond 6.5
to highly structured haystacks, and good algorithms can exploit this structure.

Since each instance of an NP problem corresponds to a witness-checking program that we can run in polynomial time, a better analogy is the following. I have a box with clear sides. I can see a fairly simple mechanism inside it, with a series of gears, ratchets, and so on, which is connected to the input at one end and a bell at the other. The question is whether I can "reverse engineer" this mechanism, and tell whether there is an input that causes the bell to ring. Our belief that $\mathsf{P} \neq \mathsf{NP}$ is the belief that some boxes of this type—those corresponding to NP-complete problems—might as well be black.

6.3 Diagonalization and the Time Hierarchy

> I shall show you the upward abyss of Intelligences, of which you are the bottom, whereas I stand just a little higher than you, though I am separated from the unknown heights by a series of barriers of irrevocable transitions.
>
> Stanislaw Lem, *Golem XIV, Lecture XLIII—About Itself*

Proving that a particular problem can't be solved in polynomial time seems very challenging. However, we can *construct* artificial problems which are outside P, by making sure that any polynomial-time algorithm gets them wrong in at least one case. By showing that, on the other hand, these problems are in some class C, we can prove that $\mathsf{P} \neq \mathsf{C}$.

In this section, we will use this approach to prove that polynomial time is less powerful than exponential time, i.e., that $\mathsf{P} \subset \mathsf{EXPTIME}$. More generally, we will show that $\mathsf{TIME}(g(n)) \subset \mathsf{TIME}(f(n))$, whenever $f(n)$ is large enough compared to $g(n)$. Then, in the next section, we will discuss whether similar techniques can prove that $\mathsf{P} \neq \mathsf{NP}$.

Let's choose a particular time complexity function $f(n)$. We start by constructing a problem that captures all the problems we can solve in $f(n)$ time. The idea is somewhat similar to our NP-complete problem WITNESS EXISTENCE, which asks whether there is an input that makes a given program Π return "yes." In this case, we ask how Π will respond to a given input x of length n, if it is allowed to run for up to $f(n)$ steps.

PREDICT(Π, x)
Input: A program Π and an input x
Output: If Π halts within $f(\|x\|)$ steps when given x as its input, return its output $\Pi(x)$. Otherwise, answer "don't know."

Note that the function f is fixed, rather than being part of the input—different choices of f give different versions of PREDICT. Note also that PREDICT predicts Π's behavior for *precisely* $f(|x|)$ steps or less, as opposed to some constant times $f(|x|)$. For this to be well-defined, we have to choose a specific programming language, so we know what counts as a single step. But let's assume that we have done this.

Of course, we can solve PREDICT in about $f(|x|)$ steps, by simulating Π and seeing what it does. But it is conceivable, *a priori*, that there is some other way to solve PREDICT—some shortcut that lets us see forwards in time and predict Π's output without actually running it. We will prove that this is impossible, by setting up a paradox where a program tries to predict itself.

First we modify PREDICT by applying a program to its own source code, and "twisting" the result:

> CATCH22(Π)
>
> Input: A program Π
>
> Output: If Π halts within $f(|\Pi|)$ steps when given its own source code as input, return the *negation* of its output, $\overline{\Pi(\Pi)}$. Otherwise, answer "don't know."

Note that when we write things like Π(Π), we are using the symbol Π to mean two different things: the function that Π computes, and its source code. Feeding a program to itself may sound like an infinite regress. But its source code is simply a finite string, and it can be fed to itself like any other string.

We will prove that CATCH22 can't be solved in $f(n)$ steps or less. Assume the contrary—that it can be solved by some program Π_{22} which runs on inputs x in $f(|x|)$ steps or less. In particular, if given its own source code, it will halt within $f(|\Pi_{22}|)$ steps. But then it must return the negation of its own output,

$$\Pi_{22}(\Pi_{22}) = \overline{\Pi_{22}(\Pi_{22})}.$$

The only way out of this contradiction is to conclude that there is no such program.

Another way to put this is that for any program Π which runs in $f(n)$ steps, we have

$$\text{CATCH22}(\Pi) = \overline{\Pi(\Pi)}.$$

Thus for at least one input—namely, its own source code—Π and CATCH22 give opposite answers. Therefore, no such Π can solve CATCH22.

Since CATCH22 is just a special case of PREDICT, and since it takes at most one more step to negate the result, this also means that PREDICT can't be solved in less than $f(n)$ steps. In other words, there is no magic shortcut that lets us predict Π's output—we just have to run it, and see what happens. To draw an analogy, suppose that with the help of a high-resolution scan of your brain and a fancy quantum particle-dynamics simulator, I can predict the next hour of your behavior in only 59 minutes. Then in the last minute of the hour, you could prove the simulator wrong by doing the opposite of what it simulated. There is no way to predict arbitrary physical systems faster than waiting to see what they do, and the same is true of arbitrary programs.

The cultured reader will recognize this as a classic technique called *diagonalization.* As we will see in Chapter 7, it goes back to Cantor's proof that the reals are uncountable, Gödel's proof that there are unprovable truths, and Turing's proof that the Halting Problem is undecidable. Indeed, it is really just a time-bounded version of Turing's proof, where PREDICT and CATCH22 only attempt to see $f(n)$ steps into the future instead of to the end of time.

Now that we know that CATCH22 can't be solved in $f(n)$ steps, how hard is it? Clearly we can solve it by simulating Π for $f(|\Pi|)$ steps and seeing what happens. At first, this seems like a contradiction. However, we need to keep in mind that this simulation has to be carried out by some kind of interpreter, or universal program, which takes Π's source code as input and runs it.

This interpreter has to consult Π's source code at each step to determine what to do. It also has to run some sort of "alarm clock" which goes off when Π has taken $f(|\Pi|)$ steps. These things require some overhead, so it takes the interpreter more than t steps to carry out t steps of Π. Let's use $s(t)$ to denote the amount of time an interpreter needs to simulate t steps of Π, while running a clock that keeps track of t. Then our diagonalization argument shows that $s(t) > t$.

The form of $s(t)$ depends on the details of our programming language or our model of computation. For instance, a Turing machine (see Chapter 7) has only local access to its memory. So, it needs to carry a copy of Π's source code and the clock around with it, or shuttle back and forth between them and the workspace where it carries out the simulation. This makes the simulation relatively slow, and the best known method takes $s(t)=O(t\log t)$ time.

A random-access machine, on the other hand, can access Π's source code and the clock directly. If we can increment the clock in $O(1)$ time per tick, then running the clock adds a total of $O(t)$ time to the simulation, and $s(t)=O(t)$.

6.6 We have made one additional assumption—namely, that $f(n)$ can be computed in $O(f(n))$ time. In other words, $f(n)$ is *time constructible* in the sense that we introduced in Section 6.1.2. That way, we can calculate $f(n)$ and "wind" the alarm clock, while adding a total of $O(f(n))$ to the running time.

Under these assumptions, CATCH22 can be solved in $s(f(n))+O(f(n))=O\big(s(f(n))\big)$ time. But since it can't be solved in exactly $f(n)$ steps, it can't be solved in $O(g(n))$ steps for any $g(n)$ such that $g(n)=o(f(n))$. (Strictly speaking, to prove this we need a version of CATCH22 that differs on an infinite number of inputs from every Π which runs in $f(n)$ steps—see Problem 6.8.) Thus it is in $\mathsf{TIME}\big(s(f(n))\big)$ but not in $\mathsf{TIME}(g(n))$, and we have proved the following:

Theorem 6.1 (Time Hierarchy Theorem) *Assume that our programming language and model of computation allows us to simulate t steps of an arbitrary program Π, while running an alarm clock that goes off after t steps, in $s(t)$ time. Then if $f(n)$ is time constructible and $g(n)=o(f(n))$,*

$$\mathsf{TIME}(g(n))\subset\mathsf{TIME}\big(s(f(n))\big).$$

The Time Hierarchy Theorem shows us that we really can compute more if we have more time. For instance, it shows that exponential time is more powerful than polynomial time,

$$\mathsf{P}\subset\mathsf{EXP}\subset\mathsf{EXPEXP}\subset\cdots.$$

But it makes much finer distinctions than that. For instance, even if $s(t)=O(t\log t)$ instead of $s(t)=O(t)$, it shows that some polynomials are more powerful than others—that for any real numbers a,b with $1\le a<b$,

$$\mathsf{TIME}(n^a)\subset\mathsf{TIME}(n^b).$$

Proving a hierarchy theorem for *nondeterministic* time requires more cleverness, since we can't just run an algorithm and negate the result. However, as Problem 6.9 shows, we can also prove that

$$\mathsf{NP}\subset\mathsf{NEXP}\subset\mathsf{NEXPEXP}\subset\cdots.$$

Thus deterministic and nondeterministic time classes both form rich hierarchies, where each level is stronger than the one below. It looks like we are very good at separating complexity classes from each other. Why can't we separate P from NP?

The problem is that we don't know how to compare these hierarchies to each other. Because of its circular nature, diagonalization is a powerful tool for comparing apples to apples, or oranges to oranges—but it doesn't help us compare one type of computational resource to another, such as deterministic vs. nondeterministic time, or time vs. memory. Since $\mathsf{P}\subset\mathsf{EXP}$, we know that at least one of the inclusions in

$$\mathsf{P}\subseteq\mathsf{NP}\subseteq\mathsf{EXP}$$

is proper, but we can't prove this about either one.

There is still hope, you say. Perhaps we can use a different style of diagonalization, and show that some CATCH22-like function is in NP but not in P? Sadly, we will see in the next section that no argument remotely like this one can prove that $\mathsf{P} \neq \mathsf{NP}$.

6.4 Possible Worlds

> The optimist proclaims that we live in the best of all possible worlds; and the pessimist fears this is true.
>
> James Branch Cabell, *The Silver Stallion*

Many of the things we believe to be true, we believe *must* be true. We believe that it would be nonsensical for them to be false—that they are true, not just in our world, but in all possible worlds. But if $\mathsf{P} = \mathsf{NP}$ in some worlds and $\mathsf{P} \neq \mathsf{NP}$ in others, then no type of reasoning that applies equally well in every world can resolve which one of these is true. As we will see in this section, this rules out many potential attacks on the P vs. NP question, including diagonalization.

In computer science, we can contemplate possible worlds in which we are given access to new computational powers. Suppose there is an oracle, such as the one shown in Figure 6.3, who will solve any instance of a problem A we ask her to. Obviously, having access to this oracle might allow us to compute things, or verify witnesses, which we could not before. In such a world, our familiar complexity classes become the *relativized* classes P^A, NP^A, and so on.

For instance, P^{SAT} is the class of problems we can solve in polynomial time if we have access to an oracle who can instantly tell us whether or not a given SAT formula is satisfiable. Since SAT is NP-complete, this is just as good as having an oracle who can answer any question in NP.

$$\mathsf{P}^{\text{SAT}} = \mathsf{P}^{\mathsf{NP}}.$$

Although we didn't call it by this name, we have seen P^{NP} before. Problems 1.10 and 6.10 show that we can find Hamiltonian paths if they exist, or solve optimization problems like finding the smallest vertex cover, by asking an oracle to solve a polynomial number of instances of the corresponding decision problem. Thus these problems are in P^{NP}.

Note that P^A is not the same as "what P would be if A were in P." These two versions of P correspond to two different possible worlds. In the first world, our programming language is augmented with a magical new command that solves any instance of A in a single step. However, we don't know how this command works, just as we cannot see inside the oracle's thought processes. In the second world, where A is in P, there is a program that solves A in polynomial time, written with just the *ordinary* commands of our programming language. This is a program like any other, which we can understand and modify as we like.

For instance, for any program Π we can translate the statement $\Pi(w) =$ "yes" into a SAT formula. Then in P^{NP}, we can ask the oracle whether there is a witness w such that $\Pi(w)$ returns "yes," and thus solve any problem in NP. But we can only do this if Π itself doesn't make any calls to the oracle. Each SAT problem has a single layer of quantifiers, and if we try to nest one SAT problem inside another, we get problems with more layers than the oracle can handle.

In contrast, if SAT is in P then we can replace calls to the SAT oracle with a subroutine written in the ordinary version of our programming language. We can then translate the statement that the resulting program returns "yes" into another SAT formula, run our SAT subroutine on it, and so on, nesting as many

FIGURE 6.3: The King of Athens receiving advice from Pythia, the Oracle of Delphi.

levels deep as we like. As we showed in Section 6.1.1, in this case we can absorb any constant number of quantifiers, and P includes the entire polynomial hierarchy.

If you like, $\mathsf{P}^{\mathsf{SAT}}$ is what P would be if we were given a supply of magic boxes that solve SAT problems. These boxes are black, and we can't take them apart and look at them. But if we can build these boxes out of earthly components, and nest them inside each other, then $\mathsf{P} = \mathsf{NP}$ and the polynomial hierarchy collapses.

To make this more precise, suppose we are trying to verify a statement of the following form:

$$\forall x_1, x_2, \ldots, x_n : \exists y_1, y_2, \ldots, y_n : \phi(x_1, \ldots, x_n, y_1, \ldots, y_n). \tag{6.4}$$

Is this problem in P^{NP}? That is, can we solve it in polynomial time if we have access to a SAT oracle? For each truth assignment $x_1, \ldots, x_n$, we can ask her whether there is an assignment $y_1, \ldots, y_n$ such that ϕ is true. But since there are 2^n ways to set the x_i, we would have to call on her an exponential number of times. We would like to ask her the question "is there an assignment $x_1, \ldots, x_n$ such that you will say no?" But that would be asking the oracle a question about herself, and that's cheating.

Thus we believe that P^{NP} is smaller than $\Pi_2\mathsf{P}$, the level of the polynomial hierarchy where problems like (6.4) live. Specifically, Problem 6.11 shows that

$$\mathsf{P}^{\mathsf{NP}} \subseteq \Sigma_2\mathsf{P} \cap \Pi_2\mathsf{P}.$$

This is the class of problems where the quantifiers can be written in either order, ∀∃ or ∃∀. On the other hand,

$$\mathsf{NP}^{\mathsf{NP}} = \Sigma_2\mathsf{P} \quad \text{and} \quad \mathsf{coNP}^{\mathsf{NP}} = \Pi_2\mathsf{P}\,.$$

If we are willing to contemplate dizzying concatenations of oracles, $\mathsf{P}^{\mathsf{NP}^{\mathsf{NP}}} = \mathsf{P}^{\Sigma_2\mathsf{P}}$ is contained in the third level of the polynomial hierarchy, and so on.

Now that we have defined these relativized worlds, let's say that a proof technique *relativizes* if it applies in all of them. For instance, given a nondeterministic program that uses the **goto both** command from Section 4.3.3, we can simulate it deterministically by trying all possible computation paths. This simulation works for programs that make calls to an oracle, if the simulator can call the oracle too. Therefore, the fact that $\mathsf{NP} \subseteq \mathsf{EXP}$ relativizes, and $\mathsf{NP}^A \subseteq \mathsf{EXP}^A$ for all oracles A.

Most such simulation arguments relativize. In particular, if there were some way to simulate nondeterminism deterministically with only a polynomial increase in the running time—such that calls to the oracle could be treated as calls to a subroutine, and passed through to the simulator—then we would have, not just $\mathsf{P} = \mathsf{NP}$, but $\mathsf{P}^A = \mathsf{NP}^A$ for all A.

The diagonalization argument we used in Section 6.3 also relativizes. We can define the problems PREDICT and CATCH22 in the presence of any oracle, and assuming that CATCH22 can be solved in $f(n)$ time leads to the same contradiction as before. Thus the Time Hierarchy Theorem holds in every relativized world. Similarly, if there were a proof using diagonalization that $\mathsf{P} \neq \mathsf{NP}$, then we would have $\mathsf{P}^A \neq \mathsf{NP}^A$ for all A as well.

The following theorem—which was the first major result on the P vs. NP question—spells doom for both these techniques.

Theorem 6.2 *There exist oracles A and B such that*

$$\mathsf{P}^A = \mathsf{NP}^A \quad \textit{but} \quad \mathsf{P}^B \neq \mathsf{NP}^B\,.$$

In other words, there are possible worlds where $\mathsf{P} = \mathsf{NP}$, and others where $\mathsf{P} \neq \mathsf{NP}$. Therefore, no proof technique that relativizes can hope to resolve which one is true.

Each of the oracles in Theorem 6.2 gives us additional computational power, but in very different ways. While A adds a kind of power that erases the distinction between P and NP, B adds a kind of power that enhances it. Designing these oracles helps teach us about the P vs. NP problem, and why it might be difficult to resolve in our own unrelativized world.

To find an oracle A for which $\mathsf{P}^A = \mathsf{NP}^A$, recall once again that NP problems are P problems with a single ∃ in front of them. If adding this quantifier gives us no more power than we had already, A must be a fixed point with respect to adding quantifiers. To construct such an oracle, we consider *quantified Boolean formulas*, i.e., formulas of the form

$$\mathsf{Q}z_1 : \mathsf{Q}z_2 : \cdots : \mathsf{Q}z_n : \phi(z_1, z_2, \ldots, z_n)\,,$$

where $\phi(z_1, z_2, \ldots, z_n)$ is a Boolean formula and where each quantifier Q is ∀ or ∃. The problem QUANTIFIED SAT asks whether a given formula of this kind is true.

We will give a precise characterization of QUANTIFIED SAT's complexity in Chapter 8. For now, we note that it appears to be outside the entire polynomial hierarchy, since each level of PH only allows a fixed number of quantifiers, while for QUANTIFIED SAT the number of quantifiers can grow with n. For instance, in QUANTIFIED SAT we can express the claim that the first player has a winning strategy in a game which will last a polynomial number of moves, while in each level of PH the number of moves is a constant.

Now suppose that A is an oracle for QUANTIFIED SAT. Since an instance of QUANTIFIED SAT with a $\exists$ in front of it is simply another instance of QUANTIFIED SAT, we claim that $\mathsf{P}^A = \mathsf{NP}^A$. In the presence of such an oracle, adding one more quantifier does nothing to increase our (already enormous) computational power, and finding solutions is just as easy as checking them.

Now for the other half of the theorem. Intuitively, an oracle B such that $\mathsf{P}^B \neq \mathsf{NP}^B$ would be one that truly is a black box—one with finicky and idiosyncratic tastes, such that finding an input to which she says "yes" really is like finding a needle in a haystack.

The simplest way to construct such an oracle is to do so randomly. For each $n \geq 0$, we flip a coin. If it comes up heads, we choose a random string $w_n \in \{0,1\}^n$ by flipping n more coins. Then B will say "yes" to w_n and "no" to all other strings of length n. If it comes up tails, then B will say "no" to all strings of length n. Note that we make these random choices once and for all, and B's responses are fixed thereafter.

Now, consider the following problem:

FINICKY ORACLE
Input: An integer n in unary
Question: Is there a string w_n of length n to which B says "yes"?

This problem is clearly in NP^B, since we can verify the witness w_n by asking B about it. On the other hand, we claim that with probability 1 it is not in P^B.

The reason is that any polynomial-time program Π only has time to ask B about poly(n) strings of length n. For large n, this is a vanishingly small fraction of the 2^n possible strings. No matter how Π chooses which strings to ask about, with overwhelming probability B will say "no" to every one, leaving Π none the wiser about whether a w_n exists. Thus whatever response Π gives is correct with probability exponentially close to 1/2. Since B's responses are chosen independently for each n, with probability 1 there will be some value of n at which Π fails.

It may seem odd to construct a random oracle and then pose a question about it. Another way to put this is that for any set S containing at most one string of each length, the problem "is there a string of length n in S?" is in NP^B for the oracle B who tells us whether a string is in S. On the other hand, the vast majority of such sets have no discernible pattern—indeed, only a set of measure zero has any pattern at all. Thus if S is chosen randomly, the strings in S truly are needles in a haystack. Even if we can test each blade of hay by consulting B, we have no hope of finding the needles in polynomial time.

Since many of the basic arguments one can make about programs and computations relativize, what type of arguments does Theorem 6.2 leave us? For many years, simulation and diagonalization were virtually the only tools we had for proving inclusions or separations of complexity classes. We now know of a few non-relativizing techniques for proving that two complexity classes are the same or different. In particular, in Section 11.2 we will use *arithmetization* to prove that IP = PSPACE, where IP is the class of problems for which there are *interactive proofs* of a certain kind. However, as we will discuss in the next
6.7 section, there is another major barrier to resolving the P vs. NP question.

6.5 Natural Proofs

An intuitive way to prove that $\mathsf{P} \neq \mathsf{NP}$ is to invent some property, some kind of "complicatedness," that no problem in P can possess—and then show that some problem A in NP possesses it. Of course, the first property that comes to mind is the property of lacking any polynomial-time algorithm, but this just restates the question. What we would like is a concrete mathematical property—say, some combinatorial property of the set of yes-instances of A—that separates NP from P.

If we go far enough down the complexity hierarchy, we can in fact separate some complexity classes from others in exactly this way. To see an example of this, let's look deep inside P, at problems that can be solved by very simple Boolean circuits.

6.5.1 *Circuit Complexity Classes*

Recall that a Boolean circuit has an input layer, followed by successive layers of AND, OR, and NOT gates, where each gate receives its input from the layer above it. It ends with an output layer, consisting of one or more bits. The *depth* of a circuit is the number of layers, its *width* is the largest number of gates in any one layer, and its *size* is the total number of gates.

We are interested in understanding what problems can be solved by circuits with a given width and depth. Since solving a problem means, by definition, solving instances of any size n, we need to consider families of circuits, with a circuit with n inputs for each value of n. We can then define complexity classes by limiting how the depth, width, and size of these circuits scale with n.

Let's consider one of the lowest possible complexity classes we could define in this way—those whose depth is constant, and whose width is polynomial. If our AND and OR gates are binary, i.e., if each one has just two inputs, then it would take $\log_2 n$ layers just to get an output bit that depends on all the inputs. So, we will allow our ANDs and ORs to have arbitrary *fan-in*—that is, each gate can have as many inputs as we like. We show an example of such a circuit in Figure 6.4. This gives us the following class:

> AC^0 is the class of problems that can be solved by families of Boolean circuits of constant depth and poly(n) width, where the AND and OR gates have arbitrary fan-in.

We can think of AC^0 and classes like it as simple models of parallel computation. The width is the number of processors, each layer is one step of computation, and the depth is the computation time. We assume, unrealistically, that every processor can send a message to every other one in a single step. For instance, a processor computing an AND gate sends out `false` if it receives `false` from any of its inputs, and otherwise sends out `true`. Then AC^0 is the class of problems that can be solved by a polynomial number of processors in constant time.

It is important to note that our definition of AC^0 does not demand that the circuits for different values of n have anything in common with each other. It simply states that a circuit family with constant depth and polynomial width exists. Such complexity classes are called *nonuniform*, and they can be somewhat counterintuitive. For instance, consider the following class:

> P/poly is the class of problems that can be solved by families of circuits of polynomial size.

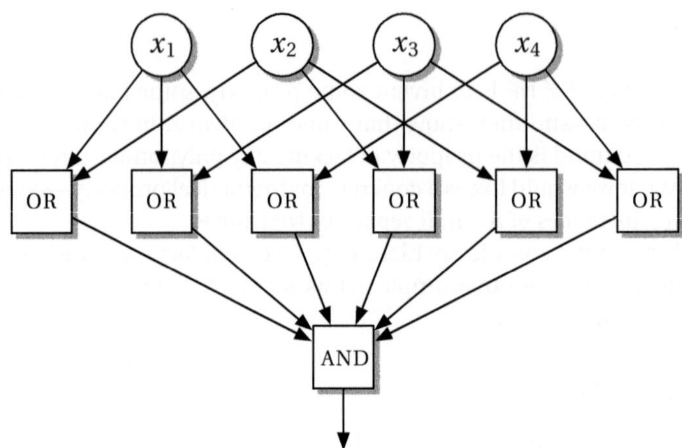

FIGURE 6.4: A circuit that checks that at most one of its four input bits is `false`. For general n, this problem can be solved with a circuit with constant depth and $O(n^2)$ width. So, the problem AT MOST ONE FALSE BIT is in AC^0.

At first blush, this class sounds just like P. Indeed, it includes P, since any polynomial-time computation can be carried out by a circuit of polynomial size. However, P/poly is larger than P because there is no way, in general, to generate the nth circuit in the family in poly(n) time. The circuits can differ in arbitrary ways from one value of n to the next, and the entire family may not even have a finite description. In contrast, P is the class of problems for which there is a *uniform* family of circuits—that is, one that can be finitely described by a single program that works for all n.

6.8 Similarly, we can define a class called *uniform* AC^0, consisting of problems that can be solved by circuit families of constant depth and polynomial width, where there is a polynomial-time program that outputs a blueprint of the nth circuit when given n as input (in unary). Alternately, uniform AC^0 is the set of problems that can be solved by a program, written in some programming language for parallel computers, in constant time and a polynomial number of processors.

As the following exercise shows, in the absence of uniformity, the constraint that the width of an AC^0 circuit is polynomial is quite important.

Exercise 6.2 *Show that any function whatsoever can be computed by a family of Boolean circuits of* $2^{O(n)}$ *width, with constant depth if we allow gates with arbitrary fan-in, and* poly(n) *depth if the gates are binary.*

6.5.2 PARITY *is Not in* AC^0

Our main motivation for considering AC^0 is that we can prove that certain functions are outside it, just as we dream of doing for P. Given n bits $x_1,\ldots,x_n$, let PARITY be the problem of telling whether an even number of the x_i are true. Obviously PARITY is in P. However, by defining a kind of "complicatedness" that no problem in AC^0 can have but PARITY does, we can show that it is not in AC^0.

The idea is to ask what happens to a function when we restrict it by fixing the values of some of its inputs. PARITY is "complicated" in the sense that it is sensitive to every variable, no matter how the others are set. Therefore, if we fix any subset of the variables to particular values, the parity still depends on the remaining variables. In the same sense, the AND function is rather "uncomplicated," since setting even a single input `false` causes the output to be the constant `false`.

To state this sort of thing quantitatively, consider the following definition:

Definition 6.3 *Let $f(x_1,\ldots,x_n)$ be a function of n variables, and let $\gamma(n)$ be some function of n. We will say that f is* γ-complicated *if, for any subset $S \subseteq \{x_1,\ldots,x_n\}$ where $|S| = \gamma(n)$, any setting of the variables outside S to Boolean values results in a non-constant function of the variables in S.*

Note that making $\gamma(n)$ smaller makes this property stronger. In particular, PARITY is γ-complicated for any $\gamma(n) \geq 1$, since it does not become a constant until we set every variable. On the other hand, the following theorem states that functions in AC^0 cannot be very complicated in this sense.

Theorem 6.4 *If $\gamma = n^{o(1)}$, i.e., if $\gamma = o(n^\varepsilon)$ for all $\varepsilon > 0$, then no function in AC^0 is γ-complicated.*

Therefore, PARITY is not in AC^0. This implies that AC^0 is smaller than $\mathsf{P/poly}$, and uniform AC^0 is smaller than P.

Proof The proof of Theorem 6.4 is quite clever. First, note that by adding dummy gates if necessary, we can always think of an AC^0 circuit as alternating between layers of AND and OR gates. And, using De Morgan's laws, we can move NOT gates up through the circuit, until they simply negate some of the variables at the input layer.

Now, given a function f in AC^0, we set most of its variables to random values, leaving about $\sqrt{n}$ of them unset. Intuitively, this restriction fixes the outputs of many of the AND and OR gates in the top layers of f's circuit to constants, causing them to disappear—especially those with large fan-in.

In particular, suppose that f can be computed by a circuit of d layers, alternating between ANDs and ORs. Problems 6.12 and 6.13 show that after this random restriction, with high probability we can switch the top two layers while increasing the width of the circuit by just a constant factor. Since the second and third layers then consist of gates of the same type, we can merge them into a single layer, obtaining a circuit of depth $d-1$ for the restricted function.

We then set most of these $\sqrt{n}$ variables, leaving only about $n^{1/4}$ unset and reducing the depth to $d-2$, and so on. By the time there are $n^{1/2^d}$ variables left, the circuit has depth zero, in which case the function on these remaining variables is a constant. Thus f cannot be n^ε-complicated for any $\varepsilon < 1/2^d$. Since each f in AC^0 can be computed by circuits of depth d for some constant d, no such f can be $n^{o(1)}$-complicated, proving Theorem 6.4. □

6.9

Once we know that PARITY is outside AC^0, we can show that other problems are outside AC^0 by reducing PARITY to them. These include MAJORITY, the problem of telling whether a majority of the x_i are true, as well as REACHABILITY and binary multiplication. We discuss these in Problems 6.17 and 6.18.

Of course, AC^0 is such a low-lying complexity class that separating it from P may not strike the reader as a great achievement. But structurally speaking, this seems like the kind of proof we are looking for. Can we use a similar proof to show that $\mathsf{P} \neq \mathsf{NP}$? As we will see in the next section, the answer is probably no—assuming that there are good pseudorandom generators in P.

6.5.3 *Pseudorandom Functions and Natural Proofs*

Obviously, to prove that $\mathsf{P} \neq \mathsf{NP}$ we would need some definition of complicatedness that is much more subtle than the property we used for AC^0. However, there are several criteria that we might expect these properties to meet.

We say that a property C is *common* if it is true of a random function with reasonable probability. Specifically, suppose we choose the value of a function $f : \{0,1\}^n \to \{0,1\}$ on each of its 2^n possible inputs uniformly and independently from $\{0,1\}$, so that f's truth table consists of 2^n random bits. Then C is common if such an f has property C with probability $2^{-O(n)}$. Note that there are 2^{2^n} possible functions, so even a $2^{-O(n)}$ fraction of them constitutes a very large set.

Second, we say that C is *constructive* if there is an algorithm that tells whether or not it is true for a function $f : \{0,1\}^n \to \{0,1\}$ in $2^{O(n)}$ time, given access to f as an oracle or subroutine. Alternately, we can think of this algorithm as being given f's truth table all at once, in the form of a gigantic string of 2^n bits. Since $2^{O(n)} = \text{poly}(2^n)$, this means that C can be computed in polynomial time, as a function of the size of f's entire truth table. Thus, in a bizarre sense, we can say that C is constructive if it is in P.

Finally, we say that C is *useful* against a complexity class X if it is false for every function in X. Then we can show that $\mathsf{X} \neq \mathsf{Y}$ if we can show that it is true for some function in Y. If C is common and constructive, we call this a *natural proof* that $\mathsf{X} \neq \mathsf{Y}$.

Problem 6.15 shows that γ-complicatedness is constructive, and Problem 6.16 shows that it is common whenever γ is sufficiently greater than $\log_2 n$. Thus it provides a natural proof that AC^0 is smaller than P. But do we expect other notions of complicatedness to be common and constructive as well? Does this definition of "natural" deserve its name?

To justify the requirement of commonality, it seems reasonable to say that random functions are complicated—that simplicity is rare, and complicatedness is common. Of the 2^{2^n} possible functions on n variables, the vast majority have no rhyme or reason. Their truth tables have no compact description, and they depend on their variables in arbitrarily complicated ways.

As a justification for constructivity, we point out that most concrete properties of a function that one can think of are constructive. For instance, suppose $\phi(x_1,\ldots,x_n)$ returns the value of some Boolean formula ϕ. Then the property that ϕ is satisfiable is simply the property that there is at least one 1 in its truth table, and the property that a string of 0s and 1s contains a 1 is certainly in P. Of course, we are not saying that SAT is in P—just that it can be solved in time which is polynomial as a function of 2^n, the length of ϕ's truth table.

Exercise 6.3 *If $\phi(x_1,\ldots,x_n)$ is a Boolean function, show that the property that the* QUANTIFIED SAT *formula $\exists x_1 : \forall x_2 : \exists x_3 : \cdots : \phi$ is true is a constructive property of ϕ. That is, it is in* P *as a function of ϕ's truth table.*

In addition, even if we initially prove a lower bound using an "unnatural" property C, which is not constructive or not common, in all known cases we can "naturalize" the proof by replacing C with a natural property C'. For instance, in our proof that PARITY is outside AC^0, we could have said that a function f is complicated if flipping any input negates its output—a property that only PARITY and its negation have. Instead, we used the more general property that f is not constant even when restricted to a small number of inputs—a property shared by almost all functions.

If we accept this definition of "natural," can there be a natural proof that $\mathsf{P} \neq \mathsf{NP}$? For now, let's focus on the slightly different question of whether there is a natural proof that NP is not contained in P/poly. We will prove that, under reasonable assumptions, no such proof is possible. Specifically, we will prove that if there are polynomial-time functions that "look random," then for every common, constructive property C, there are functions in P/poly for which C is true—because, by the definition of commonality, C is true for random functions with reasonable probability.

First we need to define a pseudorandom function. This is an algorithm that takes a "seed" s of length k and an input x of length n, and produces an output $f_s(x)$. The idea is that if s is chosen randomly from $\{0,1\}^k$, then $f_s(x)$ looks like a completely random function of x. In other words, the truth table of f_s looks as if all 2^n bits were independently random.

We formalize the idea that f_s "looks random" as follows. A *tester* is an algorithm B which is allowed to call f as a subroutine, and which returns "yes" if it thinks that f is random. We focus on testers which run in time polynomial in the size of the truth table, $2^{O(n)} = \text{poly}(2^n)$. Then we say that f_s is *pseudorandom* if the probability that any such tester returns "yes," when s is chosen randomly, is almost identical to what it would be if the entire truth table were chosen randomly:

$$\left| \Pr_{s\in\{0,1\}^k}[B(f_s) = \text{“yes”}] - \Pr_{f\in\{0,1\}^{2^n}}[B(f) = \text{“yes”}] \right| = 2^{-\omega(n)}, \tag{6.5}$$

where $2^{-\omega(n)}$ means less than 2^{-cn} for any constant c.

If the seed length k is polynomial in n, and if $f_s(x)$ can be computed from s and x in polynomial time, we say that f_s is a *polynomial-time pseudorandom function.* In that case, if we choose a family of seeds s, one for each input size n, then we can hard-wire them into a family of polynomial-size circuits, and $f_s(x)$ is in P/poly as a function of x.

Now suppose someone comes forward and announces that some property C gives a natural proof that $\mathsf{NP} \subset \mathsf{P/poly}$. Since C is useful against P/poly, then C is false for any polynomial-time function $f_s(x)$. But since C is common, it is true of truly random functions with probability $2^{-O(n)}$. Finally, if C is constructive, then there is an algorithm B which runs in $2^{O(n)}$ time which tells us whether C is true.

But this means that B is a tester such that can tell the difference between f_s and a random function with probability $2^{-O(n)}$. That is,

$$\Pr_{s\in\{0,1\}^k}[B(f_s) = \text{“yes”}] = 0 \quad \text{but} \quad \Pr_{f\in\{0,1\}^{2^n}}[B(f) = \text{“yes”}] = 2^{-O(n)}.$$

This contradicts the proposition (6.5) that f_s is pseudorandom.

Thus any property which is common, constructive, and useful against P/poly yields a tester that distinguishes random functions from those in P/poly. If polynomial-time pseudorandom functions exist, there can be no natural proof that some function in NP—or, for that matter, any function at all—is outside P/poly.

So do pseudorandom functions exist? We will explore the machinery of pseudorandomness in Chapter 11. For now, recall that a *one-way function* is a function which is in P, but whose inverse is not. For instance, while $g(x) = a^x \bmod p$ is in P, its inverse DISCRETE LOG seems hard. As the following theorem states, if we assume that DISCRETE LOG requires circuits of size $2^{\omega(n^\varepsilon)}$ for some $\varepsilon > 0$—or that some other one-way function is this hard—then we can use g to generate pseudorandom functions:

Theorem 6.5 *Suppose there is a one-way function g and a constant $\varepsilon > 0$ such that no circuit family of size $2^{O(n^\varepsilon)}$ can invert g with probability $2^{-O(n^\varepsilon)}$. Then there exist polynomial-time pseudorandom functions, and there is no natural proof that* $\mathsf{NP} \subset \mathsf{P/poly}$.

We will prove this theorem in Problems 11.16 and 11.17.

The concept of a natural proof is due to Alexander Razborov and Steven Rudich. As they wrote in their 1993 paper, natural proofs have a self-defeating character. If there is a natural proof that some function is harder than P/poly, then there are circuits of size $2^{o(n^\varepsilon)}$, for arbitrarily small $\varepsilon > 0$, that break any pseudorandom generator and invert any one-way function. Thus if we can prove that some problems are hard, other problems become easy. For instance, there can be no natural proof that DISCRETE LOG requires circuits of 2^{n^ε} size for some $\varepsilon > 0$, since such a proof would imply that it doesn't!

6.5.4 *Where To Next?*

Given our belief that there are one-way functions and polynomial-time pseudorandom generators, Theorem 6.5 is a powerful barrier to proving that P $\neq$ NP. What kinds of proof are still available to us? It leaves us with the following three options:

1. *Use uniformity.* Theorem 6.5 leaves open the possibility that there is a natural property which is false for every function in P, even though it holds for some in P/poly. However, it is hard to imagine how such a property would work. It would have to involve, not how the amount of computation scales with n, but how the computation we do varies from one value of n to the next. In terms of Boolean circuits, we need to ask not how large they are, but whether the structure of the entire family can be captured by a unified description.

2. *Don't be common.* We might be able to prove that a function is outside P by showing that it is complicated in some rare, specific way, which it shares neither with functions in P nor with random functions. In a sense, we have already seen an example of this in the diagonalization proof of Section 6.3 where the function CATCH22, owing to its clever circular nature, disagrees with every possible polynomial-time algorithm.

3. *Violate constructivity.* We argued that most natural properties of a function are in P as a function of the truth table. However, this is not a terribly convincing argument. For instance, we could imagine a property C of the following form: C is true of a function f if there is no function g that has some relationship with f. In this case, g would be a witness that f is uncomplicated. If f and g are both functions on $\{0,1\}^n$ and we can confirm this relationship in poly(2^n) time, then C would be in coNP as a function of f's truth table.

As mathematics advances, it seems possible that some subtle property will emerge which can truly separate P from NP. At the time we write this, this hope still seems fairly distant. But at least the concepts of relativization and natural proofs give us some guidance about where to look.

6.10

6.6 Problems in the Gap

There is far more going on inside NP than a binary distinction between P and NP-completeness. We have already seen problems, such as FACTORING, which seem to lie in between these two extremes. In this section and the next, we look inside NP and survey its internal structure.

Suppose that P $\neq$ NP. Is it possible that every problem in NP is either easy, or maximally hard? In other words, that every problem is either in P or is NP-complete? We have argued that problems in NP $\cap$ coNP such as SMALL FACTOR are probably not NP-complete, since that would cause the polynomial hierarchy to collapse. But this is not a proof.

Here we show that if $\mathsf{P} \neq \mathsf{NP}$, then there are problems in the middle ground between P and NP-completeness. In fact, NP contains an infinite number of gradations in difficulty, where each one is harder, with respect to polynomial-time reductions, than the one below.

We warn the reader in advance that the only problems that we can actually *prove* are in this middle ground are highly artificial. They are constructed using a kind of diagonalization, which forces them to avoid NP-completeness by carefully shooting themselves in the foot. Nevertheless, their existence shows that, unless $\mathsf{P} = \mathsf{NP}$ and the entire edifice collapses, NP has a rich internal structure. They also bolster our belief that natural problems such as FACTORING lie in this limbo as well.

Without further ado, let us prove the following theorem.

Theorem 6.6 *If* $\mathsf{P} \neq \mathsf{NP}$, *then there are problems which are outside* P *but which are not* NP*-complete.*

Proof We start by choosing an NP-complete problem. Any one will do, but we will use SAT. For notational convenience, we define a function SAT(x) as

$$\text{SAT}(x) = \begin{cases} \texttt{true} & \text{if } x \text{ is a yes-instance of SAT} \\ \texttt{false} & \text{if } x \text{ is a no-instance of SAT}. \end{cases}$$

Similarly, we define the yes- and no-instances of our constructed problem A as follows,

$$A(x) = \begin{cases} \text{SAT}(x) & \text{if } f(|x|) \text{ is even} \\ \texttt{false} & \text{if } f(|x|) \text{ is odd}, \end{cases} \tag{6.6}$$

where f is some function from $\mathbb{N}$ to $\mathbb{N}$. We will carefully design f such that

1. $f(n)$ is computable in poly(n) time, so that A is in NP;
2. no polynomial-time program Π solves A, so that A is not in P; and
3. no polynomial-time program Π reduces SAT to A, so that A is not NP-complete.

We will do this using a lazy kind of diagonalization. First, let Π_i denote the ith program written in our favorite programming language. If this sounds strange, recall that programs are just finite strings, and finite strings written in an alphabet consisting of b symbols can be interpreted as integers in base b. Then Π_i is simply the program whose source code is the integer i. Of course, the vast majority of integers are complete nonsense when interpreted as programs, and we assume that these Π_i return "don't know."

We want to focus on those Π_i which run in polynomial time. In general, picking out these programs is very hard. To make it easier, we will demand that each Π_i has a "header" which declares explicitly how long it is allowed to run—say, by giving constants c and k such that its running time on an input x should be limited to $t = c|x|^k$. We keep a clock running while Π_i runs, and if it does more than t steps we force it to stop and return "don't know."

We can then rewrite conditions 2 and 3 above as

2′. For all i, there is a y such that $\Pi_i(y) \neq A(y)$, and

3′. For all i, there is a y such that $\text{SAT}(y) \neq A(\Pi_i(y))$.

```
f(n)
begin
    if f(n − 1) = 2i then
        for at most n steps of computation do
            for all y in lexicographic order do
                if Π_i(y) ≠ A(y) then return f(n − 1) + 1 ;
            end
        end
        if we run out of time before finding such a y then return f(n − 1) ;
    end
    else if f(n − 1) = 2i + 1 then
        for at most n steps of computation do
            for all y in lexicographic order do
                if SAT(y) ≠ A(Π_i(y)) then return f(n − 1) + 1 ;
            end
        end
        if we run out of time before finding such a y then return f(n − 1) ;
    end
end
```

FIGURE 6.5: The definition of the function $f(n)$ used in Theorem 6.6.

We will define $f(n)$ so that it increases very slowly as a function of n, making sure that these two conditions hold for each Π_i. Specifically, when $f(n-1)$ is even, we search for an instance y such that 2′ holds, proving that Π_i fails to solve A where $f(n-1)=2i$. Similarly, when $f(n)$ is odd, we search for an instance y such that 3′ holds, proving that Π_i fails to reduce SAT to A where $f(n)=2i+1$.

For each value of n, we spend n steps of computation searching for instances y that defeat Π_i in whichever sense we are working on. If we find a suitable y within that time, we set $f(n)=f(n-1)+1$, moving on to the next case of 2′ or 3′ we need to prove. Otherwise, we set $f(n)=f(n-1)$, so that next time we will devote $n+1$ steps to the search instead of n.

Note that for each y, we need to compute $\Pi_i(y)$, which takes poly($|y|$) time. We also need to compute SAT(y) and $A(y)$, which presumably takes exponential time. However, we keep the clock running during these computations, and give up as soon as n steps have passed. In particular, this means that $f(n)$ only depends on $f(|y|)$ where $|y|=O(\log n)$, so there is no danger of circularity in our definition. It also means that $f(n)$ grows extremely slowly. But that doesn't matter—all we need is to define it in such a way that every Π_i fails at some point.

We can express all this with the pseudocode shown in Figure 6.5. The time it takes to compute $f(n)$ is the time it takes to compute $f(n-1)$ plus $O(n)$ steps, so $f(n)$ can be computed in poly(n) time.

Now if $f(n)$ increases without bound, then clearly conditions 2′ and 3′ are true for all i. In that case, A is outside P since no Π_i solves it, and A is not NP-complete since no Π_i reduces SAT to it. But what happens if we get stuck, and f becomes constant at some point?

This is where the cleverness of our definition (6.6) of A comes in. First, suppose that f gets stuck at an even value, i.e., that there is an n_0 such that $f(n)=2i$ for all $n \geq n_0$. This means that $\Pi_i(y)=A(y)$ for all y, so Π_i is a polynomial-time algorithm for A. But according to (6.6), it also this means that A and SAT only differ on instances of size less than n_0. There are a finite number of such instances, so we can put all their answers in a lookup table of constant size. This gives us a polynomial-time algorithm for SAT: given an instance x of size n, if $n < n_0$ we look it up in the table, and if $n \geq n_0$ we run $\Pi_i(x)$. Since SAT is NP-complete, this would contradict our assumption that $\mathsf{P} \neq \mathsf{NP}$.

Similarly, suppose that f gets stuck at an odd value, so that $f(n_0) = 2i+1$ for all $n \geq n_0$. That means that $\text{SAT}(y) = A(\Pi(y))$ for all y, so Π_i is a polynomial-time reduction from SAT to A. But according to (6.6), it also means that $A(x) = \texttt{false}$ for all instances of size $n \geq n_0$. This means that A only has a finite number of yes-instances—but then A is in P, since we can put all the yes-instances in a lookup table of constant size. Since SAT is reducible to A, this would imply that SAT is in P, and again contradict our assumption that $\mathsf{P} \neq \mathsf{NP}$. □

As Problem 6.19 shows, for any two problems B, C where B is easier than C, we can use similar techniques to construct a problem A in between them. In other words, if $B < C$, by which we mean $B \leq C$ but $C \not\leq B$, there is an A such that $B < A < C$. We can then construct problems D and E such that $B < D < A < E < C$, and so on. Therefore, unless $\mathsf{P} = \mathsf{NP}$, there are infinite hierarchies of problems within NP, where each one is strictly harder than the one below it.

We can also, as Problem 6.20 shows, construct pairs of problems that are *incomparable*—that is, such that neither one can be reduced to the other. Both problems are hard, but in different ways. Being able to solve either one would not help us solve the other.

These existence proofs are very interesting. But are there "naturally occurring" problems with these 6.11 properties? Most problems we encounter in practice are either in P or NP-complete. However, there are a few that appear to be in between. We have already mentioned FACTORING, or rather decision problems based on it such as SMALL FACTOR, which lie in $\mathsf{NP} \cap \mathsf{coNP}$.

As we will see in Section 11.1, another candidate is GRAPH ISOMORPHISM, the problem of telling whether two graphs are topologically identical. While it is not known to be in coNP, there is a kind of *interactive proof* for no-instances which works with high probability, and the polynomial hierarchy would collapse if it were NP-complete. Moreover, we know of no way to reduce FACTORING to GRAPH ISOMORPHISM or vice versa, and they seem to have rather different characters. So, perhaps they are incomparable.

6.7 Nonconstructive Proofs

> Taking the principle of excluded middle from the mathematician would be the same, say, as proscribing the telescope to the astronomer or to the boxer the use of his fists.
>
> David Hilbert

We have already seen that there is an intimate relationship between computational complexity and the nature of proof. The class NP consists of problems where, whenever the answer is "yes," there is a simple proof of that fact. Can we refine this relationship further? Are there subclasses of NP corresponding to different kinds of proof? In particular, most NP problems ask whether something exists. But even if we have a proof that it does, does that mean we can find it?

In mathematics, there are two types of proof of existence. A *constructive* proof shows us that x exists by showing us how to find or construct it. In modern terms, it gives us an algorithm that produces x. In contrast, a *nonconstructive* proof proves that x exists, but it gives us no information at all about where it is. Ultimately, nonconstructive proofs rely on the *principle of excluded middle*, which goes back to Aristotle: either x exists or it doesn't. If assuming that it doesn't exist leads to a contradiction, then it must exist.

There are mathematicians, the *constructivists*, who do not accept such arguments, and argue that proofs by contradiction aren't proofs at all. In a sense, they believe that mathematical objects don't exist unless there is an algorithm that finds them. While this point of view seems a bit extreme, it's reasonable to say that the more insight into a problem we have, the more constructive we can be. As we said in Chapter 3, problems are in P precisely when there is some insight that allows us to find their solutions quickly, rather than searching blindly for them.

In this section, we will see that each family of nonconstructive proofs, such as the pigeonhole principle, parity arguments, and fixed point theorems, corresponds to a subclass of NP, or more precisely, the analog of NP for search problems. So, rather than engaging in a philosophical debate about whether nonconstructive proofs count, we can use computational complexity to explore the power of these arguments, and the gap between nonconstructive and constructive proofs—the gap between knowing that something exists, and finding it.

6.7.1 *Needles, Pigeons, and Local Search*

Let's start by defining an NP-like class of search problems, or *function problems* as they are often called. Rather than just asking whether a witness exists, let's demand to see it:

> FNP is the class of functions $A(x)$ of the form
>
> $$A(x) = \begin{cases} w \text{ such that } B(x,w) \text{ and } |w| = \text{poly}(|x|) \\ \text{undefined if there is no such } w\,, \end{cases}$$
>
> where B is in P.

Equivalently, a function $A(x)$ is in FNP if the statement that $A(x) = w$ can be checked in polynomial time. For instance, the function that takes a Boolean formula and returns one of its satisfying assignments is in FNP. Note that FNP includes multi-valued functions, since in general there is more than one witness w for a yes-instance. It also includes *partial* functions, i.e., those which are undefined on some of their inputs, since there is no witness w if x is a no-instance.

Now let's define TFNP as the subclass of FNP consisting of *total* functions—that is, those where a witness always exists. For instance, FACTORING is in TFNP because every integer has a unique prime factorization. The proof that this factorization exists is rather specific to the primes, but many other problems are in TFNP because of some more general type of existence proof. We can capture the power of each type of proof by defining a corresponding subclass of TFNP.

Let's start with the pigeonhole principle. If f is a function from a set S to a smaller set T, then there must be some pair x, y such that $f(x) = f(y)$. After all, if $|S|$ pigeons nest in $|T| < |S|$ holes, then at least one pair of pigeons must nest together. Moreover, if f is in P, then finding such a pair is in TFNP.

If we define PPP, for "polynomial pigeonhole principle," as the class of problems of this form, then the following problem is PPP-complete:

PIGEONHOLE CIRCUIT
Input: A Boolean circuit with n inputs and n outputs, which computes a function $f:\{0,1\}^n \to \{0,1\}^n$ such that $f(x) \neq (0,\dots,0)$ for all x
Output: A pair x, y such that $f(x) = f(y)$

Here the pigeons are the 2^n possible inputs x, and the holes are the $2^n - 1$ possible values of $f(x)$ other than $(0,\dots,0)$. This is a *promise problem*, where we promise that the input circuit doesn't map anything to $(0,\dots,0)$. If we would rather not make that promise, we can ask either for a pair of pigeons, or a witness that the promise is broken:

PIGEONHOLE CIRCUIT
Input: A Boolean circuit with n inputs and n outputs, which computes a function $f:\{0,1\}^n \to \{0,1\}^n$
Output: A pair x, y such that $f(x) = f(y)$, or an x such that $f(x) = (0,\dots,0)$

Exercise 6.4 *Show that the following problem is in* PPP. *(Compare Problem 4.12.)*

PIGEONHOLE SUBSET SUM
Input: A list of ℓ integers $S = \{x_1,\dots,x_\ell\}$
Output: Two distinct subsets $A, B \subseteq S$ whose totals are equivalent mod $2^\ell - 1$

Since SUBSET SUM is NP-complete, one might expect that PIGEONHOLE SUBSET SUM is PPP-complete. At the time we write this, however, this is not known.

Another type of nonconstructive proof is the following. Any function f defined on a graph G must have a local minimum, i.e., a vertex x for which $f(x) \leq f(y)$ for all neighbors y of x. Suppose that G is exponentially large—for instance, that the set of vertices is $\{0,1\}^n$. To keep the problem of finding a local minimum in TFNP, we assume that G has polynomial degree, and that there is a polynomial-time algorithm $N(x)$ that lists all the neighbors of a given vertex x. Then we can check that x is a local minimum by computing $f(y)$ for all its neighbors.

The class of problems that ask for a local minimum in this setting is called PLS for "polynomial local search." Several interesting problems are known to be PLS-complete, including finding a local minimum of the energy of a spin glass (see Problem 5.30), or a local optimum of the TRAVELING SALESMAN problem with respect to certain edge-swapping heuristics (see Section 9.10.1).

Of course, there is a simple greedy algorithm for solving problems in PLS, modeled after gradient descent: at each step, check all the neighbors of the current vertex, and move to the one that minimizes f. As Problem 6.22 shows, we can speed this algorithm up by sampling random vertices, and starting our descent from the lowest one. However, it still takes exponential time in general.

6.7.2 *Parity and Hamiltonian Paths*

Another family of nonconstructive proofs comes from parity arguments, such as the fact that any graph has an even number of vertices with odd degree. Suppose we again have a graph G of exponential size

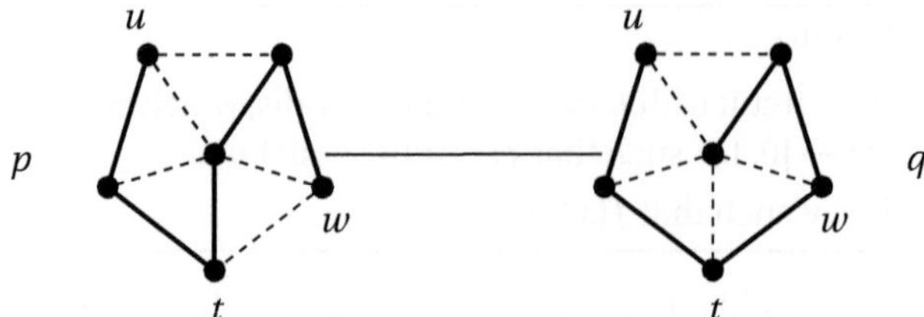

FIGURE 6.6: Two Hamiltonian paths p,q differ on two edges if and only if they are related by this kind of move. To change p to q, we connect its endpoint w to one of its neighbors t, and then disconnect t from that part of the path.

whose maximum degree is polynomial, defined implicitly by a polynomial-time algorithm $N(x)$ which returns a list of x's neighbors. Then, given a vertex of odd degree, we can ask for another one.

The class of problems of this form is called PPA for "polynomial parity argument." It contains a number of charming problems, including the following:

> ANOTHER HAMILTONIAN CYCLE
> Input: A graph H where every vertex has odd degree, and a Hamiltonian cycle c in H
> Output: Another Hamiltonian cycle in H

Thus while finding one Hamiltonian cycle is NP-complete, finding a *second* one is in PPA, at least in a graph with odd degrees. The fact that there always is another Hamiltonian cycle in such a graph is given by the following lemma.

Lemma 6.7 *Let H be a graph where every vertex has odd degree. For each edge (u,v), the number of Hamiltonian cycles passing through (u,v) is even.*

Proof Hamiltonian cycles that pass through (u,v) are in one-to-one correspondence with Hamiltonian paths that start at u and end at v. We will show that there are an even number of such paths.

We define a graph G whose vertices correspond to Hamiltonian paths in H, and where two paths p,q are neighbors if they differ on just two edges. As Figure 6.6 shows, if p starts at u and ends at w, we can change it to q by adding the edge connecting w to one of its neighbors t, and removing the edge that previously connected t to the part of p that includes w.

Now define G in this way, but only with the paths that start at u and do not go through (u,v). If p's endpoint w has odd degree, then w has an even number of neighbors t that we could connect it to in order to form q. Thus p has even degree in G.

The only exception is when $w=v$, since then we are not allowed to add the edge connecting w to u. Therefore, the odd-degree vertices in G are exactly the paths that start at u and end at v, and there are an even number of them. We show an example in Figure 6.7. □

Note that the graph H is of polynomial size by definition, since it is given as part of the input. While G is typically of exponential size, its maximum degree is at most the maximum degree of H, which is

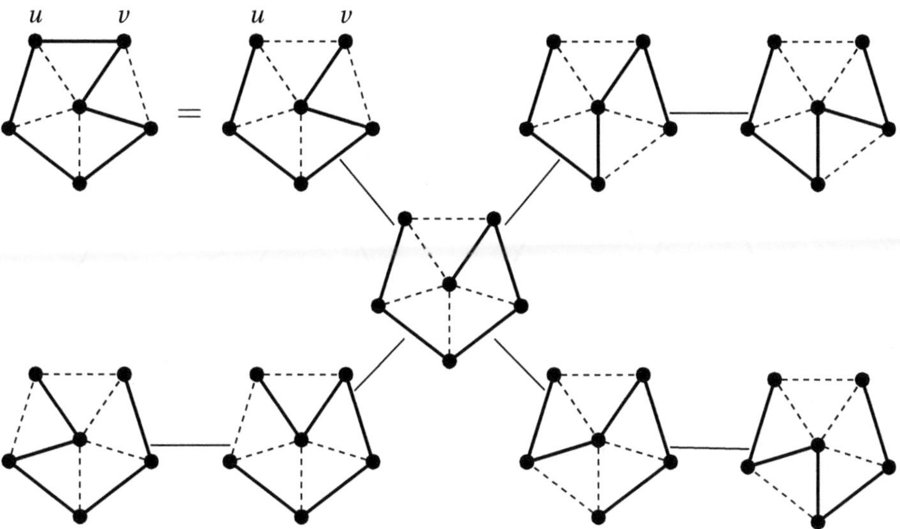

FIGURE 6.7: If we form a graph G consisting of Hamiltonian paths that start at a vertex u and do not go through an edge (u,v), where two paths are neighbors if they differ on two edges, then its odd-degree vertices are the paths that start at u and end at v. These paths are in one-to-one correspondence to Hamiltonian cycles that pass through (u,v), so there are an even number of such cycles.

polynomial. Moreover, the proof of Lemma 6.7 gives a polynomial-time algorithm $N(p)$ that computes the list of neighbors of any vertex p in G, i.e., the list of paths that differ from p on two edges. Thus ANOTHER HAMILTONIAN CYCLE is in PPA.

There is an important special case of PPA: namely, where no vertex in G has degree more than 2. In that case, G is the disjoint union of a set of paths or cycles (in G, not in H) and our parity argument simply says that the total number of endpoints of these paths is even. Given one of these endpoints, we can ask for another one—not necessarily the other endpoint of the same path.

A priori, the class of such problems might be smaller than PPA. However, it turns out that any problem in PPA can be translated into one of this form. In other words, this kind of parity argument is just as powerful, and can prove that the same things exist, as the more general parity argument where G has polynomial degree.

The reason is shown in Figure 6.8. Given a graph G, we define a new graph G' as follows, For each pair of vertices x,y of G with an edge between them, G' has two vertices associated with the ordered pairs (x,y) and (y,x). This splits each vertex x of degree d into d vertices (x,y).

Now we need to describe the edges of G'. For each x, we define a perfect matching between its neighbors, giving each y a partner y'. If x has odd degree, we define a perfect matching between all but one of its neighbors, leaving one unmatched. For instance, we could list its neighbors in order $y_1,\ldots,y_d$, and match y_i with $y' = y_{i+1}$ whenever i is odd. Then we connect each vertex (x,y) of G' to (y,x) and (x,y') as shown in Figure 6.8.

Every vertex of G' has degree 1 or 2, so G' consists of a disjoint union of paths and cycles. Moreover, G' has exactly one vertex of degree 1 for each vertex x of odd degree in G—namely (x,y) where y is x's

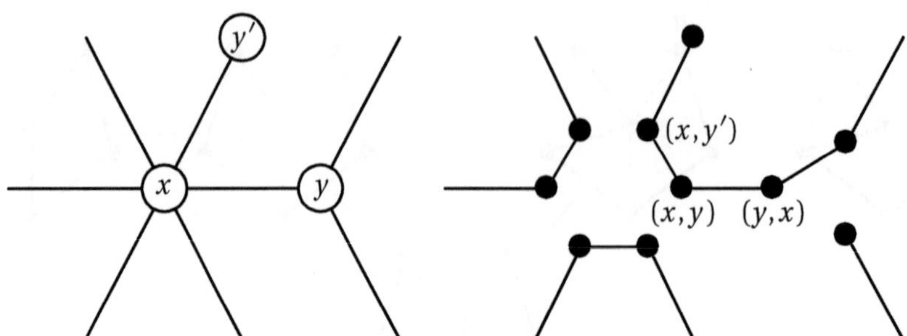

FIGURE 6.8: Splitting vertices of degree d into d vertices of degree at most 2, by defining a perfect matching on the neighbors of each vertex x.

unmatched neighbor. Thus if we can find another endpoint in G', we can find another vertex of odd degree in G.

For another problem in PPA, recall that the complement of a graph $H = (V, E)$ is $\overline{H} = (V, \overline{E})$. In other words, u and v are adjacent in $\overline{H}$ if and only if they are not adjacent in H. Then consider the following problem:

ANOTHER HAMILTONIAN PATH
Input: A graph H and a Hamiltonian path p
Output: Another Hamiltonian path in H, or a Hamiltonian path in $\overline{H}$

The following lemma shows that there always is such a path.

Lemma 6.8 *For any graph H, the total number of Hamiltonian paths in H and $\overline{H}$ is even.*

Proof We will define a bipartite graph G of exponential size, and let you finish the job. On one side, the vertices of G correspond to the Hamiltonian paths p of the complete graph defined on H's vertices. On the other side, they correspond to spanning subgraphs of H, i.e., subsets of H's edges. Thus if H has n vertices and m edges, G has $n!/2$ vertices on one side and 2^m vertices on the other.

The edges of G are defined as follows: a path p is connected to a subgraph $K \subseteq H$ if and only if $K \subset p$. That is, K is a subgraph of p other than p itself. We show an example in Figure 6.9. In Problem 6.23, we ask you to show that the only odd-degree vertices in G are the Hamiltonian paths in H and in $\overline{H}$. Therefore, there are an even number of them. □

However, we have not quite shown that ANOTHER HAMILTONIAN PATH is in PPA. The problem is that G contains vertices x of exponentially large degree, so we can't produce a list of x's neighbors with a polynomial-time algorithm $N(x)$. We would like to fix this by splitting vertices as in Figure 6.8, creating a new graph G' whose maximum degree is 2. However, there's a little problem. If x has exponentially many neighbors, we don't have time to list them all in some order and match adjacent pairs.

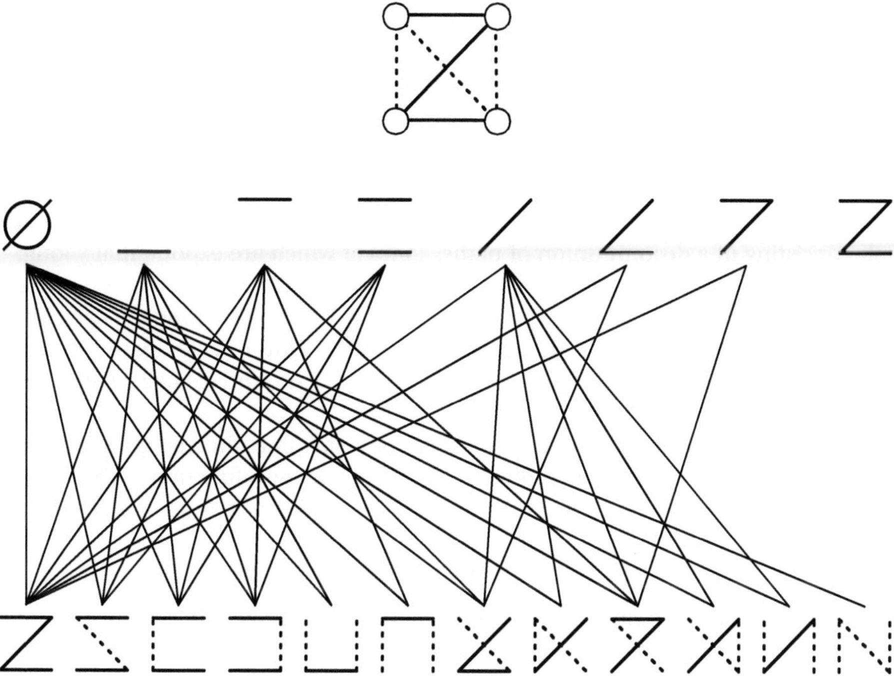

FIGURE 6.9: At the top, a graph H and its complement $\overline{H}$, drawn with solid and dotted edges respectively. Below, the bipartite graph G defined in the text. On top, we have the subgraphs of H, and below we have all the Hamiltonian paths on the complete graph $H \cup \overline{H}$. The only vertices with odd degree are the Hamiltonian paths of H and of $\overline{H}$, namely the Z on the bottom left and the N on the bottom right.

All is not lost. All we need is a perfect matching of x's neighbors that we can compute in polynomial time—in other words, a polynomial-time function ϕ such that $\phi(x,y) = y'$ and $\phi(x,y') = y$. In Problem 6.24, we ask you to find such a function for the graph G. This completes the proof that ANOTHER HAMILTONIAN PATH is in PPA.

6.7.3 *Sources, Sinks, and Fixed Points*

We can define yet another complexity class using parity arguments on *directed* graphs. Suppose that G is an exponentially large graph, where each vertex x has a polynomial number of edges pointing in or out of it, given by a pair of polynomial-time functions $N_{\text{in}}(x)$ and $N_{\text{out}}(x)$. Then, given an unbalanced vertex—that is, one whose out-degree and in-degree are different—we can ask for another one. The set of search problems of this kind is called PPAD, where D stands for "directed."

Using a similar trick as in Figure 6.8, we can focus on the case where the in-degree and out-degree of each vertex is at most one, so that each vertex has at most one predecessor and at most one successor. Then the graph is a disjoint union of directed paths, and the unbalanced vertices are the sources and sinks at which these paths start and end. Given a source or a sink, we can ask for another source or sink—again, not necessarily the one at the other end of that path.

If we remove the arrows from the edges of this graph, we get an undirected graph of maximum degree 2. This turns a PPAD problem into one in PPA: namely, given a vertex of degree 1, find another one. Thus

$$\mathsf{PPAD} \subseteq \mathsf{PPA}\,.$$

However, in general it seems hard to convert PPA problems into PPAD ones. To see why, imagine an undirected graph consisting of a disjoint union of paths, some of which are exponentially long. There is no obvious way to compute, in polynomial time, an orientation for each edge that is consistent throughout each path. If we try to do this using local information, there will be points along the path where the orientation reverses, creating vertices with in-degree or out-degree 2. Thus directed parity arguments seem to demand more structure from the problem than undirected ones do, and we believe that PPAD is smaller than PPA.

On the other hand, we can also convert the search for a sink in a directed graph with maximum in-degree and out-degree 1 to a pigeonhole problem. We simply define a function $f(x)$ as x's successor y if it has one, and otherwise $f(x)=x$. Then $f(x)=f(y)$ if y is a sink and x is its predecessor, or vice versa. Thus we also have

$$\mathsf{PPAD} \subseteq \mathsf{PPP}\,.$$

For an example of a problem in PPAD, suppose we divide a triangle into smaller triangles. We will assume here that these triangles form a regular lattice, but what we are about to say applies to any triangulation. We color the vertices of the resulting graph with three colors. This coloring doesn't have to be proper in the GRAPH 3-COLORING sense. Instead, we require the three outer corners to have three different colors—say cyan, yellow, and magenta. In addition, we require the vertices on the edge between the cyan corner and the yellow corner to be colored cyan or yellow, and analogously on the other two edges. Let's call such a coloring *legal*. Now consider the following lemma:

Lemma 6.9 (Sperner's Lemma) *Any legal coloring contains at least one trichromatic triangle.*

We can prove this lemma using an undirected parity argument, shown in Figure 6.10. We define a graph G by drawing the boundaries between the colors—that is, the edges of the dual lattice that separate vertices of different colors. As we travel along an edge of the triangle, from one outer corner to another, the color must alternate an odd number of times. Thus G has an odd number of vertices, all of which have degree 1, along each edge. Since there are three edges, the total number of such vertices is odd—but the number of odd-degree vertices must be even, so G must also have an odd number of odd-degree vertices in the interior. These vertices have degree 3, and correspond to trichromatic triangles. Thus there must be at least one trichromatic triangle, and in fact an odd number of them.

However, there is also a *directed* parity argument, shown in Figure 6.11. We choose a pair of colors, such as yellow and magenta. We define a directed graph G consisting of the edges of the dual lattice which have yellow on one side and magenta on the other, and orient them so that yellow is on their left. Paths in G can neither merge nor branch, so G consists of a disjoint union of paths and cycles. Moreover, if we add an additional row of yellow vertices as shown in the figure, then there is only one place where a path enters the lattice, and nowhere where it can leave.

Therefore, there must be a sink in the interior, and this corresponds to a trichromatic triangle with the same handedness as the coloring of the three outer corners. In addition, there can be pairs of trichromatic

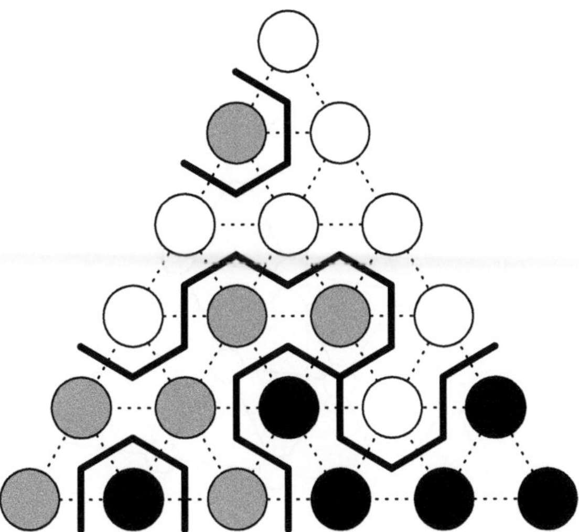

FIGURE 6.10: The undirected proof of Sperner's Lemma. We take a subset of the edges of the dual lattice, consisting of edges that separate vertices of different colors. There are an odd number of vertices of degree 1 on the outside of the triangle. Thus there must be an odd number of odd-degree vertices in the interior as well, and these correspond to trichromatic triangles.

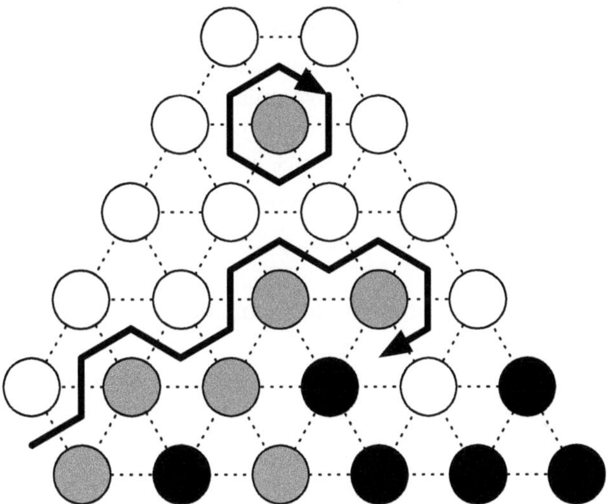

FIGURE 6.11: The directed proof of Sperner's Lemma. If we choose a pair of colors, start at a corner, and steer in between these two colors—in this case, keeping yellow vertices on our left and magenta vertices on our right—then this path must end at a trichromatic triangle. Note the additional row of yellow vertices, which keep us from floating off into space.

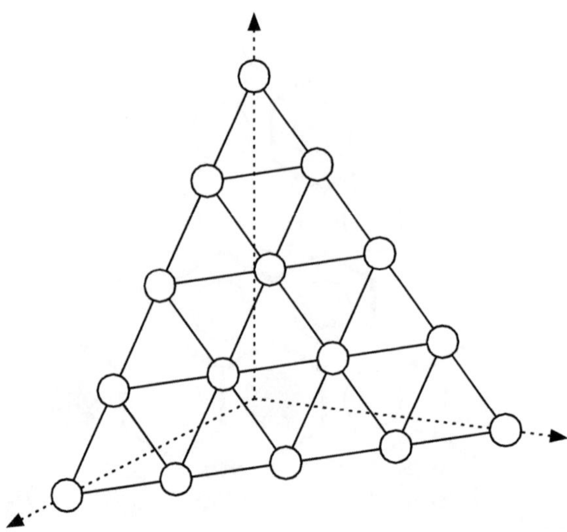

FIGURE 6.12: We can describe a triangular lattice of exponential size as the set of points (x_1, x_2, x_3) such that $x_1 + x_2 + x_3 = 2^n$.

triangles of opposite handedness, connected by the same kind of directed paths. Note that the undirected argument is insensitive to the triangles' handedness.

To turn Sperner's Lemma into a problem in PPAD, we make this lattice exponentially large. We can represent it as the lattice of nonnegative integer points (x_1, x_2, x_3) such that $x_1 + x_2 + x_3 = 2^n$ as shown in Figure 6.12. The coloring is given by some polynomial-time function c from the lattice to $\{1,2,3\}$. It is legal if $c(x_1, x_2, x_3) \in \{1,2\}$ along the edge where $c(x_3) = 0$, and analogously along the other edges. A trichromatic triangle is a triplet of neighboring points, $\{(x_1, x_2, x_3), (x_1 + 1, x_2 - 1, x_3), (x_1 + 1, x_2, x_3 - 1)\}$, to which c assigns all three colors.

> SPERNER
>
> Input: A Boolean circuit that computes a function $c(x_1, x_2, x_3)$, defined on the triplets such that $x_1 + x_2 + x_3 = 2^n$
>
> Output: Either a triplet (x_1, x_2, x_3) where c is illegal, or a trichromatic triangle

We can generalize this lemma to any number of dimensions, where we subdivide a d-dimensional simplex and color its corners with $d+1$ different colors, and restrict the colors along any face to those of its d corners. We then look for a *panchromatic* simplex, i.e., one whose vertices have all $d+1$ colors. While the lattices become slightly more complicated, we can similarly define a problem SPERNER, and in three or more dimensions this problem is PPAD-complete.

Sperner's Lemma can also be viewed as a discrete version of one of the most famous nonconstructive arguments in mathematics, Brouwer's fixed point theorem. Let B^d be an d-dimensional ball, i.e., the set of points in $\mathbb{R}^d$ whose distance from the origin is less than or equal to 1. Topologically speaking, any

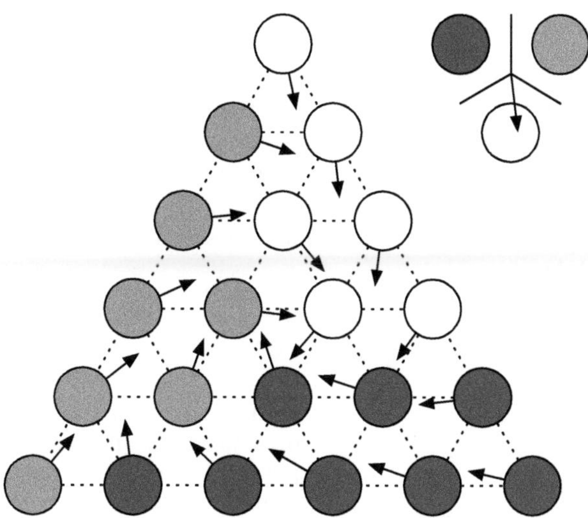

FIGURE 6.13: Reducing BROUWER to SPERNER by coloring each point according to the direction of $f(x) - x$. The center of a trichromatic triangle is an approximate fixed point.

simply-connected n-dimensional region is equivalent to B^d. For instance, a triangle, including both its boundary and its interior, is equivalent to B^2. Brouwer proved the following:

Theorem 6.10 *Let f be a continuous function from B^d to B^d. Then f has a fixed point, i.e., an $x \in B^d$ such that $f(x) = x$.*

We can define BROUWER as the problem of finding a fixed point of a given continuous function f to within a given precision ε.

We can reduce BROUWER to SPERNER, as shown for the two-dimensional case in Figure 6.13. At each point x in the lattice, there is a nonzero vector $f(x) - x$, unless x is a fixed point. If we color x according to the direction of this vector, assigning each of the three colors to a sector of width $2\pi/3$ as shown in the figure, then the center of any trichromatic triangle is a nearly-fixed point in the sense that $|f(x) - x|$ is small. By subdividing the triangle at finer and finer resolutions, we can find a sequence of these centers that converge to a fixed point.

We close this section by mentioning another problem: the problem of finding a Nash equilibrium, i.e., a set of strategies for players in a game such that no player has an incentive to change their strategy. While giving the full definition of a Nash equilibrium would take us too far afield, NASH can be phrased as an instance of BROUWER, and is therefore in PPAD. In 2006, it was shown to be PPAD-complete, even for two-player non-zero-sum games. This has led to a flurry of activity at the boundary between economics and computer science. 6.12

6.7.4 *Nonconstructivism and Nondeterminism*

These subclasses of TFNP are another example of how the theory of computational complexity can capture deep mathematical ideas. Just as P $\neq$ NP if finding solutions is harder than checking them, PPP, PPA, and PPAD are outside P if and only if these proof techniques are truly nonconstructive—if we cannot turn these arguments that something exists into a polynomial-time algorithm to find it.

Of course, if P = NP then all these classes collapse to P. Then there is no distinction between constructive and nonconstructive proofs—any proof that x exists can be made constructive, as long as we can check x in polynomial time.

6.8 The Road Ahead

> The P versus NP problem deals with the central mystery of computation. The story of the long assault on this problem is our Iliad and our Odyssey; it is the defining myth of our field.
>
> Eric Allender

We have seen that if P = NP, then finding everything from mathematical proofs to scientific theories is easier than we think it is. Much of the complexity hierarchy would collapse, and the internal structure of NP would be crushed out of existence. On the other hand, we have also seen that proving P $\neq$ NP seems very difficult, and that many of the arguments that we would like to use are doomed to failure.

Can we let ourselves off the hook? Is it possible that P = NP could be true in some way that avoids these terrible consequences, or that the P vs. NP question is impossible to resolve? Let's explore several possible loopholes and gray areas.

First, it could be that P = NP, but that the fastest algorithms for 3-SAT and SHORT PROOF take, say, $\Theta(n^{100})$ time. This would certainly make it impractical to replace the search for truth with a mechanical algorithm. Even so, the difference between checking proofs and finding them would be just one of degree, rather than one of kind. While mathematicians would stay employed for a while, there would be a fixed size of proof beyond which creativity is no longer needed.

On the other hand, this possibility also drives home the point that, counter to its usual motivation, P is not the class of "tractable" problems, that can be solved efficiently in some real sense—it is much larger than that. We know from the Time Hierarchy Theorem that there are problems in P that require $\Theta(n^c)$ time for arbitrarily large c, and we can invent artificial problems like "does G have a clique of size 100?" for which this is probably the case. We like to think that no "naturally occurring" problems have this property—that the problems we really care about are either outside P, or solvable in $O(n^c)$ time where $c \leq 6$, say. However, this may fit our past experience simply because we humans can only conceive of relatively simple algorithms.

Second, there might be a nonconstructive proof that P = NP. That is, there might be a proof that 3-SAT is in P, but which, like Robertson and Seymour's proof that certain graph properties are in P (see Note 2.11) gives no hint about what the algorithm is. But even if we don't know what it is, the algorithm is still out there. It is just a finite program, and it can be written in your favorite programming language.

One final possibility is that the P vs. NP question cannot be resolved using our standard logical axioms—that it is *formally independent*, and that neither P = NP nor P $\neq$ NP is provable. If P = NP, this would create a bizarre situation. There would be a specific program, in our favorite programming language, that solves 3-SAT in polynomial time. But it would have the odd property that we cannot prove that it works, even if someone shows its source code to us.

There are good reasons to think that this last possibility is unlikely. To see why, consider Fermat's Last Theorem—ignoring Wiles' recent proof—stating that there are no integers $x, y, z > 0$ and $n > 2$, such that $x^n + y^n \neq z^n$. If it were false, there would be a simple proof that it is false—namely, a counterexample x, y, z, n. Therefore, if it were independent of some system of logic powerful enough to check counterexamples, it would have to be true.

Now consider the following statement, which we call "3-SAT is hard":

> For all $n \geq 1000$, there is no Boolean circuit with $n^{\log_{10} n}$ or fewer gates that solves all 3-SAT instances of size n.

If $\mathsf{P} = \mathsf{NP}$, this statement is clearly false, since if 3-SAT is in P then such circuits exist for all sufficiently large n. On the other hand, if $\mathsf{P} \neq \mathsf{NP}$, it is reasonable to believe that this statement is true. It could only fail if NP is contained in $\mathsf{TIME}(n \log n)$ or its nonuniform counterpart, or if the complexity of 3-SAT oscillates bizarrely, making it easy for some values of n even though it is hard for others. Barring these possibilities, we can take "3-SAT is hard," or something like it, as a proxy for $\mathsf{P} \neq \mathsf{NP}$.

But "3-SAT is hard" has the same logical structure as Fermat's Last Theorem. If it is false, then there is a finite proof of that fact—such as a circuit C with only a billion gates which solves all 3-SAT instances of size 1000. Checking that C works on all such instances would take a long, but finite, time. Therefore, if "3-SAT is hard" is independent, then it must be true, and $\mathsf{P} \neq \mathsf{NP}$. This suggests that the P vs. NP question really does have an answer, and that one possibility or the other will eventually be proved.

The P vs. NP question will be with us for a long time. In fact, we would wager that it will turn out 6.13
to be the hardest of the Millennium Prize Problems. We have faith that $\mathsf{P} \neq \mathsf{NP}$, and that this will be proved someday. But this proof will require truly new mathematics, and will use ideas that are currently far beyond our reach.

Problems

> If you don't work on important problems, it's not likely that you'll do important work.
>
> Richard Hamming

6.1 This far and no farther. Consider the version of INDEPENDENT SET given at the end of Section 4.2.4:

> INDEPENDENT SET (exact version)
>
> Input: A graph $G = (V, E)$ and an integer k
>
> Question: Does G's largest independent set have size k?

Or similarly, the problem CHROMATIC NUMBER from Problem 4.22:

CHROMATIC NUMBER
Input: A graph G and an integer k
Question: Is k the smallest integer such that G is k-colorable?

Show that for both these problems, the property $A(x)$ that x is a yes-instance can be expressed as follows, where B and C are in P:

$$A(x) = (\exists y : B(x,y)) \wedge (\forall z : C(x,z)) .$$

Equivalently, show that there are two properties A_1, A_2 in NP such that $A(x)$ holds exactly if $A_1(x)$ is true but $A_2(x)$ is false. Thus A can be thought of as the difference between two NP properties, and the class of such properties is called DP. Now show that

$$\mathsf{DP} \subseteq \Sigma_2\mathsf{P} \cap \Pi_2\mathsf{P} ,$$

and give some intuition for why this inclusion should be proper.

6.2 Mate in k. Argue that "mate in k moves" problems in Generalized Chess are in Σ_kP even on a board of *exponential* size, as long as there are only a polynomial number of pieces. Hint: you just need to check that the moves can be described with a polynomial number of bits, and that we can check in polynomial time, given the initial position and the sequence of moves, whether White has won.

6.3 Constructible time. Prove that the function $f(n) = 2^n$ is time constructible. Prove also that if $f(n)$ and $g(n)$ are time constructible, then so are $f(n) + g(n)$, $f(n) \times g(n)$, and $f(g(n))$, where for the last one we assume that $f(n) \geq n$.

6.4 More padding. Generalize the padding argument of Section 6.1.2 to show that, for any functions f_1, f_2, and g where $g(n) \geq n$, if $\mathsf{NTIME}(f_1(n)) \subseteq \mathsf{TIME}(f_2(n))$ then $\mathsf{NTIME}(f_1(g(n))) \subseteq \mathsf{TIME}(f_2(g(n)))$.

Use this to show that if EXP = NEXP, then EXPEXP = NEXPEXP; or more concretely, that for any constant $c > 0$,

$$\mathsf{TIME}(2^{2^{O(n^c)}}) = \mathsf{NTIME}(2^{2^{O(n^c)}}) .$$

6.5 In the stratosphere, nondeterminism doesn't matter. Define $2 \uparrow^k n$ as $2^{\cdot^{\cdot^{2^n}}}$, where the tower is k levels high (see also Note 7.12). Now consider the following rather powerful complexity class:

$$\mathsf{ZOWIE} = \bigcup_k \mathsf{TIME}(2 \uparrow^k n) .$$

Show that ZOWIE = NZOWIE. So, if our computation time is stratospheric enough—if our class of time complexity functions has strong enough closure properties—then nondeterminism makes no difference.

6.6 Better than exhaustive search. Figure 6.14 shows an algorithm for 3-SAT, which we call `MS` for Monien–Speckenmeyer. It relies on the following fact: suppose a 3-SAT formula ϕ contains a clause $(x_1 \vee x_2 \vee x_3)$, where each x_i is either a variable or its negation. Then either $x_1 = \texttt{true}$, or $x_1 = \texttt{false}$ and $x_2 = \texttt{true}$, or $x_1 = x_2 = \texttt{false}$ and $x_3 = \texttt{true}$. We use $\phi[x_i = \texttt{true}]$, for instance, to denote the formula resulting from setting $x_i = \texttt{true}$. Finally, we use `2-sat` to denote a polynomial-time algorithm for 2-SAT, such as the one given in Section 4.2.2.

Let $T(n)$ be the running time of this algorithm on a 3-SAT formula with n variables and poly(n) clauses, measured in the number of recursive function calls. Show that $T(n)$ is at worst

$$T(n) = T(n-1) + T(n-2) + T(n-3) .$$

Conclude that the running time of this algorithm is $c^n \operatorname{poly}(n)$, where $c = 1.839...$ is the unique real root of the cubic equation

$$c^3 - c^2 - c - 1 = 0 .$$

```
MS
input: a 3-SAT formula φ
output: whether φ is satisfiable or not
begin
    if φ contains no 3-clauses then return 2-SAT(φ) ;
    let c = (x1, x2, x3) be the first 3-clause in φ ;
    if MS(φ[x1 = true]) = "yes" then return "yes";
    else if MS(φ[x1 = false, x2 = true]) = "yes" then return "yes";
    else if MS(φ[x1 = false, x2 = false, x3 = true]) = "yes" then
        return "yes" ;
    else return "no" ;
end
```

FIGURE 6.14: The Monien–Speckenmeyer algorithm for 3-SAT.

Generalize this to larger k, and show that k-SAT can be solved in time $c_k^n \operatorname{poly}(n)$ where c_k is the largest real root of

$$c^k(c-2)+1=0.$$

When k is large, c_k approaches $2-2^{-k}$. Better algorithms are known, but this at least shows that we can achieve $c_k < 2$ for any k.

6.5

6.7 Ticking clocks. We claim in the text that a random-access machine can run a clock in $O(1)$ time per tick. This is certainly true if our variables can be arbitrarily large integers, and we can increment them in a single step. However, what if each variable can only hold a constant number of bits? To increment the time t from $2^\ell - 1$ to 2^ℓ, we have to flip $\ell+1$ bits, so some ticks take more than $O(1)$ time. However, show that the total number of bit flips we need to do in order to count from 0 to t is at $O(t)$.

6.8 An infinite number of paradoxes. For the version of CATCH22 we defined in the text, all we know is that it differs from any program Π that runs in $f(n)$ steps on at least one input—namely, Π itself. If CATCH22 and Π differ on only a finite number of inputs, there is some danger that we can fix Π by "patching" it on these inputs, and thus solve CATCH22 in $o(f(n))$ time. To fix this, consider the following version of CATCH22:

> CATCH22$'(\Pi,\ell)$
>
> Input: A program Π and an integer ℓ given in unary
>
> Output: If Π halts within $f(|\Pi|+\ell)$ steps when given (Π,ℓ) as input, return $\overline{\Pi(\Pi,\ell)}$. Otherwise, answer "don't know."

Show that for any program Π that runs in $f(n)$ steps, CATCH22$'$ disagrees with Π on an infinite number of inputs. Use this to show that CATCH22$'$ is not in TIME$(g(n))$ if $g(n)=o(f(n))$. Hint: show that $g(n) < f(n)$ for all but a finite number of values of n.

6.9 A hierarchy for nondeterministic time. How can we use diagonalization against a nondeterministic complexity class? We can't simply run a witness-checking program Π on all possible witnesses and negate its result, since this would take exponential time. But we can use a trick called "lazy diagonalization" to define a range of inputs, and make sure that our problem differs from Π on at least one input in this range.

```
NCatch22((Π, y), w)
begin
    if y < 2^b then
        if |w| ≤ f(|(Π, y)|) then run Π((Π, y), w) for f(|(Π, y)|) steps ;
        if Π((Π, y), w) returns "yes" then return "yes" ;
        else return "no" ;
    else if y = 2^b then
        for all strings z with |z| ≤ f(|(Π, 0)|) do
            run Π((Π, 0), z) for f(|(Π, 0)|) steps ;
            if Π((Π, 0), z) returns "yes" then return "no" ;
        return "yes" ;
    else return "no" ;
end
```

FIGURE 6.15: The witness-checking program for NCATCH22, our nondeterministic version of CATCH22. If $y < 2^b$ then (Π, y) is a yes-instance of NCATCH22 if and only if $(\Pi, y+1)$ is a yes-instance of Π, while $(\Pi, 2^b)$ is a yes-instance of NCATCH22 if and only if $(\Pi, 0)$ is a no-instance of Π.

As before, fix a function $f(n)$. We want to diagonalize against all witness-checking programs $\Pi(x, w)$ with the property that, if x is a yes-instance, there is a witness w consisting of $|w| \leq f(|x|)$ bits such that $\Pi(x, w)$ returns "yes" in $f(|x|)$ steps or less. One type of input x that we could give to Π is a pair (Π, y), consisting of Π's own source code and an integer y. If there is a w such that $\Pi((\Pi, y), w)$ returns "yes," then (Π, y) is a yes-instance of Π.

Let's say that y lies in the range $0 \leq y \leq 2^b$, where b is a parameter that we will choose below. We define a witness-checking program NCATCH22 with a clever twisted loop through this range. If $y < 2^b$, then NCATCH22 runs Π on $(\Pi, y+1)$ with the witness w. If $\Pi((\Pi, y+1), w)$ returns "yes," then NCATCH22 does too.

On the other hand, if $y = 2^b$, NCATCH22 cycles around to $y = 0$. In this case it ignores w and scans through all possible witnesses z, running $\Pi((\Pi, 0), z)$ for each one, and determining whether $(\Pi, 0)$ is a yes-instance of Π or not. It then negates this result, returning "yes" if and only if there is no such z. We show the pseudocode of NCATCH22, complete with bounds on Π's running time and witness size, in Figure 6.15.

Thus for any $y < 2^b$, (Π, y) is a yes-instance of NCATCH22 if and only if $(\Pi, y+1)$ is a yes-instance of Π. On the other hand, $(\Pi, 2^b)$ is a yes-instance of NCATCH22 if and only if $(\Pi, 0)$ is a no-instance of Π. Show that this implies there is some y such that (Π, y) is a yes-instance of NCATCH22 but a no-instance of Π, or vice versa. Thus there is no witness-checking program for NCATCH22 whose running time and witness size is $f(n)$.

On the other hand, show that if b is sufficiently large as a function of $|\Pi|$, then the running time of NCATCH22 is $O(s(f(n)))$ as a function of its input size $n = \left|(\Pi, y)\right|$, where $s(f(n))$ is the time it takes the interpreter to simulate $f(n)$ steps. Feel free to assume that f is time constructible and that $s(n) = O(n \log n)$, say. Then conclude, for instance, that

$$\mathsf{NTIME}(n^c) \subset \mathsf{NTIME}(n^d),$$

for any constants $1 < c < d$. Hint: checking all possible witnesses z for $(\Pi, 0)$ takes exponential time as a function of $f(|(\Pi, 0)|)$, but we only do this when the input size is $n = \left|(\Pi, 2^b)\right|$.

6.10 Optimization with an oracle. Show that the optimization problems of finding the largest independent set, the smallest vertex cover, or the largest clique, are all in P^{NP}. Hint: if you have access to an oracle for the decision versions of these problems as defined in Section 4.2.4, first find the size of these sets. Then use their self-reducibility, as you did for SAT in Problem 1.10, to find the sets themselves.

6.11 The oracle says yes or no. Let's explore the class P^{NP} in more detail. First, show that P^{NP} contains both NP and coNP. Then, show that $\mathsf{P}^{\mathsf{NP}} \subseteq \Sigma_2\mathsf{P} \cap \Pi_2\mathsf{P}$ and also that $\mathsf{NP}^{\mathsf{NP}} = \Sigma_2\mathsf{P}$.

Hint: let Π be the polynomial-time program that makes calls to the NP oracle. An input x is a yes-instance if there is a run of Π which returns "yes," which receives "yes" and "no" answers from the oracle at the appropriate places.

6.12 Switching ANDs and ORs. In this problem, we will guide you through the proof of the following lemma. The proof uses some machinery from discrete probability, for which we refer you to the Appendix.

Lemma 6.11 (The Switching Lemma) *Consider a circuit with n inputs $x_1,\ldots,x_n$ consisting of a layer of* OR *gates which feed into a single* AND *gate, where both types of gates have arbitrary fan-in. Now suppose that we randomly restrict this circuit as follows. For each variable x_i, with probability $1-n^{-1/2}$ we set x_i randomly to* `true` *or* `false`*, and with probability $n^{-1/2}$ we leave it unset. For any constant a, if there are $O(n^a)$* OR *gates, then with probability $1-o(n^{-a})$ the resulting circuit can be expressed as a layer of a constant number of* AND *gates, which feed into a single* OR *gate.*

We will split the restriction into two stages. At the first stage, we set each variable with probability $1-p$, where $p = n^{-1/4}$. Then in the second stage, we set each of the remaining variables with probability $1-p$. In each case we set the variable `true` or `false` with equal probability $(1-p)/2$. Then the probability that a variable is still unset after both stages is $p^2 = n^{-1/2}$. The purpose of these two stages is to give us two bites at the apple—we use the first stage to eliminate gates of high fan-in, and the second stage to simplify the remaining gates.

Now, each OR gate could have fan-in up to n. But if any of its inputs are set to `true`, it disappears. Thus it survives and depends on c variables if exactly c of its inputs are unset, and the others are set to `false`. Show that there is a constant c such that, with probability $1-o(n^{-a})$, after the first stage of the restriction all surviving OR gates depend on at most c inputs. You may find the bound (A.10) from Appendix A.2.5 useful.

So, assume that after the first stage, all surviving OR gates have fan-in at most c. We now move on to the second stage, which is a bit tricky. If there are t OR gates, let $H_1,\ldots,H_t$ denote the sets of variables feeding into them. If the H_i are disjoint, then the restriction affects each OR gate independently. The AND gate disappears if any of the ORs have their output set to `false`. Show that this happens to a given OR with constant probability.

The problem is that the H_i are not in general disjoint, so a more careful analysis is needed. We will show by induction on c that, for each constant c, there is a constant h_c with the property that if every OR gate on the first level has fan-in at most c, then with probability $1-o(n^{-a})$ after the second stage, the AND gate depends on at most h_c of the remaining variables.

For the base case, when $c=1$ the AND gate depends directly on the x_i. In this case, show using an argument similar to the one in the first stage that, for some constant h_1, with probability $1-o(n^{-a})$ the AND gate either disappears, or depends on at most h_1 unset variables.

For the induction step, let $I \subset \{H_1,\ldots,H_t\}$ be the largest family of OR gates such that the H_i are disjoint. Let $H' = \bigcup_{i\in I} H_i$ denote the set of variables they depend on, and let $\ell = |I|$. Show that there is a constant r such that the AND survives with probability $o(n^{-a})$ if $|I| > r \log n$. Then show that if $\ell = O(\log n)$, then with high probability only c' of the variables in H' remain unset after the second stage, for some constant c'. Finally, use the fact that the AND only depends on a variable outside H' if it would do so for some setting of these c' variables.

Finally, show that if the AND gate only depends on a constant number of variables, then we can write its output as the OR of a constant number of ANDs.

6.13 Reducing the depth. Now suppose that a function f can be computed by an AC^0 circuit of depth d and width $O(n^a)$ consisting of alternating layers of AND and OR gates. We can ignore NOT gates, since as discussed in the text we can push them up through the circuit until they simply negate some of the inputs.

Show that if we randomly restrict f as in Problem 6.12, then with probability $1-o(1)$ the resulting function can be computed by an AC^0 circuit of depth $d-1$ whose width is still $O(n^a)$. Finally, use the Chernoff bound (see Appendix A.5.1) to show that, with high probability, $\Omega(n^{1/2})$ variables remain unset.

6.14 You can't be complicated without being deep. Use the Switching Lemma and Problem 6.13 to prove Theorem 6.4, showing that no function in AC^0 is γ-complicated if $\gamma(n) = n^{o(1)}$. Then sketch a proof that PARITY requires circuits of depth $\Omega(\log\log n)$.

6.15 Complicatedness is constructive. Finally, show that under reasonable assumptions about the complexity of computing $\gamma(n)$, the property of being γ-complicated is in P as a function of the truth table of f. In fact, with no assumptions about γ, it is in AC^0. Thus it is "constructive" in the sense of Section 6.5.3.

6.16 Complicatedness is common. Show that if $\gamma = A\log_2 n$ for any constant $A > 1$, then a uniformly random function is γ-complicated with high probability. Thus it is "common" in the sense of Section 6.5.3. Hint: compute the expected number of subsets S and settings of the variables outside S such that this restriction of f is constant, and use the first moment method of Appendix A.3.2.

6.17 Majority is deep. Show that PARITY is AC^0-reducible to MAJORITY. In other words, show that if, in addition to AND and OR gates, we have a supply of MAJORITY gates, which output `true` if a majority of their inputs are `true`, then we can compute PARITY with a circuit of constant depth and polynomial width. We can write this as

$$\text{PARITY} \le_{\mathsf{AC}^0} \text{MAJORITY}.$$

Conclude that since PARITY is not in AC^0, neither is MAJORITY. Hint: prove the more general fact that MAJORITY gates let us compute any symmetric function, i.e., any function that depends just on the number of true inputs.

6.18 More deep problems. In the same sense as the previous problem, show that PARITY is AC^0-reducible to REACHABILITY and to MULTIPLICATION. In other words, show that if we have gates which take the adjacency matrix of a graph as input and return whether there is a path between, say, the 1st and nth vertices—or which take two binary integers and return their product—then we can compute PARITY in constant depth and polynomial width. Conclude that, like MAJORITY, neither of these problems is in AC^0.

6.19 Splitting the difference. Show how to generalize Theorem 6.6 to prove the following: for any pair of problems B, C in NP where $B \le C$ but $C \not\le B$, there is a problem A such that

$$B \le A \le C \text{ but } C \not\le A \not\le B.$$

Hint: Theorem 6.6 is the special case where B is in P and C is NP-complete. By iterating this process, we can show that, unless P = NP, there is an infinite hierarchy of problems in NP, each of which is reducible to the one above it but not to the one below.

6.20 Incomparable problems. Show that if P $\ne$ NP, then there is a polynomial-time computable function f such that, if we define two problems A_{even} and A_{odd} as follows,

$$A_{\text{even}}(x) = \text{SAT}(x) \wedge (f(|x|) \text{ is even})$$
$$A_{\text{odd}}(x) = \text{SAT}(x) \wedge (f(|x|) \text{ is odd}),$$

then A_{even} and A_{odd} are incomparable. That is, neither one is reducible to the other.

Hint: as in the proof of Theorem 6.6, look for instances on which each polynomial-time program Π_i fails to reduce A_{even} to A_{odd} and vice versa, as well as failing to solve either one, or reduce SAT to either one. Work on some of these when $f(n)$ is odd, and on others when $f(n)$ is even, in a way that contradicts our assumption that P $\ne$ NP if we get stuck and $f(n)$ becomes constant.

6.21 Bounded nondeterminism. Another way to generate complexity classes between P and NP-completeness is to bound the amount of nondeterminism. For a given function $f(n)$ which is at most polynomial in n, let $\mathsf{NONDET}(f(n))$ be the class of problems for which yes-instances have witnesses of length $O(f(n))$ which can be checked in polynomial time. In particular, let's consider witnesses of polylogarithmic length. For each integer k, define

$$\mathsf{B}_k = \mathsf{NONDET}(\log^k n).$$

We showed in Problem 4.19 that $\mathsf{B}_1 = \mathsf{P}$. More generally, show that $\bigcup_k \mathsf{B}_k$ is in the class QuasiP defined in Problem 2.20.

It seems likely that the B_k form a hierarchy of distinct classes. As evidence for this, show that if $\mathsf{B}_j = \mathsf{B}_k$ for some $j < k$, then SAT problems on n variables can be solved deterministically in $2^{O(n^{j/k})}$ time. Hint: using a padding argument to put a version of SAT in B_k.

6.22 Randomized search. Let's consider local search problems on the n-dimensional hypercube. We want to find a local minimum of a function f, i.e., an $x \in \{0,1\}^n$ such that $f(x) \le f(y)$ for all y which differ from x on a single bit. Rather than requiring that f is in P as in Section 6.7.1, we assume that f is a black-box function, designed by an adversary. We will define the running time T of an algorithm as the number of queries it makes to this black box—that is, the number of points x in the hypercube at which it evaluates $f(x)$.

Show that for any deterministic algorithm, the adversary can design an f such that $T = 2^n$—that is, the algorithm is forced to evaluate f everywhere. On the other hand, show that there is a randomized algorithm that achieves $T = O(2^{n/2}\sqrt{n})$ with constant probability. Hint: start by evaluating f at t random places, and then pursue a greedy algorithm from the one with the smallest value of f. Show that this gives $T \le 2^n n/t + t$ with constant probability, and optimize t. 6.14

6.23 Paths and parity. Prove Lemma 6.8 by showing that the only odd-degree vertices in the bipartite graph G are the Hamiltonian paths in H and in $\overline{H}$.

6.24 Matching paths. Define a matching function for the neighbors of each vertex in the bipartite graph G defined in Lemma 6.8, and complete the proof that ANOTHER HAMILTONIAN PATH is in PPA.

In other words, define a polynomial-time computable functions ϕ and ψ with the following properties. Let K, p be a neighboring pair in G. That is, let $K \subseteq H$ and let p be a Hamiltonian path in $H \cup \overline{H}$ such that $K \subset p$. Then for each K, we have $\phi(K,p) = p'$ and $\phi(K,p') = p$ where p' is another path such that $K \subset p'$. Similarly, for each p we have $\psi(p,K) = K'$ and $\psi(p,K') = K$ where K' is another subgraph such that $K' \subset p$—unless p is a Hamiltonian path of H or $\overline{H}$, in which case it has a single neighbor K which is unmatched or matched to itself.

6.25 Parity mod p. Let p a positive integer. Prove that in any bipartite graph, if there is a vertex whose degree is not a multiple of p, then there is another one. This can be used to define yet another complexity class, $\mathsf{PPA}[p]$, based on this method of nonconstructive proof.

6.26 Permutations and pigeons. A *one-way permutation* is a one-to-one function $f : \{0,1\}^n \to \{0,1\}^n$ such that f is in P but f^{-1} is not. For instance, $f(x) = a^x \bmod p$ for prime p is believed to be one such function if we restrict to the domain $x \in \{0,\ldots,p-2\}$, since DISCRETE LOG appears to be outside P. Show that if PPP = P then no one-way permutations exist, by reducing the problem of inverting a given function f to PIGEONHOLE CIRCUIT.

6.27 Almost-natural proofs. It's natural to ask how far we can push the parameters of commonality and constructiveness in our definition of natural proofs. The Razborov–Rudich result still works if we call a property "common" if it holds for a random function with probability $2^{-\mathrm{poly}(n)} = 2^{-2^{O(\log n)}}$. However, if we reduce this probability to $2^{-2^{O(\log^c n)}}$ for some constant $c > 1$, then there are properties that are useful against P/poly, subject to the assumption that SAT is exponentially hard.

Specifically, let m be a function of n, and let C be the following property of f. If the last $n - m$ bits of x are zero, then $f(x) = 1$ if and only if the first m bits encode a yes-instance of SAT, and $f(x) = 0$ otherwise. If any of the last $n - m$ bits of x are nonzero, then $f(x)$ can take any value. Show that if f is a random function, then C holds with probability 2^{-2^m}.

Now suppose that there is some constant $\varepsilon > 0$ such that no family of circuits of size $2^{O(n^\varepsilon)}$ can solve SAT. Show that if $m = \log^c n$ for some $c > 1/\varepsilon$, then C is useful against P/poly. On the other hand, show that C is constructive, in that it can be checked in time polynomial (indeed, linear) as a function of the length of f's truth table.

6.15

6.28 Sparse sets. A set of strings of length n is called *sparse* if it has only poly(n) members of each length n. Show that if a problem with a sparse set of yes-instances is NP-complete, then NP $\subseteq$ P/poly.

Notes

6.1 Problems for the next century. The Clay Mathematics Institute has offered a $1,000,000 prize for each of their seven Millennium Prize Problems. Of course, if you prove that P = NP, you might be able to prove all the others as well. The Fields medalist Steven Smale also lists P vs. NP as one of the key mathematical challenges for the 21st century [762].

6.2 The polynomial hierarchy. The classes Σ_kP and Π_kP first appeared in Meyer and Stockmeyer [575]. Stockmeyer [776] and Wrathall [835] developed it further, and gave a number of problems which are complete for various levels of the hierarchy. In particular, if ϕ is a 3-SAT formula, verifying statements of the form

$$\underbrace{\exists x_{1,1} : \cdots : \exists x_{1,n_1}}_{\exists} : \underbrace{\forall x_{2,1} : \cdots : \forall x_{2,n_2}}_{\forall} : \underbrace{\exists x_{3,1} : \cdots : \exists x_{3,n_3}}_{\exists} : \cdots : \phi(x_{1,1}, \ldots, x_{k,n_k}),$$

where there are k alternating blocks of quantifiers, is Σ_kP-complete. Similarly, they are Π_kP-complete if their outermost quantifier is a $\forall$.

6.3 Gödel's letter. We are very much indebted to Hartmanis [370], Sipser [758] and Aaronson [1] for their discussion's of Gödel's letter. Surprisingly, and contrary to our current intuition, Gödel expresses some optimism that proofs can be found in polynomial time. He points out that exhaustive search can be avoided in certain number-theoretic problems, like calculating a quadratic residue, and expresses curiosity about the hardness of PRIMALITY. If only we could discuss computational complexity with Gödel today!

Gödel's footnote is also important. One can argue that the choice of axioms is a major part of the creative process; this is certainly true in the theory of infinite sets, for instance. On the other hand, one could argue that the problem of finding short, elegant sets of axioms, that imply all the mathematical truths we are used to, and do not lead to a contradiction, is also in NP. So, even the process of selecting axioms might be easily automated if P = NP.

6.4 Artificial Intelligence. By decrying mechanical solutions rather than creative ones, we do not mean to raise the canard of machine vs. human intelligence. Neither of us see any reason why machines cannot be just as creative and intuitive as humans are, and filled with just as much true understanding of mathematical and physical problems. Our point is that it would be disappointing if all the problems we care about can be solved by a machine that *lacks* these qualities—a machine that has more in common with a coffee grinder than it does with us.

Finally, one could argue that the deepest scientific problems cannot be described as searching for explanations in a preordained space, however large, of possible theories. Einstein's insight that space and time are not fixed feels more like a radical expansion of the space of theories, rather than a clever focusing on one part of it. Of course, this may be because the search space is wider than we can currently conceive.

6.5 Doing better than exhaustive search. There is a long history of exponential-time algorithms for NP-complete problems that do far better than a naive exhaustive search. The 3-SAT algorithm of Problem 6.6 is due to Monien and Speckenmeyer [598]. They also gave an improved version, which solves k-SAT in time $O(c_{k-1}^n \operatorname{poly}(n))$. For $k = 3$, this gives a running time that grows as c_2^n where $c_2 = \varphi = 1.618...$, the golden ratio. For more algorithms of this kind, both randomized and deterministic, see Section 10.3, Note 10.3, and David Eppstein [267].

6.6 Hierarchy Theorems. The original Time Hierarchy Theorem was proved by Hartmanis and Stearns [372]. They show that the overhead of simulation on a Turing machine is $s(t) = O(t^2)$, and therefore that $\mathsf{TIME}(g(n)) \subseteq \mathsf{TIME}(f(n))$ if $g(n)^2 = o(f(n))$. This was improved to $s(t) = O(t \log t)$ by Hennie and Stearns [397]. For multi-tape Turing machines, tighter results are possible; see Fürer [305].

As we state in the text, there are RAM models and programming languages where $s(t) = O(t)$, and where the constant in O does not depend on Π. For such models there is a Time Hierarchy Theorem with constant factors. In other words, there is a constant C such that we can do more in $Cf(n)$ time than in $f(n)$ time. In fact, for some languages and models this constant can be $C = 1 + \varepsilon$ where ε is arbitrarily small. See Jones [446], Sudborough and Zalcberg [781], and Ben-Amram [96].

Hierarchy theorems were given by Cook [197] for nondeterministic time, by Hartmanis, Lewis, and Stearns for deterministic space [773], and by Ibarra [415] for nondeterministic space.

6.7 Oracles and relativized classes. The fact that there exist oracles A and B such that $\mathsf{P}^A = \mathsf{NP}^A$ but $\mathsf{P}^B \neq \mathsf{NP}^B$ was shown in 1975 by Baker, Gill, and Solovay [74]. While we use a random oracle for B in our discussion, they showed that it is possible to create such an oracle deterministically, using a clever type of diagonalization.

Bennett and Gill [103] showed that, in the presence of random oracle A, many of the relationships we believe in hold with probability 1. In particular, for almost all oracles A, we have

$$\mathsf{LOGSPACE}^A \subset \mathsf{P}^A \subset \mathsf{NP}^A \subset \mathsf{PSPACE}^A \subset \mathsf{EXP}^A,$$

as well as $\mathsf{NP}^A \neq \mathsf{coNP}^A$, where LOGSPACE and PSPACE are the classes of problems we can solve with $O(\log n)$ or $\operatorname{poly}(n)$ memory (see Chapter 8). We also have $\mathsf{BPP}^A = \mathsf{P}^A$, where BPP is the class of problems we can solve with a randomized polynomial-time algorithm (see Section 10.9). All of these relations are exactly what we believe to be true in the real world.

This inspired the *random oracle hypothesis*, which claimed that for any two complexity classes $\mathsf{C}_1, \mathsf{C}_2$, $\mathsf{C}_1 = \mathsf{C}_2$ if and only if $\mathsf{C}_1^A = \mathsf{C}_2^A$ for almost all A. However, in 1990 Chang et al. [152] showed that $\mathsf{IP}^A \neq \mathsf{PSPACE}^A$ for almost all A. As we will show in Section 11.2 using the non-relativizing technique of arithmetization, $\mathsf{IP} = \mathsf{PSPACE}$, and so the random oracle hypothesis is false.

6.8 Nonuniform circuit families and complexity with advice. We will see nonuniform complexity classes like P/poly again in our discussion of pseudorandom generators in Section 11.4.4. In order to be useful for cryptography and derandomization, pseudorandom generators need to have the property that they "fool" nonuniform algorithms as well as uniform ones—in other words, that no family of small circuits can detect any pattern that distinguishes their output from truly random bits. In that chapter we will describe P/poly as the class of problems solvable in polynomial time given a polynomial amount of *advice*, i.e., a string of $\operatorname{poly}(n)$ bits that depends only on n. This is equivalent to the circuit definition given here, since we can hardcode the advice into the circuit, or describe the structure of the circuit with the advice.

The class P/poly includes some undecidable problems, simply because any function of n can be included in the advice. However, Karp and Lipton [463] showed that if $\mathsf{NP} \subseteq \mathsf{P/poly}$, then $\Sigma_2\mathsf{P} = \Pi_2\mathsf{P}$ and the polynomial hierarchy collapses to its second level. The proof goes like this. First, if $\mathsf{NP} \subset \mathsf{P/poly}$, then there is a family of polynomial-size circuits that tell whether a SAT formula is satisfiable. Since SAT is self-reducible (see Problem 4.1), this means that there is a family of polynomial-size circuits that output a satisfying assignment if there is one. Now consider a property A in Π_2P. It can be expressed in the form

$$A(x) = \forall y : \exists z : \phi(x, y, z)$$

for some SAT formula ϕ. For a given x and y, $\phi(x, y, z)$ is a SAT formula depending only on z. So, if $A(x)$ is true, there exists a circuit C that takes x and y as input and returns a z such that $\phi(x, y, z)$ holds, and we can write

$$A(x) = \exists C : \forall y : \phi(x, y, C(x, y)).$$

Thus any Π_2P property can be written as a Σ_2P property, and $\Pi_2\text{P} = \Sigma_2\text{P}$. This means that we can switch the two innermost quantifiers in any quantified Boolean formula. But since we can merge like quantifiers, we can absorb any constant number of quantifiers into these two, and $\text{PH} = \Sigma_2\text{P} = \Pi_2\text{P}$. (Compare Exercise 6.1, where we showed that if NP = coNP then PH collapses to its first level.)

Results like these suggest that nonuniform complexity classes are not that much more powerful than their uniform counterparts—at least until we get to circuits of size $O(2^n)$, which as Exercise 6.2 shows can compute any function at all simply by hardcoding its truth table into the circuit.

6.9 The Switching Lemma and AC^0. The Switching Lemma of Problem 6.12 and the fact that PARITY $\notin \text{AC}^0$ was proved by Furst, Saxe, and Sipser [307]. Independently, Ajtai [27] proved PARITY $\notin \text{AC}^0$ using tools from first-order logic. Our presentation in Problem 6.12 is close to that in [307], but we are also indebted to lecture notes by Madhu Sudan. Johan Håstad [377] gave stronger versions of the switching lemma, and proved that any circuit of poly(n) width that computes PARITY must have depth $\Omega(\log n / \log\log n)$. Håstad received the Gödel Prize for this work in 1994.

6.10 Natural proofs and beyond. Razborov and Rudich gave their natural proofs result in [686], for which they received the Gödel Prize in 2007. They showed that essentially all the lower bound proofs known at the time can be *naturalized*, in the sense that they can be phrased in terms of common, constructive properties even if they are not initially stated that way. There is a nice review by Rudich in [717].

Since then, yet another barrier to proving that P $\neq$ NP has emerged. As we will see in Section 11.2, we can learn a lot by *arithmetizing* a Boolean function, and extending it to a polynomial over a finite field such as the integers mod p. Aaronson and Wigderson [5] define *algebrization* as a generalized version of relativization, in which we are allowed to query oracles in this extended sense, and show that no proof technique which algebrizes can separate P from NP.

In a series of recent papers, Mulmuley and Sohoni [618] have proposed a sophisticated approach to complexity lower bounds using algebraic geometry and representation theory. They rely on the algebraic properties of specific functions like the permanent, and their methods arguably avoid both commonality and constructivity. For a relatively accessible review of their approach, see Regan [687].

If the reader would like to gaze upon the wreckage of those who have dashed themselves against this problem, see Gerhard Woeginger's page at www.win.tue.nl/~gwoegi/P-versus-NP.htm.

6.11 Problems in the gap. Theorem 6.6 was proved in 1975 by Ladner [515], and is often called Ladner's Theorem. It mimics earlier results in recursive function theory, which showed that there are problems that are undecidable but not "Turing complete"—that is, such that the Halting Problem is not reducible to them. In the same paper, he showed a number of other curious results: for instance, if P $\neq$ NP then there are two problems, both of which are outside P, such that any problem which is reducible to both of them is in P.

A nice treatment of diagonalization and Ladner's Theorem can be found in the book [75] by Balcázar, Díaz, and Gabarró, including generalizing Problem 6.20 to show that there are infinite sets of problems in NP that are incomparable to each other. We also note an alternate unpublished proof of Ladner's Theorem due to Impagliazzo, where we pad SAT questions so they have length $f(n)$ instead of n. We can then make $f(n)$ grow fast enough that these padded questions are not NP-complete, but not so fast that they are in P.

The classes B_k of problems with witnesses of polylogarithmic size, and the result of Problem 6.21 that they are distinct unless SAT can be solved in time $2^{O(n^\alpha)}$ for $\alpha < 1$, are discussed by Díaz and Torán in [236].

6.12 Nonconstructive proofs and subclasses of TFNP. For a cute example of a nonconstructive proof, let's show that there exist irrational numbers x, y such that x^y is rational. Let $z = \sqrt{2}^{\sqrt{2}}$, and note that $z^{\sqrt{2}} = 2$. Then z is either rational or irrational, and in either case we're done.

The classes PPP, PPA, and PPAD, based on the pigeonhole principle and parity arguments, were defined by Papadamitriou [647]; see also Megiddo and Papadimitriou [564]. These papers showed that ANOTHER HAMILTONIAN CYCLE and ANOTHER HAMILTONIAN PATH are in PPA. Lemma 6.7 is known as Smith's Lemma, and the proof given here is from Thomason [788]. These papers also showed that SPERNER, BROUWER, and NASH are in PPAD. For a reduction from NASH to BROUWER, see Geanakoplos [321].

Ironically, Brouwer was a fierce opponent of nonconstructive arguments. He rejected the Law of the Excluded Middle, and founded a school of thought called the Intuitionists. They insisted that the only sets S which exist are those for which there is a finite procedure—what we would now call an algorithm—that determines whether a given x is in S. Of course, for problems in TFNP there is a finite procedure: namely, an exhaustive search through the possible witnesses. Thus we might follow [647] and call these proofs *inefficiently* constructive, as opposed to nonconstructive.

The PPAD-completeness of NASH was shown first for three-player games by Daskalakis, Goldberg, and Papadimitriou [216], and then for two-player games by Chen and Deng [158]. For zero-sum two-player games, Nash equilibria can be found in polynomial time using linear algebra. For an introduction to connections between economics, game theory, and computer science, we recommend the book *Algorithmic Game Theory*, edited by Nisan, Roughgarden, Tardos, and Vazirani [637].

The class PLS based on local search was defined by Johnson, Papadimitriou, and Yannakakis [442]. The PLS-completeness of finding local optima of TRAVELING SALESMAN and local minima in spin glasses (often referred to in computer science as the Hopfield model of neural networks) was shown by Papadimitriou, Schäffer, and Yannakakis [648].

Another natural directed parity argument is the following: in a directed graph where no vertex has in-degree or out-degree more than 1, given a source, find a sink. This gives yet another class, PPADS, which contains PPAD and is contained in PPP. Beame et al. [91] showed that, in a relativized world with a generic oracle, these classes are related just as we expect: PPA and PPADS are incomparable and therefore PPAD is a proper subclass of each of them, PPADS is a proper subclass of PPP, and PPA and PPP are incomparable.

Note that finding the particular local minimum where the gradient descent algorithm will end up from a given initial vertex is harder, *a priori*, than potentially harder than finding an arbitrary local minimum. In fact, this version of the problem is PSPACE-complete (see Chapter 8 for the definition of PSPACE). The same is true for finding the other end of a path that starts at a given vertex in a graph or directed graph of maximum degree 2.

Are there TFNP-complete problems? It seems difficult to define a problem which is to TFNP as WITNESS EXISTENCE is to NP, since given a witness-checking program Π we would have to first prove, or promise, that there is a witness for every instance. Such classes are called *semantic*, since they are defined in terms of the meaning of a program or circuit rather than its surface structure. One of the attractions of PPP, PLS, PPA and so on is that they can be defined *syntactically*, with promises that are easy to check. Thus they have complete problems, like SPERNER or the second version of PIGEONHOLE CIRCUIT, which make no promises about their input, and give the problem an easy way out if the promise is broken. This was one of the motivations of defining these classes in [647].

6.13 Formal independence. For a more thorough exploration of the possibility that P vs. NP is formally independent, we highly recommend the paper by Aaronson [1]. Ben-David and Halevi [97] show that if P $\neq$ NP can be proved to be formally independent using any of our current techniques, then the gap between P and NP is incredibly small. Specifically, every problem in NP would be solvable by circuits of size $n^{\alpha(n)}$ where $\alpha(n)$ is the inverse of the Ackermann function—which grows more slowly than $\log n$, $\log\log n$, or even $\log^* n$, the number of times we need to iterate the logarithm to bring n below 1.

6.14 Local search. Problem 6.22 is from Aldous [28]. Using ideas from Markov chains, he showed that this simple algorithm is essentially optimal—that no randomized algorithm can find a local minimum with $2^{(1/2-\varepsilon)n}$ queries for any $\varepsilon > 0$. Aaronson [3] tightened this lower bound to $2^n/\text{poly}(n)$ using quantum arguments, and this was generalized to other graphs by Dinh and Russell [242].

6.15 Almost-natural proofs. The result in Problem 6.27 on almost-natural proofs is due to Chow [165]. The variation we use here was found by Vadhan.

Chapter 7

The Grand Unified Theory of Computation

> It may be desirable to explain, that by the word *operation*, we mean *any process which alters the mutual relation of two or more things*, be this relation of what kind it may. This is the most general definition, and would include all subjects in the universe. In abstract mathematics, of course operations alter those particular relations which are involved in the considerations of number and space... But the science of operations, as derived from mathematics more especially, is a science of itself, and has its own abstract truth and value; just as logic has its own peculiar truth and value, independently of the subjects to which we may apply its reasonings and processes.
>
> Augusta Ada, Countess of Lovelace, 1842

We have spent the first part of this book asking what problems can be solved in polynomial time, exponential time, and so on. But what if our time is unlimited? Let's call a problem *decidable*, or *computable*, if there is an algorithm that solves it in some finite amount of time. We place no bounds whatsoever on how long the algorithm takes—we just know that it will halt eventually. In some sense, the class Decidable of such problems is the ultimate complexity class. Is there anything outside it? Is every problem decidable, and every function computable, if we give our computer enough time?

For this question to make sense, we have to wrestle with an issue that is partly mathematical and partly philosophical: what is computation anyway? In Chapter 2 we argued that the class P is robust—that the running time of an algorithm doesn't depend very much on our model of computation or our choice of programming language. Based on this claim, we have allowed ourselves to define our algorithms using an informal "pseudocode." We have assumed that an algorithm's behavior is essentially the same regardless of the details of how we implement it, or what hardware we run it on.

But do the details really not matter? *A priori*, it might be that each programming language, and each computing technology, is capable of solving a different set of problems—and therefore that there are as many definitions of computability and decidability as there are models of computation. Is there a universal definition of computation, with many equivalent forms, that captures every reasonable model?

Many of the greatest moments in the history of science have been *unifications*: realizations that seemingly disparate phenomena are all aspects of one underlying structure. Newton showed that apples and planets are governed by the same laws of motion and gravity, creating for the first time a single explanatory framework that joins the terrestrial to the celestial. Maxwell showed that electricity and magnetism can be explained by a single electromagnetic field, and that light consists of its oscillations. This quest for unity drives physics to this day, as we hunt for a Grand Unified Theory that combines gravity with quantum mechanics.

In 1936, computer science had its own grand unification. In the early 20th century, driven by the desire to create a firm underpinning for mathematics, logicians and mathematicians invented various notions of an "effective calculation" or "mechanical procedure"—what we would call a program or algorithm today. Each of their definitions took a very different attitude towards what it means to compute. One of them defines functions recursively in terms of simpler ones; another unfolds the definition of a function by manipulating strings of symbols; and, most famously, the Turing machine reads, writes, and changes the data in its memory by following a set of simple instructions.

Turing, Kleene, Church, Post, and others showed that each of these models can simulate any of the others. What one can do, they all can do. This universal notion of computation, and the fact that it transcends what programming language we use, or what technology we possess, is the topic of this chapter.

We begin by describing Babbage's vision of a mechanical device which can be programmed to carry out any calculation, and Hilbert's dream of a mechanical procedure that can prove or refute any mathematical claim. We discuss the fact that any sufficiently powerful model of computation can talk about itself—programs can run other programs, or even run themselves.

The self-referential nature of computation has some surprising consequences, including the fact that some problems are *undecidable*—there is no program that solves them in any finite amount of time. Chief among these is the Halting Problem, which asks whether a given program will halt or run forever. We will see how these undecidable problems imply Gödel's Incompleteness Theorem that no formal system is capable of proving all mathematical truths. Thus, while Babbage's vision has now been realized, Hilbert's vision can never be.

We then describe the three great models of computation proposed in the early 20th century—partial recursive functions, the λ-calculus, and Turing machines—and show that they are all equivalent to each other. We argue for the *Church–Turing Thesis*, which states that these models are strong enough to carry out any conceivable computation. Finally, we cast our net wide, and find universal computation and undecidability nearly everywhere we look: in products of fractions, tilings of the plane, and the motions of a chaotic system.

7.1 Babbage's Vision and Hilbert's Dream

There are two fundamental concepts at the heart of computer science. One is the *algorithm* or *program*: a sequence of instructions, written in some general but precise language, that carries out some calculation or solves some problem. The other is the *universal computer*: a programmable device that can carry out any algorithm we give to it.

These concepts emerged in the 19th and early 20th centuries, in forms that are surprisingly recognizable to us today. Regrettably, they appeared in two separate threads that, as far as we know, never interacted—the visions of an inventor, and the dream of an unshakeable foundation for mathematics. Let's go back in time, and watch as our forebears set computer science in motion.

7.1.1 *Computing with Steam*

> As soon as an Analytical Engine exists, it will necessarily guide the future course of the science.
>
> Charles Babbage, *Passages from the Life of a Philosopher*

> What shall we do to get rid of Mr. Babbage and his calculating machine? Surely if completed it would be worthless as far as science is concerned?
>
> British Prime Minister Sir Robert Peel, 1842

For centuries, printed tables of trigonometric functions, logarithms, and planetary orbits were used as computational aids by scientists, merchants, and navigators. Being calculated by human computers, these tables were subject to human fallibility. But errors, especially those in navigational tables, can have severe consequences.

One day in 1821, while checking astronomical tables and discovering numerous errors, Charles Babbage declared "I wish to God that these calculations had been executed by steam!" Babbage, being not only a mathematician and philosopher, but also an inventor and mechanical engineer, decided to develop a machine that could calculate and print mathematical tables, automatically and free of errors.

Babbage knew that many functions can be closely approximated by polynomials of low degree. For example, the 4th-order Taylor series for the cosine,

$$\cos x \approx 1 - \frac{x^2}{2} + \frac{x^4}{24},$$

has an error of at most 0.02 for $0 \le x \le \pi/2$. In order to compute functions such as $\cos x$ and $\ln x$, Babbage designed his first machine, the Difference Engine, to tabulate polynomials. He proposed to do this in an ingenious way, called the *method of differences.*

If $f(x)$ is a polynomial of degree n then its difference $\Delta f(x) = f(x+1) - f(x)$ is a polynomial of degree $n-1$, and so on, until its nth difference $\Delta^n f$ is a constant. Each of these differences was stored in a column of wheels as shown in Figure 7.1, where each wheel stored a single digit. The user set the initial values $f(0), \Delta f(0), \ldots, \Delta^n f(0)$ by turning these wheels, and the machine then computed $f(x), \Delta f(x), \ldots$ for successive values of x by adding these differences to each other. Thus the Difference Engine could be "programmed" to calculate any polynomial whose degree is less than the number of columns, and approximate a wide variety of functions.

The Difference Engine was already a very innovative concept. But before it was built, Babbage turned 7.1
his attention to a far more ambitious machine—a truly universal computer, which could perform any conceivable calculation. In his autobiography, he wrote:

> The whole of arithmetic now appeared within the grasp of mechanism. A vague glimpse even of an Analytical Engine at length opened out, and I pursued with enthusiasm the shadowy vision.

Babbage spent most of the rest of his life turning this "shadowy vision" into a detailed blueprint.

FIGURE 7.1: Charles Babbage's difference engine: 19th century woodcut (left) and 20th century construction from Meccano™ parts (right).

The Analytical Engine was designed with a "store" that could keep a thousand 50-digit numbers, each digit being represented by the position of a cogwheel. The arithmetical heart of the machine—the "mill"—could perform addition, subtraction, multiplication, and division, using clever mechanical assemblies. To control the mill, Babbage proposed a technique that was then being used to direct mechanical looms, and that survived in the programming of electronic computers until the 1970s. In his own words:

> It is known as a fact that the Jacquard loom is capable of weaving any design which the imagination of man may conceive. It is also the constant practice for skilled artists to be employed by manufacturers in designing patterns. These patterns are then sent to a peculiar artist, who, by means of a certain machine, punches holes in a set of pasteboard cards in such a manner that when those cards are placed in a Jacquard loom, it will then weave upon its produce the exact pattern designed by the artist.

Babbage's design included "variable" cards that copied numbers from the store into the mill and vice versa, and "operation" cards that instructed the mill to carry out particular operations on its contents.

Babbage realized that it was not enough to carry out a fixed sequence of operations. He also included "combinatorial" cards that governed the motion of the variable and operation cards, allowing for conditional execution, skipping forwards or backwards in the sequence, or for a cycle of cards to be repeated a certain number of times. In other words, he invented **if–then** and **goto** statements, and **for** loops. He even proposed that the machine could punch its own "number" cards, recording data from previous calculations, giving it an effectively unlimited memory.

Babbage's goals for the Analytical Engine consisted solely of numerical calculations. It was Augusta Ada, Countess of Lovelace (and daughter of Lord Byron), who realized that the Analytical Engine could do much more than number crunching. She wrote:

> The bounds of *arithmetic* were however outstepped the moment the idea of applying the cards had occurred; and the Analytical Engine does not occupy common ground with mere "calculating machines." In enabling mechanism to combine together *general* symbols in successions of unlimited variety and extent, a uniting link is established between the operations of matter and the abstract mental processes of the most abstract branch of mathematical science...

Just as we do today, Ada thought of computation as a manipulation of *symbols*, not just digits. And, she realized that these symbols can encode many different kinds of information:

> Again, it might act upon other things besides number, were objects found whose mutual fundamental relations could be expressed by those of the abstract science of operations... Supposing, for instance, that the fundamental relations of pitched sounds in the science of harmony and of musical composition were susceptible of such expression and adaptations, the engine might compose elaborate and scientific pieces of music of any degree of complexity or extent.

Ada also wrote a program for the Analytical Engine, with two nested **for** loops, to compute a sequence of fractions called the Bernoulli numbers. For this reason, she is often referred to as the first computer programmer.

Sadly, Babbage and Lovelace were almost one hundred years ahead of their time. The first truly programmable computers were built in the 1940s, with electric relays instead of cogwheels. But their basic 7.2
architecture—a central processing unit, governed by a program, which reads and writes data stored in memory—would have been perfectly recognizable to Babbage. Moreover, the idea of a universal computer played another important role in the early 20th century: as a thought experiment, to understand the ultimate mathematical power of algorithms.

7.1.2 *The Foundations of Mathematics*

> We hear within us the perpetual call: There is the problem. Seek its solution. You can find it by pure reason, for in mathematics there is no *ignorabimus*.
>
> David Hilbert

Mathematicians from Euclid to Gauss have been thinking about algorithms for millennia. But the idea of algorithms as well-defined mathematical objects, worthy of investigation in and of themselves, didn't emerge until the dawn of the 20th century.

In 1900, David Hilbert delivered an address to the International Congress of Mathematicians. He presented a list of 23 problems which he saw as being the greatest challenges for 20th-century mathematics. These problems became known as Hilbert's problems, and inspired ground-breaking work across the mathematical landscape. His 10th problem was as follows:

> Specify a procedure which, in a finite number of operations, enables one to determine whether or not a given Diophantine equation [a polynomial equation with integer coefficients] with an arbitrary number of variables has an integer solution.

He felt sure that there was such an algorithm—the only problem was to find it.

Hilbert showed even more optimism about the power of algorithms in 1928, when he challenged his fellow mathematicians with the *Entscheidungsproblem* (in English, the decision problem):

> The Entscheidungsproblem is solved if one knows a procedure that allows one to decide the validity of a given logical expression by a finite number of operations.

In other words, he asked for an algorithm that will take any statement written in a formal language—for instance,

$$\neg\exists x, y, z, n \in \mathbb{N} - \{0\} : (n \geq 3) \wedge (x^n + y^n = z^n),$$

which expresses Fermat's last theorem—and determine, in a finite amount of time, whether it is true or false. Such an algorithm would constitute nothing less than the mechanization of mathematics.

7.3 Let's compare this with our discussion in Section 6.1.3. If $\mathsf{P} = \mathsf{NP}$, we can tell in poly(ℓ) time whether a given statement has a proof of length ℓ or less. Hilbert wasn't concerned with the distinction between polynomial and exponential time—he wanted a procedure that halts after a finite amount of time, regardless of how long the shortest proof or disproof of the statement is.

Hilbert didn't spell out what he meant by a "procedure." He didn't have computing devices in mind—he was unaware of Babbage's work, and electronic computers were decades away. He did, however, want this procedure to be mechanical, in the sense that it could be carried out by a human mathematician according to a clear sequence of operations. This gave rise to the question of what functions are "computable" or "effectively calculable."

At the time, many mathematicians were struggling to build an axiomatic foundation for mathematics. Their goal was to reduce all of mathematics to set theory and logic, creating a formal system powerful enough to prove all the mathematical facts we know and love. But at the turn of the century, several paradoxes shook these foundations, showing that a naive approach to set theory could lead to contradictions.

The best known of these paradoxes is *Russell's paradox*, which deals with sets. Sets can be elements of other sets—for instance, consider the set of all intervals on the real line, each of which is a set of real numbers. So, it seems reasonable to ask which sets are elements of themselves. Once we can do that, we can define the set of sets that are *not* elements of themselves,

$$R = \{S : S \notin S\}.$$

But now we're in trouble. By definition, R is an element of itself if and only if it isn't:

$$R \in R \Leftrightarrow R \notin R.$$

It would be intolerable to have this kind of contradiction at the heart of mathematics. Our only escape is banish R from the mathematical universe—to conclude that, for some reason, R is not a well-defined mathematical object.

The problem with R is the "loopiness" of its definition—it is a loop with a twist in it, like a Möbius strip. We can avoid this kind of loop by insisting on a strictly stratified mathematical universe, where each set can only contain sets from lower levels, until we reach "atoms" like the empty set. In such a universe the idea of a set containing itself is, in Russell's words, "noise without meaning." Thus we eliminate the paradox by dismissing the definition of R as nonsensical. 7.4

This stratified picture of set theory paved the way for an analogous approach to computation. We start with a handful of atomic functions, and build more complicated functions out of them. This lets us build up a universe of computable functions, each of which is inductively defined in terms of simpler functions, defined at an earlier stage of the process.

However, there are many particular ways that we could do this—many notions of what constitutes an atomic operation, and many ways to combine them to make higher-level functions. In the first decades of the 20th century, several definitions of computability along these lines were proposed, but it was not clear whether any of them captured the universe of all conceivable computations.

In 1936, the *annus mirabilis* of computation, a series of results showed that all these definitions are equivalent. This gave rise to the *Church–Turing Thesis*, that these models capture anything that could be reasonably called a computation—in Hilbert's words, any procedure with a finite number of operations. We sketch these models, and their equivalence, in Sections 7.3, 7.4, and 7.5.

At the same time, Turing proposed his Halting Problem, which asks whether a given program will ever halt. Using an argument similar in form to Russell's paradox, we will show in Section 7.2 that if there were a program that solved this problem, we could create a paradox by asking it about itself. Therefore, no such program exists, and the Halting Problem problem is undecidable.

Since the Halting Problem is a well-defined mathematical question, this shows that the Entscheidungsproblem cannot be solved—there is no algorithm that can take an arbitrary mathematical statement as input, and determine whether it is true or false. In 1970, it was shown that telling whether a given Diophantine equation has integer roots is just as hard as the Halting Problem. So, Hilbert's 10th problem is undecidable as well. 7.5

Hilbert may have found the phenomenon of undecidability disappointing, but we find it liberating. Universal computation is everywhere, even in seemingly simple mathematical questions. It lends those questions an infinite richness, and proves that there is no mechanical procedure that can solve them once and for all. Mathematics will never end.

Now that the stage is set, let's explore the self-referential nature of computation, and how it leads to undecidable problems and unprovable truths.

7.2 Universality and Undecidability

There's no getting round a proof... even if it proves that something is unprovable!

Apostolos Doxiadis and Christos H. Papadimitriou, *Logicomix*

The most basic fact about modern computers is their universality. Like the Analytical Engine, they can carry out any program we give to them. In particular, *there are programs that run other programs.* A computer's operating system is a program that runs and manages many programs at once. Even web browsers run programs on virtual machines that they simulate inside their own workspace.

In any programming language worth its salt, one can write an *interpreter* or *universal program*—a program that takes the source code of another program as input, and runs it step-by-step, keeping track of its variables and which instruction to perform next. Symbolically, we can define this universal program like this:

$$U(\Pi, x) = \Pi(x). \tag{7.1}$$

That is, U takes a program Π and an input x, and returns the same results that Π would give if we ran Π on x. Note that we are using the symbol Π in two different ways. On the left, we mean Π's source code, which can be fed to U like any other finite string. On the right, we mean Π in its active form, taking an input x and returning an output.

The fact that universal programs exist has enormous consequences. We can set up self-referential situations—what Douglas Hofstadter calls "strange loops"—in which programs run themselves, or modified versions of themselves. These loops can lead to paradoxes in which the only way out is to accept that
7.6 some problems cannot be solved—that they are *undecidable*, and beyond the ability of any algorithm.

Let's start by showing that, in any programming language powerful enough to interpret itself, there are programs that will run forever and never return any output. Then, we will show that there is no program that can tell whether or not a given program, with a given input, will ever halt.

7.2.1 *Diagonalization and Halting*

Let $U(\Pi, x)$ be a universal program as defined in (7.1). Consider the special case where $x = \Pi$. Then U simulates what Π would do given its own source code:

$$U(\Pi, \Pi) = \Pi(\Pi).$$

Now suppose, for simplicity, that the programs in question returns a Boolean value, true or false. Then we can define a new program V which runs Π on itself, and negates the result:

$$V(\Pi) = \overline{\Pi(\Pi)}.$$

There is nothing mythical about V—it is a concrete program, that really can be written in your favorite programming language. But if we feed V its own source code, an apparent contradiction arises, since

$$V(V) = \overline{V(V)}.$$

The only way to resolve this paradox is if $V(V)$ is undefined. In other words, when given its own source code as input, V runs forever, and never returns any output. This shows that any programming language powerful enough to express a universal program possesses programs that never halt, at least when given certain inputs. In brief,

Universality implies non-halting programs.

Thus any reasonable definition of computable functions includes *partial* functions, which are undefined for some values of their input, in addition to *total* ones, which are always well-defined.

This kind of argument, where we feed things to themselves and add a twist, is known as *diagonalization.* It originated with Georg Cantor, who used it to show that some infinities are bigger than others. Specifically, let $\mathbb{N} = \{0,1,2,3,\ldots\}$ denote the set of natural numbers, and let $\wp(\mathbb{N})$ denote its *power set*—that is, the set $\{S : S \subseteq \mathbb{N}\}$ of all subsets of $\mathbb{N}$. If these sets were the same size, it would be possible to put them in one-to-one correspondence. In other words, there would be a way to list all subsets as S_0, S_1, S_2, and so on, such that for every $S \subseteq \mathbb{N}$ we have $S = S_i$ for some $i \in \mathbb{N}$.

Cantor proved that no such correspondence exists. If one did, we could imagine listing all the subsets in a table, where the ith row contains a series of bits marking which integers are in S_i. For instance, let's say that $S_0 = \emptyset$ is the empty set, $S_1 = \mathbb{N}$, S_2 is the even numbers, S_3 is the odd ones, S_4 is the primes, and so on. Then the beginning of this table would look like this:

	0	1	2	3	4	$\cdots$
S_0	0	0	0	0	0	
S_1	1	1	1	1	1	
S_2	1	0	1	0	1	
S_3	0	1	0	1	0	
S_4	0	0	1	1	0	
$\vdots$						$\ddots$

Cantor's idea is to take the diagonal of this table, and flip all its bits. This gives us a new set T:

	0	1	2	3	4	$\cdots$
S_0	**0**	0	0	0	0	
S_1	1	**1**	1	1	1	
S_2	1	0	**1**	0	1	
S_3	0	1	0	**1**	0	
S_4	0	0	1	1	**0**	
$\vdots$						
T	**1**	**0**	**0**	**0**	**1**	$\cdots$

Since $i \in T$ if and only if $i \notin S_i$, we have $T \neq S_i$ for all i. Therefore, T was not included in our table of subsets, and our table was not complete. Any correspondence between $\mathbb{N}$ and $\wp(\mathbb{N})$ leaves at least one subset out. So, while both $\mathbb{N}$ and $\wp(\mathbb{N})$ are infinite, $\wp(\mathbb{N})$ is bigger.

We say that $\mathbb{N}$ is *countably* infinite, and we denote its cardinality $\aleph_0$. In contrast, $\wp(\mathbb{N})$ is *uncountably* infinite—its cardinality is a larger infinity. For any set S, the cardinality of its power set is $\left|\wp(S)\right| = 2^{|S|}$, since for each $x \in S$ we can choose whether or not to include x in a subset T. In these terms, the diagonal argument shows that $2^{\aleph_0} > \aleph_0$, and more generally that $2^x > x$ even when x is infinite. The cardinality of $\wp(\wp(\mathbb{N}))$ is an even larger infinity, $2^{2^{\aleph_0}}$, and so on.

7.7 We can phrase our argument about V in diagonal form as follows. Let Π_i denote the ith program in some convenient order. If programs always halt, Π_i gives a well-defined output when given Π_j as input, for all i and j. If we imagine a table of all these outputs, then V consists of taking the diagonal and negating it:

	Π_1	Π_2	Π_3	$\ldots$
Π_1	$\Pi_1(\Pi_1)$	$\Pi_1(\Pi_2)$	$\Pi_1(\Pi_3)$	
Π_2	$\Pi_2(\Pi_1)$	$\Pi_2(\Pi_2)$	$\Pi_2(\Pi_3)$	
Π_3	$\Pi_3(\Pi_1)$	$\Pi_3(\Pi_2)$	$\Pi_3(\Pi_3)$	
$\vdots$				
V	$\overline{\Pi_1(\Pi_1)}$	$\overline{\Pi_2(\Pi_2)}$	$\overline{\Pi_3(\Pi_3)}$	

Once again, we have a contradiction. The difference is that now V *is* somewhere in the table—if a universal program exists then so does V, so $V = \Pi_j$ for some j. So, this time around, we escape the contradiction by admitting that $\Pi_i(\Pi_j)$ is not well-defined for all i and j, and that some programs never halt on some inputs. In particular, V cannot differ with itself, so $V(V)$ must not be well-defined. Conversely, if a programming language guarantees, for some reason, that its programs always halt, then it is not powerful enough to interpret itself.

Exercise 7.1 *In one of the first diagonal statements, Epimenides the Cretan stated that all Cretans are liars. Show that this is not in fact a contradiction, by giving a consistent interpretation of Epimenides' statement and its truth or falsehood.*

Exercise 7.2 *Phrase Russell's paradox as a diagonal argument. Hint: imagine a vast table which lists whether or not $S \in T$ for every pair of sets S, T.*

7.2.2 *The Halting Problem*

Since some programs halt and others don't, it would be nice to be able to tell which is which. Consider the following problem:

> HALTING
>
> Input: A program Π and an input x
>
> Question: Will Π halt when given x as input? Equivalently, is $\Pi(x)$ well-defined?

We will prove that this problem is *undecidable*—that no program can solve it in any finite amount of time.

In 1936, the fact that undecidable problems exist came as quite a surprise. We have spent much of this book discussing classes of problems that can be solved in polynomial time, exponential time, and so on. Just as we defined complexity classes like P and EXP, we can define a class Decidable as the class of all problems that can be solved in *any finite amount of time*:

```
Fermat
begin
   t := 3;
   repeat
      for n = 3 to t do
         for x = 1 to t do
            for y = 1 to t do
               for z = 1 to t do
                  if x^n + y^n = z^n then return (x, y, z, n);
               end
            end
         end
      end
      t := t + 1;
   until forever;
end
```

FIGURE 7.2: This program does not halt. Andrew Wiles has discovered a truly marvelous proof of this, which this caption is too narrow to contain.

> Decidable is the class of problems A for which there is a program Π that halts after a finite amount of time, and which, given any input x, answers whether or not $A(x)$ is true.

Hilbert believed that Decidable includes *all* well-defined mathematical problems. But the Halting Problem shows that it does not—that some problems are beyond the reach of any algorithm.

At first, it seems that we can solve HALTING by running the universal program U, simulating what Π would do given x, and seeing what happens. But if $\Pi(x)$ doesn't halt, neither will U's simulation. To solve HALTING, we have to be able to see into the future, and predict, in finite time, whether or not $\Pi(x)$ will ever halt. We will show that no program can do this—since if it could, it could create a contradiction by trying to predict itself.

Before we prove that no program can solve HALTING, let's think about what other problems we could solve if such a program existed—a program with the following behavior:

$$\texttt{Halts}(\Pi, x) = \begin{cases} \texttt{true} & \text{if } \Pi \text{ halts on input } x \\ \texttt{false} & \text{otherwise} \end{cases} \tag{7.2}$$

Such a program would be very handy for debugging software. It could tell us whether a program will hang, and never get back to the user. However, it would have much more profound uses as well. Consider the program `Fermat` in Figure 7.2. It looks for a counterexample to Fermat's Last Theorem—that is, positive integers x, y, z, n with $n \geq 3$, such that

$$x^n + y^n = z^n, \tag{7.3}$$

It does this by exhaustively checking all quadruples $x, y, z, n \leq t$, with t ranging up to infinity.

Of course, we don't actually want to run `Fermat`—we only want to feed it to `Halts`. It took the mathematical community 358 years to prove that no such quadruples exist, and that `Fermat` will run forever. Obviously having `Halts` would save us a lot of time.

Exercise 7.3 *According to a conjecture formulated by Christian Goldbach in 1771, every even integer greater than 2 can be written as the sum of two primes, such as* $10=3+7$ *or* $12=5+7$. *Write a program that could be fed to* `Halts`, *if it existed, to prove or disprove Goldbach's Conjecture.*

There is a truly marvelous proof that the program `Halts` cannot exist, and the proof fits into a few lines. Once again, we use diagonalization. Suppose that `Halts` existed. Then we could call it as a subroutine, set $x=\Pi$, and build the following program:

```
Catch22(Π)
begin
   if Halts(Π,Π) = true then loop forever;
   else return true;
end
```

We have designed `Catch22`(Π) so that it halts if and only if $\Pi(\Pi)$ runs forever. But now diagonalization strikes again. If we run `Catch22` on itself, `Catch22(Catch22)` halts if and only if it doesn't. The only escape from the contradiction is to conclude that there is no such program as `Halts`. So, HALTING is undecidable.

7.8 Of course, this does not show that HALTING can never be solved. If we restrict our attention to a particular program Π and define a version of HALTING whose only input is x, then for some choices of Π this problem is clearly decidable—for instance, if Π immediately halts without doing anything with its input. But for some other choices of Π, such as the universal program U, this version of the Halting Problem is just as hard as the general one, and so is undecidable.

Now that we have one undecidable problem, we can prove that other problems are undecidable by reducing HALTING to them. For instance, consider the following problem:

> FORTY-TWO
>
> Input: A program Π
>
> Question: Is there an x such that $\Pi(x)$ halts and returns 42?

Given a program Π and input x, we can convert them to a program Π' which ignores its input, runs $\Pi(x)$ instead, and return 42 if it halts:

```
Π′(x′)
begin
   run Π(x);
   return 42;
end
```

If $\Pi(x)$ halts, then $\Pi'(x')$ returns 42 for all x'. If $\Pi(x)$ doesn't halt, then neither does Π', no matter what input x' we give it. Thus if FORTY-TWO were decidable, HALTING would be too. But we know that HALTING is undecidable, so FORTY-TWO must be undecidable as well.

This recipe for Π' is a reduction analogous to the ones we used for NP-completeness—it maps instances of HALTING to instances of FORTY-TWO with the same answer. It shows that FORTY-TWO is at least as hard as HALTING, or

$$\text{HALTING} \leq \text{FORTY-TWO}\,.$$

Just as we used polynomial-time reductions for exploring P and NP, here we use *computable* reductions. That is, a reduction can be any function from instances of A to instances of B that we can compute in finite time. In that case, $A \leq B$ implies that if B is decidable then A is decidable, and conversely, if A is undecidable then B is undecidable. As Problems 7.3 and 7.4 show, we can prove that virtually any nontrivial question about a program's long-term behavior is undecidable by reducing HALTING to it—typically, with a simple reduction that modifies the program's source code.

It's worth noting that if the only thing we had to worry about was falling into an endless loop, the Halting Problem would be decidable:

Exercise 7.4 *Suppose that a program Π has the following property: for any input x, either $\Pi(x)$ halts, or it falls into a periodic loop, where all its variables return again and again to the same values. Show that in that case Π's Halting Problem is decidable.*

Thus the undecidability of the Halting Problem stems from the fact that programs can run forever, but in complicated ways—such as searching for larger and larger counterexamples to Fermat's Last Theorem.

The astute reader may object that we haven't really proved that the Halting Problem is undecidable. All we have proved is that no programming language can answer *its own Halting Problem*, unless it is so weak that it cannot interpret its own programs. To show that the Halting Problem—and therefore, the Entscheidungsproblem, of which the Halting Problem is a special case—is truly undecidable, we need to show that there are particular programming languages, or models of computation, that capture the power of all possible algorithms. This is, to some extent, a philosophical claim, rather than a mathematical one, and we will return to it in Section 7.5.

It's also important to realize that we are making some very important assumptions about how programs can be built out of other programs. For instance, we assume that if a program for `Halts`(Π, x) exists, then so does a program that calls `Halts` as a subroutine with the special case $x = \Pi$, and then uses an **if–then** statement to do one thing or another depending on its results. Most fundamentally, we assume that each program has a source code that can be read by other programs, or even by itself. Living, as we do, long after the events of 1936, we have the luxury of thinking in terms of high-level programming languages which let us take these things for granted. But at the dawn of computer science, these concepts had to be built from scratch. As we will see later in this chapter, carrying out the diagonal argument took much more work back then than it does now.

But first, let's use the Halting Problem as a springboard to explore the upper reaches of the complexity hierarchy. In the land of polynomial time, we defined classes like P, NP, and the polynomial hierarchy to understand the relationship between a problem's logical structure and its computational complexity. How does this landscape look when we are allowed any finite amount of time to do our computation? And how is the undecidability of the Halting Program related to the other great upheaval of the 1930s—Gödel's Theorem that no formal system can provide a complete foundation for mathematics?

7.2.3 Recursive Enumerability

While the Halting Problem is undecidable, it has a kind of one-sided decidability. If the answer is "yes," then we can learn that fact in a finite amount of time, by simulating Π until it halts. To put this differently, we can write

$$\texttt{Halts}(\Pi,x) = \exists t : \texttt{HaltsInTime}(\Pi,x,t),$$

where $\texttt{HaltsInTime}(\Pi,x,t)$ is the property that Π, given x as input, halts on its tth step. Since we can tell if this is true by running Π for t steps, HaltsInTime is decidable. Thus Halts is a combination of a decidable problem with a single "there exists."

In some ways, this is analogous to the relationship between P and NP. Recall that a property A is in NP if it can be written $A(x) = \exists w : B(x,w)$ where B is in P. In other words, x is a yes-instance of A if some witness w exists, and the property $B(x,w)$ that w is a valid witness for x can be checked in polynomial time. Here t is a witness that Π halts, and we can check the validity of this witness in finite time.

There is an interesting alternate way to describe this kind of problem. Typically, we give a program an input x, and ask it to tell whether x is in some set S of yes-instances. If such a program exists, we say that S is decidable. Equivalently, we say that a problem is decidable if its set of yes-instances is decidable.

But we can also hook a program Π up to a printer—with an infinite amount of paper, of course—and ask it to print out a list of all the elements of S. We demand that every $x \in S$ appears at some point in this list, and that no $x \notin S$ ever will. If S is infinite, Π will run forever, but any particular $x \in S$ appears on the printer within a finite amount of time. Note that we allow Π to print the $x \in S$ in any order, and even repeat them, as long as each one appears eventually. If there is such a program, we say that Π *enumerates* S, and that S is *recursively enumerable.*

Now let RE denote the class of problems for which the set of yes-instances is recursively enumerable. We claim that these are exactly the problems of the form $A(x) = \exists w : B(x,w)$ where B is decidable. In one direction, suppose that Π enumerates the yes-instances of a problem A, and let $\texttt{Prints}(\Pi,x,t)$ denote the property that x is the tth string that Π produces. Then Prints is decidable, and we can express the claim that x is a yes-instance as $\exists t : \texttt{Prints}(\Pi,x,t)$.

In the other direction, suppose that $A(x) = \exists w : B(x,w)$. For simplicity, we treat x and w as integers. Our first stab at a program that prints out all yes-instances x would look like this:

```
begin
  x := 0 ;
  repeat
    w := 0 ;
    while ¬B(x, w) do w := w + 1 ;
    print x ;
    x := x + 1 ;
  until forever;
end
```

The problem with this program, of course, is that if x is a no-instance, the inner **while** loop gets stuck looking for a nonexistent witness w. To avoid this, we need to use an approach similar to `Fermat` in Figure 7.2. Here is an improved program:

```
begin
    t := 0 ;
    repeat
        for x = 0 to t do
            for w = 0 to t do
                if B(x, w) then print x ;
            end
        end
        t := t + 1 ;
    until forever;
end
```

This technique, where an inner loop goes through all pairs (x, w) with $x, w \le t$ and an outer loop increases t, is called *dovetailing*. Note that this program might print a small yes-instance much later than a large one, if its witness is much larger. But it prints each yes-instance in finite time, so A is in RE.

Exercise 7.5 *Show that the Halting Problem is in* RE. *In other words, describe a program that prints out the set of yes-instances of the Halting Problem, i.e., the set of all pairs* (Π, x) *such that* Π *halts when given* x *as input.*

Exercise 7.6 *Show that the Halting Problem is* RE*-complete. That is, show that any problem in* RE *can be expressed as an instance of the Halting Problem.*

Just as we defined coNP as the set of problems for which no-instances have witnesses, we can define coRE as the set of problems for which the set of no-instances is recursively enumerable. Equivalently, these are the problems that can be expressed with a single "for all" quantifier, such as the statement

$$\texttt{Hangs}(\Pi, x) = \forall t : \overline{\texttt{HaltsInTime}(\Pi, x, t)},$$

that $\Pi(x)$ will run forever.

Unlike in the polynomial-time world, where the P vs. NP question remains unresolved, we know that RE, coRE, and Decidable are all distinct. Since the Halting Problem is undecidable, Exercise 7.5 tells us that Decidable $\subset$ RE. Then the following exercise shows that RE and coRE are different:

Exercise 7.7 *Show that*

$$\text{Decidable} = \text{RE} \cap \text{coRE}. \qquad (7.4)$$

In other words, show that if both S *and* $\overline{S}$ *are recursively enumerable then* S *is decidable.*

We can conclude from this that

$$\text{RE} \neq \text{coRE},$$

since otherwise (7.4) would imply that Decidable = RE. In contrast, the question of whether NP $\neq$ coNP is still open.

7.2.4 A Tower of Gods

> You jest with me. So I jest with you. Probably Koshchei jests with all of us. And he, no doubt—even Koshchei who made things as they are—is in turn the butt of some larger jest... and there is merriment overhead, but it is very far away.
>
> James Branch Cabell, *Jurgen*

The classes RE and coRE are just the beginning. Analogous to the polynomial hierarchy, we can define an infinite tower of complexity classes, Σ_k and Π_k, consisting of problems with k quantifiers alternating between $\exists$ and $\forall$. In particular,

$$\mathsf{RE} = \Sigma_1\,, \quad \mathsf{coRE} = \Pi_1\,, \quad \text{and} \quad \Sigma_0 = \Pi_0 = \Sigma_1 \cap \Pi_1 = \mathsf{Decidable}\,.$$

For an example at the second level of this hierarchy, the property that a program Π halts on all inputs is in Π_2, since it has a $\exists$ nested inside a $\forall$:

$$\forall x : \exists t : \texttt{HaltsInTime}(\Pi, x, t)\,,$$

The union of all these classes, $\bigcup_{k=0}^{\infty} \Sigma_k = \bigcup_{k=0}^{\infty} \Pi_k$, is known as the *arithmetical hierarchy*. Unlike the polynomial hierarchy, we know that the levels of the arithmetical hierarchy are distinct—that $\Sigma_k \subset \Sigma_{k+1}$ and $\Pi_k \subset \Pi_{k+1}$ for all k. To see why, imagine a "hypercomputer" with access to an oracle that can answer the Halting Problem in finite time. As Problem 7.12 shows, such a computer can solve any problem in $\Sigma_2 \cap \Pi_2$. However, by the same diagonalization argument as before, this hypercomputer cannot solve its own Halting Problem—and this hyper-Halting Problem is complete for Σ_2.

Thus each time we add another layer of quantification, we get a more powerful class, containing problems at higher and higher orders of undecidability. Climbing this ladder leads into a cosmic realm, populated by gods with increasingly infinite computational power. Let's return to the world of finite computations, and connect the two great impossibility results of the 1930s: the undecidability of the Halting Problem, and the incompleteness of formal systems.

7.2.5 From Undecidable Problems to Unprovable Truths

> God exists since mathematics is consistent, and the Devil exists since we cannot prove it.
>
> André Weil

> It cannot be a complete coincidence that several outstanding logicians of the twentieth century found shelter in asylums at some time in their lives.
>
> Gian-Carlo Rota, *Indiscrete Thoughts*

A *formal system* has a finite set of axioms, including rules of inference such as *modus ponens*, "A and $(A \Rightarrow B)$ implies B." A *theorem* is a statement that can be proved, with some finite chain of reasoning,

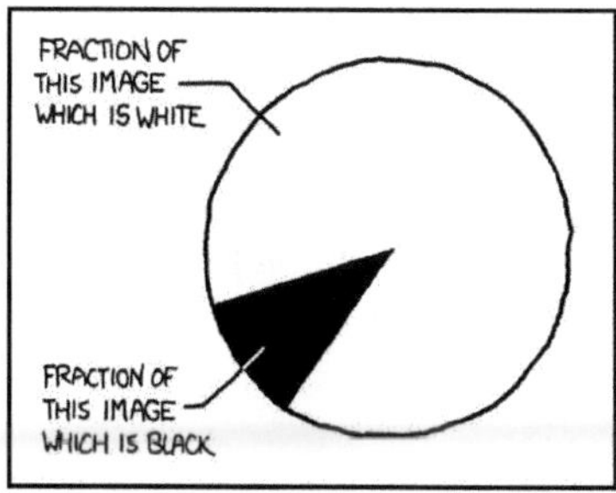

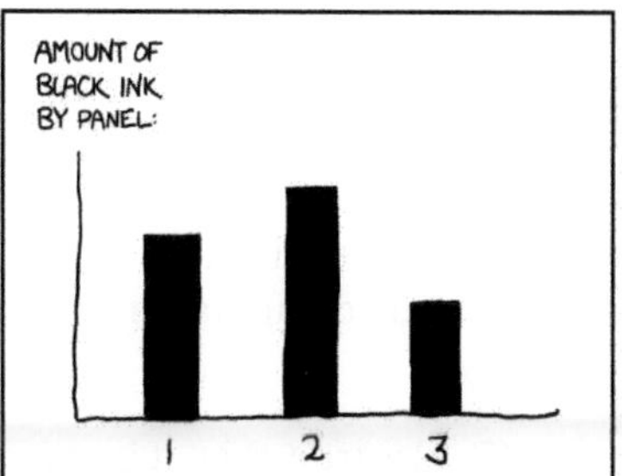

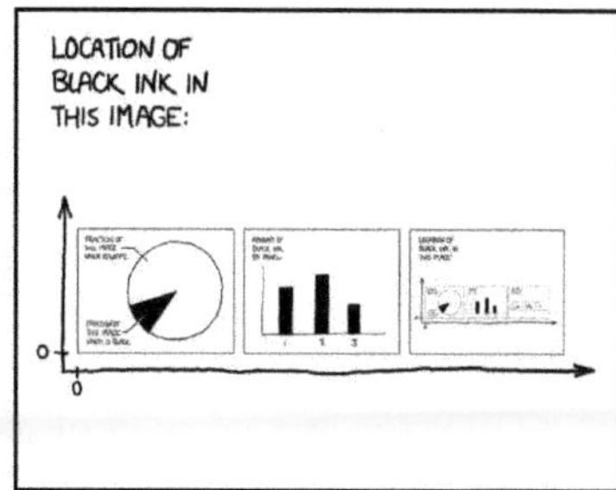

FIGURE 7.3: Above, some self-referential cartoons. Below, this caption.

from the axioms. We say that a formal system is *consistent* if there is no statement T such that both T and $\overline{T}$ are theorems, and *complete* if, for all T, at least one of T or $\overline{T}$ is a theorem.

We can define a statement as true or false by interpreting the symbols of the formal system in some standard way—treating $\exists$ as "there exists," $\wedge$ as "and," and so on—and assuming that its variables refer to a specific set of mathematical objects, such as the integers. Then the ideal formal system would be consistent and complete, in that all its theorems are true, and all true statements are theorems. Such a system would fulfill Hilbert's dream of an axiomatic foundation for mathematics. It would be powerful enough to prove all truths, and yet be free from paradoxes.

In 1931, Kurt Gödel dashed Hilbert's hopes. He proved the astonishing fact that *no sufficiently powerful formal system is both consistent and complete.* He did this by constructing a self-referential statement which can be interpreted as

This statement cannot be proved.

If this were false, it could be proved—but this would violate consistency by proving something false. Therefore, it must be true, and hence unprovable—and so there are truths that cannot be proved.

If all that Gödel had done was write down this sentence in English (or German), it would be a paradox 7.9
about language, not mathematics. After all, natural language makes self-reference very easy, as the figure at the top of this page illustrates. He did much more than that. He showed, amazingly, that this kind of self-reference can happen even in formal systems that talk about the integers. In other words, he constructed a sentence that, on one level, claims that an integer with some property does not exist—but that, on another level, can be interpreted as referring to itself, and asserting that no proof of it exists.

Rather than following Gödel's original proof, we will prove his Incompleteness Theorem using the undecidability of the Halting Problem. In other words, we will trade sentences that talk about themselves for programs that run themselves. Ultimately, these two forms of self-reference are the same—but what Gödel had to build from scratch, universal computation gives us for free.

We start by discussing the property $\texttt{Theorem}(T)$ that a statement T is provable. We can state it like this:

$$\texttt{Theorem}(T) = \exists P : \texttt{Proof}(P, T),$$

where $\texttt{Proof}(P, T)$ is the property that P is a valid proof of T. As we discussed in Section 6.1.3, `Proof` is decidable, since we can check P step-by-step and see if each line follows from the previous ones according to the axioms. Thus the set of theorems is recursively enumerable—we can go through all possible proofs, in order of increasing length, and print out all the theorems we find.

We are interested in formal systems that are powerful enough to talk about computation. First of all, we assume that they include the usual logical operators, including quantifiers like $\exists$ and $\forall$. Secondly, we assume that they can express claims like $\texttt{Halts}(\Pi, x)$, and that their axioms are strong enough to derive each step of a computation from the previous one.

Now let's ask whether such a system can prove all truths about halting or non-halting programs. If $\Pi(x)$ halts on its tth step, then a transcript of its computation is a proof of this fact, about t lines long. Therefore,

If $\texttt{Halts}(\Pi, x)$ is true, then it is provable.

But what if Π doesn't halt? Suppose that all true statements of the form $\overline{\texttt{Halts}(\Pi, x)}$ were provable. Then we could solve the Halting Problem, by doing two things in parallel: running $\Pi(x)$ to see if it halts, and looking for proofs that it won't. Since the Halting Problem is undecidable, there must exist statements of the form $\overline{\texttt{Halts}(\Pi, x)}$ that are true, but are not provable. Thus, for some programs and inputs, the fact that they will never halt is an unprovable truth. In that case, neither $\texttt{Halts}(\Pi, x)$ nor $\overline{\texttt{Halts}(\Pi, x)}$ is a theorem—it is formally independent of our axioms.

Of course, we can always add $\overline{\texttt{Halts}(\Pi, x)}$ to our list of axioms. Then the fact that Π doesn't halt on input x becomes, trivially, a theorem of the system. But then there will be another program Π', and another input x', for which $\overline{\texttt{Halts}(\Pi', x')}$ is true but not provable in this new system, and so on. No finite set of axioms captures all the non-halting programs—for any formal system, there will be some truth that it cannot prove.

7.10

Exercise 7.8 *Suppose that a formal system is consistent, and powerful enough to discuss computation. Show that the set of theorems is recursively enumerable but not decidable. Equivalently, the set of non-theorems is not recursively enumerable.*

Exercise 7.9 *Theorems that are easy to state can be very hard to prove, and the shortest proof can be much longer than the statement of the theorem. Let* $\texttt{ProofLength}(\ell)$ *be the maximum, over all theorems* T *of length* ℓ*, of the length of the shortest proof of* T*. Show that* $\texttt{ProofLength}$ *grows faster than any computable function.*

7.3 Building Blocks: Recursive Functions

> The fact that we can build a computer out of parts… is due to the recursive dependence of the basic operations between natural numbers.
>
> Rósza Péter, *Recursive Functions in Computer Theory*

How can we give a clear mathematical definition of computation? Intuitively, a function is computable if it can be defined in terms of simpler functions which are computable too. These simpler functions are defined in turn in terms of even simpler ones, and so on, until we reach a set of primordial functions for which no further explanation is necessary. These primordial functions form the "atoms" of computation. In terms of programming, they are the elementary operations that we can carry out in a single step.

Modern programming languages let us build up these definitions in a number of ways. Functions can call each other as subroutines; they can call themselves, reducing the problem to a simpler instance, until they reach an easy base case; and they can repeat a set of operations many times by forming **for** and **while** loops. Each of these basic structures can be traced back to the theory of *recursive functions*, which was one of the three main models of computation leading to the unification of 1936. In this section, we take a look at this theory—its atoms and chemistry—and learn what ingredients we really need for universal computation.

7.3.1 Building Functions from Scratch

To build a set of computable functions recursively—in modern terms, to define a programming language—we need to define a set of primordial functions and a set of schemes by which functions can be combined. Aesthetically, we would like both these sets to be as small and simple as possible, while ensuring that we can build any conceivable computation.

As we will see, there is no need to include arithmetic operations like addition and multiplication among our primordial functions. In order to get arithmetic off the ground, all we need is the number zero, i.e., the constant function

$$0(x) = 0\,,$$

and the *successor* function,

$$S(x) = x + 1\,.$$

Strictly speaking, we also need to include the identity function $I(x) = x$, and more generally functions like $I_2^3(x, y, z) = y$ that pull numbers out of a list. Hopefully everyone agrees that these are computable.

Now we need some schemes by which we can construct new functions from old ones. The first one is composition. If f and g are already defined, we can define a new function $h = f \circ g$ by

$$h(\mathbf{x}) = f(g(\mathbf{x}))\,, \tag{7.5}$$

where $\mathbf{x} = (x_1, \ldots, x_k)$ represents an arbitrary number of variables. In terms of programming, composition lets us call previously defined functions as subroutines, using the output of one as the input of the other.

More generally, we allow functions to access each of their variables, and to call a subroutine on any subset of them. For instance, if $f(x_1, x_2)$, $g(x_1, x_2)$ and $g'(x_1, x_2)$ are already defined, we can define

$$h(x_1, x_2, x_3) = f\big(g(x_1, x_2), g'(x_3, x_1)\big)\,.$$

The second scheme is called *primitive recursion*. It allows us to define a function in terms of itself in a specific way: for one of its variables y, we can define its value at $y+1$ in terms of its value at y, until we get to the base case $y = 0$. Thus if f and g are defined, we can define h as

$$\begin{aligned} h(\mathbf{x}, 0) &= f(\mathbf{x}) \\ h(\mathbf{x}, y+1) &= g(\mathbf{x}, y, h(\mathbf{x}, y))\,. \end{aligned} \tag{7.6}$$

As Figure 7.4 shows, we can turn this into modern-looking pseudocode in two different ways. We can use recursion, as in the definition, or use a **for** loop to iterate through the values of y. Thus primitive recursion adds **for** loops to our programming toolbox.

Note that when we start evaluating a function defined using primitive recursion, we know to what depth it will recurse, since evaluating $h(\mathbf{x}, y)$ takes y levels of recursion. Equivalently, whenever we enter

```
h(x, y)
begin
    if y = 0 then return f(x);
    else return g(x, y, h(x, y − 1));
end
```

```
h(x, y)
begin
    z := f(x);
    for y′ = 0 to y − 1 do
        z := g(x, y′, z);
    end
    return z;
end
```

FIGURE 7.4: Two ways to program the primitive recursive function $h(\mathbf{x}, y)$ defined in (7.6). On the left, we use recursion. On the right, we use a **for** loop.

the **for** loop of Figure 7.4, we know that it will run y times and then halt. Of course we disallow nasty tricks like changing the value of y from inside the loop.

Using primitive recursion, we can define addition as

$$\begin{aligned}\texttt{add}(x,0)&=x\\ \texttt{add}(x,y+1)&=S\big(\texttt{add}(x,y)\big),\end{aligned}$$

or, in plainer language,

$$x+0=x \text{ and } x+(y+1)=(x+y)+1.$$

Once we have addition, we can define multiplication and then exponentiation:

$$\begin{aligned}\texttt{mult}(x,0)&=0\\ \texttt{mult}(x,y+1)&=\texttt{add}\big(\texttt{mult}(x,y),x\big),\end{aligned}$$

$$\begin{aligned}\texttt{exp}(x,0)&=1\\ \texttt{exp}(x,y+1)&=\texttt{mult}\big(\texttt{exp}(x,y),x\big).\end{aligned}$$

Exercise 7.10 *Express the following functions in primitive recursive fashion: predecessor and subtraction,*

$$\texttt{pred}(x)=\begin{cases}0 & \text{if } x=0\\ x-1 & \text{if } x>0\end{cases}\qquad \texttt{sub}(x,y)=\begin{cases}0 & \text{if } x<y\\ x-y & \text{if } x\geq y\end{cases},$$

positivity and less-or-equal,

$$\texttt{pos}(x)=\begin{cases}0 & \text{if } x=0\\ 1 & \text{if } x>0\end{cases}\qquad \texttt{leq}(x,y)=\begin{cases}1 & \text{if } x\leq y\\ 0 & \text{if } x>y\end{cases},$$

and parity and division by two,

$$\texttt{even}(x)=x \bmod 2=\begin{cases}0 & \text{if } x \text{ is odd}\\ 1 & \text{if } x \text{ is even}\end{cases}\qquad \texttt{div}_2(x)=\lfloor x/2\rfloor.$$

Functions that can be generated from the primordial functions 0 and S by using composition and primitive recursion are called *primitive recursive.* In addition to the basic arithmetic functions we just defined, this includes many "higher" functions, such as `prime`(x) which is 1 if x is prime and 0 otherwise. Indeed, most functions we encounter in everyday life are primitive recursive—we can program them using subroutines, **if–then–else**, and **for** loops.

We claim, however, that not all computable functions are primitive recursive. Why? Because we can 7.11
compute the value of any primitive recursive function, given its definition, with a finite number of steps. In other words, the universal function $U(f,x) = f(x)$ is computable, where by the first f we mean f's "source code"—a finite string defining f in terms of simpler functions, and defining those, all the way down to 0 and S.

If U were primitive recursive, the diagonalization argument of Section 7.2.1 would show that some primitive recursive functions are only partially defined—that, for some inputs, the corresponding program never halts. But since the number of times that each **for** loop runs is bounded, primitive recursive programs always halt, and their outputs are always defined. Therefore, they cannot interpret themselves—there is no primitive recursive function whose behavior includes that of all the others.

There is another way to show that the primitive recursive functions are a limited subset of the computable functions. We can define a function that is computable, but whose explosive growth outstrips that of any primitive recursive function. This is what we will do next.

7.3.2 Primitive Programming with BLOOP*, and a Not-So-Primitive Function*

To explore the power of primitive recursion, it helps to imagine writing programs in a fictional programming language, which we follow Douglas Hofstadter in calling BLOOP. The hallmark of BLOOP is that it has only bounded loops, such as **for** or **repeat** y **times** loops, where the number of iterations is fixed when we enter the loop.

Figure 7.5 shows BLOOP programs for addition, multiplication, and exponentiation. Instead of calling the earlier functions as subroutines, we write them out in the source code, so you can see how each level of primitive recursion creates another nested loop. The differences in the initial values, namely x for addition, 0 for multiplication, and 1 for exponentiation, are unimportant.

It seems that addition, multiplication, and exponentiation are the first three in a series of operations. What happens if we continued this series? Consider the following definitions:

$$A_1(x,y) = x + y = x + \underbrace{1 + \cdots + 1}_{y \text{ times}}$$

$$A_2(x,y) = x \times y = \underbrace{x + x + \cdots + x}_{y \text{ times}}$$

$$A_3(x,y) = x^y = \underbrace{x \times x \times \cdots \times x}_{y \text{ times}}$$

$$A_4(x,y) = x^{x^{\cdot^{\cdot^{x}}}} = \underbrace{x\wedge(x\wedge(\cdots\wedge x))}_{y \text{ times}}.$$

You see the pattern: $A_n(x,y)$ is defined by applying x to itself y times using A_{n-1}. For $n \geq 2$, we can write

```
add(x,y)
begin
    z1 := x;
    repeat y times do
        z1 := z1 + 1;
    end
    return z1;
end

mult(x,y)
begin
    z2 := 0;
    repeat y times do
        z1 := x;
        repeat z2 times do
            z1 := z1 + 1;
        end
        z2 := z1;
    end
    return z2;
end

exp(x,y)
begin
    z3 := 1;
    repeat y times do
        z2 := 0;
        repeat z3 times do
            z1 := x;
            repeat z2 times do
                z1 := z1 + 1;
            end
            z2 := z1;
        end
        z3 := z2;
    end
    return z3;
end
```

FIGURE 7.5: BLOOP programs for addition, multiplication, and exponentiation, with one, two, and three levels of nested loops.

this recursively as

$$A_n(x,y) = \begin{cases} x \times y & \text{if } n = 2, \\ 1 & \text{if } n > 2 \text{ and } y = 0, \\ A_{n-1}\big(x, A_n(x, y-1)\big) & \text{if } y > 0. \end{cases} \tag{7.7}$$

As n increases, $A_n(x,y)$ grows astronomically, as the following exercise will help you appreciate.

Exercise 7.11 *List all pairs (x,y) for which $A_4(x,y)$ does not exceed 2^{64}. Compare $A_4(3,4)$ to the number of atoms in the visible universe, which is roughly 10^{80}. Which is larger, $A_5(4,2)$ or $A_4(5,2)$?*

The function $A_n(x,y)$ is called the *Ackermann function.* For each fixed n, it is a primitive recursive function of x and y, since, as Figure 7.6 shows, we can calculate it with a BLOOP program that has n levels of nested **for** loops. But what if n is a variable rather than a constant, giving a three-variable function $A(n,x,y) = A_n(x,y)$? Any particular BLOOP program has a fixed number of nested loops, so no such program can suffice for all n. This suggests that A is not primitive recursive.

In Problem 7.17, we lead you through a proof of this intuition. Specifically, we ask you to show that, for any primitive recursive function $f(y)$, there is an n such that A_n *majorizes* f in the following sense: for all $y \geq 3$,

$$f(y) < A_n(2,y). \tag{7.8}$$

```
A_n (x,y)
begin
    z_n := 1;
    repeat y times do
        z_{n-1} := 1;
        repeat z_n times do
            ⋱      (n nested loops)
                z_1 := x;
                repeat z_2 times do
                    z_1 := z_1 + 1;
                end
                z_2 := z_1;
            ⋰
        end
        z_n := z_{n-1};
    end
    return z_n;
end
```

FIGURE 7.6: BLOOP program for the Ackermann function $A_n(x,y)$ for fixed n.

In essence, n is the number of nested **for** loops in the BLOOP program for f, or the number of levels of primitive recursion it takes to define f from S and 0. Now, if $A(n,x,y)$ were primitive recursive, so would be the special case $f(y) = A_y(2,y)$. But then for some n we would have

$$A_y(2,y) < A_n(2,y).$$

If we set $y = n$, which is yet another form of diagonalization, we get a contradiction.

We can certainly compute $A(n,x,y)$ in finite time, since applying the recursive equation (7.7) repeatedly will eventually reach a base case. However, unless n is constant, there is no way to express (7.7) in primitive recursive form, i.e., with a constant number of equations of the form (7.6)—it is a more complicated type of recursion.

Thus $A(n,x,y)$ is a concrete example of a function that is computable but not primitive recursive. What kind of programming language do we need to compute it? What is missing from BLOOP?

7.12

7.3.3 *Run Until You're Done: **while** Loops and μ-Recursion*

> But what has been said once can always be repeated.
>
> Zeno of Elea

We can always program the Ackermann function by translating (7.7) directly into pseudocode. This gives a recursive algorithm, as shown on the left side of Figure 7.7, in which $A(n,x,y)$ calls itself on smaller

```
A(n,x,y)
begin
    if n = 2 then
        return mult(x,y);
    else if y = 0 then
        return 1;
    else
        return A(n − 1,x,A(n,x,y − 1));
end
```

```
A(n,x,y)
begin
    push(n);
    while stack not empty do
        pop(n);
        if n = 2 then
            y := mult(x,y);
        else if y = 0 then
            y := 1;
        else
            y := y − 1;
            push(n − 1);
            push(n);
    end
    return y;
end
```

FIGURE 7.7: We can implement the Ackermann function recursively, or iteratively using a **while** loop and a stack. For convenience we assume that $n \geq 2$ and use $A(2,x,y) = \texttt{mult}(x,y)$ as our base case.

values of n. But what if we want to "unfold" this recursion as in Figures 7.5 and 7.6? What kind of loop do we need to calculate $A(n,x,y)$ iteratively instead of recursively?

We can carry out a recursive algorithm by using a *stack* to keep track of the tasks and subtasks along the way. We kick the process off by pushing the number n on the stack. Then, at each step, we pop the top number off the stack to see what value of n we are working on. Recall from (7.7) that

$$A(n,x,y) = A(n-1,x,A(n,x,y-1)).$$

This tells us that, in order to calculate $A(n,x,y)$, we have to calculate $A(n,x,y-1)$ first. To do this, we decrement y, and push another copy of n on the stack. But first we push $n-1$ on the stack, in order to remember that after we calculate $y' = A(n,x,y-1)$, we will feed it into $A(n-1,x,y')$.

We continue in this way, pushing subtasks onto the stack, until we reach a base case where we can compute $A(n,x,y)$ directly. To keep things simple, in Figure 7.7 we define our base case as $n = 2$, at which point we call the `mult` function on x and y. Once we have computed $A(n,x,y)$, we use the variable y to pass its value back to the subtask below it on the stack. Thus we push and pop subtasks until the stack is empty, our task is done, and we return the answer.

It is a simple matter to encode an entire stack as a single integer, and to write primitive recursive functions `push`, `pop`, and `isempty` to manipulate it. For one method, see Problem 7.18. So each step of this process is primitive recursive. But how many times do we need to repeat it? If we knew the running time in advance, we could put the entire process inside a **for** loop. Unfortunately, the running time is essentially $A(n,x,y)$ itself. So we can't use a bounded loop, since the bound is the function we're trying to calculate in the first place.

What we need is a **while** loop, which runs as long as some condition is satisfied—in this case, as long as there are still tasks left to be done on the stack. In the theory of recursive functions, we represent these loops with an operator called *μ-recursion*. Given a function $f(\mathbf{x}, y)$ which is already defined, μ-recursion lets us define a new function $h(\mathbf{x})$ as follows:

$$h(\mathbf{x}) = \mu_y f(\mathbf{x}, y) = \min\{y : f(\mathbf{x}, y) = 0\}. \tag{7.9}$$

Thus $h(\mathbf{x})$ returns the smallest solution y to the equation $f(\mathbf{x}, y) = 0$. This corresponds to the following **while** loop:

```
h(x)
begin
    y := 0;
    while f(x, y) ≠ 0 do y := y + 1;
    return y;
end
```

Of course, there may be no such y, and this **while** loop might never halt. In that case, $h(\mathbf{x})$ is undefined. Thus when we add μ-recursion and **while** loops to our toolbox, we gain a great deal of computational power, but at a cost—we must now include partial functions, which might be undefined for some values of their input. Indeed, when we apply μ to partial functions, we require that $f(\mathbf{x}, y')$ is defined for all the y' we need to check:

$$\mu_y f(\mathbf{x}, y) = \min\{y : f(\mathbf{x}, y) = 0 \text{ and } f(\mathbf{x}, y') \text{ is defined for all } y' \leq y\}. \tag{7.10}$$

Functions that can be generated from the primordial functions 0 and S using composition, primitive recursion, and μ-recursion are called *partial recursive*. Since the corresponding programming language allows *free* loops in addition to bounded ones, we follow Hofstadter in calling it FLOOP. Unlike BLOOP, FLOOP contains all the basic constructs that appear in modern programming languages. So, we can just as well define the partial recursive functions as those you can compute in C, FORTRAN, or whatever your favorite programming language is.

We can consider the subclass of *total* recursive functions, i.e., those which are defined for all values of their input. Since there exist functions, such as the Ackermann function, that are total but not primitive recursive, the primitive recursive functions form a proper subclass of the total recursive ones:

$$\text{primitive recursive} \subset \text{total recursive} \subset \text{partial recursive}.$$

In terms of programming,

$$\text{BLOOP programs} \subset \text{FLOOP programs that always halt} \subset \text{FLOOP programs}.$$

7.3.4 *A Universal Function*

At the end of Section 7.3.1, we pointed out that the universal function $U(f, x) = f(x)$ is not primitive recursive, even when f is restricted to the primitive recursive functions. In other words, BLOOP is not strong enough to interpret itself—it is not flexible enough to bite its own tail.

FLOOP, on the other hand, does have this self-referential power. In other words, there is a partial recursive function U, such that $U(f, x) = f(x)$ for any partial recursive function f and any x, with the understanding that $U(f, x)$ is well-defined if and only if $f(x)$ is.

We can justify this claim by writing U in FLOOP. Since we don't know f's running time, we use a **while** loop to run it until it halts:

```
U(f,x)
begin
   initialize f with input x ;
   while f hasn't halted yet do
      run a step of f ;
   return f's output ;
end
```

For now, we beg the reader's indulgence on the line "run a step of f," since there is a more serious issue to deal with first. By definition, partial recursive functions act on integers, not on functions. How can we feed one function to another as input?

In the modern age, the answer is simple. Each partial recursive function f has a source code—a FLOOP program, or a series of definitions using composition, primitive recursion, and μ-recursion. In either case, this source code is a finite string. If it is written in an alphabet of, say, 50 symbols, we can write it out and treat it as an integer written in base 50. Thus every program is an integer that describes the function it computes, and we can feed this integer to U.

In the early days of computation, however, this question threw people for a loop. The fact that a function can be encoded as an integer, which can in turn be acted on by other functions—in other words, that functions can act both as programs and as data—came as a revelation.

This idea first appeared in the work of Gödel. In order to feed them to other functions, he devised a way to give each function an integer description—what we now call a *Gödel number*. Let $\overline{h}$ denote the Gödel number of a partial recursive function h. We can start by defining $\overline{h}$ for the zero, successor, and identity functions—the atoms of our language—and then work upward from there. One possible scheme might look like this:

$$\overline{h} = \begin{cases} 7 & \text{if } h = 0 \\ 11 & \text{if } h = S \\ 13 & \text{if } h = I \\ 2 \cdot 3^{\overline{f}} \cdot 5^{\overline{g}} & \text{if } h = f \circ g \\ 2^2 \cdot 3^{\overline{f}} \cdot 5^{\overline{g}} & \text{if } h \text{ is defined from } f \text{ and } g \text{ using primitive recursion} \\ 2^3 \cdot 3^{\overline{f}} & \text{if } h = \mu_y f \,. \end{cases} \tag{7.11}$$

Note that these descriptions are unambiguous, since the prime factorization of an integer is unique. The Gödel number of our addition function would then be

$$\overline{\texttt{add}} = 2^2 \cdot 3^{13} \cdot 5^{11} = 311\,391\,210\,937\,500 \,.$$

We need to add some bells and whistles to deal with functions of several variables, but you get the idea. The Gödel number of h is defined inductively from the Gödel numbers of simpler functions, until we get down to elementary functions that don't have to be defined. Thus it is much like a program in a modern programming language, albeit a little less readable.

Now we claim that, using this Gödel numbering or any other reasonable definition of a function's source code, the universal function $U(f,x) = f(x)$ is partial recursive. The proof of this claim is rather technical, and with a thankful nod to the pioneers of computation, we relegate its details to history. But the idea is simple. There are primitive recursive functions that "read" integers, either in a format like (7.11) or as a string of symbols, and parse it as a string of instructions. Then there is a primitive recursive function `step` that executes a single instruction, updates the variables that keep track of intermediate results, and keeps track of what instruction it should execute next. Thus the line "run a step of f" in our pseudocode for U is primitive recursive, and U is partial recursive.

Exercise 7.12 Kleene's Normal Form Theorem. *Sketch a proof that there exist primitive recursive functions f and g such that every partial recursive function h can be written in the form*

$$h(\mathbf{x}) = g\big(\mathbf{x}, \mu_y f(p,\mathbf{x},y)\big)$$

for some integer p. Therefore, any partial recursive function can be written with a single use of μ-recursion. Equivalently, any partial recursive function can be computed by a FLOOP *program with a single **while** loop.*

We have seen that the partial recursive functions are very rich, and that they can simulate and interpret themselves just as today's computers can. Nevertheless, we can still ask whether the set of partial recursive functions is synonymous with the set of computable ones. Do composition, primitive recursion, and μ-recursion cover every reasonable way in which functions can be defined inductively in terms of simpler ones? Equivalently, can all possible computations be carried out with basic arithmetic, along with **for** and **while** loops?

With the benefit of hindsight, the answer is yes—otherwise, there is something missing from all the programming languages we use today. However, if we put ourselves in the shoes of our counterparts in 1936, we have no reason to be convinced of this. Let's look at another approach to computation, which at first seems totally different from the partial recursive functions, but which turns out to be exactly equivalent to them.

7.4 Form is Function: the λ-Calculus

> You say that I pay too much attention to form. Alas! it is like body and soul: form and content to me are one; I don't know what either is without the other.
>
> Gustave Flaubert

Traditional mathematics is a rigidly stratified society. At the bottom of the ladder we have numbers. Above them are functions, which act on numbers and return numbers. On the level above that are more abstract operators, such as composition, primitive recursion, or the derivative, which take one or more functions as input and return functions as output.

This strict hierarchy is perfectly fine for almost all of mathematics, and as we discussed in Section 7.1.2, it banishes a host of paradoxes based on self-reference. But for computer science, it is too confining. We need to be able to make statements such as $U(f,x) = f(x)$ where objects live on two levels at once—where

f plays two roles, as both a function and a piece of data that can be fed to other functions. Level-crossing statements like these are our bread and butter, including "loopy" ones like $U(f,f)=f(f)$ where we feed a function to itself.

If we have to, we can close these loops using Gödel numbers, where each function f has a corresponding integer $\overline{f}$, but this involves a lot of technical work. In the modern world, programs and data are both simply finite strings, and we can feed any string to any program without breaking any laws. Is there an elegant mathematical model which is similarly democratic?

The logician Alonzo Church invented a system, called the *λ-calculus*, in which functions at all levels of abstraction can interact with each other as equals. At first, the notation used in this system is quite strange, and even forbidding. However, it has its own charms which will reward the persistent reader. It represents an attitude towards computation that is very different from other models, and which still survives today in functional programming languages such as LISP and SCHEME. Because it makes no distinction between functions and data, it makes diagonal arguments, like the undecidability of the Halting Problem, elegant and immediate.

7.4.1 Functions are Forms

The format of the λ-calculus takes a little getting used to. First, rather than writing

$$f(x)=x^2\,,$$

we will write

$$f=\lambda x\,.\,x^2\,.$$

In other words, $\lambda x\,.\,x^2$ is the function that takes a variable x and returns x^2. If we place a number to the right of this function, it substitutes that number for the variable x:

$$(\lambda x\,.\,x^2)7=49\,.$$

So far, this is just a strange shift in notation. But the way the λ-calculus treats the *process* of computation is also very different. Consider the addition function, which we write

$$\lambda xy\,.\,x+y\,.$$

Let's apply this to 3 and 5. Rather than writing

$$(\lambda xy\,.\,x+y)(3,5)\,,$$

we write

$$((\lambda xy\,.\,x+y)3)5=(\lambda y\,.\,3+y)5=8\,.$$

Halfway through this computation, we have the object $\lambda y\,.\,3+y$. In other words, having fixed $x=3$, we have a function on the other variable y, which adds 3 to it.

This is a subtle but powerful shift in how we think about performing a computation. Rather than modifying variables and passing their values up to functions, we evaluate functions from the top down, modifying what they will do with the rest of their data when they get it. The notation λxy is really a shorthand, which works like this:

$$\lambda xy\,.\,x+y=\lambda x\,.(\lambda y\,.\,x+y)\,.$$

Thus addition is a function on *one* variable x, which, when given x, returns the function that adds x to its input y.

The purest form of the λ-calculus doesn't have arithmetic operations like addition, or even the concept of the integers. The basic objects in the λ-calculus are strings, or *combinators*, which take the next string to their right and substitute it for each appearance of the first variable marked with λ. In other words, if x is a variable and U and V are strings, then

$$(\lambda x.U)V = U[x \to V], \tag{7.12}$$

where $U[x \to V]$ denotes the string we get from U by replacing each appearance of x with V. Each such step is called a *reduction*. A computation then consists of a chain of reductions, such as

$$(\lambda xy\,.\,yxy)ab = (\lambda y\,.\,yay)b = bab\,.$$

Note that we evaluate strings from left to right, and that in the absence of parentheses, we interpret the string zab as $(za)b$. The operation of reduction is nonassociative, meaning that $z(ab) \neq (za)b$ in general. For instance, grouping a and b together in this example gives

$$(\lambda xy\,.\,yxy)(ab) = \lambda y\,.\,y(ab)y\,.$$

This process of substituting strings for symbols is all the λ-calculus ever does. Each combinator simply provides a form for strings to plug into. And yet, these forms give us all the function we need. As we will prove below, the λ-calculus is just as powerful as our other models of computation.

To see how things might get interesting, let's look at some examples where we feed one combinator to another. Consider the following three combinators:

$$\begin{aligned} \mathbf{T} &= \lambda xy\,.\,x \\ \mathbf{F} &= \lambda xy\,.\,y \\ \mathbf{I} &= \lambda x\,.\,x \end{aligned} \tag{7.13}$$

The combinators **T** and **F** take two inputs and return the first or second one. Since **I** is the identity, returning its input unchanged, we have **IT** = **T** and **IF** = **F**. But if we combine these in the opposite order, something funny happens:

$$\begin{aligned} \mathbf{TI} &= (\lambda xy\,.\,x)(\lambda x\,.\,x) \\ &= (\lambda xy\,.\,x)(\lambda z\,.\,z) \\ &= \lambda y\,.\,(\lambda z\,.\,z) \\ &= \lambda yz\,.\,z \\ &= \mathbf{F}. \end{aligned}$$

In the second line, we changed the variable in **I** from x to z to avoid ambiguity, and we renamed variables again in the last line to identify $\lambda yz\,.\,z$ with **F**. These renamings are allowed because $\lambda x\,.\,x$ and $\lambda z\,.\,z$ are both the combinator **I**, and $\lambda xy\,.\,y$ and $\lambda yz\,.\,z$ are both the combinator **F**, just as $f(x) = x^2$ and $f(y) = y^2$ are both the function that returns the square of its input.

Some products, such as **TI**, reach a final state, or *normal form*, in which no more reductions are possible. Either there are no λs left, or there are no strings to the right of the first combinator to plug into it. However, this is not true of all expressions. Consider the combinator $\mathbf{Q} = \lambda x.xx$, which takes x as input and returns xx. If we feed $\mathbf{Q}$ to itself, we get

$$\begin{aligned}\mathbf{QQ} &= (\lambda x.xx)(\lambda x.xx) = (\lambda x.xx)(\lambda x.xx)\\ &= (\lambda x.xx)(\lambda x.xx)\\ &= \cdots.\end{aligned}$$

The chain of reductions never halts. We keep plugging the second $\lambda x.xx$ into the λx of the first, and the expression never reaches a normal form. Even worse is the following expression, which grows without bound as we try to reduce it:

$$\begin{aligned}(\lambda x.xx)(\lambda y.yyy) &= (\lambda y.yyy)(\lambda y.yyy)\\ &= (\lambda y.yyy)(\lambda y.yyy)(\lambda y.yyy)\\ &= (\lambda y.yyy)(\lambda y.yyy)(\lambda y.yyy)(\lambda y.yyy)\\ &= \cdots.\end{aligned}$$

These are examples of infinite loops in the λ-calculus. Like partial recursive functions, λ-expressions may never return an answer, and the functions they compute may be undefined for some values of their input. Telling which expressions reach a normal form and which ones don't is λ-calculus's version of the Halting Problem, and we will prove below that it is undecidable.

As Problem 7.21 shows, we can also think of $\mathbf{QQ} = (\lambda x.xx)(\lambda x.xx)$ as a "quine"—a program that prints out its own source code. We can even use it as a recipe to create quines in other programming languages.

7.13

7.4.2 Numbers are Functions Too

So far, so good. But how does all this string manipulation actually compute anything? For an example of how λ-calculus can perform simple logical operations, take a look at the following exercise:

Exercise 7.13 *Consider the combinators* **T** *and* **F** *defined in* (7.13). *As their names imply, let's think of these as representing* `true` *and* `false` *respectively. Show that we can then represent Boolean functions as follows:*

$$\begin{aligned}\mathbf{not} &= \lambda pxy.pyx\\ \mathbf{and} &= \lambda pq.pqp\\ \mathbf{or} &= \lambda pq.ppq.\end{aligned}$$

For instance, show that

$$\mathbf{and\,T\,F} = (\mathbf{and\,T})\,\mathbf{F} = \mathbf{I\,F} = \mathbf{F}.$$

Of course, there is no inherent connection between **T** and **F** and the truth values `true` and `false`. Indeed, λ-calculus allows us to manipulate strings in such general ways that nearly any two strings could play these roles.

How about arithmetic? There are many ways to represent the natural numbers $0, 1, 2, 3 \ldots$ in λ-calculus. We will use the *Church numerals*, defined as follows:

$$\begin{aligned}
\overline{0} &= \lambda f x . x\,, \\
\overline{1} &= \lambda f x . f x\,, \\
\overline{2} &= \lambda f x . f(f x)\,, \\
\overline{3} &= \lambda f x . f(f(f x))\,,
\end{aligned}$$

and so on. Given a function f, the Church numeral $\overline{n}$ returns the function f^n, i.e., f iterated n times:

$$\overline{n} f = \lambda x . f^n x\,. \tag{7.14}$$

To flex our muscles, let's define combinators that carry out arithmetic operations on these numerals. First, consider

$$\mathbf{succ} = \lambda m f x . f(m f x)\,. \tag{7.15}$$

If $\overline{m}$ is a numeral, $\mathbf{succ}\,\overline{m}$ applies f one more time than $\overline{m}$ would. Thus **succ** acts as the successor function. Let's check this, reducing one step at a time:

$$\begin{aligned}
\mathbf{succ}\,\overline{n} &= (\lambda m f x . f(m f x))(\lambda g y . g^n y) \\
&= \lambda f x . f((\lambda g y . g^n y) f x) \\
&= \lambda f x . f((\lambda y . f^n y) x) \\
&= \lambda f x . f(f^n x) \\
&= \lambda f x . f^{n+1} x \\
&= \overline{n+1}\,.
\end{aligned}$$

In contrast to traditional numbers, Church numerals are not simply passive data. They are active functions, which create iterated versions of other functions. This lets us write inductive definitions of functions very compactly. For instance, suppose we want a combinator **add** that adds two numerals together. One way to do this is

$$\mathbf{add} = \lambda n m f x . n f(m f x)\,. \tag{7.16}$$

Since $f^m(f^n(x)) = f^{m+n}(x)$, we have

$$\mathbf{add}\,\overline{m}\,\overline{n} = \overline{m+n}\,.$$

But we can also describe adding n as iterating the successor function n times. So alternately, we could write

$$\mathbf{add}' = \lambda n . n\,\mathbf{succ}\,. \tag{7.17}$$

Problem 7.22 shows that **add** and **add**$'$ are equivalent, at least when applied to Church numerals.

For multiplication, we use the fact that if we iterate f^n m times we get $(f^n)^m = f^{nm}$. Thus if

$$\mathbf{mult} = \lambda n m f . n(m f)\,,$$

we have

$$\mathbf{mult}\,\overline{m}\,\overline{n} = \overline{m n}\,.$$

Note that all these combinators are happy to take inputs n and m that are not Church numerals, but in that case their behavior may have nothing to do with arithmetic.

Finally, what happens if we apply one numeral to another? Applying $\overline{n}$ to $\overline{m}$ means iterating, n times, the operator that changes f to f^m. This may seem a bit abstract, but this is the sort of thing that λ-calculus is good at. The answer is given by the following exercise.

Exercise 7.14 *Show that if*

$$\mathbf{exp} = \lambda mn.nm$$

then

$$\mathbf{exp}\,\overline{m}\,\overline{n} = \overline{m^n}\,.$$

Writing the successor function for Church numerals was easy, and this let us construct combinators for addition, multiplication, and exponentiation. But what about the predecessor? Is there a combinator **pred** such that $\mathbf{pred}\,\overline{n} = \overline{n-1}$ if $n > 0$, and $\mathbf{pred}\,\overline{0} = \overline{0}$? Constructing **pred** is a little tricky, and Church himself believed for a while that it was impossible. His student Stephen Kleene found a clever solution, which we invite you to work out in Problem 7.23.

7.4.3 The Power of λ

> When one first works with the λ-calculus it seems artificial and its methods devious; but later one appreciates its simplicity and power.
>
> Robin Gandy

As we have seen, λ-calculus is strong enough to do basic arithmetic. Can it do more? Let us say that a function $f : \mathbb{N} \to \mathbb{N}$ is *λ-definable* if there is a combinator ϕ such that

$$\phi\,\overline{n} = \overline{f(n)}$$

for any Church numeral $\overline{n}$, as long as $f(n)$ is defined. How expressive is the λ-calculus compared to the recursive functions? Are all partial recursive functions λ-definable, or vice versa?

As described in Section 7.3, the partial recursive functions are defined inductively from zero and the successor function using composition, primitive recursion, and μ-recursion. We already have zero and the successor. We can compose two functions with the following combinator,

$$\mathbf{comp} = \lambda fgx.f(gx), \tag{7.18}$$

which takes two functions f, g as input and returns $f \circ g$. In fact, this is the same combinator **mult** that we used to multiply Church numerals above. If we can also perform primitive recursion and μ-recursion in λ-calculus, then by induction all partial recursive functions are λ-definable.

One way to handle primitive recursion is to treat it as a **for** loop. We can easily write a **for** loop that runs n times using the Church numeral $\overline{n}$. We ask you to do this for the factorial function in Problem 7.24, and for primitive recursion in general in Problem 7.25. This shows that any primitive recursive function is λ-definable.

However, a **for** loop is often not the most elegant way to define a function. We would like to be able to use recursive definitions like this,

$$n! = \begin{cases} 1 & \text{if } n = 0 \\ n(n-1)! & \text{if } n > 0\,. \end{cases} \tag{7.19}$$

This requires us to know when we have reached the base case $n = 0$, and to act accordingly. We can do this using the following exercise:

Exercise 7.15 *Show that if*

$$\mathbf{iszero} = \lambda n\,.\,n(\lambda x.\mathbf{F})\mathbf{T}$$

where $\mathbf{F}$ *and* $\mathbf{T}$ *are defined as in* (7.13), *then*

$$\mathbf{iszero}\,\overline{n}\,x\,y = \begin{cases} x & \text{if } n = 0 \\ y & \text{if } n > 0\,. \end{cases}$$

Using **iszero**, **mult**, and the predecessor **pred**, it seems that we can write the factorial as

$$\mathbf{fac} = \lambda n\,.\,\mathbf{iszero}\,n\,\overline{1}\,(\mathbf{mult}\,n\,(\mathbf{fac}(\mathbf{pred}\,n)))\,.$$

However, this definition doesn't quite work. Like any other combinator, **fac** must be a finite string of symbols. We are free to use previously-defined functions as shorthand, but defining it directly in terms of itself would lead to an infinite regress.

This makes it seem that functions in λ-calculus cannot be defined recursively in terms of themselves. Yet we don't want to lose the expressive elegance of recursion. How can we arrange for λ-expressions like **fac** to refer to themselves indirectly, rather than directly?

The idea is to treat recursive definitions like (7.19) as equations that functions must satisfy. Let's define an operator **R** as follows:

$$(\mathbf{R}f)(n) = \begin{cases} 1 & \text{if } n = 0 \\ n f(n-1) & \text{if } n > 0\,, \end{cases} \tag{7.20}$$

or, in λ-style,

$$\mathbf{R} = \lambda f n\,.\,\mathbf{iszero}\,n\,\overline{1}\,(\mathbf{mult}\,n\,(f(\mathbf{pred}\,n)))\,. \tag{7.21}$$

The effect of **R** is to modify a function f by applying a single step of the recursion (7.19). Then $f(n) = n!$ is the unique function such that

$$f = \mathbf{R}f\,.$$

In other words, f should be a *fixed point* of **R**.

But how can we know that **R** has a fixed point? And if it exists, how can we find it? In traditional mathematics, functions may or may not have fixed points, and for those that do, it is generally impossible to calculate them explicitly. One of the most beautiful results of λ-calculus is that such fixed points always exist, and that we can actually construct them.

Theorem 7.1 (Fixed Point Theorem) *For any λ-expression* **R**, *there is at least one* f *such that*

$$\mathbf{R}f = f\,.$$

Moreover, one such f can be written

$$f = \mathbf{YR},$$

where

$$\mathbf{Y} = \lambda R.\big(\lambda x.R(xx)\big)\big(\lambda x.R(xx)\big). \tag{7.22}$$

The combinator $\mathbf{Y}$ is called a *fixed point combinator.* It seems quite magical—given any combinator $\mathbf{R}$, it returns a fixed point of $\mathbf{R}$. Skeptical? Here's the proof:

$$\begin{aligned}\mathbf{YR} &= \lambda R.\big(\lambda x.R(xx)\big)\big(\lambda x.R(xx)\big)\,\mathbf{R}\\ &= \big(\lambda x.\mathbf{R}(xx)\big)\big(\lambda x.\mathbf{R}(xx)\big)\\ &= \mathbf{R}\Big(\big(\lambda x.\mathbf{R}(xx)\big)\big(\lambda x.\mathbf{R}(xx)\big)\Big)\\ &= \mathbf{R}(\mathbf{YR}).\end{aligned}$$

Nor is $\mathbf{Y}$ the only combinator with this property—there are an infinite number of fixed point combinators. We give two other examples in Problem 7.26.

With $\mathbf{Y}$'s help, we can define the factorial as $\mathbf{YR}$ where $\mathbf{R}$ is defined as in (7.21). At first, this definition seems like abstract nonsense. But it is perfectly concrete, and $\mathbf{fac} = \mathbf{YR}$ really does compute $n!$ in a finite number of steps. As we reduce it, it repeatedly applies the recursion $\mathbf{R}$,

$$\mathbf{fac}\,\overline{n} = \mathbf{R}(\mathbf{fac})\,\overline{n} = \mathbf{R}(\mathbf{R}(\mathbf{fac}))\,\overline{n} = \cdots,$$

or, in more familiar notation,

$$n! = n(n-1)! = n(n-1)(n-2)! = \cdots,$$

until it reaches the base case $n = 0$ and returns the answer.

As we show in Problem 7.27, this ability to construct fixed points is not limited to the λ-calculus. In any programming language capable of universal computation, if $\mathbf{R}$ is a program that modifies other programs, there is a program f such that $\mathbf{R}(f)$ computes the same function that f does—and there is a program $\mathbf{Y}$ such that $\mathbf{Y}(\mathbf{R})$ is one such f. But without the loopy power of the λ-calculus, in which functions can act on functions that act on other functions, it would be difficult to juggle the different levels of abstraction at play here.

Our proof that all partial recursive functions are λ-definable will be complete if we can also handle μ-recursion. First, recall its definition. If $f(x,y)$ is partial recursive, we can define a function $h(x)$ as the smallest y such that $f(x,y) = 0$:

$$h(x) = \mu_y f(x,y) = \min\{y : f(x,y) = 0\}, \tag{7.23}$$

assuming that $f(x,y)$ is defined for all $y \leq h(x)$. As we discussed in Section 7.3.3, μ-recursion is like a **while** loop. We don't know in advance how many times it will repeat, or if it will ever halt at all. This is similar in spirit to iterating a function until we reach a fixed point, and we can express $h(x)$ in this style.

First let ϕ be a function which takes a pair (x,y) as input. It explores larger and larger values of y until it finds one such that $f(x,y) = 0$, which it returns:

$$\phi(x,y) = \begin{cases} y & \text{if } f(x,y) = 0\\ \phi(x,y+1) & \text{otherwise.}\end{cases}$$

Then we can write

$$h(x) = \phi(x,0).$$

If f is λ-definable, we can write the recursive definition

$$\phi = \lambda xy.\,\mathbf{iszero}(f\,x\,y)\,y\,(\phi\,x(\mathbf{succ}\,y)).$$

Thus ϕ is the fixed point of the function $\mathbf{H}f$, where

$$\mathbf{H} = \lambda\phi xy.\,\mathbf{iszero}(f\,x\,y)\,y\,(\phi\,x(\mathbf{succ}\,y)),$$

and we can write

$$h = \lambda x.\,\phi\,x\,\overline{0} \text{ where } \phi = \mathbf{YH}.$$

In practice, this function works like this:

$$\begin{aligned} hx &= \phi\,x\,\overline{0} = \overline{0} \text{ if } f(x,0) = 0, \text{ or...} \\ &= \mathbf{H}(\phi\,x\,\overline{0}) = \phi\,x\,\overline{1} = \overline{1} \text{ if } f(x,1) = 0, \text{ or...} \\ &= \mathbf{H}(\mathbf{H}(\phi\,x\,\overline{0})) = \phi\,x\,\overline{2} = \overline{2} \text{ if } f(x,2) = 0, \text{ or...} \end{aligned}$$

If $f(x,y) \neq 0$ for all y, this chain of reductions goes on forever, exactly as the corresponding **while** loop would. On the other hand, if $f(x,y) = 0$ for some y, and if $f(x,y')$ is defined for all $y' < y$, then we arrive at a normal form after a finite chain of reductions, and $h(x)$ is defined.

Thus if f is λ-definable, so is h. By stepping up one more level of abstraction, we can even write μ-recursion itself as a combinator in the λ-calculus, which takes f and returns h:

$$\mu = \lambda f x.\,\mathbf{Y}(\mathbf{H}'f)\,x\,\overline{0} \text{ where } \mathbf{H}' = \lambda f\phi xy.\,\mathbf{iszero}(f\,x\,y)\,y\,\phi(x\,\mathbf{succ}\,y).$$

This completes the proof that all partial recursive functions are λ-definable.

It is not too hard to see, at least intuitively, that the converse is true as well—all λ-definable functions are partial recursive. Like any other finite string of symbols, we can think of a λ-expression as an integer. Then there is a primitive recursive function that carries out a single reduction step, by searching for strings of digits in an integer and substituting them according to (7.12). We can iterate this function, and find its fixed point if one exists, using μ-recursion. Therefore, the partial function **eval** which maps each expression to its normal form, and which is undefined if the reduction process never halts, is partial recursive. This was proved by Kleene in 1936, showing that the λ-calculus and the partial recursive functions are exactly equivalent in their computational power.

7.4.4 *Universality and Undecidability*

In Section 7.3, we saw that in order to define a universal function $U(f,x) = f(x)$ in the partial recursive functions, we had to go through a somewhat complicated process of assigning a Gödel number to each function. In the λ-calculus, in contrast, functions can already take other functions as input, so no such translation is needed. In a sense, the universal function is simply the identity:

$$\mathbf{U} = \lambda f x.\,f x.$$

This same ability to handle abstraction makes it very easy to prove in the λ-calculus that the Halting Problem is undecidable. That is, we can prove that there is no combinator **halts** that tells whether a given λ-expression $\alpha\beta$ reaches a normal form after a finite series of reductions:

$$\textbf{halts}\,\alpha\beta = \begin{cases} \textbf{T} & \text{if } \alpha\beta \text{ reduces to a normal form} \\ \textbf{F} & \text{otherwise}, \end{cases}$$

where $\mathbf{T} = \lambda xy\,.\,x$ and $\mathbf{F} = \lambda xy\,.\,y$ as before.

The proof exactly mirrors the one we gave in Section 7.2.2, but in the λ-calculus it is even more compact. Suppose that **halts** exists. Then let **normal** be some λ-expression, such as $\lambda x\,.\,x$, which is already in normal form—and let **never** be an an expression such as $(\lambda x\,.\,xx)(\lambda x\,.\,xx)$ which never halts.

Now define the combinator

$$\mathbf{C}_{22} = \lambda\alpha\,.\,(\textbf{halts}\,\alpha\alpha)\;\textbf{never}\;\textbf{normal},$$

where **C** stands for "catch." Then $\mathbf{C}_{22}\,\alpha$ evaluates to normal form if and only if $\alpha\alpha$ never does. Feeding this beast to itself gives a contradiction, since the λ-expression

$$\mathbf{C}_{22}\,\mathbf{C}_{22}$$

reaches normal form if and only if it doesn't. We are forced to conclude that the combinator **halts** does not exist. Thus the λ-calculus cannot solve its own Halting Problem. Equivalently, it cannot tell whether a partial recursive function α is defined on a given input β.

7.14

7.5 Turing's Applied Philosophy

> What does a Turing machine look like? You can certainly imagine some crazy looking machine, but a better approach is to look in a mirror.
>
> Charles Petzold, *The Annotated Turing*

The partial recursive functions and the λ-calculus are clearly very powerful, and we have sketched a proof that they are equivalent. But why should we believe that they encompass the power of all possible computations? Are the particular forms of recursion and substitution that they offer general enough to carry out anything that Hilbert would have called a finite procedure?

Before the work of Alan Turing, this was genuinely unclear. Several other ways to define recursive functions had been proposed, and Gödel himself found the λ-calculus "thoroughly unsatisfactory." It was not until Turing's proposal of a simple model of computation, and the proof that all three of these models are equivalent, that a consensus was established about the definition of computability.

7.5.1 Meet the Computer

Turing's machine begins as a model of a *human* computer like those shown in Figure 7.8, not an electronic one—a human carrying out a mathematical procedure according to a fixed set of rules. He distills this process down to its bare essentials, until it consists of elementary steps in which we read and write single symbols. In his own words:

FIGURE 7.8: Computers in the Hamburg observatory in the 1920s.

> Computing is usually done by writing certain symbols on paper. We may suppose this paper is divided into squares like a child's arithmetic book... The behavior of the computer at any moment is determined by the symbols which he is observing, and of his "state of mind" at that moment.
>
> Let us imagine the operations performed by the computer to be split up into "simple operations" which are so elementary that it is not easy to imagine them further divided. Every such operation consists of some change of the physical system consisting of the computer and his tape... We may suppose that in a simple operation not more than one symbol is altered. Any other changes can be split up into simple changes of this kind.

While paper is two-dimensional, Turing points out that a one-dimensional paper tape will do, with one symbol written in each square on the tape. Since we can only distinguish a finite number of different symbols from each other, he requires that the "alphabet" of symbols that can be written on the tape be finite. He makes the same argument about "states of mind," saying that these form a finite set as well. One argument for this is the finiteness of our brains. But he gives another argument as well:

> It is always possible for the computer to break off from his work, to go away and forget all about it, and later to come back and go on with it. If he does this he must leave a note of instructions (written in some standard form) explaining how the work is to be continued. This note is the counterpart of the "state of mind." We will suppose that the computer works

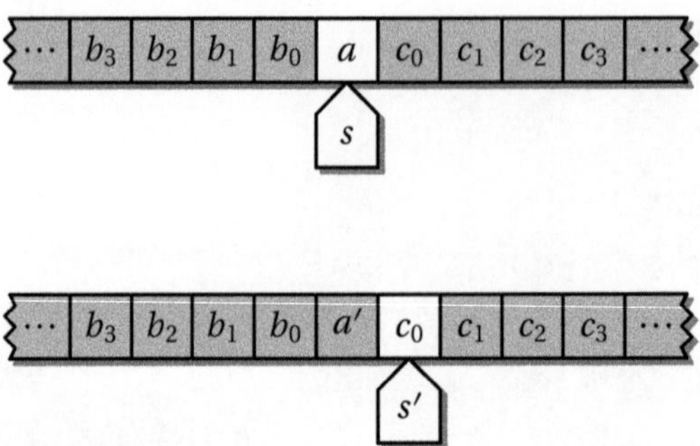

FIGURE 7.9: One step of a Turing machine. It changes the tape symbol from a to a', changes its internal state from s to s', and moves one step to the right.

> in such a desultory manner that he never does more than one step at a sitting. The note of instructions must enable him to carry out one step and write the next note. Thus the state of progress of the computation at any stage is completely determined by the note of instructions and the symbols on the tape.

In other words, a finite procedure is, by definition, something where we can record our current state, and our progress so far, with a finite string of symbols. Note that Turing doesn't claim here that his machine is an adequate model for all of human cognition, let alone consciousness—that is a separate philosophical question. He simply claims that it can model the type of cognition we engage in when we carry out a computation according to a fixed procedure.

At the end of this distillation, we have the following mathematical model. A *Turing machine* has a "head" which can be in a finite number of internal states, and a "tape" where each square contains a symbol drawn from a finite alphabet. In each step of computation, it observes the symbol at its current location on the tape. Based on this symbol and its internal state, it then

1. updates the symbol at its current location on the tape,
2. updates its internal state, and
3. moves one step left or right on the tape.

If we let A denote the alphabet of symbols, and let S denote the set of internal states, we can write the machine's behavior as a transition function,

$$F : A \times S \to A \times S \times \{\pm 1\}.$$

For instance, $F(a,s) = (a', s', +1)$ would mean that if the machine sees the symbol a on the tape when it is in state s, then it changes the tape symbol to a', changes its state to s', and moves to the right. We show this in Figure 7.9.

We start the machine by writing the input string x on the tape, with a special blank symbol ␣ written on the rest of the tape stretching left and right to infinity. We place the head at one end of the input, and start it in some specific initial state. The machine uses the tape as its workspace, reading and writing symbols representing its intermediate results until the output is all that remains. At that point, it enters a special HALT state to announce that it's done.

Exercise 7.16 *Write a Turing machine that computes the successor function $S(x) = x + 1$. In other words, let $A = \{0, 1, \sqcup\}$, and design a Turing machine that starts at the least significant bit of x, changes x to $x + 1$, ends up back at the least significant bit, and halts. Hint: let the set of states be $S = \{\text{CARRY}, \text{RETURN}, \text{HALT}\}$, and start it in the* CARRY *state.*

We can think of a Turing machine's transition function as its "source code." Moreover, we can write out a table of its values as a finite string of symbols, and thus feed one Turing machine to another. We can then ask whether there is a universal machine U, which takes a pair (M, x) as input—where M is the source code of a Turing machine, and x is an input string, written next to each other on U's tape—and simulates M's action on the input x.

Turing gave an explicit construction of just such a machine. It uses marker symbols to keep track of M's simulated state and location, and runs over to M's source code after each step to look up what transition it should perform next. If its simulation of M halts, then U returns $U(M, x) = M(x)$. Note that U can simulate Turing machines with an arbitrary number of states and tape symbols, even though it only has a fixed number of them itself. His original construction was fairly large, but since then clever people have found universal Turing machines with just a handful of states and symbols.

Turing then applied the diagonal argument to show that the Halting Problem is undecidable. Suppose 7.15
there were a machine $H(M, x)$ that could tell, in finite time, whether another machine M, given input x, will ever halt. It would be easy to modify H, and make a machine $C_{22}(M)$ that halts if $M(M)$ doesn't halt, and enters an infinite loop if it does. Then if we fed C_{22} to itself, $C_{22}(C_{22})$ would halt if and only if it wouldn't, and we are forced to conclude that H does not exist. Analogous to Problem 7.4, virtually any question about the long-term behavior of a Turing machine is undecidable.

In the minds of most computer scientists, in 1936 and today, Turing succeeded in finding the definitive model of computation. One reason that his work is so convincing is that he did not proceed by inventing a particular formalism, such as the partial recursive functions or the λ-calculus, and arguing that certain kinds of recursion or string substitution are sufficient. Instead, he arrived at his machine by contemplating the process of computation in general, with what Emil Post—who independently invented a very similar model—called "psychological fidelity." Once we accept that a computation is something that we carry out with pencil and paper, by reading and writing symbols until we arrive at the result, it becomes clear that the Turing machine leaves nothing out. As Gödel said in 1946,

> Tarski has stressed in his lecture (and I think justly) the great importance of the concept of general recursiveness (or Turing's computability). It seems to me that this importance is largely due to the fact that with this concept one has for the first time succeeded in giving an absolute definition of an interesting epistemological notion, i.e., one not depending on the formalism chosen.

Turing's model was also the first in which a universal function can be defined concretely and easily, without the technical hurdle of Gödel numbers or the abstraction of the λ-calculus. Gödel referred to the ease with which Turing machines can simulate each other—and, therefore, the ease with which the diagonal argument can be applied—as "a kind of miracle."

The Turing machine is also extremely robust to changes in its definition. If we try to make it more elaborate and powerful, we find that it can always be simulated by the original model. Multiple heads? Represent their positions with additional symbols on the tape, and have a single head shuttle back and forth between them. Multiple tapes? Interleave them, or combine their symbols in a larger alphabet, and write them all on one tape. A two-dimensional grid, where the head can move north, south, east, or west? Draw a spiral on the grid and unroll it into a one-dimensional tape, and so on. There are certainly things we can do to the Turing machine to cripple it, such as forcing it to move to the right on every step. But there is nothing we can do to make it stronger.

Finally, as we will see next, Turing's model of computation is equivalent to the partial recursive functions and the λ-calculus. Thus both psychological fidelity and mathematical unity suggest that it is the right definition.

7.16

Exercise 7.17 *Define a version of the Turing machine that, in addition to moving left and right, can insert and delete squares on the tape. Sketch a proof that such a machine can be simulated by a standard Turing machine.*

Exercise 7.18 *A* one-sided Turing machine *is one that can never move to the left of its initial position, so that its tape stretches from* 0 *to* $+\infty$ *rather than from* $-\infty$ *to* $+\infty$*. Sketch a proof that such a machine can simulate a standard Turing machine. Hint: fold the tape in half.*

Exercise 7.19 *Sketch a proof that two tape symbols suffice—in other words, that any Turing machine can be simulated by one where* $A=\{0,1\}$*, possibly with a larger number of states.*

Exercise 7.20 *Show that* BLANK TAPE HALTING*, the problem of whether a given Turing machine will halt if started on a blank tape, is undecidable.*

Exercise 7.21 *The* busy beaver function *$BB(n,m)$ is the maximum, over all Turing machines with n states and m tape symbols that halt when given a blank tape as input, of the number of steps it takes them to do so. Show that $BB(n,m)$ grows faster, as a function of n and m, than any computable function. Hint: compare to Exercise 7.9.*

7.17

7.5.2 The Grand Unification

All the prophets of God proclaim the same faith.

Baha'u'llah

As we discussed in Sections 7.3 and 7.4, the partial recursive functions and the λ-calculus are equivalent in their computational power—a function is partial recursive if and only if it is λ-definable. Let's complete the triumvirate by showing that they are equivalent to Turing machines as well. In what follows, we say that a partial function $f(x)$ is *Turing-computable* if there is a Turing machine that, given input x, halts with $f(x)$ on its tape if $f(x)$ is defined, and runs forever if $f(x)$ is undefined.

First, we sketch a proof that Turing-computable functions are partial recursive. There is a simple way to "arithmetize" the tape—namely, break it into two halves and convert each half to an integer, with the least significant digit closest to the head. Using the notation of Figure 7.9, this gives

$$B=\sum_{i=0} 2^i b_i \quad \text{and} \quad C=\sum_{i=0} 2^i c_i\,, \tag{7.24}$$

where we assume for simplicity that the tape alphabet is $A = \{0, 1\}$, and that 0 also acts as the blank symbol. Note that B and C are finite, since all but a finite number of squares on the tape are blank. We represent the tape symbol at the head's current position as an additional integer a.

With this representation, it's easy to carry out a step of the Turing machine's computation. For instance, to change the tape symbol from a to a' and move one step to the right as in Figure 7.9, we update B, C, and a as follows:

$$\begin{aligned} B &\to 2B + a' \\ C &\to \lfloor C/2 \rfloor \\ a &\to C \bmod 2\,. \end{aligned} \tag{7.25}$$

We showed in Exercise 7.10 that all of these functions are primitive recursive. The internal state s is just another integer, and we can update it—as well as determine a' and what direction to move the head—by checking a finite number of cases.

Thus we claim that, for any given Turing machine M, there is a primitive recursive function `step` which performs a single step of M's computation, updating the vector of variables (B, C, a, s). We give M an input x by setting, say, $B = x$ and $C = a = s = 0$. Then, since primitive recursion allows us to iterate the `step` function t times, the state (B, C, a, s) of the machine after t steps is a primitive recursive function of x and t. Finally, we can use μ-recursion to find the t, if there is one, such that M enters the HALT state on its tth step, and return its state at that point. This shows that the partial function computed by M is partial recursive.

Conversely, recall that the partial recursive functions are those that can be defined from zero and the successor function $S(x) = x+1$ using composition, primitive recursion, and μ-recursion. Turing machines can certainly compute zero, and Exercise 7.16 shows that they can compute $S(x)$. We just need to show that the Turing-computable functions are closed under these three ways of defining new functions from old ones. For composition, for instance, we need to show that if $f(x)$ and $g(x)$ are Turing-computable, then so is $f(g(x))$. But this is easy—simply combine the two machines into one by taking the union of their sets of states, and identify the HALT state of g's machine with the initial state of f's.

Similarly, there are Turing machines that compute functions formed from f and g by primitive recursion or μ-recursion. The details are tedious, but the ideas are simple—the machine applies these functions repeatedly, while incrementing or decrementing counters to keep track of its work. If a function $h(\mathbf{x})$ defined by μ-recursion is undefined because $f(\mathbf{x}, y) \neq 0$ for all y, then the machine runs forever, trying larger and larger values of y, just as the corresponding FLOOP program would. Thus all partial recursive functions are Turing-computable.

As an alternate proof, it is not hard to see that Turing machines can scan through a string for some symbol, and substitute another string at each place where that symbol appears. Thus they can simulate reductions in the λ-calculus, halting if they reach normal form. Conversely, the λ-calculus can represent the two halves of a Turing machine's tape as strings. With a little work we can represent the `step` function as a combinator, which reaches a normal form if the machine halts. Thus a function is Turing-computable if and only if it is λ-definable.

7.5.3 *The Church–Turing Thesis*

We have seen that all our definitions of computability are equivalent:

$$\text{partial recursive} = \lambda\text{-definable} = \text{Turing-computable}\,.$$

For good measure, we can add

= programmable in FLOOP, or your favorite programming language.

This brings us to the philosophical stance that has structured computer science ever since 1936. The *Church–Turing Thesis* states that Turing computability, in any of its equivalent forms, is the right definition of computability—that it captures anything that could reasonably be called a procedure, or a computation, or an algorithm. The universal Turing machine is capable of simulating, not just any Turing machine, but any computing device that can be finitely described, where each step accesses and modifies a finite amount of information. In the words of John von Neumann:

> [A Turing machine] is able to imitate any automaton, even a much more complicated one... if it has reached a certain minimum level of complexity. In other words, a simpler thing will never perform certain operations, no matter what instructions you give it; but there is a very definite finite point where an automaton of this complexity can, when given suitable instructions, do anything that can be done by automata at all.

The Church–Turing Thesis allows us to think intuitively about computation, without worrying unduly about the details of our model. In a world of bounded resources, such as polynomial time, there seem to be differences between what, say, classical and quantum computers can do. But in the unbounded world of computability and decidability, these differences disappear completely.

7.18 The Thesis has another consequence as well. If Turing machines can carry out any computation, so can any system that can simulate a Turing machine. This suggests that there is a large class of systems that are *computationally universal*, and that each one can simulate all the others—just as any NP-complete problem can simulate, or express, every other one. In the next section, we will see just how widespread this phenomenon is.

7.6 Computation Everywhere

> Computers made of water, wind and wood... may be bubbling, sighing or quietly growing without our suspicion that such activities are tantamount to a turmoil of computation whose best description is itself.
>
> A. K. Dewdney, *Computer Recreations*

If computation is software, it can run on many different kinds of hardware—any physical or mathematical system whose interactions allow it to store and process information in sufficiently complicated ways. In this section, we will see that many surprisingly simple systems are computationally universal, and therefore that even simple questions about these systems are undecidable.

If anything, computational universality, and undecidability, are the rule rather than the exception. Once we start looking for them, we see them everywhere.

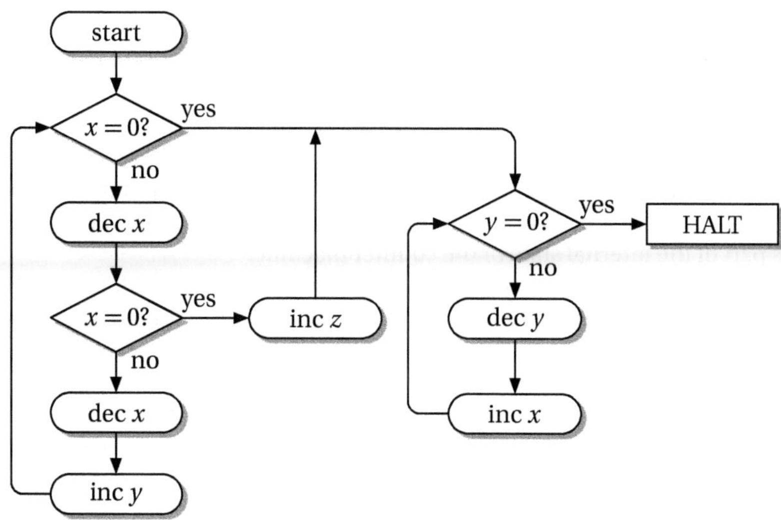

FIGURE 7.10: A counter machine with three counters. Given an initial value of x, where $y = z = 0$, it sets $x \to \lfloor x/2 \rfloor$ and $z \to x \bmod 2$. It uses y as temporary storage for $\lfloor x/2 \rfloor$. Each internal state of the machine, or line of the program, corresponds to a node in the flowchart.

7.6.1 *Computing with Counters*

> "Can you do Addition?" the White Queen asked. "What's one and one and one and one and one and one and one and one and one and one?"
> "I don't know," said Alice. "I lost count."
>
> Lewis Carroll, *Through The Looking-Glass*

If a machine only has a finite number of states, with no access to anything but its own thoughts, then it is doomed to halt or fall into a periodic loop after a finite number of steps. But if we give it access to a memory with an infinite number of possible states, then it is capable of universal computation. Turing showed that this is the case even if this memory has a simple structure, and the machine can only observe and modify it through a very narrow window—namely, the symbol at its current location on the tape. Can we make the memory even simpler, and this window even narrower?

A *counter machine* has a finite number of internal states, and access to a finite number of integer counters. The only thing it can do to these counters is increment or decrement them, and the only question it can ask about them is whether or not they are zero. The machine receives its input through the initial state of one of these counters, and like the Turing machine, it enters a HALT state when it is done.

We can think of a counter machine as a flowchart or a program in a miniature programming language. Each internal state corresponds to a node in the flowchart or a line of the program, and the only allowed instructions are **inc** (increment), **dec** (decrement), and conditionals like **if** $x = 0$. For instance, the flowchart in Figure 7.10 uses three counters, x, y, and z. Assuming that their initial values are $(x, 0, 0)$, it transforms them to $(\lfloor x/2 \rfloor, 0, x \bmod 2)$. It uses y as temporary storage for $\lfloor x/2 \rfloor$, first incrementing y for every two times it decrements x, and then incrementing x and decrementing y at the same rate. Similarly, we can compute $2x$ by incrementing y twice each time we decrement x, and then feeding y back into x.

But if we can compute $\lfloor x/2 \rfloor$, $2x$, and $x \bmod 2$, we can simulate a Turing machine arithmetically as in Section 7.5.2. Specifically, we can transform any Turing machine into a counter machine with just three counters: one for each half of the tape, B and C as in (7.25), and a temporary one in which we store, say, $2B$ or $\lfloor C/2 \rfloor$ before feeding it back into B and C. We include the current tape symbol, and the head's internal state, as part of the internal state of the counter machine.

Thus the Halting Problem for three-counter machines is undecidable. Moreover, three-counter machines can compute any partial recursive function f, by starting with x in one counter and ending with $f(x)$ in that counter when they halt.

Exercise 7.22 *Design a counter machine with two different* HALT *states, called* YES *and* NO, *such that it ends in one or the other depending on whether x is a power of 2, and such that the final value of another counter is* $\lfloor \log_2 x \rfloor$.

We can reduce the number of counters even further, First, let's consider *multiplicative* counter machines, where each step can multiply or divide a counter by a constant, or check to see whether a counter is a multiple of some constant. Now even a *single* counter suffices. This is because we can encode three additive counters as a single multiplicative one:

$$w = 2^x 3^y 5^z .$$

We can increment or decrement y, say, by multiplying or dividing w by 3. Similarly, we can test whether $y \neq 0$ by asking whether w is a multiple of 3.

But a one-counter multiplicative machine, in turn, can be simulated by a two-counter additive one. If we have an additional counter u, we can multiply or divide w by 2, 3, or 5, or check to see whether w is a multiple of any of these, using loops similar to that in Figure 7.10. To divide w by 3, for instance, we decrement w three times for each time we increment u. If w becomes zero partway through a group of three decrements, we know that w is not a multiple of 3.

Thus even two counters suffice for universal computation. Specifically, for any partial recursive function f, there is a two-counter machine which, given the initial value $w = 2^x$, ends with $w = 3^{f(x)}$, or runs forever if $f(x)$ is undefined.

For yet another simple but universal automaton, consider a *two-dimensional finite automaton*, which wanders on the northeast quadrant of the infinite plane (Figure 7.11). It has a finite number of states, and at each step it moves north, south, east, or west. It cannot write anything on the squares it touches, and it cannot see. All it knows about its position is whether it is touching the southern or western edge. But this is a two-counter machine in disguise, whose counters are its x and y coordinates. Therefore, it is undecidable to tell whether such an automaton will ever halt—or ever touch the origin—or just wander off to infinity.

It's worth noting that these counter machines are extremely inefficient. If there are n symbols on the tape, it takes the three-counter machine about 2^n steps to move the head left or right. The two-counter machine is even worse, taking as many as 5^{2^n} steps. Thus an algorithm that a Turing machine can run in polynomial time could take doubly-exponential time. But where computational universality

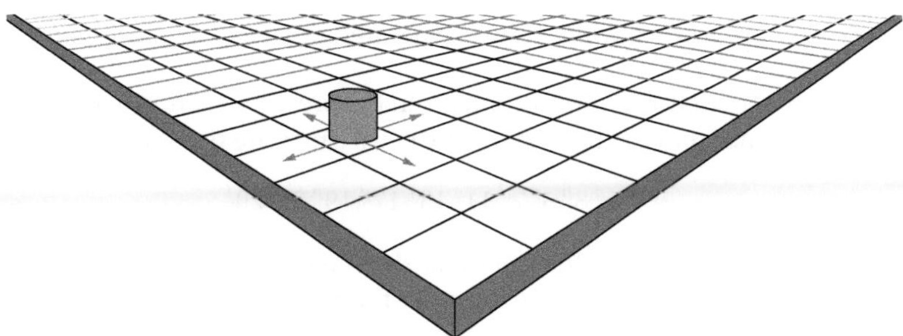

FIGURE 7.11: This automaton moves blindly on a quadrant of the infinite plane. It can only sense whether it is touching one of the boundaries, and it only has a finite number of internal states—and yet it is computationally universal.

is concerned, all that matters is that these machines can carry out any computation if we give them long enough. What's exponential time when you have eternity?

Exercise 7.23 *Argue that any counter machine with a constant number of counters needs exponential time to simulate a Turing machine. Hint: how long does it take to read or write n bits of information from or to an integer counter?*

7.6.2 *Fantastic Fractions*

If computers can come in many forms, then so can programs. Here is a program written in a rather odd programming language called FRACTRAN, consisting of nothing more than a list of fractions:

$$\frac{17}{91} \quad \frac{78}{85} \quad \frac{19}{51} \quad \frac{23}{38} \quad \frac{29}{33} \quad \frac{77}{29} \quad \frac{95}{23} \quad \frac{77}{19} \quad \frac{1}{17} \quad \frac{11}{13} \quad \frac{13}{11} \quad \frac{15}{14} \quad \frac{15}{2} \quad \frac{55}{1}$$

The state of the computer consists of an integer n. At each step, we run through the list until we find the first fraction p/q such that $(p/q)n$ is an integer—that is, the first one whose denominator q divides n. Then we multiply n by p/q, and start again at the beginning of the list. If there is no such p/q in the list, the program halts.

This particular program never halts, but it does something rather surprising. If we start it with the initial value $n=2$, it produces the sequence

$$2, \; \frac{15}{2}\times 2 = 15, \; \frac{55}{1}\times 15 = 825, \; \frac{29}{33}\times 825 = 725 \ldots,$$

until, after 19 steps, $n=4=2^2$. After 50 more steps, $n=2^3$, and 211 steps after that, $n=2^5$. The steps where n is a power of 2 are shown in Table 7.1. As you can see, this program produces the prime numbers x, in increasing order, in the form 2^x.

step	19	69	280	707	2363	3876	8068	11319	...
n	2^2	2^3	2^5	2^7	2^{11}	2^{13}	2^{17}	2^{19}	...

TABLE 7.1: Prime powers of two, generated by a FRACTRAN program.

At first glance, this is extremely mysterious. But given the previous section, we know that this program is a kind of multiplicative counter machine. In this case, we can write the state in the form

$$n = 2^a\, 3^b\, 5^c\, 7^d\, 11^e\, 13^f\, 17^g\, 19^h\, 23^i\, 29^j\,,$$

since these are the primes appearing in the fractions' denominators. Thus we have 10 counters, which we can increment and decrement by multiplying n by fractions. For instance, the first fraction in the program is $17/91 = 17/(7\times 13)$. So, if both d and f are nonzero, we decrement them and increment g. If either $d=0$ or $f=0$, then we move on to $78/85 = 78/(5\times 17)$ and check to see if both c and g are nonzero, and so on.

We already know that additive counter machines can simulate Turing machines. To convert such a machine into a FRACTRAN program, we just need to do a little extra work. First, we modify the machine so that whenever it checks to see if a counter is zero, it decrements it if it is nonzero. We can always increment the counter afterward with an additional state.

We need to express the machine's internal state as part of the integer n. We can do this by associating each state s with an additional prime p_s. Then, for each kind of step that we could take from the state s, we include p_s in the denominator of the corresponding fraction, and include $p_{s'}$ for the new state s' in the numerator. Since the numerator and denominator must be relatively prime, we need to ensure that $s' \neq s$. We can do this by adding more states, if necessary replacing a self-loop with a pair of states transitioning to each other. We ask the reader to work out some examples in Problem 7.32 and 7.33.

Thus, like its namesake FORTRAN, FRACTRAN is computationally universal. There are FRACTRAN problems that simulate other FRACTRAN programs, or counter machines, or Turing machines. Its Halting Problem is undecidable, and for any computable function $f(x)$ there is a program that, if given 2^x as input, returns $3^{f(x)}$ as output whenever $f(x)$ is well-defined.

Exercise 7.24 *Use* FRACTRAN *to prove that any partial computable function $f(x)$ can be written in the form*

$$f(x) = g\left(\mu_y f(x,y)\right),$$

where f and g are primitive recursive. See also Exercise 7.12 on page 249.

FRACTRAN might seem like a gimmick, but it is related to deep problems in number theory. Consider the function

$$g(x) = \begin{cases} x/2 & \text{if } x \text{ is even} \\ 3x+1 & \text{if } x \text{ is odd}. \end{cases} \tag{7.26}$$

The *Collatz conjecture*, also known as the $3x+1$ problem, states that the sequence

$$x, g(x), g^2(x) = g(g(x)), \ldots$$

ends in the cycle $4,2,1,4,2,1,\ldots$, no matter what the initial value of x is. This conjecture has been verified for all x up to 10^{18}, and there are convincing heuristic arguments for it. However, there is still no proof, and resolving it seems quite difficult. According to the great mathematician Paul Erdős, "mathematics is not yet ready for such problems."

The $3x+1$ function belongs to a family of functions whose behavior depends on $x \bmod q$ for some q:

$$g(x)=\frac{a_i}{q}(x-i)+b_i \quad \text{where} \quad x \equiv_q i\,. \tag{7.27}$$

Here a_i and b_i are integer coefficients for $0 \le i < q$. Note that $g(x)$ is an integer for any integer x. For any function of this form, we can ask whether a given initial x will reach some y after some number of iterations:

> COLLATZ
>
> Input: Integers x, y, and q, and integer coefficients a_i, b_i for $0 \le i < q$
>
> Question: Is there an integer t such that $g^t(x)=y$?

Given the current state n of a FRACTRAN program, let $g(n)$ denote its state after the next step. If Q is the lowest common multiple of the program's denominators, it is easy to see that $g(n)=a\,n$ where a only depends on $n \bmod Q$. Thus we can express a single step of a FRACTRAN program as a Collatz function—indeed one where $b_i=0$ for all i.

Since we can arrange for the FRACTRAN program to arrive a particular value y if it halts, we can convert its Halting Program to an instance of COLLATZ. This gives us the reduction

$$\text{HALTING} \le \text{COLLATZ}\,,$$

and proves that COLLATZ is undecidable.

This does not show that the original $3x+1$ problem is undecidable. But it does show that there is no way of addressing it that is general enough to solve all problems of this form. It is tempting to believe that this is part of the reason why this particular one is so difficult.

7.19

7.6.3 *The Game of Tag*

In 1921, the young logician Emil Post was studying the axioms and rules of inference in *Principia Mathematica*, the system of mathematical logic proposed by Bertrand Russell and Alfred North Whitehead. His idea was to approach formal systems abstractly as string-rewriting systems, not unlike the λ-calculus that Church would invent a decade later. Post was one of the first to think of formal systems in a combinatorial sense, as sets of strings produced by formal rules, regardless of the meaning that we assign to them.

Post showed, remarkably, that the axioms of the *Principia* can be reduced to a single initial string, and that its rules of inference can be replaced with rules of the following *normal form*:

$$gs \Rightarrow sh\,,$$

where g and h are fixed strings and s is arbitrary. In other words, for any string that begins with g, we can remove g from its head and append h to its tail. This is a beautiful early example of a proof that a simple system is universal, and capable of simulating much more general systems—even though Post thought of these as formal systems, rather than models of computation.

At first, Post hoped to solve the *Entscheidungsproblem* by finding an algorithm to determine whether a given string can be produced from an initial string, representing the system's axioms, using a given set of rules $\{(g_i, h_i)\}$ in normal form. We can think of the production process as a computation, where each step consists of applying one of the rules. In general, this computation might be nondeterministic, since we might have a choice of which rule to apply to a given string. For instance, with the rules $\{(a, b), (ab, c)\}$, we could change abc to either bcb or cc.

As a warm-up, Post considered the case where all the g_i have the same length v for some constant v, and where each h_i is determined by g_i's first symbol a. Thus, at each step, if a string begins with the symbol a, we remove its first v symbols including a, and append the string $h(a)$. This process is deterministic, so each initial string leads to a fixed sequence. If this sequence reaches a string whose length is less than v, it halts.

These are called *tag systems*, and Post initially expected them to be an easy special case. However, he found that even tag systems possess, in his words, "bewildering complexity." To get a sense of how complicated they can be, consider the following tag system, where $v = 2$ and the alphabet is $\{a, b, c\}$:

$$\begin{aligned} h(a) &= bc \\ h(b) &= a \\ h(c) &= aaa\,. \end{aligned}$$

Let's start with the initial string $a^3 = aaa$ and see what happens.

$$\begin{array}{l} aaa \\ \quad abc \\ \qquad cbc \\ \qquad\quad caaa \\ \qquad\qquad aaaaa = a^5\,. \end{array}$$

The string seems to be growing, and six iterations later, it becomes a^8. But then it becomes, in short order, a^4, a^2, and then a. At that point its length is less than $v = 2$, so it halts.

After exploring a few examples, the reader might ask whether the initial string a^x eventually halts for all $x \geq 1$. We agree that this is an excellent question, especially considering the following exercise.

Exercise 7.25 *Show that this tag system maps a^x to $a^{g(x)}$, where*

$$g(x) = \begin{cases} x/2 & \text{if } x \text{ is even} \\ (3x+1)/2 & \text{if } x \text{ is odd.} \end{cases} \tag{7.28}$$

This is the $3x + 1$ function from (7.26), except that it performs an extra iteration if x is odd. Thus the question of whether a^x will halt is exactly the question of whether x will ever reach the cycle $\{4, 2, 1\}$ in the $3x + 1$ problem.

As Problem 7.34 shows, it is not too hard to convert any Collatz function of the form (7.27) into a tag system, and thus reduce COLLATZ to the problem of telling whether one string in a tag system will lead to another. We can construct a similar reduction for the problem of whether the string will shrink to a length less than v and halt. Thus the following problem is undecidable:

TAG SYSTEM HALTING
Input: An integer v, a finite alphabet A, a string $h(a)$ for each $a \in A$, and an initial string x
Question: Does x lead to a string of length less than v?

Since tag systems are a special case of normal form, the Halting Problem for systems in normal form is undecidable as well:

NORMAL FORM HALTING
Input: A list of pairs of strings $\{(g, h)\}$ and an initial string x
Question: By starting with x and applying the rules $gs \Rightarrow sh$, is it possible to derive a string y to which no rule can be applied, i.e., which does not begin with any g in the list?

We have shown that this is undecidable even for systems which are deterministic, or as Post called them, *monogenic*—that is, where only one sequence of strings can be derived from the initial string.

Exercise 7.26 *Show that, given a system in normal form, an initial string x, and a string y, it is undecidable tell whether or not y can be derived from x.*

We can use tag systems to prove that another seemingly simple problem is undecidable. In Post's *correspondence problem*, we have a list of pairs of strings,

$$\left\{ \binom{s_1}{t_1}, \binom{s_2}{t_2}, \ldots, \binom{s_k}{t_k} \right\}.$$

We then ask whether there is any finite, nonempty string that can be written both as the concatenation of some sequence of s_i and that of the corresponding t_i, where a given i can appear any number of times in this sequence. For instance, if we have

$$\binom{s_1}{t_1} = \binom{a}{abab}, \binom{s_2}{t_2} = \binom{bab}{ba}, \binom{s_3}{t_3} = \binom{aa}{a},$$

then the answer is "yes," since

$$s_1 s_2 s_2 s_3 = t_1 t_2 t_2 t_3 = ababbabaa.$$

Formally, we state this problem as follows.

CORRESPONDENCE
Input: A list of pairs of strings $\{\binom{s_1}{t_1}, \binom{s_2}{t_2}, \ldots, \binom{s_k}{t_k}\}$
Question: Is there a finite sequence $i_1, i_2, \ldots, i_\ell$ with $\ell \geq 1$ such that $s_{i_1} s_{i_2} \cdots s_{i_\ell} = t_{i_1} t_{i_2} \cdots t_{i_\ell}$?

We will prove that CORRESPONDENCE is undecidable by reducing TAG SYSTEM HALTING to it. Given a tag system, we add three symbols to its alphabet, $\langle$, $|$, and $\rangle$. If the initial string is $x = x_1 x_2 \cdots x_m$, we include the pair

$$\binom{s_x}{t_x} = \binom{\langle}{\langle x_1 | x_2 | \cdots | x_m} .$$

Then, for each string $g = g_1 g_2 \cdots g_v$ of length v, if $h(g_1) = h_1 h_2 \cdots h_n$, we include the pair

$$\binom{s_g}{t_g} = \binom{g_1 | g_2 | \cdots | g_v |}{| h_1 | h_2 | \cdots | h_n} .$$

Finally, for each string $y = y_1 y_2 \cdots y_t$ of length $t < v$, we include the pair

$$\binom{s_y}{t_y} = \binom{y_1 | y_2 | \cdots | y_t \rangle}{\rangle} .$$

The pairs $\binom{s_x}{t_x}$ and $\binom{s_y}{t_y}$ force us to start with x and end with y, and each pair $\binom{s_g}{t_g}$ forces us to carry out one step of the tag system. For instance, if the initial string is $x = aaaa$, the $3x+1$ tag system described above gives the pairs

$$\binom{\langle}{\langle a|a|a|a}, \binom{a|a|}{|b|c}, \binom{a|b|}{|b|c}, \binom{b|c|}{|a}, \binom{c|a|}{|a|a|a}, \binom{c|b|}{|a|a|a}, \binom{a\rangle}{\rangle},$$

where we exclude pairs corresponding to strings g that never occur. The tag system halts after the computation

$$\begin{array}{l} aaaa \\ \quad aabc \\ \qquad bcbc \\ \qquad\quad bca \\ \qquad\qquad aa \\ \qquad\qquad\quad bc \\ \qquad\qquad\qquad a, \end{array}$$

which corresponds to the string

$$\begin{aligned} w &= \langle a|a|a|a|b|c|b|c|a|a|b|c|a\rangle \\ &= s_x s_{aa} s_{aa} s_{bc} s_{bc} s_{aa} s_{bc} s_a \\ &= t_x t_{aa} t_{aa} t_{bc} t_{bc} t_{aa} t_{bc} t_a . \end{aligned}$$

Any finite string that can be written as a concatenation of these pairs is either w, or w repeated some number of times. Thus x halts if and only if a corresponding string exists. This completes the reduction

$$\text{TAG SYSTEM HALTING} \le \text{CORRESPONDENCE},$$

7.20 and proves that CORRESPONDENCE is undecidable.

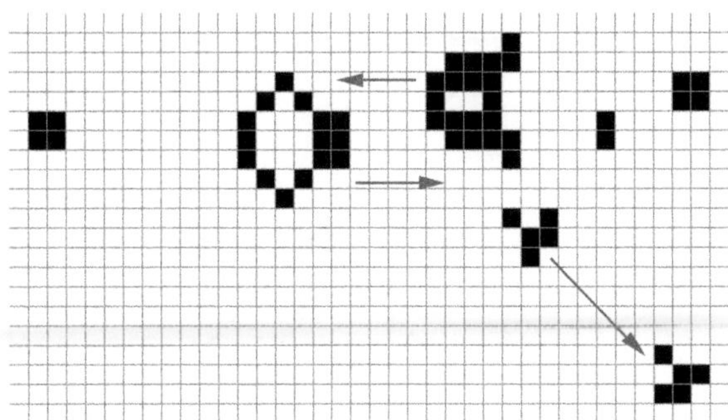

FIGURE 7.12: The glider gun, the first pattern found in Life where the number of live cells increases without bound. Its two central components shuttle back and forth, producing a glider every 30 steps.

7.6.4 The Meaning of Life

In Section 5.5.3, we encountered *cellular automata*—discrete universes whose laws of physics update each cell of a lattice simultaneously. Like our universe, where signals are limited by the speed of light, their dynamics is *local*. That is, the new state of each cell depends only on its own current state, and those of its neighbors. The physics of our universe is rich enough to build universal computers. Is this true of these toy universes as well?

One famous rule in which we can build computers is Conway's "Game of Life." In Figure 5.21 on page 159, we saw a *glider*—an elementary particle that travels one step diagonally every four time steps. There is also a *glider gun*, shown in Figure 7.12. It produces a glider every 30 steps, emitting a steady stream of them into the vacuum.

Depending on its precise timing, a collision between two gliders can make one of them bounce back the way it came, or annihilate both of them. We can use these collisions to perform Boolean operations, where we represent `true` and `false` as glider streams which are turned on or off. For instance, Figure 7.13 shows a NOT gate, where we intersect a stream A with the stream from a glider gun in a way that annihilates both. This produces a stream $\overline{A}$ which is on if and only if A is off.

Exercise 7.27 *By drawing a diagram in the spirit of Figure 7.13, show how to convert two glider streams A and B to Boolean combinations such as $A \vee B$ and $A \wedge B$, using glider guns and collisions that annihilate both streams if they are present. Can you use a similar approach to perform* fanout, *i.e., to copy the truth value of a stream to two others, or do you need another kind of collision to do this?*

Gliders are just the beginning. A dedicated community of "Life hackers" has invented oscillators, spaceships, puffer trains, buckaroos, breeders, switches, latches, adders, and memories. By putting these and other components together, it is possible to build an entire Turing machine in the Game of Life—a finite state control which sends and receives signals from a tape, updating its symbols and shifting it left and right. We show this astonishing construction in Figure 7.14.

In order to perform universal computation, this Turing machine must be equipped with an infinite tape, stretching to the horizon in both directions. Its initial state can be described with a finite number of bits, since a blank tape consists of a finite unit repeated periodically. However, since we are used to

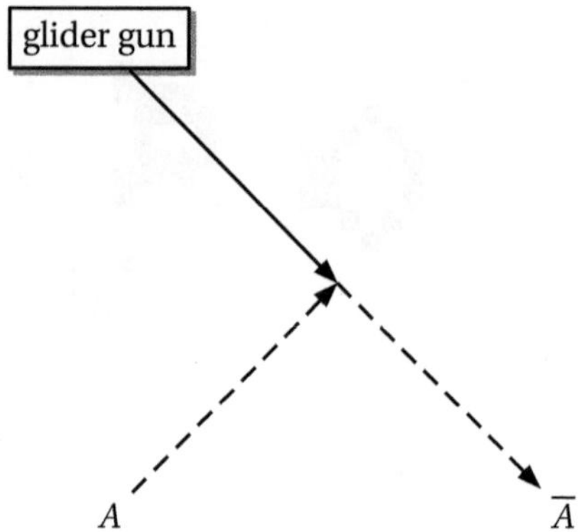

FIGURE 7.13: A schematic of a NOT gate in Life. The two streams of gliders annihilate if they are both turned on, so $\overline{A}$ is on if and only if A is off.

computers of finite physical extent, we might prefer a computer whose initial state consists of only a finite number of "live" cells, surrounded entirely by zeros.

There are two possible ways to do this. One is to design a flotilla of spaceships that move out into the lattice, firing gliders back toward the tape that collide and combine to form new tape cells. While this construction has not yet been completed, it seems perfectly possible, since glider collisions have been designed to build nearly anything.

Another type of universal computer, which can be built in Life with a finite number of live cells, is the counter machine discussed in Section 7.6.1. We can represent the value of each counter as the position of a stable 2×2 block. We can then push and pull these blocks, incrementing and decrementing the counter, by sending carefully timed sets of gliders at them.

In either of these constructions, we can arrange for a particular cell to turn on if and only if the Turing machine, or the counter machine, enters its HALT state. Therefore, the following problem is undecidable:

> LIFE PREDICTION
>
> Input: An initial configuration C and integers x, y
>
> Question: If we start with C, will the cell at (x, y) ever turn on?

Thus the eventual fate of an initial configuration in Life is impossible to predict.

7.21 The Game of Life shows that we can perform universal computation in a simple two-dimensional universe. Is even one dimension enough? Turing machines are one-dimensional objects, and since the head only moves one step at a time, their dynamics is already local. So, designing one-dimensional CAs that simulate Turing machines is easy. One method is to combine the internal state with the tape symbol at the head's current location. If the machine has tape symbols A and internal states S, this gives the CA the set of states $Q = A \cup (A \times S)$. We then define the transition function f of the CA,

$$q_{x,t+1} = f(q_{x-1,t}, q_{x,t}, q_{x+1,t}),$$

so that it simulates the Turing machine at each step.

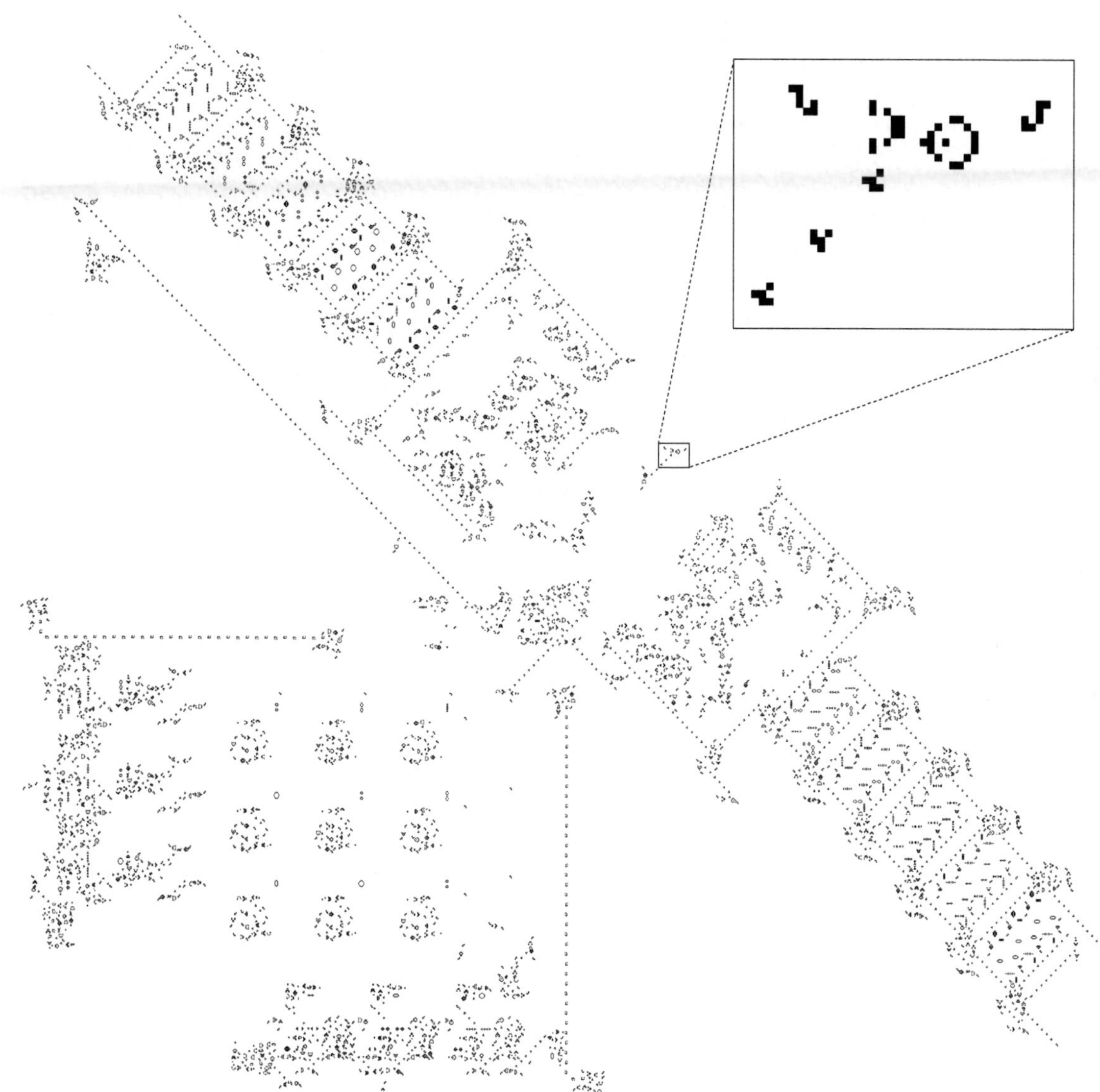

FIGURE 7.14: Constructing a Turing machine in the Game of Life. The head with its finite-state control is on the lower left, and the stack stretches from upper left to lower right. When animated, it is a truly impressive sight. We magnify one of its components, a glider gun, to give a sense of its scale.

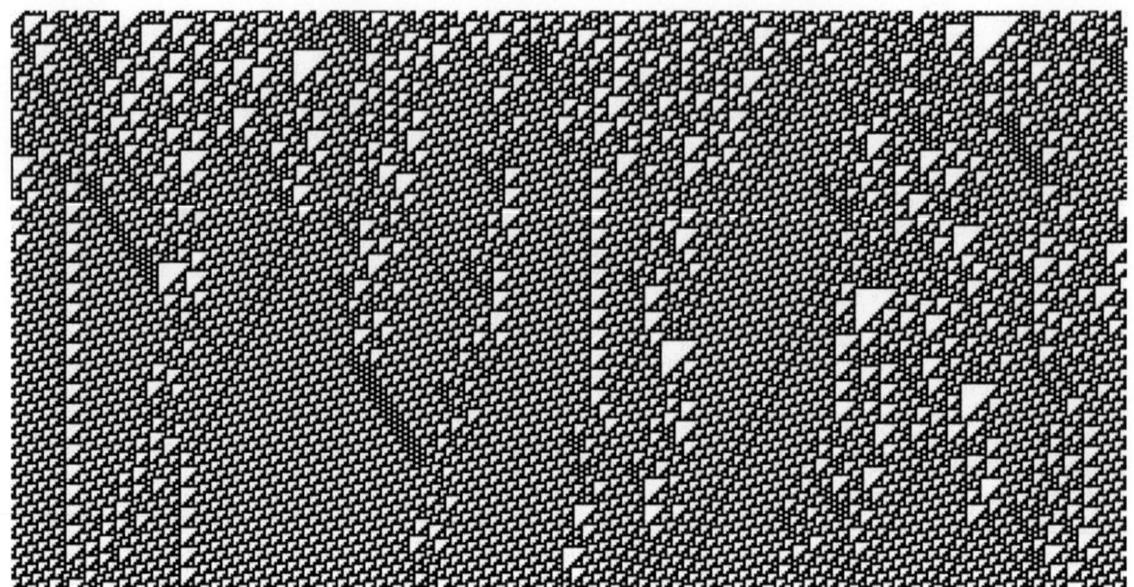

FIGURE 7.15: The evolution of elementary CA rule 110 from a random initial state, showing collisions between several types of gliders and periodic structures. Time increases downward, and each row of the diagram is a single step.

For instance, suppose that when the Turing machine is in state s and the tape symbol is b, it writes b' on the tape, changes its state to s', and moves to the right. Then we would write, for any $a,c,d \in A$,

$$f(a,(b,s),c)=b' \quad \text{and} \quad f((b,s),c,d)=(c,s').$$

We also define $f(a,b,c)=b$ for any $a,b,c \in A$, since the tape symbols stay fixed if the head isn't around to modify them. Then one step of this CA would look like

$\cdots$	a	(b,s)	c	d	$\cdots$
$\cdots$	a	b'	(c,s')	d	$\cdots$

moving the head to the right and updating the tape. Of course, the parallel nature of a cellular automaton allows us to simulate many Turing machine heads simultaneously. We haven't specified here what happens when two heads collide.

If we start with a small universal Turing machine, i.e., one with a small number of states and symbols, this construction gives a universal CA with a fairly small number of states. But what is more surprising is that even "elementary" CAs—those with just two states and nearest-neighbor interactions—are capable of universal computation. Consider the following rule:

111	110	101	100	011	010	001	000
0	1	1	0	1	1	1	0

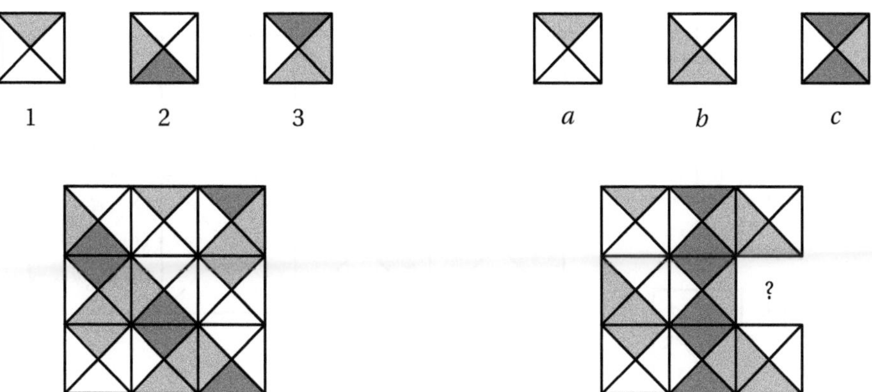

FIGURE 7.16: The set $\{1,2,3\}$ of Wang tiles allows us to tile the infinite plane, but the set $\{a,b,c\}$ does not. Telling which of these is the case is undecidable. Note that we don't, in general, allow tiles to be rotated or reflected.

This CA is called rule 110, since 01101110 is 110 in binary.

As Figure 7.15 shows, rule 110 has a wide variety of gliders and periodic structures. By using their collisions in exceedingly clever ways, we can simulate a kind of tag system, whose rules and initial string are given as part of the initial state. Like the Turing machine in Figure 7.14, the initial state of this simulation consists of a finite string, with spatially periodic patterns repeated to its left and right. Hence rule 110 is capable of universal computation, and the question of whether a given string of bits will ever appear in the lattice, given an initial state of this kind, is undecidable.

In turn, the smallest known universal Turing machines work by simulating rule 110. However, like the 7.22
initial state of the CA, the initial tapes of these Turing machines are filled with periodic patterns to the left and right of the input string, as opposed to being filled with blanks. They never halt, but the question of whether a given string of bits will ever appear on their tape, immediately to the left, say, of their head, is undecidable.

Life and rule 110 show that even systems with very simple rules can perform universal computation, 7.15
and thus generate undecidable behavior. Indeed, it's hard to imagine a system whose "laws of physics" are simpler without being trivial. We can build computers out of virtually anything, as long as their physics allows for particles that can carry information from place to place, and nontrivial interactions between these particles that let us process this information.

7.6.5 Tiling the Plane

The TILING problem asks whether we can cover the plane with tiles of a given set of types. In Section 5.3.4, we saw that for finite regions, this problem is NP-complete. Here we will show that for the *infinite* plane, it is undecidable—by showing how tiles can perform universal computation in space, rather than time.

We can fit tiles together according to shape or color. *Wang tiles* have a color along each edge, and the colors of neighboring tiles are required to match. Figure 7.16 shows two sets of these tiles. The set on the

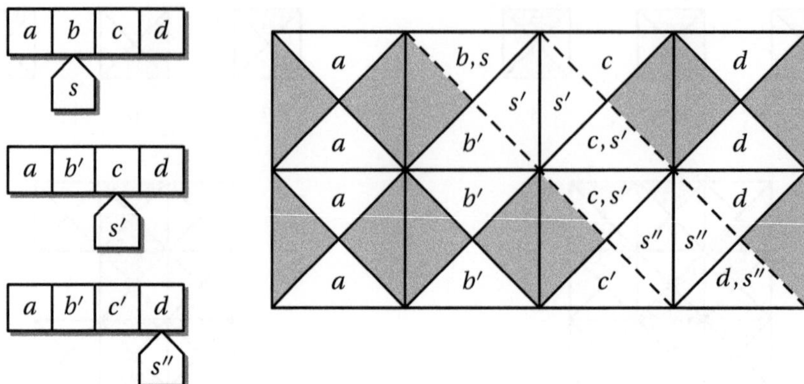

FIGURE 7.17: Simulating a Turing machine with Wang tiles. Plain tape symbols have gray on their sides, and transmit their symbols from top to bottom. Tiles representing the head transmit its state to the left or right, and also modify the tape symbol. The dashed lines show the head's path.

left can tile the infinite plane, but the set on the right always produces a gap that cannot be filled. The following problem asks which is which:

> TILING
> Input: A set of T Wang tiles
> Question: Can we tile the infinite plane with tiles from T?

Given a Turing machine, it's easy to design a set of Wang tiles such that each row corresponds to a step of the machine's computation. As Figure 7.17 shows, the tiles correspond to tape symbols, with special tiles marking the head's location and state. Each color corresponds to a tape symbol a, a head state s, or a pair (a, s). Matching these colors transmits a tape symbol from each tile to the one below it, and transmits the head's state to the left or right.

Now suppose that there are no tiles in our set colored with the Turing machine's HALT state. In that case, if the machine halts, there is no way to complete the tiling. If we specify the tiles on one row, starting the machine in its initial state and giving it an input string, then we can fill in the tiles below that row, all the way out to infinity, if and only if the machine never halts. In fact, since the Halting Problem is undecidable even for blank tapes (see Exercise 7.20), it suffices to specify a single tile at the origin corresponding to the machine's head. This shows that the following variant of TILING is undecidable:

> TILING COMPLETION
> Input: A set T of Wang tiles, and specified tiles at a finite set of positions
> Question: Can this tiling be completed so that it covers the infinite plane?

However, this construction doesn't show that TILING is undecidable. After all, it's easy to tile the plane with the tiles in Figure 7.17—just fill the plane with tape symbols, with no Turing machine head around to modify them. Without specifying at least one initial tile, how can we make sure that, somewhere in the plane, there is actually a Turing machine performing a computation?

Let's pause for a moment to discuss an issue that may seem unrelated at first. The Wang tiles on the left of Figure 7.16 tile the plane periodically. In fact, the unique tiling they admit consists of the 3×3 block shown in the figure, repeated forever. Now consider the following exercise:

Exercise 7.28 *Suppose that, for any set T of Wang tiles, either T allows a periodic tiling of the plane, or cannot tile the plane at all. Show that if this were true then* TILING *would be decidable.*

Therefore, if TILING is undecidable, there must be a set of tiles that can tile the plane, but not periodically. We call such a set of tiles *aperiodic*.

Finding an aperiodic set of tiles takes some thought, but such sets do exist. One of them, consisting of six shapes and their rotations and reflections, is shown in Figure 7.18. If we decorate these tiles with stripes of two different colors, say black and gray, and if we require neighboring tiles to match both according to color and shape, they yield a fractal pattern of squares of size 2×2, 4×4, 8×8, and so on.

We can use these squares as scaffolds on which to force a Turing machine to perform computations of increasing length. Specifically, we expand the set of tiles so that each one carries a tape symbol or a state-symbol pair. We then require the tile at the upper-left corner of each square to contain a head in its initial state, and the other tiles along the square's upper edge to encode a blank tape. By requiring these tiles to match like the Wang tiles described above, we force the Turing machine to run for 2 steps in each 2×2 square, 4 steps in each 4×4 square, and so on. (Using Exercise 7.18, we can assume that the machine will never move to the left of its initial position.) Since there are squares of arbitrarily large size, we can only tile the infinite plane if the machine runs forever.

We have a little more work to do. The problem is that these computations will overlap—since each tile is a part of an infinite number of squares of increasing size, which computation should it be a part of? We can avoid this problem with a clever construction shown in Figure 7.19. First, we focus on the $1/4$ of the tiles which protrude at their corners, i.e., the leftmost two shapes in Figure 7.18. Within each black square, we then pick out a subset of these tiles, namely the ones at the intersections of the rows and columns that don't cross a smaller black square.

In each black square of side 2^{2k}, this gives us a $2^k \times 2^k$ "board" of tiles. The tiles in each board are not adjacent; they are connected by the dotted lines in Figure 7.19. But we can simulate adjacency with additional tiles that transmit states and symbols along these dotted lines. These boards lie inside each other, but they don't overlap—so we can use each one to force the Turing machine to run for 2^k steps.

This reduces BLANK TAPE HALTING, or rather its complement, to TILING, and proves that the latter is undecidable. 7.23

7.6.6 *From Chaos to Computation*

Computation can also occur in continuous dynamical systems, where we start with an initial point and repeatedly apply a map f to it. Consider the following function, defined on the unit square $[0,1]\times[0,1]$:

$$f(x,y)=\begin{cases}(x/2,2y) & \text{if } y<1/2\\ (x/2+1/2,2y-1) & \text{if } y\geq 1/2.\end{cases}$$

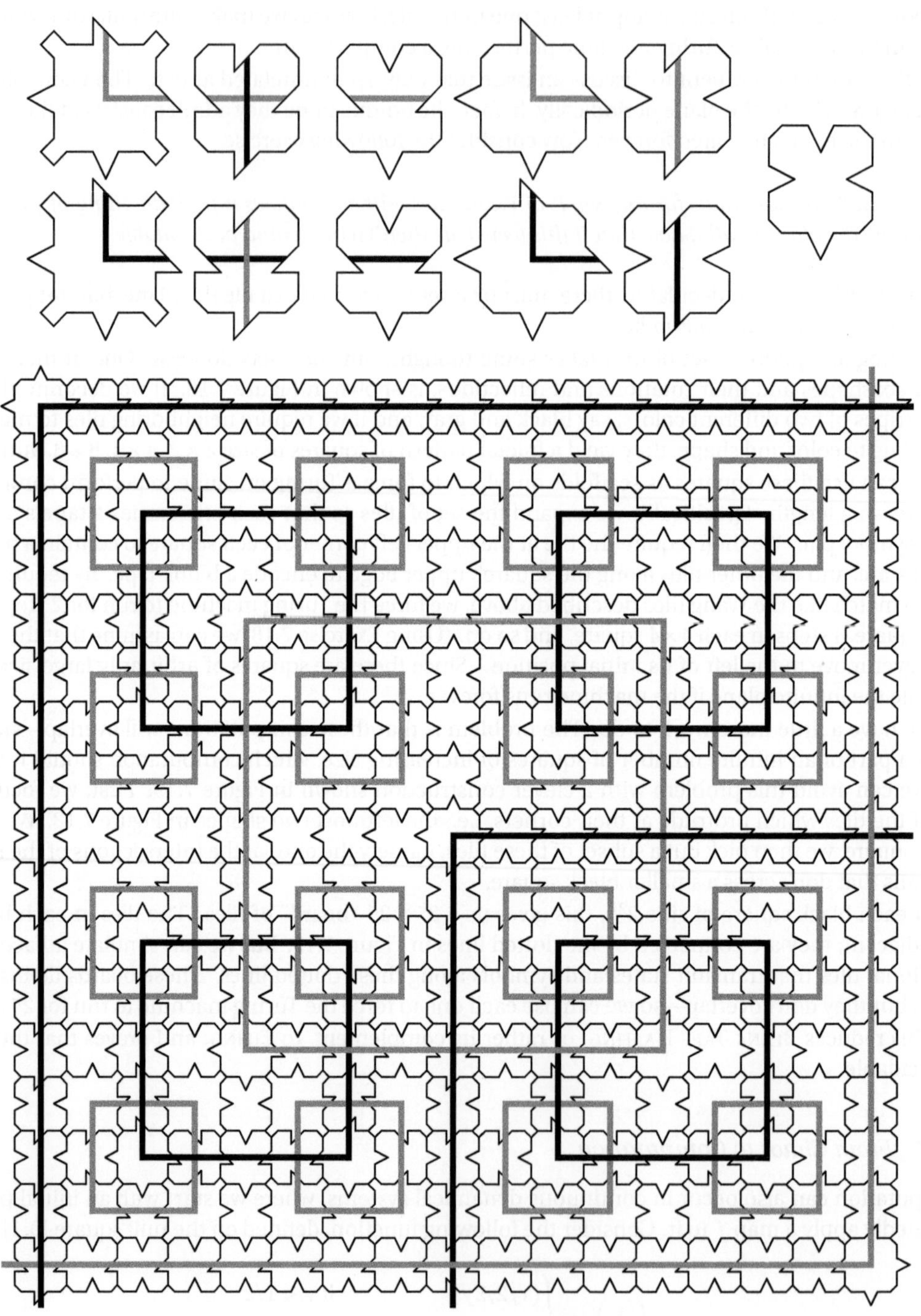

FIGURE 7.18: Robinson's aperiodic tiles. When decorated with black and gray stripes, they form a fractal hierarchy of squares whose sides are powers of 2. We then use these squares as scaffolds for Turing machine computations.

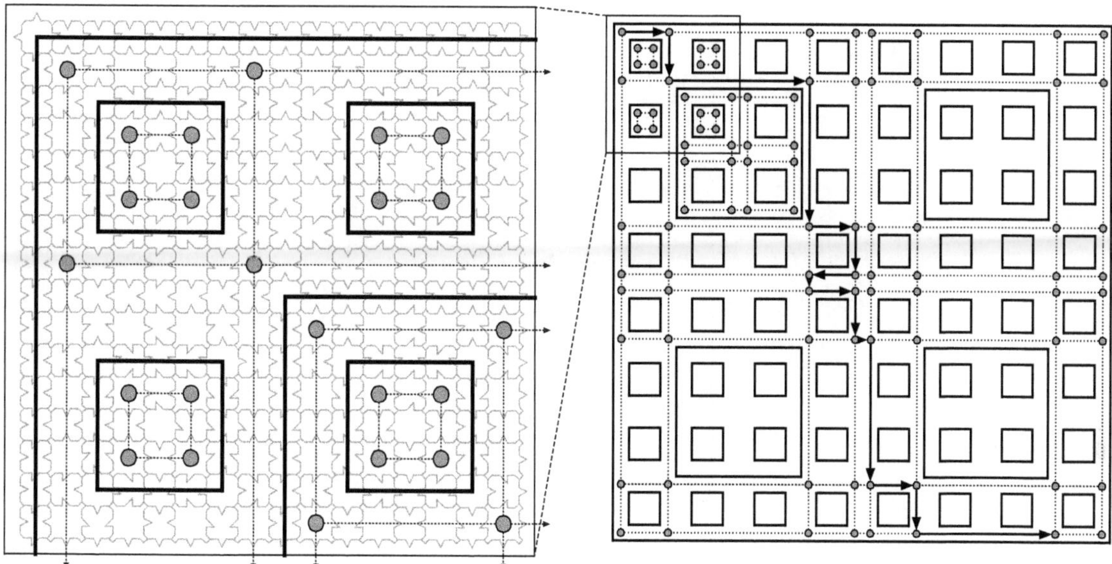

FIGURE 7.19: Each black square contains a "board" consisting of the tiles with protruding corners whose rows and columns are not interrupted by smaller black squares. Here these tiles are marked with gray circles, connected by dotted lines. On the left, the region shown in Figure 7.18. On the right, we zoom out, and show how black squares of side 4×4, 16×16, and 64×64 contain boards of size 2×2, 4×4, and 8×8. We use additional types of tiles to transmit the machine's states and symbols along the dotted lines, so that each $\ell \times \ell$ board simulates ℓ steps of the Turing machine. Shown in bold is a possible path that the machine might take as part of an 8-step simulation.

This is called the *Baker's Map*. It stretches the unit square, cuts it, and then places the two stretched halves together, like a baker stretching and kneading dough. We show its action in Figure 7.20.

What does this map do in digital terms? Let's write the binary expansions of x and y as $x = 0.x_1x_2x_3\ldots$ and $y = 0.y_1y_2y_3\ldots$ respectively, where

$$x = \sum_{i=1}^{\infty} 2^{-i} x_i \quad \text{and} \quad y = \sum_{i=1}^{\infty} 2^{-i} y_i\,.$$

Then f shifts y's bits towards the decimal point, shifts x's bits away from it, and moves the most significant bit of y to x:

$$f(0.x_1x_2x_3\ldots, 0.y_1y_2y_3\ldots) = (0.y_1x_1x_2\ldots, 0.y_2y_3y_4\ldots)\,.$$

Suppose we start with two different initial points, (x,y) and (x',y'), and apply f repeatedly to both of them. If y and y' differ in their ith bit then this bit gets shifted towards the decimal point, doubling the distance $|y - y'|$ at each step. After i steps, y and y' disagree on their most significant bits, and the two points are in completely different parts of the square.

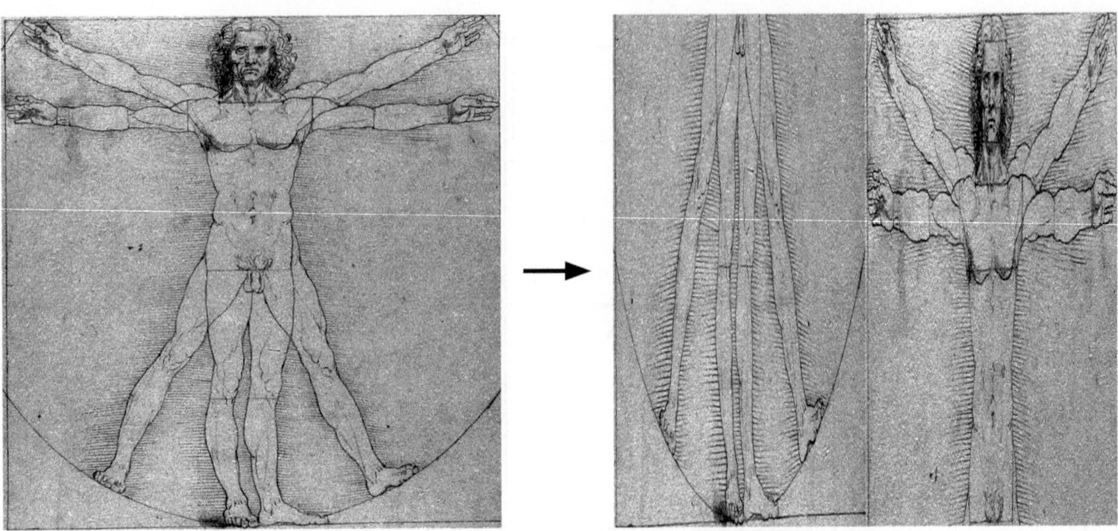

FIGURE 7.20: The Baker's Map f on the unit square. It doubles y, halves x, and moves y's most significant bit to x. We can think of the coordinates (x,y) as a tape full of bits, $\ldots y_3 y_2 y_1 . x_1 x_2 x_3 \ldots$, where $0.x_1 x_2 x_3 \ldots$ and $y = 0.y_1 y_2 y_3 \ldots$ are the binary expansions of x and y respectively. Then f moves the decimal point, representing the head of a Turing machine, one step to the left.

This makes the Baker's Map a classic example of a chaotic dynamical system. It quickly magnifies any perturbations or uncertainties in its initial conditions, until it becomes completely unpredictable. If we only know the initial conditions to t bits of accuracy, we cannot predict the state of the system more than t steps in the future.

Exercise 7.29 *Show that the Baker's Map has two unique fixed points and a unique period-2 orbit, i.e., a pair of points (x,y) and (x',y') such that $f(x,y)=(x',y')$ and $f(x',y')=(x,y)$. What are their digit sequences, and where are they in the unit square? How many period-3 orbits are there? What can you say about period-t orbits for larger t?*

How can maps like this simulate Turing machines? If we write the bits of y in reverse, we can think of (x,y) as an infinite tape whose tape symbols are 0 and 1,

$$\ldots y_3 y_2 y_1 . x_1 x_2 x_3 \ldots ,$$

where the decimal point represents the machine's head. In that case, we have

$$f(\ldots y_3 y_2 y_1 . x_1 x_2 x_3 \ldots) = f(\ldots y_3 y_2 . y_1 x_1 x_2 x_3 \ldots),$$

so f moves the head one step to the left. Note the similarity to the arithmetization of the Turing machine on page 262, where we treated each half of the tape as the bit sequence of an integer. Both there and here, we move the machine's head by halving one of these numbers and doubling the other.

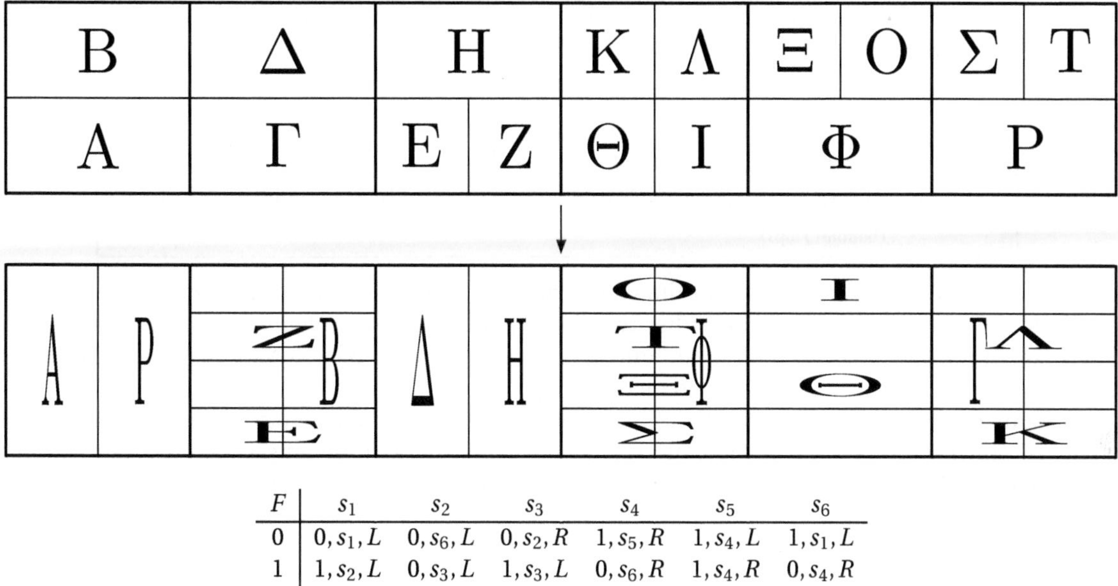

F	s_1	s_2	s_3	s_4	s_5	s_6
0	$0, s_1, L$	$0, s_6, L$	$0, s_2, R$	$1, s_5, R$	$1, s_4, L$	$1, s_1, L$
1	$1, s_2, L$	$0, s_3, L$	$1, s_3, L$	$0, s_6, R$	$1, s_4, R$	$0, s_4, R$

FIGURE 7.21: Above, an iterated map f on the rectangle $[0,6] \times [0,1]$ equivalent to a universal Turing machine with 6 states and 2 symbols. Each large square corresponds to one of the machine's states, and the current tape symbol is the most significant bit of y. Stretching vertically or squashing horizontally corresponds to moving the machine's head left or right on the tape. Below, the machine's transition function. Each triplet gives the new symbol, the new state, and whether the head moves left or right.

To complete the simulation, let's say that the tape symbol at the machine's current location is y_1, which is 0 or 1 if we are in the lower or upper half of the square respectively. At each step, we can change y_1 by shifting up or down by 1/2. We then move the head left or right on the tape by stretching vertically or squashing horizontally.

Finally, we define our function on a set of unit squares, one for each of the machine's internal states, and update the state by mapping pieces of one square to another. If the machine has s states, and we put these squares next to each other, we get a piecewise-continuous function f from the rectangle $[0,s] \times [0,1]$ to itself. This function divides this rectangle into a finite number of pieces, stretches or squashes each one, and maps it back somewhere inside the rectangle.

In Figure 7.21, we carry out this construction for a universal Turing machine with 6 states and 2 tape symbols. This machine simulates cellular automaton rule 110, so the question of whether a particular finite string of bits will ever appear on its tape, at its head's current location, is undecidable. In our map, this corresponds to x and y lying in a pair of finite intervals, where their binary expansions start out with particular finite strings. Thus, given an initial point (x,y), the question of whether it will ever land in a particular rectangle is undecidable.

As we discussed at the end of Section 7.6.4, the initial tape of this Turing machine is filled with periodic patterns to the left and right of its input. Such a tape corresponds to an initial point (x,y) with rational coordinates, since a real number is rational if and only if its binary expansion becomes periodic after a finite number of bits. So, even if we know the initial point *exactly*, questions about f's dynamics like the following are undecidable:

> ITERATED MAP PREDICTION
>
> Input: Rational numbers $x, y, x_{\text{left}}, x_{\text{right}}, y_{\text{bottom}}, y_{\text{top}}$
>
> Question: Is there an integer t such that $f^t(x,y)$ lies in the rectangle $[x_{\text{left}}, x_{\text{right}}] \times [y_{\text{bottom}}, y_{\text{top}}]$?

Indeed, essentially any question about the long-term trajectory of such maps is undecidable, such as whether a given point is periodic, whether two initial points will ever collide, and so on. Thus these maps are unpredictable in a much more fundamental way than the Baker's Map is. Their behavior is computationally universal, as opposed to merely chaotic.

7.24 To sum up, iterated maps can simulate Turing machines, which simulate rule 110, which simulate tag systems, which simulate Collatz functions, which simulate FRACTRAN programs, which simulate counter machines, which can in turn simulate other Turing machines... a fitting chain of simulations with which to end our chapter.

Problems

> As long as a branch of science offers an abundance of problems, so long is it alive; a lack of problems foreshadows extinction or the cessation of independent development.
>
> David Hilbert, 1900

7.1 Computability is rare. Prove that the number of programs is countably infinite, but the number of functions $f:\mathbb{N}\mapsto\mathbb{N}$ or $f:\mathbb{N}\to\{\texttt{true},\texttt{false}\}$ is uncountably infinite. Therefore, almost all functions are uncomputable.

7.2 Goldbach's other conjecture. The mathematician Christian Goldbach (1690–1764) loved to invent mathematical conjectures, a passion he shared with his friend Leonhard Euler. His most famous conjecture, the one known today as *Goldbach's Conjecture*, states that every even number greater than or equal to 4 is the sum of two primes. In another conjecture, Goldbach claimed that every odd number can be written as the sum of a prime p (including $p=1$ which was considered a prime back then) and two times the square of an integer $a \geq 0$, like $11=3+2\times 2^2$ or $23=5+2\times 3^2$. Euler himself checked that this conjecture holds for all odd numbers up to 2500, but he couldn't find a proof.

Write a computer program that halts if and only if Goldbach's other conjecture is false. Then run the program.

7.3 More undecidable problems. Show that it is undecidable, given the source code of a program Π, to tell whether or not any of the following is true:

1. Π halts on input 0.
2. Π computes a total function—that is, $\Pi(x)$ halts for all x.
3. $\Pi(x)=\texttt{true}$ for all x.
4. The set of x on which Π halts is finite.
5. There is an x such that $\Pi(x)=x$.

6. Π is equivalent to some other program Ψ. That is, even though Π and Ψ have different source codes, they compute the same partial function—for all x, either $\Pi(x)$ and $\Psi(x)$ both halt and return the same answer, or neither halts. Show that for any program Ψ, this problem is undecidable.

Prove each of these by reducing Halting to them. That is, show to how convert an instance (Π, x) of Halting to an instance of B. For instance, you can modify Π's source code, or write a new program that calls B as a subroutine. Note that your reduction does not necessarily have to map yes-instances to yes-instances and no-instances to no-instances—all that matters is that if you could solve B, you could solve Halting.

7.4 Rice's Theorem. We saw some examples of reductions from Halting in the previous problem. Generalize these reductions and prove the following theorem:

Theorem 7.2 (Rice's Theorem) *Let P be a property of programs that depends only on the partial functions that they compute. Assume that there is at least one program Π_1 for which P is true, and at least one program Π_2 for which P is false. Then show that it is undecidable, given a program Π, to tell whether P is true for Π or not.*

Properties like these, that depend only on a program's output and not on its internal structure or dynamics, are called *semantic.*

Hint: show that for any such property P, we can reduce Halting to the question of whether or not P is true. Note that in some cases your reduction may have to map yes-instances to no-instances and vice versa.

7.8

7.5 It's hard to tell if the only loops are simple. In Exercise 7.4 you showed that the Halting Problem is decidable for any program Π with the property that, for each x, $\Pi(x)$ either halts or falls into an endless loop. Now prove that telling whether a given program Π has this property is itself undecidable.

7.6 P vs. NP. We saw in Section 6.1.2 that if $\mathsf{P} = \mathsf{NP}$ then $\mathsf{TIME}(G) = \mathsf{NTIME}(G)$ for any class of functions $G(n)$ that is closed under composition with polynomials. This is certainly true if G is the class of all computable functions. On the other hand, we argued in Section 7.2.3 that Decidable and RE are analogous to P and NP respectively, and we proved using the Halting Problem that Decidable $\neq$ RE. Why isn't this a proof that $\mathsf{P} \neq \mathsf{NP}$?

7.7 Domains and ranges. Given a partial function $f(x)$, we define its *domain* $\mathrm{Dom}\, f$ as the set of x for which $f(x)$ is well-defined. In particular, if f is computed by some program, its domain is the set of inputs for which that program halts. Similarly, its *range* is the set of all its well-defined outputs, i.e., $\{f(x) : x \in \mathrm{Dom}\, f\}$. Show that the following conditions are equivalent properties of a set S:

1. S is recursively enumerable.
2. S is the domain of some computable partial function.
3. S is the range of some computable partial function.

7.8 Enumerating in order. Consider a set $S \subseteq \mathbb{N}$ such that there is a program that prints out all the elements of S in increasing order. Show that S is decidable.

7.9 Decidable subsets of enumerable sets. Show that if S is infinite and recursively enumerable, it has an infinite decidable subset. Hint: consider Problem 7.8.

7.10 Total functions can't be enumerated. Show that the set of total computable functions is not recursively enumerable. In other words, show that there is no computable function U_{tot} such that $U_{\text{tot}}(i, x)$ is well-defined for all i and all x, with the property that, for every total computable function $f(x)$, there is an i such that $f(x) = U_{\text{tot}}(i, x)$ for all x. To put this differently, there is no program that prints out a list of all the programs that always halt.

7.11 We can't print the ones that don't halt. Show that the set of programs Π that halt when given themselves as input is recursively enumerable, but that its complement is not.

7.12 Hypercomputers. Show that if we have access to an oracle for the Halting Problem, the class of problems we can solve in finite time becomes $\Sigma_2 \cap \Pi_2$.

Hint: note that a problem is in $\Sigma_2 \cap \Pi_2$ if both its yes-instances and its no-instances can be described in the form $\exists y : \forall z : B(x,y,z)$ for some decidable B. (Compare Problem 6.11, which shows that $\mathsf{P}^{\mathsf{NP}} \subseteq \Sigma_2\mathsf{P} \cap \Pi_2\mathsf{P}$.)

7.13 Infinite sets of axioms. Show that Gödel's Incompleteness Theorem holds for formal systems with an infinite number of axioms, as long as the set of axioms is recursively enumerable.

7.14 Primitive logarithms. Starting with the results of Exercise 7.10, show that the function $\texttt{log}_2(x) = \lfloor \log_2 x \rfloor$ is primitive recursive (with the convention $\texttt{log}_2(0) = 0$). More generally, show that for any primitive recursive functions $g(y)$ and $t(x)$, the function

$$f(x) = \max\{y : y \le t(x) \text{ and } g(y) \le x\}$$

is primitive recursive. Hint: Note that $\texttt{log}_2(x)$ is the inverse of 2^y, and the function $f(x)$ is the generalized inverse of of $g(y)$.

7.15 Pairing up. In modern computer science, it is trivial to think of any finite collection of integers as a finite sequence of bits, and therefore as a single integer. In the world of recursive functions, we can do this with the functions $\texttt{pair} : \mathbb{N} \times \mathbb{N} \to \mathbb{N}$, $\texttt{left} : \mathbb{N} \to \mathbb{N}$, and $\texttt{right} : \mathbb{N} \to \mathbb{N}$ such that if $p = \texttt{pair}(x,y)$, then $\texttt{left}(p) = x$ and $\texttt{right}(p) = y$. One such pairing works as follows:

p	0	1	2	3	4	5	6	7	8	9	…
x	0	1	0	2	1	0	3	2	1	0	…
y	0	0	1	0	1	2	0	1	2	3	…

Show that `pair`, `left`, and `right` are primitive recursive. Hint: first figure out what $\texttt{pair}(x,0)$ is. Then use the technique of Problem 7.14 to define `left` and `right`.

7.16 Recursion through iteration. Using the results of the previous problem, show that when defining primitive recursive functions, we really only need to think about functions of a single variable. In particular, show that a function $f(x)$ is primitive recursive if and only if it can be generated from 0, S, `pair`, `left`, and `right`, using composition and *iteration*, where iteration is defined as follows: if $f(x)$ and $g(x)$ are already defined, we can define a new function

$$h(y) = g^y(f(0)). \tag{7.29}$$

In fact, we can simplify this further and just take $h(y) = g^y(0)$.

7.17 Ackermann explodes. In this problem, we will prove that the Ackermann function grows more quickly than any primitive recursive function. We will focus on the following family of functions of a single variable:

$$G_n(y) = A(n,2,y).$$

In particular,

$$G_1(y) = y+2,\ G_2(y) = 2y,\ G_3(y) = 2^y,\ G_4(y) = \underbrace{2^{2^{\cdot^{\cdot^{2}}}}}_{y},\ \ldots$$

For $n \ge 3$, we can write G_n iteratively as

$$G_n(y) = G_{n-1}^y(1) = G_{n-1}(G_n(y-1)).$$

Show that these functions *majorize* the primitive recursive functions in the following sense: for any primitive recursive function $f(y)$, there is an n such that $f(y) < G_n(y)$ for all $y \geq 3$. It follows that $G_y(y) = A(y, 2, y)$ grows faster than any primitive recursive function.

Hint: use the definition of the primitive recursive functions given by Problem 7.16. Clearly 0, S, `pair`, `left`, and `right` are all majorized by $G_n(y)$ for some n. Then show that if two functions f and g are majorized, so are the functions formed from them by composition and iteration. In other words, if $f(y) < G_n(y)$ and $g(y) < G_m(y)$ for all $y \geq 3$, then there is some k, ℓ such that $f(g(y)) < G_k(y)$ and $g^y(f(0)) < G_\ell(y)$. You may find it useful to prove that, for all $n \geq 2$ and $y \geq 3$,

$$G_n(y) \geq G_{n-1}(y+1). \tag{7.30}$$

7.18 Packing stacks into numbers. Show that we can represent a stack of arbitrary depth, where each cell contains an arbitrary integer, as a single integer—and that we can manipulate this state with functions `push`, `pop`, and `isempty` that are primitive recursive. Hint: represent the stack as a pair, consisting of the top integer and everything else.

7.19 Metaprograms, metametaprograms... Imagine computing a function $f(x)$ in the following way. We give an input x to a BLOOP program Π, which prints out another BLOOP program Π_x. We then run Π_x, which produces $f(x)$. Show that the class of functions that can be computed this way is larger than the primitive recursive functions. Does it include all computable functions? What about programs that print programs that print programs, and so on?

Along the same lines, show that the Ackermann function $A(n, x, y)$ can be implemented as a BLOOP program provided that we can call a universal interpreter for BLOOP programs as a subroutine. Why does this not prove that $A(n, x, y)$ is primitive recursive?

7.20 Primitive recursive time. Sketch a proof that a function $f(n)$ is primitive recursive if and only if it can be calculated in $t(n)$ time for some primitive recursive function $t(n)$.

7.21 Quines. Let's think of the combinator $\mathbf{Q} = \lambda x.xx$ in a slightly different way: as a program that takes a program Π and prints the source code—not the output—of the program $\Pi(\Pi)$. That is, if Π is a program that expects an input x, then **Q**(Π) is a version of Π which takes no input, and instead assumes that x is the source code of Π.

Show that **Q**(**Q**) is a *quine*—a program that, when we run it, prints out a copy of its own source code. If you are a good programmer, write **Q** in your favorite language. (Actually dealing with input and output conventions, line breaks, and so on can be quite tricky.) Then feed it to itself, run the resulting program **Q**(**Q**), and see if it works.

7.13

7.22 Adding in λ. Show that the λ-expressions **add** and **add**$'$ defined in (7.16) and (7.17) are equivalent when applied to Church numerals. That is, show that for any $\overline{i}, \overline{j}$,

$$\mathbf{add}\,\overline{i}\,\overline{j} = \mathbf{add}'\,\overline{i}\,\overline{j}.$$

This is intuitively clear, but proving it takes some work. Hint: use induction. Start with the base case $\overline{i} = \overline{0}$, and then show that

$$\mathbf{add}(\mathbf{succ}\,\overline{i})\,\overline{j} = \mathbf{succ}(\mathbf{add}\,\overline{i}\,\overline{j}) \text{ and } \mathbf{add}'(\mathbf{succ}\,\overline{i})\,\overline{j} = \mathbf{succ}(\mathbf{add}'\,\overline{i}\,\overline{j})$$

where **succ** is the successor function defined in (7.15).

7.23 The wisdom teeth trick. Express the predecessor of a Church numeral in λ-calculus, i.e., find a combinator **pred** such that

$$\mathbf{pred}\,\overline{n} = \overline{n-1} \quad (n > 0)$$
$$\mathbf{pred}\,\overline{0} = \overline{0}.$$

As legend has it, Stephen Kleene solved this problem while having his wisdom teeth pulled, so his solution is sometimes referred to as the "wisdom teeth trick." Here are some hints to help you find it without the use of novocaine.

First, we need a way to keep track of a pair of Church numerals without having them act on each other. The following combinators let us form pairs and extract their components:

$$\mathbf{pair} = \lambda xyf.fxy\,,\ \mathbf{first} = \lambda p.pT\,,\ \mathbf{second} = \lambda p.pF\,,$$

where T and F are defined as in (7.13). These are analogous to the primitive recursive functions of Problem 7.15, but in the λ-calculus we don't need to use arithmetic to do this.

Now write a combinator Φ that changes the pair (i,j) to $(j,j+1)$. What happens if we apply Φ n times to $(0,0)$?

7.24 Factorials in λ. Using the same pairing combinators as in the previous problem, write a λ-expression for the factorial function that doesn't use the Y-combinator.

7.25 Primitive recursion in λ. Now generalize the previous problem to show that if f and g are λ-definable, so is the function h defined from them using primitive recursion (7.6). Then use the λ-calculus's power of abstraction to show that the primitive recursion operator, which takes f and g as input and returns h, can itself be written as a combinator.

7.26 More fixed point combinators. Here are two other fixed-point combinators. One, discovered by Turing, is

$$\Theta = \eta\eta \quad \text{where} \quad \eta = \lambda xy.y(xxy).$$

Another, discovered by John Tromp, is

$$\mathbf{T} = \mathbf{T}_1\mathbf{T}_2 \quad \text{where} \quad \mathbf{T}_1 = \lambda xy.xyx \quad \text{and} \quad \mathbf{T}_2 = \lambda yx.y(xyx).$$

Prove that $\Theta\,\mathbf{R} = \mathbf{R}(\Theta\,\mathbf{R})$ for any $\mathbf{R}$, and similarly for $\mathbf{T}$.

7.27 Fixed point programs. Following the style of Problem 7.21, show how to interpret the fixed point combinator $\mathbf{Y}$ as a program that takes a program-modifying program $\mathbf{R}$ as input, and prints out the source code of a program f which is a fixed point of $\mathbf{R}$. We don't necessarily mean that $\mathbf{R}$ leaves f's source code unchanged—we mean that it is a *functional* fixed point, in that f and $\mathbf{R}(f)$ compute the same partial function.

7.28 Two stacks suffice. As we have seen in several contexts, a *stack* is a data structure that works like a stack of plates. Each plate has a symbol written on it, which belongs to a finite alphabet A. We can "push" a symbol onto the stack, adding another plate on top; we can "pop" a symbol off the stack, removing the top plate and reading the symbol on it; or we can check to see if the stack is empty. But like a Turing machine's tape, we can only observe a stack in a restricted way. Only the top symbol is visible to us, and we have to pop it off if we want to see the symbols below it.

Consider the following kind of automaton. It has a finite set S of internal states, and access to k stacks. At each step, it can update its own internal state, and push or pop a symbol on each stack. It chooses these actions based only on its own internal state and the top symbols on the stacks, or whether or not they are empty. We can give such an automaton an input by putting a string of symbols on one of its stacks, and read its output from another stack when it enters a HALT state.

Show that if $k \geq 2$, such an automaton can simulate a Turing machine, and hence is computationally universal. Furthermore, show that if we are willing to accept an inefficient simulation, even a one-symbol alphabet $A = \{a\}$ suffices. Note that in that case, all that matters about a stack's state is its depth.

7.29 One queue suffices. Instead of stacks, suppose our automaton has access to data structures called *queues*. The only visible symbol is the one at the front of the queue, which is the one that gets removed if we pop a symbol. When we push a symbol, it gets added at the rear of the queue. We again have an automaton with a finite set S of internal states, which can push or pop a symbol on each queue or check to see if they are empty. Show that even one queue suffices to simulate a Turing machine.

7.30 Two, three, and four dimensions. Consider two-dimensional finite automata which are trapped inside some region in the planar grid. At each step, they can move one step in one of the cardinal directions based on their current state, and then update their state. Like the automata discussed in Section 7.6.1, they cannot write anything on the cells they visit, and the only way they can sense their current position is to tell when they bump into a wall. Thus their transition function can be written

$$F : \{\text{wall}, \text{not wall}\} \times S \to S \times \{\uparrow, \downarrow, \leftarrow, \rightarrow\},$$

where S is their finite set of states. We assume that whenever they bump into a wall, they bounce off it, back into the interior of the region.

Sketch definitions of two-dimensional automata that can determine

1. whether the region they are trapped in is a rectangle;
2. whether or not they are in a square;
3. whether or not they are in a square whose side is a power of 2.

Similarly, sketch a definition of a three-dimensional automaton that can tell whether it is in a cube whose side is a prime—and a four-dimensional automaton that can tell whether it is in a hypercube whose side is a perfect number, i.e., a number which is the sum of its divisors.

7.31 One-counter machines. Show that the Halting Problem for one-counter additive machines is decidable. Hint: first show that if the machine has n states and the counter reaches zero more than n times in the course of a computation, it will run forever.

7.32 FRACTRAN practice. Write a non-halting FRACTRAN program that generates the Fibonacci sequence in the following sense: if the initial value is $n = 6 = 2^1 3^1$, it generates $2^1 3^2$, $2^2 3^3$, $2^3 3^5$, and so on, so that the exponents of 2 and 3 are adjacent Fibonacci numbers whenever n has no other factors. The shortest such program we have found is

$$\frac{33}{14} \quad \frac{21}{22} \quad \frac{13}{7} \quad \frac{13}{11} \quad \frac{26}{85} \quad \frac{34}{65} \quad \frac{1}{13} \quad \frac{1}{17} \quad \frac{10}{3} \quad \frac{7}{1}$$

How does this program use the exponents of 2, 3, and 5, and larger primes to encode the state? Can you find a shorter program?

Then write a FRACTRAN program that, given the input 2^k in the form of its initial value n, halts if k is a power of 2, and otherwise falls into an infinite loop. Hint: you can arrange an infinite loop by putting a pair of fractions $p/q, q/p$ at the head of the program.

7.33 Euclid's algorithm in FRACTRAN. Write a FRACTRAN program that, given input $2^a 3^b 5$, halts with output $2^{\gcd(a,b)}$. Hint: start by implementing a (very short) FRACTRAN program for the map

$$2^a 3^b 5 \mapsto 2^{a-\min(a,b)} 3^{b-\min(a,b)} 7^{\min(a,b)} 5 .$$

Extend this program to deal with the cases $a > b$, $a < b$, and $a = b$. Then iterate it and move the counters around so that leads to the state $2^b 3^{a \bmod b} 5$. Finally, arrange the program so that it strips off the 5 and terminates if $b = 0$.

7.34 Collatz with tags. Let g be a Collatz function with coefficients a_i and b_i defined as in (7.27). Show that there is a tag system with $v = q$ on the following set of $2q + 2$ symbols,

$$\{\alpha, \beta_0, \beta_1, \ldots, \beta_{q-1}, \gamma, \delta_1, \delta_2, \ldots, \delta_{q-1}\},$$

such that, for any x, the string $\alpha\gamma^x$ is mapped to $\alpha\gamma^{g(x)}$.

Hint: start with the rules

$$h(\alpha) = \beta_{q-1} \cdots \beta_1 \beta_0$$
$$h(\gamma) = \delta_{q-1} \cdots \delta_1 \delta_0.$$

Notice that if $x \equiv i \bmod q$, we will meet strings beginning with β_i and δ_i. Then devise rules of the form

$$h(\beta_i) = \gamma^r \alpha \gamma^s$$
$$h(\delta_i) = \gamma^t,$$

where r, s, and t depend on a_i and b_i for each $0 \le i < q$.

7.35 Tagging Turing. Give another proof that NORMAL FORM HALTING is undecidable, by using a tag system in normal form to simulate Turing machines directly. Hint: pretend the tape is circular.

7.36 Rotating tiles. Show that TILING with Wang tiles remains undecidable even if rotations are allowed. If we change the matching rule so that colors come in complementary pairs, and neighboring tiles have to have complementary colors along their shared edge, show that TILING is undecidable even if we allow rotations and reflections.

7.37 Stripes imply rectangles. Suppose that a set of Wang tiles can tile the plane with horizontal periodicity. In other words, there is some integer ℓ such that the tiling remains the same if we shift the lattice ℓ sites to the left. Equivalently, the tiling consists of an infinite vertical strip of width ℓ, repeated endlessly to the left and right. Show that in this case, the same set of Wang tiles can tile the plane periodically, with a finite rectangle repeated forever.

Notes

7.1 The Difference Engine. Babbage's Difference Engine was designed to evaluate a polynomial $p(x)$ at a sequence of equidistant values $x, x+h, x+2h, \ldots$ for some step size h. As stated in the text for the case $h = 1$, if f is a polynomial of order n, its difference $\Delta f(x) = f(x+h) - f(x)$ is of order $n-1$, until its nth-order difference is a constant. After manually setting the initial values $f(0), \Delta f(0), \ldots, \Delta^n f(0)$, we update them as follows:

$$f(x+\delta) = f(x) + \Delta f(x)$$
$$\Delta f(x+\delta) = \Delta f(x) + \Delta^2 f(x)$$
$$\vdots$$
$$\Delta^{n-1} f(x+\delta) = \Delta^{n-1} f(x) + \Delta^n f.$$

Note that the machine never needs to multiply. If $\Delta^i f(x)$ is stored in the ith column, at each step it simply adds the current value of the ith column to the $(i-1)$st. The nth column stays constant, and the 0th column contains $f(x)$.

If Babbage had gotten his machine working, it is doubtful whether he would have simply used Taylor series. In many cases, we can obtain much better approximations by using low-degree polynomials whose coefficients are tuned to achieve the smallest possible error in a given interval. For instance, the approximation

$$\cos x \approx 1 - 0.4967x^2 + 0.03715x^4$$

has an error of less than 10^{-3} on the interval $0 \le x \le \pi/2$, much smaller than that of the 4th-order Taylor series.

Babbage first proposed the Difference Engine in 1822. In 1847–1849 he designed an improved machine, "Difference Engine No. 2," which would have evaluated polynomials up to order 7 with a precision of 31 digits. None of Babbage's machines were completed during his lifetime, but in 1989, the London Science Museum began to build a working Difference Engine No. 2 according to Babbage's design. It was completed in 1991, two hundred years after Charles Babbage's birth, and worked perfectly.

7.2 Other pioneers of computing. Percy Ludgate was an Irish accountant, who independently of Babbage (at least at first) spent his spare time designing mechanical computers. His scheme for multiplication was quite idiosyncratic. Instead of storing a multiplication table, he assigned "index numbers"—or as they were called in a 1909 article about his work [133], "Irish logarithms"—to all products of one-digit numbers:

x	0	1	2	3	4	5	6	7	8	9	10	12	14	15	16	18	20	$\cdots$
$\log x$	50	0	1	7	2	23	8	33	3	14	24	9	34	30	4	15	25	$\cdots$

These numbers look quite strange, but they obey $\log xy = \log x + \log y$ for all $0 < x, y \le 9$. Ludgate's proposed machine could multiply large numbers quickly by computing all these double-digit products, and then carrying the tens in parallel. Zero is a special case, and corresponds to all logarithms greater than or equal to 50.

Ludgate's scheme for division was also very original. To compute p/q, he first multiplied the numerator and denominator by a number close to $1/q$, giving $p/q = p'/(1+\varepsilon)$ for a small ε. He then used the Taylor series

$$\frac{1}{1+\varepsilon} = 1 - \varepsilon + \varepsilon^2 - \varepsilon^3 + \cdots,$$

to approximate p/q to high precision. In modern terms, he found a reduction from DIVISION to MULTIPLICATION.

Another early figure was Leonardo Torres y Quevedo, a Spanish engineer. In the 1890s he designed a solver for algebraic equations, and he was also famous for building funiculars, airships, remote-controlled boards, and a machine that played a perfect Chess endgame of a king and rook against a human king. He knew of Babbage's work, and in his *Ensayos sobre automática* he proposed building an Analytical Engine with floating-point arithmetic using electromechanical components. In 1920, he presented an "Electromechanical Arithmometer" in Paris, which allowed a user to input arithmetic problems through a keyboard, and computed and printed the results. In a *Scientific American* article about Torres' creations, he had this to say about artificial intelligence:

> There is no claim that [the chess player] will think or accomplish things where thought is necessary, but its inventor claims that the limits within which thought is really necessary need to be better defined, and that an automaton can do many things that are popularly classed with thought.

See Randell [684] for more discussion of this history, along with Vannevar Bush's early work on analog and digital computers.

7.3 The Entscheidungsproblem. Hilbert's Entscheidungsproblem was originally published in a textbook with Ackermann [401]. It's worth noting that Hilbert asked for a procedure which would determine whether a statement is true or false, as opposed to asking for a proof or disproof of it. The distinction between truth and provability took some time to emerge, culminating in Gödel's result that there is an unbridgeable gap between them.

7.4 Sets, barbers, and adjectives. Russell's paradox shows that so-called *naive set theory*, in which we are allowed to use any property to define the set of objects with that property, can lead to contradictions. He described his paradox in a letter to Gottlob Frege in 1902, just as Frege's book—which used naive set theory to lay the foundations of arithmetic—was going to print. Frege hastily added an Appendix to his book, in which he wrote

> Hardly anything more unfortunate can befall a scientific writer than to have one of the foundations of his edifice shaken after the work is finished. This was the position I was placed in by a letter of Mr. Bertrand Russell, just when the printing of this volume was nearing its completion.

While adopting a stratified set-theoretic universe gets rid of paradoxes like Russell's, it has other consequences as well. For instance, we do not speak of the set of all sets, since it would have to contain itself. Instead, we kick it upstairs by calling it a "proper class."

Russell's paradox has many equivalent forms. As he pointed out in [719], R is like the barber who shaves every man who doesn't shave himself, although this version of the paradox goes away if the barber doesn't shave at all. Another form concerns adjectives that describe themselves. For instance, "polysyllabic" is polysyllabic, "recondite" is recondite, "short" is short, and so on. Let's call such words *autological*, and let's call non-self-describing words like "monosyllabic", "quotidian", and "long" *heterological*. Is "heterological" heterological or not?

For another diagonal paradox, consider Berry's paradox. Define an integer x as follows: "the smallest integer that cannot be defined in less than thirteen words." From this we have to conclude that there is a difference between language and metalanguage—that "definability" is not a definable notion.

Similarly, let's say that 0 and 1 are interesting because they are the identities of addition and multiplication, 2 is interesting because it is the only even prime, and so on. At some point we get to the smallest uninteresting number—but surely that makes it interesting!

For a wonderful exploration of Russell's life, and the anguish and passion of his search for foundations, we recommend *Logicomix* by Doxiadis and Papadimitriou [247].

7.5 Hilbert's 10th Problem. When David Hilbert proposed his list of problems in 1900, probably every mathematician alive would have agreed that the 10th problem must have a solution. But after the appearance of undecidable problems in 1936, people began to wonder whether the solvability of Diophantine equations might be undecidable as well. In 1944, Emil Post declared that Hilbert's 10th problem "begs for an unsolvability proof," but it took decades for this proof to appear.

In 1961, Julia Robinson, Martin Davis, and Hilary Putnam [224] showed the undecidability of *exponential* Diophantine equations, i.e., integer equations where exponentiation and polynomials are allowed. Building on earlier work of Robinson [706], their proof worked by encoding the sequence of states of a Turing machine in an enormous integer, so that an integer solution exists if and only if the computation halts.

The challenge was then to show that exponentiation was not required—or rather, that it could be represented with purely polynomial Diophantine equations. In 1970, Yuri Matiyasevič [559] found a system of polynomials that encode the Fibonacci numbers. Using the fact that they grow exponentially (see Problem 2.4) he was able to complete the proof. A history and exposition can be found in [560].

Since then, the proof of undecidability has been simplified to some extent. The simplest proof, that of Jones and Matiyasevič [443, 444], builds integers that encode trajectories of the counter machines of Section 7.6.1.

7.6 Hilbert, Turing, Church: A Hilarious Tour of Computation. The phrase "strange loop," and the names BLOOP and FLOOP for the programming languages corresponding to primitive recursive and partial recursive functions, were introduced by Douglas Hofstadter in his Pulitzer Prize winning opus *Gödel, Escher, Bach: An Eternal Golden Braid* [407]. Both lighter and deeper than any textbook, this gorgeous book discusses self-reference and paradox in music, art and mathematics. In particular, it gives the clearest explanation of Turing's Halting Problem, and Gödel's Incompleteness Theorem, that we have seen. Buy it for your children and read it to them at bedtime.

7.7 The Continuum Hypothesis. Since real numbers, like subsets of the integers, can be represented as infinite binary strings, $2^{\aleph_0}$ is also the cardinality of the reals. Cantor conjectured that $2^{\aleph_0}$ is the smallest cardinal greater than $\aleph_0$, which is denoted $\aleph_1$. Proving or disproving this hypothesis was the first of Hilbert's problems in 1900. However, in 1940 Kurt Gödel showed that it cannot be proved using the standard Zermelo–Fraenkel axioms of set theory, and in 1963 Paul Cohen showed that it cannot be *disproved* either. Thus it is formally independent of these axioms.

A variety of other axioms have been put forward that would imply the Continuum Hypothesis or its negation. But until these axioms gain wide acceptance, its truth or falsity is a doctrinal question in the theology of infinite sets, on which reasonable mathematicians can disagree.

7.8 The Halting Problem and undecidability. The original Halting Problem was posed by Turing [797] for Turing machines, and by Church [173] for the existence of a normal form in the λ-calculus. A modern presentation, similar to the "software-level" one we give here—including the role played by an interpreter and the fact that BLOOP (by another name) cannot interpret itself—was given in 1972 by Hoare and Allison [403].

The fact that all nontrivial semantic properties of programs are undecidable, as stated in Problem 7.4, was proved by Rice [703]. Here "semantic" means a property of the function that a program computes, and "nontrivial" means that it is neither true for all computable partial functions, nor false for all of them.

As Problem 7.5 shows, some "dynamical" questions about programs are also undecidable, such as whether the only way that Π fails to halt is by falling into an endless loop. Other examples include whether Π contains "dead code," i.e., lines that are never executed—or whether there is a shorter program that computes the same function that Π does.

7.9 Gödel's Second Incompleteness Theorem. The fact that no sufficiently powerful formal system is both consistent and complete—that is, that there are statements S such that neither S nor $\overline{S}$ is a theorem, or (even worse) they both are—is Gödel's First Incompleteness Theorem. His Second Incompleteness Theorem is, in some ways, even more severe: no formal system can prove its own consistency. In other words, it is impossible to prove, within the axioms of the system, that there is no statement S such that both S and $\overline{S}$ are theorems. Thus no formal system strong enough to do the mathematics we care about can prove, about itself, that it is free of paradoxes.

According to the rules of propositional logic, if one can prove a contradiction then one can prove anything. In other words, for any statements S and T, we have $(S \wedge \overline{S}) \Rightarrow T$. It follows that all a system needs to do in order to prove its own consistency is prove that some T is not a theorem. Since the set of theorems is undecidable (see Exercise 7.8), and since finite computations correspond to finite proofs, this rules out the possibility that the system can prove, for all nontheorems T, that T is a nontheorem. However, the Second Incompleteness Theorem implies the stronger statement that the system cannot do this for *even one* nontheorem.

7.10 Formal independence. If it is true, but unprovable, that Π never halts, then neither $\texttt{Halts}(\Pi, x)$ nor $\overline{\texttt{Halts}(\Pi, x)}$ is a theorem. In other words, the question of whether or not $\Pi(x)$ halts cannot be resolved by the axioms either way.

As a consequence, we have the freedom to add *either one* of these statements to our list of axioms, without causing an inconsistency. If we choose to believe that Π halts, even though it doesn't, we have the following strange situation. Since we can run Π for any finite amount of time, each of the statements

$$\overline{\texttt{HaltsInTime}(\Pi, x, 1)}$$
$$\overline{\texttt{HaltsInTime}(\Pi, x, 2)}$$
$$\overline{\texttt{HaltsInTime}(\Pi, x, 3)}$$
$$\vdots$$

is a theorem—and yet so is

$$\exists t : \texttt{HaltsInTime}(\Pi, x, t).$$

Such a system is called ω*-inconsistent*. It believes that there exists an integer t with a certain property, even though it knows that each particular t fails to qualify. Paraphrasing Scott Aaronson [1], if we ask such a system to tell us what t is, it cheerfully replies "why, the t such that $\Pi(x)$ halts after t steps, of course!"

7.11 Recursion and primitive recursion. The idea of defining functions recursively goes back to Dedekind and Peano in the late 1880s. In an effort to build an axiomatic foundation for number theory, they described how to define addition, multiplication, and so on inductively using the successor function. The phrase *primitive recursion* was coined by Rózsa Péter in 1935, who proved many important results about the class of primitive recursive functions. She also defined a hierarchy of more elaborate types of recursion, roughly equivalent to the hierarchy described in Problem 7.19. The equivalence between primitive recursive functions and programs with bounded loops—that is, programs written in BLOOP—was demonstrated by Meyer and Ritchie [574].

In the 1930s, Herbrand and Gödel defined recursive functions more generally, as solutions to systems of equations: for each x, the value $h(x)$ is defined if there is a unique y such that the statement $h(x) = y$ can be derived from the system. This derivation consists of a finite number of steps, where each step substitutes the right-hand side of one of the equations wherever its left-hand side appears. At the time, however, Gödel was not convinced that this concept of recursion "comprises all possible recursions." In 1936, Kleene [488] showed that that this definition is equivalent to what we now call partial recursive functions, where we use the μ operator to solve equations of the form $f(\mathbf{x}, y) = 0$, and Exercise 7.12 is called Kleene's Normal Form Theorem. Some of this history appears in Kleene [489].

7.12 Exploding functions. The Ackermann function, and a proof that it is not primitive recursive, was published by Hilbert's student Wilhelm Ackermann in 1928 [19]. A similar function was published concurrently in a little-known paper by another of Hilbert's students, Gabriel Sudan [780]; see [146].

Other recursive functions with simple proofs that they are not primitive recursive were given by Rózsa Péter [657] and by Robinson [707]. The proof we give in Problem 7.17 that the Ackermann function majorizes every primitive recursive function essentially follows [707]. The kind of iterative implementation of Ackermann's function which we give in Section 7.3, using a **while** loop and a stack, appears not to have been published before 1988 [355].

A popular way to write the sequence of binary operations starting with addition, multiplication, and exponentiation comes from Knuth. If we define $x \uparrow^1 y = x + y$ and

$$x \uparrow^{n+1} y = \underbrace{x \uparrow^n (x \uparrow^n (\cdots x))}_{n \text{ copies of } x},$$

then $A(n, x, y) = x \uparrow^n y$. The operation $x \uparrow^4 y = \underbrace{x^{x^{\cdot^{\cdot^{x}}}}}_{y \text{ times}}$ is sometimes called "tetration." This notation allows us to express very large numbers very compactly.

7.13 Quine and "quines." The philosopher and logician Willard Van Orman Quine made numerous contributions to epistemology, set theory, and the philosophy of science. He also studied indirect versions of self-reference. Consider the following paradoxical sentence:

> "Yields a falsehood when preceded by its quotation" yields a falsehood when preceded by its quotation.

This sentence manages to assert that it is false, without using the words "this sentence." The first half acts as data and the second half acts as a function on this data, and together they expand to create another copy of the original sentence. In *Gödel, Escher, Bach,* Douglas Hofstadter draws numerous analogies between this kind of self-reference and self-reproduction. For instance, our genome (the first half) is the data that our ribosomes (the second half) read, translate into proteins, and unfold into another copy of ourselves.

In Quine's honor, a computer program that "self-reproduces," in the sense that it prints out a copy of its own source code, is called a *quine.* You can find quines in virtually every programming language ever invented—some readable, and some deliberately obfuscated—at Gary Thompson's Quine Page, www.nyx.net/~gthompso/quine.htm.

7.14 The λ-calculus and combinatory logic. The notion of a combinator, and the realization that they allow functions to act directly on each other, was originally given by Moses Ilyich Schönfinkel in 1924 [731]. It was reinvented and extended by Haskell Curry [208], after whom the functional programming language HASKELL is named.

Schönfinkel and Curry showed that we can generate any combinator whatsoever from these two,

$$\mathbf{S} = \lambda xyz\,.(xz)(yz) \text{ and } \mathbf{K} = \lambda xy\,.x\,.$$

(Note that **K** was called **T** in Section 7.4.) For instance, the identity is $\lambda x\,.x = \mathbf{SKK}$, and the composition combinator **comp** from (7.18) is

$$\lambda fgx\,.f(gx) = \mathbf{S}(\mathbf{KS})\mathbf{K}.$$

Thus **S** and **K** form one of the most compact systems capable of universal computation. If you think two combinators is still too many, you can generate them both from this one:

$$\mathbf{X} = \lambda x\,.x\mathbf{KSK},$$

since

$$\mathbf{K} = (\mathbf{XX})\mathbf{X} \text{ and } \mathbf{S} = \mathbf{X}(\mathbf{XX}).$$

In addition to the basic substitution rule, which is called *β-reduction* in the λ-calculus, Schönfinkel used the *axiom of extensionality* which states that two combinators are equivalent if they always carry out the same function on their inputs. However, this axiom, which is called *η-reduction*, isn't really necessary.

Alonzo Church proposed the λ-calculus in 1932 [171, 172] as a foundation for logic. His students Kleene and Rosser modified its original definition, and played crucial roles in its development. Church and Rosser [174] showed that, if a normal form exists, it is unique and any chain of reductions will lead to it. Many of the combinators for arithmetic on Church numerals are due to Rosser; those for predecessor and subtraction are due to Kleene, as is the proof that all partial recursive functions are λ-definable [487].

In the same year as Turing's work, Church [173] showed that the problem of whether a λ-expression has a normal form—or equivalently, whether a partial recursive function is defined for a given input—is undecidable. The fixed point combinator **Y** was discovered by Curry, and there are an infinite number of others.

The standard reference for the λ-calculus is Barendregt [79]; for a lighter introduction, see the lecture notes by Barendregt and Barendsen [80]. Smullyan gives a lovely exposition of combinatory logic in terms of birds in a forest responding to each others' calls in *To Mock a Mockingbird* [763].

The programming language LISP was invented by John McCarthy in 1958 [563], who later received the Turing Award. Inspired by the λ-calculus, it treats programs and data as the same kind of object, and lets the user define functions from a handful of simple primitives. Along with its descendants, such as HASKELL, SCHEME, and ML, LISP is considered a *functional* programming language because it computes by unfolding the definition of a function until it reaches a normal form. In contrast, *imperative* programming languages such as FORTRAN, C, and JAVA work primarily by ordering the computer to change the values of its variables, like a Turing machine updating its tape. Because of the grand unification of 1936, we know that these two types of programming language are equally powerful.

7.15 Small universal Turing machines. Turing's universal machine had a fairly large number of states and symbols. But since then, clever people have found much smaller ones—for instance, with just 6 states and 4 symbols. As Figure 7.22 shows, there is a tradeoff between states and symbols. Their product is essentially the size of the table of the transition function, and this product is roughly constant along the universal curve.

The smallest universal machines on record are those of Rogozhin [714], Kudlek and Rogozhin [507], and Neary and Woods [626]. Rather than simulating other Turing machines directly, these machines simulate tag systems like those described in Section 7.6.3. In some cases, a small amount of postprocessing is needed to extract the output from the tape.

If we modify our definitions a little, allowing the initial tape to be filled with periodic patterns to the left and right of the input word instead of with blanks, we can make these machines even smaller—with 3 states and 3 symbols, 2 states and 4 symbols, or 6 states and 2 symbols. We used the last of these for our iterated map in Section 7.6.6. These so-called *weakly* universal machines are also due to Neary and Woods [627], and they work by simulating cellular

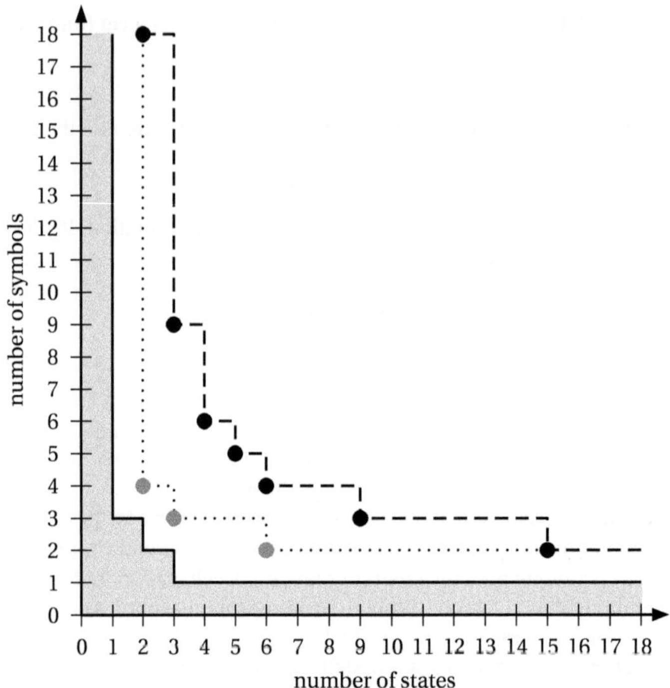

FIGURE 7.22: The smallest known universal Turing machines are indicated by black dots. Thus, everywhere at or above the dashed line, there is at least one universal Turing machine with that number of states and symbols. Weakly universal machines, where the tape is filled with a periodic pattern to the left and right of a finite initial word, rather than blanks, are indicated by gray dots. No universal machines exist at or below the solid line, because their Halting Problems are decidable.

automaton rule 110. They never halt, but the question of whether or not they will ever produce a certain string on their tape is undecidable.

7.16 Alan Turing and his machine. Our quotes are from Turing's 1936 paper "On computable numbers, with an application to the *Entscheidungsproblem*" [797]. It is reprinted, along with the other seminal papers of 1936 by Church, Kleene, and Post, in the collection of Davis [221]. For a guided tour through Turing's paper we recommend Petzold [658]. We also recommend Gandy [314], who describes the various strands of thought that came together in 1936. Gödel's quote is from his remarks to the Princeton Bicentennial Conference on Problems in Mathematics [221], and von Neumann's quote is from [815].

During World War II, Turing played a key role in the British effort to break the Nazi's Enigma code—a success which was, arguably, as important to the outcome of the war as the development of radar or the atomic bomb. He and his collaborators designed Colossus, the first large-scale electronic computer, slightly predating John von Neumann's work on ENIAC in the United States.

Turing also wrote presciently on the subject of artificial intelligence. In [798] he proposed the so-called Turing Test, in which a computer tries to fool an interviewer into believing that it is human. He began by describing a game in which a man tries to convince an interviewer that he is a woman. He argued that computers would someday pass this test, that "at the end of the century the use of words and general educated opinion will have altered so much that one will be able to speak of machines thinking without expecting to be contradicted," and that "machines will eventually compete with men in all purely intellectual fields."

Turing believed that we would create such machines, not by writing a fixed program, but by setting a process of learning or evolution in motion. He described a simulated neural network, trained by giving it positive and negative feedback. He wrote, with a poignant tongue in his cheek,

> It will not be possible to apply exactly the same teaching process to the machine as to a normal child. It will not, for instance, be provided with legs, so that it could not be asked to go out and fill the coal scuttle... however well these deficiencies might be overcome by clever engineering, one could not send the creature to school without the other children making excessive fun of it.

In a little-known 1948 report [800], Turing laid out several kinds of search, including "genetical or evolutionary search," presaging the field of genetic algorithms. He also did foundational work on pattern formation, proposing and simulating nonlinear reaction-diffusion models for biological systems such as embryogenesis and phyllotaxis [799].

Alan Turing had the misfortune of being born into a time, and a society, in which homosexual relationships were illegal. After being convicted of "gross indecency," he was stripped of his security clearance, thus ending his work on cryptography, and sentenced to injections of estrogen. Two years later, he committed suicide. Far too late, in 2009, British Prime Minister Gordon Brown apologized for the shameful treatment that Turing received, and in 2013 he was pardoned by the Queen. It is poignant to imagine how great a role Turing might have played in the further development of computer science if it had not been for this tragedy.

7.17 Busy beavers. The busy beaver function of Exercise 7.21 was invented by Tibor Rado [680]. He defined two functions, and proved that both are uncomputable: the maximum number of moves $S(n,m)$ a machine makes before halting, and the maximum number $\Sigma(n,m)$ of 1s it can write on the tape before halting. Even for small machines, these numbers are astronomical. For $n=6$ states and $m=2$ symbols, for instance, Pavel Kropitz has found a Turing machine showing that $S(n,m) \geq 7.4 \times 10^{36534}$.

7.18 The Church–Turing Thesis. Church and Turing were concerned with capturing the class of functions that are "effectively computable"—those for which there is an algorithm, each step of which could be carried out by a human with pencil and paper. Their thesis states that this class is synonymous with the Turing-computable functions. This is equivalent to von Neumann's statement that a Turing machine can simulate any automaton, as long as each step of the automaton is something that a human could do.

The *physical Church–Turing Thesis* is a very different statement—namely, that any function which can be computed by a physical device can also be computed by a Turing machine, and more generally, that any physical system can be simulated by a Turing machine to any desired accuracy. This could be false, for instance, if the laws of physics allow for exotic processes that perform an infinite amount of computation in a finite time. We briefly address this thesis in Chapter 15 as part of our discussion of quantum computation.

The Church–Turing Thesis is often misunderstood as the claim that digital computers can simulate human intelligence. Of course, if the physical Church–Turing Thesis is true, the brain can be simulated by a computer since it is a finite physical system. The question of whether this simulation would be conscious in the same way that we are opens up an additional can of worms, such as whether a computer can "really" enjoy strawberries and cream, or "just simulate" that experience—and whether this distinction makes any sense. Turing deliberately sidestepped this question in [798] by defining the Turing test only in terms of a computer's outward behavior, not its inner life.

In any case, while it is certainly connected to these questions, the original thesis does not address intelligence in general. It focuses on the set of decidable problems, those for which there is an algorithm—in Turing's words, those "which can be solved by human clerical labour, working to fixed rules, and without understanding."

7.19 Counter machines, FRACTRAN, and the Collatz Problem. Our definition of additive and multiplicative counter machines, and the fact that two-counter machines are universal, follows Minsky's book [583]. FRACTRAN was introduced and shown to be computationally universal by John Horton Conway [195]; see Guy [358] and Havil [382] for additional examples and discussion. Conway's proof that FRACTRAN is universal builds on his earlier proof [194] of universality for generalized Collatz problems.

The Collatz problem is named after its inventor Lothar Collatz. Among small numbers, one of the initial values of x that takes the longest to reach 1 is 27. It takes 111 steps, and reaches 9232 before it comes back down. One heuristic argument for the conjecture, which we take from Jeffrey Lagarias [516], is to consider the number of times we divide by 2 in between odd values of x. Clearly $3x+1$ is even. If we assume that it is divisible by 4 with probability 1/2, by 8 with probability 1/4, and so on, just as if it were a random even integer, then we will divide by 2 once with probability 1/2, by 4 with probability 1/4, and so on, before multiplying by 3 the next time. The geometric mean of the ratio between the new value and the old one is then

$$\left(\frac{3}{2}\right)^{1/2}\left(\frac{3}{4}\right)^{1/4}\left(\frac{3}{8}\right)^{1/8}\cdots=\frac{3}{4}<1.$$

This suggests that, "on average," $g(x)<x$, and therefore that the trajectory of x cannot diverge. However, like many problems in mathematics, the question is whether the values of x can really be treated as random in this way.

The two-, three-, and four-dimensional finite automata of Problem 7.30 are explored in [458, 534].

7.20 Tag systems and Post's anticipation. Even though it took place in 1921, Emil Post's work on normal form and tag systems did not appear in print until 1943 [664]. He struggled with a particular example, where $v=3$, $h(0)=00$, and $h(1)=1101$. Its general behavior is still not understood—there is neither an algorithm for its halting problem nor a proof that it is undecidable.

Minsky [582] proved that tag systems are universal using the counter simulation of Turing machines with tape alphabet $\{0,1\}$, and Cocke and Minsky [184] showed this even in the case $v=2$. The tag system which computes the $3x+1$ function and the proof of universality using Collatz functions in Problem 7.34 are from Liesbeth De Mol [593], who explains the role that tag systems played in Post's thinking in [592].

After realizing that even tag systems possessed extremely complicated behavior, Post reversed his thinking, and used the diagonal method (Problem 7.11) to show that systems in normal form cannot prove all true propositions about themselves. Thus he proved—in essence—Gödel's and Turing's results a decade before they did. See Stillwell [775] for a discussion of Post's approach, which was arguably simpler than Gödel's.

In 1936, Post invented a machine essentially identical to the Turing machine [663], although he did not construct a universal machine or prove that the Halting Problem is undecidable. Later, he invented the notion of recursive enumerability [665]. His correspondence problem appeared in [666], and the undecidability proof we give here is from Minsky [583]. CORRESPONDENCE has been used to show that many other problems, ranging from automata theory to abstract algebra, are undecidable as well—including the problem of whether two four-dimensional manifolds are homeomorphic.

Owing to his anticipation of Gödel's and Turing's more famous works, Post is something of an unsung hero in logic and computer science. We agree with his sentiments when he says

> ...it seemed to us to be inevitable that these developments will result in a reversal of the entire axiomatic trend of the late 19th and early 20th centuries, with a return to meaning and truth.

7.21 The Game of Life. Like FRACTRAN, Life was invented (or should we say discovered?) by John Horton Conway. It first appeared in 1970, in Martin Gardner's "Mathematical Games" column in *Scientific American.* The first simulations were carried out on graph paper and Go boards, but it quickly became a favorite topic of the emerging Hacker community. That year, Bill Gosper won a $50 prize from Conway by finding the glider gun in Figure 7.12, thus proving that the number of live sites can grow without bound.

The Turing machine shown in Figure 7.14 was constructed by Paul Rendell, who described it in [20]. Sliding block memories, which simulate counters by moving a 2×2 block, were invented by Dean Hickerson. A universal counter machine, consisting of a finite state control coupled to two or more of these blocks, was developed by Paul Chapman.

7.22 One-dimensional cellular automata. The universality of rule 110 was conjectured by Stephen Wolfram and proved by Matthew Cook [196]. Like Minsky's counter machines, Cook's original construction was exponentially inefficient. However, Neary and Woods [625] showed that the relevant tag systems, and therefore rule 110, can simulate a Turing machine with polynomial slowdown.

Wolfram [832] has conjectured that every cellular automaton is either easy to predict and analyze, or is computationally universal. Another candidate for universality among the "elementary" one-dimensional rules, with two states per site and nearest-neighbor interactions, is rule 54. See Hanson and Crutchfield [366] for a study of its gliders and their interactions.

7.23 Tilings. Wang tiles are named after Hao Wang, who showed in 1961 that TILING COMPLETION is undecidable [817]. He conjectured that any set of tiles can tile the plane periodically or not at all, and pointed out that in that case TILING would be decidable. In 1966, his student Robert Berger found a (very large) aperiodic set of tiles, and proved that TILING is in fact undecidable [105]. The much smaller set of tiles we show in Figure 7.18 was designed by Raphael Robinson [708].

Interestingly, the problem of whether a given set of Wang tiles can tile the plane *periodically* is also undecidable, since we can design the tiles so that a periodic tiling exists if and only if a given Turing machine halts. See Gurevich and Koriakov [357].

7.24 Iterated maps. The fact that simple piecewise-continuous maps of the plane can simulate Turing machines was shown by Moore [601, 602]. He showed that a number of questions about their behavior are undecidable, including whether a given point lies on a periodic orbit or whether or not its trajectory is chaotic. He also showed that these two-dimensional maps can be made infinitely differentiable, and that they can be embedded in continuous flows in three dimensions.

Similar results were obtained independently by Reif, Tygar, and Yoshida [694] for optical ray-tracing, Siegelmann and Sontag [752] for neural networks with piecewise linear activation functions, and Cosnard, Garzon, and Koiran [204] for maps of the plane.

The universal 6-state, 2-symbol Turing machine we simulate in Figure 7.21 is due to Neary and Woods; see Note 7.15.

Chapter 8

Memory, Paths, and Games

> We cross our bridges when we come to them and burn them behind us, with nothing to show for our progress except a memory of the smell of smoke, and a presumption that once our eyes watered.
>
> Tom Stoppard, *Rosencrantz and Guildenstern Are Dead*

> It is hard enough to remember my opinions, without also remembering my reasons for them!
>
> Friedrich Nietzsche

So far, we have focused on time as our measure of computational complexity. But in the real world, both time and memory are limited resources. How does the complexity landscape change if we ask what we can achieve with a given amount of memory instead?

The fundamental difference between time and memory is that *memory can be reused.* (If only that were true of time as well.) This causes memory to behave very differently as a computational resource, and leads to some highly nonintuitive results. In particular, where memory is concerned, the gap between determinism and nondeterministic computation—between finding a solution and checking one—is much smaller than we believe it is with time. Even more bizarrely, in a world limited by memory rather than by time, there isn't much difference between proving that something exists and proving that it doesn't. Proofs of existence and of nonexistence are equally easy to check.

In this chapter we will meet complexity classes of problems that can be solved, deterministically or nondeterministically, with various amounts of memory. Just as the concept of NP-completeness illuminates the hardness of HAMILTONIAN PATH, these classes help us understand how hard it is to find a path through an enormous graph, or find a winning strategy against a clever opponent. They will help reveal the logical structure of these problems, the kinds of witnesses they require, and how hard it is to check their witnesses. By the end of this chapter, we will see how computational complexity lurks in mazes, wooden puzzles, and in board games thousands of years old.

8.1 Welcome to the State Space

Had we but world enough, and time...
Andrew Marvell, *To His Coy Mistress*

At the time we write this, a typical laptop has roughly 10^{10} bits of memory, or 10^{12} if we include its hard drive. At each moment, these bits describe the computer's entire internal state, including the values of all its variables, the contents of its arrays and data structures, and so on. Since each of these bits can be true or false, the total number of possible states the computer can be in is

$$2^{10^{12}} .$$

Let's call this vast, but finite, set of all possible states the computer's *state space*. As the program proceeds, it follows a path through this space, where each elementary step takes it from one state to the next. If we are concerned with running time, we limit the *length* of these paths. But if we care about memory, what matters is the size of the space in which they wander.

Let's be a little more precise about what things we keep in memory. Where do we keep the program? Since the length of any given program is just a constant, it doesn't really matter whether we think of it as stored in memory or coded directly into the computer's hardware. However, we do need to use the memory to keep track of what instruction of the program the computer is currently running, in order to know what to do next. For instance, one part of the memory could contain the fact that we are currently running line 17 of the program, which tells us to increment the variable x_3, store the result in the variable y, and then go to line 18.

More importantly, where do we put the input? We normally think of loading a program's input into the computer's memory, so that its bits can be read and modified alongside the computer's other bits. However, many interesting problems can be solved with an amount of memory that is much smaller than the input itself. This may seem odd, but we live in a data-rich age, where satellites and radio telescopes produce terabytes of images every day. Our computer cannot hope to hold this much data at once, but it can search these images for patterns by looking at a little at a time.

To model this kind of situation, we separate the n bits of the input from the $f(n)$ bits of memory the computer uses to process it. To keep it from cheating and using these bits as additional memory, we give it *read-only access* to the input: it can look up any bit of the input it wants, but it cannot modify these bits or overwrite them with new data. We can think of the input as being stored in some capacious but unwritable medium, such as a stack of optical disks. In contrast, the computer can both read and write the bits in its own memory.

How does our computer request a given bit of the input, or of its own memory? We will assume that it is equipped with a *random access memory* (RAM), which allows it to access any bit by specifying its location. So, we will allow it to carry out instructions like "increment the ith element of the array a" or "set y to the ith bit of the input" in a single step. We will assume throughout the chapter that we have at least $\log n$ bits of memory, since it takes $\log_2 n$ bits just to specify which bit of an n-bit input we want.

8.1 We are now in a position to define "space-bounded" complexity classes, consisting of problems we can solve with a given amount of memory:

```
input: a string x of length n
output: is x a palindrome?
begin
    i = 1 ;
    j = n ;
    while i < j do
        if x_i ≠ x_j then return "no" ;
        i := i + 1 ;
        j := j − 1 ;
    end
    return "yes" ;
end
```

FIGURE 8.1: A program that checks whether x is a palindrome using $O(\log n)$ memory, showing that this problem is in L. It has read-only access to x, so it only needs to keep track of i and j.

> SPACE($f(n)$) is the class of problems for which a program exists that solves instances of size n using read-only access to the input and $O(f(n))$ bits of memory.

In particular, much of this chapter will focus on what we can do with logarithmic, or polynomial, memory:

$$\mathsf{L} = \mathsf{SPACE}(\log n)$$

$$\mathsf{PSPACE} = \mathsf{SPACE}(\text{poly}(n)) = \bigcup_{k>0} \mathsf{SPACE}(n^k)$$

What sorts of problems live in these two classes? If a problem is in L, we can solve it even though our memory is far too small to hold the entire input. We can look at any part of the input we like, but we can only see a few bits of it at a time, and we can't remember very much about what we've seen so far. Typically, problems in L are "local" rather than "global," in the sense that we can examine the input piece by piece and confirm that some simple property holds at each place.

For example, consider the problem of telling whether a string is a palindrome, such as

saippuakivikauppias,

which means "soapstone seller" in Finnish. We just need to make sure that each letter matches its mirror image, and we can do this with the program shown in Figure 8.1. Since we have read-only access to x, the only things we need to keep track of in memory are the integer counters i and j. Since each of these is at most n, we can encode them with $\log_2 n$ bits each. The total amount of memory we need is $2\log_2 n = O(\log n)$, so this problem is in L.

For an example of a problem in PSPACE, suppose we have some physical system in which each state can be described with poly(n) bits. At each time step we update the state according to some dynamical rule. We would like to know whether the initial state is part of a *periodic orbit*—that is, if there is some t

FIGURE 8.2: A periodic orbit of cellular automaton rule 30, with time running from left to right. Each column is a state with n bits, where 0 is white and 1 is black. The transition rule is illustrated above, showing how the current values of a bit and its neighbors determine its value on the next time step. We use cyclic boundary conditions so that the lattice wraps around. Here $n = 7$, the initial state is 0000101, and the period is $t = 63$. While the period may be exponentially long, we only need n bits to remember the current state, so telling whether the initial state lies on a periodic orbit is in PSPACE.

such that the system returns to its initial state after t steps. For instance, Figure 8.2 shows a periodic orbit of the cellular automaton discussed in Section 5.5.3.

We can solve this problem simply by simulating the system, and checking at each point to see if the current state is identical to the initial one. Since there are $2^{\text{poly}(n)}$ possible states, and since the dynamics of the cellular automaton could cycle through a large fraction of them, it could take an exponential amount of time for this to happen, if it ever does. However, it only takes $\text{poly}(n)$ bits to keep track of the current state, so this problem is in PSPACE.

How do classes like L and PSPACE relate to the time-bounded complexity classes we have already seen? Let's set down some basic relationships between the amount of memory a program uses and the time it takes to run.

First of all, if a program uses m bits of memory then it has 2^m possible states. Therefore, no path in the state space can be more than 2^m steps long, unless it returns to a state it has already visited. But if this happens, the computer will perform exactly the same steps again, and get stuck in an endless loop. Thus a program that uses m bits of memory needs at most 2^m time to complete its task:

$$\mathsf{SPACE}(f(n)) \subseteq \mathsf{TIME}(2^{O(f(n))}) \tag{8.1}$$

In particular, $2^{O(\log n)} = \text{poly}(n)$. For instance, $2^{c \log_2 n} = n^c$. Therefore, a program that runs in $O(\log n)$ memory only has a polynomial number of states, and only needs polynomial time:

$$\mathsf{L} \subseteq \mathsf{P}. \tag{8.2}$$

Conversely, suppose a program runs in time t. If each elementary step reads or writes a single bit of memory—or even a block of memory of constant size, such as a 64-bit word—then such a program only has time to read or write $O(t)$ bits. Even a random-access machine with large memory can only access $O(t)$ different bits in t steps, and Problem 8.1 shows that we can then simulate it with only $O(t)$ bits memory. Thus

$$\mathsf{TIME}(f(n)) \subseteq \mathsf{SPACE}(f(n)). \tag{8.3}$$

In particular, $\mathsf{P} \subseteq \mathsf{PSPACE}$.

Now suppose a problem can be solved nondeterministically in time t. That is, suppose that there is a witness-checking program Π that runs in time t, such that x is a yes-instance if and only if there is a witness such that Π returns "yes." Assuming that Π reads one bit per step, it only has time to read t bits. Therefore this witness, if it exists, has length at most t.

As we discussed in Section 4.3, we can solve such a problem deterministically by generating all possible 2^t witnesses of length t and running Π on each one. The total time we need is $2^t\, t$, and setting $t = O(f(n))$ gives

$$\mathsf{NTIME}(f(n)) \subseteq \mathsf{TIME}(2^{f(n)} f(n)) \subseteq \mathsf{TIME}(2^{O(f(n))}).$$

This exhaustive search takes exponential time—but its memory use isn't so bad, since we can reuse the same t bits for all 2^t witnesses. Moreover, since Π runs in time t it only needs $O(t)$ bits of memory for its own work, so the total memory we need is just $t + O(t) = O(t)$. This strengthens (8.3) to

$$\mathsf{NTIME}(f(n)) \subseteq \mathsf{SPACE}(f(n)), \tag{8.4}$$

and in particular,

$$\mathsf{NP} \subseteq \mathsf{PSPACE}.$$

Let's summarize what we know so far. The relations (8.1) and (8.4), along with the fact that $\mathsf{TIME}(f(n)) \subseteq \mathsf{NTIME}(f(n))$, give a chain of inclusions climbing up the complexity hierarchy, alternating between time-bounded and space-bounded classes:

$$\mathsf{L} \subseteq \mathsf{P} \subseteq \mathsf{NP} \subseteq \mathsf{PSPACE} \subseteq \mathsf{EXPTIME} \subseteq \mathsf{NEXPTIME} \subseteq \mathsf{EXPSPACE} \subseteq \cdots$$

Are these classes different? In Section 6.3, we discussed the Time Hierarchy Theorem, which shows that $\mathsf{TIME}(g(n)) \subset \mathsf{TIME}(f(n))$ whenever $g(n) = o(f(n))$. There is a similar Space Hierarchy Theorem, which shows that

$$\mathsf{L} \subset \mathsf{PSPACE} \subset \mathsf{EXPSPACE} \subset \cdots$$

So, we know that at least one of the inclusions in

$$\mathsf{L} \subseteq \mathsf{P} \subseteq \mathsf{NP} \subseteq \mathsf{PSPACE}$$

is proper, and we believe that they all are. However, we don't have a proof for any of them.

There's something missing. How do we define nondeterministic versions of L and PSPACE? What kind of a witness do we need to find our way through state space? For the answer, dear reader, turn to the following section.

Exercise 8.1 *A function is* polylogarithmic *if it is $O(\log^c n)$ for some constant c. Consider the class of problems solvable in polylogarithmic space,*

$$\mathsf{SPACE}(\text{polylog}(n)) = \bigcup_{c>0} \mathsf{SPACE}(\log^c n).$$

How much time might it take to solve such problems? Do you believe that this class is contained in P*?*

```
Integer Partitioning Witness Checker
input: a list x_1,...,x_ℓ of integers and a witness w_1,...,w_ℓ
begin
    r = 0 ;
    for i = 1 to ℓ do
        if w_i = +1 then r := r + x_i ;
        if w_i = −1 then r := r + x_i ;
    if r = 0 then return "yes" ;
end
```

FIGURE 8.3: A witness-checking program for INTEGER PARTITIONING.

8.2 Show Me The Way

> And see ye not yon bonny road
> That winds about the fernie brae?
> That is the Road to fair Elfland,
> Where thou and I this night maun gae.
>
> Anonymous, *Thomas the Rhymer*

Imagine watching a clever friend solve a mechanical puzzle. She has been playing with it for a long time, sliding the pieces back and forth. Finally, she reaches the final state and holds it up in triumph. You can't remember the entire sequence of moves, but she has proved to you that a solution exists—that there is a path, through the space of possible states, from the initial state to the final one.

This is the essence of nondeterministic computation when memory is limited. As we did for problems in NP, let's imagine that a Prover claims that some input is a yes-instance. She can prove this claim to you, the Verifier, by providing a witness that describes a path through your state space. You check this witness by following this path, reaching a state where you agree with her proof and return "yes."

The only problem with this kind of witness is that it's far too large for you to hold in your memory. After all, if you have m bits of memory then the size of your state space, and the possible length of the path, is 2^m. So, rather than giving you the witness all at once, the Prover takes you by the hand, and shows it to you one step at a time, telling you which move to make next in the state space.

To make this more concrete, consider the problem INTEGER PARTITIONING from Chapter 4. Given a list of integers $\{x_1,\ldots,x_\ell\}$, we want to know if we can separate them into two subsets with the same total. Equivalently, we want to know whether there is a sequence of bits, or signs $w_1,\ldots,w_\ell \in \{+1,-1\}$, such that $\sum_{i=1}^{\ell} w_i x_i = 0$. To prove that the answer is "yes," the Prover gives you a witness consisting of ℓ bits, telling you for each i whether w_i is $+1$ or -1. You respond by adding or subtracting x_i from a running total r, and you return "yes" if you end with $r = 0$. We show this witness-checking program in Figure 8.3.

Figure 8.4 shows a small portion of the state space of this program. At each step, all you need to hold in your own memory are the current values of i and r, so there is one state for each combination (i,r). These states form a directed graph in which each edge corresponds to a single step of computation. Each state has two edges leading out of it, corresponding to the different actions you can take based on the

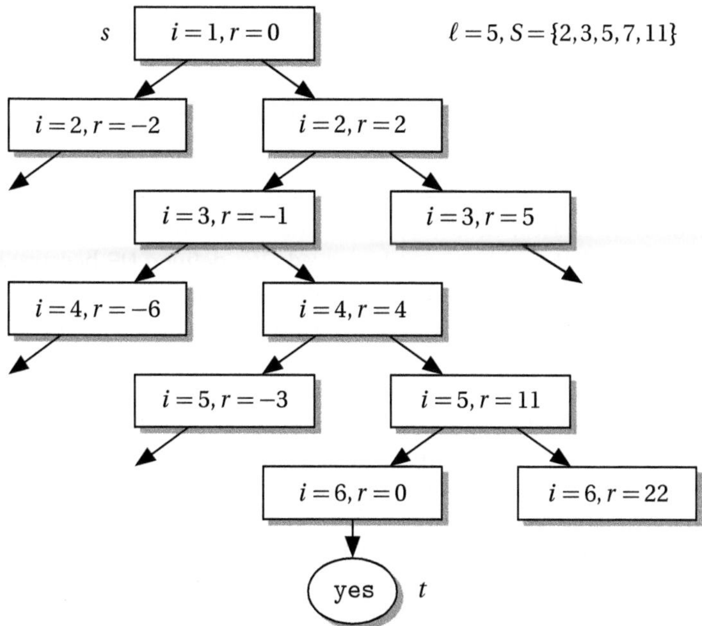

FIGURE 8.4: A tiny piece of the state space of the witness-checking program for INTEGER PARTITIONING, with the input $S=\{2,3,5,7,11\}$. There is one state for each combination of i and r. At each step the witness tells you whether to change r by $+x_i$ or $-x_i$, until $r=0$ and you return "yes." The path shown corresponds to the witness $+1,-1,+1,+1,-1$, putting $\{2,5,7\}$ on one side of the partition and $\{3,11\}$ on the other. Note that the state space depends on the instance.

next bit of the witness. If the Prover can guide you from the initial state $s=(1,0)$ to the final "yes" state $t=(\ell,0)$, you agree that there is a solution.

But as with the mechanical puzzle, there is no need for you to hold the entire witness in your mind at once. The Prover can give it to you one bit at a time, letting you read it from left to right, and thus telling you at each step whether to add or subtract x_i. Inspired by this example, we give the following definition of NSPACE complexity classes.

> NSPACE($f(n)$) is the class of problems with the following property. There is a program Π which is given read-only access to the input x and the witness w, and which uses $O(f(n))$ memory, such that x is a yes-instance if and only if there exists a w such that Π returns "yes." Moreover, Π is only allowed to read w from left to right.

In particular, we can define the nondeterministic versions of L and PSPACE:

$$\mathsf{NL} = \mathsf{NSPACE}(\log n)$$
$$\mathsf{NPSPACE} = \mathsf{NSPACE}(\mathrm{poly}(n)) = \bigcup_{k>0} \mathsf{NSPACE}(n^k)$$

What's the point of these classes? As always, we define complexity classes in order to express fundamental ideas about the way that problems can be solved. We defined NP because for many problems, solutions are easy to check, even if they are hard to find. In the same way, classes like NL and NPSPACE capture the fact that for some problems, their solutions are very long, but we can check them step-by-step with a small amount of memory.

At this point, the reader might ask why we don't simply define NSPACE($f(n)$) as the set of problems with witnesses that can be checked with $O(f(n))$ memory, just as NTIME($f(n)$) is the set of problems for which witnesses can be checked in $O(f(n))$ time. Good question! Why not just give the Verifier read-only access to the witness and the input, and limit its working memory?

The reason is that such Verifiers are a little too powerful. For instance, suppose that I claim that a 3-SAT formula is satisfiable, and I give you a satisfying assignment as a witness. Given read-only access to both the formula and the assignment, it is easy for you to check that each clause is satisfied. The only things you need to store in memory are a handful of counters that keep track of which clause you are checking, which variable you want to look up next, and so on. Since the values of these counters are poly(n), you can store them with $O(\log n)$ bits. So, if we used this definition, NL would include NP-complete problems.

This is all well and good, and we can define things however we like. But our goal in defining NL and other NSPACE classes is to give the Verifier just enough access to the witness to check a proposed path through state space. We do this with the restriction that it can only read the witness from left to right. This keeps it from going back and looking at previous bits of the witness, while leaving it enough power to check a path one step at a time.

As another definition of NSPACE, we can pretend that the Verifier guesses the path as it goes along, rather than having the Prover show it the way. In other words, we can consider nondeterministic programs like those in Section 4.3.3, which have a magical **goto both** command that splits the computation into two possible paths. This gives yet another definition, which is actually the more traditional one:

NSPACE($f(n)$) is the class of problems for which a nondeterministic program exists that uses $O(f(n))$ memory, such that the input is a yes-instance if and only if there exists a computation path that returns "yes."

All this talk of paths through state space should remind you of a problem we saw long ago—our old friend REACHABILITY from Section 3.4:

REACHABILITY
Input: A directed graph G and two vertices s, t
Question: Is there a path from s to t?

Now we claim that

FIGURE 8.5: A sliding-block puzzle known as Hua Rong Dao, Hakoiri Musume (Daughter in the Box), L'Âne Rouge, Khun Chang Khun Phaen, and Klotski. Can you slide the large block at the top out through the slot at the bottom? The shortest solution has 81 moves. Telling whether an $n \times n$ generalization of this puzzle can be solved is a REACHABILITY problem on the exponentially large graph of all possible states. Since each state can be described with a picture consisting of poly(n) bits, it is in NPSPACE.

REACHABILITY is in NL.

The witness is simply the path from s to t. Given read-only access to G, say in the form of its adjacency matrix, you can follow the path and check that each step is along a valid edge. If G has n vertices, the path might be n steps long, but you only need $O(\log n)$ bits of memory to keep track of the current vertex.

We can also consider REACHABILITY problems on graphs that are exponentially large, but where each state, or vertex, can be described with a polynomial number of bits. For instance, imagine that your friend is solving a puzzle made of blocks that slide back and forth in an $n \times n$ box, like the one in Figure 8.5. There could easily be an exponential number of possible states, and the solution could be exponentially long. But you only need poly(n) bits to describe the current state, and it is easy to check whether each move is allowed. The problem of telling whether such a puzzle has a solution is thus in NPSPACE, and is equivalent to a REACHABILITY problem on the exponentially large graph of puzzle states.

More generally, any problem in an NSPACE class is a REACHABILITY problem in the state space of some program Π. (Note that as in Figure 8.4, the graph G depends on the input x.) So we can say that

> Any problem in NSPACE($f(n)$) is equivalent to a REACHABILITY problem on a graph G of size $N = 2^{O(f(n))}$.

This also gives us an upper bound on NSPACE in terms of time complexity. Since REACHABILITY is in P, we can solve NSPACE($f(n)$) problems in an amount of time that is polynomial in $N = 2^{O(f(n))}$, the size of the state space. Therefore, we have

$$\mathsf{NSPACE}(f(n)) \subseteq \mathsf{TIME}(\mathrm{poly}(N)) \subseteq \mathsf{TIME}(2^{O(f(n))}), \tag{8.5}$$

which strengthens (8.1) to the nondeterministic case. In particular, since the state space of a problem in NL has size $N = \mathrm{poly}(n)$, we have

$$\mathsf{NL} \subseteq \mathsf{P}.$$

In a traditional REACHABILITY problem, we are given G's adjacency matrix as part of the input. If G is an exponentially large state space, writing down this kind of explicit map is impossible—but it is also unnecessary, since G has a very succinct description. We can describe G *implicitly*, in the form of the puzzle and its rules, or the program Π and its input x. Each time an algorithm for REACHABILITY wants to know whether G has an edge between two states u and v, we can find out by simulating Π, and telling whether we can get from u to v in a single step of computation. Thus the algorithm can act just as if it had read-only access to G's adjacency matrix.

We will use this relationship between space-bounded computation and REACHABILITY repeatedly in the rest of this chapter. By looking at REACHABILITY from several points of view—solving it with different kinds of algorithms, or expressing its logical structure in a compact way—we will gain a deep understanding of how nondeterminism works in the space-bounded setting. We will start by proving that REACHABILITY is to NL as SAT is to NP—in other words, that it is NL-complete.

Exercise 8.2 *By reviewing the witness-checking program given above, invent a restricted version of* INTEGER PARTITIONING *that is in* NL.

8.3 L and NL-Completeness

Now that we have defined L and NL, we can ask a question analogous to P vs. NP. Can we always find our own way through state space, or are there problems where we need a Prover to show us the path? Just as we used NP-completeness to illuminate the P vs. NP question, we can define a notion of NL-completeness, so that L = NL if and only if some NL-complete problem is in L.

We have already seen that REACHABILITY is in NL, and conversely, that any problem in NL is equivalent to a REACHABILITY problem on a graph of polynomial size. In this section, we will tighten what we mean by "equivalent" here, and show that REACHABILITY is indeed NL-complete. Thus the L vs. NL question boils down to whether or not REACHABILITY can be solved deterministically with only $O(\log n)$ memory. While this question is not quite as earth-shaking as P vs. NP, it remains unsolved.

8.3.1 One Bit at a Time: Log-Space Reductions

Recall our definition of NP-completeness:

> A problem B in NP is NP-*complete* if, for any problem A in NP, there is a polynomial-time reduction from A to B.

We use polynomial-time reductions for two reasons. First, this definition allows us to say that $\mathsf{P} = \mathsf{NP}$ if and only if some NP-complete problem is in P. Secondly, unless $\mathsf{P} = \mathsf{NP}$, it makes the process of transforming A to B relatively easy compared to solving either one. This gives our reductions enough power to translate questions, but not enough to answer them.

Analogously, let's define NL-completeness as follows.

> A problem B in NL is NL-*complete* if, for any problem A in NL, there is a log-space reduction from A to B.

A log-space reduction is a function f that takes an instance x of A and returns an instance $f(x)$ of B, such that $f(x)$ is a yes-instance if and only if x is. Moreover, we can compute $f(x)$ using $O(\log n)$ memory and read-only access to x. If such a reduction exists from A to B, we write

$$A \le_{\mathsf{L}} B\,.$$

But where do we put $f(x)$ once we've computed it? If $f(x)$ is roughly as large as x—say, if they are both n bits long—we can't hold it in our memory. This is confusing, since the whole idea of a reduction is to solve A by transforming it to B, and then solving the resulting instance using our favorite algorithm for B. How can we feed $f(x)$ to the algorithm for B, if we can't even write it down?

The answer is that we don't have to, or at least not all at once. The algorithm for B is supposed to have read-only access to $f(x)$, but if it is broken down into sufficiently simple steps, it only accesses one bit of $f(x)$ at a time. So, we will generate each bit of $f(x)$ on demand, whenever the algorithm for B asks for it:

> A function $f(x)$ is in L if there is a program Π that uses $O(\log n)$ memory which, when given read-only access to x and i, produces the ith bit of $f(x)$. We also require that $|f(x)| = \mathrm{poly}(n)$, so i can be encoded in $O(\log n)$ bits.

We can think of a log-space function as a device that simulates read-only access to $f(x)$, but doesn't actually store $f(x)$ anywhere. The algorithm for B doesn't mind, since each bit of $f(x)$ is there when it needs it. From its point of view, it might as well be looking at $f(x)$'s bits directly.

We can chain several such devices together, where each one simulates read-only access to the input of the next one. In other words, if $f(x)$ and $g(y)$ are log-space functions then their composition $g(f(x))$ is as well. Let Π_f and Π_g denote the programs that compute them. Each time Π_g asks for a bit of $y = f(x)$, we call Π_f as a subroutine to compute it using read-only access to x. Since we can reuse Π_f's memory each time we call it, and since $|f(x)| = \mathrm{poly}(n)$, our total memory use is $O(\log|f(x)|) + O(\log n) = O(\log n)$.

This means that log-space reducibility, like polynomial-time reducibility, is transitive:

$$\text{if } A \le_{\mathsf{L}} B \text{ and } B \le_{\mathsf{L}} C, \text{ then } A \le_{\mathsf{L}} C.$$

In particular, if B is NL-complete and $B \le_{\mathsf{L}} C$ for some problem C in NL, then C is NL-complete as well. Just as for NP-complete problems, we can show that a problem is NL-complete by reducing to it from any problem we already know to be NL-complete. We don't have to start from scratch.

Note that we typically recalculate each bit of $f(x)$ many times, once for each time it is requested. As a result, the algorithms resulting from log-space reductions can be quite slow. We could avoid this slowdown by keeping a table of the bits of $f(x)$ we have already calculated, as in the dynamic programming algorithms of Section 3.3, but this would take too much memory. This tradeoff between memory use and running time is a common phenomenon—we can save time by using more memory, or vice versa.

How powerful are log-space reductions? In fact, they include many of the reductions we saw in Chapter 5. For instance, consider the reduction from NAE-3-SAT to GRAPH 3-COLORING in Section 5.3.2. It converts a formula ϕ to a graph G, where each clause becomes a gadget consisting of a few vertices. Now suppose you have read-only access to ϕ, and I ask you for a single bit of G's adjacency matrix—that is, I ask you whether a given pair of vertices is connected. If we give each vertex a name $O(\log n)$ bits long, such as "vertex 3 of the gadget corresponding to clause 226," you can easily answer my question by looking at the corresponding clauses. Thus this kind of "local gadget replacement" can be carried out in L, and in even lower complexity classes.

8.2

8.3.2 REACHABILITY *is* NL*-Complete*

Our first NP-complete problem, WITNESS EXISTENCE, essentially repeated the definition of NP by asking if there is a witness that a given witness-checking program will accept. An analogous problem for NL is:

> NL WITNESS EXISTENCE
>
> Input: A program Π, an input x, and an integer k given in unary
>
> Question: Is there a witness w such that $\Pi(x, w)$ returns "yes," given access to $\log_2 k$ bits of memory, read-only access to x, and read-only, left-to-right access to w?

We can check a given witness w for this problem by running Π and seeing what it does. Moreover, since k is given in unary, Π uses $\log_2 k \leq \log_2 n$ bits of memory where n is the total size of the input. Thus we can run Π in $O(\log n)$ memory, and NL WITNESS EXISTENCE is in NL. Conversely, any problem in NL can, by definition, be phrased as an instance of NL WITNESS EXISTENCE.

We wish to show that NL WITNESS EXISTENCE is log-space reducible to REACHABILITY. As discussed in the previous two sections, there is a witness w for a given Π, x, and k if and only if there is a path from the initial state to the "yes" state in the directed graph G describing Π's state space. Since Π has access to $\log_2 k$ bits of memory, G has $k \leq n$ vertices.

The definition of log-space reduction demands that we be able to generate each bit of G on demand. If we represent G as its adjacency matrix A, this means calculating A_{ij} for a given i and j—in other words, telling whether there is an edge from i to j. We can answer this question simply by simulating Π and seeing whether a single step of computation would take it from state i to state j, or whether such a step is possible if i is a state in which Π consults the witness. The amount of memory we need to do this simulation is just the amount that Π needs to run, which is $O(\log n)$. Therefore

$$\text{NL WITNESS EXISTENCE} \leq_{\mathsf{L}} \text{REACHABILITY},$$

and REACHABILITY is NL-complete.

So, is REACHABILITY in L? Equivalently, is L = NL? Here's another way to put this: how much memory do you need to find your way through a maze with n locations? In a planar maze with no loops, like the one shown in Figure 8.6, all you need to keep track of is your current location, since you can follow the

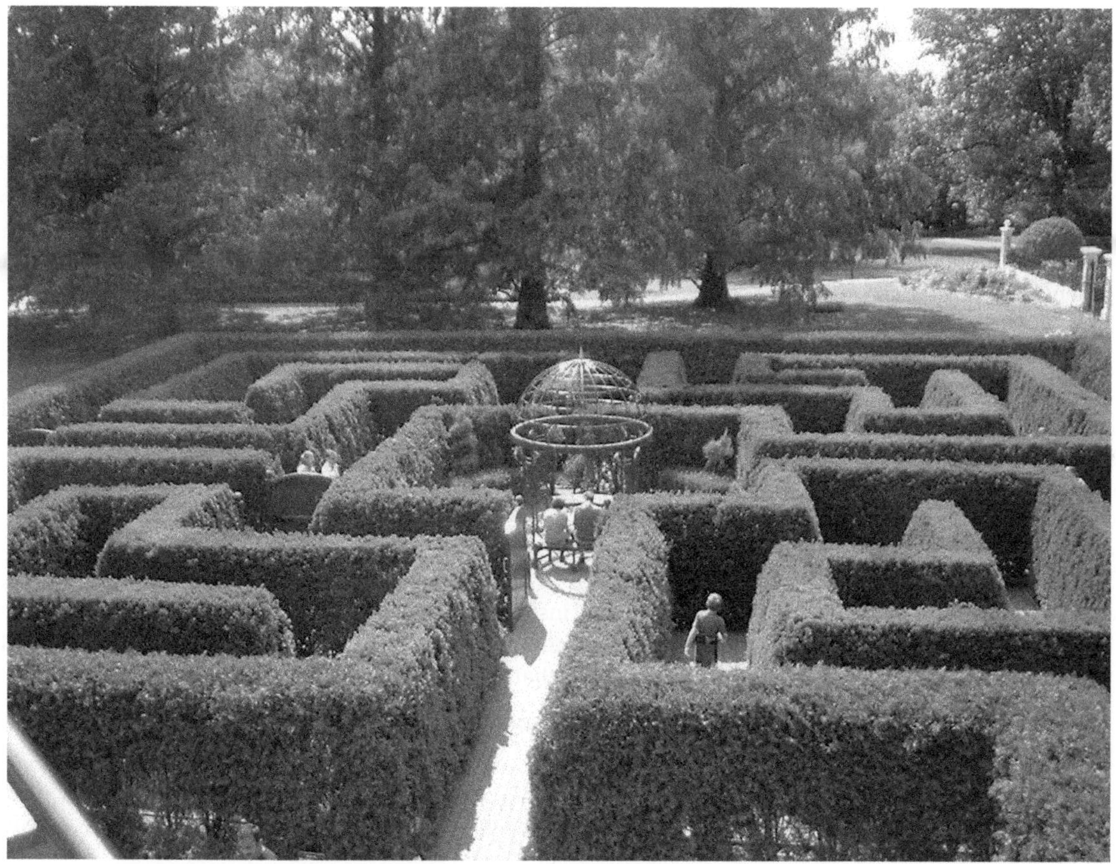

FIGURE 8.6: This REACHABILITY problem is in L; just keep your hand on the right-hand wall. But what if the maze were not planar, or if it had loops?

classic advice of keeping your hand on the right-hand wall. At the end of this chapter, we will also discuss the recent discovery that REACHABILITY for *undirected* graphs is in L.

In general, however, it seems that to find your way through a directed graph you either need a guide, or an amount of memory significantly larger than the $O(\log n)$ it takes to remember your current location. In the next section, we will derive an upper bound on how much memory you need to do it on your own.

Exercise 8.3 *Let* ONE-OUT REACHABILITY *be the variant of* REACHABILITY *on a directed graph G where each vertex has at most one outgoing edge. Show that* ONE-OUT REACHABILITY *is in* L*. Do you think that it is* L*-complete under some reasonable kind of reduction?*

8.4 Middle-First Search and Nondeterministic Space

> In the middle of the journey of our life, I came to myself
> within a dark wood, where the straight way was lost.
>
> Dante Alighieri, *The Divine Comedy*

What is the gap between deterministic and nondeterministic computation? How hard is it to find solutions, as a function of the time it takes to check them? Where time is concerned, we believe that we must essentially perform an exhaustive search, so that the inclusion

$$\mathsf{NTIME}(f(n)) \subseteq \mathsf{TIME}(2^{O(f(n))})$$

is the best we can do. Where memory is concerned, however, the gap between nondeterministic and deterministic computation is much smaller. In this section, we will prove the following:

Theorem 8.1 (Savitch's Theorem) *For any* $f(n) \geq \log n$*,*

$$\mathsf{NSPACE}(f(n)) \subseteq \mathsf{SPACE}(f(n)^2).$$

In other words, solving problems deterministically instead of nondeterministically only requires us to expand our memory quadratically, instead of exponentially. In particular, since the square of a polynomial is just a polynomial with a larger degree, this means that

$$\mathsf{PSPACE} = \mathsf{NPSPACE},$$

in contrast with our belief that $\mathsf{P} \neq \mathsf{NP}$.

The proof works by constructing a very memory-efficient algorithm for REACHABILITY. Namely, we will show that

$$\text{REACHABILITY on a graph with } n \text{ vertices is in } \mathsf{SPACE}(\log^2 n).$$

For standard REACHABILITY problems where G's adjacency matrix is given explicitly as part of the input, REACHABILITY is in $\mathsf{SPACE}(\log^2 n)$. Since REACHABILITY is NL-complete, this implies Savitch's theorem in the special case where $f(n) = \log n$,

$$\mathsf{NL} \subseteq \mathsf{SPACE}(\log^2 n).$$

But our algorithm also works for exponentially large graphs that are described implicitly by a program. By simulating read-only access to G's adjacency matrix, we can scale our algorithm up to state spaces of size $N = 2^{O(f(n))}$, solving their REACHABILITY problems in $O(f(n)^2)$ space.

How can we solve REACHABILITY with only $O(\log^2 n)$ memory? This is not so obvious. If we explore the graph using depth-first or breadth-first search as in Section 3.4.1, we need $\Theta(n)$ bits of memory—one bit per vertex—to keep track of where we have already been.

Instead, we will use a strategy that we first saw in Section 3.4.2: namely, *middle-first search.* Suppose we wish to know whether we can get from i to j in ℓ or fewer steps. As Figure 8.7 shows, this is true if there is a midpoint k such that we can get from i to k, and from k to j, in at most $\ell/2$ steps each. We can then look for midpoints for each of these paths, recursively dividing ℓ by 2 until $\ell = 1$. At that point, we use our read-only access to G's adjacency matrix A_{ij} to tell if there is an edge from i to j.

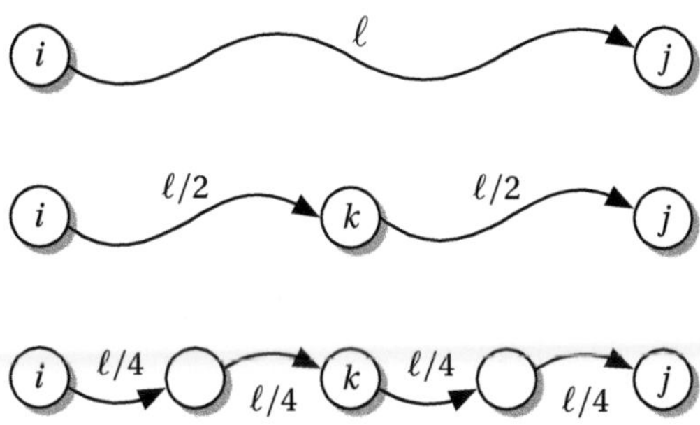

FIGURE 8.7: Middle-first search: to find a path from i to j of length ℓ, look for a midpoint k such that there are paths of length $\ell/2$ from i to k and from k to j. Recurse until the length of each path is 1 or less.

`MiddleFirst`(i,j,ℓ)
input: an adjacency matrix A, two vertices i,j, and a length ℓ
output: is there a path of length ℓ or less from i to j?
begin
  if $\ell=1$ **then**
    if $i=j$ **or** $A_{ij}=1$ **then return** "yes" ;
    else return "no" ;
  for $k=1$ **to** n **do**
    if `MiddleFirst`$(i,k,\ell/2)$ **and** `MiddleFirst`$(k,j,\ell/2)$ **then return** "yes" ;
  return "no" ;
end

FIGURE 8.8: A recursive algorithm for middle-first search. We assume for simplicity that ℓ is a power of 2.

This gives the recursive algorithm shown in Figure 8.8. Initially we set $\ell = n$, since this is the largest number of steps we could need. We assume for simplicity that n is a power of 2 so that ℓ is always even.

In Section 3.4.2, we pointed out that middle-first search is similar to squaring the adjacency matrix $\log_2 n$ times, giving a matrix that describes all the pairs of vertices that can be reached from each other. If we keep track of the matrix at each stage, remembering what reachable pairs we have discovered so far, we obtain a polynomial-time algorithm. However, this requires n^2 bits of memory, far more than the $O(\log^2 n)$ we are aiming for.

So, this time around we simply call `MiddleFirst` recursively, without keeping track of our previous results. This makes our algorithm very slow, since it will recalculate `MiddleFirst`(i,j,ℓ) for a given triple i,j,ℓ many times. In fact, Problem 8.11 shows that its running time is now superpolynomial. However, it is very efficient in its use of memory.

To analyze `MiddleFirst`'s memory use, we have to think about how much memory it takes to run a recursive algorithm. We need to keep track, not just of what "incarnation" of `MiddleFirst` we are currently running, but also of the previous incarnations that called this one, all the way back to the original one.

`MiddleFirst(4,3,1)`	Is there a path from 4 to 3 of length 1 or less?
`MiddleFirst(7,3,2)`	Is there a path from 7 to 3 of length 2 or less?
`MiddleFirst(7,2,4)`	Is there a path from 7 to 2 of length 4 or less?
`MiddleFirst(1,2,8)`	Is there a path from 1 to 2 of length 8 or less?
⋮	
`MiddleFirst`(s,t,n)	Is there a path from s to t of length n or less?

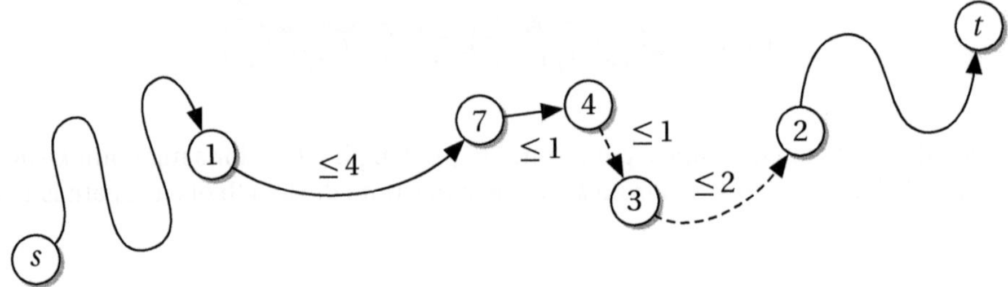

FIGURE 8.9: The current stack of recursive function calls, and the path that `MiddleFirst` is currently checking. Solid parts of the path have already been confirmed, and dashed ones are still being checked. At the moment, the algorithm is consulting G to see if there is an edge from vertex 4 to vertex 3.

As in Section 7.3.3, the structure on which we keep track of all these tasks and subtasks is called a *stack*. The task we are currently doing is on the top of the stack, its parent task is the next one down, and so on. Whenever we call the algorithm recursively, we "push" the child task onto the stack, and make it our top priority. Whenever we complete a task, we "pop" it off the top of the stack, and continue working on its parent. This continues until we have completed the entire nested sequence of subtasks, pop all the way back out to the original problem, and end with an empty stack.

For instance, consider the situation in Figure 8.9. We are currently running `MiddleFirst`$(4,3,1)$ and checking whether there is an edge from vertex 4 to vertex 3. We are doing this in order to check the second half of a path of length 2 from vertex 7 to vertex 3, as part of running `MiddleFirst`$(7,3,2)$; we are doing this, in turn, in order to check the first half of a path of length 4 from vertex 7 to vertex 2, as part of running `MiddleFirst`$(7,2,4)$; and so on, all the way back out to `MiddleFirst`(s,t,n).

How much memory does the entire stack take up? Since we divide ℓ by 2 at each stage of recursion, it takes $\log_2 n$ stages to go from $\ell = n$ to the base case $\ell = 1$. Therefore, the depth of the stack is $O(\log n)$. At each level of the stack, in order to record that we are currently running `MiddleFirst`(i,j,ℓ), we have to record i, j, and ℓ, and these take $O(\log n)$ digits each. The total size of the stack is thus $O(\log^2 n)$ bits, so REACHABILITY is in SPACE$(\log^2 n)$.

Now let's scale this result up to the state space G of a witness-checking program Π that uses $O(f(n))$ bits of memory. The number of states is $N = 2^{O(f(n))}$, so we can solve the resulting case of REACHABILITY deterministically with $O(\log^2 N) = O(f(n)^2)$ bits of memory—assuming we have read-only access to G.

Of course, we don't have enough memory to write down G's adjacency matrix explicitly. But, as in Section 8.3.1, we don't have to. `MiddleFirst` only needs one bit of G's adjacency matrix at a time, when it reaches its base case $\ell = 1$. We can generate this bit whenever `MiddleFirst` requests it, by simulating Π on state i and seeing if the next step could lead to state j. It takes just $O(f(n))$ bits to simulate

Π, so the total memory use is still $O(f(n)^2)$. Thus $\mathsf{NSPACE}(f(n)) \subseteq \mathsf{SPACE}(f(n)^2)$, and our proof of Savitch's Theorem is complete.

An interesting question is whether Savitch's Theorem is optimal. Is there an $\alpha < 2$ such that 8.3

$$\mathsf{NSPACE}(f(n)) \subseteq \mathsf{SPACE}(f(n)^\alpha),$$

or is $\alpha = 2$ the best we can do? To put this differently, do we really need $\Theta(\log^2 n)$ memory to solve REACHABILITY deterministically? So far, no one has been able to come up with a better approach, but space-bounded complexity is full of surprises.

8.5 You Can't Get There From Here

As we discussed in Chapter 4, nondeterministic computation is a fundamentally asymmetric notion. I can prove to you that a graph has a Hamiltonian path by showing you one, but there seems to be no way to prove that no such path exists without doing some kind of search.

Recall that coNP is the class of problems where there is a simple proof if the answer is "no," just as NP is the class of problems where there is a simple proof if the answer is "yes." Unless some NP-complete problem has easily checkable proofs in both cases, these classes are fundamentally different. As Exercise 6.1 showed, if coNP = NP then the polynomial hierarchy collapses to its first level. Thus our belief that coNP ≠ NP is almost as strong as our belief that P ≠ NP.

In this section, we show once again that nondeterminism is radically different in the space-bounded case. There is a kind of proof that something doesn't exist that can be checked with about the same amount of memory as a proof that it does. In terms of complexity classes,

Theorem 8.2 (The Immerman–Szelepcsényi Theorem) *For any $f(n) \geq \log n$,*

$$\mathsf{coNSPACE}(f(n)) = \mathsf{NSPACE}(f(n)).$$

In particular,

$$\mathsf{coNL} = \mathsf{NL}.$$

The proof works by showing that NON-REACHABILITY is in NL, or equivalently that REACHABILITY is in coNL. Since REACHABILITY and NON-REACHABILITY are NL-complete and coNL-complete respectively, this also means that

NON-REACHABILITY is NL-complete.

Thus any problem in NL can be reduced to one in coNL and vice versa.

Ponder this for a moment. We defined NL and $\mathsf{NSPACE}(f(n))$ as the set of problems for which the answer is "yes" if there exists a path through state space. The Immerman–Szelepcsényi Theorem tells us that we can define these classes just as well in terms of the *nonexistence* of these paths.

To show that NON-REACHABILITY is in NL, we use the witness-checking definition of NL given in Section 8.2. How can I prove to you, the Verifier, that there is no path from s to t? The idea is to establish, through a clever induction, the number of vertices reachable from s in 0 steps, 1 step, 2 steps, and so on. Once we agree on the total number r of vertices that are reachable in n or fewer steps, I show you paths to r different vertices. If none of these are t, we're done.

To see how this works, let R_ℓ denote the set of vertices that can be reached from s in ℓ or fewer steps, and let $r_\ell = |R_\ell|$ denote the number of such vertices. I can prove to you that a given vertex v is in R_ℓ by showing you a path from s to v of length ℓ or less. You check each step of this path using read-only access to G's adjacency matrix, while maintaining a counter to make sure that the number of steps is at most ℓ.

But if you and I agree on the value of r_ℓ, I can also prove to you that v is *not* in R_ℓ, by showing you r_ℓ other vertices u which *are* in R_ℓ. To keep me from cheating and showing you the same $u \in R_\ell$ more than once, you force me to show you the elements of R_ℓ in increasing order. You keep track of the last u that I showed you, so I can't go back and show you a u that I showed you before.

It's important to realize that you don't have to remember which $u \in R_\ell$ I have shown you, or how to get to any of them. You just need to count how many such u I have shown you so far. When your counter gets to r_ℓ, you know that I have shown you all of R_ℓ, and therefore that v is not reachable in ℓ steps.

Now, assuming that we already agree on r_ℓ, we can establish $r_{\ell+1}$ as follows. I go through the vertices u in order from 1 to n. For each u, I either claim that $u \in R_{\ell+1}$ or that $u \notin R_{\ell+1}$—either that u can be reached in $\ell+1$ or fewer steps, or that it can't. If I claim that $u \in R_{\ell+1}$, I show you a path from s to u of length $\ell+1$ or less, and you increment $r_{\ell+1}$. On the other hand, if I claim that $u \notin R_{\ell+1}$, I have to prove that none of its incoming neighbors are in R_ℓ. So as before I show you, in increasing order, r_ℓ vertices w in R_ℓ, giving you a path from s to each w, and pointing out that none of these w are neighbors of u.

We start with the base case $r_0 = 1$, since the only vertex reachable in 0 steps is s itself. We then inductively establish the values of r_1, r_2, and so on. Once we agree on $r_n = r$, I again go through the vertices in increasing order, showing you paths to r reachable vertices other than t. The cleverness of this inductive scheme is that I, the Prover, cannot cheat. There is no way for me to hide from you the fact that some vertex is reachable.

Just as log-space algorithms save on memory by recalculating the same bit many times, I remind you over and over again which vertices are reachable, and how to reach them. This redundancy allows you to check the witness with a small amount of memory. All you need are a handful of counters to keep track of the vertices we are discussing, the length of the path I have shown you so far, the number of vertices in R_ℓ that I have shown you so far, and so on.

In total, the things you need to remember during the stage of the witness-checking process when we are establishing the value of $r_{\ell+1}$ are the following:

- ℓ and r_ℓ,
- the number of vertices $u \in R_{\ell+1}$ that I have shown you,
- the current vertex u that I claim is in $R_{\ell+1}$ or not,
- during a proof that $u \in R_{\ell+1}$:
  - the length of the path from s to u that I have shown you so far, and its current location.
- during a proof that $u \notin R_{\ell+1}$:
  - the vertex w that I am currently proving is in R_ℓ,
  - the number of $w \in R_\ell$ that I have shown you,
  - the length of the path from s to w that I have shown you so far, and its current location.

Each of these is an integer between 0 and n, so the total amount of memory you need is just $O(\log n)$. This completes the proof that NON-REACHABILITY is in NL, and therefore that NL = coNL.

To prove Theorem 8.2 in general, we apply the same argument to NON-REACHABILITY on the state space of a witness-checking program Π that uses $O(f(n))$ memory. As in the proof of Savitch's Theorem, you can check each individual claim that there is or isn't an edge from one state to another by simulating one computation step of Π. The integers ℓ, r_ℓ, u and so on now range from 0 to $N = 2^{O(f(n))}$, so they can be stored in $O(\log N) = O(f(n))$ memory. Thus a NON-REACHABILITY problem of this kind is in NSPACE($f(n)$), and coNSPACE($f(n)$) = NSPACE($f(n)$).

Now that we have explored determinism and nondeterminism, and existence and nonexistence, in a 8.4
world where memory is limited but time is not, let's look at problems with deeper logical structures—at games between a Prover who thinks that something is true, and a Skeptic who thinks that it is false.

8.6 PSPACE, Games, and Quantified SAT

Throughout this book, we have argued that computational complexity theory is the study of the fundamental nature of problems. For the most part, we have addressed this question from an *algorithmic* point of view, asking how much time or memory it takes to solve a given problem. But we can also classify problems according to their *logical* structure—what kinds of claims they make, and what type of logical language is needed to express these claims.

These two points of view are deeply related. Many complexity classes are best viewed as the set of problems that can be expressed in a particular logical form. We saw in Section 4.3 that NP consists of problems with a single "there exists" in front of a polynomial-time property. If A is in NP, the claim that x is a yes-instance is of the form

$$A(x) = \exists w : B(x, w),$$

where $B(x, w)$ is the property that w is a valid witness for x. To prove a claim of this form to you, all I need to do is provide you with w, whereupon you can verify in polynomial time that B is true.

The quantifier $\exists$ corresponds exactly to the process of searching for a solution, or the choices a nondeterministic machine can make, or the conversation between a Prover and a Verifier. By adding more quantifiers, we rise up through the levels of the polynomial hierarchy, which we discussed in Section 6.1.1.

Is there a similar characterization of space-bounded complexity classes? Indeed there is, and it turns out to be very beautiful. In this section and the next, we will explore the class PSPACE from a computational and logical point of view. We will see that just as NP problems claim the existence of a single object of polynomial size, PSPACE problems can make claims with a much richer logical structure—such as the claim that the first player has a winning strategy in a game. We start by looking once again at the canonical problem associated with space-bounded computation, REACHABILITY.

8.6.1 *Logic and the* REACHABILITY *Game*

At first glance, the claim that the answer to a REACHABILITY problem is "yes" has just one "there exists" in it: namely, there exists a path from s to t. But what if the path in question is extremely long? For instance, suppose we start with a problem in PSPACE, and convert it to a REACHABILITY problem on its state space. The corresponding graph has size $N = 2^{\text{poly}(n)}$, and the length of the path—that is, the running time of the corresponding program—could be exponentially large.

The claim that such a path exists is fine as far as logic is concerned, but it seems like a rather large object to postulate all at once. We want quantifiers like $\exists$ to tell an algorithmic story, in which I present an object to you, and you check that it works. If the object I claim exists is exponentially large, there is no way for me to show it to you, or for you to verify it, in a reasonable amount of time.

To give our logical statements an algorithmic meaning, let's adopt a language in which each quantifier only refers to an object of polynomial size. In particular, let's restrict ourselves to *first-order* quantifiers, where each one refers to a single vertex of the graph. Our first attempt to express REACHABILITY with such quantifiers might look like this:

$$\exists i_1 : \exists i_2 : \cdots : \exists i_{N-1} : \text{edge}(s, i_1) \wedge \text{edge}(i_1, i_2) \wedge \cdots \wedge \text{edge}(i_{N-1}, t), \tag{8.6}$$

where edge(i, j) denotes the claim that either $i = j$ or there is an edge from i to j. Each individual quantifier now postulates just a single vertex. But if N is exponentially large then so is the number of variables we are postulating, and the set of properties that they are supposed to satisfy. All we have done is write out "there exists a path" one vertex at a time. We haven't avoided the path's exponential size.

Can we express REACHABILITY with a small number of first-order quantifiers? The answer is yes—but only if we are willing to accept a deeper logical structure. Once again, we use the idea of middle-first search. Let $R(i, j, \ell)$ denote the claim that we can get from i to j in ℓ or fewer steps. Then $R(i, j, \ell)$ holds only if there exists a midpoint k such that we can get from i to k, and from k to j, in $\ell/2$ steps each. We can write this as

$$R(i, j, \ell) = \exists k : R(i, k, \ell/2) \wedge R(k, j, \ell/2). \tag{8.7}$$

In a graph with N vertices, the claim that we can get from s to t is $R(s, t, N)$. By applying (8.7) repeatedly, we can break this claim down as follows:

$$\begin{aligned} R(s, t, N) &= \exists k : R(s, k, N/2) \wedge R(k, t, N/2) \\ &= \exists k, k', k'' : R(s, k', N/4) \wedge R(k', k, N/4) \wedge R(k, k'', N/4) \wedge R(k'', t, N/4) \\ &= \cdots \end{aligned}$$

However, each time we iterate (8.7), replacing $R(i, j, \ell)$ with $R(i, k, \ell/2) \wedge R(k, j, \ell/2)$, the formula doubles in size. Taking $\log_2 N$ steps down this road gives us N quantifiers, and N claims of the form $R(i, j, 1) =$ edge(i, j). This is just (8.6) again, in which we postulate each vertex in the chain. How can we avoid this exponential blowup?

Just as $\exists$ is a compact way to write a long string of ORs, we can use a "for all" quantifier $\forall$ to write an AND more compactly. If $S(z)$ is a statement about a Boolean variable z, we can abbreviate

$$S(0) \wedge S(1)$$

as

$$\forall z \in \{0, 1\} : S(z).$$

If S is a complicated property, this saves a lot of ink. Let's use a similar trick to combine the two claims on the right-hand side of (8.7), and write

$$R(i, j, \ell) = \exists k : \forall (i', j') \in \{(i, k), (k, j)\} : R(i', j', \ell/2). \tag{8.8}$$

In other words, there is a vertex k such that, for any pair of vertices (i', j') which is either (i, k) or (k, j), we can get from i' to j' in $\ell/2$ steps.

This may seem needlessly technical, but something very important is going on. Just as the middle-first algorithm reuses memory, the expression (8.8) reuses the symbols $R(i', j', \ell/2)$ for both halves of the path, rather than devoting separate symbols to them as (8.7) does. Now, if we iterate (8.8), the claim that we can get from s to t becomes

$$\begin{aligned}
R(s,t,N) &= \exists k_1 : \forall (i_1, j_1) \in \{(s, k_1), (k_1, t)\} : R(i_1, j_1, N/2) \\
&= \exists k_1 : \forall (i_1, j_1) \in \{(s, k_1), (k_1, t)\} : \\
&\qquad \exists k_2 : \forall (i_2, j_2) \in \{(i_1, k_2), (k_2, j_1)\} : R(i_2, j_2, N/4) \\
&= \exists k_1 : \forall (i_1, j_1) \in \{(s, k_1), (k_1, t)\} : \\
&\qquad \exists k_2 : \forall (i_2, j_2) \in \{(i_1, k_2), (k_2, j_1)\} : \\
&\qquad\qquad \ddots \\
&\qquad\qquad\quad \exists k_m : \forall (i_m, j_m) \in \{(i_{m-1}, k_m), (k_m, j_{m-1})\} : \mathrm{edge}(i_m, j_m) .
\end{aligned}$$

This leads to a formula with $m = \log_2 N = \mathrm{poly}(n)$ layers of quantifiers, alternating between $\exists$ and $\forall$, surrounding a single claim of the form $\mathrm{edge}(i, j)$. Thus the total length of the formula, and the number of vertices we discuss, is only $\mathrm{poly}(n)$.

Now that we have expressed REACHABILITY in a compact logical form, even when the state space is exponentially large, what story does it tell? What kind of algorithmic process does this alternation between "there exists" and "for all" correspond to?

It's a game! There are two players: a Prover who claims there is a path from s to t, and a Skeptic who claims there isn't. At each point in time, the state of play is a pair of vertices which may or may not have a path between them of length ℓ or less. In each round, the Prover chooses a midpoint, and the Skeptic then chooses which half of the path to focus on for the rest of the game. Here is an example, corresponding to Figure 8.10:

PROVER: I claim we can get from s to t in N steps, through the midpoint k_1.
SKEPTIC: I don't believe you. How can you get from s to k_1 in $N/2$ steps?
PROVER: By going through k_2.
SKEPTIC: Oh yeah? How can you get from k_2 to k_1 in $N/4$ steps?
PROVER: By going through k_3.

...and so on.

This continues for $m = \mathrm{poly}(n)$ moves, at which point we arrive at the endgame where $\ell = 1$. The Prover wins if there is an edge from i_m to j_m, and the Skeptic wins if there isn't. We can tell who won by simulating a single step of the program, and seeing if it takes us from i_m to j_m. There is a path from s to t if and only if the Prover has a *winning strategy* in this game—if we end up with a valid edge in state space, no matter which part of the path the Skeptic focuses on.

Something quite marvelous has happened here. We started with a problem for which the witness, i.e., the path, is much too large for a Prover to describe, or for a Verifier to check, in a single round of communication. But by having a conversation between a Prover and a Skeptic, and allowing the Skeptic to challenge the Prover by "zooming in" on any part of the witness she desires, we can, in essence, verify

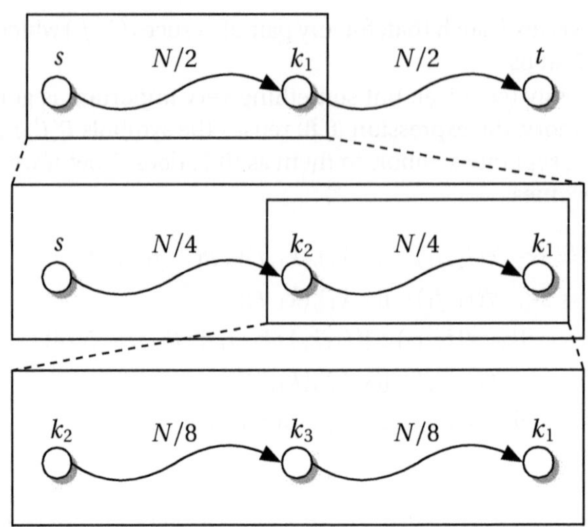

FIGURE 8.10: The REACHABILITY game. The Prover names a midpoint, and then the Skeptic chooses which half of the path to zoom in on. There is a path from s to t if and only if the Prover has a winning strategy, which always ends with a single valid edge.

an exponentially large witness with a conversation that is only polynomially long—if by "verify" we mean that the Prover can *always* win, no matter what moves the Skeptic makes.

Just as one round of conversation between the Prover and the Verifier is enough to capture NP, we have shown that games with a polynomial number of moves can capture all of PSPACE. Next we will make this connection more explicit, by showing that games involving SAT formulas, and many games that people actually play, are in fact PSPACE-complete.

8.6.2 PSPACE-*Completeness and the* SAT *Game*

The previous section showed how we can treat any PSPACE problem as a formula with poly(n) alternating quantifiers, and that this in turn can be thought of as a two-player game poly(n) moves long. In this section we prove the converse by showing that a two-player version of SAT is PSPACE-complete. In Section 8.7 we will use this problem as a jumping-off point to prove that some of our favorite board games are PSPACE-complete or even harder.

Here is our definition of PSPACE-completeness:

> A problem $B \in$ PSPACE is PSPACE-*complete* if, for any problem A in PSPACE, there is a polynomial-time reduction from A to B.

We use polynomial-time reductions so that if any PSPACE-complete problem is in P, then P = PSPACE—which would be even more astonishing than P = NP. If we want to compare PSPACE to L, say, we can use log-space reductions instead. But we will stick to this definition for simplicity.

8.2

As we did for NP and NL, we start with a problem which is PSPACE-complete by definition:

> PSPACE PROGRAM
>
> Input: A program Π, an input x, and an integer m given in unary
>
> Question: Will $\Pi(x)$ return "yes" using m bits of memory or less?

We give m in unary so that we can simulate Π in memory less than or equal to the total input size, so PSPACE PROGRAM is in PSPACE. Note that since NPSPACE = PSPACE, we are free to focus on deterministic programs which interact only with the input x, rather than witness-checking ones. Now consider the following problem:

> TWO-PLAYER SAT
>
> Input: A SAT formula $\phi(x_1,\ldots,x_n)$
>
> Question: Two players, the Prover and the Skeptic, take turns setting the x_i. The Prover sets x_1, then the Skeptic sets x_2, and so on. After all the variables are set, the Prover wins if ϕ is true, and the Skeptic wins if ϕ is false. Does the Prover have a winning strategy?

Since the Prover's and Skeptic's moves correspond to $\exists$s and $\forall$s respectively, TWO-PLAYER SAT is equivalent to the following:

> QUANTIFIED SAT
>
> Input: A SAT formula $\phi(x_1,\ldots,x_n)$
>
> Question: Let $\Phi = \exists x_1 : \forall x_2 : \exists x_3 : \cdots : \phi(x_1,\ldots,x_n)$. Is Φ true?

We will prove that these problems are PSPACE-complete by reducing PSPACE PROGRAM to QUANTIFIED SAT in polynomial time.

The reduction uses the REACHABILITY game of the previous section. Since Π runs in $m = \text{poly}(n)$ bits of memory, the size of its state space is $N = 2^m$, so there are $m = \log_2 N$ moves in the game. The "endgame" is the claim edge(i_m, j_m) that Π would go from state i_m to state j_m in a single step of computation.

Assuming as always that each of Π's instructions is reasonably simple, this claim can be evaluated in polynomial time. Specifically, each state i is a string of m bits, and we can simulate a single step of Π in poly(m) time. For that matter, if Π is written in a sufficiently stripped-down programming language, given read-only access to i and j we can tell whether edge(i, j) is true in $O(\log m)$ space, suggesting that we can do all this with a log-space reduction. However, since we use polynomial-time reductions in our definition of PSPACE-completeness, this stronger claim isn't necessary to our argument.

One technical issue is that in each move of the REACHABILITY game, the Prover chooses a state k, and the Skeptic chooses a pair of states (i, j). Since each state is a string of m bits, how can we encode such moves in TWO-PLAYER SAT, where each move sets just a single Boolean variable? The solution is easy: the player chooses these vertices, one bit at a time, over the course of m or $2m$ moves. We intersperse these with "dummy" moves, where the other player sets a variable which doesn't appear anywhere in ϕ. The total number of moves is then $O(m^2)$, but this is still poly(n).

Another issue is that the Skeptic is required to choose a pair of vertices corresponding to either the first half or the second half of the current path. Her tth move is of the form

$$\forall (i_t, j_t) \in \{(i_{t-1}, k_t), (k_t, j_{t-1})\},$$

i.e., she has to choose the pair (i_{t-1}, k_t) or the pair (k_t, j_{t-1}). To impose this constraint, we give the Prover an automatic win if the Skeptic tries to cheat—that is, if there is a t such that (i_t, j_t) is not in the set $\{(i_{t-1}, k_t), (k_t, j_{t-1})\}$. Thus the Prover wins if the following property holds,

$$W = \text{edge}(i_m, j_m) \vee \left(\bigvee_{t=1}^{m} \left((i_t \neq i_{t-1}) \vee (j_t \neq k_t)\right) \wedge \left((i_t \neq k_t) \vee (j_t \neq j_{t-1})\right) \right).$$

Clearly W is a polynomial-time property of all the bits the players have chosen. Just as in Section 5.2 when we proved that 3-SAT is NP-complete, we can convert it into a 3-SAT formula. We write it out as a Boolean circuit with a polynomial number of gates which outputs `true` if W is true, add internal variables corresponding to the wires, and convert each gate to a handful of 3-SAT clauses. This gives a 3-SAT formula ϕ, and converts W to the statement that there is some way to set the internal variables so that ϕ is true. The Prover sets these variables with poly(m) additional moves, alternating with dummy moves for the Skeptic.

Thus we can convert our instance of PSPACE PROGRAM to an instance of QUANTIFIED SAT with poly(m) = poly(n) variables. Moreover, this reduction can be carried out in poly(n) time. However, to complete the proof that TWO-PLAYER SAT and QUANTIFIED SAT are PSPACE-complete, we need to show that they are in PSPACE. We will prove this in the next section, along with a much more general fact: for any game that lasts for a polynomial number of moves, telling whether a player has a winning strategy is in PSPACE.

8.6.3 Evaluating Game Trees

> Now and then this or that move, provided in the text with an exclamation or a question mark (depending upon whether it had been beautifully or wretchedly played), would be followed by several series of moves in parentheses, since that remarkable move branched out like a river and every branch had to be traced to its conclusion before one returned to the main channel.
>
> Vladimir Nabokov, *The Defense*

In the Prologue, we talked about how in a game like Chess, the choices you and your opponent make create a branching tree of possibilities. Each line of play is a path through the tree, and each leaf is a win, loss, or a draw for you. You and your opponent take turns deciding which step to take down the tree, and you have a winning strategy if you can always guide the game to a winning leaf.

Let's call a position *winning* if you have a winning strategy starting from there. If it's your move, the current position is winning if at least one of your possible moves results in a winning position. If it's your opponent's move, the position is winning if *every* move your opponent could make leads to a winning position. These alternating conditions—that there exists a move to a winning position, or that all moves lead to one—are exactly the quantifiers $\exists$ and $\forall$ we saw in the REACHABILITY game and in QUANTIFIED SAT.

We can also think of the game tree as a Boolean circuit. It alternates between OR gates and AND gates, corresponding to $\exists$ and $\forall$ respectively. If we label each leaf `true` or `false` depending on whether you

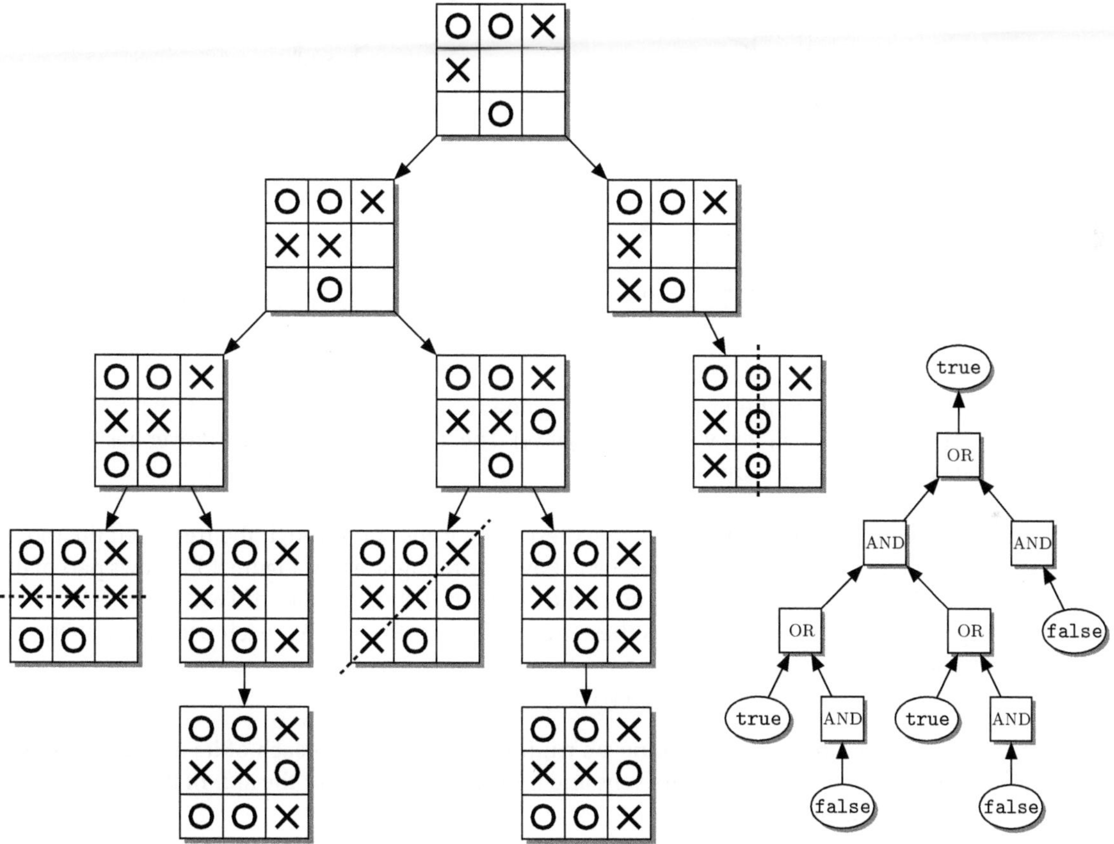

FIGURE 8.11: The game tree, and the corresponding AND-OR tree, resulting from a position in Tic-Tac-Toe. Positions are labeled `true` if they lead to a win for X, and `false` if they lead to a draw or a win for O. At the root, it is X's turn, and X has a winning strategy. Some suboptimal lines of play are not shown.

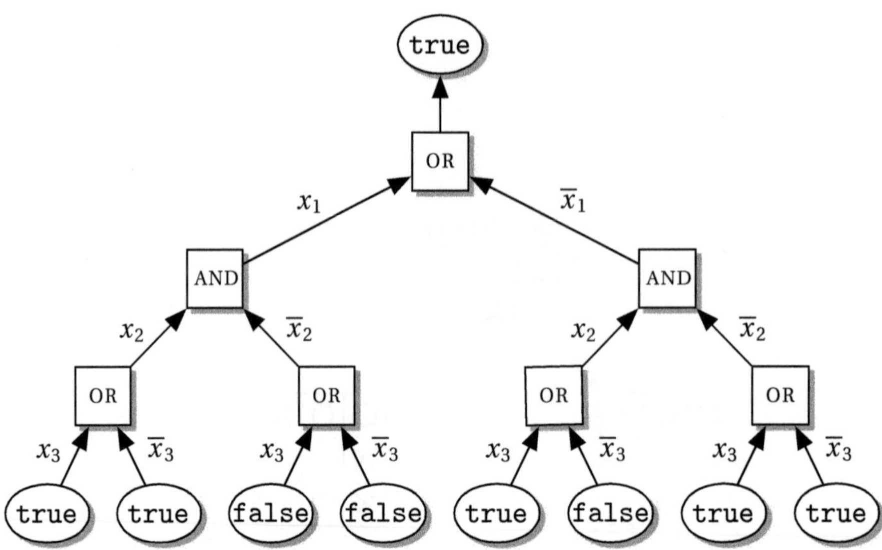

FIGURE 8.12: The game tree for the TWO-PLAYER SAT game corresponding to the QUANTIFIED SAT formula $\exists x_1 : \forall x_2 : \exists x_3 : (\overline{x}_1 \vee x_2 \vee x_3) \wedge (\overline{x}_1 \vee x_2 \vee \overline{x}_3) \wedge (x_1 \vee \overline{x}_2 \vee x_3)$. The formula is true, so the Prover has a winning strategy. What is his winning first move?

win the corresponding endgame, and propagate these truth values up through the tree, then the root is `true` or `false` depending on whether or not you have a winning strategy. Figure 8.11 shows part of the game tree for Tic-Tac-Toe, and Figure 8.12 shows the game tree for an instance of TWO-PLAYER SAT.

Let's design an algorithm that evaluates these trees. For TWO-PLAYER SAT, each position is a QUANTIFIED SAT formula

$$\Phi = Q_1 x_1 : Q_2 x_2 : \cdots : \phi(x_1, x_2, \ldots, x_n)$$

where each Q_i is either $\exists$ or $\forall$. To determine Φ's truth value, we set x_1 to each of its possible truth values, evaluate the resulting formulas recursively, and return either the OR or the AND of their truth values depending on whether Q_1 is $\exists$ or $\forall$.

Setting $x_1 = \texttt{true}$, say, gives a formula on the remaining variables, which we denote

$$\Phi[x_1 \to \texttt{true}] = Q_2 x_2 : \cdots : \phi(\texttt{true}, x_2, \ldots, x_n).$$

We continue until every variable is set and the final position of the game is decided. Then Φ is simply ϕ for some truth assignment of its variables, and the Prover wins if ϕ is true. This gives the recursive algorithm shown in Figure 8.13.

Now suppose we have a game between two players, Black and White. Given a position p, we label its children p_1, p_2, and so on, one for each of the current player's possible moves. We recursively evaluate positions p_i until we reach an endgame, giving a win, loss, or draw for Black. This gives the algorithm shown in Figure 8.14. It carries out a depth-first exploration of the game tree, pursuing each line of play to its conclusion, and then backtracking and exploring other branches.

```
Evaluate(Φ)
input: a quantified formula Φ = Q₁x₁ : Q₂x₂ : ··· : φ(x₁, x₂, ..., xₙ)
output: is Φ true, so that the Prover has a winning strategy?
begin
    if there are no quantifiers then return φ ;
    if Q₁ = ∃ then return OR(Evaluate(Φ[x₁ → true]), Evaluate(Φ[x₁ → false])) ;
    if Q₁ = ∀ then return AND(Evaluate(Φ[x₁ → true]), Evaluate(Φ[x₁ → false])) ;
end
```

FIGURE 8.13: A recursive algorithm to evaluate QUANTIFIED SAT formulas.

```
Evaluate(p)
input: a position p in a board game
output: is p winning for Black?
begin
    if p is an endgame then
        if Black has won then return true ;
        else return false ;
    end
    if it is Black's turn then
        v = false ;
        for all of Black's moves i do v = OR(v, Evaluate(p_i)) ;
        return v ;
    end
    if it is White's turn then
        v = true ;
        for all of White's moves i do v = AND(v, Evaluate(p_i)) ;
        return v ;
    end
end
```

FIGURE 8.14: A depth-first algorithm for evaluating game trees.

How much memory does this algorithm need? As we discussed in Section 8.4, to run a recursive algorithm we have to maintain a stack of tasks and subtasks. In this case, the stack encodes the line of play we are currently exploring—in other words, the chain of positions leading to the current one. The maximum depth of the stack is the depth of the tree, which in turn is the maximum length of the game.

So, if no game lasts for more than a polynomial number of moves, and if each position can be encoded with a polynomial number of bits, `Evaluate` will run in polynomial space. For Two-Player SAT, each game lasts for n moves, and each position is a setting of some of the variables, which is a string of at most n bits, so the total memory we need is $O(n^2) = \text{poly}(n)$.

Thus Two-Player SAT and Quantified SAT are in PSPACE. Since we showed in Section 8.6.2 that any problem in PSPACE can be reduced to them, they are PSPACE-complete.

8.5 Note that if we fix the number of alternating quantifiers to a constant k—as in a "mate in k moves" problem in Chess—then the resulting version of Quantified SAT is in Σ_kP, the kth level of the polynomial hierarchy (see Section 6.1.1). Thus we have also shown that PSPACE contains the entire polynomial hierarchy PH. Since we believe that PH consists of an infinite tower of complexity classes, each one of which is more powerful than the one below it, we can take this as evidence that PSPACE is much more powerful than P or NP.

What about board games? If there are a finite number of types of pieces, we can encode a position on an $n \times n$ board with $O(n^2)$ bits. But how long do games last? In Tic-Tac-Toe, each move fills in a square. So, on an $n \times n$ board the game can last for at most n^2 moves, and telling whether the current player has a winning strategy is in PSPACE. On the other hand, in Chess, Checkers, and Go—or, rather, their generalizations to $n \times n$ boards—the length of a game can be exponentially large, raising their computational complexity to EXPTIME or even EXPSPACE.

We end this section by discussing an important generalization of `Evaluate`, which is much closer to how real game-playing programs work. For most of the games we care about, it is computationally infeasible to explore the game tree completely. Depending on our computational resources, we can only see ahead to a "horizon" a certain number of moves away. Once we reach this horizon, we assign a value `Points`(p) to each position. For instance, in Chess it is common to assign 9 points to the queen, 5 to each rook, 3 for knights and bishops, and 1 to each pawn, with additional points for controlling the center of the board and having pawns support each other.

If we assign positive values to Black and negative ones to White, the value of a position is the maximum of its children's values if it is Black's turn, and the minimum if it is White's turn. This gives the algorithm shown in Figure 8.15, known as *minimax search.* Deciding what features of a position to value and how much to value them, and how deeply to explore a given line of play, are important and difficult problems—but they belong to the realm of Artificial Intelligence, and we will speak no more of them here.

8.7 Games People Play

In the previous section, we saw two abstract games—Two-Player SAT and the Reachability game on the state space of a program with poly(n) memory—and showed that they are PSPACE-complete. Many real games that we and our computer counterparts enjoy also turn out to be PSPACE-complete, and some are even higher up the complexity hierarchy. In this section, we give two examples: the word game Geography, and the venerable Asian game of Go.

```
Minimax(p)
input: a position p in a board game
output: our approximate value for p, where Black is positive
begin
    if we have reached our search horizon then return Points(p);
    else if it is Black's turn then
        v = −∞;
        for all of Black's moves i do v = max(v, Evaluate(p_i));
        return v;
    end
    if it is White's turn then
        v = +∞;
        for all of White's moves i do v = min(v, Evaluate(p_i));
        return v;
    end
end
```

FIGURE 8.15: The minimax search algorithm. When we reach the horizon of our search tree, we estimate how far ahead a given player is using some `Points` function.

8.7.1 *Geography*

Habeam, geographer of wide reknown,
Native of Abu-Keber's ancient town,
In passing thence along the river Zam
To the adjacent village of Xelam,
Bewildered by the multitude of roads,
Got lost, lived long on migratory toads,
Then from exposure miserably died,
And grateful travelers bewailed their guide.

Ambrose Bierce

There are many ways to pass the time during a long journey. One of them is a word game called Geography. The first player names a place, say Magdeburg. The second player replies with a place whose name starts with the same letter the previous place ended with, such as Genoa. The first player could then respond with Albuquerque, and so on. Each place can only be named once, and the first player who cannot make a move loses.

To make a mathematical version of this game, let's assume that there is a finite list of agreed-upon places, and a directed graph describing the legal moves. We start at a particular vertex, and you and I take turns deciding which step to take next. Whoever ends up in a dead end, where every outgoing edge points to a vertex we have already visited, loses. This gives us the following problem, which we put in small capitals to distinguish it from the game:

GEOGRAPHY
Input: A directed graph G and a starting vertex v
Question: Is v a winning position for the first player?

What is the computational complexity of this problem? If G has n vertices then each position in the game can be encoded with n bits marking which of the vertices have already been visited, plus $O(\log n)$ bits to keep track of the current vertex, for a total of $O(n)$. More importantly, no game can last for more than n moves. By the argument of Section 8.6.3, it takes just $O(n^2)$ memory to explore the entire game tree. Thus GEOGRAPHY is in PSPACE.

We will prove that GEOGRAPHY is PSPACE-complete by showing that

$$\text{QUANTIFIED SAT} \leq \text{GEOGRAPHY}.$$

To explain the reduction, let's start with an unquantified SAT formula ϕ, where the truth values of its variables $x_1,\ldots,x_n$ are already set. We can decide whether ϕ is true by playing a little game. The Skeptic chooses a clause c, and the Prover chooses one of c's variables x. The Prover wins if x's value satisfies c, and the Skeptic wins if it doesn't. Clearly ϕ is true if the Prover has a winning strategy—if he can exhibit a satisfying variable for any clause the Skeptic chooses.

We can implement this in GEOGRAPHY as shown in Figure 8.16. There are vertices for each variable and its negation, and a vertex for each clause. The Skeptic goes first, following an edge to one of the clause vertices, and the Prover replies by moving to one of the variable vertices. The variables' truth values are represented by marking some of their vertices as already visited, so that the Prover cannot use them in his reply. For instance, if we have already visited x_1 but $\overline{x}_1$ is still available, then x_1 is false.

The variable vertices are dead ends, so the Prover wins if he can move to one that satisfies the Skeptic's clause. On the other hand, if the clause is dissatisfied by all its variables—that is, if all its variables are marked as visited—then the clause vertex is a dead end, and the Prover loses. Thus ϕ is satisfied if the Prover has a winning strategy.

To complete the reduction from QUANTIFIED SAT, we need to arrange for the players to take turns setting the variables' truth values. We do this with a string of choice gadgets as shown in Figure 8.17, which let the current player set x_i true by visiting the $\overline{x}_i$ vertex or vice versa. The Prover and the Skeptic choose the values of x_i for i odd and even respectively, so the Prover has a winning strategy if and only if the QUANTIFIED SAT formula $\exists x_1 : \forall x_2 : \exists x_3 : \cdots : \phi(x_1,\ldots,x_n)$ is true. This reduction can clearly be carried out in polynomial time—or, for that matter, logarithmic space—and GEOGRAPHY is PSPACE-complete.

8.6 This reduction was easy to construct because the process of playing a game, where the players take turns deciding which branch to follow in the game tree, is already a lot like GEOGRAPHY. Indeed, just as a game tree can be thought of as a Boolean circuit whose layers alternate between AND and OR gates, Problem 8.16 shows that any Boolean circuit can be converted into an instance of GEOGRAPHY, where the first and second players try to prove that the circuit's output is `true` or `false` respectively.

Every two-player game can be thought of as battle between a Prover and Skeptic, where the Prover wishes to prove that he has a winning strategy and the Skeptic begs to differ. However, for most games it is not so obvious how we can encode arbitrary logical relationships in the game's dynamics. In the next section, we show how to do this for one of the oldest and most popular games in the world.

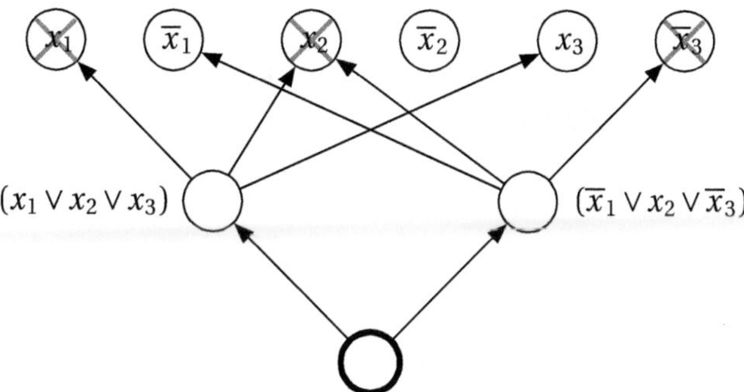

FIGURE 8.16: Evaluating SAT formulas with GEOGRAPHY. We start at the bold vertex at the bottom. The Skeptic goes first and chooses which clause to examine, and the Prover chooses a variable which satisfies that clause. Vertices that have already been visited, which force the Prover to use the other truth value for those variables, are marked with an X. Here $x_1 = \texttt{false}$, $x_2 = \texttt{false}$, and $x_3 = \texttt{true}$, and the formula $\phi = (x_1 \vee x_2 \vee x_3) \wedge (\overline{x}_1 \vee x_2 \vee \overline{x}_3)$ is satisfied.

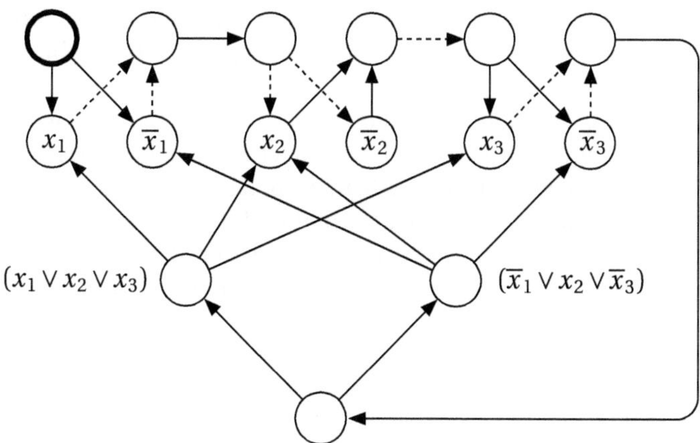

FIGURE 8.17: Adding choice gadgets lets the players alternately set the truth values of the variables, giving a GEOGRAPHY instance that encodes the QUANTIFIED SAT formula $\exists x_1 : \forall x_2 : \exists x_3 : (x_1 \vee x_2 \vee x_3) \wedge (\overline{x}_1 \vee x_2 \vee \overline{x}_3)$. We start at the bold vertex at the upper left, and the Prover goes first. As a visual aid, the Skeptic's moves are shown as dashed lines. What is the Prover's winning strategy?

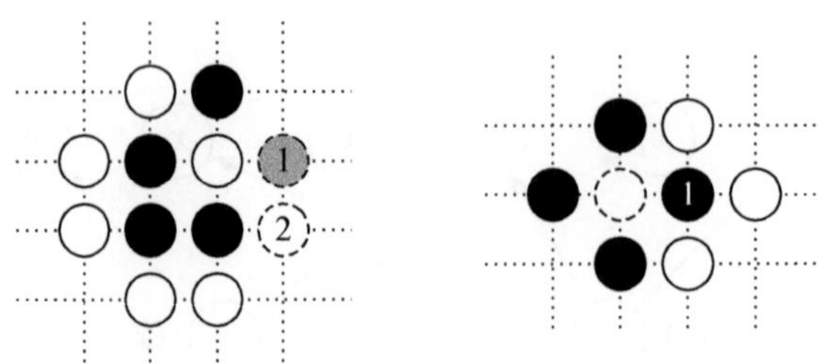

FIGURE 8.18: On the left, two potential captures. Black can take one white stone by playing at 1, but if it is White's turn she can play at 2 and take three black stones. On the right, a ko. Black has just captured White's stone by playing at 1. White is not allowed to immediately recapture Black's stone, since this would return the board to its previous position and create an endless loop.

8.7.2 Go

> In the depth of night
> even the ceiling
> becomes a go board.
>
> Senryu

Go is an exceedingly elegant board game, invented in China at least 2500 years ago. While its rules are very simple, its strategies are very deep. We will see in this section that the computational complexity of Go exceeds even PSPACE-completeness. In fact, it belongs to some of the highest complexity classes we have discussed.

While there are a number of variations, Go's rules are easy to state. Players take turns placing a stone of their color on the board, and they may also pass. If a connected group of stones is completely surrounded by stones of the other color as shown in Figure 8.18, they are taken prisoner and removed from the board. The end of the game is determined by agreement, when both players pass consecutively. At that point, each player's score is the size of their territory, i.e., the number of empty spaces surrounded by their stones, minus the number of their stones that were taken prisoner during the game. The player with the highest score wins.

There is essentially only one other rule. The right-hand side of Figure 8.18 shows a common situation called a *ko*, the japanese word for eternity and for threat. Black has just taken one of White's stones. White could retaliate by playing that stone again, and taking Black's stone. But then Black could play as he did before, and the game could go back and forth in an endless loop. To prevent such a loop, the ko rule prohibits White from immediately replying in this way. Instead, she has to play elsewhere on the board, hopefully forcing Black to respond there so that she can return to the ko fight and save her stone. In its basic form, the ko rule simply states that no move may return the board to its previous position.

How computationally complex is Go? Traditionally, it is played on a 19×19 board. There are only finitely many positions, so we could prepare a lookup table giving the right move in every possible situation. This lookup table only takes $O(1)$ memory, so Go is quite easy. Unfortunately, when we say $O(1)$ we mean $3^{19^2} \approx 10^{172}$. Since this answer is not very helpful, let's generalize Go to boards of arbitrary size:

8.7

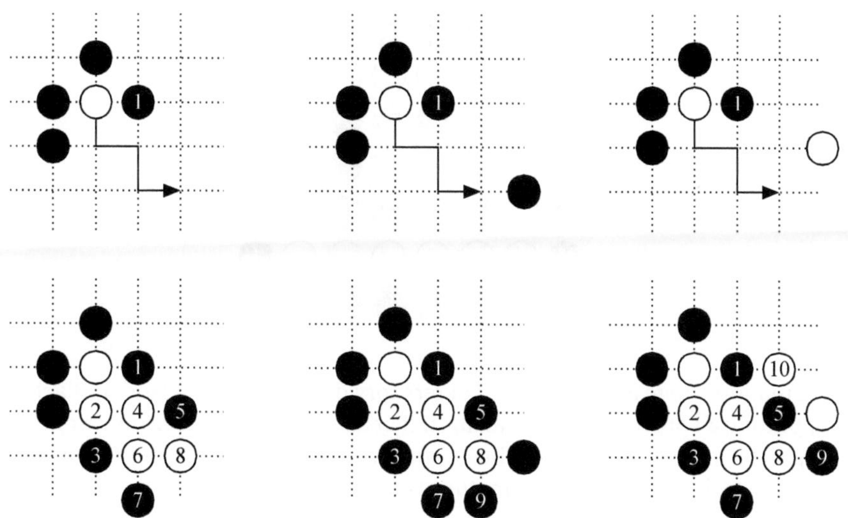

FIGURE 8.19: A ladder in Go. On the left, Black keeps White in atari, and White tries to escape along a zig-zag path. In the center, this path meets a black stone and Black wins. On the right, the path meets a white stone. If Black tries to continue the ladder, White captures the stone at 5 and escapes.

> GO
>
> Input: A position on an $n \times n$ board
>
> Question: Does Black have a winning strategy from this position?

We will show that

$$\text{QUANTIFIED SAT} \le \text{GO}.$$

This implies that GO is PSPACE-hard, i.e., that any problem in PSPACE can be reduced to it. If GO were in PSPACE, it would be PSPACE-complete. However, as we will see, it is even harder than that.

We focus on a particular type of play called a *ladder*, shown in Figure 8.19. The ladder starts when Black plays at 1 and threatens to capture the white stone. White can survive by playing at 2, 4, 6 and so on, but Black keeps the pressure on with moves at 3, 5, and 7. Throughout this process, Black keeps White in *atari*, an immediate threat of capture. Just like a Chess player who is kept in check, White must respond immediately or choose to lose the battle.

Once the ladder begins, the fate of White's stone depends on what the ladder runs into first. If it runs into a black stone or the edge of the board, then White's entire group is lost. But if it runs into a white stone instead, this reinforcement allows White to escape as shown on the right of Figure 8.19.

Now suppose that Black's ladder attacks a large group of white stones rather than a single one. Furthermore, suppose that if Black fails to keep White in a constant state of atari, White will have time to make a crucial move that eliminates Black's threat and secures the white stones. An example of such a situation is shown in Figure 8.20. If White's group is large enough, the entire game depends on whether White escapes the ladder or not.

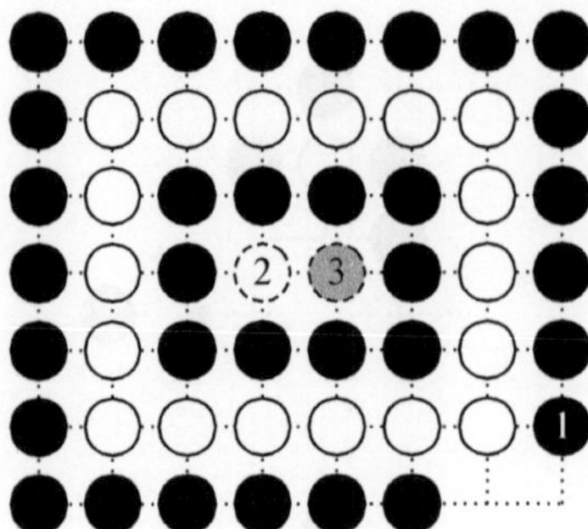

FIGURE 8.20: A situation where a large amount of territory depends upon a ladder. Black starts the ladder by playing at 1, putting all of White's stones in atari. If at any point White is no longer in atari, White plays at 2, threatening to take the inner group of black stones. If Black takes 2 by playing at 3, White plays at 2 again and takes the black stones, whereupon White's territory is secure.

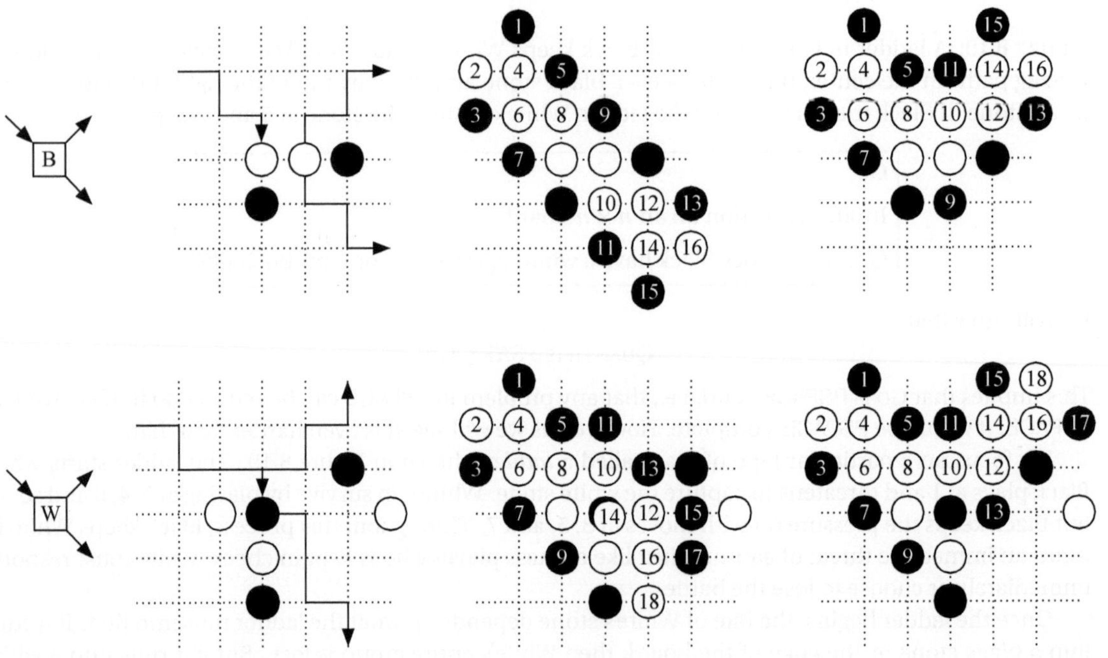

FIGURE 8.21: The choice gadgets. In each case a ladder enters from upper left, and one of the players controls whether it goes to the upper right or lower right. Black's gadget depends on whether she plays 9 above or below the original white stones. White's gadget depends on whether she captures the black stone with 12 and replaces it with 14, or plays 12 next to 10.

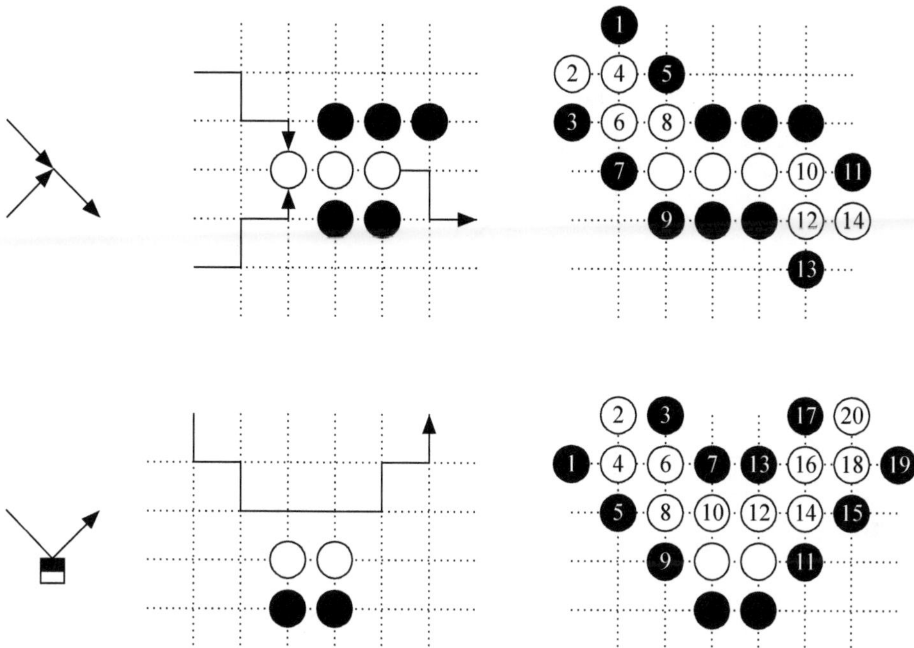

FIGURE 8.22: Above, the merge gadget, which merges two possible ladder paths into one. Regardless of whether a ladder enters from the upper left (the line of play shown) or the lower left, it exits to the lower right. Below, the mirror gadget, and a ladder bouncing off it.

In most real Go games, the outcome of a ladder is easily predictable, so neither player has any incentive to follow it to its conclusion. But with some clever gadgets, we can let Black and White choose where the ladder goes, and cause its eventual fate to depend on whether their choices satisfy a SAT formula. The key gadgets are shown in Figure 8.21, where Black and White can send the ladder in one direction or another. Two other useful gadgets are shown in Figure 8.22. These let us merge two possible ladder paths into one, or change a ladder's direction by bouncing it off a mirror.

We implement a QUANTIFIED SAT formula with a position like that shown in Figure 8.23. Each variable is set when a player uses a choice gadget to send the ladder one way or the other around a white stone. Later, Black will send the ladder towards one of these stones. At that point, Black will win if the ladder collides with the black stones on the outside of its previous path, and White will win if it hits the white stone instead.

After the variables are set, White, playing the Skeptic, chooses a clause c, and Black, playing the Prover, bounces the ladder towards one of c's variables x. The reduction is arranged so that the ladder then collides with its previous path if x satisfies c, and hits the white stone if it doesn't. Thus Black has a winning strategy if and only if the corresponding QUANTIFIED SAT formula is true.

This completes the proof that GO is PSPACE-hard. Why isn't it in PSPACE? The reason is that, unlike GEOGRAPHY, there is no guarantee that the game will last for only a polynomial number of moves. While

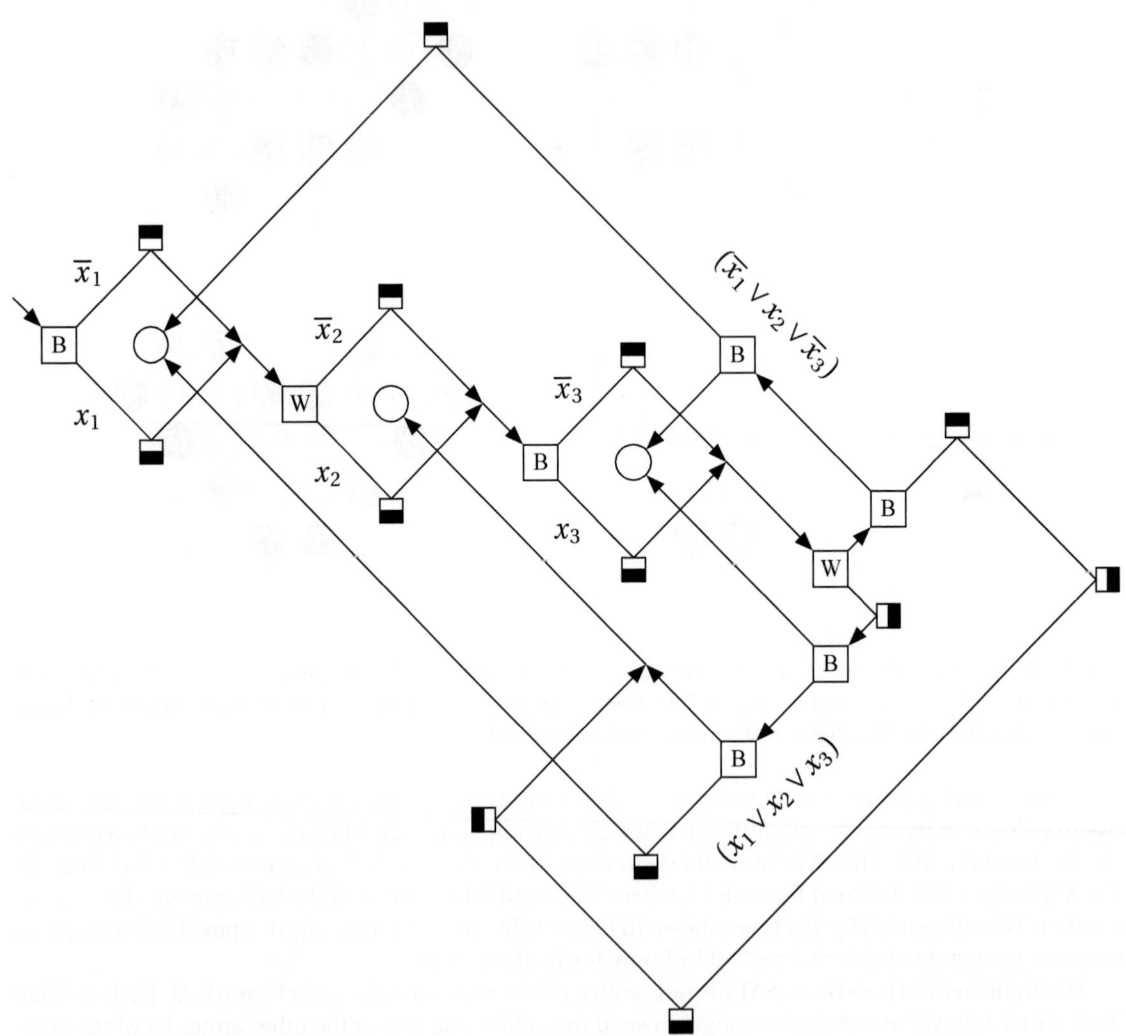

FIGURE 8.23: A GO position corresponding to the QUANTIFIED SAT formula $\exists x_1 : \forall x_2 : \exists x_3 : (x_1 \vee x_2 \vee x_3) \wedge (\overline{x}_1 \vee x_2 \vee \overline{x}_3)$. The players set x_i true or false by sending the ladder one way or the other around a white stone. White chooses which clause to check, and Black chooses which variable to satisfy it with. White wins if the ladder hits a white stone, and Black wins if it collides with its earlier path.

there are only n^2 places on an $n \times n$ board, pieces can be captured and removed, so players could place a stone at the same location many times. In particular, complicated sets of ko fights, in which players retake one ko by using another ko as a threat, can lead to games of exponential length. Thus the game tree can be exponentially deep, and there is no obvious way to explore it with only polynomial memory.

If Go is not in PSPACE, where is it? Problem 8.24 shows that it is in EXPTIME, and while we will not present the proof here, it turns out to be EXPTIME-complete. This is under the simple ko rule where we cannot return to the previous position. However, some variants of Go have a "superko rule," in which we cannot revisit any position that has occurred earlier in the game—similar to a rule in modern Chess, stating that no position may be repeated three times.

The superko rule turns Go into a kind of GEOGRAPHY on the exponentially large graph of possible positions. This puts it in EXPSPACE, and it might even be EXPSPACE-complete—making it one of the hardest problems that humans ever try to solve.

Human aficionados of Go might argue that these exponential-length games are unrealistic, and have little to do with why these games are deep and subtle in practice. After all, real Go games between good players involve a small number of captures, and a few short ko fights. The longest Go game on record was only 411 moves long. However, the fact that Go is EXPTIME-complete is a proof that, at least in some artificial situations, the optimal line of play really does lead to games of exponential length. Perhaps gods play Go very differently from the way humans do; see Figure 8.24.

8.8

8.7.3 Games One Person Plays

Where PSPACE-completeness is concerned, does it take two to tango? On the contrary, even *one-player* games—that is, games of solitaire and puzzles—can be PSPACE-complete. In Section 8.2, we discussed sliding block puzzles. Let's define the following decision problem:

> SLIDING BLOCKS
>
> Input: A starting configuration of labeled blocks in an $n \times n$ grid, with desired ending positions for one or more blocks
>
> Question: Is there a sequence of moves that solves the puzzle?

As Problem 8.27 shows, the shortest solution to such puzzles can consists of an exponentially large number of moves. Unless there is some way to compress this solution, or a proof of its existence, into a polynomial number of bits, SLIDING BLOCKS is not in NP. However, as we said before it is a REACHABILITY problem where each state can be described with poly(N) bits, so it is in PSPACE.

It turns out that these puzzles, and many others similar in spirit, are PSPACE-complete. We will not give the full proof, but Figure 8.25 shows the kinds of gadgets that are involved. They represent AND and OR gates, since an "output" block can move only if both or either "input" blocks can move. We can build circuits by tiling these gadgets together. Then, as the blocks move back and forth, the puzzle executes a kind of reversible computation.

We showed in Section 5.3.4 that TILING, the "static" problem of telling whether we can fill a region with tiles of a certain shape, is NP-complete. If you like, it is the "dynamic" nature of SLIDING BLOCKS that raises its computational complexity from NP to PSPACE.

8.9

FIGURE 8.24: Two gods playing Go inside a persimmon.

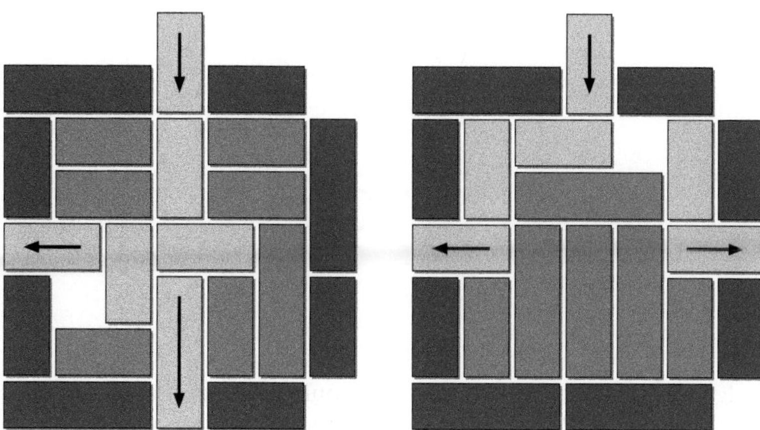

FIGURE 8.25: AND and OR gadgets in the proof that sliding block puzzles are PSPACE-complete. On the left, the north block cannot move until both the west and south blocks do. On the right, the north block can move if either the west or east blocks do. The shaded blocks remain fixed.

8.8 Symmetric Space

Throughout this chapter, we have thought of computations as paths through state space, and vice versa. If a vertex has multiple outgoing edges, the computation can consult the witness to decide which way to go, or branch nondeterministically. Thus REACHABILITY for general directed graphs is NL-complete. On the other hand, in the state space of a deterministic computation, each vertex has only one outgoing edge. As Exercise 8.3 suggests, ONE-OUT REACHABILITY is L-complete under an appropriate kind of reduction.

What about UNDIRECTED REACHABILITY, the case of REACHABILITY on undirected graphs? It doesn't seem NL-complete, since as Figure 8.26 shows, without arrows on the edges it can be difficult to make sure that we follow a legal computation path. On the other hand, it seems harder than ONE-OUT REACHABILITY, since paths in an undirected graph can certainly branch.

UNDIRECTED REACHABILITY seems to be a problem in search of a complexity class. Surely it's complete for something! If we wish, we can define such a class by fiat:

> SL is the class of problems that are log-space reducible to UNDIRECTED REACHABILITY.

The S here stands for "symmetric", since edges go both ways.

More generally, we can consider symmetric programs, where every computation step is also allowed in reverse. Then the state space is an undirected graph, where a path exists from the initial state to a state that returns "yes" if and only if the input is a yes-instance. The set of problems solvable by such programs with $O(f(n))$ memory is called SSPACE($f(n)$), and SL = SSPACE($\log n$).

Since UNDIRECTED REACHABILITY is a special case of REACHABILITY, we know that SL is contained in NL. In addition, Problem 8.29 shows that ONE-OUT REACHABILITY is in SL. Thus we have

$$\mathsf{L} \subseteq \mathsf{SL} \subseteq \mathsf{NL}\,.$$

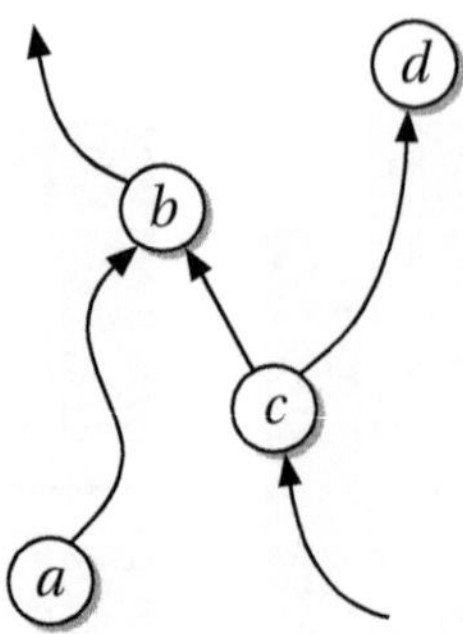

FIGURE 8.26: If we ignore the arrows in state space, we could go from a to b, travel backwards from b to c, and then follow the other branch forwards to d, even though there is no legal computation path from a to d. Thus it seems unlikely that UNDIRECTED REACHABILITY is NL-complete.

Is the power of SL really intermediate between L and NL? Or does it coincide with one of them?

In Chapter 12, we will see that we can solve UNDIRECTED REACHABILITY with a *randomized* algorithm—namely, by taking a random walk. Suppose an undirected graph G has n vertices, and that there is a path from s to t. Then if we start at s and move to a random neighbor at each step, after $O(n^3)$ steps we will hit t with constant probability.

Since we only need to keep track of our current location, we can perform this algorithm with just $O(\log n)$ memory. This puts UNDIRECTED REACHABILITY inside yet another complexity class—RL, the class of problems that can be solved, with high probability, by a randomized algorithm that takes poly(n) time and $O(\log n)$ memory. However, it seems hard to *derandomize* this algorithm and make it deterministic.

In 2005, a wonderful breakthrough occurred. Omer Reingold proved that

UNDIRECTED REACHABILITY is in L,

and therefore

$$\mathsf{SL} = \mathsf{L}.$$

More generally, SSPACE($f(n)$) = SPACE($f(n)$) for all $f(n) \geq \log n$. Thus if memory is our limited resource, "symmetric nondeterminism" is no more powerful than deterministic computation. To put this more poetically, we can find our way out of an undirected maze without keeping track of much more than our current location.

To get a sense of how this proof works, suppose that a graph has diameter D and maximum degree d. We can solve UNDIRECTED REACHABILITY, telling whether there is a path from s to t, by starting at s and trying all possible paths of length D. Using depth-first search, we can do this with a stack of depth D. Moreover, we can record which way the path goes with $\log_2 d$ bits at each level of the stack. So, the total amount of memory we need is $O(D \log d)$. This shows that UNDIRECTED REACHABILITY is in L for any graph whose diameter D is $O(\log n)$, and whose maximum degree d is $O(1)$. The problem, of course, is that a graph with n vertices can have diameter $\Theta(n)$.

We can reduce the diameter from D to $D/2$ by "squaring" the graph, defining a new graph G^2 where two vertices are adjacent if there is a path of length 2 or less between them in G. However, squaring also increases the degree from d to d^2. Thus, by the time we square the graph enough times to reduce its diameter to $O(\log n)$, its degree could be quite large. What we would like is a way to square a graph, in some sense, without increasing its degree. If we can bring its diameter down to $O(\log n)$ while keeping its degree to $d = O(1)$, we can solve UNDIRECTED REACHABILITY with $O(\log n)$ memory.

The solution to this problem is very beautiful. It relies on the spectral theory of graphs to turn G into an *expander*—a graph where the set of vertices we can reach in ℓ steps grows exponentially with ℓ—while keeping its degree constant. We will learn about this machinery in Chapter 12, and discuss the proof that UNDIRECTED REACHABILITY is in L in Section 12.9.6. For now, we leave the reader basking in the warm surprises of space-bounded computation.

8.10

Problems

> It is the duty of all teachers, and of teachers of
> mathematics in particular,
> to expose their students to problems
> much more than to facts.
>
> Paul Halmos

8.1 Simulating a big memory. We claim in Section 8.1 that a program that runs in t steps only has time to access $O(t)$ bits of memory. This is trivial if our model of computation is a Turing machine, since in t steps it can only visit t locations on its tape. But if our machine has a random-access memory, it can read or write the ith bit of memory in a single step. In t steps we could set $i = 2^t$, say, and this suggests that we might need 2^t bits of memory. Show that, nevertheless, we can simulate any such program with one that uses only $O(t)$ bits of memory.

8.2 Matched parentheses. Suppose we are given a string of parentheses. Show that telling whether they are properly matched and nested, such as (( ) ( )) or ((( ))) ( ) but not ( )) or ( )) (( ), is in L.

8.3 More matched parentheses. (Considerably trickier than the previous problem.) Now suppose there are two types of parentheses, round ones and square brackets. We wish to know if they are properly nested: for instance, ([ ]) [ ] and [( ) [ ]] are allowed, but ([ )] is not. Show that this problem is in L. Hint: prove by induction that it suffices to check some relatively simple property of each substring.

8.4 How much do we need to remember? Following up on the previous two problems, suppose we restrict the computer's access to the input further, so that it must read the input from left to right. Show that in this case, Problem 8.2 can still be solved in $O(\log n)$ memory, but that Problem 8.3 and the palindrome example from Section 8.1 each require $\Theta(n)$ memory. Hint: when we have read halfway through the input, how much information do we need to remember about it?

8.5 Addition, one bit at a time. Show that binary addition is in L in the one-bit-at-a-time sense discussed in the text. In other words, show that the problem of computing the ith bit of $x+y$, where x, y, and i are n-bit integers, is in L.

8.6 A pebble puzzle. Consider the following variant of REACHABILITY. I give you a graph G and the initial and final locations of k pebbles, i.e., vertices $s_1, \ldots, s_k$ and $t_1, \ldots, t_k$. I ask you whether we can move each pebble from its initial location to its final one. In each step, we may move a pebble from its current vertex to a neighboring one, but we can never have two pebbles at the same vertex.

How does the amount of memory you need to solve this problem grow as a function of k? Show that this problem is in NL for any constant k, but that it could be quite hard if k is, say, $\Theta(n)$.

8.7 From everywhere to everywhere else. A directed graph G is *strongly connected* if for every pair of vertices u, v there is a path from u to v. Show that STRONG CONNECTEDNESS is in NL. Then show that it is NL-complete by giving a log-space reduction from REACHABILITY to STRONG CONNECTEDNESS.

8.8 2-SAT is to NL as 3-SAT is to NP. Show that 2-SAT is NL-complete. Hint: first use the discussion in Section 4.2.2 to reduce REACHABILITY to the problem 2-UNSAT, the problem of whether a 2-SAT formula is unsatisfiable, and then use the Immerman–Szelepcsényi Theorem. Can you invent a restricted case of 2-SAT which is L-complete?

8.9 Cyclic or acyclic? Consider the problem ACYCLIC GRAPH of telling whether a directed graph is acyclic. Show that this problem is in NL, and then show that it is NL-complete. Hint: first, change a directed graph into an acyclic one by keeping track of the length of the walk.

8.10 Finite-state automata. A *deterministic finite-state automaton* (DFA) consists of a finite set of states S, an initial state $s_0 \in S$, a subset $S_{\text{yes}} \subseteq S$, an alphabet of input symbols Σ, and a transition function $\delta : S \times \Sigma \to S$. It starts in the state s_0, and reads a word $w = a_1 a_2 \cdots a_t$ from left to right, where each symbol a_i is in Σ. At each step, it updates its state from s_{i-1} to $s_i = \delta(s_{i-1}, a_i)$. We say the automaton *accepts* w if, after reading all of w, its final state s_t is in S_{yes}.

A *nondeterministic finite-state automaton* (NFA) is defined similarly, except that now the transition function is multi-valued, returning a subset of S and allowing the new state to be any one in that set. This creates a tree of possible paths, and we say that an NFA accepts if any of these paths do.

Now consider the following problem:

> DOES THIS NFA ACCEPT ANY WORDS?
>
> Input: An NFA A specified by $S, s_0, S_{\text{yes}}, \Sigma$, and δ
>
> Question: Does there exist a word w, with symbols in Σ, that A accepts?

Show that this problem is NL-complete. Is the analogous problem for DFAs in L, or is it still NL-complete? If you believe it is NL-complete, invent a variant of this problem that is L-complete—specifically, to which ONE-OUT REACHABILITY can be reduced using local gadget replacement.

8.11 Middle-first search is slow. We stated in the text that, because it revisits the same questions many times rather than saving its previous results à la dynamic programming, `MiddleFirst` takes more than polynomial time. Let $T(n, \ell)$ be its worst-case running time, measured in the number of times it calls itself recursively. Show that

$$T(n,\ell) = 2n\,T(n,\ell/2).$$

Then solve this equation, assuming the base case $T(n,1) = 1$ since when $\ell = 1$ the algorithm can look directly at the adjacency matrix. Show that $T(n,n)$ is a superpolynomial—specifically, quasipolynomial—function of n,

$$T(n,n) = n^{\Theta(\log n)}.$$

8.12 Space and time. Define NSPACETIME($f(n), g(n)$) as the class of problems that can be solved nondeterministically by an algorithm that uses $O(f(n))$ memory and runs in $O(g(n))$ time. First explain why this is, *a priori*, a stronger property than being in both NSPACE($f(n)$) and NTIME($g(n)$). In other words, explain why

$$\mathsf{NSPACETIME}(f(n), g(n)) \subseteq \mathsf{NSPACE}(f(n)) \cap \mathsf{NTIME}(g(n))$$

might generally be a proper inclusion, and similarly for the deterministic case.

Now prove the following generalization of Savitch's Theorem:

Theorem 8.3 $\mathsf{NSPACETIME}(f(n), g(n)) \subseteq \mathsf{SPACE}(f(n) \log g(n))$.

For instance, if a problem can be solved nondeterministically by an algorithm that uses linear space and polynomial time, it can be solved deterministically in $O(n \log n)$ space. What is the running time of this algorithm?

8.13 Reachable and unreachable. Consider the following restatement of the Immerman–Szelepcsényi Theorem:

Theorem 8.4 *There is a log-space reduction which takes a graph G and a pair of vertices s, t, and returns a graph G' and a pair of vertices s', t', such that there is a path from s to t in G if and only if there is no path from s' to t' in G'.*

If G has n vertices, how many vertices does G' have?

8.14 Two-player HAMILTONIAN PATH. In the Prologue, we discussed Hamilton's Icosian Game, in which we try to find a Hamiltonian path on the edges of a dodecahedron. In fact, his original game involved two players, where the first player chose the first five steps. The second player wins if she is able to complete a Hamiltonian path, and the first player wins if she cannot. We can generalize this game as follows:

> HAMILTON'S GAME
>
> Input: A directed graph G with n vertices and an integer $m < n$
>
> Question: White tries to complete a Hamiltonian path after Black chooses the first m steps. Does White have a winning strategy?

Say as much as you can about the computational complexity of this problem.

8.15 A hierarchy in logarithmic space? In analogy with the polynomial hierarchy (see Section 6.1.1), consider the logarithmic space hierarchy, where $\Sigma_k \mathsf{L}$ and $\Pi_k \mathsf{L}$ consist of properties in L surrounded by k alternating of quantifiers. Show that this hierarchy collapses to its first level. In other words, for all $k \geq 1$ we have

$$\Sigma_k \mathsf{L} = \Pi_k \mathsf{L} = \mathsf{NL}.$$

8.16 From circuits to Geography. Show how to convert an arbitrary Boolean circuit, where the truth values of its inputs are fixed, to an instance of GEOGRAPHY such that the first player has a winning strategy if and only if the output is `true`. Hint: start at the output and move towards the inputs, so that each gate is a winning position for the current player if it is `true`. When we reach an AND or an OR gate, who should decide which of its inputs to move to? What about a NOT gate?

8.17 Geography on the edges. EDGE GEOGRAPHY is to GEOGRAPHY as Eulerian paths are to Hamiltonian ones. In other words, vertices can be visited multiple times, but each directed edge can only be used once. Show, using a reduction almost identical to the one given in Section 8.7.1, that EDGE GEOGRAPHY is PSPACE-complete.

8.18 Planar Geography. Consider the gadget in Figure 8.27. Show that if one player enters from the left, both players will guide the path through to the right in order to avoid a quick loss. Similarly, if a player enters from below, they will play through and exit above. Therefore, it acts as a crossover gadget analogous to Figure 4.3 for GRAPH 3-COLORING, allowing two edges to cross—as long as no more than one of these edges would ever be used in a game. Show that all the crossings in Figure 8.17 are of this type, and conclude that PLANAR GEOGRAPHY is PSPACE-complete.

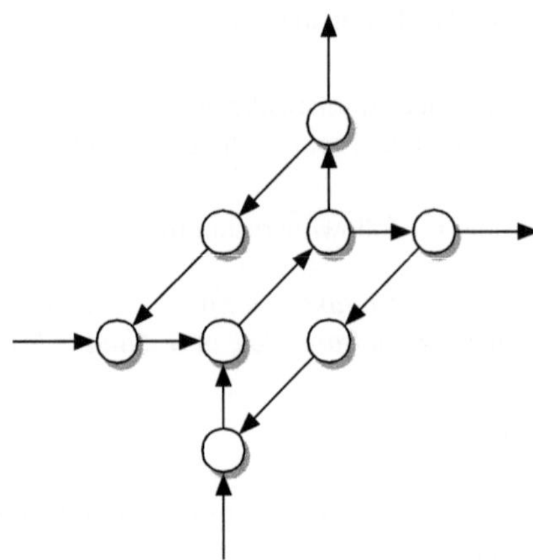

FIGURE 8.27: A crossover gadget for Geography.

8.19 Undirected Geography. What happens if we play Geography on an undirected graph? Prove the following fact: the first player has a winning strategy if and only if every maximal matching includes the starting vertex. Using the fact that we can find such paths in polynomial time, even for non-bipartite graphs (see Note 5.6) show that UNDIRECTED GEOGRAPHY is in P.

Hint: recall Theorem 5.2 from Section 5.5.2 that a matching M is maximal if and only if there is no alternating path, i.e., a path that alternates between edges in M and $\overline{M}$ and that starts and ends at isolated vertices.

8.20 Alphabetic Geography. In traditional Geography, a move from a city u to a city v is allowed if v starts with the same letter that u ends with. Let's say that a directed graph G is *alphabetic* if it of this form: in other words, if for some finite alphabet A, each vertex u can be assigned a pair $f(u), \ell(u) \in A$ such that there is an edge from u to v if and only if $\ell(u) = f(v)$. Give a small example of a graph that is not alphabetic. Then show that ALPHABETIC GEOGRAPHY is PSPACE-complete, by converting any graph to an alphabetic one (albeit with an alphabet that grows with the number of vertices).

8.21 Good old-fashioned Geography. Now consider *really* traditional Geography, where the alphabet is limited to, say, 26 symbols. The input now consists of a list of n cities. Show that the corresponding version of GEOGRAPHY is in P. How does its complexity grow with the size of the alphabet?

8.22 Geography with repeat visits. Suppose we play GEOGRAPHY without the restriction that we can only visit each vertex once. However, the graph is still directed, and you lose if you find yourself at a vertex with no outgoing edges. Show that the problem of telling whether the first player has a winning strategy in this case is in P. As a corollary, show that GEOGRAPHY is in P if played on an acyclic directed graph. Hint: find an iterative way to label each vertex as a win, loss or draw for whichever player moves there.

8.23 Cast a hex. Hex is a simple and beautiful game which was invented independently in the 1940s by Piet Hein and John Nash. It is played on a lozenge-shaped board with hexagonal cells. Black and White take turns placing stones of their color on the board. Black wins if she forms a path from upper left to lower right, and White wins if she forms a path from upper right to lower left. The left side of Figure 8.28 shows a game on a 5×5 board where White has won.

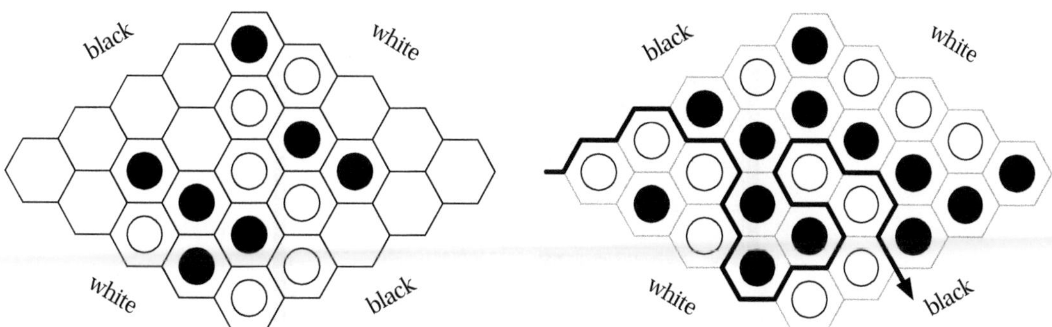

FIGURE 8.28: The game of Hex. On the left, a game where Black went first, but White won after blocking Black's path in the lower corner. On the right, we can prove that there are no draws in Hex by drawing a path that starts at the left corner and keeps Black and White on its left and right respectively.

Show that there are no draws in Hex. In other words, show that if the board is filled with black and white stones, that either Black or White has won, but not both. You may find the right side of Figure 8.28 helpful—also, compare Sperner's Lemma from Section 6.7.3.

Intuitively, having an extra stone on the board can never hurt. Confirm this intuition by proving that the first player has a winning strategy, starting from the initial position of an empty board. Hint: show that if the second player had a winning strategy, the first player could "steal" it and win by pretending to be the second player, giving a contradiction. Why does this proof not work in other games where the first player appears to have an advantage, such as Chess or Go?

8.11

> If this were go
> I'd start a ko fight
> and surely live,
> but on the road to death
> there's no move left at all.
>
> Sansa (1623)

8.24 Simple ko. As a follow-up to Problem 8.22, consider a board game, such as Go or Chess, where the state space of possible positions on an $n \times n$ board is exponentially large and there is no guarantee that the game only lasts for a polynomial number of moves. Assume for the moment that there is no restriction on visiting the same position twice. Show that telling whether the first player has a winning strategy is in EXPTIME.

Now consider the simple ko rule where we are prohibited from revisiting the immediately previous position. Show that telling whether the first player has a winning strategy is still in EXPTIME. What goes wrong with the "superko rule," where we are prohibited from revisiting any previous position? What type of ko rule would keep this problem in EXPTIME?

8.25 Let's call it a draw. Modern Chess has a "50 move rule" in which a player can claim a draw if, for 50 consecutive moves, no piece has been captured and no pawn has been moved. Suppose that we generalize this to $n \times n$ boards with a function $f(n)$, such that there is a draw if no captures or pawn moves have occurred in $f(n)$ consecutive moves. Show that if $f(n) = \text{poly}(n)$, this version of Chess is in PSPACE.

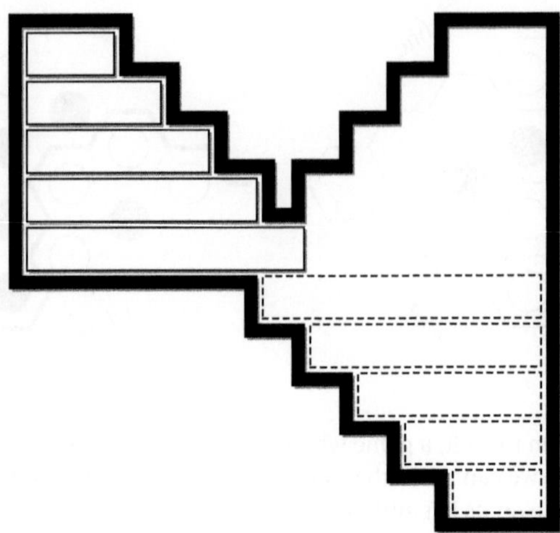

FIGURE 8.29: A sliding-block puzzle called Junk's Hanoi, designed by Toshi "Junk" Kato, which takes an exponential number of moves to solve.

8.26 A hill of beans. Consider the following game. There is a pile of n beans. In each move, one of us removes a certain number of beans from the pile. We each have a set $S, T \subseteq \{1,\ldots,n\}$ describing which numbers of beans we can remove. For instance, if my set is $S = \{1,2,3\}$ and your set is $T = \{2,4,6\}$, then on my move I can remove 1, 2, or 3 beans, and on your move you can remove 2, 4, or 6. You cannot pass, the number of beans you remove has to be exactly in your allowed set, and whoever first empties the pile wins. If the pile is smaller than the current player's smallest legal move, so that no move is possible, the game is a draw. For instance, in this example if $n = 8$ and it is my turn, I can win by removing 3 beans. There are 5 beans left: you have to remove 2 or 4, leaving me with 3 or 1 respectively, and in either case I can then empty the pile.

Now consider the problem BEANPILE where the input is n, S, and T, and the question is who has a winning strategy. Say as much as you can about the computational complexity of this problem.

8.27 Sliding block puzzles can take a long time. Consider the sliding-block puzzle shown in Figure 8.29, and its generalization to n blocks. Prove that the shortest solution consists of $f(n) = 2\cdot 3^{n-1} - 1$ moves, where moving a block from one place to another counts as a single move. For instance, here $n = 5$ and $f(n) = 161$.

8.28 Printing the solution. Show that there is a PSPACE program that prints out the entire sequence of moves for the shortest solution to a sliding block puzzle, or one of the shortest solutions if they are not unique. Note that if the solution is exponentially long, as Problem 8.27 shows it could be, this program doesn't have enough memory to hold it all at once. Instead, assume that the program has access to a printer, on which it prints out one move at a time.

8.29 Erasing the arrows. Suppose G is a directed graph in which each vertex has at most one outgoing edge. Suppose that s and t are vertices, and that t is a "sink": that is, t has no outgoing edges. Now change G into an undirected graph G' by erasing the arrows on the edges, so that we can move in either direction. Show that there is an undirected path from s to t in G' if and only if there is a directed path from s to t in G. Conclude that ONE-OUT REACHABILITY can be reduced to UNDIRECTED REACHABILITY using local gadget replacement. This is part of the original argument that $\mathsf{L} \subseteq \mathsf{SL}$.

8.30 Graph 2-Coloring is to L as Graph 3-Coloring is to NP. Show that Graph 2-Coloring is coSL-complete by showing that Undirected Reachability can be reduced to No Graph 2-Coloring using local gadget replacement, and vice versa. In other words, give a simple procedure that transforms an undirected graph G to an undirected graph G' with a pair of vertices s, t such that G is 2-colorable if and only if there is no path in G' from s to t, and another simple procedure that does the reverse. Given that SL = L, conclude that Graph 2-Coloring is L-complete—a nice counterpart to the fact that Graph 3-Coloring is NP-complete.

Notes

8.1 Models of space-bounded computation. Most books on computational complexity use the Turing machine as their model of computation. The benefit of the Turing machine is that it makes the notion of an "elementary step" completely explicit: each step only changes the machine's internal state and one symbol on the tape, so telling whether two states are connected by a single step is easily seen to be in L (and in even lower complexity classes). Here we have chosen to use the model of programs with random access memory, trusting that, even without specifying precisely what steps are considered "elementary" in our programming language, most readers will find it familiar and intuitive.

The programming model is perfectly adequate for defining NL and L. However, the farther one goes down the complexity hierarchy, the more our model of computation matters. For instance, if we want to think about space-bounded computation with $o(\log n)$ memory, we have to abandon the RAM model of memory access; since we no longer have enough memory to store an index i ranging from 1 to n, we don't even have enough memory to keep track of which bit of the input we want. Down at this microscopic level, the precise details of how our computer is structured, and what operations it can carry out in each step, can make a large difference in what it can do.

8.2 Log-space, and lower-level, reductions. For each complexity class X, deciding what type of reduction we will allow in our definition of X-completeness is partly a matter of taste. If our motivation is to understand whether or not X is more powerful than a subclass Y, such as whether P $\neq$ NP or L $\neq$ NL, we should use Y-reductions, since then X = Y if and only if some X-complete problem is in Y.

On the other hand, demanding simpler reductions gives a finer-grained picture of computational complexity. For instance, the fact that 3-SAT and Graph 3-Coloring can be reduced to each other using very simple reductions shows that they really are almost exactly as hard as each other, as opposed to the weaker property of being within polynomial time of each other. These "local gadget replacement" reductions don't even require the full power of L, since they can be carried out in parallel by very shallow Boolean circuits—for instance, by the AC^0 circuits of Section 6.5.1. Most known NP-complete problems are still NP-complete even if we only allow log-space reductions, and this is a stronger property than NP-completeness under polynomial-time reductions.

8.3 Savitch's Theorem and Reachability. Savitch's Theorem, and the generalization in Problem 8.12, was proved in [726], although the type of recursion used to show that Reachability is in SPACE($\log^2 n$) appeared earlier in Lewis, Stearns, and Hartmanis [530]. Savitch introduces the general connection between Reachability and nondeterministic computation by discussing "threadable mazes" in which each room corresponds to a state of a computer and each corridor corresponds to a single computation step. However, it wasn't until several years later that Jones [445] formalized the notions of NL-completeness and L-completeness, and showed that Reachability and One-Out Reachability are NL- and L-complete. Problem 8.10 is also from [445].

8.4 The Immerman–Szelepcsényi Theorem. The fact that NL = coNL, and more generally that NSPACE($f(n)$) = coNSPACE($f(n)$), was proved independently by Neil Immerman [416] and Róbert Szelepcsényi [783], for which they shared the Gödel Prize in 1995. Most textbooks describe it in terms of a nondeterministic log-space algorithm (see e.g. [759]) but we feel that this log-space checkable witness is easier to understand.

Just before the Immerman–Szelepcsényi Theorem was proved, Lange, Jenner, and Kirsig [518] showed that the logarithmic space hierarchy defined in Problem 8.15 collapses to its second level. That is, they showed that $\Pi_2\mathsf{L} = \text{co}{-}\Pi_2\mathsf{L} = \Sigma_2\mathsf{L}$, and therefore that $\Pi_k\mathsf{L} = \Sigma_k\mathsf{L} = \Pi_2\mathsf{L}$ for any $k \geq 2$.

8.5 Quantified SAT and alternation. Quantified 3-SAT was formulated, and shown to be PSPACE-complete, by Stockmeyer and Meyer [777]. One way to view strings of $\exists$ and $\forall$s is as a generalization of nondeterministic computation. Go back and look at Figure 4.8, where we describe NP problems as a tree of computation paths which returns "yes" if any of its branches does. We can think of this tree as an enormous Boolean circuit, with an OR gate at each **goto both**. If instead we allow two kinds of **goto both** instruction, one which returns "yes" if either branch does, and another which demands that both branches do, we get a mixture of AND and OR gates.

Chandra, Kozen and Stockmeyer [151] defined complexity classes $\mathsf{ATIME}(f(n))$ and $\mathsf{ASPACE}(f(n))$ consisting of those problems that can be solved by these rather exotic programs, where no computation path takes more than $O(f(n))$ time or space respectively. The fact that Quantified SAT is PSPACE-complete then shows that

$$\mathsf{AP} = \mathsf{PSPACE}$$

and more generally that, for any $f(n) \geq n$,

$$\mathsf{ATIME}(f(n)) \subseteq \mathsf{SPACE}(f(n)) \subseteq \mathsf{NSPACE}(f(n)) \subseteq \mathsf{ATIME}(f(n)^2).$$

This quadratic increase from $f(n)$ to $f(n)^2$ is a generalization of Savitch's Theorem. Similarly, Problem 8.22 can be used to show that

$$\mathsf{AL} = \mathsf{P}$$

and more generally that, for any $f(n) \geq \log n$,

$$\mathsf{ASPACE}(f(n)) = \mathsf{TIME}(2^{O(f(n))}).$$

8.6 Geography. Geography, or rather the variant Edge Geography given in Problem 8.17, was shown to be PSPACE-complete by Schaefer [728]. Problem 8.19, showing that Undirected Geography is in P, appeared as Exercise 5.1.4 in [121]. Fraenkel, Scheinerman and Ullman [291] studied Undirected Edge Geography, and showed that some cases of it are in P.

In [728], Schaefer proved that a number of other simple games are PSPACE-complete as well. One of these is Generalized Kayles, which is a sort of two-player Independent Set: players take turns marking vertices that aren't adjacent to any marked vertex, and the first person unable to do this loses. Another is a version of Two-Player SAT in which players take turns setting the value of any variable that hasn't already been set. This is a useful starting point for other PSPACE-completeness proofs, since it removes the requirement that the players set the variables $x_1, x_2, \ldots$ in a rigid order.

8.7 Counting Go positions. Readers concerned about the size of the lookup table for Go will be relieved to know that only 1% of positions on the 19×19 board can actually arise legally [795]. This reduces the size of our lookup table from 10^{172} to about 10^{170}.

8.8 Go, Chess, and Checkers. Go was first shown to be PSPACE-hard by Lichtenstein and Sipser [531]. The proof using ladders we give here, which we feel is considerably more elegant, is from Crâşmaru and Tromp [206]. Using the combinatorial game theory of Berlekamp, Conway, and Guy [106], Wolfe [830] gave another proof that Go endgames are PSPACE-hard, even when the board is separated into independent battles where each one has a game tree of polynomial size.

The EXPTIME-completeness of Go under the simple ko rule was shown by Robson [709] using positions with many simultaneous ko fights. Robson [710] also proved that certain combinatorial games are EXPSPACE-complete under the "superko rule" that no position can be visited twice, but it is still unknown whether this is true of Go.

Chess was shown to be PSPACE-hard by Storer [779], where we generalize to $n \times n$ boards by allowing an arbitrary number of pieces of each kind except the King. If we use the generalized "50 move rule" as in Problem 8.25 then Chess is in PSPACE, so it is PSPACE-complete. In the absence of the 50 move rule, there is no guarantee that a game will last for only poly(n) moves, and Fraenkel and Lichtenstein [290] showed that in that case Chess is EXPTIME-complete. On the other hand, if we also include the rule that no position may be repeated three times, Chess might be EXPSPACE-complete, just as Go under the superko rule might be. Of course, just as for Go, these exponentially long games are very different from how humans play—the longest Chess game on record lasted for just 269 moves.

Checkers was shown to be PSPACE-hard by Fraenkel et al. [289] and EXPTIME-complete by Robson [711], although this also depends partly on the rules for a draw.

8.9 PSPACE-complete puzzles. The PSPACE-completeness of motion puzzles began with Reif [692], who considered the "Generalized Mover's Problem" of moving a jointed object past a set of obstacles. Sliding-block puzzles were shown to be NP-hard by Spirakis and Yap [772], and the "Warehouseman's Problem," in which each block must be moved to a specific location, was shown to be PSPACE-complete by Hopcroft, Schwartz, and Sharir [410].

The complexity of sliding block puzzles in which just one block must be moved to a particular place remained open until Hearn and Demaine [391] showed they are PSPACE-complete, even when all the blocks are dominoes. Their approach builds on an earlier PSPACE-completeness result of Flake and Baum [285] for a family of puzzles called "Rush Hour."

Hearn and Demaine [392] developed a model of computation they call nondeterministic constraint logic, which makes it possible to prove completeness results with just a few simple gadgets like those in Figure 8.25. This allowed them to give greatly simplified proofs of PSPACE-completeness for a number of puzzles, such as Sokoban, which had previously been shown to be PSPACE-complete by Culberson [207].

An excellent history of sliding block puzzles can be found in Hordern [411], which is unfortunately out of print.

8.10 Symmetric space. Symmetric space was defined by Lewis and Papadimitriou [529]. They showed that symmetric space contains deterministic space, and showed that UNDIRECTED REACHABILITY and GRAPH NON-BIPARTITENESS (or equivalently, NO GRAPH 2-COLORING) are SL-complete. They also showed that if SL = L, then symmetric and deterministic space are identical for larger amounts of space as well—in other words, that SSPACE($f(n)$) = SPACE($f(n)$) for all $f(n) \geq \log n$.

Using methods completely unlike the Immerman–Szelepcsényi theorem, Nisan and Ta-Shma [639] showed that SL = coSL. This causes a hierarchy of classes defined by Reif [693] to collapse to SL. A compendium of SL-complete problems can be found in Àlvarez and Greenlaw [39].

The fact that SL = L was shown by Reingold [696] in 2005, using the *zig-zag product* of Reingold, Vadhan and Wigderson [698]. We discuss the zig-zag product in Section 12.9.5. See Note 12.20 for additional references.

8.11 Hex and friends. Like the proof of Sperner's Lemma in Section 6.7.3, the proof of Problem 8.23 that the first player has a winning strategy in Hex is completely nonconstructive. It tells us nothing about what the winning strategy might be. On small boards, playing in the center is a decisive first move, but it is not known whether this is true on larger boards. To eliminate the first player's advantage, it is common to give the second player the option of switching sides after the first move—rather like having one player cut a cake, and then letting the other choose which slice to take.

While going first in Chess is generally believed to give an advantage, being the current player isn't always helpful. For instance, if a piece is currently protecting a friend or pinning an enemy, any move might break this alignment. Positions in Chess like this, where the current player would rather pass than make a move, are called *zugzwang*.

Since each move fills in a cell, a game of Hex on an $n \times n$ board lasts at most n^2 moves, and HEX is in PSPACE. Even and Tarjan [273] proved PSPACE-completeness for a generalized version of Hex, which is essentially a two-player version of REACHABILITY. There is a graph G and a pair of vertices s, t. Players take turns marking vertices with their color. Black wins if she connects s and t with a path of black vertices, and White wins if she creates a cut between s and t. With some additional gadgets, Reisch [700] then proved PSPACE-completeness for standard Hex. Of course, we start from an arbitrary initial position, since we know the first player has a win on the empty board.

Many other board games are now known to be PSPACE-complete, including Othello [424], Gobang [699], Amazons [308, 390] and the traditional Hawaiian game of Konane [390].

Chapter 9

Optimization and Approximation

> It's a funny thing about life; if you refuse to accept anything but the best, you very often get it.
>
> W. Somerset Maugham

We have talked quite a bit about the difficulty of finding a needle in a haystack. What happens if we have a whole stack of needles, but we want to find the longest one, or the shortest, or the sharpest?

We have already several problems like this. Some of them, like MINIMUM SPANNING TREE, MAX FLOW, and MIN CUT, have a structure that lets us find the optimal solution in polynomial time—building it from scratch, or improving it steadily until it becomes as good as possible. But for others, like MAX-SAT and VERTEX COVER, even the decision problem of whether a solution exists above or below a certain threshold is NP-complete.

We start this chapter by exploring the relationships between these decision problems and their optimization versions. We will see that, for most problems, we can find the optimal solution in polynomial time if and only if we can tell whether a solution with a given a quality exists.

If finding the best solution is computationally intractable, we can ask instead for a solution that isn't too bad—one whose quality is within some factor of the best possible one. We will study approximation algorithms for the famous TRAVELING SALESMAN problem, and see how a salesman can find solutions of varying quality depending on the geometry of his turf. We will see that for some problems, we can get as close as we like to the optimum with a reasonable amount of computation. But for some other problems, we can get to a particular factor times the optimum, but no closer unless $\mathsf{P} = \mathsf{NP}$.

We then look at some large families of optimization problems that can be solved in polynomial time, such as LINEAR PROGRAMMING and SEMIDEFINITE PROGRAMMING, where we try to maximize a function subject to a set of constraints. We will see that the duality between MAX FLOW and MIN CUT is no accident—that LINEAR PROGRAMMING problems come in pairs, where the variables of each one correspond to the constraints of the other. We will learn to solve these problems by climbing to the top of a many-faceted jewel, floating a balloon to the top of a cathedral, and cutting away the space of solutions until only the optimum remains, like Michelangelo finding David in the stone.

We then turn to LINEAR PROGRAMMING's more difficult cousin INTEGER LINEAR PROGRAMMING. If we demand that the variables take integer values, then even telling whether or not a solution exists is NP-complete. But we can obtain good approximations for many problems by relaxing this constraint, using LINEAR PROGRAMMING to find a "fuzzy" solution, and then turning this fuzzy solution into a real one by rounding each variable to the nearest integer. Moreover, if the constraints have special properties, we can solve INTEGER LINEAR PROGRAMMING in polynomial time after all. This lets us build a powerful framework for optimization algorithms, where we understand the complexity of a problem in terms of the geometry of its set of solutions.

Finally, we look at optimization problems from a practical point of view. Even if we are faced with an exponentially large search tree of possible solutions, we can prune its branches by proving upper and lower bounds on the quality of the solutions they lead to. With the help of these techniques, many problems that are exponentially hard in theory can be solved quickly in practice.

9.1 Three Flavors of Optimization

How hard is it to find the longest needle? For instance, suppose we want to find the truth assignment that maximizes the number of satisfied clauses in a SAT formula:

> MAX-SAT (optimization)
>
> Input: A SAT formula ϕ
>
> Output: A truth assignment that satisfies as many of ϕ's clauses as possible

This is a search problem, or as we called it in Chapter 2, a function problem. It asks for the best possible truth assignment, or for one of them if there are multiple assignments that are equally good. Another version of MAX-SAT asks for the value of the optimum, without demanding the optimum itself:

> MAX-SAT (value)
>
> Input: A SAT formula ϕ
>
> Output: The largest number ℓ of clauses that can be satisfied at once.

Finally, in our discussion of NP-completeness, we considered the decision problem of whether a solution exists at or above a certain threshold:

> MAX-SAT (threshold)
>
> Input: A SAT formula ϕ and an integer ℓ
>
> Question: Is there a truth assignment that satisfies ℓ or more clauses in ϕ?

What is the relationship between these problems? We showed in Section 5.4.2 that the threshold version of MAX-SAT is NP-complete. In particular, MAX-k-SAT is NP-complete for any $k \geq 2$. What can we say about the complexity of the other two versions?

If we can find the optimum assignment or compute its value, we can immediately decide whether an assignment exists above a certain threshold. Thus

$$\text{MAX-SAT (threshold)} \leq \text{MAX-SAT (optimization)}\,,\ \text{MAX-SAT (value)}\,.$$

Since the threshold version is NP-complete, we can reduce any problem in NP to the other two versions as well. We call problems with this property NP-*hard*. They are at least as hard as any problem in NP, and possibly harder. Similarly, the optimization versions of INDEPENDENT SET, VERTEX COVER, CLIQUE, and MAX CUT are NP-hard, since we showed in Sections 5.3.3 and 5.4.1 that their decision versions are NP-complete.

Moreover, in a certain sense, the search and value versions are harder than the threshold version. To prove that a solution exists above a certain threshold, all I have to do is exhibit one, so the threshold version is in NP. On the other hand, suppose I show you a solution x and claim that it is the best possible one. This is equivalent to the claim that all other solutions are at most as good as x:

$$\forall x' : \text{value}(x') \leq \text{value}(x) .$$

Similarly, suppose I claim that ℓ is the best value we can achieve—that there is a solution with value ℓ, but there are no solutions with value greater than ℓ. This claim involves both a "there exists" and a "for all":

$$\left(\exists x : \text{value}(x) = \ell\right) \wedge \left(\forall x' : \text{value}(x') \leq \ell\right) .$$

A priori, these claims cannot be expressed with just a "there exists," so they do not appear to lie in NP. They belong to other levels of the polynomial hierarchy that we defined in Section 6.1.1, namely Π_1P and DP $\subseteq \Pi_2$P $\cap\, \Sigma_2$P respectively (see Problem 6.1 for the definition of DP).

On the other hand, suppose that we had an oracle that can solve the threshold decision problem. In that case, we could solve the other versions by calling on this oracle a polynomial number of times. If ϕ has m clauses, the largest number of clauses we could possibly satisfy is m. We then find the optimal value by binary search. We ask the threshold oracle whether there is a solution with value at least $m/2$. If it answers "yes" or "no," we ask it if there is a solution with value at least $3m/4$ or at least $m/4$ respectively, and so on. This determines the optimal value with $\log_2 m$ calls to the oracle.

If we can determine the optimal value, can we find the optimal solution? For most natural problems, including MAX-SAT, the answer is yes. The idea is to set one variable at a time. Given a SAT formula $\phi(x_1, x_2, \ldots, x_n)$, setting x_1 true or false satisfies some of ϕ's clauses, shortens others, and yields a SAT formula on the remaining variables, $\phi' = \phi[x_1 \to \texttt{true}]$ or $\phi' = \phi[x_1 \to \texttt{false}]$. One of these settings leads to the optimal assignment. Thus if we let opt(ϕ) denote the optimal value of ϕ,

$$\text{opt}(\phi) = \max\left(\text{opt}(\phi[x_1 \to \texttt{true}]), \text{opt}(\phi[x_1 \to \texttt{false}])\right) .$$

Note that opt(ϕ') counts the clauses of ϕ that are satisfied by our setting of x_1. We compute the optimal value opt(ϕ') for both values of x_1, and give x_1 the truth value such that opt(ϕ') = opt(ϕ). We then try both settings of x_2, and so on.

We can visualize this as in Figure 9.1. We start at the root of a search tree, where no variables are set, and move downward to the leaves, which represent complete assignments. If we can compute the optimal value of any formula, we can tell which branch leads to the optimal leaf and get there without any backtracking. If ϕ has n variables, the total number of times we run the optimal value algorithm is at most $2n$.

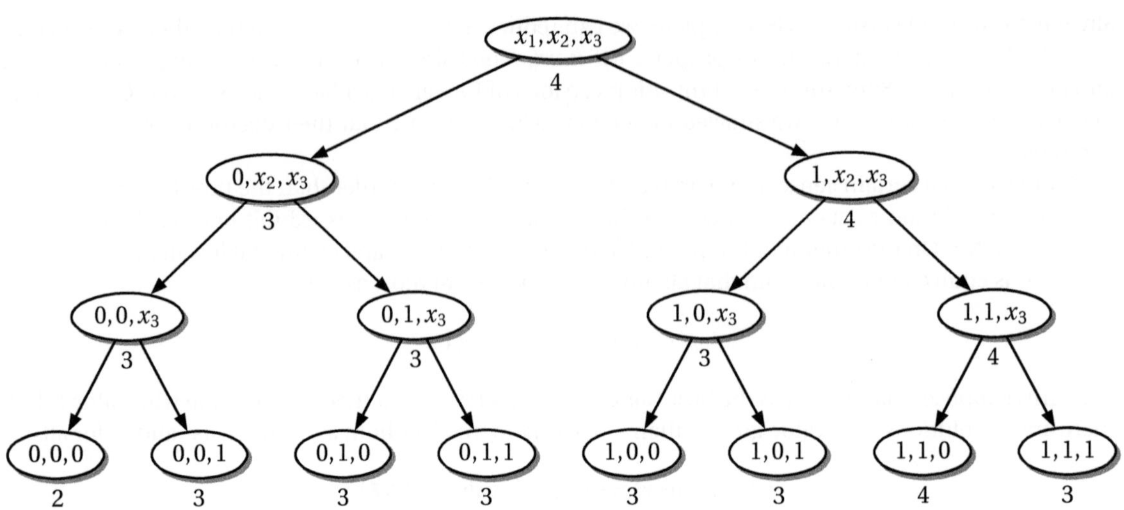

FIGURE 9.1: The search tree of the MAX-2-SAT problem $\phi = (x_1 \vee x_2) \wedge (x_1 \vee \overline{x_2}) \wedge (\overline{x}_1 \vee \overline{x}_3) \wedge (x_2 \vee x_3)$. Each node corresponds to a partial assignment, and a SAT formula on the unset variables. We show the number of satisfied clauses below each leaf, and the optimal number below each node. We can find our way to the optimal leaf by following the branch with the largest optimal value.

This gives us the following chain of reductions,

$$\text{MAX-SAT (optimization)} \leq \text{MAX-SAT (optimal value)} \leq \text{MAX-SAT (threshold)}.$$

It's important to note that these reductions are more general than the ones we used for NP-completeness in Chapter 5. There we wrote $A \leq B$ for two decision problems A, B if we could map single instances of A to single instances of B, mapping yes-instances to yes-instances and no-instances to no-instances. Here $A \leq B$ denotes a *Turing reduction* as defined in Note 5.2. That is, if we have an oracle for B, we can solve A with a polynomial number of calls to that oracle.

This kind of reduction implies that if B is in P, then so is A. So if any of these versions of MAX-SAT is in P, they all are, and P = NP. We can also say that the optimization and optimal value versions are in P^{NP}, the class of problems that we can solve in polynomial time if we have access to an oracle for problems in NP (see Section 6.4).

Our strategy for finding the optimal assignment given an algorithm for computing the optimal value, relies crucially on the fact that MAX-SAT is *self-reducible*: setting a variable in a MAX-SAT problem leads to MAX-SAT problem on the remaining variables. This is not true of haystacks in general. Even if I give you a magical device that tells you the length of the longest needle in the haystack, there is no obvious way to use this device to find the needle.

However, problems in NP are not unstructured haystacks, and most natural optimization problems are self-reducible. Moreover, Problem 9.1 shows that an optimization problem is Turing-reducible to its threshold version whenever the latter is NP-complete.

9.2 Approximations

> It is the mark of an educated mind to rest satisfied with the degree of precision which the nature of the subject admits and not to seek exactness where only an approximation is possible.
>
> Aristotle

If an optimization problem is NP-hard, we can't expect to find an efficient algorithm that finds the optimal solution. But we might hope for an algorithm that finds a *good* solution—one that is guaranteed to be not much worse than the optimum.

9.1

We define "not much worse" in terms of the ratio between the quality of the optimal solution and that of the solution returned by the algorithm. For an algorithm A and an instance x, let $A(x)$ denote the quality of the solution that A produces, and let $\mathrm{OPT}(x)$ be the quality of the optimal solution. We say that A is a ρ*-approximation* for a maximization problem if, for all instances x,

$$\frac{A(x)}{\mathrm{OPT}(x)} \geq \rho \, .$$

In this case $\rho \leq 1$, and we would like ρ to be as large as possible. For a minimization problem, we demand that, for all x,

$$\frac{A(x)}{\mathrm{OPT}(x)} \leq \rho \, ,$$

in which case $\rho \geq 1$ and we would like ρ to be as small as possible. In other words, an approximation algorithm guarantees that the solution it finds is, say, at least half as good—or at most twice as bad—as the optimum.

9.2

Where finding the precise optimum is concerned, NP-complete problems—or, rather, their optimization versions—are all created equal. But when we ask whether we can approximate them within various factors ρ, we find that some are harder than others. As we will see, some can be approximated within a factor arbitrarily close to 1, others can be approximated within a certain constant factor, but no closer, and still others cannot be approximated within any polynomial factor—unless, of course, $\mathsf{P} = \mathsf{NP}$.

9.2.1 *Pebbles and Processors*

We begin with two problems that have very simple approximation algorithms. Recall from Section 4.2.4 that a vertex cover of a graph G is a set S of vertices such that every edge has at least one endpoint in S. Then MIN VERTEX COVER asks for the smallest possible vertex cover of G. Since VERTEX COVER is NP-complete, MIN VERTEX COVER is NP-hard.

Let's play a game in which we cover G's vertices with pebbles. We start with all vertices unpebbled. At each step, we choose an edge whose endpoints are both unpebbled, and we pebble both of them. We continue until there are no such edges. At that point, the set S of pebbled vertices is a vertex cover.

In fact, S is not much larger than the optimal vertex cover S_{opt}. To see this, note that $|S| = 2k$ where k is the number of edges that our algorithm pebbled. Since these k edges have no vertices in common, and

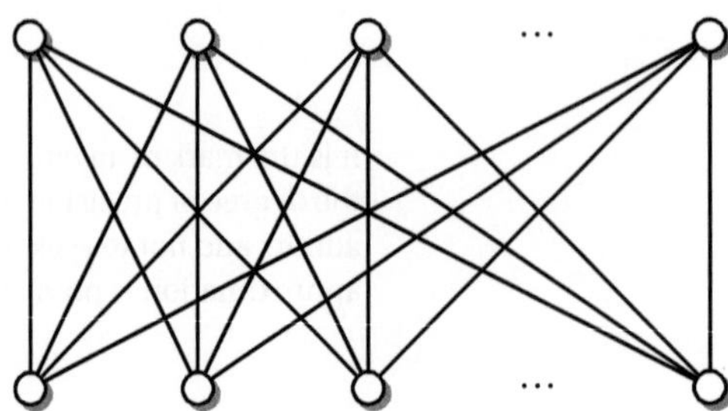

FIGURE 9.2: The pebble algorithm covers all $2n$ vertices in the complete bipartite graph, whereas covering all the vertices on one side gives a vertex cover of size n.

since each one has at least one endpoint in S_{opt}, we have $|S_{\text{opt}}| \geq k$. Thus S is at most twice as large as S_{opt},

$$\rho = \frac{|S|}{|S_{\text{opt}}|} \leq \frac{2k}{k} = 2,$$

and our pebble game is a 2-approximation for MIN VERTEX COVER. This upper bound $\rho \leq 2$ on the pebble game's performance is tight, as the example of Figure 9.2 shows.

Note the two ingredients that went into our proof that $\rho \leq 2$. We need both an upper bound on the performance of our algorithm, and a lower bound on the optimum. We will see this pattern again and again in this chapter—reversed, of course, for maximization problems.

Exercise 9.1 *Does our 2-approximation algorithm for* MIN VERTEX COVER *give a 1/2-approximation for* MAX INDEPENDENT SET*? Why or why not?*

Can we do better? Is there a ρ-approximation for MIN VERTEX COVER for some constant $\rho < 2$? The pebble game achieves the ratio $\rho = 2$ because we put pebbles on both endpoints of each edge we choose. This seems wasteful, since each edge only needs one pebble to cover it. Moreover, we choose these edges in no particular order, without any regard to how many other edges they touch.

Let's be greedier, and place each new pebble so that it covers as many uncovered edges as possible. If we define the degree of a vertex as the number of uncovered edges it is a part of—that is, the number of uncovered neighbors it has—then each step of the greedy algorithm chooses a vertex v of maximal degree and covers it. Equivalently, we add v to S and remove v's edges from the graph.

An algorithm like this that tries to solve an optimization problem by using a reasonable rule-of-thumb is called a *heuristic*, from the Greek εὑρίσκω for "I discover." But before we cry "Eureka!" we had better see how well our heuristic works.

Surprisingly, while this greedy heuristic seems much smarter than the pebble game, it is actually much worse. In fact, it is not a ρ-approximation for any constant ρ. The best approximation ratio it can guarantee grows as $\Theta(\log n)$.

To see why, consider the following bipartite graph, which we call B_t. On one side, there is a set of t vertices $W = \{w_1, \ldots, w_t\}$. On the other side, for each $1 \leq i \leq t$, there is a set of $\lfloor t/i \rfloor$ vertices, $U_i = \{u_{i,1}, \ldots, u_{i,\lfloor t/i \rfloor}\}$. Thus the total number of vertices is

$$n = |W| + \sum_{i=1}^{t} |U_i| = t + \lfloor t/1 \rfloor + \lfloor t/2 \rfloor + \cdots + \lfloor t/t \rfloor \geq t \ln t\,.$$

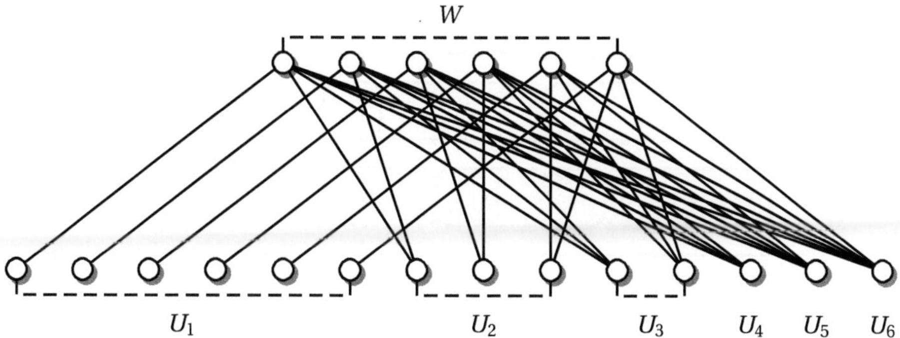

FIGURE 9.3: The bipartite graph B_t for $t = 6$. The smallest vertex cover is W, but the greedy algorithm covers all the vertices in $U = \bigcup_{i=1}^{t} U_i$, and $|U|/|W| = \Theta(\log n)$.

We connect u_{ij} to w_k if and only if $\lceil k/i \rceil = j$. Thus each vertex in U_i has degree i. Moreover, each vertex in W is connected to at most one vertex in U_i for each i, so each vertex in W has degree at most t. We show all this in Figure 9.3.

How does the greedy algorithm perform on B_t? The single vertex in U_t is a vertex of maximum degree t. If we cover it, the degree of each vertex in W decreases by one, and the vertex in U_{t-1} now has maximum degree $t-1$. Covering this vertex leaves us with U_{t-2}, and so on. After we have covered all of U_j for all $j > i$, the vertices in U_i still have degree i, and the vertices in W have degree at most i. So, through a series of unfortunate but legal decisions, the greedy algorithm covers every vertex in $U = \cup_{i=1}^{t} U_i$.

Exercise 9.2 *If you think we're being unfair to the greedy algorithm by breaking ties in favor of covering vertices in U_i instead of vertices of equal degree in W, show how to modify B_t so that the degrees of the vertices in U_i are strictly greater.*

Thus the greedy algorithm produces a vertex cover S of size

$$|S| = \sum_{i=1}^{t} |U_i| \geq t \ln t - t \,.$$

But since W is also a vertex cover, the size of the optimal one is at most

$$\left|S_{\text{opt}}\right| \leq |W| = t \,.$$

Therefore, the approximation ratio is

$$\rho \geq \frac{t \ln t - t}{t} = \Theta(\log n) \,.$$

So, in the worst case, being greedy can give us a vertex cover $\Omega(\log n)$ worse than the optimum.

The good news is that this is as bad as it gets. The greedy algorithm achieves an approximation ratio of $\rho = O(\log n)$ on any graph, and the analogous algorithm performs this well on more general problems

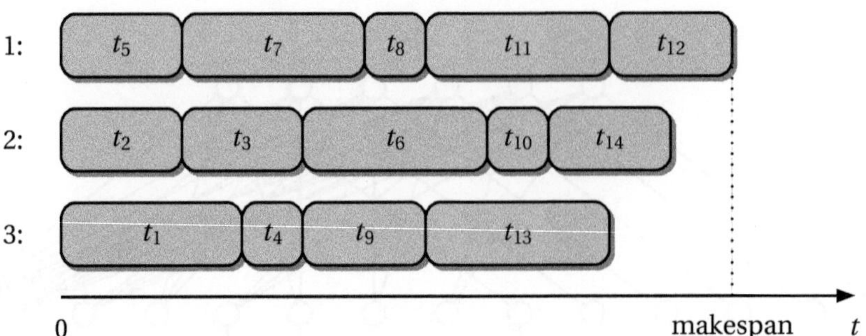

FIGURE 9.4: Scheduling $n = 14$ tasks on $p = 3$ machines.

as well (see Problem 9.8). But it is a little shocking that such an intuitive algorithm fails to achieve any constant ρ, especially given how well the naive pebble algorithm does.

We should take this as a cautionary tale, showing that seemingly obvious and natural heuristics can be easy prey for the adversary. Nevertheless, heuristics are important tools for solving real-world instances of hard optimization problems. In Section 9.10 we will see that, in practice, the right combination of heuristics and rigorous bounds can find solutions much better, and much faster, than the worst-case bounds from theory would suggest.

Let's look at another problem with a simple approximation algorithm. Imagine that you run a computer center with p identical machines, or a parallel computer with p processors, for some constant p. You are given n jobs with running times $t_1, t_2, \ldots, t_n$. Your task is to find a schedule—a function $f : \{1,\ldots,n\} \to \{1,\ldots,p\}$ that assigns the jth job to the $f(j)$th machine.

The *load* on the jth machine is the total time it takes to run the jobs assigned to it,

$$L_j = \sum_{i:f(i)=j} t_i .$$

Since the machines run in parallel, the time it takes us to finish all the jobs is the maximum load on any machine,

$$M = \max_{1 \le j \le p} L_j .$$

We call this time the *makespan*, and your goal is to minimize it. This gives the following problem, illustrated in Figure 9.4:

> MAKESPAN
>
> Input: A set of n running times $t_i \ge 0$, $i = 1, \ldots, n$
>
> Output: A function $f : \{1,\ldots,n\} \to \{1,\ldots,p\}$ that minimizes the makespan $M = \max_{1 \le j \le p} \sum_{i:f(i)=j} t_i$

The following exercise shows that MAKESPAN is NP-hard:

Exercise 9.3 *Reduce* INTEGER PARTITIONING *to the case of* MAKESPAN *where* $p = 2$.

There is, however, a simple algorithm that comes with a quality guarantee. The idea is to divide tasks into two groups, "large" and "small," and schedule the large tasks first. Let $T_{\text{tot}} = \sum_{j=1}^{n} t_j$ be the total running time of all the tasks. We choose some constant $\eta > 0$, say $\eta = 1/10$. We call the jth task *large* if its running time is at least a fraction η of the total, $t_j \geq \eta T_{\text{tot}}$, and *small* if $t_j < \eta T_{\text{tot}}$.

The algorithm then proceeds in two phases:

1. Find a schedule that minimizes the makespan of the large tasks, using exhaustive search.
2. Distribute the small tasks by giving each one to the machine with the smallest load so far, regardless of that task's running time.

How long does this algorithm take? Since there are at most $1/\eta$ large tasks, there are $p^{1/\eta}$ ways to schedule them. Since p and η are constants, this gives a constant number of possible schedules to explore in phase 1, and we can check all of them in polynomial time using exhaustive search. Phase 2 clearly runs in polynomial time as well.

Now let M denote the makespan of the schedule produced by this algorithm. How does it compare to M_{opt}? Let M_{large} denote the maximum load of a machine after phase 1, i.e., the minimum makespan of the large tasks. Since adding the small tasks can only increase the makespan, we have $M_{\text{large}} \leq M_{\text{opt}}$. If the algorithm manages to fit all the small tasks in phase 2 without increasing the makespan, putting them on machines with load less than M_{large}, then $M = M_{\text{large}}$. But since $M_{\text{opt}} \leq M$, in this case $M = M_{\text{opt}}$ and the algorithm performs perfectly.

So, we will focus on the harder case where the makespan increases in phase 2, so that $M > M_{\text{large}}$ and our schedule for the small tasks might not be optimal. Let L_j denote the load of machine j after phase 2. Without loss of generality, we can assume that $L_1 \geq L_2 \geq \cdots \geq L_p$. Since $L_1 = M > M_{\text{large}}$, we know that at least one small task was added to machine 1.

Let t be the running time of the last such task. Since our strategy greedily assigns small tasks to the machines with the smallest load, the load of machine 1 must have been less than or equal to the others just before this task was added. This means that, for all $2 \leq j \leq p$,

$$M - t \leq L_j \leq M .$$

Since $t < \eta T_{\text{tot}}$, we also have

$$T_{\text{tot}} = M + \sum_{j=2}^{p} L_j \geq pM - (p-1)t \geq pM - \eta(p-1)T_{\text{tot}} .$$

Rearranging this gives an upper bound on M,

$$M \leq \frac{T_{\text{tot}}}{p}\left(1 + \eta(p-1)\right) . \tag{9.1}$$

On the other hand, the best schedule we could hope for would be perfectly balanced, keeping all the machines busy all the time. Thus the optimum makespan is bounded below by

$$M_{\text{opt}} \geq \frac{T_{\text{tot}}}{p} .$$

Combining this lower bound on M_{opt} with the upper bound (9.1) on M gives an approximation ratio

$$\rho \le \frac{M}{M_{\text{opt}}} \le 1+\eta(p-1). \tag{9.2}$$

By setting $\eta = \varepsilon/(p-1)$, we can approximate M_{opt} within a factor of $\rho = 1+\varepsilon$ for arbitrarily small $\varepsilon > 0$. In fact, we can define an algorithm that takes both the problem instance and ε as inputs, and returns a $(1+\varepsilon)$-approximation. For each fixed ε, this algorithm runs in polynomial time. Such an algorithm is called a *polynomial-time approximation scheme* or PTAS.

On the face of it, a PTAS seems to be sufficient for all practical purposes. It allows us to find solutions within 1%, 0.1%, 0.01%, ... of the optimum, each in polynomial time. However, improving the quality of the approximation comes at a cost, since the running time increases as ε decreases. Our PTAS for MAKESPAN, for example, has to explore $p^{1/\eta} = p^{(p-1)/\varepsilon}$ possible schedules in phase 1, so its running time grows exponentially as a function of $1/\varepsilon$. Even for $p = 2$ and $\varepsilon = 0.01$, this number is 2^{100}—hardly a constant that we can ignore.

We would prefer an algorithm whose running time is polynomial both in n and in $1/\varepsilon$. Such an algorithm is called a *fully polynomial-time approximation scheme*, or FPTAS for short. There is, in fact, an FPTAS for MAKESPAN, which we will not describe here. However, as the following exercise shows, the idea
9.3 of an FPTAS only makes sense for problems where the optimum can be real-valued or exponentially large.

Exercise 9.4 *Let A be the optimization version of a problem in* NP, *such as* MIN VERTEX COVER, *where the optimum is a nonnegative integer bounded by some polynomial in n. Show that there can be no FPTAS for A unless* P = NP.

9.2.2 The Approximate Salesman

> I don't say he's a great man. Willy Loman never made a lot of money. His name was never in the paper. He's not the finest character that ever lived. But he's a human being, and a terrible thing is happening to him. So attention must be paid.
>
> Arthur Miller, *Death of a Salesman*

In this section, we explore approximation algorithms for what is probably the most famous optimization problem—the TRAVELING SALESMAN problem, or TSP for short.

Given a collection of cities and the cost of travel between each pair of them, TSP asks for the cheapest way to visit each city exactly once and return to your starting point. Thus if we number the cities arbitrarily from 1 to n, an instance of the problem is specified by a $n \times n$ matrix d, where d_{ij} is the cost of traveling from city i to city j.

9.4 We can represent this tour as a permutation $\pi : \{1 \ldots n\} \to \{1 \ldots n\}$ where $\pi(i)$ denotes i's successor. Since our salesman has to visit all n cities on a single round trip, we demand that π strings them all together in a single cycle, $1 \to \pi(1) \to \pi(\pi(1)) \to \cdots \to 1$. We then define TSP as follows,

> TRAVELING SALESMAN
>
> Input: An $n \times n$ matrix $d_{ij} \ge 0$
>
> Output: A cyclic permutation π that minimizes $c(\pi) = \sum_{i=1}^{n} d_{i,\pi(i)}$

Of course, we can also define a threshold version of this problem,

> TRAVELING SALESMAN (threshold)
> Input: An $n \times n$ matrix $d_{ij} \geq 0$ and a number ℓ
> Question: Is there a cyclic permutation π such that $c(\pi) \leq \ell$?

TSP is a weighted version of HAMILTONIAN CYCLE, so we will prove that TSP is NP-hard by reducing HAMILTONIAN CYCLE to it. Given a graph $G = (V, E)$ with n vertices, we define an instance of TSP as

$$d_{ij} = \begin{cases} 1 & \text{if } (i,j) \in E \\ 100 & \text{if } (i,j) \notin E. \end{cases} \tag{9.3}$$

Thus traveling along the edges of G is cheap, but jumping from i to j without such an edge is expensive. Then G has a Hamiltonian cycle if and only if the corresponding TSP has a tour of cost n. If G is not Hamiltonian, the shortest tour has length at least $n+99$. Thus the threshold version of TSP is NP-complete, and the optimization version is NP-hard.

Can we hope for an approximation algorithm for TSP that achieves some reasonable approximation ratio ρ? In general we can't, unless $\mathsf{P} = \mathsf{NP}$. By increasing the cost of jumping from i to j without an edge, we can make the gap between Hamiltonian and non-Hamiltonian graphs as large as we want. If we define the distances as

$$d_{ij} = \begin{cases} 1 & \text{if } (i,j) \in E \\ 1 + \rho n & \text{if } (i,j) \notin E, \end{cases}$$

then the length of the shortest tour is n if G is Hamiltonian and at least $(\rho + 1)n$ if it isn't. If we had a ρ-approximation algorithm for TSP, we could distinguish these two cases from each other, and thus solve HAMILTONIAN CYCLE in polynomial time.

However, there is an important case of TSP for which approximation algorithms do exist. We say that the distances d_{ij} form a *metric* if they are symmetric, i.e., $d_{ij} = d_{ji}$, and they obey the triangle inequality,

$$d_{ij} \leq d_{ik} + d_{kj} \quad \text{for all } i, j, k.$$

We call this case of TSP the metric TSP, or ΔTSP for short.

Exercise 9.5 *Modify the reduction* (9.3) *from* HAMILTONIAN CYCLE *to show that* ΔTSP *is* NP*-hard.*

The best-known metric is the Euclidean distance in the plane, where the distance between (x_i, y_i) and (x_j, y_j) is

$$d_{ij} = \sqrt{(x_i - x_j)^2 + (y_i - y_j)^2}.$$

The corresponding case of ΔTSP is called EUCLIDEAN TSP. Even EUCLIDEAN TSP is NP-hard, but here the reduction from HAMILTONIAN CYCLE is much more involved.

In the remainder of this section, we will give a 2-approximation, and then a (3/2)-approximation, 9.5
for ΔTSP. Of course, these approximation algorithms also work for EUCLIDEAN TSP. One approach that *doesn't* work is being greedy—moving, at each step, to the nearest city we haven't visited yet. This can lead us down a primrose path, forcing us to return to cities that we could have visited earlier by going just

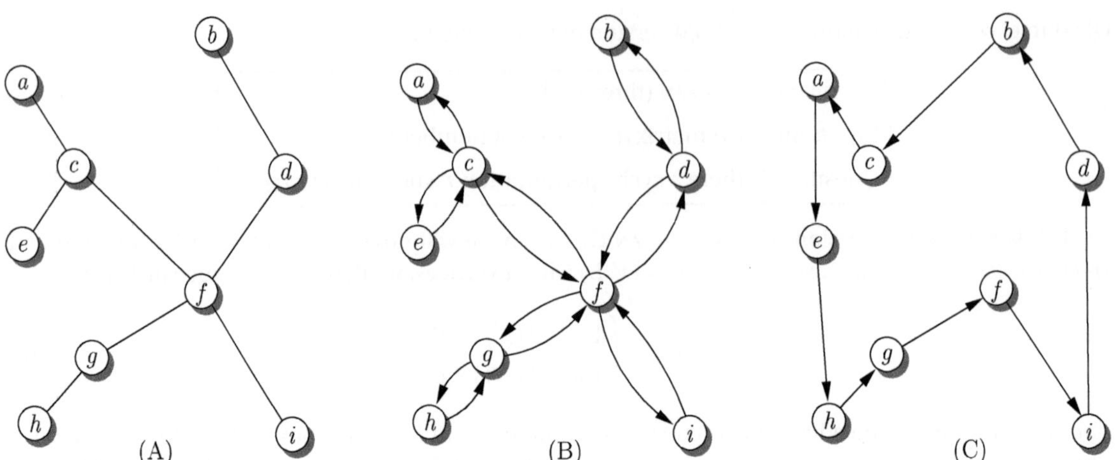

FIGURE 9.5: Constructing a tour from a spanning tree. (A) The minimum spanning tree T, which we can find in polynomial time. (B) Doubling each edge makes T Eulerian. (C) Skipping vertices we have already visited turns the Eulerian tour into a Hamiltonian one, suitable for Mr. Loman.

a little out of our way. Problem 9.10 shows that the greedy tour can be as much as $\Theta(\log n)$ longer than the optimal tour.

Instead of greed, the basic ingredients of our approximation algorithms are Eulerian spanning graphs. A spanning graph is a graph that connects all of the cities. As we saw in the Prologue, if all its vertices have even degree, it has an Eulerian tour—one which traverses each edge once.

We can change an Eulerian tour to a Hamiltonian one simply by skipping vertices we have already visited. If the triangle inequality holds, this never increases the length of the tour—if we have already visited k, skipping directly from i to j incurs a cost $d_{ij} \leq d_{ik} + d_{kj}$. So if we can find an Eulerian spanning graph G whose edges have total length $\ell(G)$, we can find a Hamiltonian tour of length $\ell \leq \ell(G)$.

The shortest spanning graph of all is the minimum spanning tree, which we can find in polynomial time as in Section 3.5.1. Let's denote it T. In general, T has vertices of odd degree, but we can "Eulerify" it simply by doubling each edge as in Figure 9.5. The resulting Eulerian spanning graph has twice the total length of the minimum spanning tree, so skipping vertices we have already visited gives a tour of length at most

$$\ell \leq 2\ell(T). \tag{9.4}$$

On the other hand, if we remove an edge from the optimal cyclic tour, we get a spanning tree consisting of a path of $n-1$ edges. This path has length at most ℓ_{opt} and it is at least as long as the minimum spanning tree, so

$$\ell(T) \leq \ell_{\text{opt}}. \tag{9.5}$$

Combining (9.4) and (9.5) gives $\ell \leq 2\ell_{\text{opt}}$, so we have found a 2-approximation for ΔTSP.

Can we improve this factor of 2? It was the price we paid to Eulerify the spanning tree by doubling all its edges. But only the odd-degree vertices prevent a graph from being Eulerian, so all we have to do is add one edge to each of these vertices. Since there are an even number of odd-degree vertices (see Problem 1.1) we can do this by connecting them in pairs—that is, according to a perfect matching.

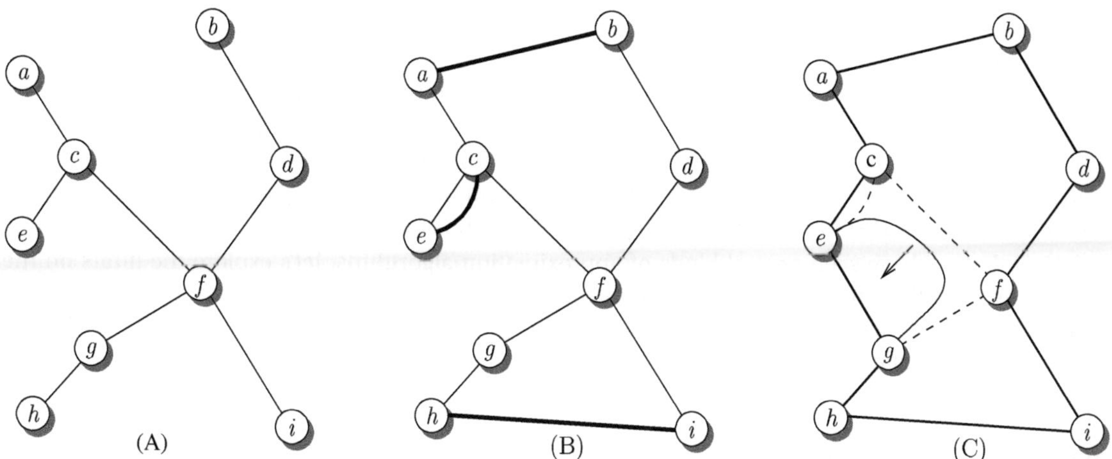

FIGURE 9.6: Christofides' algorithm, for the same cities as in Figure 9.5. (A) The minimum spanning tree T. (B) We Eulerify T by adding a perfect matching M of the odd-degree vertices (the bold edges). (C) We again follow an Eulerian tour, skipping cities we have already visited.

This is the idea behind Christofides' algorithm, which we show in Figure 9.6. We again start with the minimum spanning tree T. Let U denote the set of vertices that have odd degree in T, and let M be the perfect matching on U whose edges have the smallest total length $\ell(M)$. We then take an Eulerian tour of $T \cup M$, skipping the cities we have already visited. This gives a tour of length

$$\ell \leq \ell(T) + \ell(M).$$

What approximation ratio does this algorithm achieve? As before, $\ell(T) \leq \ell_{\text{opt}}$. We claim also that

$$\ell(M) \leq \frac{\ell_{\text{opt}}}{2}, \tag{9.6}$$

and therefore that Christofides' algorithm gives a (3/2)-approximation,

$$\ell \leq \frac{3}{2}\ell_{\text{opt}}.$$

To prove (9.6), consider a salesman who only has to visit the cities in U. He can do this by following the optimal tour but skipping the other cities. By the triangle inequality, the resulting tour has length $\ell_U \leq \ell_{\text{opt}}$. But this tour contains two perfect matchings on U, namely

$$M_1 = \{(u_1, u_2), (u_3, u_4), \ldots, (u_{t-1}, u_t)\} \text{ and } M_2 = \{(u_2, u_3), (u_4, u_5), \ldots, (u_t, u_1)\},$$

where we number U's vertices $u_1, u_2, \ldots, u_t$ according to their order in this tour. Since

$$\ell(M_1) + \ell(M_2) = \ell_U \leq \ell_{\text{opt}},$$

either $\ell(M_1) \leq \ell_{\text{opt}}/2$ or $\ell(M_2) \leq \ell_{\text{opt}}/2$. In either case, the shortest possible perfect matching on U has length at most $\ell_{\text{opt}}/2$, proving (9.6).

We can find the minimum-length perfect matching M in polynomial time using a variant of Edmonds' "paths, trees, and flowers" algorithm (see Note 5.6). Thus Christofides' algorithm is a polynomial-time 3/2-approximation. Despite its simplicity, it is still the best known approximation for ΔTSP in general.

9.6 On the other hand, the triangle inequality is a very weak assumption about the distances d_{ij}. For particular metrics such as the EUCLIDEAN TSP, we can do much better. By dividing the plane into squares, solving the TSP for the subset of cities in each square, and then connecting the resulting tours to each other, we can get a $(1+\varepsilon)$-approximation in polynomial time for any fixed $\varepsilon > 0$—that is, a PTAS.

Now that we have seen several kinds of approximation algorithms, let's explore the limits on their
9.7 power. As we will see in the next section, for some NP-complete problems there are precise thresholds α such that we can achieve an α-approximation ratio in polynomial time, but no ρ-approximation with $\rho < \alpha$ is possible unless P = NP.

9.3 Inapproximability

> All of us failed to match our dreams of perfection. So I rate us on the basis of our splendid failure to do the impossible.
>
> William Faulkner

How far can approximation algorithms go? Is there a ρ-approximation for MIN VERTEX COVER with $\rho < 2$, or for ΔTSP with $\rho < 3/2$? How close can ρ get to 1 for these problems?

We have already seen a few cases where approximation algorithms can only go so far. We showed that, in the absence of the triangle inequality, we cannot approximate the TSP within any constant factor ρ, since doing so would let us solve HAMILTONIAN CYCLE. Similarly, Exercise 9.4 showed that MIN VERTEX COVER cannot have an FPTAS—that is, we cannot achieve $\rho = 1 + \varepsilon$ with a running time that is only polynomial in $1/\varepsilon$. However, we could still hope for a PTAS, whose running time grows superpolynomially in $1/\varepsilon$, but which is polynomial in n for any fixed $\varepsilon > 0$.

In this section, we will show that for many NP-complete problems there is a constant ρ beyond which no approximation is possible unless P = NP. By drawing on deep results on interactive proofs—which we will get a taste of in Chapter 11—we will show that it is NP-hard to ρ-approximate MIN VERTEX COVER for any $\rho < 7/6$, or to ρ-approximate MAX-3-SAT for any $\rho > 7/8$. For some problems, including MAX-SAT, we will determine the precise threshold beyond which polynomial-time approximation is impossible.

9.3.1 *Mind the Gap*

Let's rephrase the fact that the threshold version of VERTEX COVER is NP-complete. Given a graph G and an integer k, it is NP-complete to tell which of the following is true: whether the size of the smallest vertex cover is at most k, or at least $k+1$.

What happens if we increase the gap between k and $k+1$? Consider the following problem, where α and β are real numbers such that $0 \le \beta < \alpha \le 1$:

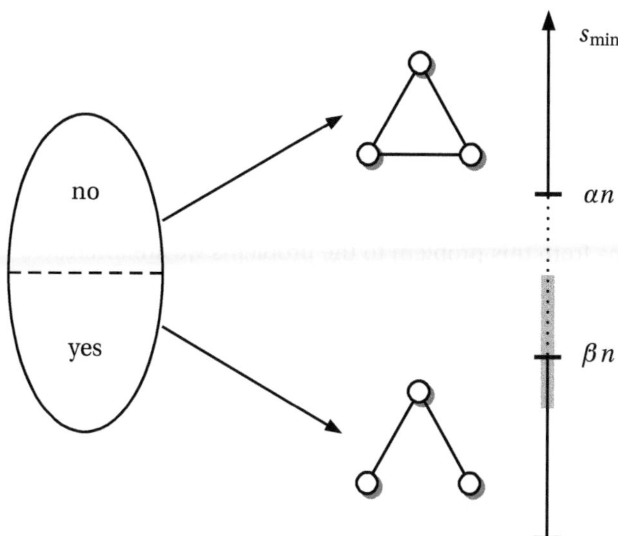

FIGURE 9.7: A reduction from an NP-complete problem to GAP VERTEX COVER. It maps yes-instances to graphs with vertex covers of size $s_{\min} \leq \beta n$, and no-instances to graphs where $s_{\min} \geq \alpha n$. If $\rho < \alpha / \beta$a ρ-approximation to MIN VERTEX COVER (the gray line) would be able to distinguish between these two cases, implying P = NP.

GAP VERTEX COVER
Input: A graph G with n vertices whose smallest vertex cover has size $s_{\min}$, where either $s_{\min} \leq \beta n$ or $s_{\min} \geq \alpha n$
Question: Is $s_{\min} \leq \beta n$?

This is a *promise problem*, in which we are assured that the input has a given property. If this promise is broken—that is, if the size of G's smallest vertex cover is between αn and βn—then the algorithm is allowed to return any answer it likes.

Clearly GAP VERTEX COVER is NP-complete if $\alpha = \beta + 1/n$. But intuitively, it gets easier as the gap between α and β gets wider. If $\alpha > 2\beta$ it is in P, since our 2-approximation algorithm for MIN VERTEX COVER will give a vertex cover of size $s \leq 2\beta n < \alpha n$ if and only if $s_{\min} \leq \beta n$. In general, if there is a polynomial-time ρ-approximation for VERTEX COVER, then GAP VERTEX COVER is in P whenever $\alpha/\beta > \rho$.

Now suppose that we can reduce some NP-complete problem to GAP VERTEX COVER as shown in Figure 9.7, where yes- and no-instances are mapped to graphs where $s_{\min} \leq \beta n$ and $s_{\min} \geq \alpha n$ respectively. Then unless P = NP, there can be no polynomial-time ρ-approximation for MIN VERTEX COVER for any $\rho < \alpha / \beta$The same goes for maximization problems with $\rho > \beta / \alpha$

Unfortunately, the NP-complete problems we have seen so far come with asymptotically vanishing gaps. All it takes to turn a yes-instance into a no-instance is a single unsatisfied clause in MAX-SAT, or a single edge in MAX CUT. Thus in a formula with m clauses, or a graph with m edges, the largest gap that we can promise is just $\alpha/\beta = 1 + \varepsilon$ where $\varepsilon \geq 1/m = 1/\text{poly}(n)$.

To prove the NP-hardness of a ρ-approximation for some $\rho > 1$, we need a *gap-amplifying* reduction that widens the gap between yes- and no-instances from $1/\text{poly}(n)$ to $\Theta(1)$ while preserving NP-completeness. It is far from obvious how to design such reductions, or even whether they exist. In fact they do, but they rely on sophisticated ideas from algebra and graph theory. They lie at the heart of the PCP Theorem, one of the most celebrated results in theoretical computer science, and we will meet them in Section 11.3.

Once we know a single NP-complete problem that comes with a sufficiently large gap, all we need are *gap-preserving* reductions from this problem to the problems we are interested in. It turns out that the problem that works best as a starting point for this type of reduction is MAX-XORSAT.

Like SAT, XORSAT is a Boolean constraint satisfaction problem with clauses and literals, but an XORSAT clause combines literals through exclusive OR rather than OR. The exclusive OR of x and y is written as $x \oplus y$. Here is an example of a 3-XORSAT formula:

$$(x_1 \oplus \overline{x}_3 \oplus x_4) \wedge (\overline{x}_2 \oplus x_3 \oplus \overline{x}_5) \wedge (\overline{x}_1 \oplus x_2 \oplus x_5).$$

If we map `true` and `false` to 1 and 0 respectively, $\oplus$ becomes addition mod 2 and an XORSAT formula becomes a system of linear equations. The example above corresponds to the system

$$\begin{aligned} x_1 + x_3 + x_4 &\equiv_2 0 \\ x_2 + x_3 + x_5 &\equiv_2 1 \\ x_1 + x_2 + x_5 &\equiv_2 0. \end{aligned}$$

Note how we got rid of negations by moving them to the right-hand side of the equation. We can solve such systems in polynomial time, since algorithms like Gaussian elimination work perfectly well in the finite field of integers mod 2. Hence the problem XORSAT, of telling whether we can satisfy *every* clause of a XORSAT formula, is in P.

If we cannot satisfy all the clauses, however, the problem of telling how many of them we can satisfy is NP-complete. Analogous to MAX-SAT, we define

> MAX-k-XORSAT
> Input: An k-XORSAT formula ϕ and an integer ℓ
> Question: Is there a truth assignment that satisfies ℓ or more clauses in ϕ?

Exercise 9.6 *Show that* MAX-k-XORSAT *is* NP*-hard for* $k \geq 2$. *Hint: reduce from* MAX CUT.

To serve as a starting point for gap-preserving reductions, we focus on instances of XORSAT where we are promised that the maximum number of satisfiable clauses is either at most βm or at least αm, where m is the total number of clauses. The problem is to decide on which side of this gap the instance lies:

> GAP-k-XORSAT
> Input: A k-XORSAT formula with m clauses such that the maximum number of satisfiable clauses is either at most βm or at least αm
> Question: Is there an assigment that satisfies at least αm clauses?

We want GAP-XORSAT to be NP-complete, and we want the gap to be as large as possible. Obviously α has to be smaller than 1, since the case $\alpha = 1$ is in P.

On the other hand, β has to be larger than 1/2, because we can *always* satisfy at least $m/2$ clauses. Consider a random truth assignment $x = (x_1, \ldots, x_n)$ where we choose each $x_i \in \{0,1\}$ by flipping a coin. The probability that x satisfies a given clause is 1/2. By linearity of expectation (see Appendix A.3.2), $m/2$ clauses are satisfied on average. Now either every assignment satisfies precisely half of the clauses, or some satisfy less and others more. In either case, there is at least one assignment that satisfies at least half of the clauses—and as we discuss below, we can find one in polynomial time.

So in order to make GAP-XORSAT hard, we need $\alpha < 1$ and $\beta > 1/2$. Surprisingly, these bounds are enough to ensure NP-completeness. As we will discuss in Section 11.3, the strongest form of the PCP Theorem implies the following:

Theorem 9.1 (Inapproximability of MAX-XORSAT) *For any $k \geq 3$ and any α, β such that $1/2 < \beta < \alpha < 1$, GAP-k-XORSAT is NP-complete. Therefore, it is NP-hard to ρ-approximate MAX-XORSAT for any $\rho > 1/2$.*

Now, just as we used SAT as a starting point for NP-completeness, we can use MAX-XORSAT to prove inapproximability for many other problems by applying gap-preserving reductions. Let's start by reducing MAX-3-XORSAT to MAX-3-SAT. Any 3-XORSAT clause is the conjunction of four 3-SAT clauses, which forbid the four truth assignments with the wrong parity. For example,

$$x_1 \oplus x_2 \oplus x_3 = (x_1 \vee x_2 \vee x_3) \wedge (x_1 \vee \overline{x}_2 \vee \overline{x}_3) \wedge (\overline{x}_1 \vee x_2 \vee \overline{x}_3) \wedge (\overline{x}_1 \vee \overline{x}_2 \vee x_3).$$

If the XORSAT clause is satisfied, then all four 3-SAT clauses in this gadget are satisfied. If it isn't, then exactly three of them are.

Applying this reduction converts a 3-XORSAT formula ϕ with m clauses to a 3-SAT formula ϕ' with $m' = 4m$ clauses. Moreover, we can satisfy αm clauses of ϕ if and only if we can satisfy $\alpha' m'$ clauses of ϕ', where

$$\alpha' = \frac{3+\alpha}{4}.$$

This shows that GAP-3-SAT is NP-complete whenever $\alpha < 1$ and $\beta > 7/8$. Therefore, it is NP-hard to ρ-approximate MAX-3-SAT for any $\rho > 7/8$.

Exercise 9.7 *Generalize this to show that MAX-k-SAT is NP-hard to approximate for any $\rho > 1 - 2^{-k}$.*

On the other hand, we can satisfy 7/8 of the clauses of a 3-SAT formula just as we satisfied 1/2 the clauses of an XORSAT formula—by choosing a random truth assignment. Since each clause only forbids one of the eight possible truth assignments of its variables, this assignment satisfies it with probability 7/8. Again using linearity of expectation, if there are m clauses then the expected number of satisfied clauses is $(7/8)m$.

This gives a randomized approximation algorithm for MAX-3-SAT—just flip n coins and output the result. Since the best the optimum could do is satisfy every clause, this algorithm achieves an approximation ratio of $\rho = 7/8$ on average. We can do the same for MAX-k-SAT, approximating it within $\rho = 1 - 2^{-k}$.

Many of the simplest and most beautiful approximation algorithms are randomized, and we will see more of them in Chapter 10. But an algorithm like this one could have bad luck, repeatedly choosing a worse-than-average assignment. If you would prefer a deterministic algorithm, we show you how to

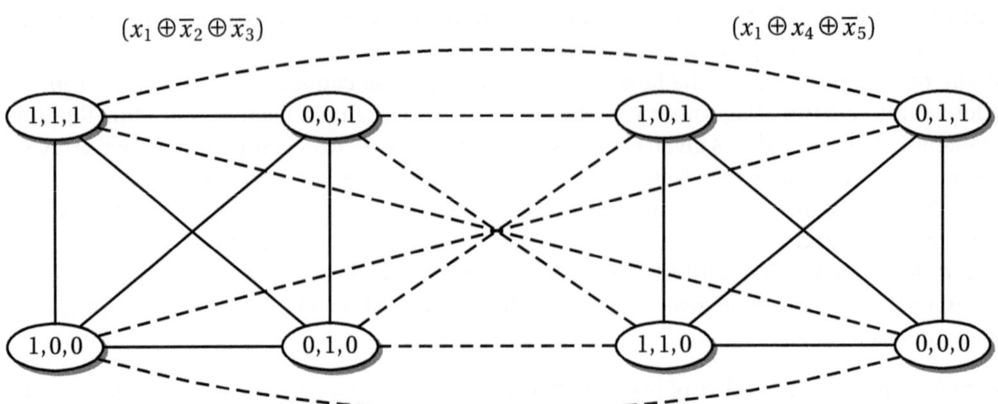

FIGURE 9.8: Reducing 3-XORSAT to INDEPENDENT SET. Each clause gadget is a fully connected graph (solid lines), in which each vertex represents a satisfying assignment of that clause's variables. Additional edges (dashed lines) run between the clause gadgets, representing conflicting assignments (here for the variable x_1). An independent set S in this graph corresponds to an assignment that satisfies at least $|S|$ clauses.

derandomize it in Problem 9.9, deterministically finding a truth assignment that is at least as good as a random one. This technique lets us approximate MAX-k-SAT within a factor of $1-2^{-k}$, and MAX-XORSAT within a factor of $1/2$, in polynomial time.

For MAX-XORSAT and MAX-k-SAT, we have found a precise bound on how well we can approximate them in polynomial time—a constant ρ such that there there is an efficient ρ-approximation, but where ρ'-approximation is NP-hard for any $\rho' > \rho$. Such an *optimal inapproximability result* tells us exactly what we can and can't achieve in polynomial time. It would be nice to have similar sharp results for other optimization problems.

9.3.2 *Cover Me:* MIN VERTEX COVER

Let's turn our attention to MIN VERTEX COVER. We claimed at the beginning of this section that it is NP-hard to ρ-approximate MIN VERTEX COVER for any $\rho < 7/6$. To prove this, we will construct a gap-preserving reduction from 3-XORSAT, through INDEPENDENT SET, to VERTEX COVER.

We convert an 3-XORSAT formula ϕ to a graph G as follows. For each clause c we include a complete graph with four vertices, where each vertex corresponds to one of the four satisfying truth assignments of c's variables. We then add edges between these clause gadgets, connecting any two vertices that correspond to conflicting truth assignments as shown in Figure 9.8. Now consider the following exercise:

Exercise 9.8 *Show that we can satisfy at least ℓ of ϕ's clauses if and only if G has an independent set of size at least ℓ.*

Since this reduction maps the number of satisfied clauses directly to the size of the independent set, the ratio between yes- and no-instances remains exactly the same. If we could approximate MAX INDEPENDENT SET for some $\rho > \beta / \alpha$ we could distinguish the case where G's largest independent set has size at least αm from that where it is most βm, and thus solve the corresponding GAP-3-XORSAT problem. Hence ρ-approximating MAX INDEPENDENT SET is NP-hard for any $\rho > 1/2$.

Now, at some cost to the gap, we reduce to VERTEX COVER. Since G has $n = 4m$ vertices, it has an independent set of size αm if and only if it has a vertex cover of size $(4-\alpha)m = (1-\alpha/4)n$, and similarly for β. This reduces GAP-3-XORSAT to GAP VERTEX COVER where

$$\alpha' = 1 - \beta/4 \quad \text{and} \quad \beta' = 1 - \alpha/4 .$$

Since GAP-XORSAT is NP-complete whenever $\alpha < 1$ and $\beta > 1/2$, this shows that GAP VERTEX COVER is NP-complete whenever $\alpha < 7/8$ and $\beta > 3/4$. Therefore, approximating MIN VERTEX COVER is NP-hard for any $\rho < 7/6$.

At this point, our best approximation algorithm for MIN VERTEX COVER has $\rho = 2$, and we have an inap- 9.8
proximability result for $\rho < 7/6$. Contrary to the situation for MAX-SAT, this leaves us with a gap between the best known approximation and the best known inapproximability bound. Is there an approximation algorithm with $\rho < 2$? This question is still open—it is conjectured, but not known, that $\rho = 2$ is the best possible approximation ratio unless P = NP. Reductions from MAX-XORSAT have yielded inapproximability results for many other problems as well, including ΔTSP.

9.3.3 *Münchhausen's Method:* MAX INDEPENDENT SET

What about MAX INDEPENDENT SET? We showed above that ρ-approximating it is NP-hard for $\rho > 1/2$. Can we meet that bound, and approximate it within a factor of $1/2$?

We will show here that we can't. In fact, unless P = NP, we can't ρ-approximate MAX INDEPENDENT SET for any constant $\rho > 0$. The reason is a property called *self-improvability.* Just as Baron Münchhausen escaped from a swamp by pulling himself up by his own bootstraps, we can improve any ρ-approximation for MAX INDEPENDENT SET until $\rho > 1/2$. To prevent the Baron from proving that P = NP—the sort of claim that he is likely to make—we can't have $\rho > 0$ in the first place.

Our bootstraps are compositions of graphs, as shown in Figure 9.9. The composition $G_1[G_2]$ of G_1 and G_2 is a graph with vertices $V = V_1 \times V_2$. We construct it by replacing each vertex of G_1 with a copy of G_2 and replacing each edge (u, v) of G_1 with a complete bipartite graph, joining every vertex in the copy of G_2 at u to every vertex of the copy of G_2 at v.

Let $s_{\max}(G)$ denote the size of the maximum independent set of G. The following exercise shows that composing two graphs multiplies their values of $s_{\max}$:

Exercise 9.9 *Show that, for any two graphs G_1, G_2,*

$$s_{\max}(G_1[G_2]) = s_{\max}(G_1)\, s_{\max}(G_2) .$$

Now suppose that a polynomial-time algorithm A_1 is a ρ-approximation for MAX INDEPENDENT SET. That is, it returns an independent set of size

$$A_1(G) \geq \rho\, s_{\max}(G) .$$

Then we can define a new algorithm $A_2(G) = \sqrt{A_1(G[G])}$ which is a $\sqrt{\rho}$-approximation, since

$$A_2(G) = \sqrt{A_1(G[G])} \geq \sqrt{\rho\, s_{\max}(G[G])} = \sqrt{\rho}\, s_{\max}(G) .$$

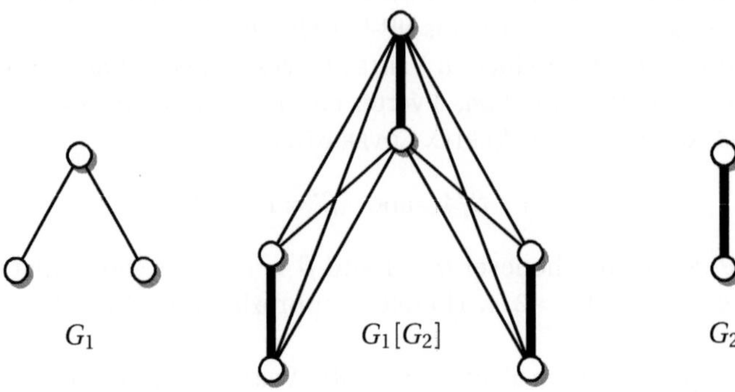

FIGURE 9.9: The composition of two graphs.

More generally, for any integer $k \geq 0$ we can define an algorithm A_k as

$$A_k(G) = A_1(G^k)^{1/k} \quad \text{where} \quad G^k = \underbrace{G[G[\cdots[G]]\cdots]}_{k \text{ times}},$$

and then

$$A_k(G) \geq \rho^{1/k}\, s_{\max}(G).$$

While the size of G', and therefore the running time of A_k, grows exponentially as a function of k, for any fixed k it is polynomial in n. Therefore, A_k is a $\rho^{1/k}$-approximation.

But this leads to a contradiction. Whatever approximation ratio $\rho > 0$ we assume that our original algorithm A_1 can achieve, there is some constant k such that $\rho^{1/k} > 1/2$, and A_k is a ρ'-approximation where $\rho' > 1/2$. Since we showed above that this is impossible, A_1 can't have been a ρ-approximation in the first place. Thus, like TSP, we cannot approximate MAX INDEPENDENT SET within any constant factor unless $\mathsf{P} = \mathsf{NP}$.

9.9

9.4 Jewels and Facets: Linear Programming

> There are hardly any speculations in geometry more useful or more entertaining than those which relate to maxima and minima.
>
> Colin Maclaurin

In the next few sections, we focus on classes of optimization problems that ask us to maximize or minimize some function subject to a set of constraints. The crucial property that allows us to solve these problems efficiently is their convexity—the fact that any local optimum is the global optimum. In the landscape analogy of Section 3.5.3, their landscapes consist of one large mountain or valley, albeit a high-dimensional one.

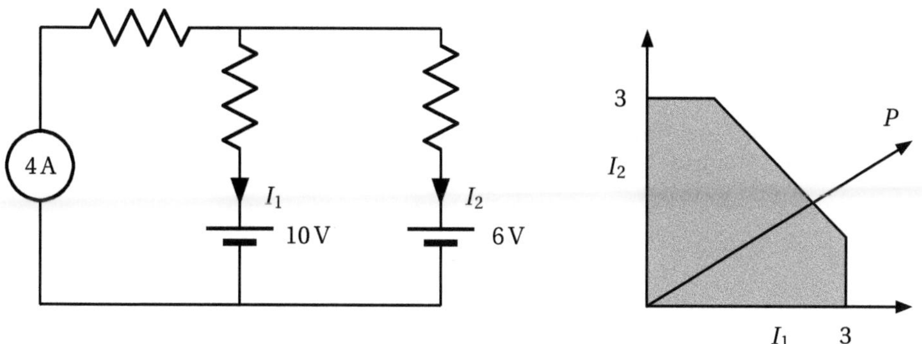

FIGURE 9.10: On the left, a circuit for charging two batteries. We want to maximize the total power $P = 10I_1 + 6I_2$, subject to the constraints $I_1 \leq 3$, $I_2 \leq 3$, and $I_1 + I_2 \leq 4$. On the right, this corresponds to moving as far out as possible parallel to the vector P, while staying within the feasible polygon.

We start with LINEAR PROGRAMMING—a name that, surprisingly, does not refer to writing computer programs line by line. It refers to the problem of finding the minimum or maximum of a linear function, where the variables are subject to linear inequalities. It was born in the 1950s when very few electronic computers were around, and "programming" meant planning or scheduling.

LINEAR PROGRAMMING was motivated by practical planning problems from industry and economics, and even today it is used for a wide variety of practical optimization problems. Our motivation for discussing LINEAR PROGRAMMING is not its practical relevance, however, but its beauty and versatility. It has a very intuitive geometrical nature, and it allows us to reformulate many other optimization problems, both easy and hard, in this geometrical setting. This will foster our understanding of optimization in general, and the struggle to achieve a high-quality solution while working within a set of constraints.

We will look at several types of algorithms for LINEAR PROGRAMMING, including crawling along the 9.10
surface of allowed solutions and approaching the optimum from within. Along the way, we will see that the duality between MAX FLOW and MIN CUT that we studied in Section 3.6 is just an example of a larger phenomenon, and that it has an intuitive explanation in terms of the equilibrium of physical forces. Later on in this chapter, we will see how to use LINEAR PROGRAMMING to narrow the search space for NP-hard optimization problems, and to find approximate solutions to them—finding good solutions even in the worst case, and often finding the best solution in practice.

9.4.1 *Currents and Flows*

We begin our discussion of LINEAR PROGRAMMING with a somewhat prosaic example. For better or worse, we have become inseparable from our electronic companions—our laptops, mobile phones, and so on. But when we travel we have to carry a charger for each device, and this is quite annoying. Let's build a charger that can be used for all our devices, and that will charge the batteries as fast as possible.

Let's assume that we want to charge two batteries whose voltages are 10 volts and 6 volts respectively. We plug a power supply into a circuit like that shown in Figure 9.10. Our goal is to determine the currents I_1 and I_2 in a way that maximizes the total power delivered to the batteries, while respecting the constraints on the power supply and the batteries. Since the power is $P = VI$, the total power is

$$P = 10I_1 + 6I_2 .$$

Suppose, however, that we will burn out the batteries if we give either one more than 3 amperes of current. Suppose also that the power supply has enough voltage to power the batteries, but that it can only deliver a current of up to 4 amperes. Finally, we don't want to discharge the batteries, so neither I_1 nor I_2 can be negative. This gives us a system of constraints

$$\begin{aligned} I_1 &\le 3 \\ I_2 &\le 3 \\ I_1 + I_2 &\le 4 \\ I_1 &\ge 0 \\ I_2 &\ge 0 . \end{aligned} \tag{9.7}$$

We call a solution (I_1, I_2) *feasible* if it satisfies these constraints. In the (I_1, I_2) plane, each constraint becomes a half-plane. The feasible set is their intersection, which is the five-sided polygon shown on the right of Figure 9.10.

Maximizing $P = 10I_1 + 6I_2$ means moving as far as possible in the direction of the vector $(10, 6)$ while staying inside the feasible polygon. It is not hard to see that the optimal solution is the corner of the polygon at $(I_1, I_2) = (3, 1)$.

The problem we have just solved is an instance of LINEAR PROGRAMMING: maximizing a linear function of real-valued of variables, subject to a set of constraints in the form of linear inequalities. To formulate it in general, we express the variables and constraints with vectors and matrices. We adopt the convention that $\mathbf{x}$ is a column vector, so that its transpose $\mathbf{x}^T$ is a row vector. We denote the inner product of two vectors in $\mathbb{R}^d$ as $\mathbf{c}^T\mathbf{x} = \sum_{i=1}^{d} c_i x_i$. We also write $\mathbf{a} \le \mathbf{b}$ to mean that $a_i \le b_i$ for all i.

The general definition of LINEAR PROGRAMMING, or LP for short, is this:

> LINEAR PROGRAMMING
>
> Input: An $m \times d$ matrix $\mathbf{A}$ and vectors $\mathbf{b} \in \mathbb{R}^m$ and $\mathbf{c} \in \mathbb{R}^d$
>
> Output: A vector $\mathbf{x} \in \mathbb{R}^d$ that maximizes $\mathbf{c}^T\mathbf{x}$, subject to the m constraints
>
> $$\mathbf{A}\mathbf{x} \le \mathbf{b} \tag{9.9}$$

In our battery-charging example, there are $d = 2$ variables and $m = 5$ constraints. The total power is $\mathbf{c}^T\mathbf{x}$ where

$$\mathbf{c} = \begin{pmatrix} 10 \\ 6 \end{pmatrix} \quad \text{and} \quad \mathbf{x} = \begin{pmatrix} I_1 \\ I_2 \end{pmatrix} .$$

The five constraints (9.7) on the currents can be expressed as $\mathbf{A}\mathbf{x} \le \mathbf{b}$, where

$$\mathbf{A} = \begin{pmatrix} 1 & 0 \\ 0 & 1 \\ 1 & 1 \\ -1 & 0 \\ 0 & -1 \end{pmatrix} \quad \text{and} \quad \mathbf{b} = \begin{pmatrix} 3 \\ 3 \\ 4 \\ 0 \\ 0 \end{pmatrix} .$$

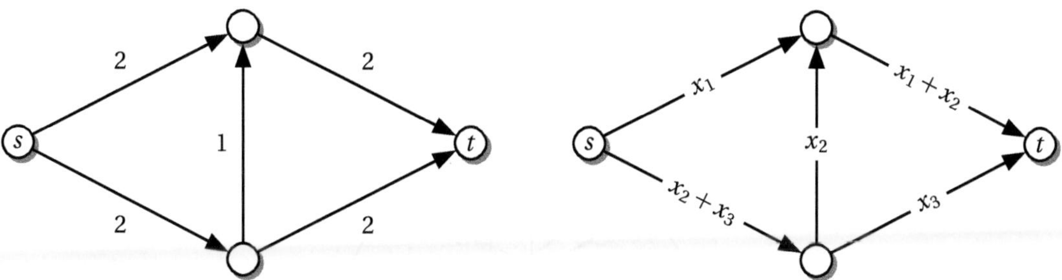

FIGURE 9.11: An instance of MAX FLOW. On the left, the capacities. On the right, we can describe the flow with three variables.

The types of constraints we have allowed in our definition of LP are somewhat arbitrary. For instance, we could replace the inequalities $\mathbf{Ax} \le \mathbf{b}$ with the equation $\mathbf{Ax} + \mathbf{s} = \mathbf{b}$, and simply require that $\mathbf{s} \ge \mathbf{0}$. Conversely, we can express linear equations as a pair of inequalities. We have chosen the form (9.9) to simplify our discussion, but we will sometimes use other forms. In particular, we will often separate the constraints $\mathbf{x} \ge \mathbf{0}$ from other constraints of the form $\mathbf{Ax} \le \mathbf{b}$.

There are two ways in which an instance of LP could be degenerate. First, the inequalities $\mathbf{Ax} \le \mathbf{b}$ could be contradictory, so that the feasible set is empty. Second, the feasible set could be infinite, and could include points arbitrarily far out along the $\mathbf{c}$ direction, so that $\mathbf{c}^T\mathbf{x}$ is unbounded. For now, we will assume that the feasible set is nonempty but bounded, so that the optimum $\max_{\mathbf{x}} \mathbf{c}^T\mathbf{x}$ is well-defined.

LINEAR PROGRAMMING derives much of its importance from its versatility—the fact that a wide range of optimization problems can be expressed as maximizing a linear function subject to linear inequalities. For example, consider the MAX FLOW problem of Section 3.6. The flow through the network shown in Figure 9.11 can be characterized by the vector $\mathbf{x}^T = (x_1, x_2, x_3)$. In order to describe the flow with just three independent variables, we have built the constraints that the total flow in and out of each vertex is zero directly into our definition. The other constraints—that the flow on each edge is at most its capacity, and that the flow along each edge is nonnegative—can be written as $\mathbf{Ax} \le \mathbf{b}$ where

$$\mathbf{A} = \begin{pmatrix} 1 & 0 & 0 \\ 1 & 1 & 0 \\ 0 & 1 & 0 \\ 0 & 1 & 1 \\ 0 & 0 & 1 \\ -1 & 0 & 0 \\ 0 & -1 & 0 \\ 0 & 0 & -1 \end{pmatrix} \quad \text{and} \quad \mathbf{b} = \begin{pmatrix} 2 \\ 2 \\ 1 \\ 2 \\ 2 \\ 0 \\ 0 \\ 0 \end{pmatrix}.$$

We wish to maximize the value of the flow, $x_1 + x_2 + x_3 = \mathbf{c}^T\mathbf{x}$ where $\mathbf{c}^T = (1,1,1)$. Figure 9.12 shows the feasible region—it is a three-dimensional polyhedron, where each facet is the set of points where one of the constraints holds exactly.

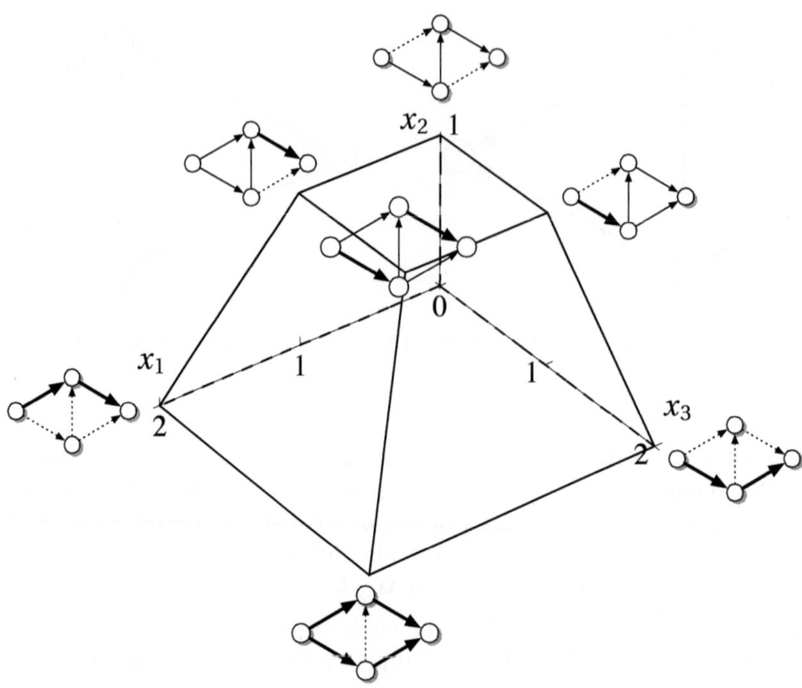

FIGURE 9.12: The polytope of feasible solutions in our instance of MAX FLOW. Dotted, solid, and bold lines correspond to flows of 0, 1 and 2 respectively. The optimal flow is at the vertex $(x_1, x_2, x_3) = (2, 0, 2)$ at the bottom of the figure.

9.4.2 The Feasible Polytope

LINEAR PROGRAMMING has an intuitive geometrical interpretation. The set of feasible solutions, i.e., the set of vectors

$$\left\{\mathbf{x} \in \mathbb{R}^d : \mathbf{A}\mathbf{x} \leq \mathbf{b}\right\},$$

forms a d-dimensional jewel or *polytope*. Each of its facets is a $(d-1)$-dimensional hyperplane, corresponding to one of the constraints being satisfied exactly. In two or three dimensions, this jewel is a just a polygon or a polyhedron, and sparkles as much as the average engagement ring. But when d and m are large, it becomes a far more beautiful and complicated object—typically with an exponential number of vertices, even if the number of constraints is polynomial.

On the surface, LP is an optimization problem with continuous variables. However, as in the battery charging and MAX FLOW examples, the function $\mathbf{c}^T\mathbf{x}$ is always maximized at a vertex of the feasible polytope. Hence LP is really a combinatorial problem, where the set of possible solutions is discrete.

To see that the optimum is always a vertex, observe that the set of points in $\mathbb{R}^d$ with a given value of $\mathbf{c}^T\mathbf{x}$ is also a $(d-1)$-dimensional hyperplane—namely, one that is perpendicular to the vector $\mathbf{c}$. Imagine moving such a hyperplane "down" from infinity until it touches the polytope. Clearly this happens first at a vertex—or simultaneously at several vertices, and hence on the edge or facet they span.

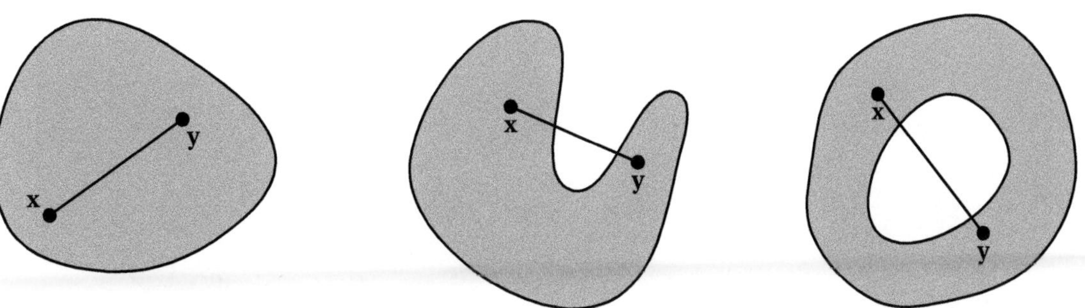

FIGURE 9.13: A convex set (left) and two non-convex sets (center and right).

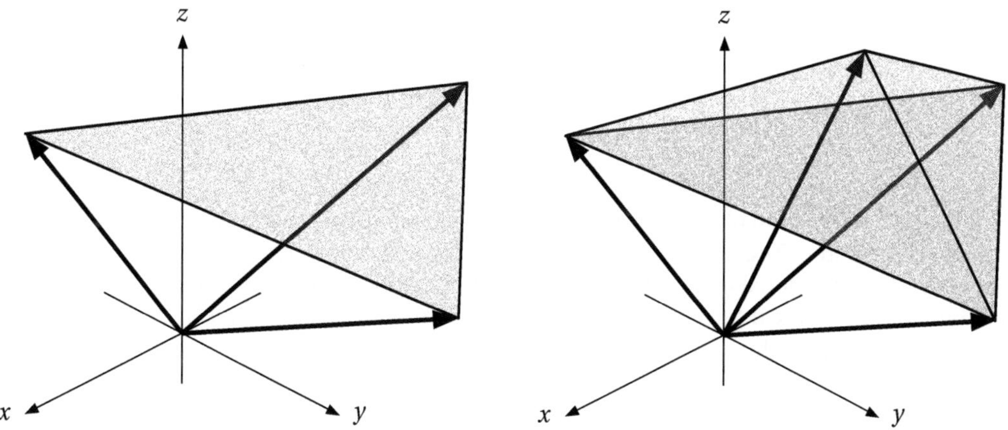

FIGURE 9.14: Convex hulls of three and four vectors.

It's instructive to go through a proof of this fact. We start by noting a key property of the feasible polytope. A set $S \subseteq \mathbb{R}^d$ is *convex* if, for any pair $\mathbf{x}, \mathbf{y} \in S$, we have $\lambda \mathbf{x} + (1-\lambda)\mathbf{y} \in S$ for all $0 \le \lambda \le 1$. In other words, any straight line between two points in S lies in S as in Figure 9.13.

Exercise 9.10 *Show that for any instance of* LP*, the feasible polytope is convex.*

More specifically, a polytope is the *convex hull* of its vertices. If we denote the vertices $\mathbf{v}_1, \ldots, \mathbf{v}_s$, their convex hull is defined as

$$\operatorname{conv}\{\mathbf{v}_1, \ldots, \mathbf{v}_s\} = \left\{ \sum_{i=1}^{s} \lambda_i \mathbf{v}_i : \sum_{i=1}^{s} \lambda_i = 1\,,\, \lambda_i \ge 0 \right\}.$$

The convex hull of two vectors is the straight line that runs between the endpoints of the vectors. As shown in Figure 9.14, the convex hull of three linearly independent vectors is a triangle, and the convex hull of four vectors is a tetrahedron.

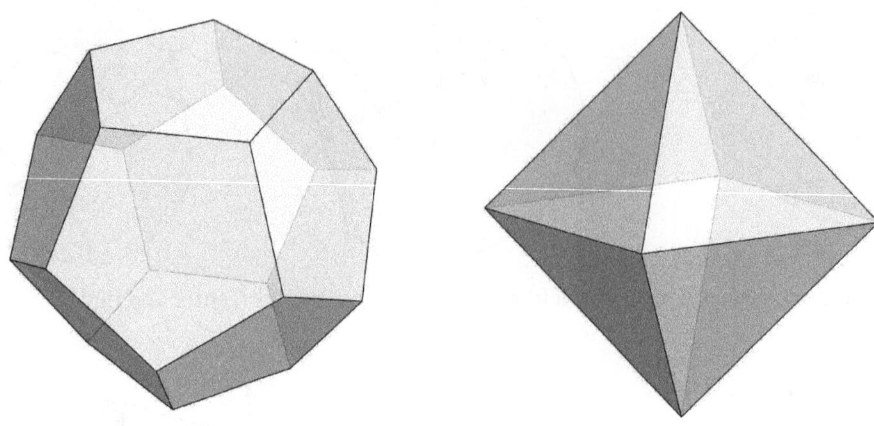

FIGURE 9.15: The intersection of k facets is a $(d-k)$-dimensional hyperplane. Here $d=3$ and each facet is a 2-dimensional plane. The intersection of two facets is a 1-dimensional edge, and the intersection of three facets is a 0-dimensional vertex. As in the octahedron, a vertex can be the intersection of more than d facets, but it is always the intersection of at least d of them. Thus at any vertex of the feasible polytope, at least d constraints are tight.

Thus any feasible $\mathbf{x}$ is a *convex combination* of the vertices $\mathbf{v}_i$,

$$\mathbf{x}=\sum_{i=1}^{s}\lambda_i\mathbf{v}_i\,.$$

Since $\mathbf{c}^T\mathbf{x}$ is a linear function of $\mathbf{x}$, this means that the quality of any $\mathbf{x}$ in the polytope is a weighted average of the quality of the vertices:

$$\mathbf{c}^T\mathbf{x}=\sum_{i=1}^{s}\lambda_i\,\mathbf{c}^T\mathbf{v}_i\,.$$

But the weights λ_i are nonnegative and sum to 1, so there must be at least one $\mathbf{v}_i$ such that $\mathbf{c}^T\mathbf{v}_i\geq\mathbf{c}^T\mathbf{x}$. Thus either the optimum is a vertex, or several vertices are equally good. In the latter case, any vertex in the convex hull of the optimal vertices—any point on the edge or facet that they span—is also optimal.

Now let's think about the combinatorics of the feasible polytope. As stated above, each facet is a $(d-1)$-dimensional hyperplane, corresponding to the set of points $\mathbf{x}$ that satisfy one of the constraints exactly. We will say that this constraint is *tight* on this facet.

As Figure 9.15 shows, the intersection of 2 facets is a $(d-2)$-dimensional hyperplane, as long as the facets aren't parallel. More generally, the intersection of k facets is $(d-k)$-dimensional hyperplane, as long as the corresponding rows of $\mathbf{A}$ are linearly independent. Each 1-dimensional edge is the intersection of $d-1$ facets, and each 0-dimensional vertex is the intersection of d facets. Thus at any vertex, at least d of the constraints are tight.

This gives us an algorithm for LP. There are m constraints, so the feasible polytope has at most m facets. Since each vertex is the intersection of d facets, there are at most $\binom{m}{d}$ vertices. We can find each one by taking the intersection $\mathbf{x}$ of the hyperplanes corresponding to d constraints, which is just an exercise in linear algebra. We check to see if each such $\mathbf{x}$ is feasible—that is, if it satisfies all the other constraints as well. Finally, we compute $\mathbf{c}^T\mathbf{x}$ for each vertex, and return the maximum.

Exercise 9.11 *Show that this brute-force approach solves* LP *in polynomial time if d is constant.*

Unfortunately, the number of vertices can be exponentially large even if the number of dimensions and facets is polynomial. The simplest example is the d-dimensional hypercube,

$$\left\{\mathbf{x} \in \mathbb{R}^d : 0 \le x_i \le 1 \text{ for all } i\right\}.$$

This polytope has $m = 2d$ facets but 2^d vertices, so the brute-force approach takes exponential time.

This leaves us with the question of whether LP is in P. In the 1970s, most common problems were quickly proved either to be in P or to be NP-hard—but a handful of problems, including FACTORING, GRAPH ISOMORPHISM, and LP, resisted this simple classification. FACTORING and GRAPH ISOMORPHISM are still floating in limbo, and as far as we know are neither in P nor are NP-complete. But after several years of uncertainty, LP turned out to be in P after all.

In the next few sections, we will explore several kinds of algorithms for LP, including some that provably run in polynomial time. These come in three families: climbing the surface of the feasible polytope, approaching the optimal vertex from inside the polytope, or using hyperplanes to cut away infeasible solutions until only the optimum remains.

9.4.3 *Climbing the Polytope: the Simplex Algorithm*

When one sees LP, a simple greedy algorithm immediately suggests itself. Starting at one vertex of the polytope, we move from vertex to vertex along its edges, always heading farther "up" along the $\mathbf{c}$ direction. The following exercise shows that this algorithm always ends at an optimal vertex:

Exercise 9.12 *Show that any vertex $\mathbf{x}$ that is a local optimum is in fact a global optimum. In other words, if none of its edges leads to a vertex with a larger value of $\mathbf{c}^T\mathbf{x}$, then it maximizes $\mathbf{c}^T\mathbf{x}$. Hint: show that the set of feasible $\mathbf{x}$ with a given value of $\mathbf{c}^T\mathbf{x}$ is convex.*

So as soon as we reach a vertex for which $\mathbf{c}^T\mathbf{x}$ cannot be increased, we are done. This strategy is called the *simplex algorithm*. We illustrate it in Figure 9.16.

Where should we start? As we will see, finding even a single feasible point can be as hard as solving LP. But for now, we will assume that we can always find an initial vertex at which to start our climb. In problems like MAX FLOW or our charger example, where the constraints include $\mathbf{x} \ge \mathbf{0}$ and the other entries of $\mathbf{A}$ and $\mathbf{b}$ are nonnegative, the vertex $(0, \ldots, 0)$ is a fine place to start.

We have seen similar algorithms before. In the Ford–Fulkerson algorithm for MAX FLOW (see Section 3.6) we increase the flow along some path from the source to the sink, until the flow on one of the edges reaches its capacity—that is, until we run into one of the facets of the polytope of feasible flows. At least for some networks, the Ford–Fulkerson algorithm is just the simplex algorithm in disguise. Moving along an edge of the feasible polytope corresponds to increasing the flow along some path—including, if this path includes the reverse edges defined in Section 3.6, canceling some of the existing flow.

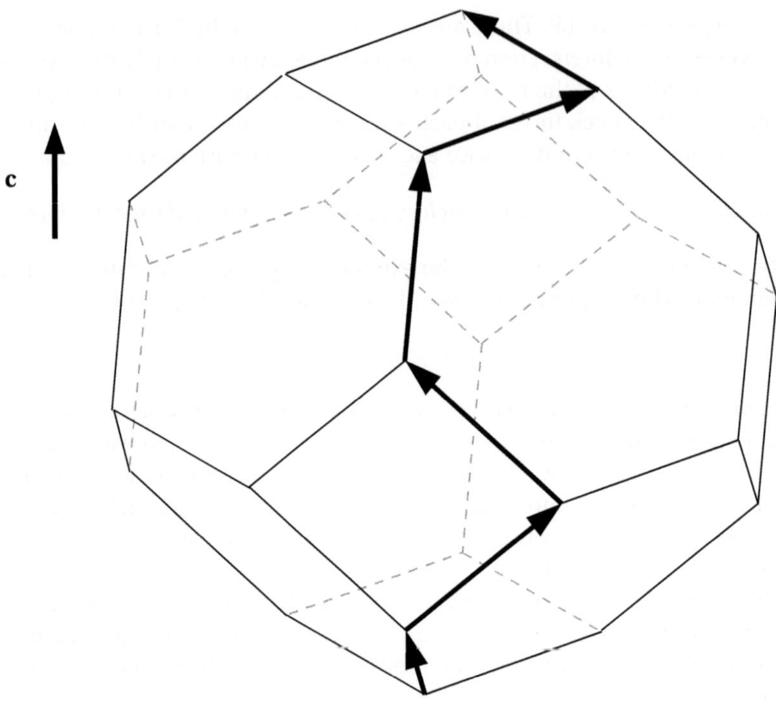

FIGURE 9.16: The simplex algorithm on its way to the top.

Look again at Figure 9.12, where we show the flow corresponding to each vertex of the feasible polytope. The following exercise asks you to find the correspondence between edges of the polytope and moves of the Ford–Fulkerson algorithm:

Exercise 9.13 *For each of the* 12 *edges of the polytope shown in Figure 9.12, find the path along which the flow is being increased, including possibly along reverse edges of the residual graph. Show that for those edges that increase the value* $x_1 + x_2 + x_3$, *this path leads from the source to the sink. For instance, which edge of the polytope corresponds to the improvement shown in Figure 3.19 on page 69?*

Exercise 9.14 *Find an example of* MAX FLOW *where some moves of the Ford–Fulkerson algorithm lead to flows that do not correspond to vertices of the feasible polytope. Hint: increase the capacities of the outer four edges in our example.*

Typically, there are many different ways that the Ford–Fulkerson algorithm could increase the flow. If it chooses badly, it can take exponential time. On the other hand, if it chooses its improvements wisely—for instance, as Problems 3.44 and 3.45 show, if it increases the flow along the shortest or "fattest" path from the source to the sink—then it always succeeds in polynomial time.

Similarly, the simplex algorithm is really a family of algorithms. There are typically many different neighboring vertices we could move to that would increase $\mathbf{c}^T\mathbf{x}$. The question is which move we make—what *pivot rule* we use. One possible rule is to be greedy and move to the neighboring vertex with the largest value of $\mathbf{c}^T\mathbf{x}$. Another is to choose randomly from among all the neighbors that increase $\mathbf{c}^T\mathbf{x}$. Is there a pivot rule such that the simplex algorithm always works in polynomial time?

9.4.4 *How Complex is Simplex?*

> My intution told me that the procedure would require too many steps wandering from one adjacent vertex to the next. In practice it takes a few steps. In brief, one's intuition in higher dimensional space is not worth a damn!
>
> George Dantzig

The simplex algorithm is extremely effective in practice, and it is still one of the most popular methods for solving LINEAR PROGRAMMING in the real world. But does it run in polynomial time in the worst case?

Let's start by showing that each step of the simplex algorithm takes polynomial time. In a generic polytope, each vertex is the intersection of d facets, corresponding to d tight constraints that are linearly independent. Each move consists of relaxing one of these constraints and moving away from the corresponding facet. We slide along the edge formed by the intersection of the other $d-1$ facets until we bump into another facet, which becomes the new tight constraint. There are d possible moves, and we choose one that increases $\mathbf{c}^T\mathbf{x}$—that is, a move along an edge such that $\mathbf{c}^T\mathbf{r} > 0$, where $\mathbf{r}$ denotes the vector pointing along that edge.

What do we do if more than d facets meet at the current vertex? As the following exercise shows, we can't simply check all the edges of the polytope that meet at that vertex, since there could be exponentially many of them:

Exercise 9.15 *Show that there are d-dimensional polytopes with $m = \text{poly}(d)$ facets, in which some vertex has an exponential number of edges incident to it. Hint: where would you bury a d-dimensional pharaoh?*

We can get around this problem by regarding a "vertex" as a set of d facets, even if many such sets correspond to the same point $\mathbf{x}$. Then there are just d possible moves as before. Some of these moves don't actually change $\mathbf{x}$, but they change which facets we use to define $\mathbf{x}$. If you like, they move infinitesimally, from one subset of $\mathbf{x}$'s facets to another. As long as we always "move" in a direction $\mathbf{r}$ such that $\mathbf{c}^T\mathbf{r} > 0$, there is no danger that we will fall into an endless loop.

Since each step is easy, the running time of the simplex algorithm is essentially the number of vertices it visits. But as its inventor George Dantzig complained, in high dimensions it can be difficult to understand the possible paths it can take.

As the following exercise shows, the simplex algorithm can easily handle some situations where the 9.11
polytope has an exponential number of vertices:

Exercise 9.16 *Suppose the feasible polytope is the d-dimensional hypercube. Show that the simplex algorithm visits at most d vertices no matter what pivot rule it uses, as long as it always moves to a better vertex.*

However, this is a little too easy. The constraints $0 \le x_i \le 1$ defining the hypercube are independent of each other, so the feasible polytope is simply the Cartesian product of d unit intervals.

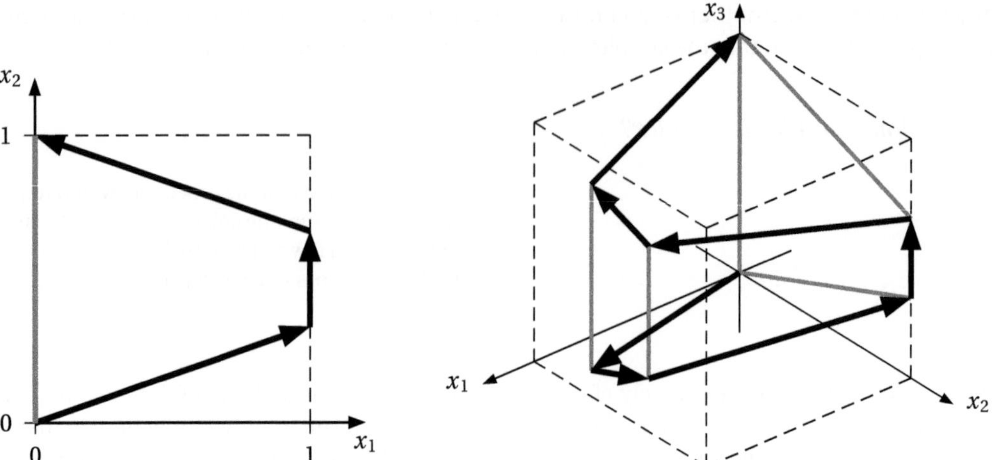

FIGURE 9.17: Klee–Minty cubes for $\varepsilon = 1/3$ and $d = 2,3$. If it starts at the origin, the simplex algorithm could visit all 2^d vertices on its way to the optimum.

Things get more interesting when the variables interact with each other. Let ε be a real number such that $0 \le \varepsilon < 1/2$, and consider the d-dimensional polytope defined by the inequalities

$$\begin{aligned} 0 &\le x_1 \le 1 \\ \varepsilon x_1 &\le x_2 \le 1 - \varepsilon x_1 \\ &\vdots \\ \varepsilon x_{d-1} &\le x_d \le 1 - \varepsilon x_{d-1}. \end{aligned} \tag{9.10}$$

This is called the *Klee–Minty cube.* It is a perturbation of the d-dimensional hypercube, since setting $\varepsilon = 0$ gives the hypercube itself. As in the hypercube, it has 2^d vertices where each variable either saturates its upper or its lower bound. However, now the bounds for x_i depend on the values of x_j for all $j < i$. We show two- and three-dimensional Klee–Minty cubes in Figure 9.17.

Suppose we are trying to find the vertex with the largest value of x_d—namely, the vertex $(0, \ldots, 0, 1)$. If we start at the origin $(0, \ldots, 0)$, the simplex algorithm could visit all 2^d vertices, following a Hamiltonian path as shown in Figure 9.17. In particular, suppose it uses the following "lowest index" pivot rule: look for the smallest i such that sliding along the (tilted) x_i-axis, flipping x_i from its lower bound to its upper bound or vice versa, will increase x_d. For $d = 2$, this gives the sequence of vertices

$$(0,0) \to (1,\varepsilon) \to (1, 1-\varepsilon) \to (0,1).$$

Exercise 9.17 *Show that the lowest-index pivot rule visits all 2^d vertices of the Klee–Minty cube when $d = 3$.*

We invite you to prove, in Problem 9.12, that the same thing happens for any d. Don't worry, we'll wait here for you.

The Klee–Minty cube shows that, at least for some instances of LP and some pivot rules, the simplex algorithm takes exponential time. You might object that its dismal performance in this case is just an artifact of our pivot rule, and you would be right. The greedy pivot rule, for instance, finds the optimum with a single jump from the origin.

But at the time we write this, no pivot rule is known that always succeeds in polynomial time. And for most of them, worst-case instances are known that force them to visit an exponential number of vertices. This includes the greedy pivot rule, although here the worst-case instance is more complicated than the Klee–Minty cube.

9.12

9.4.5 *Theory and Practice: Smoothed Analysis*

The fact that the simplex algorithm requires exponential time in the worst case is in sharp contrast to its performance in practice. Various pivot rules are routinely applied to instances of LP with thousands of variables and constraints, and the number of vertices they visit is typically just *linear* in the number of constraints. In practice, simplex just works.

This discrepancy between empirical and theoretical complexity poses a real challenge to theoretical computer science. In 2004, Dan Spielman and Shang-Hua Teng met this challenge by proposing a new tool called *smoothed analysis.*

9.13

In worst-case analysis, we propose an algorithm, and then the adversary chooses an instance x. In smoothed analysis, chance steps in, adding a small amount of noise and producing a perturbed instance x'. If you think of x and x' as vectors in the space of instances, then x' is chosen uniformly from a ball of radius ε around x. The adversary can decide where this ball is centered, but he cannot control where in the ball the perturbed instance x' will be.

We then define the running time as the average over the perturbed instances, maximized over the adversary's original instance:

$$T_{\text{smoothed}} = \max_x \left(\mathop{\mathbb{E}}_{x':|x'-x|\le\varepsilon} T(x') \right).$$

If the adversary has to carefully tune the parameters of an instance to make it hard, adding noise upsets his plans. It smooths out the peaks in the running time, and makes the problem easy on average.

Spielman and Teng considered a particular pivot rule called the *shadow vertex rule.* It projects a two-dimensional shadow of the polytope, forming a two-dimensional polygon, and tries to climb up the outside of the polygon. This rule performs poorly if the shadow has exponentially many sides, with exponentially small angles between them. They showed that if we add noise to the constraints, perturbing the entries of $\mathbf{A}$ and $\mathbf{b}$, then the angles between the facets—and between the sides of the shadow polygon—are $1/\text{poly}(m)$ with high probability. In that case, the shadow has $\text{poly}(m)$ sides, and the shadow vertex rule takes polynomial time. We illustrate this in Figure 9.18.

Smoothed analysis provides a sound theoretical explanation for the suprisingly good performance of the simplex algorithm on real instances of LP. But it still leaves open the question of whether LP is in P—whether it can can be solved in polynomial time even in the worst case, when the adversary has the last word. We will answer that question soon. But first we explore a remarkable property of LP that generalizes the relationship between Max Flow and Min Cut, and which says something about LP's worst-case complexity as well.

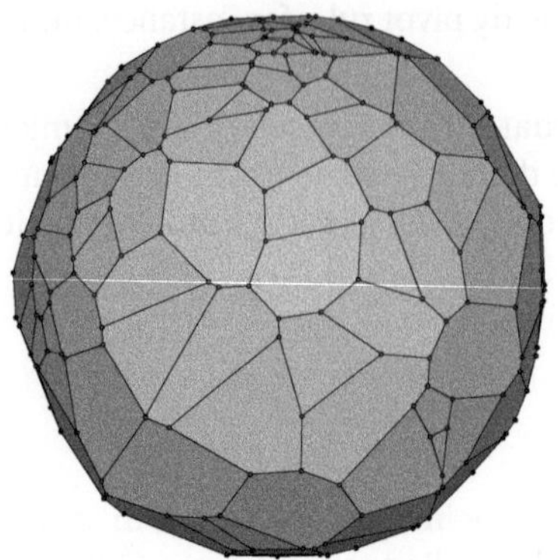

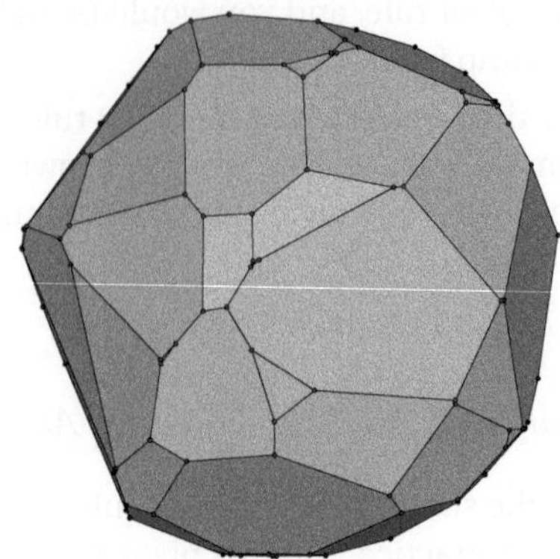

FIGURE 9.18: The polytope on the left has 200 facets and 396 vertices. On the right, we perturb this polytope by adding 5% random noise to the right-hand sides of the inequalities, reducing the number of facets and vertices to 87 and 170. The angles between the facets have become sharper, so the polytopes' shadows, i.e., the two-dimensional polygons that form the outlines of these pictures, have fewer sides.

9.5 Through the Looking-Glass: Duality

Every explicit duality is an implicit unity.
Alan Watts

We saw in Section 3.6 that two seemingly disparate problems, MAX FLOW and MIN CUT, have the same answer. The weight of any cut is an upper bound on MAX FLOW, and the value of any flow is a lower bound on MIN CUT. These bounds meet at each others' optima—the minimum cut is a witness for the maximum flow, and vice versa.

In this section, we will see that the duality between MAX FLOW and MIN CUT is just an example of a much more general phenomenon—that instances of LINEAR PROGRAMMING come in mirror-image pairs. One maximizes, the other minimizes, and the variables of each problem correspond to the constraints of the other one. A feasible solution for either problem gives a bound on the other one, and the optimum of each one is a witness for the optimum of the other.

9.5.1 *The Lowest Upper Bound*

We start by writing LP in a slightly different way. In the battery charger and MAX FLOW examples in Section 9.4.1, we had the constraints $\mathbf{x} \geq \mathbf{0}$ that the currents are nonnegative. Here we write those constraints separately, rather than including them in $\mathbf{A}$ and $\mathbf{b}$. We want to solve the following maximization problem:

$$\max_{\mathbf{x}} \left(\mathbf{c}^T \mathbf{x}\right) \quad \text{subject to} \quad \mathbf{x} \geq \mathbf{0} \quad \text{and} \quad \mathbf{A}\mathbf{x} \leq \mathbf{b}. \tag{9.11}$$

As before, $\mathbf{x}, \mathbf{c} \in \mathbb{R}^d$, $\mathbf{b} \in \mathbb{R}^m$. and $\mathbf{A}$ is an $m \times d$ matrix. Thus there are d variables and a total of $d + m$ constraints. We will call this the *primal* problem.

Now multiply the constraints $\mathbf{Ax} \leq \mathbf{b}$ on the left by a vector $\mathbf{y} \in \mathbb{R}^m$ where $\mathbf{y} \geq \mathbf{0}$. This gives

$$\mathbf{y}^T \mathbf{Ax} \leq \mathbf{y}^T \mathbf{b}.$$

If we also assume that

$$\mathbf{c}^T \leq \mathbf{y}^T \mathbf{A}, \tag{9.12}$$

then we have

$$\mathbf{c}^T \mathbf{x} \leq \mathbf{y}^T \mathbf{Ax} \leq \mathbf{y}^T \mathbf{b}. \tag{9.13}$$

Thus for any $\mathbf{y}$ such that $\mathbf{y} \geq \mathbf{0}$ and $\mathbf{c}^T \leq \mathbf{y}^T \mathbf{A}$, the inner product $\mathbf{y}^T \mathbf{b}$ is an upper bound on $\mathbf{c}^T \mathbf{x}$—that is, an upper bound on the optimum of the primal problem.

To make this upper bound as tight as possible, we want to minimize it. This gives us the following minimization problem, where for aesthetic reasons we write $\mathbf{b}^T \mathbf{y}$ instead of $\mathbf{y}^T \mathbf{b}$:

$$\min_{\mathbf{y}} \left(\mathbf{b}^T \mathbf{y} \right) \quad \text{subject to} \quad \mathbf{y} \geq \mathbf{0} \quad \text{and} \quad \mathbf{A}^T \mathbf{y} \geq \mathbf{c}, \tag{9.14}$$

where we got the second constraint by taking the transpose of (9.12).

We call (9.14) the *dual* problem. It looks much like the primal (9.11), but we have swapped $\mathbf{b}$ with $\mathbf{c}$, max with min, and $\leq$ with $\geq$ in the constraints. It has m variables $\mathbf{y}$, one for each constraint of the primal problem. There are a total of $m + d$ constraints on $\mathbf{y}$, namely m from $\mathbf{y} \geq \mathbf{0}$, and d from $\mathbf{A}^T \mathbf{y} \geq \mathbf{c}$. Each column of $\mathbf{A}$ corresponds to a constraint on $\mathbf{y}$, just as each row of $\mathbf{A}$ corresponds to a constraint on $\mathbf{x}$.

According to (9.13), any feasible solution of the dual problem gives an upper bound on the primal optimum and vice versa. In particular, the dual optimum is greater than or equal to the primal optimum:

$$\max_{\mathbf{x}} \left(\mathbf{c}^T \mathbf{x} \right) \leq \min_{\mathbf{y}} \left(\mathbf{b}^T \mathbf{y} \right). \tag{9.15}$$

This is called *weak duality*—weak because (9.15) doesn't tell us how close these optima are to each other. If there is a large gap between them, solving the dual doesn't tell us much about the primal or vice versa.

The good news is that there is no gap. The primal and dual problem have exactly the same optimal value,

$$\max_{\mathbf{x}} \left(\mathbf{c}^T \mathbf{x} \right) = \min_{\mathbf{y}} \left(\mathbf{b}^T \mathbf{y} \right). \tag{9.16}$$

In the degenerate cases, the value of the primal is unbounded if and only if the feasible polytope of the dual is empty, and vice versa. Thus the solutions to the two problems are either both undefined, or both well-defined and equal.

This fact (9.16) is known as *strong* duality. Since we already have weak duality, to prove (9.16) it suffices to exhibit a single $\mathbf{y}$ such that $\mathbf{b}^T \mathbf{y} = \mathbf{c}^T \mathbf{x}$ where $\mathbf{x}$ is the primal optimum. We will show that such a $\mathbf{y}$ exists using a simple physical argument.

Imagine standing inside a cathedral. Its floor and walls are simple and flat, but the ceiling that vaults the nave is made of elaborately painted facets. The tour guide explains that the paintings were made to distract the eye in order to hide the highest point of the ceiling, the point nearest to heaven. No mortal has ever located this point.

Luckily, a vendor sells helium-filled balloons on the plaza in front of the cathedral. In a flash of inspiration, you buy one, smuggle it inside, and let it go. Driven by its buoyancy, the balloon moves straight

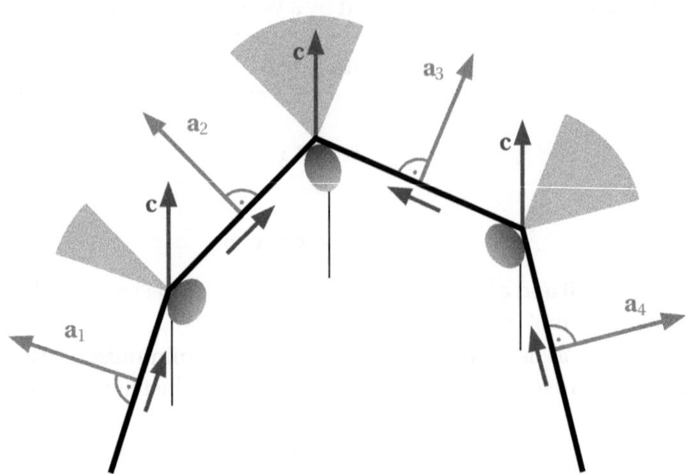

FIGURE 9.19: Proof of strong duality by balloon.

up until it hits the ceiling. It then slides along a series of edges and facets, until it finally comes to rest at the highest point.

The balloon's position is $\mathbf{x}$, and its buoyancy is $\mathbf{c}$. The facets at which it comes to rest are the constraints which are tight at the optimum. We will assume that all of $\mathbf{x}$'s components are positive—that it touches some facets of the "ceiling" formed by the inequalities $\mathbf{Ax} \le \mathbf{b}$, but none of the "walls" formed by $\mathbf{x} \ge \mathbf{0}$.

Let $S \subseteq \{1, \dots, m\}$ denote the set of indices j such that the constraint corresponding to the jth row of $\mathbf{A}$ is tight. We denote this row $\mathbf{a}_j^T$, using the T to remind you that it's a row vector. Then

$$S = \left\{ j : \mathbf{a}_j^T \mathbf{x} = b_j \right\}. \tag{9.17}$$

The balloon presses against each of these facets with some force, and each facet presses back with an equal and opposite force. In order for the balloon to be in equilibrium, the sum of these forces must exactly cancel its buoyancy.

In the absence of friction, the force on each facet is perpendicular to its surface—that is, along the normal vector $\mathbf{a}_j$ as shown in Figure 9.19. If we let $y_j \ge 0$ denote the magnitude of the force on the jth facet, we have

$$\sum_{j \in S} \mathbf{a}_j y_j = \mathbf{c}. \tag{9.18}$$

The balloon can't push on any facet it isn't touching, so $y_j = 0$ for all $j \notin S$. This defines a vector $\mathbf{y} \in \mathbb{R}^m$ such that

$$\mathbf{A}^T \mathbf{y} = \mathbf{c}. \tag{9.19}$$

Along with the fact that $\mathbf{y} \ge \mathbf{0}$, this shows that $\mathbf{y}$ is a feasible solution of the dual problem. This is the key to strong duality.

All that remains is to show that the value of $\mathbf{y}$ in the dual problem is equal to the value of $\mathbf{x}$ in the primal. Taking the transpose of (9.19) and multiplying it on the right by $\mathbf{x}$ gives

$$\mathbf{y}^T \mathbf{A} \mathbf{x} = \mathbf{c}^T \mathbf{x}. \tag{9.20}$$

And, from (9.17) we have

$$(\mathbf{Ax})_j = \mathbf{a}_j^T \mathbf{x} = b_j \quad \text{for all } j \in S.$$

Since $y_j = 0$ for $j \notin S$, this gives

$$\mathbf{y}^T \mathbf{A} \mathbf{x} = \mathbf{y}^T \mathbf{b} = \mathbf{b}^T \mathbf{y}.$$

Together with (9.20), this yields

$$\mathbf{c}^T \mathbf{x} = \mathbf{b}^T \mathbf{y},$$

proving strong duality.

It's worth asking which constraints of the dual problem are tight. In the simplest case, where exactly d facets of the primal problem meet at its optimum $\mathbf{x}$, we have $y_j = 0$ for the other $m - d$ facets, so $m - d$ constraints of the form $\mathbf{y} \geq 0$ are tight. On the other hand, (9.19) shows that all d constraints of the form $\mathbf{A}^T \mathbf{y} = \mathbf{c}$ are tight. Thus when the primal problem has exactly d tight constraints at $\mathbf{x}$, the dual has exactly m tight constraints at $\mathbf{y}$.

Exercise 9.18 *Say that an instance of* LP *is* overdetermined *if more than d facets meet at the optimal vertex, i.e., if more than d constraints are tight. Similarly, say that it is* underdetermined *if it has more than one optimal vertex. Show that if the primal problem is overdetermined then the dual is underdetermined, and vice versa.*

In our discussion, we made the simplifying assumption that all of $\mathbf{x}$'s components are positive. You can handle the case where some of the constraints $\mathbf{x} \geq \mathbf{0}$ are tight, so that the balloon touches both the "ceiling" and the "walls," in Problem 9.14.

9.14

9.5.2 *Flows and Cuts, Again*

Let's look again at our Max Flow example from Section 9.4.1 Since we are now writing the constraints $\mathbf{x} \geq \mathbf{0}$ separately, the primal problem maximizes $\mathbf{c}^T \mathbf{x}$ where $\mathbf{c}^T = (1,1,1)$, subject to $\mathbf{Ax} \leq \mathbf{b}$ where

$$\mathbf{A} = \begin{pmatrix} 1 & 0 & 0 \\ 1 & 1 & 0 \\ 0 & 1 & 0 \\ 0 & 1 & 1 \\ 0 & 0 & 1 \end{pmatrix} \quad \text{and} \quad \mathbf{b} = \begin{pmatrix} 2 \\ 2 \\ 1 \\ 2 \\ 2 \end{pmatrix}.$$

Thus the dual problem asks us to minimize

$$\mathbf{b}^T \mathbf{y} = 2y_1 + 2y_2 + y_3 + 2y_4 + 2y_5, \tag{9.21}$$

subject to the constraints $\mathbf{y} \geq \mathbf{0}$ and $\mathbf{A}^T \mathbf{y} \geq \mathbf{c}$, or

$$\begin{aligned} y_1 + y_2 &\geq 1 \\ y_2 + y_3 + y_4 &\geq 1 \\ y_4 + y_5 &\geq 1. \end{aligned} \tag{9.22}$$

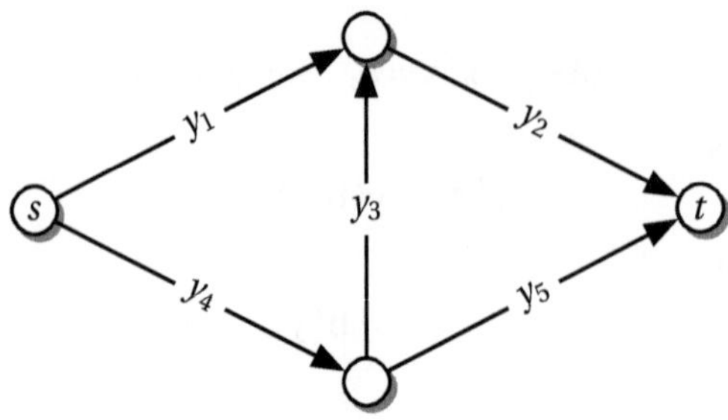

FIGURE 9.20: The variables of the dual problem in our example, each of which corresponds to including an edge in the minimum cut.

What does this dual problem mean? The variables $\mathbf{y}$ correspond to the capacity constraints on the flow, with one y_j for each edge of the network as shown in Figure 9.20. Suppose that we have a subset C of the edges, and that each y_j is 0 or 1 depending on whether or not the corresponding edge is in C. Then the constraints (9.22) demand that each of the three possible paths through the network be blocked by at least one edge in C. In other words, C is a cut.

Moreover, the components of $\mathbf{b}$ are the capacities of the edges of the network, so $\mathbf{b}^T\mathbf{y}$ is simply C's total weight. Hence the dual problem is exactly the corresponding instance of MIN CUT. Our proof of strong duality recovers, and generalizes, the fact that the weight of the MIN CUT equals the value of the MAX FLOW.

Our interpretation of the dual problem as MIN CUT hinges on the assumption that each the y_j take integer values at the optimum $\mathbf{y}$. This condition is called *integrality*. In Section 3.6, we used the Ford–Fulkerson algorithm to show that any instance of MAX FLOW with integer capacities does indeed have an integral optimum, and one can use the same approach to prove integrality for the instance of LP corresponding to MIN CUT. In Section 9.8.3, we will give a general condition under which both the primal and the dual have integral optima.

In addition to being a beautiful mathematical fact, duality gives us new kinds of algorithms. If we want to solve an instance of LP, we are free to solve either the primal or the dual, and one might be easier than the other. For instance, most algorithms for MIN CUT work by solving the corresponding instance of MAX FLOW. An interesting family of "primal-dual" algorithms work simultaneously on the primal and the dual, using each one as the guide for the other.

9.5.3 *Duality and Complexity*

We pause here to point out that duality has some consequences for the complexity of LINEAR PROGRAMMING. Consider the following version of the problem:

LINEAR PROGRAMMING (threshold)
Input: An $m \times d$ matrix $\mathbf{A}$, vectors $\mathbf{b} \in \mathbb{R}^m$ and $\mathbf{c} \in \mathbb{R}^d$, and a number $\ell \in \mathbb{R}$
Question: Is there a vector $\mathbf{x} \in \mathbb{R}^d$ such that $\mathbf{A}\mathbf{x} \leq \mathbf{b}$ and $\mathbf{c}^T\mathbf{x} \geq \ell$?

We can certainly answer this question if we can find the optimum vertex. Conversely, if we can solve this threshold problem then we can find the maximum of $\mathbf{c}^T\mathbf{x}$ by binary search, just as we did for MAX-SAT in Section 9.1.

In this case, $\mathbf{c}^T\mathbf{x}$ is real-valued, so *a priori* our binary search never ends—it just narrows our search to smaller and smaller intervals. However, suppose we start with some initial interval of width E that includes the maximum, and that we want to find the maximum within an error ε. Since each step divides the interval in half, this takes $\log_2(E/\varepsilon)$ steps.

Happily, as Problem 9.18 shows, we can bound $\mathbf{c}^T\mathbf{x}$ within an interval of width $E = 2^{O(n)}$ where n is the total number of bits it takes to express the problem instance—that is, the total number of bits of the entries of **A**, **b**, and **c**. The same problem shows that we can determine the optimal vertex exactly when $\varepsilon = 2^{-O(n)}$, by checking to see which constraints are nearly tight. Thus we can solve the optimization problem with $\log_2(E/\varepsilon) = O(n)$ calls to an oracle for the threshold version. The two versions are Turing-reducible to each other, and each one is in P if and only if the other is.

The threshold version of LINEAR PROGRAMMING is clearly in NP, since any $\mathbf{x}$ that satisfies these inequalities is a witness that the answer is yes. But by duality it is also in coNP, because any feasible vector $\mathbf{y}$ of the dual problem such that $\mathbf{b}^T\mathbf{y} < \ell$ is a witness that the answer is no.

As we discussed in Section 6.1.1, if any problem in NP$\cap$coNP is NP-complete, then NP = coNP and the polynomial hierarchy collapses to its first level. Thus duality gives us strong evidence that the threshold version of LINEAR PROGRAMMING is not NP-complete, and that its optimization version is not NP-hard.

Enough suspense. It is time now to show that LINEAR PROGRAMMING is, in fact, in P. In the next section we will again use our balloon, with some additional forces on it, to solve LINEAR PROGRAMMING in polynomial time.

9.6 Solving by Balloon: Interior Point Methods

> Lift my spirits so that I may be of use—
> And trap the wicked with the works of their own hands.
>
> Christine Robinson, *Psalms for a New World*

Because the simplex algorithm crawls from vertex to vertex, it can be forced to visit an exponential number of vertices by constructions like the Klee–Minty cube. There is another class of algorithms called *interior point methods* that approach the optimal vertex from *inside* the feasible polytope. By moving freely and smoothly in the polytope's interior, these algorithms can avoid getting caught in long sequences of moves on its surface.

9.6.1 *Buoyancy and Repulsion*

The interior point method we will explore here works much like the balloon we released in the cathedral to prove strong duality. However, if we just let the balloon go, it will bump against a $(d-1)$-dimensional facet, then the $(d-2)$-dimensional intersection of two facets, and so on until it gets stuck on an edge. From then on the balloon will follow the simplex algorithm, and the cathedral designer can trap it in a Klee–Minty maze.

To avoid this, we introduce repulsive forces that push the balloon away from the facets—so that instead of getting caught in the polytope's edges, it floats gently past them on its way up to the optimum. Of course, these repulsive forces push the balloon down and away from the optimum as well. We will reduce the strength of the repulsion over time so that the balloon's equilibrium position can converge to the optimum.

From physics we know that most forces can be written as the gradient of a potential energy $V(\mathbf{x})$,

$$\mathbf{F} = -\nabla V \quad \text{or} \quad F_i = -\frac{\partial V}{\partial x_i}.$$

Such forces tend to drive objects "downhill" towards lower values of V, so that they come to rest at a minimum of V. To give our balloon a buoyancy $\mathbf{F} = \mathbf{c}$ we define its potential as $V(\mathbf{x}) = -\mathbf{c}^T\mathbf{x}$, so that its energy decreases as it floats up.

We need to add a term to the potential that keeps the balloon away from the facets of the polytope. Recall that the jth facet is the set

$$\left\{\mathbf{x} : \mathbf{a}_j^T \mathbf{x} = b_j\right\},$$

where $\mathbf{a}_j^T$ is the jth row of $\mathbf{A}$. One choice of potential, called the *logarithmic barrier function*, is this:

$$V_{\text{facets}}(\mathbf{x}) = -\sum_{j=1}^{m} \ln\left(b_j - \mathbf{a}_j^T \mathbf{x}\right). \tag{9.23}$$

This potential is well-defined in the interior of the polytope where $\mathbf{A}\mathbf{x} < \mathbf{b}$, and approaches $+\infty$ as we approach any of the facets. If r is the distance to the nearest facet, V_{facets} scales as $-\ln r$. Thus the repulsive force scales as $1/r$, getting stronger as we approach the nearest facet.

The potential V_{facets} is a strictly convex function of $\mathbf{x}$. Thus if we assume as usual that the polytope is bounded, V_{facets} has a unique minimum $\mathbf{x}_0$ in the interior. This is the equilibrium position of the balloon if the only forces on it are the facets' repulsion, and we will use it as our starting point.

Now let's give the balloon a buoyancy of λ in the direction $\mathbf{c}$. The equilibrium position $\mathbf{x}_\lambda$ is the minimum of the total potential,

$$V(\mathbf{x}) = V_{\text{facets}}(\mathbf{x}) - \lambda \mathbf{c}^T \mathbf{x}.$$

Alternately, $\mathbf{x}_\lambda$ is the place where the forces on the balloon cancel,

$$-\nabla V(\mathbf{x}_\lambda) = -\nabla V_{\text{facets}}(\mathbf{x}_\lambda) + \lambda \mathbf{c} = 0,$$

or

$$\nabla V_{\text{facets}}(\mathbf{x}_\lambda) = \lambda \mathbf{c}. \tag{9.24}$$

As λ increases the balloon rises, pushing harder against the repulsion and coming to rest closer and closer to the optimal vertex.

Suppose we increase λ very slowly so that the balloon is in equilibrium at each moment. How does its position $\mathbf{x}_\lambda$ change as function of λ? If we increase λ just a little bit, we have

$$\mathbf{x}_{\lambda+\mathrm{d}\lambda} = \mathbf{x}_\lambda + \frac{\mathrm{d}\mathbf{x}_\lambda}{\mathrm{d}\lambda}\,\mathrm{d}\lambda.$$

To find $\mathrm{d}\mathbf{x}_\lambda/\mathrm{d}\lambda$ we differentiate (9.24) with respect to λ, giving

$$\mathbf{H}\frac{\mathrm{d}\mathbf{x}_\lambda}{\mathrm{d}\lambda} = \mathbf{c}. \tag{9.25}$$

Here $\mathbf{H}$ is the *Hessian* of V_{facets}, or the matrix of its second derivatives:

$$H_{ij} = \frac{\partial^2 V_{\text{facets}}}{\partial x_i \, \partial x_j}.$$

Multiplying (9.25) by the inverse of $\mathbf{H}$ then gives us our equation for $\mathrm{d}\mathbf{x}_\lambda/\mathrm{d}\lambda$,

$$\frac{\mathrm{d}\mathbf{x}_\lambda}{\mathrm{d}\lambda} = \mathbf{H}^{-1}\mathbf{c}. \tag{9.26}$$

As Problem 9.15 shows, we can write an explicit formula for $\mathbf{H}$ as a function of $\mathbf{A}$, $\mathbf{b}$, and $\mathbf{x}$, and compute it in polynomial time. Thus one way to solve LINEAR PROGRAMMING is to integrate the differential equation (9.26) numerically, starting at the initial point $\mathbf{x}_0$, increasing λ until $\mathbf{x}$ is as close to the optimum as we like. We willfully ignore several important issues, including what step size we should use for the numerical integration and how to find $\mathbf{x}_0$ in the first place, but we claim that the time it takes to compute $\mathbf{x}_\lambda$ is polynomial in m and λ.

Finally, since the repulsive force a distance r from a facet scales as $1/r$, and since at most m facets meet at the optimum, the distance between $\mathbf{x}_\lambda$ and the optimal vertex is $r = O(m/\lambda)$. Therefore, we can get within ε of the optimum by integrating the equation up to $\lambda = O(m/\varepsilon)$.

A differential equation like (9.26) may not look like the kind of algorithm we are used to. But many interior point methods are essentially discrete-time approximations of this continuous process. The most famous of these, and the first which was proved to work in polynomial time, is *Karmarkar's algorithm*. It updates $\mathbf{x}$ by jumping in the direction $\mathrm{d}\mathbf{x}/\mathrm{d}\lambda$.

$$\mathbf{x} \to \mathbf{x} + \Delta\lambda\, \mathbf{H}^{-1}\mathbf{c}.$$

The step size $\Delta\lambda$ is set so that $\mathbf{x}$ moves halfway, say, from its current position to the boundary of the polytope. Each of these steps multiplies the buoyancy λ by a constant, so we can get within ε of the minimum in $O(\log(m/\varepsilon))$ steps. Problem 9.18 shows that we can determine the optimal vertex exactly once $\varepsilon = 2^{-O(n)}$, where n is the total number of bits in the problem instance. Thus the number of steps is $O(n)$, giving us a polynomial-time algorithm for LINEAR PROGRAMMING.

9.15

9.6.2 *The Newton Flow*

The differential equation (9.26) describes how the balloon's equilibrium position changes as its buoyancy increases. Let's look at its trajectory in another way—as running a minimization algorithm in reverse. This process has the charming property that it maximizes $\mathbf{c}^T\mathbf{x}$ for any $\mathbf{c}$ we like. All we have to do is point it in the right direction.

Suppose that our balloon is somewhere in the polytope when it suddenly loses most of its helium, leaving just enough to counteract its own weight. Then $\lambda = 0$ and the balloon is driven towards the center by the facets' repulsion, until it reaches the point $\mathbf{x}_0$ where V_{facets} is minimized.

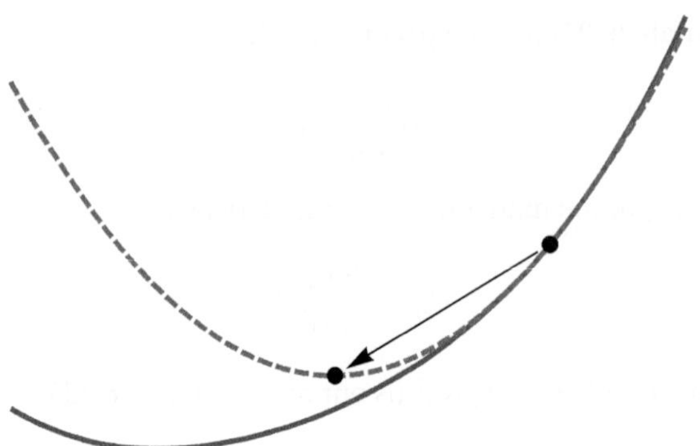

FIGURE 9.21: Newton's method. To minimize V (solid) we use a second-order Taylor series (dashed), based on the first and second derivatives at our current position x, to estimate the jump δ such that $V(x+\delta)$ is minimized.

If the balloon follows Newton's laws, the force on it is the gradient $F=-\nabla V$. But rather than using Newton's laws, let's use *Newton's method*—a method for finding the minimum of V which is much faster than simple gradient descent.

The idea is to compute the first and second derivatives at our current position x, approximate V with a second-order Taylor series, and use that Taylor series to estimate where V is minimized. We can then jump from x to our estimated minimum $x+\delta$ as in Figure 9.21. If $V(x)$ is actually quadratic so that the Taylor series is exact, this works in a single step. If not, we can repeat the process, and hopefully converge to the true minimum.

For functions of one dimension, the second-order Taylor series around x is

$$V(x+\delta) \approx V(x) + V'\delta + \frac{1}{2}V''\delta^2 .$$

Minimizing the right-hand side suggests that V is minimized at $x+\delta$, where

$$\delta = -\frac{1}{V''}V' .$$

In higher dimensions, we have

$$V(\mathbf{x}+\boldsymbol{\delta}) \approx V(\mathbf{x}) + (\nabla V)^T\boldsymbol{\delta} + \frac{1}{2}\boldsymbol{\delta}^T\mathbf{H}\boldsymbol{\delta} .$$

Minimizing the right-hand side, using the fact that the Hessian $\mathbf{H}$ is symmetric, gives

$$\boldsymbol{\delta} = -\mathbf{H}^{-1}\nabla V .$$

Rather than jumping discretely from $\mathbf{x}$ to $\mathbf{x}+\boldsymbol{\delta}$, we can push $\mathbf{x}$ toward the minimum in continuous time by setting its derivative equal to $\boldsymbol{\delta}$,

$$\frac{d\mathbf{x}}{dt} = \boldsymbol{\delta} . \tag{9.27}$$

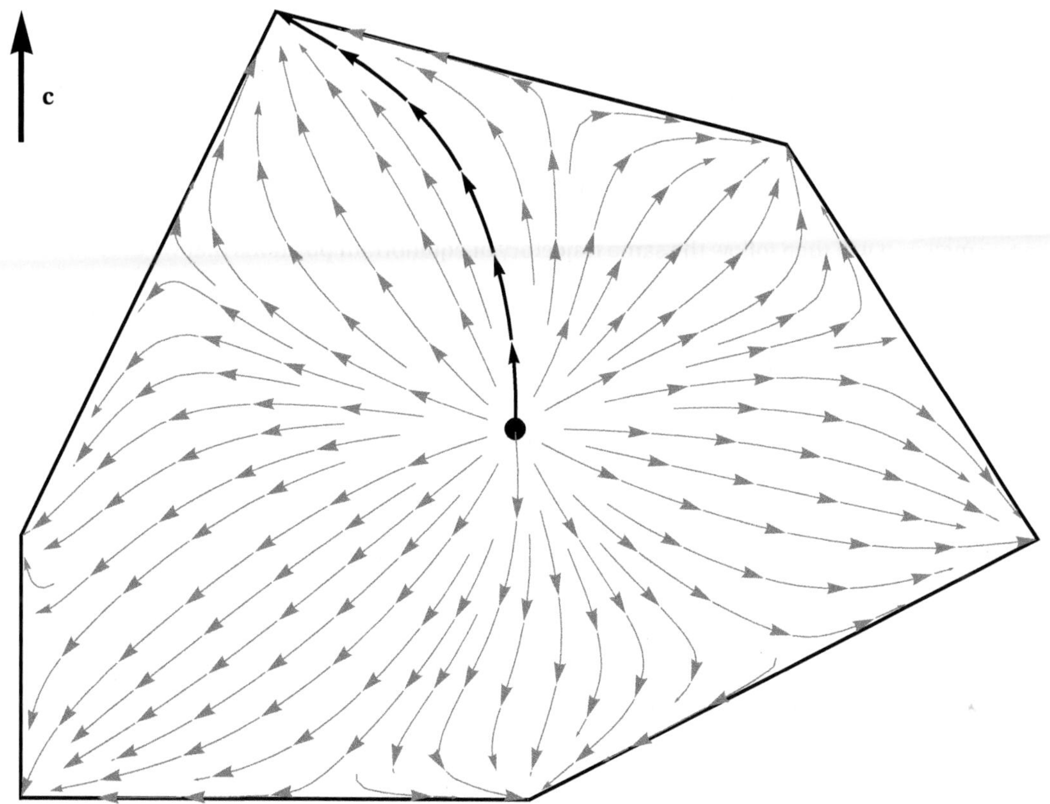

FIGURE 9.22: If we start at $\mathbf{x}_0$ and give ourselves a nudge in the direction $\mathbf{c}$, the reverse Newton flow (9.28) takes us to the optimal vertex.

If we integrate (9.27), $\mathbf{x}$ tends towards $\mathbf{x}_0$ no matter where it starts in the polytope. Conversely, if we run this process backwards, $\mathbf{x}_0$ becomes an unstable fixed point, and the flow pushes $\mathbf{x}$ out towards the vertices of the polytope. This gives

$$\frac{d\mathbf{x}}{dt} = -\boldsymbol{\delta} = \mathbf{H}^{-1}\nabla V_{\text{facets}}. \tag{9.28}$$

This is called the *reverse Newton flow*, and we illustrate it in Figure 9.22.

For every point $\mathbf{x}$ in the polytope, there is some buoyant force $\lambda\mathbf{c}$, along some direction $\mathbf{c}$, such that $\mathbf{x}$ would be the equilibrium position of the balloon. As in (9.24), in order for the forces to cancel we must have $\nabla V_{\text{facets}} = \lambda\mathbf{c}$, and (9.28) becomes

$$\frac{d\mathbf{x}}{dt} = \lambda\mathbf{H}^{-1}\mathbf{c}.$$

Except for the factor of λ, this is exactly the differential equation (9.26) for the balloon's trajectory that we wrote before.

What does this mean? Each trajectory of the reverse Newton flow tends to the vertex that maximizes $\mathbf{c}^T\mathbf{x}$ for some $\mathbf{c}$, where $\mathbf{c}$ depends only on its initial direction as it leaves $\mathbf{x}_0$. We can see this in Figure 9.22, where the initial direction determines which vertex we converge to. To select the $\mathbf{c}$ we want, we just have to give the balloon a nudge in the direction $\mathbf{H}^{-1}\mathbf{c}$, starting it at

$$\mathbf{x}_0 + \lambda \mathbf{H}^{-1}\mathbf{c}$$

for some small λ. It will then follow the same trajectory of equilibrium positions that we described in the previous section, just as if its buoyancy λ was increasing.

9.7 Hunting with Eggshells

> Directions for hunting a lion in the Sahara: we fence all of the Sahara, we divide it into two halves by another fence, and we detect one half that has no lion in it. Then we divide the other half by a fence, and we continue in this manner until the fenced piece of ground is so small that the lion cannot move and so is caught.
>
> Mathematical folklore, as told by Jiří Matoušek and Bernd Gärtner [561]

Each instance of LINEAR PROGRAMMING has a feasible polytope consisting of the set of all vectors $\mathbf{x}$ such that $\mathbf{A}\mathbf{x} \le \mathbf{b}$. What if we just want to know whether this polytope is nonempty—that is, whether a feasible solution exists? This gives us the following variant of LINEAR PROGRAMMING:

LINEAR PROGRAMMING FEASIBILITY
Input: An $m \times d$ matrix $\mathbf{A}$ and a vector $\mathbf{b} \in \mathbb{R}^m$
Question: Is there a vector $\mathbf{x} \in \mathbb{R}^d$ such that $\mathbf{A}\mathbf{x} \le \mathbf{b}$?

As we commented earlier, for some kinds of instances this problem is easy. For instance, if $\mathbf{A}$ and $\mathbf{b}$ have nonnegative entries, $\mathbf{x} = (0,\ldots,0)$ is always feasible.

But in general, FEASIBILITY is just as hard as the optimization version of LINEAR PROGRAMMING. In particular, FEASIBILITY includes the threshold version of LINEAR PROGRAMMING that we defined in Section 9.5.3, since $\mathbf{c}^T\mathbf{x} \ge \ell$ is just another linear inequality that we can add to the polytope. If we can solve FEASIBILITY, we can use binary search to find the largest value of ℓ such that a feasible $\mathbf{x}$ with $\mathbf{c}^T\mathbf{x} \ge \ell$ exists. We can determine ℓ to within n digits of accuracy in poly(n) time, and thus solve LINEAR PROGRAMMING.

In this section, we will show how to hunt feasible lions in polynomial time, fencing them in a d-dimensional Sahara. To mix our metaphors a little more, we will fence our lions in with a series of eggshells. The volumes enclosed by these eggshells will decrease until the lion cannot move, or until they are too small to contain one, proving that there was no lion in the first place.

This is called the *ellipsoid method*. Historically, it was the first polynomial-time algorithm for LINEAR PROGRAMMING, although it is significantly less efficient than the interior-point methods we discussed in the previous section. More importantly, it applies to many other optimization problems as well, including many with nonlinear constraints. It works whenever the feasible set is convex, as long as we can tell which half of the desert lacks a lion.

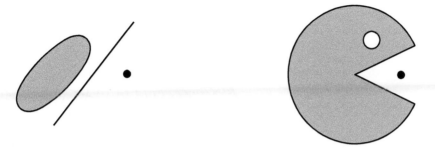

FIGURE 9.23: A point outside a convex set can always be separated from that set by a hyperplane (left). For non-convex sets (right), this isn't always the case.

9.7.1 *Convexity, Hyperplanes, and Separation Oracles*

It's easy to get the impression that LINEAR PROGRAMMING can be solved in polynomial time because of *linearity.* Since each constraint is linear, and since there are only a polynomial number of constraints, the feasible set is a polytope with a polynomial number of facets. This makes it easy to carry out each step of the simplex algorithm, and lets us define and calculate the repulsive potential we used for our interior point method.

But for our lion-hunting strategy, the key property is not linearity but *convexity.* As Figure 9.23 shows, if a point $\mathbf{x}$ is outside a convex set P, there is a *separating hyperplane* that puts $\mathbf{x}$ on one side and all of P on the other. This hyperplane tells us how to fence the desert, assuming we haven't already found a feasible feline.

The ellipsoid method doesn't care what kind of optimization problem we're trying to solve, or what kinds of inequalities define the feasible set. It just needs to be able to tell whether a given point is feasible, and to find a separating hyperplane if it isn't. In other words, it needs to be able to call on the following kind of oracle:

> A *separation oracle* for a set P takes a point $\mathbf{x}$ as input. If $\mathbf{x} \in P$ it replies that x is feasible. If $\mathbf{x} \notin P$ then it provides a separating hyperplane, i.e., a set of the form
>
> $$\left\{\mathbf{y} : \mathbf{a}^T\mathbf{y} = b\right\},$$
>
> such that $\mathbf{a}^T\mathbf{x} \geq b$ but $\mathbf{a}^T\mathbf{y} \leq b$ for all $\mathbf{y} \in P$.

For LINEAR PROGRAMMING, we can do the oracle's job in polynomial time—we just have to check the m constraints $\mathbf{Ax} \leq \mathbf{b}$. If they are all satisfied then $\mathbf{x}$ is feasible. If the jth one is violated, the corresponding facet $\{\mathbf{y} : \mathbf{a}_j^T\mathbf{y} = b_j\}$ is a separating hyperplane.

But there are many other kinds of constraints for which the separation oracle can be implemented in polynomial time. For instance, consider the constraint that $\mathbf{x}$ lies inside the unit sphere:

$$|\mathbf{x}|^2 \leq 1.$$

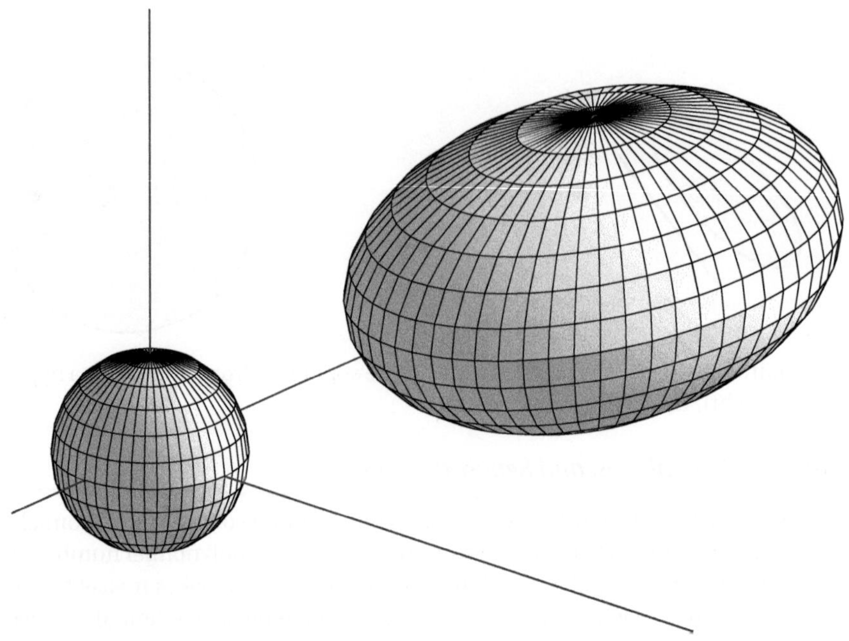

FIGURE 9.24: An ellipsoid is an affine transformation of the unit sphere.

This constraint is quadratic, rather than linear, in the components of **x**. We can certainly check to see if it holds. If it doesn't, the vector **x** pokes out through the sphere somewhere. The hyperplane tangent to the sphere at that point, namely

$$\left\{\mathbf{y} : \mathbf{x}^T\mathbf{y} = |\mathbf{x}|\right\},$$

is a separating hyperplane.

This ellipsoid leads to a rich family of optimization problems. By completing the square, we can express any quadratic constraint on **x** in the form

$$(\mathbf{x}-\mathbf{r})^T\mathbf{Q}(\mathbf{x}-\mathbf{r}) \le 1, \tag{9.29}$$

where $\mathbf{Q}$ is a symmetric $d \times d$ matrix. If $\mathbf{Q}$ is *positive definite*, i.e., if all its eigenvalues are positive, then (9.29) defines an ellipsoid like that shown in Figure 9.24. It is centered at **r**, and its principal axes are the eigenvectors of **Q**.

Such an ellipsoid is the image of the unit sphere under an *affine transformation*, where we apply a linear transformation and then add a constant,

$$\mathbf{x} \to \mathbf{M}^{-1}\mathbf{x} + \mathbf{r},$$

for some $d \times d$ matrix **M**. In that case we have

$$\mathbf{Q} = \mathbf{M}^T\mathbf{M}.$$

Indeed, any positive definite $\mathbf{Q}$ can be written in this form. We say that a symmetric matrix $\mathbf{Q}$ is *positive semidefinite*, and write $\mathbf{Q} \succeq 0$, if it meets any one of the following equivalent conditions:

1. $\mathbf{x}^T\mathbf{Q}\mathbf{x} \geq \mathbf{0}$ for all $\mathbf{x} \in \mathbb{R}^d$.
2. All the eigenvalues of $\mathbf{Q}$ are nonnegative.
3. $\mathbf{Q} = \mathbf{M}^T\mathbf{M}$ for some $d \times d$ matrix $\mathbf{M}$.

We ask you to prove that these conditions are equivalent in Problem 9.20.

Exercise 9.19 *Suppose that* $\mathbf{x}$ *violates the constraint* (9.29), *where* $\mathbf{Q}$ *is positive definite. Show that the set*

$$\left\{\mathbf{y} : (\mathbf{x}-\mathbf{r})^T\mathbf{Q}(\mathbf{y}-\mathbf{r}) = \sqrt{(\mathbf{x}-\mathbf{r})^T\mathbf{Q}(\mathbf{x}-\mathbf{r})}\right\}$$

is a separating hyperplane tangent to the ellipsoid.

We can also get convex sets by bounding convex functions. We say that a function $f(\mathbf{x})$ is convex if, for any $\mathbf{x}, \mathbf{y} \in \mathbb{R}^d$ and all $0 \leq \lambda \leq 1$,

$$f(\lambda\mathbf{x} + (1-\lambda)\mathbf{y}) \leq \lambda f(\mathbf{x}) + (1-\lambda) f(\mathbf{y}). \tag{9.30}$$

If $f(x)$ is a twice-differentiable function of one dimension, it is convex if its second derivative is non-negative. Recall that in higher dimensions, the Hessian of $f(\mathbf{x})$ is its matrix of second derivatives, $H_{ij} = \partial^2 f/\partial x_i \partial x_j$. Then $f(\mathbf{x})$ is convex if and only if its Hessian is positive semidefinite.

Exercise 9.20 *Suppose that the Hessian of* $f(\mathbf{x})$ *is positive definite, i.e.,* $\mathbf{x}^T\mathbf{H}\mathbf{x} > 0$ *for all* $\mathbf{x} \in \mathbb{R}^d$. *Then show that* f *has a unique minimum in any convex set. Hint: given* $\mathbf{x}, \mathbf{y}$, *compute the second derivative of* $f(\lambda\mathbf{x} + (1-\lambda)\mathbf{y})$ *with respect to* λ.

As Figure 9.25 shows, if $f(\mathbf{x})$ is convex then for any ℓ the set

$$P_\ell = \{\mathbf{x} : f(\mathbf{x}) \leq \ell\}$$

is convex. If the gradient ∇f can be computed in polynomial time, we can use it to find separating hyperplanes for such sets. Specifically, suppose that $f(\mathbf{x}) > \ell$. Then the hyperplane that passes through $\mathbf{x}$ and which is perpendicular to ∇f,

$$\left\{\mathbf{y} : (\nabla f)^T\mathbf{y} = (\nabla f)^T\mathbf{x}\right\}, \tag{9.31}$$

is a separating hyperplane. It is tangent to the contour surface $\{\mathbf{y} : f(\mathbf{y}) = f(\mathbf{x})\}$ on which $\mathbf{x}$ lies, and P_ℓ is contained inside this surface.

Exercise 9.21 *Show that if* $f(\mathbf{x})$ *is convex, it is bounded below by its first-order Taylor series. That is, for any* $\mathbf{x}, \mathbf{y}$, *we have*

$$f(\mathbf{y}) \geq f(\mathbf{x}) + (\nabla f)^T(\mathbf{y}-\mathbf{x}).$$

Use this to prove that the hyperplane (9.31) *separates* $\mathbf{x}$ *from* P_ℓ.

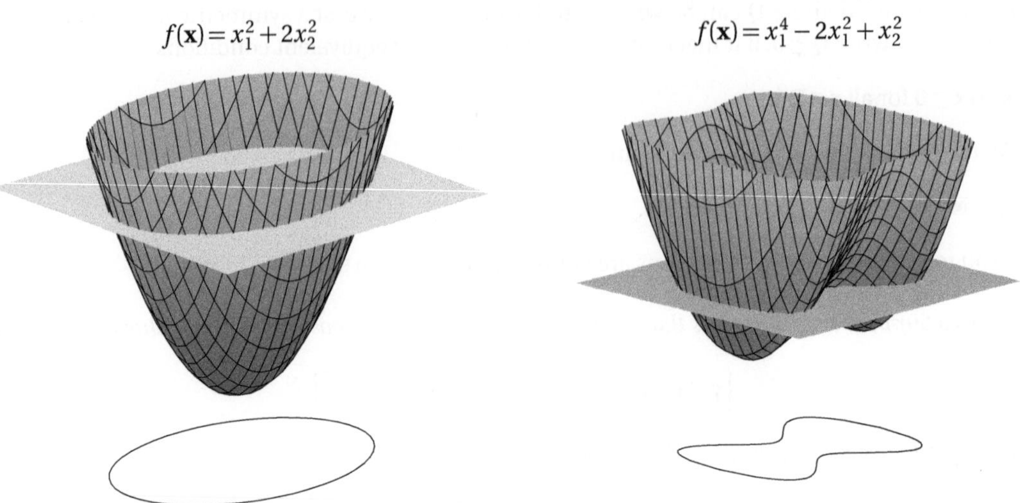

FIGURE 9.25: For a convex function $f(\mathbf{x})$ (left), the set of $\mathbf{x}$ such that $f(\mathbf{x}) \leq \ell$ is convex. For a non-convex function (right), this set can be non-convex.

FIGURE 9.26: The Gömböc, a convex body with a single point of stable equilibrium with respect to gravity.

These examples suggest that we can find separating hyperplanes for nearly any convex set we might come across—whether it is bounded by flat facets like the feasible polytope of LINEAR PROGRAMMING, or by curved surfaces like the one shown in Figure 9.26. If $f(\mathbf{x})$ is convex, we can add the constraint $f(\mathbf{x}) \le \ell$, and use binary search to find the smallest ℓ such that the feasible set is nonempty. Thus we can minimize convex functions—or maximize concave functions—inside convex sets.

But to do all this, we need to show how to turn these hyperplanar fences into a polynomial-time algorithm for FEASIBILITY. This is where ellipsoids come in.

9.7.2 *Cutting Eggs in Half and Fitting Them in Smaller Eggs*

If we take our lion-hunting strategy literally, it seems we have everything we need. At each step, we ask the separation oracle whether the point in the center of the fenced-in area is feasible. If it isn't, she gives us a separating hyperplane. We then build a fence along this hyperplane, making the area smaller, and so on. In essence, we perform a d-dimensional generalization of binary search.

There is some danger, however, that as we add more fences, the fenced-in area could become a complicated polytope—indeed, just as complicated as the feasible polytope in LINEAR PROGRAMMING. In that case, even telling whether there is anything left inside the fence could be just as hard as solving FEASIBILITY in the first place. What we need is a type of fence that can be specified with just a few parameters, whose size and shape can be updated to take the separating hyperplane into account, and whose volume decreases reliably at each step. Ellipsoids fill the bill admirably.

Suppose the feasible set is contained in an ellipsoid E. As in (9.29), E is defined by its center $\mathbf{r}$ and a positive definite matrix $\mathbf{Q}$,

$$E = \left\{\mathbf{x} : (\mathbf{x}-\mathbf{r})^T \mathbf{Q} (\mathbf{x}-\mathbf{r}) \le 1\right\}.$$

We ask the separation oracle if the center $\mathbf{r}$ is feasible. If it is, we're done. If not, the oracle returns a separating hyperplane,

$$\left\{\mathbf{y} : \mathbf{a}^T \mathbf{y} = b\right\}.$$

We can cut E along the hyperplane parallel to this one, but which passes through $\mathbf{r}$. Then we know that the feasible set is contained in one half of E,

$$\left\{\mathbf{y} \in E : \mathbf{a}^T \mathbf{y} \le \mathbf{a}^T \mathbf{r}\right\}.$$

But in order to keep our fences simple, we don't want to deal with half-ellipsoids, quarter-ellipsoids, and so on. Instead, we update our fence to a new ellipsoid E', with center $\mathbf{r}'$ and matrix $\mathbf{Q}'$, which circumscribes this half-ellipsoid.

As Figure 9.27 shows, there are many ellipsoids that circumscribe a given half-ellipsoid. We want E' to be the one with the smallest volume, so that the volume of the fenced-in area decreases as much as possible at each step. We can find E' by solving this minimization problem in the case where E is the unit sphere, and then applying the affine transformation from the sphere to the ellipsoid. Problems 9.22 and 9.23 guide you through this derivation, giving explicit formulas for $\mathbf{Q}'$ and $\mathbf{r}'$ as a function of $\mathbf{Q}, \mathbf{r}$, and $\mathbf{a}$. Here they are, in all their glory:

$$\begin{aligned} \mathbf{Q}' &= \left(1 - \frac{1}{d^2}\right)\mathbf{Q} + \frac{2(d+1)}{d^2} \frac{\mathbf{a} \otimes \mathbf{a}^T}{\mathbf{a}^T \mathbf{Q}^{-1} \mathbf{a}} \\ \mathbf{r}' &= \mathbf{r} - \frac{1}{d+1} \frac{\mathbf{Q}^{-1}\mathbf{a}}{\sqrt{\mathbf{a}^T \mathbf{Q}^{-1} \mathbf{a}}} . \end{aligned} \tag{9.32}$$

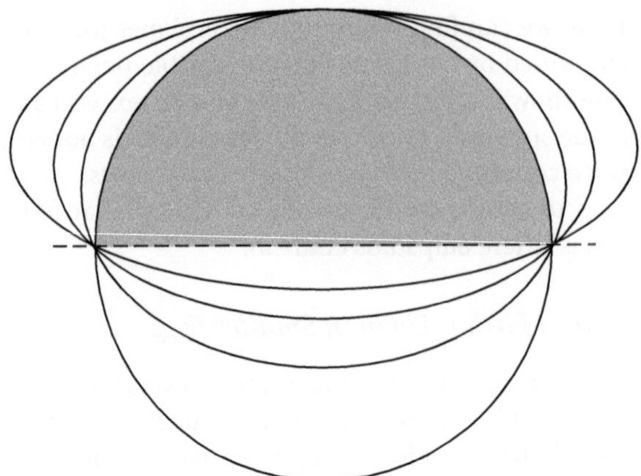

FIGURE 9.27: If we cut away half of an ellipsoid along a separating hyperplane, there is a family of ellipsoids that contain the remaining half. The ellipsoid method uses the one with the smallest volume.

Here $\mathbf{a} \otimes \mathbf{a}^T$ denotes the outer product of $\mathbf{a}$ with itself, i.e., the matrix

$$\left(\mathbf{a} \otimes \mathbf{a}^T\right)_{ij} = a_i a_j .$$

Updating $\mathbf{Q}$ and $\mathbf{r}$ repeatedly yields a sequence of ellipsoids that enclose the feasible set more and more tightly. Figure 9.28 shows two steps of this process. Moreover, at each step the volume decreases by

$$\frac{\text{volume}(E')}{\text{volume}(E)} = \left(\frac{d}{d+1}\right)\left(\frac{d^2}{d^2-1}\right)^{(d-1)/2} \leq \mathrm{e}^{-1/(2d+2)} . \tag{9.33}$$

So we can reduce the volume of the fenced-in area exponentially, say by a factor of 2^{-n}, in $O(d\,n)$ steps.

To start our hunt, we need to fence the entire Sahara—that is, we need an ellipsoid that is guaranteed to contain the feasible set. For LINEAR PROGRAMMING, we showed in Problem 9.17 that the coordinates of any vertex are $2^{O(n)}$ in absolute value, where n is the total number of bits in the instance. So we can start with a sphere of radius $2^{O(n)}$ centered at the origin.

On the other hand, in Problem 9.19 we give a lower bound of $2^{-O(dn)}$ on the volume of the feasible polytope if it is nonempty, ignoring the degenerate case where it is squashed flat and has volume zero. If the volume of the ellipsoid becomes smaller than that, we know that the feasible set is empty. Putting this together, we only need to iterate the ellipsoid method $O(d^2 n)$ times before we either find a feasible point, or prove that there is none.

This gives us another proof that LINEAR PROGRAMMING FEASIBILITY, and therefore LINEAR PROGRAMMING, is in P. But it applies equally well to nearly any convex optimization problem. We just need two ingredients: (1) a separation oracle that we can implement in polynomial time, and (2) exponential upper and lower bounds—$2^{\text{poly}(n)}$ and $2^{-\text{poly}(n)}$—on the volume of the feasible set.

9.16

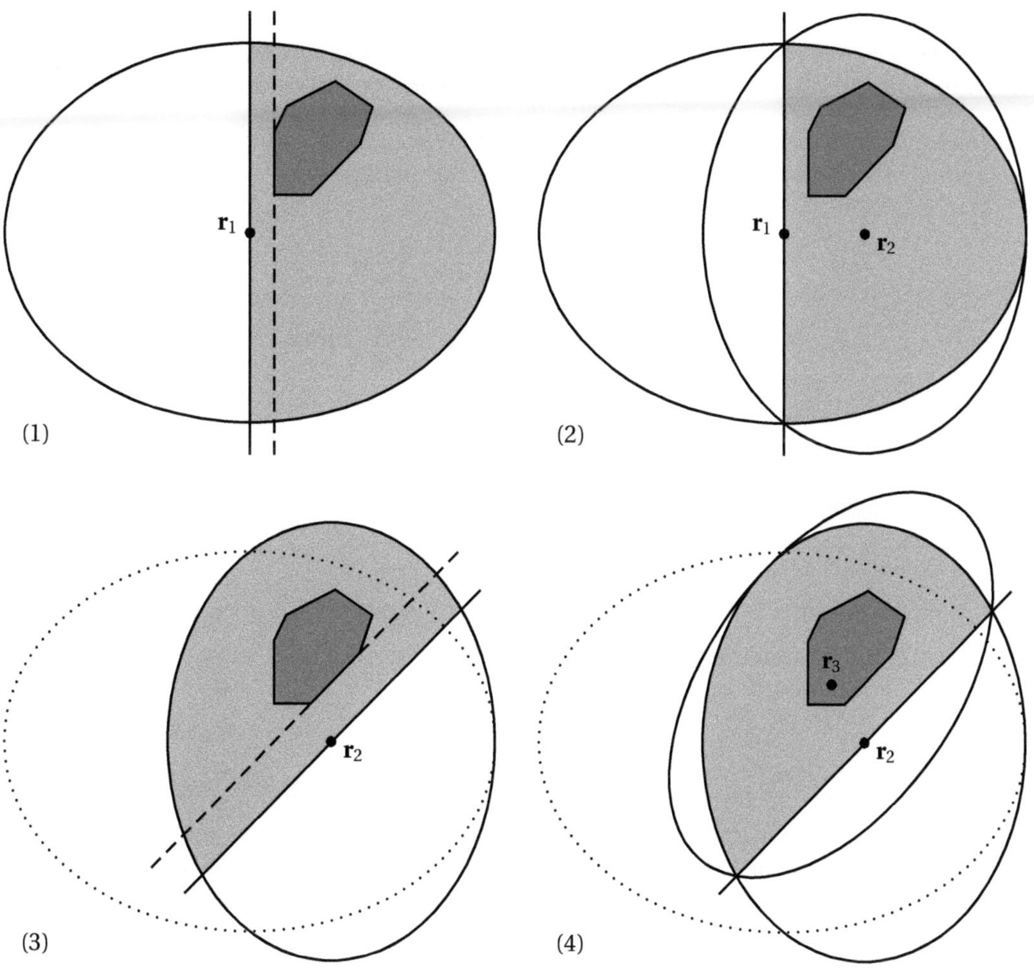

FIGURE 9.28: The ellipsoid method in action. Dashed lines are separating hyperplanes—in this case, violated facets of an instance of LINEAR PROGRAMMING. In each step we divide the ellipsoid E into two halves, along the line through the center $\mathbf{r}$ parallel to the separating hyperplane, and then bound the remaining half in a new ellipsoid E'.

9.7.3 *Semidefinite Programming*

A particularly important kind of nonlinear optimization problem places constraints, not on a vector $\mathbf{x}$, but on a matrix $\mathbf{X}$. Consider the constraint that $\mathbf{X}$ is positive semidefinite, which as before we denote $\mathbf{X} \succeq 0$. This constraint is convex, as the following exercise shows:

Exercise 9.22 *Prove that the set of positive semidefinite $d \times d$ matrices is a convex subset of the real symmetric matrices. Hint: review the three equivalent definitions of positive semidefiniteness in Section 9.7.1.*

We can also place linear constraints on the entries of $\mathbf{X}$. We can write such constraints compactly as the inner product of $\mathbf{X}$ with a matrix $\mathbf{A}$, which is the trace of the matrix product $\mathbf{A}^T\mathbf{X}$:

$$\langle \mathbf{A}, \mathbf{X} \rangle = \operatorname{tr} \mathbf{A}^T \mathbf{X} = \sum_{i,j=1}^{d} A_{ij} X_{ij} \le b \,.$$

Given these constraints, we can maximize a linear function of $\mathbf{X}$'s entries,

$$\langle \mathbf{C}, \mathbf{X} \rangle = \operatorname{tr} \mathbf{C}^T \mathbf{X} = \sum_{i,j=1}^{d} C_{ij} X_{ij} \,.$$

This gives us the following problem:

SEMIDEFINITE PROGRAMMING

Input: A list of $d \times d$ matrices $\mathbf{A}_1, \ldots, \mathbf{A}_m$, a vector $\mathbf{b} \in \mathbb{R}^m$, and a $d \times d$ matrix $\mathbf{C}$

Output: A real symmetric $d \times d$ matrix $\mathbf{X}$ that maximizes $\langle \mathbf{C}, \mathbf{X} \rangle$ subject to the constraints

$$\begin{aligned} \langle \mathbf{A}_i, \mathbf{X} \rangle &\le b_i \text{ for } i = 1, \ldots, m \\ \mathbf{X} &\succeq \mathbf{0} \end{aligned} \tag{9.35}$$

Can we implement a separation oracle for SEMIDEFINITE PROGRAMMING in polynomial time? We can handle the linear constraints just as we did for LINEAR PROGRAMMING, checking each one to see if it is satisfied. If the jth constraint is violated, we can use the corresponding facet

$$\left\{ \mathbf{Y} : \langle \mathbf{A}_j, \mathbf{Y} \rangle = b_j \right\}$$

as a separating hyperplane.

To find a separating hyperplane for positive semidefiniteness, first note that we can express $\mathbf{X} \succeq 0$ as an infinite number of linear constraints on $\mathbf{X}$. Namely, for all $\mathbf{v} \in \mathbb{R}^d$ we have

$$\mathbf{v}^T \mathbf{X} \mathbf{v} \ge 0 \,.$$

We can also write $\mathbf{v}^T \mathbf{X} \mathbf{v}$ as the inner product of $\mathbf{X}$ with the outer product $\mathbf{v} \otimes \mathbf{v}^T$, so

$$\langle \mathbf{v} \otimes \mathbf{v}^T, \mathbf{X} \rangle \ge 0 \,.$$

Using Gaussian elimination, we can diagonalize **X** and thus find its eigenvalues and eigenvectors in polynomial time. If none of its eigenvalues are negative, **X** is positive semidefinite. On the other hand, suppose **X** has an eigenvector **v** with a negative eigenvalue, so that

$$\langle \mathbf{v} \otimes \mathbf{v}^T, \mathbf{X} \rangle < 0 .$$

Then the set of real symmetric matrices **Y** such that $\mathbf{Y}\mathbf{v} = 0$, or equivalently

$$\left\{ \mathbf{Y} : \langle \mathbf{v} \otimes \mathbf{v}^T, \mathbf{Y} \rangle = 0 \right\} ,$$

is a separating hyperplane. Note that this hyperplane consists of singular matrices, i.e., matrices **Y** with zero determinant.

As for LINEAR PROGRAMMING, it's possible to prove exponential upper and lower bounds on the volume of the feasible set as a subset of the real symmetric matrices. Together with the separation oracle, these bounds let us solve SEMIDEFINITE PROGRAMMING in polynomial time using the ellipsoid method.

We hasten to add that there are also interior point methods for SEMIDEFINITE PROGRAMMING analogous to those for LINEAR PROGRAMMING that we described in Section 9.6. As for LINEAR PROGRAMMING, these methods are more efficient than the ellipsoid method, and they are simple as well—one can define a potential energy $V(\mathbf{X})$ that pushes **X** away from the singular matrices, and compute its gradient and Hessian efficiently. We introduce such a potential in Problem 9.24.

SEMIDEFINITE PROGRAMMING has many applications. In Section 10.4 we will use it for a randomized approximation algorithm for MAX CUT, and in Problem 15.53 we will use it to find the optimal measurement to identify quantum states. Perhaps its most marvelous application is Problem 9.44, which shows that whenever the chromatic number of a graph equals the size of its largest clique, we can use SEMIDEFINITE PROGRAMMING to find both in polynomial time.

9.17

9.7.4 *Convexity and Complexity*

The ellipsoid method lets us solve nearly any optimization problem, as long as the feasible set is convex, and the function we are trying to minimize or maximize is convex or concave respectively. We end this section by showing that in the absence of convexity, optimization becomes NP-hard even for functions with a simple analytic form.

Consider the following problem:

> QUADRATIC PROGRAMMING
>
> Input: A $d \times d$ matrix **Q**, an $m \times d$ matrix **A**, and vectors $\mathbf{b} \in \mathbb{R}^m$ and $\mathbf{r} \in \mathbb{R}^d$
>
> Output: A vector $\mathbf{x} \in \mathbb{R}^d$ that minimizes the function
>
> $$f(\mathbf{x}) = (\mathbf{x} - \mathbf{r})^T \mathbf{Q} (\mathbf{x} - \mathbf{r}) ,$$
>
> subject to the constraints $\mathbf{A}\mathbf{x} \leq \mathbf{b}$.

The Hessian of f is $2\mathbf{Q}$. So if **Q** is positive semidefinite, f is convex and QUADRATIC PROGRAMMING is in P.

If $\mathbf{Q}$ is not positive semidefinite, however, this problem is NP-hard. To see why, consider an instance of QUADRATIC PROGRAMMING where $\mathbf{Q} = -(4/d)\mathbb{1}$ and $\mathbf{r} = (1/2, \ldots, 1/2)$. Then

$$f(\mathbf{x}) = -\frac{1}{d}\sum_{i=1}^{d}(2x_i - 1)^2 .$$

If we impose the constraints

$$0 \leq x_i \leq 1 \quad \text{for all } i , \tag{9.36}$$

then f is minimized at the corners of the hypercube $\{0,1\}^d$ where $f(\mathbf{x}) = -1$.

This gives us an easy reduction from 3-SAT. Given a 3-SAT formula ϕ with Boolean variables $x_1, \ldots, x_d$, we translate each clause into a linear constraint, such as

$$(x_1 \vee x_2 \vee \overline{x_3}) \quad \to \quad x_1 + x_2 + (1 - x_3) \geq 1 .$$

If we add these constraints to (9.36), the resulting instance of QUADRATIC PROGRAMMING has a solution $\mathbf{x}$ with $f(\mathbf{x}) = -1$ if and only if ϕ is satisfiable. Thus QUADRATIC PROGRAMMING is NP-hard in general—even though it only involves quadratic functions and linear inequalities.

9.8 Algorithmic Cubism

> If a man is at once acquainted with the geometric foundation of things and with their festal splendor, his poetry is exact and his arithmetic musical.
>
> Ralph Waldo Emerson

In the early 20th century, painters like Pablo Picasso and George Braque founded cubism—a movement that revolutionized European painting and sculpture. The Cubists believed, as did post-impressionist Paul Cézanne, that artists should treat nature "in terms of the cylinder, the sphere and the cone."

A few decades later, cubism entered optimization. Motivated by the success of the simplex algorithm, people began to treat more and more optimization problems in terms of polytopes. We have already seen that MAX FLOW and MIN CUT can be written as linear programs, but the expressive power of linear inequalities goes far beyond that. In this section and the next, we treat a variety of optimization problems as maximizing a function on the surface of a polytope, ranging from easy problems like MINIMUM SPANNING TREE to hard ones like VERTEX COVER and TSP.

First we face a constraint that we haven't dealt with yet in this chapter. Namely, for many problems we are only interested in integral solutions. We have already seen how integrality can turn harmless polynomial equations into NP-complete problems (the quadratic Diophantine equations of Section 5.4.4) or even make them undecidable (Hilbert's 10th problem, which we briefly discussed in Section 7.1.2). Similarly, we will see that demanding integral solutions turns LINEAR PROGRAMMING into its NP-hard cousin INTEGER LINEAR PROGRAMMING.

Even when integrality makes a problem NP-hard, we can sometimes find a good approximation to it by *relaxing* and *rounding*. First we relax the integrality constraint, pretending that we don't mind fractional solutions, and solve the resulting instance of LINEAR PROGRAMMING. We then obtain a feasible solution to the original problem by rounding each variable x_i to a nearby integer. For some problems, such as MIN VERTEX COVER, this gives a solution within a constant factor of the optimum—in the language of Section 9.2, a ρ-approximation for some constant ρ.

On the other hand, we will see that when the problem has a certain special form—for instance, when it shares some of the mathematical structure of MINIMUM SPANNING TREE or MAX FLOW—then the polytope's vertices are integral in the first place, and the integrality constraint has no effect. In that case, we can solve the corresponding instance of INTEGER LINEAR PROGRAMMING in polynomial time simply by treating it as an instance of LINEAR PROGRAMMING.

9.8.1 *Integer Linear Programming*

Let's start with the cubist view of MIN VERTEX COVER. Suppose a graph $G=(V,E)$ has n vertices and m edges. A subset $S \subseteq V$ of vertices can be specified by its characteristic vector $\mathbf{x} \in \{0,1\}^n$, where

$$x_v = \begin{cases} 1 & \text{if } v \in S \\ 0 & \text{if } v \notin S. \end{cases}$$

By definition, S is a vertex cover if it includes at least one endpoint of every edge, so we demand that

$$x_u + x_v \geq 1 \quad \text{for all } (u,v) \in E.$$

We can write this as $\mathbf{Ax} \geq \mathbf{1}$ where $\mathbf{1}$ denotes the column vector $(1,\ldots,1)^T$ and $\mathbf{A}$ is the following $m \times n$ matrix,

$$a_{ij} = \begin{cases} 1 & \text{if vertex } j \text{ is an endpoint of edge } i \\ 0 & \text{otherwise.} \end{cases} \tag{9.37}$$

This $\mathbf{A}$ is called the *incidence matrix* of G. To solve MIN VERTEX COVER we want to minimize $|S|$, which we can write as $\mathbf{1}^T\mathbf{x}$.

Thus we want to minimize a function of the form $\mathbf{c}^T\mathbf{x}$, subject to constraints of the form $\mathbf{Ax} \geq \mathbf{b}$. This is just an instance of LINEAR PROGRAMMING, except that we want $\mathbf{x}$ to have entries in $\{0,1\}$ rather than varying continuously. If we add the constraints $\mathbf{x} \geq \mathbf{0}$ then it suffices to demand that $\mathbf{x}$ is integral, since minimizing $|S|$ will push each x_v down to 0 or 1.

Asking for the best possible integral solution of a linear program is known as INTEGER LINEAR PROGRAMMING, or ILP for short. The general definition reads

> INTEGER LINEAR PROGRAMMING (ILP)
>
> Input: A matrix $\mathbf{A} \in \mathbb{Z}^{m \times d}$ and vectors $\mathbf{b} \in \mathbb{Z}^m$ and $\mathbf{c} \in \mathbb{Z}^d$
>
> Output: A vector $\mathbf{x}$ that maximizes $\mathbf{c}^T\mathbf{x}$, subject to the constraints
>
> $$\begin{aligned} \mathbf{Ax} &\leq \mathbf{b} \\ \mathbf{x} &\in \mathbb{Z}^d \end{aligned} \tag{9.39}$$

We have shown that

$$\text{MIN VERTEX COVER} \le \text{ILP},$$

and therefore that ILP is NP-hard. In fact, it is easy to express a wide variety of optimization problems as integer linear programs, and thus reduce them to ILP without the need for fancy gadgets. This makes ILP one of the most versatile NP-hard problems, and one of the most important in practice.

Exercise 9.23 *Reduce* MAX-SAT *to* ILP.

9.8.2 *Fuzzy Covers*

The fact that ILP is so close to its polynomial-time cousin LP suggests a general strategy for finding approximate solutions: *relaxation* and *rounding*. First we relax the problem by ignoring the integrality constraint, producing an instance of LP. We can solve this relaxed problem in polynomial time, but the optimal vertex **x** will usually be fractional. We then try to round **x** to an integral solution, replacing each of its components with the nearest integer. Hopefully, this gives a feasible solution to the original problem, and one which is not much worse than the optimum.

The relaxed version of MIN VERTEX COVER lets each x_v take fractional values between zero and one,

$$x_v \in \{0,1\} \quad \rightarrow \quad 0 \le x_v \le 1 .$$

A feasible solution **x** of this relaxed problem is a "fuzzy vertex cover," where each vertex v is in S to some extent x_v, and $x_u + x_v \ge 1$ for every edge (u,v). The "size" of **x** is $\mathbf{1}^T\mathbf{x} = \sum_v x_v$, and we can find the fuzzy vertex cover of minimum size in polynomial time.

If this fuzzy cover happens to be a genuine cover, with $x_v \in \{0,1\}$ for each v, it is the smallest vertex cover. We show a graph for which this is the case on the left of Figure 9.29. In general, however, the LP relaxation of MIN VERTEX COVER produces fractional solutions. For instance, on a triangle the smallest fuzzy cover is $(1/2, 1/2, 1/2)$ and has size $3/2$, while the smallest genuine vertex cover has size 2. What good are these fuzzy solutions?

First of all, relaxing a constraint can only improve the quality of the optimum. Every genuine cover is also a fuzzy cover, so the size $k_{\text{relaxed}} = \mathbf{1}^T\mathbf{x}$ of the smallest fuzzy cover **x** is less than or equal to the size $k_{\text{opt}} = |S_{\text{opt}}|$ of the smallest genuine one:

$$k_{\text{relaxed}} \le k_{\text{opt}} . \tag{9.40}$$

Having a lower bound on k_{opt} that we can compute in polynomial time can be very handy. As we will see in the next section, we can use bounds like these to prune the branches of a search tree, greatly reducing the time it takes to find the optimum.

On the other hand, we can change a fuzzy vertex cover to a genuine one by rounding each x_v to the nearest integer, 0 or 1, rounding up if $x_v = 1/2$. This gives the set

$$S_{\text{rounded}} = \{v \in V : x_v \ge 1/2\} .$$

Exercise 9.24 *Show that* $S_{rounded}$ *is a vertex cover.*

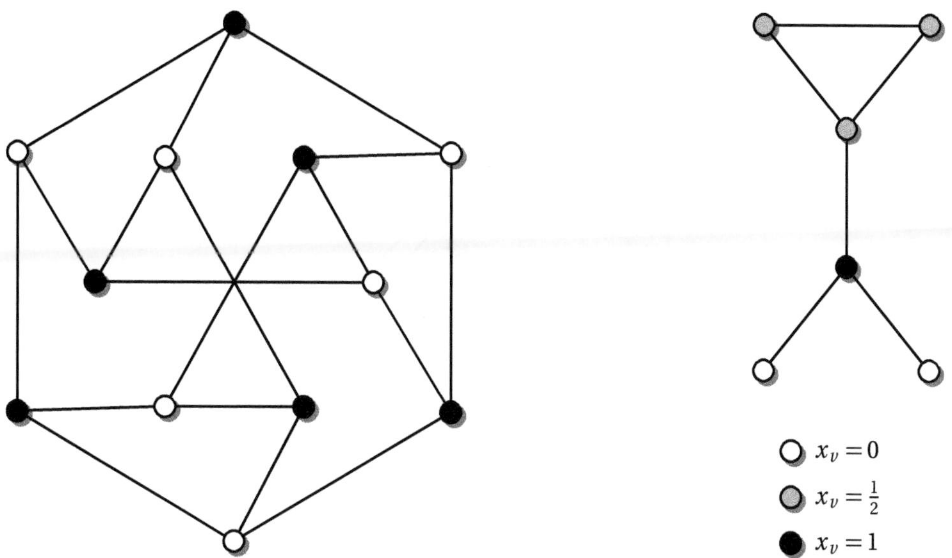

FIGURE 9.29: The LP relaxation of MIN VERTEX COVER, and the fuzzy covers it produces. For some graphs, like the Franklin graph (left), the optimal fuzzy cover is the smallest vertex cover; shown is one of the two optimal vertices of the polytope. For others, such as the graph on the right, the smallest fuzzy cover is fractional. In fact, Problem 9.28 shows that $x_v \in \{0, 1/2, 1\}$ for all v.

How large is S_{rounded}? Rounding at most doubles the size, since the worst case occurs if $x_v = 1/2$ for all v. Thus $k_{\text{rounded}} = |S_{\text{rounded}}|$ is bounded above by

$$k_{\text{rounded}} \leq 2k_{\text{relaxed}}, \tag{9.41}$$

and combining this with (9.40) gives

$$k_{\text{rounded}} \leq 2k_{\text{opt}}. \tag{9.42}$$

Thus, like the pebble game of Section 9.2.1, relaxation and rounding gives a 2-approximation for MIN VERTEX COVER. The following exercise shows that, in general, it doesn't do any better:

Exercise 9.25 *Give a family of graphs with n vertices, for arbitrarily large n, such that $x_v = 1/2$ for all v in the minimal fuzzy cover, and for which the approximation ratio achieved by the relax-and-round algorithm approaches* 2 *as n goes to infinity.*

The LP relaxation of MIN VERTEX COVER has a particularly nice property. The relaxed optimum is *half-integral*—that is, $x_v \in \{0, 1/2, 1\}$ for all v. The proof is easy, and we leave it to you in Problem 9.28. Moreover, Problem 9.29 shows that there is always a minimal vertex cover that includes all vertices v with $x_v = 1$, and excludes all v with $x_v = 0$. So even when the minimal fuzzy cover is fractional, it can give us some information about the minimal vertex cover.

It's tempting to think that the relax-and-round strategy works for ILP in general. Unfortunately, this is not the case. As Figure 9.30 shows, the optimum of the relaxed problem might sit at the top of a long,

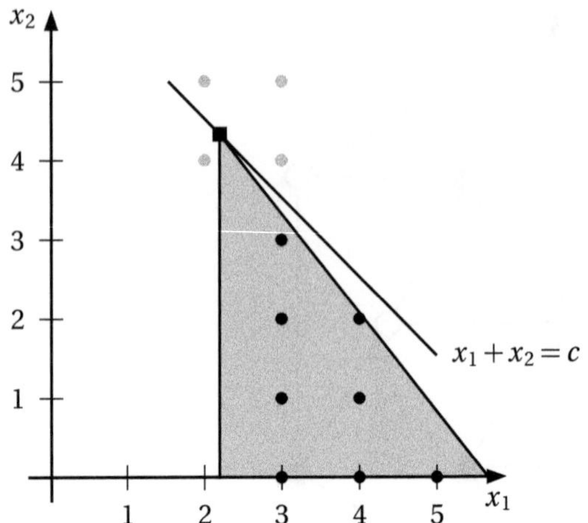

FIGURE 9.30: Relaxation and rounding doesn't always work. In some cases, none of the integral vectors near the relaxed optimum are feasible.

narrow corner of the polytope that fits between the lattice of integral vectors. In that case, none of the nearby integral vectors are feasible. Even if we can find a feasible integral solution, its quality might be much worse than that of the relaxed optimum, giving the relax-and-round algorithm an arbitrarily poor approximation ratio.

9.8.3 *Easy Polyhedra: Unimodularity*

We know that INTEGER LINEAR PROGRAMMING is NP-hard in general. But for some cases of LINEAR PROGRAMMING, adding the constraint of integrality makes no difference, because the vertices of the feasible polytope—including the optimal vertex—are already integral. In that case, we can solve the corresponding case of INTEGER LINEAR PROGRAMMING in polynomial time. We just ignore the integrality constraint, and solve the relaxed version using balloons or eggshells.

We have already seen one example of this phenomenon. Namely, MAX FLOW has integral solutions whenever the capacities of the edges are integers. We proved this in Section 3.6 using the Ford–Fulkerson algorithm. But something more general is true: if we define the polytope of feasible flows as in Section 9.4.1, then *all* of its vertices are integral.

As another example, for the graph on the left side of Figure 9.29, the optimal fuzzy vertex cover—the solution to the LP relaxation of MIN VERTEX COVER—is integral. We will see below that this graph has the property that the polytope of fuzzy covers has integral vertices. For any such graph, we can solve MIN VERTEX COVER in polynomial time.

This offers us a new kind of algorithmic insight, and a new family of polynomial-time algorithms: reduce your problem to ILP, and check to see whether the integrality constraint is redundant. The problem is then how to tell whether a given set of linear inequalities describes a polytope with integral vertices. Our goal in this section is to give some sufficient conditions for integrality, in terms of the underlying linear algebra of these inequalities.

Recall that any vertex of the polytope $\{\mathbf{x} \in \mathbb{R}^d : \mathbf{A}\mathbf{x} \le \mathbf{b}\}$ is the intersection of some set of d linearly independent inequalities. That is, $\mathbf{x}$ is a solution of $\mathbf{A}'\mathbf{x} = \mathbf{b}'$, where $\mathbf{A}'$ is a nonsingular $d \times d$ submatrix of $\mathbf{A}$ and $\mathbf{b}'$ consists of the corresponding elements of $\mathbf{b}$. As in Problem 9.17 we can then invoke Cramer's rule, giving

$$x_i = \frac{\det \mathbf{A}'_i}{\det \mathbf{A}'},$$

where $\mathbf{A}'_i$ is the matrix formed by replacing the ith column of $\mathbf{A}'$ with $\mathbf{b}'$. If $\mathbf{A}'$ and $\mathbf{b}'$ are integral, the numerator is integral. If the denominator $\det \mathbf{A}'$ is ± 1, then $\mathbf{x}$ is integral as well.

An integer matrix $\mathbf{A}'$ is called *unimodular* if its determinant is ± 1, or equivalently if its inverse is also integer-valued. Thus one sufficient condition for the vertices of the feasible polytope to be integral is if $\mathbf{b}$ is integral, and if every nonsingular $d \times d$ submatrix of A is unimodular. Alternately, we can say that every $d \times d$ submatrix of A has determinant $+1$, -1, or 0.

As we saw in Section 9.4, the constraints on $\mathbf{x}$ often include simple upper and lower bounds. For instance, in MAX FLOW we demand that the flow on each edge is at least zero and at most the edge's capacity. In general, we might want to impose constraints of the form $\mathbf{u} \le \mathbf{x} \le \mathbf{w}$, in addition to more complex constraints of the form $\mathbf{A}\mathbf{x} \le \mathbf{b}$. We can express these constraints all at once in the form

$$\hat{\mathbf{A}}\mathbf{x} \le \hat{\mathbf{b}} \quad \text{where} \quad \hat{\mathbf{A}} = \begin{pmatrix} \mathbf{A} \\ -\mathbb{1} \\ \mathbb{1} \end{pmatrix} \quad \text{and} \quad \hat{\mathbf{b}} = \begin{pmatrix} \mathbf{b} \\ -\mathbf{u} \\ \mathbf{w} \end{pmatrix}. \tag{9.43}$$

If $\mathbf{u}$ and $\mathbf{w}$ are integral, the vertices of the resulting polytope are integral whenever every nonsingular $d \times d$ submatrix of $\hat{\mathbf{A}}$ is unimodular. What property does $\mathbf{A}$ need to have for this to be the case?

We say that $\mathbf{A}$ is *totally unimodular* if all of its square submatrices have determinant ± 1 or 0. That is, if we choose any k rows and any k columns of $\mathbf{A}$, the $k \times k$ matrix formed by their intersections is either singular or unimodular. In particular, taking $k = 1$ implies that each of $\mathbf{A}$'s entries is ± 1 or 0. Problem 9.30 shows that $\hat{\mathbf{A}}$ is totally unimodular if and only if $\mathbf{A}$ is totally unimodular. Thus if $\mathbf{A}$ is totally unimodular, the polytope described by (9.43) has integral vertices, for any integral vectors $\mathbf{u}$ and $\mathbf{w}$.

Total unimodularity is not necessary for the feasible polytope to have integral vertices. However, it is sufficient, and it holds for many problems. It can be used to show that a network with integer capacities has an integral MAX FLOW without resorting to the Ford–Fulkerson algorithm (Problem 9.31), or that we can minimize the itinerary for a group of traveling salesmen in polynomial time (Problem 9.33). In addition, it is often easy to prove, since there are quite a few easily checkable properties of $\mathbf{A}$ that imply total unimodularity (Problem 9.30).

We content ourselves here with proving total unimodularity for a particular family of matrices: the 9.18
incidence matrices of bipartite graphs. Since the graph on the left of Figure 9.29 is bipartite, this explains why its optimal fuzzy vertex cover is integral. More generally, total unimodularity lets us prove that MIN VERTEX COVER, and some other problems, are in P for bipartite graphs.

Without further ado, let's prove the following theorem.

Theorem 9.2 *A graph G is bipartite if and only if its incidence matrix* $\mathbf{A}$, *defined as in* (9.37), *is totally unimodular.*

Proof Assume that G is bipartite, and let $\mathbf{B}$ be a $k \times k$ submatrix of $\mathbf{A}$. We will show that $\det \mathbf{B}$ is ± 1 or 0 by induction on k. For $k = 1$, this is trivial, since the each entry of $\mathbf{A}$ is 0 or 1. For $k > 1$, recall that the rows of **A**, and therefore of **B**, correspond to edges of the graph, while columns correspond to vertices. Thus each row of **B** has at most two 1s, and we can distinguish three cases:

1. **B** has a row of 0s. Then $\det \mathbf{B} = 0$.

2. **B** has a row with exactly one 1. By permuting rows and columns, we can write

$$\mathbf{B} = \begin{pmatrix} 1 & \mathbf{0}^T \\ \mathbf{b} & \mathbf{B}' \end{pmatrix}$$

   for some vector **b** and some $(k-1) \times (k-1)$ matrix $\mathbf{B}'$. But then $\det \mathbf{B} = \det \mathbf{B}'$, which by induction is ± 1 or 0.

3. Every row of **B** contains exactly two 1s. In that case, the vertices corresponding to **B**'s columns must include both endpoints of the edge corresponding to each row of **B**. That means that **B** is the incidence matrix of a subgraph of G. Like G, this subgraph is bipartite, and we can divide its vertices into two groups L, R such that edges only run between L and R. Thus by permuting the columns, we can write

$$\mathbf{B} = \begin{pmatrix} \mathbf{B}_L & \mathbf{B}_R \end{pmatrix},$$

   where each row of $\mathbf{B}_L$ contains exactly one 1, and the same is true for $\mathbf{B}_R$. Adding up all the columns in $\mathbf{B}_L$ gives **1**, and the same is true for $\mathbf{B}_R$. Hence the columns of **B** are linearly dependent, and $\det \mathbf{B} = 0$.

Conversely, suppose that G is not bipartite. Then G contains a cycle of length k for some odd k, and **A** has a $k \times k$ submatrix

$$\begin{pmatrix} 1 & 1 & & & & \\ & 1 & 1 & & & \\ & & 1 & & & \\ & & & \ddots & & \\ & & & & 1 & 1 \\ 1 & & & & & 1 \end{pmatrix},$$

where matrix entries not shown are zero. For any odd k, this matrix has determinant 2, so **A** is not totally unimodular. □

Now recall that MIN VERTEX COVER is an instance of ILP, namely

$$\min_{\mathbf{x}} (\mathbf{1}^T \mathbf{x}) \quad \text{subject to} \quad \mathbf{x} \geq \mathbf{0} \quad \text{and} \quad \mathbf{A}\mathbf{x} \geq \mathbf{1}, \tag{9.44}$$

where **A** is the incidence matrix of G. Theorem 9.2 tells us that for bipartite graphs, the feasible polytope, i.e., the polytope of fuzzy vertex covers,

$$\{\mathbf{x} : \mathbf{x} \geq \mathbf{0} \text{ and } \mathbf{A}\mathbf{x} \geq \mathbf{1}\},$$

has integral vertices. Therefore, BIPARTITE VERTEX COVER is in P.

Theorem 9.2 applies to another old friend: MAX BIPARTITE MATCHING from Section 3.8. To write MAX BIPARTITE MATCHING as a linear program, we encode a set of edges $F \subseteq E$ by the characteristic vector

$$y_e = \begin{cases} 1 & \text{if } e \in F \\ 0 & \text{if } e \notin F. \end{cases}$$

The set F is a matching if each vertex is incident to at most one edge in F—but this means that $\mathbf{A}^T\mathbf{y} \le \mathbf{1}$ where $\mathbf{A}$ is again G's incidence matrix. We want to maximize $|F| = \mathbf{1}^T\mathbf{y}$. Thus MAX BIPARTITE MATCHING is another instance of ILP,

$$\max_{\mathbf{y}}(\mathbf{1}^T\mathbf{y}) \quad \text{subject to} \quad \mathbf{y} \ge \mathbf{0} \quad \text{and} \quad \mathbf{A}^T\mathbf{y} \le \mathbf{1}. \tag{9.45}$$

Now if $\mathbf{A}$ is unimodular, so is $\mathbf{A}^T$. Thus if G is bipartite, the polytope of "fuzzy matchings,"

$$\left\{\mathbf{y} : \mathbf{y} \ge \mathbf{0} \text{ and } \mathbf{A}^T\mathbf{y} \le \mathbf{1}\right\},$$

also has integral vertices, and we can solve MAX BIPARTITE MATCHING in polynomial time. Note that we omitted the constraint $\mathbf{y} \le \mathbf{1}$, since it is already implied by $\mathbf{A}^T\mathbf{x} \le \mathbf{1}$.

In fact, looking at (9.44) and (9.45), we see these two problems are precisely each others' duals as defined in Section 9.5. As a consequence, the size of the minimum vertex cover equals the size of the maximum matching. We can prove this directly using combinatorial arguments, as perhaps the reader did in Problem 4.16. But the cubist view of these problems makes their relationship immediate, and reveals that it is an example of a deeper phenomenon. In the next section, we will see what else this point of view can offer us.

Exercise 9.26 *Consider the weighted versions of* BIPARTITE VERTEX COVER *and* MAX BIPARTITE MATCHING, *where each vertex v or edge e has a weight c_v or b_e, and we wish to minimize $\mathbf{c}^T\mathbf{x}$ or maximize $\mathbf{b}^T\mathbf{y}$. Show that both these problems are in* P. *Note that* $\mathbf{c}$ *and* $\mathbf{b}$ *do not have to be integral.*

9.9 Trees, Tours, and Polytopes

In MIN VERTEX COVER we optimize over all subsets $S \subseteq V$ of the vertices of a graph that form a vertex cover of G. In MAX BIPARTITE MATCHING, we optimize over subsets $F \subseteq E$ of the edges that form a matching. This is a prevalent theme in combinatorial optimization—we have a base set such as V or E, and we want to optimize over all subsets of this base set that have a certain property.

We have already seen how viewing these subsets as the vertices of a polytope can give us insight into these problems. Let's look at two more problems this way, an easy one and a hard one: MINIMUM SPANNING TREE, where we optimize over subsets $F \subseteq E$ that form a spanning tree, and TRAVELING SALESMAN, where $F \subseteq E$ forms a Hamiltonian cycle.

As before, we encode a subset $F \subseteq E$ by its characteristic vector $\mathbf{x} \in \{0,1\}^{|E|}$,

$$x_e = \begin{cases} 1 & \text{if } e \in F \\ 0 & \text{if } e \notin F. \end{cases}$$

Let $\mathbf{c}$ denote the vector of weights c_e of the edges $e \in E$. Then we can write MINIMUM SPANNING TREE and TRAVELING SALESMAN as

$$\min_{\mathbf{x}}(\mathbf{c}^T\mathbf{x}) \quad \text{subject to} \quad \mathbf{x} \in \mathscr{F},$$

where the set $\mathcal{F}$ of feasible solutions is given by those characteristic vectors that represent spanning trees or Hamiltonian cycles.

Now $\mathcal{F}$ is a discrete set of vectors, but it is also the set of vertices of a polytope—namely, its convex hull conv($\mathcal{F}$). We can think of $\mathcal{F}$ as the polytope of "fuzzy spanning trees" or "fuzzy tours," with the genuine trees and tours at its vertices. Since $\mathbf{c}^T\mathbf{x}$ is always minimized at a vertex, we can recast MINIMUM SPANNING TREE and TRAVELING SALESMAN as linear programs:

$$\min_{\mathbf{x}}(\mathbf{c}^T\mathbf{x}) \quad \text{subject to} \quad \mathbf{x} \in \text{conv}(\mathcal{F}).$$

If $\mathcal{F}$ can be described by a polynomial number of linear inequalities then MINIMUM SPANNING TREE and TSP are simply instances of LINEAR PROGRAMMING, and we can solve them accordingly. Even if the number of inequalities is exponential, we may still be able to implement a separation oracle in polynomial time, in which case we can use the ellipsoid method.

Thus our goal is to understand as much as we can about the geometric structure of conv($\mathcal{F}$) in each case. This is complementary to our approach in the preceding section. For MIN VERTEX COVER and MAX BIPARTITE MATCHING, we started with a set of linear inequalities, and checked whether the resulting polytope has integral vertices. Here we start with a set of integral vertices, and we ask for the linear inequalities that describe their convex hull.

This gives us a geometric picture of computational complexity, where the complexity of the problem is deeply related to the complexity of the polytope. For instance, we will see that both MINIMUM SPANNING TREE and TSP correspond to polytopes with an exponential number of facets—but for MINIMUM SPANNING TREE, we can nevertheless implement a separation oracle in polynomial time. For TSP, in contrast, no such separation oracle is known. Thus the fundamental difference in complexity between these problems shows up when we try to find an explicit description of the corresponding polytopes and their facets.

9.9.1 Spanning Trees

We start with MINIMUM SPANNING TREE. If $G = (V, E)$, a set of edges $F \subseteq E$ defines a spanning subgraph—that is, a subgraph (V, F) on the vertices of G. This subgraph is a spanning tree if and only if it is connected and has exactly $|V| - 1$ edges. A graph is connected, in turn, if and only if its minimum cut is greater or equal to one—that is, for any proper subset $S \subset V$ there is at least one edge in F connecting S to the rest of the graph. Hence the set of spanning trees is given by

$$\mathcal{F}_{\text{tree}} = \left\{ \mathbf{x} \in \{0,1\}^{|E|} : \sum_{e \in E} x_e = |V| - 1, \sum_{e \in \delta(S)} x_e \geq 1 \text{ for all } \emptyset \neq S \subset V \right\},$$

where $\delta(S)$ denotes the set of edges with one endpoint in S and the other in $V - S$. Since there are $2^{|V|} - 2$ proper subsets $S \subset V$, this definition includes an exponential number of constraints.

The convex hull of $\mathcal{F}_{\text{tree}}$ is called the *spanning tree polytope*. What inequalities describe it? Our first attempt might be simply to relax the integrality constraint, letting each x_e range over the unit interval

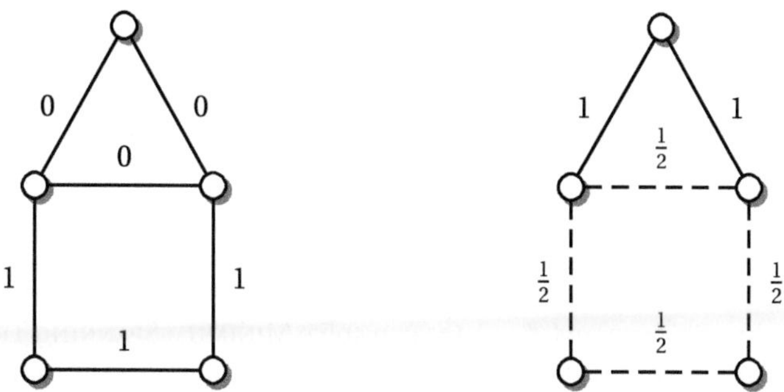

FIGURE 9.31: A weighted graph in which the minimum spanning tree has weight 2, but the minimum over the cut polytope $P_{\text{tree}}^{\text{cut}}$ has weight 3/2. Therefore, $P_{\text{tree}}^{\text{cut}}$ is generally larger than the spanning tree polytope $\text{conv}(\mathcal{F}_{\text{tree}})$. On the left, the edge weights. On the right, the optimal vertex in $P_{\text{tree}}^{\text{cut}}$.

[0,1], and keeping the cut constraints unchanged. This gives a polytope

$$P_{\text{tree}}^{\text{cut}} = \left\{ \mathbf{x} \in [0,1]^{|E|} : \sum_{e \in E} x_e = |V|-1, \sum_{e \in \delta(S)} x_e \geq 1 \text{ for all } \emptyset \neq S \subset V \right\}.$$

We call $P_{\text{tree}}^{\text{cut}}$ the *cut polytope.*

Clearly $P_{\text{tree}}^{\text{cut}}$ contains $\text{conv}(\mathcal{F})$, but in general $P_{\text{tree}}^{\text{cut}}$ is larger. As Figure 9.31 shows, there are graphs for which $P_{\text{tree}}^{\text{cut}}$ contains fractional vertices that do not correspond to spanning trees, and which have weights smaller than any genuine spanning tree can have. Thus while minimizing $\mathbf{c}^T\mathbf{x}$ over $P_{\text{tree}}^{\text{cut}}$ would give us a lower bound on the weight of the minimum spanning tree, it is not the right description of $\text{conv}(\mathcal{F}_{\text{tree}})$.

Let's try another characterization of the spanning tree polytope. A spanning tree can also be defined as a spanning subgraph with exactly $|V|-1$ edges that has no cycles. Because a cycle through a set of vertices S has $|S|$ edges, the no-cycle condition can be written as

$$\sum_{e \in E(S)} x_e \leq |S|-1 \quad \text{for all } \emptyset \neq S \subset V, \tag{9.46}$$

where $E(S)$ denotes the set of all edges with both endpoints in S. Equation (9.46) is called the *subtour elimination constraint* because, in TSP, a cycle that visits some proper subset $S \subset V$ is called a subtour.

Using subtour elimination and relaxing integrality gives a polytope

$$P_{\text{tree}}^{\text{sub}} = \left\{ \mathbf{x} \in [0,1]^{|E|} : \begin{array}{ll} \sum_{e \in E} x_e = |V|-1, & \\ \sum_{e \in E(S)} x_e \leq |S|-1 & \text{for all } \emptyset \neq S \subset V \end{array} \right\}, \tag{9.47}$$

which we call the *subtour polytope.* Note that the constraints $x_e \leq 1$ are included in the subtour elimination constraints, since we can take S to be e's endpoints.

Exercise 9.27 *Show that P_{tree}^{sub} is contained in P_{tree}^{cut}. Hint: show that if $\sum_{e \in \delta(S)} x_e$ is too small, then $\sum_{e \in E(S)} x_e$ or $\sum_{e \in E(\overline{S})} x_e$ is too large. Then, by examining Figure 9.31, show that P_{tree}^{sub} is in fact a proper subset of P_{tree}^{cut}.*

Again, it's obvious that $P_{\text{tree}}^{\text{sub}}$ contains $\text{conv}(\mathscr{F}_{\text{tree}})$, but this time both polytopes are the same. In other words, along with the constraints that the weights are nonnegative and that the total weight is $|V|-1$, the subtour elimination constraints give a complete characterization of the spanning tree polytope:

$$P_{\text{tree}}^{\text{sub}} = \text{conv}(\mathscr{F}_{\text{tree}}). \tag{9.48}$$

The integrality of $P_{\text{tree}}^{\text{sub}}$ can be traced back to the same insight that underlies the greedy algorithms for MINIMUM SPANNING TREE from Section 3.5.1. Problem 9.40 guides you through a proof of this.

Now that we know that $P_{\text{tree}}^{\text{sub}} = \text{conv}(\mathscr{F}_{\text{tree}})$, we can solve MINIMUM SPANNING TREE, in principle, by minimizing $\mathbf{c}^T\mathbf{x}$ on $P_{\text{tree}}^{\text{sub}}$. Is this easy to do? On one hand, there are about $2^{|V|}$ subtour elimination constraints, one for each subset $S \subset V$. Thus we can't simply treat this as an instance of LINEAR PROGRAMMING.

On the other hand, we saw in Section 9.7 that we can solve optimization problems on any convex set using the ellipsoid method, as long as we can implement a separation oracle in polynomial time. Thus we can handle even an exponential number of constraints, as long as there is a polynomial-time algorithm that either confirms that they are all satisfied or finds one that is violated.

Let's go back to $P_{\text{tree}}^{\text{cut}}$ for a moment. Even though it is not the polytope we want, it will give us a clue for how to implement a separation oracle for $P_{\text{tree}}^{\text{sub}}$. The constraint $\sum_{e\in E} x_e = |V|-1$ is easy to check, so we focus on the cut constraints. To check these, we find the set $S_{\min}$ with the smallest cut $\sum_{e\in\delta(S)} x_e$ separating it from $V - S_{\min}$. This is an instance of MIN CUT, which we can solve in polynomial time. Since for all S we have

$$\sum_{e\in\delta(S)} x_e \geq \sum_{e\in\delta(S_{\min})} x_e \,,$$

if $\sum_{e\in\delta(S_{\min})} x_e \geq 1$ then all the cut constraints are satisfied. Otherwise, $S_{\min}$ corresponds to a violated constraint, and the set

$$\left\{ \mathbf{y} : \sum_{e\in\delta(S_{\min})} y_e = 1 \right\}$$

is a separating hyperplane in $\mathbb{R}^{|E|}$.

Now let's turn to $P_{\text{tree}}^{\text{sub}}$. We rewrite the subtour constraints (9.46) as

$$\min_{\emptyset\neq S\subset V} \left(|S| - \sum_{e\in E(S)} x_e \right) \geq 1 \,. \tag{9.49}$$

If this is not the case, the set S that minimizes this expression corresponds to a violated subtour constraint. While the expression we need to minimize is a little more complicated than the cut, we can reduce this problem to MIN CUT by modifying the graph as described in Problem 9.42. Thus we can confirm the subtour constraints, or find one that is violated, in polynomial time.

Of course, this isn't how we solve MINIMUM SPANNING TREE in practice. Kruskal's and Prim's algorithms, which we saw in Chapter 3, are far more efficient and easier to implement. Yet this geometric perspective gives us some fresh insight into MINIMUM SPANNING TREE. It tells us, for example, that maximizing the weight is just as easy as minimizing it, since we can maximize or minimize $\mathbf{c}^T\mathbf{x}$ for any vector $\mathbf{c}$. We can even give some edges positive weights and others negative weights, expressing a preference for some and trying to avoid others.

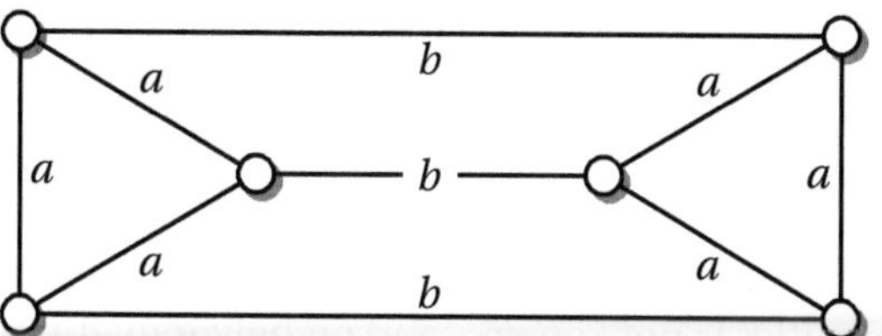

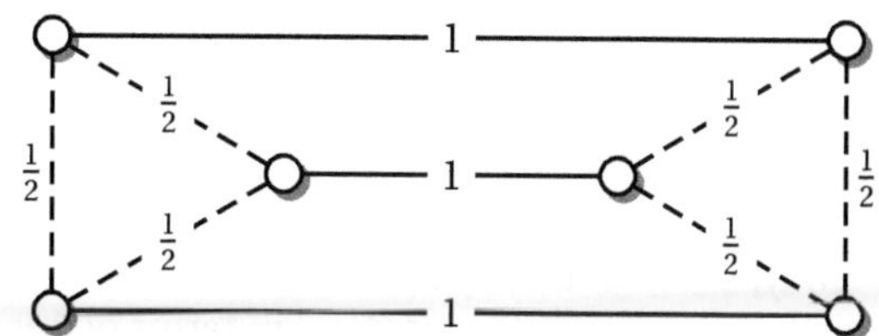

FIGURE 9.32: Every tour in the left graph has length $4a+2b$. The fractional vertex of the cut polytope $P_{\text{tour}}^{\text{cut}}$ shown on the right has weight $3a+3b$, which is smaller than any genuine tour when $a>b$.

In contrast, finding the minimum spanning tree whose weight is greater than or equal to a given w is a fundamentally different problem. Intersecting $P_{\text{tree}}^{\text{sub}}$ with the constraint $\mathbf{c}^T\mathbf{x}\geq w$ creates new vertices which are not necessarily integral. In fact, this problem is NP-hard, as can easily be shown by reducing from EXACT SPANNING TREE (Section 5.4.3).

Now let's see what this geometric perspective can tell us about a more demanding problem—TRAVELING SALESMAN.

9.9.2 *The Relaxed Salesman*

If we want to define a polytope whose vertices are salesmen's tours, we need a characterization of tours in terms of inequalities. First, Hamiltonian cycles are *2-regular*: every vertex has degree exactly two. Secondly, they are *2-connected*: for any proper subset $S\subset V$, we need to cut at least two edges to separate S from $V-S$. Thus we can define the set of tours as

$$\mathscr{F}_{\text{tour}}=\left\{\mathbf{x}\in\{0,1\}^{|E|}:\begin{array}{ll}\sum_{e\in\delta(v)}x_e=2 & \text{for all } v\in V,\\ \sum_{e\in\delta(S)}x_e\geq 2 & \text{for all } \emptyset\neq S\subset V\end{array}\right\},$$

where $\delta(v)$ is the set of edges incident to v. If c_e is the length of the edge e, we can think of TSP as minimizing the weight $\mathbf{c}^T\mathbf{x}$ in the convex hull of this set, conv($\mathscr{F}_{\text{tour}}$). Thus our goal is to understand the geometry of this polytope.

As for MINIMUM SPANNING TREE, we can start by relaxing the integrality constraint and keeping the cut and degree constraints. Thus gives us a polytope analogous to $P_{\text{tree}}^{\text{cut}}$,

$$P_{\text{tour}}^{\text{cut}}=\left\{\mathbf{x}\in[0,1]^{|E|}:\begin{array}{ll}\sum_{e\in\delta(v)}x_e=2 & \text{for all } v\in V,\\ \sum_{e\in\delta(S)}x_e\geq 2 & \text{for all } \emptyset\neq S\subset V\end{array}\right\}. \tag{9.50}$$

Clearly $P_{\text{tour}}^{\text{cut}}$ contains conv($\mathscr{F}_{\text{tour}}$), but the example in Figure 9.32 shows that $P_{\text{tour}}^{\text{cut}}$ is larger. Thus minimizing $\mathbf{c}^T\mathbf{x}$ over $P_{\text{tour}}^{\text{cut}}$ gives us a lower bound on the length of the shortest tour, but it isn't generally tight.

For MINIMUM SPANNING TREE, we obtained a complete characterization of the polytope conv($\mathscr{F}_{\text{tree}}$) by switching our constraints from cuts to subtours. We can try the same thing for TSP, since an alternate definition of a Hamiltonian cycle is a 2-regular subgraph that does not contain any subtours. This gives us the polytope

$$P_{\text{tour}}^{\text{sub}}=\left\{\mathbf{x}\in\{0,1\}^{|E|}:\begin{array}{ll}\sum_{e\in\delta(v)}x_e=2 & \text{for all } v\in V,\\ \sum_{e\in E(S)}x_e\leq |S|-1 & \text{for all } \emptyset\neq S\subset V\end{array}\right\}. \tag{9.51}$$

Unfortunately, this time around the subtour elimination constraints don't bring us any closer to a full characterization of conv($\mathscr{F}_{\text{tour}}$), as the following exercise shows.

Exercise 9.28 *Prove that $P^{sub}_{tour} = P^{cut}_{tour}$. Hint: show that in the presence of the degree constraints $\sum_{e\in\delta(v)} x_e = 2$, the cut and subtour constraints are equivalent for each $S \subset V$.*

Maybe we should try to characterize the tour polytope with yet another set of constraints. Alas, an explicit description of conv($\mathscr{F}_{\text{tour}}$) in terms of linear inequalities is not known—and no polynomial-time separation oracle for it exists unless $\mathsf{P} = \mathsf{NP}$. Nevertheless, by settling for the larger polytope $P^{\text{sub}}_{\text{tour}}$ we can obtain lower bounds on the shortest possible tour, and use these bounds to narrow our search for the optimum. This is the key to solving large real-world instances of TSP, as we will see next.

9.10 Solving Hard Problems in Practice

If you can't get rid of the skeleton in your closet, you'd best teach it to dance.

George Bernard Shaw

The fact that an optimization problem is NP-hard doesn't make it disappear. We still want to solve it—preferably, in a reasonable amount of time.

In this section we will explore some of the techniques that have been developed to tackle large optimization problems in practice. The idea is to bound the optimum of a given instance from above and below. For a minimization problem, we get an upper bound from any feasible solution—the one produced by an approximation algorithm or heuristic, or simply the best solution we have found so far. On the other hand, we get a lower bound from the optimum of a relaxed version of the problem, which we can solve efficiently.

The goal is to bring these bounds as close together as possible. If they match, we have found the optimum and we're done. If not, we at least know how good our current solution is compared to the optimum, and we know that certain branches of the search tree are not worth following.

We illustrate this approach with the TRAVELING SALESMAN problem, using the 42-city instance shown in Figure 9.33. But the underlying ideas are applicable to many other problems as well.

9.10.1 Heuristics and Local Search

To solve this instance of TSP, we start by applying a heuristic to get a reasonable solution. We could use an approximation algorithm, such as Christofides', which is guaranteed to give a solution within a certain ratio ρ of the optimum. But we are equally free to try heuristics with no such guarantees, and use whichever one works best for the instance at hand.

Let's start with the greedy algorithm, which always moves to the nearest city it hasn't visited yet. Starting in Santa Fe (21), our greedy salesman follows a reasonable route until in Portland (15) he realizes that he has left out a couple of cities on the East coast. He flies down to Florida, then up to New England, and finally back to Santa Fe as shown in Figure 9.34. This is a valid tour, but not a very good one.

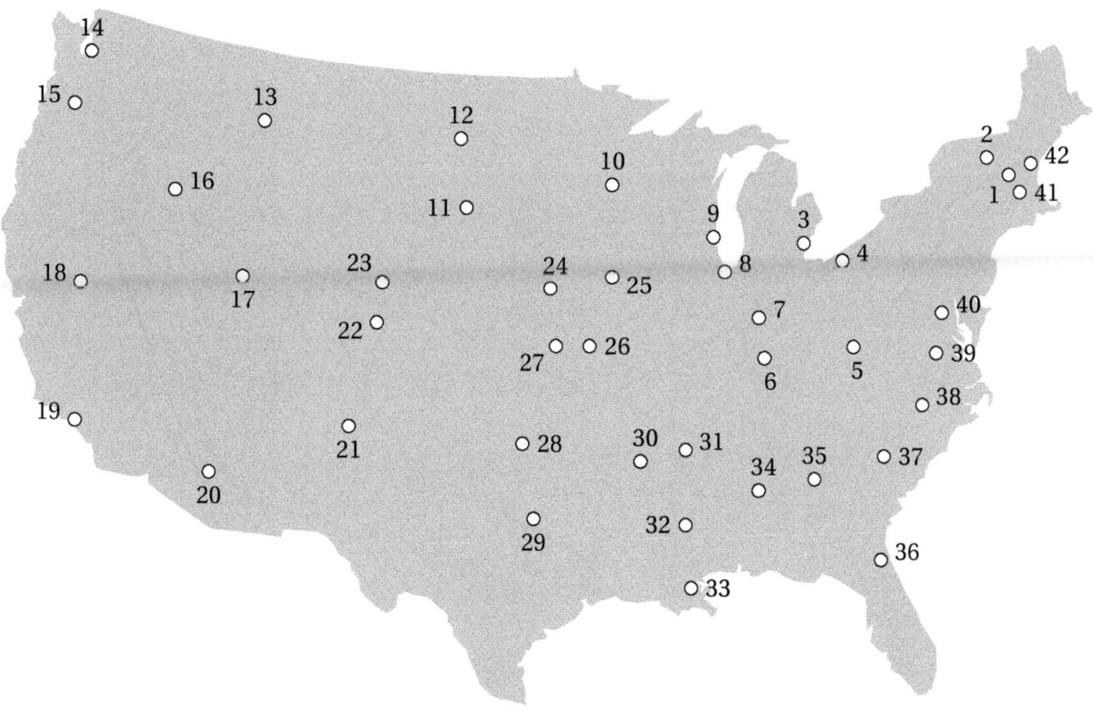

FIGURE 9.33: A challenge: find the shortest itinerary through these 42 cities.

Exercise 9.29 *Consider a tour for the Euclidean* TSP *that crosses itself. Show that this tour is not optimal, and describe how it can be shortened.*

We can improve our tour by exploring its neighborhood, making local alterations in an effort to decrease its length. Figure 9.35 shows two types of local move that transform one tour into another. A *vertex insertion* move takes a vertex out of the tour and places it between two other vertices. A 2-*opt* move cuts two edges and swaps the connections between the resulting paths.

Each of these two types of move defines a neighborhood of $O(n^2)$ tours, since there are $O(n^2)$ moves we could apply. Since this neighborhood is of polynomial size, we can explore it in polynomial time. We then perform a greedy local search. At each step we take the current tour T, check out its neighborhood by trying out all possible moves of a given type, and move to the shortest neighboring tour. We iterate this process until T is the shortest in its neighborhood—that is, until it is a local optimum with respect to these moves.

Figure 9.36 shows the local optimum with respect to 2-opt moves that we reach if we start with the greedy tour. All the edge crossings are gone, and the whole tour looks quite reasonable. Its total length is 12217 miles, considerably shorter than the greedy tour. But how good is it compared to the global optimum? To answer that question, we need a lower bound on the length of the shortest tour.

We can derive a simple lower bound from the degree constraints of Section 9.9.2: namely, in any tour, each vertex is connected to exactly two other vertices. If we take the distances of a vertex to its two nearest neighbors, sum up these distances for all vertices and divide by two, the result is a lower bound for the

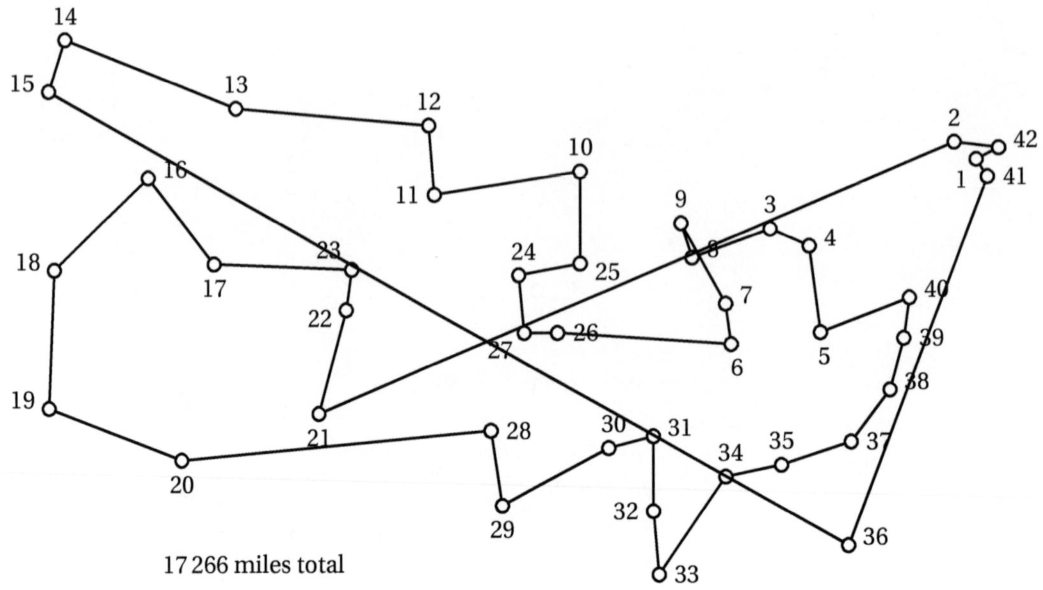

FIGURE 9.34: A greedy tour that starts in Santa Fe (21).

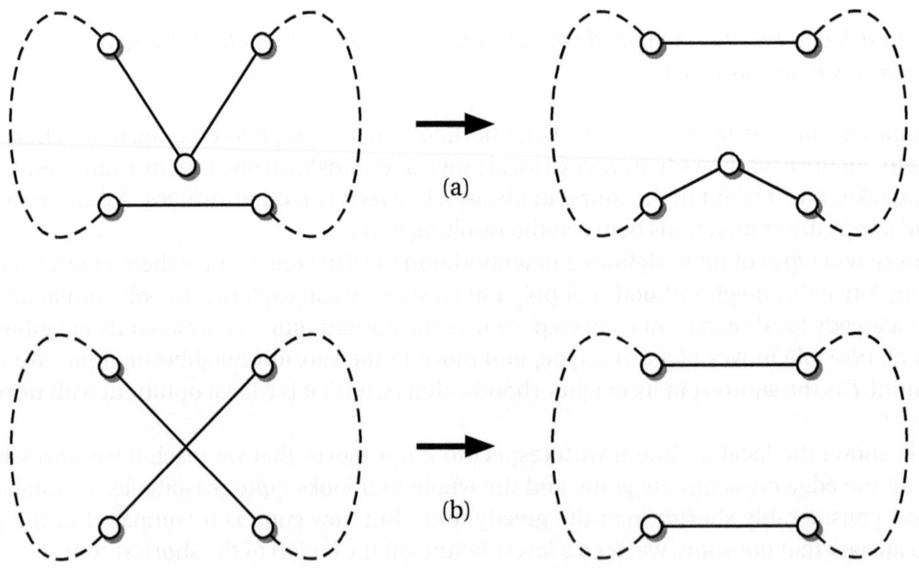

FIGURE 9.35: Local moves in TSP: (a) vertex insertion and (b) 2-opt moves.

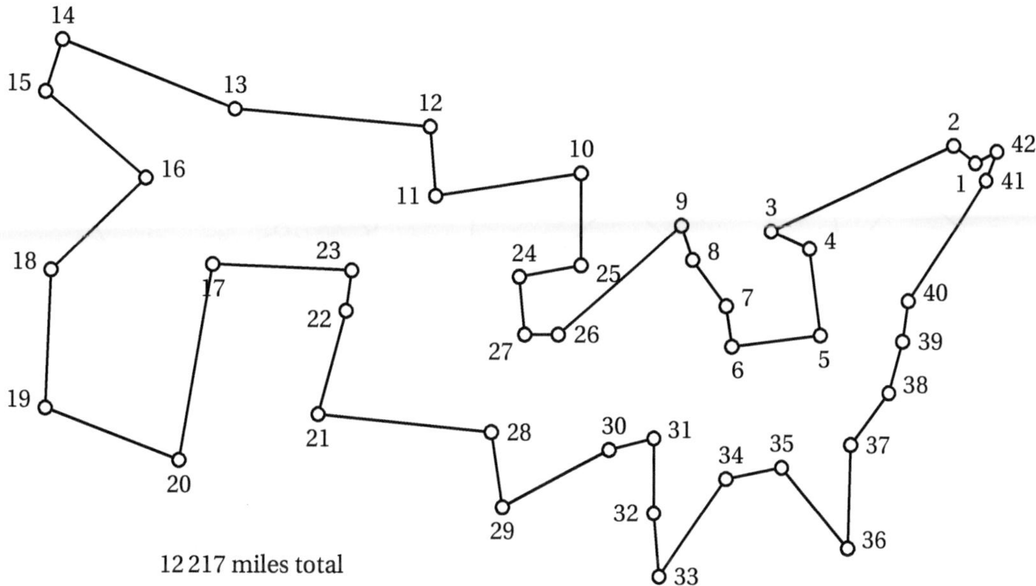

FIGURE 9.36: Local optimum with respect to 2-opt moves.

shortest tour. For our instance, this gives a lower bound of 10 092 miles. Hence we know that our 2-opt tour exceeds the optimal tour by at most 21%. The 2-opt tour is actually much closer to the optimum than this simple bound suggests—but to see this, we need a better lower bound.

There are various ways to get lower bounds, but they all rely on the same idea: relaxation. By removing some of the constraints, we get a minimization problem which we can solve in polynomial time. Relaxation also enlarges the set of solutions, including subgraphs that are not Hamiltonian cycles, or "fuzzy tours" which have fractional weights on some edges. Thus the optimum of the relaxed problem is not usually a genuine tour. However, the genuine tours are also feasible solutions of the relaxed problem, so the relaxed optimum is a lower bound on the length of the shortest tour.

9.10.2 The Relaxed Salesman: Tree Bounds

We start by considering a relaxation for TSP based on spanning trees. A tour is a spanning graph with exactly one loop, so one possible relaxation of TSP is to allow other graphs with one loop. This is where *one-trees* come in.

Fix an arbitrary vertex $v \in V$. A one-tree is a spanning tree on the other vertices $V - v$, plus two edges incident to v. Thus a one-tree has exactly one loop, which contains v, and v has degree 2. To find the *minimum* one-tree, we compute the minimum spanning tree on $V - v$, and then close the loop by adding the two shortest edges incident to v.

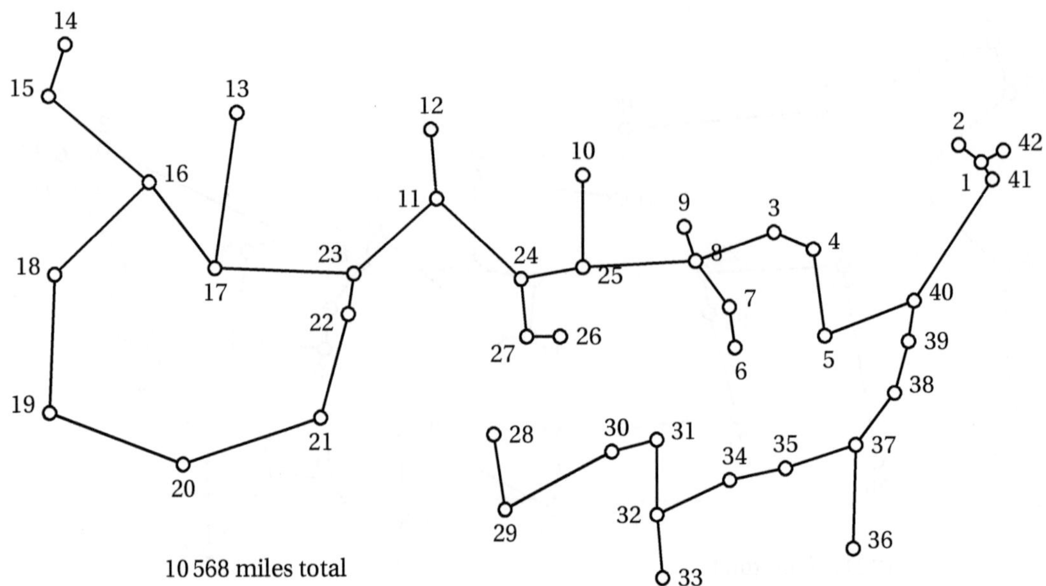

FIGURE 9.37: The minimum one-tree, where the vertex v at which we close the loop is Santa Fe (21).

Let $\mathbf{c}$ denote the vector of edge lengths, and let OneT($\mathbf{c}$) and TSP($\mathbf{c}$) denote the lengths of the minimum one-tree and the shortest tour respectively. Since a tour is also a one-tree, we have

$$\text{OneT}(\mathbf{c}) \leq \text{TSP}(\mathbf{c}).$$

Figure 9.37 shows the minimum one-tree for our instance, where v is Santa Fe. Its length is 10 568 miles, telling us that our 2-opt tour from Figure 9.36 is at most 16% longer than the shortest tour. However, it looks more like a tree than a tour, suggesting that this lower bound is not very tight. We could try to increase this lower bound by closing the loop at some other vertex v, and maximizing over all v. But there is something much better we can do.

To make a one-tree look more like a tour, we need to increase the number of vertices of degree 2. We can do this by penalizing the one-tree for visiting vertices of degree greater than 2, and rewarding it for visiting vertices of degree 1. In general, we can impose a penalty π_v for entering or leaving each vertex v by defining a set of modified edge lengths $\mathbf{c}^\pi$,

$$c_{uv}^\pi = c_{uv} + \pi_u + \pi_v \,.$$

For any vector of penalties π, the length of the minimum one-tree with weights $\mathbf{c}^\pi$ gives a lower bound on the shortest tour with distances $\mathbf{c}^\pi$. Thus

$$\text{OneT}(\mathbf{c}^\pi) \leq \text{TSP}(\mathbf{c}^\pi) = \text{TSP}(\mathbf{c}) + 2\sum_v \pi_v \,,$$

where the equality follows since every vertex on a tour has degree 2. This gives the lower bound

$$w(\pi) \leq \text{TSP}(\mathbf{c}),$$

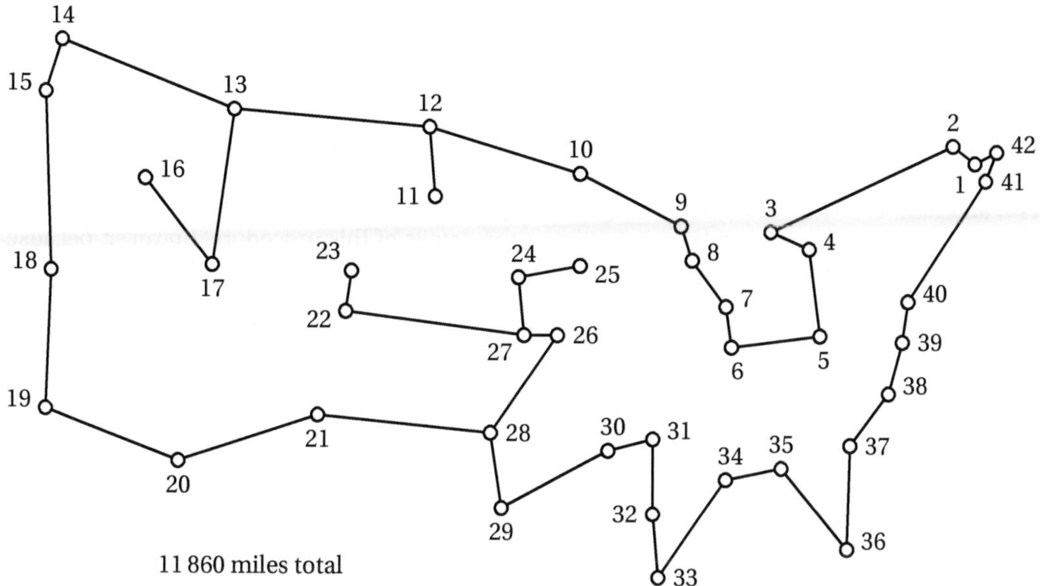

FIGURE 9.38: The one-tree corresponding to the Held–Karp bound.

where

$$w(\pi) = \text{OneT}(\mathbf{c}^{\pi}) - 2\sum_{v} \pi_v . \tag{9.52}$$

The greatest possible lower bound on TSP($\mathbf{c}$) of the form (9.52) is the maximum of $w(\pi)$ over all penalty vectors π,

$$\max_{\pi} w(\pi) \leq \text{TSP}(\mathbf{c}) .$$

This is called the *Held–Karp bound.*

Figure 9.38 shows the one-tree corresponding to the Held–Karp bound. It still has a few vertices with degree other than 2, but it looks much more like a tour than the one-tree in Figure 9.37. Its total length is 11 860 miles, so our locally optimal 2-opt tour is at most 3% longer than the global optimum.

Conversely, the Held–Karp bound is at most 3% below the optimum—so it is quite sharp, at least in this instance. However, we haven't explained how to calculate it yet. To maximize $w(\pi)$ we rewrite (9.52) as follows,

$$w(\pi) = \min_{T} \left[c(T) + \sum_{v} (d_v^T - 2)\pi_v \right] . \tag{9.53}$$

Here we minimize over all one-trees T, $c(T)$ is the weight of T with the original edge distances $\mathbf{c}$, and d_v^T is the degree of vertex v in T.

For any fixed one-tree T, the right-hand side of (9.53) is a linear function of π. Thus for small changes of π, the T that minimizes it does not change. Hence $w(\pi)$ is a piecewise linear function with gradient

$$\frac{\partial w(\pi)}{\partial \pi_v} = d_v^T - 2 ,$$

and we can maximize $w(\pi)$ by gradient ascent. Given the current penalties π, we calculate new penalties π' according to

$$\pi'_v = \pi_v + \Delta(d_v^T - 2),$$

for some step size $\Delta > 0$. This is intuitive—we increase π_v if $d_v^T > 2$ and decrease π_v if $d_v^T < 2$, encouraging T to remove edges from vertices of degree greater than 2 and add edges to vertices of degree 1.

Each time we update π, we compute the new minimum one-tree T', use its degrees $d_v^{T'}$ to define the gradient, and iterate. This procedure converges in polynomial time to the maximum of $w(\pi)$, provided the step size Δ is chosen properly. We won't discuss the details of this procedure, however, because in the next section we will learn how to calculate the Held–Karp bound by more familiar means. We will see that the Held–Karp bound is independent of the vertex v where we close the one-tree's loop, and that additional linear inequalities can bring us even closer to the optimum.

9.10.3 *The Relaxed Salesman: Polyhedral Bounds*

Returning to our instance, there is still a 3% gap between the Held–Karp lower bound and the length of our 2-opt tour. To close this gap, we will improve our lower bounds using an LP relaxation of TSP.

Specifically, we will consider "fuzzy tours" $\mathbf{x} \in [0,1]^{|E|}$ in the cut polytope $P_{\text{tour}}^{\text{cut}}$ defined in Section 9.9.2, and minimize their total "length" $\mathbf{c}^T\mathbf{x} = \sum_e c_e x_e$:

$$\min_{\mathbf{x}} \Big(\sum_e c_e x_e\Big) \qquad \text{subject to}$$

$$\sum_{e \in \delta(v)} x_e = 2 \qquad \text{for all } v \in V, \text{ and} \tag{9.54a}$$

$$\sum_{e \in \delta(S)} x_e \geq 2 \qquad \text{for all } \emptyset \neq S \subset V. \tag{9.54b}$$

Like the other polytopes we discussed in Section 9.9, this polytope has an exponential number of facets, since it has a cut constraint (9.54b) for every proper subset S. However, we can again implement a separation oracle in polynomial time by solving MIN CUT. Given a vector $\mathbf{x}$, we construct an undirected graph with edge weights x_e. If the minimum cut that splits this network into two pieces has weight at least 2 then all the cut constraints are satisfied. Otherwise, the minimum cut gives us a subset $S \subset V$ for which (9.54b) is violated.

One way to use this separation oracle is to start by ignoring the cut constraints, solve the resulting LINEAR PROGRAMMING problem, and then check to see which cut constraints are violated. We then add one of the violated constraints to the problem, solve it again, and so on, until we get a solution that satisfies all the cut constraints.

We start this process by solving the LP relaxation with just the degree constraints,

$$\begin{aligned} &\min_{\mathbf{x}} \Big(\sum_e c_e x_e\Big) && \text{subject to} \\ &\sum_{e \in \delta(v)} x_e = 2 && \text{for all } v \in V. \end{aligned} \tag{9.55}$$

The solution to this problem is shown in Figure 9.39. It has some fractional edges where $x_e = 1/2$. Moreover, it violates several cut constraints, including for the vertices $S = \{1,2,42,41\}$, since the cut between

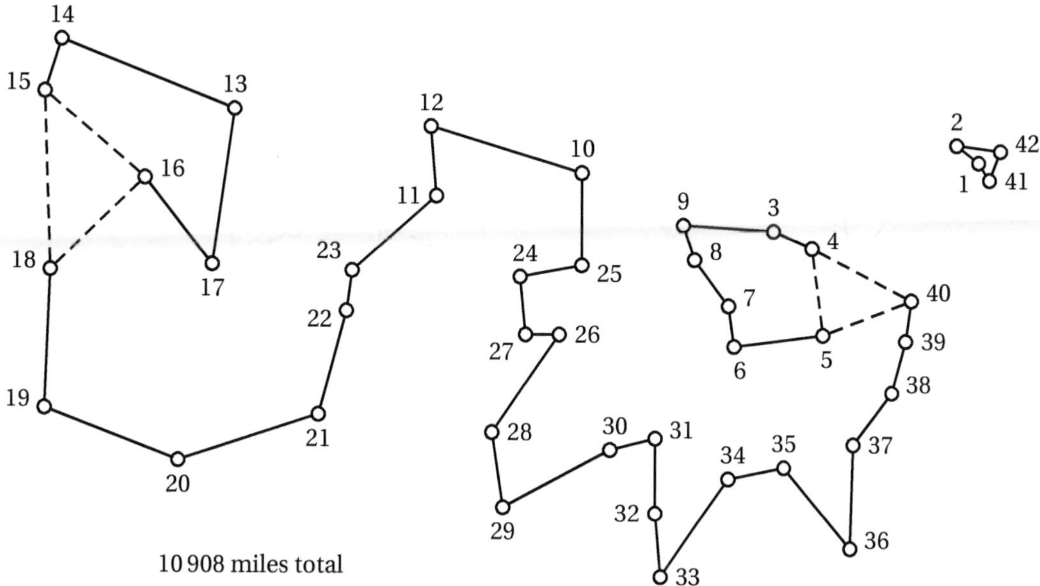

FIGURE 9.39: Solution of the LP relaxation (9.55) with just the degree constraints. Dashed lines indicate fractional edges where $x_e = 1/2$. The vertices $S = \{1,2,41,42\}$ form a subtour, and correspond to a violated cut constraint.

them and the rest of the graph is zero. As Exercise 9.28 showed, this is equivalent to the fact that they form a subtour. Thus we add the constraint

$$\sum_{e \in \delta(S)} x_e \geq 2 \qquad \text{for } S = \{1,2,41,42\},$$

and run the LP solver again. The resulting solution contains a new subtour, so we add the corresponding cut constraint and solve again, and so on.

After adding three cut constraints, we get the solution shown in Figure 9.40. It is connected, but there are still violated cut constraints. Consider the set $S = \{13,14,15,16,17\}$: it is connected to the rest of the graph through two edges of weight $1/2$, so its cut is just 1. With three more constraints we get a solution that is 2-connected, i.e., a solution that satisfies *all* the cut constraints and has no subtours, shown in Figure 9.41.

Thus the fuzzy tour shown in Figure 9.41 is the minimum of the LP relaxation (9.54). Remarkably, its length is exactly the Held–Karp bound—that is, the total length of the minimum one-tree shown in Figure 9.38. This is no accident. The Held–Karp bound is the maximum, over the penalties π, of a lower bound on a minimization problem. Does this ring a bell? As the clever reader will guess, the Held–Karp bound is the dual of the linear program (9.54). Problem 9.43 guides you through a proof of this fact.

The duality between these two approaches tells us that we can derive a lower bound on the shortest tour using either one-trees with penalties, or fuzzy tours bounded by inequalities. But the latter approach

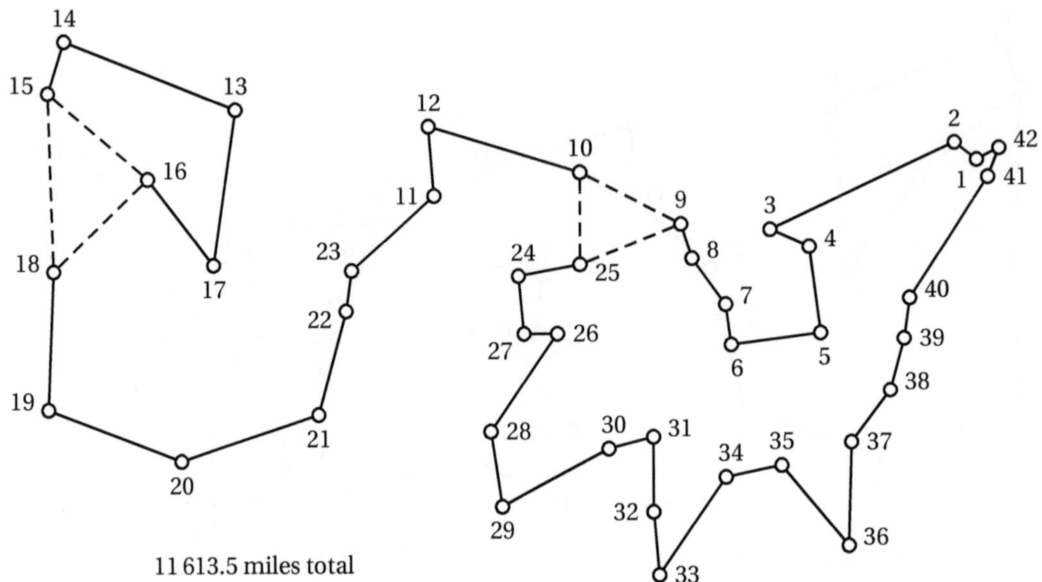

FIGURE 9.40: Solution of the LP relaxation with cut constraints on $\{1,2,41,42\}$, $\{3,4,5,\ldots,9\}$, and $\{24,25,26,27\}$. The solution is connected, but not 2-connected. For instance, the vertices $\{13,14,15,16,17\}$ form a subtour, since they are connected the rest of the graph by two fractional edges with total weight 1.

suggests that we can improve our results by coming up with additional inequalities, constraining the LP relaxation further, and increasing our lower bound. And improve we must, since we still haven't found the shortest tour.

Even though we have satisfied all the cut/subtour constraints, there are other linear constraints that are satisfied by any valid tour but violated by the solution in Figure 9.41. Consider the triangle $S_0 = \{15,16,18\}$ and the sets $S_1 = \{14,15\}$, $S_2 = \{16,17\}$ and $S_3 = \{18,19\}$. The sum of their cuts is

$$\sum_{i=0}^{3} \sum_{e \in \delta(S_i)} x_e = 3+2+2+2 = 9\,.$$

On the other hand, we claim that for any valid tour,

$$\sum_{i=0}^{3} \sum_{e \in \delta(S_i)} x_e \geq 10\,. \tag{9.56}$$

This is called a *comb constraint* because the Venn diagram of the sets S_i resembles a comb with three teeth, as shown in Figure 9.42. Why does it hold?

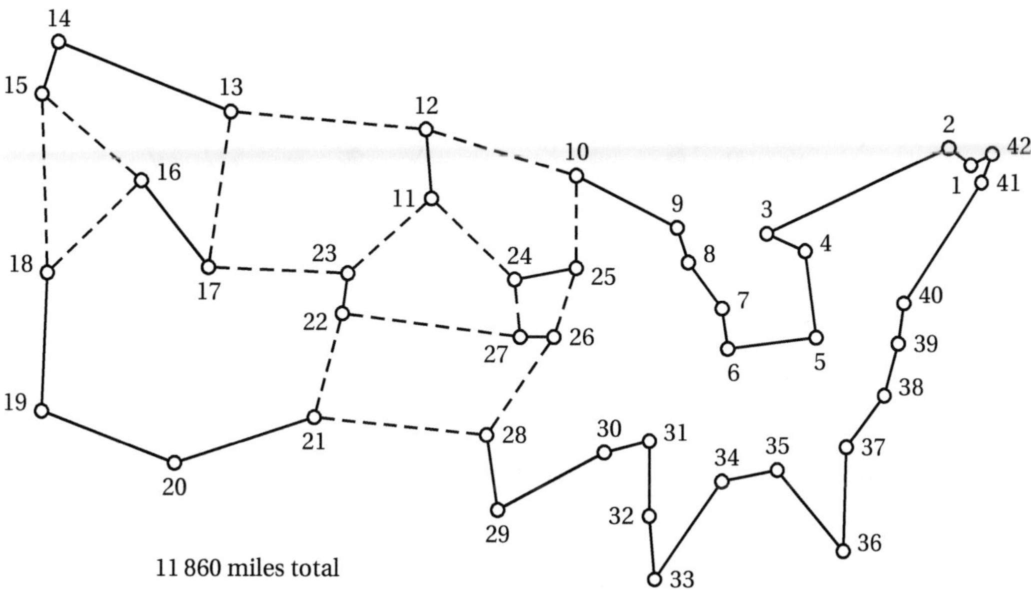

FIGURE 9.41: Solution of the LP relaxation with additional cut constraints on $\{13,14,15,16,17\}$, $\{13,14,\ldots,23\}$, $\{10,11,12\}$, and $\{11,12,\ldots,23\}$. Now all the cut constraints are satisfied, and there are no subtours.

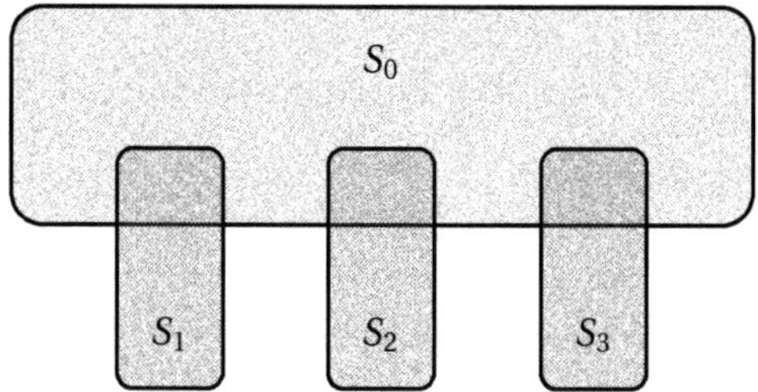

FIGURE 9.42: Venn diagram of vertex sets for comb constraints.

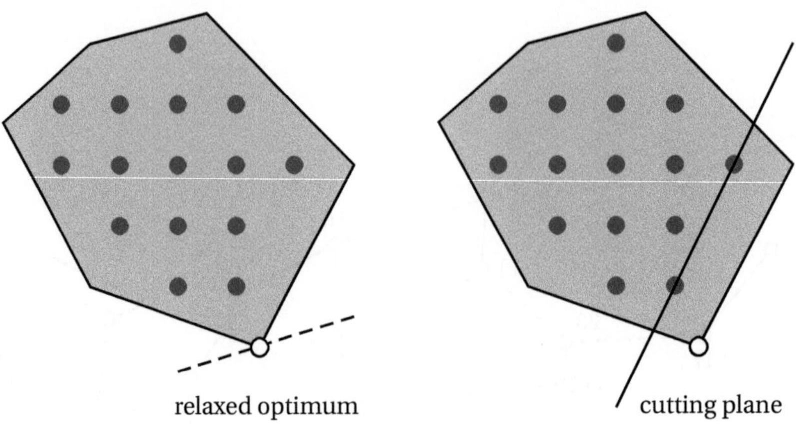

FIGURE 9.43: A cutting plane brings us closer to an integer solution.

Comb constraints arise from the same insight that Euler applied to the Königsberg bridges—what comes in must go out. For a tour this means that the cut size of any proper subset $S \subset V$ must be even:

$$\sum_{e \in \delta(S)} x_e \in \{2,4,6,\ldots\}.$$

A cut size of 2 means that the tour enters S, visits *all* the vertices in S, and then leaves S. Now consider the following exercise:

Exercise 9.30 *A* comb with k teeth *is a family of subsets $S_0, S_1, \ldots, S_k \subset V$ such that $S_1, \ldots, S_k$ are disjoint and each one is incomparable with S_0. That is, $S_i \cap S_j = \emptyset$ for all $1 \le i < j \le k$, and $S_0 \cap S_i \ne \emptyset$ and $S_0 \cap \overline{S}_i \ne \emptyset$ for all $1 \le i \le k$. Prove that if k is odd, the sum of the cut sizes of all these sets is bounded by*

$$\sum_{i=0}^{k} \sum_{e \in \delta(S_i)} x_e \ge 3k+1. \tag{9.57}$$

Hint: show that if one of the teeth S_i has cut size 2, it contributes at least 1 to the cut size of the handle S_0.

Applying (9.57) with $k = 3$ gives (9.56).

If we add this particular comb constraint to our linear program, the relaxed optimum becomes a better lower bound. It is still not a tour, but we can look for more comb constraints, and indeed adding just one more comb constraint gives a flawless tour. At this point, we have found a relaxed problem with an integral optimum, which must in fact be the shortest tour of these 42 cities.

This method of adding more and more constraints to the LP relaxation of an integer program is known as the *cutting plane* strategy, illustrated in Figure 9.43. If the relaxed optimum has fractional components, we try to find a constraint that is satisfied by any integer solution, but violated by the current relaxed optimum. We add this constraint to the relaxed problem, cutting its feasible polytope along the corresponding hyperplane. Solving this improved LP relaxation gives an optimum that is hopefully less fractional.

The difficulty of this approach is that we have to identify increasingly complicated constraints and devise polynomial-time separation oracles for them. This can take considerable creativity, and even if we succeed, the corresponding improvement in the relaxed solution is often quite small. In order to solve large instances of INTEGER LINEAR PROGRAMMING, we need to combine the cutting plane strategy with another powerful tool: branch and bound.

9.10.4 Closing in: Branch and Bound

The problem with our relaxed solution in Figure 9.41 is that it contains fractional edges where $x_e = 1/2$. To push it towards a genuine tour, we can choose one of these edges and force it to be either present or absent, by setting x_e to 1 or 0. Each of these corresponds to a constrained instance of TSP, giving our search algorithm two branches to explore.

Figure 9.44 shows the effects of forcing the edge $(12,13)$ to be present or absent. Forcing it to be present gives a valid tour of length 11 894. This is the best tour we have seen so far, but to make sure it's optimal we have to explore the other branch.

Forcing $(12,13)$ to be absent creates a relaxed solution with a new violated cut constraint, in this case for $S = \{24,26,27\}$. If we had to, we could explore this branch further by adding this constraint to our relaxation and solving again. This may yield a 2-connected but fractional solution, in which case we would branch on another fractional edge, and so on.

But in this case no further exploration is necessary. Adding more constraints can only increase the length of a solution, so the length 11 928 of this relaxed solution is a lower bound on any solution that that branch could lead to—that is, a lower bound on *any* solution where $(12,13)$ is absent. Since 11 928 is greater than the length 11 894 of the valid tour where $(12,13)$ is present, the latter must be the shortest tour possible. Thus the tour shown on the top of Figure 9.44 is the optimum, and we have solved our instance of TSP—the shortest tour is 11 894 miles long.

This general idea is called *branch and bound*. We divide the set of all solutions into disjoint sets by choosing some variable and giving it all possible settings. Continuing this branching process generates a search tree of exponential size—but if an entire branch leads to solutions of greater length than a valid solution we have already found, there is no need to explore it.

At each branch of our search tree, we derive lower bounds on the solutions it could lead to, by solving the best possible LP relaxation we can find. And we derive upper bounds on the optimum by finding the best feasible solution we can, using approximation algorithms or heuristics. Whenever these bounds cross each other, we know that the shortest solution on that branch is longer than the optimum, so we can prune that branch off the tree.

Even after we prune away all the branches we can, the search tree may still be exponentially large in the worst case. But in practice it is far smaller. In our 42-city instance of TSP, the tree got pruned all the way down to a single branch. In general, the size of the search tree depends on the quality of the bounds and on the branching rule, i.e., which fractional variable we try to set next. The sharpest bounds are usually obtained by adding cutting planes to the LP relaxation, in which case the method is called *branch and cut*.

Large instances of TSP have been solved with branch and cut algorithms, including the shortest tours through all 24 978 cities of Sweden and all 15 112 cities in Germany. For even larger instances, such as 1 904 711 populated locations around the world, branch and cut methods have yielded a solution that is provably within 0.05% of the optimum. 9.19

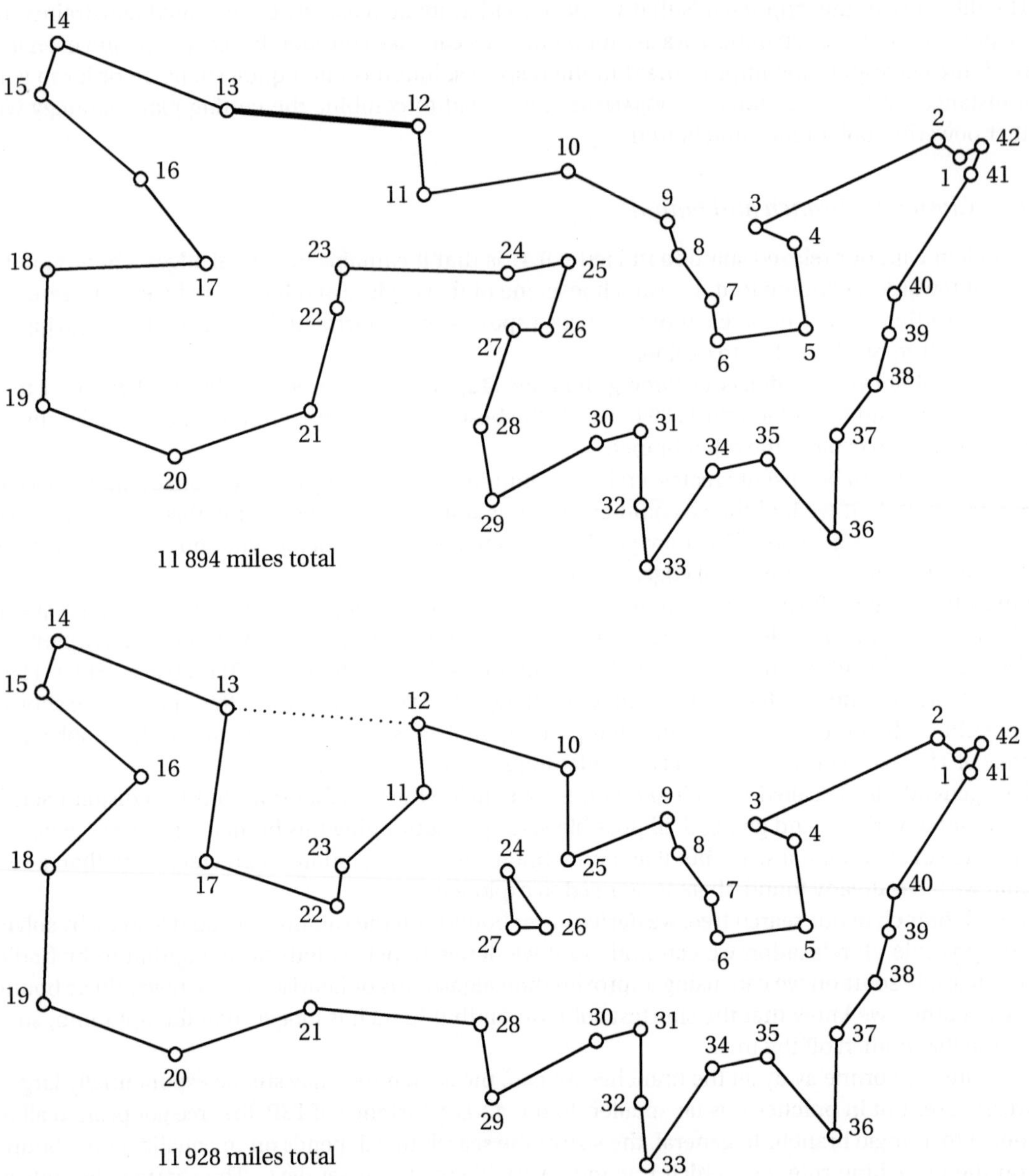

FIGURE 9.44: Forcing the edge (12,13) to be present (top) gives a valid tour 11 894 miles long. Forcing (12,13) to be absent (bottom) gives a solution 11 928 miles long. It violates a cut constraint since $S = \{24,26,27\}$ is a subtour, but adding this constraint can only increase its length, which already exceeds that of the valid tour at the top. Thus there is no need to search further—the tour at the top is the shortest possible, and we have found the optimal solution.

These numbers tell us that even if a problem is NP-hard, we should still long for insight. The better we understand a problem—in particular, the tighter bounds we can place on its optimum—the faster we can solve it, by narrowing our search to parts of the haystack where the optimum must lie.

Problems

> Nothing is more attractive to intelligent people than an honest, challenging problem, whose possible solution will bestow fame and remain as a lasting monument.
>
> Johann Bernoulli

9.1 When two flavors are the same. The claim that an optimization problem can be reduced to its threshold verification may seem counterintuitive. After all, the optimum solution $\mathbf{x}$ generally contains more information than its value $f(\mathbf{x})$. However, show that if the threshold version of an optimization problem A is NP-complete, then its optimization version is reducible to its threshold version.

Hint: each solution $\mathbf{x}$ can be described with a sequence of bits. Consider questions of the form "is there an optimal solution whose bit sequence starts out with s?" where s is some initial subsequence.

9.2 Embedding the solution in its value. Suppose that each possible solution $\mathbf{x}$ to an optimization problem can be described with n bits, and that we are trying to optimize some integer-valued function $f(\mathbf{x})$. Show how to modify f, giving a new quality function f', such that the solution $\mathbf{x}$ that maximizes f' is unique, and is the lexicographically first solution out of all those that optimize f. Moreover, show that finding the maximum $f'(\mathbf{x})$ immediately tells us $\mathbf{x}$.

For instance, consider MIN-WEIGHT VERTEX COVER, where each vertex has an integer weight w_v, and we want to find the vertex cover $S \subset V$ with the smallest total weight $w(S) = \sum_{v \in S} w_v$. Show how to define new weights w'_v such that, if there are multiple vertex covers with the same minimum weight $w(S)$, then the lexicographically first such vertex cover is the unique minimum of $w'(S)$.

9.3 Three flavors of TSP. Suppose we have n cities and that the distances between them are given as m-bit numbers. Show how to find the length of the optimal tour, and the optimal tour itself, with a polynomial number of calls to an oracle for the threshold version of TSP. In other words, show that TSP is self-reducible.

9.4 Dynamic salesman. A naive search algorithm for TSP takes $O(n!)$ time time to check all tours. Use dynamic programming to reduce this to a simple exponential, i.e., to solve TSP in $O(2^n \text{poly}(n))$ time as in Figure 9.45. Hint: look at Problem 3.31.

9.5 Pack it in. Every backpacker knows this problem: you can't carry all the objects that you would like to take with you on your trip. You have to choose a subset of objects, and since objects have different values, you want to maximize the total value while not exceeding the total weight you can carry. If we can carry a total weight W and there are ℓ objects with sizes $w_1, \dots, w_\ell$ and values $v_1 \dots, v_\ell$, this gives us the KNAPSACK problem:

KNAPSACK
Input: Positive integers $W, w_1, \dots, w_\ell$ and $v_1, \dots, v_\ell$
Output: A subset $S \subseteq \{1, \dots, \ell\}$ that maximizes $\sum_{i \in S} v_i$ subject to $\sum_{i \in S} w_i \le W$

Show that KNAPSACK is NP-hard. Hint: reduce from SUBSET SUM.

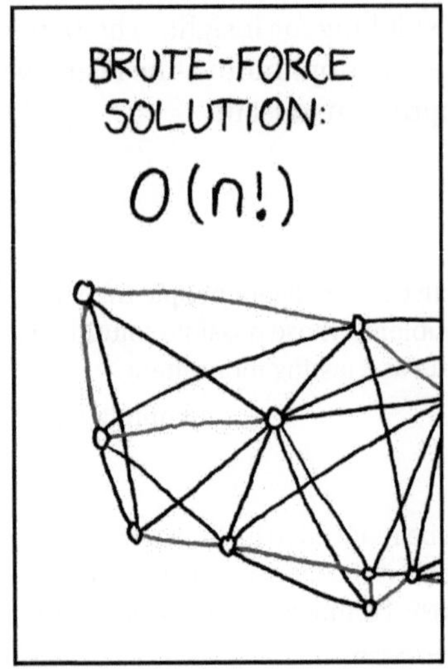

FIGURE 9.45: Solving TSP in the modern age.

9.6 Dynamic knapsacks. Let $v_{\max} = \max(v_1, \dots, v_\ell)$ denote the value of the most valuable object in an instance of KNAPSACK. Then $\ell v_{\max}$ is an upper bound for the value we can achieve. Use this fact to devise an algorithm that solves KNAPSACK in $O(\ell^2 v_{\max})$ time. Why doesn't this prove that $\mathsf{P} = \mathsf{NP}$?

Hint: for each $1 \le j \le \ell$ and each $1 \le v \le \ell v_{\max}$, let $S_{j,v}$ be the smallest-weight subset of $\{j, \dots, \ell\}$ whose value is exactly v. Show that we can compute the weight of $S_{i,v}$ using dynamic programming, as we did for SUBSET SUM in Section 4.2.3.

9.7 Approximate knapsack packing. The preceding problem showed that we can solve KNAPSACK in polynomial time if the values v_i are small, i.e, if they are bounded by a polynomial in n. Exploit this fact to devise a FPTAS for KNAPSACK.

Hint: replace each v_i with a scaled and truncated value, $\lfloor v_i / K \rfloor$ for some scaling factor K, and run the algorithm of Problem 9.6 on the resulting instance. Then analyze how the value of the resulting set compares with the optimum, and figure out what K you need to make this a $(1-\varepsilon)$-approximation. Finally, show that with this value of K, the running time is polynomial in n and $1/\varepsilon$.

9.8 Greedy set covers. In VERTEX COVER, we need to cover every edge, and each vertex covers the set of edges incident to it. Let's generalize this problem as follows:

> MIN SET COVER
> Input: A family of subsets $S_1, S_2, \dots, S_\ell \subseteq \{1, \dots, m\}$
> Output: The smallest subset $A \subseteq \{1, \dots, \ell\}$ such that $\bigcup_{i \in A} S_i = \{1, \dots, m\}$?

This is often described as recruiting a team. There is a set of m skills that we need to have, and each person i possesses a set of skills S_i. The problem is then to find the smallest team that covers all m skills. VERTEX COVER is the case where skills are edges, people are vertices, and each skill is possessed by exactly two people.

The obvious greedy strategy is to add people to the team one at a time, at each step adding the one that covers the largest number of additional skills. This is simply the generalization of our greedy algorithm for VERTEX COVER.

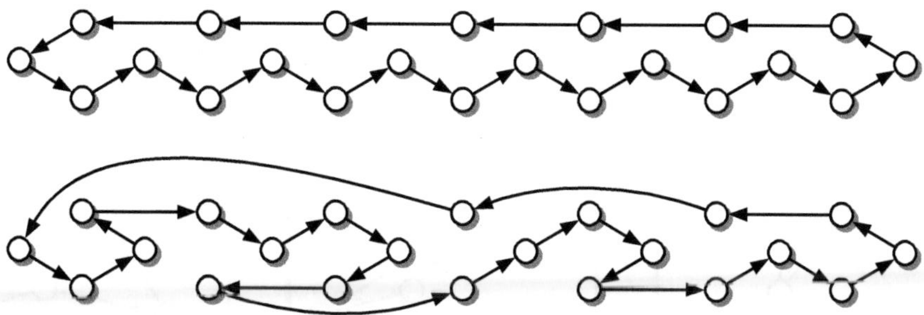

FIGURE 9.46: The shortest tour (top) versus the greedy one (bottom).

Formally, we start with $A = \emptyset$, and at each step we update A to $A' = A \cup \{j\}$, where j maximizes the number of newly covered elements $\left|S_j - \cup_{i \in A} S_i\right|$.

Show that this algorithm achieves an approximation ratio $\rho = O(\log m)$. Specifically, show that $\rho \leq \ln m$. Hint: suppose that the optimal cover has size k. Then show that each step of the greedy algorithm reduces the number of uncovered elements of $\{1, \ldots, m\}$ by a factor of $1 - 1/k$ or less.

9.9 As good as random. How can we deterministically choose a truth assignment that satisfies at least the expected number of clauses? First consider a mixed SAT formula ϕ with clauses of various lengths. Show how to compute the expected number of satisfied clauses $\mathbb{E}[c]$ given a uniformly random truth assignment. Then show how to compute the conditional expectations $\mathbb{E}[c \mid x_1 = \texttt{true}]$ and $\mathbb{E}[c \mid x_1 = \texttt{false}]$—in other words, the expected number of satisfied clauses if x_1 is given a particular value and the other variables are set randomly. Show that at least one of these is at least as big as the overall expectation,

$$\mathbb{E}[c \mid x_1 = \texttt{true}] \geq \mathbb{E}[c] \quad \text{or} \quad \mathbb{E}[c \mid x_1 = \texttt{false}] \geq \mathbb{E}[c].$$

Then describe how to travel down the search tree of truth assignments, setting one variable at a time, until all the variables are set and we reach a leaf that satisfies at least the expected number of clauses.

This technique of finding a deterministic object which is just as good as a random one is called the *method of conditional expectation*, and lets us derandomize some randomized algorithms. In this case, it gives a deterministic ρ-approximation for MAX-k-SAT with $\rho = 1 - 2^{-k}$, or for MAX-XORSAT with $\rho = 1/2$.

9.10 The short-sighted salesman. Show that there are graphs where the greedy algorithm for ΔTSP, which moves to the nearest city that it hasn't already visited, yields tours whose length is $\Omega(\log n)$ times the optimum. Hint: consider Figure 9.46. You may need to alter the distances minutely in order to break ties.

9.11 Insertion heuristics. The insertion heuristic for ΔTSP grows a tour by adding one vertex at a time until the tour includes all the vertices. Suppose that τ is a partial tour, i.e., a tour through a subset of the cities. Let k be a city not on τ, and let $\text{Tour}(\tau, k)$ be the partial tour obtained by inserting k into τ as follows:

1. If τ contains only one city i, then $\text{Tour}(\tau, k)$ is the two-city tour consisting of edges (i, k) and (k, i).
2. If τ passes through more than one city, let (i, j) be the edge in τ that minimizes $d_{ik} + d_{kj} - d_{ij}$. Then $\text{Tour}(\tau, k)$ is constructed by deleting edge (i, j) and adding edges (i, k) and (k, j).

In other words, the new city k is added such that the length of the tour increases as little as possible. This greedy strategy still gives us the freedom to choose which city to add next, and one particular choice is to always add the city that is nearest to a city on the tour (Figure 9.47).

Show, using the triangle inequality, that this *nearest insertion heuristic* is a 2-approximation for ΔTSP. Hint: this heuristic can be seen as a variant of the 2-approximation in Section 9.2.2, where constructing the minimum spanning tree, Eulerification, and tour building are lumped together in one incremental process.

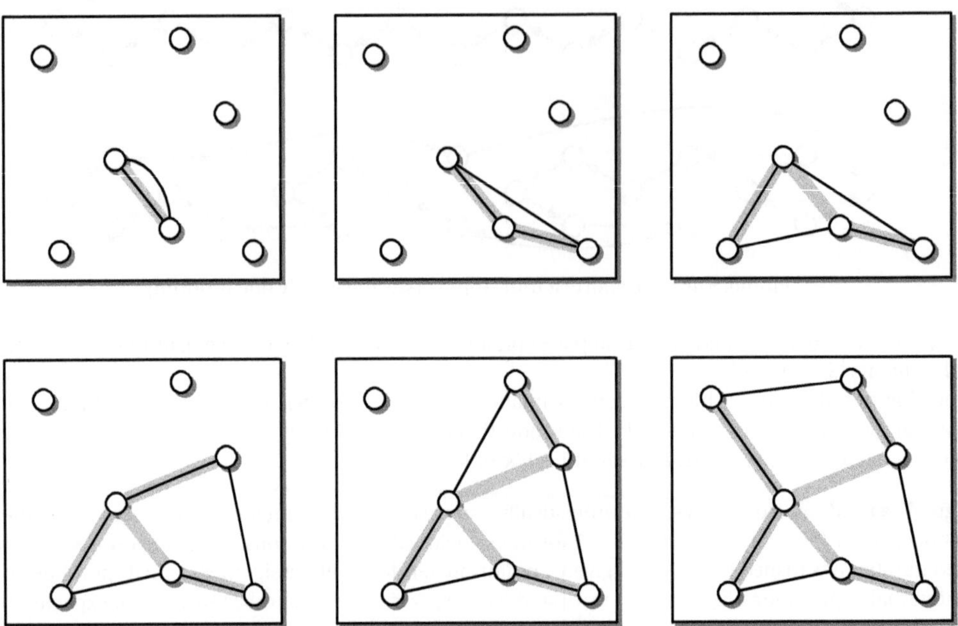

FIGURE 9.47: Nearest insertion heuristics in action.

9.12 The Klee–Minty cube. Prove that the simplex method with the "lowest index" pivot rule visits every vertex of the Klee–Minty cube in both the following cases: (a) it starts at the origin and maximizes x_d, or (b) it starts at $(0,\ldots,0,1)$ and minimizes x_d. Hint: use induction over d, and focus on the variable x_d.

9.13 The Hirsch conjecture. Define the diameter of a polytope in the graph-theoretic sense, i.e., the number of steps in the shortest path through the graph of edges, maximized over all pairs of vertices. Let $\Delta(d,m)$ denote the largest possible diameter of any d-dimensional polytope with no more than m facets for $m > d \geq 2$. The Hirsch conjecture claims that

$$\Delta(d,m) \leq m-d. \tag{9.58}$$

In this problem, we will prove the Hirsch conjecture for 0-1 polytopes, i.e., those whose vertices are corners of the hypercube $\{0,1\}^d$. This type of polytope arises frequently in LP relaxations of NP-hard problems.

9.11 Let $\mathrm{diam}(P)$ denote the diameter of the polytope P. Show that for any d-dimensional 0-1 polytope

$$\mathrm{diam}(P) \leq d, \tag{9.59}$$

with equality if and only if P is the hypercube. For $m \geq 2d$, this implies the Hirsch conjecture. The case $m < 2d$ can be reduced to the case $m = 2d$ since

$$\Delta(d,m) \leq \Delta(m-d, 2(m-d)) \qquad \text{for } m < 2d, \tag{9.60}$$

which is true for general polytopes. Prove this too.

Hints: prove (9.59) by induction on d, and assume that there are two vertices $\mathbf{u},\mathbf{v} \in P$ with distance $\delta(\mathbf{u},\mathbf{v}) \geq d$. You can assume that $\mathbf{v} = \mathbf{0}$ (why?). Show that this implies $\mathbf{u} = \mathbf{1} = (1,\ldots,1)$. Use the fact that any two vertices with

$u_i = v_i$ for some i share a common facet, i.e., are connected by a path along a $(d-1)$-dimensional 0-1 polytope. To prove (9.60), use the fact that for $m < 2d$, any two vertices of a d-dimensional polytope share at least one facet.

9.14 Touching the walls. In Section 9.5, we assumed that the components of the optimal vertex $\mathbf{x}$ are all positive. Now consider the possibility that $x_i = 0$ for some $i \in \{1,\ldots,d\}$. Generalize (9.18) and (9.19) as follows: if $x_i > 0$, the constraint corresponding to the ith column of $\mathbf{A}$ is tight at the dual optimum $\mathbf{y}$. Therefore, if

$$R = \left\{i : \mathbf{a}_i^T \mathbf{y} = c_i\right\} \subseteq \{1,\ldots,d\},$$

where now $\mathbf{a}_i$ denotes the ith column of $\mathbf{A}$, we have $x_i = 0$ for all $i \notin R$. This is the dual of the statement that $y_j = 0$ for all $j \notin S$. These two statements together are called *complementary slackness*: $y_j = 0$ unless the jth constraint on $\mathbf{x}$ is tight, and $x_i = 0$ unless the ith constraint on $\mathbf{y}$ is tight.

9.15 Repulsive facets. Show that the gradient of the potential V_{facets} as defined in (9.23) is

$$\nabla V_{\text{facets}}(\mathbf{x}) = \sum_{j=1}^{m} \frac{\mathbf{a}_j^T}{b_j - \mathbf{a}_j^T \mathbf{x}}.$$

Then, defining $\mathbf{D}$ as the diagonal matrix with entries

$$D_{jj} = b_j - \mathbf{a}_j^T \mathbf{x},$$

show that the Hessian of V_{facets} is

$$\mathbf{H}(\mathbf{x}) = \mathbf{A}^T \mathbf{D}^{-2} \mathbf{A} = \sum_{j=1}^{m} \frac{\mathbf{a}_j \otimes \mathbf{a}_j^T}{(b_j - \mathbf{a}_j^T \mathbf{x})^2}.$$

Here $\mathbf{u} \otimes \mathbf{v}^T$ denotes the outer product of two vectors, i.e., the matrix $(\mathbf{u} \otimes \mathbf{v}^T)_{ij} = u_i v_j$. Note that $\mathbf{H}$ is real and symmetric, so it can be diagonalized.

Now suppose that the feasible polytope $\{\mathbf{x} : \mathbf{A}\mathbf{x} \leq \mathbf{b}\}$ is bounded, and assume as always that it has nonzero volume, i.e., it is not squashed flat into a lower-dimensional subspace. Show that $\mathbf{A}$ has full rank—that is, that the rows $\mathbf{a}_j$ span $\mathbb{R}^d$. Finally, show that $\mathbf{H}$ is then positive definite, i.e., all of its eigenvalues are positive. In particular, $\mathbf{H}$ is invertible, and V_{facets} has a unique minimum inside the polytope.

9.16 Jumpstart for LP. How do we find an initial feasible solution for the simplex algorithm? By solving another linear program for which we know a feasible solution! As we outlined in Section 9.4.1, we can transform the set of inequalities $\mathbf{A}\mathbf{x} \leq \mathbf{b}$ into a set of equalities $\mathbf{A}'\mathbf{x}' = \mathbf{b}'$ by adding "slack" variables $\mathbf{s} \geq 0$. Hence we can assume that our linear program is of the form

$$\max_{\mathbf{x}} \left(\mathbf{c}^T \mathbf{x}\right) \quad \text{subject to} \quad \mathbf{A}\mathbf{x} = \mathbf{b} \quad \text{and} \quad \mathbf{x} \geq \mathbf{0}. \tag{9.61}$$

By adding more variables, construct yet another linear program that has a trivial feasible solution, and whose optimal solution is a feasible solution of (9.61). Hint: if $\mathbf{b} \in \mathbb{R}^m$, you need m additional variables. Without loss of generality, you may assume that $\mathbf{b} \geq \mathbf{0}$.

9.17 Rational vertices. Suppose we have an instance of LINEAR PROGRAMMING where the feasible polytope is $P = \{\mathbf{x} \in \mathbb{R}^d : \mathbf{A}\mathbf{x} \leq \mathbf{b}\}$. Assume that $\mathbf{A}$ and $\mathbf{b}$ have integer entries, and let n be the total number of bits of all these entries. Show that the vertices of P are rational—that is, each vertex $\mathbf{v}$ can be written

$$\mathbf{v} = \frac{\mathbf{u}}{D}, \tag{9.62}$$

where $\mathbf{u}$'s entries and the denominator D are integers. Moreover, show that $D \leq 2^n$ and $|u_i| \leq 2^n$ for all i.

Hint: use *Cramer's rule*, which states that the inverse of any $d \times d$ matrix $\mathbf{M}$ is given by

$$(\mathbf{M}^{-1})_{ij} = (-1)^{i-j} \frac{\det \mathbf{M}^{(ji)}}{\det \mathbf{M}}, \tag{9.63}$$

where $\mathbf{M}^{(ji)}$ is the $(d-1) \times (d-1)$ matrix formed by removing the ith column and jth row of M. Therefore, if $\mathbf{M}\mathbf{x} = \mathbf{y}$ so that $\mathbf{x} = \mathbf{M}^{-1}\mathbf{y}$, we have

$$x_i = \frac{1}{\det \mathbf{M}} \sum_{j=1}^{d} (-1)^{i-j} y_j \det \mathbf{M}^{(ji)} = \frac{\det \mathbf{M}'_i}{\det \mathbf{M}}, \tag{9.64}$$

where $\mathbf{M}'_i$ is the matrix formed by replacing the ith column of $\mathbf{M}$ with $\mathbf{y}$.

9.18 Range and resolution. Consider an instance of LINEAR PROGRAMMING where $\mathbf{A}$, $\mathbf{b}$, and $\mathbf{c}$ have integer entries with a total of n bits. Show that for any vertex, including the optimum, we have $|\mathbf{c}^T\mathbf{x}| \leq 2^{O(n)}$. Show also that if a given constraint $\mathbf{a}_j^T \mathbf{x} \leq b_j$ is not tight at a given vertex $\mathbf{v}$ then it is loose by at least 2^{-n}, i.e.,

$$b_j - \mathbf{a}_j^T \mathbf{v} \leq 2^{-n}.$$

Similarly, show that if $\mathbf{v}$ is a non-optimal vertex, then $\mathbf{c}^T\mathbf{v}$ is less than the optimal value of $\mathbf{c}^T\mathbf{x}$ by at least 2^{-2n}.

Use these results to show that there is an $\varepsilon = 2^{-O(n)}$ such that, if we can determine the optimal vertex $\mathbf{x}$, or even just its quality $\mathbf{c}^T\mathbf{x}$, within ε, then we can find the optimal vertex exactly. This completes the argument that LP is Turing-reducible to its threshold version using binary search. Hint: use the bounds derived in Problem 9.17.

9.19 Minimum volume of the feasible polytope. As in the previous two problems, suppose we have an instance of LINEAR PROGRAMMING where $\mathbf{A}$ and $\mathbf{b}$ have integer entries with a total of n bits. Show that unless it is empty or squashed flat (i.e., contained in a lower-dimensional subspace) the feasible polytope P has volume at least $2^{-O(dn)}$.

Hint: show that P must contain the convex hull S of some set of $d+1$ vertices $\mathbf{v}_0, \ldots, \mathbf{v}_d$, where the d vectors $\mathbf{v}_j - \mathbf{v}_0$ for $1 \leq j \leq d$ are linearly independent. This hull is a d-dimensional simplex, defined by

$$S = \left\{ \sum_{i=0}^{d} \lambda_i \mathbf{v}_i : 0 \leq \lambda_i \leq 1, \sum_{i=0}^{d} \lambda_i = 1 \right\}.$$

Show that its volume is proportional to the determinant of a $(d+1) \times (d+1)$ matrix whose jth column is 1 followed by $\mathbf{v}_j$,

$$\text{volume}(S) = \frac{1}{d!} \left| \det \begin{pmatrix} 1 & 1 & \cdots & 1 \\ \mathbf{v}_0 & \mathbf{v}_1 & \cdots & \mathbf{v}_d \end{pmatrix} \right|. \tag{9.65}$$

Then use the results from Problem 9.17 to give a lower bound on this determinant, assuming that it isn't zero.

9.20 Positive semidefinite matrices. Let $\mathbf{Q}$ be a real symmetric $d \times d$ matrix. Show that the following three conditions are equivalent to each other:

1. $\mathbf{x}^T\mathbf{Q}\mathbf{x} \geq 0$ for all $\mathbf{x} \in \mathbb{R}^d$.
2. All the eigenvalues of $\mathbf{Q}$ are nonnegative.
3. $\mathbf{Q} = \mathbf{M}^T\mathbf{M}$ for some $d \times d$ matrix $\mathbf{M}$.

Hint: what do you know about the eigenvalues and the eigenvectors of a real symmetric matrix?

9.21 Volume of an ellipsoid. Show that the volume of the d-dimensional hypersphere of radius r, i.e., $S_r^d = \{\mathbf{x} \in \mathbb{R}^d : |\mathbf{x}|^2 \leq r\}$, is

$$\text{volume}(S_r^d) = \frac{\pi^{d/2}}{\Gamma(\frac{d}{2}+1)}\, r^d .$$

where $\Gamma(x) = \int_0^\infty t^{x-1}\,\mathrm{e}^{-t}\,\mathrm{d}t$ is the Gamma function. Hint: evaluate the Gaussian integral

$$I = \int_{-\infty}^{\infty} \mathrm{e}^{-(x_1^2+x_2^2+\cdots+x_d^2)}\,\mathrm{d}x_1 \cdots \mathrm{d}x_d$$

twice, first by direct evaluation, and then by using radial coordinates.

Then use the affine transformation from the unit sphere to the ellipsoid

$$E = \{\mathbf{x} \in \mathbb{R}^d : (\mathbf{x}-\mathbf{r})^T \mathbf{Q}(\mathbf{x}-\mathbf{r}) \leq 1\},$$

to show that the volume of the latter is

$$\text{volume}(E) = \frac{\text{volume}(S_1^d)}{\sqrt{\det \mathbf{Q}}} .$$

9.22 Updating ellipsoids, part I. Let $E = \{\mathbf{y} : |\mathbf{y}|^2 \leq 1\}$, and find the ellipsoid E' with the smallest volume that contains the hemisphere

$$\{\mathbf{y} \in E : y_1 \geq 0\} .$$

Then show that

$$\frac{\text{volume}(E')}{\text{volume}(E)} = \left(\frac{d}{d+1}\right)\left(\frac{d^2}{d^2-1}\right)^{(d-1)/2} \leq \mathrm{e}^{-1/(2d+2)} .$$

Hint: as Figure 9.27 suggests, E' is centered at $(r, 0, \ldots, 0)$ for some $0 \leq r \leq 1/2$. Compute its height along the y_1-axis, and what its width along the other $d-1$ axes needs to be to contain the hemisphere. Show that its volume is then given by

$$\frac{\text{volume}(E')}{\text{volume}(E)} = (1-r)\left(1 - \left(\frac{r}{1-r}\right)^2\right)^{-(d-1)/2}$$

and that this expression is minimized at

$$r = \frac{1}{d+1} .$$

Finally, bound the ratio between the volumes with the inequality $1 + x \leq \mathrm{e}^x$.

9.23 Updating ellipsoids, part II. Show that the minimum-volume ellipsoid E' that contains the half-ellipsoid

$$\{\mathbf{y} \in E : \mathbf{a}^T \mathbf{y} \leq \mathbf{a}^T \mathbf{r}\}$$

is given by (9.32). Hint: first use the results of Problem 9.22 to show that if E is the unit sphere and $\mathbf{a}$ is a unit vector,

$$\begin{aligned} \mathbf{Q}' &= \left(1 - \frac{1}{d^2}\right)\mathbb{1} + \frac{2(d+1)}{d^2}\,\mathbf{a} \otimes \mathbf{a}^T \\ \mathbf{r}' &= -\frac{\mathbf{a}}{d+1} . \end{aligned} \tag{9.66}$$

The outer product $\mathbf{a} \otimes \mathbf{a}^T$ is the projection operator onto the subspace parallel to $\mathbf{a}$. In particular, if $\mathbf{a}$ is the unit vector $(1, 0, \ldots, 0)$ then $\mathbf{a} \otimes \mathbf{a}^T$ is the matrix with a 1 in its upper-left corner and 0s everywhere else.

In the general case, apply the affine transformation $\mathbf{x} \to \mathbf{M}(\mathbf{x}-\mathbf{r})$ to map E to the unit sphere, compute the resulting unit vector perpendicular to the separating hyperplane, use (9.66), and transform back. Finally, note that the ratio volume(E')/volume(E) is invariant under affine transformations.

9.24 Logarithmic barriers for semidefinite programming. To design an interior-point method for SEMIDEFINITE PROGRAMMING, we have to define a potential energy that pushes $\mathbf{X}$ away from the singular matrices, i.e., those with a zero eigenvalue. Consider the following choice:

$$V(\mathbf{X}) = -\ln\det\mathbf{X}.$$

Show that the resulting repulsive force is

$$F = -\nabla V = (\mathbf{X}^{-1})^T.$$

In other words, show that for any $\mathbf{Y}$, to first order in ε we have

$$V(\mathbf{X}+\varepsilon\mathbf{Y}) = V(\mathbf{X}) + \varepsilon\langle\nabla V, \mathbf{Y}\rangle = V(\mathbf{X}) - \varepsilon\,\mathrm{tr}(\mathbf{X}^{-1}\mathbf{Y}).$$

Hint: first show that, to first order, $\det(\mathbb{1}+\varepsilon\mathbf{Z}) = \varepsilon\,\mathrm{tr}\,\mathbf{Z}$.

By the way, this is also a proof of Cramer's rule (9.63), which we used in Problem 9.17. Prove that too.

9.25 Even local optima can be hard. In this problem we construct a non-convex quadratic optimization problem where it is NP-hard even to verify that a given vector is a local optimum. We start with an instance of 3-SAT. Let ϕ be a 3-SAT formula with n variables x_i and m clauses. Our problem will have $n+1$ variables $y_0, y_1, \ldots, y_n$, and for each clause in ϕ we add a linear constraint like

$$(x_1 \vee x_2 \vee \overline{x}_3) \quad\rightarrow\quad y_1 + y_2 + (1-y_3) + y_0 \geq \frac{3}{2}. \tag{9.67}$$

We also require that $\left|y_i - 1/2\right| \leq y_0$ for each $1 \leq i \leq n$:

$$\frac{1}{2} - y_0 \leq y_i \leq \frac{1}{2} + y_0. \tag{9.68}$$

Finally, we require that

$$0 \leq y_0 \leq 1/2.$$

The polytope of feasible solutions defined by these constraints is convex. But now we ask to minimize a non-convex function, namely

$$f(\mathbf{y}) = p(\mathbf{y}) + q(\mathbf{y}),$$

where

$$p(\mathbf{y}) = \sum_{i=1}^{n}\left(y_0^2 - \left(y_i - \frac{1}{2}\right)^2\right) \quad\text{and}\quad q(\mathbf{y}) = -\frac{1}{2n}\sum_{i=1}^{n}\left(y_i - \frac{1}{2}\right)^2.$$

Show that $\mathbf{y}^\star = (0, 1/2, \ldots, 1/2)^T$ is a local minimum if and only if ϕ is not satisfiable.

Hint: note that $\mathbf{y}^\star$ is a vertex of the feasible polytope and that $p(\mathbf{y}^\star) = 0$. Show that for fixed y_0, the minimum of f is less or equal to $-y_0^2/2$ if ϕ is satisfiable. Then show that for fixed y_0, the minimum is greater or equal to $(7/18)y_0^2$ if ϕ is not satisfiable. If a clause is unsatisfied, the corresponding variables y_i or $1-y_i$ are all at most $1/2$. On the other hand, the clause constraint (9.67) implies that at least one of them is greater than or equal to $1/2 - y_0/3$.

9.26 Maximum entropy. The principle of maximum entropy states that among all probability distributions that are consistent with a set of observations, the most likely distribution is the one that maximizes the entropy. Let p_i denote the probability of an event i. We want to maximize the Gibbs–Shannon entropy

$$H(\mathbf{p}) = -\sum_i p_i \log p_i$$

subject to the constraints $\mathbf{p} \geq 0$ and $\sum_i p_i = 1$.

By making observations, we impose additional constraints $\mathbf{A}\mathbf{p} = \mathbf{b}$. For instance, we might observe the average of some quantity f, $\sum_i p_i f_i = \mathbf{f}^T\mathbf{p}$. In Problem 12.3, for example, we constrain the average energy of a physical system. Show that maximizing the entropy subject to these constraints is a convex optimization problem, and solve it analytically using Lagrange multipliers.

9.27 Integer programming without inequalities. Let $\mathbf{A}$ be an integer $m \times d$ matrix and let $\mathbf{b}$ be an integer vector. Show that the set of integer solutions $\mathbf{x}$ of $\mathbf{Ax} = \mathbf{b}$ forms a lattice, and describe how to find this lattice in polynomial time. Then explain why this implies a polynomial time solution to the integer linear program that maximizes $\mathbf{c}^T\mathbf{x}$ subject to $\mathbf{Ax} = \mathbf{b}$ and $\mathbf{x} \in \mathbb{Z}^d$.

9.28 Half-integral vertex covers. Prove that the LP relaxation of MIN VERTEX COVER is half-integral, i.e., it yields a vector $\mathbf{x}$ with $x_v \in \{0, 1/2, 1\}$ for all v. Hint: assume that there is a vertex $\mathbf{x}$ that is not half-integral. Use this vertex to construct two feasible solutions $\mathbf{y}, \mathbf{z} \neq \mathbf{x}$ such that $\mathbf{x} = (\mathbf{y} + \mathbf{z})/2$. But a convex combination of two feasible solutions can't be a vertex, giving a contradiction.

9.29 The unfuzzy parts of fuzzy covers. Let $G = (V, E)$ be a graph and let $\mathbf{x}$ be the solution of the LP relaxation of MIN VERTEX COVER on G. Let $V_1 = \{v : x_v = 1\}$, $V_{1/2} = \{v : x_v = 1/2\}$, and $V_0 = \{v : x_v = 0\}$. Show that there exists a minimum vertex cover $S_{\min}$ of G that contains all of V_1 and none of V_0. This is called the *Nemhauser–Trotter Theorem.*

Hint: first show that if $S_{\min}$ contains none of V_0 then it has to contain all of V_1. Then show that there exists an $S_{\min}$ that contains none of V_0. Both proofs work by contradiction. Consider the sets $V_0 \cap S_{\min}$, $V_0 - S_{\min}$, and $V_1 - S_{\min}$. Can there be edges between these sets?

9.30 Total unimodularity. How can we check whether a given matrix $\mathbf{A}$ is totally unimodular? Prove that each of the following conditions is sufficient. Throughout, we assume that $\mathbf{A}$ has entries in $\{0, \pm 1\}$.

(a) $\mathbf{A}$ is totally unimodular if and only if any of the following matrices is totally unimodular: $-\mathbf{A}$, $\mathbf{A}^T$, $(\mathbf{A}, \mathbb{1})$, $(\mathbf{A}, \mathbf{A})$.

(b) If each column of $\mathbf{A}$ has at most one $+1$ and at most one -1 then $\mathbf{A}$ is totally unimodular. Hint: first show that any matrix where every column has one $+1$ and one -1, and all other entries are zero, has zero determinant.

(c) If $\mathbf{A}$ has no -1 entries and if the rows of $\mathbf{A}$ can be permuted so that the 1s in each column appear consecutively, then A is totally unimodular. Hint: every square submatrix $\mathbf{B}$ of $\mathbf{A}$ inherits this "consecutive ones" property. Subtract rows of $\mathbf{B}$ from each other such that each column of $\mathbf{B}$ contains at most one $+1$ and one -1.

(d) If both of the following conditions hold, then $\mathbf{A}$ is totally unimodular:

- Each column of $\mathbf{A}$ contains at most two nonzero entries.
- The rows of $\mathbf{A}$ can be partitioned into two sets $\mathbf{A}_1$ and $\mathbf{A}_2$ such that two nonzero entries in a column are in the same set of rows if they have different signs, and in different sets of rows if they have the same sign.

Hint: show that if every column of $\mathbf{A}$ has exactly two nonzero entries, and if the second condition holds, then $\mathbf{A}$ has zero determinant.

(e) The signed incidence matrix of a directed graph, i.e., the $|E| \times |V|$ matrix $\mathbf{A}$ where

$$a_{ev} = \begin{cases} +1 & \text{if } e \text{ is an outgoing edge from } v, \\ -1 & \text{if } e \text{ is an incoming edge to } v, \\ 0 & \text{otherwise,} \end{cases}$$

is totally unimodular.

(f) Let $\mathbf{B}$ and $\mathbf{C}$ be totally unimodular. Then the matrices

$$\begin{pmatrix} \mathbf{B} & 0 \\ 0 & \mathbf{C} \end{pmatrix} \quad \text{and} \quad \begin{pmatrix} 0 & \mathbf{B} \\ \mathbf{C} & 0 \end{pmatrix}$$

are totally unimodular.

9.31 Integral flows. Write MAX FLOW as a linear program and show that the constraint matrix is totally unimodular. If the edge capacities are integers then this program has an integral optimum, giving another proof that there is an integral maximum flow. Hint: consider the flow in a directed network.

9.32 Stretching the shortest path. Write SHORTEST PATH as a linear program and show that its constraint matrix is unimodular. Then write down the dual problem and give an interpretation of its variables.

Hint: treat the graph as directed by replacing each edge (u,v) by two directed edges $u \to v$ and $v \to u$ of equal capacity c_{uv}, and then treat the path as a flow. To interpret the dual problem, think of a physical network in which vertices are beads, edges are strings, and where beads i and j are joined by a piece of string of length c_{ij}. Imagine grabbing beads s and t and pulling them apart...

9.33 Traveling sales force. In times of global warming and high unemployment, we decide to reduce the mileage of our sales force by hiring more salesmen and dividing the country among them. As in our definition of TSP (Section 9.2.2) we define a permutation π of the cities, where the sales route goes from city i to city $\pi(i)$, then to $\pi(\pi(i))$, and so on. However, we now drop the requirement that π consists of a single cycle. This gives us the following problem:

> ASSIGNMENT
>
> Input: An $n \times n$ matrix with entries $d_{ij} \geq 0$
>
> Output: A permutation π that minimizes $\sum_{i=1}^{n} d_{i,\pi(i)}$

The name ASSIGNMENT reflects another interpretation, where we want to assign n tasks to n workers in a way that minimizes the total cost, and where d_{ij} is the cost of worker i handling task j. In our setting, each cycle of π is a subtour to which we assign one of our salesmen. Note that if $d_{ii} = 0$ for all i, the optimal solution is to assign a stationary salesman to each city. We can prevent this, forcing each salesman to leave his home town, by setting $d_{ii} = \infty$.

Prove that ASSIGNMENT can be solved in polynomial time. Hint: we want to minimize $\sum_{ij} d_{ij} M_{ij}^{(\pi)}$ where $M^{(\pi)}$ is a *permutation matrix*, with one 1 in each row and each column and 0s elsewhere:

$$M_{ij}^{(\pi)} = \begin{cases} 1 & \text{if } j = \pi(i) \\ 0 & \text{otherwise}. \end{cases}$$

The polytope of "fuzzy permutations" is the set of doubly stochastic matrices, i.e., nonnegative matrices whose rows and columns each sum to 1:

$$A_{ij} \geq 0 \text{ for all } i,j\,, \quad \sum_{i=1}^{n} A_{ij} = 1 \text{ for all } i\,, \quad \text{and} \quad \sum_{j=1}^{n} A_{ji} = 1 \text{ for all } i\,, \tag{9.69}$$

Prove that this polytope is integral and that its vertices are exactly the permutation matrices $M^{(\pi)}$. Do this by showing that the row and column constraints on A are totally unimodular, by interpreting their coefficients as the incidence matrix of a bipartite graph.

Note that this gives an alternate proof that MAX MATCHING for bipartite graphs is in P. More generally, suppose that we have a weighted bipartite graph, with n vertices on the left and n on the right, and that each edge has a weight w_{ij}. Show that MAX-WEIGHT PERFECT MATCHING, the problem of finding the perfect matching π with the highest total weight $\sum_{i=1}^{n} w_{i,\pi(i)}$, is in P.

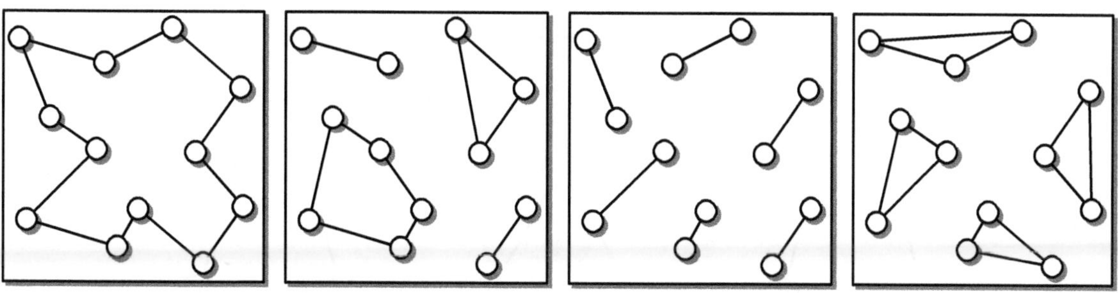

FIGURE 9.48: Same instance, different problems. Left to right: TSP(hard), ASSIGNMENT (easy), MIN-2-CYCLE COVER (easy) and MIN-3-CYCLE COVER (hard).

9.34 Doubly stochastic. The previous problem showed that the polytope of doubly stochastic matrices is integral. We can rephrase this as *Birkhoff's Theorem*, which states that any doubly stochastic matrix is a convex combination of permutation matrices. That is,

$$A = \sum_{\pi} a_\pi M^{(\pi)} \quad \text{where} \quad M_{ij}^{(\pi)} = \begin{cases} 1 & \text{if } j = \pi(i) \\ 0 & \text{otherwise,} \end{cases}$$

where π ranges over all $n!$ permutations and $a_\pi \geq 0$ for each π. (This is one of many theorems named after Birkhoff—another one is about black holes.)

In Problem 9.33 we proved Birkhoff's Theorem using the unimodularity of the constraints on the rows and columns of A. Now give a second proof along the following lines. First show that it suffices to prove that there exists a permutation matrix in the support of A, i.e., there is a π such that $A_{i,\pi(i)} > 0$ for all i. Then prove this fact using Hall's Theorem from Problem 3.47.

Hint: given a set $S \subseteq \{1, \ldots, n\}$, consider the characteristic vector $\mathbf{v}$ where $v_i = 1$ if $i \in S$ and $v_i = 0$ if $i \notin S$. Then consider the support of $A\mathbf{v}$, i.e., the set of nonzero components.

9.35 Triply stochastic. Inspired by the previous problem, let's call an $n \times n \times n$ array A of nonnegative numbers *triply stochastic* if any two-dimensional slice sums to 1. That is,

$$A_{ijk} \geq 0 \text{ for all } i, j, k$$
$$\sum_{j,k=1}^{n} A_{ijk} = 1 \text{ for all } i\,, \quad \sum_{i,k=1}^{n} A_{ijk} = 1 \text{ for all } j\,, \quad \sum_{i,j=1}^{n} A_{ijk} = 1 \text{ for all } k\,.$$

Show that if the polytope P of triply-transitive arrays has integer vertices, then $\mathsf{P} = \mathsf{NP}$. Then show that it doesn't, even for $n = 2$. Thus there is no analog of Birkhoff's Theorem in three dimensions.

9.36 Why is TSP hard? TSP can be viewed as ASSIGNMENT (Problem 9.33) with the additional constraint that the permutation π consists of a single cycle. There are other ways we could constrain π. Consider the case where n is even and we have $n/2$ salesmen, and we want each one to commute between a pair of cities. In this case we have to minimize $\sum_i d_{i,\pi(i)}$ over all π, with the restriction that π is composed of $n/2$ two-cycles (also called transpositions). This problem is equivalent to MIN-WEIGHT PERFECT MATCHING, which is in P even for non-bipartite graphs.

More generally, we can hire n/k salesmen and ask each one to tour k cities. A *cycle cover* of a graph is a set of cycles such that every vertex is part of exactly one cycle. For our n/k salesmen we need to solve MIN-k-CYCLE COVER. We already know that MIN-2-CYCLE COVER is in P, but for $k \geq 3$, MIN-k-CYCLE COVER is NP-hard. From this perspective (Figure 9.48), the hardness of TSP is another example of the "two vs. three" phenomenon that we discussed in Section 5.5. 5.6

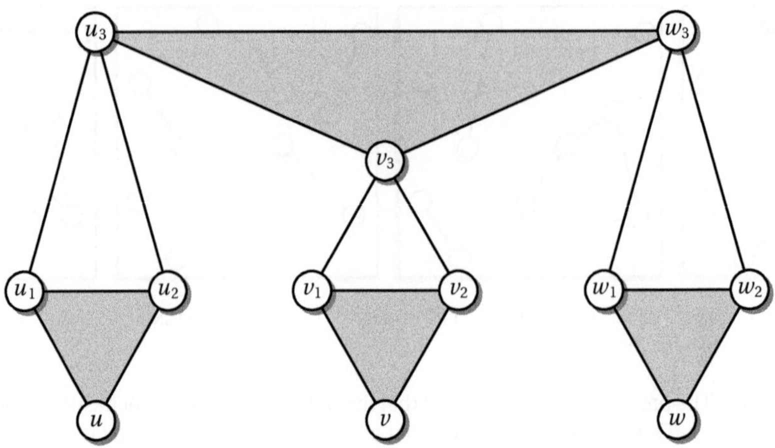

FIGURE 9.49: Gadget to reduce PERFECT 3-MATCHING to 3-CYCLE COVER.

9.20 Prove that MIN-3-CYCLE COVER is NP-hard by reducing from the NP-complete problem PERFECT 3-MATCHING from Section 5.5.2 to the following decision problem:

3-CYCLE COVER
Input: A graph $G(V,E)$ with $|V|=3n$
Question: Does G have a 3-cycle cover?

Hint: Figure 9.49 shows a possible gadget for the reduction from PERFECT 3-MATCHING to 3-CYCLE COVER. The vertices u, v, w are elements of the sets U, V, W in the definition of PERFECT 3-MATCHING.

9.37 Submodular Functions. The next few problems deal with the following kind of function, and the polytopes and optimization problems associated with them. Let E be a finite set, and let f be a function that maps subsets of E to nonnegative numbers. We call f *submodular* if

$$f(S)+f(T)\geq f(S\cap T)+f(S\cup T) \tag{9.70}$$

for all $S,T\subseteq E$. We will usually assume that $f(\emptyset)=0$. In addition, we call f *nondecreasing* if for all $S\subseteq T$,

$$f(S)\leq f(T). \tag{9.71}$$

Submodular functions are discrete analogs of concave functions. If we define a "derivative" as the increment we get when adding an element $e\in E$,

$$\delta_e(S)=f(S\cup\{e\})-f(S),$$

then δ_e is nonincreasing, i.e.,

$$\delta_e(S)\geq\delta_e(T) \tag{9.72}$$

for all $S\subset T$. Of course, $\delta_e(T)=0$ if $e\in T$.

Prove the equivalence of (9.70) and (9.72). Hint: apply (9.70) to $S\cup\{e\}$ and T. For the converse direction, transform $f(T)-f(S\cap T)$ into $f(S\cup T)-f(S)$ by adding the elements of $S-T$ one by one to both arguments.

9.38 Graphical rank. The rank r of a graph $G(V,E)$ is a function that maps subsets $F \subseteq E$ of edges to integers via

$$r(F) = |V| - \text{number of connected components of } G(V,F). \tag{9.73}$$

Show that the rank is nondecreasing and submodular. Hint: what are the possible values of $r(F \cup \{e\}) - r(F)$? Then check that (9.72) holds.

9.39 Submodular polytopes, greediness and integrality. Let E be a finite set, and let f be a submodular, nondecreasing function mapping subsets of E to nonnegative numbers as in Problem 9.37, with $f(\emptyset) = 0$. If we impose a subtour-like constraint that no subset S has a total weight greater than $f(S)$, we get the polytope

$$P_f = \left\{ \mathbf{x} \in \mathbb{R}^{|E|} : \mathbf{x} \geq 0, \sum_{j \in S} x_j \leq f(S) \text{ for all } S \subseteq E \right\}. \tag{9.74}$$

We call such a polytope submodular. We can then consider the linear optimization problem

$$\max_{\mathbf{x}}(\mathbf{c}^T \mathbf{x}) \quad \text{subject to} \quad \mathbf{x} \in P_f. \tag{9.75}$$

Without loss of generality, we can assume that

$$c_1 \geq c_2 \geq \cdots \geq c_k > 0 \geq c_{k+1} \geq \cdots \geq c_{|E|}.$$

Let $S^j = \{1, \ldots, j\}$ for $j \in E$ be the sequence of subsets of E obtained by adding the elements of E one by one in the order of decreasing c_j. Show that the solution to (9.75) is the following vector $\mathbf{x}$,

$$x_j = \begin{cases} f(S^j) - f(S^{j-1}) & \text{for } 1 \leq j \leq k, \\ 0 & \text{for } j > k. \end{cases} \tag{9.76}$$

Use this to show that if f is integer valued then P_f is integral. Why do we call $\mathbf{x}$ a greedy solution?

Hint: first show that $\mathbf{x} \in P_f$. Then prove that the following vector $\mathbf{y}$ is a feasible solution of the dual of (9.76),

$$y_S = \begin{cases} c_j - c_{j+1} & \text{for } S = S^j, 1 \leq j < k, \\ c_k & \text{for } S = S^k, \\ 0 & \text{otherwise}, \end{cases}$$

and use this to show that $\mathbf{x}$ is the primal optimum. Then use the fact that a polytope P is integral if $\max\{\mathbf{c}^T\mathbf{x} : \mathbf{x} \in P\}$ has an integral solution $\mathbf{x}$ for every vector $\mathbf{c}$.

9.40 Integrality of the forest polytope. The *forest polytope* of a graph $G = (V,E)$ is defined as

$$P_{\text{forest}} = \left\{ \mathbf{x} \in \mathbb{R}^{|E|} : \mathbf{x} \geq \mathbf{0}, \sum_{e \in E(S)} x_e \leq |S| - 1 \text{ for all } 0 \neq S \subseteq V \right\}.$$

Show that P_{forest} is integral, and use this result to show that $P^{\text{sub}}_{\text{tree}}$ (9.47) is integral.

Hint: for the integrality of P_{forest}, show that

$$P_{\text{forest}} = \left\{ \mathbf{x} \in \mathbb{R}^{|E|} : \mathbf{x} \geq \mathbf{0}, \sum_{e \in E'} x_e \leq r(E') \text{ for all } E' \subseteq E \right\},$$

where r is the rank of the graph $G(V,E)$. Then use Problems 9.38 and 9.39. For the integrality of $P^{\text{sub}}_{\text{tree}}$, use the fact that if a hyperplane H is tangent to a polytope P, then $H \cap P$ is a polytope whose vertices are a subset of those of P.

9.41 Submodularity and greed redux. Now let's try to maximize a submodular function, conditioned on the size of the subset. Suppose f is submodular and nondecreasing as defined in Problem 9.37, and that $f(\emptyset)=0$. Let $K\geq 0$ be an integer. Then we can ask for the set T that maximizes $f(T)$, among all the subsets of size at most K.

An obvious greedy algorithm suggests itself: start with $S_0=\emptyset$ and add elements one at a time, setting $S_{t+1}=S_t\cup\{e_t\}$ where e_t maximizes the added value $\delta_e(S_t)$. If we write $\eta_t=\delta_{e_t}(S_t)$, this algorithm obtains a subset S_K with

$$f(S_K)=\sum_{t=0}^{K-1}\eta_t\,.$$

If $\eta_t=0$ for some $t<K$ the algorithm can stop early, giving a subset of size $K^\star=t<K$.

Prove that this is a ρ-approximation, where

$$\rho=1-\left(1-\frac{1}{K}\right)^K\geq 1-\frac{1}{\mathrm{e}}\,.$$

Furthermore, prove that if the greedy algorithm stops with a subset of size $K^\star<K$ then its solution is optimal.

Hint: let $f(S_{\mathrm{opt}})$ denote the optimum. Use submodularity to prove that, for all $0\leq t<K$,

$$f(S_{\mathrm{opt}})\leq\sum_{i=0}^{t-1}\eta_i+K\eta_t\,. \tag{9.77}$$

This gives us a set of K inequalities. If we fix η_0, we have a linear program where $\eta_1,\ldots,\eta_{K-1}$ are variables. Show that $f(S_{\mathrm{opt}})$ is maximized at a vertex where all K inequalities intersect, and where the η_i take a special form. Then calculate $f(S_{\mathrm{opt}})$ and $f(S_K)$ at that point and compare them.

9.42 A separation oracle for the spanning tree polytope. To confirm that the subtour elimination constraints are satisfied, or find one which is violated, we need to find the set $\emptyset\neq S\subset V$ which minimizes (9.49). After adding $\sum_{e\in E}x_e=|V|-1$ to both sides of (9.49), this gives

$$\min_{\emptyset\neq S\subset V}\left(|S|+\sum_{e\in E}x_e-\sum_{e\in E(S)}x_e\right)=\min_{\emptyset\neq S\subset V}\left(|S|+\sum_{e\notin E(S)}x_e\right). \tag{9.78}$$

The subtour elimination constraints are satisfied if and only if this minimum is at least $|V|$. We will show how to solve this minimization problem in polynomial time by reducing it to s-t MIN CUT.

Given $G=(V,E)$ and the vector $\mathbf{x}$, we define a weighted directed graph G' as follows. It has a vertex w_v for each $v\in V$, a vertex w_e for each $e\in E$, and additional vertices s,t. As shown in Figure 9.50, we connect s to each w_e with weight x_e, and connect each w_v to t with weight 1. Finally, for each $e=(u,v)\in E$, we connect w_e to w_u and to w_v with infinite weight—that is, we prohibit these edges from being cut.

Show that the minimum cut in G' that separates s from t has weight $|S|+\sum_{e\notin E(S)}x_e$, where S is the set of all $v\in V$ such that w_v is on the s-side of the cut. Hint: let S be the set of v such that we cut the edge from w_v to t. Show that we have to cut the edge from s to w_e for any $e=(u,v)$ such that either u or v is not in S. How do we deal with the case $S=\emptyset$?

9.43 Held–Karp bound for TSP. Show that value of the LP relaxation (9.54b) of TSP equals the Held–Karp bound from Section 9.10.2. Hint: take (9.54), single out a vertex v_1, replace the cut constraints on $V-v_1$ by subtour elimination constraints, and introduce dual variables for the degree constraints $\sum_{e\in\delta(v)}x_e=2$ for all $v\neq v_1$. Write down the corresponding dual problem and compare it with (9.47).

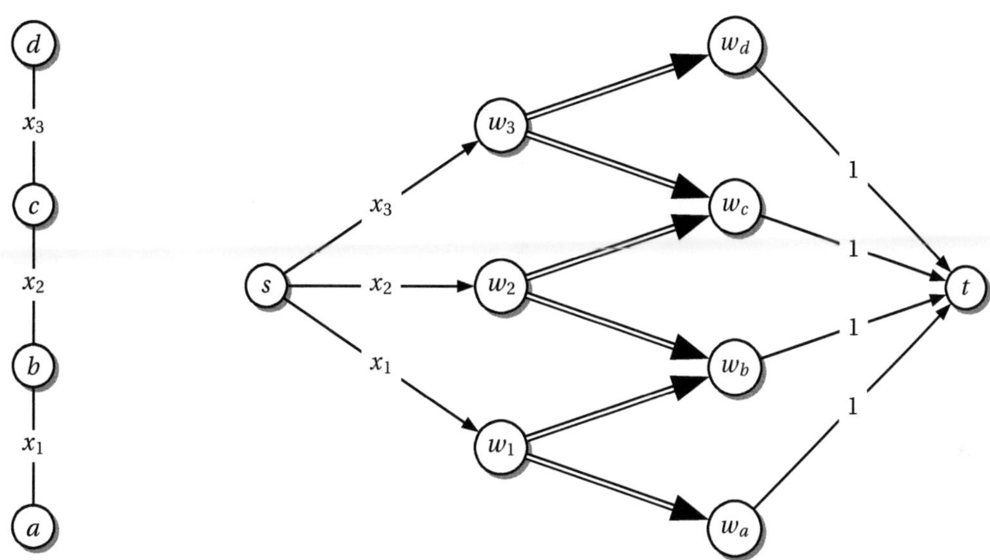

FIGURE 9.50: The separation oracle for the subtour elimination constraints. On the left, the original graph $G=(V,E)$, with edge weights x_e. On the right, a weighted directed graph G', which has vertices w_e for each $e \in E$ and w_v for each $v \in V$. Double arrows have infinite weight.

9.44 The Lovász theta function and a useful sandwich. In this problem we will use SEMIDEFINITE PROGRAMMING to compute an interesting function of a graph G, denoted $\vartheta(G)$. As in Problem 5.16, we define the clique number $\omega(G)$ as the size of G's largest clique, and $\chi(G)$ as its chromatic number. Then $\vartheta(G)$ is sandwiched between these two:

$$\omega(G) \le \vartheta(G) \le \chi(G). \tag{9.79}$$

This gives us the wonderful result that if $\omega(G)=\chi(G)$, both are computable in polynomial time.

Given $G=(V,E)$ where $|V|=n$, let S denote the following set of $n \times n$ symmetric matrices A: 9.21

$$S = \{A : A_{vv} = 1 \text{ for all } v \in V, \text{ and } A_{uv} = 1 \text{ for all } (u,v) \in E\}\,.$$

All other entries, $A_{uv} = A_{vu}$ where $u \neq v$ and $(u,v) \neq E$, are unconstrained. Then we define $\vartheta(G)$ as the minimum, over all $A \in S$, of the maximum eigenvalue $\lambda_{\max}$ of A. Equivalently,

$$\vartheta(G) = \min_{A \in S} \max_{w} \frac{w^T A w}{|w|^2}\,,$$

where w ranges over all n-dimensional vectors. Note that since A is symmetric, its eigenvalues are real. Note also that we want its largest eigenvalue $\lambda_{\max}$, not the eigenvalue with the largest absolute value.

First prove that $\vartheta(G)$ can be computed to any desired precision in polynomial time by reducing it to SEMIDEFINITE PROGRAMMING. Hint: what can you say about $\lambda_{\max}\mathbb{1} - A$?

Then prove (9.79). For the lower bound $\omega(G) \le \vartheta(G)$, construct a vector w such that $w^T A w \ge \omega(G)|w|^2$ for any $A \in S$. For the upper bound $\vartheta(G) \le \chi(G)$, first show that adding vertices or edges can only increase $\vartheta(G)$. Given a k-coloring c of G, add vertices so that there are an equal number of each color, and add an edge between every pair of vertices of different colors. Then consider the adjacency matrix B of the complement $\overline{G}$,

$$B_{uv} = \begin{cases} 1 & u \neq v \text{ and } c(u) = c(v) \\ 0 & \text{otherwise}. \end{cases}$$

Let J denote the matrix where every entry is 1, and note that $J + tB \in S$ for any t. Show that J and B commute, so they have the same eigenvectors. Finally, compute the eigenvalues of $J + tB$ directly, and find the t for which $\lambda_{\max}$ is minimized.

9.45 The clique polytope. In the spirit of Section 9.9, let's define "fuzzy cliques" on a graph G in two different ways. First, for any clique $C \subseteq V$, define its characteristic vector as

$$x_v = \begin{cases} 1 & \text{if } v \in C \\ 0 & \text{if } v \notin C. \end{cases}$$

Let $\mathscr{C}$ denote the set of all such vectors, and conv($\mathscr{C}$) denote their convex hull. This turns MAX CLIQUE into instance of LP, maximizing $\mathbf{1}^T\mathbf{x}$ in conv($\mathscr{C}$).

Alternately, we can define a fuzzy clique as a nonnegative vector whose inner product with any independent set is at most 1,

$$\mathscr{C}_{\text{ind}} = \left\{\mathbf{x} \in \{0,1\}^{|V|} : \textstyle\sum_{v \in S} x_v \leq 1 \text{ for all independent sets } S \subseteq V\right\}.$$

Show that conv($\mathscr{C}$) $\subseteq \mathscr{C}_{\text{ind}}$. Then show that if these two polytopes are the same, the clique number and the chromatic number are equal. In that case, Problem 9.44 shows that we can compute them in polynomial time—giving another case of LP which is in P even though the relevant polytope can have exponentially many facets. Hint: use duality.

Notes

9.1 Further reading. We recommend the outstanding book *Approximation Algorithms* by Vijay Vazirani [810]. In addition, Ausiello et al. [60] includes a compendium documenting the approximability of more than 200 optimization problems.

9.2 Absolute versus relative approximations. We defined ρ-approximations in terms of their ratio with the optimum, i.e., in terms of their "relative" or multiplicative error. Why didn't we discuss "absolute" or additive performance guarantees?

One reason is that most optimization problems are rescalable. Given an instance, one can easily construct an equivalent instance in which the quantity we are trying to optimize has been multiplied by some constant λ. In TRAVELING SALESMAN we can scale all the distances by a factor of λ, in MAX-SAT we can repeat each clause λ times, and so on. An algorithm that promises to get within an additive constant c of the optimum value can then be used to find the exact optimum, if the function we are trying to optimize is integer-valued, simply by rescaling the problem by a factor $\lambda > c$. Thus for rescalable problems, the best we can hope for are relative performance guarantees—unless P = NP, of course.

9.3 Approximability of MAKESPAN. The PTAS in Section 9.2.1 was found by R. L. Graham in 1969, long before the term PTAS was established [349]. Graham's algorithm is exponential in both $1/\varepsilon$ and p. This was improved by Sahni [722], who reduced the dependence of the running time on $1/\varepsilon$ to a polynomial. This qualifies as FPTAS if we fix the value of p. In 1987, Hochbaum and Shmoys [405] published a PTAS that is polynomial in p but not in $1/\varepsilon$.

9.4 Guidebooks. TRAVELING SALESMAN is by far the most famous optimization problem, and there is a vast literature devoted to it. We mention only the recent monograph of Applegate et al. [50] and the classic volume of Lawler et al. [520]. The book of Reinelt [695] focuses on algorithms for the TSP.

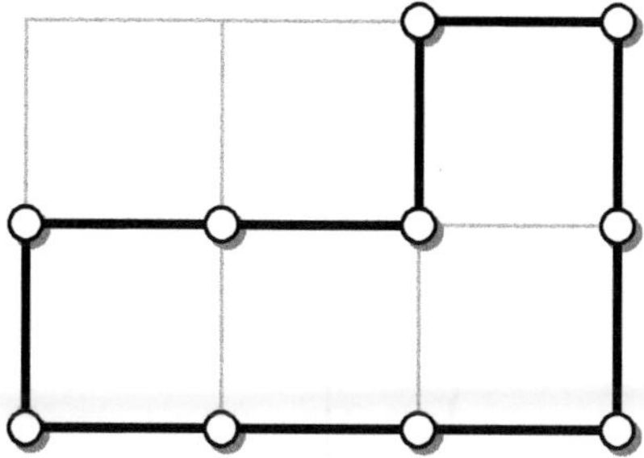

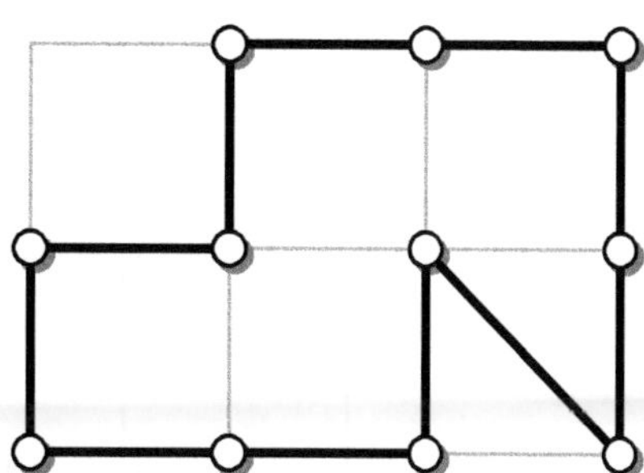

FIGURE 9.51: EUCLIDEAN TSP on grids: The shortest tour through n cities has length n if the underlying grid graph has a Hamiltonian cycle (left). If it has no Hamiltonian cycle, the shortest tour is strictly longer than n because of the diagonal moves (right).

9.5 Hardness of the EUCLIDEAN TSP. One way to prove that EUCLIDEAN TSP is NP-hard is to reduce from HAMILTONIAN CYCLE on *grid graphs*—subgraphs of the integer lattice $\mathbb{Z}^2$ induced by some finite set of vertices. Two vertices of a grid graph are connected by an edge only if they are nearest neighbors on the grid as shown in Figure 9.51. If a grid graph with n vertices has a Hamiltonian cycle, this cycle corresponds to a tour of length n in the Euclidean plane. If it has no Hamiltonian cycle then any tour has to use diagonal links of length greater than one, so the shortest tour has length strictly larger than n. Hence an algorithm for EUCLIDEAN TSP can used to solve HAMILTONIAN CYCLE on grid graphs, but the latter is NP-complete [422].

Note that this reduction also shows that TSP is NP-hard for distances based on the "Manhattan metric," $d_{ij} = |x_i - x_j| + |y_i - y_j|$. The original proof of the hardness of EUCLIDEAN TSP employed a more complicated reduction from SET COVER [643, 318].

9.6 Approximating the TSP. Insertion heuristics for ΔTSP were analysed by Rosenkrantz, Stearns, and Lewis [715]. They proved that "nearest insertion" as defined in Problem 9.11, as well as "cheapest insertion" where you select the city that can be inserted at minimum cost, are both 2-approximations. These methods correspond to Prim's algorithm for MINIMUM SPANNING TREE. Similarly, insertion heuristics that handle several partial tours and merge them to form larger tours correspond to Kruskal's algorithm, and form another class of 2-approximations for ΔTSP.

Christofides' algorithm was invented in 1976 by Nicos Christofides [166]. As we state in the text, it is still the best known approximation when the only restriction on the distances d_{ij} is the triangle inequality. However, a recent breakthrough by Gharan, Saberi, and Singh [327] shows that the ratio 3/2 can be improved for *graphical* metrics, i.e., those that can be described as shortest-path distances in a graph. The family of instances of EUCLIDEAN TSP shown in Problem 9.10, for which the greedy algorithm yields a tour $\Theta(\log n)$ longer than the optimal one, is taken from [414].

9.7 Gated communities: a PTAS for EUCLIDEAN TSP. In 1996, Sanjeev Arora and Joseph Mitchell independently found polynomial-time approximation schemes for EUCLIDEAN TSP [52, 587], for which they jointly received the Gödel Prize in 2010. Both their algorithms recursively divide the problem into weakly interacting subproblems, and then solve these subproblems using dynamic programming. We sketch Arora's algorithm here. The division is geometric: as shown in Figure 9.52, the square that contains all n cities is divided into 4 squares, each of these is divided into 4 smaller squares, and so on until each square contains no more than one city.

The interaction between a pair of adjacent squares at the same level of this construction is controlled by a set of "portals" on the boundaries between them. A "well-behaved" tour is one that crosses the boundary between two squares only through these portals. While the shortest well-behaved tour is longer than the optimal tour, we can make this discrepancy arbitrarily small by increasing the number of portals. Specifically, for any $\varepsilon > 0$, the shortest

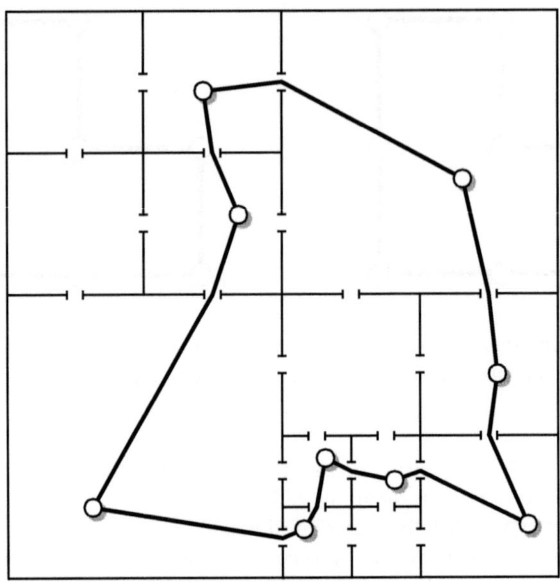

FIGURE 9.52: Recursive partitioning of the plane into squares used in the PTAS for EUCLIDEAN TSP.

well-behaved tour is at most $(1+\varepsilon)$ longer than the shortest tour if the number of portals between each pair of squares is $m = O(\log n/\varepsilon)$.

The triangle inequality implies (exercise!) that the shortest well-behaved tour uses each portal at most 3 times. Hence the total number of times the tour crosses a given boundary is $O(3^m) = O(n^{1/\varepsilon})$. Starting with the smallest squares, the shortest well-behaved tour can then be constructed in $O(n^{1/\varepsilon})$ time by dynamic programming. This running time can be made almost linear, $O(n(\log n)^{1/\varepsilon})$. However, the dependence on $1/\varepsilon$ is still exponential, so this is not an FPTAS.

9.8 Sometimes planarity helps. Many NP-hard problems on graphs remain NP-hard on planar graphs. For instance, we saw in Section 4.2.1 that GRAPH 3-COLORING is just as hard in the planar case as it is for general graphs. Where approximability is concerned, however, planarity often makes a big difference.

For example, MIN VERTEX COVER has a PTAS on planar graphs. The reason is that planar graphs cannot have "too many" edges: by Euler's formula, $|E| \leq 2(|V|-3)$. This lets us decompose the graph recursively into subgraphs that interact only through very few shared vertices, and to organize a dynamic programming algorithm around this decomposition [73]. Similarly, when the distances between cities are given by the shortest paths in a planar graph, a decomposition into weakly interacting subgraphs constitutes the core of a PTAS for this case of the TSP [55].

9.9 Approximation limited. Most known optimal inapproximability results rely on the PCP Theorem, which we will meet, and partly prove, in Section 11.3. The fact that it is NP-hard to approximate MAX-3-XORSAT within a factor better than $\rho = 1/2$ is an immediate consequence of Theorem 11.4. This strong version of the PCP Theorem, due to Johan Håstad [378], essentially states that proofs or witnesses of polynomial size can be written "holographically," in the sense that we can reject a false proof with probability at least $1/2$ by looking at just three of its bits. Read Section 11.3 and Note 11.4 to gain enlightenment about this seemingly magical result.

The proof that TSP cannot be approximated within any constant factor is due to Sahni and Gonzales [721]. This is a piece of cake compared to proving approximability bounds for ΔTSP: by reducing from GAP-XORSAT, Papadimitriou and Vempala [649] showed that approximating ΔTSP within a ratio of $\rho = 117/116 \approx 1.009$ is NP-hard, and Karpinski, Lampis, and Schmied [467] improved this to $75/74 \approx 1.014$. It is generally believed that the Held–Karp relaxation achieves an approximation ratio of $\rho = 4/3$ for ΔTSP.

Another source for strong inapproximability results is the Unique Games Conjecture (see Note 10.4), which states that approximating a certain kind of coloring problem within a certain gap is NP-hard. If this conjecture is true, it provides another starting point from which inapproximability results can be derived by gap-preserving reductions. It would imply, for example, that approximating MIN VERTEX COVER within any factor less than 2—the factor achieved by the pebble game and the LP relaxation—is NP-hard.

Finally, we note that in the original story, Baron Münchhausen pulled himself up by his hair rather than his bootstraps.

9.10 Linear programming. Systems of linear inequalities have been studied since the 19th century, most notably by Fourier in 1826 [288]. The idea of casting planning problems as linear programs was first published by the Russian mathematician Leonid Kantorovich in 1939, but his paper [452] was ignored. This changed in the late 1950s, after linear programming had been established as a major field of mathematical economics. Kantorovich shared the 1975 Nobel Prize in economics with T. C. Koopmans, who coined the term "linear programming."

George B. Dantzig, who independently discovered linear programming in 1947 [214], was not included as a Nobel laureate. But he received his share of fame as the inventor of the simplex algorithm, the first method to systematically solve linear programs. Fueled by the increasing availability of the first digital computers, the simplex algorithm sparked a whole new industry of modeling and solving real-world problems as linear programs, and this industry has been very active ever since then. In 2000, the simplex method was listed among the ten algorithms with the greatest influence on the development and practice of science and engineering in the 20th century [622].

According to [669], "the subject of linear programming is surrounded by notational and terminological thickets. Both of these thorny defenses are lovingly cultivated by a coterie of stern acolytes who have devoted themselves to the field." We couldn't agree more. Our exposition is intended to give a painless introduction to the key ideas without explaining how to handle the many exceptions and special cases that can occur. We owe a great deal to the books of Dasgupta, Papadimitrioiu and Vazirani [215], Bertsimas and Tsitsiklis [109], and Matou š ek and Gärtner[561], and an expository article by Clausen [178]. The battery-charging example was taken from [130].

9.11 Lean polytopes: the Hirsch Conjecture. One way to understand the suprisingly good performance of the simplex method is to consider the topological distance between the vertices of a polytope P—that is, the number of steps in the shortest path between the two vertices along the edges of P. The *diameter* of P is the maximum distance over all pairs of vertices. If our pivot rule were perfectly clairvoyant, the simplex method would never take more steps than the diameter of P.

This is where the Hirsch conjecture comes in. Let $\Delta(d, m)$ denote the maximum diameter of a d-dimensional polytope with at most m facets for $m > d \geq 2$. In 1954, Warren Hirsch claimed that

$$\Delta(d, m) \leq m - d\,. \tag{9.58}$$

If this conjecture were true, every vertex on the polytope would be at most $m - d$ steps away from the optimum. Of course, we would still need to find this path, but it would be possible in principle for some pivot rule to succeed in this many steps.

As Problem 9.13 shows, the Hirsch conjecture is true for 0-1 polytopes, i.e., for polytopes whose vertices are corners of the hypercube $\{0, 1\}^d$. It is also true for arbitrary polytopes in dimension $d \leq 3$.

The Hirsch conjecture was open for decades, but in 2010, Francisco Santos found a counterexample [725]. His counterexample has $d = 43$ dimensions and $m = 86$ facets, and its diameter is $86 - 43 + 1$. Although this disproves the Hirsch conjecture, the algorithmically relevant question of whether the diameter of a polytope is bounded by a polynomial function of m and d is still open. The sharpest result so far is due to Kalai and Kleitman [451], who proved the subexponential bound $\Delta(d, n) < n^{\log_2 d + 2}$.

The first proof of (9.59) in Problem 9.13 was given by Naddef [620]. Our solution follows the proof given in Ziegler's book *Lectures on Polytopes* [848], who also added the fact that equality in (9.59) only holds on the hypercube. Ziegler is an excellent source on the Hirsch conjecture and polytopes in general.

9.12 Worst-case instances for simplex. The Klee–Minty cube is named after Victor Klee and George Minty [486], and was the first exponential-time instance for a simplex algorithm. While the lowest-index rule may seem artificial, similar pivot rules such as Bland's anti-cycling rule are often used in practice.

Shortly after Klee and Minty's work, Jeroslow [430] found a corresponding instance for the greedy pivot rule. Since then, exponential-time instances have been found for many other pivot rules. See Amenta and Ziegler [45] for a general framework for constructing them. It is much harder to construct hard instances for randomized pivot rules, but recently Friedmann, Hansen, and Zwick [299] constructed instances for which simple randomized rules take superpolynomial time on average.

9.13 Smoothed analysis. The idea of smoothed analysis was introduced in 2004 by Spielman and Teng [770]. They received the 2008 Gödel Prize and the 2009 Fulkerson Prize for their work, and in 2010 Spielman won the Nevanlinna prize as well. Since then, smoothed analysis has been extended and applied to a number of other algorithms and problems. See [771] for a survey.

Earlier, researchers such as Borgwardt [127] and Smale [761] had proved that the simplex algorithm runs in polynomial time on average. However, these results hold for completely random instances of LINEAR PROGRAMMING, such as those whose constraints are chosen from a spherically symmetric distribution.

9.14 More on duality. We derived the dual of a linear program in terms of the forces on a balloon. Another derivation uses an economic metaphor. Suppose I want you to find the optimal solution $\mathbf{x}$ to the primal problem. However, we live in a market society rather than a dictatorship, so rather than forcing you to satisfy the inequalities $\mathbf{Ax} \le \mathbf{b}$, I will give you an incentive to do so. Namely, for each i, I charge you a price $y_i \ge 0$ per unit for violating the ith constraint $\sum_j A_{ij} x_j \le b_i$. Thus in addition to gaining the value $\mathbf{c}^T\mathbf{x}$, you pay a total price of

$$\sum_i y_i \left(\sum_j A_{ij} x_j - b_i \right) = \mathbf{y}^T(\mathbf{Ax} - \mathbf{b}),$$

Note that this price can be negative, so you can make a profit if some constraints hold with room to spare.

This leaves you with a maximization problem, in which you wish to find

$$\max_{\mathbf{x}} \left(\mathbf{c}^T\mathbf{x} - \mathbf{y}^T(\mathbf{Ax} - \mathbf{b}) \right), \tag{9.80}$$

subject only to the positivity constraints $\mathbf{x} \ge 0$. You can avoid paying a price, and even make an additional profit, by staying within the feasible region $\mathbf{Ax} \le \mathbf{b}$. Therefore, the optimum of this relaxed problem is greater than or equal to the optimum of the original primal problem.

As the price-setter, my goal is to set $\mathbf{y}$ so that the maximum (9.80) is minimized. To do this, I need to set the prices y_i high enough so that it is not worth it for you to violate the constraints. In particular, the total profit you gain by moving out along the x_j direction, say, must be zero or negative,

$$c_j - \sum_i y_i A_{ij} \le 0. \tag{9.81}$$

Requiring this for every j gives the dual constraints we know and love,

$$\mathbf{y}^T\mathbf{A} \ge \mathbf{c}^T.$$

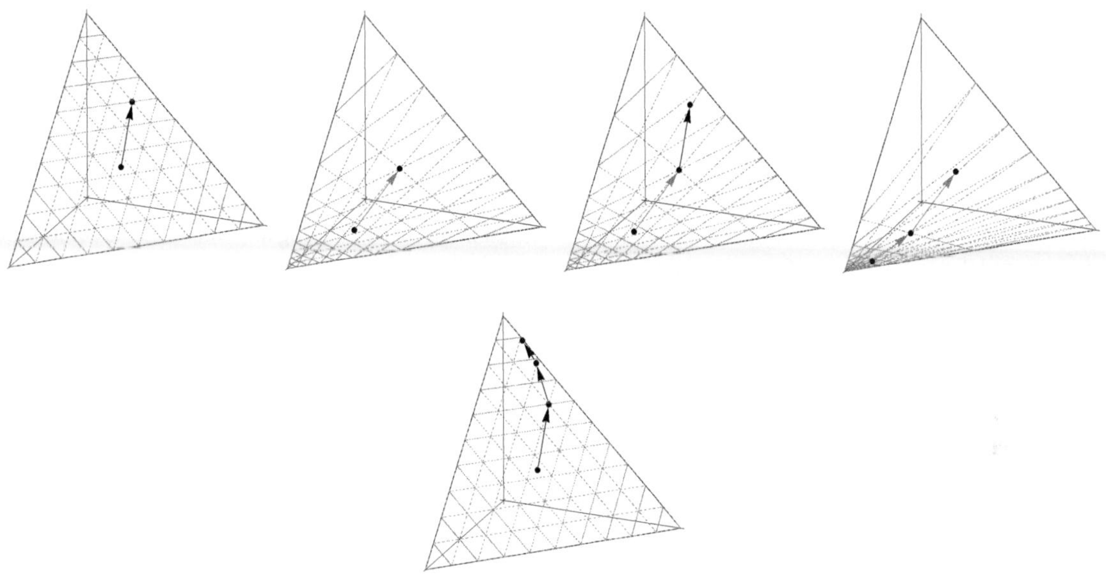

FIGURE 9.53: Karmarkar's algorithm starts at the center of the polytope and takes a step in the direction **c**. It then applies a projective transformation that maps its position back to the center, takes another step, projects again, and so on (top, left to right). In the original coordinates (bottom) it converges rapidly to the optimal vertex.

Now, rewriting (9.80) as

$$\mathbf{y}^T\mathbf{b} + \max_{\mathbf{x}} \left((\mathbf{c}^T - \mathbf{y}^T\mathbf{A})\mathbf{x} \right), \tag{9.82}$$

we see that you are forced to set $x_j = 0$ for each j such that (9.81) holds as a strict inequality. Thus the maximum value you can obtain for the relaxed problem is simply $\mathbf{y}^T\mathbf{b}$, and my price-setting problem is exactly the dual problem—that is, minimizing $\mathbf{y}^T\mathbf{b}$ subject to the constraints $\mathbf{y}^T\mathbf{A} \geq \mathbf{c}^T$ and $\mathbf{y} \geq 0$.

For those familiar with the method of Lagrange multipliers for constrained optimization problems, we offer one more picture of duality. The dual optimum $\mathbf{y}$ consists of the Lagrange multipliers for the tight constraints, where $y_j = 0$ if the jth constraint is not tight. For this reason (9.80) is also called the Lagrangean relaxation.

The concept of duality was introduced by John von Neumann in 1928 [814] through his famous minimax theorem, which we discuss in the context of randomized algorithms in Section 10.5.2. In a conversation in 1947, von Neumann and Dantzig discovered that the game-theoretic duality is closely related to the duality of linear programs; see Problem 10.15. The first published proof of strong duality was given by Gale, Kuhn, and Tucker [311] in 1951.

The dual description of SHORTEST PATH that we asked for in Problem 9.32 is from Minty [585]. We learned about it from Dasgupta, Papadimitriou, and Vazirani [215].

9.15 Interior point methods. Interior point methods were popular for nonlinear optimization in the 1960s, primarily using barrier functions like our potential energy V to repel the solution from the boundary of the feasible region. By the early 1980s, barrier methods had gone out of fashion, but interest in them was revived in 1984 when Narendra Karmarkar [459] published a polynomial-time interior point method for LINEAR PROGRAMMING.

Unlike the ellipsoid method, Karmarkar's algorithm is efficient in both theory and practice. In its original form, it starts at the center of the polytope and takes a large step in the direction of **c**. It then applies a *projective transforma-*

tion, mapping its current position back to the center of the polytope as shown in Figure 9.53. It then takes another step, projects back to the center, and so on. Projecting back to the center ensures that it has room to make a large jump, so that in the original coordinates it converges rapidly to the optimum. Essentially, the projective transformations push away the facets that get too close, so that it has plenty of room to move.

Shortly after Karmarkar's seminal paper, it became clear that his algorithm is equivalent to classical barrier methods, including the Newton flow; see Gill et al. [331] and Bayer and Lagarias [87]. Our presentation of the Newton flow is inspired by Anstreicher [47]. For a complete discussion of the Newton flow approach to LINEAR PROGRAMMING, including the proof of polynomial complexity and discussion on how to finding an initial feasible solution, we recommend Renegar [701]. See Gonzaga [345] for a nice presentation of Karmarkar's original algorithm.

9.16 The ellipsoid method. The ellipsoid method was introduced as an algorithm for convex optimization by N. Z. Shor [749] and Nemirovsky and Yudin [630] in the 1970s, but it was Leonid Khachiyan in 1979 [477] who showed that it solves LINEAR PROGRAMMING in polynomial time. Although it is far less efficient than Karmarkar's algorithm and other interior point methods, Khachiyan's result proved for the first time that LINEAR PROGRAMMING—and, as we describe in the text, any convex optimization problem with a polynomial-time separation oracle—is in P.

In our discussion we reduced optimization problems to FEASIBILITY using binary search. However, there is a more direct method. Suppose we trying to minimize a convex function f inside a convex set. Once we have an ellipsoid whose center $\mathbf{r}$ is feasible, we cut it along the hyperplane defined in (9.31), $\{\mathbf{x} : (\nabla f)^T \mathbf{x} = (\nabla f)^T \mathbf{r}\}$, which passes through $\mathbf{r}$ and is tangent to the contour surface of f. We then use (9.32) to compute a smaller ellipsoid that contains the minimum, and so on.

9.17 Semidefinite programming. SEMIDEFINITE PROGRAMMING may seem like a rather special convex optimization problem, but it includes many important optimization problems as special cases. It contains LINEAR PROGRAMMING since $\mathbf{x} \geq \mathbf{0}$ can be written as $\text{diag}(\mathbf{x}) \succeq 0$ where $\text{diag}(x_1, \ldots, x_d)$ denotes the diagonal matrix with entries $x_1, \ldots, x_d$. It also contains convex QUADRATIC PROGRAMMING, since the constraint

$$(\mathbf{Ax}+\mathbf{b})^T(\mathbf{Ax}+\mathbf{b}) - \mathbf{c}^T\mathbf{x} - \mathbf{d} \leq \mathbf{0}$$

can be written as

$$\begin{pmatrix} \mathbb{1} & \mathbf{Ax}+\mathbf{b} \\ (\mathbf{Ax}+\mathbf{b})^T & \mathbf{c}^T\mathbf{x}+\mathbf{d} \end{pmatrix} \succeq 0 .$$

The literature on SEMIDEFINITE PROGRAMMING gives many applications; see e.g. [807, 295] and Problem 9.44. A number of NP-hard optimization problems have convex relaxations that are semidefinite programs, and rounding these relaxations yields good approximation algorithms. The canonical example is MAX CUT, for which we give a randomized approximation algorithm in Section 10.4.

9.18 Unimodularity. In Section 9.8.3 we showed that the vertices of a polytope are integral if the constraints are given by a totally unimodular matrix. Conversely, an integer matrix $\mathbf{A}$ is totally unimodular if and only if the vertices of the polytope $P = \{\mathbf{x} : \mathbf{Ax} \leq \mathbf{b}, \mathbf{x} \geq 0\}$ are integral for any integer vector $\mathbf{b}$. Note that even if $\mathbf{A}$ is not totally unimodular, P can still have *some* integral vertices for *some* vectors $\mathbf{b}$.

The original proof of the relation between unimodularity and integrality goes back to Hoffman and Kruskal [406]. A simpler proof by Veinott and Dantzig [811] can be found in [110].

The fact that solutions to the LP relaxation of VERTEX COVER are half-integral, and the theorem of Problem 9.29, are from Nemhauser and Trotter [629].

9.19 Solving large instances of TRAVELING SALESMAN. The 42-city instance that we used throughout Section 9.10 is a historic example—it was the first instance of TSP that was solved using LP relaxations. In 1954, the year that Dantzig, Fulkerson and Johnson solved it [213], 42 cities was considered a large-scale instance. The actual distance matrix can be found online under the name `dantzig42.tsp` as part of the TSPLIB, a library of sample instances for the TSP. Ask your favorite search engine.

If you run your own experiments on `dantzig42.tsp` you should note that the file contains distances scaled in a curious way, $(d_{ij} - 11)/17$. Dantzig et al. wanted each distance to range from 0 to 255 so it would fit into a single byte, despite the fact that they performed their calculations by hand. We used the original road distances in miles, rounded to the nearest integer.

The size of the largest solved instance of TSP has grown dramatically since then. The current record is the shortest tour that connects 85 900 locations in a VLSI application. See www.tsp.gatech.edu for current and past milestones in TSP-solving, including software packages and many more exciting facts about TSP.

9.20 Exclusive cycles. In Problems 9.33 and 9.36 we learned that finding minimum-weight cycle covers is easy if we allow cycles of any length, but becomes hard if we are restricted to k-cycles for any $k \geq 3$.

In a complementary setting we are allowed to have cycles of various lengths $L \subseteq \{3, 4, 5, \ldots\}$. The corresponding existence problem is known as L-CYCLE COVER. Surprisingly, the complexity of L-CYCLE COVER depends crucially on the lengths that are *excluded* from L: specifically, it is NP-hard if there is any $\ell > 4$ such that $\ell \notin L$. This result, which was proved by Hell, Kirkpatrick, Kratochvíl, and Kříž [394], implies that optimizing a traveling sales force becomes NP-hard if, for example, we exclude cycles of length 13.

9.21 The Lovász theta function and perfect graphs. Lovász first introduced the function $\vartheta(G)$ in [540] as an upper bound for the capacity of an error-correcting code based on a graph. It is often defined in terms of the complement graph, $\vartheta(\overline{G})$, in which case it is bounded below by the size of the largest independent set (i.e., the size of the largest clique of $\overline{G}$).

The theta function has many equivalent definitions, including one given in terms of a kind of vector coloring of the graph where the vectors on neighboring vertices are required to be orthogonal; see Knuth [494] for a review. Our proof in Problem 9.44 is from Lovász's book [541].

As Problem 5.16 showed, it's NP-complete to tell whether $\omega(G) = \chi(G)$. Graphs with the stronger property that $\omega(G') = \chi(G')$ for any induced subgraph G', i.e., any subgraph defined by a subset of the vertices, are called *perfect*. Lovász [538] proved that G is perfect if and only if $\overline{G}$ is perfect. Along with Problem 9.44, this shows that for perfect graphs, the problems GRAPH COLORING, CLIQUE, VERTEX COVER, and INDEPENDENT SET are all in P.

Many special classes of graphs are perfect. These include bipartite graphs, giving an alternate proof of Problem 4.16; chordal graphs, for which any cycle has an internal edge connecting two of its vertices so that its largest induced cycle is a triangle; and the interval graphs of Problem 3.27.

An induced subgraph that is a cycle of odd length 5 or greater is called a *hole*. No perfect graph can contain a hole, since a hole has clique number 2 but chromatic number 3. An induced subgraph that is the complement of a hole is called an *antihole*. In 1960, Claude Berge conjectured that a graph is perfect if and only if it contains no holes or antiholes, and such graphs are called *Berge graphs* in his honor. In 2002, just before his death, this conjecture was proved by Maria Chudnovsky, Neil Robertson, Paul Seymour, and Robin Thomas [168], who shared the 2009 Fulkerson prize for their work. In addition, Chudnovsky et al. gave a polynomial-time algorithm for determining whether a given graph is perfect [167].

In Problem 9.45, we showed that if the clique polytope $\text{conv}(\mathscr{C})$ coincides with $\mathscr{C}_{\text{ind}}$ then $\omega(G) = \chi(G)$. More to the point, Fulkerson [303] and Chvátal [175] showed that these two polytopes coincide if and only if G is perfect.

Chapter 10

Randomized Algorithms

—O! many a shaft, at random sent,
Finds mark the archer little meant!
And many a word, at random spoken,
May soothe or wound a heart that's broken!

Sir Walter Scott, *Lord of the Isles*

Whether you are a mouse escaping from a cat, a cryptographer evading a spy, or an algorithm designer solving hard problems, the best strategy is often a random one. If you commit to a deterministic strategy, the adversary can exploit it—in algorithmic terms, he can choose an instance on which your algorithm will perform as poorly as possible. But if you make unpredictable choices, you may be able to keep him off balance, and do well with high probability regardless of what instance he throws at you.

In this chapter, we get a taste of the variety and power of randomized algorithms. We will meet algorithms that find the smallest cut in a graph by merging random vertices until only two remain; that find satisfying assignments by wandering randomly through the space of possible solutions; that solve problems by raising them up to a high-dimensional space, and then casting their shadow back down along a random direction; and that play games by searching the tree of possible moves in random order. In many cases, these algorithms are as good or better than the best known deterministic algorithms, and they are often simpler and more beautiful.

Along the way, we will learn how to check that the software on a space probe is uncorrupted, even if it is too far away to exchange more than a few bits with us, how to change a puzzle with many solutions into one where the solution is unique, and how to confirm that two functions are equal by sampling them at a few random places. We will develop tools like random hash functions and polynomial identity testing, which will be important ingredients of interactive proofs in Chapter 11.

We will conclude by delving more deeply into the nature of the primes. We will see a series of randomized algorithms for PRIMALITY based on different number-theoretic ideas. Finally, we will sketch the deterministic algorithm, found in 2004, which establishes that PRIMALITY is in P. 10.1

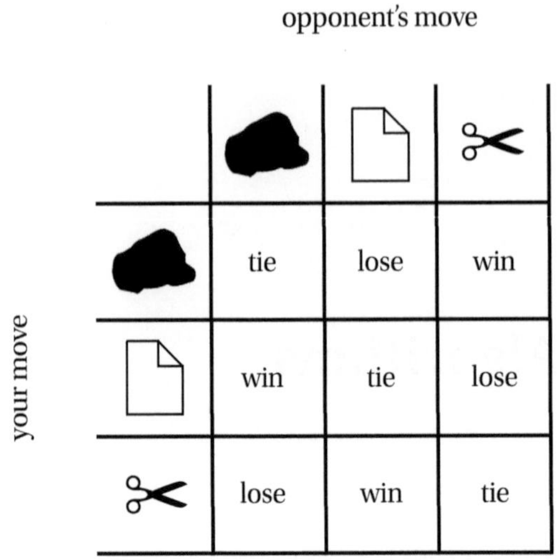

opponent's move

your move

	tie	lose	win
	win	tie	lose
	lose	win	tie

FIGURE 10.1: The classic game of Rock, Paper, Scissors (or Jan-Ken-Pon, or Ca-Chi-Pun, or Ching Chong Cha, or Kawi Bawi Bo...)

10.1 Foiling the Adversary

> Those who are skilled in producing surprises will win. Such tacticians are as versatile as the changes in heaven and earth.
>
> Sun Tzu, *The Art of War*

Readers from many cultures will recognize the game shown in Figure 10.1. When playing this game with a friend, your goal is to predict her moves as accurately as possible, while making your own moves as unpredictable as possible.

Now imagine that we are playing against our old bête noire, the adversary. He possesses enormous computational power, and complete understanding of whatever strategy we plan to use. If we commit to any deterministic strategy, no matter how sophisticated, he can predict our moves and win every time. Our only recourse is to play randomly, in which we choose from the three possible moves with equal probability. If we have a source of genuinely random bits—a Geiger counter, say, or a fair die—then the adversary cannot predict our moves no matter how intelligent he is.

Consider the analogy to computation. If we commit ourselves to a deterministic algorithm, the adversary can choose the instance on which that algorithm performs as poorly as possible. But we can keep the adversary off guard by using an algorithm that flips coins and makes random choices. Even if the adversary knows what algorithm we are using, he doesn't know which way these coin flips will come up. No matter how computationally powerful he is, he cannot predict what actions our algorithm will take, and he cannot construct a hard instance in advance.

To see this concretely, let's revisit the classic algorithm `Quicksort` from Chapter 3. We sort a list by choosing a pivot element, separating the elements into two sublists according to whether they are less than or greater than this pivot, and then recursively sorting these sublists. The average running time of this algorithm is $\Theta(n \log n)$. However, as we discussed in Section 3.2.1, the word "average" here can have two very different meanings.

First, "average" could mean that the algorithm is deterministic but the instance is random, so that all $n!$ permutations are equally likely. However, this is naive in a world where the instance is chosen by an adversary. If we choose the pivot deterministically, the adversary can choose an order for the instance such that the pivot is always the smallest or largest element of the list. For instance, if we use the first element as the pivot, the adversary can give us an instance that is already sorted, or that is sorted in reverse order. In that case, `Quicksort` takes $\Theta(n^2)$ time instead of $\Theta(n \log n)$.

In fact, most real-world instances of sorting are already mostly sorted, so this poor performance can happen even without the adversary's intervention. For better or worse, real instances are very far from random—they have rich statistical structure that may cause a given algorithm to perform much better, or much worse, than it would on random instances.

On the other hand, "average" could mean that the adversary gets to choose the instance, but that the *algorithm* is random, and that we average over all possible sequences of choices the algorithm could make. In randomized `Quicksort`, we choose the pivot randomly at each step, and the average running time is $\Theta(n \log n)$ no matter what order the adversary puts the instance in. Indeed, as far as the algorithm is concerned, every instance looks just like a random one.

Of course, we might still choose bad pivots at each step, and again take $\Theta(n^2)$ time. However, instead of happening deterministically for a few instances, this happens with small probability on every instance. Our algorithm can still perform poorly, but only if we are very unlucky.

To take another example, suppose we are trying to estimate the integral $\int_0^1 f(x)\,\mathrm{d}x$ of some function $f(x)$, or equivalently, its average in the unit interval. The standard way to do this is to sample the function at a series of points $x_1, \ldots, x_n$, and use the average of these samples as our estimate:

$$\mathbb{E}[f] = \int_0^1 f(x)\,\mathrm{d}x \approx \frac{1}{n} \sum_{i=1}^{n} f(x_i).$$

For most functions, sampling $f(x)$ at a periodic set of points such as $x = 1/n, 2/n, \ldots$ would give perfectly good answers. But if the adversary challenges us to estimate the average of $f(x) = \sin^2 n\pi x$, this method gives 0 rather than the correct value $1/2$.

We could spend a great deal of time trying to design a set of sample points that avoids every kind of periodicity we expect to encounter. But if the adversary knows what points we plan to use, he can always choose a function that is zero at those points but has a nonzero average. On the other hand, if we choose the sample points randomly, then as long as f is bounded, our estimate will converge to $\mathbb{E}[f]$ as $n \to \infty$ unless we are extremely unlucky and most of our samples fall in atypical places.

In both these cases, by adopting a randomized algorithm instead of a deterministic one, we can trade a small number of bad instances for a small probability of bad performance. The deterministic algorithm works well on most instances, while the randomized algorithm works well, for all instances, most of the time. In a sense, randomization "spreads out" the badness as shown in Figure 10.2. Rather than throwing bad instances at us, the adversary can only gnash his teeth and hope that we get unlucky.

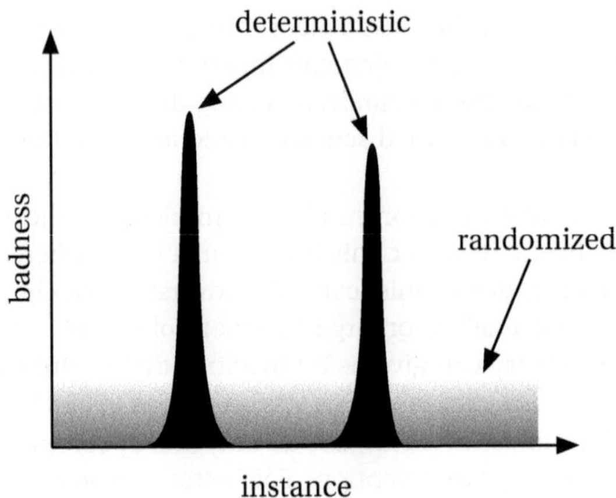

FIGURE 10.2: A cartoon of how randomized algorithms spread out the badness. For a deterministic algorithm, there are a few bad instances, but the adversary gets to choose the instance. For a randomized algorithm, there is a small chance of bad performance on any given instance, but for all instances the likely performance is good.

In addition, both these algorithms are *simple*. The best possible pivot for `Quicksort` is the median, which would take a while to find, but a random pivot does just fine. It's hard to design a deterministic set of sample points that is far from periodic in an appropriate sense, but a random set of points possesses this property with high probability. We will see below that naive choices like these work surprisingly well for many problems.

10.2 The Smallest Cut

What is life but a series of inspired follies?
George Bernard Shaw, *Pygmalion*

In this and the next two sections, we will look at randomized algorithms for three classic problems: MIN CUT, k-SAT, and MAX CUT. These algorithms are very different from each other, and each one illustrates a simple but beautiful way in which randomness can help us find optimal or near-optimal solutions.

There are many settings in which we want to measure the reliability, or vulnerability, of a network. Given a graph G, we call a subset C of its edges a *cut* if cutting them would cause it to fall apart into two pieces. Then one natural measure of G's reliability is given by the following problem:

> MIN CUT
> Input: A graph $G = (V, E)$
> Output: The size of the smallest cut

```
Karger Min-Cut
input: a graph G = (V, E)
output: a cut C ⊆ E
begin
    repeat
        choose a random edge e ;
        contract e and merge its endpoints into a single vertex ;
    until there are only two vertices a, b left;
    let C be the set of edges between a and b ;
    return C ;
end
```

FIGURE 10.3: Karger's randomized algorithm for MIN CUT.

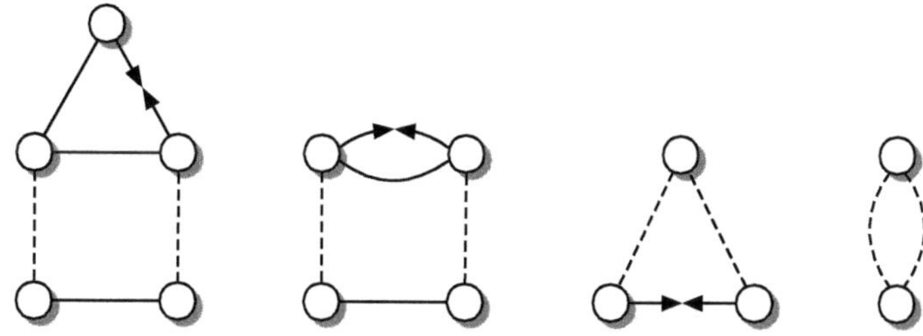

FIGURE 10.4: After three contractions, a particular minimum cut $C_{\min}$ consisting of the two dashed edges survives. What is the probability that this occurs?

Note that this differs from the version of MIN CUT we discussed in Section 3.7 in several ways. First, we no longer demand that the cut separate two specific vertices s and t from each other—we simply ask that it divides the graph into two parts. Second, here G is undirected and its edges all have equal weight, so the size of the cut is just the number of edges that cross from one piece to the other.

One way to solve this version of MIN CUT would be to choose a fixed s, find the minimum s-t cut for each vertex t, and take the minimum over t. We can do this using the duality between MAX FLOW and s-t MIN CUT and our favorite polynomial-time algorithm for MAX FLOW. Indeed, generalizations of this duality give us polynomial-time algorithms for many problems of this kind.

However, we focus here on an elegant randomized algorithm invented by David Karger. As shown in Figure 10.3, at each step we choose a random edge (u, v) and contract it, merging its endpoints u, v into a single vertex. This vertex inherits all of u's and v's edges, other than the one between them.

As the graph shrinks, it becomes a multigraph, where some pairs of vertices have multiple edges between them. We give each of these edges an equal probability of being chosen for contraction. For instance, in the second part of Figure 10.4, the probability that we merge the top two vertices is $2/5$, since there are two edges between them.

We repeat this process until only two vertices a, b are left. At that point, let C be the set of surviving edges connecting these two vertices. Clearly C is a cut, since it separates G into the two pieces that contracted to form a and b respectively. What is surprising is that, with reasonably high probability, C is as small as possible.

To see this, suppose that the smallest cut $C_{\min}$ has size k. If the smallest cut is not unique, let $C_{\min}$ denote a particular one of the smallest cuts. Then the algorithm finds $C_{\min}$ if and only if its k edges survive until only two vertices remain, as shown in Figure 10.4 for a particular $C_{\min}$ of size $k=2$. What is the probability that none of the edges in $C_{\min}$ ever get contracted?

Consider the step of the algorithm when there are t vertices left. Each vertex must have at least k edges, since otherwise cutting its edges would give a cut of size less than k. Thus there are at least $tk/2$ edges left, and the probability that none of the k edges in $C_{\min}$ are chosen for contraction on that step is

$$1-\frac{k}{\#\text{ of edges}} \geq 1-\frac{k}{tk/2}=\frac{t-2}{t}.$$

If G has n vertices to start with, the probability p that $C_{\min}$ survives until the end is the product of this probability over all $n-2$ steps. This gives the lower bound

$$p \geq \prod_{t=3}^{n} \frac{t-2}{t}=\frac{1\cdot 2\cdot 3\cdots(n-2)}{3\cdot 4\cdot 5\cdots n}=\frac{2}{n(n-1)}. \tag{10.1}$$

Thus the probability of success is $\Omega(1/n^2)$. Note that if the minimum cut is not unique, this lower bound applies to the probability that *any particular one* survives.

Exercise 10.1 *Show that a connected graph with n vertices can have at most* $n(n-1)/2=\binom{n}{2}$ *minimum cuts. Can you think of a graph that has exactly this many?*

Succeeding with probability $\Omega(1/n^2)$ may not seem like a great achievement. However, we can increase the probability of success by trying the algorithm multiple times. Since each attempt is independent, if we make $1/p=O(n^2)$ attempts, the probability that none of them finds $C_{\min}$ is

$$(1-p)^{1/p} \leq 1/\mathrm{e},$$

where we used the inequality $1-x \leq \mathrm{e}^{-x}$. Thus the probability that at least one of these $1/p$ attempts succeeds is at least $1-1/\mathrm{e} \approx 0.63$. This gives an algorithm that succeeds with constant probability, i.e., with probability $\Omega(1)$.

If we want an algorithm that succeeds with *high* probability, i.e., with probability $1-o(1)$, we can make a somewhat larger number of attempts. If we try the algorithm $(1/p)\ln n = O(n^2 \log n)$ times, say, the probability that every attempt fails is

$$(1-p)^{(1/p)\ln n} \leq \mathrm{e}^{-\ln n}=1/n=o(1).$$

Thus we can raise the probability of success from $\Omega(1)$ to $1-O(1/n)$ with just a $\log n$ increase in the running time. Increasing the number of attempts by a factor of n makes the probability of failure exponentially small,

$$(1-p)^{n/p} \leq \mathrm{e}^{-n},$$

bringing the probability of success exponentially close to 1.

This "boosting" technique is a simple but important theme in randomized algorithms. If each attempt succeeds with probability p, the average number of attempts before we succeed is $1/p$ (see Exercise A.15 in Appendix A.3.4). We can succeed with constant probability by making $1/p$ attempts, and with high probability by making a few more. If each attempt takes polynomial time and $p = 1/\text{poly}(n)$, the total running time is poly(n). In this case, we get a total running time comparable to the best known deterministic algorithms.

10.2

Exercise 10.2 *Show how to use Karger's algorithm to find the* MIN CUT *in a weighted graph where edges have nonnegative integer weights, as long as these weights are* poly(n). *What can we do if the weights are exponentially large?*

10.3 The Satisfied Drunkard: WalkSAT

Keep on going and the chances are you will stumble on something, perhaps when you are least expecting it. I have never heard of anyone stumbling on something sitting down.

Charles F. Kettering

For our next example of a randomized algorithm, let's return to the canonical NP-complete problem: 3-SAT. We can always tell if a 3-SAT formula is satisfiable by performing an exhaustive search. But another approach is to take a random walk in the space of truth assignments, starting in a random place and then flipping variables until we stumble on a solution. This gives a randomized algorithm for 3-SAT that is astonishingly simple. And, while it takes exponential time, it is one of the fastest algorithms known.

Here's the idea. At any point in time, our current truth assignment will satisfy some clauses and dissatisfy others. We simply choose one of the unsatisfied clauses and satisfy it by flipping one of its variables. This could, of course, upset some clauses that were satisfied before. Nevertheless, we blithely continue on our way, flipping variables until every clause is satisfied, or until we run out of time and give up. We give the pseudocode for this algorithm, expressed for k-SAT in general, in Figure 10.5.

We will show that if ϕ is satisfiable, `WalkSAT` finds a solution with a reasonably large probability after taking a fairly small number of steps. So, if after running it many times we still haven't found a solution, we can conclude that ϕ is probably unsatisfiable.

How can we analyze `WalkSAT`'s performance? Each move satisfies the chosen clause c by flipping x to the value that c wants, but this also dissatisfies any clauses that were satisfied only by x's previous value. Thus the number u of unsatisfied clauses follows a kind of random walk. We would like to show that if ϕ is satisfiable then this walk reaches the origin $u = 0$ in a reasonable amount of time and with reasonable probability. However, the statistical properties of this walk seem very complicated, and no one knows how to carry out the analysis this way.

We can indeed treat `WalkSAT` as a random walk, but with a different measure of its distance from a solution. In our analysis of Karger's algorithm, we focused our attention on a particular minimal cut, and showed that the algorithm finds that one with reasonable probability. Similarly, if ϕ is satisfiable we can focus our attention on a particular satisfying assignment A. Then let d be the *Hamming distance* between A and B, i.e., the number of variables on which our current truth assignment B differs from A.

```
WalkSAT
input: a k-SAT formula φ
output: a satisfying assignment or "don't know"
begin
    start at a uniformly random truth assignment B ;
    repeat
        if B satisfies φ then return B ;
        else
            choose a clause c uniformly from among the unsatisfied clauses ;
            choose a variable x uniformly from among c's variables ;
            update B by flipping x ;
        end
    until we run out of time;
    return "don't know" ;
end
```

FIGURE 10.5: The WalkSAT algorithm for k-SAT.

Then WalkSAT performs a random walk on the d-axis, where d ranges from 0 to n. We succeed if we reach the origin, where $d = 0$ and $B = A$, or if we run out of unsatisfied clauses first, in which case we have hit upon some other satisfying assignment.

How does d change at each step? Flipping a variable on which A and B disagree brings B closer to A, in which case $\Delta d = -1$. But flipping a variable on which they agree drives B farther away, so $\Delta d = +1$. How often do each of these things happen?

Let c be the chosen clause. Since A is a satisfying assignment, A agrees with c on at least one of its variables. On the other hand, since c is currently unsatisfied, B disagrees with c on all k of its variables. In the worst case, A only agrees with c, and disagrees with B, on one of its variables, and A and B agree on the other $k-1$. Since x is chosen uniformly from these k variables, flipping x decreases or increases d with the probabilities

$$\Pr[\Delta d = -1] = \frac{1}{k}, \quad \Pr[\Delta d = +1] = \frac{k-1}{k}.$$

If we start at a typical value of d and move left or right with these probabilities, how long will it take us to reach the origin $d = 0$?

Let's explore the case $k = 2$ first. Then even in the worst case the random walk is *balanced*, i.e., it is equally likely to move toward or away from the origin. As we show in Appendix A.4.4, the expected time it takes us to reach the origin is $O(n^2)$. Therefore, for 2-SAT, WalkSAT succeeds in expected time $O(n^2)$ if the formula is satisfiable. Of course, we already showed in Section 4.2.2 that 2-SAT can be solved in polynomial time, but it's nice that WalkSAT works in polynomial time as well.

Now let's consider the case $k \geq 3$, where k-SAT is NP-complete. Now $\Pr[\Delta d = +1] > \Pr[\Delta d = -1]$ and the random walk is biased away from the origin. What is the probability P_{success} that we will reach $d = 0$, and thus find A, in a reasonable amount of time?

It is convenient to imagine that d ranges all the way from zero to infinity. This is pessimistic since the maximum Hamming distance is n, but it gives a lower bound on P_{success} that is very close to the correct value. So imagine performing this walk on the infinite line, and let $p(d)$ denote the probability that we will *ever* reach the origin, if we start at an initial distance d. Since we will reach the origin from our current position if and only if we will do so from our next position, $p(d)$ obeys the equation

$$p(d) = \frac{1}{k}\,p(d-1) + \frac{k-1}{k}\,p(d+1). \tag{10.2}$$

We have $p(0) = 1$, since if $d = 0$ we're already at the origin. But if $k > 2$, the farther out from shore we start our swim, the greater the probability that we will be swept out to sea—so $\lim_{d\to\infty} p(d) = 0$ whenever $k > 2$. The following exercise asks you to solve for $p(d)$ with these boundary conditions.

Exercise 10.3 *Using* (10.2) *and assuming that* $p(0) = 1$ *and* $p(\infty) = 0$, *show that*

$$p(d) = (k-1)^{-d}.$$

Since our initial truth assignment B is chosen randomly, d will itself be a random variable. To compute the probability P_{success} that a given attempt succeeds, we need to average $p(d)$ over d. If we flip a coin for each variable, B will agree with A on $n/2$ variables on average. We might think, therefore, that the probability that `WalkSAT` succeeds is

$$P_{\text{success}} = p(n/2) = (k-1)^{-n/2}.$$

However, this is unduly pessimistic. While it is exponentially unlikely that our initial distance is, say, $n/3$, the probability of success in this case is exponentially greater. Thus we can get lucky either by starting out unusually close to A, or by decreasing d on many steps of our walk. To calculate P_{success} correctly, we have to sum over all the ways that these two kinds of luck can combine.

This sum turns out to be quite simple. There are 2^n possible initial truth assignments B, each of which is chosen with probability 2^{-n}. There are $\binom{n}{d}$ of them at distance d, and each of these succeeds with probability $p(d)$. Thus

$$\begin{aligned}
P_{\text{success}} &= 2^{-n}\sum_B \Pr[\text{we succeed starting from } B] \\
&\geq 2^{-n}\sum_{d=0}^{n}\binom{n}{d}p(d) \\
&= 2^{-n}\sum_{d=0}^{n}\binom{n}{d}(k-1)^{-d} \\
&= \left(\frac{2(k-1)}{k}\right)^{-n}.
\end{aligned} \tag{10.3}$$

In the last line, we used the binomial theorem $\sum_{i=0}^{n}\binom{n}{i}a^i b^{n-i} = (a+b)^n$ with $a = 1/(k-1)$ and $b = 1$.

How many steps should we take in each attempt before we give up? We defined $p(d)$ as the probability that we *ever* reach $d = 0$ on the infinite line. But it turns out that if we will ever reach the origin, we will do so in the first $O(n)$ steps. Problem 10.7 shows that for 3-SAT, just $3n$ steps suffice. So if we haven't found a solution after $3n$ steps, we should try again with a new random initial assignment.

These "random restarts" are much more effective than simply continuing the random walk. Rather than prolonging our search in the same region of solution space, it makes sense to start over somewhere new. As we show in Problem 10.11, random restarts are generally a good idea whenever the time it takes for an algorithm to succeed has a heavy-tailed probability distribution.

Each of these restarts gives us a fresh chance of succeeding with probability P_{success}. As in Karger's algorithm, if we make $1/P_{\text{success}}$ attempts we succeed with constant probability. If we increase the number of attempts by a factor of n, we succeed with probability exponentially close to 1. Since we take $O(n)$ steps per attempt, the running time of each attempt is poly(n). Thus the total running time is

$$\frac{1}{P_{\text{success}}}\,\text{poly}(n) = \left(\frac{2(k-1)}{k}\right)^n \text{poly}(n).$$

For 3-SAT, this gives

$$\left(\frac{4}{3}\right)^n \text{poly}(n).$$

While $(4/3)^n$ is still exponentially large, it is far faster than the 2^n running time of a naive exhaustive search. Moreover, when this analysis of `WalkSAT` was carried out in 1999, it was faster than *any known deterministic algorithm*. At the time we write this, the best known algorithm is only slightly faster, working in time 1.3^n instead of $(4/3)^n$.

10.3 Given its simplicity and the simplicity of its analysis, `WalkSAT` remains one of the gems of randomized algorithms. It is also the stepping-off point for a vast family of local search algorithms. Rather than simply flipping a random variable in a random unsatisfied clause, we can choose our moves with varying amounts of greed, or using physics-inspired rules such as simulated annealing. Finally, it introduces us to the idea of a random walk through solution space. Such walks, more technically known as *Markov chains*, are so important that we devote the entirety of Chapter 12 to them.

10.4 Solving in Heaven, Projecting to Earth

> To them, I said, the truth would be literally nothing but the shadows of the images.
>
> Plato, *The Republic*, Book VII

Our third example of a randomized algorithm uses a simple but audacious idea. Imagine taking an optimization problem, and lifting it up to one in a much higher-dimensional space—where, paradoxically, it becomes easier to solve. Then you map this high-dimensional solution back down to a solution of the original problem, by casting its shadow along a random direction. In this section, we show how this idea leads to one of the most sophisticated and beautiful families of randomized approximation algorithms.

Recall the Max Cut problem:

> Max Cut
>
> Input: A weighted graph G with nonnegative edge weights w_{ij}
>
> Output: The weight $W_{\text{MAX CUT}}$ of the maximum cut

We showed in Chapter 5 that the decision version of MAX CUT is NP-complete, so this optimization problem is NP-hard. Suppose our goal is to get a good cut, rather than the best one—say, a cut whose weight is at least some fraction of the optimum. Can we obtain a cut which is at least α as weighty as the maximum, for some α? And what is the best α we can achieve in polynomial time?

One obvious approach is a greedy algorithm. We start by painting some vertex black. We then go through the vertices in an arbitrary order, painting vertex i black or white so that its color differs from the majority of its neighbors that have been painted so far. When determining the majority, we give the vote of each painted neighbor j weight w_{ij}.

Exercise 10.4 *Show that the greedy algorithm obtains a cut whose weight is at least half the total weight of all the edges.*

Therefore, this algorithm gives a cut whose weight is at least half the optimum, giving $\alpha = 1/2$. In terms of Section 9.2, this algorithm is a 1/2-approximation. Is this the best we can do?

As we saw in Section 9.8.2, we can often get good approximations for optimization problems by *relaxing* them to a less constrained problem, and then *rounding* the resulting solution to get a solution of the original problem. For instance, we found approximate solutions for VERTEX COVER by treating it as an instance of INTEGER LINEAR PROGRAMMING, relaxing it to an instance of LINEAR PROGRAMMING by allowing the variables to be reals instead of integers, and finally rounding each variable to the nearest integer. This produced vertex covers that are no more than twice as large as the optimal one.

In 1994, Michel Goemans and David Williamson found a beautiful way to relax the MAX CUT problem, and obtain cuts whose weight is at least 0.878 times the optimum. It uses beautiful ideas from linear algebra and convex optimization, and inaugurated a large family of randomized approximation algorithms. How does it work?

First we rephrase the problem by giving each vertex i a value $s_i = \pm 1$ depending on which side of the cut it is on. Then $s_i s_j = +1$ if i and j are on the same side of the cut, and $s_i s_j = -1$ if they are on opposite sides. We can then write the weight of the cut as

$$W = \sum_{i<j} w_{ij} \frac{1 - s_i s_j}{2},$$

where $w_{ij} = 0$ if i and j are not adjacent. Next, we rewrite this as

$$W = \frac{1}{2}\left(\sum_{i<j} w_{ij} - E\right), \tag{10.4}$$

where

$$E = \sum_{i<j} w_{ij} s_i s_j .$$

Maximizing W is equivalent to minimizing E. To a physicist, E is the energy of an antiferromagnetic Ising model with interaction strengths w_{ij} (see Problem 5.30), but our non-physicist readers should feel free to ignore this connection.

The idea of the relaxation is this. Rather than limiting the s_i to the values ± 1, we will let them be d-dimensional unit vectors $\mathbf{s}_i$. Curiously, if d is large enough, this problem actually becomes easier.

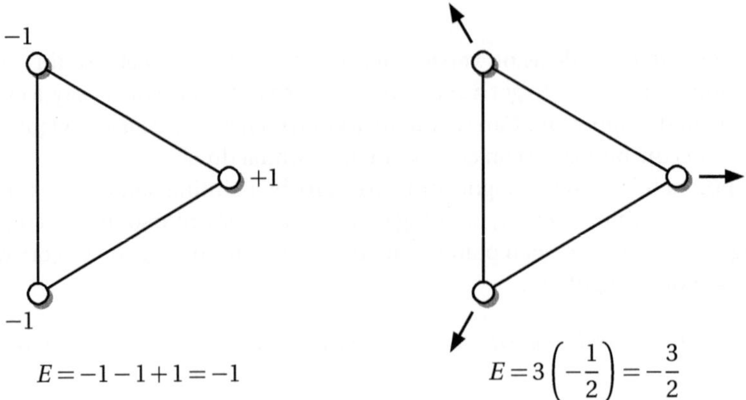

FIGURE 10.6: Relaxing the MAX CUT problem for a triangle, where all edges have unit weight. On the left, the original problem has a minimum energy $E=-1$, which gives a cut of weight $W=(1/2)(3-E)=2$. On the right, allowing the s_i to be vectors gives a solution with $E=-3/2$, or a "vector cut" of weight 9/4.

We generalize the "energy" by defining it as the weighted sum of the inner products of neighboring vectors,

$$E=\sum_{i<j} w_{ij}\, \mathbf{s}_i^T \mathbf{s}_j\,.$$

If all the $\mathbf{s}_i$ are parallel or antiparallel, lying at $+1$ or -1 along the same axis, this is the same E we had before. But if we let the $\mathbf{s}_i$ rotate freely, then E can be smaller, and W can be greater, than any solution to the original problem.

For instance, suppose G is a triangle as shown in Figure 10.6. If all the edges have unit weight, the optimum of the original MAX CUT problem is achieved by setting $s_i=+1$ for one vertex and $s_i=-1$ for the other two, or vice versa, in which case $E=-1$. However, in the relaxed problem, the $\mathbf{s}_i$ can be vectors with angles $2\pi/3$ between them, in which case $\mathbf{s}_i^T\mathbf{s}_j=-1/2$ and $E=-3/2$. Plugging these values of E into (10.4) gives a cut of weight $W=2$ in the original problem, and a "vector cut" of weight 9/4 in the relaxed problem.

Let's call MAX VECTOR CUT the problem of finding the vector cut of maximum weight, and denote its weight by W_{VECTOR}. Cuts are also vector cuts, so as always the optimum of the relaxed problem is at least as good as that of the original problem:

$$W_{\text{MAX CUT}} \le W_{\text{VECTOR}}\,.$$

Our goal will be to round the vector cut to produce a genuine cut, whose weight W_{ROUNDED} is not too much less than W_{VECTOR}. This will pin the true optimum between these two values,

$$W_{\text{ROUNDED}} \le W_{\text{MAX CUT}} \le W_{\text{VECTOR}}\,.$$

The approximation ratio will then be bounded by

$$\frac{W_{\text{ROUNDED}}}{W_{\text{MAX CUT}}} \ge \frac{W_{\text{ROUNDED}}}{W_{\text{VECTOR}}}\,, \tag{10.5}$$

and we will show that $W_{\text{ROUNDED}}/W_{\text{VECTOR}}$ is at least 0.878.

First we have to show how to solve MAX VECTOR CUT. The reader might think that allowing the $\mathbf{s}_i$ to be vectors rather than scalars only makes our life more complicated. After all, we are forcing ourselves to solve a much higher-dimensional problem. However, when we let these vectors be *as high-dimensional as they want*, something wonderful happens that lets us solve this problem in polynomial time.

If G has n vertices, any particular set of n vectors $\mathbf{s}_i$ lies in an n-dimensional space. Therefore, to achieve the best possible vector cut, it suffices to let the $\mathbf{s}_i$ be n-dimensional. In that case, we can write out the list of vectors $\mathbf{s}_i$ as an $n \times n$ matrix $\mathbf{S}$, where $\mathbf{s}_i$ is the ith column of $\mathbf{S}$. If we multiply this matrix by its transpose, we get a symmetric matrix $\mathbf{S}^T\mathbf{S}$ whose entries are the inner products $\mathbf{s}_i^T\mathbf{s}_j$:

$$\begin{pmatrix} \mathbf{s}_1^T \\ \hline \mathbf{s}_2^T \\ \hline \vdots \\ \hline \mathbf{s}_n^T \end{pmatrix} \cdot \left(\begin{array}{c|c|c|c} \mathbf{s}_1 & \mathbf{s}_2 & \cdots & \mathbf{s}_n \end{array}\right) = \begin{pmatrix} \mathbf{s}_1^T\mathbf{s}_1 & \mathbf{s}_1^T\mathbf{s}_2 & \cdots & \mathbf{s}_1^T\mathbf{s}_n \\ \mathbf{s}_2^T\mathbf{s}_1 & \mathbf{s}_2^T\mathbf{s}_2 & & \\ \vdots & & \ddots & \\ \mathbf{s}_n^T\mathbf{s}_1 & & & \mathbf{s}_n^T\mathbf{s}_n \end{pmatrix},$$

or more compactly,

$$(\mathbf{S}^T\mathbf{S})_{ij} = \mathbf{s}_i^T\mathbf{s}_j\,.$$

If we let $\mathbf{W}$ denote the matrix of weights w_{ij} where the diagonal entries w_{ii} are zero, we can write

$$E = \frac{1}{2}\sum_{ij} w_{ij}\,(\mathbf{S}^T\mathbf{S})_{ij} = \frac{1}{2}\operatorname{tr}\mathbf{W}\mathbf{S}^T\mathbf{S},$$

where tr denotes the trace. If we denote $\mathbf{S}^T\mathbf{S}$ as a single matrix $\mathbf{A}$, then

$$E = \frac{1}{2}\operatorname{tr}\mathbf{W}\mathbf{A},$$

and E is a linear function of the entries of $\mathbf{A}$. This is all well and good. But we want to maximize this linear function subject to a complicated set of quadratic constraints—namely, that $\mathbf{A} = \mathbf{S}^T\mathbf{S}$ for some matrix $\mathbf{S}$.

At this point, we recall a fact that we learned in Section 9.7.1. If $\mathbf{A}$ is a real symmetric matrix, then there is a matrix $\mathbf{S}$ such that $\mathbf{A} = \mathbf{S}^T\mathbf{S}$ if and only if $\mathbf{A}$ is positive semidefinite, i.e., all its eigenvalues are nonnegative. Denoting this $\mathbf{A} \succeq \mathbf{0}$ and adding the linear constraint $A_{ii} = |\mathbf{s}_i|^2 = 1$ for each i, we get the following optimization problem:

$$\max_{\mathbf{A}} \operatorname{tr}\mathbf{W}\mathbf{A} \quad \text{subject to} \quad A_{ii} = 1 \text{ for all } i \quad \text{and} \quad \mathbf{A} \succeq \mathbf{0}. \tag{10.6}$$

This is an instance of SEMIDEFINITE PROGRAMMING. As we discussed in Section 9.7.3, we can solve it in polynomial time using the ellipsoid method or with interior point methods. Thus we can find the maximum vector cut in polynomial time.

Next, we need to convert this vector cut into a good old-fashioned cut that puts some vertices on one side and some on the other. In Section 9.8.2, we converted a "fuzzy vertex cover" into a real vertex cover by rounding fractional variables to the nearest integer. How can we "round" an n-dimensional vector $\mathbf{s}$ so that it becomes $+1$ or -1?

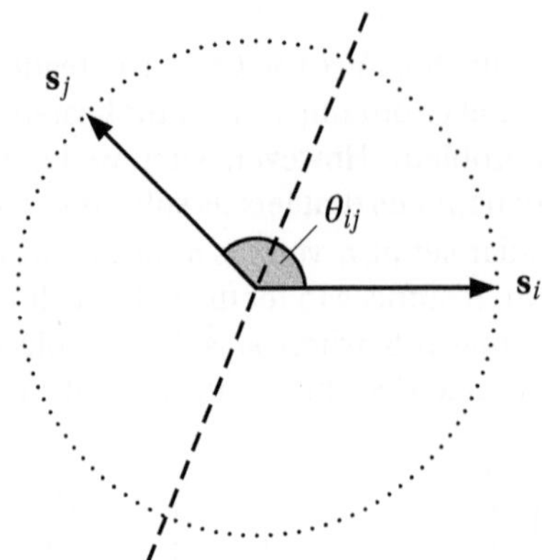

FIGURE 10.7: If we draw a random hyperplane, the probability that $\mathbf{s}_i$ and $\mathbf{s}_j$ are on opposite sides is θ_{ij}/π where θ_{ij} is the angle between them.

We would like the rounding process to preserve as much of the geometric information in the vector cut as possible. In particular, we would like $\mathbf{s}_i$ and $\mathbf{s}_j$ to be placed on opposite sides of the cut if the angle between them is large. It's not easy to do this deterministically, but here is a simple way to do it randomly. Divide the n-dimensional space into two halves by drawing a random hyperplane through the origin, and set $s_i = +1$ or -1 depending on which side $\mathbf{s}_i$ is on. Equivalently, choose a random n-dimensional vector $\mathbf{r}$ (the normal to the hyperplane) and set s_i according to whether $\mathbf{s}_i^T \mathbf{r}$ is positive or negative.

What expected weight $\mathbb{E}[W_{\text{ROUNDED}}]$ does this process achieve? For each edge (i, j), the contribution that $\mathbf{s}_i$ and $\mathbf{s}_j$ make to W_{ROUNDED} is w_{ij} if they fall on opposite sides of the hyperplane, and 0 if they don't. Thus their expected contribution is w_{ij} times the probability that this happens. By linearity of expectation (see Appendix A.3.2) the expected weight is the sum of these expectations:

$$\mathbb{E}[W_{\text{ROUNDED}}] = \sum_{i<j} w_{ij} \Pr[\mathbf{s}_i \text{ and } \mathbf{s}_j \text{ are on opposite sides}]. \tag{10.7}$$

What is the probability that $\mathbf{s}_i$ and $\mathbf{s}_j$ are on opposite sides? From the point of view of $\mathbf{s}_i$ and $\mathbf{s}_j$, drawing a random hyperplane amounts to drawing a random line through the origin in the two-dimensional plane they span. As Figure 10.7 shows, the probability that they end up on opposite sides of this line is θ_{ij}/π where θ_{ij} is the angle between them. Thus (10.7) becomes

$$\mathbb{E}[W_{\text{ROUNDED}}] = \sum_{i<j} w_{ij} \frac{\theta_{ij}}{\pi}. \tag{10.8}$$

How much does this rounding process cost us? In other words, how much smaller than W_{VECTOR} is $\mathbb{E}[W_{\text{ROUNDED}}]$? Since $\mathbf{s}_i^T \mathbf{s}_j = \cos\theta_{ij}$,

$$W_{\text{VECTOR}} = \sum_{i<j} w_{ij} \frac{1-\cos\theta_{ij}}{2}. \tag{10.9}$$

Since the ratio between two sums with nonnegative terms is at least the smallest ratio between any pair of corresponding terms, we can bound the ratio between (10.9) and (10.8) as follows:

$$\frac{\mathbb{E}[W_{\text{ROUNDED}}]}{W_{\text{VECTOR}}} = \frac{\sum_{i<j} w_{ij}\,\theta_{ij}/\pi}{\sum_{i<j} w_{ij}(1-\cos\theta_{ij})/2} \geq \min_{i<j} \frac{\theta_{ij}/\pi}{(1-\cos\theta_{ij})/2} \geq \alpha,$$

where

$$\alpha = \min_{\theta\in[0,\pi]} \frac{\theta/\pi}{(1-\cos\theta)/2} = 0.878\ldots$$

Thus the expected ratio between W_{ROUNDED} and W_{VECTOR} is at least 0.878. Plugging this back into (10.5), the expected ratio between W_{ROUNDED} and the optimum $W_{\text{MAX CUT}}$ is bounded by

$$\frac{\mathbb{E}[W_{\text{ROUNDED}}]}{W_{\text{MAX CUT}}} \geq \frac{\mathbb{E}[W_{\text{ROUNDED}}]}{W_{\text{VECTOR}}} \geq 0.878.$$

This bounds the *expected* ratio obtained by each run of the algorithm. To show that running the algorithm repeatedly gives, with high probability, a solution arbitrarily close to this ratio or better, we refer the reader to Problem 10.12.

The idea behind the Goemans–Williamson algorithm—of lifting a problem to a higher-dimensional space and then randomly projecting the solution back down—has produced approximation algorithms for a wide variety of problems. Moreover, there are good reasons to believe that it gives the best possible approximation algorithm for MAX CUT unless P = NP.

On the other hand, if we move away from the worst case, the Goemans–Williamson algorithm does 10.4
much better. As Problem 10.14 shows, if the MAX CUT includes a large fraction of the edges then the approximation ratio improves, and approaches 1 as the graph becomes bipartite.

10.5 Games Against the Adversary

> As you like to play so much, you shall play against Captain Najork and his hired sportsmen. They play hard games and they play them jolly hard. Prepare yourself.
>
> Russell Hoban, *How Tom Beat Captain Najork and His Hired Sportsmen*

Now that we have seen several examples of randomized algorithms, let's build further on the analogy with which we began this chapter: computation is a game where we choose the algorithm and the adversary chooses the instance. By thinking about the strategies available to both sides in this game, we can sometimes prove both upper and lower bounds on randomized computation, and even prove that some randomized algorithms are optimal. To illustrate this, in this section we will see how a simple randomized algorithm can help us analyze a two-player game—and then show that no algorithm can do any better.

10.5.1 AND-OR *Trees and* NAND *Trees*

Suppose you and I are playing a game, and it's your move. How can we tell if you have a winning strategy? As we discussed in Chapter 8—and for that matter, all the way back in the Prologue—having a winning strategy means that you have a move, such that no matter how I reply, you have a move, and so on... until the game ends in a winning position for you.

In an idealized game where there are two possible moves in each position, and where every game lasts the same number of moves, this gives us a binary tree like the one we showed for tic-tac-toe on page 325. The nodes of the tree alternate between AND and OR gates. If it is your move, you win if either possible move leads to a win, and if it is my move, you win if *both* of my possible moves lead to a win for you. The leaves of the tree are the endgame positions, which we mark `true` or `false` if they are wins for you or me respectively. Finally, you have a winning strategy if the entire tree evaluates to `true`.

We can get rid of the AND-OR alternation, and make every level of the tree look the same, by complementing the truth values on every other level as shown in Figure 10.8. Now the value at each node corresponds to whether that position is a win for the *current* player. This is true if at least one move leads to a loss for the other player, so each node outputs `true` if at least one of its inputs is `false`, or equivalently if not all its inputs are `true`. Thus each node outputs the complement of the AND, also known as the NAND, of its inputs. We will call this a NAND tree.

How can we evaluate such a tree? The most obvious approach is a depth-first algorithm, that evaluates each node y by recursively evaluating its children y_{left} and y_{right}, until we reach the base case where we evaluate leaves directly. This algorithm is shown in Figure 10.9. It is just a more compact version of the algorithm on page 327.

We can make this algorithm more efficient by taking advantage of the following fact. As soon as we see a child who evaluates to `false`—that is, a winning move for the current player, which puts their opponent in a losing position—we can immediately return `true` without evaluating the other child. This gives the improved algorithm shown in Figure 10.10, which only evaluates both children if the first one evaluates to `true`. Any `true` node has at least one `false` child, and if it's the one on the left this algorithm will save some time.

But what happens if the game is designed by an adversary? Consider the following exercise:

Exercise 10.5 *Assign truth values to the leaves in the* NAND *tree of Figure 10.8 so that even the improved* `Evaluate` *algorithm is forced to evaluate every leaf.*

As Problem 10.17 shows, if we use *any* deterministic scheme to decide what subtree or leaf to evaluate, the adversary can choose the values of the leaves in a way that forces us to evaluate every one. If we define the running time of the algorithm as the total number of leaves it evaluates, the worst-case running time is then 2^k for a tree of depth k, since there are $N = 2^k$ leaves.

The answer, of course, is to evaluate the two children in *random* order. Then no matter how the adversary chooses the values of the leaves, if a node has a `false` child we will see it first with probability 1/2. How long does the resulting algorithm take average? How many leaves does it have to evaluate?

Consider the two cases where the output is `true` or `false`, and denote their average running times as $f_{\texttt{true}}(k)$ and $f_{\texttt{false}}(k)$ respectively. If the output is `true`, in the worst case one child is `true` and the other is `false`. With probability 1/2 we only evaluate the `false` one, and with probability 1/2 we evaluate them both, giving

$$f_{\texttt{true}}(k) = f_{\texttt{false}}(k-1) + \frac{1}{2} f_{\texttt{true}}(k-1). \tag{10.10}$$

On the other hand, if the output is `false` then both children are `true`, and we are forced to evaluate both of them, giving

$$f_{\texttt{false}}(k) = 2 f_{\texttt{true}}(k-1). \tag{10.11}$$

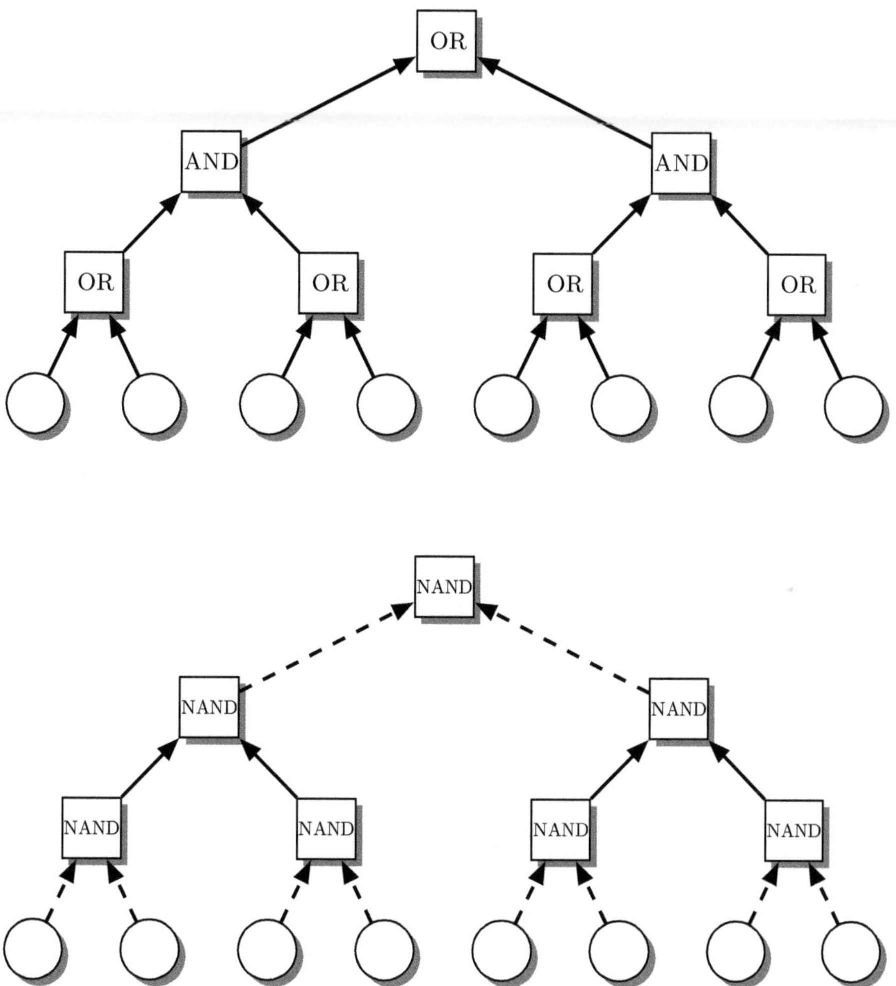

FIGURE 10.8: An AND-OR tree. Nodes are `true` if they are a win for you. It's your turn, and you have a winning strategy if the entire tree evaluates to `true`. In the tree above, it's your move. Below, by complementing the values at every other level (the dashed arrows), we obtain a NAND tree, where a node is `true` if it is a win for the current player.

```
Evaluate(y)
input: a node y
output: is y a win for the current player?
begin
    if y is a leaf then return y's value ;
    else return NAND(Evaluate(y_left), Evaluate(y_right)) ;
end
```

FIGURE 10.9: A recursive algorithm for evaluating NAND trees.

```
Evaluate(y)
begin
    if y is a leaf then return y's value ;
    else if Evaluate(y_left) = false then return true ;
    else if Evaluate(y_right) = false then return true ;
    else return false ;
end
```

FIGURE 10.10: An improved algorithm that only evaluates the second child if it has to.

Combining these two gives the following equation for $f_{\texttt{true}}$ in the worst case, where every `true` node has only one `false` child:

$$f_{\texttt{true}}(k) = \frac{1}{2} f_{\texttt{true}}(k-1) + 2 f_{\texttt{true}}(k-2). \tag{10.12}$$

The solution to this Fibonacci-like equation is given by the following exercise.

Exercise 10.6 *Show that the solution to* (10.12) *is*

$$f(k) = \Theta(\alpha^k),$$

where

$$\alpha = \frac{1+\sqrt{33}}{4} = 1.686...$$

is the positive root of the quadratic equation $\alpha^2 - \alpha/2 - 2 = 0$. Then substitute $N = 2^k$ to show that the average number of leaves the randomized algorithm evaluates grows as

$$N^{\log_2 \alpha} = N^{0.753...}.$$

When N is exponentially large, $N^{0.753}$ is an exponentially small fraction of N. Thus we succeed in evaluating the entire tree while only evaluating, on average, a vanishingly small fraction of the leaves. Combined with Problem 10.17, which shows that for any deterministic algorithm the adversary can force us to evaluate every leaf, this shows that randomized algorithms are provably better than deterministic ones for this problem.

Are there other algorithms that are even more efficient? Is 0.753 the best power of N we can achieve? In the next section we will see how to prove that this is optimal, by thinking about our choice of algorithm—and the adversary's choice of values for the leaves—as a game we are playing against him.

10.5.2 The Minimax Theorem and Yao's Principle

Let's use the language of game theory to describe our conflict with the adversary. Imagine an enormous matrix M_{ij} where i ranges over all possible deterministic algorithms and j ranges over all possible instances. The entry M_{ij} is the running time—or the probability of error, or some other measure of poor performance—of algorithm i on instance j. We choose i, and the adversary chooses j. The adversary's score, or *payoff*, is M_{ij}, and ours is $-M_{ij}$. This is a *zero-sum* game, in which our gain is the adversary's loss and vice versa. He wants to maximize M_{ij}, and we want to minimize it.

If we go first, committing to a deterministic algorithm, the adversary can choose the instance on which our algorithm performs the worst. But if he goes first, revealing the instance, we can tailor our algorithm to it. Thus for a typical payoff matrix M,

$$\min_i \max_j M_{ij} > \max_j \min_i M_{ij} . \tag{10.13}$$

Note how the order of these quantifiers corresponds to the order in which we move. On the left-hand side, we go first, choosing the algorithm i. The adversary then chooses whichever instance j maximizes M_{ij}. Our best move is the i that minimizes $\max_j M_{ij}$—that is, the algorithm with the best worst-case behavior. On the right-hand side, the situation is reversed—he chooses the instance j, and we choose the algorithm i that minimizes M_{ij}.

Equation (10.13) tells us that where deterministic computation is concerned, each player would rather move second. As in Rock, Paper, Scissors, whoever shows their hand first is at a grave disadvantage. Unfortunately, whether we like it or not, we have to go first. If we want to make worst-case guarantees about an algorithm's performance, we have to reveal it to the world, and let anyone challenge us with any instance they like.

In game-theoretic terms, a deterministic algorithm is a *pure strategy*, corresponding to a single row of M. A randomized algorithm corresponds to a *mixed strategy*, i.e., a probabilistic distribution over the rows. Consider the sequence of coin flips that a randomized algorithm A uses to choose random vertices, take random steps, or whatever. If we flip all these coins in advance, A's behavior becomes deterministic. Thus we can think of A as a probability distribution of deterministic algorithms, one for each possible sequence of coin flips.

Of course, most probability distributions over the universe of deterministic algorithms cannot be expressed succinctly in the form of a randomized algorithm. But to be generous to ourselves, we will assume that our mixed strategy is an arbitrary distribution. We denote it as a column vector $\mathbf{a}$, where a_i is the probability that we carry out the ith deterministic algorithm.

Similarly, the adversary can pursue a mixed strategy, choosing his instance according to some probability distribution. We denote this as a column vector $\mathbf{b}$, where b_j is the probability he gives us the jth instance. His average payoff is then

$$\mathbf{a}^T \mathbf{M} \mathbf{b} = \sum_{i,j} a_i M_{ij} b_j .$$

We choose $\mathbf{a}$, and he chooses $\mathbf{b}$. He wants to maximize $\mathbf{a}^T \mathbf{M} \mathbf{b}$, and we want to minimize it.

We still have to go first. But for mixed strategies, "going first" doesn't mean announcing a deterministic algorithm or an instance. Instead, it means announcing the probability distribution $\mathbf{a}$, the randomized algorithm we plan to use, but not the coin flips that will drive its random choices. It's like revealing to our opponent that we will play Rock, Paper, and Scissors according to the probability distribution $(1/2, 1/3, 1/6)$.

Fascinatingly, for mixed strategies the expected payoff the players can achieve no longer depends on the order in which they move, assuming that both players play optimally. In other words, there are optimal strategies $\mathbf{a}^\star$ and $\mathbf{b}^\star$, defined as

$$\mathbf{a}^\star = \text{argmin}_{\mathbf{a}} \left[\max_{\mathbf{b}} \mathbf{a}^T \mathbf{M} \mathbf{b} \right] \text{ and } \mathbf{b}^\star = \text{argmax}_{\mathbf{b}} \left[\min_{\mathbf{a}} \mathbf{a}^T \mathbf{M} \mathbf{b} \right], \tag{10.14}$$

such that

$$\min_{\mathbf{a}} \max_{\mathbf{b}} \mathbf{a}^T \mathbf{M} \mathbf{b} = \max_{\mathbf{b}} \min_{\mathbf{a}} \mathbf{a}^T \mathbf{M} \mathbf{b} = \mathbf{a}^{\star T} \mathbf{M} \mathbf{b}^\star . \tag{10.15}$$

10.5 This is the *Minimax Theorem*, first proved by John von Neumann.

Exercise 10.7 *Prove that the optimal mixed strategy in Rock, Paper, Scissors for both players is* $\mathbf{a}^\star = \mathbf{b}^\star = (1/3, 1/3, 1/3)$.

There are many proofs of the minimax theorem. One of them, which we give in Problem 10.15, shows that if we treat the two sides of (10.15) as LINEAR PROGRAMMING problems, they are dual in the sense of Section 9.4. In other words, their relationship is analogous to that between MAX FLOW and MIN CUT. Another proof, given in Problem 10.16, uses Brouwer's Fixed-Point Theorem to show that if each player tries to update their strategy to adapt to the other's, they reach a fixed point where neither has any incentive to change.

Now suppose we go first, and announce the randomized algorithm $\mathbf{a}$ we plan to use. The only instances j which the adversary has any reason to include in his distribution $\mathbf{b}$ are those which maximize $\mathbf{a}$'s running time. But in that case, he might as well just choose one of these worst-case instances. Similarly, if the adversary goes first and announces his distribution of instances $\mathbf{b}$, our reply $\mathbf{a}$ should only include deterministic algorithms i that minimize the average running time, and we might as well just use one of them.

Therefore, in both cases the second player might as well respond with a pure strategy rather than a mixed one, and we can rewrite (10.15) as

$$\min_{\mathbf{a}} \max_{j} \sum_i a_i M_{ij} = \max_{\mathbf{b}} \min_{i} \sum_j M_{ij} b_j . \tag{10.16}$$

What does (10.16) mean in terms of computation and running times? The left-hand side is the average running time of the best possible randomized algorithm, when run on worst-case instances. The right-hand side is the average running time, maximized over all instance distributions, of the best deterministic algorithm. Note the two different uses of the word "average" here: on the left, it's the average over the runs of the randomized algorithm, while on the right, it's the average over the possible instances.

We can turn (10.16) into a lower bound on the running time by fixing a distribution of instances. For any particular $\mathbf{b}$, the expected running time is at most the maximum over all $\mathbf{b}$, so

$$\text{For any } \mathbf{b},\ \min_{\mathbf{a}} \max_{j} \sum_i a_i M_{ij} \ge \min_{i} \sum_j M_{ij} b_j . \tag{10.17}$$

We can state this more concretely as follows:

Theorem 10.1 (Yao's principle) *For any distribution* **b** *on instances, the average running time of the best deterministic algorithm, when run on instances chosen according to* **b**, *is a lower bound on the average running time of the best randomized algorithm, when run on worst-case instances.*

We urge the reader to take the time to parse this statement, and appreciate what it can do for us. We are allowed to consider any distribution on instances we like, and can design this distribution in order to make our analysis easy. Any distribution yields a lower bound on the running time of the best randomized algorithm. If we choose the right distribution, we might be able to prove that our favorite randomized algorithm is actually optimal.

Of course, it's generally very hard to reason about the best possible deterministic algorithm—otherwise, it wouldn't be so hard to prove that $\mathsf{P} \neq \mathsf{NP}$. However, we can often do this in *black-box* problems, where our only access to the input is through a series of queries. For instance, in Section 6.2 we showed that if a sorting algorithm can only access its input by asking which of two elements is larger, it needs $\Omega(n \log n)$ comparisons to sort a list of size n. Similarly, in the NAND tree, we can only access the input by asking, one at a time, for the truth values of the leaves. In the next section, we will see that we can then apply Yao's principle, and prove that the randomized algorithm of Section 10.5.1 is the best possible.

10.5.3 Adversarial Inputs on the NAND *Tree*

To prove that the algorithm of Section 10.5.1 is optimal, we want to design a probability distribution on the inputs so that no deterministic algorithm can do better than our randomized one. The idea is to force the algorithm to evaluate both children just as often, on average, as our randomized algorithm does.

Our first impulse might be to flip a coin, perhaps with some bias towards `true` or `false`, and choose the truth value of each leaf independently. As Problem 10.19 shows, this gives a fairly good lower bound if we tune the bias of the coin correctly. However, this distribution is a little too easy for the following reason: fairly often, a `true` node has two `false` children, and in this case we never need to evaluate more than one child.

In order to get a lower bound that matches our algorithm, we need to avoid this easy case, and force the deterministic algorithm to look at both children of a `true` node half the time. So, rather than choosing the truth values independently, we build our instance from the top down. We start by flipping a coin and setting the root to `true` or `false` with equal probability. Then we descend the tree, determining the values of the nodes at each level. If a node is `false`, we set both its children to `true`; if a node is `true`, we set one of its children to `true` and the other to `false`, flipping a fair coin to decide which is which. Then the truth values of the nodes are correlated in such a way that the easy case never happens.

Exercise 10.8 *Show that if we assign truth values to nodes this way, the expected fraction of true leaves approaches* $2/3$ *as the depth increases.*

Now suppose that we have a deterministic, depth-first algorithm, and suppose it is working on a `true` node. No matter how it decides which child to evaluate first, our instances are chosen so that the first child is `true` with probability $1/2$. Therefore, it is forced to evaluate the other child with probability $1/2$, just as for the randomized algorithm of Section 10.5.1.

Combining this with the fact that we always need to evaluate both children of a `false` node, we see that the expected running time of the deterministic algorithm is governed by exactly the same equations,

(10.10) and (10.11), as that of the randomized algorithm. The solution is again $\Theta(\alpha^k) = \Theta(N^{0.753\ldots})$, giving a lower bound that matches the randomized algorithm exactly. The only difference is that this probability now comes from randomness in the instances, instead of randomness in the algorithm's behavior.

This completes the proof that the randomized algorithm is optimal—almost. The problem is that we limited our analysis to depth-first algorithms that proceed recursively from the root down to the leaves, evaluating entire subtrees before looking anywhere else. For NAND trees, the optimal algorithm is indeed of this type, but we omit this (much more difficult) part of the proof. In fact, as Problem 10.21 shows, for some other types of trees we can get a better algorithm by skipping a generation, sampling several grandchildren in order to decide which children to evaluate first.

10.6

10.6 Fingerprints, Hash Functions, and Uniqueness

At the next stop in our journey through randomized algorithms, we will meet an important new tool: *hash functions*, which map a large set to a small one in a random, or nearly random, way.

Hash functions have many uses. They let us compress information, generating a short "fingerprint" of a much longer string. They can help two people solve a problem together while exchanging just a few bits. They give efficient randomized algorithms for searching for one string in another, or checking that a calculation is correct. They let us add random constraints to a problem, so that its solution becomes unique. In Chapter 11, we will see that hash functions also play a role in *interactive proofs*, where a verifier probes a prover with random questions.

In this section, we will also meet the Isolation Lemma. This is a clever probabilistic analysis of problems like MINIMUM SPANNING TREE, MIN-WEIGHT MATCHING, and MAX-WEIGHT INDEPENDENT SET, where the score of a solution is the total weight of the edges or vertices it contains. The Isolation Lemma shows how to make the optimal solution to such a problem unique by choosing these weights randomly.

10.6.1 Slinging Hash across the Solar System

"Heavens, Ginger! There must be something in this world that you wouldn't make a hash of."

P. G. Wodehouse, *The Adventures of Sally*

Consider the following predicament. We have sent a space probe beyond Neptune's orbit, to explore Kuiper belt objects named after Arctic deities. We have loaded it with a gigabyte of software to help it make autonomous decisions as it navigates from one planetoid to another.

Worryingly, a sensor on the probe notices that the software may have been corrupted in some way. We would like to check that its copy of the software still matches the original. However, at this great distance, the probe can only send and receive information at a slow rate. Asking it to send us the entire gigabyte is out of the question. How can we check its software in a way that only requires the probe to exchange a small amount of information with us?

The idea is to ask the probe to send us a kind of summary, or *fingerprint*, of its copy of the software. Let h be a *hash function*, i.e., a function that maps strings of n bits to strings of ℓ bits for some $\ell \ll n$. Let x denote the original version of the software, and let y denote the version now occupying the probe's memory. Then we ask the probe to send us $h(y)$, and compare $h(x)$ with $h(y)$. If $h(x) \neq h(y)$, we know that the probe has been corrupted.

But perhaps the probe's software has been modified deliberately, by shy, crafty creatures who wish to remain undiscovered. In this case, they may know which hash function h we plan to use, and they may have carefully chosen a string $y \neq x$ such that $h(y) = h(x)$. But we know how to foil such adversaries. Rather than committing ourselves to a particular h, we will choose h randomly from a *family* of hash functions. Our goal is then to find a family such that it is very unlikely that $h(y) = h(x)$ unless $y = x$.

A common type of hash function is to interpret x as a large integer, and define $h_p(x) = x \bmod p$ where p is a prime. Such a function distinguishes x from y unless $x \equiv_p y$, or equivalently, unless p divides $x - y$. We will show that, if $x \neq y$, this is very unlikely as long as p is chosen from a large enough set of primes.

First note that if x and y have at most n bits, $x - y$ has at most n bits as well. Long ago, in Exercise 4.21, we showed that an n-bit integer has at most n distinct prime factors. Therefore, as long as we choose p uniformly from a set containing many more than n primes, with high probability p does not divide $x - y$ unless $x - y = 0$.

Now, the *Prime Number Theorem* (see Problem 10.24) states that the number of primes less than a given number t is asymptotically

$$\pi(t) \sim \frac{t}{\ln t}.$$ 10.7

The probability that a random prime p between 1 and t divides $x - y$ is then at most

$$\frac{n}{\pi(t)} \sim \frac{n \ln t}{t}.$$

This is $o(1)$ if $t = n^2$, say, or indeed if $t = n^{1+\varepsilon}$ for any $\varepsilon > 0$. So, we choose p randomly from the primes between 1 and t, send p to the probe, and ask it to send $h_p(y) = y \bmod p$ back to us.

This allows us to detect any difference between x and y with probability $1 - o(1)$, while exchanging just $O(\log t) = O(\log n)$ bits. In contrast, Problem 10.29 shows that no deterministic communication strategy can ensure that $x = y$ unless we and the probe exchange at least n bits. Thus in the context of communication, randomized strategies can be exponentially more powerful than deterministic ones.

The idea of fingerprinting has many uses. Problem 10.30 uses fingerprinting to search rapidly for 10.8
a pattern in a string, Problem 10.31 uses it to check matrix multiplication, and in Section 10.8 we will see a kind of fingerprinting in the polynomial-time algorithm for PRIMALITY. But first, let's meet another interesting use of hash functions: changing many solutions to one.

10.6.2 Making Solutions Unique 10.9

> Out of the blue, as promised, of a New York
> Puzzle-rental shop the puzzle comes—
> A superior one, containing a thousand hand-sawn,
> Sandal-scented pieces.
>
> James Merrill, *Lost in Translation*

Suppose I promise you that a search problem has exactly one solution. *A priori*, this might make it easier to solve. For instance, in a tiling puzzle like those we discussed in Section 5.3.4, there may be many ways to cover a given region with the tiles, and these many possibilities create a difficult search problem. In a jigsaw puzzle, on the other hand, there are more constraints on how the pieces fit together. These constraints make the solution unique, and help us find it.

Of course, there are two kinds of uniqueness at work here. There is *local* uniqueness, where the shapes of the pieces and the pictures on them allow only one piece to fit in a given place, and *global* uniqueness, where there is only one way to complete the entire puzzle. Clearly, local uniqueness makes problems easier, since it allows us to solve the puzzle piece-by-piece without any backtracking. But does global uniqueness help? Suppose I give you an instance of your favorite NP-complete problem, and promise that it has a unique solution or no solution at all. Are such instances easier to solve than general ones?

In this section we will show that, in general, the promise of uniqueness does not make the search any easier. Specifically, we will give a simple randomized algorithm that turns any satisfiable SAT formula into one with a single satisfying assignment. Let UNIQUE SAT denote the special case of SAT where the number of solutions is zero or one. Then this is a *randomized polynomial-time reduction* from SAT to UNIQUE SAT, which we denote

$$\text{SAT} \leq_{\text{RP}} \text{UNIQUE SAT}\,.$$

This shows that, in a world where algorithms have access to random bits, UNIQUE SAT is just as hard as SAT. In particular, if there is a randomized polynomial-time algorithm for UNIQUE SAT, there is one for SAT as well.

The idea of the reduction is extremely simple. Given a satisfiable SAT formula, we will add new constraints to it, forbidding most of its solutions. The question is what kind of constraints we should add, and how many, and how often exactly one solution remains. As in our conversation with the space probe, the essential tool will be to apply a random hash function—but we need to choose it from a family with an important statistical property.

Suppose ϕ is a SAT formula. If h is a hash function that maps truth assignments to integers in some range, we can define a new formula ϕ' by adding the constraint that h is zero:

$$\phi'(x) = \phi(x) \wedge (h(x) = 0)\,.$$

Presumably $h(x)$ can be computed in polynomial time. Since SAT is NP-complete, we can then translate the statement $h(x) = 0$ into a polynomial number of SAT clauses, so that ϕ' is a SAT formula of polynomial size. We could do the same with any other NP-complete constraint satisfaction problem.

If S is the set of solutions to ϕ, the set of solutions to ϕ' is

$$S' = \{x \in S : h(x) = 0\}\,.$$

Our goal is to design h so that $|S'| = 1$ with reasonably large probability. Of course, if the adversary knows what h we will use, he could design an instance in which $h(x) = 0$ for all $x \in S$ or for none of them. Once again, we will foil him by choosing h randomly from some family.

At a minimum, this family should have the property that any given x is mapped to a random place. In particular, every x should have an equal probability of being mapped to zero. If the range of h is $\{0, L-1\}$,

$$\text{for all } x,\ \Pr_h[h(x) = 0] = \frac{1}{L}\,. \tag{10.18}$$

However, this is not enough. We also need to limit the correlations between the events $h(x) = 0$ and $h(y) = 0$. For example, one extremely bad way to choose h would be to choose z randomly from 0 to $L-1$, and then set $h(x)=z$ for all x. This satisfies (10.18), but either $S'=S$ or $S'=\emptyset$, and in neither case do we make the solution unique.

What we would really like is for h to be a completely random function from the set of possible solutions to $\{0,\ldots,L-1\}$. Then the events $h(x)=0$ are independent, and $|S'|$ is binomially distributed, like $|S|$ flips of a biased coin that comes up heads $1/L$ of the time (see Appendix A.4.1). The probability that exactly one flip comes up heads is then

$$\Pr_h\left[\left|S'\right|=1\right]=\frac{|S|}{L}\left(1-\frac{1}{L}\right)^{|S|-1}. \tag{10.19}$$

If $L=|S|$ this is at least $1/\mathrm{e}$, and it is bounded below by a constant whenever L is within a constant factor of $|S|$. As long as L is a reasonably good approximation of $|S|$, we get a unique solution with constant probability.

However, a completely random h is too much to ask. If ϕ is a SAT formula on n variables, the set of possible solutions is $\{0,1\}^n$. There are L^{2^n} different functions $h:\{0,1\}^n\to\{0,\ldots,L-1\}$, and each one consists of an arbitrary table of 2^n values of h. Describing such a table takes exponentially many bits, and choosing one randomly would require us to flip exponentially many coins. Since a randomized algorithm that runs in time t can flip at most t coins, and since we want our reduction from SAT to UNIQUE SAT to run in polynomial time, we have to choose h from a family of hash functions that can be described succinctly—that is, with $\mathrm{poly}(n)$ bits—so that we can choose one of them by flipping $\mathrm{poly}(n)$ coins.

Is there something between total randomness and determinism that is sufficient for our purposes? Indeed there is. Let us say that a family of hash functions is *pairwise independent* if, for any distinct pair x,y, they get mapped to any given pair of values with probability $1/L^2$, just as if the function were completely random:

$$\text{for all } x,y,z,w \text{ with } x\neq y,\ \Pr_h[h(x)=w \text{ and } h(y)=z]=\frac{1}{L^2}. \tag{10.20}$$

In particular, the probability that both x and y are mapped to zero is

$$\Pr_h[h(x)=0 \text{ and } h(y)=0]=\frac{1}{L^2}.$$

Another way to put this is that the probability that $h(y)=0$ doesn't depend on whether $h(x)=0$ or not. Thus the conditional probability that $h(y)=0$ given that $h(x)=0$ is $1/L$, just as if we didn't know $h(x)$:

$$\text{for all } x,y \text{ with } x\neq y,\ \Pr_h[h(y)=0\mid h(x)=0]\ =\ \Pr_h[h(y)=0]\ =\ \frac{1}{L}. \tag{10.21}$$

Leaving aside for the moment how to satisfy this property, let's see what it would do for us. We have $|S'|=1$ if and only if there is some $x\in S$ such that $h(x)=0$ and $h(y)\neq 0$ for all other $y\in S$. These events are disjoint for different x, so their total probability is the sum

$$\begin{aligned}\Pr\left[\left|S'\right|=1\right]&=\sum_{x\in S}\Pr[h(x)=0]\cdot\Pr\left[h(y)\neq 0 \text{ for all } y\neq x\mid h(x)=0\right]\\&=\sum_{x\in S}\Pr[h(x)=0]\cdot\left(1-\Pr\left[h(y)=0 \text{ for some } y\neq x\mid h(x)=0\right]\right).\end{aligned}$$

According to the union bound (see Appendix A.3.1), the probability that $h(y)=0$ for some $y \neq x$ is at most the sum of this probability over all y. Applying pairwise independence (10.21) then gives

$$\Pr\left[|S'|=1\right] \geq \sum_{x\in S} \Pr[h(x)=0] \left(1 - \sum_{y\in S: y\neq x} \Pr\left[h(y)=0 \mid h(x)=0\right]\right)$$
$$= \frac{|S|}{L}\left(1-\frac{|S|-1}{L}\right) \geq \frac{|S|}{L}\left(1-\frac{|S|}{L}\right). \tag{10.22}$$

Note the similarity between this expression and (10.19). When $|S|$ and L are large, (10.19) is to (10.22) as $x\mathrm{e}^{-x}$ is to $x(1-x)$, which is the approximation we would get if we took the Taylor series of $x\mathrm{e}^{-x}$ to second order. In essence, pairwise independence means independence to second order—no correlations appear until we look at correlations between three or more events.

Now let's suppose, as before, that L is a reasonably good estimate of $|S|$. In fact, we want it to be a slight overestimate, in the following sense:

$$2\,|S| \leq L \leq 4\,|S|\,. \tag{10.23}$$

In that case, $|S|/L$ lies between $1/4$ and $1/2$, and (10.22) tells us that we get a unique solution with probability at least $3/16$. Unfortunately, we have no idea what $|S|$ is. How can we choose L so that (10.23) holds reasonably often?

Estimating $|S|$ seems hard, since it could lie anywhere between 1 and 2^n. So, we will estimate $\log_2 |S|$ instead. Specifically, if we choose m randomly from $\{1,\ldots,n\}$, with probability $1/n$ we get lucky and choose the m such that

$$2^{m-1} \leq |S| \leq 2^m\,.$$

Then (10.23) holds if we set $L=2^\ell$ where $\ell = m+1$. In this case, we can think of $\{0,\ldots,L-1\}$ as the set $\{0,1\}^\ell$ of ℓ-bit strings.

Let's put this all together. If we have a SAT formula ϕ on n variables, we choose ℓ randomly from $\{2,\ldots,n+1\}$, and then choose h randomly from a pairwise independent family of hash functions from n-bit strings to ℓ-bit ones. If h can be computed in polynomial time, we can express the constraint $h(x)=0$ with a SAT formula of polynomial size. This gives a new formula $\phi' = \phi \wedge (h(x)=0)$, such that if ϕ is satisfiable, ϕ' has a unique solution with probability $\Omega(1/n)$. Thus if we have an algorithm that can solve instances of UNIQUE SAT, we can solve the original SAT problem with $O(n)$ attempts on average.

It remains to show that there are, in fact, pairwise independent families of hash functions that can be described succinctly and computed efficiently. We describe one such family in the next section.

10.6.3 Affine Functions and Pairwise Independence

If we want to map n-bit strings to ℓ-bit ones, one approach that comes to mind is to treat strings as vectors. We can map an n-dimensional vector $\mathbf{x} \in \{0,1\}^n$ to an ℓ-dimensional one by multiplying it by an $\ell \times n$ matrix, $\mathbf{A} \in \{0,1\}^{\ell\times n}$. Since $\mathbf{Ax}=0$ whenever $\mathbf{x}=0$, we mix things up a little more by adding an ℓ-dimensional vector $\mathbf{b} \in \{0,1\}^\ell$. Thus we define $h(\mathbf{x}) = \mathbf{Ax}+\mathbf{b} \bmod 2$, or more explicitly,

$$h(\mathbf{x})_i = \sum_{j=1}^{n} A_{ij} x_j + b_i \bmod 2\,.$$

Functions like these, where we apply a linear transformation and then add a constant, are called *affine.* In the rest of this section, we will take it as given that all our arithmetic is done mod 2, but our arguments apply equally well if work with integers mod p for any prime p.

Now consider the family of hash functions h where **A** and **b** range over all $\ell \times n$ matrices and all ℓ-dimensional vectors. We will show that if **A** and **b** are chosen uniformly at random, this family is pairwise independent.

First note that for any $\mathbf{x} \in \{0,1\}^n$ and any $\mathbf{w} \in \{0,1\}^\ell$, the probability that $\mathbf{Ax}+\mathbf{b}=\mathbf{w}$ is $1/2^\ell$. This follows from the randomness of **b**, since whatever **Ax** is, adding a uniformly random vector **b** to it makes it uniformly random in $\{0,1\}^\ell$.

Next we need to show that $h(\mathbf{x})$ and $h(\mathbf{y})$ are independent whenever $\mathbf{y} \neq \mathbf{x}$. In other words, even if we know that $\mathbf{Ax}+\mathbf{b}=\mathbf{w}$, $\mathbf{Ay}+\mathbf{b}$ is still uniformly random. This will follow from the randomness of **A**.

First we claim that, for any nonzero vector **u**, if **A** is uniformly random then so is **Au**. To see this, think of **Au** as the sum, over j, of u_j times the jth column of **A**. Then if we focus on any nonzero u_j, the randomness of the jth column is enough to make **Au** uniformly random.

Now we write

$$h(\mathbf{y}) = \mathbf{Ay}+\mathbf{b} = \mathbf{Ax}+\mathbf{b}+\mathbf{A}(\mathbf{y}-\mathbf{x}) = h(\mathbf{x})+\mathbf{A}(\mathbf{y}-\mathbf{x}).$$

If $\mathbf{y} \neq \mathbf{x}$, setting $\mathbf{u}=\mathbf{y}-\mathbf{x}$ shows that $\mathbf{A}(\mathbf{y}-\mathbf{x})$ is uniformly random. Therefore, whatever $h(\mathbf{x})$ is, $h(\mathbf{y})$ is uniformly random, and $h(\mathbf{x})$ and $h(\mathbf{y})$ are independent.

Finally, we note that **A** and **b** can be described with $\ell(n+1)=O(n^2)$ bits, and therefore we can choose them randomly by flipping $O(n^2)$ coins. Moreover, we can clearly calculate h in polynomial time. Therefore, we can use this family of hash functions to complete the argument of the previous section, and reduce SAT to UNIQUE SAT in randomized polynomial time.

The larger story here is that randomness is expensive—but in many cases, we don't need as much of it as we thought. A completely random function from $\{0,1\}^n$ to $\{0,1\}^\ell$ costs $2^{n+\ell}$ coin flips. But for the reduction from SAT to UNIQUE SAT to work, all we need is pairwise randomness, and we can get that with only poly(n) coin flips.

Pairwise independent hash functions have many other uses as well. We will see in Problems 11.2 and 11.3 that we can use them to distill purely random bits from slightly random ones, and even "recycle" random bits we have used before. This lets us run a randomized algorithm many times, making many nearly independent attempts, while using just a few truly random bits.

10.6.4 It's Lonely at the Top (or Bottom)

We close this section by learning another technique for making solutions unique—and whose analysis makes a seemingly thorny problem delightfully easy.

Suppose we have a graph G with n vertices and m edges. Typically, G has exponentially many spanning trees. If each edge e has a random integer weight $w(e)$, independently chosen from some interval, then there are spanning trees of many different weights. How wide do we need to make this interval if we want the MINIMUM SPANNING TREE to be unique with large probability?

For instance, the 4×4 grid has 24 edges and 100352 spanning trees. (We will learn how to do such calculations in Chapter 13.) If the edge weights are random, the total weight of the trees follows a roughly Gaussian distribution as shown in Figure 10.11. However, this distribution is full of complicated correlations, since the weights of two spanning trees are correlated whenever they have an edge in common.

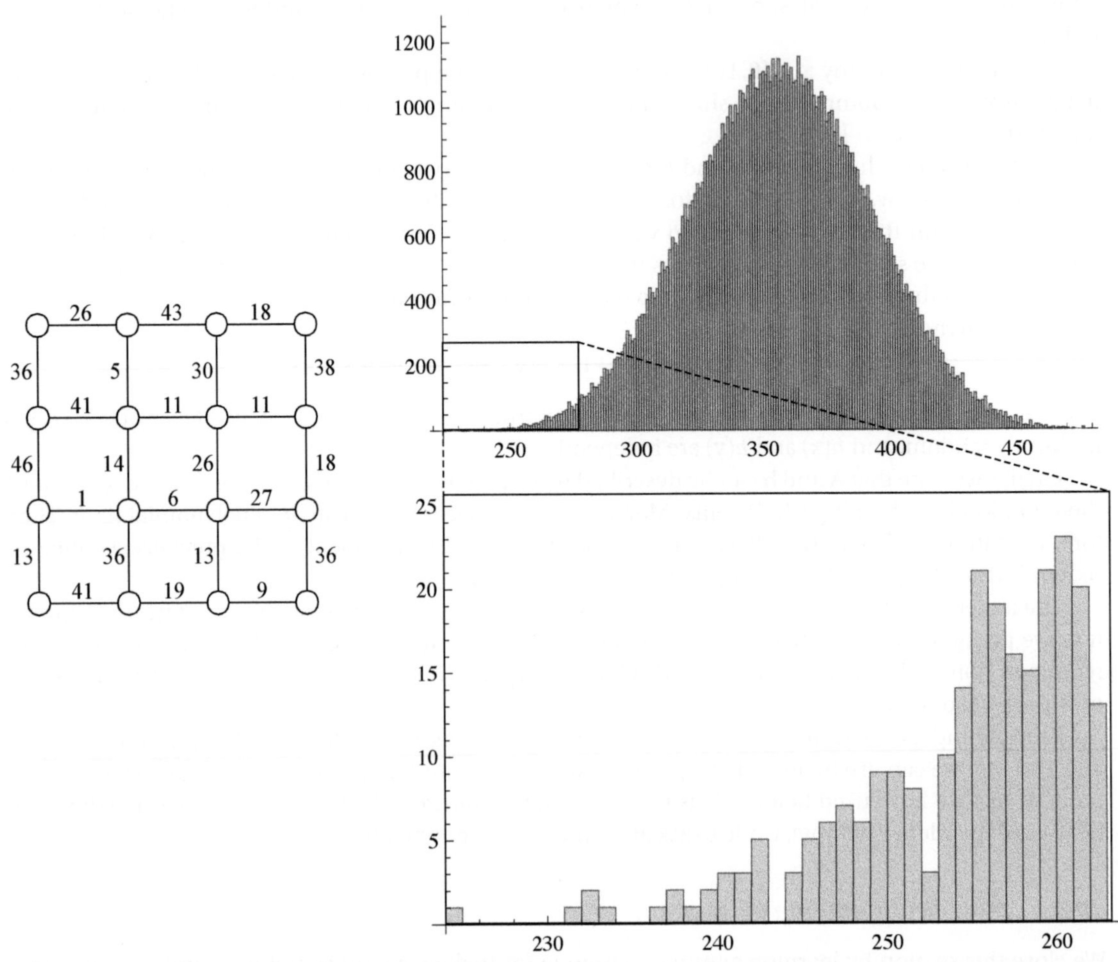

FIGURE 10.11: On the left, a graph G with 24 edges, each of which has a random weight $1 \leq w(e) \leq 48$. On the upper right, the histogram of the weights of all 100352 spanning trees. On the lower right, we zoom in on the 250 lightest spanning trees. The minimum spanning tree, whose weight is 224, is unique.

It seems difficult to reason about such a distribution, and especially about its extreme values—the lowest and highest weights in its tails. However, there is a clever probabilistic argument that shows that if each $w(e)$ is chosen uniformly from a sufficiently large set, such as the interval $\{1,\ldots,r\}$ where $r = \alpha m$ for some $\alpha > 1$, there is a significant chance that the lightest spanning tree is unique.

Let's call an edge e *optional* if, among the lightest spanning trees, there are some that include e and some that don't. Obviously, if the lightest spanning tree is not unique, there is at least one optional edge. We will bound the probability that this occurs, by computing the expected number of optional edges.

Imagine varying the weight of a particular edge e, while keeping the other edges' weights fixed. Let $t(e)$ be the largest weight that e can have while still appearing in one of the lightest trees. If we reduce $w(e)$ below this threshold, the trees containing e get even lighter—so if $w(e) < t(e)$ then e appears in all the lightest trees. On the other hand, if $w(e) > t(e)$, by definition e doesn't appear in any of them. Therefore, e can only be optional if $w(e) = t(e)$.

Now observe that $t(e)$ is only a function of the weights of the *other* edges, and that $w(e)$ is chosen independently of them. Thus, whatever $t(e)$ is, the probability that $w(e) = t(e)$ is at most $1/r$. Since there are m edges, the expected number of optional edges is at most m/r. By Markov's inequality, this is also an upper bound on the probability that there are any optional edges, so

$$\Pr[\text{lightest spanning tree is unique}] \geq 1 - \frac{m}{r} = 1 - \frac{1}{\alpha}.$$

Thus if $\alpha = 2$ and $r = 2m$, as we chose in Figure 10.11, the lightest spanning tree is unique with probability at least $1/2$. By symmetry, the same is true of the heaviest one, although not necessarily at the same time.

As the reader will probably have noticed, this argument has nothing whatsoever to do with spanning trees. It applies equally well to any problem where we want some kind of subset of the vertices or edges of a graph—or any other set, for that matter—whose total weight is minimized or maximized. To state it as generally as possible,

Lemma 10.2 (The Isolation Lemma) *For any set $S = \{e_1,\ldots,e_m\}$ of size m, and any family $\{T_1,\ldots,T_N\}$ of subsets of S, choosing weights $w(e_i)$ independently and uniformly from $\{1,\ldots,\alpha m\}$ makes the minimum-weight T_j unique with probability at least $1 - 1/\alpha$.*

The original application of this lemma was to give a fast parallel algorithm for PERFECT MATCHING. By picking out a unique matching of minimal weight, and connecting matchings with matrix determinants in a way that we describe in Section 10.7.2, it makes it possible for multiple processors to determine, in parallel, which edges appear in that matching.

10.10

10.7 The Roots of Identity

In this section we look at another important building block of randomized algorithms—*identity testing*, in which we confirm that two polynomials are identical by testing them at a few random places. We will use this to build an elegant randomized algorithm for PERFECT MATCHING, and in Chapter 11 it will give us an interactive proof that one player has a winning strategy in a game.

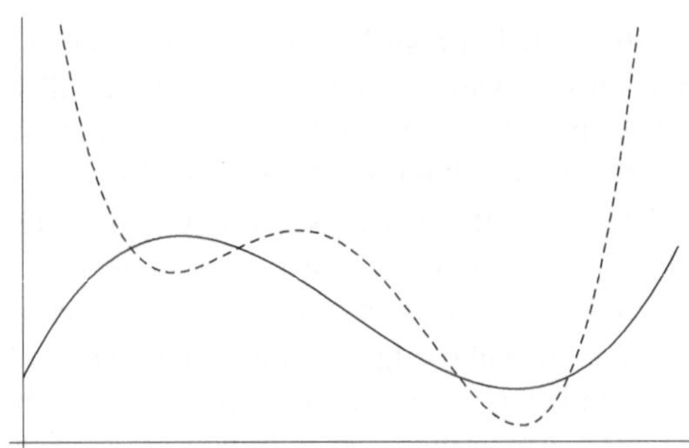

FIGURE 10.12: Two polynomials $f(x)$, $g(x)$ of degree d can coincide at at most d values of x. Here $d = 4$.

10.7.1 Digging at the Roots

Suppose I wish to convince you that two functions f and g are identical—that is, that $f(x) = g(x)$ for all x. With profuse apologies, I tell you that it's a little too complicated for me to prove this analytically. But I offer to play a game with you. You can challenge me with any value of x you like, and I will respond by showing you that $f(x) = g(x)$ for that particular value.

Of course, showing that $f(x) = g(x)$ for a particular x is hardly a proof that $f(x) = g(x)$ everywhere. But under the right circumstances, it can be very convincing. Specifically, suppose that f and g belong to a family of functions with the following property: for any two functions in the family, either they are identical, or they coincide at just a few values of x. Then if I am being dishonest and f and g are different, you will only be fooled if you choose one of the few values of x at which they coincide. You can make this very unlikely by choosing x randomly, so a few rounds of this game should give you a high degree of confidence.

The most important, and common, families of this kind consist of polynomials. If f and g are polynomials of degree d, then $f(x) = g(x)$ if and only if x is a root of the polynomial $\phi(x) = f(x) - g(x)$. Since ϕ also has degree d, it has at most d roots unless it is identically zero—that is, unless $\phi(x) = 0$ for all x. Therefore, if you choose x randomly from a set S, the probability that $f(x) = g(x)$ is at most $d/|S|$ unless f and g really are identical. We illustrate this in Figure 10.12,

In the process of evaluating $f(x)$ and $g(x)$, we might find the integers involved getting too large to handle. In that case, we can carry out identity testing mod p for some n-digit prime p, rather than on the entire set of integers. Like the real or complex numbers, the integers mod p form a field whenever p is prime (see Appendix A.7.4). Therefore, a polynomial of degree d can have at most d roots mod p.

So we let p be a prime much larger than d, and we choose x randomly from $\mathbb{F}_p$. If $\phi(x) \equiv_p 0$, this confirms with high probability that $\phi \equiv_p 0$ everywhere, or equivalently, that $\phi(x) = f(x) - g(x)$ is a multiple of p for all x. But, as in the space probe example of Section 10.6.1, ϕ can't have too many prime divisors. Therefore, if p is chosen randomly from a reasonably large set of primes, this implies with high probability that $\phi = 0$ and $f = g$ everywhere.

Polynomial identity testing is one of the few problems for which we have a randomized algorithm, but for which no deterministic algorithm is known. Let's turn now to a nice, and unexpected, application: telling whether a graph has a perfect matching.

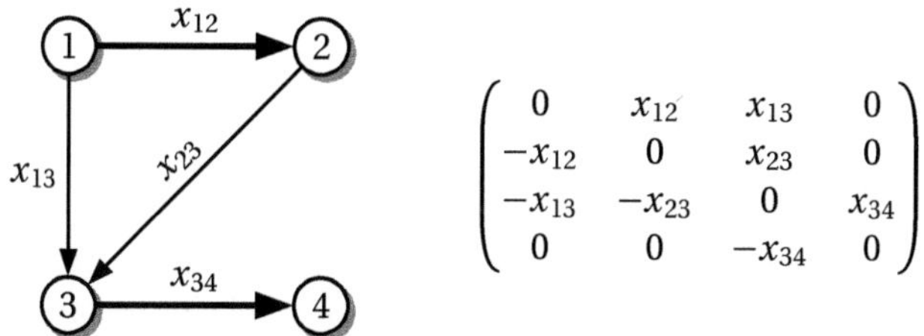

FIGURE 10.13: A graph G with an arbitrary orientation, and the matrix $A(\mathbf{x})$ defined in (10.24). Its determinant is $x_{12}^2 + x_{34}^2$, which corresponds to the perfect matching shown in bold.

10.7.2 From Roots to Matchings

The uses of polynomial identity testing are not limited to algebra. In this section, we will show that it yields a charming randomized algorithm for a graph-theoretic problem: PERFECT MATCHING, which we met in the bipartite case in Section 3.8.

First let's generalize identity testing to polynomial functions of more than one variable. The following lemma, which we ask you to prove in Problem 10.37, shows that sampling at a few random points works for these polynomials as well:

Lemma 10.3 (The Schwartz–Zippel Lemma) *Let $\phi(\mathbf{x}) = \phi(x_1, \ldots, x_n)$ be a multivariate polynomial where the largest total degree of any term is d. Suppose that each x_i is chosen independently from a set S. Then*

$$\Pr[\phi(\mathbf{x}) = 0] \le d/|S|$$

unless ϕ is identically zero.

Now suppose we have an undirected graph G with n vertices. Choose an arbitrary orientation of its edges, so that $i \to j$ or $j \to i$ for each edge (i, j). For each edge (i, j), define a variable x_{ij}. Denote the vector of all these variables $\mathbf{x}$, and consider the matrix $A(\mathbf{x})$ defined as follows:

$$A_{ij}(\mathbf{x}) = \begin{cases} x_{ij} & \text{if } i \to j \\ -x_{ji} & \text{if } j \to i \\ 0 & \text{if } i \text{ and } j \text{ are not adjacent.} \end{cases} \tag{10.24}$$

This is a weighted, antisymmetric version of G's adjacency matrix as shown in Figure 10.13. Its determinant is given by

$$\det A(\mathbf{x}) = \sum_{\pi} (-1)^{\pi} \prod_{i=1}^{n} A_{i,\pi(i)}.$$

Here π runs over all $n!$ permutations of the vertices, and $(-1)^{\pi}$ means $+1$ or -1 if π has even or odd parity—that is, if π can be written as a product of an even or odd number of swaps of two vertices each.

The heart of our algorithm is given by the following exercise:

Exercise 10.9 *Show that $\det A(\mathbf{x})$ is identically zero if and only if G has no perfect matchings. In one direction, show that any matching $\{(u_1, v_1), (u_2, v_2), \ldots, (u_{n/2}, v_{n/2})\}$ contributes a term $\prod_{i=1}^{n} x_{u_i, v_i}^2$.*

In the other direction, recall that any permutation can be written as a product of a set of cycles. Show that the contributions from permutations with odd cycles cancel out. Conclude that if $\det A \neq 0$*, there is at least one way to cover the graph with cycles of even length, and therefore at least one perfect matching.*

We will explore the relationship between determinants and matchings more deeply in Chapter 13.

The determinant $\det A(\mathbf{x})$ is a polynomial in the variables x_{ij}. While its degree is just n, in general it has $n!$ terms, so it would take too long to write it out explicitly. On the other hand, if we choose particular values for the x_{ij}, we can calculate $\det A$ in polynomial time using Gaussian elimination (see Section 13.1.3).

If we use integer values for the x_{ij}, these integers might become extremely large during the process of Gaussian elimination. So we use fingerprinting, and calculate $\det A(\mathbf{x}) \bmod p$ for some prime p. Exercise 10.9 shows that each perfect matching contributes a term with coefficient 1. Therefore, if G has any perfect matchings, $\det A \bmod p$ is not identically zero. If we choose each x_{ij} uniformly from $\mathbb{F}_p$ for some prime $p > n^2$, Lemma 10.3 tells us that $\det A(\mathbf{x}) \not\equiv_p 0$ with probability at least $1 - n/p > 1 - 1/n$ unless G has no perfect matchings.

Let's compare this with deterministic algorithms for this problem. If G is bipartite, we can tell whether it has a perfect matching by reducing BIPARTITE MATCHING to MAX FLOW as discussed in Section 3.6. However, in the case of general graphs, the standard algorithm for PERFECT MATCHING is considerably more complicated: namely, Edmond's "paths, trees, and flowers" algorithm (see Note 5.6). Once again, the
10.11 randomized algorithm is far simpler.

10.8 Primality

> The problem of distinguishing prime numbers from composite numbers, and of resolving the latter into their prime factors, is known to be one of the most important and useful in arithmetic... Further, the dignity of the science itself seems to require that every possible means be explored for the solution of a problem so elegant and so celebrated.
>
> Carl Friedrich Gauss, *Disquisitiones Arithmeticae* (1801)

One of the most fundamental number-theoretic problems is PRIMALITY. Given an n-digit integer p, is it prime? This problem has a rich history, starting with Fermat and Gauss and going on up through the modern age. A series of randomized algorithms developed in the 20th century finally led, in 2004, to a deterministic polynomial-time algorithm, proving that PRIMALITY is in P. Here we give a glimpse of this history, and of the number theory underlying it.

10.8.1 The Little Theorem That Could: the Fermat Test

Fermat's Little Theorem states that if p is prime, for all a we have

$$a^{p-1} \equiv_p 1\,.$$

We have met the Little Theorem twice: in Section 4.4, where we proved that PRIMALITY is in NP, and in Section 5.4.4, where we showed that modular Diophantine equations are NP-complete. We give a modern proof of it, using the fact that $\mathbb{Z}_p^* = \{1, 2, \ldots, p-1\}$ forms a group under multiplication, in Appendix A.7.1.

```
Fermat(p)
begin
    choose a randomly from {1,2,...,p−1} ;
    if gcd(a,p) ≠ 1 then return "composite" ;
    else if a^(p−1) ≢_p 1 then return "composite" ;
    else return "possibly prime" ;
end
```

FIGURE 10.14: The Fermat test. It catches most composite numbers, but it thinks that Carmichael numbers are prime.

Since any a that violates Fermat's Little Theorem is a witness that p is composite, one idea for a randomized algorithm for PRIMALITY is to choose a random a and check to see if indeed $a^{p-1} \equiv_p 1$. We answer "possibly prime" if it is, and "composite" if it isn't. The question is what fraction of the possible values of a violate the Little Theorem, and hence with what probability this algorithm returns "composite," if p is indeed composite.

To explore this, let us introduce a little more notation. Let p be an integer which may or may not be prime. Then an integer a has a *multiplicative inverse*, i.e., an a^{-1} such that $a^{-1}a \equiv_p 1$, if and only if a and p are mutually prime. So if $\mathbb{Z}_p^*$ denotes the set of nonzero integers mod p which are mutually prime to p, then $\mathbb{Z}_p^*$ forms a group under multiplication. For instance, $\mathbb{Z}_{15}^* = \{1,2,4,7,8,11,13,14\}$.

Exercise 10.10 *Write the multiplication table of $\mathbb{Z}_p^*$ for $p = 15$. For each element a, identify its inverse a^{-1}.*

Now consider the subset of $\mathbb{Z}_p^*$ consisting of those a which satisfy the Little Theorem:

$$H = \{a \in \mathbb{Z}_p^* : a^{p-1} = 1\}.$$

This is the set of a which fool the Fermat test—that is, which fail to serve as witnesses that p is composite. (Note that we write $=$ instead of $\equiv_p$ for elements of $\mathbb{Z}_p^*$.)

If a and b are in H, their product ab is in H as well since $(ab)^{p-1} = a^{p-1}b^{p-1} = 1$. Therefore, H is a *subgroup* of $\mathbb{Z}_p^*$. The size of a subgroup divides the size of the group (see Appendix A.7.1). Thus if H is a *proper* subgroup, i.e., if H is not all of $\mathbb{Z}_p^*$, then $|H|/|\mathbb{Z}_p^*| \le 1/2$. In other words, unless every choice of a satisfies the Little Theorem, at most half of them do. Since a is chosen randomly, the Fermat test will then return "composite" with probability at least $1/2$.

Exercise 10.11 *Find the subgroup $H \subseteq \mathbb{Z}_p^*$ for $p = 15$.*

This suggests the algorithm shown in Figure 10.14, known as the Fermat test. We choose a randomly from among the integers less than p. First we check that a and p are mutually prime, since if they share a factor we immediately know that p is composite. We then check whether $a^{p-1} \equiv_p 1$. We can compute $\gcd(a,p)$ using Euclid's algorithm, and compute $a^{p-1} \bmod p$ using repeated squaring as in Section 3.2.2, so both these steps can be carried out in polynomial time.

The only problem with the Fermat test is that there are composite numbers for which $a^{p-1} \equiv_p 1$ for all a, so that $H = \mathbb{Z}_p^*$. These are called *Carmichael numbers*, and the smallest one is $561 = 3 \times 11 \times 17$. The Fermat test will always return "possibly prime" for these numbers, so we have to tighten it somehow. 10.12

```
Miller-Rabin(p)
begin
    choose a randomly from {1,2,...,p−1} ;
    if gcd(a,p) ≠ 1 then return "composite" ;
    if a^{p−1} ≢_p 1 then return "composite" ;
    else begin
        let k = p−1 ;
        while a^k ≡_p 1 and k is even do k = k/2 ;
        if a^k ≢_p 1 and a^k ≢_p p−1 then return "composite ;
        else return "possibly prime" ;
    end
end
```

FIGURE 10.15: The Miller-Rabin test, which catches all composite numbers with probability at least 1/2.

10.8.2 *Square Roots: the Miller–Rabin Test*

We can tighten the Fermat test, and catch all the composite numbers, by using another question from number theory: what are the square roots of 1 mod p? That is, for what $x \in \mathbb{Z}_p^*$ is $x^2 = 1$? There are at least two such x for any p, namely 1 and $p-1 \equiv_p -1$. But if there are any others, p must be composite.

To see this, assume that $x^2 \equiv_p 1$. Then there is a k such that

$$x^2 - 1 = (x+1)(x-1) = kp\,. \tag{10.25}$$

Now assume that $x \not\equiv_p \pm 1$. Since neither $x+1$ nor $x-1$ is a multiple of p, the prime factors of p must be divided between $x+1$ and $x-1$. Thus $\gcd(x+1,p)$ and $\gcd(x-1,p)$ are each proper divisors of p,

Exercise 10.12 *Find the four square roots of* 1 *in* $\mathbb{Z}_{35}^*$, *and show how those other than* ± 1 *yield* 35*'s divisors.*

This argument shows that any square root of 1 other than 1 or $p-1$ is a witness that p is composite. Conversely, if p is divisible by two or more distinct odd primes, Problem 10.39 shows that there at least four square roots of 1 in $\mathbb{Z}_p^*$.

It would be nice if we could simply compute the square roots of 1, but since this would give us proper divisors of p, it is just as hard as FACTORING. Instead, we start by applying the Fermat test. Once we have an a that satisfies the Little Theorem, we consider the following sequence of numbers:

$$x_0 = a^{p-1} \equiv_p 1,\; x_1 = a^{(p-1)/2},\; x_2 = a^{(p-1)/4}, \ldots \tag{10.26}$$

Each of these numbers is a square root of the previous one. We continue dividing the power of a by 2 until it is odd, at which point we can take no more square roots. Thus $0 \le i \le t$ where 2^t is the largest power of 2 that divides $p-1$.

Consider the first x_i in this series such that $x_i \ne 1$, if there is one. Then $x_i^2 = x_{i-1} = 1$, so x_i is a square root of 1 other than 1. If $x_i \ne p-1$, we have found our witness and we return "composite." On the other hand, if $x_i = p-1$, or if $x_t = 1$ so that there is no $x_i \ne 1$, we return "possibly prime." This gives the algorithm shown in Figure 10.15, known as the *Miller–Rabin* test.

Just as we did for the Fermat test, let's consider the set of a that fool the Miller–Rabin test, and cause it to return "possibly prime" even when p is composite. This is the set of a which pass the Fermat test, but where either $x_t = 1$ or $x_i = p - 1$:

$$F = \left\{ a \in \mathbb{Z}_p^* : a^{p-1} = 1 \text{ and } (x_t = 1 \text{ or } x_i = p-1) \right\}. \tag{10.27}$$

While F is not necessarily a subgroup of $\mathbb{Z}_p^*$, Problem 10.40 shows that it is contained in a subgroup K. Moreover, if p is divisible by two or more distinct odd primes then K is a proper subgroup, and $|F|/|\mathbb{Z}_p^*| \le |K|/|\mathbb{Z}_p^*| \le 1/2$. As we argued for the Fermat test above, the Miller–Rabin test then returns "composite" with probability at least $1/2$ if a is chosen randomly.

This lets us catch all the composite numbers except for those which are only divisible by one odd prime: namely, those of the form q^k or $2^\ell q^k$ where q is prime. Happily, Problem 10.41 shows that it's easy to check whether p is a power of a prime, and it is certainly easy to check whether p is even. Moreover, it can be shown that such numbers cannot be Carmichael numbers, so they fail the Fermat test anyway. Therefore, for any composite number p, the Miller–Rabin test correctly identifies it as such at least half the time.

10.8.3 Derandomization and the Riemann Hypothesis

Let's review the Fermat and Miller–Rabin tests. In both of them, our goal is to find a witness a that p is composite. The sets H and F of non-witnesses, i.e., the values of a that fool these tests and make p seem prime, are contained in subgroups of $\mathbb{Z}_p^*$. If these subgroups are proper, they contain at most half the elements of $\mathbb{Z}_p^*$. So, at least half of the possible values of a are witnesses, and if we choose a uniformly we find a witness with probability at least $1/2$.

It's a little frustrating, however, that in a world where half the possible values of a are witnesses, we can't find one deterministically. Instead of choosing a randomly, can't we just scan through some set of polynomial size—say the a ranging from 1 to b for some $b = \text{poly}(n)$—and see if any of them work?

The answer is yes, if a celebrated conjecture in mathematics is true. The *Extended Riemann Hypothesis* (ERH) is a statement about the zeros of certain functions in the complex plane, or equivalently about the density of primes in certain series. It has the following interesting consequence: 10.13

Theorem 10.4 *If the ERH holds, then there is a constant C such that if a subgroup $K \subseteq \mathbb{Z}_p^*$ contains the set $\{1, 2, \ldots, C(\ln p)^2\}$ then $K = \mathbb{Z}_p^*$.*

Therefore, if the ERH holds, we don't need to sample randomly from all of $\mathbb{Z}_p^*$. Instead, we can just try the first $O(\log^2 p) = O(n^2)$ values of a. If they all satisfy the Miller–Rabin test, we know that p is prime. In fact, this version of the algorithm came first—Miller proposed it in 1976, and Rabin then proposed the randomized version in 1980.

This is an important kind of *derandomization*, in which we can sample deterministically from a set of polynomial size, rather than randomly from an exponentially large one. As we will see in Section 11.4, if sufficiently good pseudorandom generators exist, we can do this for any randomized algorithm. However, given our current state of knowledge, constructing these small sample sets, and proving that they work for particular problems, often involves sophisticated mathematics.

10.8.4 *Polynomial Identities Again, and the AKS Algorithm*

> It is in the nature of the problem that any method will become more complicated as the numbers get larger. Nevertheless, in the following methods the difficulties increase rather slowly... The techniques that were previously known would require intolerable labor even for the most indefatigable calculator.
>
> Carl Friedrich Gauss, *Disquisitiones Arithmeticae* (1801)

We conclude this section by presenting one more randomized algorithm for PRIMALITY. We then sketch how it can be derandomized to give a deterministic polynomial-time algorithm, thus finally proving that PRIMALITY is in P.

Consider the following polynomial identity,

$$(x+1)^p \equiv_p x^p + 1\,, \tag{10.28}$$

or equivalently

$$P(x) \equiv_p 0 \text{ where } P(x) = (x+1)^p - x^p - 1\,. \tag{10.29}$$

Note that this holds for all x if and only if all of P's coefficients are equivalent to zero mod p. Problem 10.42 shows that this is the case if and only if p is prime.

We might like to use the technique of polynomial identity testing, described in Section 10.7, to confirm that (10.28) holds. However, that technique relies on the fact that a polynomial of degree d can have at most d roots in a finite field—and $\mathbb{Z}_p$ is only a field if p is prime, which is what we're trying to prove. Moreover, $P(x)$ has degree p, so *a priori* its roots could include any subset of $\mathbb{Z}_p$.

How about using the fingerprinting idea of Section 10.6.1? By the Chinese Remainder Theorem (see Appendix A.7), we have $P \not\equiv_p 0$ if and only if $P \not\equiv_q 0$ for some prime q that divides p. But if we knew p's prime factors, we would certainly already know whether it is prime. So computing $P(x)$ modulo a random prime doesn't help us either.

It turns out that another kind of fingerprinting does work. Instead of computing $P(x)$ modulo a prime, we will compute it modulo a *polynomial.* The set of polynomials over x with coefficients in $\mathbb{Z}_p$ forms a *ring* (see Appendix A.7.4) denoted $\mathbb{Z}_p[x]$. Given two polynomials P and Q, we can define $P \bmod Q$ as the remainder after dividing P by Q, and obtaining the quotient with the largest possible leading term. For instance,

$$(x+1)^3 \bmod (x^2+1) = 2x - 2\,,$$

since

$$(x+1)^3 = (x+3)(x^2+1) + 2x - 2\,.$$

Returning to (10.29), if $P \equiv_p 0$ then $P \bmod Q \equiv_p 0$ for all Q. For instance, for $p = 3$ we have

$$P(x) \bmod (x^2+1) = \left((x+1)^3 - x^3 - 1\right) \bmod (x^2+1) = 3x - 3 \equiv_3 0\,.$$

If we can find a Q such that $P \bmod Q \not\equiv_p 0$—that is, such that Q is not a divisor of P in the ring $\mathbb{Z}_p[x]$—we know that $P \not\equiv_p 0$, and that p is composite.

Thus a sensible test would consist of choosing a random polynomial Q, and checking whether $P \bmod Q \equiv_p 0$. Specifically, we will choose Q from the set of all polynomials in $\mathbb{Z}_p[x]$ of degree d, whose leading term is x^d. Denote this set S, and note that $|S| = p^d$. We need to make d large enough so that at least $1/\text{poly}(n)$ of the polynomials in S do not divide P.

How large a degree d can we handle? If $d = \text{poly}(n)$ then we can compute $P \bmod Q$ in polynomial time, even though P has exponentially large degree, using the repeated-squaring algorithm for modular exponentiation in Section 3.2.2. For instance, we can compute $(x+1)^{2^n} \bmod Q$ by starting with $x+1$, squaring n times, and taking the result mod Q each time.

How many coins do we need to flip to choose Q? Since Q has d coefficients other than its leading term x^d, we need $d \log_2 p = O(dn)$ random bits. Since we only have time to flip $\text{poly}(n)$ coins, we can again handle $d = \text{poly}(n)$. On the other hand, if d is superpolynomial in n, we don't even have time to choose Q or write it down, nor can we compute $P \bmod Q$. So hopefully $d = \text{poly}(n)$ suffices.

Happily, we will show that $d = O(n)$ is enough. The analysis is similar in spirit to that in Section 10.6.1. There, we used the fact that a nonzero n-bit integer has at most n prime factors, and that these are a small fraction of the primes between 1 and n^2. Now we need to bound the fraction of polynomials of degree d that divide P, and show that these are not too large a fraction of S.

Just as integers have unique prime factorizations, if p is prime then polynomials in $\mathbb{Z}_p[x]$ can be factored uniquely into *irreducible* polynomials—those which cannot be factored into products of smaller ones. For instance, the polynomial x^3+x+1 is irreducible in $\mathbb{Z}_2[x]$, while in $\mathbb{Z}_3[x]$ it can be factored uniquely as $(x^2+x+2)(x+2)$.

A result analogous to the Prime Number Theorem tells us that roughly $1/d$ of the p^d polynomials in S are irreducible. On the other hand, if P is a nonzero polynomial of degree p, it can have at most p/d irreducible factors of degree d. Recall that $|S| = p^d$, and assume for simplicity that exactly $1/d$ of these are irreducible. Then the number of irreducible polynomials in S that do not divide P is at least

$$\frac{p^d}{d} - \frac{p}{d}.$$

Therefore, if we choose Q uniformly from S, we get a non-divisor of P with probability at least

$$\frac{1}{p^d}\left(\frac{p^d}{d} - \frac{p}{d}\right) = \frac{1}{d}\left(1 - \frac{1}{p^{d-1}}\right).$$

Unfortunately, when p is composite—which is the case we're trying to find a witness for—the factorization of polynomials in $\mathbb{Z}_p[x]$ into irreducibles is not unique. However, Problem 10.43 shows that unless p is a prime power, $P \not\equiv_q 0$ for any prime q that divides p. So, we can carry out a similar counting argument by asking how often $Q \bmod q$ is a non-divisor of $P \bmod q$ instead. We don't need to know what q is—we're just using it to perform our analysis.

Let S_q denote the set of polynomials in $\mathbb{Z}_q[x]$ of degree d with leading term x^d. Then if Q is uniformly random in S, $Q \bmod q$ is uniformly random in S_q. There are q^d polynomials in S_q, and we again assume that $1/d$ of them are irreducible. On the other hand, using the unique factorization of polynomials in $\mathbb{Z}_q[x]$, at most p/d of them can be factors of $P \bmod q$. Thus the probability that $Q \bmod q$ does not divide $P \bmod q$ is at least

$$\frac{1}{q^d}\left(\frac{q^d}{d} - \frac{p}{d}\right) = \frac{1}{d}\left(1 - \frac{p}{q^d}\right).$$

Since $q \ge 2$, this probability is $\Theta(1/d)$ whenever $d > \log_2 p = n$. So setting $d = 2n$, say, lets us find a non-divisor of P, and thus a proof that p is composite, with probability $\Theta(1/n)$. By repeating the algorithm n times, we learn that p is composite with constant probability.

This gives us a probabilistic algorithm for PRIMALITY based on rather different ideas from the Fermat and Miller–Rabin tests. Three years after this algorithm was published, Manindra Agrawal, Neeraj Kayal, and Nitin Saxena found a way to derandomize it—in other words, to restrict Q to a set of polynomial size, so that we can try all possible Q in polynomial time. In essence, they showed that it suffices to check the slightly different identity

$$(x+a)^p \equiv_p x^p + a$$

mod $Q(x) = x^r - 1$, where a ranges from 1 to poly($\log p$) = poly(n), and r is a particular integer which is also poly(n). This is just a polynomial number of different identities, and we can check every one mod $Q(x)$ in polynomial time. This breakthrough finally established, after decades of work—or, if you start with Fermat, centuries—that PRIMALITY is in P.

10.14

10.9 Randomized Complexity Classes

We conclude this chapter by defining several complexity classes, consisting of problems we can solve with various kinds of randomized algorithms in polynomial time. Each of these classes will correspond to a certain kind of guarantee, and certain kinds of errors, that the randomized algorithm can make.

Where decision problems are concerned, a *Monte Carlo* algorithm is one that gives the right answer a clear majority of the time. The class BPP, for "bounded-error probabilistic polynomial time," consists of the problems we can solve with a Monte Carlo algorithm in polynomial time:

> BPP is the class of decision problems for which there is a randomized algorithm A, which always runs in poly(n) time and which gives the correct answer with probability at least $2/3$. That is, for all x, we have
>
> $$\Pr[A \text{ returns "yes"}] \geq 2/3 \text{ if } x \text{ is a yes-instance}$$
> $$\Pr[A \text{ returns "yes"}] \leq 1/3 \text{ if } x \text{ is a no-instance}.$$

The constants $1/3$ and $2/3$ in this definition are completely arbitrary. As Problem 10.46 shows, if we run a BPP algorithm multiple times and take the majority of the results, we can amplify the probability of getting the right answer until it is exponentially close to 1.

BPP algorithms are allowed to have *two-sided* error. In other words, their answer can be wrong whether x is a yes-instance or a no-instance. An algorithm with *one-sided* error is slightly stronger, in that it guarantees its answer is correct in one of these cases. In particular, RP is the class of problems for which there is a randomized algorithm that is correct whenever it says "yes":

> RP is the class of decision problems for which there is a randomized algorithm A, which always runs in poly(n) time and which returns "yes" or "don't know" such that, for all x,
>
> $$\Pr[A \text{ returns "yes"}] \geq 1/2 \text{ if } x \text{ is a yes-instance}$$
> $$\Pr[A \text{ returns "yes"}] = 0 \text{ if } x \text{ is a no-instance}.$$

Of course, if we run an RP algorithm t times on a yes-instance, the probability that at least one run returns "yes" is greater than or equal to $1-2^{-t}$.

While BPP, like P, is symmetric under complementation, RP might not be. We can define its mirror image coRP as the class of problems for which there is an algorithm A that returns "no" or "don't know," such that A returns "no" with probability at least $1/2$ on a no-instance, and never on a yes-instance. For instance, since the Miller–Rabin algorithm returns "composite" at least half the time if p is composite, and never returns "composite" if p is prime, PRIMALITY is in coRP.

Las Vegas algorithms have the additional property that their answers can be checked, so we can confirm that they are correct. Equivalently, they either answer the question correctly or return "don't know." Then ZPP, for "zero-error probabilistic polynomial time," is the class of problems for which there is a Las Vegas algorithm whose running time is polynomial, and that answers the question a reasonable fraction of the time:

> ZPP is the class of decision problems for which there is a randomized algorithm A, which always runs in poly(n) time and which either returns the correct answer or returns "don't know." Moreover, for all inputs x, A returns the correct answer with probability at least $1/2$.

There is a simple tradeoff between uncertainty about the running time and uncertainty about the answer. Consider the following alternate definition of ZPP:

> ZPP is the class of decision problems for which there is a randomized algorithm that always returns the correct answer, and whose expected running time is poly(n).

Exercise 10.13 *Show that these two definitions are equivalent. In one direction, use Markov's inequality (see Appendix A.3.2) to show that if we run an algorithm for twice its expected running time, it returns the right answer with probability at least* $1/2$. *In the other direction, simply run the algorithm until it succeeds.*

The next two exercises show some simple relationships between these classes:

Exercise 10.14 *Show that*

$$\mathsf{RP} \subseteq \mathsf{BPP} \text{ and } \mathsf{coRP} \subseteq \mathsf{BPP}.$$

Hint: run an RP *algorithm twice.*

Exercise 10.15 *Show that*

$$\mathsf{ZPP} = \mathsf{RP} \cap \mathsf{coRP}\,.$$

Hint: if a problem is in RP∩coRP*, it has an* RP *algorithm and a* coRP *algorithm. Run them both. Compare Exercise 7.7.*

How are these classes related to P and NP? They certainly all contain P, since deterministic algorithms are just randomized algorithms that don't flip any coins, and that always give the right answer. On the other hand, any string of coin flips that causes an RP algorithm to return "yes" is a witness that x is a yes-instance, since it is always correct in this case. Therefore,

$$\mathsf{RP} \subseteq \mathsf{NP}\,,$$

and similarly coRP ⊆ coNP.

However, it seems likely that RP is smaller than NP. There is an obvious randomized algorithm for NP-complete problems: namely, just try a random witness, and see if it works. However, if there are yes-instances with just a few witnesses—such as a graph that can only be colored in a few ways—then the probability that this algorithm works is exponentially small. If there is only one needle, dipping randomly into the haystack won't work very often. In contrast, RP is the class of problems where, if x is a yes-instance, half the haystack consists of needles.

Is BPP contained in NP? This is not so clear. For a BPP problem, no single run of the program proves that x is a yes-instance, since both "yes" and "no" answers can be wrong. The claim that x is a yes-instance is now a claim that 2/3 of the possible runs of the program return "yes"—but since a randomized algorithm that runs in time t can flip t coins, the number of possible runs of a BPP algorithm is $2^{\text{poly}(n)}$. Therefore, it is not obvious how to prove this claim deterministically without checking an exponential number of possible runs, and it is logically possible that BPP and NP are incomparable.

Exercise 10.16 *Suppose I promise you that the fraction of a haystack consisting of needles is either less than* 1/3 *or greater than* 2/3*. To learn which of these is true, show that you need to examine* 2/3 *of the haystack in the worst case.*

Now we come to the following question. Where algorithms are concerned, does randomness help at all? Are these complexity classes larger than P? Many randomized algorithms can be derandomized, and replaced with deterministic versions. However, as we saw in Section 10.8, doing this often involves sophisticated mathematics. For instance, we can derandomize the Miller–Rabin algorithm for PRIMALITY using the Extended Riemann Hypothesis, assuming that it holds. We can also derandomize `WalkSAT` and the Goemans–Williamson algorithm for MAX CUT, but both of these require clever ideas from error-correcting codes.

One way we could derandomize all randomized algorithms at once is to find a pseudorandom generator—a deterministic process that produces bits that seem random to any polynomial-time algorithm. If such a generator exists, then ZPP, RP, and BPP are all equal to P. Such generators are one of the chief topics of the next chapter.

10.15

Problems

> Fortune can, for her pleasure, fools advance,
> And toss them on the wheels of Chance.
>
> Juvenal

10.1 Worst cases for `Quicksort`. A deterministic version of `Quicksort` uses the first element of the list as the pivot. Show that there are 2^{n-1} ways the adversary could order the input list so that this algorithm does the worst possible number of comparisons.

10.2 Decision trees on average. In Section 6.2 we used the "Twenty Questions" argument to show that any comparison-based sorting algorithm needs to make at least $\log_2 n! \approx n \log_2 n$ comparisons to sort a list of n elements. Specifically, we viewed the algorithm as a decision tree, and used the fact that any binary tree with ℓ leaves has depth at least $\log_2 \ell$. However, this argument only shows that *some* leaves are $\log_2 \ell$ steps from the root. We might hope that the *average* number of comparisons can be smaller. Dash this hope by showing that if a binary tree has ℓ leaves, the average over all leaves of the depth is at least $\log_2 \ell$.

Hint: there are two nice ways to solve this problem. One is to use induction on subtrees. The other is to consider a random walk from the root of the tree, going left or right with equal probability at each branch, so that we arrive at a leaf at depth d with probability 2^{-d}. Then use Jensen's inequality (see Appendix A.3.6), which states that $\mathbb{E}[f(x)] \geq f(\mathbb{E}[x])$ for any convex function f, i.e., any function whose second derivative is nonnegative.

10.3 `Quicksort`, pair by pair. We showed in Section 3.2.1 that the average number of comparisons that `Quicksort` does on a list of n items approaches $2n \ln n$ when n is large. Here is another way to do this. Let X_{ij} be the indicator random variable for the event that the ith smallest item in the list ever gets compared with the jth smallest one: in other words, $X_{ij} = 1$ if this occurs and 0 otherwise. Then the total number of comparisons is

$$X = \sum_{1 \leq i < j \leq n} X_{ij} .$$

By linearity of expectation (see Section A.3.2) we have

$$\mathbb{E}[X] = \sum_{1 \leq i < j \leq n} \mathbb{E}[X_{ij}] ,$$

and $\mathbb{E}[X_{ij}]$ is the probability that these two items are compared. Show that this gives

$$\mathbb{E}[X] = 2 \sum_{j=1}^{n} (H_j - 1) ,$$

where $H_j = \sum_{k=1}^{j} 1/k$ is the jth harmonic number. Check that this gives exactly the right answer for a few values of n; for instance, for $n = 3$ we have $\mathbb{E}[X] = 8/3$, and for $n = 4$ we have $\mathbb{E}[X] = 29/6$. Then recover our earlier result $\mathbb{E}[X] \approx 2n \ln n$ by writing $H_j \approx \ln j$ and replacing the sum over j with an integral.

Hint: every item eventually gets chosen as a pivot, if only in the base case where the sublist consists just of that item. Whether or not the ith item is ever compared with the jth one depends entirely on which item in the range $\{i, \ldots, j\}$ is chosen first as the pivot. If it is between i and j, these two items are separated into two sublists, and will never be compared; but if it is i or j, we compare i and j.

10.4 Bad running times are rare. We know that `Quicksort`'s running time is $O(n \log n)$ on average. Let's prove the stronger statement that this is true with high probability. First, review the recursion tree shown in Figure 3.5. Since each level of recursion takes $O(n)$ time to compare the pivot with each item in the sublist, we wish to show that the depth of this tree is $O(\log n)$ with high probability.

Focus on a particular item x in the list, and show that choosing a random pivot cuts down the expected size of the sublist containing x—which we define as zero if x has already been chosen as a pivot—by a constant factor. Therefore, the expected size of this sublist decreases exponentially as a function of depth. Then use the first moment method (see Appendix A.3.2) to show that with high probability x is chosen as a pivot at a depth at most $b \log n$ for some constant b. Use the union bound (see Appendix A.3.1) to bound the probability that this fails to be true for any of the n items. Finally, conclude that, for any constant a, there is a constant b such that the probability the running time exceeds $b n \log n$, and therefore that the depth exceeds $b \log n$, is at most n^{-a}.

10.5 Kruskal and Karger. Recall from Section 3.5.1 that Kruskal's algorithm for MINIMUM SPANNING TREE works by adding edges to a forest until the forest becomes one big tree. At each step, we add the lightest edge that doesn't complete a cycle, i.e., whose endpoints lie in different trees.

Suppose all edges have equal weight, and that we choose randomly from all the edges which don't complete a cycle. If we stop one edge short of a spanning tree, so that we have two trees left, they define a cut. Show that this is equivalent to Karger's algorithm for MIN CUT.

10.2

10.6 When Karger doesn't work. Suppose we want to solve s-t MIN CUT as defined in Section 3.7, in which we want to separate a specific pair of vertices s and t from each other. We might try a modification of Karger's algorithm, in which at each step we choose randomly from the edges, except those that connect s to t directly, and contract it, continuing until s and t are the only two vertices left. Show that this approach does not work. Specifically, construct a family of graphs or multigraphs for which the probability that this algorithm finds the minimum s-t cut is exponentially small.

10.7 $3n$ steps suffice. Show that if we start a distance d from the origin and follow a biased random walk, where d increases with probability $2/3$ and decreases with probability $1/3$, then the probability that we are exactly at $d = 0$ after $t = 3d$ steps is $\Theta(2^{-d}/\sqrt{d})$. Pretend that d is allowed to be negative—if this happens it just means that we reach $d = 0$ earlier.

Thus the probability that `WalkSAT` finds the satisfying assignment after $3d$ steps is within a polynomial factor of the probability that it ever will if we let it walk forever. Hint: write this probability as the product of a binomial and an exponential term, and use Stirling's approximation $n! \approx n^n \mathrm{e}^{-n} \sqrt{2\pi n}$.

10.8 Which distances matter? By maximizing the summand in the third line of (10.3), find which initial value of the Hamming distance d from the satisfying assignment contributes the most to `WalkSAT`'s overall probability of success. Is it the typical value $n/2$, or some other constant times n?

10.9 A random walk of colorings. Design a `WalkSAT`-like algorithm for GRAPH 3-COLORING that chooses a random edge whose endpoints are the same color, and recolors one of its endpoints. Show that if G is a 3-colorable graph with n vertices, it finds a coloring for G with high probability in $(3/2)^n \operatorname{poly}(n)$ time.

Hint: define the Hamming distance between two colorings as the number of vertices on which they differ, fix some proper 3-coloring C, and describe the change in the distance between the current coloring and C as a biased random walk. Finally, average the probability that this walk will touch the origin over all initial colorings a distance d from C as in (10.3).

10.10 Forbidding colors. Here is another simple randomized algorithm for GRAPH 3-COLORING. For each vertex v, choose randomly from the three colors, and restrict v's color to the other two. Show that the question of whether we can 3-color the graph with these colors is equivalent to an instance of 2-SAT. (You may already have proved this in

Problem 5.6.) Then show that if G is a 3-colorable graph with n vertices, we can find a coloring with high probability in $(3/2)^n \operatorname{poly}(n)$ time by choosing this restriction randomly.

10.11 Random restarts and heavy tails. When do random restarts make sense? Suppose that the probability $P(t)$ that an algorithm will succeed on the tth step is exponentially distributed with mean τ:

$$P(t) = \frac{1}{\tau} \mathrm{e}^{-t/\tau}.$$

Now suppose that we have been running the algorithm for t_0 steps, and it still hasn't succeeded. Show that the remaining time we need to run the algorithm has exactly the same distribution $P(t)$ that it did before, so that in particular the expected time we have to wait is τ. This is analogous to radioactive decay in physics, where the half-life of a particle is the same regardless of how long it has been in existence. Thus there is no advantage to starting over, and we might as well simply continue.

On the other hand, suppose that the running time has a heavy-tailed or power-law distribution, such as

$$P(t) \propto t^{-\alpha}.$$

Assume that $\alpha > 2$, since otherwise the expected time diverges. Show that in this case, the expected waiting time gets worse as t_0 increases—so the longer we've been waiting, the longer we have to wait. Therefore, if the algorithm doesn't succeed fairly quickly, it's better to start over rather than continuing to run it.

10.12 Randomized ratios. Suppose a randomized algorithm gives an approximate solution to a maximization problem, such that the expected ratio between the algorithm's solution and the optimum is α. Then show that for any $\varepsilon > 0$, the algorithm produces solutions for which this ratio is at least $\alpha - \varepsilon$ with probability $p(\varepsilon) \geq \varepsilon$. Indeed, show that when ε is small, $p(\varepsilon)$ is closer to $\varepsilon/(1-\alpha)$. Then, use the boosting techniques at the end of Section 10.2 to show that we achieve this ratio with constant probability after running the algorithm $1/\varepsilon$ times.

10.13 Vector cuts. Suppose G is a complete graph with n vertices, where all $\binom{n}{2}$ edges have unit weight. One way to design a vector cut for G is to divide the vertices into k groups of n/k each, and associate each group of vertices with the vector of length 1 pointing toward one of the corners of a $(k-1)$-dimensional simplex centered at the origin. For instance, if $k = 4$ these vectors point to the corners of a tetrahedron. The inner product of any two such vectors, if they are distinct, is

$$\mathbf{s}_i^T \mathbf{s}_j = -\frac{1}{k-1}.$$

Use this fact to show that, for any k that divides n, the resulting vector cut has weight $n^2/4$. This is precisely the weight of the MAX CUT of such a graph, which corresponds to the case $k = 2$.

10.14 Better cuts are easier to approximate. The best approximation ratio that the Goemans–Williamson algorithm can achieve for worst-case graphs is 0.878..., but if the MAX CUT is large it does much better. For simplicity, consider the case where G is unweighted. Suppose that G has m edges, and that its MAX CUT has weight $W = qm$. Show that if $q \geq q_0 = 0.844...$, the Goemans–Williamson algorithm produces a cut whose expected weight is at least $\alpha(q)W$, where

$$\alpha(q) = \frac{\cos^{-1}(1-2q)}{\pi q}.$$

Note that $\alpha(q)$ is minimized at q_0, where $\alpha(q_0)$ is the worst-case ratio $\alpha = 0.878...$ we calculated in the text. Then conclude that if G is almost bipartite, so that $q = 1 - \varepsilon$ for some small ε, the Goemans–Williamson algorithm gives a cut of size $1 - \varepsilon'$ where

$$\varepsilon' \leq \frac{2}{\pi}\sqrt{\varepsilon}.$$

Hint: show that if we average over all the edges, the average contribution to the weight of the MAX VECTOR CUT is

$$\mathbb{E}\left[\frac{1-\cos\theta_{ij}}{2}\right] \geq q\,.$$

Then use Jensen's inequality from Appendix A.3.6.

10.15 Minimax from duality. Here is one proof of von Neumann's Minimax Theorem. First, review the definition of LINEAR PROGRAMMING in Section 9.4. Then, given a payoff matrix $\mathbf{M}$, show that

$$v_1 = \min_{\mathbf{a}} \max_{\mathbf{b}} \mathbf{a}^T \mathbf{M} \mathbf{b}$$

can be written as the following instance of LINEAR PROGRAMMING. Our variables are v and the components a_i of the mixed strategy. We minimize v subject to the following constraints:

$$\forall i : a_i \geq 0$$
$$\sum_i a_i = 1$$
$$\forall j : \sum_i a_i M_{ij} \leq v\,.$$

Hint: use the form of the Minimax Theorem given in (10.16), where the second player might as well respond with a pure strategy. Then, show that

$$v_2 = \max_{\mathbf{b}} \min_{\mathbf{a}} \mathbf{a}^T \mathbf{M} \mathbf{b}$$

is the solution to the dual maximization problem as defined in Section 9.5. Conclude, using the fact that the primal and dual problems have the same optimum, that $v_1 = v_2$.

10.16 Minimax from fixed points. Here is another proof of the Minimax Theorem. Suppose that we have a two-player game with payoff matrix $\mathbf{M}$, and that the players currently have mixed strategies $\mathbf{a}$ and $\mathbf{b}$. In an effort to improve her payoff, the first player could increase the probability a_i of any move i that does better against $\mathbf{b}$ than her current strategy. So, define

$$c_i = \begin{cases} \sum_j M_{ij} b_j - \mathbf{a}^T \mathbf{M} \mathbf{b} & \text{if } \sum_j M_{ij} b_j > \mathbf{a}^T \mathbf{M} \mathbf{b} \\ 0 & \text{otherwise}, \end{cases}$$

and define a new strategy $\mathbf{a}'$ as follows:

$$a_i' = \frac{a_i + c_i}{1 + \sum_k c_k}\,,$$

where the denominator maintains the normalization $\sum_i a_i' = 1$. Define a strategy $\mathbf{b}'$ for the second player similarly.

Brouwer's Fixed Point Theorem (see Section 6.7.3) states that, given any continuous map f from a closed, simply connected set into itself, there is an $\mathbf{x}$ for which $f(\mathbf{x}) = \mathbf{x}$. The set of strategy pairs, i.e., vectors $(\mathbf{a}, \mathbf{b})$ such that $\sum_i a_i = \sum_j b_j = 1$, is closed and simply connected. Show that the map from $(\mathbf{a}, \mathbf{b})$ to $(\mathbf{a}', \mathbf{b}')$ defined above is continuous. Then show that if $(\mathbf{a}, \mathbf{b})$ is a fixed point of this map, i.e., if $\mathbf{a}' = \mathbf{a}$ and $\mathbf{b}' = \mathbf{b}$, then (10.15) holds.

10.17 Adversarial trees. Suppose we follow a deterministic strategy for the NAND tree in Section 10.5.1: in other words, a deterministic algorithm that takes the values of the leaves we have seen so far, and determines which leaf to evaluate next. This algorithm might proceed in a top-down manner, by completely evaluating a subtree, or it might skip around among the leaves in an arbitrary way. Show that for any such algorithm, the adversary can choose truth values for the leaves that force us to evaluate every leaf. Hint: show by induction that he can make the value of any subtree depend on the last leaf in it that we look at.

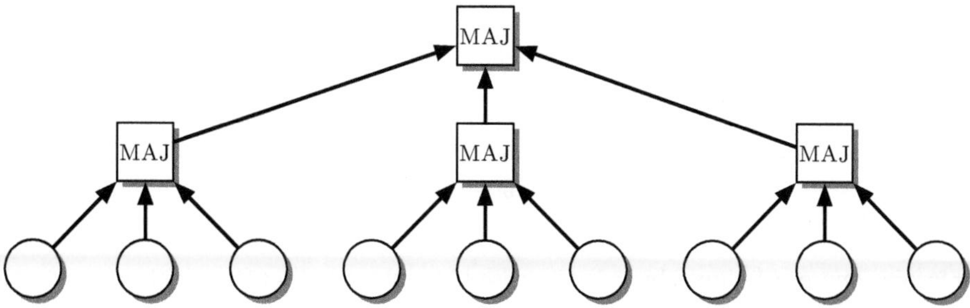

FIGURE 10.16: A majority tree of depth 2.

10.18 I will meet you by the witness tree. The smallest possible proof, or witness, of the value of an NAND tree consists of a single `false` child of each `true` node, and both children of each `false` node, until we have a set of leaves that determine the value of the root. Show that in any NAND tree of depth k and $N = 2^k$ leaves, this witness consists of exactly $2^{k/2} = \sqrt{N}$ leaves if k is even. Of course, this witness might be hard to find, but this does give a simple lower bound on the number of leaves we need to evaluate in order to evaluate the tree.

10.19 Independent leaves. Suppose we choose the truth values of the leaves of a NAND tree independently, setting each one `true` with probability p and `false` with probability $1-p$. Show that if

$$p = \frac{\sqrt{5}-1}{2} = \varphi - 1 = 0.618...$$

where φ is the golden ratio, then each node of the tree is `true` with probability p, and the nodes on a given level are independent of each other. Then argue that any depth-first algorithm is forced to evaluate both children of each node with probability p. Finally, modify the analysis of Section 10.5.1 to show that, if the tree has depth k and $N = 2^k$ leaves, the expected number of leaves it evaluates is

$$\varphi^k = N^{\log_2 \varphi} = N^{0.694...}.$$

This is a good lower bound, but it doesn't quite show that our $N^{0.753...}$ algorithm is optimal.

10.20 Trees of larger arity. Suppose we have a tree where each node has d children, so that a tree of depth k has $N = d^k$ leaves. Each node has a NAND gate, which outputs `true` if at least one of its children is `false`, and outputs `false` if all of its children are `true`. Such a tree corresponds to a game where there are d possible moves at each point. By generalizing the analysis of Section 10.5.1, show that by evaluating subtrees in random order, we can evaluate the tree by looking at an average of $\alpha^k = N^{\log_d \alpha}$ leaves, where

$$\alpha = \frac{d-1+\sqrt{d^2+14d+1}}{4}.$$

10.21 Majority trees. Suppose we have a tree where each node has three children, so a tree of depth k has $N = 3^k$ leaves. Place a MAJORITY gate at each node which outputs the majority of its children's truth values as shown in Figure 10.16. Show that by evaluating the subtrees in random order, we can evaluate the entire tree while only looking at

$$(8/3)^k = N^{\log_3(8/3)} = N^{0.892...}$$

leaves on average. Then use Yao's principle to show that this is the optimal depth-first algorithm. Hint: with what probability do the first two inputs of a majority gate determine its output? What is the worst type of input the adversary can throw at us?

Interestingly, unlike the NAND tree, in this case the depth-first algorithm is not optimal. Rather than evaluating an entire subtree before moving on to the next, we can skip down to a grandchild of the current node, and use that to guess which children we should evaluate first. This gives a slightly better algorithm, which looks at $2.654^k = N^{0.888}$ leaves. It is not known whether this is optimal.

10.6

10.22 Fewer factors. Improve the upper bound of Exercise 4.21 to show that any integer x has at most $O(\log x / \log\log x)$ distinct prime factors. Hint: if we list the primes in increasing order, the ith prime is at least i.

10.23 An infinity of primes. There are many proofs that there are an infinite number of primes, and most books give a classic proof by contradiction from Euclid. Here is a very fast and easy proof due to Erdős. First show that any integer x can be written in the form

$$x = s^2 \prod_{\text{prime } p \le x} p^{z_p}, \tag{10.30}$$

where $z_p = 0$ or 1 for each p. In other words, any integer can be written as the product of a square and a *square-free* integer, where no prime factor appears twice. Conclude from this that the number of primes less than t is bounded below by

$$\pi(t) \ge \frac{1}{2}\log_2 t .$$

This is nowhere near the right answer, but it does show that $\pi(t)$ goes to infinity.

10.24 A physicist's proof of the Prime Number Theorem. Here is a "back of the envelope" argument (not a proof!) for the Prime Number Theorem. An integer t is prime if it is not a multiple of 2, or 3, or 5, and so on. Let's pretend that these events are random and independent. Then the probability that t is prime is

$$P(t) = \left(1-\frac{1}{2}\right)\left(1-\frac{1}{3}\right)\left(1-\frac{1}{5}\right)\left(1-\frac{1}{7}\right)\cdots = \prod_{\text{prime } p < t}\left(1-\frac{1}{p}\right).$$

Take the logarithm of both sides, cavalierly use the Taylor series $\ln(1-x) \approx -x$ and replace differences with derivatives to obtain

$$\frac{\mathrm{d}\ln P(t)}{\mathrm{d}t} = -\frac{P(t)}{t}.$$

Solve this differential equation and conclude that, at least when t is large,

$$P(t) \approx \frac{1}{\ln t}.$$

Finally, integrate P and conclude that the number of primes less than t is roughly

$$\pi(t) \approx \int_2^t \frac{\mathrm{d}x}{\ln x} = \frac{t}{\ln t} + O\!\left(\frac{t}{\ln^2 t}\right). \tag{10.31}$$

10.25 An infinity of primes, redux. Here is a lovely proof from Euler that there are an infinite number of primes, which also gives a strong hint as to their density. First prove the following identity, which is at the heart of much of analytic number theory:

$$\prod_{\text{prime } p}\left(1-\frac{1}{p}\right)^{-1} = \sum_{n=1}^{\infty}\frac{1}{n}. \tag{10.32}$$

Now since the harmonic series on the right-hand side diverges, the left-hand side must diverge as well. But this means that the product must contain an infinite number of terms, so the number of primes is infinite.

This equation can tell us something about the density of the primes, rather than just their infinitude. If the reciprocals $1/p$ of the primes went to zero too quickly—that is, if the density of primes were too low—then the product on the left-hand side of (10.32) would converge, even with an infinite number of terms. To pursue this line of reasoning, first prove that

$$\prod_{\text{prime } p \le t} \left(1-\frac{1}{p}\right)^{-1} \ge \sum_{n=1}^{t} \frac{1}{n} > \ln t\,.$$

Then, approximating $1-1/p$ as $\mathrm{e}^{-1/p}$, argue that the sum of the reciprocals of the primes grows at least as fast as this,

$$\sum_{\text{prime } p \le t} \frac{1}{p} \ge \ln\ln t - O(1)\,.$$

and therefore, in some average sense, the probability that t is prime is at least $1/\ln t$. Proving a matching upper bound is not easy—one is provided by Mertens' Theorem (no relation!) that

$$\prod_{\text{prime } p \le t} \left(1-\frac{1}{p}\right)^{-1} = \mathrm{e}^{\gamma} \ln n\,, \tag{10.33}$$

where γ is Euler's constant.

10.26 Mutually prime. Suppose that two integers x, y are chosen randomly, say from the set $\{1,2,\ldots,t\}$ from some large t. Assuming that the events that x and y are divisible by a given prime p are independent, and using an approach similar to (10.32), argue that as t goes to infinity, the probability x and y are mutually prime is

$$P = 1 \bigg/ \sum_{n=1}^{\infty} \frac{1}{n^2}\,.$$

Using the Riemann zeta function, we can write $P = 1/\zeta(2)$. Euler showed that $\zeta(2) = \pi^2/6$, so 10.13

$$P = \frac{6}{\pi^2} = 0.6079...$$

For instance, 6087 out of the 10 000 pairs of integers between 1 and 100 are mutually prime.

10.27 Even fewer factors. Assuming again that the events that a given integer t is divisible by various primes are independent, give a heuristic argument that if t is randomly chosen, its number of distinct prime factors is Poisson-distributed with mean $\ln\ln t$. (See Appendix A.4 for the definition of the Poisson distribution.) This was shown rigorously by Erdős and Kac [268]. Thus the upper bounds of Exercise 4.21 and Problem 10.22 are very generous in almost all cases.

10.28 Another Frenchman. Mersenne primes are of the form $2^n - 1$, such as 3, 7, 31, 127, and 8191. Fermat primes are of the form $2^{2^k} + 1$, such as 3, 5, 17, 257, and 65 537. Give heuristic arguments that there are an infinite number of Mersenne primes, but only a finite number of Fermat primes. Neither of these things are known!

10.29 Communication. Suppose x and y are n-bit strings. I possess x and you possess y. We want to work together to compute the equality function, $f(x,y) = 1$ if $x = y$ and 0 if $x \neq y$. We use some communication protocol where I send you a message, you send one back, and so on until we reach a conclusion. Each message we send can depend on the string in our possession and all the messages we have received so far. Prove that any deterministic protocol requires us to exchange a total of at least n bits—in which case I might as well send you all of x.

Hint: for each string x, there is some set of messages exchanged that confirms that $f(x,x)=1$. Show that if the total length of these messages is less than n for all x, there is some pair of distinct strings x,y such that you and I send each other exactly the same messages when calculating $f(x,x)$ and $f(y,y)$. But in that case, we would send each other these same messages when calculating $f(x,y)$ or $f(y,x)$, and would incorrectly conclude that $x=y$.

10.30 String searching with fingerprints. Suppose I have a string u of length m, such as `strep`, and a string w of length n, such as `obstreperous`. I wish to find all appearances of u in w. Let $v_i = w_i w_{i+1} \cdots w_{i+m-1}$ denote the string of length m starting at the ith letter of w. A naive algorithm for this problem would compare u to v_i for each i, which would take $\Theta(mn)$ time.

There is a simple randomized algorithm whose running time is essentially $O(n+m)$. The idea is to compute a fingerprint $h(u)$ and compare it with $h(v_i)$ for each i. Specifically, we interpret u and v_i as integers, written in base b if the alphabet they are written in has b different symbols. Then we let $h(u)=u \bmod p$ for some randomly chosen prime p. If $h(u)=h(v_i)$, we check to see if $u=v_i$.

Show that for any constant p, if we have already computed $h(v_i)$ then we can compute $h(v_{i+1})$ in constant time, so that the total running time is $O(n+m)$. Then estimate how large p needs to be in order for the total probability of a false positive, where there is an i such that $h(u)=h(v_i)$ but $u \neq v_i$, to be $o(1)$.

10.31 Using fingerprints to check matrix multiplication. As we discussed in Note 2.5, the best known algorithms for multiplying two $n \times n$ matrices take time $O(n^\alpha)$ for some $\alpha > 2$. However, there is a simple randomized algorithm for *verifying* matrix products with high probability in $O(n^2)$ time.

Suppose I claim that $\mathbf{AB}=\mathbf{C}$ where $\mathbf{A}$, $\mathbf{B}$, and $\mathbf{C}$ are $n \times n$ matrices. You can confirm this by applying both sides of this equation to a random vector $\mathbf{v}$, and checking that

$$\mathbf{A}(\mathbf{B}\mathbf{v}) = \mathbf{C}\mathbf{v}. \tag{10.34}$$

Calculating each side of this equation takes $O(n^2)$ time.

Show that if we simply compute the fingerprint of both sides, i.e., if we compute both sides mod p for a random prime p, and if $\mathbf{v}$ is chosen uniformly at random from $\{0,1,\ldots,p-1\}^n$, then the probability that (10.34) holds is at most $1/p$ if $\mathbf{AB} \neq \mathbf{C}$. If p is not a prime, show that (10.34) holds with probability at most $1/2$ if $\mathbf{AB} \neq \mathbf{C}$. Hint: (10.34) only holds if $\mathbf{v}$ is an element of the *kernel* of the matrix $\mathbf{M}=\mathbf{AB}-\mathbf{C}$, i.e., if $\mathbf{Mv}=0$.

10.32 Affine functions with fewer coins. A *Toeplitz matrix* is one whose entries are a constant along each diagonal: that is,

$$\mathbf{A} = \begin{pmatrix} a_0 & a_1 & a_2 & \\ a_{-1} & a_0 & a_1 & \cdots \\ a_{-2} & a_{-1} & a_0 & \\ & \vdots & & \ddots \end{pmatrix}.$$

Show that the family of hash functions $h(\mathbf{x})=\mathbf{Ax}+\mathbf{b}$ described in Section 10.6.3 remains pairwise independent even if we restrict $\mathbf{A}$ to be a Toeplitz matrix, choosing each a_i uniformly from $\mathbb{F}_p=\{0,1,\ldots,p-1\}$ for some prime p. How does this reduce the number of coins we need to flip to reduce SAT to UNIQUE SAT?

10.33 Affine lines. Fix a prime p, and consider the family of functions $h_{a,b}(x)=ax+b \bmod p$ from $\mathbb{F}_p$ to $\mathbb{F}_p$. Show that if a and b are chosen uniformly at random from $\mathbb{F}_p$, this family is pairwise independent. Therefore, another way to reduce SAT to UNIQUE SAT is to let p be a prime bigger than 2^n, guess $L=2^\ell$ as in Section 10.6.2, and add the constraint that $h_{a,b}(x) \leq p/L$. For that matter, the family $h_{a,b}$ works over any finite field, so we can treat $\{0,1\}^n$ as the finite field of size 2^n (see Appendix A.7.4) in which case the family $\{h_{a,b}\}$ has size 2^{2n}.

10.34 Triply independent. Let h be chosen randomly from the family of hash functions described in Section 10.6.3. That is, let $\mathbb{F}_2=\{0,1\}$ denote the field with addition and multiplication mod 2. Then let $h(\mathbf{x}):\mathbb{F}_2^n\to\mathbb{F}_2^\ell$ be defined as $h(\mathbf{x})=\mathbf{A}\mathbf{x}+\mathbf{b}$, where $\mathbf{A}$ and $\mathbf{b}$ are uniformly random in $\mathbb{F}_2^{\ell\times n}$ and $\mathbb{F}_2^\ell$ respectively. Show that h is 3*-wise independent.* In other words, show that if $\mathbf{x},\mathbf{y},\mathbf{z}\in\mathbb{F}_2^n$ are distinct then $h(\mathbf{z})$ is uniformly random in $\mathbb{F}_2^\ell$, regardless of the values of $h(\mathbf{x})$ and $h(\mathbf{y})$. Is h 4-wise independent?

Then consider the case where we do all this over $\mathbb{F}_p$ for some prime $p>2$. Show that in this case, $h(\mathbf{x})$, $h(\mathbf{y})$, and $h(\mathbf{z})$ are 3-wise independent if $\mathbf{x}$, $\mathbf{y}$, and $\mathbf{z}$ are linearly independent. Why is the case $p=2$ special?

10.35 Almost triply independent. Fix a prime p. Given a two-dimensional vector $\begin{pmatrix}x\\y\end{pmatrix}$, we can define its slope $r\in\mathbb{F}_p\cup\{\infty\}$as $r=yx^{-1}$, where x^{-1} is the inverse of x mod p and $r=\infty$ if $x=0$. Show that if we multiply a vector of slope r by a matrix $\mathbf{M}=\begin{pmatrix}a&b\\c&d\end{pmatrix}$, its slope becomes

$$h(r)=\frac{a+br}{c+dr}.$$

Now suppose that $\mathbf{M}$ is chosen uniformly at random from all invertible matrices whose entries are integers mod p. Show that h is almost 3-wise independent in the following sense: for any distinct triple $r,s,t\in\mathbb{F}_p\cup\{\infty\}$ the triple $h(r),h(s),h(t)$ is equally likely to be any of the $(p+1)p(p-1)$ possible distinct triples.

10.36 Unique cliques. Use the Isolation Lemma of Section 10.6.4 to give a randomized reduction from CLIQUE to UNIQUE CLIQUE. In other words, give a procedure that takes an instance (G,k) of CLIQUE and returns an instance (G',k'), at most polynomially larger, with the following properties:

1. If G does not have a clique of size k, then G' does not have a clique of size k'.
2. If G has a clique of size k, then G' has a clique of size k' with probability $\Omega(1/n)$.

Note that the vertices of G' are unweighted.

10.37 The Schwartz–Zippel Lemma. Prove Lemma 10.3. Hint: use induction on the number of variables, with the fact that a univariate polynomial $\phi(x_1)$ of degree d has at most d roots as your base case. Then write

$$\phi(x_1,\ldots,x_n)=\sum_{i=0}^{d}\phi_i(x_1,\ldots,x_{n-1})\,x_n^i,$$

where ϕ_i is a polynomial of degree $d-i$, and take the largest i for which ϕ_i is not identically zero. If we fix the values of $x_1,\ldots,x_{n-1}$ then ϕ becomes a polynomial $\psi(x_n)$ of degree i.

10.38 Secret sharing. Suppose an organization has k trustees, and their bylaws state that any ℓ members form a quorum. They wish to share a secret s among them—say, the combination to a vault—in such a way that if any subset consisting of ℓ trustees get together, they can deduce the secret, but any subset of size less than ℓ can recover no information about it at all.

There is a delightful way to do this with polynomials over a finite field. Let p be some prime greater than k, and let P be a degree-d polynomial $P(z)=\sum_{i=0}^{d}a_iz^i$ defined on $\mathbb{F}_p$. We set the constant term $P(0)=a_0$ to the secret s, and choose the other d coefficients $a_1,\ldots,a_d$ uniformly from $\mathbb{F}_p$. We then give the ith trustee $P(i)$ for each $1\le i\le k$.

Show that if $d+1$ trustees get together and share their values of P, they can deduce s. In other words, given any $d+1$ values of a degree-d polynomial, we can obtain the entire polynomial by interpolation: it takes three points to determine a parabola, four to determine a cubic, and so on.

Hint: show that for any set $\{z_1,\ldots,z_{d+1}\}$ where the z_i are distinct, the following $(d+1)\times(d+1)$ matrix, known as the *Vandermonde matrix*, is invertible:

$$\mathbf{V}=\begin{pmatrix} 1 & z_1 & z_1^2 & & z_1^d \\ 1 & z_2 & z_2^2 & \cdots & z_2^d \\ 1 & z_3 & z_3^2 & & z_3^d \\ & \vdots & & \ddots & \\ 1 & z_{d+1} & z_{d+1}^2 & & z_{d+1}^d \end{pmatrix}.$$

Specifically, prove that its determinant is

$$\det \mathbf{V}=\prod_{1\le i<j\le d+1}(z_j-z_i),$$

and that this is nonzero mod p.

Conversely, show that $P(0)$ is uniformly random in $\mathbb{F}_p$, even if we condition on the values $P(z_i)$ for any set of d or fewer z_i. Thus if we set $d=\ell-1$, no subset smaller than a quorum can learn anything about s.

10.11

10.39 Four square roots. Show that if p has two distinct odd prime factors q and r, there are at least 4 square roots of 1 in $\mathbb{Z}_p^*$. Hint: use the Chinese Remainder Theorem (see Appendix A.7) and consider elements x such that $x\equiv_q \pm 1$ and $x\equiv_r \pm 1$. More generally, if p has t distinct odd prime factors, show that $\mathbb{Z}_p^*$ has 2^t square roots.

10.40 Fooling Miller–Rabin. In Section 10.8.2 we defined the set $F\subseteq\mathbb{Z}_p^*$ of those a that fool the Miller–Rabin test. Here we define a subgroup K which contains F. Let j be the smallest integer such that there exists an $a\in\mathbb{Z}_p^*$ for which $a^{p-1}=1$ but $a^{(p-1)/2^j}\neq 1$. Now consider the following set:

$$K=\left\{a\in\mathbb{Z}_p^* : a^{(p-1)/2^j}\in\{1,p-1\}\right\}. \tag{10.35}$$

Show that K is a subgroup of $\mathbb{Z}_p^*$ and that $F\subseteq K$. Then show that if p has two distinct odd prime factors, K is a proper subgroup.

Hint: suppose that p is divisible by odd primes q and r. Consider an $a\in K$ such that $a^{(p-1)/2^j}=p-1$. As in the previous problem, by the Chinese Remainder Theorem there is an a' such that $a'\equiv_q a$ but $a'\equiv_r 1$. Show that $a'\notin K$.

10.41 Prime powers. Show that we can tell in poly(n) time whether an n-digit number p is a prime power. More generally, we can tell if $p=q^k$ for integers q and $k>1$, regardless of whether or not q is prime. Hint: how large can k be?

10.42 Binomials mod p. Complete the proof that the identity $(x+1)^p\equiv_p x^p+1$ for all x if and only if p is prime. Hint: use the binomial theorem to write

$$(x+1)^p=\sum_{k=0}^{p}\binom{p}{k}x^k$$

and then show that, if and only if p is prime, $\binom{p}{k}$ is divisible by p for all k except 0 and p. Conversely, if p is not prime, let q^t be the largest power of some prime q that divides q, and show that $\binom{p}{k}$ is not divisible by p.

10.43 Binomials mod q. Let p have two or more distinct prime factors, and let q be one of them. Reasoning as in the previous problem, show that $(x+1)^p\not\equiv_q x^p+1$ for some x.

10.44 Factoring random numbers. While FACTORING seems to be hard, there is an elegant randomized algorithm that generates a uniformly random integer y in the range $1 \leq y \leq N$ along with its prime factorization. If N is an n-bit integer, the expected running time of this algorithm is poly(n).

The algorithm is astonishingly simple. We generate a sequence of integers $x_1, x_2, \ldots$ as follows. First we choose x_1 uniformly from $\{1, \ldots, N\}$, then we choose x_2 uniformly from $\{1, \ldots, x_1\}$, and so on. At each step we choose x_{i+1} uniformly from $\{1, \ldots, x_i\}$, until $x_i = 1$ and we stop.

Now, for each $x \in \{2, \ldots, N\}$, let t_x denote the number of times that x appears in this sequence. We claim that the t_x are independent of each other, and that each one is distributed as

$$P(t_x) = \frac{1}{x^{t_x}} \left(1 - \frac{1}{x}\right). \tag{10.36}$$

In other words, if we have a biased coin that comes up heads with probability $1/x$ and tails with probability $1 - 1/x$, then t_x is distributed exactly like the number of heads before the first tail. We ask you to prove this claim below.

Next we check these x_i for primality, and define y as the product of the prime ones. This gives

$$y = \prod_{\text{prime } x_i} x_i = \prod_{\text{prime } p \leq N} p^{t_p}.$$

Applying (10.36), the probability of generating any given y is

$$P(y) = \prod_{\text{prime } p \leq N} \frac{1}{p^{t_p}} \left(1 - \frac{1}{p}\right) = \frac{1}{y} \prod_{\text{prime } p \leq t} \left(1 - \frac{1}{p}\right) = \Theta\left(\frac{1}{y \log N}\right), \tag{10.37}$$

where we used Mertens' Theorem (10.33) in the last equality.

If $y > N$, we throw the whole thing out and start over. If $y \leq N$, we even out the probability distribution (10.37) by printing out y with probability y/N, and starting over with probability $1 - y/N$. Then each $y \leq N$ is printed with the same probability $\Theta(1/(N \log N))$, the total probability that we output some $y \leq N$ instead of starting over is $\Theta(1/\log N)$, and the expected number of trials is $\Theta(\log N) = O(n)$.

Finally, show that the expected number of primality tests we need to run in each trial is

$$\mathbb{E}[\#\text{ of distinct } x_i > 1 \text{ in the sequence}] = \sum_{i=2}^{N} \frac{1}{i} < \ln N = O(n). \tag{10.38}$$

Thus the expected total number of primality tests we need to run in the entire process is $O(n^2)$. Since PRIMALITY is in P, the expected running time is poly(n).

Hint: first observe the following alternate method to choose a random integer in the range $\{1, \ldots, N\}$. For each x, suppose we have a coin as described above, which comes up heads with probability $1/x$ and tails with probability $1 - 1/x$. Flip these in decreasing order, with $x = N$, $x = N-1$, and so on, and set x according to the first one which comes up heads. Now analyze the entire process in terms of these coins, and use this picture to prove (10.36) and (10.38).

10.14

10.45 Counting a stream. Randomization can help us solve problems with a surprisingly small amount of memory. Consider the following *data streaming* problem: we are watching a very long string of events, such as packets of data flying by on the Internet. We want to count the number of events, but the number N we are likely to see is so large that we cannot fit it in our memory. So we maintain an estimate of $k = \log_2 N$ instead, which only takes $\log_2 k = O(\log \log N)$ bits to store. We do this by starting with $k = 0$ and incrementing k with probability 2^{-k} each time another packet goes by. Show that the final value of k is a good estimate of $\log_2 N$ in the sense that for any $\varepsilon > 0$, with high probability we have

$$(1 - \varepsilon) \log_2 N < k < (1 + \varepsilon) \log_2 N.$$

Hint: we can drive this algorithm by choosing a random real number r uniformly from $[0,1]$ at each of the N steps, and incrementing k if $r < 2^{-k}$. Show that if $k = (1-\varepsilon)\log_2 N$, there are probably at least k such steps, while for $k = (1+\varepsilon)\log_2 N$ there are probably none. (You may find the Chernoff and union bounds from Appendices A.5.1 and A.3.1 helpful.) What does this mean for the final value of k?

10.46 Amplifying randomized algorithms. For any pair of probabilities p, q, define $\mathsf{C}[p,q]$ as the class of problems for which a randomized polynomial-time algorithm of some kind exists such that

$$\Pr[A \text{ returns "yes"}] \geq p \text{ if } x \text{ is a yes-instance}$$
$$\Pr[A \text{ returns "yes"}] \leq q \text{ if } x \text{ is a no-instance}.$$

Assume that for any algorithm A in C, there is an algorithm A' in C that runs A a polynomial number of times and returns "yes" if the number of times that A returns "yes" is above a certain threshold. Formally, we say that C is *closed under majority*. This is certainly true of BPP.

Show that for any constants p, q with $p > q$, we have $\mathsf{C}[p,q] = \mathsf{C}[2/3, 1/3]$. In fact, show that this is true even if $p(n)$ and $q(n)$ are functions of n and the gap between them is polynomially small, such as if $p(n) = 1/2 + 1/n^c$ and $q(n) = 1/2 - 1/n^c$ for some constant c.

At the other extreme, show that we can make $p(n)$ and $q(n)$ exponentially close to 1 and 0. In other words, if $q(n) = \mathrm{e}^{-\mathrm{poly}(n)}$ and $p(n) = 1 - \mathrm{e}^{-\mathrm{poly}(n)}$ we have $\mathsf{C}[p(n), q(n)] = \mathsf{C}[2/3, 1/3]$. Thus the definitions of classes like AM and BPP are quite insensitive to what values of p and q we use.

Hint: use Chernoff bounds (see Section A.5.1) to show that we can amplify the gap between p and q by running A a polynomial number of times and setting the right threshold for the number of times it returns "yes." Why does this not work if the initial gap between p and q is exponentially small?

10.47 Pairwise independence and mixing. For a given function $h: \{0,1\}^\ell \to \{0,1\}^\ell$ and a pair of sets $A, B \subseteq \{0,1\}^\ell$, let $p_{A,B}(h)$ be the fraction of $\{0,1\}^\ell$ that h maps from A to B:

$$p_{A,B}(h) = \Pr_{x \in \{0,1\}^\ell} [x \in A \text{ and } h(x) \in B] .$$

Show that if h is drawn uniformly from a pairwise independent family of hash functions, $p_{A,B}(h)$ is usually close to its expectation—a fact that we'll use to construct a pseudorandom generator in Problem 11.18. Namely, for any A, B and any $\varepsilon > 0$,

$$\Pr_h \left[\left| p_{A,B}(h) - \alpha\beta \right| > \varepsilon \right] \leq \frac{\alpha\beta}{2^\ell \varepsilon^2} \leq \frac{1}{2^\ell \varepsilon^2}, \tag{10.39}$$

where $\alpha = |A|/2^\ell$ and $\beta = |B|/2^\ell$. Hint: use pairwise independence to show that the variance $\mathrm{Var}_h\, p_{A,B}(h)$ is exactly what it would be if h were uniformly random among *all* functions from $\{0,1\}^\ell$ to $\{0,1\}^\ell$, and then use Chebyshev's inequality from Appendix A.3.2. (In fact, you'll find you can prove a slightly tighter bound.)

Notes

10.1 Books. Standard texts on randomized algorithms include Motwani and Raghavan [617] and Mitzenmacher and Upfal [590]. Where PRIMALITY is concerned, we recommend the marvelous book *Prime Numbers: A Computational Perspective* by Crandall and Pomerance [205], as well as *Primality Testing in Polynomial Time* by Dietzfelbinger [237]. An introduction to the classes BPP and ZPP can be found in Balcázar, Díaz, and Gabarró [75].

10.2 MIN CUT. Karger's algorithm appeared in [456]. If G has n vertices and m edges, it has a running time of $O(mn^2 \log^2 n)$. This was improved to $O(n^2 \log^3 n)$ time by Karger and Stein [457]. Thus it is competitive with the best known deterministic algorithms, whose running times are essentially $O(nm)$. These log factors, of course, depend on the details of the data structures we use.

We learned about the relationship between Kruskal's and Karger's algorithms, given in Problem 10.5, from [215].

10.3 `WalkSAT` and local search. The first analysis of `WalkSAT` was given by Papadimitriou [645], who showed that for 2-SAT it succeeds in $O(n^2)$ time. The analysis we give here is a simplified version of Schöning's 1999 paper [734]. Since then a number of improvements have been found. Dantsin et al. [212] gave a deterministic version of `WalkSAT` whose running time is slightly worse, $(2k/(k+1))^n$ instead of $(2(k-1)/k)^n$. The 1.3^n algorithm we quote (actually, 1.307^n) is from Hertli [399], and builds on a previous algorithm of Paturi, Pudlák, Saks, and Zane [651] for SAT problems with a unique solution.

`WalkSAT` has many variants. Once we choose an unsatisfied clause, rather than choosing randomly from its three variables, we could act more greedily. For instance, we could flip the variable that minimizes the total number of unsatisfied clauses, or that causes the smallest number of currently satisfied clauses to become unsatisfied [739]. In practice, we run the algorithm with a "noise parameter" p, where we make random moves with probability p and greedy ones with probability $1-p$. We can then use experimental data to tune p to the instances we care about [562], or invent adaptive mechanisms that tune it on the fly [409].

Other possibilities include *tabu search* [333], in which we refuse to flip variables that have been flipped recently, and *simulated annealing* [481], in which we interpret the number of unsatisfied clauses as the energy of a physical system at a certain temperature, and then make moves according to the laws of statistical mechanics. All of these techniques are designed to help us escape local optima and avoid getting stuck in the same region of solution space.

We owe Problem 10.10 to Beigel and Eppstein [93], who showed how to solve GRAPH 3-COLORING in $O(1.3289^n)$ time. To our knowledge this is currently the best algorithm known.

10.4 MAX CUT and unique games. The Goemans–Williamson algorithm, and the improved approximation ratio of Problem 10.14 for graphs with a large MAX CUT, appeared in [335]. For a review of the power of random projection, where we project sets in a high-dimensional space into a random low-dimensional subspace, see the excellent monograph of Vempala [812].

An exciting recent conjecture would imply that the Goemans–Williamson algorithm, and many of our other favorite approximation algorithms, are optimal unless $\mathsf{P}=\mathsf{NP}$. A *unique game* is a kind of graph coloring problem. We have k colors, for each edge (i,j) there is a permutation π_{ij} acting on the k colors, and we say an edge is satisfied if the colors of its endpoints are related by $c_j=\pi_{ij}(c_i)$. The *Unique Games Conjecture* (UGC), formulated by Subhash Khot [478], states that the analogous MAX-SAT problem is hard to approximate in the following sense: for every constant $\delta>0$, there is a k such that it is NP-hard to tell the difference between the case where only a fraction δ of the edges can be satisfied, and the case where $1-\delta$ or more can be satisfied.

Khot, Kindler, Mossel and O'Donnell [479] showed that if the UGC is true then approximating MAX CUT any better than the Goemans–Williamson algorithm, i.e., with any ratio better than $\alpha=0.878...$, is NP-hard. Khot and Regev [480] also showed under the UGC that it is NP-hard to approximate VERTEX COVER with any ratio better than 2, suggesting that the simple pebble algorithm discussed in Section 9.2.1 is optimal.

On the other hand, in 2010, Arora, Barak, and Steurer [54] presented an algorithm that solves unique games in $O(2^{n^\varepsilon})$ time for any $\varepsilon>0$. Although this subexponential algorithm doesn't disprove the UGC, it casts some doubt on its validity.

10.5 The Minimax Theorem. The Minimax Theorem first appeared in a 1928 paper by John von Neumann [814]. The proof we give in Problem 10.16, using Brouwer's Fixed Point Theorem, is due to John Nash [623]. Much more general forms of the minimax theorem exist, including where the payoff to the players is a convex function of their strategies, rather than the simple bilinear function $\mathbf{a}^T\mathbf{M}\,\mathbf{b}$. The text by Luce and Raiffa [547] gives a number of different proofs and interpretations. Yao's principle was formulated by Andrew Chi-Chih Yao, and appeared in [838].

10.6 Game trees. The analysis we give here of the randomized algorithm for NAND trees, and the proof using Yao's principle that it is the optimal depth-first algorithm, is taken from Saks and Wigderson [723]. We omit the more difficult part of their proof: namely, that the optimal algorithm is depth-first. Earlier, Snir [764] used a simpler argument to give the upper bound $3^{k/2} = N^{\log_4 3} = N^{0.792\ldots}$. The lower bound of Problem 10.19 was given by Pearl [652, 653] and Tarsi [785].

In [723], the authors credit Boppana with the $(8/3)^k$ algorithm for the majority tree problem of Problem 10.21, and comment that in this case the optimal algorithm is not depth-first. The improved algorithm mentioned in Problem 10.21 is from Jayram, Kumar, and Sivakumar [428], who also proved a $(7/3)^k$ lower bound without the depth-first assumption.

10.7 The Prime Number Theorem. The Prime Number Theorem was conjectured in the late 1700s by Legendre and by Gauss. Gauss conjectured the approximation (10.31) for $\pi(t)$ given in Problem 10.24; this is called the *logarithmic integral*, and is denoted $\mathrm{Li}(t)$.

The theorem was proved in the late 1800s by Hadamard and by de la Vallée Poussin, who applied complex analysis to the Riemann zeta function (see Note 10.13 below). Elementary proofs were found in the 20th century by Selberg and Erdős, and were the subject of an unfortunate priority dispute.

The back-of-the-envelope calculation we give in Problem 10.24 is similar to a passage in Havil's lovely book *Gamma: Exploring Euler's Constant* [381]. We learned Euler's proof of the infinity of the primes, given in Problem 10.25, from [205].

10.8 Communication. Suppose Alice has a string x, Bob has a string y, and they are trying to calculate some function $f(x,y)$. The *communication complexity* of f is the minimum number of bits of information they need to exchange. Communication complexity was introduced by Andrew Chi-Chih Yao in 1979 [839], who was honored for this and other work with the Turing Award in 2000. It has since grown into a rich subfield of computer science, not least because randomized communication protocols can be exponentially more efficient than deterministic ones. We refer the interested reader to Arora and Barak [53, Chapter 13].

10.9 Fingerprinting. The string-matching algorithm of Problem 10.30 is called the Rabin–Karp algorithm [465]. The algorithm for checking matrix multiplication given in Problem 10.31 is from Freivalds [294]. We are indebted to Umesh Vazirani for the space probe example of fingerprinting, although his probe only went to the moon.

10.10 Unique solutions, hash functions, and the Isolation Lemma. The reduction from SAT to UNIQUE SAT was given by Valiant and Vazirani [804]. In terms of complexity classes, we can say that the class UP, consisting of NP problems where the solution is unique if one exists, is just as hard as NP under randomized polynomial-time reductions. Therefore, if UP $\subseteq$ RP then NP $=$ RP. One motivation for UP comes from cryptography, since decrypting a message corresponds to finding the unique cleartext it conveys.

Since 1 is odd, this reduction also shows that telling whether the number of solutions to an NP-complete problem is even or odd is NP-hard under RP reductions. This is a key ingredient in showing that counting the number of solutions to problems in NP is as hard as the polynomial hierarchy; see Note 13.2.

The pairwise independent families of hash functions described in Section 10.6.3 and Problem 10.33 were found by Carter and Wegman [147].

The Isolation Lemma is due to Mulmuley, Vazirani, and Vazirani [619], who used it to give the fast parallel algorithm for PERFECT MATCHING that we alluded to in the text. Since then, it has found many applications, including in digital watermarking and bounds on circuit complexity.

10.11 Identity testing. The Schwartz–Zippel Lemma is named after Schwartz [736] and Zippel [849]. A slightly weaker version was found earlier by DeMillo and Lipton [226]. Zippel's book [850] is an excellent survey of algorithms, both randomized and deterministic, for identity testing and other problems involving polynomials.

Exercise 10.9, showing that a graph has a perfect matching if and only if an antisymmetric version of its adjacency matrix has nonzero determinant, is from Tutte [801]. The idea of sampling the determinant at a random value of **x**, thus obtaining a randomized algorithm for PERFECT MATCHING, is due to Lovász [539]. We will use determinants and permanents in Chapter 13 to calculate the number of spanning trees and perfect matchings.

The secret sharing scheme of Problem 10.38 was invented by Shamir [743].

10.12 Carmichael numbers. If there were only a finite number of Carmichael numbers, we could patch the Fermat test by checking for each one. However, it was shown in 1994 by Alford, Granville, and Pomerance [35] that there are infinitely many of them.

10.13 The Riemann Hypothesis. The Riemann zeta function is defined as

$$\zeta(z)=\sum_{n=1}^{\infty}\frac{1}{n^z}\,.$$

This sum converges if $\mathrm{Re}\,z > 1$, and we define $\zeta(z)$ for other values of z by analytic continuation. It has zeros at $z=-2,-4,-6,\ldots$ The Riemann Hypothesis states that except for these, all its zeros have real part 1/2: that is, that they lie on the line $1/2 \pm \imath y$ in the complex plane where y is real.

The zeta function is connected to the density of primes through Euler's relation, which is a generalization of (10.32):

$$\zeta(z)=\prod_{p\text{ prime}}\left(1-\frac{1}{p^z}\right)^{-1}$$

Through this connection, it turns out that the Riemann Hypothesis is equivalent to the following statement, which quantifies how closely the logarithmic integral of Problem 10.24 approximates the number of primes less than t:

$$\left|\pi(t)-\int_2^t\frac{\mathrm{d}x}{\ln x}\right|=O\!\left(\sqrt{t}\log t\right).$$

For an enjoyable popular introduction to the Riemann Hypothesis, see [713].

To define the *Extended Riemann Hypothesis* (ERH), we first need to define a *Dirichlet character*. This is a complex-valued function χ defined on the natural numbers which is *multiplicative*, i.e., $\chi(xy)=\chi(x)\chi(y)$ for all x,y. Furthermore, $\chi(x)=\chi(x+k)$ for all x, i.e., χ is periodic with period k, and $\chi(x)\neq 0$ if and only if x and k are mutually prime. For example, if $k=5$ we could have

$x \bmod 5$	0	1	2	3	4
$\chi(x)$	0	1	$\imath$	$-\imath$	-1

For readers familiar with group theory, χ is a homomorphism from $\mathbb{Z}_k^*$ to $\mathbb{C}$.

The Dirichlet L-function is the following generalization of the zeta function,

$$L_\chi(z)=\sum_{n=1}^{\infty}\frac{\chi(n)}{n^z}\,,$$

so that $L_\chi(z)=\zeta(z)$ if $\chi=1$. Then the ERH states that, for any Dirichlet character χ, the zeros of L_χ with positive real part all have real part 1/2. Ankeny [46] proved Theorem 10.4, and Bach [70] showed that the constant C is at most 2. Therefore, if the ERH holds, any subgroup of $\mathbb{Z}_p^*$ that includes every integer from 1 to $2\ln^2 p$ in fact includes all of $\mathbb{Z}_p^*$.

10.14 Primality. Miller's algorithm appeared in [581], and Rabin's randomized version appeared in [676]. Ultimately, if the ERH turns out to be true, we will have had a deterministic polynomial-time algorithm for PRIMALITY since 1976. Another important randomized algorithm for PRIMALITY, based on a number-theoretic function called the Jacobi symbol, was given by Solovay and Strassen [766]; it can also be made deterministic assuming the ERH is true.

The randomized algorithm described in Section 10.8.4, in which we calculate $P_p \bmod Q$ for random polynomials Q of degree n, was found by Agrawal and Biswas in 1999 [22]. The deterministic version was found by Agrawal, Kayal, and Saxena in 2002 [23], who shared both the Gödel Prize and the Fulkerson Prize in 2006. This is even more notable because Kayal and Saxena were undergraduates at the time they did most of the work!

Lenstra and Pomerance [524] improved the AKS algorithm so that it runs in time $O(\log^6 p) = O(n^6)$, ignoring $\text{polylog}(n)$ factors. Excellent reviews of these and other algorithms for PRIMALITY can be found in Granville [350] and the book by Crandall and Pomerance [205].

It should be noted that for practical purposes, randomized algorithms like the Miller–Rabin test are still considerably faster than the AKS algorithm. As we emphasized in Section 2.5, however, the real impact of the AKS algorithm is not that PRIMALITY is a "tractable" problem: it is a fundamental new insight into the nature of the primes, which shows us that PRIMALITY is not as complex a property as we thought it was.

The randomized algorithm given in Problem 10.44 for producing random numbers along with their factorizations is due to Adam Kalai [450],

10.15 Complexity classes. Monte Carlo algorithms have their origins in estimating high-dimensional integrals in statistical physics by random sampling, a subject we treat extensively in Chapter 12. The classes RP, ZPP, and BPP were defined by Gill [330]. The term *Las Vegas* algorithm was coined by Babai in [61]. To our knowledge, no one has proposed a class of Atlantic City algorithms. Presumably they would be like Las Vegas algorithms, but a little seedier.

Note that BPP is defined as a semantic class rather than a syntactic one (see Note 6.12). In other words, if we try to define a BPP-complete problem analogous to WITNESS EXISTENCE, where we are given a program Π as part of the input, we have to promise that Π returns "yes" with probability greater than $2/3$ or less than $1/3$. Of course, if BPP = P then this promise can be removed.

Chapter 11

Interaction and Pseudorandomness

I saw mage Merlin, whose vast wit
And hundred winters are but as the hands
Of loyal vassals toiling for their liege.
And near him stood the Lady of the Lake,
Who knows a subtler magic than his own—
Clothed in white samite, mystic, wonderful.

Alfred, Lord Tennyson, *Idylls of the King*

In the previous chapter, we saw how randomness can give us simple, efficient, and beautiful algorithms. But it changes the landscape of computation in many other ways as well. Problems in NP are those where, if the answer is "yes," the Prover can prove this to the Verifier. But what happens if the Verifier is willing to accept that something is *probably* true, instead of certainly true?

In this chapter the Verifier and Prover return in the form of Arthur and Merlin. We will watch Merlin convince Arthur that two graphs are topologically different, and see how Arthur can keep Merlin honest by asking him random questions. We will see mysterious *zero-knowledge* proofs, in which Merlin convinces Arthur that a graph is colorable, while telling him nothing at all about how to color it. We will see how a mysterious Chess player can convince us that White has a winning strategy without playing a single game. And we will meet one of the most celebrated results in computer science, the PCP Theorem, which shows that problems in NP have proofs that can be checked by looking at just a few bits.

We will then turn to another one of complexity theory's deep questions: whether randomized algorithms can be *derandomized*. Our computers are deterministic—unless we hook them up to a Geiger counter, they don't have access to truly random numbers. Instead, they have functions that purport to generate pseudorandom numbers. But are there pseudorandom number generators that can fool any polynomial-time algorithm into thinking that they are truly random? If so, then as far as polynomial-time algorithms are concerned, we don't really need random numbers—we can use pseudorandom ones instead, and simulate our algorithms deterministically. In terms of the complexity classes introduced at the end of Chapter 10, this would imply that BPP = P.

Pseudorandom generators turn out to be closely related to *one-way functions*, which are easy to compute but hard to invert, and to secure cryptosystems where an encrypted message is indistinguishable from random noise. We describe several pseudorandom generators, and prove that they can fool any polynomial-time algorithm as long as problems like DISCRETE LOG are as hard as we think they are. We end by proving a general connection between hardness and randomness—that if there exist functions that are exponentially hard to compute in a certain sense, all randomized polynomial-time algorithms can be derandomized.

11.1 The Tale of Arthur and Merlin

We can think of NP problems as conversations between the Prover and the Verifier, where the Verifier asks for a proof and the Prover responds with one. What if we let these conversations go on a little longer, and let the Verifier ask the Prover a series of questions?

If the Verifier chooses his questions deterministically, the Prover can construct the entire conversation in advance. In that case, he can provide it to the Verifier as a proof, and we just have NP again. But if the Verifier can ask *random* questions, we get an entirely new kind of proof—one where the Prover can give satisfactory answers to the Verifier's questions, with high probability, if and only if the answer is "yes."

In this section, we will meet the godlike Merlin and the polynomial-time Arthur, and see how Merlin can prove to Arthur that certain things are almost certainly true using this kind of conversation. We will see how Arthur and Merlin can establish trust by asking each other random questions, so that these interactive proofs work even if one of them tries to cheat—if Merlin tries to deceive Arthur, or if Arthur tries to learn something other than what Merlin is trying to prove.

11.1.1 In Which Arthur Scrambles a Graph

Consider the two graphs in Figure 11.1. Are they topologically identical? That is, does the graph on the right have the same structure as the one on the left, except that its vertices have been rearranged? If so, we say that they are *isomorphic*, and write $G_1 \cong G_2$. Formally:

> GRAPH ISOMORPHISM
> Input: Two graphs $G_1 = (V_1, E_1), G_2 = (V_2, E_2)$
> Question: Is there a permutation $\pi : V_1 \to V_2$ such that $\pi(G_1) = G_2$, i.e., $(u, v) \in E_1$ if and only if $(\pi(u), \pi(v)) \in E_2$?

Clearly GRAPH ISOMORPHISM is in NP, since if $G_1 \cong G_2$ I can prove it to you by showing you the isomorphism π. But is it in coNP? Equivalently, is GRAPH NONISOMORPHISM in NP? Of course, I can prove to you that $G_1 \not\cong G_2$ if they differ in some obvious way, such as not having the same number of vertices of a given degree. But if two graphs are nonisomorphic, is there always a proof of this fact which can be verified in polynomial time?

In a sense there is, if we relax our definition of "proof" and "verify" a little. To make our story a little grander, let's welcome two more characters to the stage: Arthur and Merlin. Merlin is the Prover—possessed of godlike intelligence, and capable of solving any problem that can be solved. Arthur is the Verifier—like us, a mere mortal, and bound within polynomial time.

Now suppose G_1 and G_2 are nonisomorphic. Merlin wants to convince Arthur of this, but he can only use arguments that Arthur can check with his limited computational powers. But as far as we know,

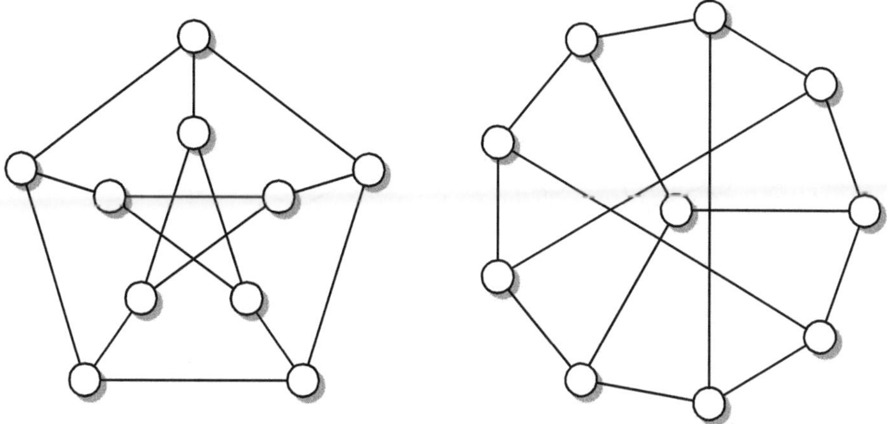

FIGURE 11.1: Are these two graphs isomorphic?

FIGURE 11.2: Merlin giving Arthur sage advice.

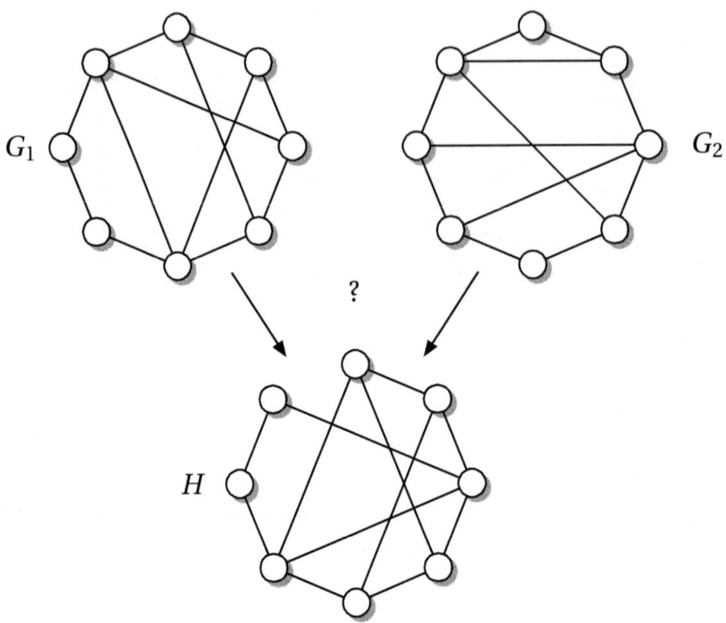

FIGURE 11.3: Arthur generates H by applying a random permutation to either G_1 or G_2. If Merlin can consistently tell which graph Arthur started with (can you?) then G_1 and G_2 are nonisomorphic.

GRAPH NONISOMORPHISM is not in NP, so Merlin cannot simply provide a witness that Arthur can check once and for all. So, Merlin suggests that they play a game.

In each round, Arthur chooses one of the two graphs, $G = G_1$ or $G = G_2$. He then chooses a permutation ϕ, permutes the vertices of his chosen graph, and hands the result $H = \phi(G)$ to Merlin. Arthur then challenges Merlin to tell him which G he used to generate H. If $G_1 \ncong G_2$, Merlin can use his vast computational powers to tell which one is isomorphic to H, and answer Arthur's question correctly. We invite you to play the role of Merlin in Figure 11.3.

If $G_1 \cong G_2$, on the other hand, Merlin is helpless. Arthur's optimal strategy is random: he flips a coin and chooses G uniformly from $\{G_1, G_2\}$, and chooses ϕ uniformly from the $n!$ possible permutations. Then $H = \phi(G)$ has exactly the same probability distribution whether $G = G_1$ or $G = G_2$. No matter how computationally powerful Merlin is, he has no information he could use to tell which graph Arthur started with, and whatever method he uses to respond will be correct half the time. If we play the game for t rounds, the probability that Merlin answers correctly every time is only $1/2^t$.

This is our first example of an *interactive proof*, or an *Arthur–Merlin game*: a game where Merlin tries to convince Arthur of a certain claim, such that Merlin can win with high probability if and only if his claim is true. In this case, Merlin wins if he answers all of Arthur's questions correctly. More generally, we determine the winner by applying some polynomial-time algorithm to a transcript of the entire exchange between them. If this algorithm returns "yes," then Merlin is declared the winner, and Arthur declares himself convinced.

11.1.2 In Which Merlin Sees All

When we analyzed the interactive proof for GRAPH NONISOMORPHISM in the previous section, we made an important assumption: namely, that Arthur's random choices are *private*, so that Merlin doesn't know which graph he started with, or what permutation he applied. What happens if Merlin is not just computationally powerful, but omniscient? If he can see Arthur's coin flips then obviously he can cheat, and convince Arthur that he can tell which G Arthur chose even if $G_1 \cong G_2$.

Is there a kind of interactive proof for GRAPH NONISOMORPHISM that Arthur can trust even if Merlin can see Arthur's random choices? That is, even if Arthur's coins are *public*, and flipped in full view where all parties can see them? This seems counterintuitive. Surely, in the presence of a deceptive demigod, a mortal's private thoughts are his only resource.

There is indeed such a proof. However, it relies on ideas rather different from those in the previous section. If S is the set of all graphs that are isomorphic to either G_1 or G_2, then S is larger in the case where $G_1 \ncong G_2$ than if $G_1 \cong G_2$. So, we will construct an interactive proof for a more abstract problem, in which Merlin tries to prove to Arthur that a set S is at least a certain size M.

We will assume that if Merlin is lying, S is significantly smaller than he says it is. Specifically, we assume that there is an M and a constant $c < 1$ such that either $|S| \geq M$ or $|S| \leq cM$, and Merlin claims that the former is true. We also assume that membership in S is in NP—that for any $x \in S$, Merlin can prove that $x \in S$ to Arthur.

One way Merlin could prove that $|S| \geq M$ is simply to show Arthur M elements of S, but this would take too long if M is exponentially large. On the other hand, if S is a subset of some larger set U, such as the set of all strings of a certain length, then Arthur could estimate the fraction $|S|/|U|$ by sampling random elements of U. However, if $|S|/|U|$ is exponentially small, it would take an exponential number of samples for Arthur to find even a single element of S.

Thus neither of these strategies work if M is exponentially large, but $M/|U|$ is exponentially small. To handle this case, we will construct an interactive proof using one of our tools from Section 10.6—random hash functions.

Suppose Arthur chooses a random function h that maps U to $\{0,\ldots,M-1\}$. There is no need to keep h secret—indeed, Arthur needs to show h to Merlin. So Arthur chooses h by flipping a set of public coins. Arthur then challenges Merlin to show him an $x \in S$ such that $h(x) = 0$. The expected number of such x is $|S|/M$, so the probability that Merlin can meet this challenge is much larger if $|S| \geq M$ than if $|S| \leq cM$. We illustrate this in Figure 11.4.

The first kind of hash function that comes to mind is a completely random one, where $h(x)$ is independent and uniform in $\{0,\ldots,M-1\}$ for each $x \in U$. But since U is exponentially large, Arthur would have to flip exponentially many coins to choose such a function. We need to restrict ourselves to a *pseudorandom* family of hash functions instead—a family that only requires us to flip a polynomial number of coins. Such a family cannot offer us complete independence, but perhaps some weaker property is good enough for our purposes.

The following lemma shows that, just as in Section 10.6.2, a family of hash functions with *pairwise* independence is good enough. Note that our hash function maps U into $\{0,\ldots,L-1\}$ for some L, rather than $\{0,\ldots,M-1\}$. When we apply this lemma below, we will set L to be a constant times M.

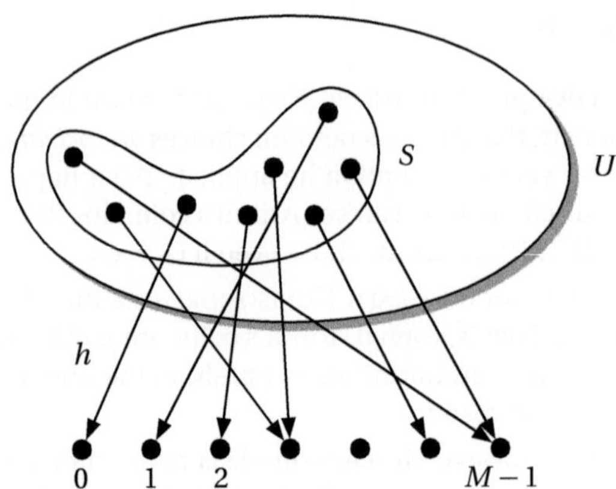

FIGURE 11.4: An interactive proof that $|S| \geq M$ using public coins. Arthur chooses a random hash function h from U to $\{0,\ldots,M-1\}$, and challenges Merlin to show him an $x \in S$ such that $h(x)=0$. Merlin has a better chance of doing this if $|S| \geq M$ than if $|S| \leq cM$ for some constant $c < 1$.

Lemma 11.1 *Let $S \subseteq U$, and let h be chosen uniformly from a pairwise independent family of hash functions from U to $\{0,\ldots,L-1\}$. The probability P that there is at least one $x \in S$ such that $h(x)=0$ satisfies*

$$\frac{|S|}{|S|+L} \leq P \leq \frac{|S|}{L}.$$

Proof Let Z be the number of $x \in S$ such that $h(x)=0$, so that $P = \Pr[Z \geq 1]$. The first moment method (see Appendix A.3.2) gives an upper bound on P in terms of Z's expectation:

$$\Pr[Z \geq 1] \leq \mathbb{E}[Z].$$

Since each $x \in S$ is mapped to 0 with probability $1/L$, by linearity of expectation we have $\mathbb{E}[Z] = |S|/L$, proving the upper bound.

The second moment method (see Appendix A.3.5) gives a lower bound on P in terms of Z's expectation and its second moment, i.e., the expectation of Z^2:

$$\Pr[Z \geq 1] \geq \frac{\mathbb{E}[Z]^2}{\mathbb{E}[Z^2]}. \tag{11.1}$$

The second moment $\mathbb{E}[Z^2]$ is the sum over all ordered pairs (x,y) where $x,y \in S$ of the probability that both x and y are mapped to 0. Using pairwise independence, we can write $\mathbb{E}[Z^2]$ as

$$\begin{aligned}\mathbb{E}[Z^2] &= \sum_{x\in S}\left(\Pr[h(x)=0]\sum_{y\in S}\Pr[h(y)=0 \mid h(x)=0]\right)\\ &= \frac{|S|}{L}\left(1+\frac{|S|-1}{L}\right) \leq \frac{|S|}{L}\left(1+\frac{|S|}{L}\right).\end{aligned}$$

Combining this with (11.1) and $\mathbb{E}[Z] = |S|/L$ yields the lower bound. □

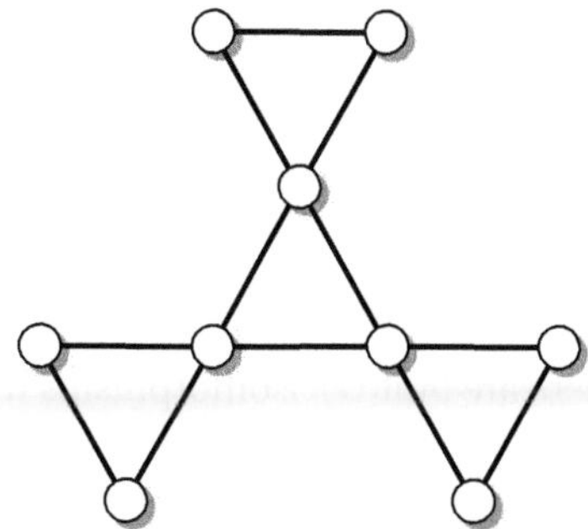

FIGURE 11.5: A graph G. We can rotate and reflect the central triangle, and each of the outer triangles can be independently flipped. So, its automorphism group $\mathrm{Aut}(G)$ consists of 48 permutations.

In Section 10.6.3, we saw how to construct a pairwise independent family of hash functions for any L which is a power of 2. By choosing L to be a power of 2 somewhere between cM and M, and making c small enough, we can make sure that Merlin can win with much higher probability if $|S| \geq M$ than if $|S| \leq cM$. Specifically, the reader can check that Lemma 11.1 gives the following, where $c = 1/16$:

Corollary 11.2 *Let L be the largest power of 2 less than or equal to $M/2$, and let h be chosen uniformly from a pairwise independent family of hash functions from U to $\{0,\ldots,L-1\}$. Then the probability that there is an $x \in S$ such that $h(x)=0$ is at least $2/3$ if $|S| \geq M$, and is at most $1/4$ if $|S| \leq M/16$.*

Now let's apply this approach to GRAPH NONISOMORPHISM. Given a graph G, let $I(G)$ denote the set of graphs isomorphic to G, or equivalently the image of G under all possible permutations:

$$I(G) = \{H : H \cong G\} = \{H : H = \pi(G) \text{ for some } \pi\}.$$

Then the set S of graphs isomorphic to G_1 or G_2 is

$$S = I(G_1) \cup I(G_2) = \{H : H \cong G_1 \text{ or } H \cong G_2\}.$$

If $G_1 \cong G_2$ then $S = I(G_1) = I(G_2)$. On the other hand, if $G_1 \ncong G_2$, then $I(G_1)$ and $I(G_2)$ are disjoint, and S is larger than either one.

However, while we know that S is bigger in one of these cases than the other, we don't know how big it is in either case. This is because $|I(G)|$ depends on how symmetric G is. As an extreme case, if G is a complete graph then $|I(G)| = 1$, since every permutation leaves G unchanged. We would like to compensate for the amount of symmetry each graph has, so that $|S|$ takes one value if G_1 and G_2 are isomorphic, and another value if they are not.

We can do this using a little group theory. An *automorphism* of a graph G is a permutation α of its vertices that maps edges to edges, so that $\alpha(G) = G$. If α_1 and α_2 are automorphisms, so is their composition. Therefore, the set of G's automorphisms, which we denote

$$\mathrm{Aut}(G) = \{\alpha : \alpha(G) = G\},$$

forms a subgroup of the permutation group S_n. We call Aut(G) the *automorphism group* of G. For the graph G shown in Figure 11.5, Aut(G) has 48 elements.

Now consider the following exercise:

Exercise 11.1 *Let ϕ, ϕ' be permutations. Show that $\phi'(G) = \phi(G)$ if and only if $\phi' = \phi \circ \alpha$ for some $\alpha \in$ Aut(G), where $\circ$ denotes composition.*

It follows that for each $H \in I(G)$, there are $|\mathrm{Aut}(G)|$ isomorphisms from G to H. The number of different graphs H we can get by applying all $n!$ permutations to G is then

$$|I(G)| = \frac{n!}{|\mathrm{Aut}(G)|}.$$

Formally, for each $H \in I(G)$ the set of ϕ such that $\phi(G) = H$ is a *coset* of Aut(G), and there are $n!/|\mathrm{Aut}(G)|$ different cosets (see Appendix A.7).

To cancel this factor of $|\mathrm{Aut}(G)|$, we define a new set $I'(G)$ where each element consists of a graph $H \cong G$ and one of H's automorphisms:

$$I'(G) = \{(H, \alpha) : H \cong G \text{ and } \alpha \in \mathrm{Aut}(H)\}.$$

Then for any graph G with n vertices, $|I'(G)| = n!$. If we define

$$S' = I'(G_1) \cup I'(G_2),$$

we have

$$|S'| = \begin{cases} n! & \text{if } G_1 \cong G_2 \\ 2n! & \text{if } G_1 \ncong G_2. \end{cases}$$

Clearly Merlin can prove to Arthur that a given pair (H, α) is in S' by showing him an isomorphism between H and either G_1 or G_2.

We are almost there. We have constructed a set whose size differs by a factor $c = 1/2$ depending on whether $G_1 \cong G_2$ or not. In order to use Corollary 11.2, we have to amplify this factor to $c = 1/16$. We can do this by using the set of ordered 4-tuples of elements of S',

$$S'' = S' \times S' \times S' \times S'.$$

Then

$$|S'| = \begin{cases} (n!)^4 & \text{if } G_1 \cong G_2 \\ 16(n!)^4 & \text{if } G_1 \ncong G_2. \end{cases}$$

Putting all this together, Merlin claims that $|S''| \geq 16(n!)^4$. Arthur chooses a random hash function h from a pairwise independent family, and challenges Merlin to show him an $x \in S''$ such that $h(x) = 0$. By Corollary 11.2, Merlin can win with probability at least 2/3 if $G_1 \ncong G_2$, and at most 1/4 if $G_1 \cong G_2$. Thus even when Arthur's coins are public, there is an interactive proof for GRAPH NONISOMORPHISM.

This technique, where Merlin convinces Arthur of a lower bound on the size of a set, can be used to replace private coins with public ones in *any* interactive proof. Roughly speaking, Merlin convinces Arthur that he could answer most of his questions correctly by proving that the set of questions he could answer correctly is large. So as far as talking with Merlin is concerned, public coins are just as good as private ones.

Moreover, if Arthur's coins are public, he might as well just send the sequence of coin flips to Merlin, since Merlin can figure out for himself what question Arthur *would* have asked if he had used those coin flips in his question-generating algorithm. Therefore, if there is an Arthur–Merlin game for a given problem, there is one where Arthur's questions are simply random sequences of bits.

Let's phrase what we have learned so far about interactive proofs in terms of complexity classes. We have shown that GRAPH NONISOMORPHISM is in the following class:

> AM is the set of problems for which there is an Arthur–Merlin game consisting of a constant number of rounds, exchanging a total of poly(n) bits, and where Arthur's coin flips are public, such that
>
> $$\Pr[\text{Merlin can win}] \geq 2/3 \text{ if } x \text{ is a yes-instance}$$
> $$\Pr[\text{Merlin can win}] \leq 1/3 \text{ if } x \text{ is a no-instance}.$$

Just as for the complexity classes BPP and RP defined in Section 10.9, the constants 2/3 and 1/3 are arbitrary. By playing the game multiple times and taking the majority of the results, we can *amplify* any pair of probabilities with a gap between them as in Problem 10.46, and make the probability that Merlin can win arbitrarily close to 1 and 0 for yes-instances and no-instances respectively. On the other hand, the fact that the number of rounds is constant is critical, and we will show below that interactive proofs with a polynomial number of rounds are far more powerful.

Since GRAPH ISOMORPHISM is in NP, this means that

$$\text{GRAPH ISOMORPHISM is in NP} \cap \text{coAM}.$$

We believe that AM is "slightly larger" than NP. Therefore this result is just a bit weaker than if GRAPH ISOMORPHISM were in NP∩coNP. In particular, we believe that NP-complete problems do not have interactive proofs for no-instances. Indeed, it can be shown that if NP $\subseteq$ coAM then the polynomial hierarchy would collapse, much as it would if NP = coNP (see Section 6.1).

For this reason, while we have not yet found a polynomial-time algorithm for GRAPH ISOMORPHISM, we believe that it is not NP-complete either. We illustrate this belief in Figure 11.6.

11.1

11.1.3 In Which Arthur Learns Nothing but the Truth

Let's look now at another variation on interactive proofs. Is there a way for Merlin to prove something to Arthur, while conveying no information other than the fact that it is true?

If this question strikes you as odd, suppose you are buying something online. Your goal is to prove that you are who you say you are, while conveying no other information. In particular, you don't want the merchant, or anyone who is eavesdropping on the conversation, to learn how to impersonate you in the future. Our current methods of online shopping are not this secure, but they should be. And as we will see in this section and the next, they could be.

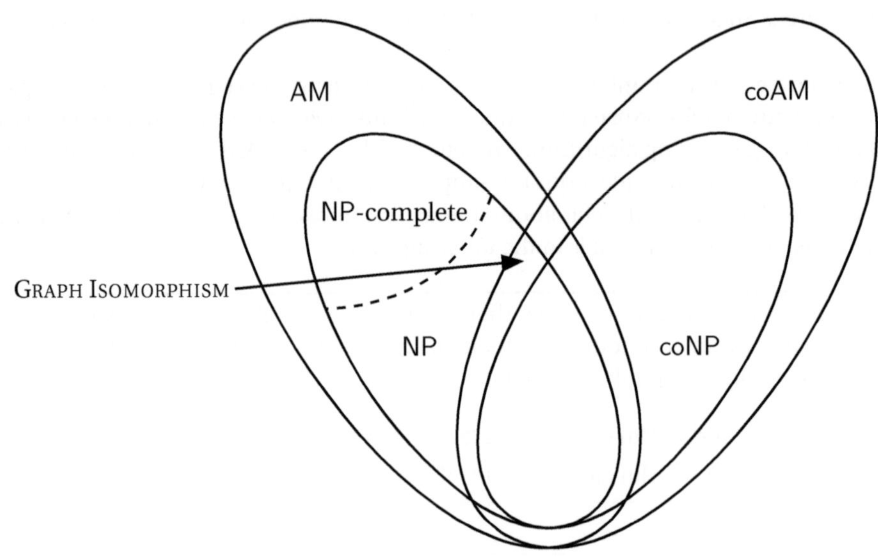

FIGURE 11.6: What we believe about the classes NP, coNP, AM, and coAM, and what we know about GRAPH ISOMORPHISM so far.

To explore these issues, let's return to our first interactive proof for GRAPH NONISOMORPHISM. If both parties "play fair," then Arthur chooses G randomly from $\{G_1, G_2\}$, applies a random permutation ϕ, and asks Merlin which of G_1 and G_2 is isomorphic to $H = \phi(G)$. If Merlin cheats and claims that $G_1 \ncong G_2$ when in fact $G_1 \cong G_2$, Arthur will catch him in the act with probability 1/2.

However, there is a sense in which Arthur can cheat as well. Suppose Arthur has a third graph G_3 that he knows is isomorphic to either G_1 or G_2, but he doesn't know which. If Arthur gives Merlin G_3 and pretends that he generated it from G_1 or G_2, Merlin will gladly tell him which one it is isomorphic to. Thus Arthur can learn more from this protocol than we originally intended. Is there a way for Merlin to convince Arthur that $G_1 \ncong G_2$, without revealing any additional information?

Interactive protocols of this kind exist, and are called *zero-knowledge proofs.* As our first example, let's switch from nonisomorphism to isomorphism. Suppose that $G_1 \cong G_2$. Then Merlin can prove this fact to Arthur, while revealing nothing whatsoever about the isomorphism between them, in the following way.

First Merlin, rather than Arthur, chooses G uniformly from $\{G_1, G_2\}$, applies a uniformly random permutation ϕ, and shows Arthur $H = \phi(G)$. Arthur then chooses $G' \in \{G_1, G_2\}$ and challenges Merlin to show him an isomorphism ϕ' from G' to H. If $G_1 \cong G_2$ then Merlin can always provide ϕ'. But if $G_1 \ncong G_2$ and Arthur chooses G' by flipping a coin, Merlin can only provide such a ϕ' with probability 1/2.

Intuitively, Arthur learns nothing from this interaction other than the fact that G_1 and G_2 are probably isomorphic. For each H, he only learns one of the two isomorphisms in the equation $G_1 \cong H \cong G_2$. He never gets to compose them to form the isomorphism from G_1 to G_2. Moreover, since ϕ is uniformly random, it is easy to see that ϕ' is uniformly random as well.

```
Simulator #1
in each round begin
    choose G′ ∈ {G1, G2} uniformly ;
    choose φ′ ∈ Sn uniformly ;
    H := φ′(G′) ;
    return (H, G′, φ′) ;
end
```

FIGURE 11.7: Our first attempt at a simulator for the zero-knowledge protocol for GRAPH ISOMORPHISM. It produces the same distribution of conversations that talking with Merlin would—if Arthur plays fair.

```
Simulator #2
in each round begin
    repeat
        choose G′ ∈ {G1, G2} uniformly ;
        choose φ′ ∈ Sn uniformly ;
        H := φ′(G′) ;
    until A(H) = G′;
    return (H, G′, φ′) ;
end
```

FIGURE 11.8: An improved simulator that works even if Arthur cheats—that is, regardless of what algorithm $A(H)$ Arthur uses to choose G'.

But in that case, Arthur *might as well be generating the entire conversation himself,* by choosing $G' \in \{G_1, G_2\}$ and ϕ' uniformly, and setting $H = \phi'(G')$. He could be having this conversation in a dream, in which Merlin's answers are generated by his own subconscious. Intuitively, this means that Arthur learns nothing from this protocol that he doesn't already know.

Formalizing this intuition involves some subtlety. Let's define a *simulator* as a randomized algorithm that produces exactly the same probability distribution of conversations as a real interaction with Merlin would, assuming that Merlin's claim is true. Since Arthur can run this simulator himself, the conversation with Merlin doesn't help Arthur compute anything that he couldn't compute already.

Our first attempt at a simulator for this protocol looks like Figure 11.7. Each round of conversation is a triple (H, G', ϕ'). As we suggested earlier, Arthur simulates a round by choosing G' and ϕ' uniformly, and setting $H = \phi'(G')$.

This simulator works perfectly as long as Arthur is honest, and flips a coin to decide which isomorphism to demand. But this is not enough. Suppose Arthur cheats and generates his questions in some other way, running some algorithm A on Merlin's graph H to choose G'. Then he will be able to tell the difference between the dream conversation and a real one, since in the dream he plays fair.

In order for the proof to be zero-knowledge, we need a simulator that can produce realistic conversations, no matter how Arthur generates his questions. This seems like a lot to ask. Happily, like Arthur's subconscious, the simulator knows what algorithm Arthur uses, and can use this knowledge to produce a convincing conversation.

We show an improved simulator in Figure 11.8. It repeatedly generates triples (H, G', ϕ') until it finds one which is consistent with Arthur's algorithm, i.e., such that $G' = A(H)$. These triples have exactly the same probability distribution as if Merlin generated H and Arthur then applied A to decide which G' to ask about. Note that A can be randomized, in which case by $A(H) = G'$ we mean that A returns G' on this particular run. Note also that if $G_1 \cong G_2$, the simulator is just as likely to generate a given H from G_1 as it is from G_2. Thus $G' = A(H)$ with probability $1/2$ no matter what $A(H)$ is, and the expected number of times we need to run the **repeat** loop is 2.

This simulator produces realistic conversations, in polynomial time, with probability exponentially close to 1. No matter what algorithm Arthur uses to generate his questions, his conversation with Merlin teaches him nothing that he could not generate himself, with high probability, with a randomized polynomial-time algorithm—assuming that Merlin's claim $G_1 \cong G_2$ is true. Thus the protocol we described above is a zero-knowledge interactive proof for GRAPH ISOMORPHISM.

Exercise 11.2 *Show that Merlin doesn't need superhuman powers to play his part in this protocol for* GRAPH ISOMORPHISM. *If he knows an isomorphism π from G_1 to G_2, he can generate his responses with a randomized polynomial-time algorithm.*

Now that we have a zero-knowledge proof for GRAPH ISOMORPHISM, what about GRAPH NONISOMORPHISM? Here we will use similar ideas to make sure that Arthur, as well as Merlin, is being honest.

As in our first protocol for GRAPH NONISOMORPHISM, Arthur sends Merlin a graph H and asks him which $G \in \{G_1, G_2\}$ is isomorphic to it. However, before he answers, Merlin turns the tables and demands an interactive proof that Arthur generated H honestly from G_1 or G_2, and that he knows an isomorphism from one of them to H.

This proof consists of a series of n rounds. In each one, Arthur generates an ordered pair of graphs (K_1, K_2) by applying some pair of permutations to G_1 and G_2. If he wishes, he can then switch K_1 and K_2. He then sends the resulting ordered pair to Merlin.

Merlin then has two options. He can demand that Arthur show him a pair of isomorphisms from $\{G_1, G_2\}$ to $\{K_1, K_2\}$, forcing Arthur to prove that each K_i is isomorphic to some G_j. Alternately, he can demand an isomorphism from H to one of the K_i, allowing Arthur to choose which one. If Arthur generated H and (K_1, K_2) honestly then he can pass either test easily, since he knows an isomorphism from some $G \in \{G_1, G_2\}$ to H and the isomorphisms between the G_i and the K_i.

If Arthur fails in any round, Merlin exclaims "you're cheating!", hits him with a stick, and ends the conversation. But if Arthur succeeds for n rounds then Merlin consents, and sends some $G_i \in \{G_1, G_2\}$ to Arthur. If Arthur did indeed generate H from G_i, Merlin wins.

Now that we know the rules of the game, what are the players' optimal strategies? When choosing which test to give Arthur, Merlin should simply flip a coin. Then if Arthur generated either H or the pair (K_1, K_2) dishonestly, Arthur will fail with probability $1/2$ in each round, and the probability that he fools Merlin for all n rounds is 2^{-n}.

Similarly, Arthur's optimal strategy is to generate the K_i by applying independently random permutations to G_1 and G_2, and switching K_1 and K_2 with probability $1/2$. In that case, each (K_1, K_2) is uniformly random among all possible ordered pairs that can be generated from $\{G_1, G_2\}$, so this part of the conversation gives Merlin no information about which G_i Arthur used to generate H. Thus if $G_1 \cong G_2$, Merlin can only win with probability $1/2$, just as in the original protocol.

In Problem 11.4, we ask you, dear reader, to prove that this protocol is zero-knowledge by writing a simulator for it. However, this simulator is not perfect. For instance, suppose Arthur cheats by generating

H from some third graph G_3, but generates the K_i honestly. With probability 2^{-n}, Merlin goes all n rounds without demanding an isomorphism from H to one of the K_i, and doesn't realize that Arthur is cheating. At that point, the simulator cannot generate Merlin's reply—which is to say that Arthur has gotten away with it, and has learned something from Merlin that he wasn't supposed to.

As a consequence, the simulator produces conversations with a probability distribution that is exponentially close, but not identical, to the that of a real interaction with Merlin. Such protocols are called *statistically* zero-knowledge proofs. In contrast, our protocol for GRAPH ISOMORPHISM is *perfectly* zero-knowledge—the simulator produces exactly the same probability distribution as a real conversation.

11.1.4 One-Way Functions, Bit Commitment, and Dinner Plates

We have seen zero-knowledge interactive proofs for GRAPH ISOMORPHISM and GRAPH NONISOMORPHISM. For what other problems do such proofs exist? We end this section by showing that—assuming there are secure cryptosystems—this class includes all of NP. We do this by giving a zero-knowledge proof for an NP-complete problem, namely GRAPH 3-COLORING.

Suppose that Merlin knows a 3-coloring C of a graph G. He wants to convince Arthur of this fact while revealing nothing at all about C. He can do this as follows. While Arthur is out of the room, Merlin chooses a permutation π uniformly from the $3! = 6$ permutations of the 3 colors, and colors G according to $\pi(C)$. Merlin then covers G's vertices with dinner plates. Arthur is invited back into the room, and is allowed to choose any pair of neighboring vertices. Merlin lifts the plates off those two vertices, and shows Arthur that they have different colors. Arthur then leaves the room, and Merlin chooses a new permutation π for the next round.

Since Merlin applies a fresh permutation in each round, Arthur has no opportunity to build up information about the coloring C. To be more precise, all Arthur sees in each round is a random pair of different colors, where all 6 such pairs are equally likely. No matter how he chooses which vertices to look at, a simulator could produce these colors just as well as a conversation with Merlin could.

On the other hand, if G is not 3-colorable, then no matter how Merlin colors the vertices, at least one pair of neighbors have the same color. Therefore, if G has n vertices and m edges and Arthur chooses from them uniformly, he will catch Merlin cheating with probability at least $1/m = 1/O(n^2)$ in each round. If they play, say, n^3 rounds, the probability that Merlin succeeds on every round is exponentially small.

This gives an excellent interactive proof for GRAPH 3-COLORING—as long as we have some paint and dinner plates. But what does this mean computationally? We need a scheme in which Merlin gives Arthur the color of each vertex in some encrypted form, and then shows Arthur how to decrypt the colors of the two vertices he chooses. This scheme must have two properties. First, Arthur must be unable to decrypt the colors on his own—that is, he cannot see through the plates. Second, Merlin must be forced to *commit* to a color for each vertex—he must be unable to cheat by changing the colors after Arthur chooses which pair to look at. In other words, there must be a unique way to decrypt each encrypted color. Such a scheme is called *bit commitment*.

One way to perform bit commitment is for Merlin to apply a *one-way function*—a function f that can be computed, but not inverted, in polynomial time. Since f is hard to invert, it acts as a dinner plate, which is opaque to Arthur's polynomial-time vision. Let's see how this works in the case where Merlin commits to, and encrypts, a single bit.

Let $f(x)$ be a one-to-one function from n-bit strings to n-bit strings, and let b be a polynomial-time function that maps each string x to a single bit $b(x)$. To commit to a bit c, Merlin chooses uniformly from all strings x such that $b(x) = c$, and gives Arthur $f(x)$. Then, to reveal c, Merlin gives Arthur x. At that point, Arthur can check that Merlin gave him $f(x)$ before, and calculate $c = b(x)$.

For this scheme to be secure, f has to be one-way in the following strong sense. Not only is Arthur incapable finding the string x from $f(x)$—he cannot even find the single bit $b(x)$. Specifically, we want the problem of finding x to be reducible, in polynomial time, to the problem of finding $b(x)$, so that finding $b(x)$ is just as hard as finding all of x. In that case b is called a *hard-core bit* for f.

To make all this concrete, let's consider a particular scheme. Let p be an n-bit prime, and recall that $\mathbb{Z}_p^* = \{1,2,\ldots,p-1\}$ is the group of integers mod p with multiplication. As in Section 4.4.1, let a be a *primitive root*, so that for every $z \in \mathbb{Z}_p^*$ we have $z \equiv_p a^x$ for some $0 \le x < p-1$. Let $f(x) = a^x \bmod p$. We showed in Section 3.2.2 that we can calculate $f(x)$ in polynomial time. However, we believe that f is a one-way function—in other words, that the problem DISCRETE LOG of determining $x = f^{-1}(z) = \log_a z$ from $z = f(x)$ is hard.

Now let $b(x)$ denote the "most significant bit" of x,

$$b(x) = \begin{cases} 1 & \text{if } x \ge (p-1)/2 \\ 0 & \text{if } x < (p-1)/2. \end{cases}$$

In Problem 11.8, you will show that $b(x)$ is a hard-core bit for $f(x)$—if we can compute the most significant bit of the discrete log, we can compute all of it. So unless there is a randomized polynomial-time algorithm for DISCRETE LOG, there is no randomized polynomial-time algorithm for finding $b(x)$ from $f(x)$, or even guessing $b(x)$, with probability noticeably better than chance.

Back to GRAPH 3-COLORING. For each vertex v, Merlin commits to two bits that describe v's color. For each bit c, he chooses x uniformly from those x were $b(x) = c$, i.e., from $\{0,\ldots,(p-1)/2-1\}$ or $\{(p-1)/2,\ldots,p-2\}$ if $c = 0$ or $c = 1$ respectively. He then sends Arthur $f(x) = a^x \bmod p$. If Arthur can solve DISCRETE LOG for any of these inputs then some of the dinner plates will be transparent, and Merlin's proof that the graph is 3-colorable will no longer be zero-knowledge.

It's important to note that for cryptographic purposes like these, we need a stronger notion of "hardness" than we have used before. When we defined NP-completeness, we called a problem hard if there *exist* hard instances, so that it is hard in the worst case. For a cryptosystem to be secure, on the other hand, it's not enough for there to exist secret keys that are hard to break—we need almost all keys to have this property. Therefore, worst-case hardness is not enough. We need problems that are hard on average, and indeed for almost all instances—that is, for almost all integers x that Merlin could use in this protocol.

Establishing that a problem is hard for most instances, rather than just the worst ones, requires different kinds of reductions than the ones we used to prove NP-completeness. Problem 11.10 shows that DISCRETE LOG has the interesting property that we can reduce arbitrary instances to random ones. In other words, if there is a polynomial-time algorithm that finds $\log_a z$ for a $1/\text{poly}(n)$ fraction of the possible values of z, then there is a randomized polynomial-time algorithm that works for all z with probability $1/\text{poly}(n)$, and by running this algorithm $\text{poly}(n)$ times we can make it work with high probability. Thus if DISCRETE LOG is easy on average, it is easy even in the worst case—and if it is hard in the worst case, it is hard on average.

This gives us a zero-knowledge proof for GRAPH 3-COLORING, as long as DISCRETE LOG is as hard as we think it is. Even if it isn't, there are schemes for bit commitment based on other one-way functions, pseudorandom generators, or secure cryptosystems—see, for instance, Problem 11.14. As we will discuss in Section 11.4, we strongly believe that all these things exist. If we are right, all problems in NP have zero-knowledge proofs similar to this one.

However, we stress that these proofs are *computationally* zero-knowledge, as opposed to the perfectly or statistically zero-knowledge proofs discussed in the previous section. Arthur is prevented from seeing through the dinner plates because of his limited computational powers—the information is there in front of him, but he is unable to extract it in polynomial time.

11.2

11.2 The Fable of the Chess Master

An odd character with a flowing white beard—looking suspiciously like someone in Figure 11.2—appears in Washington Square Park in New York City, a favorite hangout of Chess players. He sets up a folding wooden table, and sits down at it. He then announces to everyone within earshot that White has a winning strategy in Chess, and can always win within 300 moves.

You approach him and ask him to prove this claim. Oddly, he doesn't provide a Chess board, and he doesn't challenge you to a game. After all, what would you learn by having him beat you? Unless you played optimally—which is far beyond your limited computational powers—you would always harbor a nagging doubt that you might have won if you had chosen a different line of play.

Instead, he offers to answer a series of random questions *about integer polynomials*. After a few thousand of these questions, and a little arithmetic on your part, you are forced to admit that his claim is almost certainly true—that if he were lying, you would have caught him with probability very close to 1.

Of course, you are playing the part of Arthur, and the Chess Master is revealed to be an out-of-work Merlin. We saw in Section 11.1 that by answering questions that Arthur chooses randomly, Merlin can convince him that things like GRAPH NONISOMORPHISM are almost certainly true. In this section, we will see that interactive proofs are incredibly powerful—that a conversation with Merlin can convince Arthur of things far above NP in the complexity hierarchy.

As in Section 11.1, we define an Arthur–Merlin game as a conversation between the two, where Merlin is declared the winner or not by applying a polynomial-time algorithm to the transcript of the conversation. Then we can define the complexity class IP, which stands for Interactive Proof, as follows:

> IP is the set of problems for which there is an Arthur–Merlin game consisting of a polynomial number of rounds, such that Merlin can win with probability at least 2/3 if the input is a yes-instance, and at most 1/3 if it is a no-instance.

We restrict the number of rounds and the total number of bits exchanged to poly(n) in order to keep the interaction within Arthur's ability to participate.

Clearly NP $\subseteq$ IP, since the conversation could consist just of Merlin showing Arthur a witness, and AM $\subseteq$ IP since AM is the class of problems for which a constant number of rounds of conversation suffice. What things can Merlin convince Arthur of in a polynomial number of rounds?

In this section, we will show that interactive proofs of this kind are exactly as powerful as giving Arthur a polynomial amount of memory, and an unlimited amount of time. In other words,

$$\mathsf{IP} = \mathsf{PSPACE}.$$

In one direction, IP $\subseteq$ PSPACE follows from the fact we can go through every possible conversation between Arthur and Merlin, and check that Merlin wins in at least 2/3 of them. Since the length of the conversation is poly(n), it only takes poly(n) memory to do this using a depth-first search.

To show the converse, PSPACE $\subseteq$ IP, we will describe an interactive proof for QUANTIFIED SAT. Recall from Section 8.6 that this problem is equivalent to telling who has a winning strategy in TWO-PLAYER SAT, a game in which the players take turns setting the variables of a Boolean formula, one player tries to make it true, and the other tries to make it false. We showed there that QUANTIFIED SAT is PSPACE-complete, and that any reasonable two-player board game with a polynomial bound on the length of the game can be reduced to it.

The reader might object that Chess games can be exponentially long. However, in our fable, Merlin has provided an upper bound by claiming that White can win in just 300 moves. If Merlin makes claims about generalized Chess on $n \times n$ boards, saying that White can win from a certain starting position within t moves, his claim becomes a QUANTIFIED SAT formula whose size is polynomial in n and t.

The interactive proof for QUANTIFIED SAT involves a profound transformation, called *arithmetization*, from Boolean formulas to polynomials, which has had an enormous impact on theoretical computer science. Let's sit at Merlin's table and see what he has to say.

11.2.1 From Formulas to Polynomials

"All right then, my liege—sorry, you remind me of an old friend," Merlin says. "Surely you're familiar with the fact that we can treat the players' moves as choices of Boolean variables, and write the proposition that White has won as a SAT formula. The claim that White has a winning strategy then becomes a quantified formula, where White's and Black's moves correspond to 'there exists' and 'for all' respectively. This formula is true if White has a move, such that for all possible replies Black could make, White has a move, and so on, until White wins."

"Of course," you reply. "I've been to high school, where I read Chapter 8 of this book."

"Good," says Merlin. "Now, with the magic of arithmetization, I will translate the claim that such a formula is true into a series of claims about integer polynomials." He presents the following SAT formula as an example:

$$\phi = (x_1 \vee x_2 \vee x_3) \wedge (\overline{x}_1 \vee \overline{x}_2 \vee \overline{x}_3). \tag{11.2}$$

If we replace `true` and `false` by the integers 1 and 0, then $x \wedge y$ can be written xy and $\overline{x}$ can be written $1-x$. Similarly, $x \vee y$ can be written $1-(1-x)(1-y)$ using De Morgan's law $x \vee y = \overline{\overline{x} \wedge \overline{y}}$. This turns ϕ into a polynomial,

$$\phi(x_1, x_2, x_3) = \big(1-(1-x_1)(1-x_2)(1-x_3)\big)(1-x_1x_2x_3), \tag{11.3}$$

where we abuse notation by using ϕ to represent both a Boolean formula and the corresponding polynomial. You can check, for instance, that if $x_1, x_2, x_3 \in \{0,1\}$, then $\phi = 1$ if at least one of the x_i is 1 and at least one is 0.

We can handle quantifiers in the same way. Applying $\forall$ or $\exists$ to x_3, for instance, is just an AND or an OR of the formulas we get by setting x_3 to 0 or 1. This gives

$$\begin{aligned} \forall x_3 : \phi(x_1,x_2,x_3) &= \phi(x_1,x_2,0)\,\phi(x_1,x_2,1) \\ \exists x_3 : \phi(x_1,x_2,x_3) &= 1-\big(1-\phi(x_1,x_2,0)\big)\big(1-\phi(x_1,x_2,1)\big). \end{aligned} \tag{11.4}$$

Now consider the following quantified formula:

$$\begin{aligned}\phi &= \forall x_1 : \exists x_2 : \forall x_3 : \psi(x_1, x_2, x_3) \\ &= \forall x_1 : \exists x_2 : \forall x_3 : (x_1 \vee x_2 \vee x_3) \wedge (\overline{x}_1 \vee \overline{x}_2 \vee \overline{x}_3).\end{aligned} \tag{11.5}$$

We will use ϕ_i to denote the intermediate formula inside the ith quantifier. That is,

$$\begin{aligned}\phi_3 &= \psi \\ \phi_2 &= \forall x_3 : \phi_3 \\ \phi_1 &= \exists x_2 : \phi_2 \\ \phi_0 &= \forall x_1 : \phi_1 = \phi .\end{aligned}$$

Applying the rules (11.4) gives

$$\begin{aligned}\phi_3(x_1, x_2, x_3) &= \big(1 - (1 - x_1)(1 - x_2)(1 - x_3)\big)(1 - x_1 x_2 x_3) \\ \phi_2(x_1, x_2) &= \phi_3(x_1, x_2, 0)\, \phi_3(x_1, x_2, 1) \\ \phi_1(x_1) &= 1 - \big(1 - \phi_2(x_1, 0)\big)\big(1 - \phi_2(x_1, 1)\big) \\ \phi_0 &= \phi_1(0)\, \phi_1(1).\end{aligned}$$

Then the quantified formula ϕ is true if and only if applying these sums and products gives $\phi_0 = 1$.

Exercise 11.3 *Translate ϕ into English and confirm that it is true. Then calculate ϕ_0 and confirm that $\phi_0 = 1$.*

Our interactive proof starts with Merlin claiming that ϕ_0, the arithmetization of a quantified SAT formula, is equal to 1. To back up his claim, Merlin gives Arthur the polynomial $\phi_1(x_1)$. Arthur can check that $\phi_1(x_1)$ is what Merlin says it is using the technique of polynomial identity testing from Section 10.7. That is, he sets x_1 to a random integer in a certain range, and demands a proof that the value of $\phi_1(x_1)$ is correct. Merlin proves this claim, in turn, by giving Arthur $\phi_2(x_1, x_2)$ as a polynomial of x_2, and so on.

At the ith stage of the proof, the variables $x_1, \ldots, x_{i-1}$ have been set by Arthur's choices, and Merlin claims that $\phi_i(x_1, \ldots, x_{i-1}, x_i)$ is a certain polynomial of x_i. Arthur checks this claim by setting x_i randomly. Since ϕ_i is defined in terms of ϕ_{i+1} according to one of the rules (11.4), Merlin responds by giving $\phi_{i+1}(x_1, \ldots, x_{i+1})$ as a polynomial in x_{i+1}, and so on. Finally, after all the variables have been set, Arthur translates the SAT formula inside all the quantifiers into a polynomial ϕ_n himself, and checks that it gives Merlin's stated value.

Let's use our formula ϕ as an example. The conversation, between Merlin and a somewhat truculent Arthur, could go like this:

Merlin: I claim that $\phi_0 = 1$, and therefore that ϕ is true.

Arthur: Why should I believe that $\phi_0 = 1$?

Merlin: Because $\phi_0 = \phi_1(0)\, \phi_1(1)$, and I claim that $\phi_1(x_1) = x_1^2 - x_1 + 1$.

Arthur: I'm not convinced. If that claim were true, setting $x_1 = 3$ would give $\phi_1(3) = 7$. Can you prove that to me?

Merlin: Certainly. Since $\phi_1(3) = 1 - \big(1 - \phi_2(3, 0)\big)\big(1 - \phi_2(3, 1)\big)$, this follows from my claim that $\phi_2(3, x_2) = 6x_2^2 - 11x_2 + 3$.

Arthur: Hmm. If that claim were true, setting $x_2 = 2$ would give $\phi_2(3,2) = 5$. Why is that?

Merlin: As you know, $\phi_2(3,2) = \phi_3(3,2,0)\,\phi_3(3,2,1)$. I claim that $\phi_3(3,2,x_3) = -12x_3^2 + 8x_3 - 1$.

Arthur. If that were true, setting $x_3 = 5$ would give—give me a moment—$\phi_3(3,2,5) = -261$. Why should I believe that?

Merlin: Because, my dear King, even you are capable of translating the SAT formula ψ at the heart of ϕ into the polynomial $\phi_3 = \psi(x_1, x_2, x_3)$ given in (11.3). If you would care to substitute the values $x_1 = 3$, $x_2 = 2$, and $x_3 = 5$ into ϕ, you will see that indeed $\phi(3,2,5) = -261$.

Arthur: Let me see—yes, that works. But perhaps you got lucky, and fooled me about one of these polynomials. Shall we play again, with different x_i?

Merlin: [sighs] Of course, my liege. I would be happy to.

Of course, for all this to work, these polynomials have to be within Arthur's abilities to read and test them. For instance, if they have an exponential number of terms then they are too large for Merlin to tell Arthur about, and too large for Arthur to check. Since a polynomial of degree d can have $d+1$ terms, we would like all these polynomials to have degree poly(n).

Unfortunately, according to (11.4), each quantifier doubles the degree, so d could be exponentially large as a function of n. For instance, the formula

$$\phi(x) = \forall y_1 : \forall y_2 : \cdots : \forall y_n : x \vee (y_1 \wedge y_2 \wedge \cdots \wedge y_n),$$

which is true if and only if x is true, becomes the polynomial

$$\phi(x) = \prod_{y_1=0,1} \prod_{y_2=0,1} \cdots \prod_{y_n=0,1} \left(1 - (1-x)(1 - y_1 y_2 \cdots y_n)\right) = x^{2^n - 1}. \tag{11.6}$$

Of course, both x^{2^n-1} and x are arithmetizations of the Boolean formula stating that x is true, since they both yield 0 or 1 if $x = 0$ or 1 respectively. But as polynomials over the integers, they are very different.

Exercise 11.4 *Prove* (11.6) *by induction on* n.

11.2.2 Reducing the Degree

We need to modify our arithmetization scheme somehow to keep the degree of the polynomials low. One way to do this is to introduce an additional operator, which we call Я for "reduce." We will write it as a quantifier, although it operates on polynomials rather than formulas.

For any polynomial ϕ, we define Я$x : \phi$ as the polynomial we get if, for all $i > 0$, we replace every occurrence of x^i in ϕ with x. Equivalently, we can write

$$(\text{Я} x : \phi)(x) = \phi(0) + \left(\phi(1) - \phi(0)\right) x. \tag{11.7}$$

If the other variables are fixed, this changes ϕ, as a function of x, into a straight line through its values at $x = 0$ and $x = 1$. Thus Я reduces ϕ's degree while keeping its value on Boolean variables the same. If we apply Я to each of ϕ's variables, it turns ϕ into a multilinear function, i.e., a sum of terms where each x_i appears with power at most 1 in each term.

Exercise 11.5 *In our example, where* $\psi(x_1,x_2,x_3)$ *is given by* (11.3)*, show that*

$$\mathsf{R}x_1 : \mathsf{R}x_2 : \mathsf{R}x_3 : \psi(x_1,x_2,x_3) = x_1 + x_2 + x_3 - x_1x_2 - x_1x_3 - x_2x_3\,,$$

and that this coincides with ψ *on Boolean values.*

Now suppose we interleave our original quantifiers with Rs, and make each polynomial linear before it gets fed to a $\forall$ or $\exists$. In our example, this produces the following:

$$\begin{aligned}\phi = {} & \forall x_1 : \mathsf{R}x_1 : \\ & \exists x_2 : \mathsf{R}x_1 : \mathsf{R}x_2 : \\ & \forall x_3 : \mathsf{R}x_1 : \mathsf{R}x_2 : \mathsf{R}x_3 : \psi(x_1,x_2,x_3)\,. \end{aligned} \tag{11.8}$$

As before, let's define ϕ_i as the polynomial inside the ith quantifier, where now i ranges from 0 to 9. Merlin again tries to convince Arthur that $\phi_0 = 1$. However, in addition to claims regarding the $\forall$s and $\exists$s, Merlin now also makes claims about the Rs.

Let's take $\phi_4(x_1,x_2) = \mathsf{R}x_2 : \phi_5(x_1,x_2)$ as an example. Suppose that Arthur has already chosen the values $x_1 = 3$ and $x_2 = 2$, and Merlin claims that $\phi_4(3,2) = -7$. When challenged to prove this claim, Merlin responds that $\phi_5(3,x_2) = 6x_2^2 - 11x_2 + 3$. By plugging the values $x_2 = 0$ and $x_2 = 1$ into ϕ_5 and using (11.7), Arthur confirms that $\phi_4(3,x_2) = 3 - 5x_2$ and therefore $\phi_4(3,2) = -7$ as Merlin claimed. Arthur then chooses a new random value for x_2 and challenges Merlin to prove that $\phi_5(3,x_2)$ is what he says it is.

How large are the degrees of the ϕ_i? Once we apply R, the degree of each variable is at most 1, and each $\forall$ or $\exists$ increases its degree to at most 2. The innermost polynomial, corresponding to the SAT formula ψ, has degree less than or equal to ψ's length, since each variable appears in ψ with degree equal to the total number of times it appears in the formula. So, for a formula with n variables and length ℓ, the degree d of each ϕ_i is at most ℓ. Thus the degrees of the ϕ_i are now at most linear, as opposed to exponential, in the length of the formula.

Recall from Section 10.7 that two polynomials of degree d can only agree at d places unless they are identical. So if Arthur chooses each x_i from a range much larger than d, he will almost certainly catch Merlin cheating, unless Merlin is indeed telling the truth about each ϕ_i. Specifically, if Arthur chooses each x_i from a set of size $n^3\ell$, or from the integers mod p for some prime $p > n^3\ell$, then the chance Merlin fools him about a particular ϕ_i is at most $d/(n^3\ell) \le 1/n^3$.

There are a total of $n + n(n+1)/2 = O(n^2)$ quantifiers, including the Rs, in a formula like (11.8). Taking a union bound over all $O(n^2)$ stages of the proof, the probability that Merlin succeeds in cheating at any stage is $O(1/n)$. By choosing the x_i from a larger set, or playing this game multiple times, we can make the probability that Merlin succeeds in cheating exponentially small.

This gives an interactive proof, with a polynomial number of rounds, for QUANTIFIED SAT. Since QUANTIFIED SAT is PSPACE-complete, we have shown that IP = PSPACE. In theological terms, a polynomial-time conversation with a god is equivalent to polynomial memory, and all the time in the world.

Mathematically, this proof is not too hard to follow. But it may leave the reader a little bewildered. First of all, in some sense Arthur already knows the polynomials that Merlin is telling him about. In our example, if we ignore the Rs, Arthur can translate the formula ϕ directly into

$$\phi_0 = \prod_{x_1=0,1}\left(1 - \prod_{x_2=0,1}\left(1 - \prod_{x_3=0,1}\psi(x_1,x_2,x_3)\right)\right).$$

However, expanding all these products would create intermediate formulas with exponentially many terms. Unfolding this highly compact form to get the answer $\phi_0 = 1$ does not seem to be something Arthur can do in polynomial time. If it is, then Arthur has no need of Merlin's advice, and PSPACE = P.

Secondly, by exchanging these challenges and replies, Arthur and Merlin are making moves in a game, which Merlin wins if Arthur becomes convinced. But what does this game have to do with Chess?

There are two important differences between Chess and the game that Arthur and Merlin are playing. First, Arthur sets the variables for every quantifier, while in a real game, the variables are chosen by one player at the $\exists$s and the other player at the $\forall$s.

More importantly, while the central formula ψ inside the quantifiers still corresponds to White placing Black in checkmate, we are giving this formula a very different kind of input. While moves in Chess or TWO-PLAYER SAT correspond to setting the x_i to Boolean values, Arthur sets them to integers over a large range. Rather like extending real numbers to complex ones, Arthur and Merlin are playing some kind of imaginary Chess, where most moves cannot be interpreted as moving pieces on the board. By extending Chess to this much larger set of moves, we obtain a game which Merlin will lose almost all the time, unless White can win the original Chess game every time.

11.3

11.3 Probabilistically Checkable Proofs

> I have discovered a truly marvelous proof of this theorem, which this margin is too narrow to contain.
>
> Pierre de Fermat

Suppose I claim to have proved a marvelous new theorem, and I send you a purported proof of it. As we discussed in Section 6.1.3, given the axioms of a formal system, you can check this proof in polynomial time as a function of its length. But as every mathematician knows, going through this process can be tedious. If your attention falters for a moment, you might miss the one flaw in a thousand steps of reasoning, and think that the proof is valid even if it isn't.

How lovely it would be if this proof were *holographic*—if every part of it were a good indication of the validity of the whole. In such a proof, if any flaws exist, they are somehow spread throughout the entire proof rather than being localized. Then by checking the proof in just a few places, you could conclude that it is almost certainly correct.

As we will describe in this section, proofs of this kind really do exist. For any problem in NP, there is a proof in which you can detect any flaw with probability 1/2 by checking a constant number of bits of the proof. In fact, even *three* bits are enough. Once again, computational complexity has transformed, and expanded, our notion of proof.

11.3.1 A Letter from Merlin

The seeds of this idea are already present in our first interactive proof for GRAPH NONISOMORPHISM, where Merlin has to send Arthur just a few bits to convince him that $G_1 \not\cong G_2$.

This time, Merlin has been called away on important business (see Figure 11.9), so he is unable to have a conversation with Arthur. However, he has written out a long scroll, with one bit for each of the

FIGURE 11.9: Merlin is too busy to have a conversation with Arthur. He will send a proof instead.

$2^{\binom{n}{2}}$ possible graphs H on n vertices. The Hth bit in this scroll is 0 or 1 if H is isomorphic to G_1 or G_2 respectively. If H is not isomorphic to either graph, Merlin can set this bit either way.

Arthur chooses G_1 or G_2 as before, applies a random permutation ϕ to get H, and looks up the Hth bit in the scroll. If $G_1 \ncong G_2$, this bit tells Arthur which graph he started with, just as Merlin would if he were there. On the other hand, if $G_1 \cong G_2$, then no matter what bits Merlin wrote on the scroll, the Hth bit is wrong half the time.

Thus GRAPH NONISOMORPHISM has a *probabilistically checkable proof*, or PCP for short. For a yes-instance, there is a proof that Arthur will accept with probability 1, but for a no-instance, there is no proof that Arthur will accept with probability more than 1/2. We say that this PCP has *completeness* 1 and *soundness* 1/2. This terminology is a little awkward, since a good PCP has small soundness, but we will stick to it.

In essence, a probabilistically checkable proof is like a conversation with a consistent Merlin—one who is willing to write down, in advance, how he would reply to any of Arthur's questions, and who sticks to those replies. In an interactive proof, in contrast, we have to guard against the possibility that Merlin changes his story as the game goes on.

PCPs have three important parameters: the length of the proof, the number of bits that Arthur is allowed to examine, and the number of coins he needs to flip. The proof of GRAPH NONISOMORPHISM that we just described is exponentially long. Arthur looks at a single bit, and he flips poly(n) coins to choose which bit to look at—one coin to choose between G_1 or G_2, and $\log_2 n! = \Theta(n \log n)$ coins to choose what permutation ϕ to apply.

What happens if we scale this down? What problems have PCPs of polynomial length, where Arthur flips just $O(\log n)$ coins, and where he is only allowed to check a constant number of bits of the proof? In the 1990s, a series of breakthroughs showed that this class includes all of NP:

Theorem 11.3 (PCP Theorem) *Every problem in* NP *has probabilistically checkable proofs of length* $\mathrm{poly}(n)$, *completeness* 1 *and soundness* 1/2, *where the verifier flips* $O(\log n)$ *coins and looks at* $O(1)$ *bits of the proof.*

Amazingly, even three bits is enough, at an infinitesimal cost to the completeness and soundness:

Theorem 11.4 *For any constant* $\delta > 0$, *every problem in* NP *has probabilistically checkable proofs of length* $\mathrm{poly}(n)$, *where the verifier flips* $O(\log n)$ *coins and looks at three bits of the proof, with completeness* $1-\delta$ *and soundness* $1/2+\delta$.

Moreover, the conditions under which Arthur accepts are extremely simple—namely, if the parity of these three bits is even.

This theorem has powerful consequences for inapproximability. If we write down all the triplets of bits that Arthur could look at, we get a system of linear equations mod 2. If Merlin takes his best shot at a proof, the probability that Arthur will accept is the largest fraction of these equations that can be simultaneously satisfied. Thus even if we know that this fraction is either at least $1-\delta$ or at most $1/2+\delta$, it is NP-hard to tell which. As we discussed in Section 9.3, this implies that it is NP-hard to approximate MAX-XORSAT within any constant factor better than 1/2.

Proving these theorems in full is beyond our scope. However, we can get a good feel for them by proving the following version:

Theorem 11.5 *Every problem in* NP *has probabilistically checkable proofs of length* $2^{\mathrm{poly}(n)}$, *where the verifier flips* $\mathrm{poly}(n)$ *coins and looks at* $O(1)$ *bits of the proof, with completeness* 1 *and soundness* 1/2.

Unlike Theorems 11.3 and 11.4, these proofs are exponentially long. But it is still very surprising that we can check them by looking at a constant number of bits.

Even proving this weaker version of the PCP Theorem takes a fair amount of technical work. However, the reader will be well-rewarded, with a tour of polynomial equations, error-correcting codes, and Fourier analysis—tools which have found many other applications in computer science. After completing the proof, we sketch a recent proof of Theorem 11.3 that works by amplifying the gap between satisfiability and unsatisfiability.

11.3.2 Systems of Equations

As with many of the topics in this chapter, our approach to Theorem 11.5 is algebraic. We start by formulating an NP-complete problem regarding systems of quadratic equations. Suppose we have n variables $k_1, \ldots, k_n$. We can represent any quadratic function of the k_i in the following way:

$$\begin{aligned} q(k_1,\ldots,k_n) &= \sum_{i,j=1}^{n} A_{ij} k_i k_j + \sum_{i=1}^{n} b_i k_i + c \\ &= \mathbf{k}^T \mathbf{A} \mathbf{k} + \mathbf{b}^T \mathbf{k} + c . \end{aligned} \tag{11.9}$$

Here $\mathbf{A}$ is an $n \times n$ matrix, $\mathbf{b}$ and $\mathbf{k}$ are n-dimensional vectors, and c is an integer. Given a set of m quadratic functions $q_1, \ldots, q_m$, we can then ask whether there is a solution $\mathbf{k}$ to the system of equations $q_1(\mathbf{k}) = 0$, $q_2(\mathbf{k}) = 0$, and so on.

We will focus on the case where all our arithmetic is done mod 2. To save ink, we will write $q(\mathbf{k}) = 0$ rather than $q(\mathbf{k}) \equiv_2 0$.

> MOD-2 QUADRATIC EQUATIONS
>
> Input: A set of mod-2 quadratic functions $q_1(\mathbf{k}), \ldots, q_m(\mathbf{k})$ of the form (11.9), $q_\ell(\mathbf{k}) = \mathbf{k}^T \mathbf{A}_\ell \mathbf{k} + \mathbf{b}_\ell^T \mathbf{k} + c_\ell$, where $\mathbf{A}_\ell \in \{0,1\}^{n \times n}$, $\mathbf{b}_\ell \in \{0,1\}^n$ and $c_\ell \in \{0,1\}$ for each ℓ.
>
> Question: Is there a $\mathbf{k} \in \{0,1\}^n$ such that $q_\ell(\mathbf{k}) = 0$ for all $1 \le \ell \le m$?

Exercise 11.6 *Prove that* MOD-2 QUADRATIC EQUATIONS *is* NP*-complete by reducing* 3-SAT *to it. Hint: use the k_i to represent Boolean variables, and arithmetize the clauses as in Section 11.2.1. This gives a system of cubic polynomials. Then reduce the degree from* 3 *to* 2 *by adding additional variables and equations.*

Now consider an instance of MOD-2 QUADRATIC EQUATIONS, and suppose that the answer is "yes." If we follow our usual approach to NP problems, Merlin would simply send Arthur a vector $\mathbf{k}$—but Arthur would have to look at all n bits of $\mathbf{k}$ to confirm that it is a solution. Can Merlin provide a holographic proof instead, that Arthur can check by looking at just a constant number of bits?

Indeed he can. To achieve this holographic character, Merlin's proof encodes the solution $\mathbf{k}$ in a highly redundant way, or claims to do so. He sends Arthur a scroll containing a string s_1 of 2^n bits, giving the value of every possible linear function of $\mathbf{k}$—which, mod 2, means the parity of every possible subset of the k_i. Equivalently, this string is the truth table of the function

$$s_1(\mathbf{x}) = \sum_{i=1}^{n} k_i x_i = \mathbf{k}^T \mathbf{x}. \tag{11.10}$$

Merlin's scroll also includes a string s_2 of 2^{n^2} bits, containing every possible homogeneous quadratic function of $\mathbf{k}$—that is, those with no linear or constant terms. This is the truth table of the following function on $n \times n$ matrices $\mathbf{M}$,

$$s_2(\mathbf{M}) = \sum_{i,j=1}^{n} M_{ij} k_i k_j = \mathbf{k}^T \mathbf{M} \mathbf{k}. \tag{11.11}$$

In the next section, we will see how Arthur can use s_1 and s_2 to confirm that $\mathbf{k}$ is a solution to the system of equations.

As we will see, designing our PCP so that Arthur accepts a valid proof is easy. The tricky part is making sure that he often rejects an invalid one. There are two ways that Merlin can be dishonest: he can send strings s_1, s_2 that encode a vector $\mathbf{k}$ as in (11.10) and (11.11), or are very close to strings that do, but where $\mathbf{k}$ is not a solution. Or, Merlin can send strings s_1, s_2 that are nowhere near strings of this form, and don't consistently encode any $\mathbf{k}$ at all. We have to make sure that Arthur can detect Merlin's deception in both cases, and that the probability he mistakenly accepts Merlin's proof is not too large.

11.3.3 Random Subsets of the System

Let's assume for now that Merlin is telling the truth, at least insofar as s_1 and s_2 actually encode a vector $\mathbf{k}$ according to (11.10) and (11.11). How can Arthur check that $\mathbf{k}$ is a solution? Given a particular quadratic function,

$$q(\mathbf{k}) = \mathbf{k}^T \mathbf{A} \mathbf{k} + \mathbf{b}^T \mathbf{k} + c = s_2(\mathbf{A}) + s_1(\mathbf{b}) + c\,,$$

Arthur can confirm that $q(\mathbf{k}) = 0$ by checking s_1 at $\mathbf{b}$ and s_2 at $\mathbf{A}$. But since there are m polynomials q_ℓ, how can he confirm that $q_\ell(\mathbf{k}) = 0$ for all of them, while checking just a constant number of bits?

The answer is to compute the sum of a random subset of the q_ℓ, or, equivalently, the inner product of the vector $\mathbf{q}(\mathbf{k}) = (q_1(\mathbf{k}), \ldots, q_m(\mathbf{k}))$ with a random vector $\mathbf{r} \in \{0,1\}^m$. For each ℓ, Arthur flips a coin, choosing $r_\ell \in \{0,1\}$ uniformly. He then computes

$$q(\mathbf{k}) = \mathbf{r}^T \mathbf{q}(\mathbf{k}) = \sum_{\ell=1}^{m} r_\ell q_\ell(\mathbf{k})\,.$$

He can obtain $\mathbf{r}^T \mathbf{q}(\mathbf{k})$ from s_1 and s_2 by writing

$$q(\mathbf{k}) = \mathbf{k}^T \mathbf{A} \mathbf{k} + \mathbf{b}^T \mathbf{k} + c = s_2(\mathbf{A}) + s_1(\mathbf{b}) + c\,, \tag{11.12}$$

where $\mathbf{A}$, $\mathbf{b}$, and c are defined as

$$\mathbf{A} = \sum_{\ell=1}^{m} r_\ell \mathbf{A}_\ell\,, \quad \mathbf{b} = \sum_{\ell=1}^{m} r_\ell \mathbf{b}_\ell \quad \text{and} \quad c = \sum_{\ell=1}^{m} r_\ell c_\ell\,. \tag{11.13}$$

If $q_\ell(\mathbf{k}) = 0$ for all ℓ, then clearly $q(\mathbf{k}) = 0$. On the other hand, we claim that if $q_\ell(\mathbf{k}) \neq 0$ for any ℓ, then $q(\mathbf{k}) \neq 0$ with probability 1/2. Here we rely on the following exercise:

Exercise 11.7 *Let $\mathbf{q} \in \{0,1\}^m$ be a fixed m-dimensional vector. Show that if $\mathbf{q} \neq (0,\ldots,0)$, then if $\mathbf{r} \in \{0,1\}^m$ is uniformly random,*

$$\Pr_{\mathbf{r}}\left[\mathbf{r}^T \mathbf{q} \neq 0\right] = 1/2\,.$$

Hint: focus on a particular $q_\ell \neq 0$, and use the fact that the components of $\mathbf{r}$ are chosen independently of each other.

So if s_1 and s_2 are of the form (11.10) and (11.11), Arthur rejects a non-solution $\mathbf{k}$ with probability 1/2 each time he performs this test. But what if s_1 and s_2 are not exactly of this form, but only close to them?

11.3.4 Error Correction

Let's say that two functions s, s' from $\{0,1\}^n$ to $\{0,1\}$ are *δ-close* if they differ on at most a fraction δ of their possible inputs. That is, if $\mathbf{x}$ is chosen uniformly from $\{0,1\}^n$,

$$\Pr_{\mathbf{x}}\left[s(\mathbf{x}) = s'(\mathbf{x})\right] \geq 1 - \delta\,.$$

To make our PCP sound, so that Arthur rejects any invalid proof with probability 1/2, we will force a deceptive Merlin to make a difficult choice. He can give Arthur strings s_1, s_2 that are δ-close to $\mathbf{k}^T\mathbf{x}$ and

$\mathbf{k}^T\mathbf{M}\mathbf{k}$, but where $\mathbf{k}$ is not a solution—or he can give Arthur strings that are not δ-close to functions of this form. We will show that if Merlin takes the first route, and if δ is small enough, Arthur can reject $\mathbf{k}$ as in Section 11.3.3. Then in Section 11.3.5, we will show how Arthur can detect the second form of deception.

How small must δ be for this to work? The following exercise shows that if $\delta < 1/4$, then s_1 determines $\mathbf{k}$ uniquely:

Exercise 11.8 *Suppose that the function $s_1(\mathbf{x})$ is δ-close to $\mathbf{k}^T\mathbf{x}$ for some $\mathbf{k}$. Show that if $\delta < 1/4$, there can only be one such $\mathbf{k}$. Hint: show that if $\mathbf{k}_1 \neq \mathbf{k}_2$, the functions $\mathbf{k}_1^T\mathbf{x}$ and $\mathbf{k}_2^T\mathbf{x}$ differ on half their inputs, and use the triangle inequality.*

This suggests that we can recover a perfectly linear function from a nearly linear one. But how? If Arthur wants to know $\mathbf{k}^T\mathbf{x}$ for a particular $\mathbf{x}$, he can't simply look at $s_1(\mathbf{x})$. After all, this might be one of the $\mathbf{x}$ on which s_1 differs from $\mathbf{k}^T\mathbf{x}$.

There is a simple but clever way for Arthur to spread this difference out, and deduce $\mathbf{k}^T\mathbf{x}$ by comparing two different values of s_1:

Exercise 11.9 *Suppose that the function $s_1(\mathbf{x})$ is δ-close to $\mathbf{k}^T\mathbf{x}$. Show that, for any fixed $\mathbf{x}$, if $\mathbf{y}$ is chosen uniformly,*

$$\Pr_{\mathbf{y}}\left[s_1(\mathbf{x}+\mathbf{y}) - s_1(\mathbf{y}) = \mathbf{k}^T\mathbf{x}\right] \geq 1 - 2\delta\,. \tag{11.14}$$

Hint: use the union bound.

We could write $s_1(\mathbf{x}+\mathbf{y}) + s_1(\mathbf{y})$ instead of $s_1(\mathbf{x}+\mathbf{y}) - s_1(\mathbf{y})$, since we are doing our arithmetic mod 2. However, here and in a few places below, we write '$-$' instead of '$+$' to keep the ideas behind the expression clear.

Exercise 11.9 lets Arthur recover $\mathbf{k}^T\mathbf{x}$ a majority of the time whenever $\delta < 1/4$. In other words, because it represents the n-bit string $\mathbf{k}$ redundantly with 2^n bits, the string $s_1 = \mathbf{k}^T\mathbf{x}$ is an *error-correcting code.* Even if Merlin adds errors to s_1, flipping some of its bits, Arthur can correct these errors as long as there are fewer than $n/4$ of them.

Similarly, suppose that s_2 is δ-close to $\mathbf{k}^T\mathbf{M}\mathbf{k}$. That is, if $\mathbf{M}$ is chosen uniformly from all $n \times n$ matrices,

$$\Pr_{\mathbf{M}}\left[s_2(\mathbf{M}) = \mathbf{k}^T\mathbf{M}\mathbf{k}\right] \geq 1 - \delta\,.$$

Then for any fixed $\mathbf{M}$, if $\mathbf{N}$ is chosen uniformly,

$$\Pr_{\mathbf{N}}\left[s_2(\mathbf{M}+\mathbf{N}) - s_2(\mathbf{N}) = \mathbf{k}^T\mathbf{M}\mathbf{k}\right] \geq 1 - 2\delta\,. \tag{11.15}$$

This gives Arthur an error-correcting way to test whether $\mathbf{k}$ is a solution. He chooses a random vector $\mathbf{r}$ as before, and defines $\mathbf{A}$, $\mathbf{b}$, and c in order to compute $q(\mathbf{k}) = \mathbf{r}^T\mathbf{q}(\mathbf{k})$. But rather than assuming that $q(\mathbf{k}) = s_2(\mathbf{A}) + s_1(\mathbf{b}) + c$ as in (11.12), he chooses $\mathbf{y}$ and $\mathbf{N}$ uniformly, and computes the following surrogate for $q(\mathbf{k})$:

$$q'(\mathbf{k}) = s_2(\mathbf{A}+\mathbf{N}) - s_2(\mathbf{N}) + s_1(\mathbf{b}+\mathbf{y}) - s_1(\mathbf{y}) + c\,, \tag{11.16}$$

He then accepts $\mathbf{k}$ as a solution if $q'(\mathbf{k}) = 0$. With what probability can Merlin fool him?

Lemma 11.6 *Suppose that s_1 and s_2 are δ-close to $\mathbf{k}^T\mathbf{x}$ and $\mathbf{k}^T\mathbf{M}\mathbf{k}$ respectively, where $\mathbf{k}$ is not a solution to the system of equations. Arthur chooses $\mathbf{r}$ uniformly, and computes $q'(\mathbf{k})$ according to* (11.16), *with $\mathbf{A}$, $\mathbf{b}$, and c defined as in* (11.13). *Then the probability that $q'(\mathbf{k}) = 0$, and therefore that Arthur accepts $\mathbf{k}$ as a solution, is at most $1/2 + 2\delta$.*

Proof The probability that $q'(\mathbf{k}) \neq q(\mathbf{k})$ is at most the sum of the probabilities that $s_2(\mathbf{A}+\mathbf{N}) - s_2(\mathbf{N}) \neq s_2(\mathbf{A})$ or $s_1(\mathbf{b}+\mathbf{y}) - s_1(\mathbf{y}) \neq s_1(\mathbf{b})$. According to (11.14) and (11.15) each of these probabilities is at most 2δ, so $q'(\mathbf{k}) = q(\mathbf{k})$ with probability at least $1-4\delta$. If $q'(\mathbf{k}) = q(\mathbf{k})$ but $\mathbf{k}$ is not a solution, Exercise 11.7 tells us that Arthur rejects with probability 1/2. Thus the probability that he rejects is at least $(1-4\delta)/2 = 1/2 - 2\delta$, and he accepts with probability at most $1/2 + 2\delta$. □

If Arthur repeats this test t times, he accepts a non-solution $\mathbf{k}$ with probability at most $(1/2+2\delta)^t$. For any $\delta < 1/4$, he can make this probability as small as he likes by setting t to a suitable constant.

This shows that if Merlin tries to cheat by giving Arthur strings s_1 and s_2 that are δ-close to $\mathbf{k}^T\mathbf{x}$ and $\mathbf{k}^T M \mathbf{k}$, but where $\mathbf{k}$ is not a solution, then Arthur can catch him. Our next goal is to show how to catch Merlin if he cheats the other way—if s_1 and s_2 are not δ-close to any functions of these forms.

11.3.5 Linearity Testing and Fourier Analysis

A function $s(\mathbf{x})$ is of the form $\mathbf{k}^T\mathbf{x}$ if and only if $s(\mathbf{x})$ is *linear*. That is,

$$s(\mathbf{x}+\mathbf{y}) = s(\mathbf{x}) + s(\mathbf{y}), \tag{11.17}$$

for all $\mathbf{x}, \mathbf{y} \in \{0,1\}^n$. An intuitive way to test for linearity is to choose $\mathbf{x}$ and $\mathbf{y}$ randomly, check s at $\mathbf{x}$, $\mathbf{y}$, and $\mathbf{x}+\mathbf{y}$, and see if (11.17) is true.

The next lemma shows that if s is far from linear, then (11.17) is often false. Therefore, if testing (11.17) on random pairs $\mathbf{x}, \mathbf{y}$ succeeds repeatedly, s is probably close to a linear function.

Lemma 11.7 *Suppose that s is not δ-close to any linear function. That is, for all $\mathbf{k}$, if $\mathbf{x}$ is chosen uniformly then*

$$\Pr_{\mathbf{x}}\left[s(\mathbf{x}) = \mathbf{k}^T\mathbf{x}\right] < 1-\delta .$$

Then if $\mathbf{x}$ and $\mathbf{y}$ are chosen uniformly and independently, the linearity test (11.17) *fails with probability at least δ:*

$$\Pr_{\mathbf{x},\mathbf{y}}\left[s(\mathbf{x}+\mathbf{y}) = s(\mathbf{x}) + s(\mathbf{y})\right] < 1-\delta . \tag{11.18}$$

Proof It's convenient to change our perspective a bit, and focus on the function

$$f(\mathbf{x}) = (-1)^{s(\mathbf{x})} .$$

Then the linearity test (11.17) for s becomes

$$f(\mathbf{x}+\mathbf{y}) f(\mathbf{x}) f(\mathbf{y}) = 1 . \tag{11.19}$$

Let $g(\mathbf{x},\mathbf{y})$ denote the function $f(\mathbf{x}+\mathbf{y})f(\mathbf{x})f(\mathbf{y})$. Then the probability that (11.19) holds for a random pair $\mathbf{x}$, $\mathbf{y}$ is

$$\Pr_{\mathbf{x},\mathbf{y}}\left[g(\mathbf{x},\mathbf{y}) = 1\right] = \frac{1}{2}\left(1 + \mathop{\mathbb{E}}_{\mathbf{x},\mathbf{y}}\left[g(\mathbf{x},\mathbf{y})\right]\right) . \tag{11.20}$$

We will bound this probability using Fourier analysis. We can decompose any function $f : \{0,1\}^n \to \mathbb{R}$ as a linear combination of Fourier basis functions:

$$f(\mathbf{x}) = \sum_{\mathbf{k}\in\{0,1\}^n} (-1)^{\mathbf{k}^T\mathbf{x}}\, \tilde{f}(\mathbf{k}) . \tag{11.21}$$

Here $\mathbf{k}$ is the frequency vector, and

$$\tilde{f}(\mathbf{k}) = \mathop{\mathbb{E}}_{\mathbf{x}} \left[(-1)^{\mathbf{k}^T\mathbf{x}} f(\mathbf{x})\right]$$

is the Fourier coefficient of f at $\mathbf{k}$.

The Fourier transform preserves the sum of the squares of f, in the following sense:

$$\sum_{\mathbf{k}} \left|\tilde{f}(\mathbf{k})\right|^2 = \mathop{\mathbb{E}}_{\mathbf{x}} \left[\left|f(\mathbf{x})\right|^2\right]. \tag{11.22}$$

This equation is often called Parseval's identity, after Marc-Antoine Parseval—no relation, as far as we know, to the Parsifal in King Arthur's court. It follows from the fact that the Fourier transform is unitary, as in Section 3.2.3 where we discussed the Fourier transform on the integers mod n, and that unitary operators preserve inner products and the lengths of vectors. If we regard f and $\tilde{f}$ as 2^n-dimensional vectors, then with our choice of normalization, their lengths squared are $\mathbb{E}_{\mathbf{x}}\left|f(\mathbf{x})\right|^2$ and $\sum_{\mathbf{k}}\left|\tilde{f}(\mathbf{k})\right|^2$ respectively.

Our assumption is that s differs from any linear function, and therefore that f differs from any Fourier basis function $(-1)^{\mathbf{k}^T\mathbf{x}}$, on at least δ of its possible inputs $\mathbf{x}$. This places an upper bound on f's Fourier coefficients. Namely, for all $\mathbf{k}$,

$$\tilde{f}(\mathbf{k}) = 1 - 2 \Pr_{\mathbf{x}} \left[s(\mathbf{x}) \neq \mathbf{k}^T\mathbf{x}\right] < 1 - 2\delta. \tag{11.23}$$

The next exercise gives the expectation of g, and therefore the probability that the linearity test succeeds, in terms of f's Fourier transform.

Exercise 11.10 *Show that*

$$\mathop{\mathbb{E}}_{\mathbf{x},\mathbf{y}} \left[g(\mathbf{x},\mathbf{y})\right] = \sum_{\mathbf{k}} \tilde{f}(\mathbf{k})^3. \tag{11.24}$$

Hint: write f as its Fourier sum (11.21), *and use the fact that*

$$\mathop{\mathbb{E}}_{\mathbf{x}}(-1)^{\mathbf{k}^T\mathbf{x}} = \begin{cases} 1 & \mathbf{k} = \mathbf{0} \\ 0 & \mathbf{k} \neq \mathbf{0}. \end{cases}$$

By separating out one factor of $\tilde{f}(\mathbf{k})$ from the sum (11.24) and bounding it with the maximum of $\tilde{f}(\mathbf{k})$ over all $\mathbf{k}$, we get

$$\begin{aligned} \mathop{\mathbb{E}}_{\mathbf{x},\mathbf{y}} \left[g(\mathbf{x},\mathbf{y})\right] &\leq \left(\max_{\mathbf{k}} \tilde{f}(\mathbf{k})\right) \sum_{\mathbf{k}} \left|\tilde{f}(\mathbf{k})\right|^2 \\ &= \left(\max_{\mathbf{k}} \tilde{f}(\mathbf{k})\right) \mathop{\mathbb{E}}_{\mathbf{x}}\left[\left|f(\mathbf{x})\right|^2\right] \\ &< (1-2\delta) \mathop{\mathbb{E}}_{\mathbf{x}}\left[\left|f(\mathbf{x})\right|^2\right] \\ &= 1 - 2\delta. \end{aligned}$$

Here we used Parseval's identity in the second line, the bound (11.23) in the third, and the fact that $\left|f(\mathbf{x})\right|^2 = 1$ in the fourth. Combining this with (11.20) completes the proof of (11.18). □

With Lemma 11.7 in hand, we can now describe the first part of the PCP for MOD-2 QUADRATIC EQUATIONS. Arthur wishes to establish that s_1 is δ-close to $\mathbf{k}^T\mathbf{x}$ for some $\mathbf{k}$. He does this by performing the linearity test with t_1 independently random pairs $\mathbf{x},\mathbf{y}$. If $s_1(\mathbf{x}+\mathbf{y})=s_1(\mathbf{x})+s_1(\mathbf{y})$ for all these pairs, he proceeds to the next stage of the PCP. If not, he rejects the proof and throws Merlin's scroll out the window.

Lemma 11.7 shows that the probability that Arthur will be deceived at this point is at most $(1-\delta)^{t_1}$. Henceforth we, and Arthur, will assume that s_1 is indeed δ-close to $\mathbf{k}^T\mathbf{x}$. Next, he has to check that s_2 is δ-close to $\mathbf{k}^T\mathbf{M}\mathbf{k}$ for the same $\mathbf{k}$.

11.3.6 Quadratic Consistency

Now that Arthur has checked that s_1 represents the linear functions of some $\mathbf{k}$, his next task is to check that s_2 represents the quadratic functions of the same $\mathbf{k}$.

First note that while $\mathbf{k}^T\mathbf{M}\mathbf{k}$ is quadratic in $\mathbf{k}$, it is linear in the matrix $\mathbf{M}$ of coefficients. So Arthur starts by establishing that $s_2(\mathbf{M})$ is close to a linear function of $\mathbf{M}$. If we "flatten" $\mathbf{M}$ and treat it as an n^2-dimensional vector, rather than an $n\times n$ matrix, we can rewrite (11.11) as

$$s_2(\mathbf{M})=(\mathbf{k}\otimes\mathbf{k})^T\mathbf{M}. \tag{11.25}$$

Here $\mathbf{k}\otimes\mathbf{k}$ is the n^2-dimensional vector

$$(\mathbf{k}\otimes\mathbf{k})_{ij}=k_i k_j\,,$$

and we treat the pair (i,j) as a single index running from 1 to n^2.

Arthur then applies the linearity test to s_2, just as he did to s_1. After a suitable number of trials, he concludes that s_2 is δ-close to some linear function. In other words, for some n^2-dimensional vector $\mathbf{w}$,

$$\Pr_M\left[s_2(\mathbf{M})=\mathbf{w}^T\mathbf{M}\right]>1-\delta\,.$$

Now Arthur has to check Merlin's claim that $\mathbf{w}=\mathbf{k}\otimes\mathbf{k}$. How can he do this?

If Merlin is being honest, then for any pair $\mathbf{x},\mathbf{y}\in\{0,1\}^n$ we have

$$s_2(\mathbf{x}\otimes\mathbf{y})=s_1(\mathbf{x})\,s_1(\mathbf{y})\,. \tag{11.26}$$

This follows since both sides represent a quadratic function of $\mathbf{k}$, which can be factored as the product of two linear functions:

$$(\mathbf{k}\otimes\mathbf{k})^T(\mathbf{x}\otimes\mathbf{y})=\sum_{ij}k_i k_j x_i y_j=\left(\sum_i k_i x_i\right)\left(\sum_j k_j x_j\right)=(\mathbf{k}^T\mathbf{x})(\mathbf{k}^T\mathbf{y}). \tag{11.27}$$

Thus an intuitive test is to choose $\mathbf{x}$ and $\mathbf{y}$ independently and uniformly, and check to see whether (11.26) is true. How well does this test work? That is, if s_1 and s_2 are both linear but $\mathbf{w}\neq\mathbf{k}\otimes\mathbf{k}$, for what fraction of pairs $\mathbf{x}$, $\mathbf{y}$ does (11.26) hold?

Let's rewrite (11.27) a little:

$$\begin{aligned}s_2(\mathbf{x}\otimes\mathbf{y})-s_1(\mathbf{x})\,s_1(\mathbf{y})&=\sum_{ij}(w_{ij}-k_i k_j)x_i y_j\\&=\left(\mathbf{w}-(\mathbf{k}\otimes\mathbf{k})\right)^T(\mathbf{x}\otimes\mathbf{y}).\end{aligned} \tag{11.28}$$

We need one more notational move. If we treat $\mathbf{w}$ and $\mathbf{k}\otimes\mathbf{k}$ as $n\times n$ matrices rather than n^2-dimensional vectors, then (11.28) becomes

$$s_2(\mathbf{x}\otimes\mathbf{y}) - s_1(\mathbf{x})\,s_1(\mathbf{y}) = \mathbf{x}^T\mathbf{Q}\,\mathbf{y}\,,$$

where

$$\mathbf{Q} = \mathbf{w} - \mathbf{k}\otimes\mathbf{k}^T\,.$$

The tensor product in $\mathbf{k}\otimes\mathbf{k}$ has become the outer product $\mathbf{k}\otimes\mathbf{k}^T$. It means the same thing,

$$Q_{ij} = w_{ij} - k_i k_j\,,$$

but where i and j index a row and a column respectively.

Now consider the following exercise, analogous to Exercise 11.7:

Exercise 11.11 *Suppose* $\mathbf{Q}$ *is a nonzero* $n\times n$ *matrix. Show that if* $\mathbf{x},\mathbf{y}\in\{0,1\}^n$ *are uniformly random,*

$$\Pr_{\mathbf{x},\mathbf{y}}\left[\mathbf{x}^T\mathbf{Q}\mathbf{y}\neq 0\right] \geq 1/4\,.$$

Hint: use Exercise 11.7 twice.

Thus if s_1 and s_2 are linear but $\mathbf{w}\neq\mathbf{k}\otimes\mathbf{k}$, Arthur will be deceived with probability at most 3/4 each time he does this test. He can reduce this probability to $(3/4)^{t_2}$ by repeating the test t_2 times.

There is a problem with this test, however, if s_2 is close to linear, rather than exactly linear. The reason is that it only checks s_2 at matrices of the form $\mathbf{x}\otimes\mathbf{y}$. A vanishingly small fraction of $n\times n$ matrices are of this form—namely, those of rank at most 1. If s_2 is δ-close to $\mathbf{w}^T\mathbf{M}$, it is entirely possible that these are the matrices on which s_2 and $\mathbf{w}^T\mathbf{M}$ differ.

We can fix this problem by using the error-correcting approach of Exercise 11.9. In addition to choosing $\mathbf{x}$ and $\mathbf{y}$ uniformly, Arthur chooses a uniformly random matrix $\mathbf{M}$. Then he checks whether

$$s_2(\mathbf{x}\otimes\mathbf{y}+\mathbf{M}) - s_2(\mathbf{M}) = s_1(\mathbf{x})\,s_1(\mathbf{y})\,. \tag{11.29}$$

We call this the *quadratic consistency test*. Combining Exercises 11.9 and 11.11 gives the following:

Lemma 11.8 *Suppose that the functions* $s_1(\mathbf{x})$ *and* $s_2(\mathbf{M})$ *are* δ*-close to* $\mathbf{k}^T\mathbf{x}$ *and* $\mathbf{w}^T\mathbf{M}$ *respectively. Show that if* $\mathbf{w}\neq\mathbf{k}\otimes\mathbf{k}$, *and if* $\mathbf{x}$, $\mathbf{y}$, *and* $\mathbf{M}$ *are chosen independently and uniformly, then* (11.29) *holds with probability at most* $3/4+\delta$.

Arthur runs the quadratic consistency test t_2 times. If s_1 and s_2 have already passed the linearity test, but they fail to represent an assignment $\mathbf{k}$ consistently, Arthur will be deceived with probability at most $(3/4+\delta)^{t_2}$.

11.3.7 *Putting It Together*

We can finally describe the entire PCP for MOD-2 QUADRATIC EQUATIONS. It has three stages:

1. Arthur applies the linearity test to s_1 and s_2, running t_1 trials of the test for each one, and concluding that they are δ-close to linear functions $\mathbf{k}^T\mathbf{x}$ and $\mathbf{w}^T\mathbf{M}$.

2. Arthur then runs t_2 trials of the quadratic consistency test, concluding that $\mathbf{w}=\mathbf{k}\otimes\mathbf{k}$, so that $s_1(\mathbf{x})$ and $s_2(\mathbf{M})$ are δ-close to $\mathbf{k}^T\mathbf{x}$ and $\mathbf{k}^T\mathbf{M}\mathbf{k}$ respectively.

3. Finally, Arthur checks that $\mathbf{k}$ is a solution to the system of equations, using the error-correcting test described in Section 11.3.4. He does this t_3 times.

Clearly, if we are dealing with a yes-instance of MOD-2 QUADRATIC EQUATIONS, Merlin can provide a proof that Arthur accepts at every stage. Thus our PCP has completeness 1.

To show that it has soundness 1/2, let's bound the probability that Arthur accepts an invalid proof. For any $\delta < 1/4$, if there really is no solution $\mathbf{k}$, then

1. either s_1 or s_2 is not δ-close to a linear function;

2. or they both are, but $\mathbf{w}\neq\mathbf{k}\otimes\mathbf{k}$; or

3. $\mathbf{w}=\mathbf{k}\otimes\mathbf{k}$, but $\mathbf{k}$ is not a solution.

Thus if Merlin is being deceptive, Arthur has a good chance of rejecting the proof in at least one of the three stages. Combining Lemmas 11.7, 11.8, and 11.6, the probability that Arthur accepts the proof is at most

$$\max((1-\delta)^{t_1}, (3/4+\delta)^{t_2}, (1/2+2\delta)^{t_3}) .$$

If we set $\delta = 1/8$ and $t_1 = t_2 = 6$ and $t_3 = 3$, this probability is at most $(7/8)^6 < 1/2$. Thus our PCP is sound.

To generate all these **x**s, **y**s, **M**s, and **r**s, Arthur needs to flip n, n^2, or m coins each. Since he performs $O(1)$ tests, the total number of coins he needs to flip is poly(n) as promised.

Finally, how many bits of the proof does Arthur look at? A little bookkeeping shows that this is

$$6t_1 + 4t_2 + 4t_3 = 72 .$$

Thus Arthur can probabilistically check proofs for MOD-2 QUADRATIC EQUATIONS, rejecting false proofs with probability at least 1/2, by checking only 72 bits of the proof. The number of bits he needs to look at remains constant, *no matter how large the instance of* MOD-2 QUADRATIC EQUATIONS *is*—regardless of the number n of variables, or the number m of equations. This completes the proof of Theorem 11.5.

We pause to note two things. First, somewhat like the zero-knowledge proofs we studied in Section 11.1.3, Arthur becomes convinced that the system of equations has a solution, without ever learning what it is. At most, he learns $\mathbf{k}^T\mathbf{x}$ and $\mathbf{k}^T\mathbf{M}\mathbf{k}$ for a finite number of **x**s and **M**s, and this is not enough to determine **k**.

Second, every test that Arthur applies in this three-stage procedure consists of checking the sum mod 2, or the parity, of some constant number of bits. We can think of each such test as an XORSAT clause. For instance, we can rewrite the linearity test as

$$s_1(\mathbf{x}+\mathbf{y}) \oplus s_1(\mathbf{x}) \oplus s_1(\mathbf{y}) = 0 ,$$

where $\oplus$ denotes XOR or addition mod 2.

Now imagine writing down an exponentially long list of all the tests that Arthur might choose. We can treat this list as an enormous XORSAT formula, whose variables are the $2^n + 2^{n^2}$ bits of s_1 and s_2. If we started with a yes-instance of MOD-2 QUADRATIC EQUATIONS, we can satisfy every clause in this formula—there is a choice of s_1 and s_2 such that all the linearity, consistency, and solution tests come out right.

On the other hand, if the input is a no-instance then a constant fraction of the clauses are violated no matter how s_1 and s_2 are chosen. If Arthur checks a random clause, he will reject with constant probability. This is very different from the usual situation—even if a SAT or XORSAT formula is unsatisfiable, it might be possible to satisfy all but one of its clauses.

To put this differently, suppose we start with a 3-SAT formula ϕ. We convert it to an instance of MOD-2 QUADRATIC EQUATIONS, and convert this to an XORSAT formula whose clauses correspond to Arthur's tests. Finally, we convert each XORSAT clause to a constant number of 3-SAT clauses. This produces a new 3-SAT formula ϕ', which is satisfiable if and only if ϕ is.

But something important has happened. If ϕ is even a little unsatisfiable then ϕ' is *very* unsatisfiable, in the sense that a constant fraction of its clauses are always violated. If Merlin's proof is wrong, it is wrong in many places.

In the weak version of the PCP theorem we proved here, ϕ' is exponentially larger than ϕ. However, this idea of amplifying the number of unsatisfied clauses is also the key to proving the stronger version, Theorem 11.3. We sketch that proof in the next section.

11.3.8 Gap Amplification

We have shown that the NP-complete problem MOD-2 QUADRATIC EQUATIONS, and therefore all problems in NP, have probabilistically checkable proofs of exponential size, where Arthur only needs to look at a constant number of bits. This result is fantastic—but how do we bring the size of the proof down to poly(n), and the number of coins that Arthur needs to flip down to $O(\log n)$, as stated in Theorem 11.3?

Let's step back for a moment, and note that constraint satisfaction problems such as SAT already possess a weak kind of PCP. Suppose Merlin claims that a 3-SAT formula ϕ is satisfiable. He sends a scroll to Arthur, purporting to give a satisfying assignment. Arthur chooses a random clause, looks up its variables' truth values on the scroll, and checks to see if it is satisfied. If ϕ is not satisfiable, there is at least one unsatisfied clause no matter what truth assignment Merlin wrote on his scroll. Thus if G has m clauses, Arthur will reject Merlin's claim with probability at least $1/m$.

Unfortunately, even if ϕ is unsatisfiable, it might be possible to satisfy all but one of its clauses. Then if Arthur checks a random clause, he will be fooled most of the time—just as if he were checking a proof where most of the steps are correct.

Our goal is to amplify this gap between satisfiability and unsatisfiability. Let us say that a formula ϕ is *δ-unsatisfiable* if, for any truth assignment, at least a fraction δ of ϕ's clauses are violated. Then consider the following promise problem, similar to those we saw in Section 9.3.1:

> δ-GAP-3-SAT
>
> Input: A 3-SAT formula ϕ that is either satisfiable or δ-unsatisfiable.
>
> Question: Which is it?

If he checks a random clause, Arthur will be fooled by a δ-unsatisfiable formula with probability at most $1-\delta$, and he can improve this to $(1-\delta)^t$ by checking t independently random clauses. For any constant $\delta > 0$ there is a constant t such that $(1-\delta)^t < 1/2$, so Arthur only needs to look at a total of $O(1)$ bits. Thus another way to state Theorem 11.3 is the following:

Theorem 11.9 *There is a constant $\delta > 0$ such that δ-GAP-3-SAT is NP-hard.*

It's clear that GAP-3-SAT is NP-hard if $\delta = 1/m$, so the challenge is pumping the gap δ up to a constant. In Section 9.3, we used *gap-preserving* reductions to relate approximation problems to each other. To prove Theorem 11.9, we need a *gap-amplifying* reduction—one that makes unsatisfiable formulas even less satisfiable.

This strategy was carried out by Irit Dinur in 2005. She described a reduction that increases the gap by a constant factor, while increasing the size of the formula by another constant:

Lemma 11.10 *There are constants $A > 1$, $B > 1$, and $\delta_{\max} > 0$, and a polynomial-time algorithm that transforms* 3-SAT *formulas ϕ with m clauses to new formulas ϕ' with m' clauses, such that*

1. *if ϕ is satisfiable, so is ϕ',*
2. *if ϕ is δ-unsatisfiable then ϕ' is δ'-unsatisfiable, where $\delta' = \min(A\delta, \delta_{\max})$, and*
3. *$m' \leq Bm$.*

By applying this lemma $k = O(\log m)$ times, we can amplify the gap from $1/m$ up to $\delta_{\max}$. Unlike the exponentially long XORSAT formulas we described in the previous section, the final formula has $B^k m = B^{O(\log m)} m = \mathrm{poly}(m)$ clauses, so Arthur only needs to flip $O(\log m) = O(\log n)$ coins to choose a random clause. Finally, the entire process can be carried out in polynomial time. This gives a polynomial-time reduction from 3-SAT to $\delta_{\max}$-GAP-3-SAT, showing that the latter is NP-hard and proving Theorem 11.3.

We will sketch a proof of Lemma 11.10. First we define a graph G whose vertices are the variables x_i of ϕ, where two variables are neighbors if they appear together in a clause. We will assume that the maximum degree of G is some constant d. If this is not the case, there is a simple way to "preprocess" G, reducing its degree to a constant, which only decreases the gap by a constant factor.

For each variable x_i, consider the sphere S_i of variables less than ℓ steps away from x_i, where ℓ is a constant we will choose later. Each of these spheres contains at most d^ℓ variables. Our goal is to increase the gap, and make the proof more holographic, by using the fact that each violated clause appears in many overlapping spheres.

To do this, we define a new constraint satisfaction problem. For each x_i, this new problem has a "supervariable" X_i, which represents a truth assignment of all the variables in the sphere S_i. Thus X_i can take up to 2^{d^ℓ} different values. We impose two kinds of constraints on these variables. For each pair of overlapping spheres S_i, S_j we impose a *consistency* constraint, demanding that X_i and X_j agree on the truth values of the variables in $S_i \cap S_j$. And, we impose a *satisfaction* constraint on each X_i, demanding that all the clauses inside S_i are satisfied. Then if one of ϕ's clauses is violated, we must either violate the satisfaction constraints of the spheres containing it, or violate the consistency constraints between them.

Let's suppose first that the X_i are consistent with each other, so that they correspond to some truth assignment of the original variables x_i. Then if U is the set of violated clauses in ϕ, the number of violated constraints in our new problem is the number of spheres of radius ℓ that overlap with U.

In Section 12.9, we will discuss *expanders*: graphs where, for any subset $U \subset V$ of their vertices with $|U| \leq |V|/2$, taking a step in the graph connects U to at least $\alpha|U|$ vertices outside U for some constant $\alpha > 0$. If G is an expander, the number of spheres that overlap with U grows as $(1+\alpha)^\ell\,|U|$. This amplifies the gap by a factor of $(1+\alpha)^\ell$, and we can make this factor as large as we want by choosing ℓ large enough.

If G is not already an expander, we can overlay an expander H on the same set of vertices, adding H's edges to those of G. These new edges have nothing to do with the original formula—they are simply a way to ensure that the spheres grow fast enough, so that the proof becomes sufficiently holographic.

The case where the X_i are not consistent with any underlying truth assignment is more subtle. Suppose that Merlin tries to cheat by giving some variables different truth values in different spheres. One can show that one of two things happens. Either most spheres agree with the majority values of these variables, and their satisfaction constraints are violated—or, due to the expander property, many overlapping pairs of spheres have different values for these variables, and their consistency constraints are violated.

So this new problem has a much larger gap than ϕ has. However, this comes at a cost. We also have a larger "alphabet" of states, since each variable can take 2^{d^ℓ} different values instead of 2. We can map this problem back to a 3-SAT formula on Boolean variables by writing each X_i as a string of d^ℓ bits. However, if we translate the constraints on the X_i naively into 3-SAT clauses, it could be that only one of the d^ℓ clauses in a given X_i is violated, or that two overlapping X_i, X_j only disagree on one of the d^ℓ variables in their intersection. Thus the gap could go back down by a factor of $O(d^\ell)$, undoing all our progress.

Happily, we saw in the proof of Theorem 11.5 how to build 3-SAT formulas with a constant gap. For each overlapping pair X_i, X_j, we express their consistency and satisfaction constraints as an instance of MOD-2 QUADRATIC EQUATIONS. We then transform this into a 3-SAT formula $\phi_{i,j}$ with gap ζ, so that if any of the constraints on X_i and X_j are violated, at least ζ of the clauses of $\phi_{i,j}$ are violated as well.

Our 3-SAT formula ϕ' consists of the conjunction of all these $\phi_{i,j}$. If ϕ is δ-unsatisfiable, then at least $(1+\alpha)^\ell \delta$ of the constraints on the X_i are violated, and for each such constraint, at least ζ of the clauses in the corresponding $\phi_{i,j}$ are violated. Thus ϕ' is δ'-unsatisfiable, where

$$\delta' = \zeta(1+\alpha)^\ell \,\delta = A\delta .$$

Since ζ is a fixed constant, we can make $A > 1$ by taking ℓ to be large enough. We are ignoring several other constant factors here, including the cost of preprocessing G to reduce its maximum degree to d, but this gives the idea.

How large a formula ϕ' does this construction give? If ϕ has n variables, there are at most $d^{2\ell} n$ overlapping pairs X_i, X_j. For each one, the number of clauses in $\phi_{i,j}$ is exponential as a function of d^ℓ. But since d and ℓ are constants, this is just another constant. Thus the size of ϕ' is a (rather large) constant B times that of ϕ, as promised in Lemma 11.10. Finally, applying this lemma $O(\log n)$ times, as we discussed above, completes our proof sketch for Theorem 11.3.

Looking back at our journey so far, we have seen many ways to transform one kind of mathematical truth into another—from the simple gadget reductions of Chapter 5, to arithmetization, polynomials over finite fields, and Fourier analysis. The combination of algebraic, analytic, and combinatorial ideas in these proofs makes the PCP theorem one of the great triumphs of theoretical computer science. As we move forward into the 21st century, we can expect new mathematical tools to have similarly unexpected consequences in computational complexity.

11.4

11.4 Pseudorandom Generators and Derandomization

> How dare we speak of the laws of chance? Is not chance the antithesis of all law?
>
> Joseph Bertrand, *Calcul des probabilités*, 1889

Like time and memory, randomness is a limited resource. If there is any real randomness in the world, it comes from chaotic or quantum-mechanical processes to which most computers don't have access. Just

11.5

as we ask how much time or memory we need to solve a problem, we should ask how much randomness we need—how many coins we need to flip, as a function of the size of the input.

When we run a randomized algorithm in practice, we use *pseudorandom* numbers produced by an oxymoronic "deterministic random number generator." How do we know whether these numbers are random enough? And what does that mean?

We have seen situations where true randomness is essential, such as communicating with as few bits as possible (Section 10.6.1), interactive proofs, and probabilistically checkable proofs. On the other hand, we have seen cases like PRIMALITY where a randomized algorithm was *derandomized* and replaced by a deterministic one. Where polynomial-time algorithms are concerned, it is not known whether we *ever* need truly random numbers. If pseudorandom numbers are good enough, in some sense, then randomized polynomial-time algorithms are no more powerful than deterministic ones.

In this section, we define exactly what we mean by a pseudorandom generator, and what it would take for such a generator to be good enough to derandomize all polynomial-time algorithms. We will see that the existence of such generators is intimately connected to other complexity-theoretic ideas, including one-way functions and the building blocks of cryptography. Good pseudorandom generators are much like secure cryptosystems, which create a random-seeming message from an underlying seed.

11.4.1 Pseudorandom Generators

> Anyone who considers arithmetical methods of producing random digits is, of course, in a state of sin.
>
> John von Neumann

Let A be a randomized algorithm that returns "yes" or "no." Suppose it uses ℓ random bits, where ℓ is some function of the input size n. When can we replace these random bits with pseudorandom ones, without perceptibly changing A's behavior?

We can think of A as a *deterministic* algorithm that takes two inputs: the instance x it is trying to solve, and the string r of bits, random or pseudorandom, that it uses to drive the choices it makes. To focus on how A depends on r, we will fix x and define $A(r) = A(x, r)$. There is a subtle issue lurking behind the phrase "fix x" which we will address in Section 11.4.4. But for now, we will think of A as a deterministic algorithm whose only input is r.

We will define a *pseudorandom generator* as a function $g : \{0,1\}^k \to \{0,1\}^\ell$ for some $k < \ell$. We think of g as taking a *seed* s of length k and expanding it to a string $g(s)$ of length ℓ. The idea is that if s is truly random, then $g(s)$ looks random in some sense. Thus g takes k random bits, and "stretches" them into ℓ pseudorandom ones.

What happens if we run A with pseudorandom bits instead of random ones? If r is uniformly random in $\{0,1\}^\ell$, the probability that A returns "yes" is

$$\Pr_{r \in \{0,1\}^\ell}[A(r) = \text{"yes"}] .$$

On the other hand, if we choose s uniformly from $\{0,1\}^k$ and then give A the string $g(s)$, the probability that it returns "yes" is

$$\Pr_{s \in \{0,1\}^k}[A(g(s)) = \text{"yes"}] .$$

The following definition expresses the claim that $g(s)$ seems just like a string of random bits as far as A is concerned. For simplicity, we use ℓ rather than n as our basic parameter. Since ℓ is at most the running time, we have $\ell = \text{poly}(n)$ anyway.

Definition 11.11 *Let g be a function from $\{0,1\}^k$ to $\{0,1\}^\ell$ where $k < \ell$ is some function of ℓ. Let $A(r)$ be a deterministic algorithm that takes a string r of ℓ bits as input. We say that g* fools *A if, for all constants $c > 0$,*

$$\left| \Pr_{s\in\{0,1\}^k}[A(g(s)) = \text{“yes”}] - \Pr_{r\in\{0,1\}^\ell}[A(r) = \text{“yes”}] \right| = o(\ell^{-c}).$$

In other words, A gives essentially the same results if, instead of flipping ℓ coins, we flip just k coins and use g to turn the resulting seed s into ℓ pseudorandom bits. Since this changes the probability that A returns "yes" by a superpolynomially small amount, it would take a superpolynomial number of experiments to distinguish the bits generated by g from truly random ones.

Let's turn this around. What happens if g fails to fool some algorithm A? In that case, we distinguish g's output from random strings by running A a polynomial number of times and measuring the probability that it returns "yes." We will call such an algorithm a *tester*.

We say that g *is a strong pseudorandom generator* if it fools all polynomial-time algorithms. That is, no polynomial-time tester can distinguish the distribution of pseudorandom strings $g(s)$, where s is uniform in $\{0,1\}^k$, from the uniform distribution on $\{0,1\}^\ell$. This a much stronger property than simple statistical measures of pseudorandomness, such as a lack of correlations or the ability to fool a particular algorithm we care about. It states that g's output has *no pattern in it that any polynomial-time algorithm can discover.*

11.4.2 Pseudorandomness and Cryptography

Pseudorandom generators, if they exist, can give us excellent cryptosystems. Suppose I want to send you a secret message. The most secure way to encrypt it is with a *one-time pad*, like that shown in Figure 11.10.

Periodically, you and I meet under a bridge and exchange a book filled with random bits. Whenever you want to send me a message m, you XOR it with a string of bits r from the book, flipping the ith bit of your message if the ith bit in the book is 1. You send me the resulting string, which we denote $m \oplus r$. I decrypt it by flipping the same bits back, obtaining $(m \oplus r) \oplus r = m$. Then we both burn those pages of the book and never use them again.

If r is uniformly random then so is $m \oplus r$, no matter what m is. Therefore, if the bits of r are truly random, and we never reuse them, the encrypted messages are completely random as well. If an eavesdropper overhears our message, there is no way they can decode it, or even distinguish it from random noise, no matter how computationally powerful they are.

Thus this cryptosystem is completely secure. However, it requires us to generate, and share, one bit of secret key for each bit of message we want to send. This makes it expensive, both in terms of random bits and the physical security it takes to exchange them. In fact, the United States was able to decrypt some Soviet messages during the 1940s because the KGB reused some pages of their one-time pads.

Now suppose that $g : \{0,1\}^k \to \{0,1\}^\ell$ is a strong pseudorandom generator. Instead of using a random string r of length ℓ as our pad, we can use $g(s)$ where s is a secret key of length k. We still need to meet under the bridge occasionally, but now we only need to share k bits of secret key for each ℓ bits of message.

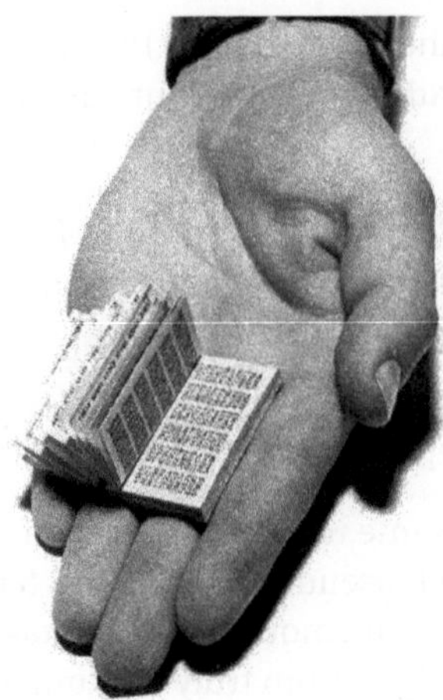

FIGURE 11.10: A one-time pad from the Cold War era.

If $k = o(\ell)$, then asymptotically we hardly ever need to meet. And if the eavesdropper can't tell $g(s)$ from a random string r, they can't distinguish our encrypted messages from random noise. This is a *private-key* or *shared-key* cryptosystem, as opposed to the public-key cryptosystems we will discuss in Section 15.5, but the amount of key sharing we need to do is greatly reduced.

As we will see next, pseudorandom generators are closely related to the same building blocks that we used for bit commitment in Section 11.1.4—namely, one-way functions and hard-core bits. Given a one-way function f, we will devise a generator g such that discovering a pattern in g's output is tantamount to inverting f. If f is sufficiently hard to invert, then g can fool any polynomial-time algorithm into thinking that its bits are truly random.

11.4.3 From One-Way Functions to Pseudorandom Generators

The generation of random numbers is
too important to be left to chance.

Robert R. Coveyou

Here we will show that if one-way functions exist, we can take a random seed of length k and expand it into a pseudorandom string of length $\ell = \text{poly}(k)$. Turning this around, we can simulate ℓ random bits with a seed of length $k = \ell^{\varepsilon}$ for arbitrarily small ε.

Let's start with a precise definition of one-way functions and hard-core bits. Note that we denote the length of the input as k rather than n, since we will be applying f to seeds of length k.

Definition 11.12 *Let* $f : \{0,1\}^k \to \{0,1\}^k$ *and* $b : \{0,1\}^k \to \{0,1\}$ *be computable in* $\text{poly}(k)$ *time. We say that* f *is a* one-way function with hard-core bit b *if, for all polynomial-time randomized algorithms* A *and all constants* c,

$$\mathop{\mathbb{E}}_{x\in\{0,1\}^k} \Pr\left[A(f(x)) = b(x)\right] = \frac{1}{2} + o(k^{-c}). \tag{11.30}$$

In other words, no algorithm can invert f, taking $f(x)$ as input and returning x, or even the single bit $b(x)$, with probability much better than chance.

This definition says that inverting f is hard, not just in the worst case, but on average over all possible x. To put this differently, for any algorithm A, the fraction of inputs x on which A returns $b(x)$ a clear majority of the time, or even $1/2+1/\text{poly}(k)$ of the time, must be exponentially small.

As we discussed in Section 11.1.4, we believe that one such function is modular exponentiation, $f(x)=a^x \bmod p$ where p is a k-bit prime. We can take $b(x)$ to be the most significant bit of x, since finding $b(x)$ from $f(x)$ is as hard as finding all of x and solving the DISCRETE LOG problem.

If a one-way function is one-to-one, such as $f(x)=a^x \bmod p$ where a is a primitive root, it is called a *one-way permutation.* We will start by showing how to use a one-way permutation to turn k random bits into $k+1$ pseudorandom ones. Then we will show that we can iterate this operation a polynomial number of times, turning a k-bit seed into a pseudorandom string of length $\ell=\text{poly}(k)$.

Given a one-way permutation $f:\{0,1\}^k \to \{0,1\}^k$ with a hard-core bit b, our function from k bits to $k+1$ bits is simply:

$$g(x)=(f(x),b(x)). \tag{11.31}$$

Thus g applies f to x, and tacks $b(x)$ on to the result. Intuitively, if f is one-way and b is hard-core, so that it is hard to calculate $b(x)$ from $f(x)$, the new bit $b(x)$ looks random to any algorithm. The following theorem confirms this intuition, and shows that g is indeed a strong pseudorandom generator.

Theorem 11.13 *Let f be a one-way permutation with hard-core bit b. Then the function $g=(f(x),b(x))$ is a strong pseudorandom generator where $\ell=k+1$.*

Proof The proof is by contradiction. We will show that if any tester A can tell the difference between $g(x)$ and a random string, then we can use A to "break" our one-way function, and calculate $b(x)$ from $f(x)$ with $1/\text{poly}(k)$ probability.

Suppose there is a polynomial-time algorithm A that is not fooled by g. That is, A behaves differently on $g(x)$ for a random seed x than it does on a random string r of $k+1$ bits. If we write r as a k-bit string y followed by a bit b', then

$$\left| \Pr_{x\in\{0,1\}^k}\left[A\big(f(x),b(x)\big)=\text{"yes"}\right] - \Pr_{\substack{y\in\{0,1\}^k\\ b'\in\{0,1\}}}\left[A\big(y,b'\big)=\text{"yes"}\right] \right| = \Omega\left(k^{-c}\right). \tag{11.32}$$

for some constant c. We assume without loss of generality that the difference between these probabilities is positive, since otherwise we can modify A by switching "yes" and "no."

The careful reader will raise an objection to our use of Ω in (11.32). Definition 11.11 says that g fools A if the difference between these probabilities is $o(k^{-c})$ for all constants c. The complement of this statement is that, for some constants $C,c>0$, this difference is at least Ck^{-c} for an infinite set of values of k. This is weaker than saying that the difference is $\Omega(k^{-c})$, since the latter means that some such inequality holds for all sufficiently large k.

This distinction is just a reflection of how strong we want our pseudorandom generators to be, and how hard our one-way functions are to invert. We can vary either one, and our proof will work just fine if we vary the other one correspondingly. If there is a tester that works for all k, there is an algorithm that inverts f for all k. Similarly, if there is a tester that works on some sparse but infinite set, such as when

k is a power of 2, there is an algorithm that inverts f on the same set. For simplicity, we will stick here to the claim that the tester succeeds with probability $\Omega(k^{-c})$—that is, with probability $1/\text{poly}(k)$.

It's convenient to write $A = 1$ or $A = 0$ if A returns "yes" or "no" respectively. Then we can write the probabilities in (11.32) as expectations:

$$\mathop{\mathbb{E}}_{x\in\{0,1\}^k}\left[A(f(x),b(x))\right] - \mathop{\mathbb{E}}_{\substack{y\in\{0,1\}^k\\ b'\in\{0,1\}}}\left[A(y,b')\right] = \Omega(k^{-c}).$$

If f is one-to-one and x is uniformly random in $\{0,1\}^k$, then $f(x)$ is uniformly random in $\{0,1\}^k$. Therefore, $\mathbb{E}[A(y,b')]$ stays the same if we set $y = f(x)$, and use the same random x in both expectations. Thus

$$\mathop{\mathbb{E}}_{x\in\{0,1\}^k}\left[A(f(x),b(x)) - \mathop{\mathbb{E}}_{b'\in\{0,1\}}\left[A(f(x),b')\right]\right] = \Omega(k^{-c}).$$

The expectation over b' is just the average of the cases $b' = b(x)$ and $b' = \overline{b(x)}$, whatever $b(x)$ is. Subtracting this average from $A(f(x),b(x))$ gives

$$\frac{1}{2}\mathop{\mathbb{E}}_{x\in\{0,1\}^k}\left[A(f(x),b(x)) - A(f(x),\overline{b(x)})\right] = \Omega(k^{-c}). \tag{11.33}$$

Thus if we run A on a pair $(f(x),b')$, the probability that it returns "yes" is polynomially greater, on average, if $b' = b(x)$ than if $b' \neq b(x)$.

This suggests a randomized polynomial-time algorithm to calculate $b(x)$ from $f(x)$. Given $f(x)$, we run $A(f(x),b')$ for both values of b'. If A returns "yes" for one value of b' and "no" for the other, we return the value of b' for which A returned "yes." If A returns "yes" for both or "no" for both, we flip a coin and return 0 or 1 with equal probability.

If we call this algorithm B, the probability that it returns the right answer is

$$\Pr\left[B(f(x)) = b(x)\right] = \frac{1}{2} + \frac{1}{2}\left(A(f(x),b(x)) - A(f(x),\overline{b(x)})\right).$$

Then (11.33) shows that, on average over all x, B is right polynomially better than chance:

$$\mathop{\mathbb{E}}_{x\in\{0,1\}^k}\Pr\left[B(f(x)) = b(x)\right] = \frac{1}{2} + \Omega(k^{-c}).$$

This contradicts (11.30), and shows that either f is not one-way, or b is not hard-core. By contradiction, g must fool every polynomial-time algorithm. □

Theorem 11.13 shows that by applying a one-way permutation f and a hard-core bit b, we can generate a single pseudorandom bit from our seed, stretching our string by one. By iterating this method, we can generate polynomially many pseudorandom bits, stretching a random seed of length k out to a pseudorandom string of length $\ell = \text{poly}(k)$.

The idea is to keep track of a variable x. Its initial value is the seed s, and at each step we update x by applying $f(x)$. However, rather than revealing all of x, we reveal just the hard-core bit $b(x)$. If we do this for ℓ steps, we produce a string of ℓ pseudorandom bits:

$$g_\ell(s) = \left(b(s), b(f(s)), b(f^2(s)), \ldots, b(f^{\ell-1}(s))\right), \tag{11.34}$$

If the seed s is k bits long, this defines a function $g_\ell : \{0,1\}^k \to \{0,1\}^\ell$ called the *Blum–Micali generator*. One nice property of this generator is that it can produce one pseudorandom bit at a time, whenever the algorithm needs it. Indeed, this is how many pseudorandom generators work in practice.

Our next result shows that g_ℓ fools any polynomial-time algorithm that uses ℓ random bits, as long as ℓ is polynomial as a function of k. In other words, for any constant a, we can stretch a random seed of length k out to a pseudorandom string of length $\ell = k^a$. The proof uses a clever "hybrid argument," where we interpolate between completely random strings and those produced by g_ℓ. Each step in this interpolation replaces one of the bits of $g_\ell(s)$ with a random bit. If any of these one-bit changes make a difference in the behavior of an algorithm A, we can use that difference to break our one-way function.

Theorem 11.14 *Let f be a one-way permutation with hard-core bit b, and let $\ell = \mathrm{poly}(k)$. Then the Blum–Micali generator g_ℓ defined in* (11.34) *is a strong pseudorandom generator.*

Proof As in Theorem 11.13, our proof is by contradiction. Suppose there is a tester A that is polynomially more likely to return "yes" if given $g_\ell(s)$ than if given a random string r of length ℓ. Since $\ell = \mathrm{poly}(k)$, a function is polynomial in ℓ if and only if it is polynomial in k. So for some constant c we have

$$\mathop{\mathbb{E}}_{s\in\{0,1\}^k}[A(g_\ell(s))] - \mathop{\mathbb{E}}_{r\in\{0,1\}^\ell}[A(r)] = \Omega(k^{-c}), \tag{11.35}$$

where, as before, we define $A=1$ or $A=0$ if A returns "yes" or "no" respectively.

We can move from one of these probability distributions to the other by replacing $g(s)$, one bit at a time, with uniformly random bits. At the ith stage of this process, where $0 \le i \le \ell$, the first i bits are random, and the remaining $\ell - i$ bits are given by $g(s)$. If $g_i = b(f^{i-1}(s))$ denotes the ith bit of $g_\ell(s)$ and r_i denotes the ith bit of a random string r, then at the ith stage we imagine running A on the hybrid string

$$(r_1, r_2, \ldots, r_i, g_{i+1}, \ldots g_\ell)\,.$$

The probability (11.35) that A distinguishes $g(\ell)$ from r is the expected total difference between the 0th stage and the ℓth stage. We can write this as a telescoping sum of the expected difference between each stage and the next:

$$\begin{aligned}
\mathop{\mathbb{E}}_{s\in\{0,1\}^k}[A(g_\ell(s))] - \mathop{\mathbb{E}}_{r\in\{0,1\}^\ell}[A(r)] &= \mathop{\mathbb{E}}_{s,r}\Big[A(g_1, g_2, \ldots, g_\ell) - A(r_1, r_2, \ldots, r_\ell)\Big] \\
&= \mathop{\mathbb{E}}_{s,r}\Big[A(g_1, g_2, \ldots, g_\ell) - A(r_1, g_2, \ldots, g_\ell) \\
&\qquad + A(r_1, g_2, \ldots, g_\ell) - A(r_1, r_2, g_3, \ldots, g_\ell) \\
&\qquad\quad \vdots \\
&\qquad + A(r_1, \ldots, r_{\ell-1}, g_\ell) - A(r_1, \ldots, r_{\ell-1}, r_\ell))\Big] \\
&= \sum_{i=1}^{\ell} \mathop{\mathbb{E}}_{s,r}\Big[A(\ldots, r_{i-1}, g_i, g_{i+1}, \ldots) - A(\ldots, r_{i-1}, r_i, g_{i+1}, \ldots)\Big] = \Omega(k^{-c}).
\end{aligned}$$

Since this sum has only a polynomial number of terms, its average term must be polynomially large. That is, if we choose i uniformly from $\{1, \ldots, \ell\}$, the average difference in A's behavior between the $(i-1)$st and

ith stages is the total difference divided by ℓ. Since $\ell = O(k^a)$ for some constant a, this gives

$$\mathop{\mathbb{E}}_{s,r,i}\left[A(\ldots,r_{i-1},g_i,g_{i+1},\ldots) - \mathop{\mathbb{E}}_{b'\in\{0,1\}} A(\ldots,r_{i-1},b',g_{i+1},\ldots)\right] = \Omega(k^{-c-a}) \,.$$

At the ith stage, we have $x = f^{i-1}(s)$ and $g_i = b(x)$. Since f is one-to-one, if s is uniformly random in $\{0,1\}^k$ then so is $f(s)$, and by induction on i we can assume that x is uniformly random. Analogously to (11.33) we can rewrite this as

$$\mathop{\mathbb{E}}_{x,r,i}\left[A(\ldots,r_{i-1},b(x),b(f(x)),\ldots) - A(\ldots,r_{i-1},\overline{b(x)},b(f(x)),\ldots)\right] = \Omega(k^{-c-a}) \,. \tag{11.36}$$

As in Theorem 11.13, we can now define an algorithm B that takes $f(x)$ and a bit b' as input, and tries to tell if $b' = b(x)$ or not. We choose i uniformly from $\{1,\ldots,\ell\}$ and flip $i-1$ coins $r_1,\ldots,r_{i-1}$. Since f and b can be computed in polynomial time, given $f(x)$ we can compute $b(f(x))$, $b(f^2(x))$, and so on. We then run A on the following input:

$$B(f(x),b') = A\big(r_1,\ldots,r_{i-1},\, b',\, b(f(x)),\ldots,b(f^{\ell-i-1}(x))\big) \,.$$

By (11.36), we know that B is polynomially more likely to return "yes" if $b' = b(x)$ than if $b' \neq b(x)$. Since b is a hard-core bit, we can then invert f with polynomial probability. This contradicts our assumption that f is a one-way function. So no tester A can exist, and g is a strong pseudorandom generator. □

This theorem shows that, if one-way permutations exist, we can fool any polynomial-time algorithm with a pseudorandom generator whose seed length is $k = \ell^\varepsilon$ for arbitrarily small $\varepsilon > 0$. Since $\ell = \text{poly}(n)$, we can also say that $k = n^\varepsilon$ for arbitrarily small ε. While the construction is considerably more complicated, we can do the same thing with any one-way function, whether or not it is one-to-one.

Conversely, if pseudorandom generators exist, they themselves are one-way functions from $\{0,1\}^k$ to $\{0,1\}^\ell$. Otherwise, given a string r, we could try to invert g and find an s such that $g(s) = r$. If r is random then such an s exists with probability $2^{k-\ell}$, so this gives us a tester that distinguishes $g(s)$ from a truly random r. Thus one-way functions exist if and only if pseudorandom generators do.

Now that we know how to stretch k random bits out to poly(k) pseudorandom ones—assuming that
11.6 one-way functions exist—how does this help us reduce the number of random bits we need to run a randomized algorithm? To what extent does it let us derandomize them, and simulate them deterministically? And what kind of pseudorandom generator would we need to derandomize them all the way down to deterministic polynomial time?

11.4.4 *Derandomizing* BPP *and Nonuniform Algorithms*

Recall the complexity class BPP that we defined in Section 10.9, of problems solvable by randomized algorithms that return the right answer at least 2/3 of the time. Let's quantify how pseudorandom generators help us simulate such algorithms deterministically.

Let A be a BPP algorithm that uses ℓ random bits. Depending on whether the input is a yes-instance or a no-instance, the probability that A returns "yes" is either greater than 2/3 or less than 1/3. We can tell deterministically which of these is true by running A on all 2^ℓ possible strings of random bits. Since each of these runs take polynomial time, we have the following:

> If a BPP algorithm A uses ℓ bits of randomness, then there is a deterministic version of A that takes $2^\ell \operatorname{poly}(n)$ time.

Since $\ell = \operatorname{poly}(n)$, this tells us that $\mathsf{BPP} \subseteq \mathsf{EXPTIME}$.

Now suppose there is a pseudorandom generator g with seed length k that fools A. In that case, it suffices to run A on $g(s)$ on all 2^k possible seeds s. This lets us reduce the factor of 2^ℓ to 2^k:

> If a BPP algorithm A is fooled by a pseudorandom generator with seeds of length k, then there is a deterministic version of A that takes $2^k \operatorname{poly}(n)$ time.

In the previous section, we saw that if one-way functions exist then we can fool polynomial-time algorithms with a seed length n^ε for ε as small as we like. This gives us the following result:

> If one-way functions exist, then $\mathsf{BPP} \subseteq \mathsf{TIME}(2^{n^\varepsilon})$ for any $\varepsilon > 0$.

This is somewhat weaker than the statement BPP = P. For instance, there could be BPP problems for which the fastest deterministic algorithm takes, say, $\Omega(n^{\log n})$ time. But it would show that the exponential dependence of the running time can be made arbitrarily mild—say, exponential in $n^{1/100}$.

However, if we are really to derandomize BPP to this extent, we need to address the subtlety we alluded to at the beginning of Section 11.4.1. In order to fool a BPP algorithm, we have to fool it on every instance. In other words, our pseudorandom generator g has to have the property that

$$\left| \Pr_{s\in\{0,1\}^k}[A(x,g(s)) = \text{“yes”}] - \Pr_{r\in\{0,1\}^\ell}[A(x,r) = \text{“yes”}] \right|$$

is vanishingly small for all x. If A is not fooled by g, that means that there exists an instance x for which these probabilities differ significantly.

But this does not quite mean that there is a deterministic tester $A'(r) = A(x,r)$, whose only input is r, that g fails to fool in the sense of Definition 11.11. The reason for this is that A' might not be able to construct x on its own. In other words, there might be instances x on which A is not fooled by g, but these instances *might be very hard to find.* Thus the ability to fool algorithms whose only input is r, as in Definition 11.11, is not enough to derandomize BPP algorithms on all instances.

Instead, we need to consider the stronger property that g fools algorithms that are given access to additional information—say, in the form of a lookup table—that they can use to distinguish $g(s)$ from a random string. We can think of the contents of this lookup table as advice from a helpful god, designed to help the algorithm achieve its goals. This advice depends only on the size n of the input, and is limited to a certain number of bits. In this case, in order to make a tester $A'(r)$ that distinguishes $g(s)$ from a random string, we give A' the instance x of size n on which A is not fooled by g.

Note that this advice can change in arbitrary ways from one value of n to the next, and there may be no finite description that ties all the values of n together. Unlike what we usually mean by an algorithm, A' can pursue very different strategies on inputs of different sizes. For this reason, algorithms with advice are called *nonuniform.*

We saw these nonuniform algorithms in Section 6.5 in their incarnation as families of Boolean circuits, where there is one circuit for each input size n, but the circuits for different n may have nothing in common. For instance, P/poly, which we defined there as the class of problems that can be solved by families of circuits with poly(n) gates, is also the class of problems solvable by nonuniform algorithms that run in poly(n) time and receive poly(n) bits of advice.

Thus, to establish that $\mathsf{BPP} \subseteq \mathsf{TIME}(2^{n^\varepsilon})$, we need a pseudorandom generator that fools all nonuniform testers—that is, all nonuniform polynomial-time algorithms with a polynomial amount of advice, or equivalently all families of Boolean circuits of polynomial size. Happily, the proofs of Theorems 11.13 and 11.14 go through just as before. All we have to do is strengthen our definition of one-way functions analogously, so that even nonuniform algorithms cannot invert $f(x)$, or compute the hard-core bit $b(x)$, with probability polynomially better than chance.

This issue of nonuniformity may seem like a technicality. But as we will see next, it plays a crucial role in the strongest known results on derandomization. If there are functions that are *exponentially* hard to compute, even for nonuniform algorithms, then we can construct pseudorandom generators whose seeds have only a logarithmic number of bits—and derandomize BPP all the way down to P.

11.4.5 Logarithmic Seeds and BPP = P

Suppose there is a pseudorandom generator which creates pseudorandom strings of length ℓ from seeds of length $k = O(\log \ell)$. Then the number of possible seeds is just $2^k = \mathrm{poly}(\ell) = \mathrm{poly}(n)$. By trying all of them, we can simulate any BPP algorithm deterministically in polynomial time, and BPP = P.

However, if we want seeds of only logarithmic length, we need to weaken our definition of "fooling" somewhat. After all, there is a simple polynomial-time algorithm that distinguishes $g(s)$ from a random string r—just try all poly(n) possible seeds s and check if $r = g(s)$ for any of them. For that matter, we can find s with probability $1/\mathrm{poly}(n)$ simply by guessing it randomly. Thus there is no hope of fooling all polynomial-time algorithms as in Definition 11.11, where the change in $\Pr[A(r) = \text{"yes"}]$ has to be superpolynomially small.

On the other hand, if our goal is to derandomize a BPP algorithm A, there's nothing wrong if A's behavior changes significantly. We just have to make sure that it still gives the right answer a clear majority of the time. For instance, if A is correct with probability at least 2/3, as in our definition of BPP, any change in $\Pr[A(r) = \text{"yes"}]$ bounded below 1/6 is acceptable. With this in mind, consider the following definition:

Definition 11.15 *An* exponentially strong pseudorandom generator *is a function* $g : \{0,1\}^k \to \{0,1\}^\ell$ *where* $k = O(\log \ell)$ *that can be computed in* poly(ℓ) *time, such that, for any nonuniform algorithm A whose running time and advice is* $O(\ell)$,

$$\left| \Pr_{s\in\{0,1\}^k}[A(g(s)) = \text{"yes"}] - \Pr_{r\in\{0,1\}^\ell}[A(r) = \text{"yes"}] \right| = o(1).$$

There are several things to note about this definition. First, g's running time can be *exponential* as a function of the seed length k, since it is polynomial in ℓ.

Secondly, we can defeat simple testers that try to go through all possible seeds by making the constant in $k = O(\log \ell)$ large enough. If $k = 10 \log_2 \ell$, say, there are $2^k = \ell^{10}$ possible seeds—but an algorithm that runs in $O(\ell)$ time and has $O(\ell)$ advice can check at most $O(\ell)$ of them. Thus it is plausible that no such algorithm can distinguish $g(s)$ from a random string with probability greater than $O(\ell^{-9})$.

Finally, suppose we are trying to fool a BPP algorithm A whose running time is $t = \mathrm{poly}(n)$. We can set $\ell = t$, even if A doesn't need this many random bits. The corresponding nonuniform algorithm only needs $n < \ell$ bits of advice, so both the running time and advice are $O(\ell)$ by definition. We have $k = O(\log \ell) =$

$O(\log n)$, so we can simulate A deterministically in poly(n) time by trying all $2^k = \text{poly}(n)$ seeds. Thus we can conclude the following:

> If exponentially strong pseudorandom generators exist, then $\mathsf{BPP} = \mathsf{P}$.

Can we construct pseudorandom generators that are exponentially strong? Unfortunately, the techniques of Section 11.4.3 are not powerful enough to do this. However, there is another way to show that such generators exist, subject to a plausible assumption—that some functions are exponentially hard to compute, even given an exponential amount of advice. We describe this construction in the next section.

11.4.6 Hardness and Randomness

The results of Section 11.4.3 show that if certain problems are hard, such as inverting a one-way function, then certain generators are pseudorandom. This ability to trade hardness for randomness creates a pleasing give and take. If we have a *lower* bound on the complexity of problems like DISCRETE LOG, then we get an *upper* bound on the amount of randomness we need, and therefore on the complexity of simulating randomized algorithms deterministically.

How far can we push this idea? Are there kinds of hard problems whose existence would imply the existence of exponentially strong pseudorandom generators, and thus push BPP all the way down to P? What kind of hardness would these problems have to possess?

We end this chapter by describing a construction that yields an exponentially strong pseudorandom generator from any function which is exponentially hard in a certain sense. Here's the idea. If we have a function $f : \{0,1\}^n \to \{0,1\}$, we can use it to generate an ℓ-bit string from a k-bit seed in the following way. First choose a family of ℓ subsets $S_i \subset \{1,\ldots,k\}$ where $1 \le i \le \ell$, such that $|S_i| = n$ for all i. Then given a seed $s \in \{0,1\}^k$, define the ith bit of $g(s)$ as

$$g(s)_i = f\left(s_{j_1}, s_{j_2}, \ldots, s_{j_n}\right) \quad \text{where} \quad S_i = \{j_1, j_2, \ldots, j_n\}.$$

Thus $g(s)_i$ is f applied to the n bits of s picked out by S_i. We will abuse notation below and write

$$g(s)_i = f(S_i).$$

This is called the *Nisan–Wigderson generator*, and we illustrate it in Figure 11.11. In our application, we will take $k = bn$ and $\ell = 2^{\delta n}$ for some constants $b > 1$ and $\delta > 0$, so that $k = O(\log \ell)$. We will also ensure that the sets S_i have a small overlap with each other—that for some constant $\alpha < 1$,

$$\left|S_i \cap S_j\right| \le \alpha n \text{ for all } i \ne j\,.$$

As Problem 11.15 shows, such families of sets do exist, and can be found in poly(ℓ) time.

To make the Nisan–Wigderson generator exponentially strong, we will apply it using a function f that is exponentially hard in the following sense.

Definition 11.16 *A function is* exponentially hard *if there is a constant $\varepsilon > 0$ such that, for any nonuniform algorithm A whose running time and advice is $O(2^{\varepsilon n})$,*

$$\Pr_{x \in \{0,1\}^n}[A(x) = f(x)] = \frac{1}{2} + O(2^{-\varepsilon n})\,.$$

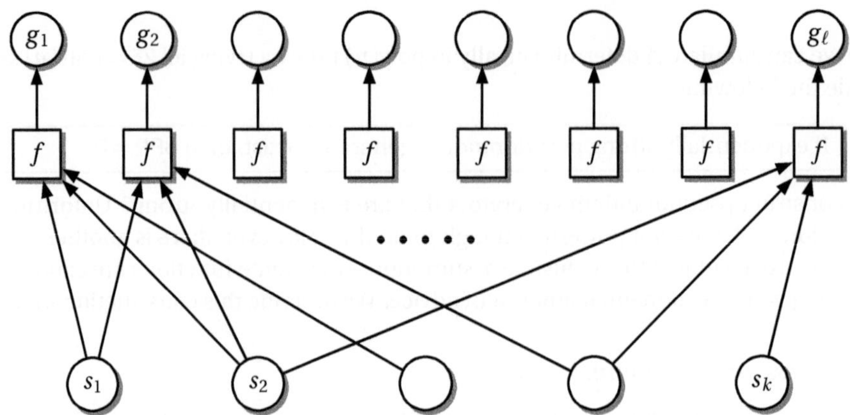

FIGURE 11.11: The Nisan–Wigderson generator. Given a seed s of length k, we generate a pseudorandom string $g(s)$ of length ℓ by applying f to ℓ different subsets S_i, each containing n bits of the seed. In this illustration, $k=5$, $n=3$, and $\ell=8$.

Compare this to our definition of one-way functions in Section 11.4.3. There we demanded that no polynomial-time algorithm can invert f with probability $1/\text{poly}(n)$ better than a random coin flip. Now we demand that even a nonuniform algorithm that has $2^{\varepsilon n}$ time and advice—or equivalently, a nonuniform family of Boolean circuits with $2^{\varepsilon n}$ gates—can only do $2^{-\varepsilon n}$ better than a coin flip.

We are now in a position to prove that we can derandomize BPP completely if there are functions in EXPTIME that are exponentially hard in this sense. The proof is actually quite simple. What matters is the conceptual leap to nonuniform algorithms, and the fact that many things that we don't know, or that may be hard to compute, can simply be included in the advice that the algorithm receives.

Theorem 11.17 *Suppose there is a function $f:\{0,1\}^n \to \{0,1\}$ that can be computed in $O(2^{\theta n})$ time for some constant θ, but which is exponentially hard. Then the Nisan–Wigderson generator is exponentially strong and* BPP = P.

Proof Once again, the proof is by contradiction. If $g(s)$ is not exponentially strong, there is some nonuniform tester A that runs in $O(\ell)$ time and receives $O(\ell)$ advice, such that

$$\mathop{\mathbb{E}}_{s\in\{0,1\}^k}[A(g(s))] - \mathop{\mathbb{E}}_{r\in\{0,1\}^\ell}[A(r)] \geq C$$

for some constant $C>0$. Our goal is to turn this tester into a nonuniform algorithm that computes f with higher probability than it ought to, given that f is exponentially hard.

We use the hybrid argument as in Theorem 11.14. If we replace the ℓ bits of $g(s)$, one at a time, with random bits, one of these replacements must make a difference of C/ℓ in A's behavior. In other words, for some i we have

$$\mathop{\mathbb{E}}_{s,r}\Big[A(\ldots,f(S_{i-1}),f(S_i),r_{i+1},\ldots) - A(\ldots,f(S_{i-1}),r_i,r_{i+1},\ldots)\Big] \geq C/\ell\,. \tag{11.37}$$

Let $x\in\{0,1\}^n$ denote the n bits of s contained in S_i, and let $y\in\{0,1\}^{k-n}$ denote the other bits of s. Then the average over s is the average over x and y, and we can write $\mathbb{E}_{s,r}$ as $\mathbb{E}_{x,y,r}$.

But if an inequality holds on average over all x, y, and r, there must exist some particular choice of y and r for which it holds if we just average over x. If we fix y and r to these values then we can give them to a nonuniform algorithm B as part of its advice, and define

$$B(x,b)=A\big(f(S_1),\ldots,f(S_{i-1}),b,r_{i+1},\ldots,r_\ell\big).$$

Then (11.37) becomes

$$\mathop{\mathbb{E}}_x\left[B(x,f(x))-\mathop{\mathbb{E}}_{b\in\{0,1\}}B(x,b)\right]\geq C/\ell\,,$$

so B is $\Omega(1/\ell)$ more likely to return "yes" if $b=f(x)$ than if $b\neq f(x)$.

As in Theorems 11.13 and 11.14, by running B on both values of b we obtain an algorithm that computes $f(x)$ with probability $\Omega(1/\ell)$ better than chance. We will show that, if ℓ, k, and the intersections between the S_j are bounded appropriately, this violates the assumption that f is exponentially hard.

Of course, B has to calculate the values $f(S_1),\ldots,f(S_{i-1})$ in order to give them to A. So saying that B is a good algorithm for calculating $f(x)=f(S_i)$ seems a bit circular. But this is where we use nonuniformity, along with our assumption that $|S_i\cap S_j|\leq \alpha n$ for all i,j.

Once the bits of y are fixed, $f(S_j)$ depends only on the bits of S_j that are part of the input x—in other words, those in $S_i\cap S_j$. Thus for each j, $f(S_j)$ really only depends on at most αn bits, so it can be specified completely with a lookup table of size $2^{\alpha n}$. We give B these lookup tables so that it can simply look each $f(S_j)$ up instead of having to calculate it. With the help of all this advice, the running time of B is essentially that of A, namely $O(\ell)$.

Now recall that $k=bn$ and $\ell=2^{\delta n}$ for suitably chosen constants b and δ. The total package of advice we give B then consists of

1. The $O(\ell)$ bits of advice on which A relies,
2. $k-n\leq bn=O(n)$ bits giving y, the bits of the seed s other than x,
3. $\ell=2^{\delta n}$ bits giving r, the random string that A can distinguish from $g(s)$, and
4. $2^{\alpha n}\,i<2^{\alpha n}\,\ell=O(2^{(\alpha+\delta)n})$ bits giving lookup tables for $f(S_j)$ for each $j<i$ as a function of x.

The total amount of advice is $O(2^{(\delta+\alpha)n})$, the running time is $O(\ell)=O(2^{\delta n})$, and on average, we calculate $f(x)$ with probability $2^{-\delta n}$ better than chance.

If $\delta+\alpha<\varepsilon$, this violates the assumption that f cannot be calculated in $O(2^{\varepsilon n})$ time, with $O(2^{\varepsilon n})$ advice, $O(2^{-\varepsilon n})$ better than chance. By contradiction, no such tester A can exist, and g is exponentially strong. This completes the proof. □

Do exponentially hard functions exist? This is not obvious. On one hand, the Time Hierarchy Theorem of Section 6.3 shows that if $\varepsilon<\theta$, there are functions computable in time $O(2^{\theta n})$ but not in time $O(2^{\varepsilon n})$. On the other hand, exponential advice is a very powerful resource. Any function f from $\{0,1\}^n$ to $\{0,1\}$ can be calculated using 2^n bits of advice, since this is enough to provide a lookup table for $f(x)$ on all 2^n possible inputs x. For that matter, any function can be calculated with probability $1/2+2^{-n/2}/2$ using $2^{n/2}$ advice—just provide a lookup table for $2^{n/2}$ of the possible inputs, and flip a coin if x isn't in the table.

However, this naive approach can only go so far. There is no obvious way to calculate $f(x)$ with probability $2^{-n/4}$, say, with a lookup table of size only $2^{n/4}$. Moreover, if no exponentially hard functions exist, this means that any function computable in exponential time can actually be computed in $2^{o(n)}$ time, $2^{-o(n)}$ better than chance, with only $2^{o(n)}$ bits of advice. As is so often the case, we have no proof that this is impossible, but it does not seem very likely. So unless advice is far more helpful to algorithms than we think it is, we can derandomize all polynomial-time algorithms, and BPP = P.

11.7

11.4.7 *Possible Worlds*

We have seen that there is an intimate relationship between one-way functions, pseudorandom generators, and secure cryptosystems. These are marvelous computational objects, assuming that they exist. How strongly do we believe in them?

Inverting a polynomial-time function, inferring the seed of a pseudorandom generator from the string it produces, and decrypting a polynomial-time cryptosystem are all problems in NP. Thus none of these things exist if P = NP.

On the other hand, even if P ≠ NP it's not clear that these things exist. When I encrypt a message, or apply a one-way function, I am creating a special kind of hard problem—a puzzle that's hard for you to solve, but where I know the solution. The question is whether there are polynomial-time algorithms that create such puzzles. We need more than hard problems—we need problems for which hard instances are easy to construct.

Russell Impagliazzo describes the situation in terms of five possible worlds. In *Algorithmica*, P = NP, or perhaps BPP = NP. As we discussed in Section 6.1, this would mean not only that all our favorite search problems are tractable, but that there is no need for creativity or intuition in the process of discovering mathematical proofs or scientific knowledge.

In *Heuristica*, P ≠ NP but hard instances of NP-complete problems are very rare. There are polynomial-time algorithms that work on almost all instances, where "almost all" refers to any probability distribution that can be sampled in polynomial time as in Problem 11.13. Thus the problem of finding hard instances of NP-complete problems is itself an intractable problem.

In *Pessiland*, there are NP-complete problems that are hard on average, but there are no one-way functions. That is, for any polynomial-time computable function f, the problem of inverting f is in P. In such a world, it is easy to create hard problems, but hard to create hard problems for which the creator knows the solution. In other words, designing hard puzzles is just as hard as solving them. In particular, there is no easy way to encrypt a message that makes it hard to decrypt.

In *Minicrypt*, there are one-way functions, but no public-key cryptography. In particular, there is no way to establish a shared secret with a stranger by communicating over a public channel, as we believe that Diffie–Hellman key exchange allows us to do (see Section 15.5.6). However, since there are one-way functions, we have bit commitment and zero-knowledge proofs as described in Section 11.1.4. We also have pseudorandom generators, and private-key cryptosystems as described in Section 11.4.2.

Finally, in *Cryptomania*, the entire apparatus of pseudorandomness and public-key cryptography is available to us. We can exchange secret keys over a public channel, and there are "trapdoor functions" that are hard to invert, but become easy to invert if one possesses a secret key. This lets us communicate securely using schemes like RSA cryptography (see Section 15.5.1).

Our current fond belief is that we live in Cryptomania. However, as Impagliazzo points out, essentially the only reason we have for believing that DISCRETE LOG and FACTORING are hard is our lack of success in finding polynomial-time algorithms for them. While there would be strong consequences for cryptography if they turn out to be in P, there would not be very strong consequences for complexity theory. And as PRIMALITY shows, there is a lot of room for new mathematical and algorithmic insights for number-theoretic problems like these.

On the other hand, even if DISCRETE LOG and FACTORING turn out to be easy, there are other candidates 11.8
for trapdoor cryptosystems waiting in the wings. These alternatives are also of interest since, as we will see in Chapter 15, quantum computers can solve DISCRETE LOG and FACTORING in polynomial time.

Problems

> In mathematics the art of proposing a question must be held of higher value than solving it.
>
> George Cantor

11.1 Randomness as advice. Prove that $\mathsf{BPP} \subseteq \mathsf{P/poly}$. Hint: first use Problem 10.46 to amplify the probability that a BPP algorithm A returns the right answer to $1-o(2^{-n})$. Then show that there exists a sequence r of $\mathrm{poly}(n)$ coin flips, which we can give to the algorithm as advice, such that using r causes A to return the right answer on every instance of length n. Thus advice is a powerful enough resource to replace randomness.

11.2 The Leftover Hash Lemma. In this problem, we will show how to distill, or *extract*, a nearly random string of bits from a partly random one. We need a few definitions. As in Section 10.6.2, a *pairwise independent family of hash functions* is a set of functions $h:\{0,1\}^n \to \{0,1\}^\ell$ such that, if h is chosen uniformly from the family, then for any $x,y \in \{0,1\}^n$ with $x \neq y$ and any $w,z \in \{0,1\}^\ell$,

$$\Pr_h[h(x)=w \text{ and } h(y)=z]=2^{-2\ell}.$$

Suppose that this family of hash functions is of size 2^k, so that it takes a "seed" $s \in \{0,1\}^k$ to choose h. Then we can describe the entire family as a function $H:\{0,1\}^k \times \{0,1\}^n \to \{0,1\}^\ell$ such that $h_s(x)=H(s,x)$:

$$\Pr_s[H(s,x)=w \text{ and } H(s,y)=z]=2^{-2\ell}.$$

We say that a probability distribution Q on $\{0,1\}^n$ has *min-entropy* b if $Q(x) \le 2^{-b}$ for all $x \in \{0,1\}^n$. Our goal is to convert such a Q into a nearly uniform distribution on $\{0,1\}^{n'}$ for some $n' < n$, extracting n' random bits from n partly-random ones. We will do this by stirring k truly random bits into the mix, using these as the seed for the hash function, and including this seed in our output. This gives a function

$$f(s,x)=(s,H(s,x)).$$

If $x \in \{0,1\}^n$ is chosen according to Q and $s \in \{0,1\}^k$ is uniformly random, then $f(s,x)$ is distributed according to a probability distribution P on $\{0,1\}^{n'}$ where $n'=k+\ell$.

We will prove that, as long as ℓ is not too large, P is close to uniform in the following sense. We say that P is ε*-close* to the uniform distribution U if $\|P-U\|_1 \le \varepsilon$:

$$\sum_{x\in\{0,1\}^{n'}} \left|P(x)-2^{-n'}\right| \le \varepsilon.$$

As we discuss in Section 12.2.3, this implies that no statistical experiment can distinguish P from the uniform distribution with probability greater than $\varepsilon/2$. In particular, if we run a randomized algorithm with a pseudorandom string of n' bits sampled according to P as opposed to a uniformly random string, this changes the probability that the algorithm returns "yes" by at most $\varepsilon/2$.

Given these definitions, prove the Leftover Hash Lemma:

Lemma 11.18 *Let Q be a distribution on $\{0,1\}^n$ with min-entropy b, and let $H: \{0,1\}^k \times \{0,1\}^n \to \{0,1\}^\ell$ be a family of pairwise independent hash functions. If $x \in \{0,1\}^n$ is chosen according to Q and $s \in \{0,1\}^k$ is chosen uniformly, then the distribution P on $\{0,1\}^{n'}$ with $n' = k + \ell$ defined by $f(s,x) = (s, H(s,x))$ is ε-close to uniform, where*

$$\varepsilon = 2^{-(b-\ell)/2}.$$

Hint: use the fact that $\|P\|_2^2$ is the *collision probability* of P, i.e., the probability that drawing from it twice produces the same result. Then use the Cauchy–Schwarz inequality $\|v\|_1 \le \sqrt{N}\|v\|_2$ (see Appendix A.2.3) and the fact that $\|u - v\|_2^2 = \|u\|_2^2 + \|v\|_2^2 - 2u^T v$ for any vectors u, v.

11.3 Recycling randomness. We can use the Leftover Hash Lemma to run a randomized algorithm many times, amplifying the probability that it returns the correct answer as in Problem 10.46, while using far fewer random bits than it would take to make each run completely independent.

Let A be a randomized algorithm that uses n random bits. Assume for simplicity that, given a yes-instance, A returns "yes" and "no" with probability 2/3 and 1/3 respectively. First show that if we run A using a uniformly random string $x_1 \in \{0,1\}^n$, and condition on the answer that A returns, the resulting distribution Q on $\{0,1\}^n$ has min-entropy at least $n - \log_2 3$.

Now, rather than running A again with a fresh string of random bits, we choose a random seed $s \in \{0,1\}^k$ for a pairwise independent family H of hash functions, and map x_1 to $f(s,x_1) = (s, H(s,x_1)) \in \{0,1\}^{n'}$. We then flip $n - n'$ coins, adding them to this string, to produce a new string $x_2 \in \{0,1\}^n$. Show that x_2 is within ε of the uniform distribution on $\{0,1\}^n$, with ε defined as in Problem 11.2.

We then run A using x_2, map x_2 to $f(x_2) = (s, H(s,x_2))$, flip $n - n'$ coins to produce x_3, and so on. However, we keep using the same seed s we chose initially, so this only costs $n - n'$ random bits per iteration. Show by induction that after t steps, the probability distribution on A's responses is within $t\varepsilon$ of what it would be if we used purely random strings at each step.

By choosing the parameters k, ℓ, and ε, show that we can increase the probability that the majority of these responses are correct up to $1 - 2^{-\Omega(t)}$ with a total of $O(n + t^2)$ random bits rather than the $O(tn)$ we would need if each x_i were chosen independently. Hint: because we include the seed s as part of the pseudorandom string $(s, H(s,x))$, and since this string is close to uniform at each stage, we only need to choose the hash function once—we don't need a new seed at each iteration.

11.4 Simulated conversations. Write a simulator for the zero-knowledge protocol for GRAPH NONISOMORPHISM given in Section 11.1.3, which generates a probability distribution of conversations exponentially close to that of a real interaction with Merlin. Specifically, if Arthur uses an honest strategy, which generates H and the pair (K_1, K_2) from $\{G_1, G_2\}$, this distribution should match the real one exactly. If Arthur cheats, the simulator should match the real distribution except in the exponentially unlikely event that Arthur fools Merlin on every round.

11.5 Random squares. For a nice illustration of why the simulator for a zero-knowledge proof needs to be able to handle the case where Arthur cheats and uses a non-random strategy, consider the following scenario. Fix a prime p. Arthur sends Merlin an integer $x \in \mathbb{Z}_p^* = \{1, 2, \ldots, p-1\}$. Then, Merlin sends Arthur a y such that $x \equiv_p y^2$, or replies "x is not a square" if there is no such y. Clearly, Arthur can learn quite a bit from this protocol, so it is not zero-knowledge. Show, on the other hand, that if Arthur chooses x uniformly from $\mathbb{Z}_p^*$ it is easy to simulate this conversation in polynomial time, so that Arthur learns nothing at all.

11.6 It's hip not to be square. In the group $\mathbb{Z}_m^*$ of integers mod m which are mutually prime to m, not all elements have a square root. Those that do are called *quadratic residues*. Consider the following problem.

> QUADRATIC RESIDUE
>
> Input: An integer m and an $x \in \mathbb{Z}_m^*$
>
> Question: Is x a quadratic residue? That is, there a $y \in \mathbb{Z}_m^*$ such that $x = y^2$?

Clearly QUADRATIC RESIDUE is in NP. Show that it is in NP ∩ coAM by providing an interactive proof—first with private coins, and then with public ones—for its complement QUADRATIC NONRESIDUE.

Hint: think of QUADRATIC RESIDUE as analogous to GRAPH ISOMORPHISM, and follow the approaches of Section 11.1. Note that xy^2 is a residue if and only if x is. In particular, the product of two residues is a residue, so the set of residues forms a subgroup, $R = \{y^2 : y \in \mathbb{Z}_m^*\}$. Feel free to assume that Merlin can prove to Arthur that R is a certain size.

11.7 Zero knowledge, square or hip. Following up on the previous problem, provide zero-knowledge interactive proofs for both QUADRATIC RESIDUE and QUADRATIC NONRESIDUE.

11.8 The hard end of the log. Let p be a prime and let a be a primitive root. Show that the case of DISCRETE LOG of finding $x = \log_a z$ is reducible in polynomial time to the problem of finding its most significant bit $b(x)$,

> DISCRETE LOG MSB
>
> Input: A prime p, a primitive root a, and a $z \in \mathbb{Z}_p^*$ where $z = a^x \bmod p$
>
> Question: Is $x \geq (p-1)/2$?

It follows that $b(x)$ is a hard-core bit for inverting $f(x) = a^x \bmod p$. Using reasoning similar to Problem 10.46, show that if there is a randomized polynomial-time algorithm that guesses $b(x)$ correctly with probability at least $1/2 + \varepsilon$ where $\varepsilon = 1/\text{poly}(n)$, there is a randomized polynomial-time algorithm that finds x with high probability.

Note that this is a Turing reduction as defined in Note 5.2, so you might need to call a subroutine for DISCRETE LOG MSB a polynomial number of times. Hint: if $p = 13$, then in binary

$$\frac{\log_2 9}{12} = 0.101010\ldots$$

11.9 The other end of the log. As a counterpoint to the previous problem, show that in $\mathbb{Z}_p^*$ for prime p, finding the least significant bit of the DISCRETE LOG—that is, whether $\log_a z$ is even or odd—is in P. As a consequence, QUADRATIC RESIDUE (see Problem 11.6) is in P if the modulus is prime.

Hint: use Fermat's Little Theorem. Interestingly, if p is composite then even the least significant bit is hard core, assuming that FACTORING is hard. 11.2

11.10 Reducing the worst case to a random case. Let's show that if we can solve DISCRETE LOG on a reasonably large fraction of instances, we can solve it probabilistically on all of them. Fix an n-digit prime p and a primitive root a. Suppose that there is a polynomial-time algorithm for DISCRETE LOG which returns $\log_a z$ whenever $z \in S$, for some set $S \subseteq \mathbb{Z}_p^*$. Show that there is a randomized polynomial algorithm that, for any z, returns $\log_a z$ with probability $|S|/|\mathbb{Z}_p^*|$. If this fraction is $1/\text{poly}(n)$, DISCRETE LOG can be solved in randomized polynomial time. Conclude that, unless DISCRETE LOG is in RP, it can only be solved on a fraction of inputs that is superpolynomially small—that is, smaller than $1/n^c$ for any constant c. Hint: how can you change an arbitrary z into a uniformly random one?

11.11 Interactive proof of the permanent. Suppose I have an $n \times n$ matrix A with integer entries a_{ij}. Its *permanent* is defined as

$$\text{perm}\, A = \sum_{\pi} \prod_{i=1}^{n} a_{i,\pi(i)},$$

where π ranges over all $n!$ permutations. This is just like the determinant, except we don't have the usual sign ± 1 for even and odd permutations. In Chapter 13, we will see that for matrices of 0s and 1s, the problem PERMANENT is complete for #P, the class of problems that count solutions for problems in NP. We believe that this class lies outside the entire polynomial hierarchy, so PERMANENT is very hard.

However, there is a simple interactive proof for PERMANENT, which was a key precursor to the interactive proof for QUANTIFIED SAT. In particular, it shows that IP contains $\mathsf{P}^{\#\mathsf{P}}$, an important step towards showing that it contains PSPACE. First note that

$$\operatorname{perm} A = \sum_{i=1}^{n} a_{1,i} \operatorname{perm} A^{(1,i)},$$

where $A^{(1,i)}$ is the $(n-1)\times(n-1)$ matrix formed by deleting the top row and the ith column. There is a unique $(n-1)\times(n-1)$ matrix of polynomials $M(x)$ of degree $n-1$ such that $M(i) = A^{(1,i)}$ for all $i = 1, \ldots, n$, which Arthur can compute in polynomial time. Merlin's proof will work by giving Arthur the polynomial $Q(x) = \operatorname{perm} M(x)$.

Merlin starts by claiming that $\operatorname{perm} A$ has a certain value, and sending Arthur the coefficients of Q. If Merlin is honest, Arthur can confirm the value of $\operatorname{perm} A$ using

$$\operatorname{perm} A = \sum_{i=1}^{n} a_{1,i}\, Q(i).$$

But Arthur needs to challenge Merlin to prove that he gave him Q's true coefficients. Describe an interactive proof where Arthur chooses x randomly from a sufficiently large range, and challenges Merlin to prove the value of $Q(x)$. This is a claim about the permanent of an $(n-1)\times(n-1)$ matrix $M(x)$, which Arthur and Merlin discuss in the same way, and so on. What is Q's degree? And what range does Arthur need to choose x from so that the total probability that Merlin deceives him, at any stage of this process, is $o(1)$?

11.12 One bit is not enough. Consider probabilistically checkable proofs of length poly(n), where the verifier flips $O(\log n)$ coins and looks at just one bit of the proof. Show that for any soundness and completeness $0 < s < c \le 1$, the set of problems with such proofs for yes-instances is simply P. Hint: show that we can compute, in polynomial time, the proof that maximizes the probability that the verifier will accept.

11.13 Fooling randomized algorithms. Suppose we modify Definition 11.11 of a good pseudorandom generator by requiring that, for all constants c,

$$\left| \Pr_{s\in\{0,1\}^k}[B(g(s)) = \text{“yes”}] - \Pr_{r\in\{0,1\}^\ell}[B(r) = \text{“yes”}] \right| = o(n^{-c}),$$

for all *randomized* polynomial-time algorithms B, where Pr denotes probability both over B's input and B's own internal coin flips. While this still does not ensure that g fools a randomized polynomial-time algorithm A on every instance, show that it does fool A *on average* in the following sense.

If P is a probability distribution on $\{0,1\}^n$, we say that P is *efficiently samplable* if there is a polynomial-time randomized algorithm that produces outputs $x \in \{0,1\}^n$ with probability $P(x)$. Then show that if instances x are chosen according to any such distribution, the average difference in A's output probabilities is very small:

$$\mathbb{E}_{P(x)} \left| \Pr_{s\in\{0,1\}^k}[A(x, g(s)) = \text{“yes”}] - \Pr_{r\in\{0,1\}^\ell}[A(x, r) = \text{“yes”}] \right| = o(n^{-c}).$$

11.14 Bit commitment from pseudorandom generators. In this problem, we will show that any good pseudorandom generator can be used to give a secure protocol for bit commitment. In other words, Alice can send Bob a bit b in encrypted form, and then show Bob how to decrypt it. Before Alice reveals it, Bob cannot determine b with probability $1/2 + 1/\text{poly}(n)$, and Alice cannot cheat by changing the value of b after the fact.

First, let's look at a scheme that doesn't work. Suppose that $g:\{0,1\}^n \to \{0,1\}^\ell$ is a good pseudorandom generator, say according to the definition of Problem 11.13, which is known to both parties. Alice chooses a seed $s \in \{0,1\}^n$ and calculates $g(s)$. She then XORs the last bit of $g(s)$ with b, flipping it if $b=1$, and sends the result to Bob. To reveal b, Alice sends Bob the seed s, so that he can calculate $g(s)$ himself and see if the last bit is flipped. Why might this scheme allow Alice to cheat, and reveal a different b than she committed to?

A better scheme is as follows. First, Bob chooses a random string $r \in \{0,1\}^\ell$ and sends it to Alice. Now, Alice chooses a seed s and calculates $g(s)$. If $b=1$ then Alice sends Bob $g(s)\oplus r$—that is, she flips the ith bit of $g(s)$ whenever $r_i = 1$. If $b=0$, she sends Bob $g(s)$. To reveal b, she again sends Bob the seed s.

Show that if $\ell = 3n$, say, it is exponentially unlikely that Alice can cheat in this case. Hint: show that if Bob's string r is chosen randomly, it is exponentially unlikely that there is a pair of seeds s, s' such that $g(s)\oplus g'(s) = r$.

11.15 Subsets for the Nisan–Wigderson generator. In this problem, we construct the family of subsets called for in the Nisan–Wigderson generator. Let $k = bn$ where $b=10$, let $\ell = 2^{\delta n}$ where $\delta = 1/30$, and let $\alpha = 1/5$. Show that when n is sufficiently large, there is a family of ℓ subsets $S_i \subset \{1,\ldots,k\}$ such that $|S_i| = n$ for all i and $\left|S_i \cap S_j\right| \le \alpha n$ for all $i \ne j$. Moreover, show that we can find such a family in poly(ℓ) time by repeating the following ℓ times: go through all $\binom{k}{n}$ possible S_j, find one that has a small intersection with all the previous S_i, and add it to the family. More generally, show that for any ε, there are constants $b, \delta, \alpha > 0$ such that this holds, and where $\alpha + \delta < \varepsilon$.

Hint: assume that each S_i is chosen uniformly and independently from all $\binom{k}{n}$ subsets of size n. Then bound the probability that $\left|S_i \cap S_j\right| > \alpha n$, and use the union bound to show that the total probability that this occurs for any pair i, j is less than 1. If choosing the S_i randomly gives a good family of subsets with nonzero probability, at least one such family must exist.

11.16 Making the Blum–Micali generator stronger. In the text, we defined a pseudorandom generator as strong if no polynomial-time tester can break it with 1/poly(n) probability. We can strengthen this definition as follows. Given a function $t(n)$, we say that a generator $g:\{0,1\}^k \to \{0,1\}^\ell$ is *$t(n)$-strong* if, for any nonuniform algorithm A whose running time and advice is $O(t(n))$, or equivalently any family of circuits of size $O(t(n))$,

$$\left| \Pr_{s\in\{0,1\}^k} [A(g(s)) = \text{“yes”}] - \Pr_{r\in\{0,1\}^\ell} [A(r) = \text{“yes”}] \right| = o(1/t(n)).$$

Now generalize the hybrid argument of Theorem 11.14 to prove the following.

Theorem 11.19 *Suppose that there is a one-way function f, a hard-core bit b, and a constant $\varepsilon > 0$ such that no nonuniform algorithm with $2^{O(n^\varepsilon)}$ running time and advice, or equivalently no circuit of size $2^{O(n^\varepsilon)}$, can compute $b(x)$ from $f(x)$ with probability $2^{-O(n^\varepsilon)}$ better than chance. Then if $\ell = 2^{O(n^\varepsilon)}$, the Blum–Micali generator g_ℓ is 2^{n^ε}-strong.*

In particular, show that this follows if no circuit family of size $2^{O(n^\varepsilon)}$ can solve DISCRETE LOG. Given our current state of knowledge, it is plausible that this is the case for $\varepsilon = 1/3$, in which case there is a pseudorandom generator that is $2^{n^{1/3}}$-strong. Show that this would imply that BPP ⊆ QuasiP, the class of problems solvable in quasipolynomial time as defined in Problem 2.20.

Unfortunately, we currently have no candidates for one-way functions that require $2^{\Omega(n)}$ time to invert. Therefore, to stretch our seed exponentially far, producing poly(n) pseudorandom bits from $O(\log n)$ random ones and completely derandomize BPP, it seems that we need something like the Nisan–Wigderson generator.

11.17 Pseudorandom functions. A *pseudorandom function generator* takes a seed s of length k and an input y of length n and produces an output bit $f_s(y)$. Equivalently, it expands s to a pseudorandom string of length 2^n, which we interpret as the truth table of the function $f_s:\{0,1\}^n \to \{0,1\}$, where $f_s(y)$ is the yth bit of f_s. If $k = \text{poly}(n)$ and $f_s(y)$ can be computed in polynomial time from s and y, we say that f_s is a polynomial-time pseudorandom function. Note that we can include s in a polynomial amount of advice, so for any fixed s, $f_s(y)$ is in P/poly.

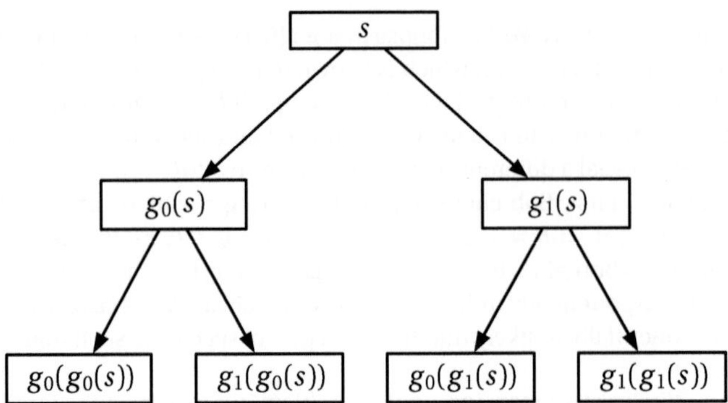

FIGURE 11.12: Building a pseudorandom function. Given a generator g that doubles the length of the seed, at each step we take the left half g_0 or the right half g_1. Doing this n times gives a binary tree of depth n, and each $y \in \{0,1\}^n$ describes a path through this tree ending at a leaf. For instance, if $n = 2$ and $y = 10$, then $G_y(s) = g_1(g_0(s))$. Finally, $f_s(y)$ is the first bit of $G_y(s)$.

In this problem we will show how to generate strong pseudorandom functions given a strong enough pseudorandom generator, and complete the argument of Razborov and Rudich from Section 6.5 that no natural proof can show that $\mathsf{NP} \not\subset \mathsf{P/poly}$. As we said there, f_s is a strong pseudorandom function if choosing s randomly from $\{0,1\}^k$ gives a function f_s which is hard to tell apart from a completely random function, i.e., one whose truth table consists of 2^n random bits—even for an algorithm whose running time is polynomial in 2^n, the size of the truth table. In other words, for any algorithm B which is given access to the truth table of f_s and whose running time is poly$(2^n) = 2^{O(n)}$,

$$\left| \Pr_{s\in\{0,1\}^k}[B(f_s) = \text{"yes"}] - \Pr_{f\in\{0,1\}^{2^n}}[B(f) = \text{"yes"}] \right| = 2^{-\omega(n)} .$$

Recall the notion of $t(n)$-strong generators from Problem 11.16. We will prove the following theorem:

Theorem 11.20 (Doubling the seed) *Suppose there is a polynomial-time generator which doubles the length of the seed, $g : \{0,1\}^k \to \{0,1\}^{2k}$, which is 2^{k^ε}-strong for some $\varepsilon > 0$. Then there is a polynomial-time pseudorandom function $f_s : \{0,1\}^n \to \{0,1\}$.*

We construct $f_s(y)$ as follows. First define the functions $g_0, g_1 : \{0,1\}^k \to \{0,1\}^k$ as the left and right halves of g respectively, so that $g(s) = (g_0(s), g_1(s))$. Now, given $y \in \{0,1\}^n$, define $G_y : \{0,1\}^k \to \{0,1\}^k$ by composing a string of n of these functions, where each function is g_0 or g_1, according to y:

$$G_y(s) = g_{y_n}\left(g_{y_{n-1}}\left(\cdots g_{y_2}\left(g_{y_1}(s)\right)\right)\right) .$$

In other words, at each step we double the length of the string by applying g, and then take the left half or the right half according to the current bit of y.

As Figure 11.12 shows, we can view this process as a path through a binary tree of depth n. Each node v has a label $x_v \in \{0,1\}^k$, and the labels of its left and right children are $g_0(x_v)$ and $g_1(x_v)$ respectively. The root's label is the seed s, and the $G_y(s)$ is the label of the yth leaf. Finally, we define $f_s(y)$ as the first bit of $G_y(s)$.

Now show that setting $n = k^\varepsilon$ proves the theorem. Hint: assume by way of contradiction that there is an algorithm B that distinguishes f_s from a random function with probability $2^{-O(n)}$. Then starting at the root, replace the labels of each node's children with uniformly random strings $x_1, x_2 \in \{0,1\}^k$. This interpolates from a pseudorandom function $G_y(s)$, where only the seed s at the root is random, to a completely random function. Show using a hybrid argument that this gives a nonuniform algorithm that distinguishes (x_1, x_2) from $g(x) = (g_1(x), g_2(x))$ with probability $2^{-O(n)} = 2^{-O(k^\varepsilon)}$, and whose running time is $2^{O(n)} = 2^{O(k^\varepsilon)}$, violating our assumption about g.

Returning to the argument of Section 6.5, we have finally proved Theorem 6.5 on page 195. In other words, if there are pseudorandom generators g that are 2^{n^ε}-strong, then no property that can be checked in time polynomial in the size of the truth table can distinguish functions in P/poly from random functions with probability $2^{-O(n)}$. Therefore, no property which is "constructive" and "common" as in Section 6.5.3 can give a natural proof that NP $\subseteq$ P/poly. In particular, Problem 11.16 shows that this follows if DISCRETE LOG requires 2^{n^ε} time for some $\varepsilon > 0$.

11.18 Nisan's generator and fooling space-bounded computation. While derandomizing polynomial-time computation is still a challenge, there is a simple and beautiful pseudorandom generator due to Noam Nisan that fools algorithms with limited memory. It uses a clever idea called *derandomized squaring*, which we will see in another form in Section 12.9.

Suppose a randomized algorithm uses s bits of memory and runs in $t \le 2^s$ time, in which case it needs at most t random bits to run. We will group these bits into $b = t/\ell$ blocks of length ℓ, and write the entire string of random bits as $r = r_1 r_2 \cdots r_b$ where $r_i \in \{0,1\}^\ell$ for each i. As in Section 11.4, we fix the instance the algorithm is running on, and think of it as a deterministic algorithm that reads r. We hope to fool it into thinking that r is uniformly random, while actually using a pseudorandom string of bits instead.

We construct r iteratively in the following way. We start by choosing $r_1 \in \{0,1\}^\ell$ uniformly. But instead of choosing r_2 independently, we choose a function $h_1 : \{0,1\}^\ell \to \{0,1\}^\ell$ and set $r_2 = h_1(r_1)$. We generate r_3 and r_4 the same way, but starting from $h_2(r_1)$ instead of r_1, for some other function h_2. Iterating this gives the following:

$$\begin{array}{cccccccc} r_1 & r_2 & r_3 & r_4 & r_5 & r_6 & r_7 & r_8 \\ r_1 & h_1(r_1) & h_2(r_1) & h_1(h_2(r_1)) & h_3(r_1) & h_1(h_3(r_1)) & h_2(h_3(r_1)) & h_1(h_2(h_3(r_1))) \end{array}$$

We will show that if these h_j are chosen uniformly from a pairwise independent family of hash functions, and if $\ell = Cs$ for a sufficiently large constant C, this pseudorandom string fools the algorithm with high probability. Since there are families of $2^{2\ell}$ hash functions on $\{0,1\}^\ell$ which are pairwise independent (see Problem 10.33), and since $j = 1, 2, \ldots, \log_2 b$, the total number of bits we need is $O(\ell \log b) = O(s \log(t/s))$ instead of t.

In particular, if $s = O(\log n)$ and $t = \mathrm{poly}(n)$, then we can get by with $O(\log^2 n)$ random bits. Thus if we define a class BPL in analogy with BPP, of problems solvable by a randomized algorithm with $O(\log n)$ memory and $\mathrm{poly}(n)$ time where the probability of acceptance is at least 2/3 or at most 1/3 for yes-instances and no-instances respectively, then BPL $\subseteq$ SPACE($\log^2 n$).

Since the algorithm has just $m = 2^s$ possible states, for each string r we can define an $m \times m$ transition matrix 11.9
$M(r)$ where $M(r)_{ij} = 1$ if reading r would take it from state i to state j, and $M(r)_{ij} = 0$ otherwise. These matrices compose when strings are concatenated: that is, $M(r_1 r_2) = M(r_1) M(r_2)$. If the algorithm starts in state i and says "yes" if it arrives at state j, then the probability it says "yes" is the expected value of M_{ij} over the random bits. The generator fools the algorithm if this expectation over r and $h_1, h_2, \ldots$ is close to what it would be if the r_i were independent and uniform. In fact, we'll prove something stronger: namely, for almost all choices of the functions $h_1, h_2, \ldots$, taking the expectation over r is enough.

Let $M^{(b)}$ and $N^{(b)}$ denote the expectations after $\log_2 b$ steps of the construction:

$$\begin{array}{rr} M^{(1)} = \mathbb{E}_{r_1} M(r_1) & N^{(1)} = \mathbb{E}_r M(r) \\ M^{(2)} = \mathbb{E}_{r_1, r_2} M(r_1 r_2) & N^{(2)} = \mathbb{E}_r M(r h_1(r)) \\ M^{(4)} = \mathbb{E}_{r_1, r_2, r_3, r_4} M(r_1 r_2 r_3 r_4) & N^{(4)} = \mathbb{E}_r M(r h_1(r) h_2(r) h_1(h_2(r))) \end{array}$$

We have $M^{(2b)} = (M^{(b)})^2$, since $r_1 \cdots r_b$ and $r_{b+1} \cdots r_{2b}$ are independent. Our goal is to show that $N^{(2b)} \approx (N^{(b)})^2$ in some sense, and thus show by induction that $N^{(b)} \approx M^{(b)}$ for all b. Hence the name *derandomized squaring*: we get to double the length of the sequence without doubling the number of random bits we need.

We will measure the distance between $M^{(b)}$ and $N^{(b)}$ using the matrix norm $\|A\| = \max_v \|vA\|_1 / \|v\|_1$, where v ranges over all row vectors and $\|v\|_1$ is the 1-norm $\sum_i |v_i|$. Show that $\|A\|$ is equal to the maximum of the 1-norms of A's rows, $\max_j \sum_i |A_{ij}|$. Show also that for any matrices A, B we have the triangle inequality $\|A+B\| \le \|A\| + \|B\|$, as well as $\|AB\| \le \|A\| \, \|B\|$.

Since the algorithm reads r deterministically, we have $\|M(r)\| = 1$ for all r, and therefore $\|\mathbb{E}_r M(r)\| = 1$ for any probability distribution over r. Show that for any two matrices M, N with $\|M\| = \|N\| = 1$, squaring them at most doubles the distance between them: $\left\|M^2 - N^2\right\| \le 2\,\|M-N\|$. Thus we have, for all b,

$$\left\|M^{(2b)} - (N^{(b)})^2\right\| \le 2\left\|M^{(b)} - N^{(b)}\right\| . \tag{11.38}$$

We start the induction with $M^{(1)} = N^{(1)}$. To get going, show that if h_1 is drawn from a pairwise independent family, then

$$\left\|N^{(2)} - (N^{(1)})^2\right\| \le \delta \quad \text{with probability at least} \quad 1 - \frac{m^7}{2^\ell \delta^2} . \tag{11.39}$$

Hint: write $M(r_1 r_2)_{ij} = \sum_k M(r_1)_{ik} M(r_2)_{kj}$. Then apply Problem 10.47 to the sets $A_{ik} = \{r : M(r)_{ik} = 1\}$ and $B_{kj} = \{r : M(r)_{kj} = 1\}$ with an appropriate value of ε, and apply the union bound to the set of all triples i, k, j. Similarly, if we fix h_1 and draw h_2 from the same family, then

$$\left\|N^{(4)} - (N^{(2)})^2\right\| \le \delta \quad \text{with probability at least} \quad 1 - \frac{m^7}{2^\ell \delta^2} .$$

Hint: the idea is the same, but now where $A_{ik} = \{r : M(r h_1(r))_{ik} = 1\}$. Repeating this process, combining with (11.38), using the triangle inequality, and taking the union bound over $\log_2 b$ stages of squaring, show that

$$\left\|M^{(b)} - N^{(b)}\right\| \le (b-1)\delta \quad \text{with probability at least} \quad 1 - \frac{m^7 \log_2 b}{2^\ell \delta^2} .$$

where we assume b is a power of 2. Finally, set $\delta = o(1/b)$ so that the probability of acceptance with pseudorandom bits instead of truly random ones differs by $o(1)$, and show that we can make the probability of failure tend to zero by setting $\ell = Cs$ for a sufficiently large constant C.

Notes

11.1 Arthur, Merlin, and interactive proofs. Interactive proofs where defined concurrently by Goldwasser, Micali, and Rackoff [342], who used private coins, and by Babai [63], who defined Arthur–Merlin games and the class AM using public coins. In [63] it was shown that any Arthur–Merlin game consisting of a constant number of rounds can be simulated with a single round, at the cost of exchanging a larger number of bits. The public-coin interactive proof for Graph Nonisomorphism we give here appeared in Babai and Moran [68]. For their development of interactive proofs, the authors of [68] and [342] shared the Gödel Prize in 1993.

Papadimitriou defined "games against Nature" [644] in which Nature plays by emitting a string of random bits, and the probability that a computationally powerful player—such as Merlin—can win is greater or less than 1/2 depending on whether the input is a yes-instance or a no-instance. Nature's moves are equivalent to public coin flips, since if his coins are public Arthur might as well simply send the entire sequence of flips to Merlin. However, unlike the Arthur–Merlin games of [63], in this model the probability that Merlin can win can be arbitrarily close to 1/2.

The private-coin interactive proof for Graph Nonisomorphism was invented by Goldreich, Micali, and Wigderson [341]. The interactive proof technique of Section 11.1.2, in which Merlin proves a lower bound on the size of a set, was given by Goldwasser and Sipser [343] building on earlier work of Sipser [757]. They used it to show that public coins can be replaced with private ones in any interactive proof.

You can find an introduction to interactive proofs, pseudorandomness, and cryptography in Goldreich's book *Modern Cryptography, Probabilistic Proofs and Pseudorandomness* [338].

11.2 Hard-core bits, bit commitment, and zero knowledge. The idea of a zero-knowledge proof appeared in [342] with the example of QUADRATIC NONRESIDUE from Problem 11.7. The zero-knowledge proof for GRAPH 3-COLORING described in Section 11.1.3 is from Goldreich, Micali, and Wigderson [341]. They showed, assuming that secure cryptography exists, that all problems in NP have zero-knowledge interactive proofs.

The idea of bit commitment first appeared in Blum [114]. Blum and Micali [116] showed that the most significant bit is a hard-core bit for DISCRETE LOG. Long and Wigderson [537] strengthened this to any function of the $O(\log n)$ most significant bits.

We will see in Section 15.5.2 that if N is composite then finding square roots mod N is just as hard as factoring N. Rabin [675] used this to show that, if FACTORING is hard, the function $f(x) = x^2 \bmod N$ is a one-way function if N is the product of two primes. Alexi, Chor, Goldreich, and Schnorr [34] showed that the least significant bit is hard-core for this function.

Goldreich and Levin [340] showed that any one-way function possesses a hard-core bit, namely the parity of some subset of x's bits; a proof is given in Arora and Barak [53, Chapter 9]. The bit commitment scheme of Problem 11.14, using pseudorandom generators, was given by Naor [621].

11.3 Interactive proofs and polynomial space. Lund, Fortnow, Karloff, and Nisan [549] showed that PERMANENT is in IP. The interactive proof given in Problem 11.11 is from Babai; the one in [549] is slightly more complicated. As we will discuss in Chapter 13, PERMANENT is #P-complete, so in conjunction with Toda's theorem that $\mathsf{PH} \subseteq \mathsf{P}^{\#\mathsf{P}}$ (see Note 13.2) this implies that IP contains the polynomial hierarchy PH.

Building on the breakthrough of [549], Shamir then showed that IP = PSPACE. Independently, Babai and Fortnow [64] used arithmetization to describe #P problems in terms of a kind of arithmetic program.

Shamir controlled the degree of the polynomial by introducing dummy variables so that no variable has more than one $\forall$ between its quantifier and its appearance in the formula. Then no variable has degree more than twice its degree in the central SAT formula. The simplified proof we give here, in which we use the operator R to make the polynomials multilinear, is due to Shen [747].

We heard the fable of the Chess Master from Scott Aaronson, who heard it from Steven Rudich.

11.4 The PCP Theorem. The PCP Theorem originally grew out of the field of interactive proofs. Ben-Or, Goldwasser, Kilian, and Wigderson [99] defined *multi-prover* interactive proofs, where Arthur can challenge two wizards—say, Merlin and Nimue—who cannot talk to each other. Like a detective interviewing suspects in separate rooms, Arthur can extract far more information from these multiple provers than he can from Merlin alone.

The first hint of the PCP theorem came from Babai, Fortnow, and Lund [66], who showed that proofs of this kind are quite powerful—the class MIP of problems for which they exist equals NEXPTIME. Their result works by showing that problems in NEXPTIME have witnesses that can be probabilistically checked by looking at a polynomial number of bits. By "scaling down" this result, Babai, Fortnow, Levin, and Szegedy [65] showed that problems in NP have witnesses that can be checked by looking at a polylogarithmic number of bits.

This led to work by Feige, Goldwasser, Lovász, Safra, and Szegedy [278], Arora and Safra [57], and Arora, Lund, Motwani, Sudan, and Szegedy [56], culminating in the proof of the PCP theorem and its consequences for inapproximability. For this work, these authors shared the Gödel Prize in 2001. For his work on PCPs and error-correcting codes, Sudan also received the Nevanlinna Prize in 2002.

The weak version of the PCP theorem that we prove in Section 11.3, where the proof is exponentially long, appears as an intermediate step in [56]. The linearity test is due to Blum, Luby, and Rubinfeld [115], and the Fourier-analytic analysis we give here is from Bellare et al. [94]. Our presentation benefited from lecture notes by Rafael Pass.

Theorem 11.4, in which the verifier just checks the parity of three bits, was proved by Johan Håstad [378]. This shows that it is NP-hard to approximate MAX-3-XORSAT within a factor greater than 1/2, or MAX-3-SAT within a ratio greater than 7/8. As we discussed in Section 9.3, these are the strongest results possible, since we can achieve these ratios simply by choosing a random assignment. Håstad received the 2011 Gödel Prize for this version of the PCP theorem, and the optimal inapproximability results that follow from it. The reader can find a proof of Theorem 11.4 in Arora and Barak [53, Chapter 22].

The proof of the PCP Theorem using gap amplification was found by Irit Dinur [244] in 2006. Our sketch follows the approach of Radhakrishnan and Sudan [678], which contains several simplifications.

11.5 Really random numbers. Some webpages offer random bits generated using a radioactive source and a Geiger counter (www.fourmilab.ch/hotbits) or radios that pick up atmospheric noise (random.org). Each of these generates on the order of 10^3 random bits per second.

In 2001, the RAND Corporation re-released their 1995 classic, *A Million Random Digits with 100,000 Normal Deviates.* One reviewer on Amazon said "Such a terrific reference work! But with so many terrific random digits, it's a shame they didn't sort them, to make it easier to find the one you're looking for."

11.6 Pseudorandom generators. The Blum–Micali generator, in which we iterate a one-way permutation and reveal a hard-core bit of each iterate, appeared in Blum and Micali [116]. In particular, they proposed iterating modular exponentiation, and proved that the resulting generator fools all polynomial-time algorithms as long as DISCRETE LOG is hard. Blum, Blum, and Shub [113] suggested using the Rabin function $f(x) = x^2 \bmod N$ (see Note 11.2).

A more complicated construction of a pseudorandom generator using any one-way function, whether or not it is one-to-one, was given by Håstad, Impagliazzo, Levin, and Luby [379]. Since a pseudorandom generator is itself a one-way function, this shows that pseudorandom generators exist if and only if one-way functions do.

The "hybrid argument" of Theorem 11.14, in which we change one bit at a time from pseudorandom to random, was given by Yao [840]. It is often stated in terms of a *next-bit test*: a pseudorandom generator is good if and only if there is no order in which we can predict its next bit from all its previous ones.

The Leftover Hash Lemma of Problem 11.2 is from Impagliazzo, Levin, and Luby [418] and Impagliazzo and Zuckerman [420]. The latter paper explored its uses in amplifying BPP algorithms with fewer random bits as in Problem 11.3. We will see another method of saving random bits, by taking a random walk in an expander graph, in Section 12.9.

The pseudorandom function generator of Problem 11.17 is due to Goldreich, Goldwasser and Micali [339]. The particular version we give here appears in Razborov and Rudich [686] as part of their result on natural proofs (see Section 6.5).

11.7 The Nisan–Wigderson generator and BPP = P. Nisan and Wigderson defined their generator in [640]. Impagliazzo and Wigderson [419] proved the stronger result that BPP = P if there is some function in $\mathsf{TIME}(2^{O(n)})$ which no nonuniform algorithm can compute in $2^{o(n)}$ time and advice.

As a partial converse, Kabanets and Impagliazzo [449] showed that if RP = P, or even if polynomial identity testing à la Schwartz–Zippel can be derandomized, then either NEXPTIME $\not\subseteq$ P/poly or calculating the permanent of a matrix requires arithmetic circuits of superpolynomial size. Thus there is a two-way connection between derandomization and hardness: we can derandomize BPP by proving that certain functions are hard, or the other way around.

11.8 Possible worlds. Impagliazzo gives a tour of his five possible worlds in [417].

11.9 Nisan's generator. Nisan's technique for fooling space-bounded computation, described in Problem 11.18, appeared in [635]. While the original paper is quite readable, we are grateful to Ryan O'Donnell's lecture notes.

If we want to use Nisan's generator to solve problems in BPL deterministically, doing this naively would take $n^{O(\log n)}$ time to scan all possible values of the $O(\log^2 n)$ random bits describing the seed r_1 and the hash functions. However, since almost all hash functions allow us to randomize just the initial seed $r_1 \in \{0,1\}^\ell$, we can reduce this to poly(n) time by searching for "good" hash functions, and then ranging over just the $2^\ell = \text{poly}(n)$ possible values of r_1. As a result, BPL is contained in "Steve's Class" SC of problems solvable in polylog(n) space and poly(n) time [636].

Chapter 12

Random Walks and Rapid Mixing

> O! Immodest mortal! Your destiny is the joy of watching the ever-shifting battle.
>
> Ludwig Boltzmann

In most of this book, we have asked how hard it is to find a solution to a problem, or tell whether one exists: a satisfying assignment for a formula, a coloring of a graph, or a way to cover a region with tiles of a certain shape. But what if our goal is not just to find a solution, but to generate *random* ones?

When the space of possible states or solutions is exponentially large, and is too complicated to be grasped in its entirety, random sampling is often the only technique we know of that allows us to learn about its structure. And the best method of random sampling is often to perform a random walk, also known as a *Markov chain.* Starting with an initial state, we take a series of steps, each of which changes the state in some small way. We continue until the probability distribution spreads out throughout the entire space and the state becomes random.

Algorithms like these first appeared in statistical physics, and are used in computational experiments every day. But they also have important applications in computer science. Most importantly, as we will see in Chapter 13, they let us approximate some quantities that are extremely hard to compute exactly, such as the permanent of a matrix.

The number of steps that it takes for a Markov chain to approach equilibrium, and thus provide a good random sample of the state space, is called its *mixing time.* As we will see in this chapter, calculating the mixing time of a Markov chain requires us to think about how quickly its choices overwhelm the system's memory of its initial state, how much one part of a system influences another, and how smoothly probability flows from one part of the state space to another. To grapple with these issues, we will find ourselves applying a panoply of mathematical ideas, from combinatorics, probability, group theory, and Fourier analysis.

We begin by considering a classic example from physics: a block of iron. 12.1

12.1 A Random Walk in Physics

If I put a block of iron next to a magnet, it will become a magnet itself and retain its magnetic field long afterward. It can even magnetize itself spontaneously without any external help. But in 1895, Pierre Curie found that iron's ability to retain a magnetic field decreases as its temperature increases, and that there is a critical temperature T_c at which it loses this ability completely.

This is a classic example of a *phase transition*, in which the macroscopic properties of a physical system undergo a sudden, qualitative change when some parameter passes a critical value. In this section we will describe a simple mathematical model that reproduces this behavior, and use it as our first example of a system where we would like to sample random states.

12.1.1 The Ising Model

We can explain Curie's observations, at least qualitatively, using a toy model of magnetism called the *Ising model.* Suppose we have a square lattice where each site i has a spin $s_i = +1$ or -1, representing an atom whose magnetic field is pointed up or down. Each spin interacts with its neighbors, giving the system an overall energy

$$E = -\sum_{ij} s_i s_j\,, \tag{12.1}$$

where the sum is over pairs of sites i, j that are nearest neighbors. In physics, systems generally try to minimize their energy, like a ball rolling downhill. Since each term $-s_i s_j$ is minimized if s_i and s_j are both up or both down, this is a *ferromagnetic* model where neighboring sites would prefer to have the same spin. If we want to study the *antiferromagnetic* Ising model, where neighbors prefer to be different from each other, we just change the $-$ in front of the sum to a $+$.

If we just want to minimize E, we can point the spins in the same direction, all up or all down. But a system is not always in its lowest energy state—depending on the temperature, its energy is sometimes higher. According to the *Boltzmann distribution*, the equilibrium probability $P_{\mathrm{eq}}(s)$ that a system is in a given state s decreases exponentially as a function of its energy $E(s)$, where the severity of this decrease depends on the temperature T:

12.2

$$P_{\mathrm{eq}}(s) \propto \mathrm{e}^{-\beta E(s)} \text{ where } \beta = 1/T\,. \tag{12.2}$$

As T approaches absolute zero, $\beta \to \infty$ and $P_{\mathrm{eq}}(s)$ becomes zero for all but the lowest energy states. At the other extreme, in the limit $T \to \infty$, we have $\beta \to 0$ and all states are equally likely. If you are wondering why $P_{\mathrm{eq}}(s)$ should depend on the energy in this particular way, we applaud your curiosity, and invite you to peruse problems 12.2 and 12.3.

Let us call a state *magnetized* if almost all its spins point in the same direction, and *unmagnetized* if there are a roughly equal number of up and down spins. All else being equal, an unmagnetized state has higher energy than a magnetized one, since many neighboring pairs of sites have opposite spin. On the other hand, there are many more ways for a state to be unmagnetized than for it to be magnetized, since there are $\binom{n}{u}$ states where u spins are up and $n-u$ are down, and this binomial coefficient is sharply peaked at $u = n/2$. Thus while a given unmagnetized state is exponentially less likely than a magnetized one, the number of unmagnetized states is exponentially greater, so the *total* probability of being unmagnetized might be greater. Which of these effects wins out?

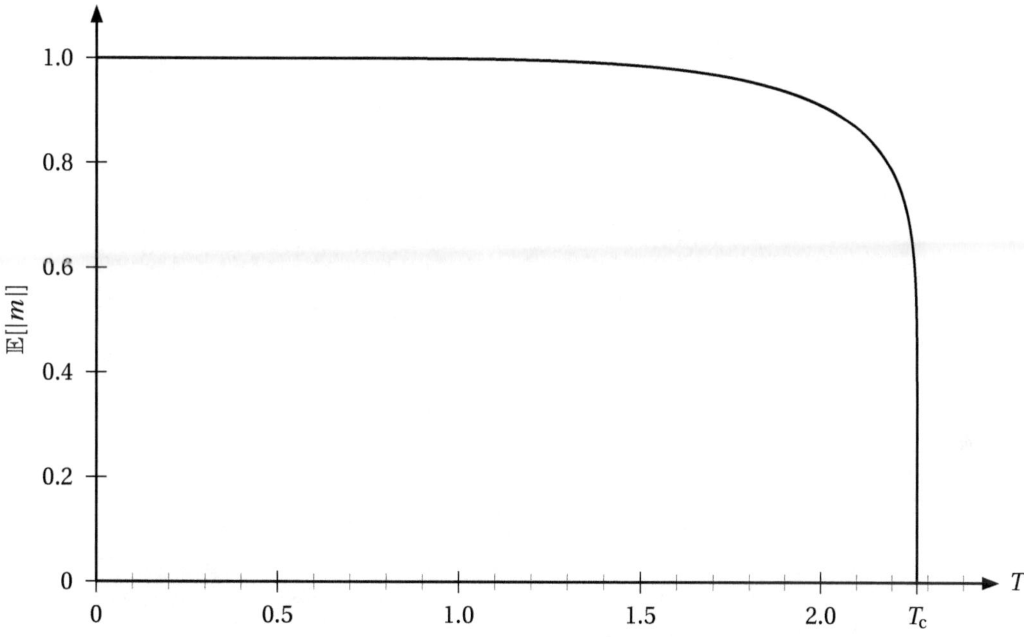

FIGURE 12.1: The typical magnetization of the two-dimensional Ising model in the limit $n \to \infty$. It drops to zero at $T_c = 2.269\ldots$

Let's lump states with the same energy together into *macrostates*. Then the total probability of being in a macrostate with energy E is proportional to

$$We^{-\beta E} = e^{S-\beta E} = e^{-\beta(E-TS)},$$

where W is the number of states in that macrostate. Its logarithm $S = \ln W$ is called the entropy. The likeliest macrostate is then the one that minimizes $E - TS$, a quantity that physicists call the *free energy*.

This creates a tug-of-war between energy and entropy, whose outcome depends on the temperature. 12.3
When T is small, $E - TS$ is minimized when E is minimized, and the system is magnetized. But when T is large enough, $E - TS$ is minimized by maximizing S. Then entropy triumphs over energy, and the system becomes unmagnetized.

Of course, the previous paragraph is just a cartoon, in which we assumed that magnetization is an all-or-nothing affair. What actually happens is shown in Figure 12.1. If we define the magnetization as the average spin, $m = (1/n)\sum_i s_i$, the expectation of its absolute value decreases continuously as T increases, and hits zero at the critical temperature $T_c = 2.269\ldots$. For the interested reader, Problem 12.4 shows how to derive this result qualitatively, using a simplified *mean field* assumption that ignores the neighborhood relationships of the spins. We will see how to compute T_c exactly in Section 13.7.3.

To get a better sense of how the Ising model behaves, consider Figure 12.2, where we show typical states above, below, and at the phase transition. When $T < T_c$ the world is covered by a sea of spins all pointing up, say, with isolated islands of spins pointing down. The fraction of islands with size s obeys a

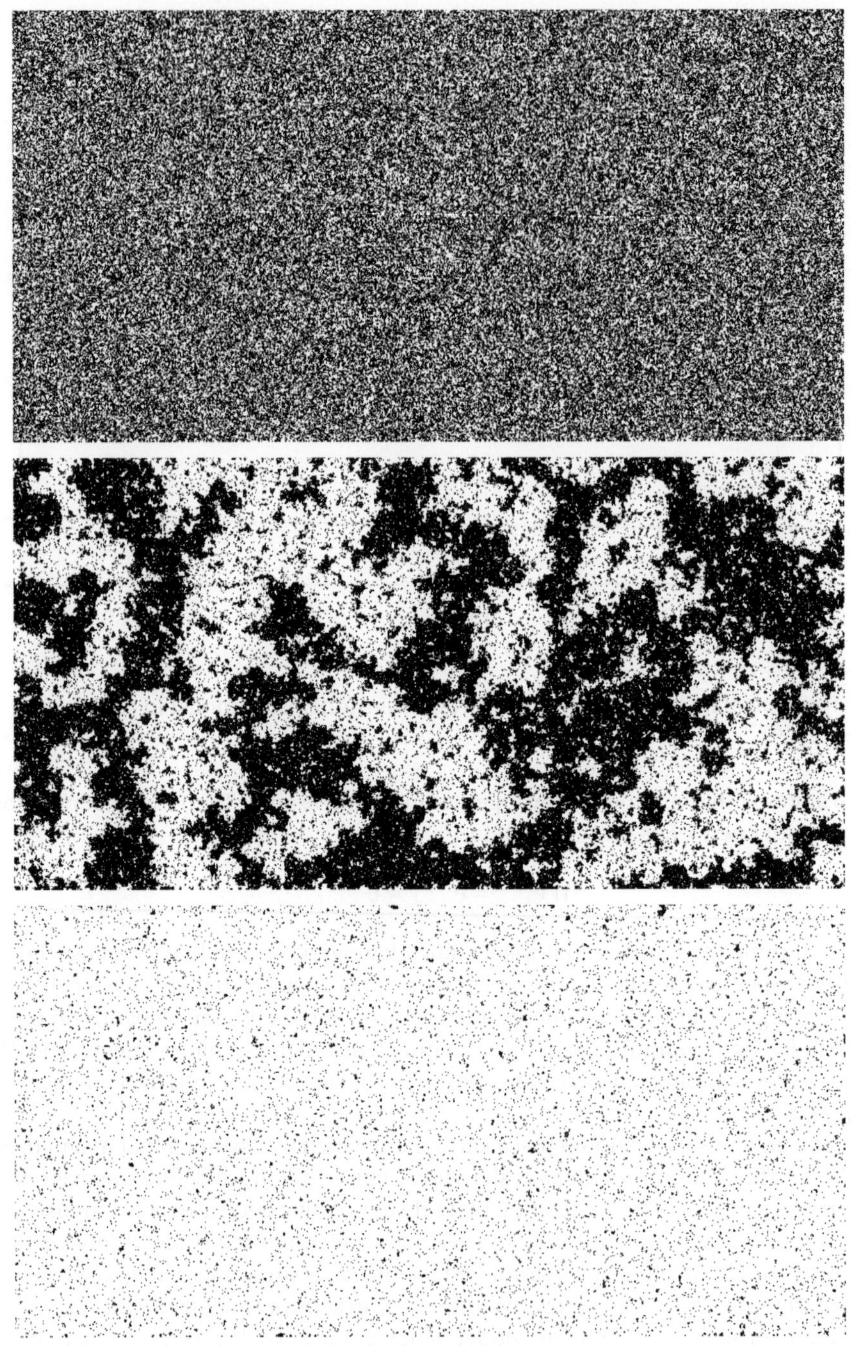

FIGURE 12.2: Typical states of a 1024×512 Ising model sampled at three different temperatures. Below T_c (bottom) there are small islands of the minority spin. Above T_c (top) there are small clumps with the same spin, but at large scales the up and down spins cancel out. At T_c (middle) there are islands and clumps at all scales, and the statistics of the model are scale-free.

distribution with an exponential tail,

$$P(s) \sim s^{-\beta} \mathrm{e}^{-s/s_0} .$$

Their average size, which is essentially s_0, is finite.

As T increases, so does s_0, and these islands grow and join each other. Finally, at the critical temperature, s_0 diverges—the islands stretch across the entire lattice, and we can no longer tell the islands from the sea. The exponential tail disappears, leaving just a power-law distribution:

$$P(s) \sim s^{-\beta} . \tag{12.3}$$

This distribution is especially interesting because it stays the same, up to normalization, if we multiply s by a constant. In other words, if we zoom out and make all the islands half as large, the distribution of their sizes looks exactly the same. More generally, if we take a large picture of the Ising model at T_c and shrink or magnify it, all its statistical properties remain the same, as long as we are too far away to see the individual pixels. This kind of *scale-free* behavior is typical of phase transitions in physics.

Above T_c, there are clumps of sites with the same spin. These clumps have finite average size, just as the islands do below T_c. If two spins are much farther apart than the typical clump size, they are effectively independent, and at large scales half the sites are up and half are down. As the temperature increases these clumps get smaller, until at $T = \infty$ even neighboring spins are completely independent of each other.

12.1.2 *Flipping Spins and Sampling States*

Now suppose that we want to generate a random state of the Ising model according to the Boltzmann distribution. There are many reasons to do this: one is to generate beautiful pictures like those in Figure 12.2, so we can gain some intuition for how the system typically behaves. But a more quantitative goal is to obtain good estimates of the average magnetization, or the correlation between spins a certain distance apart, or the average size of an island. In general, for any physical quantity X, we can estimate its expectation $\mathbb{E}[X]$ by generating a large number of random states, measuring X for each one, and averaging X over these samples. But how can we generate random states so that each one appears with the right probability?

A naive approach would be to use *rejection sampling*. Simply generate a random state by independently setting each spin to $+1$ or -1 with equal probability, calculate the energy E of this state, and then accept it as a sample with probability $P = \mathrm{e}^{-\beta(E-E_{\min})}$ (we subtract $E_{\min}$, the lowest possible energy, so that $P \le 1$). However, for almost all states P is exponentially small, so we would have to generate an exponential number of trial states in order to get a useful set of samples. Our naive trial states are constructed as if the energy doesn't matter, so they have very small probability in the Boltzmann distribution. It would be much better to take the Boltzmann factor into account during the process of constructing each state, rather than just using it to accept or reject them at the end.

Our approach is to start with an initial state—say, with all spins pointing up, or where each spin is chosen randomly—and then perform a random walk in state space, flipping one spin at a time. By defining the probabilities of these flips in the right way, we can guarantee that after taking τ steps, for some sufficiently large τ, the resulting state is effectively random and is chosen according to the Boltzmann distribution. If we take τ steps between samples, then these samples are effectively independent from each other.

The standard way to do this is called *Metropolis dynamics.* At each step, we choose a random site i, and consider what change ΔE in the energy would result if we flipped s_i. Then we flip s_i with the following probability:

$$p(\text{flip}) = \begin{cases} 1 & \text{if } \Delta E < 0 \\ \mathrm{e}^{-\beta \Delta E} & \text{if } \Delta E \geq 0. \end{cases} \tag{12.4}$$

In other words, if flipping s_i would decrease the energy, we go ahead and do it. If it would increase the energy, we do it with a probability given by the ratio between the old and new Boltzmann factors, $\mathrm{e}^{-\beta E_{\text{new}}}/\mathrm{e}^{-\beta E_{\text{old}}} = \mathrm{e}^{-\beta \Delta E}$.

12.4 This is our first example of a *Markov chain*—a process where the probability of moving from one state to another depends only on the current state, and not on the system's previous history. It has an important special property shared by most Markov chains in physics. If we denote the probability that we go from state x to state y as $M(x \to y)$, then (12.4) implies that

$$P_{\text{eq}}(x)M(x \to y) = P_{\text{eq}}(y)M(y \to x).$$

This property is called *detailed balance.* Along with the property of *ergodicity,* which we discuss in the next section, detailed balance ensures that as we continue performing random flips, we converge to the Boltzmann distribution P_{eq}.

Exercise 12.1 *Confirm that the Metropolis rule* (12.4) *satisfies detailed balance. Show that, in fact, there are an infinite number of rules that do, where* $p(\text{flip})$ *depends only on* ΔE. *Is there some sense in which the Metropolis rule is the best possible?*

So Metropolis dynamics converges to the correct equilibrium distribution. But how long does it take? In the next section, we will define the *mixing time* of a Markov chain as the number of steps we need before the probability distribution on the state space is "close enough" to P_{eq} in a certain precise sense.

12.2 The Approach to Equilibrium

In this section, we set up some general machinery. We start with the definition of a Markov chain and show how a toy example approaches equilibrium. We review the basic properties of transition matrices and their eigenvectors and eigenvalues. We then define the mixing time, and what it means for a Markov chain to mix in polynomial time.

12.2.1 Markov Chains, Transition Matrices, and Ergodicity

A *Markov chain* M is a stochastic process with no memory other than its current state. In other words, the probability of being in state y at time $t+1$ depends only on its state x at time t. We can think of a Markov chain as a random walk on a directed graph, where vertices correspond to states and edges correspond to transitions. Each edge $x \to y$ is associated with the probability $M(x \to y)$ of going from state x to state y in a single step.

If this graph is strongly connected, i.e., if for every pair of states x and y there is a path of transitions from x to y with nonzero probability, we call the Markov chain *irreducible.* We call it *aperiodic* if for every state x there is a t such that, for all $t' \geq t$, if we start at x there is a nonzero probability of returning to x in

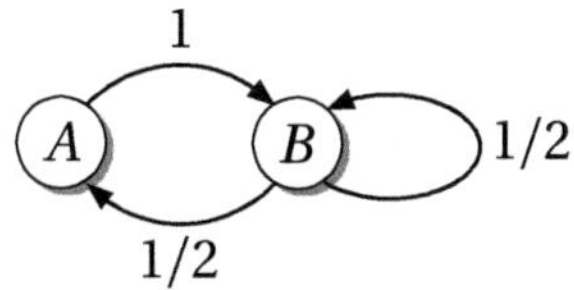

FIGURE 12.3: A Markov chain with two states.

t' steps. Aperiodicity prevents us from cycling periodically between, say, two subsets of states and never settling down. One simple way to make a Markov chain aperiodic is to add "self-loops" in which it stays in the same state with nonzero probability:

Exercise 12.2 *Suppose that a Markov chain is irreducible. Show that if there is any state x with a self-loop, i.e., such that $M_{x,x} > 0$, then it is also aperiodic.*

As we will see, any irreducible, aperiodic Markov chain with a finite number of states will converge to a unique equilibrium probability distribution P_{eq} no matter what initial state it starts in. This property is called *ergodicity*, and all the Markov chains we will consider in this chapter are ergodic.

Exercise 12.3 *Give an example of a Markov chain with an infinite number of states, which is irreducible and aperiodic, but which does not converge to an equilibrium probability distribution.*

As a toy example, consider the Markov chain in Figure 12.3. It has two states, A and B. At each step, if it is in state A, it goes to state B with probability 1, while if it is in state B, it flips a coin and goes to each state with probability 1/2. Let's write the current probability distribution as a column vector

$$P = \begin{pmatrix} P(A) \\ P(B) \end{pmatrix} .$$

When we take a step, the new probability distribution is $P' = MP$ where M is the *transition matrix*, $M_{y,x} = M(x \to y)$:

$$M = \begin{pmatrix} 0 & 1/2 \\ 1 & 1/2 \end{pmatrix} .$$

The total probability is preserved because each column of M sums to 1, and all the entries are nonnegative. Such a matrix is called *stochastic*.

If the initial probability distribution is P_0, the distribution after t steps is

$$P_t = M^t P_0 .$$

If we start in state A, the initial probability distribution is

$$P_0 = \begin{pmatrix} 1 \\ 0 \end{pmatrix}$$

and the distributions P_t we get at $t = 1,2,3,4$ are

$$\begin{pmatrix} 0 \\ 1 \end{pmatrix}, \begin{pmatrix} 1/2 \\ 1/2 \end{pmatrix}, \begin{pmatrix} 1/4 \\ 3/4 \end{pmatrix}, \begin{pmatrix} 3/8 \\ 5/8 \end{pmatrix}, \ldots$$

As t increases, P_t quickly approaches an equilibrium distribution P_{eq} such that

$$M P_{\mathrm{eq}} = P_{\mathrm{eq}} \,.$$

Thus P_{eq} is an eigenvector of M with eigenvalue 1. The only such eigenvector is

$$P_{\mathrm{eq}} = \begin{pmatrix} 1/3 \\ 2/3 \end{pmatrix} .$$

So in the limit $t \to \infty$, we visit state B twice as often as we visit state A.

How quickly do we reach equilibrium? To understand this, we need to know something about M's other eigenvectors. According to the *Perron–Frobenius Theorem*, an ergodic Markov chain with transition matrix M has a unique eigenvector P_{eq} with eigenvalue 1, and all its other eigenvectors have eigenvalues with absolute value less than 1. In this example, M's other eigenvector is

$$v = \begin{pmatrix} 1 \\ -1 \end{pmatrix},$$

and its eigenvalue is $\lambda = -1/2$. We can write the initial probability distribution P_0 as a linear combination of M's two eigenvectors. Again assuming that we start in state A, we have

$$P_0 = \begin{pmatrix} 1 \\ 0 \end{pmatrix} = P_{\mathrm{eq}} + \frac{2}{3} v \,.$$

The distribution after t steps is then

$$P_t = M^t \left(P_{\mathrm{eq}} + \frac{2}{3} v \right) = M^t P_{\mathrm{eq}} + \frac{2}{3} M^t v = P_{\mathrm{eq}} + \frac{2}{3} \lambda^t v \,,$$

The term $(2/3)\lambda^t v$ is the only memory the system retains of its initial state. As t increases, this memory fades away exponentially as $|\lambda|^t = 2^{-t}$, leaving us with the equilibrium distribution P_{eq}.

Exercise 12.4 *Show that if M is stochastic, its eigenvalues obey $|\lambda| \le 1$. Hint: for a vector v, let $\|v\|_{\max}$ denote $\max_i |v_i|$, and show that $\|Mv\|_{\max} \le \|v\|_{\max}$.*

12.2.2 Detailed Balance, Symmetry, and Walks on Graphs

Let's return to the property of detailed balance, which we introduced in Section 12.1. We repeat it here:

$$P_{\mathrm{eq}}(x) M(x \to y) = P_{\mathrm{eq}}(y) M(y \to x) \,. \tag{12.5}$$

As the next exercise shows, this implies that M's equilibrium distribution is P_{eq}.

Exercise 12.5 *Show that if M satisfies* (12.5), *then $MP_{\text{eq}} = P_{\text{eq}}$. Hint: sum* (12.5) *over all x.*

Exercise 12.6 *Check that our toy example satisfies detailed balance.*

It's worth noting that detailed balance is sufficient, but not necessary, for M to have P_{eq} as its equilibrium distribution. For instance, imagine a random walk on a cycle, where we move clockwise with probability 2/3 and counterclockwise with probability 1/3. This Markov chain converges to the uniform distribution, but it violates detailed balance.

One of the most basic Markov chains consists of a random walk on an undirected graph G. At each step, we move from the current vertex x to a uniformly random neighbor y, so that $M(x \to y) = 1/\deg(x)$ if x and y are neighbors and 0 otherwise. The following exercise shows that the equilibrium probability distribution at each vertex is proportional to its degree.

Exercise 12.7 *Show that the random walk on G is ergodic if G is connected and non-bipartite. Show that if G is connected but bipartite, it has a unique eigenvector with eigenvalue -1. Finally, show that it obeys detailed balance, and that its equilibrium distribution is*

$$P_{\text{eq}}(x) = \deg(x)/2m\,,$$

where m is the total number of edges.

We can also perform a "lazy" walk, where we stay at the current vertex with probability 1/2 and move to a random neighbor with probability 1/2.

Exercise 12.8 *Show that the lazy walk on G is ergodic if G is connected, and that it has the same equilibrium distribution P_{eq} as the basic random walk on G.*

Over the course of this chapter, we will look at Markov chains that flip bits, shuffle cards, and color graphs. Most of these Markov chains are *symmetric,* i.e., $M(x \to y) = M(y \to x)$ for all pairs of states x, y. This is a special case of detailed balance, and the equilibrium distribution P_{eq} is uniform. Thus every string of bits, permutation of the cards, or coloring of the graph will be equally likely.

12.2.3 The Total Variation Distance and Mixing Time

Let's formalize how far we are from equilibrium at a given time t. We can define the distance between two probability distributions in many ways, but for our purposes the most useful is the *total variation distance.* The total variation distance between two probability distributions P and Q is

$$\|P - Q\|_{\text{tv}} = \frac{1}{2}\sum_x |P(x) - Q(x)|\,, \tag{12.6}$$

where the sum ranges over all states x.

Exercise 12.9 *Show that the total variation distance between any two probability distributions is at most 1. Under what circumstances is it exactly 1?*

As the following two exercises show, if the total variation distance between two probability distributions is small then they give similar averages for bounded functions, including the probability that the state is in some subset of the state space. In particular, if P_t is close to P_{eq} in this sense, then sampling from P_t gives a good estimate of the equilibrium properties of the system. For instance, in the Ising model we could estimate the average magnetization, or the probability that two nearby sites have the same spin.

Exercise 12.10 *Given a subset E of the state space, define its total probability under a probability distribution P as $P(E) = \sum_{x \in E} P(x)$. Show that for any E and any pair of probability distributions P, Q we have*

$$|P(E) - Q(E)| \leq \|P - Q\|_{\text{tv}} .$$

Thus if the total variation distance between P and Q is small, any experiment that gives "yes" or "no" outcomes says "yes" with almost the same probability in both distributions. Show further that the maximum over all subsets E of this difference is exactly $\|P - Q\|_{\text{tv}}$, i.e.,

$$\max_E |P(E) - Q(E)| = \|P - Q\|_{\text{tv}} . \tag{12.7}$$

What subset E achieves this maximum?

Exercise 12.11 *Let $f(x)$ be a function defined on the state space. Given a probability distribution P, we denote the expected value of f as $\mathbb{E}_P[f] = \sum_x P(x) f(x)$. Show that the difference between the expectations resulting from two different probability distributions is bounded by*

$$\left| \mathbb{E}_P[f] - \mathbb{E}_Q[f] \right| \leq \|P - Q\|_{\text{tv}} \left(\max_x f(x) - \min_x f(x) \right) .$$

Returning to our toy example, since the probability distribution is

$$P_t = P_{\text{eq}} + \frac{2}{3} \lambda^t v = \begin{pmatrix} (1/3) + (2/3)\lambda^t \\ (2/3) - (2/3)\lambda^t \end{pmatrix} ,$$

where $\lambda = -1/2$, the total variation distance between P_t and P_{eq} is

$$\left\| P_t - P_{\text{eq}} \right\|_{\text{tv}} = \frac{2}{3} |\lambda|^t = \frac{2}{3} 2^{-t} .$$

Thus $\left\| P_t - P_{\text{eq}} \right\|_{\text{tv}}$ decreases exponentially as a function of t. How many steps do we need to take if we want the distance from equilibrium to get down to some small ε? Setting $\left\| P_t - P_{\text{eq}} \right\|_{\text{tv}}$ to ε and solving for t, we get

$$t = \log_2 \frac{2}{3\varepsilon} = O(\log \varepsilon^{-1}) .$$

This brings us to the definition of mixing time. Given an $\varepsilon > 0$, the ε*-mixing time* τ_ε is the smallest t such that, no matter what initial distribution P_0 we start in, we end up at most ε away from P_{eq}. That is,

$$\tau_\varepsilon = \min \left\{ t : \max_{P_0} \left\| P_t - P_{\text{eq}} \right\|_{\text{tv}} \leq \varepsilon \right\} .$$

Our goal is to prove upper bounds—and in some cases, lower bounds—on the mixing times of Markov chains we care about.

As our toy example suggests, τ_ε depends rather weakly on ε, with ε^{-1} appearing inside a logarithm. Once we get reasonably close to equilibrium—specifically, once $\left\|P_t - P_{\text{eq}}\right\|_{\text{tv}}$ is bounded below 1/2—the variation distance decreases exponentially, and we can make ε as small as we like with just a little extra time. For instance, Problem 12.10 shows that

$$\tau_\varepsilon \le \tau_{1/4} \log_2 \varepsilon^{-1} .$$

For this reason, we will often take ε to be a constant, and focus on how τ depends on the system size n.

In most cases, the number of states N grows exponentially as a function of n. For instance, in an Ising model with n sites we have $N = 2^n$, and in a graph with n vertices the number of 3-colorings could be as large as $N = 3^n$. To sample these state spaces in a reasonable amount of time, we would like the mixing time to be polynomial as a function of n, and therefore only polylogarithmic in N. We call this happy state of affairs *polynomial mixing*. If the mixing time scales as $\tau = O(n \log n)$, which as we will see is nearly ideal, we say that the Markov chain is *rapidly mixing*.

12.5

12.3 Equilibrium Indicators

Imagine that you are running a Markov chain, trying to produce a random state of some system. Wouldn't it be wonderful if there were an Equilibrium Indicator on the side of your computer, such that as soon as it turned green, you knew that P_t and P_{eq} were *exactly* equal, so that the current state is perfectly random? For some Markov chains, we really can define such an indicator, and prove that it will turn green with high probability within a reasonable amount of time.

What would this mean for the variation distance? If the indicator is green, we know that $P_t = P_{\text{eq}}$. If it is still red, let's say that P_t is some other distribution Q_t that we know nothing about. Then we can write

$$P_t = \Pr[\text{the indicator is green}]\, P_{\text{eq}} + \Pr[\text{the indicator is still red}]\, Q_t . \tag{12.8}$$

It's then a simple exercise to show that

$$\left\|P_t - P_{\text{eq}}\right\|_{\text{tv}} \le \Pr[\text{the indicator is still red}] . \tag{12.9}$$

Exercise 12.12 *Prove* (12.9).

Therefore, the mixing time τ_ε is at most the time it takes for the indicator to turn green with probability $1-\varepsilon$. For some Markov chains, we can use this approach to prove that they mix rapidly.

12.3.1 *Walking on the Hypercube*

Let's consider a simple sampling problem. How can we generate a random string x of n bits? Of course, we can do this by flipping n coins, and setting each bit $x_1, x_2, \ldots, x_n$ to 0 or 1 with equal probability. But we want to learn about mixing times and equilibrium indicators, so we will use a Markov chain instead—namely, a random walk on the n-dimensional hypercube.

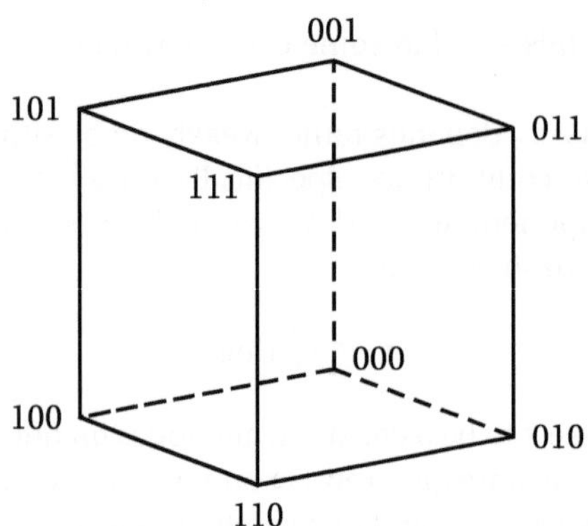

FIGURE 12.4: The 3-dimensional hypercube.

As Figure 12.4 shows, the n-dimensional hypercube has 2^n vertices, corresponding to the 2^n strings of length n. Each one has n neighbors, corresponding to the strings that differ from it at a single bit. Just as in the Ising model where we flipped one spin at a time, our Markov chain will walk along these edges and flip one bit at a time.

We will take a "lazy" random walk on this graph as defined in Section 12.2.2. At each step, with probability $1/2$ we stay where we are, and with probability $1/2$ we choose randomly from the n neighbors of our current state and move there. In terms of strings, with probability $1/2$ we do nothing, and with probability $1/2$ we choose i randomly from $\{1,\ldots,n\}$ and flip x_i.

Because we leave x_i alone or flip it with equal probability, its new value is equally likely to be 0 or 1 regardless of its previous value. So an equivalent way to describe this walk is the following: choose i randomly from $\{1,\ldots,n\}$, choose z randomly from $\{0,1\}$, and set x_i to z. This means that each bit of x becomes random whenever we touch it. Thus we can program our indicator to turn green as soon as we have have touched all the bits—that is, as soon as we have chosen all n possible values of i at least once.

This process is called the *Coupon Collector's Problem*, and we discuss it in detail in Appendix A.3.4. If each day's newspaper brings us a random coupon and there are n kinds of coupons, how many days does it take to collect at least one of each kind?

By Markov's inequality (see Appendix A.3.2) the probability that our collection is incomplete is at most the expected number of coupons still missing from it. Since the probability that we don't get a particular coupon on a particular day is $1-1/n$, the probability that it is still missing after t days is $(1-1/n)^t$. Since there are n coupons, the expected number of missing ones is $n(1-1/n)^t$. Putting all this together gives

$$\begin{aligned}\Pr[\text{the indicator is still red}] &\le \mathbb{E}[\text{\# of untouched bits}] \\ &\le n(1-1/n)^t \\ &\le n\mathrm{e}^{-t/n}, \end{aligned} \tag{12.10}$$

where we used the inequality $1-x \le \mathrm{e}^{-x}$.

Setting (12.10) equal to ε gives an upper bound on the mixing time,

$$\tau_\varepsilon < n\ln(n\varepsilon^{-1}).$$

As we show in Problem 12.37, this is an overestimate, but only by a factor of 2. Taking ε to be some constant gives

$$\tau = O(n \log n),$$

so this Markov chain mixes rapidly.

Walking around on a hypercube may seem like a very artificial thing to do, but it has a nice physical interpretation. Consider the Ising model with n sites. Its state space is precisely the n-dimensional hypercube, where x_i is 0 or 1 depending on whether the ith spin is up or down. At infinite temperature, all 2^n possible states have the same probability, the n sites are completely independent from each other, and the Metropolis dynamics discussed in Section 12.1 becomes the walk on the hypercube. This chain mixes rapidly since it takes $\Theta(n \log n)$ time to touch, and randomize, every site. As we will discuss in Section 12.10, this is true whenever the system is above its critical temperature T_c, so that sites a large distance from each other are nearly independent. Below T_c, in contrast, even distant sites are strongly correlated, and the mixing time becomes exponentially large.

12.3.2 Riffle Shuffles

Another Markov chain, which is in daily use throughout the globe—or at least at many points on its surface—is shuffling a pack of cards. We have n cards, and we want to achieve a nearly uniform distribution on the set of $n!$ possible permutations. For an ordinary deck $n = 52$ and $n! \approx 10^{68}$, but we want to understand how to shuffle decks with any value of n.

In a *riffle shuffle*, we divide the deck into two roughly equal halves and then interleave them in a random way. How many riffle shuffles do we need to get close to the uniform distribution? There is more than one plausible mathematical model of the riffle shuffle, but here we discuss a simple model of how to shuffle backwards. For each card, we flip a coin, and label the card "heads" or "tails" with equal probability. We then move all the cards labeled "heads" to the top half of the deck, leaving those labeled "tails" at the bottom, while preserving the relative order of the cards in both halves. We show an example in Figure 12.5.

Let's suppose that we write these labels on the cards, giving each card a string of Hs and Ts, with the most recent label on the left. Then, for instance, a card whose label starts with HTH is in the top half of the deck now, was in the bottom half on the previous step, the top half the step before that, and so on.

Since we preserve the relative order of the cards within each half, their relative positions must be consistent with alphabetical order. For instance, a card whose string starts with HH must be above one whose string starts with HT. If every card has a different string of labels, the permutation of the deck is completely determined—and since the labels consist of random coin flips, this permutation is uniformly random. Thus we can program our equilibrium indicator to turn green as soon as each card has a unique string of labels.

The question is how large t needs to be so that, if we choose n random strings of length t, no two strings are the same. To put it differently, if we choose n times from the set of 2^t possible strings, what is the probability that some string gets chosen twice?

This is a case of the *Birthday Problem*, which we analyze in Appendix A.3.3. If there are n people in a room and each person's birthday is chosen independently and uniformly from the y days in the year, what is the probability that two people have the same birthday? This is at most the expected number of such pairs. The number of potential pairs is $\binom{n}{2}$ and a given pair of people have the same birthday with

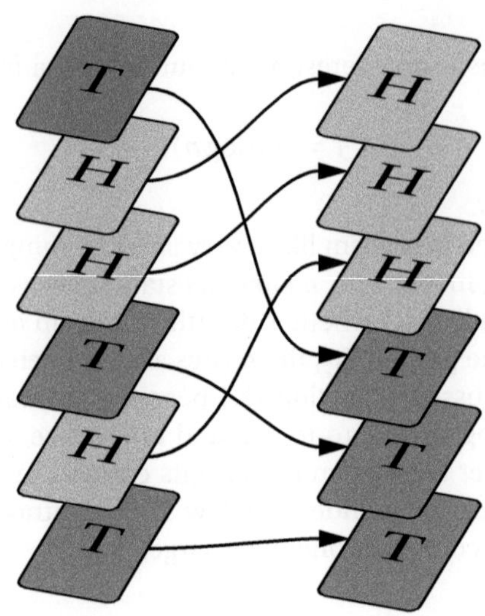

FIGURE 12.5: The riffle shuffle in reverse.

probability $1/y$, so

$$\Pr[\text{some pair has the same birthday}] \le \frac{1}{y}\binom{n}{2}.$$

Here y is the number 2^t of different strings of length t, so

$$\Pr[\text{the indicator is still red}] \le 2^{-t}\binom{n}{2} < 2^{-t}\, n^2.$$

Setting this probability to ε gives the following upper bound on the mixing time,

$$\tau_\varepsilon < \log_2(n^2\varepsilon^{-1}) = 2\log_2 n + \log_2 \varepsilon^{-1}.$$

Once again, this turns out to be correct within a constant factor. The right constant in front of $\log_2 n$ is $3/2$, which in the case $n = 52$ suggests that we should shuffle a deck of cards about 9 times.

12.6 Note that a single riffle shuffle moves all n cards. If we measure the mixing time in terms of single-card moves, we get $O(n \log n)$, the time it would take a coupon collector to touch every card at least once. In Section 12.4 and Problem 12.13 we will see that if we shuffle a deck by moving just one or two cards at a time, the mixing time is $\Theta(n \log n)$ just as the coupon collecting argument would suggest.

12.4 Coupling

While the examples of the previous section are very nice, most Markov chains do not have a simple condition that we can use to turn an equilibrium indicator on. In this section we discuss another technique, which allows us to bound the mixing times of many important Markov chains.

The idea is to run two copies of the Markov chain in parallel, starting with different initial states x_0 and x_0'. If at some time t their states x_t and x_t' are the same, we can conclude that they have "forgotten" which of these two states they started in. If this is true for all pairs x_0, x'_0, the Markov chain has forgotten its initial state completely. This means that it has reached equilibrium, and that its state is perfectly random.

If we simply ran two copies of the Markov chain independently, it would take a very long time for them to bump into each other in state space. We need to couple them together in a clever way, so that their states will be driven towards each other. A *coupling* is a way to run two copies of a Markov chain M at the same time, with the following properties:

1. If I am only allowed to see one copy, I see it evolving according to M.

2. If the two copies are in the same state, they always will be.

Thus each copy undergoes a random walk, but these two walks are correlated in such a way that if they ever coalesce to the same state x_t, they will remain together forever. The game is then to design our couplings so that they will probably coalesce within a small amount of time.

This sounds mysterious at first, so let's consider a simple example. Suppose two people are walking on the n-dimensional hypercube, and that they are currently at vertices x and x'. As in the previous section, we choose i randomly from $\{1,\ldots,n\}$ and z randomly from $\{0,1\}$. But both people use the same values of i and z, setting $x_i = z$ and $x'_i = z$. Even though each person, viewed in isolation, just took a random step, x and x' now agree on their ith bit, and will forever after.

What does coupling have to do with mixing? As you run your Markov chain, you are free to daydream that your state is one of two copies in a coupling, where the other copy is already in a completely random state. Then if the two copies have coalesced, your state is random too. Analogous to (12.9), the total variation distance from equilibrium is bounded by the probability that this hasn't happened, maximized over all pairs of initial states:

$$\left\|P_t - P_{\mathrm{eq}}\right\|_{\mathrm{tv}} \le \max_{x_0,x'_0} \Pr[x_t \ne x'_t]. \tag{12.11}$$

We give a proof of this statement in Problem 12.9.

For the hypercube, the pairs x_0, x'_0 that maximize (12.11) are complementary. They differ on every bit, so they don't coalesce until every bit has been touched—that is, until every $i \in \{1,\ldots,n\}$ has been chosen at least once. But this is the Coupon Collector's Problem again. The right-hand side of (12.11) is the probability that there is still an untouched bit, or uncollected coupon, and as in (12.10) this is at most $n\mathrm{e}^{-t/n}$. Setting this to ε gives the same bound $\tau_\varepsilon < n\ln(n\varepsilon^{-1})$ on the mixing time that we had before.

For a more general bound on the mixing time, define the *coalescence time* T_{coal} as the maximum, over all pairs of initial states x_0, x'_0, of the expected time it will take the two copies to coalesce. Problem 12.11 asks you to prove the following:

Theorem 12.1 *Suppose M is a Markov chain for which a coupling exists with coalescence time T_{coal}. Then the mixing time of M is bounded above by*

$$\tau_\varepsilon \le \mathrm{e}\, T_{\mathrm{coal}} \ln \varepsilon^{-1}. \tag{12.12}$$

For the hypercube this gives $\tau_\varepsilon \le n \ln n \ln \varepsilon^{-1}$. This is slightly weaker than the bound $n\ln(n\varepsilon^{-1})$, but both are $O(n \log n)$ when ε is fixed.

Like the walk on the hypercube, the simplest couplings consist of doing the "same thing" to both copies. However, defining what we mean by "same" sometimes takes a little thought. Here is another way to shuffle cards: at each step, choose a random number i from $\{1,\ldots,n\}$ and move the card currently in

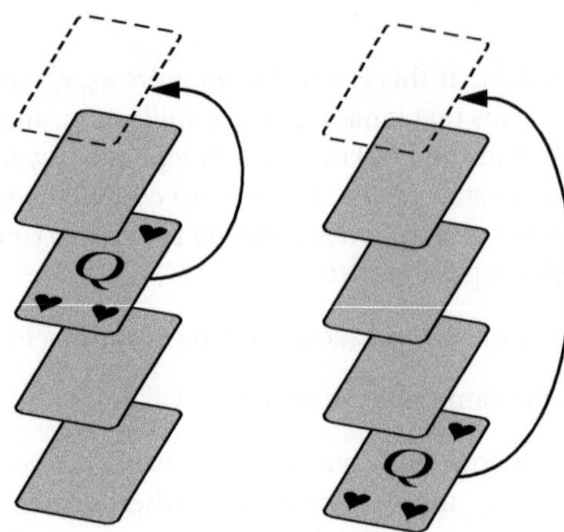

FIGURE 12.6: Moving the same card to the top in both decks.

the ith position to the top of the deck. Your card-playing partners will probably get impatient if you use this method, but let's analyze it anyway.

Now suppose we have two decks of cards, and we want to design a coupling between them. One way to do this is to use the same value of i and move, say, the 17th card in both decks to the top. But while this fits the definition of a coupling, it is completely useless to us—if the two decks start out in different permutations, they will never coalesce.

A much better approach, as the reader has probably already realized, is to choose the card by name rather than by its position, as shown in Figure 12.6. Then we move, say, the Queen of Hearts to the top in both decks. As the following exercise shows the two decks have coalesced as soon as all n cards have been chosen:

Exercise 12.13 *Show that any card that has ever been chosen is in the same position in both decks.*

Once again we are collecting coupons, and the mixing time is $O(n \log n)$.

In Problem 12.12, we consider a slightly fancier type of shuffle, in which at each step we choose two random cards in the deck and switch them. Before you look at that problem, you might want to think about what doing the "same thing" could mean, and how to design a coupling in which the two decks coalesce. The lesson here is that a single Markov chain has many different couplings, and to bound its mixing time we need to choose the right one.

Coupling may seem like a nonintuitive way to prove that a Markov chain mixes quickly. After all, a coupling is a way to drive two trajectories together, when what we want to prove is that these trajectories fly apart and spread throughout the state space. But a good coupling shows that, after a short time, the system's state is completely determined by the random moves of the Markov chain, as opposed to its initial state. Once the initial state no longer matters—once we have lost all memory of where we started—we have reached equilibrium.

Next we will see how designing clever couplings, and analyzing them in clever ways, helps us prove rapid mixing for a natural sampling problem: choosing a random k-coloring of a graph.

12.5 Coloring a Graph, Randomly

In this section, we describe one of the longest and most exciting tales in the field of Markov chains. This tale has grown in the telling, and breakthroughs are still being made every few years. Moreover, it illustrates some of the key techniques we know of for designing couplings and proving rapid mixing.

Suppose we have a graph G. We want to color its vertices with k colors, such that no neighboring vertices have the same color. This is the classic GRAPH k-COLORING problem, which we showed in Chapter 5 is NP-complete for any $k \geq 3$. But now we want to sample a *random* k-coloring from the uniform distribution, where all k-colorings are equally likely. In case you're wondering why we need Markov chains to do this, consider the following exercise:

Exercise 12.14 *The following simple algorithm generates a random coloring for $k > D$: as long as there are uncolored vertices, choose a random uncolored vertex and give it a color chosen uniformly from its available colors, i.e., one that isn't already taken by any of its neighbors. Show that this algorithm does not, in fact, generate colorings according to the uniform distribution.*

Hint: consider a chain of three vertices and let $k = 3$. Show that according to the uniform distribution, the probability that the two ends have the same color is $1/2$, whereas this naive algorithm produces such colorings with probability $4/9$.

Most work on this question has focused on a class of Markov chains known collectively as *Glauber dynamics.* Like the single-spin-flip dynamics we described in Section 12.1 for the Ising model, each move
changes the color of just one vertex. In one version, analogous to Metropolis dynamics, we choose a 12.7
random vertex v and a random color $c \in \{1, \ldots, k\}$. If c is one of v's available colors, we change v's color to c. Otherwise, we do nothing.

A slightly different Markov chain chooses a random vertex v and then chooses c uniformly from v's set of available colors. Both of these Markov chains are symmetric, so for both of them the equilibrium distribution is uniform on the set of k-colorings. They differ only in that the self-loop probability, i.e., the probability of doing nothing, is greater in Metropolis dynamics since some colors are unavailable. However, the Metropolis version will make our coupling arguments easier, so we will stick to it for now.

Of course, none of this makes sense unless G has at least one k-coloring, and unless any two colorings can be connected by a series of single-vertex moves. Let's assume that G has maximum degree D. We showed in Problem 4.4 that G is k-colorable for any $k \geq D+1$. However, as we ask you to show in Problems 12.16 and 12.17, in general we need $k \geq D+2$ for Glauber dynamics to be ergodic.

It is strongly believed that $D+2$ is also a sufficient number of colors for Glauber dynamics to mix rapidly, as the following conjecture states.

Conjecture 12.2 *For any family of graphs with n vertices and maximum degree D, if $k \geq D+2$ then Glauber dynamics mixes in time $O(n \log n)$.*

A mixing time of $O(n \log n)$ is as rapid as it possibly could be. Except in a few special cases, in order to get a random coloring we need to touch almost every vertex, and by the Coupon Collector's Problem this takes $\Theta(n \log n)$ time. Thus the heart of Conjecture 12.2 is our belief that we only have to touch each vertex a
constant number of times to get a nearly random coloring. 12.8

At the time we write this, this conjecture remains open. However, a long string of tantalizing results have gotten us closer and closer to it. We will prove rapid mixing for $k > 4D$, then $k > 3D$, and then $k > 2D$.

We will then sketch some recent ideas that have taken us down to $(1+\varepsilon)D$, within striking distance of the conjecture.

12.5.1 Coupled Colorings

We start with a proof of rapid mixing whenever $k > 4D$. Imagine that we have two colorings C, C' of the same graph. The simplest coupling would be to try the same move in both colorings: that is, to choose v and c, and try to set both $C(v)$ and $C'(v)$ to c. If c is available to v in both colorings, this move succeeds in both, and they will then agree at v. However, if c is available in one coloring but not in the other, this move only succeeds in one of the two colorings, and afterward they will disagree at v even if they agreed there before. Our task is to show that any pair of colorings are more likely to get closer together than farther apart, so that they will quickly coalesce.

Let d denote the Hamming distance between C and C', i.e., the number of vertices at which they disagree. We will call a move *good*, *bad*, or *neutral* if it causes d to decrease, increase, or stay the same. Note that any good or bad move changes the Hamming distance by exactly 1. If we denote the probability that a randomly chosen move is good or bad as p_{good} or p_{bad} respectively, the expected change in d at each step is

$$\mathbb{E}[\Delta d] = -p_{\text{good}} + p_{\text{bad}}\,. \tag{12.13}$$

There are a total of nk possible moves, one for each combination of v and c. A move is good if $C(v) \neq C'(v)$ and c is available to v in both colorings. In the worst case, v's neighbors take D distinct colors in C and another D distinct colors in C'. This leaves $k-2D$ colors that are available in both colorings, and the move is good if c is any of these. Thus for each of the d vertices v where $C(v) \neq C'(v)$ there are at least $k-2D$ good moves, and

$$p_{\text{good}} \geq \frac{d(k-2D)}{nk}\,.$$

On the other hand, a move is bad if (1) $C(v) = C'(v)$ but v has a neighbor w where $C(w) \neq C'(w)$, and (2) $c = C(w)$ or $c = C'(w)$ but none of v's other neighbors take c. In that case, c is available to v in one coloring but not the other. Thus each vertex w where $C(w) \neq C'(w)$ causes at most 2 colors to be bad for each of its neighbors v. There are d such vertices and each one has at most D neighbors, so the number of bad moves is at most $2dD$. Thus

$$p_{\text{bad}} \leq \frac{2dD}{nk}\,.$$

Then (12.13) becomes

$$\mathbb{E}[\Delta d] \leq -\left(\frac{k-4D}{nk}\right) d = -\frac{A}{n}\, d\,,$$

where

$$A = \frac{k-4D}{k}\,.$$

If $k > 4D$ then $A > 0$, and the expectation of d decreases by a factor of $(1-A/n)$ at each step. Since d's initial value is at most n, after t steps this gives

$$\mathbb{E}[d] \leq n\left(1-\frac{A}{n}\right)^t \leq n\mathrm{e}^{-At/n}\,.$$

The probability C and C' have not coalesced is exactly the probability that d is still nonzero, and by Markov's inequality (see Appendix A.3) this is at most $\mathbb{E}[d]$. Thus by (12.11) we have

$$\left\|P_t - P_{\text{eq}}\right\|_{\text{tv}} \leq \Pr[d > 0] \leq \mathbb{E}[d] \leq n\mathrm{e}^{-At/n},$$

and setting this to ε gives

$$\tau_\varepsilon \leq \frac{1}{A}\, n \ln(n\varepsilon^{-1}).$$

Fixing ε gives $\tau = O(n \log n)$, proving rapid mixing whenever $k > 4D$.

12.5.2 Shortening the Chain

In the previous section, we analyzed the effect of a coupling on arbitrary pairs of colorings, and proved that Glauber dynamics mixes rapidly whenever $k > 4D$. In this section, we prove rapid mixing for a smaller number of colors, $k > 3D$. We use the same coupling as before, but analyze it in a different way.

The idea is to focus our attention on *neighboring* pairs of colorings, i.e., pairs that differ at only one vertex as in Figure 12.7. We can think of an arbitrary pair of colorings as connected by a path of neighboring pairs. If the pairs in this path coalesce, the path gets shorter, until it shrinks to nothing and its endpoints coalesce.

Let's say that C and C' disagree at w and agree everywhere else. As in the previous section, we choose a vertex v and a color c, and attempt to color v with c in both colorings. A move is good if $v = w$ and if c is available in both colorings. But now C and C' agree on w's neighbors, so the total number of distinct colors taken by these neighbors is at most D instead of $2D$. This leaves at least $k - D$ colors available in both C and C', so the probability of a good move becomes

$$p_{\text{good}} \geq \frac{k-D}{nk}.$$

A move is bad if v is one of w's neighbors and c is either $C(w)$ or $C'(w)$, so

$$p_{\text{bad}} \leq \frac{2D}{nk}. \tag{12.14}$$

Thus (12.13) becomes

$$\mathbb{E}[\Delta d] \leq -\left(\frac{k-3D}{nk}\right) = -\frac{A}{n}\, d,$$

where

$$A = \frac{k-3D}{k}, \tag{12.15}$$

and the expected distance decreases whenever $k > 3D$.

Now that we know that two neighboring colorings are attracted to each other when $k > 3D$, what can we say about an arbitrary pair of colorings? Since Glauber dynamics is ergodic, between any pair of proper colorings there is a path of proper colorings as in Figure 12.8, where each step in the path consists of changing the color of a single vertex. Imagine running the coupling on the entire path at once. If the length of each step decreases, the total length of the path decreases as well.

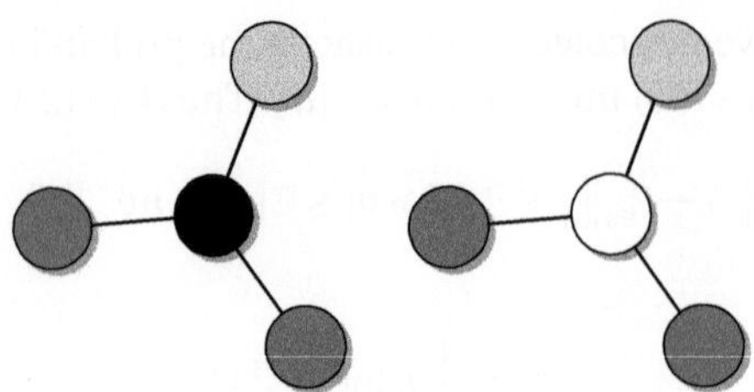

FIGURE 12.7: Two colorings of the same graph, which differ only at one vertex w.

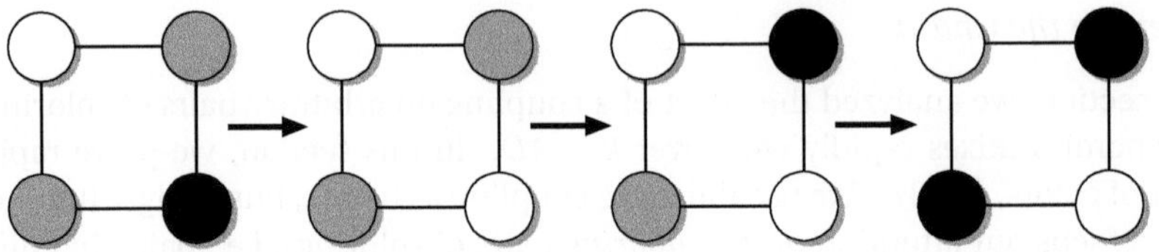

FIGURE 12.8: A path through the space of colorings. Each step in the path changes the color of one vertex.

With this in mind, let's define the distance d between two colorings as the length of the shortest path between them in state space—that is, the shortest sequence of single-vertex moves that connects them. Note that in general this is larger than the Hamming distance, since this sequence might change the color of some vertices multiple times.

For neighboring pairs, where $d = 1$, we have

$$\mathbb{E}[\Delta d] \leq -\frac{A}{n}.$$

Since the length of a path is the sum of the lengths of its steps, by linearity of expectation we have

$$\mathbb{E}[\Delta d] \leq -\frac{A}{n} d.$$

for any pair of states d steps apart. In other words, since each step in the shortest path shrinks, in expectation, by a factor of $1 - A/n$ or better, so does its overall length. As the colorings evolve, a completely different path might become even shorter, but this can only help us. By the triangle inequality, d is at most the length of any path from C to C'.

Let $d_{\max}$ denote the diameter of the state space, i.e., the maximum over all pairs of colorings C, C' of the length of the shortest path between them. Then after t steps we have

$$\mathbb{E}[d] \leq d_{\max} \left(1 - \frac{A}{n}\right)^t \leq d_{\max}\, \mathrm{e}^{-At/n}.$$

As before, the total variation distance is at most the probability that d is nonzero, which by Markov's inequality is at most $\mathbb{E}[d]$. Therefore,

$$\tau_\varepsilon \leq \frac{1}{A} n \ln(d_{\max} \varepsilon^{-1}).$$

Problem 12.16 shows that $d_{\max} = O(n)$. From (12.15) we have $A > 0$, and therefore $\tau = O(n \log n)$, whenever $k > 3D$.

This technique, where we show that a coupling brings all pairs of states closer together by analyzing its effect on neighboring pairs, is called *path coupling*. In this case, we defined two states as "neighbors" if they differ by a single step of the Markov chain. More generally, we can define a weighted graph Γ on state space where each edge (C, C') has a length $d(C, C')$, and the distance between non-neighboring pairs of states is defined as the length of the shortest path between them. Then we have the following theorem:

Theorem 12.3 (Path Coupling) *Let Γ be a weighted connected graph defined on the state space, where no edge has length less than $d_{\min}$. Define $d(C, C')$ as the length of the shortest path in Γ from C to C', and let $d_{\max} = \max_{C,C'} d(C, C')$ denote Γ's diameter. Suppose there is a coupling such that, for some $\delta > 0$,*

$$\mathbb{E}[\Delta d(C, C')] \leq -\delta d(C, C')$$

for all neighboring pairs C, C', i.e., those connected by an edge in Γ. Then the mixing time is bounded by

$$\tau_\varepsilon \leq \frac{\ln(\varepsilon^{-1} d_{\max}/d_{\min})}{\delta}.$$

The proof follows the same lines as our example above, where we had $d_{\min} = 1$ and $\delta = A/n$. Each edge in Γ shrinks by a factor of $1 - \delta \leq \mathrm{e}^{-\delta}$ at each step, and when a path of total length $d_{\max}$ has shrunk to less than $d_{\min}$ its endpoints have coalesced.

Since couplings are often much easier to analyze for neighboring pairs of states than for arbitrary pairs, path coupling is a powerful tool. As we will see next, it sometimes lets us prove rapid mixing in cases where even *defining* the coupling for non-neighboring pairs seems difficult.

12.5.3 A Clever Switch

So far, all our couplings have consisted of doing "the same thing" to both copies of the Markov chain: setting the same bit to the same value, moving the same card to the top of the deck, or trying to give the same vertex the same color. Intuitively, this is the best way to force the expected distance between the two copies to decrease as quickly as possible.

However, there are cases where this intuition is wrong. In this section, we will see that using a cleverer coupling lets us prove rapid mixing for $k > 2D$, a smaller number of colors than the simple coupling we analyzed before. Keep in mind that the Markov chain hasn't changed—we are simply proving a stronger result about it by using a better coupling.

We again consider pairs of colorings C, C' that differ at a single vertex w. As we saw in the previous section, a bad move arises when v is one of w's neighbors and c is either $C(w)$ or $C'(w)$, since—unless this color is taken by one of v's other neighbors—we succeed in changing v's color in one coloring but not the other. Wouldn't it be nice if whenever we tried to set $C(v)$ to $C(w)$, we tried to set $C'(v)$ to $C'(w)$? Then nothing would happen in either coloring, and such moves would be neutral instead of bad.

Let's consider a new coupling. We still choose a vertex v and a color c, and try to set $C(v) = c$. However, we now try to set $C'(v)$ to c' where c' might be different from c. Specifically, suppose v is a neighbor of w.

Then we define c' as follows:

$$c' = \begin{cases} C'(w) & \text{if } c = C(w) \\ C(w) & \text{if } c = C'(w) \\ c & \text{otherwise}. \end{cases}$$

This is a perfectly legal coupling—the pair (v, c'), viewed by itself, is uniformly random, so C and C' each evolve according to the original Markov chain.

In the first case, if $c = C(w)$ and $c' = C'(w)$ then v's color stays the same in both colorings, so the move is neutral. In the second case, if $c = C'(w)$ and $c' = C(w)$ then v's color might change in either or both colorings depending on the colors of its other neighbors, so the move could be bad. In all other cases, including if v and w are not neighbors, $c' = c$ as in the simple coupling of the previous sections, and these moves are neutral.

It follows that there are at most D bad moves, namely those where v is one of w's neighbors, $c = C'(w)$, and $c' = C(w)$. The probability of a bad move is thus half as large as it was in (12.14):

$$p_{\text{bad}} \leq \frac{D}{nk}.$$

As before, we get a good move if we change w to one of its available colors, so

$$p_{\text{good}} \geq \frac{k - D}{nk}.$$

Using path coupling and following our analysis as before gives

$$\tau_\varepsilon \leq \frac{1}{A} n \ln(d_{\max} \varepsilon^{-1}),$$

where now

$$A = \frac{k - 2D}{k}.$$

Thus $\tau = O(n \log n)$ for $k > 2D$.

The alert reader will have noticed something strange about this coupling. We have defined it only for pairs of colorings that differ at a single vertex. Moreover, its definition depends on the vertex w on which they differ, and how w is currently colored in each one. How can we extend this coupling to arbitrary pairs of colorings?

Here path coupling comes to the rescue. No matter how we extend this coupling to non-neighboring pairs, it still causes the shortest paths between colorings to contract, forcing them to coalesce. So, we have no obligation to analyze, or even define, this coupling for non-neighboring pairs.

We need to mention one other technical issue. If we imagine an arbitrary pair of colorings as connected by a path of proper colorings as in the previous section, where each link corresponds to a legal move of Glauber dynamics, we get into trouble for the following reason. Suppose that C and C' are one move apart. After a bad move, we can have, say, $C(w) = 1$ and $C(v) = 2$ while $C'(w) = 2$ and $C'(v) = 1$. These two colorings are now *three* moves away from each other, since we need to change w or v to some third color before we can switch their colors around. To fix this, we define the distance simply as the Hamming distance, so that each move changes it by at most one. As a consequence, we have to allow the path from C to C' to include improper colorings—but a little thought reveals that the coupling works just as well, or even better, if C or C' is improper.

The lesson here is that the best coupling might consist of a complicated correspondence between the moves we perform in the two copies. This correspondence might not even be one-to-one—a move in one copy could be associated with a probability distribution over moves in the other, as long as the total probabilities of each move match those in the original Markov chain. Finding the *optimal* coupling then becomes a LINEAR PROGRAMMING problem, and solving such problems numerically has given some computer-assisted proofs of rapid mixing in cases where simple couplings fail to coalesce.

What happens to this argument if $k = 2D$, or to the argument of the previous section if $k = 3D$? In this case, we have $\mathbb{E}[\Delta d] \leq 0$, so at worst the distance between the two colorings obeys a balanced random walk between 0 and $d_{\max}$. As we discuss in Appendix A.4.4, if we start at $d_{\max}$ and move left or right with equal probability, the expected time for us to reach the origin is $O(d_{\max}^2) = O(n^2)$. Since the probability that our move is good or bad instead of neutral is $\Omega(1/n)$, the distance changes with probability $\Omega(1/n)$ on each step, and the overall mixing time is $O(n^3)$. Thus we can prove in these cases that the mixing time is polynomial. Proving rapid mixing requires other arguments.

12.5.4 *Beyond Worst-Case Couplings*

We have seen a series of coupling arguments that prove rapid mixing for $k > \alpha D$, where the ratio α has fallen from 4, to 3, and then to 2. To prove that Glauber dynamics mixes rapidly for $k \geq D+2$, we need to get α all the way down to $1+1/D$. What kind of argument might make this possible?

One observation is that the coupling arguments we have given here are plagued by worries about the worst case. For instance, in our analysis of a pair of colorings C, C' that differ at a single vertex w, we pessimistically assume that each of w's neighbors takes a different color. This leaves just $k - D$ available colors for w, and therefore only $k-D$ good moves. In essence, we are assuming that C and C' are designed by an adversary, whose goal is to minimize the probability that they will get closer together instead of farther apart.

Let's make a wildly optimistic assumption instead: that the colors of w's neighbors are random and independent. In this case, the expected number of available colors is at least

$$k(1-1/k)^D ,$$

which, if both k and D are large, approaches

$$k\mathrm{e}^{-D/k} .$$

Since the number of bad moves is still at most D, one for each of w's neighbors, the number of good moves would then exceed the number of bad moves when

$$k\mathrm{e}^{-D/k} > D .$$

This happens whenever $k > \alpha D$, where $\alpha = 1.763...$ is the root of

$$\alpha\mathrm{e}^{-1/\alpha} = 1 .$$

As Problem 12.18 shows, we can improve α to 1.489... by using the fact that bad moves are often blocked. If v is a neighbor of w but the colors of v's other neighbors include both $C(w)$ and $C'(w)$, there is no danger we will change v's color in one coloring but not the other.

Of course, even at equilibrium w's neighbors are probably not independent. For instance, if some of them are neighbors of each other, i.e., if w is part of one or more triangles, they are required to take different colors. However, if the graph has large *girth*—that is, if there are no short cycles—then w's neighbors have no short paths connecting them except for those going through w. In this case, we might hope that they interact only weakly, and therefore that their colors are approximately independent.

There are other ways of avoiding the worst case. One is to note that after we have been running the Markov chain for a while, the neighborhood configurations for which the coupling performs poorly might be rather unlikely. Thus if we analyze the mixing process from a "warm start," i.e., an initial state chosen from a distribution which is not too far from equilibrium, the coupling might perform much better and take us the rest of the way to equilibrium.

Another approach, called "coupling with stationarity," uses the fact that we can always assume that one of the two states is random. After all, the whole point of a coupling argument is that if our initial state coalesces with a random one, then it is random too.

The most sophisticated idea so far is a *non-Markovian coupling*, in which we set up a correspondence between entire trajectories of two copies of a Markov chain, rather than a correspondence between single moves. In 2005, this approach was used to prove that Glauber dynamics mixes rapidly when $k > (1+\varepsilon)D$ for any $\varepsilon > 0$, assuming that the girth and maximum degree are large enough. While the proof is quite technical, this gives us some hope that proving Conjecture 12.2 might not be so far away.

12.9 Finally, much more is known for specific graphs such as the square lattice. In Section 12.6.4, we will see a special kind of argument for 3-colorings of the square lattice that shows that Glauber dynamics mixes in polynomial time. Other arguments, including an idea from physics called *spatial mixing* that we discuss in Section 12.10, can be used for $k \geq 6$.

12.6 Burying Ancient History: Coupling from the Past

So far, we have thought of Markov chains as processes that approach equilibrium in the limit of infinite time. We have seen many cases where it doesn't take long to get close to equilibrium. However, unless we have an equilibrium indicator, we never quite get there.

In this section, we will describe a beautiful technique for sampling *exactly* from the equilibrium distribution—in other words, generating perfectly random states in finite time. The idea is to simulate the system backwards. Rather than starting at the present and trying to approach the distant future, we ask how we could have gotten to the present from the distant past. If the system's current state is completely determined by its most recent steps then it has lost all memory of its initial state, and it is in perfect equilibrium.

We will use this approach, called *coupling from the past*, to generate random tilings, random colorings, and random states of the Ising model. As a warm-up, we illustrate the idea of time reversal by showing how to generate two kinds of random objects: spanning trees and pictures of the forest floor.

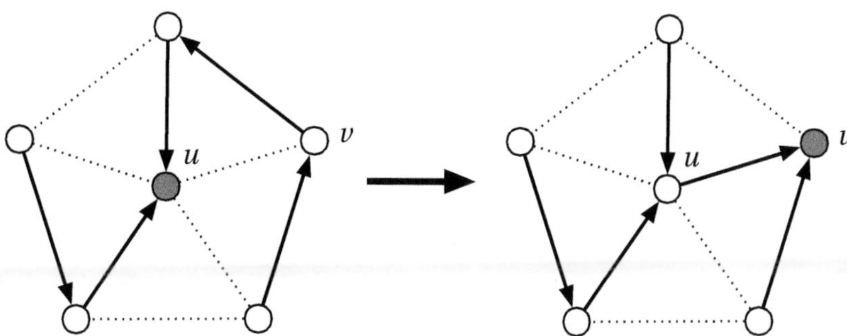

FIGURE 12.9: A step in the arrow-dragging Markov chain for rooted directed spanning trees. When the root (shown in gray) moves from u to v, we add an edge from u to v and remove v's outgoing edge. Thus each vertex w has an edge pointing to where the root went after its most recent visit to w.

12.6.1 Time Reversal, Spanning Trees and Falling Leaves

Suppose I have a graph G with n vertices. Typically, G has an exponential number of different spanning trees. How can I choose one randomly from the uniform distribution, so that each one is equally likely?

In order to generate a random spanning tree, we will actually generate a slightly fancier object: a *rooted directed spanning tree*, where one vertex is designated as the root and the edges are oriented toward it. Thus every vertex other than the root has a single outgoing edge, and all the root's edges are incoming.

Consider the following Markov chain on the set of such trees. At each step, we move the root to one of its neighbors in the graph, chosen uniformly. If its old and new locations are u and v respectively, we draw an arrow from u to v, and erase the arrow leading out of v. We show an example in Figure 12.9.

We call this the "arrow-dragging" process, since the root drags the arrows behind it. The following exercise shows that the tree T remains connected.

Exercise 12.15 *Show that the arrow-dragging process preserves the fact that T is a rooted directed spanning tree. That is, assume that every vertex had a unique directed path to u. Show that after moving the root from u to v, every vertex has a unique directed path to v.*

There are actually two random walks going on here. The root performs a random walk on the vertices of G, while the spanning tree does a random walk in the exponentially large space of all possible spanning trees. What are the equilibrium distributions of these walks? As Exercise 12.7 shows, the probability that the root is at a given vertex v is proportional to v's degree. But how likely are the various trees rooted at v? In Problem 12.22 we show the surprising fact that every spanning tree with a given root v is equally likely. Since any spanning tree can be rooted anywhere, it follows that if we remove the root and the arrows, we are left with a uniformly random spanning tree.

But when do we reach equilibrium? The key thing to notice is that the current spanning tree is a map of the root's most recent visits. Each vertex w has an outgoing arrow dating from the root's last visit to w, and pointing in the direction the root went when it left.

Now suppose that, for some t, the root has visited every vertex of G in the last t steps. Then the tree is determined by these t steps, and it has no memory of the shape it had more than t steps ago. It might

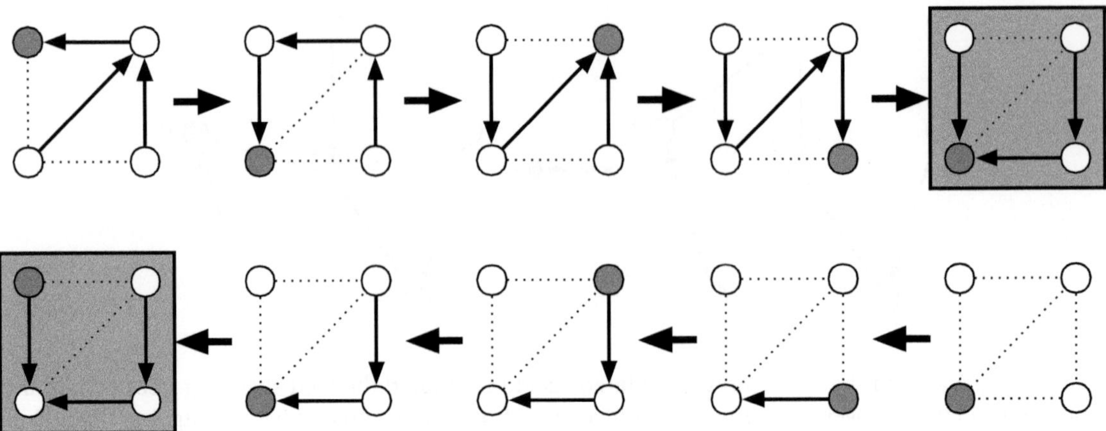

FIGURE 12.10: Sampling spanning trees by reversing time. On the top, we start with an initial spanning tree, and run the arrow-dragging process until it has visited every vertex. At this point, all memory of the initial tree has been erased. On the bottom, the root performs the same walk on the tree, but in reverse, and we add an arrow each time we visit a new vertex. These two processes generate the same uniformly random spanning tree (boxed).

as well have been running forever, with an infinite number of steps stretching back to the beginning of time. We claim that, as a consequence, the current tree is distributed exactly according to the equilibrium distribution.

Another way to see this is to run the system backwards. We start at $t = 0$ at an arbitrary vertex, and then choose the root's position at time $-1, -2$, and so on. As soon as we get to a time $-t$ where the root has visited every vertex, we can ignore all the steps farther back in time. Thus if we take any tree at time $-t$ and reverse the process, running the root back along its trajectory until $t = 0$, the resulting tree is perfectly random. Note that we don't care where the root is at $t = 0$, since every tree rooted there has the same probability.

This time-reversed process has a simple description, where we start with no edges at all and build a spanning tree from scratch. Start at any vertex you like and again follow a random walk on G, moving to a random neighbor at each step. Each time you visit a vertex you have never been to before, draw an edge from this new vertex to the old one. Now the outgoing edge of each vertex w points to where the root came from on its *first* visit to w. See Figure 12.10 for an example.

As soon as we visit the last vertex and cover the entire graph, we have generated exactly the same spanning tree that the arrow-dragging process would have if the root had visited the same sequence of vertices in the opposite order (see Figure 12.10). It follows that this "first visit" process also produces perfectly random spanning trees—a fact which would be rather hard to see without thinking of it as a Markov chain in reverse.

This argument tells us that we will reach perfect equilibrium at a certain point, but it doesn't tell us how long it will take. The expected time for a random walk on G to visit every vertex at least once is called the *cover time* t_{cover}. If G is the complete graph then covering the vertices is essentially coupon collecting, so $t_{\text{cover}} = O(n \log n)$. More generally, Problem 12.26 shows that $t_{\text{cover}} = O(n^3)$ for any graph.

FIGURE 12.11: The "falling leaves" process. Above, a forward simulation where we add leaves on top. Below, a backward simulation where we add leaves from below. As soon as the square is covered, we know that there might as well be an infinite number of leaves hidden under the ones we see, so the arrangement of leaves within this square is perfectly random.

Thus the expected time it takes for the first-visit process to generate a perfectly random spanning tree is polynomial in n. 12.10

The distinction between forward and backward processes can be quite subtle. We might naively expect to get perfectly uniform samples by running a Markov chain forwards, and stopping as soon as its history is hidden by the last t steps. However, the samples we get this way are typically far from uniform, as the following example shows.

It is autumn, and a chill is in the air. Every now and then, a leaf falls at a random location on the forest floor. We wish to sample randomly from all arrangements of leaves we could see in, say, the unit square, as shown in Figure 12.11. A forward simulation adds leaves on top, and approaches equilibrium in the limit of infinite time.

A backward simulation, on the other hand, adds leaves on the bottom, so that new leaves peek through the spaces left by old ones. As soon as the unit square is covered, we might as well be looking at the top of an infinite pile of leaves. Flipping this metaphor over, think of yourself as a small animal in a burrow, looking up at the pile from beneath. As soon as the sky is covered, you might as well be beneath an infinite pile of leaves, hibernating happily in perfect equilibrium.

The backward simulation comes with a simple equilibrium indicator—we can stop as soon as the square is covered. However, if we try to use the same indicator for the forward process, adding leaves on top and stopping the first time the square is covered, we don't get a perfect sample—the arrangements of the leaves appear with the wrong probability distribution. If you don't believe this, we invite you to try the following exercise.

Exercise 12.16 *Consider domino-shaped leaves falling on a 2×2 square. In a complete cover of the square, the number of visible leaves is either two or three. Up to symmetry, this gives us two types of arrangement, which we call A and B:*

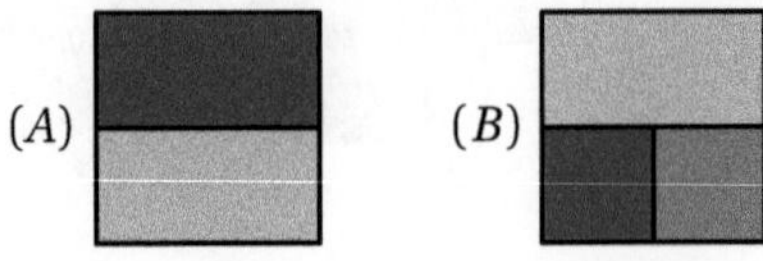

Consider the possible transitions between these two types of arrangement as we add random leaves on top. Show that at equilibrium, the total probability of a type-A arrangement is $1/3$. Show that this is also the case if we run the backward simulation, adding leaves from beneath, and stopping as soon as the square is covered. On the other hand, show that if we run the forward simulation and stop as soon as the square is covered, we get type-A arrangements with the wrong probability, $2/3$ instead of $1/3$.

12.6.2 From Time Reversal to Coupling from the Past

How can we turn the time-reversal picture of the previous section into a general algorithm? We can think of each move of a Markov chain as choosing a function f from a family $\mathscr{F}$ of possible moves, and then applying f to the current state. For instance, in the walk on the hypercube, $\mathscr{F}$ consists of $2n$ functions $f_{i,z}$, one for each combination of $i \in \{1,\ldots,n\}$ and $z \in \{0,1\}$:

$$f_{i,z}(x) = x' \text{ where } x'_j = \begin{cases} z & \text{if } j = i \\ x_j & \text{if } j \neq i\,. \end{cases}$$

In the simplest kind of coupling argument, we choose f from $\mathscr{F}$ and apply it to both copies, hoping that it brings their states closer together.

A backward simulation of the Markov chain then works as follows. We choose a series of functions f_t for $t = -1, -2$, and so on. After doing this for t steps, we ask whether these choices completely determine the current state. In other words, if we define the composition

$$\phi_{-t} = f_{-1} \circ f_{-2} \circ f_{-3} \circ \cdots \circ f_{-t}\,, \tag{12.16}$$

then we ask whether the current state

$$x_0 = \phi_{-t}(x) = f_{-1}(f_{-2}(f_{-3}(\cdots x)))$$

still depends at all on the initial state x. If it doesn't, then all memory of the initial state is lost in the depths of time. We might as well have been running the Markov chain forever, and the current state x_0 is a perfect sample from the equilibrium distribution.

Let's be a bit more formal about this. If we could, we would sample randomly from all infinite sequences of choices of f_t, going all the way back to $t = -\infty$. But for many Markov chains, we can define the family $\mathscr{F}$ of functions so that, with probability 1, the state x_0 resulting from this infinite sequence only depends on a finite subsequence of choices at the end. So if we build the sequence of moves backwards, and stop as soon as this subsequence determines x_0, then x_0 is perfectly random.

```
Coupling from the past
output: a perfectly random state x_0
begin
   φ := the identity function ;
   while x_0 = φ(x) depends on x do
      choose f randomly from 𝓕 ;
      φ := φ ∘ f ;
   end
   return x_0 ;
end
```

FIGURE 12.12: This algorithm, if we could run it, would generate perfectly random states. But how do we know that all initial states x have coalesced into a single $x_0 = \phi(x)$?

To turn this into an algorithm, note that

$$\phi_{-t} = \phi_{-(t-1)} \circ f_{-t} .$$

So, we start with the identity function ϕ_0, and compose ϕ with a randomly chosen function f at each step. Since we are going backwards in time, we compose ϕ with f on the right, so that we apply f to the state before the rest of ϕ. We halt when ϕ is a constant function, so that $x_0 = \phi(x)$ no longer depends on x. This gives the pseudocode shown in Figure 12.12.

However, this is not yet a feasible algorithm. In the coupling arguments of Section 12.4 above, we focused on just two states, and reasoned about when they coalesce. But for coupling from the past to work, we need to know that *all* initial states have coalesced, so that $\phi(x)$ no longer depends on x.

In general this would require us to run many copies of the simulation, one for each initial state, and confirm that applying the same sequence of choices f_t to each one causes them all to arrive at the same final state. Doing this for exponentially many initial states would be far more trouble than its worth. Happily, as we will see next, it sometimes suffices to keep track of just two initial states—the "highest" and the "lowest"—to tell when the entire state space has coalesced.

12.6.3 *From Above and Below: Surfaces and Random Tilings*

How can we confirm that all initial states have coalesced without simulating the Markov chain on every one? Suppose we could find a pair of "extreme" states, and show in some sense that every other state is caught between them. As these extreme states move toward each other, the entire state space contracts, and when they coalesce so do all the others. In the next few sections, we will use this approach to generate random tilings, 3-colorings of the grid, and states of the Ising model. We will start by showing a lovely correspondence between certain tilings and surfaces in three-dimensional space.

In Granada, Moorish tilings bathe the walls of the Alhambra in geometric beauty. Some distance over the French border, a baker in Aix-en-Provence is packing a hexagonal box of calissons. These are almond-flavored and rhombus-shaped, and she wishes to pack the box randomly.

Having read this book as part of her liberal education, she might use the following Markov chain. First she chooses a random vertex in the current tiling. If it is surrounded by three rhombuses as in Figure 12.13, she flips its neighborhood with probability 1/2. Otherwise, she leaves it alone.

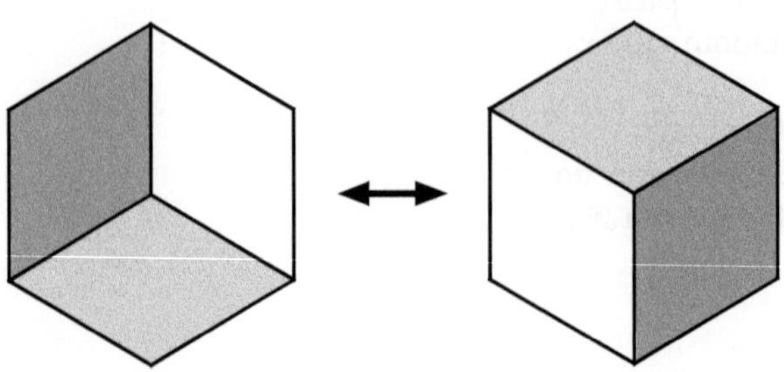

FIGURE 12.13: An elementary flip in a rhombus tiling.

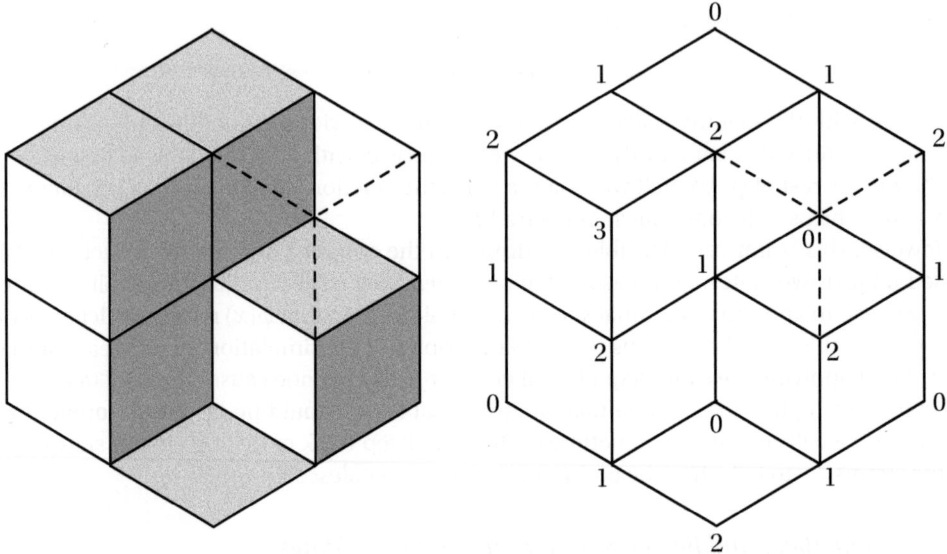

FIGURE 12.14: A rhombus tiling of a hexagon, with the heights of vertices shown on the right. The dotted lines show a place where we can perform a flip, which would change the height at that vertex from 0 to 3.

When one looks at rhombus tilings like the one shown in Figure 12.14, one tends to see a room partly filled with stacks of cubes. Applying a flip then corresponds to filling an empty space with a cube, or removing one. Let's assign a *height* to each vertex of the tiling, where moving along an edge of a tile increases or decreases the height according to whether we move towards or away from the viewer. A place where we can apply a flip is a local minimum or local maximum of the height. The flip increases or decreases the height at that vertex by ± 3, changing it from a local minimum to a local maximum or vice versa, while keeping its neighbor's heights the same.

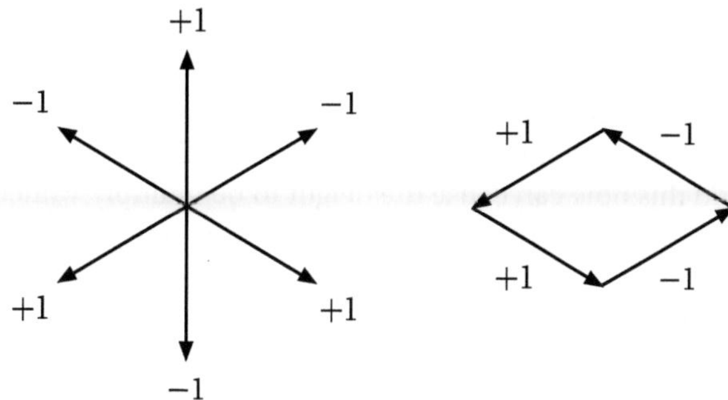

FIGURE 12.15: The proof that the height function is well-defined. Moving along the edge of a tile increases or decreases the height depending on the direction of the edge, and the total change in height around a tile is zero.

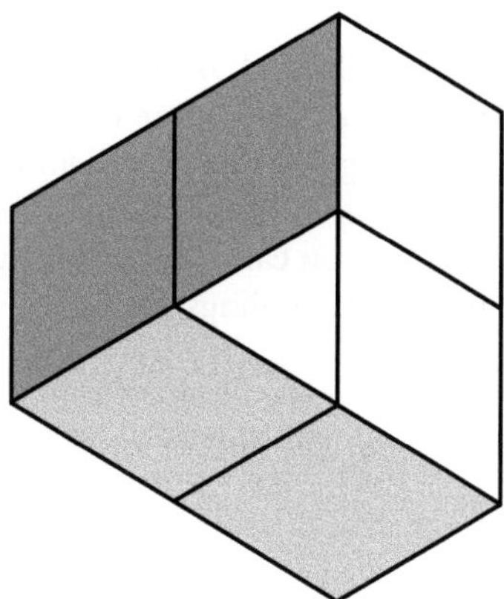

FIGURE 12.16: A tiling of a non-simply-connected region. What happens to the height when we go around the hole?

How do we know that this height is always well-defined? We have to make sure that going around a loop in the tiling always brings us back to the same height we had before, so that the height does not become multiple-valued. As Figure 12.15 shows, the total change in height around any single tile is zero. Moreover, if we stick several tiles together, the height changes along their shared edges cancel, and going around all of them gives a total change of zero. If the region we're tiling is *simply connected*, i.e., if it has no holes, then any loop can be decomposed into tiles, and this proves that the height is well-defined.

On the other hand, if a region has a hole that cannot itself be tiled with rhombuses, such as that shown in Figure 12.16, going around this hole can cause the height to be multiply-valued. Such "topological defects" are reminiscent of Escher's drawing of monks on a cyclic staircase, or Penrose's impossible triangle.

12.11 What does all this have to do with mixing? First of all, if tilings correspond to stacks of cubes, it's intuitively clear that any tiling can be converted to any other by adding and removing cubes. Equivalently, any surface can be converted to any other one by flipping local minima up and local maxima down. The following exercise asks you to turn this intuition into a theorem.

Exercise 12.17 *Prove that any two rhombus tilings of a hexagon (or more generally, any simply-connected region) can be connected by a chain of flips as in Figure 12.13, and therefore that this Markov chain is ergodic. As a corollary, prove that any rhombus tiling of a hexagon whose sides are of equal length has an equal number of rhombuses of the three orientations.*

Now that we have a mapping from tilings to surfaces, how can we use this mapping to tell when all possible initial tilings have coalesced, and thus tell when our tiling is perfectly random? Given a tiling A, let $h_A(x)$ denote the height at a vertex x. Then define the following order on the set of tilings:

$$A \preceq B \text{ if } h_A(x) \leq h_B(x) \text{ for all } x\,.$$

This is a *partial* order, since for some pairs of tilings A, B we have neither $A \preceq B$ nor $B \preceq A$.

Now consider the simplest coupling one could imagine for rhombus tilings. Given a pair of tilings A, B, choose a vertex v and one of the two neighborhood configurations in Figure 12.13, and attempt to change v's neighborhood to that configuration in both A and B. Equivalently, choose v and $\Delta h = \pm 3$, and in both tilings change v's height by Δh if you can. The following exercise asks you to show that this coupling preserves the order relationship between tilings.

Exercise 12.18 *Suppose we have a partial order $\preceq$ between states. Call a coupling* monotonic *if whenever $A_t \preceq B_t$ then $A_{t+1} \preceq B_{t+1}$ as well. Show that the coupling described here for rhombus tilings is monotonic.*

Now consider the two tilings shown in Figure 12.17. The one on the left is minimal in the partial order—its height is everywhere as low as it could possibly be—and the one on the right is maximal. Problem 12.27 shows that such tilings exist for any simply-connected region if it can be tiled at all. If we denote them as $A_{\min}$ and $A_{\max}$, then for any tiling A we have

$$A_{\min} \preceq A \preceq A_{\max}\,. \tag{12.17}$$

Since the coupling is monotonic, (12.17) also holds for the states we get if we start with $A_{\min}$, $A_{\max}$ and A and run forwards in time. Therefore, if $A_{\min}$ and $A_{\max}$ have coalesced, every other tiling A has been

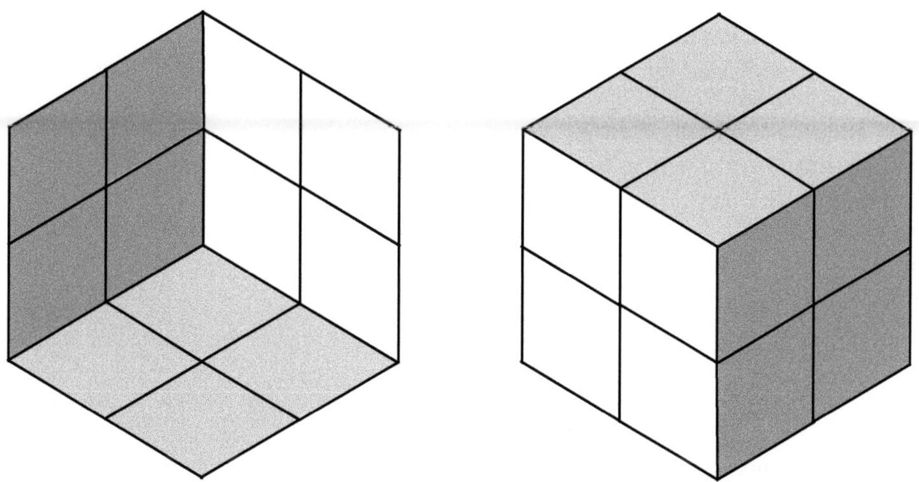

FIGURE 12.17: Minimal and maximal tilings of a hexagon.

```
Coupling from the past (monotonic version)
```
output: a perfectly random state x_0

begin

  $\phi :=$ the identity function ;

  while $\phi(A_{\min}) \neq \phi(A_{\max})$ **do**

    choose f randomly from $\mathscr{F}$;

    $\phi := \phi \circ f$;

  end

  return $x_0 = \phi(A_{\max})$;

end

FIGURE 12.18: Coupling from the past until $A_{\min}$ and $A_{\max}$ coalesce.

```
Propp-Wilson
output: a perfectly random state x_0
begin
    t := 1 ;
    φ := the identity function ;
    while φ(A_min) ≠ φ(A_max) do
        for i = 1 to t do
            choose f randomly from 𝓕 ;
            φ := φ ∘ f ;
        end
        t := 2t ;
    end
    return x_0 = φ(A_max) ;
end
```

FIGURE 12.19: The Propp–Wilson algorithm. Its expected running time is linear in the mixing time τ.

squeezed in between them and must have coalesced as well. Thus we can confirm that all initial states have coalesced by keeping track of what would happen to these two extremes, giving the algorithm shown in Figure 12.18.

Rather than prepending one random move at a time, a more efficient approach is to choose t moves of the Markov chain at once, and run these moves on $A_{\min}$ and $A_{\max}$ to see if they coalesce. If they haven't, we double t to $2t$, prepend another t random steps, and try again. This gives an algorithm whose expected running time is proportional to the mixing time τ. It is called the Propp–Wilson algorithm, and we give its pseudocode in Figure 12.19.

12.12 Running this algorithm on large hexagonal regions produces beautiful results, as shown in Figure 12.20. There are "frozen" regions near the corners, where all the rhombuses have the same orientation, and a "temperate" region in the center where the orientations vary. Amazingly, it can be shown that in the limit of large hexagons, the "arctic circle" separating these regions becomes a perfect circle.

It's important to point out several subtleties here. As in the "falling leaves" process of Exercise 12.16,
12.13 one might think that we can sample random states by running the Markov chain forward in time, and stopping the first time at $A_{\min}$ and $A_{\max}$ coalesce. Can't we use their coalescence as an equilibrium indicator? The problem is that the fact that they coalesced after t steps isn't the only thing we know—we also know that for all $t' < t$, they hadn't yet. This gives us information about the state that we shouldn't have, and Problem 12.28 gives an example where the resulting state is not uniformly random.

Another counterintuitive fact is that, each time we double t, we need to reuse the t moves we chose before as the second half of the trajectory—we can't choose $2t$ moves from scratch. Our goal is to simulate the system backwards in time, and generate a uniformly random sequence of moves f_t for $t = -1, -2$, and so on, until we find a subsequence that completely determines the state. If we choose our moves from scratch instead of reusing them in a consistent way, we unfairly bias these sequences towards those that coalesce quickly, and the resulting state is again nonrandom.

Finally, it's important to note that coupling from the past tells you how to generate random objects, but it doesn't tell you how long it will take. In fact, bounding the mixing time of the Markov chain for

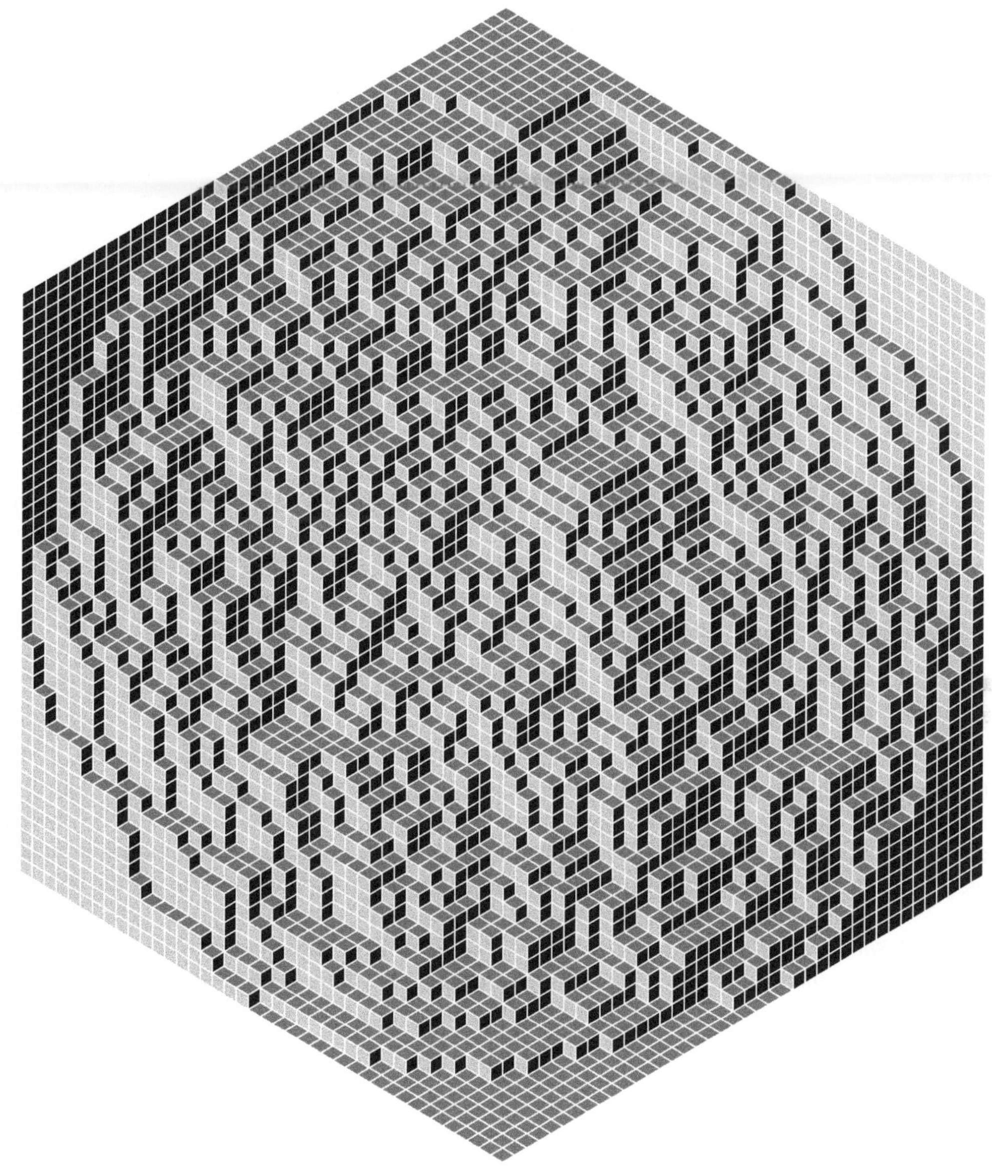

FIGURE 12.20: A random rhombus tiling of a hexagon of size 32, generated by coupling from the past. Note the "arctic circle" phenomenon.

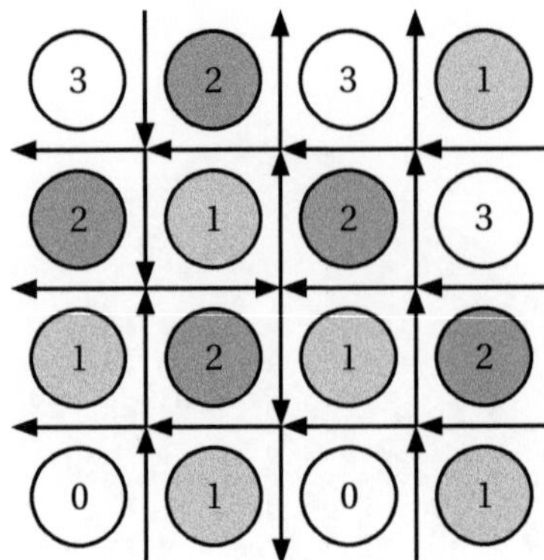

FIGURE 12.21: The height function for 3-colorings of the square lattice, where the three colors (white, light gray, and dark gray) correspond to $h \bmod 3 = 0$, 1, and 2. Shown also is the correspondence between 3-colorings and Eulerian orientations of the dual lattice.

rhombus tilings is quite tricky. As we ask you to show in Problem 12.31, a simple coupling argument doesn't work—there are many pairs of tilings where the number of bad moves exceeds the number of good moves, so naively trying the same move in both tilings tends to drive them further apart. Instead, it's necessary to consider a more complicated Markov chain in which we flip many sites at once, and then relate the mixing time of this new Markov chain to that of the original one.

12.14

12.6.4 Colorings, Ice, and Spins

If coupling from the past only worked for rhombus tilings, it would be merely a mathematical curiosity. In fact, it's possible to define height functions for a surprising number of systems, such as 3-colorings of the square lattice. As Figure 12.21 shows, for any valid 3-coloring we can assign an integer height h to each vertex, where the color 0, 1, or 2 is given by $h \bmod 3$ and where h changes by ± 1 between neighboring vertices. We can then define a partial order, and couple Glauber dynamics in a monotonic way.

Exercise 12.19 *Show that going around an elementary loop of* 4 *vertices always brings us back to the same height, and therefore that the height function is well-defined once the height of any one vertex is fixed. Show that changing the color of a single vertex corresponds to changing a local minimum in the height function to a local maximum or vice versa.*

Given a set of boundary conditions, i.e., a set of colors for the boundary sites of a lattice, we can define a pair of extreme colorings $C_{\min}$ and $C_{\max}$ analogous to the maximal and minimal rhombus tilings of the previous section. Therefore, we can use coupling from the past to generate perfectly random 3-colorings with a given boundary. Once again, the Propp–Wilson algorithm knows when to stop, but doesn't tell us how long it will take. However, a clever coupling argument shows that the time it takes $C_{\min}$ and $C_{\max}$ to coalesce, and therefore the mixing time, is polynomial in the size of the lattice.

12.14 Figure 12.21 shows that 3-colorings are equivalent to another system. By drawing arrows on the edges of the dual lattice so that the height increases whenever we step across an arrow that points to our left,

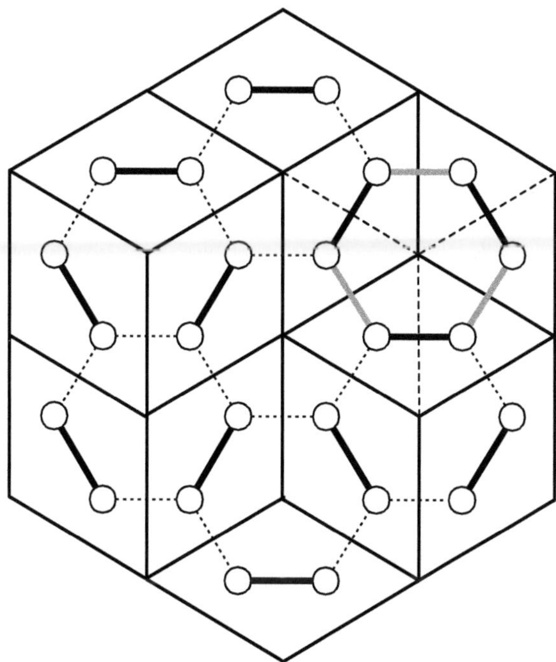

FIGURE 12.22: The correspondence between rhombus tilings and perfect matchings of the hexagonal lattice. A flip of the rhombuses corresponds to adding and removing edges around a hexagonal loop.

we can set up a correspondence between 3-colorings and orientations of the square lattice, where each vertex has two incoming arrows and two outgoing ones. Such orientations are called *Eulerian* since they describe the directions by which an Eulerian tour could enter and leave a vertex. Physicists call this the *6-vertex model.* 12.15

Exercise 12.20 *Prove that this correspondence between 3-colorings of the square lattice and Eulerian orientations works. When we update a 3-coloring according to Glauber dynamics, what happens to the corresponding Eulerian orientation?*

In another important correspondence, some tilings can be thought of as perfect matchings of a planar graph. For instance, rhombus tilings correspond to perfect matchings of a hexagonal lattice as shown in Figure 12.22. Choosing a random tiling corresponds to choosing a random perfect matching—a problem which we will address for general graphs in Chapter 13.

On the square lattice, a perfect matching corresponds to a tiling by one of the most familiar types of tile: dominoes, also known as 1×2 or 2×1 rectangles. It might not be as obvious to the eye as it is for rhombus tilings, but as Problem 12.32 shows we can define a height function here as well, and use it to show that a Markov chain based on a simple flip move is ergodic. In Figure 12.23 we show a random tiling of a region called an *Aztec diamond*, which displays the same type of "arctic circle" as rhombus tilings of the hexagon.

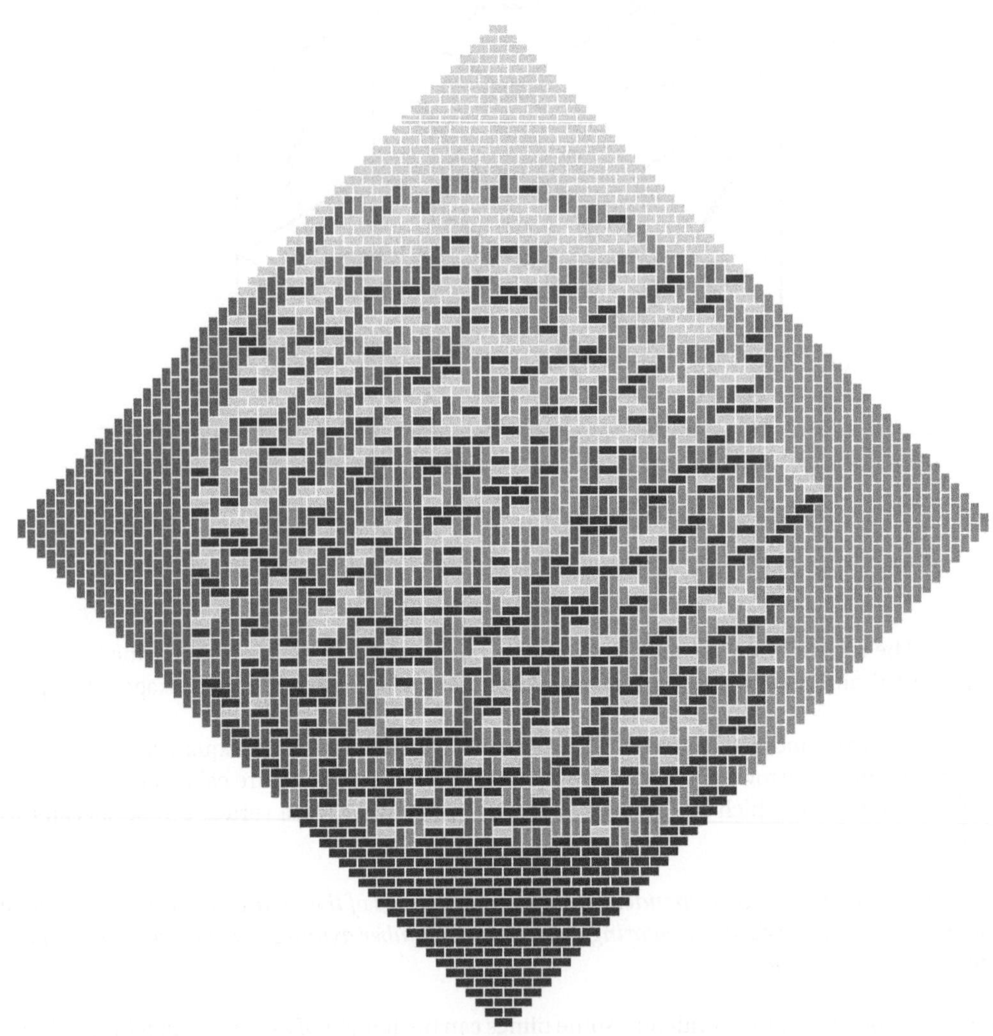

FIGURE 12.23: A random domino tiling of an Aztec diamond of size 50, generated by coupling from the past. The dominoes are colored according to their orientation and parity, so that a "brick wall" of dominoes that are staggered on alternating rows or columns is all one color. As in Figure 12.20, there is an "arctic circle" between the frozen and temperate regions.

12.6.5 The Ising Model, Revisited

We close this section by using coupling from the past to obtain perfect samples of the Ising model. First we need a partial order. Given two states x, y, say that $x \preceq y$ if $x_i \le y_i$ for all i—that is, if every spin which is up in x is also up in y. Then the maximal and minimal states are those where all the spins are all up or down respectively.

Next we need a monotonic coupling. We will use a Markov chain that is slightly different from Metropolis dynamics. At each step, we choose a random site i, and calculate the energies E_+ and E_- that the system would have if s_i were $+1$ or -1 respectively. Then we choose s_i's new value according to the Boltzmann distribution, i.e., with probabilities

$$p(+1) = \frac{e^{-\beta E_+}}{e^{-\beta E_+} + e^{-\beta E_-}} \quad \text{and} \quad p(-1) = \frac{e^{-\beta E_-}}{e^{-\beta E_+} + e^{-\beta E_-}} .$$

Since the only difference between E_+ and E_- comes from the interaction between s_i and its neighbors (see (12.1)) we have $E_+ - E_- = -2\sum_j s_j$ where the sum is over i's four neighbors j. Then we can write

$$p(+1) = \frac{1}{e^{-2\beta \sum_j s_j} + 1} \quad \text{and} \quad p(-1) = 1 - p(+1). \tag{12.18}$$

This is called "heat-bath" dynamics. It lovingly bathes s_i at temperature T until it comes to equilibrium, while holding all the other spins in the lattice fixed.

The usual way one implements a rule like (12.18) is to choose a random number r between 0 and 1. We then set s_i to $+1$ or -1 depending on whether r is less than or greater than $p(+1)$. Now define a coupling as follows: choose a site i, and set its spin according to (12.18), using the same random number r in both states. As the following exercise shows, if $x \preceq y$ then updating site i in this way will ensure that $x_i \le y_i$. Therefore, $x \preceq y$ at all times.

Exercise 12.21 *Prove that this coupling is monotonic. Is this true in the antiferromagnetic case as well?*

By starting with the all-up and all-down states, we can use coupling from the past to generate perfectly random states according to the Boltzmann distribution. How long does this take? Intuitively, the time it takes for the all-up and all-down states to coalesce depends on whether the Ising model is above or below its critical temperature. Above T_c, it is relatively easy to create islands that disagree with their surroundings. Both extreme states would naturally lose their magnetization anyway, so they have no great objection to being brought together. But below T_c, the system wants to stay magnetized—there is a strong "attractive force" towards states that are mostly up or mostly down, and neither one wants to meet the other in the middle.

As we will see in Section 12.10, this intuition is correct. There is a deep connection between the equilibrium properties of a system and the mixing times of local Markov chains, and they both undergo a phase transition at T_c. But first, let's get a deeper understanding of why mixing is fast or slow, and how this is related to the spectral properties of the transition matrix.

12.7 The Spectral Gap

In Section 12.2 we saw that the eigenvalues of a Markov chain's transition matrix play a crucial role in its approach to equilibrium. The equilibrium distribution P_{eq} is an eigenvector with eigenvalue 1, and the system approaches equilibrium as the other eigenvectors disappear. In this section, we will use this picture to derive general bounds on the mixing time, and in some cases to compute it exactly.

12.7.1 Decaying towards Equilibrium

Suppose that a Markov chain has N states, with an $N \times N$ transition matrix M. The upper-case N is meant to remind you that N is typically exponential in the size n of the underlying system—for instance, an Ising model with n sites has $N = 2^n$ possible states. We denote M's eigenvectors and eigenvalues as v_k and λ_k, with equilibrium corresponding to $v_0 = P_{\text{eq}}$ and $\lambda_0 = 1$.

Now suppose that all the other eigenvalues, or rather their absolute values, are less than or equal to $1 - \delta$ for some δ. In other words,

$$\delta = 1 - \max_{k \geq 1} |\lambda_k| .$$

This δ is called the *spectral gap*. It describes how quickly the nonequilibrium part of the probability distribution decays. If δ is large, these other eigenvalues are small, and the system mixes quickly.

We focus for now on the case where the Markov chain is symmetric and the equilibrium distribution is uniform. First let's review a nice property of symmetric matrices:

Exercise 12.22 *Let M be a real symmetric matrix. Show that its eigenvalues are real, and that if u and v are eigenvectors with different eigenvalues then u and v are orthogonal. Hint: consider the product $u^T M v$.*

The following theorem gives an upper bound on the mixing time in terms of the spectral gap. In particular, it guarantees mixing in polynomial time as long as the spectral gap is $1/\text{poly}(n)$. We handle the more general situation where M obeys detailed balance in Problem 12.35.

Theorem 12.4 *Let M be an ergodic, symmetric Markov chain with N states and spectral gap δ. Then its mixing time is bounded above by*

$$\tau_\varepsilon \leq \frac{\ln(N\varepsilon^{-1})}{\delta} . \tag{12.19}$$

Proof As in the toy example of Section 12.2, we start by writing the initial distribution as P_{eq} plus a linear combination of M's other eigenvectors:

$$P_0 = P_{\text{eq}} + P_{\text{neq}} \text{ where } P_{\text{neq}} = \sum_{k=1}^{N-1} a_k v_k , \tag{12.20}$$

where the "neq" in P_{neq} stands for "nonequilibrum." The probability distribution after t steps is

$$P_t = M^t P_0 = P_{\text{eq}} + M^t P_{\text{neq}} = P_{\text{eq}} + \sum_{k=1}^{N-1} a_k \lambda_k^t v_k . \tag{12.21}$$

We can write the total variation distance as

$$\left\|P_t - P_{\text{eq}}\right\|_{\text{tv}} = \frac{1}{2}\left\|M^t P_{\text{neq}}\right\|_1 , \tag{12.22}$$

where the 1-*norm* of a vector v (see Appendix A.2.1) is

$$\|v\|_1 = \sum_x |v(x)| .$$

To avoid an embarrassment of subscripts, we write the component of a vector v at the state x as $v(x)$ instead of v_x.

Our goal is to show that the total variation distance decays as $(1-\delta)^t$. To do this we will relate the 1-norm of $M^t P_{\text{neq}}$ to its 2-*norm*, also known as its Euclidean length:

$$\|v\|_2 = \sqrt{\sum_x |v(x)|^2} .$$

For any N-dimensional vector v, the Cauchy–Schwarz inequality (see Appendix A.2) gives

$$\|v\|_2 \le \|v\|_1 \le \sqrt{N}\,\|v\|_2 .$$

Then (12.22) implies

$$\left\|P_t - P_{\text{eq}}\right\|_{\text{tv}} \le \frac{1}{2}\sqrt{N}\left\|M^t P_{\text{neq}}\right\|_2 . \tag{12.23}$$

Since M is symmetric, by Exercise 12.22 its eigenvectors v_k are orthogonal. Then Pythagoras' Theorem, and $|\lambda_k| \le 1-\delta$ for all $k \ge 1$, gives

$$\begin{aligned}
\left\|M^t P_{\text{neq}}\right\|_2 &= \sqrt{\sum_{k=1}^{N-1} |a_k|^2\,|\lambda_k|^{2t}\,\|v_k\|_2^2} \\
&\le (1-\delta)^t \sqrt{\sum_{k=1}^{N-1} |a_k|^2\,\|v_k\|_2^2} \\
&= (1-\delta)^t \left\|P_{\text{neq}}\right\|_2 \\
&\le (1-\delta)^t .
\end{aligned} \tag{12.24}$$

Combining (12.23) and (12.24) then shows that the total variation distance decreases exponentially,

$$\left\|P_t - P_{\text{eq}}\right\|_{\text{tv}} \le \frac{1}{2}\sqrt{N}(1-\delta)^t \le \frac{1}{2}\sqrt{N}\,\mathrm{e}^{-\delta t} .$$

Finally, setting $\left\|P_t - P_{\text{eq}}\right\|_{\text{tv}} = \varepsilon$ gives an upper bound on the mixing time, $\tau_\varepsilon \le (1/\delta)\ln(\sqrt{N}/2\varepsilon)$. This implies the weaker bound (12.19) stated in the theorem. □

The spectral gap also provides a lower bound on the mixing time. The idea is that the initial probability distribution could differ from P_{eq} by a multiple of the slowest-decaying eigenvector, for which $|\lambda| = 1-\delta$. This gives the following theorem:

Theorem 12.5 *Let M be an ergodic Markov chain that obeys detailed balance and has spectral gap δ. Then for sufficiently small ε, its mixing time is bounded below by*

$$\tau_\varepsilon \geq \frac{\ln(1/2\varepsilon)}{2\delta}. \tag{12.25}$$

Setting ε to some small constant, Theorems 12.4 and 12.5 give

$$\frac{1}{\delta} \lesssim \tau \lesssim \frac{\log N}{\delta}, \tag{12.26}$$

or, in asymptotic notation, $\tau = \Omega(1/\delta)$ and $\tau = O((\log N)/\delta)$.

If the number of states N is exponential in the system size n, the upper and lower bounds of (12.26) differ by a factor of $\log N \sim n$. Thus while (12.26) might not determine the right power of n, it does show that the mixing time is polynomial in n if and only if the spectral gap is polynomially small, $\delta \sim 1/\text{poly}(n)$. Conversely, if the spectral gap is exponentially small, the mixing time is exponentially large.

12.7.2 *Walking on the Cycle*

Let's apply this machinery to a case where we can calculate the spectral gap exactly: the random walk on the cycle of N vertices. At each step, we move clockwise with probability 1/4, counterclockwise with probability 1/4, and stay where we are with probability 1/2. The transition matrix is thus (for $N=5$, say)

$$M = \begin{pmatrix} 1/2 & 1/4 & & & 1/4 \\ 1/4 & 1/2 & 1/4 & & \\ & 1/4 & 1/2 & 1/4 & \\ & & 1/4 & 1/2 & 1/4 \\ 1/4 & & & 1/4 & 1/2 \end{pmatrix}.$$

This is a *circulant* matrix, in which each row is the previous row shifted over by one. To put this algebraically, M commutes with the *shift operator* which rotates the entire cycle clockwise one step,

$$S = \begin{pmatrix} 0 & 1 & & & \\ & 0 & 1 & & \\ & & 0 & 1 & \\ & & & 0 & 1 \\ 1 & & & & 0 \end{pmatrix}.$$

Matrix elements not shown are zero.

Matrices which commute have the same eigenvectors. What are the eigenvectors of S? If v is an eigenvector with eigenvalue λ, it must be proportional to $(1, \lambda, \lambda^2, \ldots)$. But this implies that $\lambda^N = 1$, and therefore that $\lambda = \omega^k$ for some integer k, where $\omega = e^{2\pi\imath/N}$ is the Nth root of unity.

Denoting this eigenvector v_k, we have

$$v_k = \frac{1}{N}\begin{pmatrix} 1 \\ \omega^k \\ \omega^{2k} \\ \vdots \end{pmatrix}.$$

For readers familiar with the Fourier transform, v_k is simply the Fourier basis vector with frequency k. Here we have normalized it so that $\|v_k\|_1 = 1$.

Now that we know the eigenvectors v_k of M, all that remains is to find their eigenvalues. If $\mathbb{1}$ denotes the identity matrix, we can write

$$M = \frac{S}{4} + \frac{S^{-1}}{4} + \frac{\mathbb{1}}{2},$$

where these three terms correspond to moving clockwise, counterclockwise, or neither. Since $S v_k = \omega^k v_k$ we have $M v_k = \lambda_k v_k$ where

$$\lambda_k = \frac{\omega^k}{4} + \frac{\omega^{-k}}{4} + \frac{1}{2} = \frac{1 + \cos\theta_k}{2} \quad \text{where} \quad \theta_k = \frac{2\pi k}{N}.$$

Here we used the formula $\cos\theta = (e^{\imath\theta} + e^{-\imath\theta})/2$.

Setting $k = 0$ gives $\theta_0 = 0$ and $\lambda_0 = 1$, and indeed v_0 is the uniform distribution. So, what is the largest eigenvalue other than 1? Clearly λ_k is closest to 1 when θ_k is closest to zero. In other words, the eigenvectors with the lowest frequency, and the largest wavelength, take the longest to die away. The largest eigenvalue λ_k with $k \neq 0$ is

$$\lambda_1 = \frac{1}{2}\left(1 + \cos\frac{2\pi}{N}\right) = 1 - \frac{\pi^2}{N^2} + O(1/N^4),$$

where we used the Taylor series $\cos\theta = 1 - \theta^2/2 + O(\theta^4)$. Thus the spectral gap $\delta = 1 - \lambda_1$ tends to

$$\delta = \frac{\pi^2}{N^2},$$

and using (12.26) bounds the mixing time above and below as

$$N^2 \lesssim \tau \lesssim N^2 \log N.$$

Which of these bounds is tight? In the proof of Theorem 12.4, specifically, in the second line of (12.24), we took the pessimistic attitude that *every* eigenvalue other than λ_0 is as large as $1 - \delta$. The $\log N$ factor in our upper bound came from waiting for all $N - 1$ of them all to die away.

In many cases, however, there are just a few eigenvectors with eigenvalues which contribute significantly to the total variation distance. The others are considerably smaller and die away much more quickly, leaving just the components of P_{neq} parallel to the slowest eigenvectors. Physically, this corresponds to the fact that high-frequency, short-wavelength fluctuations decay quickly, leaving behind just the low-frequency modes. This is indeed the case for the walk on the cycle, and as you will show in Problem 12.36, the mixing time is $\Theta(N^2)$ rather than $\Theta(N^2 \log N)$.

From the computer science point of view, the walk on the cycle is very special. Because it has a simple periodic structure, we can diagonalize M in the Fourier basis, and write down all its eigenvalues explicitly. For most Markov chains, we cannot hope for this kind of complete analysis. Instead, we have to reason about M's eigenvalues indirectly, by thinking about how quickly probability flows from one part of the state space to another. The next section is devoted to this idea. 12.16

12.8 Flows of Probability: Conductance

The flow of probability in a Markov chain is much like the flow of electricity in a circuit, or water in a network of pipes. If there are many routes from one place to another, probability quickly spreads throughout the state space and settles down to its equilibrium level everywhere. Conversely, if there are "bottlenecks" where one region of the state space is connected to the other only by a few routes, or only by routes with low transition probabilities, it will take a long time for probability to flow from one region to the other. In this section we formalize this intuition, and show how to relate the mixing time of a Markov chain to its ability to conduct probability from one region to another.

12.8.1 Conductance

Consider a pair of states x, y that are adjacent in the state space. The flow of probability along the edge $x \to y$ is the probability of being in x, times the probability of going from x to y. We denote the equilibrium value of this flow by

$$Q(x \to y) = P_{\text{eq}}(x) M(x \to y) .$$

Note that detailed balance (12.5) is exactly the requirement that

$$Q(x \to y) = Q(y \to x) ,$$

so the net flow of probability along each edge is zero at equilibrium.

Now suppose we have a subset S of the set of states. The total flow of probability from S out to the rest of the state space is

$$Q(S \to \overline{S}) = \sum_{x \in S, y \notin S} Q(x \to y) .$$

Therefore, if we condition on being in S, the probability at equilibrium that we escape from S to the rest of the world in a single step is

$$\Phi(S) = \frac{Q(S \to \overline{S})}{P_{\text{eq}}(S)} , \tag{12.27}$$

where $P_{\text{eq}}(S) = \sum_{x \in S} P_{\text{eq}}(x)$. To put this differently, $\Phi(S)$ is the fraction of S's probability that flows out to $\overline{S}$ at each step.

Intuitively, if $\Phi(S)$ is very small, an initial distribution that starts out inside S will stay stuck in it for a long time. Roughly speaking, it will take about $1/\Phi(S)$ steps to escape S and explore the rest of the state space. Conversely, if $\Phi(S)$ is large for all S then the state space is well-connected, with many possible paths leading from everywhere to everywhere else. There are no bottlenecks, and the Markov chain will mix quickly.

The *conductance* Φ of a Markov chain is the probability of escaping from the most inescapable set, i.e., the minimum of $\Phi(S)$ over all subsets S. However, it would be silly to complain about being unable to escape from a set that contains most or all of the state space. So we take this minimum over subsets consisting of half or fewer of the states—or, more generally, containing half or less of the total probability at equilibrium. Thus

$$\Phi = \min_{S : P_{\text{eq}}(S) \le 1/2} \Phi(S) .$$

We will relate the conductance to the spectral gap δ and thus to the mixing time. It's convenient to require that M have nonnegative eigenvalues. As the following exercise shows, we can make the eigenvalues nonnegative by walking lazily, and this doesn't change the spectral gap very much.

Exercise 12.23 *Suppose that a transition matrix M is stochastic, but that some of its eigenvalues may be negative. Consider the "lazy" version of M from Exercise 12.8, which takes a step according to M with probability $1/2$ and stays put with probability $1/2$. Show that the resulting transition matrix is $M_{\text{lazy}} = (\mathbb{1} + M)/2$, that its eigenvalues $\lambda_{\text{lazy}} = (1+\lambda)/2$ are nonnegative, and that $\delta_{\text{lazy}} \geq \delta/2$.*

Then the following theorem gives upper and lower bounds for δ in terms of Φ. For simplicity, we focus again on symmetric chains.

Theorem 12.6 *Let M be an ergodic, symmetric Markov chain with nonnegative eigenvalues. Let Φ be its conductance and δ its spectral gap. Then*

$$\frac{\Phi^2}{2} \leq \delta \leq 2\Phi .$$

We ask you to prove the upper bound $\delta \leq 2\Phi$ in Problem 12.39. The idea is that if there is a bottleneck between some set S and its complement, a deviation from equilibrium that is positive on S and negative on $\overline{S}$ will take a long time to decay. The lower bound $\delta \geq \Phi^2/2$ is a little more complicated, and we relegate its proof to the Notes.

Combining Theorem (12.6) with Theorems 12.4 and 12.5 gives the following upper and lower bounds 12.17
on the mixing time, for fixed ε:

$$\frac{1}{\Phi} \lesssim \tau \lesssim \frac{\log N}{\Phi^2} . \tag{12.28}$$

Since typically $\log N \sim n$ where n is the system size, we see that the mixing time is poly(n) if and only if the conductance is $1/\text{poly}(n)$.

12.8.2 Random Walks on Graphs and UNDIRECTED REACHABILITY

One nice application of this machinery is to show that the mixing time of the random walk on any undirected graph is polynomial in the number of vertices. Recall from Exercise 12.7 that the equilibrium probability distribution of this walk is $P_{\text{eq}}(v) = \deg(v)/2m$, where $\deg(v)$ is v's degree, and m is the total number of edges. Consider the following exercise:

Exercise 12.24 *Let G be a connected undirected graph with n vertices and m edges. Show that the conductance of the lazy walk on G as defined in Exercise 12.8 is bounded by*

$$\Phi \geq \frac{1}{m} \geq 1/n^2 .$$

Conclude from (12.28) *that the spectral gap of the lazy walk on G is $\Omega(1/n^4)$ and that its mixing time is $O(n^4 \log n)$.*

This gives us a nice randomized algorithm for UNDIRECTED REACHABILITY—the question of whether there is a path from s to t in an undirected graph G. If we start at s and take, say, n^5 steps, we will be exponentially close to the equilibrium distribution on the connected component containing s. If t is in that component too, we will be at t with probability $\deg(t)/m \geq 1/n^2$. If we walk for $O(n^3)$ more steps, say, we will bump into t with probability exponentially close to 1.

To put this differently, if you wander randomly in a maze with n locations, without maintaining any memory of where you have been before, you will find the exit with high probability in poly(n) steps. In Problem 12.26 we show that even $O(n^3)$ steps suffice.

This algorithm is very memory-efficient. Suppose we have read-only access to the graph G as in Chapter 8. We only need $O(\log n)$ bits of memory to keep track of our current position, so this algorithm uses $O(\log n)$ memory and runs in poly(n) time. It returns "yes" with probability at least 1/2 if there is a path from s to t, and never returns "yes" if there isn't. This puts UNDIRECTED REACHABILITY in the class RLOGSPACE, the analog of RP for $O(\log n)$ space.

As we will see in Section 12.9.6, there is a clever way to derandomize this algorithm—to solve UNDIRECTED REACHABILITY deterministically with $O(\log n)$ bits of memory, placing it in the class LOGSPACE. To do this, we first convert G to a graph whose conductance, and therefore its spectral gap, are constant rather than $1/\text{poly}(n)$.

Next, we will explore an important technique for proving lower bounds on the conductance—by plotting out how probability can flow from one part of the state space to another.

12.8.3 *Using Flows to Bound the Conductance*

How can we prove that a Markov chain has a large conductance? Let's think of the state space as a directed graph, where each edge $e=(x,y)$ has a capacity equal to the equilibrium probability flow $Q(e)=Q(x\to y)$. Ignoring the denominator $P_{\text{eq}}(S)$ in the escape probability (12.27), finding the conductance is then like a MIN CUT problem. We look for the set S where the set of edges crossing from it to the rest of the graph has the smallest total capacity. The conductance is then proportional to the total capacity of these edges, i.e., the weight of the cut between S and $\overline{S}$.

Now recall the duality between MIN CUT and MAX FLOW we discussed in Chapter 3. The value of the MAX FLOW equals the weight of the MIN CUT. Therefore, if we can exhibit *any* flow, its value provides a lower bound on the weight of the MIN CUT. In the same vein, we can prove a lower bound on the conductance of a Markov chain by designing flows of probability in its state space. However, in this case we need to design not just a flow from one source to one sink, but a *multicommodity* flow—a family of flows from everywhere to everywhere else. Such a family, if it exists, shows that there are many paths in the state space connecting each pair of states, and therefore that the conductance is large.

Suppose that for each pair of states x,y we have a flow $f_{x,y}$ where x is the source and y is the sink. The value of $f_{x,y}$ is 1, corresponding to transporting a unit of probability from x to y. For each edge e, let $f_{x,y}(e)$ denote the current through this edge. If x and y are chosen according to the equilibrium distribution, the average flow through e is

$$F(e)=\sum_{x,y} P_{\text{eq}}(x)\,P_{\text{eq}}(y)\,f_{x,y}(e).$$

We define the *congestion* as the ratio between this average flow and the capacity $Q(e)$, maximized over all the edges:

$$\rho = \max_e \frac{F(e)}{Q(e)}. \tag{12.29}$$

This is the ratio by which we would have to scale our flows down, and hence slow down the flow of probability, for the average flow on every edge to be less than or equal to its capacity.

The following theorem shows that if the congestion isn't too large, the conductance isn't too small:

Theorem 12.7 *Let M be an ergodic Markov chain, and let $f_{x,y}$ be a family of flows with congestion ρ. Then the conductance of M is bounded below by*

$$\Phi \geq \frac{1}{2\rho}.$$

Proof We will actually bound a more symmetric version of the conductance. Define Φ' as follows,

$$\Phi' = \min_S \frac{Q(S \to \overline{S})}{P_{\text{eq}}(S)\, P_{\text{eq}}(\overline{S})}. \tag{12.30}$$

Note that, unlike $\Phi(S) = Q(S \to \overline{S})/P_{\text{eq}}(S)$, we include the probability of both S and its $\overline{S}$ in the denominator. Indeed, at the end of the day this is probably a more natural definition of conductance than Φ is.

For each pair of states x, y, let $f_{x,y}(S \to \overline{S})$ denote the total flow across the edges connecting S to $\overline{S}$. If $x \in S$ and $y \notin S$, then $f_{x,y}$ has to transport a unit of probability across these edges, so

$$f_{x,y}(S \to \overline{S}) \geq 1.$$

If x and y are chosen according to P_{eq}, the average flow $F(S \to \overline{S})$ has to transport probability from the states inside S to those outside S. This gives

$$\begin{aligned} F(S \to \overline{S}) &= \sum_{x,y} P_{\text{eq}}(x)\, P_{\text{eq}}(y)\, f_{x,y}(S \to \overline{S}) \\ &\geq \sum_{x \in S, y \notin S} P_{\text{eq}}(x)\, P_{\text{eq}}(y) \\ &= P_{\text{eq}}(S)\, P_{\text{eq}}(\overline{S}). \end{aligned} \tag{12.31}$$

By the definition of congestion, $F(e)$ is at most ρ times the capacity $Q(e)$ for any edge e, so

$$F(S \to \overline{S}) \leq \rho\, Q(S \to \overline{S}), \tag{12.32}$$

and combining (12.30), (12.31), and (12.32) gives

$$\Phi' \geq \frac{1}{\rho}.$$

We complete the proof by returning to the original definition of conductance, and noting that

$$\Phi \geq \frac{\Phi'}{2}$$

since the definition of Φ only includes sets S such that $P_{\text{eq}}(\overline{S}) \geq 1/2$. □

Combining Theorem 12.7 with (12.28), if the equilibrium distribution is uniform then the mixing time is bounded by

$$\tau = O(\rho^2 \log N). \tag{12.33}$$

Thus we can prove that a Markov chain mixes in polynomial time by defining a family of flows with polynomial congestion.

12.18 Let's again take the walk on the cycle as an example. There are N states, and $P_{\text{eq}} = 1/N$ is uniform. Since we move clockwise or counterclockwise with probability 1/4, the capacity of each directed edge is

$$Q(e) = \frac{1}{4N}.$$

Now suppose that for each pair of states x, y, we send the flow along the shortest path from x to y, so that $f_{x,y}(e) = 1$ if e is on this path and 0 otherwise. If N is even and x and y are diametrically opposite, we send 1/2 of the flow along each of the two paths of length $N/2$. We can think of this as choosing between these two paths with equal probability.

The average flow $F(e)$ is then the probability that e is on the shortest path from a random x to a random y. Since all the edges are equivalent, this is just the average length $N/4$ of a shortest path, divided by the number $2N$ of directed edges. Thus

$$F(e) = \frac{N/4}{2N} = \frac{1}{8},$$

and the congestion is

$$\rho = \frac{F(e)}{Q(e)} = \frac{1/8}{1/4N} = \frac{N}{2}.$$

Theorem 12.7 then gives a lower bound on the conductance of

$$\Phi \geq \frac{1}{2\rho} = \frac{1}{N},$$

and indeed we showed that Φ is exactly $1/N$ in the previous section.

Exercise 12.25 *Suppose that instead of always taking the shortest path from x to y, we send half the flow clockwise around the cycle and the other half counterclockwise. Show that the resulting flow has congestion $\rho = N$, leading to a weaker lower bound on the conductance of $\Phi \geq 1/2N$.*

The random walk on the hypercube is more interesting, since there are many shortest paths between most pairs of states. Suppose we divide the flow equally among them as in Figure 12.24. Then $F(e)$ is then the probability that a given edge e is included in a randomly chosen shortest path between two randomly chosen states. As for the cycle, this is the average length of a shortest path, divided by the total number of directed edges.

The length of the shortest path between two states is their Hamming distance, which on average is $n/2$. There are $n2^n$ directed edges, so

$$F(e) = \frac{n/2}{n2^n} = \frac{2^{-n}}{2}.$$

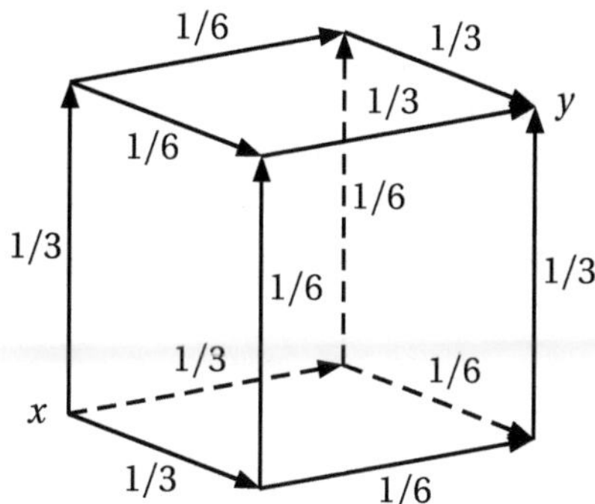

FIGURE 12.24: Dividing the flow on the hypercube equally between all shortest paths. In this case the Hamming distance between x and y is 3.

Since $P_{\mathrm{eq}} = 2^{-n}$ and we move to a given neighbor with probability $1/2n$, the capacity of each edge is

$$Q(e) = \frac{2^{-n}}{2n},$$

and the congestion is

$$\rho = \frac{F(e)}{Q(e)} = n.$$

Theorem 12.7 then gives

$$\Phi \geq \frac{1}{2\rho} = \frac{1}{2n},$$

and indeed Problem 12.40 shows that $\Phi = 1/2n$.

This calculation generalizes easily to any graph G which is sufficiently symmetric. Suppose that G is *edge transitive*, i.e., for every pair of edges e, e' there is an automorphism of G that sends e to e'. We also assume that G is d*-regular*, i.e., that every vertex x has the same degree d. We take the lazy walk where at each step we stay put with probability 1/2, and move to each neighbor with probability $1/(2d)$. The equilibrium distribution is uniform, and every edge has the same capacity $Q(e) = 1/(2dN)$.

Now if we construct a flow $f_{x,y}$ by choosing randomly from among all the shortest paths from x to y, every edge has an average flow $F(e)$ equal to the average length of a shortest path divided by the total number of edges. Let $\mathrm{diam}(G)$ denote G's diameter, i.e., the maximum over all x, y of the length of the shortest path from x to y. Since there are dN directed edges,

$$F(e) \leq \frac{\mathrm{diam}(G)}{dN},$$

and

$$\rho \leq 2\,\mathrm{diam}(G).$$

Thus if the diameter is polynomial in n, so is the mixing time.

Why doesn't a similar approach work for graph coloring, and prove that Glauber dynamics mixes in polynomial time? After all, we showed in Problem 12.16 that the graph of colorings (not to be confused with the graph we're coloring!) has diameter $O(n)$. The problem is that this is not a very symmetric graph. Not all edges are equivalent to each other, and some could act as bottlenecks. In particular, naively choosing all shortest paths with equal probability might force exponentially more flow through some edges than others, causing those edges to be exponentially congested.

12.9 Expanders

Now that we understand the role that the conductance and the spectral gap play in rapid mixing, we can present one of the most important and useful constructions in computer science: graphs that have exponential size but constant degree, on which random walks mix extremely quickly. By rapidly sampling from exponentially large sets, these graphs let us run randomized algorithms with fewer random bits. They are also at the heart of several recent breakthroughs in complexity theory, including Dinur's proof of the PCP theorem (see Section 11.3.8) and the proof that UNDIRECTED REACHABILITY is in LOGSPACE (see Section 8.8).

12.9.1 Expansion and Conductance

Let $G=(V,E)$ be a d-regular graph of size N. Given a subset $S \subseteq V$ of its vertices, let $E(S,\overline{S})$ be the set of edges crossing from S to $\overline{S}$. We say that G is an *expander* if there is some constant $\mu>0$ such that, for all S with $|S| \le N/2$,

$$\frac{\left|E(S,\overline{S})\right|}{d\,|S|} \ge \mu. \tag{12.34}$$

In other words, G has no bottlenecks—every part of G is connected to the rest of it by a large fraction of its edges. We call μ the *expansion*.

Compare this with the definition (12.27) of the conductance Φ. Given a random vertex in S, on average a fraction μ of its d neighbors are outside S. Therefore, if we take a random walk on an expander by moving to a random neighbor on each step, its conductance—the probability that we leave S—is at least μ. Conversely, if the random walk on G has conductance Φ, then G is an expander with $\mu=\Phi$.

Of course, any finite connected graph is an expander for some value of μ. What we really want is an infinite family of graphs G_N such that μ stays constant as their size N increases. Theorem 12.6 tells us this is equivalent to having a constant spectral gap. To make the eigenvalues nonnegative as that theorem demands, we use the lazy walk, whose conductance is $\Phi=\mu/2$. Then we have the following corollary:

Corollary 12.8 *Let $\{G_N\}$ be an infinite family of d-regular graphs where G_N has N vertices. The following three conditions are equivalent:*

- *There is a constant $\mu>0$ such that, for all N, G_N is an expander with expansion at least μ.*
- *There is a constant $\Phi>0$ such that, for all N, the lazy walk on G_N has conductance at least Φ.*
- *There is a constant $\delta>0$ such that, for all N, the lazy walk on G_N has spectral gap at least δ.*

We will call such a family of graphs a *constant-degree expander.*

This definition focuses on *edge expanders,* where every set has a large fraction of edges that leave it. Another common definition of an expander states that every set has a large fraction of neighbors outside it. That is, there is a constant α such that, for all $S \subset V$ with $|S| \le N/2$,

$$\frac{|\partial S|}{|S|} \ge \alpha \,,$$

where ∂S denotes the set of vertices in $\overline{S}$ that are adjacent to some vertex in S. We call families of graphs with constant $\alpha > 0$ *vertex expanders.* As the following exercise shows, these two types of expander are equivalent for graphs of constant degree.

Exercise 12.26 *Show that a family of graphs with constant degree is a vertex expander if and only if it is an edge expander by bounding α in terms of μ and vice versa. On the other hand, use Problem 12.40 to show that the n-dimensional hypercube is a vertex expander but not an edge expander.*

One important property of expanders is that their diameter is only logarithmic as a function of N:

Exercise 12.27 *Show that if G is an α-expander in the sense of Exercise 12.26, its diameter is bounded by*

$$D \le 2\log_{1+\alpha}(N/2) = O(\log N)\,.$$

Hint: how does the volume of a sphere—that is, the number of vertices reachable from a given vertex in r or fewer steps—grow with r? And how large do two spheres have to be before we know that they overlap?

For a spectral proof that expanders have diameter $O(\log N)$, see Problem 12.41.

Do constant-degree expanders exist? Many of the families of graphs that come to mind fail to qualify. In a d-dimensional lattice, the volume of a sphere of radius r grows as r^d. The expansion μ is essentially its surface-to-volume ratio r^{d-1}/r^d, which tends to zero as r tends to infinity. On the other hand, the complete graph on N vertices is an expander with $\mu = 1/2$, but its degree is $N-1$. So we are looking for graphs that are "infinite-dimensional" in the sense that the volume of a sphere grows exponentially with r, but where the degree is a constant.

It turns out that constant-degree expanders are extremely common. In fact, *almost every d-regular* graph is an expander. To be precise, we will see in Problem 14.13 that if we form a random d-regular graph by starting with N vertices, each of which has d spokes pointing out of it, and then attach these spokes to each other according to a uniformly random matching, then the result is an expander with high probability.

However, this doesn't mean that it's easy to construct expanders deterministically. Are there explicit families of graphs whose expansion, conductance, and spectral gap stay constant as their size increases? We will describe how to build such graphs in a moment. But first let's discuss one of their applications: reducing the number of coins we need to flip to get the right answer from a randomized algorithm.

12.9.2 Finding Witnesses More Quickly

In Section 10.9, we defined the class RP of problems that can be solved by a one-sided randomized algorithm A. If the answer is "yes," then A returns "yes" with probability at least $1/2$, but A never returns "yes" if the answer is "no." To put this differently, for a yes-instance at least half the potential witnesses work, while for a no-instance none of them do. For instance, the Miller–Rabin algorithm for PRIMALITY is an RP algorithm for COMPOSITENESS, since if p is composite then at least half of the elements of $\mathbb{Z}_p^*$ are a witness of that fact.

Suppose we want to improve the probability that we find a witness, if there is one. We have already discussed the most obvious method—just run the algorithm k times, choosing k independently random witnesses. This amplifies the probability that we find a witness, if there is one, to $1-2^{-k}$.

However, there is a downside. If each witness requires us to flip n coins—as in the Miller–Rabin algorithm if p is an n-bit integer—then the total number of coins we need to flip is nk. Turning this around, if we want to reduce the probability of failure to P_{fail} using this method, we have to flip $\Theta(n \log P_{\text{fail}}^{-1})$ coins. As we discussed in Chapter 11, random bits are a limited resource. Can we derandomize the process of amplification, reducing the number of random bits we need?

Instead of choosing a fresh witness for each trial of the algorithm, let's take a random walk in the space of witnesses. Let G be a d-regular expander with $N = 2^n$ vertices. We can "overlay" G on the space of witnesses, so that each witness corresponds to a vertex of G, and some witnesses become its neighbors. Because the random walk on G mixes rapidly, it takes just a few steps to get to a new witness that is effectively random. Since it takes just $\log_2 d = O(1)$ coin flips per step to decide which way to go, we should be able to save significantly on the number of coins we need.

Assume that the input is a yes-instance. Let W denote the set of valid witnesses, and keep in mind that $|W| \geq N/2$. We will show that a random walk on G hits an element of W almost as quickly as a series of independent random samples would. Our algorithm starts by flipping n coins, and choosing an initial witness w_0 from the uniform distribution. If $w_0 \in W$, it halts and returns "yes." If not, it takes t steps of the random walk on G—for some t that we will determine below—arriving at a new witness w_1. If $w_1 \notin W$, it takes another t steps, arriving at a witness w_2, and so on. If it still hasn't found a witness after k rounds of this process, it halts and returns "don't know."

What is the probability that this algorithm fails? Let Π be the projection operator which projects onto the invalid witnesses $\overline{W}$. That is, if v is a vector that assigns a probability $v(w)$ to each witness w, then Πv is the vector

$$(\Pi v)(w) = \begin{cases} 0 & \text{if } w \in W \\ v(w) & \text{if } w \notin W. \end{cases}$$

Then if M is the transition matrix of the random walk on G and $u = (1/N, \ldots, 1/N)$ is the uniform distribution, the probability of failure after k rounds is the 1-norm

$$P_{\text{fail}} = \Big\| \underbrace{\Pi M^t \Pi M^t \cdots \Pi M^t}_{k \text{ times}} \Pi u \Big\|_1 = \left\| (\Pi M^t)^k \, \Pi u \right\|_1 .$$

In other words, P_{fail} is the total probability remaining after projecting away, at the end of each round, the witnesses in W.

We want to show that P_{fail} is small for some value of t and k. We start by using the Cauchy–Schwarz inequality to bound the 1-norm with the 2-norm,

$$P_{\text{fail}} \leq \sqrt{N} \left\| (\Pi M^t)^k \, \Pi u \right\|_2 . \tag{12.35}$$

Since each round begins with a probability vector which is supported only on $\overline{W}$, i.e., such that $\Pi v = v$, it suffices to analyze how applying ΠM^t to such vectors reduces their norm. We make the following claim:

Lemma 12.9 *Let δ be the spectral gap of M. Then for any vector v such that $\Pi v = v$, we have*

$$\left\|\Pi M^t v\right\|_2 \le \left(\frac{1}{2} + (1-\delta)^t\right) \|v\|_2 .$$

We ask you to prove this lemma in Problem 12.42. The idea is to write v as $v_{\text{eq}} + v_{\text{neq}}$, where v_{eq} is proportional to the uniform distribution u, and v_{neq} is orthogonal to it. Then Π cuts down the norm of v_{eq}, since it projects away at least half the witnesses, and M^t cuts down the norm of v_{neq} by $(1-\delta)^t$.

Since $\|\Pi u\|_2 \le \|u\|_2 \le 1/\sqrt{N}$, applying Lemma 12.9 to (12.35) gives

$$P_{\text{fail}} \le \sqrt{N} \left(\frac{1}{2} + (1-\delta)^t\right)^k \|\Pi u\|_2 \le \left(\frac{1}{2} + (1-\delta)^t\right)^k . \tag{12.36}$$

Let's set t large enough so that $(1-\delta)^t \le 1/4$, say. This value of t depends on the spectral gap δ, so the better an expander G is, the fewer steps we need to take per round. Nevertheless, for any given expander, some constant t suffices. Then $P_{\text{fail}} \le (3/4)^k$, so we need $k = O(\log P_{\text{fail}}^{-1})$ rounds to reduce the probability of failure to a given P_{fail}.

How many coins do we need to flip to run this algorithm? We flip n coins to choose the initial witness. After that, in each round we flip $t \log_2 d$ coins to take t steps of the random walk—but this is just $O(1)$ coins per round, since d and t are constants. Thus the total number of coins we need to flip is

$$n + O(k) = n + O(\log P_{\text{fail}}^{-1}),$$

as opposed to the $nk = \Theta(n \log P_{\text{fail}}^{-1})$ coins we need for k independent samples.

This significantly reduces the amount of randomness we need to get the right answer out of an RP algorithm. As Problem 12.43 shows, the same approach can be used to amplify the probability of success in the class BPP, where our randomized algorithms have two-sided error.

To summarize, independent samples from a space of size $N = 2^n$ cost n random bits each. But with the help of a constant-degree expander, each sample after the first one has the astonishingly low cost of $O(1)$ random bits. Even though these samples are not truly independent, they behave statistically almost as if they were.

For this method to be efficient, we need an expander G whose size is exponential, but where the neighborhood of a given vertex w can be computed in polynomial time. We will look at several ways to construct such graphs, and then sketch the proof that UNDIRECTED REACHABILITY can be solved in logarithmic space.

12.9.3 Two Ways to Shear a Cat

As we discussed above, random d-regular graphs are expanders almost all the time. But it is often far from trivial to describe a deterministic structure that is as good as a random one—especially if we want it to be efficiently computable. Historically, the first explicit constructions of expanders came from deep insights from algebra and the Fourier analysis of nonabelian groups. With the benefit of hindsight, we can present them here in elementary terms.

FIGURE 12.25: Le Chat Domestique et Son Expansion. Each step of the random walk on the Margulis expander shears the probability distribution along the x and y axes, and adds some local diffusion as well. After just 3 steps, the cat is well on his way to the uniform distribution.

One type of expander, first studied by Margulis, is defined on an $m \times m$ square lattice. We connect each point (x,y) to four others, according to the following mappings:

$$(x,y) \mapsto \begin{cases} (x+y,y) \\ (x+y+1,y) \\ (x,x+y) \\ (x,x+y+1), \end{cases} \tag{12.37}$$

where addition is defined mod m. If we think of (x,y) as a vector then each step in G applies one of two linear operators,

$$\begin{pmatrix} 1 & 1 \\ 0 & 1 \end{pmatrix} \quad \text{or} \quad \begin{pmatrix} 1 & 0 \\ 1 & 1 \end{pmatrix}. \tag{12.38}$$

These transformations shear the lattice along the x or y axis, while the $+1$s in (12.37) perform some local diffusion as well.

Figure 12.25 shows what a random walk in G does to an initial probability distribution. To make it easier to see what's going on, we only apply the transformations in (12.37) in the forward direction, choosing each one with probability $1/4$. If you prefer an undirected graph, you can make G bipartite with $2m^2$ vertices and let these transformations send vertices in each copy of the lattice to the other, or let G consist of one copy of the lattice and add the reverse transformations as well. We prove that this family of graphs is an expander in Problem 12.45, and we can get other expanders by replacing (12.38) with nearly any set of two or more linear operators.

12.9.4 *Growing Algebraic Trees*

Another type of expander, including those that achieve the largest possible expansion μ, comes from algebra and group theory. We outline these expanders here, and explain why they work so well.

The ideal expander branches outward from each vertex, wasting as few of its edges as possible on loops. Thus it looks, locally, like the d-regular tree shown in Figure 12.26. On such a tree, the surface-to-volume ratio of a sphere stays constant, rather than tending to zero as the radius increases.

Exercise 12.28 *Show that any finite set S on the infinite d-regular tree has at least $\mu d\,|S|$ edges connecting it to $\overline{S}$, where $\mu = 1-2/d$. Hint: use induction on $|S|$.*

There is a delightful algebraic way to construct expanders with this property. Consider the following three matrices:

$$A = \begin{pmatrix} 1 & 2 \\ 0 & 1 \end{pmatrix}, \quad A^{-1} = \begin{pmatrix} 1 & -2 \\ 0 & 1 \end{pmatrix}, \quad \text{and} \quad B = B^{-1} = \begin{pmatrix} 0 & 1 \\ 1 & 0 \end{pmatrix}. \tag{12.39}$$

If we take all possible products of these matrices, they generate an infinite group G_∞ of invertible 2×2 matrices with integer entries. We can turn G_∞ into a graph by drawing an edge between two elements if they differ by one of the generators. For instance, the empty word corresponds to the identity $\mathbb{1}$, and the vertices two steps away from it are the words of length 2, namely A^2, A^{-2}, AB, $A^{-1}B$, BA, and BA^{-1}. A graph of this kind, defined by a group and a set of elements that generate it, is called a *Cayley graph*.

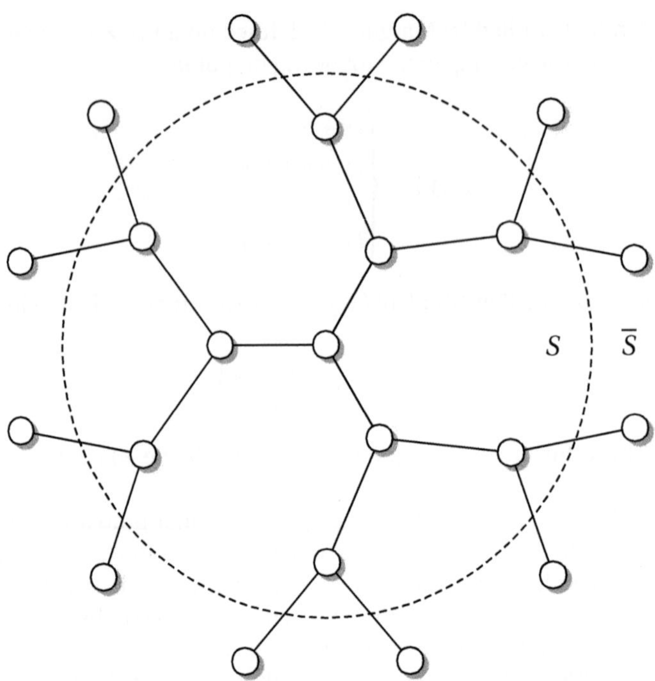

FIGURE 12.26: An ideal d-regular expander looks, locally, like a d-regular tree. Here $d = 3$. For any finite set S, starting at a random vertex in S and following a random edge takes us from S to $\overline{S}$ with probability at least $\mu = 1/3$.

The surprising thing about this particular Cayley graph, which we show in Figure 12.27, is that it is an infinite tree. There are no loops—no words that evaluate to the identity, except those such as $ABBA^{-1}$ where each matrix is canceled directly by its inverse, and we return to the identity the same way we left.

In algebraic terms, G_∞ is as noncommutative, or as *nonabelian*, as possible. If some pair of elements C, D commuted, i.e., if $CD = DC$, then the relation $CDC^{-1}D^{-1} = 1$ would correspond to a cycle of length 4. A group with no relations at all is called *free*, and this group is free except for the relation $B^2 = 1$. We offer the reader a beautiful proof of this fact in Problem 12.46.

We can fold this infinite graph up into a finite one by mapping each matrix entry to its value mod p for some prime p. This turns G_∞ into a finite group G_p—specifically, the group of all 2×2 matrices with entries in the finite field $\mathbb{F}_p$ whose determinant is ± 1. As the following exercise shows, there are $\Theta(p^3)$ such vertices, so G_p's Cayley graph has size $N = \Theta(p^3)$.

Exercise 12.29 *Show that for any prime $p > 2$, the number of invertible 2×2 matrices mod p with determinant ± 1 is $2p(p^2 - 1)$.*

A crucial fact about this construction is that, even after we fold it up in this way, the Cayley graph of G_p is still locally a tree. Its girth—that is, the length of its shortest cycle—is the length of the shortest

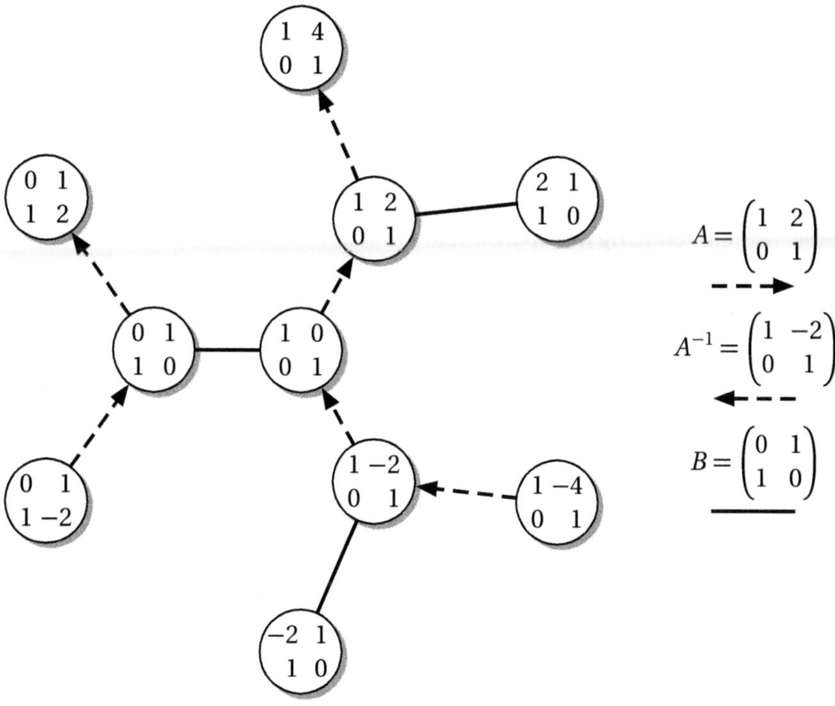

FIGURE 12.27: The Cayley graph of the group generated by the matrices A, A^{-1}, and B is an infinite tree. Equivalently, the only words formed of A, A^{-1}, and B that evaluate to the identity are those that retrace their steps, by canceling each A directly with A^{-1} and each B with another B.

word composed of A, A^{-1}, and B that evaluates to the identity mod p, not counting words that have two adjacent Bs or an A adjacent to an A^{-1}. The following exercise shows that this gives a girth of $\Omega(\log N)$, and that this is within a constant of the largest girth possible.

Exercise 12.30 *Using the fact that G_∞ is a tree, show that there is a constant $a > 0$ such that the girth of G_p is at least $a \log_2 N = O(\log N)$ for all p. Hint: how quickly can the matrix entries grow as a function of the length of the corresponding word? Conversely, show that for any d-regular graph the girth is at most $2\log_{d-1} N = O(\log N)$, and therefore that the girth of G_p is $\Theta(\log N)$.*

In a graph with girth γ, any "sphere" of radius $r < \gamma/2$ is a tree. Since G_p is 3-regular, the number of vertices in a sphere grows as $(d-1)^r = 2^r$. Thus G_p achieves the same expansion $\mu = 1 - 2/d = 1/3$ that a 3-regular tree does, up to spheres of size $2^{\gamma/2} = 2^{\Theta(\log N)} = N^\alpha$ for some $\alpha > 0$. Showing that this expansion persists all the way out to sets of size $N/2$ requires more sophisticated work, but it can be shown that any Cayley graph constructed in this way is an expander as long as the corresponding group of matrices is sufficiently free. Finally, since we can multiply matrices of n-bit integers in poly(n) time, we can compute the neighbors of any given vertex in polynomial time—even when p and N are exponentially large. Thus

12.19

G_p gives us another family of constant-degree expanders where we can carry out each step of a random walk in polynomial time.

12.9.5 The Zig-Zag Product

Next we explore the *zig-zag product*—a clever way of combining two expanders to make a larger one. To motivate it, imagine that G is a graph with good expansion, but distressingly high degree. How can we reduce its degree, without reducing its expansion too much?

One approach is to replace each vertex with a "cloud" of vertices, each one of which forms a constant-degree expander. Let G be a d_G-regular graph, and let H be a d_H-regular graph with d_G vertices. We replace each vertex of G with a cloud of d_G vertices, and connect the vertices within each cloud so that it forms a copy of H. We then connect each vertex to a unique partner in a neighboring cloud, according to an edge of G.

The zig-zag product $G \textcircled{z} H$ is defined on this set of vertices, but these are not its edges. The easiest way to define them is to describe a step of the random walk on $G \textcircled{z} H$. We start with a vertex in some cloud. We take a step within that cloud according to the random walk on H. We then move deterministically, along an edge of G, from the resulting vertex to its partner in a neighboring cloud. Finally, we take a step within the new cloud, again according to the random walk on H.

Since we have d_H choices of "zig" in our old cloud and d_H choices of "zag" in the new one, $G \textcircled{z} H$ is a d_H^2-regular graph. We show an example in Figure 12.28. Note that in order to define the edges between clouds we need a one-to-one mapping between H and the neighborhood of each vertex of G, but we will assume that we have chosen such a mapping.

The following theorem, which we ask you to prove in Problem 12.50, bounds the eigenvalue of the random walk on $G \textcircled{z} H$ in terms of those on G and H:

Theorem 12.10 *Let λ_G and λ_H denote the largest eigenvalues of the random walks on G and H respectively. Then the largest eigenvalue of the random walk on $G \textcircled{z} H$ is bounded by*

$$\lambda_{G \textcircled{z} H} \le \max(\lambda_G, \lambda_H^2) + \lambda_H . \tag{12.40}$$

Now suppose we want to improve the spectral gap of a graph G. We can do this by defining a new graph G^2 on the same set of vertices, where each edge corresponds to a path of length 2 in G. (Note that G^2 is a multigraph, since it includes self-loops corresponding to paths which double back to their starting point, and possibly multiple edges as well.) Since one step in G^2 corresponds to two steps in G, we have

$$\lambda_{G^2} = \lambda_G^2 .$$

Squaring roughly doubles the spectral gap, since $(1-\delta)^2 \approx 1-2\delta$ if δ is small. On the other hand, if G is d-regular then G^2 is d^2-regular, and squaring it repeatedly would give graphs of degree d^4, d^8, and so on. Here the zig-zag product comes to the rescue. If we possess a finite graph H with good expansion, we can reduce the degree to d_H^2 at each stage by taking the zig-zag product with H. This lets us hold the degree down to a constant, while largely keeping the benefits of squaring the graph.

Iterating the zig-zag product gives a simple construction of constant-degree expanders. Let H be a d-regular graph on d^4 vertices such that $\lambda_H \le 1/4$. We claim that such a graph exists, for some constant d, given the constructions we saw in Section 12.9.3 and 12.9.4. Now define a series of graphs G_i as follows:

$$G_1 = H^2 , \quad G_{i+1} = G_i^2 \textcircled{z} H .$$

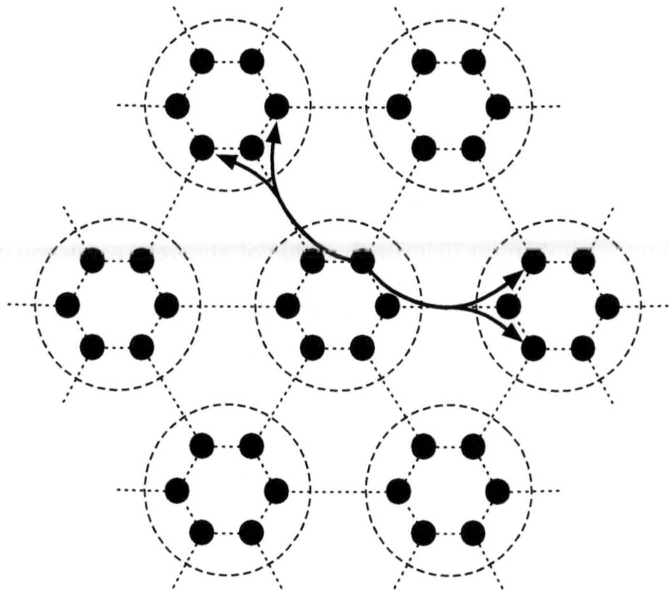

FIGURE 12.28: The zig-zag product. Each vertex of G becomes a "cloud" of d_G vertices connected as a copy of H. Each edge in $G ⓩ H$ consists of a "zig" within a cloud, then a step in G crossing to another cloud, and then a "zag" within that cloud. Here G is the triangular lattice and H is the 6-cycle.

Exercise 12.31 *Using Theorem 12.10, show that the family G_i is a constant-degree expander. Specifically, show by induction on i that G_i is a d_H^2-regular graph with $d^{4(i+1)}$ vertices and that $\lambda_{G_i} \leq 1/2$ for all $i \geq 1$.*

12.9.6 UNDIRECTED REACHABILITY

In Section 12.8.2, we showed that we can solve UNDIRECTED REACHABILITY with a randomized algorithm that uses $O(\log n)$ memory, by taking a random walk in the graph. We end this section by showing how we can use the zig-zag product to derandomize this algorithm, and solve UNDIRECTED REACHABILITY deterministically in $O(\log n)$ memory. In terms of complexity classes, this shows that $\mathsf{SL} = \mathsf{L}$ as we claimed in Section 8.8.

If an undirected graph has maximum degree d and diameter D, we can tell whether there is a path between two vertices s, t by starting at s and trying all d^D possible paths of length D or less. Using depth-first search, we can do this with $O(D \log d)$ memory—a stack of depth D, each layer of which uses $\log_2 d$ bits to record which way we went on the corresponding step. If $d = O(1)$ and $D = O(\log n)$, this is just $O(\log n)$ memory.

So we would like to be able to take an arbitrary graph G, and convert it to one where each of its connected components has logarithmic diameter and constant degree. We will do this by converting each component to a constant-degree expander, by repeatedly powering G and taking its zig-zag product with some finite expander H. We then check to see if there is a path between the clouds that corresponds to s and t.

To make this work, we need a slightly different bound on the zig-zag product's eigenvalues—specifically, on its spectral gap:

Theorem 12.11 *Let δ_G and δ_H denote the spectral gaps of the random walks on G and H respectively. Then the spectral gap of the random walk on $G \textcircled{z} H$ is bounded by*

$$\delta_{G \textcircled{z} H} \geq \delta_G \, \delta_H (1 - \delta_H/2).$$

The proof of this bound is a little more delicate than that of Theorem 12.10, and we refer the motivated reader to the Notes.

12.20 We will assume for simplicity that G is connected—otherwise, think of our comments as applying to one of its connected components. We will also assume that it is d_G-regular for some constant d_G. If it is not, we can make it 3-regular by replacing each vertex of degree d with a ring of d vertices. We can then increase the degree to any desired constant, and make it uniform throughout G, by adding multiple edges or self-loops.

Now suppose that H is a d_H-regular graph on d_G^4 vertices such that $d_H^2 = d_G$ and $\lambda_H \leq 1/2$. Then raising G to the 4th power increases the spectral gap δ by roughly 4, and zig-zagging with H reduces δ by 3/8 at worst. So if we define a series of graphs G_i as follows,

$$G_0 = G\,, \quad G_{i+1} = G_i^4 \textcircled{z} H\,,$$

then the spectral gap δ_{G_i} grows as roughly $(3/2)^i$ until it becomes a constant.

As we showed in Section 12.8.2, any connected graph G with n vertices has spectral gap $\delta_G \geq 1/\text{poly}(n)$. Thus we can turn G into a constant-degree expander by repeating this process t times for some $t = O(\log n)$. The resulting graph G_t has $n\,|V_H|^{O(\log n)} = \text{poly}(n)$ vertices, so it is still of polynomial size. Since G_t is an expander it has diameter $D = O(\log n)$ by Exercise 12.27, and it has constant degree $d_H^2 = d_G$. Thus we can try all possible paths of length D or less with just $O(\log n)$ memory

There is one part of this proof that we omit. Namely, if we only have $O(\log n)$ memory, we don't have room to construct and store G_t explicitly. Happily, given read-only access to G, it is possible to answer questions about G_t—just as if we had direct access to its adjacency matrix—with $O(\log n)$ workspace. Finally, to solve the original REACHABILITY problem on G, we check to see if there are any paths between the clouds corresponding to s and t.

In essence, the zig-zag product lets us derandomize random walks. Suppose we want to take t steps of the random walk on a d-regular graph G. If each step is independent of the previous ones, this requires us to generate $t \log_2 d$ random bits. We can effectively double t by replacing G with G^2, but this turns d into d^2, and again doubles the number of random bits we need. Like the derandomized squaring behind Nisan's pseudorandom generator in Problem 11.18, the zig-zag product lets us double the length of the path without doubling the number of random bits, so that instead of $O(t)$ random bits we only need $O(\log t)$. Then we can try all $\text{poly}(t)$ combinations of these bits, and explore G deterministically.

12.10 Mixing in Time and Space

We close this chapter by showing that *temporal* mixing, where correlations decay as a function of time, is deeply related to *spatial mixing*, where they decay as a function of distance. First, let's recall some physics. As we mentioned at the beginning of this chapter, when the Ising model is above its critical temperature T_c, the correlations between one site and another decay exponentially as a function of the distance between them. To be more precise, if two sites i and j are a distance ℓ apart, then

$$\mathbb{E}[s_i s_j] \sim e^{-\ell/\xi}.$$

Here ξ denotes the *correlation length*. Roughly speaking, it is the typical width of a blob of sites that have the same spin. If two sites are, say, 10ξ away from each other, their spins are nearly independent.

When the system is extremely hot, ξ approaches zero, and even neighboring spins are independent. On the other hand, if we cool the system then ξ increases, until we reach $T = T_c$ and ξ diverges. Below T_c, most sites agree with the overall magnetization, and $\mathbb{E}[s_i s_j] \neq 0$ even in the limit $\ell \to \infty$.

As we will see, T_c also marks a phase transition where the mixing of local Markov chains goes from fast to slow. Intuitively, if different regions of the lattice are independent of each other, we can relax each one to equilibrium by touching it a finite number of times. In this case, we would expect rapid mixing since, à la coupon collecting, after $O(n \log n)$ steps we have touched every region enough times to relax it. On the other hand, if states have global correlations—such as the entire lattice being magnetized mostly up or mostly down—then the Markov chain needs enough time to carry this information across the lattice, and the mixing time is large.

Let's consider correlations between the boundary of a lattice and its interior, rather than just between pairs of individual sites. Suppose we have a lattice L, and we fix the values of the spins—or colors, or whatever—at its boundary. While holding these boundary sites fixed, we let the interior of the lattice relax to equilibrium. For simplicity, we focus on the square lattice in two dimensions.

Now focus on a region R deep in the interior of the lattice. The sites in R will have some equilibrium probability distribution, which we call P_{eq}^R. How much does P_{eq}^R depend on the boundary conditions? Specifically, given two choices B, B' of the boundary conditions, how large can the total variation distance be between the resulting distributions, which we denote $P_{\text{eq}}^{R|B}$ and $P_{\text{eq}}^{R|B'}$?

Definition 12.12 *A system has* spatial mixing *if there are constants A, ξ such that for any lattice L and any subset $R \subset L$, and for all pairs B, B' of boundary conditions,*

$$\left\| P_{\text{eq}}^{R|B} - P_{\text{eq}}^{R|B'} \right\|_{\text{tv}} \leq A\,|\partial L|\,|R|\, e^{-\ell/\xi},$$

where ∂L denotes the set of sites on the boundary of L and ℓ denotes the smallest distance between any site in R and any site in ∂L.

Roughly speaking, if a system has spatial mixing then the boundary conditions don't matter. For instance, suppose we impose boundary conditions on the Ising model where all the boundary spins point upward. If $T < T_c$, this bias propagates all the way to the center of the lattice, and causes most of the interior spins to point upward as well. But if $T > T_c$, the boundary conditions are only felt a little way in, and the spins deep inside the lattice have no idea whether those on the boundary point up or down.

Our goal is to relate spatial mixing to the following definition of rapid mixing:

Definition 12.13 *A Markov chain has* rapid mixing *if there are constants $C, A > 0$ such that, for any lattice L with n sites, and for all boundary conditions,*

$$\left\| P_t - P_{\text{eq}} \right\|_{\text{tv}} \leq C n e^{-At/n}.$$

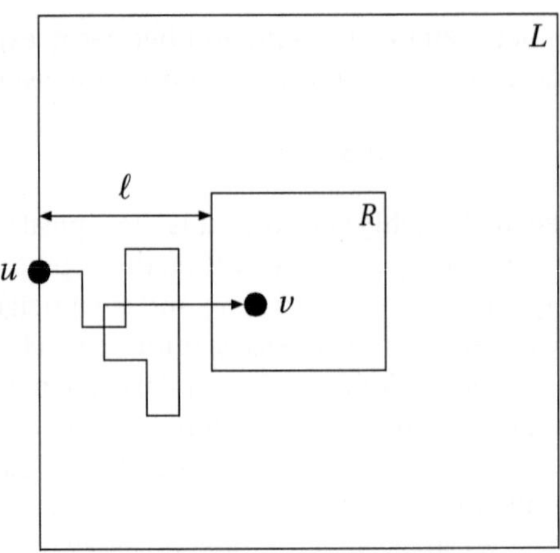

FIGURE 12.29: Counting paths along which the boundary conditions affect a region R inside the lattice.

This applies to the walk on the hypercube, the cases of graph coloring where we proved rapid mixing in Section 12.5, and so on. In particular, it implies that the mixing time is $\tau_\varepsilon = O(n \log n\varepsilon^{-1})$, or $O(n \log n)$ for fixed ε.

We will now sketch the proof of the following theorem. By a *local* Markov chain, we mean one which changes the state at a single site in each move.

Theorem 12.14 *If a system has a local Markov chain with rapid mixing as defined in Definition 12.13, then it has spatial mixing as defined in Definition 12.12.*

The idea is that the Markov chain mixes so rapidly that it has no time to propagate a difference in the boundary conditions into the heart of the lattice. Let's start by assuming that two boundary conditions B, B' differ at some boundary site u as in Figure 12.29. This difference can only affect a site v in the interior of the lattice if, by the time we reach equilibrium, a series of moves of the Markov chain carries information from u to v. This means that at a sequence of times $t_1 < t_2 < \cdots < t_m$, the Markov chain touches a sequence of sites $u_1, u_2, \ldots, u_m$ that form a path from u to v. Moreover, if u and v are a distance ℓ apart, the length m of this path must be at least ℓ.

Since each site on the square lattice has 4 neighbors, a generous upper bound on the number of paths of length m is 4^m. Moreover, the probability that a given site u_i is chosen at a given time is $1/n$, and the number of possible sequences of times $t_1 < t_2 < \cdots < t_m$ is $\binom{t}{m}$. Applying the union bound (see Appendix A.3) over all possible paths and all possible sequences of times, the probability that information has traveled from u to v is bounded by

$$p_{\text{diff}} \leq \sum_{m=\ell}^{t} \frac{4^m}{n^m} \binom{t}{m}.$$

Applying the inequality $\binom{t}{m} \le (\mathrm{e}t/m)^m$ (see Appendix A.2) then gives

$$p_{\mathrm{diff}} \le \sum_{m=\ell}^{t} \left(\frac{4\mathrm{e}t}{mn}\right)^m < \sum_{m=\ell}^{\infty} \left(\frac{4\mathrm{e}t}{\ell n}\right)^m . \tag{12.41}$$

When t/n is sufficiently small compared to ℓ, this is exponentially small. Specifically, let's set $t/n = \ell/4\mathrm{e}^2$, so that $4\mathrm{e}t/\ell n = 1/\mathrm{e}$. Then summing the geometric series in (12.41) gives

$$p_{\mathrm{diff}} < \sum_{m=\ell}^{\infty} \mathrm{e}^{-m} = A' \mathrm{e}^{-\ell},$$

where $A' = \mathrm{e}/(\mathrm{e}-1)$. Taking a union bound over all pairs u, v where u is on the boundary of L and v is in R, the probability that a difference has propagated from any vertex on the boundary to any vertex in R is at most $|\partial L|\,|R|\,p_{\mathrm{diff}}$. It follows that

$$\left\| P_t{}^{R|B} - P_t{}^{R|B'} \right\|_{\mathrm{tv}} \le A' \,|\partial L|\,|R|\,\mathrm{e}^{-\ell} . \tag{12.42}$$

Now suppose that the Markov chain mixes rapidly in the sense defined above. An important technical step, which we omit here, is to show that the probability distribution P_{eq}^{R} also mixes rapidly, so that for any boundary condition B we have

$$\left\| P_t{}^{R|B} - P_{\mathrm{eq}}^{R|B} \right\|_{\mathrm{tv}} \le C' \,|R|\,\mathrm{e}^{-A' t/n} , \tag{12.43}$$

for some constants $C', A' > 0$. The reason for this is that each site in the lattice is touched t/n times on average, and by Chernoff bounds (see Appendix A.5.1) the total number of steps that touch R is close to its expectation.

So, when $t/n = \ell/4\mathrm{e}^2$, the distributions on R induced by B and B' are close to each other—and since the Markov chain mixes rapidly, both of these are close to their respective equilibria. Combining (12.42) and (12.43) and using the triangle inequality then gives

$$\begin{aligned}
\left\| P_{\mathrm{eq}}^{R|B} - P_{\mathrm{eq}}^{R|B'} \right\|_{\mathrm{tv}} &\le \left\| P_{\mathrm{eq}}^{R|B} - P_t{}^{R|B} \right\| + \left\| P_t{}^{R|B} - P_t{}^{R|B'} \right\| + \left\| P_t{}^{R|B'} - P_{\mathrm{eq}}^{R|B'} \right\| \\
&\le 2C' \,|R|\,\mathrm{e}^{-A' t/n} + A' \,|\partial L|\,|R|\,\mathrm{e}^{-\ell} \\
&= 2C' \,|R|\,\mathrm{e}^{-(A'/4\mathrm{e}^2)\ell} + A' \,|\partial L|\,|R|\,\mathrm{e}^{-\ell} .
\end{aligned}$$

Since $|\partial L| \ge 1$, if we set $C = 2C' + A'$ and $\xi = \max(4\mathrm{e}^2/A', 1)$, this gives

$$\left\| P_{\mathrm{eq}}^{R|B} - P_{\mathrm{eq}}^{R|B'} \right\|_{\mathrm{tv}} \le C \,|\partial L|\,|R|\,\mathrm{e}^{-\ell/\xi} ,$$

establishing spatial mixing.

A similar argument applies to Markov chains that update all the sites in a patch of fixed size at each step, instead of a single site. Conversely, subject to a few conditions, spatial mixing implies the existence of a local Markov chain with rapid mixing. Thus for most systems defined by local interactions on a lattice, spatial mixing and rapid mixing are equivalent. In particular, the single-spin Markov chain for the Ising model, with either Metropolis or heat-bath dynamics, mixes rapidly if and only if $T > T_{\mathrm{c}}$.

Theorem 12.14 means that analyzing the mixing time of Markov chains can tell us something about the physical properties of a system. For instance, consider the following generalization of the Ising model, called the *antiferromagnetic Potts model.* We again have a lattice of sites, except that each s_i can be one of q colors, rather than just pointing up or down. The energy is $E = \sum_{ij} \delta_{s_i,s_j}$, where the sum is over all neighboring pairs of sites, and $\delta_{s_i,s_j} = 1$ if $s_i = s_j$ and 0 otherwise. Thus the energy is the number of edges whose endpoints are the same color. At $T = 0$, only the lowest-energy states may appear, and the Boltzmann distribution is simply the uniform distribution on all q-colorings of the lattice (assuming that at least one exists).

If we can prove that some local Markov chain mixes rapidly for q-colorings of the lattice, this establishes that the Potts model has spatial mixing even at absolute zero. In physical terms, this means that it has no phase transition—that it is usually in an unmagnetized state no matter how cold it is.

The best rigorous results to date show spatial mixing on the square lattice for $q \geq 6$ by proving rapid mixing for local Markov chains which update a finite patch of sites. On the other hand, there is overwhelming numerical evidence that spatial mixing applies for $q \geq 4$, so proving spatial mixing for $q = 4$ and $q = 5$ remains an interesting open problem. Ironically, even though a height function argument shows that Glauber dynamics mixes in polynomial time for $q = 3$ on the square lattice, we don't know that it mixes polynomially, let alone rapidly, for $q = 4$ or $q = 5$. We don't even know that increasing the number of colors decreases the mixing time, even though this seems intuitive.

12.21 When spatial mixing does not hold, and the system is globally correlated, the mixing time is typically exponentially large. For instance, suppose we have an $L \times L$ Ising model below its critical temperature T_c. In order to flip from a positively magnetized state to a negatively magnetized one, we have to pass through states with a boundary stretching across the entire lattice, separating a mostly-up region from a mostly-down one. The energy of this boundary is proportional to its length, so the Boltzmann probability of such a state is at most $e^{-\beta L}$, and below T_c the total probability of all such states is still $e^{-\Omega(L)}$. The low probability of these boundary states creates a barrier, or bottleneck, between the mostly-up and mostly-down sections of the state space, and the mixing time is $e^{\Omega(L)} = e^{\Omega(\sqrt{n})}$. This state of affairs is sometimes called *torpid* mixing.

12.22 Finally, when $T = T_c$ and the Ising model is precisely at its phase transition, spatial correlations decay according to a power law as a function of distance, $\ell^{-\gamma}$ for some constant γ, rather than exponentially. In this case, the mixing time is believed to be polynomial but not rapid, $\tau \sim n^\beta$ for some $\beta > 1$. This also occurs in rhombus tilings, 3-colorings of the square lattice, and a number of other systems with height representations. Physically, the critical temperature of these systems is $T_c = 0$.

Problems

> It is better to solve one problem five different ways than to solve five problems one way.
>
> George Pólya

12.1 Probabilities and energies. Suppose a Markov chain obeys detailed balance. Given the transition probabilities $M(x \to y)$, show how to assign an energy E to each state x so that the equilibrium distribution is the Boltzman distribution with temperature 1, i.e., $P_{eq}(x) \propto e^{-E(x)}$. What goes wrong if M doesn't obey detailed balance?

12.2 Why the Boltzmann distribution? Since physical systems tend to move towards low-energy states, it's plausible that the equilibrium probability of being in a state x should be some decreasing function of its energy $E(x)$. But why should this function take the specific form $p(x) \propto e^{-\beta E(x)}$?

Suppose that the system we care about is connected to a much larger system, called the *heat bath.* Boltzmann claimed that all states of the joint system with the same total energy are equally probable at equilibrium. In that case, the probability of our system being in a state x is proportional to the number of states of the heat bath with energy $E_{\text{hb}} = E_{\text{total}} - E(x)$.

Let W denote the number of states of the heat bath, and define its *entropy* as

$$S = \ln W.$$

Now assuming that $E_{\text{hb}} \gg E(x)$, argue that this gives

$$P(x) \propto W \propto e^{-\beta E(x)},$$

where

$$\beta = \frac{\partial S}{\partial E_{\text{hb}}}.$$

Since the definition of temperature in classical thermodynamics is $T = \partial E/\partial S$, this shows that $\beta = 1/T$ where T is the temperature of the heat bath. Moreover, if there are multiple systems all in contact with the same heat bath, they must all have the same temperature.

12.3 Boltzmann and maximum entropy. Here is another explanation of the Boltzmann distribution. Suppose we wish to maximize the Gibbs–Shannon entropy (see Problem 9.26),

$$S = -\sum_x p(x) \ln p(x), \tag{12.44}$$

subject to the constraints that the system has some fixed average energy and that the probabilities sum to 1:

$$E = \sum_x p(x) E(x)$$
$$P = \sum_x p(x),$$

where E is the average energy and $P = 1$ is the total probability. Now imagine that for each x we can vary $p(x)$ independently. The method of Lagrange multipliers tells us that there are constants α and β such that, for all x, the partial derivatives of S, E, and P with respect to $p(x)$ are related by

$$\frac{\partial S}{\partial p(x)} = \alpha \frac{\partial P}{\partial p(x)} + \beta \frac{\partial E}{\partial p(x)}.$$

Use this equation to show that

$$p(x) = \frac{1}{Z} e^{-\beta E(x)},$$

where the normalization factor is

$$Z = \sum_x e^{-\beta E(x)} = e^{\alpha+1}.$$

In fact, Z turns out to be much more than just a normalization factor. It is called the *partition function,* and we will discuss some of its uses in Chapter 13.

12.4 The mean-field Ising model. Here is a derivation of the magnetization as a function of temperature in a toy version of the Ising model. Suppose we have a lattice with n sites, and consider the "macrostate" where bn of them are up and the other $(1-b)n$ are down. The number of states in this macrostate is $W=\binom{n}{bn}$, and approximating the binomial as in Appendix A.4.3 gives the entropy

$$S=\ln W=h(b)\,n\,,$$

where $h(b)$ is (once again) the Gibbs–Shannon entropy $h(b)=-b\ln b-(1-b)\ln(1-b)$.

To estimate the energy, let's mix up the structure of the lattice. Instead of having each site interact with $d=4$ neighbors, we make a *mean field* assumption which smears the interactions over all the spins. This gives

$$E=\frac{d}{n}\sum_{i,j}s_i s_j\,,$$

where the sum ranges over all $\binom{n}{2}$ pairs of spins. Show that, when n is large, this approaches

$$E=-2dn\left(b-\frac{1}{2}\right)^2.$$

Now write $E-TS$ as a function $f(m)$ where $m=2b-1$ is the magnetization, and find the value or values of m that minimize it. Show that there is a critical temperature T_c that depends on d, such that $|m|$ is nonzero for $T<T_c$ and zero for $T>T_c$. Also derive an approximation for $|m|$ where $T=T_c-\varepsilon$ and ε is small. Hint: what is the second derivative of $f(m)$ at $m=0$?

12.5 Mean-field marginals. Here's another approach to the mean-field calculation for the Ising model. Consider a spin s with d neighbors and imagine that its neighboring spins $s_1,\ldots,s_d$ are chosen independently, setting each one to $+1$ or -1 with probability $(1+m)/2$ or $(1-m)/2$ respectively. Together, these neighbors exert a certain average magnetic field on s. Compute the resulting magnetization $m'=\mathbb{E}[s]$.

We are interested in fixed points of this process, namely self-consistent solutions where $m'=m$. Show that there is a T_c such that for $T>T_c$ the only fixed point is $m=0$, while for $T<T_c$ there are also fixed points $\pm m\neq 0$. For $T<T_c$, is there a sense in which the fixed points at $\pm m$ are stable and the one at $m=0$ is unstable?

12.6 Return visits. Justify the following claim, which may seem counterintuitive at first. Suppose I have an ergodic Markov chain in which the equilibrium distribution is uniform. Then for any pair of states x and y, if I start my walk at x, the expected number of times I visit y before returning to x is 1. For the random walk on the cycle, for instance, this is true no matter how far apart x and y are. More generally, the expected number of times we visit y before returning to x is $P_{eq}(y)/P_{eq}(x)$.

12.7 A little less lazy. Another perfectly reasonable model of a random walk on the n-dimensional hypercube is to stay put with probability $1/(n+1)$, and to move along a given axis with probability $1/(n+1)$. Show that the mixing time of this walk is also $O(n\log n)$.

Hint: show that this walk is essentially equivalent to the lazy walk discussed in the text, where we stay put with probability 1/2, except for a rescaling of time. More precisely, if we count the number s of steps in which we change the state instead of sitting still, in both cases s is given by a binomial distribution, and when these distributions have the same mean the probability distributions are nearly the same.

12.8 Riffle shuffles. Show that the reverse riffle shuffle defined in Section 12.3 is the reverse of the following model of the riffle shuffle. First choose L according to the binomial distribution $\text{Bin}(n,1/2)$, in which the probability of L is $2^{-L}\binom{n}{L}$. Then take the top L cards and put them in your left hand, and put the other $R=n-L$ cards in your right hand. Now let the cards fall one at a time, where the probability of a card falling from your left or right hand at each step is proportional to the number of cards in that hand.

Is it possible that the reverse riffle shuffle shuffles quickly, but the riffle shuffle doesn't? Piffle! Show that their mixing times are the same, at least within a polynomial factor. Hint: consider the spectral gap.

12.9 Coupling and mixing. Prove the coupling bound (12.11) on the total variation distance from equilibrium. Hint: suppose that x_0' is chosen according to P_{eq}, and consider a subset E of the state space. Show that the probability $x_t \in E$ is bounded below by

$$\Pr[x \in E] \geq P_{\text{eq}}(E) - \max_{x_0, x_0'} \Pr[x_t \neq x_t'] .$$

Then use (12.7) from Exercise 12.10.

12.10 Submultiplicativity. Let $P_t^{\,x}$ denote the probability distribution t steps after starting in an initial state x, and define

$$\delta_t = \max_{x,y} \|P_t^{\,x} - P_t^{\,y}\|_{\text{tv}} .$$

Using the triangle inequality, show that

$$\frac{\delta_t}{2} \leq \left\|P_t - P_{\text{eq}}\right\|_{\text{tv}} \leq \delta_t .$$

Now argue that δ_t has the following property, known as *submultiplicativity*:

$$\delta_{s+t} \leq \delta_s \delta_t . \tag{12.45}$$

You can prove this using the following fact: there exists a coupling for which δ_t is exactly the maximum, over all pairs x, y of initial states, of the probability that these states don't coalesce after t steps. Then note that x and y coalesce within $s+t$ steps either if they coalesce in the first s steps, or if the states they become after the first s steps coalesce in the t steps after that.

As a consequence of submultiplicativity, show that for any t_0 we have, for all $t \geq t_0$,

$$\delta_t \leq (\delta_{t_0})^{t/t_0} , \tag{12.46}$$

so that δ_t decreases exponentially. Then show that the total variation distance from equilibrium decreases exponentially once it falls below 1/2. In other words, show that for any $\varepsilon_0 < 1/2$, we have

$$\left\|P_t - P_{\text{eq}}\right\|_{\text{tv}} \leq (2\varepsilon_0)^{t/\tau_{\varepsilon_0}} ,$$

and therefore that, for any $\varepsilon \leq \varepsilon_0$,

$$\tau_\varepsilon \leq \tau_{\varepsilon_0} \log_{(2\varepsilon_0)^{-1}} \varepsilon^{-1} .$$

For instance, setting $\varepsilon_0 = 1/4$ gives, for any $\varepsilon \leq 1/4$,

$$\tau_\varepsilon \leq \tau_{1/4} \log_2 \varepsilon^{-1} .$$

12.11 Coalescence. Here we prove (12.12), bounding the mixing time in terms of the coalescence time. Suppose there is a coupling with coalescence time T_{coal}. Show using (12.11) and Markov's inequality (Appendix A.3.2) that

$$\delta_t \leq T_{\text{coal}}/t ,$$

where δ_t is defined as in the previous problem. Combine this with (12.46) to show that for any $\alpha > 1$, we have

$$\delta_t \leq \left(\alpha^{-\alpha}\right)^{t/T_{\text{coal}}} .$$

Optimize α, set $\delta_t = \varepsilon$, and solve for t.

12.12 Card swaps. The *random transposition shuffle* is another way to shuffle a deck of n cards. In each step, we choose a pair i, j of random numbers uniformly and independently from $\{1,\ldots,n\}$. If $i = j$, we do nothing. Otherwise, we swap the ith card with the jth card. Thus for any pair of i, j with $i \neq j$, we swap the ith and jth cards with probability $2/n^2$.

Now consider a coupling in which we choose the name of a card, say, the Two of Clubs, and a number, say 42. Then in both decks we swap the Two of Clubs with whichever card is in position 42. Show that the expected coalescence time, and therefore the mixing time for constant ε, is $O(n^2)$. Hint: define the distance d between the two decks as the number of cards that are in different positions, and calculate the expected time it takes for d to decrease to $d-1$.

12.13 Marking cards. Consider the random transposition shuffle from the previous problem, where i and j are chosen uniformly and independently from $\{1,\ldots,n\}$, and we swap the ith and jth card. In fact its mixing time is $O(n \log n)$, and we can show this with an interesting kind of coupon-collecting argument.

Start with all the cards unmarked. At each step, mark cards according to the following rule:

- If $i = j$, mark the ith card.
- If $i \neq j$ and one of the two cards is marked, mark the other one.
- If $i \neq j$ but neither card is marked, do nothing.

Each time you mark a card, add its name (the Queen of Hearts and so on) to a list of marked cards, and add its position to a list of marked positions.

Show that, at each step of this process, the arrangement of the marked cards at the marked positions is uniformly random. In other words, if there are k marked cards and k marked positions, all $k!$ permutations are equally likely. Hint: use induction on k.

Finally, show that the expected time for all the cards to become marked is $O(n \log n)$. Hint: if there are k marked cards so far, what is the expected number of steps it takes to mark another card?

12.14 Coupling on lines, cycles, and trees (from Tom Hayes). Consider a random walk on a line of length n, so that the position x ranges from 0 to n. We perform the lazy walk, staying put with probability $1/2$, moving left with probability $1/4$, and moving right with probability $1/4$. If it tries to move to the left of $x = 0$, or to the right of $x = n$, it does nothing. Design a coupling between two such walks in which one walker stays put whenever the other one moves. Then use the techniques of Appendix A.4.4 to show that the mixing time is $O(n^2)$.

Can you extend this argument to a similar random walk on a circle of circumference n? What about a binary tree of length n, in which at each step we either stay put, move up to our parent, or move down a random child?

12.15 Walks with momentum. In some cases, we can speed up a Markov chain by "lifting" it to one with more states. Roughly speaking, these states have both position and momentum. In other words, rather than a reversible Markov chain where we are equally likely to move forwards or backwards, we have an irreversible one which is more likely to move in the same direction that it did before.

Consider Figure 12.30. We take the cycle of size n, and double the number of states. Half of them move clockwise with probability $1 - 1/n$, the others move counterclockwise with probability $1 - 1/n$, and we switch from clockwise to counterclockwise with probability $1/n$.

Show that, neglecting parity issues, the mixing time of this lifted Markov chain is $O(n)$. More to the point, show that if we run the lifted chain for $2n$ steps and then project onto the cycle by merging pairs of states together, the resulting distribution on the cycle has a total variation distance less than $1/2$ away from the uniform distribution.

12.23 **12.16 Ergodicity of Glauber dynamics.** Suppose that we are trying to find a random q-coloring of a graph with maximum degree D. Show that Glauber dynamics is ergodic if $q \geq D + 2$. That is, show that if D is a constant we can change any q-coloring to any other with $O(n)$ moves, each of which changes the color of a single vertex. Hint: show that we can change a given vertex to any color we like by changing its neighbors first.

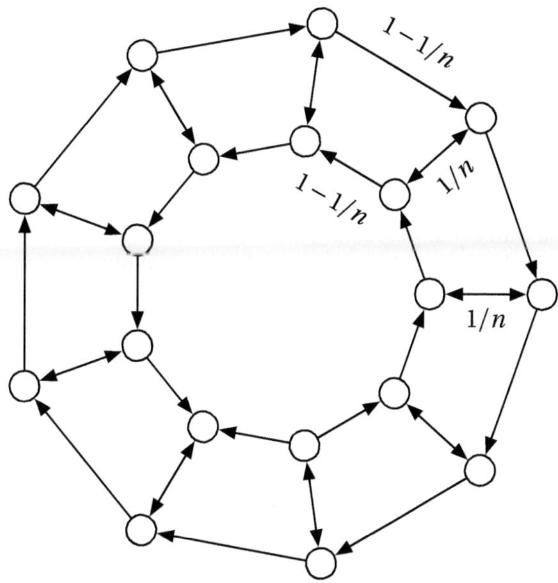

FIGURE 12.30: A lifted Markov chain, where each state corresponds to a position and direction on the cycle.

12.17 Frozen colors. Give a graph G of degree D such that Glauber dynamics is not ergodic on the set of q-colorings with $q = D+1$. In fact, give an example where every q-coloring is "frozen," so that no vertex can change its color without changing those of its neighbors as well. Hint: consider knight's moves on the square lattice.

12.18 Bad moves. Suppose we have two colorings C, C' that differ at a single vertex w. Let's say that $C(w) = 1$ and $C'(w) = 2$. Now consider a neighbor v of w. If the colors of v's other neighbors include both 1 and 2 then we can never have a bad move at v, since every color is either available to v in both colorings or in neither.

Now make the bold, and generally false, assumption that the colors of v's neighbors are random and independent. Show that when D is large, the expected number of bad moves becomes

$$D\left(1-(1-\mathrm{e}^{-D/q})^2\right).$$

Setting the number of good moves to $q\mathrm{e}^{-D/q}$ as in Section 12.5.4, show that this would imply rapid mixing for $q > \alpha D$ where $\alpha = 1.489...$ is the root of

$$\alpha \mathrm{e}^{-1/\alpha} = 1-(1-\mathrm{e}^{-1/\alpha})^2 .$$

12.19 Hard spheres. Suppose we want to sample independent sets of a graph—or as physicists call it, the *hard sphere model*, where two adjacent vertices cannot be occupied. To make larger sets more likely, we want the equilibrium probability of each set S to be proportional to $\lambda^{|S|}$ for some $\lambda > 1$. In physics, λ is called the *fugacity*.

Now consider the following Markov chain. First choose a random vertex v. With probability $1/(\lambda+1)$, remove v from S if it is already there. With probability $\lambda/(\lambda+1)$, add v to S, unless this is impossible because one of v's neighbors is already in S. Use a coupling argument to show that if the maximum degree of the graph is D, this Markov chain mixes rapidly if $\lambda < 1/(D-1)$.

12.20 Coupling on the Ising model. Consider the "heat-bath" dynamics on the Ising model described in Section 12.6.5, and analyze its effect on pairs of states x, y that differ at a single site i. By comparing the probability of good moves that set $x_i = y_i$ with the probability of bad moves that spread this difference to i's neighbors, prove that the critical temperature is bounded by $T_c \leq 4/\ln 3 = 3.641\ldots$. For comparison, the actual value is $T_c = 2.269\ldots$, as we will prove in Chapter 13.

12.21 Small spanning trees. Suppose I start at one vertex of a triangle $\{x, y, z\}$, perform a random walk where I move clockwise or counterclockwise with equal probability, and construct a spanning tree using the "first visit" process described in Section 12.6.1. Show by direct calculation that all three spanning trees are equally likely.

Hint: suppose that our first move is from x to y. From there, we might oscillate between x and y several times before finally visiting z and completing the tree. Show that it is twice as likely for our first visit to z to come from y as it is to come from x. Then summing this with the other possibility where our first step goes from x to z, show that the total probability of the three possible spanning trees is $1/3$ each.

12.22 Random spanning trees. We will show that the arrow-dragging Markov chain described in Section 12.6.1 produces uniformly random spanning trees, by proving a more general fact about its behavior in weighted graphs.

Suppose that we have a graph G where each edge from i to j is marked with a transition probability p_{ij}. For each rooted directed spanning tree T, define its weight $W(T)$ as the product of p_{ij} over all its directed edges $i \to j$. For instance, in a random walk where we move from i to a random neighbor j, we have $p_{ij} = 1/\deg(i)$. Show that in this case, since every vertex i but the root has an outgoing edge with weight $1/\deg(i)$, all trees with a given root v have the same weight, which is proportional to $\deg(v)$.

Now generalize the arrow-dragging process so that the root follows a random walk on G according to the transition probabilities p_{ij}. We will show that the equilibrium distribution of this process is proportional to $W(T)$. First suppose we currently have a tree T_v rooted at v. Every vertex w has a unique path in T_v from w to v, whose last edge passes through some neighbor of v. Denote this neighbor $u(w)$. Then show that for each neighbor w of v, T_v has one possible predecessor tree in the arrow-dragging process rooted at $u(w)$, which consists of taking T_v, adding the edge $v \to w$, and deleting the edge $u(w) \to v$. Denote this predecessor T'_w, and show that the probability the arrow-dragging process would go from T'_w to T_v is $p_{u(w),v}$. Finally, show that

$$W(T_v) = \sum_w p_{u(w),v} W(T'_w).$$

More abstractly, $MW = W$ where M is the transition matrix of the arrow-dragging process, so $P_{\text{eq}}(T)$ is proportional to $W(T)$. Conclude that if $p_{uv} = 1/\deg(u)$ and we remove the root and the arrows, all spanning trees are equally likely.

12.23 Kirchhoff's law. As a corollary to the previous problem, suppose I perform a random walk on a graph G with transition probabilities p_{ij}. Show that the equilibrium probability distribution $P_{\text{eq}}(v)$ is proportional to

$$\sum_{T_v} W(T_v) = \sum_{T_v} \prod_{(i \to j) \in T} p_{ij},$$

where the sum is over all spanning trees T_v rooted at v.

12.24 Broder's puzzle. Suppose I start at a vertex on a cycle and perform a random walk, in which I move clockwise or counterclockwise at each step with equal probability. I will eventually cross every edge, but for each edge there is some probability that it will be the last one I cross. Show that regardless of where I start, this probability is the same for every edge. Hint: consider the first-visit process of Section 12.6.1 for constructing random spanning trees.

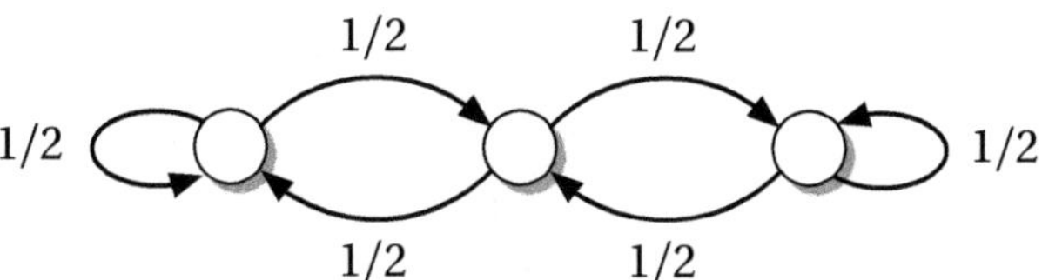

FIGURE 12.31: A random walk on a short line.

12.25 Equivalent edges. Let G be an *edge-transitive* graph with m edges, i.e., for every pair of edges e, e' there is an automorphism of G that sends e to e'. Again consider the random walk on G in which each step goes from the current vertex to a random neighbor. Suppose I start at an endpoint u of an edge $e=(u,v)$. As I walk, I will eventually reach its other endpoint v. Show that the probability I get to v without crossing e is exactly $1/m$. Hint: use Problem 12.24.

12.26 Cover times. We will show that the cover time t_{cover} of a random walk on a graph G with n vertices, i.e., the expected time it takes to visit every vertex if we move to a random neighbor at each step, is polynomial in n.

First review Exercise 12.7, which shows that if G has m edges, the equilibrium probability at a vertex i is $P_{\text{eq}}(i)=\deg(i)/2m$. Then for any two vertices i and j, let t_{ij} denote the expected time it takes a walker who starts at i to visit j for the first time. Show that t_{ii}, the expected number of steps it takes to return to i, is

$$t_{ii}=\frac{1}{P_{\text{eq}}(i)}=\frac{2m}{\deg(i)}.$$

Then, since we move to a random neighbor at each step, show that for any pair of neighboring vertices i,j we have

$$t_{ij}+t_{ji}\le 2m.$$

Now fix a spanning tree T of G. Let's say that the walker *traverses* T if for every edge $(i,j)\in T$, the walker gets from i to j and back again, although not necessarily along that edge. If we traverse T, we visit all the vertices, so the expected time it takes to traverse T is an upper bound on t_{cover}. Show that this gives

$$t_{\text{cover}}\le 2m(n-1)=O(n^3).$$

Since we only need $O(\log n)$ bits of memory to keep track of our current position, this gives another proof that UNDIRECTED REACHABILITY is in RL.

12.27 Lattices. Let S be a set with set with a partial order defined on it, such that for any two elements a,b there is a unique element $a\vee b$ which is the smallest element greater than both a and b. Formally:

$$a,b\preceq(a\vee b)$$
$$(a\vee b)\preceq c \text{ for any } c \text{ such that } a,b\preceq c.$$

Similarly, suppose that for any a,b there is a unique $a\wedge b$ which is the largest element smaller than both a and b. A set with these properties is called a *lattice*. For example, the set of points (x,y) in the plane forms a lattice if $(x,y)\preceq(x',y')$ if and only if $x\le x'$ and $y\le y'$, i.e., if (x,y) is below and to left of (x',y').

Show that any finite lattice has unique maximal and minimal elements. Then show that for any simply-connected region that can be tiled with rhombuses, the set of tilings forms a lattice.

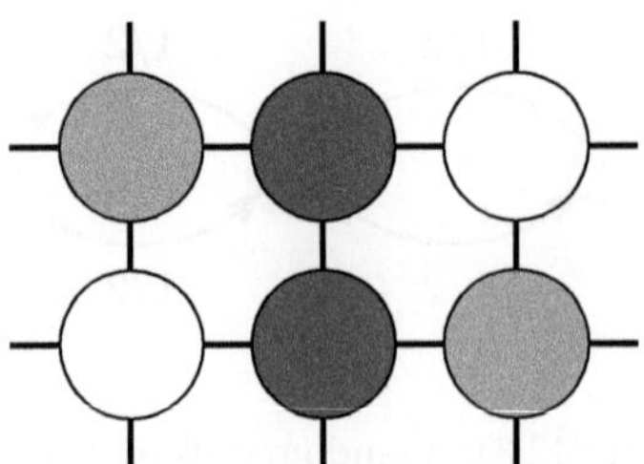

FIGURE 12.32: A topological defect in a 3-coloring of the square lattice. What happens when we try to fix it?

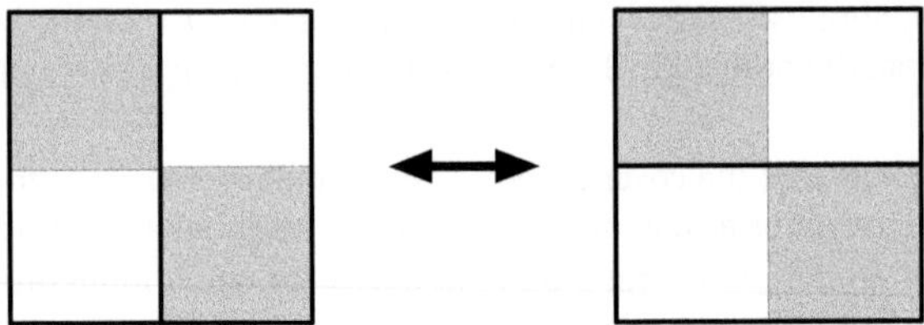

FIGURE 12.33: An elementary flip in a domino tiling.

12.28 Stopping too early. As we discussed in Section 12.6, coupling from the past does *not* give a uniform distribution if we stop at the first moment the maximal and minimal states coalesce. Consider the walk on the line of length 3 shown in Figure 12.31. At each step we choose a random direction, move in that direction if we can, and stay put if we can't. First show that the equilibrium distribution is uniform. Then suppose that we perform coupling from the past, starting one copy at the left end, the other at the right end, and attempting to move both states in the same direction at each step. Show that if we stop the moment the states first coalesce, the resulting distribution is not uniform. Indeed, the probability that we are at the middle vertex is zero.

12.29 Vortices of vertices. Suppose that we have a "topological defect" in a 3-coloring of the square lattice, in which two neighboring vertices have the same color and the surrounding vertices are colored as in Figure 12.32. If the rest of the coloring is proper, what happens if we try to fix it by changing the color of one of its vertices? What if it meets its mirror image? How does the height function behave in the vicinity of this defect?

12.30 Sperner's vortex. Use the height function for 3-colorings as an alternate proof of Sperner's Lemma (see p. 206).

12.31 When simple couplings fail. Show, for both rhombus tilings and 3-colorings of the square lattice, that a simple coupling where we try to apply the same move to the same vertex fails to coalesce. In other words, construct pairs of tilings or colorings that are more likely to get farther apart than closer together under this coupling.

12.32 Heights for dominoes. A natural Markov chain for domino tilings would be to choose a random vertex and, if it is surrounded by two horizontal or two vertical dominoes, to flip them as in Figure 12.33 with probability 1/2. Show that this Markov chain is ergodic for any simply-connected region, by designing a height function for domino tilings.

Hint: color the squares of the lattice white or black as in a checkerboard, and let the change in the height function as we move along the edge of a tile depend on the color of the square to our left. The vertex in the center of Figure 12.33 should be a local minimum or maximum of the height.

12.33 Zero probability. Let M be a stochastic matrix, which may or may not be symmetric, with a unique eigenvector $v_0 = P_{\text{eq}}$ of eigenvalue 1. Let $\mathbf{1}$ be the column vector of all 1s. Show that all of M's eigenvectors v_k other than v_0 obey $\mathbf{1}^T v_k = 0$. In other words, the total probability of each one is zero. Then assuming that M is diagonalizable, show that any probability distribution can be written in the form $P_{\text{eq}} + \sum_{k\neq 0} a_k v_k$.

12.34 Changing detailed balance to symmetry. Suppose that M obeys detailed balance. Consider the diagonal matrix T such that $T_{xx} = \sqrt{P_{\text{eq}}(x)}$, and show using (12.5) that $M' = T^{-1}MT$ is symmetric, although not necessarily stochastic. It follows that M' has the same eigenvalues as M, and therefore that its eigenvalues are real.

12.35 Bounding the mixing time. Prove the following generalization of Theorem 12.4.

Theorem 12.15 *Let M be an ergodic Markov chain with N states which obeys detailed balance and which has spectral gap δ. Let $P_{\min} = \min_x P_{\text{eq}}(x)$. Then its mixing time is bounded above by*

$$\tau_\varepsilon \le \frac{\ln(P_{\min}\varepsilon)^{-1}}{\delta}. \tag{12.47}$$

Then argue that in a physical system such as the Ising model, at any constant temperature this at worst increases the mixing time by a factor of n.

Hint: use the change of basis of Problem 12.34 so that $M' = TMT^{-1}$ symmetric and its eigenvectors are orthogonal, and bound the norm of Tv in terms of that of v. Then make an argument similar to that for the symmetric case. You may also find Problem 12.33 useful.

12.36 Tight mixing on the cycle. Let's reconsider the random walk on a cycle with N vertices. At each step, we move clockwise with probability 1/4, counterclockwise with probability 1/4, and stay where we are with probability 1/2. As in Section 12.7.2, let v_k denote the eigenvector

$$v_k = \frac{1}{N}\left(1, \omega^k, \omega^{2k}, \ldots, \omega^{(N-1)k}\right),$$

where $\omega = e^{2\imath\pi/N}$ and $k \in \{0,1,\ldots,N-1\}$ is the frequency.

Suppose that the initial distribution P_0 is concentrated at the origin $x = 0$. Show that the coefficients a_k in (12.21) are all equal to 1. (For the knowledgeable reader, this is just saying that the Fourier spectrum of the delta function is uniform.) Then applying the triangle inequality (Appendix A.2.2) gives

$$\left\|P_t - P_{\text{eq}}\right\|_{\text{tv}} \le \frac{1}{2}\sum_{k\neq 0} \lambda_k^t.$$

Sum over the eigenvalues λ_k, and use the inequality

$$\frac{1+\cos\theta}{2} \le e^{-\theta^2/4}$$

to show that the total variation distance is bounded by

$$\left\|P_t - P_{\text{eq}}\right\|_{\text{tv}} \le (1+o(1))\, e^{-(\pi^2/N^2)t}.$$

Therefore, the mixing time is at most

$$\tau_\varepsilon \le (1+o(1))\frac{N^2}{\pi^2}\ln\varepsilon^{-1} = O(N^2).$$

Along with the lower bound of (12.26), this shows that $\tau = \Theta(N^2)$.

12.37 Fourier transforming the hypercube. We showed in Section 12.3 using the Coupon Collector's argument that the mixing time for the random walk on the n-dimensional hypercube is $O(n \log n)$. However, we can compute the mixing time exactly by diagonalizing the transition matrix, as we did in Problem 12.36 for the walk on the cycle.

The state is a vector $\mathbf{x} \in \{0,1\}^n$. At each step, with probability $1/2$ we stay where we are, and with probability $1/2$ we choose a random $i \in \{1,\ldots,n\}$ and flip x_i. Show that M's eigenvectors, normalized so that $\|v_{\mathbf{k}}\|_2 = 1$, are

$$v_{\mathbf{k}}(\mathbf{x}) = \frac{1}{\sqrt{2^n}}(-1)^{\mathbf{k}^T\mathbf{x}},$$

where $\mathbf{k} \in \{0,1\}^n$ is the frequency vector. Show that their eigenvalues $\lambda_{\mathbf{k}}$ depend only on the *Hamming weight* of $\mathbf{k}$, i.e., the number of 1s. Show that the spectral gap is $\delta = 1/n$, and note that (12.26) then gives an upper bound on the mixing time of $\tau = O(n^2)$.

Then, assuming that the initial probability distribution is concentrated at the origin $\mathbf{0} = (0,\ldots,0)$, show using the Cauchy–Schwarz inequality and (12.21) that the total variation distance is bounded by

$$\left\|P_t - P_{\text{eq}}\right\|_{\text{tv}} \le \frac{1}{2}\sqrt{\sum_{\mathbf{k}\neq 0}|\lambda_{\mathbf{k}}|^{2t}} \le \frac{1}{2}\sqrt{\mathrm{e}^{n\mathrm{e}^{-2t/n}} - 1},$$

which improves our bound on the mixing time to

$$\tau_\varepsilon < \frac{1}{2} n \ln n\varepsilon^{-1}.$$

As the next problem suggests, the constant $1/2$ is exactly right.

12.38 Uncollected coupons. The previous problem shows that the mixing time of the random walk on the hypercube is at most $(1/2)n\ln n$. Returning to the Coupon Collecting picture, show that at this time there are typically still $\sqrt{n}$ bits that have not been touched, and which therefore have the same value they had initially. Explain why the walk can still be mixed even though this many bits are untouched. Then, argue that the constant $1/2$ is really the right one, and that $\tau = (1/2)n\ln n$, by sketching an argument that the total variation distance from the uniform distribution is still large at $t = a n \ln n$ for any constant $a < 1/2$.

12.24

12.39 Conductance and the gap. Prove the upper bound on the spectral gap in terms of the conductance, $\delta \le 2\Phi$, given by Theorem 12.6. Hint: use the fact that for any real, symmetric matrix M, its largest eigenvalue is given by

$$\lambda_{\max} = \max_v \frac{v^T M v}{|v|^2}.$$

Since we are interested in eigenvalues other than $\lambda_0 = 1$, we need to restrict this maximization to vectors whose total probability is zero (see Problem 12.33). Thus

$$\lambda_1 = 1 - \delta = \max_{v:\sum_x v(x)=0} \frac{v^T M v}{|v|^2}.$$

Then, given a set S of states with $|S| \le N/2$, consider the following vector:

$$v(x) = \begin{cases} 1/|S| & \text{if } x \in S \\ -1/|\overline{S}| & \text{if } x \notin S. \end{cases}$$

12.40 Surface to volume. Show by induction on n that for any set S of vertices in the n-dimensional hypercube with $|S| \le 2^{n-1}$, there are at least $|S|$ edges leading out of the set. Note that this is tight when S consists of, say, the left half of the hypercube. (This type of bound, relating the volume of a set to its surface area, is called an *isoperimetric inequality*.) Then show that the lazy walk on the hypercube has conductance $\Phi = 1/2n$. We saw in Problem 12.37 that the spectral gap is $\delta = 1/n = 2\Phi$, so in this case the upper bound of Theorem 12.6 is exact. Why is this?

12.41 Large gap, small diameter. Let G be a d-regular graph. Let M be the transition matrix of the random walk on G and suppose that M has spectral gap δ. Show that G's diameter is bounded by

$$D \le \frac{\ln N}{-\ln(1-\delta)} + 2 \le \frac{\ln N}{\delta} + 2,$$

giving a spectral proof that $D = O(\log N)$ if G is an expander. Hint: for any two vertices x, y, consider the vector v such that $v(x) = 1/2$, $v(y) = -1/2$, and $v(z) = 0$ for all other z. Show that if $\|M^t v\|_1 < 1$, the distance between x and y is at most $2t$. Then bound the 1-norm of $M^t v$ in terms of its 2-norm.

12.42 Cutting down the norm. Prove Lemma 12.9 on page 615. Hint: write $v = v_{\text{eq}} + v_{\text{neq}}$ where v_{eq} is proportional to the uniform distribution and v_{neq} is orthogonal to it, and use the triangle inequality. You may also find it useful to bound $\|v\|_1$ using the Cauchy–Schwarz inequality and the fact that v is supported only on $\overline{W}$.

12.43 Amplifying BPP with fewer random bits. In Section 10.9, we defined the class BPP of problems where a polynomial-time randomized algorithm gives the correct answer with probability at least 2/3. In other words, if the answer is "yes," then at least 2/3 of the potential witnesses work, and if the answer is "no," then at most 1/3 of them do. By running the algorithm k times and taking the majority as in Problem 10.46, we can reduce the probability P_{fail} of error to $2^{-\Theta(k)}$. But if each witness has n bits, this takes $\Theta(n \log P_{\text{fail}}^{-1})$ random bits.

Use the approach of Section 12.9.2 to show how to achieve this amplification with just $n + O(\log P_{\text{fail}}^{-1})$ random bits, by taking a walk on an expander. Hint: show that by taking a constant number of steps per round, we can reduce the probability of getting the wrong answer in a given round to $1/3 + \varepsilon$ for some small ε. Then use the union bound to show that the probability that the majority of rounds give the wrong answer is exponentially small.

12.44 The Expander Mixing Lemma. Let $G = (V, E)$ be a d-regular graph with N vertices, and let δ be the spectral gap of the simple random walk on G, where we move to each neighbor with probability $1/d$. For any subsets $A, B \subseteq V$, let $E(A, B)$ denote the number of edges connecting A to B. Then prove the following inequality, which is called the *expander mixing lemma*:

$$\left| \frac{E(A,B)}{dN} - \frac{|A||B|}{N^2} \right| \le (1-\delta)\sqrt{\frac{|A||B|}{N^2}}, \tag{12.48}$$

and then improve this slightly to

$$\left| \frac{E(A,B)}{dN} - \frac{|A||B|}{N^2} \right| \le (1-\delta)\sqrt{\frac{|A||B|(N-|A|)(N-|B|)}{N^4}}. \tag{12.49}$$

Show that if G were a random d-regular graph, the expectation of $E(A, B)$ would be $d\,|A||B|/N$. Thus if G is an expander and A and B are large enough, then $E(A, B)$ is close to what it would be in a random graph. (Compare Problem 10.47 where we achieve a similar kind of mixing with hash functions.)

Hint: let u_A be the vector such that $u_A(x) = 1$ if $x \in A$ and 0 if $x \notin A$, and similarly for u_B. Show that $u_A^T M u_B = E(A, B)/d$. Then write u_A and u_B as linear combinations of orthogonal eigenvectors of the transition matrix, and use the Cauchy–Schwarz inequality.

12.45 The Margulis expander. In this problem, we will use Fourier analysis to show that the graph G defined in Section 12.9.3 is an expander. For reasons that will be mysterious at first, we modify it as follows:

$$(x,y) \mapsto \begin{cases} (x+2y, y) \\ (x+2y+1, y) \\ (x, 2x+y) \\ (x, 2x+y+1). \end{cases} \tag{12.50}$$

In other words, each step of the random walk applies one of these two linear transformations,

$$T_1 = \begin{pmatrix} 1 & 2 \\ 0 & 1 \end{pmatrix} \quad T_2 = \begin{pmatrix} 1 & 0 \\ 2 & 1 \end{pmatrix},$$

and performs local diffusion by adding 1 to x or y with probability 1/2. Show that if the corresponding graph is an expander—that is, if this random walk has a constant spectral gap—then the graph defined by (12.37), or its undirected version, is an expander as well. Hint: take two steps.

Intuitively, the local diffusion process smooths out the high-frequency components of the probability distribution, while the shear operators T_1 and T_2 smooth out the low-frequency components. We will see that this is in fact the case. Each step of the random walk (12.50) updates a function $f : \mathbb{Z}_m^2 \to \mathbb{R}$ according to the transition matrix M,

$$Mf(x,y) = \frac{1}{4}\left(f(x+2y,y) + f(x+2y+1,y) + f(x,2x+y), f(x,2x+y+1)\right).$$

Now write f in the Fourier basis,

$$f(x,y) = \frac{1}{m}\sum_{k,\ell\in\mathbb{Z}_m} \tilde{f}(k,\ell)\,\omega^{kx+\ell y} = \frac{1}{m}\sum_{\mathbf{k}\in\mathbb{Z}_m^2} \tilde{f}(\mathbf{k})\,\omega^{\mathbf{k}^T\mathbf{x}}.$$

Here $\mathbf{x} = (x,y)$, $\mathbf{k} = (k,\ell)$, and $\omega = \mathrm{e}^{2\imath\pi/m}$ is the mth root of unity. Show that

$$\widetilde{Mf}(\mathbf{k}) = \frac{1}{2}\left(\frac{1+\omega^k}{2}\,\tilde{f}(T_2^{-1}\mathbf{k}) + \frac{1+\omega^\ell}{2}\,\tilde{f}(T_1^{-1}\mathbf{k})\right).$$

We denote the inner product of two complex-valued vectors as

$$\langle f, g\rangle = \sum_{\mathbf{x}\in\mathbb{Z}_m^2} f(\mathbf{x})^*\, g(\mathbf{x}).$$

Then writing M's largest eigenvalue as in Problem 12.39, we want to prove that, for some $\lambda < 1$,

$$\langle f, Mf\rangle \le \lambda\langle f, f\rangle,$$

or more explicitly

$$\sum_{\mathbf{x}\in\mathbb{Z}_m^2} f(\mathbf{x})^*\, Mf(\mathbf{x}) \le \lambda \sum_{\mathbf{x}\in\mathbb{Z}_m^2} \left|f(\mathbf{x})\right|^2,$$

for all f such that $\sum_{\mathbf{x}} f(\mathbf{x}) = 0$. Since the Fourier transform is unitary, it preserves inner products, so this implies

$$\langle \tilde{f}, \widetilde{Mf}\rangle = \sum_{\mathbf{k}\in\mathbb{Z}_m^2} \tilde{f}(\mathbf{k})^*\, \widetilde{Mf}(\mathbf{k}) \le \lambda \sum_{\mathbf{k}\in\mathbb{Z}_m^2} \left|\tilde{f}(\mathbf{k})\right|^2 = \langle \tilde{f}, \tilde{f}\rangle,$$

for all f such that $\tilde{f}(\mathbf{0}) = 0$ (show that this is equivalent to $\sum_{\mathbf{x}} f(\mathbf{x}) = 0$). Using the identity $\left|(1+\omega^k)/2\right| = \cos(\pi k/m)$, separating $\tilde{f}$ into its real and imaginary parts, and applying the triangle inequality, show that this would follow if

$$\frac{1}{2}\sum_{\mathbf{k}\in\mathbb{Z}_m^2} g(\mathbf{k})\left(\left|\cos\frac{\pi k}{m}\right| g(T_2^{-1}\mathbf{k}) + \left|\cos\frac{\pi\ell}{m}\right| g(T_1^{-1}\mathbf{k})\right) \le \lambda \sum_{\mathbf{k}\in\mathbb{Z}_m^2} \left|g(\mathbf{k})\right|^2, \tag{12.51}$$

for all nonnegative functions $g(\mathbf{k})$ such that $g(\mathbf{0}) = 0$.

Next, prove that, for any positive real γ, and for any a, b, we have

$$2|ab| \le \gamma|a|^2 + \frac{1}{\gamma}|b|^2. \tag{12.52}$$

To bound the left-hand side of (12.51), we will let γ depend on $\mathbf{k}$, and in a slightly different way for the two terms in the sum. Specifically, consider the following kind of function, where $\alpha < 1$ is a constant we will choose below:

$$\gamma(\mathbf{k},\mathbf{k}') = \begin{cases} \alpha & \|\mathbf{k}'\|_1 > \|\mathbf{k}\|_1 \\ 1 & \|\mathbf{k}'\|_1 = \|\mathbf{k}\|_1 \\ 1/\alpha & \|\mathbf{k}'\|_1 < \|\mathbf{k}\|_1 \,. \end{cases}$$

Here $\|\mathbf{k}\|_1 = |k| + |\ell|$, where we think of k and ℓ as ranging from $-m/2$ to $m/2-1$ instead of from 0 to $m-1$. Note that $\gamma(\mathbf{k}',\mathbf{k}) = 1/\gamma(\mathbf{k},\mathbf{k}')$.

Now apply (12.52) to the left-hand side of (12.51), using $\gamma(\mathbf{k}, T_2^{-1}\mathbf{k})$ for the first term and $\gamma(\mathbf{k}, T_1^{-1}\mathbf{k})$ for the second. Rearrange the sum and show that (12.51) holds if

$$\left|\cos\frac{\pi k}{m}\right| \left(\gamma(\mathbf{k}, T_2\mathbf{k}) + \gamma(\mathbf{k}, T_2^{-1}\mathbf{k})\right) + \left|\cos\frac{\pi \ell}{m}\right| \left(\gamma(\mathbf{k}, T_1\mathbf{k}) + \gamma(\mathbf{k}, T_1^{-1}\mathbf{k})\right) \le 4\lambda \tag{12.53}$$

holds for all $\mathbf{k} \neq \mathbf{0}$. Then show that it would be sufficient if, for each $\mathbf{k} \neq \mathbf{0}$, at least one of the following bounds is true:

$$\frac{1}{2\alpha}\left(\left|\cos\frac{\pi k}{m}\right| + \left|\cos\frac{\pi \ell}{m}\right|\right) \le \lambda\,, \tag{12.54}$$

$$\frac{1}{4}\left(\gamma(\mathbf{k}, T_2\mathbf{k}) + \gamma(\mathbf{k}, T_2^{-1}\mathbf{k}) + \gamma(\mathbf{k}, T_1\mathbf{k}) + \gamma(\mathbf{k}, T_1^{-1}\mathbf{k})\right) \le \lambda\,. \tag{12.55}$$

Note that (12.54) describes the effect of local diffusion—which we generously overestimate by a factor of $1/\alpha$—while (12.55) describes the effect of the shear operators. As we suggested above, these effects are sufficient to bound the eigenvalue in the high-frequency and low-frequency regimes respectively.

To prove this, we define the following set of "low frequency" vectors:

$$D = \{\mathbf{k} : \|\mathbf{k}\|_1 < m/4\}\,.$$

This is a diamond of radius $m/4$ centered at the origin $\mathbf{0}$. We have chosen D small enough so that T_1, T_2, and their inverses act on it just as they would on the integer lattice $\mathbb{Z}^2$, without any "wrapping around" mod m. In other words, as shown in Figure 12.34, these operators keep D inside the square where $k, \ell \in \{-m/2, \ldots, m/2-1\}$.

We are almost done. Show that, for all $\mathbf{k} = (k,\ell) \notin D$, we have

$$\frac{1}{2}\left(\left|\cos\frac{\pi k}{m}\right| + \left|\cos\frac{\pi \ell}{m}\right|\right) \le \cos\frac{\pi}{8} < 0.924\,,$$

and that for all $\mathbf{k} \in D$, we have

$$\frac{1}{4}\left(\gamma(\mathbf{k}, T_2\mathbf{k}) + \gamma(\mathbf{k}, T_2^{-1}\mathbf{k}) + \gamma(\mathbf{k}, T_1\mathbf{k}) + \gamma(\mathbf{k}, T_1^{-1}\mathbf{k})\right) \le 3\alpha + \frac{1}{\alpha}\,.$$

Hint: show that except on the k and ℓ axes and the lines $k = \pm\ell$, three out of the four operators increases $\|\mathbf{k}\|_1$ and one decreases it—and that on these lines, two of the operators keep $\|\mathbf{k}\|_1$ the same, and the other two increase it. What goes wrong if the off-diagonal entries of the shear matrices are 1 instead of 2?

Setting $\alpha = 0.95$, say, places an upper bound on the largest eigenvalue of $\lambda < 0.98$. This is larger than the correct value, but it proves that G is an expander.

12.46 Free groups and the complex plane. We can prove that the matrices A, A^{-1}, and B from (12.39) generate a free

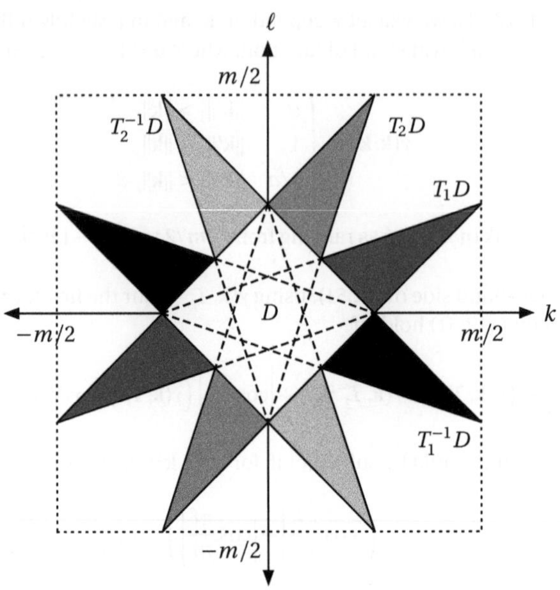

FIGURE 12.34: The diamond D, and its image under the shears T_1, T_2 and their inverses.

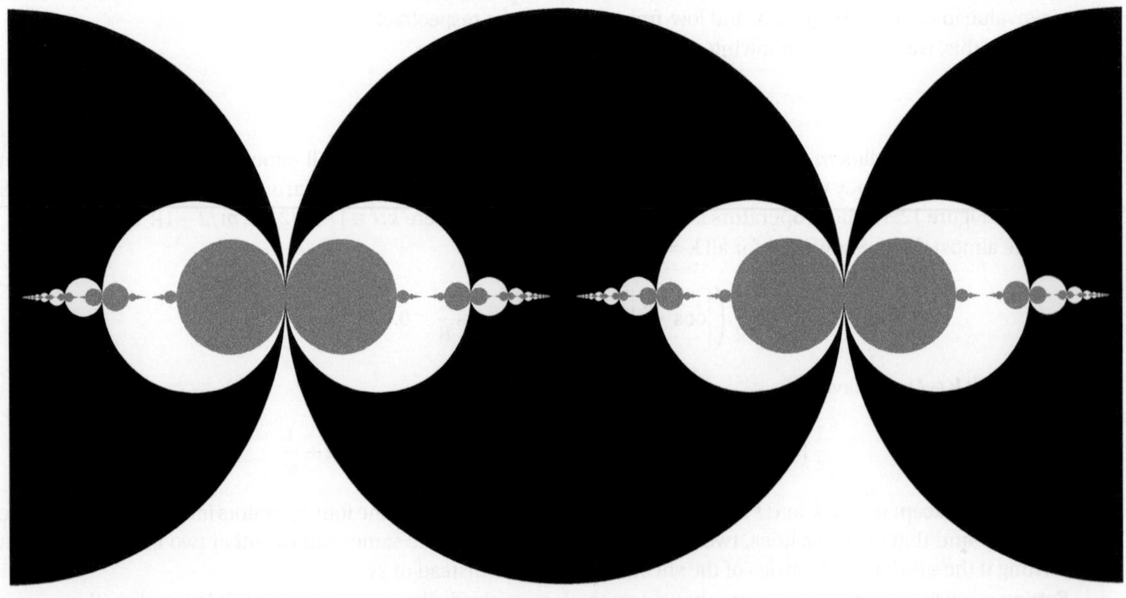

FIGURE 12.35: The action of the matrices (12.39) on the complex plane. A and A^{-1} shift it left and right, while B turns it inside out, mapping the exterior of the unit circle to its interior.

group, except for the relation $B^2 = 1$, by considering a sense in which these matrices act on the complex plane. If we multiply a vector $\begin{pmatrix} y \\ x \end{pmatrix}$ by a matrix $\begin{pmatrix} a & b \\ c & d \end{pmatrix}$, its slope $r = y/x$ is transformed to

$$r \to \frac{ar+b}{cr+d}.$$

This is called a *Möbius transformation* or *linear fractional transformation*; we saw it before in Problem 10.35. It maps circles to circles—or to straight lines, which are just circles whose radius is infinite—in the complex plane.

In our case, A and A^{-1} send r to $r \pm 2$, and B sends r to $1/r$. Show that if we start with the unit circle centered at the origin, applying a sequence of these transformations generates the circles shown in Figure 12.35. Use this picture to show that each circle in the figure corresponds to a unique word of the form

$$A^{t_1} B A^{t_2} B \cdots B A^{t_\ell},$$

for some $\ell \geq 0$ and $t_1, \ldots, t_\ell \in \mathbb{Z}$, and that no such word can evaluate to the identity.

12.47 Coin flips on trees. Let $P_{\text{return}}(t)$ be the probability that a random walk on a graph is back at its starting vertex on its tth step. Show that on an infinite d-regular tree where $d \geq 3$, $P_{\text{return}}(t)$ decreases exponentially with t.

Hint: show that if we flip t coins, each of which comes up heads with probability $1-1/d$ and tails with probability $1/d$, then P_{return} is at most the probability that no more than half of them come up heads. Then show that $P_{\text{return}}(t) \leq \mathrm{e}^{-bt}$ for some constant $b > 0$.

12.48 Generating functions and walks on trees. In this problem, we get a precise asymptotic estimate of the number of paths of length $t = 2\ell$ that return to their starting point, and therefore of $P_{\text{return}}(t)$, on an infinite d-regular tree. We will do this using the technique of generating functions. Let a_t denote the number of such paths. Then define the function

$$h(z) = \sum_{t=0}^{\infty} a_t z^t,$$

where z is a real or complex variable.

Of course, this sum might not converge. Show that if a_t grows as α^t then the value of z closest to the origin where the sum diverges, so that $h(z)$ has a singularity, is $z = 1/\alpha$. Thus the radius of convergence, within which $h(z)$ is analytic, determines the asymptotic growth of $a_t \sim \alpha^t$. Here $\sim$ hides subexponential factors, but one can determine these as well by analyzing the type of singularity. These techniques are presented with perfect clarity in Wilf's book *generatingfunctionology* [825], which no one should be without.

Let a_t be the number of paths of length t that start and end at a vertex v, and define the generating function

$$g(z) = \sum_{t=0}^{\infty} a_t z^t.$$

If we treat v as the root of the tree, v has d children. The subtree rooted at each of v's children is a tree where every vertex, including the root, has $d-1$ children. Let b_t be the number of paths of length t that start and end at the root of such a tree, and let $h(z)$ denote the corresponding generating function.

Now consider a path p of nonzero length that starts and ends at v. It starts by moving to one of v's children u. It follows a path in u's subtree that returns to u, and then moves back up to v, returning to v for the first time. After that, it can follow any additional path that returns to v. Translate this into the algebraic language of generating functions, and show that g and h obey these two quadratic equations:

$$g(z) = dz^2 g(z) h(z) + 1$$
$$h(z) = (d-1)z^2 h(z)^2 + 1.$$

Solve these equations for $g(z)$ and $h(z)$ and determine their radius of convergence. Conclude that, for even t,

$$a_t \sim \left(2\sqrt{d-1}\right)^t \quad \text{and} \quad P_{\text{return}}(t) \sim \left(\frac{2\sqrt{d-1}}{d}\right)^t .$$

As before, $\sim$ hides subexponential factors. In fact, all that is missing is a polynomial term. Derive the following exact expression for b_t, which is a lower bound for a_t:

$$b_t = (d-1)^{t/2} \frac{1}{t/2+1} \binom{t}{t/2} = \Theta\left(t^{-3/2} \left(\frac{2\sqrt{d-1}}{d}\right)^t \right) .$$

Hint: the number of paths of length 2ℓ on the integers that start and end at the origin, move one step left or right at each step, and never move to the left of the origin, is the Catalan number $\frac{1}{\ell+1}\binom{2\ell}{\ell}$.

12.49 Ramanujan graphs. How much expansion is possible in a d-regular graph? Equivalently, how small can the leading eigenvalue be? Suppose that G has N vertices and is connected and non-bipartitite. As in the previous two problems, let $P_{\text{return}}(t)$ denote the probability that a random walk on G is back at its starting point after exactly t steps, and let λ_1 be the largest eigenvalue of the transition matrix M other than 1. Show that for any even t,

$$\lambda_1 \geq \left(P_{\text{return}}(t) - \frac{1}{N-1} \right)^{1/t} .$$

Hint: consider the trace $\operatorname{tr} M^t$ and use the fact that the trace is basis-independent. Now apply the results of Problem 12.48, and conclude that, for any family of expanders, in the limit $N \to \infty$ we have

$$|\lambda_1| \geq \frac{2\sqrt{d-1}}{d} ,$$

Expanders that achieve this bound do indeed exist, and are called *Ramanujan graphs.*

12.19 **12.50 The zig-zag product.** Prove Theorem 12.10. Hint: let M be the transition matrix of the random walk on $G ⓩ H$. Write $M = BAB$ where B is the transition matrix of the random walk on H, taken simultaneously on every cloud, and A is the permutation matrix which sends each vertex to its partner in a neighboring cloud. Intuitively, the zigs and zags performed by B reduce the nonuniformity of the probability within each cloud, and A reduces the nonuniformities between clouds.

To confirm this intuition, let $\mathbf{1}$ denote the uniform distribution on $G ⓩ H$, and let $\mathbf{1}_g$ denote the uniform distribution on the cloud corresponding to $g \in V_G$, so that $\mathbf{1} = (1/|V_G|)\sum_g \mathbf{1}_g$. Consider a probability vector v defined on the vertices of $G ⓩ H$ such that v is orthogonal to $\mathbf{1}$. Write v as

$$v = v_\parallel + v_\perp ,$$

where, in each cloud, $v_\parallel$ is proportional to the uniform distribution and $v_\perp$ is orthogonal to it,

$$v_\parallel = \sum_g w(g) \mathbf{1}_g \quad \text{and} \quad \forall g : \mathbf{1}_g^T v_\perp = 0 ,$$

where $w(g)$ is some probability vector defined on G. Since $v_\parallel^T v_\perp = 0$, we have

$$\|v\|^2 = \|v_\parallel\|^2 + \|v_\perp\|^2 .$$

Now prove the following inequalities:

$$v_\parallel^T M v_\parallel \leq \lambda_G \|v_\parallel\|^2 , \quad v_\parallel^T M v_\perp \leq \lambda_H \|v_\parallel\| \|v_\perp\| , \quad v_\perp^T M v_\perp \leq \lambda_H^2 \|v_\perp\|^2 .$$

Finally, show that $v^T M v \leq \lambda \|v\|^2$ where λ is given by (12.40).

Notes

12.1 Further reading. Our exposition owes a great deal to review articles by Randall [681], Jerrum and Sinclair [434], Aldous [29], Aldous and Diaconis [31], Vazirani [809], and Dyer and Greenhill [255], as well as the book by Mitzenmacher and Upfal [590].

For further reading, we strongly recommend the book *Markov Chains and Mixing Times* by Levin, Peres, and Wilmer [526]. For a physicists' point of view, we recommend *Monte Carlo Methods in Statistical Physics* by Newman and Barkema [631], which explains the theory and practice of sampling algorithms in physics.

12.2 Boltzmann. In Chapter 7, we paid homage to the great unifying moments in physics, when very different phenomena, each with its own style of explanation, were revealed to be different aspects of the same underlying laws. One such moment was Ludwig Boltzmann's creation of statistical mechanics, which unified steam-engine thermodynamics with the microscopic physics of atoms and molecules.

Boltzmann's original expression for the entropy, $S = \ln W$ (which is engraved on his tombstone) applies when there are W states and each one is equally likely. More generally, when they are distributed according to a probability distribution $p(x)$, it is given by

$$S = -\sum_x p(x) \ln p(x). \tag{12.56}$$

For those familiar with physics, note that we ignore Boltzmann's constant k_B by measuring temperature and energy in the same units.

The expression (12.44) was found by William Gibbs, although if we restrict ourselves to the distribution of velocities in a gas it is essentially (with a minus sign) the quantity Boltzmann called H in his discussions of the Second Law of Thermodynamics. It was rediscovered half a century later by Shannon [745], who used it to describe the information capacity of a communication channel. If we use base 2 for the logarithm, S is the average number of bits of information conveyed per symbol of a message, where $p(x)$ describes the probability distribution of the symbols. We now recognize that thermodynamic entropy is really an informational quantity, representing our uncertainty about a system's microscopic state. The second law of thermodynamics is then the fact that this uncertainty can only increase as information is lost to thermal noise.

In the explanations of the Boltzmann distribution we give in Problems 12.2 and 12.3 we assume that, at equilibrium, the probability distribution is the one that maximizes the entropy subject to constraints like the conservation of energy. In particular, we assume that all states with the same energy are equally likely. Boltzmann sought a dynamical justification for this assumption, and he is credited with inventing the so-called *ergodic hypothesis* that a system's trajectory will pass through—or arbitrarily close to—every state consistent with the conservation of energy. This assumption can be proved rigorously in some simple cases using the theory of chaotic dynamical systems. We should note, however, that this is only one of the many justifications Boltzmann entertained in his writings. Others included the idea that systems are perturbed randomly by outside forces or by interactions with their environment [141, 186, 523].

12.3 Free energy. One consequence of the Second Law of Thermodynamics is that not all of a system's energy can be converted into mechanical form. The free energy $F = E - TS$ is the amount of energy that is available to run, say, a steam engine, as opposed to being locked away in the form of thermal noise. It can be written as

$$F = -\frac{1}{\beta} \ln Z,$$

where

$$Z = \sum_x \mathrm{e}^{-\beta E(x)}$$

is the partition function mentioned in Problem 12.3. In particular, we have $Z = W\mathrm{e}^{-\beta E} = \mathrm{e}^{-\beta F}$ in the simple case where every state has the same energy E.

12.4 Metropolis. Metropolis dynamics is named after Nicholas Metropolis, one of the five authors of [573], who used it in 1953 to simulate a hard-sphere model on the MANIAC computer at Los Alamos. It also goes by the name of Gibbs sampling. We look forward to the invention of Gotham City dynamics.

12.5 Rapid vs. polynomial. In the computer science literature, a mixing time τ that is polynomial in the number n of sites or vertices is often referred to as *rapid mixing*. We reserve that term for a mixing time $\tau = O(n \log n)$, which by the Coupon Collecting intuition is as fast as possible.

12.6 Card shuffling. The model of the riffle shuffle in Problem 12.8 was formalized by Gilbert and Shannon [329] and independently by Reeds. The Birthday Problem analysis of its mixing time is due to Reeds; an excellent exposition can be found in Aldous and Diaconis [31]. The more precise result of $(3/2)\log_2 n$ is from Bayer and Diaconis [88]. They give several other descriptions of riffle shuffles, including one based on chaotic maps of the unit interval. The card-marking argument in Problem 12.13 that the random transposition shuffle mixes in $O(n \log n)$ time is due to Broder; see Diaconis [231].

Even though we need about 9 shuffles to get close to the uniform distribution on the 52! permutations of a deck of cards, most people are used to distributions that result from shuffling the deck only 3 times. We are told that, consciously or subconsciously, expert Bridge players use the correlations present in the deck to their advantage.

12.7 Glauber dynamics. Glauber dynamics is named after Roy J. Glauber, who considered the evolution of the Ising model in continuous time [332]. He later won the Nobel prize for his work in quantum optics.

There is some inconsistency in the Markov chain literature about what Glauber dynamics means. Some take it to mean any Markov chain that changes the state of one site or vertex at a time, while others take it to mean a particular update rule, such as Metropolis or heat-bath dynamics.

12.8 As rapid as possible. Hayes and Sinclair [387] proved that in most natural cases, including random colorings of graphs with constant degree, the mixing time really is $\Omega(n \log n)$. However, they also point out that in some cases, the mixing time is faster than the Coupon Collecting argument would suggest. For instance, suppose we sample independent sets S with a Markov chain like that in Problem 12.19: choose a random vertex, remove it from S if it is already there, and otherwise add it if you can. On the complete graph with n vertices, there are only $n+1$ independent sets, namely the empty set and the n sets of size one. Then if we start with $S = \{v\}$, we reach $S = \emptyset$ as soon as we remove v, and the next step gives $S = \{w\}$ where w is uniformly random. Thus, on the complete graph, this chain mixes in $O(n)$ time.

12.9 Graph colorings. Let α be the best ratio for which we can prove rapid mixing for Glauber dynamics whenever the number of colors is greater than α times the maximum degree, i.e., when $q > \alpha D$. The result $\alpha = 2$ was found independently by Jerrum [431] and Salas and Sokal [724]. Bubley and Dyer [142] introduced the path coupling technique of Theorem 12.3 and used it to show polynomial-time mixing for $q = 2D$. Molloy [594] improved this to show rapid mixing for $q = 2D$.

The first breakthrough beyond $\alpha = 2$ occurred when Vigoda [813] proved that Glauber dynamics mixes in polynomial time whenever $\alpha = 11/6 = 1.833\ldots$. He did this by proving rapid mixing for a Markov chain that recolors clusters of vertices simultaneously. Dyer and Frieze [256] proved rapid mixing for $\alpha = 1.763\ldots$ by showing, as we discuss in Section 12.5.4, that the neighbors of a given vertex are essentially independent of each other as long as graph has girth and maximum degree $\Omega(\log n)$. Molloy [595] gave the argument of Problem 12.18 and improved their result to $\alpha = 1.489\ldots$. Hayes [386] showed that these results hold even for constant girth, and Dyer, Frieze, Hayes, and Vigoda [257] showed that they hold for constant degree as well.

Hayes and Vigoda [389] presented the idea of *coupling with stationarity*, in which we assume that one of the colorings is random. This provides a much simpler proof of the results of [256], and allows them to reduce the required girth to four—that is, they simply require that G has no triangles.

Although non-Markovian couplings had been discussed mathematically for some time, Hayes and Vigoda [388] were the first to use them to prove rapid mixing. They showed rapid mixing for $q > (1+\varepsilon)D$ for any $\varepsilon > 0$, assuming that $D = \Omega(\log n)$ and the girth is at least 9.

12.10 Spanning trees and time reversal. The "arrow-dragging" and "first visit" processes for spanning trees are due to Broder [137] and Aldous [30], in conversations with Diaconis. Problems 12.22, 12.23, and 12.24 are from [137]. Problem 12.23 is a form of the Matrix-Tree Theorem, which goes back to Kirchhoff and which we will see again in Chapter 13. Our hint follows a proof due to Diaconis as quoted in [137].

The argument from Problem 12.26 that the cover time in a graph of n vertices is $O(n^3)$ is from Aleliunas, Karp, Lipton, Lovász, and Rackoff [33]. The "falling leaves" model, including the viewpoint of an animal looking up from its hole, was described by Kendall and Thönnes [473]. The backward simulation of the model was also discussed by Jeulin [437].

The question of whether we get perfectly uniform spanning trees by running the forward arrow-dragging process until it covers the graph is still open. In this particular case, it may be that the forward process works just as well as the backward one, even though Exercise 12.16 shows that this isn't true in general.

12.11 Topological defects. Topological defects are of great interest in physics. Defects of opposite type like the clockwise and counterclockwise ones described in Problem 12.29 cannot be removed locally, but pairs of opposite type can annihilate when they meet [498]. Another example consists of isolated squares in domino tilings, which are "white holes" or "black holes" depending on their color on the checkerboard.

Such defects typically have attractive or repulsive forces inversely proportional to the distance between them, like those between charged particles or fluid vortices. These forces are driven purely by entropy; for instance, the number of ways to color or tile the lattice increases as a pair of opposing defects get closer together. See Fisher and Stephenson [284] for analytical results on domino tilings, and Moore, Nordahl, Minar, and Shalizi [605] for numerical experiments. In higher dimensions, such defects can take the form of strings or membranes. For instance, 3-colorings of the cubic lattice can have topological defects that are analogous to cosmic strings.

12.12 Coupling from the past. Coupling from the past, including applications to monotone couplings, prepending random moves, and doubling T until the extreme states coalesce, was first discussed in 1996 by Propp and Wilson [673]. Since then, it has been applied to a wide variety of Markov chains, including some where monotonicity is not immediately obvious.

For instance, we can sample states of the Ising model, or more generally the ferromagnetic q-state Potts model, by using a *random cluster model* found by Fortuin and Kasteleyn [287]. First we choose a random subgraph of the square lattice by adding and removing random edges, where the transition probabilities depend on the number of edges and the number of connected components. We then choose a random state for each connected component, and assign it to all the sites in that component. It turns out that for any $q \geq 1$, the Markov chain that adds and removes edges is monotone, where $x \preceq y$ if every edge present in x is also present in y. The maximal and minimal states consist of including every edge in one huge cluster (in which case every site has the same state) and no edges (in which case the sites are independent). We derive the Fortuin–Kasteleyn representation of the Ising model in Problem 13.35.

12.13 Arctic circles. For domino tilings of the Aztec diamond, Jockusch, Propp, and Shor [440] and Cohn, Elkies, and Propp [187] proved that the boundary between the frozen and non-frozen regions is asymptotically a circle. Cohn, Larsen, and Propp [189] then showed that the same is true for rhombus tilings. The reader should notice that as we go around the boundaries of these regions, the height function increases as quickly as possible along one edge, and then decreases as quickly as possible along the next one. This forces the height function to take a saddle shape in the interior—or, in the case of the hexagon, a saddle for someone with a tail.

12.14 Mixing times for tilings. The fact that the set of rhombus tilings of a hexagon is connected under the flips of Figure 12.13, and therefore that there are an equal number of rhombuses of each orientation in any tiling of a hexagon, was shown in "The Problem of the Calissons" by David and Tomei [218].

Luby, Randall, and Alistair Sinclair [545] found Markov chains with polynomial mixing times for rhombus tilings, domino tilings, and 3-colorings—or equivalently, Eulerian orientations—of the square lattice. They use Markov chains with "macromoves" which modify many sites at once.

Randall and Tetali [682] showed that the single-site Markov chains we discuss here also mix in polynomial time. They do this by proving that each macromove can be simulated by a path of single-site micromoves (i.e., the flips shown in Figures 12.13 and 12.33 for rhombus and domino tilings, or Glauber dynamics for 3-colorings) and that these paths are not too congested. It follows that if a good multicommodity flow can be constructed with macromoves, one can be constructed with micromoves as well, and so the mixing time of the single-site chains is at most polynomially larger than the macromove chains.

Finally, Wilson [827] gave somewhat tighter bounds on mixing times for rhombus tilings by analyzing the same couplings with a different notion of distance.

12.15 Height functions for magnets and ice. Height functions began in statistical physics as a way of mapping two-dimensional spin systems onto interfaces or surfaces in three-dimensional space. The height function for Eulerian orientations was first defined by van Beijeren [805]. It generalizes to Eulerian orientations of any planar graph, such as the "20-vertex model" (count them!) on the triangular lattice. The correspondence between Eulerian orientations and 3-colorings of the square lattice was pointed out by Lenard [533].

Physicists refer to Eulerian orientations as the *6-vertex model* since there are six ways to orient the edges around a given vertex. It was originally invented as a model of ice, in which each vertex corresponds to an oxygen atom and each edge corresponds to a hydrogen bond. Each edge's orientation indicates to which oxygen atom the proton in that bond is currently attached, and each oxygen atom must have exactly two protons.

The height function for the triangular antiferromagnet, and its equivalence to the surface of a set of cubes and hence to rhombus tilings, was found by Blöte and Hilhorst [112]. The height function for domino tilings seems to have appeared first in Zheng and Sachdev [847] and Levitov [528].

In a number of situations we can define a "height function" whose values are two- or more-dimensional vectors. For instance, 4-colorings of the triangular lattice have two-dimensional height functions [86, 603], so that a random coloring corresponds to a random 2-dimensional surface in 4-dimensional space! An interesting research question is whether these vector-valued height functions can be used to prove polynomial mixing times.

However, it should be noted that the existence of a height function of any kind is, in some sense, a happy accident. For 4-colorings of the square lattice, for instance, no height function exists, and as noted in Section 12.10 we currently have no proof that Glauber dynamics is rapidly mixing, despite overwhelming numerical evidence to this effect. For many simple types of tiles, such as the "right trominoes" of Section 5.3.4, we don't know of any set of local moves that is ergodic, or any other method of quickly sampling a random tiling.

12.16 Fourier analysis. The cycle is the Cayley graph of the cyclic group $\mathbb{Z}_N$, consisting of the integers mod N with addition, where the generators are $\{\pm 1\}$. Similarly, the n-dimensional hypercube is the Cayley graph of the group $\mathbb{Z}_2^n$ of n-dimensional vectors mod 2. In general, given a group G and a set of generators Γ, we can consider a random walk where each step applies a randomly chosen element of Γ. The eigenvectors of such a walk are then exactly the Fourier basis functions. For card shuffles, the group in question is the permutation group S_n. Since this group is nonabelian, i.e., multiplication is not commutative, we need to consider matrix-valued functions known as *representations*, which we discuss briefly in Section 15.6. The central reference on this technique is Diaconis [231].

12.17 High conductance, large gap. In this note, we prove the lower bound on the spectral gap in terms of the conductance, $\delta \geq \Phi^2/2$. The idea of the proof is that, if the conductance is large, probability will quickly flow "downhill" from high-probability states to low-probability ones.

Proof Let v be an eigenvector with eigenvalue $\lambda = 1-\delta$. We sort the states in descending order, so that $v(x) \ge v(y)$ if $x < y$. Since v is orthogonal to P_{eq}, we have $\sum_x v(x) = 0$. Let S be the set of states x such that $v(x)$ is positive. Without loss of generality, we can assume that $|S| \le N/2$, since otherwise we can deal with $-v$ instead of v.

It's handy to define a vector $\hat{v}$ that is zero wherever $v(x)$ is negative:

$$\hat{v}(x) = \max\left(v(x), 0\right) = \begin{cases} v(x) & x \in S \\ 0 & x \in \overline{S}. \end{cases}$$

Now consider the inner product $\hat{v}^T(\mathbb{1} - M)v$. On one hand, we have

$$(\mathbb{1} - M)v = (1-\lambda)v = \delta v,$$

so

$$\hat{v}^T(\mathbb{1} - M)v = \delta\, \hat{v}^T v = \delta \sum_x \hat{v}(x)^2. \tag{12.57}$$

On the other hand, since $\hat{v}^T M v \le \hat{v}^T M \hat{v}$, we also have

$$\begin{aligned} \hat{v}^T(\mathbb{1} - M)v &\ge \hat{v}^T v \; - \; \hat{v}^T M \hat{v} \\ &= \sum_x \hat{v}(x)^2 - \sum_{x,y} \hat{v}(x)\, \hat{v}(y) M(x \to y) \\ &= \sum_x \hat{v}(x)^2 (1 - M(x \to x)) - \sum_{x \ne y} \hat{v}(x)\, \hat{v}(y) M(x \to y). \end{aligned}$$

Since M is stochastic, $1 - M(x \to x) = \sum_{y \ne x} M(x \to y)$. So,

$$\begin{aligned} \hat{v}^T(\mathbb{1} - M)v &\ge \sum_{x \ne y} \left(\hat{v}(x)^2 - \hat{v}(x)\, \hat{v}(y)\right) M(x \to y) \\ &= \frac{1}{2} \sum_{x \ne y} \left(\hat{v}(x) - \hat{v}(y)\right)^2 M(x \to y) \\ &= \sum_{x<y} \left(\hat{v}(x) - \hat{v}(y)\right)^2 M(x \to y). \end{aligned} \tag{12.58}$$

Combining (12.57) and (12.58) gives

$$\delta \ge \frac{\sum_{x<y} \left(\hat{v}(x) - \hat{v}(y)\right)^2 M(x \to y)}{\sum_x \hat{v}(x)^2}. \tag{12.59}$$

This bounds δ in terms of the flow of probability downhill from x to y, for each pair $x < y$. To relate δ directly to the conductance, we need a bit more algebraic trickery. First, note that

$$\begin{aligned} \sum_{x<y} \left(\hat{v}(x) + \hat{v}(y)\right)^2 M(x \to y) &\le 2 \sum_{x<y} \left(\hat{v}(x)^2 + \hat{v}(y)^2\right) M(x \to y) \\ &= 2 \sum_x \hat{v}(x)^2 \sum_{y \ne x} M(x \to y) \\ &\le 2 \sum_x \hat{v}(x)^2. \end{aligned}$$

Along with (12.59), this implies

$$\delta \ge \frac{1}{2} \frac{\left(\sum_{x<y} \left(\hat{v}(x) - \hat{v}(y)\right)^2 M(x \to y)\right) \left(\sum_{x<y} \left(\hat{v}(x) + \hat{v}(y)\right)^2 M(x \to y)\right)}{\left(\sum_x \hat{v}(x)^2\right)^2},$$

and using the Cauchy–Schwarz inequality in reverse gives

$$\delta \geq \frac{1}{2}\left(\frac{\sum_{x<y}\left(\hat{v}(x)^2 - \hat{v}(y)^2\right) M(x \to y)}{\sum_x \hat{v}(x)^2}\right)^2 . \tag{12.60}$$

Next, we write the numerator of (12.60) as a telescoping sum. This lets us relate it to the equilibrium flow from S_z to $\overline{S_z}$, where S_z denotes the set of states $\{1, \ldots, z\}$ for each $z \in S$:

$$\begin{aligned}
\sum_{x<y}\left(\hat{v}(x)^2 - \hat{v}(y)^2\right) M(x \to y) &= \sum_{x<y}\left(\sum_{z:x\leq z<y} \hat{v}(z)^2 - \hat{v}(z+1)^2\right) M(x \to y) \\
&= \sum_{z\in S}\left(\hat{v}(z)^2 - \hat{v}(z+1)^2\right) \sum_{x\leq z, y>z} M(x \to y) \\
&= N \sum_{z\in S}\left(\hat{v}(z)^2 - \hat{v}(z+1)^2\right) Q\left(S_z \to \overline{S_k}\right) ,
\end{aligned} \tag{12.61}$$

since $P_{\text{eq}}(x) = 1/N$. Since $|S| \leq N/2$, for all $z \in S$ we have

$$Q\left(S_z \to \overline{S_k}\right) \geq P_{\text{eq}}(S_z)\Phi = (z/N)\Phi ,$$

so (12.61) becomes

$$\begin{aligned}
\sum_{x<y}\left(\hat{v}(x)^2 - \hat{v}(y)^2\right) M(x \to y) &\geq \Phi \sum_{z\in S} z\left(\hat{v}(z)^2 - \hat{v}(z+1)^2\right) \\
&= \Phi \sum_{z\in S} \hat{v}(z)^2 .
\end{aligned}$$

Finally, combining this with (12.60) gives

$$\delta \geq \frac{1}{2}\Phi^2 ,$$

and completes the proof. □

12.18 Conductance and flows. For random walks on graphs, the connection between expansion and conductance was laid out by Dodziuk [246] and independently by Alon [37], building on Cheeger's work bounding the eigenvalues of the Laplacian operator on continuous manifolds [155]. For this reason, the conductance is sometimes called the Cheeger constant. More general bounds relating the spectral gap of a Markov chain to its conductance were found by Sinclair and Jerrum [756]—from which our proof of Theorem 12.6 is taken almost verbatim—and independently by Lawler and Sokal [521].

By choosing "canonical paths" between pairs of states and defining a flow along these paths, Jerrum and Sinclair [432] and Jerrum, Sinclair, and Vigoda [435] proved rapid mixing for a Markov chain that counts perfect matchings and thus approximates the permanent—a major result that we will discuss in Section 13.4. Canonical paths also appeared in Diaconis and Stroock [234], and were generalized to multicommodity flows by Sinclair [755].

The bound on the mixing time given by Theorem 12.7 can be improved considerably. We can bound the mixing time directly in terms of the congestion, and for most graphs this makes τ linear in ρ rather than quadratic [755]. However, we then need to take the lengths of the paths between each pair of states into account, rather than just their capacities.

12.19 Expanders. Our discussion of expanders, especially of the zig-zag product in Section 12.9.5, owes a great deal to the review paper of Hoory, Linial, and Wigderson [408]. The technique of amplifying RP and BPP algorithms discussed in Section 12.9.2 and Problem 12.43 was discovered independently by Cohen and Wigderson [185] and

Impagliazzo and Zuckerman [420]. The bound on the diameter of an expander given in Problem 12.41 was given by Chung; see her book [169] for a survey of spectral techniques in graph theory.

Margulis [555] used group-theoretic techniques to prove that his graphs—with a slightly different set of linear operators—are expanders for some constant $\mu > 0$. Gabber and Galil [309] defined the variant we give here, and derived an explicit bound on μ by treating the probability distribution as a continuous function on the torus and using Fourier analysis. Jimbo and Maruoka [439] analyzed this and other expanders using Fourier analysis on the group $\mathbb{Z}_m^2$ instead of passing to the continuous setting.

The group generated by the matrices in (12.39) is the free product of the cyclic group $\{\ldots,A^{-1},1,A,A^2,\ldots\}$ generated by A and the group $\{1,B\}$ generated by B. We can get a completely free group, and a 4-regular expander, by taking the matrices

$$A=\begin{pmatrix}1 & 2\\ 0 & 1\end{pmatrix} \quad\text{and}\quad BAB=\begin{pmatrix}1 & 0\\ 2 & 1\end{pmatrix},$$

along with their inverses.

The lower bound on the largest eigenvalue in Problem 12.49 is due to Alon and Boppana [37]. Our proof follows Lubotzky, Phillips, and Sarnak [542], who gave the first examples of explicit Ramanujan graphs. Their construction consists of a Cayley graph in $\mathrm{SL}_2(\mathbb{F}_q)$, the group of 2×2 matrices with determinant 1 mod q, with a carefully-chosen set of generators.

12.20 The zig-zag product. The zig-zag product, including Theorems 12.10 and 12.11, was found by Reingold, Vadhan, and Wigderson [698], who won the Gödel Prize in 2009. Independent results of Martin and Randall [557] imply similar bounds on the eigenvalue of the *replacement* product of two graphs, where we replace each vertex with a cloud, but don't zig and zag quite as much. See also Rozenman and Vadhan [716], who use a somewhat different construction to square a graph without too large an increase in its degree.

The zig-zag product can also be defined for "biregular" directed graphs, namely those where every vertex has the same in-degree and the same out-degree. See Reingold, Trevisan, and Vadhan [697], who provided a simpler proof of a bound similar to Theorem 12.11, and showed that REACHABILITY for such graphs is also in LOGSPACE. More generally, Chung, Reingold, and Vadhan [170] showed that REACHABILITY is in LOGSPACE for any directed graph where the equilibrium distribution is known, where s and t each have $1/\mathrm{poly}(n)$ probability, and where the mixing time is $\mathrm{poly}(n)$.

12.21 Spatial mixing and coloring the square lattice. Our proof sketch for Theorem 12.14 follows Dyer, Sinclair, Vigoda and Weitz [259]. Another proof that spatial and temporal mixing are equivalent was given by Martinelli [558], but it uses functional analysis rather than elementary combinatorial methods.

Spatial mixing for the antiferromagnetic Potts model on the square lattice follows for $q = 8$ from the general results of Dyer and Greenhill [254], who showed that a heat-bath algorithm that updates both ends of a random edge mixes rapidly for $q \geq 2D$ on graphs of maximum degree D. Bubley, Dyer and Greenhill [143] considered an algorithm that updates a star-shaped neighborhood, and showed that it mixes rapidly for $q = 7$ on any triangle-free graph with degree 4. Spatial mixing was shown for $q = 6$ by Achlioptas, Molloy, Moore, and Van Bussell [13] using a Markov chain that updates 2×3 patches, and using different methods by Goldberg, Martin, and Paterson [337].

However, it should be noted that all existing proofs of spatial mixing for $q < 8$ are computer-assisted to some extent. In particular, the results of [143] and [13] work by numerically solving a LINEAR PROGRAMMING problem to find the optimal coupling for the heat-bath algorithm. An elementary proof of spatial mixing for $q = 6$ or 7 would still be very enlightening. We refer the reader to Ferreira and Sokal [280] and Sokal [765] for numerical evidence and physical arguments for spatial mixing in the case $q \geq 4$.

12.22 Torpid mixing. Cesi, Guadagni, Martinelli, and Schonmann [148] showed that the mixing time of the Ising model is $e^{\Omega(\sqrt{n})}$ for $T < T_c$. In the computer science community, Borgs et al. [125] proved torpid mixing for the Potts model at sufficiently low temperatures and for sampling independent sets at sufficiently high densities.

Interestingly, increasing the dimensionality of the lattice can cause mixing to become torpid. In 2007, Galvin and Randall [313] showed that when d is sufficiently large, any local Markov chain for 3-colorings of the d-dimensional cubic lattice has an exponential mixing time $e^{\Omega(L^{d-1})}$. The system "magnetizes" so that the odd or even sublattice consists mainly of one color, and to change from one magnetization to another we have to pass through states with L^{d-1} sites on the boundary between them. It is worth noting that the height function for 3-colorings can be defined on the cubic lattice in any number of dimensions, so height functions alone don't guarantee polynomial-time mixing.

12.23 Walks with momentum. The idea of speeding up a Markov chain by "lifting" it to an irreversible one on a larger number of states is due to Diaconis, Holmes, and Neal [233]. Chen, Lovász, and Pak [157] gave a lower bound of $\Omega(1/\Phi)$ on the resulting mixing time, showing that the speedup provided by this approach is at most quadratic.

12.24 The cutoff phenomenon. In a number of Markov chains, the distance from equilibrium drops very quickly from 1 to 0 at the mixing time. To be precise, there is a time τ such that for any constant $\varepsilon > 0$ the variation distance is $1 - o(1)$ for $t < (1-\varepsilon)\tau$, and is $o(1)$ for $t > (1+\varepsilon)\tau$. One example is the random walk on the hypercube, and Problem 12.38 shows that $\tau = (1/2) n \ln n$. Others include the random transposition card shuffle and the riffle shuffle. This is called the *cutoff phenomenon*, and it acts in some ways like a phase transition; see Diaconis [232] for a review.

Not all Markov chains have this property. For the random walk on the cycle, for instance, the variation distance decreases continuously as a function of t/n^2. That is, for any constant c the variation distance at $t = cn^2$ is a constant between 0 and 1.

Chapter 13

Counting, Sampling, and Statistical Physics

You cannot evade quantity. You may fly to poetry and to music, and quantity and number will face you in your rhythms and your octaves.

Alfred North Whitehead

The world is so full of a number of things
I'm sure we should all be as happy as kings.

Robert Louis Stevenson

In our travels so far, we have encountered objects ranging from Hamiltonian paths to spanning trees to satisfying assignments. We have talked about how hard it is to tell whether such objects exist, or find the best one, or construct a random one. But there is one kind of question we haven't yet discussed. Namely, how many of them are there?

If the objects in question are solutions to an NP-complete problem, counting them is clearly very hard. After all, if we can tell how many there are, we can tell whether there are any. But counting can be a subtle and difficult problem even when the corresponding existence and optimization problems are in P.

For instance, consider Figure 13.1. The 8×8 square lattice has about 13 million perfect matchings, few enough that a search algorithm could find and count them in a reasonable amount of time. On the other hand, the same lattice has roughly 1.3×10^{26} spanning trees. How can we possibly count them all?

If by "counting" a set of objects we mean counting them one by one, this is clearly infeasible. But if we mean *computing* how many of them there are, then for spanning trees it turns out that there is a lovely algorithmic shortcut, that lets us do this in polynomial time as a function of the number of vertices. Thus the problem #SPANNING TREES of finding the number of spanning trees of a given graph is in P.

In contrast, finding the number of perfect matchings in a general graph seems to be very hard, and the best known algorithms take exponential time. This is no accident. Just as 3-SAT is among the hardest search problems, #PERFECT MATCHINGS is among the hardest counting problems—at least among those where we can check easily whether a given object should count.

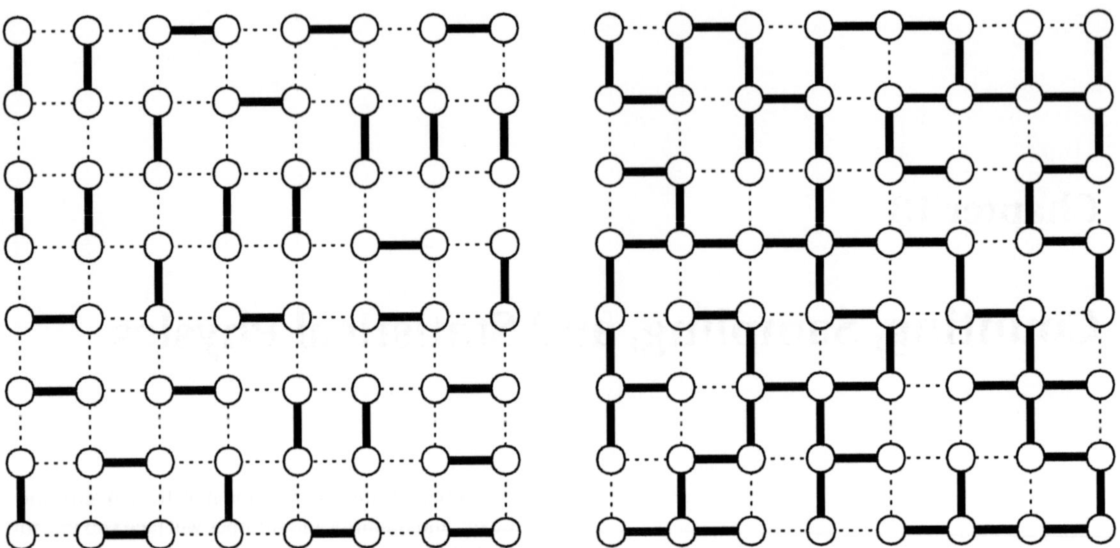

FIGURE 13.1: On the left, one of the 12988816 perfect matchings of the 8×8 lattice. On the right, one of its 126231322912498539682594816 spanning trees.

If we define #P—pronounced "sharp P"—as the class of problems that count objects satisfying some property that can be checked in polynomial time, then #PERFECT MATCHINGS is #P-complete. Thus it is just as hard as counting HAMILTONIAN PATHS or GRAPH COLORINGS. This is especially interesting since the decision problem PERFECT MATCHING, which asks if at least one perfect matching exists, is in P.

Both spanning trees and perfect matchings are simple graph-theoretic objects. At first, the fact that one of them is easy to count and the other is hard is just as mysterious as the fact that EULERIAN PATH is in P while HAMILTONIAN PATH is NP-complete. In this chapter, we will see that the difference between these two problems has deep mathematical roots. The number of spanning trees can be written as the *determinant* of a matrix, while the number of perfect matchings is the *permanent* of a matrix. We will show that computing the determinant is in P, while the permanent is #P-complete.

If we cannot compute the number of perfect matchings exactly, can we do it approximately? Counting and sampling turn out to be intimately related—if we can generate random objects then we can approximate how many of them there are, and vice versa. Using the techniques of the previous chapter, we will see how to generate random matchings, and hence count them approximately, using a Markov chain that mixes in polynomial time. This will give us a randomized polynomial-time algorithm that approximates the permanent to within any accuracy we desire.

Finally, we will see that for the special case of *planar* graphs, such as the square lattice, finding the number of perfect matchings is in P. This fact has important consequences for statistical physics, and lets us write down exact solutions for many physical models in two dimensions. In the previous chapter, we discussed the two-dimensional Ising model of magnetism, and the phase transition it undergoes at its critical temperature. We will conclude this chapter by showing how to solve it exactly.

13.1 Spanning Trees and the Determinant

Let's formalize the problem of counting the number of spanning trees of a graph.

> #SPANNING TREES
> Input: An input graph G with n vertices
> Output: The number of spanning trees of G

In this section, we will prove that this problem is in P. We will prove the *Matrix-Tree Theorem*, which states that the number of spanning trees of a graph can be written as the determinant of a version of its adjacency matrix. Then we will show that the problem of calculating the determinant, i.e.,

> DETERMINANT
> Input: An $n \times n$ matrix A with integer entries
> Output: Its determinant, $\det A$

is also in P. Thus we can count the spanning trees of a graph with n vertices in polynomial time, even though there are exponentially many of them.

13.1.1 *The Determinant and its Properties*

Let's start by reviewing the definition of the determinant. The determinant of an $n \times n$ matrix A is

$$\det A = \sum_{\pi} (-1)^{\pi} \prod_{i=1}^{n} A_{i,\pi(i)} . \tag{13.1}$$

Here the sum is over all permutations π of n objects, and $(-1)^{\pi}$ is -1 or $+1$ for permutations of odd or even parity respectively—that is, permutations made up of an odd or even number of transpositions or "swaps" of two elements. For instance, for $n = 3$ the identity and the rotations $1 \to 2 \to 3 \to 1$ and $3 \to 2 \to 1 \to 3$ are even, while the swaps $1 \leftrightarrow 2$, $1 \leftrightarrow 3$, and $2 \leftrightarrow 3$ are odd. This gives

$$\begin{pmatrix} A_{1,1} & A_{1,2} & A_{1,3} \\ A_{2,1} & A_{2,2} & A_{2,3} \\ A_{3,1} & A_{3,2} & A_{3,3} \end{pmatrix} = A_{1,1}A_{2,2}A_{3,3} + A_{1,2}A_{2,3}A_{3,1} + A_{1,3}A_{2,1}A_{3,2} - A_{1,2}A_{2,1}A_{3,3} - A_{1,3}A_{2,2}A_{3,1} - A_{1,1}A_{2,3}A_{3,2} .$$

The fundamental property of the determinant is that it is a *homomorphism* from matrices to real numbers. That is, the determinant of the product of two matrices is the product of their determinants:

$$\det AB = \det A \, \det B . \tag{13.2}$$

We will not prove this, but here's one way to think about it. As Figure 13.2 shows, if we treat A as a linear transformation on n-dimensional space, it maps the unit cube to a parallelepiped whose edges are the columns of A and whose volume is $\det A$. More generally, applying A multiplies the volume of any region by $\det A$. Applying two such transformations one after the other must multiply the volume by the determinant of the combined transformation AB.

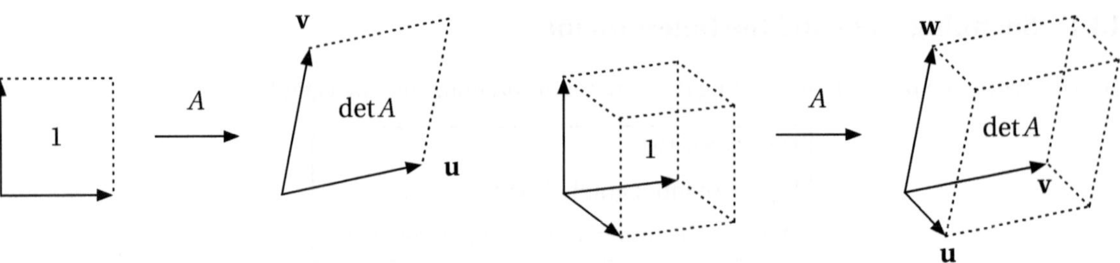

FIGURE 13.2: The linear transformation A maps the unit square to a parallelogram defined by the vectors $\mathbf{u} = (A_{1,1}, A_{2,1})$ and $\mathbf{v} = (A_{2,1}, A_{2,2})$. Its area is $\det A$, which is the component of the cross product $\mathbf{u} \times \mathbf{v}$ pointing out of the plane. In three dimensions, A maps the unit cube to a parallelepiped, defined by A's columns $\mathbf{u}, \mathbf{v}, \mathbf{w}$. Its volume is $\det A = (\mathbf{u} \times \mathbf{v}) \cdot \mathbf{w}$.

This homomorphic property also implies that the determinant is basis-independent, since for any invertible matrix U we have $\det U^{-1}AU = \det A$. In particular, this holds for the transformation matrix U which diagonalizes A. Since the determinant of a diagonal matrix is just the product of its entries, we can also write $\det A$ as the product of its eigenvalues,

$$\det A = \prod_\lambda \lambda .$$

13.1.2 The Laplacian and the Matrix-Tree Theorem

What do spanning trees have to do with the determinant? Let G be an undirected graph with n vertices, and let d_i denote the degree of each vertex i. The *Laplacian matrix* L is a modified version of G's adjacency matrix, defined as follows:

$$L_{ij} = \begin{cases} d_i & \text{if } i = j \\ -1 & \text{if there is an edge between } i \text{ and } j \\ 0 & \text{otherwise}. \end{cases}$$

More generally, if G is a multigraph we define L_{ij} as -1 times the number of edges between i and j. Note that each row and each column of L sums to zero. Thus the vector consisting of all 1s is an eigenvector with eigenvalue zero, and L has determinant zero.

In order to write the number of spanning trees as a determinant, we have to introduce one more piece of notation. Given an $n \times n$ matrix A, let $A^{(ij)}$ denote the $(n-1) \times (n-1)$ matrix formed by deleting the ith row and the jth column. These submatrices are called *minors*. The following theorem states that the minors of L give us exactly what we want:

Theorem 13.1 (Matrix-Tree Theorem) *Let G be an undirected graph or multigraph and let $T(G)$ denote the number of spanning trees in G. For any i,*

$$T(G) = \det L^{(ii)} ,$$

where L denotes the Laplacian of G. In particular, $\det L^{(ii)}$ *is the same for all i.*

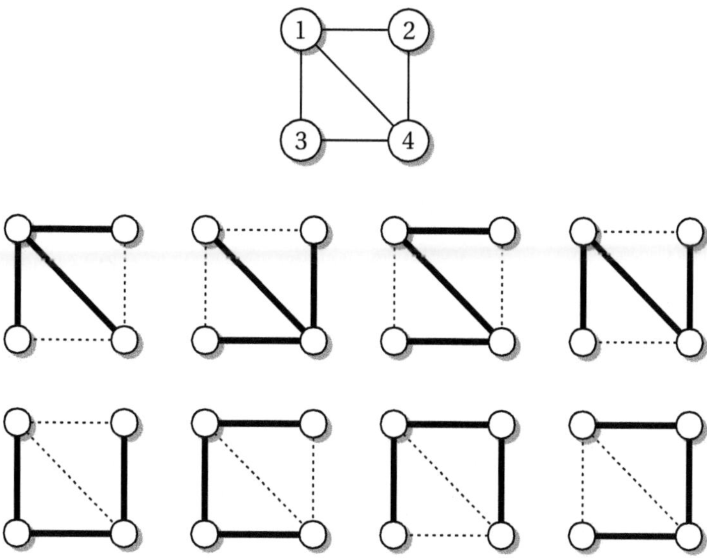

FIGURE 13.3: A graph and its 8 spanning trees.

For example, the graph shown in Figure 13.3 has 8 spanning trees. Its Laplacian matrix is

$$L = \begin{pmatrix} 3 & -1 & -1 & -1 \\ -1 & 2 & 0 & -1 \\ -1 & 0 & 2 & -1 \\ -1 & -1 & -1 & 3 \end{pmatrix}.$$

Setting $i = 1$ gives

$$\det L^{(1,1)} = \det \begin{pmatrix} 2 & 0 & -1 \\ 0 & 2 & -1 \\ -1 & -1 & 3 \end{pmatrix} = 8,$$

and we get the same result for any other choice of i.

Proof We use induction, assuming that the theorem holds for connected graphs with fewer vertices or edges. For the base case, suppose that G consists of a single vertex. Then $T(G) = 1$, and indeed the theorem holds: $L = (0)$ and $L^{(1,1)}$ is the 0×0 matrix, whose determinant is 1 by definition. If you prefer, feel free to use the multigraph consisting of a pair of vertices connected by d edges as the base case. Then

$$L = \begin{pmatrix} d & -d \\ -d & d \end{pmatrix}, \quad L^{(1,1)} = (d), \quad \text{and} \quad T(G) = d.$$

For the induction step, choose a vertex i and suppose that G has at least two vertices. If i has no edges, G has no spanning trees. In this case the theorem holds because $L^{(ii)}$ is the Laplacian of the rest of the graph and, as we pointed out above, the Laplacian has determinant zero.

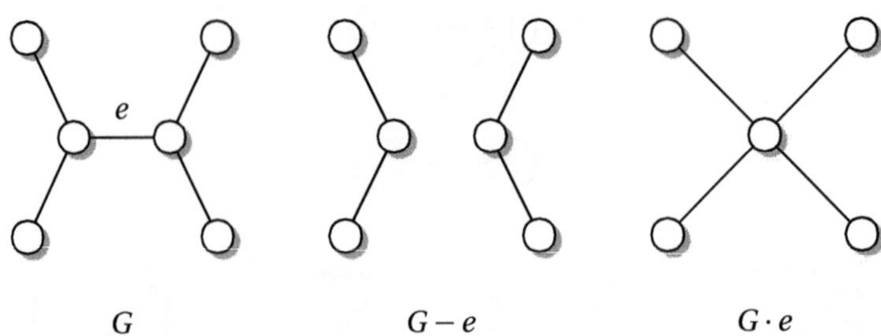

FIGURE 13.4: $G-e$ is the graph we get by cutting the edge e, and $G\cdot e$ is the graph we get by contracting e and merging its endpoints into a single vertex.

So suppose that i is connected to some other vertex j, and let e denote the edge (i,j). As shown in Figure 13.4, there are two natural ways to modify G: we can simply remove e, or we can contract it and merge i and j into a single vertex. We denote these graphs as $G-e$ and $G\cdot e$ respectively. Then we claim that $T(G)$ is given by the following recursive formula:

$$T(G)=T(G-e)+T(G\cdot e). \tag{13.3}$$

Exercise 13.1 *Prove* (13.3). *Hint: some spanning trees include the edge e, and some don't.*

Now assume by induction that the theorem holds for $G-e$ and $G\cdot e$. We are free to reorder the vertices so that i and j are the first two. Then we can write the Laplacian of G in the form

$$L_G=\left(\begin{array}{c|c|c} d_i & -1 & r_i^T \\ \hline -1 & d_j & r_j^T \\ \hline r_i & r_j & L' \end{array}\right).$$

Here r_i and r_j are $(n-2)$-dimensional column vectors describing the connections between i and j and the other $n-2$ vertices, and L' is the $(n-2)$-dimensional minor describing the rest of the graph. We can write the Laplacians of $G-e$ and $G\cdot e$ as

$$L_{G-e}=\left(\begin{array}{c|c|c} d_i-1 & 0 & r_i^T \\ \hline 0 & d_j-1 & r_j^T \\ \hline r_i & r_j & L' \end{array}\right),\quad L_{G\cdot e}=\left(\begin{array}{c|c} d_i+d_j-2 & r_i^T+r_j^T \\ \hline r_i+r_j & L' \end{array}\right).$$

To complete the induction, we wish to show that

$$\det L_G^{(ii)} = \det L_{G-e}^{(ii)} + \det L_{G\cdot e}^{(jj)}, \tag{13.4}$$

or that

$$\det \left(\begin{array}{c|c} d_j & r_j^T \\ \hline r_j & L' \end{array} \right) = \det \left(\begin{array}{c|c} d_j - 1 & r_j^T \\ \hline r_j & L' \end{array} \right) + \det L' .$$

But this follows from the fact that the determinant of a matrix can be written as a linear combination of its *cofactors*, i.e., the determinants of its minors. Specifically, for any A we have

$$\det A = \sum_{j=1}^{n} (-1)^j A_{1,j} \det A^{(1,j)} . \tag{13.5}$$

Thus if two matrices differ only in their $(1,1)$ entry, with $A_{1,1} = B_{1,1} + 1$ and $A_{ij} = B_{ij}$ for all other i, j, their determinants differ by the determinant of their $(1,1)$ minor, $\det A = \det B + \det A^{(1,1)}$. Applying this to $L_G^{(ii)}$ and $L_{G-e}^{(ii)}$ yields (13.4) and completes the proof. □ 13.1

13.1.3 *Calculating the Determinant in Polynomial Time*

The Matrix-Tree Theorem gives a reduction from #SPANNING TREES to the problem of calculating the determinant of an $n \times n$ matrix. In this section, we will show that DETERMINANT, and therefore #SPANNING TREES, are in P.

A priori, it is not at all obvious how to find the determinant of an $n \times n$ matrix in polynomial time. The definition (13.1) of the determinant is a sum over all $n!$ permutations, so evaluating it directly would take roughly $n!$ time. We could use the sum over cofactors (13.5) as a recursive algorithm, and for sparse matrices this is a good idea. For general matrices, however, the determinant of an $n \times n$ matrix involves the determinants of n minors. If $f(n)$ denotes the running time, this gives $f(n) = n f(n-1)$, causing $f(n)$ to grow like $n!$ again.

The key fact that lets us calculate the determinant in polynomial time is its homomorphic property, $\det AB = \det A \det B$. If we can write A as the product of a string of simple matrices, $A = \prod_i A_i$, then $\det A = \prod_i \det A_i$. If this string is only polynomially long, and if $\det A_i$ is easy to calculate for each A_i, this gives us a polynomial-time algorithm.

We can decompose A into just this kind of product using the classic technique of *Gaussian elimination*. We transform A to an upper-triangular matrix with a series of moves, each of which adds a multiple of one row to another or switches two rows. For instance, if $A = L^{(1,1)}$ is the minor from our spanning tree example,

$$A = \begin{pmatrix} 2 & 0 & -1 \\ 0 & 2 & -1 \\ -1 & -1 & 3 \end{pmatrix},$$

we can add half the top row to the bottom row, giving

$$\begin{pmatrix} 1 & & \\ & 1 & \\ 1/2 & & 1 \end{pmatrix} A = \begin{pmatrix} 2 & 0 & -1 \\ 0 & 2 & -1 \\ 0 & -1 & 5/2 \end{pmatrix},$$

where matrix entries not shown are zero. We then add half the second row to the bottom row,

$$\begin{pmatrix} 1 & & \\ & 1 & \\ & 1/2 & 1 \end{pmatrix} \begin{pmatrix} 1 & & \\ & 1 & \\ 1/2 & & 1 \end{pmatrix} A = \begin{pmatrix} 2 & 0 & -1 \\ 0 & 2 & -1 \\ 0 & 0 & 2 \end{pmatrix}. \tag{13.6}$$

The determinant of an upper-triangular matrix is just the product of its diagonal entries, since any permutation other than the identity hits one of the zeros below the diagonal. Similarly, the matrix corresponding to each of these moves has determinant 1. Thus taking the determinant of both sides of (13.6) yields

$$\det A = \det \begin{pmatrix} 2 & 0 & -1 \\ 0 & 2 & -1 \\ 0 & 0 & 2 \end{pmatrix} = 8.$$

How many moves does it take to transform an $n \times n$ matrix into upper-triangular form? We proceed in $n-1$ stages, where at stage j our goal is to cancel A_{ij} for all $i > j$. At the beginning of each stage, if necessary we swap the jth row with another so that $A_{jj} \neq 0$. We then subtract A_{ij}/A_{jj} times the jth row from the ith row for all $i > j$, so that the new value of A_{ij} is zero. The total number of moves is at most

$$n + (n-1) + \cdots + 3 + 2 + 1 = O(n^2).$$

Each swap has determinant -1, and the other moves have determinant 1. Then $\det A$ is the product of the determinants of these moves, and the diagonal entries of the resulting upper-triangular matrix.

This algorithm consists of poly(n) arithmetic operations. Initially A's entries have poly(n) digits, and do so all of the numerators and denominators that occur throughout the process. Thus the total running time is polynomial, and we have proved that DETERMINANT and #SPANNING TREES are in P.

13.2 Perfect Matchings and the Permanent

Now that we have learned how to count spanning trees, let's learn how to count perfect matchings. Is there a way to write the number of perfect matchings of a graph as a function of its adjacency matrix? What is to perfect matchings as the determinant is to spanning trees?

Suppose we have a bipartite graph G with n vertices on each side. We can represent it as an $n \times n$ matrix B, where $B_{ij} = 1$ if the ith vertex on the left is connected to the jth vertex on the right, and $B_{ij} = 0$ otherwise. For instance, the matrix corresponding to the graph shown in Figure 13.5 is

$$B = \begin{pmatrix} 1 & 1 & 0 & 0 \\ 0 & 1 & 0 & 1 \\ 1 & 1 & 1 & 0 \\ 0 & 0 & 1 & 1 \end{pmatrix}.$$

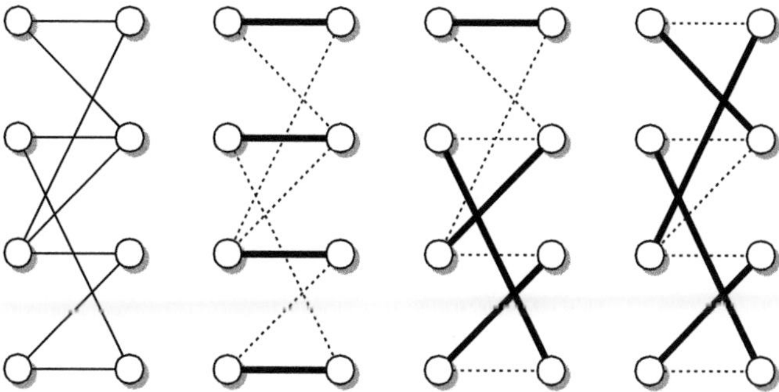

FIGURE 13.5: A bipartite graph with three perfect matchings.

Note the difference between B and the usual adjacency matrix A. While B is n-dimensional, A is $2n$-dimensional since the total number of vertices is $2n$. If we order the vertices so that the n on the left come first, followed by the n on the right, we can write A as

$$A = \begin{pmatrix} 0 & B \\ B^T & 0 \end{pmatrix}. \tag{13.7}$$

How can we express the number of perfect matchings in G in terms of these matrices? Each perfect matching is a permutation that maps the vertices on the left to their partners on the right. For instance, the first two matchings shown in Figure 13.5 correspond to the identity permutation and the permutation π such that $\pi(1) = 1$, $\pi(2) = 4$, $\pi(3) = 2$, and $\pi(4) = 3$. Each permutation π corresponds to a matching if and only if for each i there is an edge from i to $\pi(i)$, i.e., if $B_{i,\pi(i)} = 1$. Therefore, the number of matchings is given by the following quantity, which is called the *permanent* of B:

$$\operatorname{perm} B = \sum_{\pi} \prod_{i=1}^{n} B_{i,\pi(i)} \tag{13.8}$$

More generally, suppose G is a weighted graph and B_{ij} is the weight of the edge from the ith vertex on the left to the jth vertex on the right. Then if we define the weight of a matching as the *product* of the weights of its edges, $\operatorname{perm} B$ is the total weight of all the perfect matchings.

Comparing (13.8) with (13.1), we see that the permanent and the determinant differ only in the parity $(-1)^\pi$ of each permutation. This seemingly small change makes an enormous difference in its algebraic properties. The permanent is not a homomorphism. Nor is it basis-independent, although it is invariant with respect to permutations of the rows or columns. Therefore, it has no geometric interpretation analogous to the ratio by which a linear transformation changes the volume. Nevertheless, its relationship to the number of perfect matchings gives it an important combinatorial meaning.

In some cases, it is easier to deal directly with the adjacency matrix A rather than B. Since A can be written in the form (13.7), it is not hard to see that

$$\operatorname{perm} A = (\operatorname{perm} B)^2 . \tag{13.9}$$

Thus $\operatorname{perm} A$ is the square of the number of perfect matchings, or of their total weight in the case of a weighted graph.

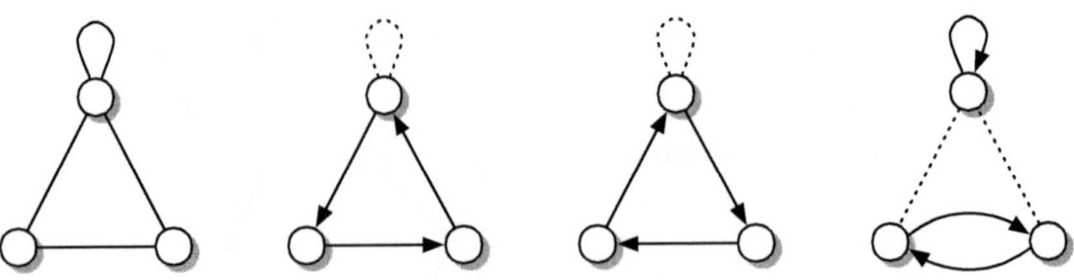

FIGURE 13.6: A graph G with three cycle covers.

Exercise 13.2 *Prove* (13.9).

There is a more constructive way to see this, which we will find useful below. A *cycle cover* of a graph G is a disjoint set of directed cycles that visit each vertex exactly once. For instance, a Hamiltonian cycle is a cycle cover consisting of a single cycle. If G is undirected, we think of it as having directed edges pointing in both directions along each edge, so a cycle can consist of a pair of adjacent vertices. A cycle can also consist of a single vertex if that vertex has a self-loop.

Each cycle cover can be thought of as a permutation π that sends each vertex to the next vertex in its cycle. Conversely, any permutation π such that there is an edge from i to $\pi(i)$ for each i can be decomposed into a disjoint set of cycles, and these form a cycle cover. Therefore if G is a graph with adjacency matrix A, each term in $\operatorname{perm} A$ corresponds to a cycle cover, and in the unweighted case we have

$$\operatorname{perm} A = \#\text{ of cycle covers}\,.$$

For instance, consider the graph in Figure 13.6. It has three cycle covers: the cycle $1 \to 2 \to 3 \to 1$, its inverse, and a cover composed of the cycles $1 \to 1$ and $2 \leftrightarrow 3$. Its adjacency matrix is

$$A = \begin{pmatrix} 1 & 1 & 1 \\ 1 & 0 & 1 \\ 1 & 1 & 0 \end{pmatrix},$$

and indeed the permanent of A is 3.

Now we claim that, in any bipartite graph,

$$\#\text{ of cycle covers} = (\#\text{ of perfect matchings})^2\,. \tag{13.10}$$

We illustrate this in Figure 13.7. It shows a "multiplication table" which takes an ordered pair of perfect matchings and returns a cycle cover. Instead of putting the vertices of the graph on the left and right, we color them black and white. Then the multiplication rule is simple: given two matchings μ_1, μ_2, we define a cycle cover π such that $\pi(i)$ is i's partner in μ_1 or μ_2 if i is black or white respectively. Thus each cycle in π follows the edges in μ_1 and μ_2 alternately.

We will prove that this rule gives a one-to-one mapping from the set of ordered pairs of perfect matchings to the set of cycle covers. Therefore, these two sets have the same size, proving (13.10).

In one direction, suppose that we have two perfect matchings μ_1, μ_2. Their *symmetric difference*, i.e., the set of edges that appear in one matching but not the other,

$$\mu_1 \oplus \mu_2 = (\mu_1 \cup \mu_2) - (\mu_1 \cap \mu_2),$$

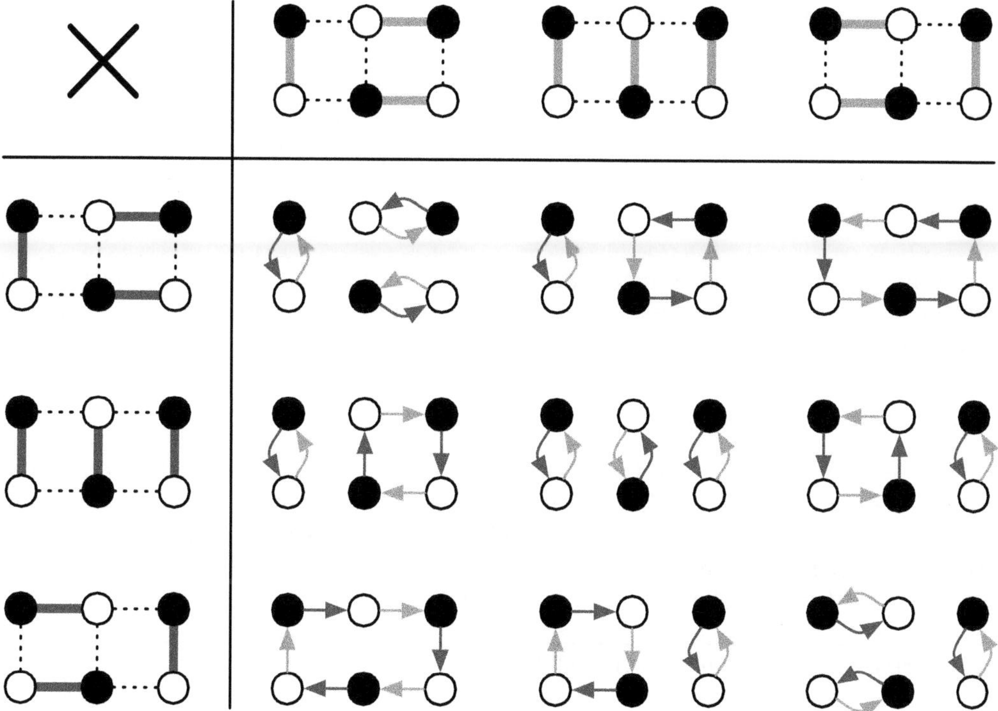

FIGURE 13.7: For any pair μ_1, μ_2 of perfect matchings, we form a cycle cover by orienting the edges in μ_1 from black to white, and the edges in μ_2 from white to black. If G is a bipartite graph, this map is one-to-one, so the permanent of its adjacency matrix is the square of the number of perfect matchings.

consists of a set of undirected cycles of even length. We can give each of these cycles an orientation as described above, pointing edges in μ_1 from black to white and edges in μ_2 from white to black. With these orientations, $\mu_1 \oplus \mu_2$ becomes a cycle cover, as long as we also add cycles of length 2 going back and forth along the shared edges in $\mu_1 \cap \mu_2$.

Conversely, suppose we are given a set σ of k even cycles, along with some isolated edges, which cover a graph. Each cycle can be oriented clockwise or counterclockwise, so σ corresponds to 2^k different cycle covers. These choices of orientation correspond to choosing, on each cycle, which edges are in μ_1 and which are in μ_2. Thus there are 2^k different pairs of matchings μ_1, μ_2 such that $\mu_1 \oplus \mu_2 = \sigma$.

Now that we know how to express the number of perfect matchings as a permanent, can we calculate it in polynomial time? As we will show in the next section, this is very unlikely. While DETERMINANT and #SPANNING TREES are in P, PERMANENT and #PERFECT MATCHINGS are among the hardest possible counting problems. Solving them is at least as hard as solving NP-complete problems, and is probably far harder.

13.3 The Complexity of Counting

The Man in the Wilderness asked me to tell
The sands in the sea, and I counted them well.
Says he with a grin, "And not one more?"
I answered him bravely, "You go and make sure!"

Traditional

When we wanted to understand how hard search problems are, and why some are so much harder than others, we defined the complexity class NP and the notion of NP-completeness. Let's use the same philosophy to help us understand the complexity of counting problems.

13.3.1 #P and #P-Completeness

Problems in NP ask whether an object with a certain property exists. We will define #P, pronounced "sharp P", as the class of problems that ask *how many* such objects exist. As in NP, we require that this property is something that we can check in polynomial time—like the property of being a proper 3-coloring, or a satisfying assignment, or a Hamiltonian path.

Here is a formal definition along the lines of the definition of NP we gave in Section 4.3:

> #P is the class of functions $A(x)$ of the form
>
> $$A(x) = |\{w : B(x,w)\}| ,$$
>
> where $B(x,w)$ is a property that can be checked in polynomial time, and where $|w| = \text{poly}(|x|)$ for all w such that $B(x,w)$ holds.

For instance, if x is a graph and $B(x,w)$ is the property that w is a proper 3-coloring of x, then $A(x)$ is the number of 3-colorings that x has. We will call this function, or equivalently the problem of calculating it, #GRAPH 3-COLORINGS. Similarly, we can define #3-SAT, #HAMILTONIAN PATHS, and so on.

In each of these cases, the objects we are counting—colorings, assignments, or paths—can be described with a polynomial number of bits as a function of the input size. So just as in the definition of NP, we require that $|w| = \text{poly}(n)$. It takes exponential time to check all $2^{\text{poly}(n)}$ possible objects w, and count how many of them work. On the other hand, we can do this with just poly(n) memory. Thus if we extend our definition of PSPACE to include functions as well as decision problems, we have #P $\subseteq$ PSPACE.

Now that we have defined a complexity class of counting problems, which are the hardest ones? How should we define #P-completeness? A first guess might be that a problem is #P-complete if it counts solutions of an NP-complete problem. However, this would be missing the mark. Just as we did for NP, we have to define a sense in which one counting problem can be reduced to another. Then #P-complete problems are those to which every other problem in #P can be reduced.

When we defined NP-completeness, all we asked of a reduction is that it map yes-instances to yes-instances and no-instances to no-instances, preserving the existence or nonexistence of solutions. In other words, it transforms an instance x of A to an instance $f(x)$ of B such that $A(x) \geq 1$ if and only if $B(f(x)) \geq 1$.

For counting problems, we want our reductions to preserve information about the number of solutions, not just whether one exists. The simplest case of this is a reduction that exactly preserves the number of solutions, so that $A(x) = B(f(x))$. Such reductions are called *parsimonious.*

More generally, we want a reduction $A \leq B$ to imply that A is easy if B is easy. Let's define a reduction between two counting problems as a polynomial-time algorithm that calls B as a subroutine on a transformed problem $f(x)$, and then uses a function g to the transform the result to the answer for $A(x)$. Formally,

If A and B are problems in #P, a *counting reduction* from A to B is a pair of functions f, g in P such that, for all x,

$$A(x) = g\big(x, B(f(x))\big) . \tag{13.11}$$

A problem B is #P*-hard* if, for every problem A in #P, there is a counting reduction from A to B. If B is in #P and is #P-hard, then B is #P-complete.

Here f is the function that transforms instances of A to instances of B, and g is the polynomial-time algorithm that converts $B(f(x))$ to $A(x)$.

By this definition, if A is reducible to B and B is in P, then A is in P. Therefore, if any #P-hard problem is in P, all of #P is contained in P. If we can count solutions then we can tell if there are any, so this would also imply that P = NP.

But this is just the tip of the iceberg. It turns out that $\mathsf{P}^{\#\mathsf{P}}$, the class of problems that we can solve in polynomial time given access to an oracle for a #P-complete problem, includes the entire polynomial hierarchy (see Section 6.1.1). Thus the power to count solutions gives us the ability, not just to tell whether any exist, but to solve problems with any number of "there exists" and "for all"s. Since we believe that each level of the polynomial hierarchy is harder than the one below, we believe that #P-complete problems are much harder than NP-complete ones.

It is easy to see that counting reductions are transitive, so we can prove that a problem is #P-complete 13.2
by reducing a known #P-complete problem to it. Just as we grew a tree of NP-complete problems by starting from WITNESS EXISTENCE, which embodies all of NP by definition, we can start with the following problem:

#WITNESSES

Input: A program Π, an input x, and an integer t given in unary

Output: The number of witnesses w of size $|w| \leq t$ such that $\Pi(x, w)$ returns "yes" in time t or less

In Section 5.2, we described how to convert the program Π and its input x to a Boolean circuit, thus reducing WITNESS EXISTENCE to CIRCUIT SAT. This reduction is parsimonious, since there is a one-to-one correspondence between witnesses and truth assignments. Therefore, the problem #CIRCUIT SAT, which counts the truth assignments that cause a circuit to output `true`, is #P-complete.

What about #3-SAT? Here we have to be a bit careful. Consider the reduction given in Section 5.2 from CIRCUIT SAT to 3-SAT. Given a Boolean circuit C, we start by writing a SAT formula ϕ with variables x_i for C's inputs, and additional variables y_j for its internal wires. Since ϕ asserts both that C returns `true` and that all its gates function properly, the truth values of the wires are determined by those of the inputs. Thus ϕ has exactly one satisfying assignment for each one that C has.

While this reduction is parsimonious, it produces a formula ϕ with a mix of 1-, 2-, and 3-variable clauses. We can convert this to a 3-SAT formula ϕ' by adding dummy variables, padding these clauses out to 3 variables each. Annoyingly, however, this changes the number of satisfying assignments. For instance, when we write the 2-variable clause $(x \vee y)$ as

$$(x \vee y \vee z) \wedge (x \vee y \vee \overline{z}),$$

the dummy variable z can be true or false, doubling the number of satisfying assignments.

We could make this reduction parsimonious by adding clauses to ϕ' that fix the values of the dummy variables, say by forcing them all to be true. But there is no need to do this. Our definition of counting reduction simply asks that we can solve #CIRCUIT SAT in polynomial time by calling #3-SAT as a subroutine. So we convert C to ϕ', count the satisfying assignments of ϕ', and divide the result by 2^d where d is the total number of dummy variables. Since d is a simple function of the original circuit—namely, of the number of gates of each type—this poses no difficulty, and #3-SAT is #P-complete.

If we look back at Chapter 5, we will see that some of our NP-completeness reductions are parsimonious and others are not. When they fail to be parsimonious, sometimes the change in the number of solutions is a simple function of the instance, as it is for #3-SAT here. A more serious problem arises if the number of solutions changes in a complicated way that depends on the solutions themselves. However, even in that case we can typically fix the reduction by modifying the gadgets, or by using a slightly different chain of reductions.

So with a little extra work, we can prove that the counting versions of our favorite NP-complete problems, including #GRAPH 3-COLORINGS and #HAMILTONIAN PATHS, are #P-complete. This comes as no surprise. Intuitively, if it is hard to tell if any solutions exist, it is very hard to count them.

What is more surprising is that counting problems can be #P-complete even when the corresponding existence problem is in P. We showed in Section 3.8 that PERFECT MATCHING is in P. Nevertheless, we will show in the next section that #PERFECT MATCHINGS, or equivalently PERMANENT, is #P-complete.

Exercise 13.3 *Show that if a counting problem #A is #P-complete with respect to parsimonious reductions—that is, if every problem in #P can be parsimoniously reduced to #A—then the corresponding existence problem A must be NP-complete. Do you think the converse is true? If A is NP-complete, is #A necessarily #P-complete? Hint: consider Section 10.6.2.*

13.3.2 The Permanent is Hard

Let's formalize the problem of computing the permanent of a matrix.

PERMANENT
Input: A $n \times n$ matrix A with integer entries
Output: Its permanent, perm A

In general, PERMANENT is not in #P, since if the permanent of a matrix is negative we can't really say that it counts a set of objects. However, the following special case is in #P:

0-1 PERMANENT

Input: A $n \times n$ matrix A of 0s and 1s

Output: Its permanent, perm$A$

We will reduce #3-SAT to PERMANENT, and thus show that PERMANENT is #P-hard. We will then reduce PERMANENT to 0-1 PERMANENT, showing that 0-1 PERMANENT is #P-complete. Since the number of perfect matchings of a bipartite graph is the permanent of its adjacency matrix (see Section 13.2), this shows that #PERFECT MATCHINGS is #P-complete as well.

To understand the reduction, first recall that if A is the adjacency matrix of a weighted graph G, then perm$A$ is the total weight of all of G's cycle covers. For each variable x, we will have a vertex with two cyclic paths passing through it, composed of edges with weight 1. To cover this vertex, each cycle cover has to include one of these paths or the other, and this choice corresponds to setting x true or false.

We then thread these paths through clause gadgets as shown in Figure 13.8. Each clause gadget has three input vertices. For each clause in which x or $\overline{x}$ appears, we place one of its input vertices on the "true" or "false" path for x. In addition to these input vertices, each clause gadget also has internal vertices and edges, and these edges have integer weights.

Now suppose we fix a truth assignment σ of the variables by including one path for each variable in the cycle cover. For each clause gadget, the input vertices corresponding to true literals are already covered. However, we still have to cover its remaining input vertices, as well as its internal vertices. The total contribution of σ to the permanent will then be the product, over all clauses, of the total weight of these internal cycle covers.

Our goal is to design the clause gadget so that this total weight is some constant C if the clause is satisfied—that is, whenever at least one of the three input vertices is already covered. If none of them are covered so that the clause is unsatisfied, we want the total weight to be 0. In addition, we want the total weight to be 0 unless, for each input vertex, its outgoing edge is covered if and only if its incoming edge is. This ensures that each variable path is entirely covered or entirely uncovered, giving that variable a consistent truth value.

If we can find a clause gadget that satisfies all these conditions, each satisfying assignment σ will contribute C^m to the permanent where m is the number of clauses. Since we can divide the permanent by C^m and recover the number of satisfying assignments, this will give a counting reduction from #3-SAT to PERMANENT, and prove that PERMANENT is #P-hard.

How can we represent these conditions algebraically? Let M be the weighted adjacency matrix representing the clause gadget's internal connections, where the first three vertices are the inputs. If the incoming and outgoing edges of vertex 1 are already covered, this removes the first row and first column from M, and the total weight of the cycle covers on the remaining vertices is the permanent of the minor $M^{(1,1)}$. Similarly, if we violate our consistency condition by covering the incoming edge of vertex 1 and the outgoing edge of vertex 2, the total weight is the permanent of $M^{(1,2)}$.

Let's generalize our earlier notation for minors as follows. If M is a $k \times k$ matrix and $S, T \subseteq \{1, \ldots, k\}$, let $M^{(S,T)}$ denote the $(k - |S|) \times (k - |T|)$ matrix resulting from removing the ith row for each $i \in S$ and the jth column for each $j \in T$. Then if we want the weight of a clause to be C if it is satisfied and 0 if it isn't,

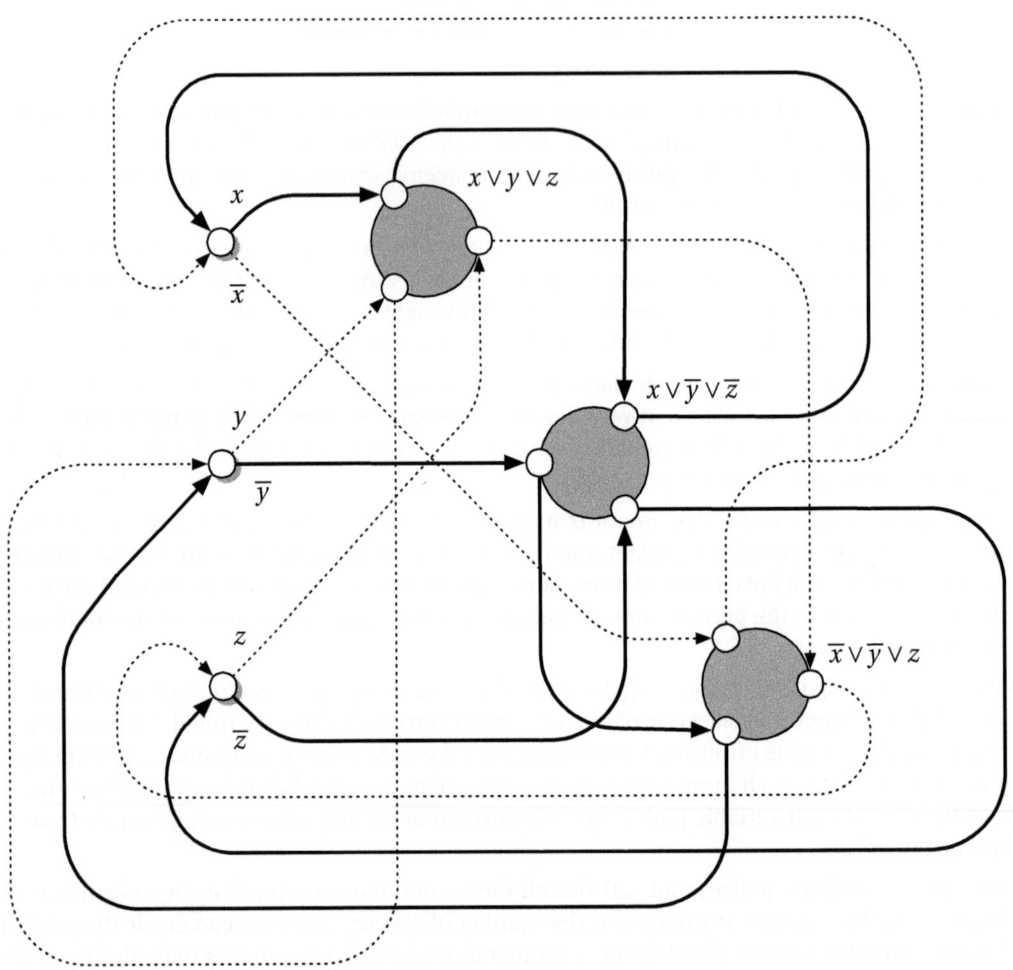

FIGURE 13.8: The reduction from #3-SAT to PERMANENT. To have a cycle cover, each variable has to be covered by its "true" path or its "false" one. The cycle cover shown in bold corresponds to the assignment $x = $ `true` and $y = z = $ `false`. The clause gadgets have the property that the total weight of their internal cycle covers is C if at least one of their inputs is covered and 0 otherwise. Thus if there are m clauses, the permanent is C^m times the number of satisfying assignments.

and 0 if the consistency condition is violated, we demand that

$$\text{for all } S,T\subseteq\{1,2,3\},\ \operatorname{perm} M^{(S,T)}=\begin{cases}0 & \text{if } S=T=\emptyset\\ C & \text{if } S=T\neq\emptyset\\ 0 & \text{if } S\neq T.\end{cases} \tag{13.12}$$

There are indeed matrices M that satisfy these conditions. Here is one with a total of $k=7$ vertices, including the 3 inputs and 4 internal vertices. The skeptical reader can check that (13.12) holds with $C=12$:

$$M=\left(\begin{array}{ccc|cccc} 0&0&-1&-1&1&1&1\\ 0&0&-1&2&-1&1&1\\ 0&0&0&-1&-1&1&1\\ \hline 1&0&0&0&2&0&0\\ 0&1&0&3&0&0&0\\ 0&0&1&1&1&2&-1\\ 0&0&1&1&1&0&1\end{array}\right)$$

This completes the proof that PERMANENT is #P-hard.

This begs the obvious question: can we do this with the determinant instead? Can we find a matrix whose determinant, and those of its minors, satisfy (13.12)? For a proof that we cannot, see Problem 13.14. Of course, this doesn't prove that #P is outside of P, or that DETERMINANT is not #P-complete. It just shows that we cannot reduce #3-SAT to DETERMINANT using the same approach.

Now we need to reduce PERMANENT to 0-1 PERMANENT. We will do this in two steps. First we reduce PERMANENT to the special case POSITIVE PERMANENT, where none of A's entries are negative. Consider the following naive bound on $\operatorname{perm} A$, where we assume that A's entries are n-bit integers:

$$\left|\operatorname{perm} A\right|\leq n!\left(\max_{i,j}\left|A_{ij}\right|\right)^n\leq n!\,2^{n^2}<2^{2n^2}.$$

Let's call this upper bound Q. Now define a matrix A' as

$$A'_{ij}=A_{ij} \bmod 2Q.$$

In other words, replace the negative entries like -1 with $2Q-1$, while leaving the positive entries unchanged. Note that Q has $O(n^2)$ bits, so A' is an instance of POSITIVE PERMANENT of polynomial size. Given a subroutine for POSITIVE PERMANENT, we can then calculate $\operatorname{perm} A \bmod 2Q=\operatorname{perm} A' \bmod 2Q$. Let's call this quantity R. Then since $\left|\operatorname{perm} A\right|<Q$, we have

$$\operatorname{perm} A=\begin{cases}R & \text{if } R<Q\\ R-2Q & \text{if } R\geq Q.\end{cases}$$

This gives a counting reduction from PERMANENT to POSITIVE PERMANENT.

To reduce POSITIVE PERMANENT to 0-1 PERMANENT, we will show how to simulate edges with positive integer weights using gadgets whose edges have weight 1. We can think of an edge with weight w as a set of w parallel edges, each with weight 1, so that there are w ways to get from one endpoint to the other.

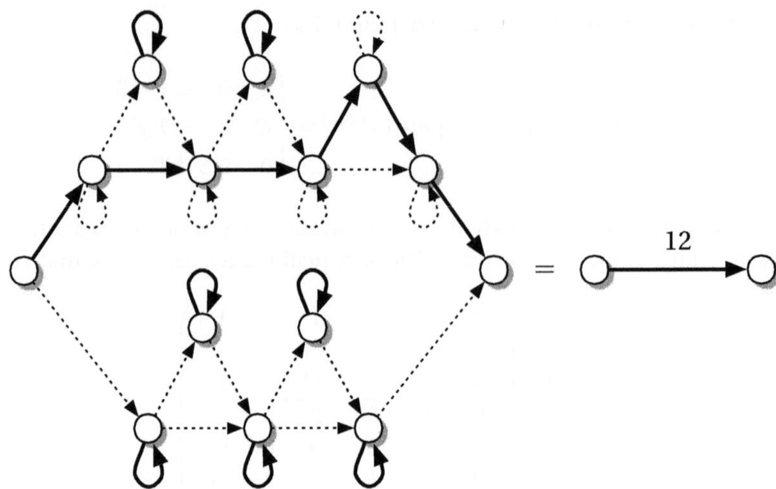

FIGURE 13.9: We can simulate an edge with weight w using edges with weight 1. For each term 2^k in w's binary expansion, we create a chain with k links, which can be covered by 2^k different paths or by its self-loops. We then wire these chains together in parallel. In this case $w = 2^3 + 2^2 = 12$, and one of the 12 ways to cover the gadget is shown in bold.

But if w is exponentially large, as an n-bit integer might be, how can be simulate these w edges with a gadget of polynomial size?

The answer is shown in Figure 13.9. We create a chain of k links, where each link can be covered in 2 different ways. Each such chain can be covered by 2^k different paths, or by its self-loops. Finally, if we attach chains of length k_1, k_2, and so on to the same endpoints, we can choose which chain to go through, and there are a total of $w = 2^{k_1} + 2^{k_2} + \cdots$ ways to pass through the gadget. If w is an integer with n bits, we can do all this with $O(n^2)$ vertices and edges.

This completes our chain of reductions,

$$\#\text{3-SAT} \le \text{PERMANENT} \le \text{POSITIVE PERMANENT} \le \text{0-1 PERMANENT},$$

and proves that 0-1 PERMANENT is #P-complete. To prove that #PERFECT MATCHINGS is #P-complete as well, we just need to remind ourselves that any $n \times n$ matrix B of 0s and 1s can be associated with a bipartite graph G with n vertices on each side. Since perm B is the number of perfect matchings of G, this gives a parsimonious reduction from 0-1 PERMANENT to #PERFECT MATCHINGS and proves that the latter is
13.3 #P-complete.

13.4 From Counting to Sampling, and Back

The fact that 0-1 PERMANENT and #PERFECT MATCHINGS are #P-complete is strong evidence that they are very hard to solve exactly. But if an exact count is too much to ask, how about an approximate one?

It turns out that there is a close relationship between counting and random sampling. For a large class of problems, if we can approximate the number of solutions then we can generate random ones with a

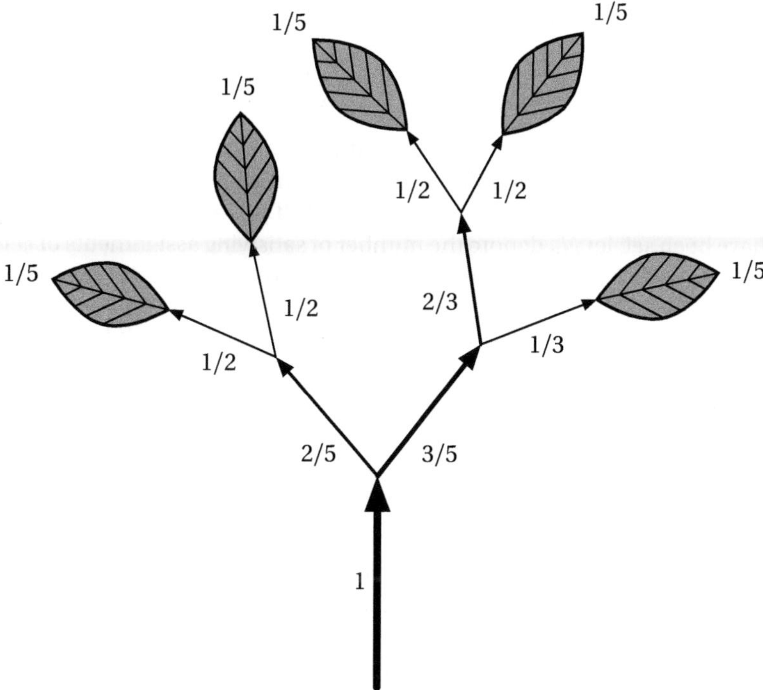

FIGURE 13.10: If you follow each branch with probability proportional to the number of leaves it leads to, you will arrive at each leaf with equal probability.

probability distribution that is almost uniform, and vice versa. In this section and the next, we will see how to use the Markov chain techniques of Chapter 12 to generate random perfect matchings, and thus approximate the permanent to any accuracy we wish.

13.4.1 The Caterpillar's Tale

Let's start with an innocent and heartwarming story, in which you play the part of a hungry caterpillar climbing a tree. At each point where a branch splits into two, you have to decide which one to follow. Finally you arrive at a leaf. How can you ensure that every leaf is equally likely to be your final destination?

Suppose that at each point where the tree branches, you can tell how many leaves there are in each subtree. If you follow each branch with probability proportional to the number of leaves on it as shown in Figure 13.10, you will reach each leaf with equal probability $1/N$ where N is the total number of leaves.

Let's describe your path through the tree step by step. We can label each path with a string p of ℓs and rs, describing a path of left and right turns from the root to a node of the tree. Let N_p denote the number of leaves on the subtree connected to that node. For instance, N is the total number of leaves, and $N_p = 1$ if p ends in a leaf. If p is your path so far, you go left or right with probability $N_{p\ell}/N_p$ or

N_{pr}/N_p respectively. If there is a leaf at the end of the path $r\ell r$, say, the probability you arrive there is

$$\frac{N_r}{N}\frac{N_{r\ell}}{N_r}\frac{N_{r\ell r}}{N_{r\ell}} = \frac{N_{r\ell r}}{N} = \frac{1}{N}. \tag{13.13}$$

Similarly, suppose you have a SAT formula ϕ, and you want to generate a random satisfying assignment. You can do this by setting one variable at a time, choosing branches in a tree of partial assignments, until you reach a leaf where all the variables have been set. If p is a string of ts and fs denoting how the first i variables have been set, let N_p denote the number of satisfying assignments of ϕ that are consistent with p. If we could compute N_p for any p, we could set x_{i+1} true or false with the following probabilities:

$$\Pr[x_{i+1} = \texttt{true}] = \frac{N_{pt}}{N_p}, \quad \Pr[x_{i+1} = \texttt{false}] = \frac{N_{pf}}{N_p}. \tag{13.14}$$

Now we invoke an important property of SAT. Namely, if we set some of ϕ's variables according to p, we get a formula ϕ_p on the remaining variables, in which some of ϕ's clauses have been satisfied and others have been shortened. Thus if we can solve #SAT, we can calculate $N_p = \#\text{SAT}(\phi_p)$ for any p. We can then choose the value of x_{i+1} according to (13.14), and each satisfying assignment will be chosen with equal probability.

This property, in which partially solving a problem leaves a smaller problem of the same kind, is called *self-reducibility*. It is shared by most natural problems—see, for instance, Problems 4.1 and 4.2. As another example, if you have solved part of a jigsaw puzzle, what you have left is a smaller puzzle in which you have to fit the remaining pieces in the remaining space.

Exercise 13.4 *Let G be a graph and e one of its edges. Show that we can modify G to produce two graphs, G_0 and G_1, such that the number of perfect matchings of G that include and exclude e is the number of perfect matchings of G_1 and G_0 respectively. Therefore, #PERFECT MATCHINGS is self-reducible.*

Thus for any self-reducible problem, if we can count its solutions then we can sample them randomly. Is the converse true? Let's put on our caterpillar costume again. If we could choose each leaf with probability $1/N$, then N would be the reciprocal of the probability with which we would arrive at any given leaf. Suppose we knew the fraction N_r/N of the leaves that lie on the right subtree, the fraction $N_{r\ell}/N_r$ of those that lie on the left subtree of the right subtree, and so on all the way down to the leaf at $r\ell r$. Then we could invert (13.13) and get

$$N = \frac{N}{N_r}\frac{N_r}{N_{r\ell}}\frac{N_{r\ell}}{N_{r\ell r}} = 1\bigg/\left(\frac{N_r}{N}\frac{N_{r\ell}}{N_r}\frac{N_{r\ell r}}{N_{r\ell}}\right).$$

We can estimate these fractions by generating random leaves. We estimate N_r/N by sampling from the entire tree, and seeing how often a random leaf lies on the right subtree. If this is a self-reducible species of tree, we can sample from that subtree's leaves, see how often they lie on its left sub-subtree, and so on. If we take enough samples, our estimates will be accurate enough to give a good approximation for N.

Both of these arguments work even if we can only count, or sample, approximately rather than perfectly. If we can approximate the number of solutions, then we can generate solutions that are almost random, in the sense that their probability distribution is almost uniform. Conversely, if we can generate almost uniform solutions, we can count approximately. We will show these things quantitatively in the next section.

13.4.2 Approximate Counting and Biased Sampling

Now that we have outlined these ideas, let's describe precisely what we mean by an algorithm that approximately counts the solutions to a problem, or which samples them with probability close to the uniform distribution. The following definition for an approximation algorithm is similar to that for optimization problems in Chapter 9.

> Let $A(x)$ be a function. We say that $A(x)$ can be *approximated in polynomial time with error* ε if there is a function $f(x)$ in P such that, for all x, we have
>
> $$(1+\varepsilon)^{-1} f(x) \le A(x) \le (1+\varepsilon) f(x) .$$
>
> If there is a randomized algorithm that takes x and ε as input and returns such an $f(x)$ with probability at least 2/3, and whose running time is polynomial in both ε^{-1} and $n = |x|$, we say that $A(x)$ has a *fully polynomial randomized approximation scheme* or FPRAS.

As always for randomized algorithms, setting the probability of success at 2/3 is arbitrary. As Problem 13.15 shows, we can make this probability exponentially close to 1 by running the algorithm multiple times and taking the median.

Similarly, let's define what we mean by sampling random solutions in a way that is almost uniform:

> Given a problem in #P, we say that we can *sample its solutions with bias* β if there is a polynomial-time randomized algorithm that produces each solution x with probability $P(x)$, such that for all x
>
> $$\frac{(1+\beta)^{-1}}{N} \le P(x) \le \frac{1+\beta}{N} , \tag{13.15}$$
>
> where N is the total number of solutions.

As the following exercise shows, this notion of approximate sampling is stronger than the notion of mixing we used in Chapter 12, where we asked that P be close to uniform in terms of the total variation distance:

Exercise 13.5 *Recall that the* total variation distance *between two probability distributions P and Q is*

$$\|P - Q\|_{\text{tv}} = \frac{1}{2} \sum_x |P(x) - Q(x)| .$$

Suppose that $P(x)$ obeys (13.15), *and let U denote the uniform distribution in which $U(x) = 1/N$ for all x. Show that $\|P - U\|_{\text{tv}} \le \beta/2$, so the distance from U goes to zero as β does. Is the converse true?*

Now that we have these definitions, we can prove a rigorous connection between approximate counting and random sampling. First we will prove that if we can count with small enough error, we can sample with small bias. The idea is to keep track of the error that accumulates when we multiply the probabilities with which we set each variable, and show that it is probably small.

Theorem 13.2 *Let A be a self-reducible problem in #P on n variables. Suppose that A can be approximated in polynomial time with error $\varepsilon = O(n^{-1})$. Then we can sample A's solutions with bias $\beta = O(n\varepsilon)$ in polynomial time as a function of n and ε^{-1}. In particular, if A has an FPRAS, we can sample its solutions with any bias $\beta > 0$ in time which is polynomial in n and β^{-1}.*

Proof First let's assume that the approximation algorithm is deterministic, i.e., that it always approximates A within error ε. In that case, the probability with which we set each variable true or false differs from the correct probability by a factor of at most $1+\varepsilon$. Since the probability $P(x)$ we arrive at a particular solution x is the product of n of these probabilities, we have

$$\frac{(1+\varepsilon)^{-n}}{N} \leq P(x) \leq \frac{(1+\varepsilon)^{n}}{N} .$$

Setting $1+\beta = (1+\varepsilon)^n$ and using the fact that $\varepsilon = O(n^{-1})$ gives $\beta = O(n\varepsilon)$.

But if the approximation algorithm is randomized, its results are not always accurate. In that case we can use Problem 13.15, taking the median of multiple trials to reduce the probability P_{fail} that it fails to ε/N. Assuming for simplicity that the n variables are Boolean, we have $N \leq 2^n$. Then we can lower the probability of failure to $P_{\text{fail}} = \varepsilon 2^{-n}$ using $t = O(\log P_{\text{fail}}^{-1}) = O(n + \log \varepsilon^{-1})$ trials.

By the union bound (see Appendix A.3.1), the probability that the approximation algorithm fails on any of the n steps is at most nP_{fail}, and this changes $p(x)$ for any particular x by at most nP_{fail}. Thus the bias β increases by at most $nP_{\text{fail}}N = n\varepsilon 2^{-n} N \leq n\varepsilon$, and we still have $\beta = O(n\varepsilon)$.

The total running time is that of the approximation algorithm, times n, times the number of trials t per step. All of these are polynomial in n and ε^{-1}. Finally, if A has an FPRAS then we can sample its solutions with bias β in polynomial time as a function of n and $\varepsilon^{-1} = O(n\beta^{-1})$, which is polynomial in n and β^{-1}. □

This sampling algorithm can be improved in a number of ways. Problem 13.16 shows that by performing a random walk that moves up and down the tree of partial solutions, rather than traveling directly from the root to a leaf, we can sample with small bias even if our counting algorithm over- or underestimates the number of solutions by a constant, or even a polynomial factor—in other words, even if $\varepsilon = O(n^d)$ for some $d > 0$.

Next, we prove a converse to Theorem 13.2. As described in the previous section, by sampling random solutions, we can estimate the fraction that lie on a given subtree, and then invert these fractions to estimate the total number of solutions. To get good estimates for these fractions, it suffices to sample a polynomial number of times from a probability distribution that is sufficiently close to the uniform distribution $U = 1/N$. For the definition of the total variation distance $\|P - U\|_{\text{tv}}$, see Section 12.2.

Theorem 13.3 *Let A be a self-reducible problem in #P on n variables. Suppose there is a polynomial-time sampling algorithm that generates solutions according to a probability distribution $P(x)$, where $\|P - U\|_{\text{tv}} \leq \varepsilon$ and $\varepsilon = O(n^{-1})$. Then there is a randomized approximation algorithm for A with error $O(n\varepsilon)$, which runs in polynomial time as a function of n and ε^{-1}.*

Proof We choose a path to a leaf, and estimate the fraction of solutions that lie in each subtree we choose. Let p denote the path we have taken so far, corresponding to setting some of the variables. Since A is self-reducible, we can sample from the N_p solutions to the subproblem on the remaining variables.

Let P_{pt} and P_{pf} denote the total probability, under the distribution P, that the next variable is true or false. Under the uniform distribution, these probabilities would be N_{pt}/N_p and N_{pf}/N_p. So, by Exercise 12.11 on page 572, we have

$$\left|P_{pt} - \frac{N_{pt}}{N_p}\right|, \left|P_{pf} - \frac{N_{pf}}{N_p}\right| \leq \varepsilon . \tag{13.16}$$

We estimate P_{pt} and P_{pf} by taking s samples. Let s_t and s_f denote the number of samples in which the next variable is true or false respectively. By the Chernoff bound (see Appendix A.5.1) if we set

$$s = n/\varepsilon^2 ,$$

then with probability $1 - \mathrm{e}^{-\Omega(n)}$ we have

$$\left|\frac{s_t}{s} - P_{pt}\right|, \left|\frac{s_f}{s} - P_{pf}\right| \leq \varepsilon . \tag{13.17}$$

Again using the union bound, the probability that this fails to be true on any of our n steps is at most $n\mathrm{e}^{-\Omega(n)}$, which is exponentially small. Then (13.16), (13.17), and the triangle inequality imply that, with probability exponentially close to 1,

$$\left|\frac{s_t}{s} - \frac{N_{pt}}{N_p}\right|, \left|\frac{s_f}{s} - \frac{N_{pf}}{N_p}\right| \leq 2\varepsilon \tag{13.18}$$

on every step of the path.

We set each variable to whichever value v agrees with a majority of the solutions, according to our estimates, of the current subproblem. With high probability, this means that $N_{pv}/N_p \geq 1/2 - O(\varepsilon)$. In that case, (13.18) implies that our estimates s_v/s differ from N_{pv}/N_p by a multiplicative factor of $1 + O(\varepsilon)$. If we multiply these factors together as in the proof of Theorem 13.2, then the product of our estimated fractions, and therefore our estimate of N, is off by a multiplicative factor of $(1 + O(\varepsilon))^n = 1 + O(n\varepsilon)$.

Finally, the total running time is that of the sampling algorithm, times n, times the number $s = n/\varepsilon^2$ of trials per step. □

Now that we know that counting and sampling are intimately related, we can use the Markov chains of Chapter 12 to approximate #P problems in polynomial time. Suppose we have a Markov chain that performs a random walk in the space of solutions. If it mixes in polynomial time, the number of steps it takes for its probability distribution to get within a distance ε from equilibrium is polynomial in n and $\log \varepsilon^{-1}$. Theorem 13.3 then gives us the following attractive connection between efficient mixing and approximate counting:

Corollary 13.4 *Let A be a problem in* #P. *If there is a Markov chain on the set of A's solutions whose equilibrium distribution is uniform and which mixes in polynomial time, then A has an FPRAS.*

In the next section, we will use exactly this approach—a Markov chain on the set of matchings on a graph, and a proof that it mixes in polynomial time—to obtain an FPRAS for 0-1 PERMANENT. 13.4

13.5 Random Matchings and Approximating the Permanent

Since it first appeared in mathematics in the early 1800s, the permanent has seemed very hard to calculate. It is a sum over exponentially many permutations, and no one knows a shortcut analogous to Gaussian elimination for the determinant. Indeed, the fact that it is #P-hard to calculate exactly is strong evidence that no such shortcut exists.

Whether we can even *approximate* the permanent remained an open question for many years. But in 2001, Mark Jerrum, Alistair Sinclair, and Eric Vigoda found a polynomial-time approximation algorithm for #PERFECT MATCHINGS and therefore for 0-1 PERMANENT. As alluded to in the previous section, it estimates the number of perfect matchings by generating uniformly random ones, using a Markov chain that mixes in polynomial time.

How can we design a Markov chain on the set of perfect matchings? What kinds of moves might it make to go from one matching to another? As we described in Section 13.2, if we have two perfect matchings μ_1 and μ_2, their symmetric difference $\mu_1 \oplus \mu_2$—that is, the set of edges which appears in one matching or the other, but not both—consists of a set of cycles of even length, where the edges on each cycle alternate between μ_1 and μ_2. We can get from μ_1 to μ_2 by flipping these cycles, adding and removing alternating edges on each one.

However, a Markov chain where we flip an entire cycle in a single move seems difficult to analyze. Given the current matching, there might be an exponential number of different cycles we could flip, and we would have to choose among these according to some well-chosen probability distribution. Thus each move would involve a potentially difficult sampling problem in itself.

Instead, our Markov chain will add or remove one edge at a time. To allow this, we will expand our state space beyond perfect matchings to include *near-perfect* ones: partial matchings with a pair of "holes," i.e., two vertices which are left unmatched. Our goal is to devise an algorithm that samples uniformly from the set of matchings of either kind. If the perfect matchings occupy a polynomial fraction of this state space—that is, if there are only poly(n) more near-perfect matchings than there are perfect ones—then we can use this algorithm to sample perfect matchings with a polynomial number of trials.

The proof that this Markov chain mixes in polynomial time involves a sophisticated use of the techniques of Chapter 12, and getting through it takes some work. But as one of the most important and celebrated uses of Markov chains in computer science, it is well worth the effort.

13.5.1 A Markov Chain and its Mixing Time

Let's fix some notation. Let $G = (V, E)$ be a bipartite graph with n vertices on each side and m edges between them. For an integer $t \geq 0$, let Ω_t denote the set of partial matchings with t pairs of holes, i.e., $2t$ unmatched vertices, and let $N_t = |\Omega_t|$. Then our state space is $\Omega = \Omega_0 \cup \Omega_1$, the set of perfect or near-perfect matchings, and its total size is $N = N_0 + N_1$.

We will assume for now that a polynomial fraction of the state space consists of perfect matchings, i.e.,

$$\frac{N_1}{N_0} = \text{poly}(n). \tag{13.19}$$

In that case, it takes poly(n) samples from Ω to get a random perfect matching. As we will discuss below, this assumption is not always true, and we need some additional ideas to handle graphs where it is false.

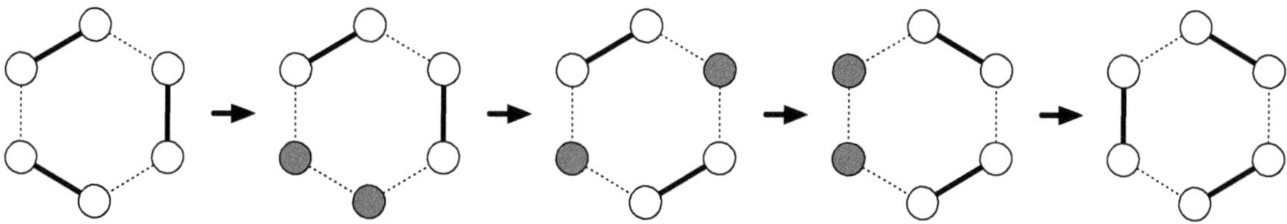

FIGURE 13.11: A sequence of moves that flip a cycle. We start by removing an edge, creating a near-perfect matching with a pair of holes (shown in gray). We then move one hole around the cycle, flipping edges as we go, until we reach the other hole and fill them both in.

If μ denotes the current matching, our Markov chain works as follows. First we choose $(u,v) \in E$ uniformly from all G's edges. We then add or remove this edge from μ, producing a new matching μ' as follows:

1. If μ is perfect and $(u,v) \in \mu$, remove (u,v) from μ.
2. If μ is near-perfect and u and v are the holes, add (u,v) to μ.
3. If μ is near-perfect, u is a hole, and $(v,w) \in \mu$ for some w, remove (v,w) and add (u,v). Similarly, if v is a hole and $(u,w) \in \mu$, remove (u,w) and add (u,v). In either case, w is now a hole.
4. Otherwise, do nothing.

This means that we do nothing quite often, especially when μ is near-perfect. We could speed up the Markov chain by choosing from among the edges where at least one endpoint is a hole, but the version we give here is easier to analyze.

Moves of type 1 or type 2 change perfect matchings to near-perfect matchings and vice versa. Type 3 moves change one near-perfect matching to another, moving a hole from u or v to w, where w is the former partner of whichever of u or v was matched before. We illustrate a sequence of these moves, showing how they flip a cycle, in Figure 13.11.

This Markov chain is symmetric—that is, $M(\mu \to \mu') = M(\mu' \to \mu)$ where M denotes the transition probability. Therefore, its equilibrium distribution P_{eq} is the uniform distribution on the set of perfect or near-perfect matchings. We will show that it mixes in polynomial time. Specifically, as in Section 12.8.3, we will show how probability can flow from every matching to every other one without overburdening any one move. This implies a lower bound on the *conductance* of the Markov chain, and hence an upper bound on its mixing time.

Let $e = (\mu, \mu')$ be an edge in state space, i.e., a move that takes us from μ to μ' according to the transition rules given above. The *capacity* of e is the flow of probability along it at equilibrium,

$$Q(e) = P_{\text{eq}}(\mu) M(\mu \to \mu').$$

Since P_{eq} is uniform, and since we choose the edge (u,v) uniformly from G's m edges, we have

$$Q(e) = \frac{1}{Nm}. \tag{13.20}$$

For every pair of matchings α, β, define a flow $f_{\alpha,\beta}$ of a unit of probability through state space from α to β. If α and β are chosen according to $P_{\rm eq}$, the average flow through e is

$$F(e) = \sum_{\alpha,\beta\in\Omega} P_{\rm eq}(\alpha)\,P_{\rm eq}(\beta)\,f_{\alpha,\beta}(e).$$

If we divide the flow equally among all the shortest paths from α to β, as we did on page 611 for the random walk on the hypercube, then

$$F(e) = \frac{1}{N^2}\sum_{\alpha,\beta\in\Omega} \Pr[\text{a random shortest path from } \alpha \text{ to } \beta \text{ goes through } e]. \tag{13.21}$$

According to (12.33) on page 610, the mixing time is

$$\tau = O(\rho^2 \log N) \tag{13.22}$$

where ρ is the *congestion*,

$$\rho = \max_e \frac{F(e)}{Q(e)}.$$

By bounding the fraction of shortest paths that go through e, we will show that ρ, and therefore τ, are polynomial in n.

Here is how we will proceed. Our key result will be to bound the average flow through e in terms of the number of matchings with two pairs of holes,

$$\sum_{\alpha,\beta\in\Omega} \Pr[\text{a random shortest path from } \alpha \text{ to } \beta \text{ goes through } e] = O(N_2). \tag{13.23}$$

Combining this with (13.20) and (13.21) will imply

$$\rho = O\left(\frac{N_2}{N}\,m\right) = O\left(\frac{N_2}{N_1}\,m\right). \tag{13.24}$$

We will then prove that, in any bipartite graph,

$$\frac{N_2}{N_1} \le \frac{N_1}{N_0}, \tag{13.25}$$

so that (13.24) becomes

$$\rho = O\left(\frac{N_1}{N_0}\,m\right).$$

Since $N_0 \le n!$ and $N_1 \le n^2 n!$, we have $\log N = O(n \log n)$. Then since $m = O(n^2)$, (13.22) will give

$$\tau = O\left(\left(\frac{N_1}{N_0}\right)^2 n^5 \log n\right).$$

Finally, our assumption (13.19) that N_1/N_0 is polynomial will imply that τ is polynomial as well.

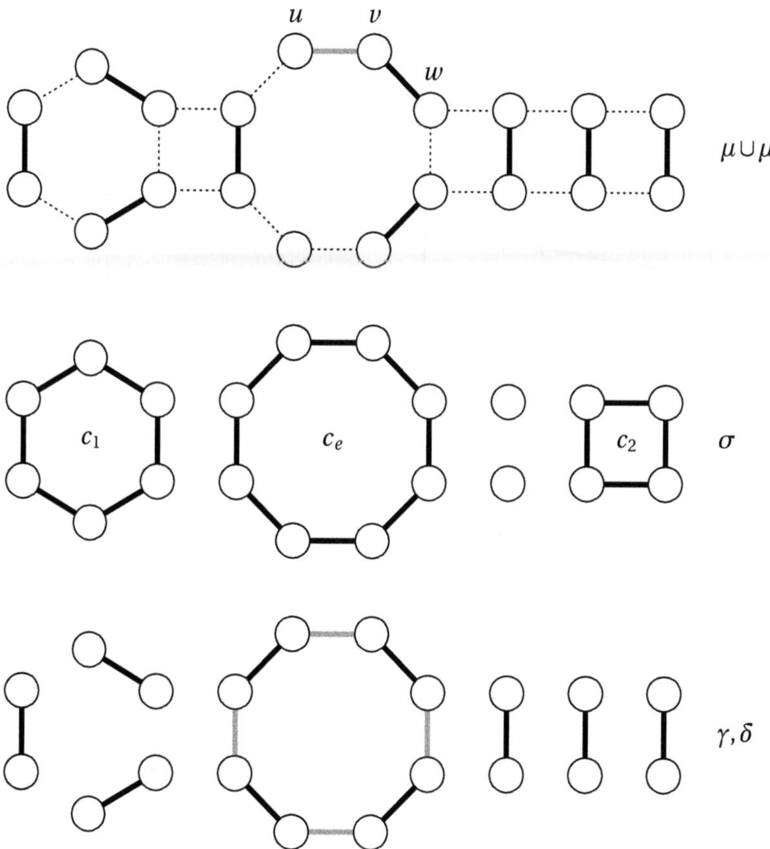

FIGURE 13.12: Above, a move e from μ to μ' in the Markov chain, which adds the edge (u,v) and removes (v,w). Middle, a set σ of cycles that is consistent with e in the sense that e could occur in the process of flipping one of them, which we call c_e. Bottom, flipping c_e from the black edges to the gray ones changes one perfect matching γ to another one, δ.

13.5.2 *Cycle Sets and Flipping Flows*

In order to prove (13.23), we need to describe how shortest paths work in the state space. Given initial and final matchings $\alpha, \beta \in \Omega$, let σ denote $\alpha \oplus \beta$. We focus for now on the case where α and β are perfect matchings, so σ consists of cycles of even length.

A shortest path from α to β consists of flipping each of the cycles in σ. If this path goes through e, then e must be one step in a sequence of moves like those shown in Figure 13.11, which flips some cycle c_e and changes one perfect matching to another. Let's call these matchings γ and δ. As Figure 13.12 shows, e removes one of γ's edges, adds one of δ's, or does both. In this case, $(v,w) \in \gamma$ and $(u,v) \in \delta$.

There are many sequences of moves which go from γ to δ, depending on where we create a pair of holes in c_e and how we move them around. However, let's pessimistically assume that e occurs in every

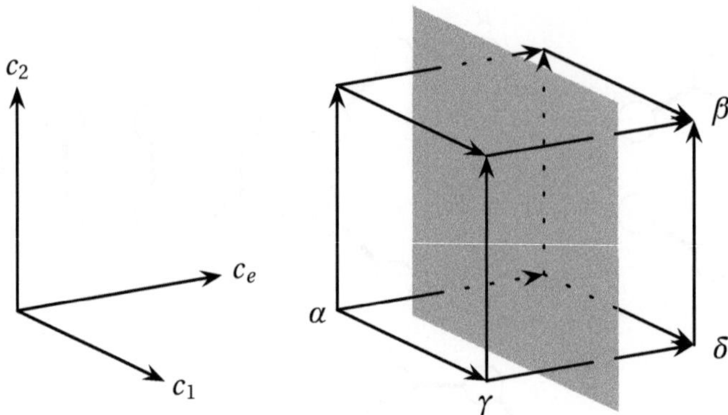

FIGURE 13.13: If σ has k cycles, α and β are opposite points on a k-dimensional hypercube, where each axis corresponds to flipping one of the cycles. At some point, we cross from one half of the hypercube to the other by flipping c_e. However, if α is chosen randomly and we choose a random shortest path from α to its antipode β, then the probability that we cross from γ to δ, and thus possibly use e, is 2^{-k}.

such sequence. Then the question is what fraction of shortest paths from α to β pass through the flip from γ to δ, and what this fraction is on average when α and β are chosen uniformly.

Suppose that $\sigma = \alpha \oplus \beta$ consists of k cycles. For each cycle, either α consists of its even edges and β consists of its odd ones, or vice versa. Thus there are 2^k pairs α, β such that $\alpha \oplus \beta = \sigma$. We can visualize α and β as opposite points on a k-dimensional hypercube as shown in Figure 13.13, where moving along an edge parallel to the ith axis corresponds to flipping the ith cycle in σ.

To get from α to β, we have to flip c_e at some point, crossing from the half of the cube where c_e agrees with α to the half where it agrees with β. There are many places where we could make this crossing, and the flip from γ to δ is one of them. If α is chosen randomly from the 2^k vertices and β is its antipode, and we choose randomly among all the shortest paths between them, the probability we will pass through the flip $\gamma \to \delta$ on the cube is 2^{-k}. To put this differently, the perfect matching we have just before we flip c_e is equally likely to include the odd or even edges of each cycle, so it is γ with probability 2^{-k}.

Let's say that σ is *consistent with e* if e could occur in the process of flipping one of σ's cycles. We have seen that each consistent σ accounts for 2^k pairs α, β such that a shortest path from α to β could pass through e. On the other hand, if α and β are chosen randomly from among these pairs, the probability that a random shortest path from α to β uses the flip $\gamma \to \delta$ is 2^{-k}, and this is an upper bound on the probability that it uses the move e. Multiplying these together, we see that each consistent σ contributes at most 1 to the sum in (13.23), so this sum is at most the number of σ that are consistent with e. Our next job is to bound how many such σ there are.

13.5.3 Counting Consistent Cycle Collections Cleverly

For a given e, how many sets of cycles σ are there that are consistent with e? Here we come to another clever idea. We can bound the number of σ by mapping them to another kind of object—namely matchings with a certain number of holes. Let Σ_e denote the set of σ consistent with e. If we can define, for

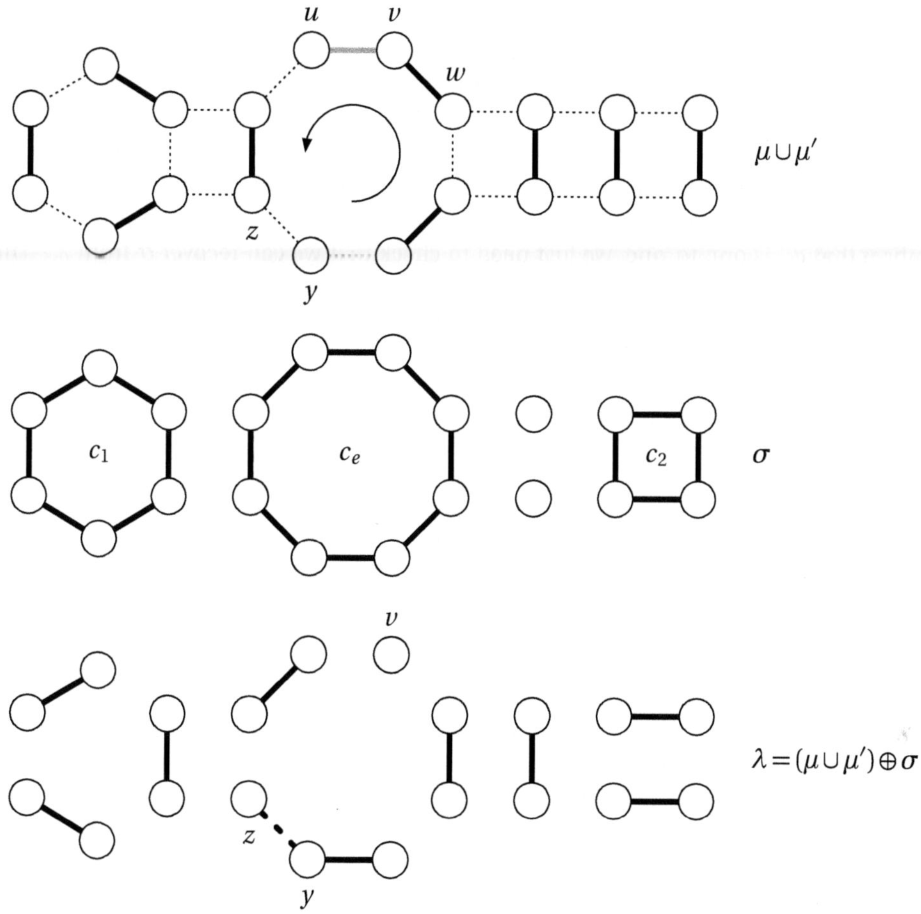

FIGURE 13.14: The mapping from the set of σs that are consistent with a given move e to the set of matchings. The set $\lambda = (\mu \cup \mu') \oplus \sigma$ is almost a matching, except that if μ and μ' are near-perfect their common hole y has two edges in λ. In this case, we use the direction of e, clockwise or counterclockwise, to decide which one of these edges (shown dashed) to remove. This gives a near-perfect matching $\lambda' = \lambda - (y, z)$ with holes at v and z.

each e, a one-to-one mapping

$$\phi_e : \Sigma_e \to \Omega_t ,$$

for some t, then $|\Sigma_e| \leq |\Omega_t| = N_t$.

For the case we have analyzed so far, where α and β are perfect matchings and σ consists of cycles, ϕ_e will map Σ_e to the set Ω_1 of near-perfect matchings. We define ϕ_e as follows. If you know e then you already know the matchings μ, μ' that e connects. To describe σ to you, it's enough to give you the edges that σ doesn't share with μ or μ'—that is, the symmetric difference $(\mu \cup \mu') \oplus \sigma$.

Let's call this difference λ. Since $\mu \cup \mu'$ includes the odd or even edges of each cycle in σ besides c_e, and a mix of odd and even edges of c_e depending on what part of c_e has already been flipped, λ is almost a matching. The only problem, as shown in Figure 13.14, is that if μ and μ' are near-perfect, their common hole y has two edges in λ.

To fix this, we take advantage of the fact that e specifies a clockwise or counterclockwise direction around c_e. If we remove whichever of y's edges (y,z) we encounter first when following this direction from (u,v), we are left with a near-perfect matching $\lambda' = \lambda - (y,z)$. Then we define $\phi_e(\sigma) = \lambda'$.

To confirm that ϕ_e is one-to-one, we just need to check that we can recover σ from λ'. This is easy, since $\sigma = (\mu \cup \mu') \oplus (\lambda' + (y,z))$, and we can determine y and z from μ, μ', and λ'. We conclude that, for each e, the number of consistent σ consisting of cycles is at most $|\Omega_1| = N_1$. This bounds the part of (13.23) coming from pairs α, β of perfect matchings:

$$\sum_{\alpha,\beta \in \Omega_0} \Pr[\text{a random shortest path from } \alpha \text{ to } \beta \text{ goes through } e] \leq N_1 . \tag{13.26}$$

Now let's consider the case where α or β is near-perfect. In addition to cycles, σ now includes one or two alternating paths as shown in Figure 13.15, as we discussed for BIPARTITE MATCHING in Section 5.5.2. To stay within the set of perfect or near-perfect matchings, we have to flip these paths at the beginning or the end of the process. Nevertheless, an argument similar to the one we used before shows that, if α, β is a random pair such that $\alpha \oplus \beta = \sigma$, the average probability that a random shortest path between α and β goes through e is at most 2^{-k} where k is the total number of cycles or paths in σ. Since there are 2^k such pairs, each σ again contributes at most 1 to to the sum in (13.23).

As before, we map σ to $\lambda = (\mu \cup \mu') \oplus \sigma$ and remove one edge from c_e if necessary. However, as Figure 13.15 shows, the resulting matching λ' sometimes has two pairs of holes. So, we have a one-to-one mapping to $\Omega_1 \cup \Omega_2$, giving the following bound on the part of (13.23) where α or β is near-perfect:

$$\sum_{\substack{\alpha,\beta \in \Omega \\ \alpha \notin \Omega_0 \text{ or } \beta \notin \Omega_1}} \Pr[\text{a random shortest path from } \alpha \text{ to } \beta \text{ goes through } e] \leq N_1 + N_2 . \tag{13.27}$$

Combining this with (13.26) and using the fact that $N_1 = O(N_2)$ (exercise!) completes the proof of (13.23).

Our next goal is to prove that $N_2/N_1 \leq N_1/N_0$ as stated in (13.25), or that going from one pair of holes to two increases the number of matchings by at most the ratio between no pairs and one. Equivalently, we will prove that

$$N_0 N_2 \leq N_1^2 . \tag{13.28}$$

The idea is to use, once again, the symmetric difference between two matchings. If α is perfect and β has two pairs of holes, then $\alpha \oplus \beta$ contains two alternating paths as shown in Figure 13.16. If we flip the parity of one of these paths, we get a pair γ, δ of near-perfect matchings. If we choose which path to flip in some deterministic way—say, by sorting the vertices in increasing order, and flipping the path whose endpoint comes first in this order—this gives a one-to-one mapping from $\Omega_0 \times \Omega_2$ to $\Omega_1 \times \Omega_1$, and proves (13.28).

This completes the proof that, under the assumption that $N_1/N_0 = \text{poly}(n)$, our Markov chain mixes in polynomial time. This allows us to sample uniformly from the set of perfect matchings, and thus approximate the permanent.

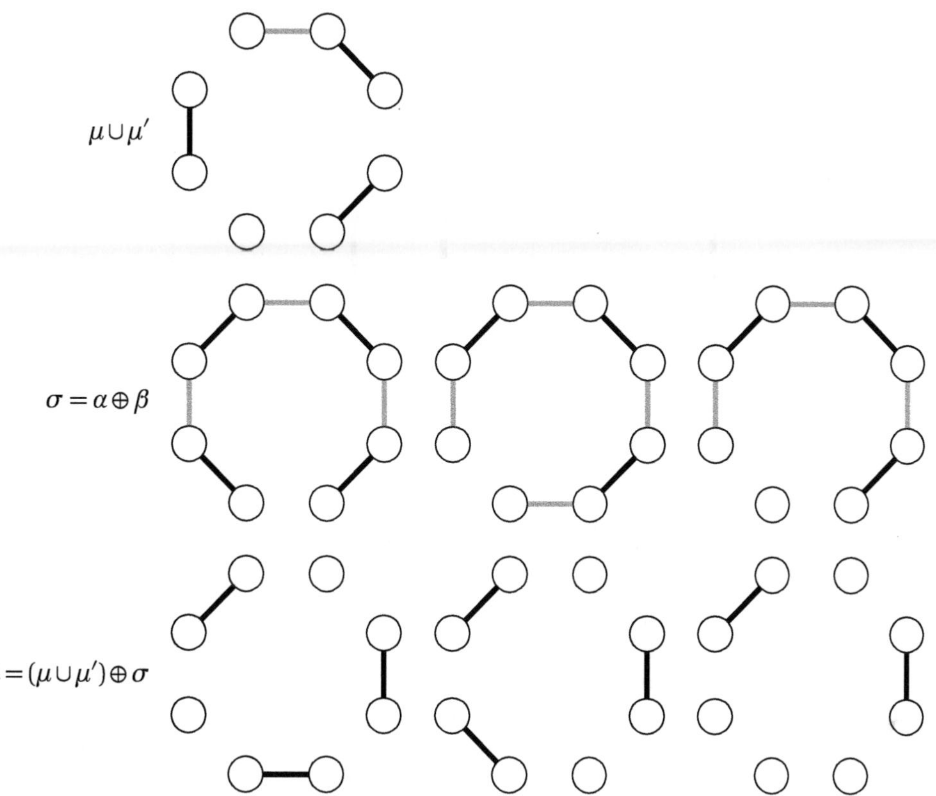

FIGURE 13.15: Extending the mapping to the case where α, β, or both are near-perfect. We have three cases, shown from left to right. Now σ contains a path whose edges alternate between α and β, and λ is a matching with one or two pairs of holes.

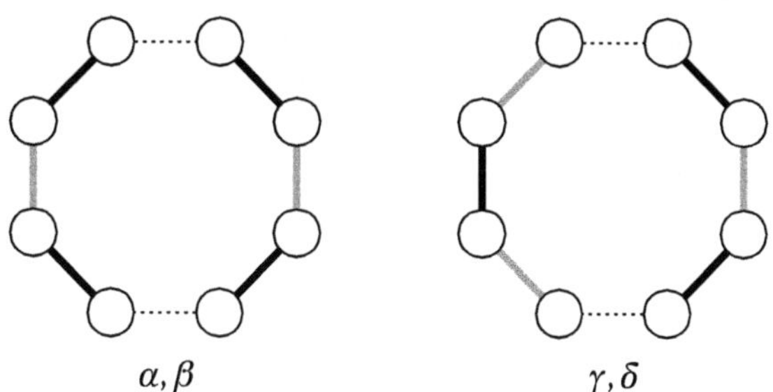

FIGURE 13.16: If α (black) is a perfect matching and β (gray) has two pairs of holes, then flipping one of the paths in $\alpha \oplus \beta$ gives two near-perfect matchings γ, δ.

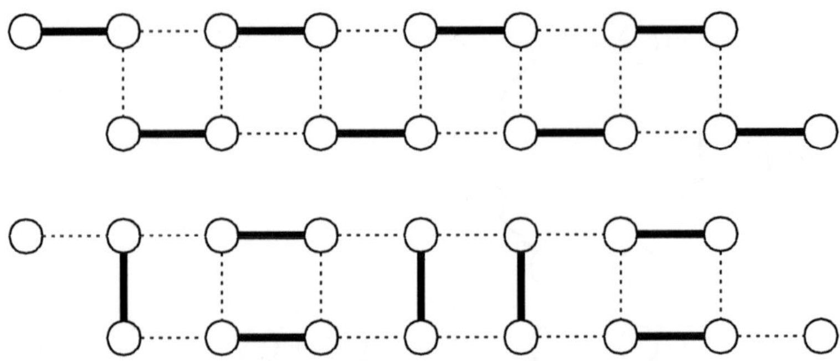

FIGURE 13.17: A type of graph with just one perfect matching (above), but exponentially many near-perfect matchings (below).

13.5.4 *Weighty Holes and Whittling Edges*

As Problem 13.17 shows, the assumption that $N_1/N_0 = \text{poly}(n)$ holds in many natural families of graphs, including lattices in any number of dimensions. However, there are graphs in which there are exponentially more near-perfect matchings than perfect ones. Consider the graph shown in Figure 13.17, for instance. For such graphs, the Markov chain might be exponentially slow. Even if it isn't, the probability N_0/N that a random sample from the state space is a perfect matching is exponentially small. Either way, sampling from the set of perfect matchings will take exponential time. What do we do then?

The answer is to compensate for the number of near-matchings by assigning weights to them. If u and v are vertices, let $\Omega_1(u,v)$ denote the set of near-perfect matchings with holes at u and v, and let $N_1(u,v) = |\Omega_1(u,v)|$. If we knew N_0 and $N_1(u,v)$, we could give each such near-perfect matching the weight

$$w(u,v) = \frac{N_0}{N_1(u,v)}.$$

The total weight of $\Omega_1(u,v)$ is then N_0. If we assign each perfect matching a weight 1, and define a probability distribution P accordingly, we have $P(\Omega_1(u,v)) = P(\Omega_0)$. Summing over all pairs u,v gives $P(\Omega_1) \le n^2 P(\Omega_0)$, so sampling from the entire state space $\Omega = \Omega_0 \cup \Omega_1$ yields a perfect matching $1/O(n^2)$ of the time. We can incorporate these weights into the Markov chain's transition probabilities, and a generalization of the argument for the unweighted version shows that it mixes in polynomial time.

The problem, of course, is that N_0 is what we are trying to calculate in the first place, and $N_1(u,v)$ is a similar quantity—so we don't know the weights $w(u,v)$. Suppose, however, that we have estimates $w_{\text{est}}(u,v)$ for them. If P_{est} is the corresponding probability distribution then we can calculate how far off these estimates are using the following equation,

$$\frac{w_{\text{est}}(u,v)}{w(u,v)} = \frac{P_{\text{est}}(\Omega_1(u,v))}{P_{\text{est}}(\Omega_0)}.$$

We can estimate these fractions by random sampling, using a Markov chain based on the weights $w_{\text{est}}(u,v)$. We then update the $w_{\text{est}}(u,v)$, refining them until they are very close to the correct weights $w(u,v)$.

To start this process off, we need estimated weights that are not too far off from the true ones. We start out with the complete bipartite graph, for which we can calculate the $w(u,v)$ exactly. We then remove one edge at a time, whittling them away until only the edges in G remain. To do this, we assign a weight to each edge, and solve the more general problem of estimating the total weight of all the matchings, where the weight of a matching is the product of the weights of its edges. At first, we give every edge weight 1. In a series of stages, we decrease the weight of the edges missing from G until they are vanishingly small, leaving the edges in G with weight 1. At that point, the total weight of the matchings is essentially the number of matchings of G.

As long as each step decreases the weight of a single edge by a constant factor, the true weights $w(u,v)$ don't change too fast. After a polynomial number of steps, we arrive at the graph G, armed with estimates $w_{\text{est}}(u,v)$ that are off by only a constant factor. This is good enough to mix quickly and give the perfect matchings a total weight of $1/O(n^2)$, and thus sample from them in polynomial time. This finally yields an FPRAS for 0-1 PERMANENT, with no assumptions about the number of perfect or near-perfect matchings.

By now, the reader should be convinced that Markov chains and random sampling are a powerful tool 13.5
for approximating hard counting problems. In fact, many #P-complete problems can be approximated using this approach. But let's return now to the question of counting perfect matchings, and calculating the permanent, exactly rather than approximately. It turns out that we can do this whenever the graph is planar—and that the same techniques will let us solve the two-dimensional Ising model exactly.

13.6 Planar Graphs and Asymptotics on Lattices

We have seen that #SPANNING TREES is a determinant, and is therefore in P, while #PERFECT MATCHINGS is a permanent and is #P-complete. However, this #P-completeness result holds for general graphs. Are there special cases where the number of matchings can be written as a determinant instead?

Indeed there are. Recall that the determinant and the permanent differ only in whether the contribution of each permutation, or equivalently each cycle cover, is multiplied by its parity $(-1)^\pi$. If we start with a bipartite graph G with adjacency matrix B, we might be able to add weights or orientations to the edges to compensate for this parity, so that the weighted adjacency matrix B' satisfies

$$\operatorname{perm} B = \det B' .$$

In this section, we will show that weights or orientations of this kind exist, in particular, whenever G is planar. This will give us a polynomial-time algorithm for exactly counting perfect matchings on any planar graph. It will also let us calculate asymptotic properties, such as how quickly the number of perfect matchings grows on a lattice as a function of its area—and this will lead us, in the next section, to an exact solution of the two-dimensional Ising model.

13.6.1 From Permanents to Determinants

We start with a planar, bipartite graph G with n vertices of each type. Let's color the two types of vertices black and white, like the squares on a checkerboard. As in Section 13.2, we define the $n\times n$ matrix B such that $B_{ij}=1$ if the ith black vertex is connected to the jth white vertex, and $B_{ij}=0$ otherwise.

Each perfect matching corresponds to a permutation π of n objects, which maps each black vertex to its white partner. The permanent of B counts all of these, but its determinant counts them weighted by their parities,

$$\det B = \sum_{\text{matchings } \pi} (-1)^\pi .$$

If we place weights $w_{ij} = \pm 1$ on the edges of G and define $B'_{ij} = w_{ij}$, then each matching π has a weight equal to the product of the weights of its edges,

$$w(\pi) = \prod_{i=1}^{n} w_{i,\pi(i)},$$

and

$$\det B' = \sum_{\text{matchings } \pi} (-1)^{\pi}\, w(\pi).$$

To write perm B as $\det B'$, we would like to design the weights w_{ij} so that $(-1)^{\pi}\, w(\pi)$ has the same sign for all π. In other words, changing the matching π should change w by -1 if π's parity changes, or by $+1$ if it stays the same. Then

$$\left|\det B'\right| = \operatorname{perm} B = \#\text{ of perfect matchings}. \tag{13.29}$$

If we prefer to work with the $2n \times 2n$ weighted adjacency matrix of the entire graph,

$$A' = \begin{pmatrix} 0 & B' \\ B'^{T} & 0 \end{pmatrix},$$

then we have

$$\det A' = (-1)^n (\det B')^2, \tag{13.30}$$

and therefore

$$\left|\det A'\right| = (\operatorname{perm} B)^2 = (\#\text{ of perfect matchings})^2. \tag{13.31}$$

Exercise 13.6 *Prove* (13.30).

If we can find weights w_{ij} such that B' and A' satisfy (13.29) and (13.31), we can compute perm B and count the perfect matchings of G in polynomial time. Do such weights exist? If so, can we construct them efficiently?

We again use the fact that the symmetric difference between any two matchings is a set of cycles whose edges alternate between them. For instance, Figure 13.18 shows two perfect matchings whose symmetric difference consists of a hexagon and a square. Flipping the edges around the square changes π by swapping vertices 4 and 5, which changes the parity. On the other hand, flipping those around the hexagon changes π by composing it with the 3-cycle $1 \to 2 \to 3 \to 1$. The parity of a 3-cycle is even since it is composed of two swaps $1 \leftrightarrow 2$ and $2 \leftrightarrow 3$, so this flip leaves π's parity unchanged.

More generally, flipping the edges around a cycle c of length $2k$ changes π by a k-cycle, and this changes π's parity if and only if k is even. To compensate for this, we want to make sure that this flip changes $w(\pi)$ by -1 or $+1$ if k is even or odd respectively. If we alternately label c's edges as red and green, the flip adds the red edges and removes the green ones or vice versa. The ratio r between the new $w(\pi)$ and the old one is the product of the weights of c's red edges, divided by the product of its green ones—and if $w_{ij} = \pm 1$, r is simply the product of all the edge weights in c. We will call c *good* if $r = -1$ and k is even, or $r = +1$ and k is odd.

Which cycles have to be good? If c appears in the difference between two matchings then the vertices in its interior have to be matched with each other, so there must be an even number of them. For instance,

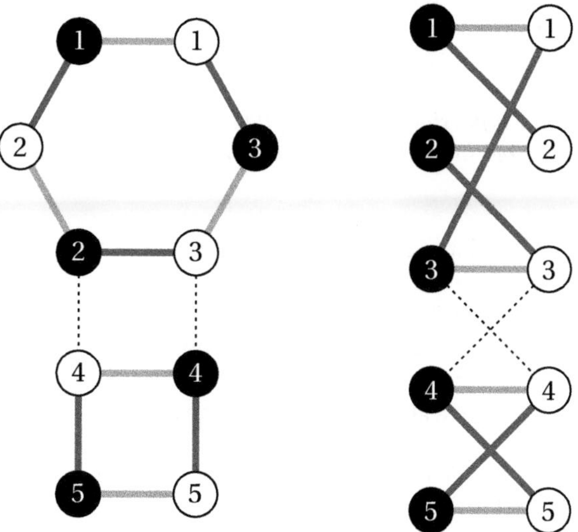

FIGURE 13.18: A planar graph, with a pair of matchings, drawn in two different ways. On the left, we can see that the difference between the two matchings consists of two cycles, a hexagon and a square. On the right, we see that flipping the hexagon changes the matching by a 3-cycle, while flipping the square changes it by a 2-cycle. Since a k-cycle has odd parity if k is even and vice versa, these matchings will make the same contribution to the determinant if these flips change the weight by $+1$ and -1 respectively.

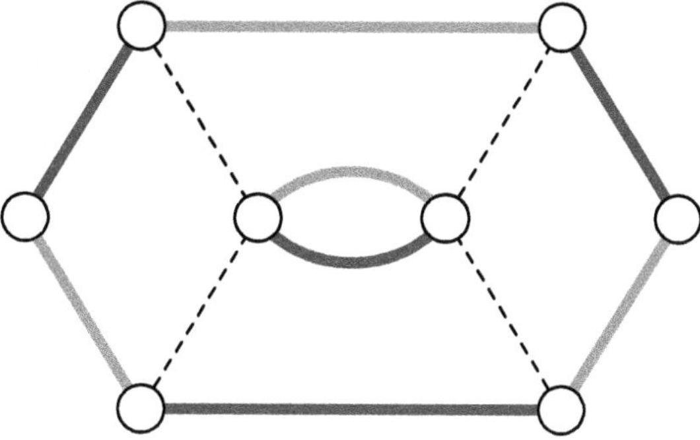

FIGURE 13.19: Two matchings that differ on a cycle of length 6, with two vertices in its interior.

Figure 13.19 shows two matchings which differ on a cycle of length 6, with two vertices inside it. Thus for the weights w_{ij} to be good, it suffices for every cycle with an even number of vertices in its interior to be good. The following exercise shows that it's enough to check the *faces* of G, i.e., cycles with no vertices inside them:

Exercise 13.7 *Let G be a planar bipartite graph, weighted in such a way that every face is good. Show that every cycle c with an even number of vertices in its interior is good. Hint: build the area enclosed by c one face at a time.*

Alternately, we can show (see Problem 13.21) that we can get from one matching to any other by a series of moves which flip around one face at a time.

Now we claim that for any planar bipartite graph, we can construct a set of weights such that every face is good. We do this by induction, growing the graph outward one face at time. Each new face has at least one edge not included in any of the previous faces. If necessary we give that edge a weight -1 in order to make that face good, and give all the other new edges a weight $+1$.

Exercise 13.8 *Construct a set of weights for the graphs in Figure 13.18 and Figure 13.19 such that all faces are good. Then check that* $|\det B'|$ *is the number of perfect matchings.*

There are polynomial-time algorithms that take a planar graph as input, find a layout in the plane, and give a list of the vertices around each face. We can then use this inductive procedure to find a good set of weights in polynomial time. Finally, since we can calculate $\det B'$ in polynomial time as well, it follows that #PERFECT MATCHINGS is in P for planar bipartite graphs. In the next section, we will look at a more general approach, which works for all planar graphs whether or not they are bipartite.

13.6.2 Pfaffian Orientations

When we solve the Ising model, we will find it useful to count perfect matchings in the case where G is planar but not bipartite. In this case, some cycle covers of G may contain cycles of odd length. Thus we can't associate them with pairs of matchings, and (13.10) is no longer true. However, we can still say that

$$\#\text{ of even cycle covers} = (\#\text{ of perfect matchings})^2 ,$$

where an *even cycle cover* is one where every cycle has even length. So, we would like to modify the adjacency matrix so that the cycle covers with one or more odd cycles cancel out, and $\det A'$ only counts the even ones.

There is an easy way to force the odd cycles to cancel out—make A' antisymmetric. If we choose an orientation for each edge, we can define

$$A'_{ij} = \begin{cases} +1 & \text{if } i \to j \\ -1 & \text{if } j \to i \\ 0 & \text{otherwise}. \end{cases} \tag{13.32}$$

Now suppose we take a cycle cover π, and reverse the edges along one of its cycles. This gives a factor of -1 for each edge in the cycle, for a total of -1 if the cycle is odd. Thus for any cycle cover with one or more

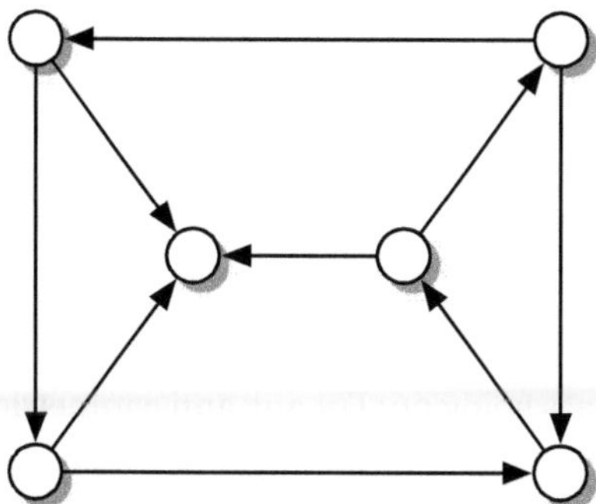

FIGURE 13.20: A non-bipartite planar graph with a Pfaffian orientation. Every face has an odd number of clockwise edges, and so does the 4-cycle around the outside with two vertices in its interior.

odd cycles, we can associate a cover with opposite weight, say by reversing the odd cycle which appears first when we write the vertices in order. These pairs of covers cancel, leaving us with

$$\det A' = \sum_{\text{even cycle covers } \pi} (-1)^\pi \, w(\pi),$$

where

$$w(\pi) = \prod_{i=1}^{n} A'_{i,\pi(i)} .$$

The diligent reader already saw this kind of adjacency matrix in Section 10.7, where we found a simple randomized algorithm to tell whether G possesses any perfect matchings. In fact, we have just solved Exercise 10.9. (Shh, don't tell!)

Now, if we can find a way to choose this orientation such that $(-1)^\pi \, w(\pi) = 1$ for every even cycle cover π, we will have

$$\det A' = \#\text{ of even cycle covers} = (\#\text{ of perfect matchings})^2 . \tag{13.33}$$

Since the parity of an even cycle is odd, $(-1)^\pi$ is simply -1 raised to the number of cycles. Thus we need to make sure that the weight of each cycle c appearing in π is -1. For each c, this means that an odd number of c's edges are oriented clockwise and an odd number are oriented counterclockwise. Then according to (13.32), going around c in either direction gives an odd number of -1s, which multiply to give $w(c) = -1$.

Since the vertices enclosed by c also form even cycles, it again suffices to consider even cycles c that have an even number of vertices in their interior. Analogous to Exercise 13.7, it suffices in turn to look at how the edges are oriented around each face:

Exercise 13.9 *Let G be a directed planar graph such that each face has an odd number of clockwise edges around it. Show that for any even cycle c with an even number of vertices in its interior, the number of clockwise edges around c is odd. Therefore, $w(c) = -1$ whether we go around c clock- or counterclockwise.*

We call such an orientation *Pfaffian*, and we show an example in Figure 13.20. Just as we constructed a set of weights that made every face good in the previous section, we can construct a Pfaffian orientation in polynomial time, by adding one face at a time and choosing the orientations of the new edges. Thus every planar graph has a Pfaffian orientation, and #PERFECT MATCHINGS is in P for planar graphs.

Exercise 13.10 *Write the* 6×6 *adjacency matrix* A' *for the directed graph shown in Figure 13.20, and check that* $\det A'$ *is the square of the number of perfect matchings.*

Here is another way to see what's going on. For any antisymmetric matrix $2n\times 2n$ matrix A,

$$\det A = (\mathrm{Pf}\,A)^2\,,$$

where the *Pfaffian* $\mathrm{Pf}\,A$ is defined as

$$\mathrm{Pf}\,A = \frac{1}{2^n\,n!}\sum_{\pi}(-1)^{\pi}\prod_{i=1}^{n} A_{\pi(2i-1),\pi(2i)}\,. \tag{13.34}$$

and the sum is over all permutations π of $2n$ objects. We ask you to prove this in Problem 13.22.

Now suppose that A corresponds to an orientation of a undirected graph G with $2n$ vertices and non-negative edge weights. Then we can write $\mathrm{Pf}\,A$ as a sum over all perfect matchings μ,

$$\mathrm{Pf}\,A = \sum_{\mu}(-1)^{\pi(\mu)}\,w(\mu)\,. \tag{13.35}$$

Here $w(\mu)$ is the product of the weights of the edges in μ, and $\pi(\mu)$ is a permutation such that μ can be written as a list of edges

$$\mu = \{(\pi(1),\pi(2)),(\pi(3),\pi(4)),\ldots,(\pi(2n-1),\pi(2n))\}\,,$$

where the ith edge is oriented from $\pi(2i-1)$ to $\pi(2i)$.

Exercise 13.11 *Prove that* (13.34) *and* (13.35) *are equivalent. Hint: show that* $(-1)^{\pi(\mu)}$ *depends only on* μ *and not on the order in which we list* μ*'s edges, and that each* μ *corresponds to* $2^n\,n!$ *different permutations* π *in* (13.34).

Now recall that we can move from any perfect matching to any other by flipping a set of even cycles, and consider the following exercise:

Exercise 13.12 *Show that if each even cycle* c *has an odd number of edges oriented with it and an odd number oriented against it, then changing the matching* μ *by flipping* c *preserves* $(-1)^{\pi(\mu)}$.

Therefore $(-1)^{\pi(\mu)}$ is the same for all perfect matchings μ, and (13.35) gives

$$|\mathrm{Pf}\,A| = \sum_{\mu} w(\mu)\,.$$

In particular, if every edge of G has weight 1, then $w(\mu)=1$ and

$$|\mathrm{Pf}\,A| = \sqrt{\det A} = \#\text{ of perfect matchings}\,.$$

13.6.3 Matchings on a Lattice

Physicists and computer scientists have two rather different notions of what a "problem" is, and what constitutes a solution to it. Consider the problem of counting perfect matchings. To a computer scientist, this is a problem that we would like to solve for arbitrary finite graphs, where we are given a precise description of the graph as our input. A "solution" is an algorithm that scales polynomially with the number of vertices, or more generally an understanding of the problem's computational complexity—for instance, a proof that it is #P-complete in general, and an efficient algorithm in the planar case.

For a physicist, on the other hand, individual graphs are not very interesting. When I give you a block of iron, I don't tell you how many atoms are in it, and I certainly don't give you a precise description of their locations. Nor do you expect these details to matter. If you have two blocks of iron of the same mass at the same temperature, you expect them to act the same. What matters is the macroscopic properties of the material they are made of—the properties that hold in the *thermodynamic limit* where n, the number of atoms, goes to infinity. After all, the typical value of n in the laboratory is 10^{24}, rather closer to infinity than our computers can reach.

Like computer scientists, physicists are interested in perfect matchings. They call them *dimer coverings*, since a dimer is a molecule composed of two atoms. But rather than asking for the exact number of matchings for particular graphs, physicists ask how the number of matchings on a graph with a particular structure, such as a square lattice, grows as a function of its size. While this is rather different from the computer science question, there are deep connections between the two. In this section, we will see how the same techniques that give us a polynomial-time algorithm for #PERFECT MATCHINGS on planar graphs help us study the "statistical physics" of this problem.

Let's suppose we have an $\ell \times \ell$ square lattice with a total of $n = \ell^2$ vertices. As we discussed in Section 12.6.4, we can think of perfect matchings on this lattice as ways to tile the plane with dominoes. How does the number M of matchings grow as the lattice gets bigger?

Each vertex has to be matched with one of its four neighbors. This gives a crude upper bound of

$$M \leq 4^n .$$

On the other hand, if we divide the lattice into 2×2 squares, each of these squares has two different perfect matchings, along the horizontal edges or along the vertical ones. This gives the lower bound

$$M \geq 2^{n/4} .$$

The true growth of M is an exponential somewhere between these two bounds. In the limit $n \to \infty$,

$$M \sim \mathrm{e}^{sn}$$

for some s such that

$$\frac{1}{4}\ln 2 \leq s \leq \ln 4 .$$

We can express s as

$$s = \lim_{n\to\infty} \frac{\ln M}{n} . \tag{13.36}$$

It is not too hard to show that this limit exists. We call s the *entropy density* of perfect matchings, and in what follows, we will calculate it exactly.

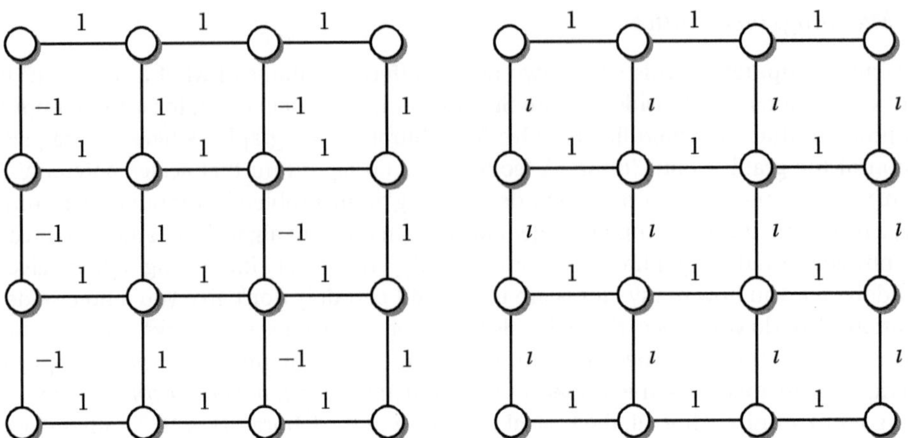

FIGURE 13.21: Two ways to put weights on the lattice so that every face is good, according to the definition of Section 13.6.1. On the left, we give half the vertical edges a weight −1. On the right, we give them all a weight $\imath$. In either case, the product of weights around any face is −1.

First let's use the approach of Section 13.6.1, in which we treat the lattice as an undirected graph. We need to put weights on the edges so that every face is good. In this case each face is a square, and we want the product of its horizontal weights to be −1 times the product of its vertical weights. There are many ways to do this. As Figure 13.21 shows, one option is to give every other vertical edge a weight −1. Another option, which we will use instead, gives every vertical edge a weight $\imath = \sqrt{-1}$.

If A' denotes the resulting weighted adjacency matrix, by (13.31) we have

$$M = \sqrt{|\det A'|}\,.$$

Since the determinant of a matrix is the product of its eigenvalues, (13.36) then becomes

$$s = \frac{1}{2}\lim_{n\to\infty} \frac{\ln|\det A'|}{n} = \frac{1}{2}\lim_{n\to\infty} \frac{1}{n}\ln\prod_{\lambda}|\lambda| = \frac{1}{2}\lim_{n\to\infty} \frac{1}{n}\sum_{\lambda}\ln|\lambda|\,, \tag{13.37}$$

where the sum ranges over all of the eigenvalues of A'.

How can we find these eigenvalues? Let's cheat a little and pretend that the lattice is cyclic, wrapping around and forming a torus. This makes the lattice non-planar, but it makes no difference to the entropy density in the limit $n\to\infty$, and the analysis is much easier. Since A' is then cyclically symmetric, its eigenvectors are Fourier basis functions, like the transition matrix of the random walk on the cycle we analyzed in Section 12.7.2. Specifically, there are $\ell^2 = n$ eigenvectors v_{jk}, one for each pair of frequencies j,k with $0 \le j,k < \ell$:

$$v_{jk}(x,y) = \mathrm{e}^{2\imath\pi(jx+ky)/\ell}\,.$$

Since A' connects each vertex (x,y) to $(x\pm 1,y)$ with weight 1, and with $(x,y\pm 1)$ with weight $\imath$, we have

$$A' v_{jk} = \lambda_{jk} v_{jk}$$

where the eigenvalue λ_{jk} is

$$\lambda_{jk} = e^{2\imath\pi j/\ell} + e^{-2\imath\pi j/\ell} + \imath e^{2\imath\pi k/\ell} + \imath e^{-2\imath\pi k/\ell} = 2\left(\cos\frac{2\pi j}{\ell} + \imath\cos\frac{2\pi k}{\ell}\right).$$

Then (13.37) becomes

$$\begin{aligned} s &= \frac{\ln 2}{2} + \frac{1}{2}\frac{1}{\ell^2}\sum_{j,k=0}^{\ell-1} \ln\left|\cos\frac{2\pi j}{\ell} + \imath\cos\frac{2\pi k}{\ell}\right| \\ &= \frac{\ln 2}{2} + \frac{1}{4}\frac{1}{\ell^2}\sum_{j,k=0}^{\ell-1} \ln\left(\cos^2\frac{2\pi j}{\ell} + \cos^2\frac{2\pi k}{\ell}\right) \end{aligned} \tag{13.38}$$

In the limit $n \to \infty$ we can change the sum over j and k into a double integral, and obtain

$$s = \frac{\ln 2}{2} + \frac{1}{4}\frac{1}{(2\pi)^2}\int_0^{2\pi}\int_0^{2\pi} \ln\left(\cos^2\theta + \cos^2\phi\right) \mathrm{d}\theta\, \mathrm{d}\phi\,. \tag{13.39}$$

As it happens, this integral can be evaluated exactly, and the entropy is

$$s = \frac{C}{\pi} = 0.29156...$$

where C is Catalan's constant $C = \sum_{z=0}^{\infty}(-1)^z/(2z+1)^2$.

Exercise 13.13 *Argue, at the same level of rigor as our discussion here, that this entropy calculation also holds for rectangles as long as their aspect ratio stays constant as $n \to \infty$. In other words, if $k/\sqrt{n}$ and $\ell/\sqrt{n}$ are positive constants, the number of matchings of a $k \times \ell$ rectangle grows as e^{sn} with the same value of s as for squares.*

13.6.4 Once More, With Pfaffians

Let's calculate the entropy of perfect matchings on the square lattice again, but this time using the Pfaffian method of Section 13.6.2. This will be a useful warm-up for solving the two-dimensional Ising model in the next section.

We wish to orient the edges of the lattice so that there are an odd number of clockwise edges on every face. Figure 13.22 shows one way to do this, where the horizontal edges are oriented to the right and the vertical edges alternate between up and down. We again let A' denote the oriented adjacency matrix, where for each edge $i \to j$ we have $A'_{ij} = 1$ and $A'_{ji} = -1$.

This orientation of the lattice is not quite cyclically symmetric, since shifting it one step to the right reverses the orientation of the vertical edges. However, we can restore this symmetry by gathering pairs of adjacent vertices into cells as shown in Figure 13.22, forming an $(\ell/2) \times \ell$ lattice of cells. On this new lattice, we again have Fourier eigenvectors which oscillate with some pair of frequencies j, k when we move from one cell to the next, where $0 \le j < \ell/2$ and $0 \le k < \ell$. The coefficients of these eigenvectors

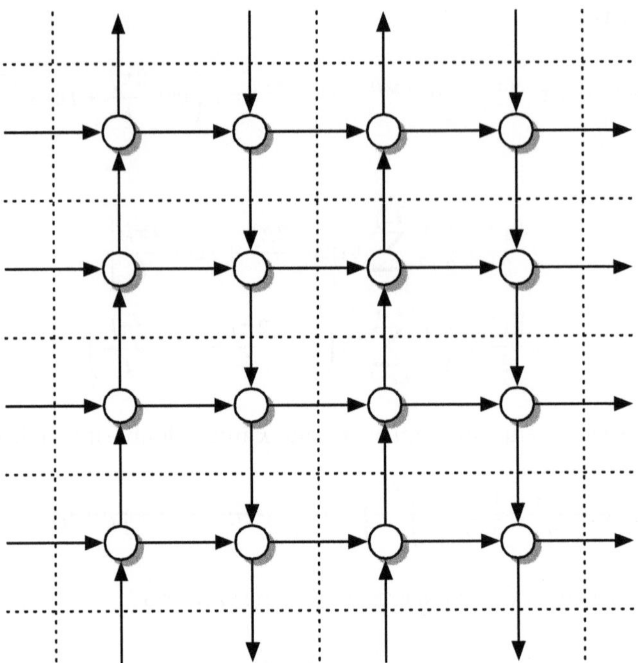

FIGURE 13.22: A Pfaffian orientation of the lattice, i.e., one where the number of clockwise edges around any face is odd. We can treat this graph as cyclically symmetric by dividing it along the dotted lines into cells of two vertices each.

are now two-dimensional vectors, whose components w_1, w_2 correspond to the left and right vertices in each cell. Thus every eigenvector can be written in the form

$$v_{jk}(x,y) = \begin{pmatrix} w_1 \\ w_2 \end{pmatrix} e^{2\imath\pi(2jx+ky)/\ell} .$$

Let's look at how A' acts on v_{jk}. For each cell, we have an internal edge from the left vertex to the right one, and three pairs of edges which connect it to the cells above, below, and to either side. Thus

$$A' v_{jk} = A'_{jk} v_{jk} ,$$

where

$$\begin{aligned} A'_{jk} &= \begin{pmatrix} 0 & 1 \\ -1 & 0 \end{pmatrix} + \begin{pmatrix} e^{\imath\phi} & 0 \\ 0 & -e^{\imath\phi} \end{pmatrix} + \begin{pmatrix} -e^{-\imath\phi} & 0 \\ 0 & e^{-\imath\phi} \end{pmatrix} + \begin{pmatrix} 0 & -e^{-2\imath\theta} \\ e^{2\imath\theta} & 0 \end{pmatrix} \\ &= \begin{pmatrix} 2\imath \sin\phi & 1-e^{-2\imath\theta} \\ -1+e^{2\imath\theta} & -2\imath\sin\phi \end{pmatrix} , \end{aligned} \tag{13.40}$$

and where $\theta = 2\pi j/\ell$ and $\phi = 2\pi k/\ell$.

For each of the $\ell^2/2$ pairs of frequencies j, k, there are two eigenvectors living in the two-dimensional space described by the vector w. The product of their eigenvalues is

$$\det A'_{jk} = 4\left(\sin^2\theta + \sin^2\phi\right),$$

and the product of all A's eigenvalues is

$$\det A' = \prod_{jk} \det A'_{jk}.$$

Using (13.37), we can then write the entropy density as

$$s = \frac{1}{2}\frac{\ln\det A'}{n} = \frac{1}{2n}\ln\prod_{jk}\det A'_{jk} = \frac{1}{2n}\sum_{jk}\ln\det A'_{jk}.$$

We again replace the sum with an integral, and get

$$\begin{aligned} s &= \frac{1}{4}\frac{1}{\ell^2/2}\sum_{j=0}^{\ell/2-1}\sum_{k=0}^{\ell-1}\ln\det A'_{jk} \\ &= \frac{\ln 4}{4} + \frac{1}{4}\frac{1}{2\pi^2}\int_0^{\pi}\mathrm{d}\theta\int_0^{2\pi}\mathrm{d}\phi\,\ln\left(\sin^2\theta + \sin^2\phi\right). \end{aligned}$$

A simple change of variables turns this into our previous expression (13.39), giving the same result $s = 0.29156....$ 13.6

13.7 Solving the Ising Model

We met the Ising model in Section 12.1, where we used it to explain Curie's observation that a block of iron suddenly loses its magnetic properties when we heat it beyond a certain point. On one hand, it is a highly simplified model, where atoms only interact if they are nearest neighbors on a lattice. On the other hand, its behavior is very rich, and reproduces—at least qualitatively—the phase transition that real magnetic materials undergo at their critical temperature. For this reason, it remains one of the most important and fundamental models in statistical physics.

In 1944, Lars Onsager found an exact solution of the Ising model, allowing us to calculate its energy and magnetization analytically as a function of temperature. This ushered in a new era of statistical physics, in which exactly solvable models allowed us to test our theoretical ideas about phase transitions, and opened up new frontiers of collaboration between physics, combinatorics, and mathematics.

We begin this section with a crash course in statistical physics, and explain how a quantity called the partition function lets us describe the properties of a system in thermal equilibrium. As a warm-up, we solve the one-dimensional Ising model using a simple inductive approach. We then map the two-dimensional Ising model onto the perfect matchings of a weighted planar graph. This allows us to use the techniques of Section 13.6 to solve it exactly, and determine its critical temperature.

13.7.1 *The Partition Function*

Suppose we have a physical system where each state S has energy $E(S)$. As we discussed in Section 12.1, if it is in equilibrium at temperature T, the probability of each state S is proportional to the Boltzmann factor $\mathrm{e}^{-\beta E(S)}$ where $\beta = 1/T$. To normalize this, we write

$$P_{\mathrm{eq}}(S) = \frac{\mathrm{e}^{-\beta E(S)}}{Z(\beta)},$$

where $Z(\beta)$ is a weighted sum over all possible states,

$$Z(\beta) = \sum_S \mathrm{e}^{-\beta E(S)}.$$

However, $Z(\beta)$ is far more than a normalization factor. It is called the *partition function*, and it encodes an enormous amount of information about the system's properties. Many physical quantities can be written in terms of Z and its derivatives. For instance, suppose that we are interested in the average energy,

$$\mathbb{E}[E] = \sum_S P_{\mathrm{eq}}(S) E(S) = \frac{1}{Z} \sum_S E(S) \mathrm{e}^{-\beta E(S)}.$$

Since

$$E \mathrm{e}^{-\beta E} = -\frac{\partial}{\partial \beta} \mathrm{e}^{-\beta E},$$

we can write

$$\mathbb{E}[E] = -\frac{1}{Z}\frac{\partial Z}{\partial \beta} = -\frac{\partial}{\partial \beta} \ln Z. \tag{13.41}$$

Similarly, if we are interested in how much the energy fluctuates around its average at equilibrium, we can calculate its variance. Since

$$E^2 \mathrm{e}^{-\beta E} = \frac{\partial^2}{\partial \beta^2} \mathrm{e}^{-\beta E},$$

the variance of the energy is

$$\mathrm{Var}\, E = \mathbb{E}[E^2] - \mathbb{E}[E]^2 = \frac{1}{Z}\frac{\partial^2 Z}{\partial \beta^2} - \frac{1}{Z^2}\left(\frac{\partial Z}{\partial \beta}\right)^2 = \frac{\partial^2}{\partial \beta^2} \ln Z. \tag{13.42}$$

The logarithm $\ln Z$ appears so often that it deserves a name of its own. Up to a factor of the temperature, physicists call it the *free energy*:

$$F = -\frac{1}{\beta} \ln Z.$$

If a system consists of n atoms, and each one can be in one of two states—such as a spin that is up or down—then Z is a sum over 2^n states. Generically, Z grows as e^{fn} for some function $f(\beta)$. Like the entropy density we defined for perfect matchings in (13.36), we can express f as the thermodynamic limit

$$f = \lim_{n \to \infty} \frac{\ln Z}{n}. \tag{13.43}$$

We call f the *free energy per site*. When a physicist says that a system is "exactly solvable," she usually means that we can calculate f, just as we calculated the entropy density of perfect matchings on the lattice in Section 13.6.3. By taking various derivatives of f, we can calculate quantities such as the expected energy per site, or its variance. In the remainder of this chapter, we will do this for the Ising model in one and two dimensions.

13.7.2 The One-Dimensional Ising Model

As we discussed in Section 12.1, each atom in the Ising model has a spin $s_i = \pm 1$ pointing up or down. In the ferromagnetic case, neighboring atoms prefer to point in the same direction. Each edge of the graph contributes -1 or $+1$ to the energy, depending on whether the spins of its endpoints are the same or different, so the total energy of a state S is

$$E(S) = -\sum_{ij} s_i s_j ,$$

where the sum is over all pairs of neighbors i, j, and the partition function is

$$Z = \sum_{\{s_i\}} e^{\beta \sum_{ij} s_i s_j} .$$

On the one-dimensional lattice in particular, this is

$$Z = \sum_{\{s_i\}} e^{\beta \sum_i s_i s_{i+1}} = \sum_{\{s_i\}} \prod_i e^{\beta s_i s_{i+1}} .$$

Our goal is to calculate Z, or at least the limit (13.43). We can do this inductively, by building the lattice one vertex at a time. Let Z_n be the partition function for a lattice of size n, i.e., a chain of n vertices. Since Z_n is a sum over all possible states, we can separate it into two parts,

$$Z_n = Z_n^+ + Z_n^- ,$$

where Z_n^+ and Z_n^- sum over all states where the last vertex in the chain points up or down respectively.

Now suppose we add another vertex. If its spin is the same as the last one, the new edge has an energy of -1, which multiplies Z by a factor of e^{β}. If it is different, it multiplies Z by $e^{-\beta}$ instead. This gives

$$\begin{aligned} Z_{n+1}^+ &= e^{\beta} Z_n^+ + e^{-\beta} Z_n^- \\ Z_{n+1}^- &= e^{-\beta} Z_n^+ + e^{\beta} Z_n^- , \end{aligned}$$

or equivalently

$$\begin{pmatrix} Z_{n+1}^+ \\ Z_{n+1}^- \end{pmatrix} = M \cdot \begin{pmatrix} Z_n^+ \\ Z_n^- \end{pmatrix} \quad \text{where } M = \begin{pmatrix} e^{\beta} & e^{-\beta} \\ e^{-\beta} & e^{\beta} \end{pmatrix} . \tag{13.44}$$

This matrix M is called the *transfer matrix*. It is conceptually similar to the transition matrix of a Markov chain, in that it describes how the system makes a transition from one vertex to the next. However, this transition is in space, rather than time, and M's entries are Boltzmann factors rather than normalized probabilities.

We can use (13.44) to derive the partition function exactly as in Exercise 13.14 below. But for our purposes, let's focus on how Z behaves asymptotically. The eigenvalues of M are

$$\lambda_1 = e^{\beta} + e^{-\beta} = 2\cosh\beta \text{ and } \lambda_2 = e^{\beta} - e^{-\beta} = 2\sinh\beta\,.$$

As we multiply by M repeatedly, the largest eigenvalue λ_1 dominates. Thus Z grows as

$$Z_n \sim \lambda_1^n\,,$$

and the free energy per site is

$$f = \lim_{n\to\infty} \frac{\ln Z}{n} = \ln \lambda_1 = \ln(2\cosh\beta)\,. \tag{13.45}$$

Exercise 13.14 *Show that the partition function for an Ising model on a chain of n vertices with* free boundary conditions, *where the two ends are not adjacent, is exactly*

$$Z = 2^n \cosh^{n-1}\beta\,.$$

On the other hand, if we use cyclic boundary conditions where the two ends join to form a circle, show that

$$Z = \operatorname{tr} M^n = 2^n \left(\cosh^n\beta + \sinh^n\beta\right)\,.$$

Note that while Z differs by a noticeable ratio between these two cases, they both obey the asymptotic statement (13.45). *Then consider what Z becomes in the limits $T \to 0$ and $T \to \infty$, or equivalently $\beta \to \infty$ and $\beta \to 0$, and check that these expressions make sense.*

We can now calculate quantities such as the average energy per site. Applying (13.41) gives

$$\lim_{n\to\infty} \frac{\mathbb{E}[E]}{n} = -\frac{\partial f}{\partial \beta} = -\frac{\partial \ln\lambda_1}{\partial\beta} = -\frac{\sinh\beta}{\cosh\beta} = -\tanh\beta\,. \tag{13.46}$$

If the system is very cold, the spins all line up, and each edge has an energy -1. The number of edges is equal to the number of vertices, and indeed (13.46) gives $\mathbb{E}[E]/n = -1$ in the limit $T \to 0$ and $\beta \to \infty$. At the other extreme where $T = \infty$ and $\beta = 0$, the spins are completely random. Thus each edge is equally likely to have an energy of $+1$ or -1, and (13.46) gives $\mathbb{E}[E]/n = 0$.

What we have done here is not limited to the Ising model. By finding the largest eigenvalue of its transfer matrix, we can solve any one-dimensional physical system with local interactions—those whose interactions are limited to neighboring pairs, or to pairs within some fixed distance of each other.

In two dimensions, on the other hand, the transfer matrix becomes infinite-dimensional. If we grow a rectangular lattice by adding a new row, the partition function depends on all of the spins in the previous row. Therefore, if the lattice has width w, the transfer matrix is 2^w-dimensional. In Problem 13.24 we use rectangles of various finite widths to derive lower bounds on the entropy of perfect matchings on the square lattice—but to get the exact answer, we have to understand the limit $w \to \infty$. Onsager solved the two-dimensional Ising model by finding the largest eigenvalue of this infinite-dimensional transfer matrix. We will use a much simpler method—a mapping from states of the Ising model to perfect matchings.

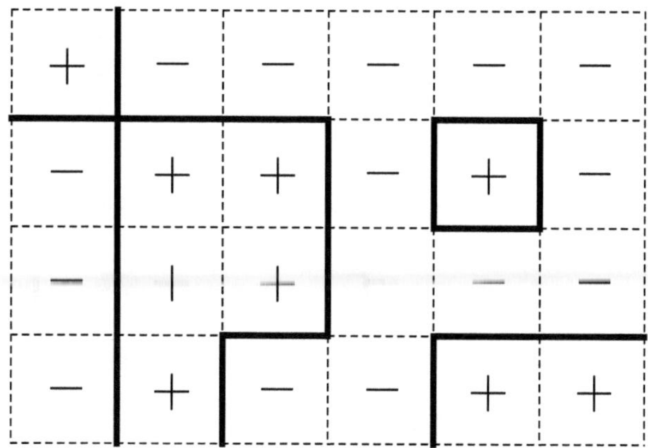

FIGURE 13.23: By drawing the boundaries between regions of up and down spins, we turn states of the Ising model into configurations of polygons where an even number of edges meet at each vertex of the dual lattice.

13.7.3 The Two-Dimensional Ising Model

We come now to the crowning achievement of this chapter—an exact solution, in the statistical physics sense, of the two-dimensional Ising model. To do this, we will rewrite the partition function as the total weight of all the perfect matchings of a certain planar graph. We will then use the Pfaffian method of Section 13.6.2 to write this sum as the square root of a determinant. Finally, we will calculate this determinant analytically in a way analogous to Section 13.6.4.

We will transform Ising states to perfect matchings in two steps. First, consider the *dual lattice* whose vertices correspond to the original lattice's faces and vice versa. Given a state of the Ising model, we draw a set of boundaries between regions of up and down spins as shown in Figure 13.23. If we ignore what happens at the edge of the world, these boundaries form a set of polygons drawn on the dual lattice, where an even number of edges meet at each vertex. Conversely, each such configuration of polygons corresponds to two Ising states, since there are two ways to assign opposite spins to neighboring regions.

If an edge in the original lattice does not cross one of C's edges, its endpoints have the same spin. Then its energy is $-\beta$ and its Boltzmann factor is e^{β}. If it crosses one of C's edges, on the other hand, its endpoints have different spins. This increases its energy by 2β to $+\beta$, and decreases its Boltzmann factor by a ratio $e^{-2\beta}$ to $e^{-\beta}$. Since a square lattice with n sites has $2n$ edges, and since each configuration corresponds to two states, the total Boltzmann factor of a configuration C is $2e^{2\beta n}\,e^{-2\beta|C|}$ where $|C|$ denotes the number of edges in C.

We can then write the partition function as a sum over all configurations C. If $g(z)$ denotes the weighted sum

$$g(z)=\sum_C z^{|C|}, \tag{13.47}$$

then

$$Z=2e^{2\beta n}\sum_C e^{-2\beta|C|}=2e^{2\beta n}\,g\left(e^{-2\beta}\right). \tag{13.48}$$

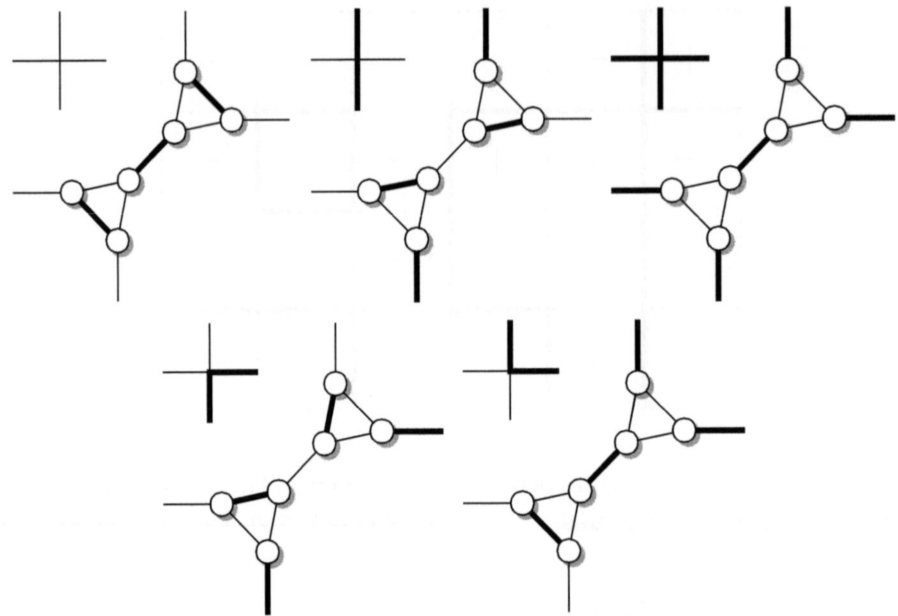

FIGURE 13.24: Replacing each vertex of the dual lattice with a gadget so that polygon configurations become perfect matchings. Each gadget has exactly one perfect matching for each way that an even number of C's edges could meet at the corresponding vertex.

Next, to transform polygon configurations to perfect matchings, we decorate the dual lattice by replacing each vertex with a gadget consisting of six vertices. As shown in Figure 13.24, this gadget has one perfect matching for each way that C's edges can meet at that vertex. In order to give each configuration C the weight $z^{|C|}$, we place a weight z on the edges between gadgets, and give the internal edges of each gadget weight 1. Then $g(z)$ is the total weight of all the perfect matchings.

To calculate $g(z)$, we give this decorated graph a Pfaffian orientation as shown in Figure 13.25. Analogous to Section 13.6.4, we divide it into cells, where each cell contains one of the six-vertex gadgets corresponding to a vertex in the dual lattice. If we pretend that the resulting lattice has cyclic boundary conditions, each eigenvector is a 6-dimensional vector times the Fourier basis function $e^{2\imath\pi(jx+ky)/\ell}$.

Taking the internal edges of each gadget into account, along with the edges connecting it to the neighboring gadgets, the oriented adjacency matrix A' acts on these vectors according to a 6×6 matrix,

$$A'_{jk}(z) = \left(\begin{array}{ccc|ccc} 0 & 1 & 1 & & -ze^{-\imath\theta} & \\ -1 & 0 & 1 & & & -ze^{-\imath\phi} \\ -1 & -1 & 0 & 1 & & \\ \hline & & -1 & 0 & 1 & 1 \\ ze^{\imath\theta} & & & -1 & 0 & 1 \\ & ze^{\imath\phi} & & -1 & -1 & 0 \end{array}\right),$$

where $\theta = 2\pi j/\ell$, $\phi = 2\pi k/\ell$, and matrix entries not shown are zero. There are 6 eigenvalues for each pair

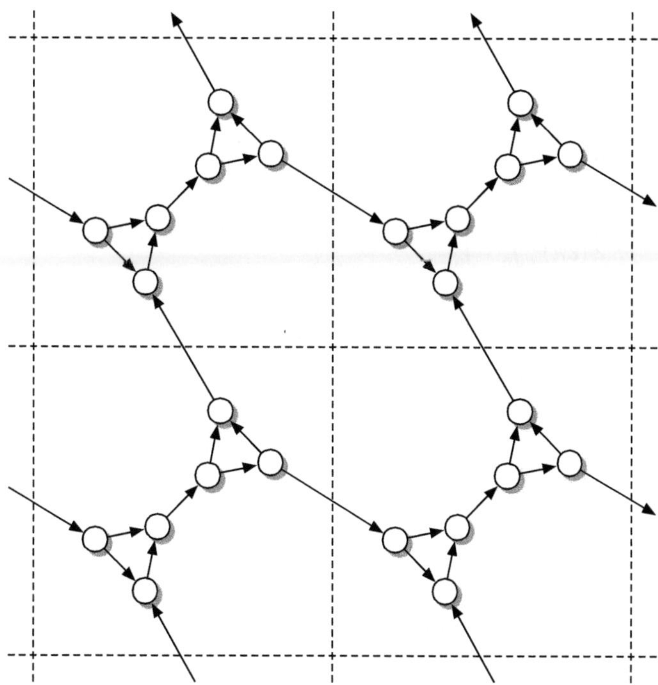

FIGURE 13.25: A Pfaffian orientation of the decorated lattice.

of frequencies j, k, and their product is

$$\det A'_{jk}(z) = (1+z^2)^2 + 2z(1-z^2)(\cos\theta + \cos\phi)\,.$$

We are almost done. We have

$$g(z) = \sqrt{\det A'(z)} = \prod_{jk} \sqrt{\det A'_{jk}(z)}\,,$$

so (13.48) gives

$$\begin{aligned} Z &= 2\mathrm{e}^{2\beta n} \prod_{j,k=0}^{\ell-1} \sqrt{\det A'_{jk}(\mathrm{e}^{-2\beta})} = 2 \prod_{j,k=0}^{\ell-1} \sqrt{\mathrm{e}^{4\beta} \det A'_{jk}(\mathrm{e}^{-2\beta})} \\ &= 2 \prod_{j,k=0}^{\ell-1} \sqrt{(\mathrm{e}^{2\beta} + \mathrm{e}^{-2\beta})^2 + 2(\mathrm{e}^{2\beta} - \mathrm{e}^{-2\beta})(\cos\theta + \cos\phi)} \\ &= 2^{n+1} \prod_{j,k=0}^{\ell-1} \sqrt{\cosh^2 2\beta + (\cos\theta + \cos\phi)\sinh 2\beta}\,. \end{aligned}$$

Writing the log of the product as a sum of the logs and replacing this sum as an integral as before, the free energy per site is

$$f = \lim_{n\to\infty} \frac{\ln Z}{n} = \ln 2 + \frac{1}{2\ell^2} \sum_{j,k=0}^{\ell-1} \ln\left(\cosh^2 2\beta + (\cos\theta + \cos\phi)\sinh 2\beta\right)$$

$$= \ln 2 + \frac{1}{2}\frac{1}{(2\pi)^2} \int_0^{2\pi}\int_0^{2\pi} \ln\left(\cosh^2 2\beta + (\cos\theta + \cos\phi)\sinh 2\beta\right) \mathrm{d}\theta\, \mathrm{d}\phi\,.$$

This integral looks rather complicated. What does it mean? As we discussed in Section 12.1, the Ising model undergoes a phase transition at a critical temperature T_c. How does this phase transition show up in our exact solution, and how can we find T_c?

If we simply plot the free energy per site as a function of temperature, it looks quite smooth. However, we can make the phase transition visible by looking at the expected energy. According to (13.41), the average energy per site is

$$\frac{\mathbb{E}[E]}{n} = -\frac{\partial f}{\partial \beta} = -\frac{1}{(2\pi)^2}\int_0^{2\pi}\int_0^{2\pi} \frac{\sinh 4\beta + (\cos\theta + \cos\phi)\cosh 2\beta}{\cosh^2 2\beta + (\cos\theta + \cos\phi)\sinh 2\beta}\, \mathrm{d}\theta\, \mathrm{d}\phi\,.$$

We plot this as a function of the temperature $T = 1/\beta$ in Figure 13.26. At $T = 0$ all the spins are aligned, and each edge has energy -1. Since there are $2n$ edges, we have $\mathbb{E}[E]/n = -2$ as expected.

Now let's heat the system up. As we discussed in Section 12.1, a typical state is mostly up, say, with small islands pointing down. The energy of these islands is proportional to their perimeter, which increases with the temperature. The steepest increase occurs at the phase transition, where the typical size of the islands diverges and they suddenly stretch across the entire lattice. Beyond the transition, $\mathbb{E}[E]/n$ approaches zero, since the spins become independent and each edge is equally likely to have an energy of $+1$ or -1.

The phase transition is even clearer if we look at the variance of the energy. According to (13.42), the variance divided by n is

$$\frac{\operatorname{Var} E}{n} = \frac{\partial^2 f}{\partial \beta^2}$$

$$= \frac{1}{(2\pi)^2}\iint \frac{4\cosh^2 2\beta + \frac{1}{2}(\cos\theta+\cos\phi)(\sinh 6\beta - 7\sinh 2\beta) - 2(\cos\theta + \cos\phi)^2}{(\cosh^2 2\beta + (\cos\theta+\cos\phi)\sinh 2\beta)^2}\, \mathrm{d}\theta\, \mathrm{d}\phi\,. \qquad (13.49)$$

The result is shown in Figure 13.27. Clearly, there is a divergence at the phase transition. What's going on?

We can think of the fluctuations around equilibrium as groups of sites whose spins flip up and down. Below T_c, these flips create and destroy small islands of the minority spin, and above T_c they flip small clumps of sites with the same spin. In both cases, the correlation length is finite, so distant parts of the system are essentially independent of each other. As a result, the energy behaves roughly like the sum of $O(n)$ independent random variables. Since the variance of a sum of independent variables is the sum of their variances, the variance is then linear in n, and $(\operatorname{Var} E)/n$ is finite.

At T_c, on the other hand, there is a scale-free distribution of islands of all sizes, ranging up to the size of the entire lattice. You can see these islands in Figure 12.2 on page 566. Fluctuations now take place at

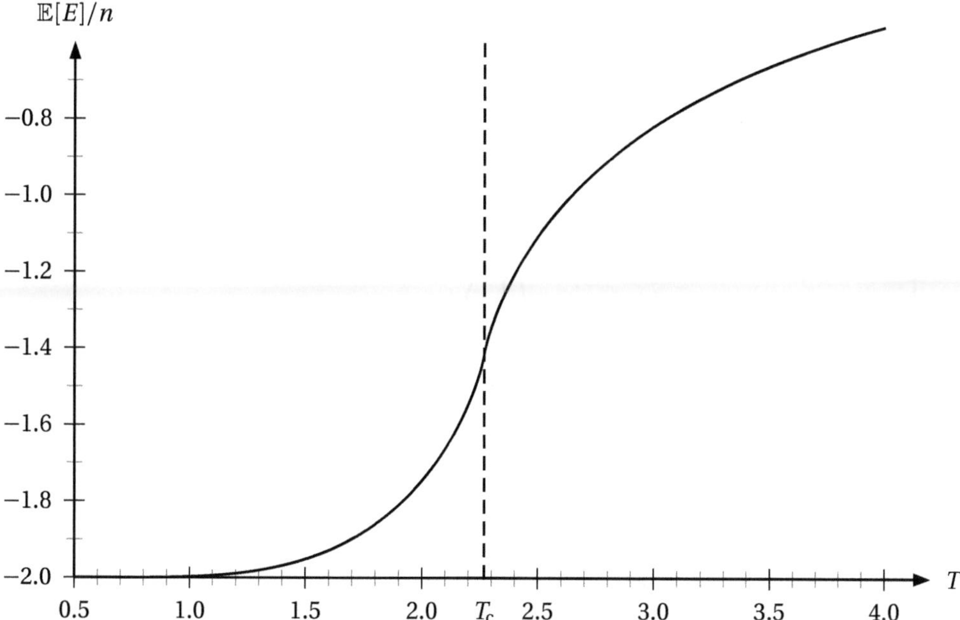

FIGURE 13.26: The expected energy per site of the two-dimensional Ising model as a function of temperature. At $T = 0$, all the spins are aligned and the energy is -1 per edge. It increases as we heat the system up, and the steepest increase occurs at the phase transition where the islands of the minority spin suddenly stretch across the lattice.

all scales, with small, medium, and large groups of vertices flipping their spins, and the energy is strongly correlated between distant parts of the lattice. As a result, $\operatorname{Var} E$ is proportional to n^γ for some $\gamma > 1$, and $(\operatorname{Var} E)/n$ diverges.

We can calculate T_c by determining where $(\operatorname{Var} E)/n$ diverges. While the integral in (13.49) is quite nasty, all that matters to us is that it diverges when the denominator of the integrand is zero for some θ, ϕ. Since $\left|\cos\theta + \cos\phi\right| \le 2$, this happens when

$$\left|\frac{\cosh^2 2\beta}{\sinh 2\beta}\right| \le 2 .$$

Using the identity $\cosh^2 x - \sinh^2 x = 1$, we find that

$$\frac{\cosh^2 2\beta}{\sinh 2\beta} = \sinh 2\beta + \frac{1}{\sinh 2\beta} .$$

If β is positive, this function is always at least 2. There is a unique β for which it is exactly 2, namely the β for which $\sinh 2\beta = 1$. Thus the critical value of β is

$$\beta_c = \frac{1}{2}\sinh^{-1} 1 = \frac{\ln(1+\sqrt{2})}{2} ,$$

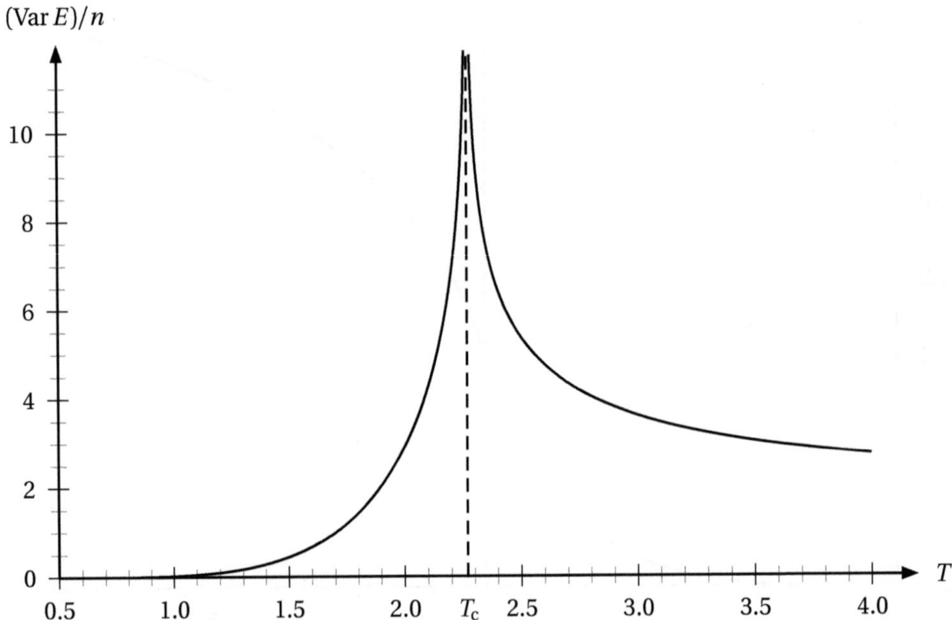

FIGURE 13.27: The variance of the energy, divided by the number of sites, in the two-dimensional Ising model. It diverges at T_c, since at the phase transition there are fluctuations away from equilibrium at all scales.

and the critical temperature is

$$T_c = \frac{1}{\beta_c} = \frac{2}{\ln(1+\sqrt{2})} = 2.269\ldots$$

Looking back at Figure 12.1 on page 565, this is indeed the temperature at which the average magnetization drops to zero.

13.7.4 *On to Three Dimensions?*

Since the two-dimensional Ising model was solved in 1944, physicists have searched in vain for an exact solution of the three-dimensional case. From the computational complexity point of view, there is some reason to be pessimistic about whether such a solution exists. Clearly the methods we used here only work for planar graphs, and Problem 13.44 shows that, for general graphs, calculating the partition function of the ferromagnetic Ising model is #P-hard.

However, it would be a fallacy to take these computational hardness results as a proof that there is no solution to the three-dimensional Ising model in the physics sense. Just because a problem is hard for arbitrary non-planar graphs does not make it impossible to calculate its asymptotic behavior on a specific type of graph. There may well be a closed-form expression for the free energy per site, or for that matter the entropy density of the perfect matchings, on the cubic lattice in any number of dimensions.

On the other hand, perhaps Nature is not quite that kind. We may have to accept that while these
13.7 limits exist, they cannot be expressed in terms of the functions and constants we are familiar with.

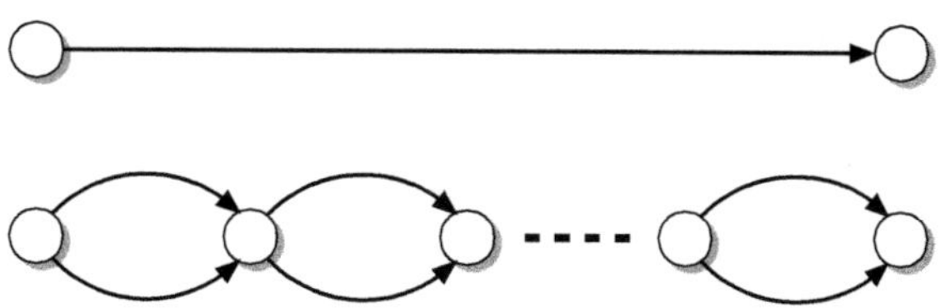

FIGURE 13.28: Replacing an edge with a chain of length k.

Problems

> Don't just read it; fight it! Ask your own questions,
> look for your own examples, discover your own proofs.
> Is the hypothesis necessary? Is the converse true? What happens in
> the classical special case? What about the degenerate cases?
> Where does the proof use the hypothesis?
>
> Paul Halmos

13.1 Random cycles. Here is a beautiful example of a sampling problem that is hard unless NP-complete problems have efficient randomized algorithms. Given a directed graph G, suppose we could generate a random cycle in polynomial time, such that every cycle, regardless of its length, is equally likely. Suppose we first modify G as shown in Figure 13.28, where we replace each edge with a chain of k vertices with a total of 2^k possible paths through them. Then show that if k is sufficiently large (but still polynomial in n) then almost all the probability belongs to the cycles that are as long as possible, and that this would give a randomized algorithm for HAMILTONIAN PATH. In terms of complexity classes, this would imply that NP $\subseteq$ RP (see Section 10.9 for the definition of RP). 13.4

13.2 Cayley and Kirchhoff. Cayley's formula states that there are n^{n-2} trees with n labeled vertices. We proved this formula in Problem 3.36 using Prüfer codes. Prove it again using the Matrix-Tree Theorem. Hint: what are the eigenvalues of the matrix $n\mathbb{1} - J$, where $\mathbb{1}$ is the identity and J is the $(n-1)$-dimensional matrix consisting of all 1s?

13.3 Minors and forests. Prove the following generalization of the Matrix-Tree Theorem. Suppose G is a graph with n vertices. Let $I \subseteq \{1,\ldots,n\}$ be a subset of its vertices, and let $L^{(I)}$ denote the $(n-|I|)$-dimensional minor of its Laplacian formed by deleting the ith row and column for each $i \in I$. Then $\det L^{(I)}$ is the number of spanning forests of G consisting of $|I|$ trees, where each tree in the forest contains exactly one of the vertices in I.

Even more generally, let $I, J \subseteq \{1,\ldots,n\}$ be two subsets of the same size, and let $L^{(I,J)}$ denote the minor formed by removing the rows corresponding to I and the columns corresponding to J. Show that, up to a sign, $\det L^{(I,J)}$ is the number of spanning forests consisting of $|I| = |J|$ trees where each tree contains exactly one vertex in I and one (possibly the same one) in J. This is called the *all-minors Matrix-Tree Theorem.*

13.4 Equilibrium and spanning trees. In Problem 12.23, we stated that the equilibrium distribution $P_{\text{eq}}(i)$ of a Markov chain at each state i is proportional to the total weight of all the spanning trees rooted at i, and in Note 12.10 we claimed this is a consequence of the Matrix-Tree Theorem. Justify this claim by proving the following generalization.

Let G be a weighted directed graph. Let A_{ij} be its adjacency matrix, where A_{ij} is the weight of the edge from j to i. Define the Laplacian matrix L as $L_{ij} = -A_{ij}$ if $i \neq j$ and $L_{jj} = \sum_{i \neq j} A_{ij}$, so that each column of L sums to zero. Given a spanning tree T rooted at a vertex i, i.e., whose edges are oriented towards i, define its weight $w(T)$ as the product of the weights of its edges. Then show that for any i and j,

$$\det L^{(i,j)} = (-1)^{i+j} \sum_{T_j} w(T_j),$$

where the sum is over all spanning trees rooted at j.

Suppose in particular that G is the state space of a Markov chain with transition matrix A, where A_{ij} is the probability of a transition from state j to state i. Show that $LP_{\mathrm{eq}} = 0$, so that the equilibrium distribution P_{eq} is an eigenvector of L with eigenvalue zero. Now given a matrix L, its *adjugate* adj L is defined as the transpose of the matrix of cofactors,

$$(\operatorname{adj} L)_{ij} = (-1)^{i+j} \det L^{(j,i)}.$$

A classic result of linear algebra is that L and adj L commute. Therefore, they have the same eigenvectors. Use this fact to prove, as we showed using Markov chains in Problems 12.22 and 12.23, that $P_{\mathrm{eq}}(i)$ is proportional to $\sum_{T_i} w(T_i)$.

13.5 Finding the minimum spanning tree with the Laplacian. Let $G = (V, E)$ be a weighted undirected graph with nonnegative edge weights w_{ij}. Define a matrix $L(\beta)$ as

$$L_{ij} = \begin{cases} \mathrm{e}^{-\beta w_{ij}} & \text{if } (i,j) \in E \\ 0 & \text{otherwise}. \end{cases}$$

Now show that

$$\text{weight of } G\text{'s minimum spanning tree} = \lim_{\beta \to \infty} \frac{\det L(\beta)^{(1,1)}}{\beta}.$$

If the w_{ij} are integers, how large does β have to be for $\det L^{(1,1)}/\beta$ to be within 1 of the weight of the minimum spanning tree?

13.6 Ryser's algorithm for the permanent. Show the following expression for the permanent of an $n \times n$ matrix:

$$\operatorname{perm} A = \sum_{S \subseteq \{1,\ldots,n\}} (-1)^{n-|S|} \prod_{i=1}^{n} \sum_{j \in S} A_{ij}.$$

This lets us calculate the permanent in $2^n \operatorname{poly}(n)$ time. This is still exponential, but is far better than summing directly over all $n!$ permutations. Hint: use the inclusion–exclusion principle of Appendix A.3.1.

13.5

13.7 Combining counters. Show that if A_1 and A_2 are in #P, then so are $A_1 + A_2$ and $A_1 A_2$. By induction, if $A_1, \ldots, A_k$ are in #P and Q is a polynomial with nonnegative integer coefficients and a polynomial number of terms, then $Q(A_1, \ldots, A_k)$ is in #P. Hint: consider pairs of objects.

13.8 If you can compute, you can count. Show that if $f(x)$ is a function in P which takes nonnegative integer values, then $f(x)$ is in #P.

13.9 Dicey derivatives. Show that calculating nth-order partial derivatives of polynomials is #P-hard. Specifically, consider the partial derivative

$$\frac{\partial^n}{\partial x_1 \, \partial x_2 \cdots \partial x_n} \prod_{i=1}^{n} \left(a_{i,1} x_1 + a_{i,2} x_2 + \cdots + a_{i,n} x_n \right) \Big|_{x_1 = 0, \ldots, x_n = 0},$$

where the $a_{i,j}$ are integer coefficients.

13.10 Don't look at the input. A more conservative definition of counting reductions, which includes parsimonious reductions, is to demand that the algorithm g look only at $B(f(x))$ and not at the original input x. In other words, we change (13.11) to

$$A(x) = g\left(B(f(x))\right).$$

How much difference does this make to our definition of #P-completeness? Hint: show how to modify B to another counting problem B' such that both x and $B(x)$ are encoded in $B'(x)$.

13.11 More #P-complete problems. Review the chain of reductions we used in Sections 5.3.1, 5.3.2, and 5.4.1 to prove that NAE-3-SAT, GRAPH 3-COLORING, and MAX CUT are NP-complete. To what extent are they parsimonious? Either use them to prove that #NAE-3-SAT, #GRAPH 3-COLORINGS, and #MAX CUTS are #P-complete, or design other reductions to do this.

13.12 Parsimonious tours. Show that the reduction from 3-SAT to HAMILTONIAN PATH given in Section 5.6 is not parsimonious. Devise a different reduction, or chain of reductions, that proves that #HAMILTONIAN PATHS is #P-complete. Hint: consider reducing from 1-IN-3 SAT instead (see Problem 5.2).

13.13 Even #2-SAT is #P-complete. The original definition of #P-completeness used *Turing reductions,* in which we say $A \leq B$ if there is a polynomial-time algorithm for A which calls B as a subroutine a polynomial number of times—in contrast to counting reductions where we can only call B once. In this problem, we will prove that even though 2-SAT is in P, #2-SAT is #P-complete with respect to Turing reductions.

First suppose that we have a polynomial of degree n with integer coefficients, $P(z) = \sum_{i=0}^{n} a_i z^i$. As we discussed in Problem 10.38, if we have $n+1$ samples $P(z_1), P(z_2), \ldots, P(z_{n+1})$ where $z_1, \ldots, z_{n+1}$ are distinct, then we can determine P's coefficients a_i in polynomial time.

Given a graph G with n vertices, let N_t denote the number of partial matchings with t "holes" or uncovered vertices. (This differs from our notation N_t below, where t is the number of pairs of holes.) Define

$$P_G(z) = \sum_{t=0}^{n} N_t z^t .$$

In particular, $P_G(0) = $ #PERFECT MATCHINGS and $P_G(1) = $ #MATCHINGS, i.e., the number of partial matchings with any number of holes.

Show that #MATCHINGS is #P-complete under Turing reductions. Hint: show how to modify G to produce a graph G_k, for any integer $k \geq 1$, such that $P_{G_k}(1) = P_G(k)$. Therefore, if we can solve #MATCHINGS for G_k where $k = 1, \ldots, n+1$, we can solve #PERFECT MATCHINGS. Finally, give a parsimonious reduction from #MATCHINGS to #2-SAT. In fact, show that we can even reduce to #POSITIVE 2-SAT, where no variables are negated.

13.14 Determinant gadgets don't work. Prove that there is no matrix M with the following properties, analogous to the conditions (13.12) on the permanent gadget we constructed in Section 13.3.2:

$$\det M = \det M^{(1,2)} = \det M^{(2,1)} = 0$$
$$\det M^{(1,1)} = \det M^{(2,2)} = \det M^{(\{1,2\},\{1,2\})} \neq 0$$

Therefore, there is no way to implement a 3-SAT clause, or even a 2-SAT clause, using the determinant in the same way that we can with the permanent.

Hint: consider the simpler property that the determinants of $M^{(1,1)}$, $M^{(2,2)}$, and $M^{(\{1,2\},\{1,2\})}$ are nonzero but not necessarily equal, while those of M, $M^{(1,2)}$, and $M^{(2,1)}$ are zero. Show that these properties are preserved if we add a multiple of the jth row to any other row, where $j>2$, and similarly for columns. Conclude that if if there were such a matrix M, there would be one of the form

$$M=\begin{pmatrix} M_2 & 0 \\ 0 & \mathbb{1} \end{pmatrix},$$

where M_2 is a 2×2 matrix. Then show that this is impossible.

13.15 The happy median. Suppose that there is a polynomial-time randomized algorithm that approximates $A(x)$ within error ε with probability at least $2/3$. Show that we can decrease the probability of failure from $1/3$ to P_{fail} by performing $O(\log P_{\text{fail}}^{-1})$ trials and taking the median.

13.16 Wandering up and down the tree. As in Section 13.4, let A be a self-reducible problem, and consider a binary tree of partial solutions. Nodes at level ℓ of the tree have ℓ variables set, ranging from the root at $\ell=0$ to the leaves, corresponding to complete solutions, at $\ell=n$. Label each node with a path p of ℓs and rs, and let N_p denote the number of leaves in the subtree rooted at p.

Now consider a Markov chain on the nodes of this tree. Let p denote the current node, and let q denote its parent. We stay at p with probability $1/2$, and move to a neighboring node—down to a subtree, or up to q—according to the following probabilities:

$$M(p\to p\ell)=\frac{1}{2}\frac{N_{p\ell}}{N_p+N_{p\ell}+N_{pr}}, \quad M(p\to pr)=\frac{1}{2}\frac{N_{pr}}{N_p+N_{p\ell}+N_{pr}}, \quad M(p\to q)=\frac{1}{2}\frac{N_p}{N_p+N_{p\ell}+N_{pr}}.$$

Note that we move up or down the tree with equal probability, so that at equilibrium we are equally likely to be at any level between 0 and n. Thus the total probability of the leaves at equilibrium is $1/(n+1)$. Moreover, show that at each level, the equilibrium probability of each node is proportional to the number of leaves on its subtree:

$$P_{\text{eq}}(p)=\frac{1}{n+1}\frac{N_p}{N}.$$

Finally, prove that the conductance of this Markov chain (see Section 12.8) is $1/O(n)$, and conclude, using (12.28) on page 607, that its mixing time is $O(n^3)$.

Now suppose that we have an approximate counting algorithm that gives estimates N_p^{est} such that $\alpha^{-1}N_p\le N_p^{\text{est}}\le \alpha N_p$. We do not demand that $N_p^{\text{est}}=N_{p\ell}^{\text{est}}+N_{pr}^{\text{est}}$. However, we do assume that $N_p^{\varepsilon}=1$ if p is a leaf, since we can check a complete assignment of the variables to see if it is a solution.

Show that if we run this Markov chain using these estimates, then at equilibrium all leaves are still equally likely, and the total probability of the leaves is $1/O(\alpha n)$. Furthermore, show that the conductance is $1/O(\alpha^2 n)$, so the mixing time is $O(\alpha^4 n^3)$. Therefore, as long as $\alpha=\text{poly}(n)$, we have a polynomial-time algorithm that generates solutions which are almost uniformly random.

Comparing this with Theorem 13.3, conclude the following marvelous fact: if a self-reducible problem A can be approximated within a polynomial factor, it has an FPRAS. In other words, if we can approximate it within a factor $O(n^d)$ for some constant $d>0$, we can do so within $1+O(n^{-c})$ for any $c>0$.

13.17 Perfect and near-perfect. Suppose that G is a d-dimensional lattice with cyclic boundary conditions, i.e., which wraps around like a torus. As in Section 13.5.4, let $\Omega_1(u,v)$ denote the number of near-perfect matchings with holes at u and v, and let Ω_0 denote the set of perfect matchings, and let $N_1(u,v)=|\Omega_1(u,v)|$ and $N_0=|\Omega_0|$. Show that for any u,v, we have $|\Omega_1(u,v)|\le|\Omega_0|$. Therefore, using the notation of Section 13.4, we have $N_1/N_0=O(n^2)$, and the unweighted Markov chain described there gives an FPRAS for the number of perfect matchings.

Hint: consider Figure 13.29, where we have shifted one of the near-perfect matchings over so that its holes are neighbors of the other matching's holes. Show that this gives a one-to-one mapping from $\Omega_1(u,v)\times\Omega_1(u,v)$ to $\Omega_0\times\Omega_0$. Generalize this argument to arbitrary *vertex-transitive* graphs, i.e., those for which, for any pair of vertices v,v', there is a one-to-one mapping from the graph to itself that preserves the edges and maps v to v'.

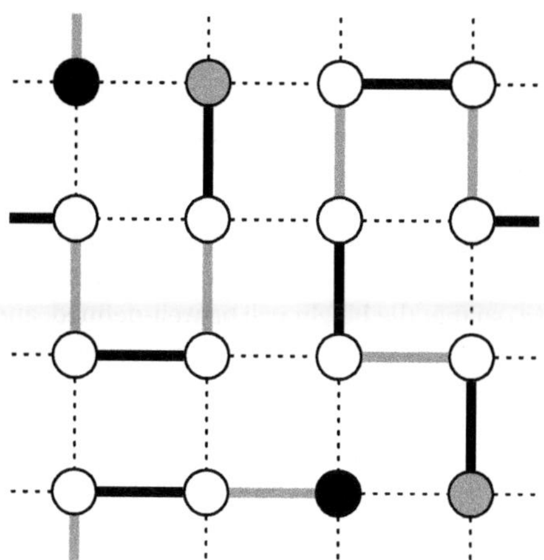

FIGURE 13.29: If we have two near-perfect matchings with the same pair of holes on a lattice with cyclic boundary conditions, we can shift one of them over, and then define a pair of perfect matchings.

13.18 Weighing the holes. In the *monomer–dimer* model, each state is a partial matching of a graph, and is given a probability proportional to λ^s where s is the number of edges it contains. Equivalently, if it has t holes, its probability is proportional to λ^{-t}. A physicist would call λ the *fugacity*, and would write $\lambda = \mathrm{e}^{\beta\mu}$ where β is the inverse temperature and μ is the chemical potential of the dimers. Alternately, we could say that each pair of holes is associated with an energy μ.

Generalize the proof of (13.28) on page 680, and show that for all $0 < t < n$ we have $N_{t+1}/N_t \le N_t/N_{t-1}$, or equivalently

$$N_{t-1}N_{t+1} \le N_t^2 .$$

This property is called *log-concavity*, since it shows that $\log N_t$ is a concave function of t. Then show that, if $N_1/N_0 = \mathrm{poly}(n)$, there is a $\lambda = \mathrm{poly}(n)$ such that the total probability of the perfect matchings in the monomer–dimer model is at least, say, $1/2$.

13.19 Estimating the permanent with random determinants. Let G be an undirected graph with m edges. Given an orientation of its edges, we can form an antisymmetric adjacency matrix A as in Section 13.6.2. Show that if we choose uniformly at random from all 2^m possible orientations,

$$\mathbb{E}[\det A] = \#\text{ of perfect matchings of } G .$$

Similarly, suppose M is a matrix of 0s and 1s. Let $M^{\pm}$ denote a matrix that results from replacing each 1 in M with $+1$ or -1. If we choose these signs independently at random, show that the expectation of $\det M^{\pm}$ is zero, but that the expectation of its square is

$$\mathbb{E}\left[(\det M^{\pm})^2\right] = \mathrm{perm}\, M .$$

Regrettably, the variance in these estimators is too large for this approach to give an efficient approximation algorithm for #PERFECT MATCHINGS or PERMANENT. 13.8

13.20 The parity of the permanent. Let ⊕PERMANENT be the decision problem of telling, given a matrix M with integer entries, whether $\mathrm{perm}\, M$ is even or odd. Show that ⊕PERMANENT is in P. It turns out that for any prime p other than 2, determining $\mathrm{perm}\, M \bmod p$ is NP-hard!

13.21 Flipping faces. Review Figure 12.22 on page 599, where we treat rhombus tilings as perfect matchings on the hexagonal lattice. In Section 12.6.3, we used a height function to argue that any two such tilings can be connected by a sequence of local moves, each of which corresponds to flipping the edges around a single face. Generalize this argument to all planar bipartite graphs. In other words, show how to convert a perfect matching to a three-dimensional surface which assigns a height to each face of the graph.

Hint: color the vertices black and white. Then move from one face to another across edges not included in the matching. When you cross an edge e, let the height increase or decrease by some amount $\delta(e)$ depending on whether the vertex to your left is black or white, analogous to the height function for domino tilings described in Problem 12.32. Derive conditions on the function $\delta(e)$ so that the height will be well-defined, and show that we can arrange for $\delta(e)$ to be positive for any edge e that appears in at least one perfect matching. Finally, show that we can flip the edges around a face if and only if that face is a local minimum or local maximum of the height function.

13.22 Proving the Pfaffian. Prove that $\det A = (\mathrm{Pf} A)^2$ for any antisymmetric matrix A, where the Pfaffian $\mathrm{Pf} A$ is defined as in (13.34). Hint: first show that we can restrict the sum $\det A = \sum_\pi (-1)^\pi \prod_{i=1}^n A_{i,\pi(i)}$ to permutations where every cycle has even length.

13.23 A little less crude. Our upper bound of 4^n for the number of perfect matchings on the square lattice is very crude indeed. The bound $2^{n/2}$ is much better, and is almost as easy to prove. Do so. Hint: describe the matching by scanning from left to right, and top to bottom, and specify the partner of each unmatched vertex you meet.

13.24 Matchings on rectangles. Show that the number of perfect matchings of a $2 \times \ell$ rectangle is the ℓth Fibonacci number, where we use the convention that $F(0) = F(1) = 1$. Conclude that another simple lower bound on the entropy density of perfect matchings on the square lattice is $s \geq (1/2)\ln \varphi = 0.24...$ where φ is the golden ratio.

Similarly, show that the number of perfect matchings of a $3 \times \ell$ rectangle, where ℓ is even, grows as b_3^ℓ where $b_3 = \sqrt{2+\sqrt{3}}$. Hint: write a transfer matrix that describes the number of ways to extend a matching from ℓ to $\ell+2$, and find its largest eigenvalue. If you're feeling up to it, show that the number of matchings on a $4 \times \ell$ rectangle grows as b_4^ℓ where $b_4 = 2.84...$, and conclude that $s \geq (1/4)\ln b_4 = 0.26....$

By continuing in this vein and analyzing perfect matchings on $w \times \ell$ rectangles, we can derive a series of lower bounds $s \geq (1/w)\ln b_w$, which will eventually converge to the true entropy $s = 0.29156....$ But as we commented in the text, the size of the transfer matrix grows exponentially as a function of the width w.

13.25 Matchings on a torus, and beyond. Several times in Sections 13.6 and 13.7, we treated the square lattice as if it had cyclic boundary conditions, wrapping around from top to bottom and from left to right so that it forms a torus. This gave the right asymptotics in the limit of large lattices. But can we calculate the number of matchings on a toroidal lattice exactly, even though it is not planar?

Suppose we have an $\ell \times \ell$ square lattice with toroidal boundary conditions where ℓ is even. Along the top of the lattice, there are ℓ vertical edges that wrap around to the bottom; along the left edge of the lattice, there are ℓ horizontal edges that wrap around to the right. Define $M_{\text{even, even}}$ as the number of perfect matchings containing an even number of vertical wraparound edges and an even number of horizontal ones, and define $M_{\text{even,odd}}$ and so on similarly. Now show that if we extend the orientation of Figure 13.25 to the entire lattice in the natural way, the Pfaffian of the resulting antisymmetric adjacency matrix A is

$$\mathrm{Pf} A = \sqrt{\det A} = M_{\text{odd,odd}} + M_{\text{odd,even}} + M_{\text{even,odd}} - M_{\text{even,even}} .$$

Now show that we can change which one of these terms is negative by reversing the orientation of the vertical wraparound edges, or the horizontal ones, or both. This gives four different adjacency matrices, which we call $A^{\pm\pm}$ where $A^{++} = A$. Show that the total number of perfect matchings can then be written as

$$M = \frac{1}{2}\left(\mathrm{Pf} A^{++} + \mathrm{Pf} A^{+-} + \mathrm{Pf} A^{-+} + \mathrm{Pf} A^{--}\right) .$$

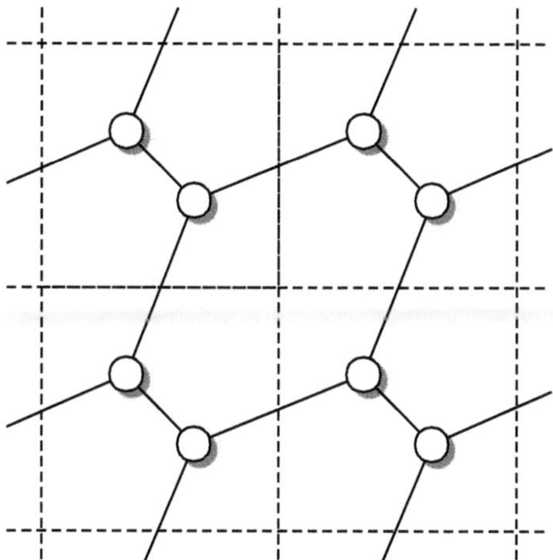

FIGURE 13.30: A hexagonal lattice is a square lattice with two vertices per cell.

This generalizes in the following way. Given a graph G, its *genus* $g(G)$ is the smallest g such that G can be drawn, without crossings, on the surface of genus g: that is, on a doughnut with g holes. For instance, planar graphs have genus 0 and the toroidal lattice has genus 1. Then the number of perfect matchings in G can be written as a linear combination of $4^{g(G)}$ Pfaffians. 13.6

13.26 Counting rhombus tilings. Let's calculate the entropy of perfect matchings on the hexagonal lattice, or equivalently, given problem 13.27, of rhombus tilings or ground states of the triangular antiferromagnet. First, argue that for the adjacency matrix A of the hexagonal lattice, where every face is a hexagon, we have

$$\operatorname{perm} A = |\det A| \,,$$

without the need for any weights on the edges. Then treat the hexagonal lattice as a square lattice with two vertices in each cell, as shown in Figure 13.30. As we did for the square lattice in Section 13.6.4, calculate the determinant of the resulting 2×2 matrix, and conclude that the number of rhombus tilings of an $\ell \times \ell$ lattice with $n = 2\ell^2$ sites grows as e^{sn} where

$$s = \frac{1}{16\pi^2} \int_0^{2\pi} \int_0^{2\pi} \ln\left(3 + 2\cos\theta + 2\cos\phi + 2\cos(\theta - \phi)\right) \mathrm{d}\theta \, \mathrm{d}\phi \,.$$

This can be simplified to

$$s = \frac{1}{\pi} \int_0^{\pi/3} \ln(2\cos\omega) \mathrm{d}\omega = 0.16153\ldots \qquad (13.50)$$

13.27 Frustrated magnets. In the antiferromagnetic Ising model, each vertex wants to have an opposite spin from its neighbors. On the triangular lattice, the three spins around each triangle can't all be opposite from each other, so they have to settle for having one opposite to the other two. A *ground state* is a state where this is true of every triangle, as shown in Figure 13.31, so the energy is as low as possible. Find a correspondence between ground states and perfect matchings on the hexagonal lattice—and, through Figure 12.22 on page 599, with rhombus tilings as well. Conclude that the entropy density of the triangular antiferromagnetic at zero temperature is given by (13.50).

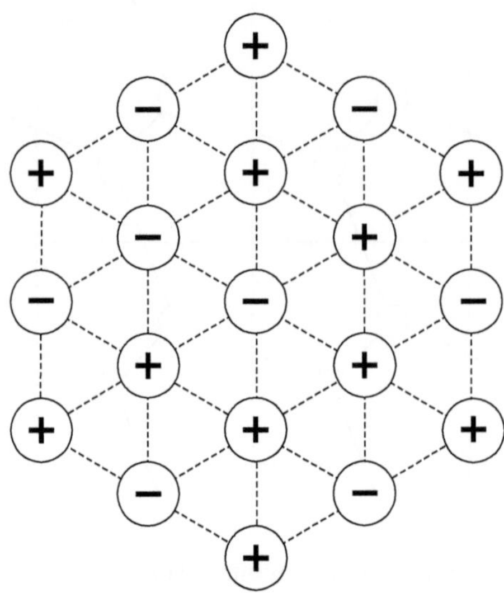

FIGURE 13.31: A ground state configuration of the antiferromagnetic Ising model on the triangular lattice.

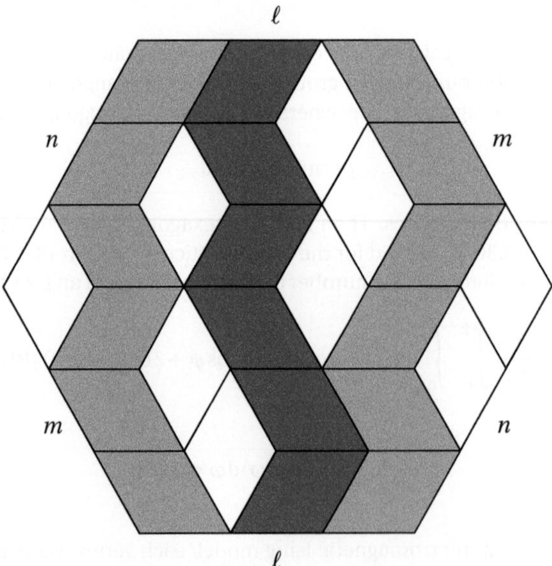

FIGURE 13.32: We can think of a rhombus tiling of a hexagon as a set of nonintersecting paths from its top edge to its bottom edge.

13.28 Rhombus tilings and nonintersecting paths. Here is another lovely way to count certain kinds of tilings. As Figure 13.32 shows, we can associate each rhombus tiling of a hexagon with a set of paths from its top edge to its bottom edge. Show that this is in fact a one-to-one mapping from rhombus tilings to families of nonintersecting paths, where the ith path leads from the ith tile on the top row to the ith tile on the bottom row. The challenge is then to count the number of such families. In particular, it seems challenging to take the constraint that the paths don't intersect each other into account.

Show that we can do this as follows. If the top and bottom edges are of length ℓ, define an $\ell \times \ell$ matrix P such that p_{ij} is the number of possible paths from the ith tile on the top to the jth one on the bottom. Then show that the number of nonintersecting families is simply $\det P$.

In particular, if we have a hexagon whose edge lengths are ℓ, m, and n, prove that no hexagon can be tiled with rhombuses unless each edge has the same length as the one opposite to it. Then show that the number of tilings is

$$T = \det P \text{ where } p_{ij} = \binom{m+n}{j-i+m}.$$

For instance, setting $\ell = m = n = 3$ as in Figure 13.32 gives $T = 980$. Combinatorial formulas like this help us understand the structure of random tilings, including the "arctic circle" phenomenon shown in Figure 12.20 on page 597. 13.9

13.29 The entropy of trees. Use the Matrix-Tree Theorem to calculate the entropy density of spanning trees on the square lattice. Show that the number of spanning trees of an $\ell \times \ell$ lattice with $n = \ell^2$ vertices grows as e^{sn}, where

$$s = \frac{1}{(2\pi)^2} \int_0^{2\pi} \int_0^{2\pi} \ln(4 - 2\cos\theta - 2\cos\phi)\, \mathrm{d}\theta\, \mathrm{d}\phi = 1.16624\ldots$$

13.30 Trees, forests, and matchings. Observe Figure 13.33. It illustrates a set of mappings between three kinds of objects: spanning trees on an $n \times n$ lattice, spanning forests on an $(n-1)\times(n-1)$ lattice where each tree is connected to the boundary, and perfect matchings on a $(2n-1)\times(2n-1)$ lattice with one corner removed. Describe these mappings, and prove that they are well-defined and one-to-one, so that for each n there are the same number of each of these three types of objects. Conclude that the spanning tree entropy is exactly 4 times the perfect matching entropy, and use the previous problem to check that this is so.

13.31 The chromatic polynomial. As in Section 13.1, given a graph G with an edge e, let $G - e$ denote G with e removed, and let $G \cdot e$ denote G with e contracted. Many quantities can be written recursively in terms of $G - e$ and $G \cdot e$, analogous to the formula (13.3) for the number of spanning trees. Show that, for any k, the number $P(G,k)$ of k-colorings of a graph G can be expressed as

$$P(G,k) = P(G-e,k) - P(G \cdot e,k),$$

along with the base case that $P(G,k) = n^k$ if G consists of n isolated vertices. Note that if e is a self-loop then $G - e$ and $G \cdot e$ are the same.

Conclude from this that if G has n vertices, $P(G,k)$ is a polynomial function of k of degree at most n. This is called the *chromatic polynomial*. It allows us to define the number of k-colorings even if k is negative, fractional, or complex! For instance, if G is the graph shown in Figure 13.34 then

$$P(G,k) = k(k-1)(k-2)^2 = k^4 - 5k^3 + 8k^2 - 4k\,.$$

What is the chromatic polynomial of the complete graph with n vertices? What about the line of length n, or the cycle of length n?

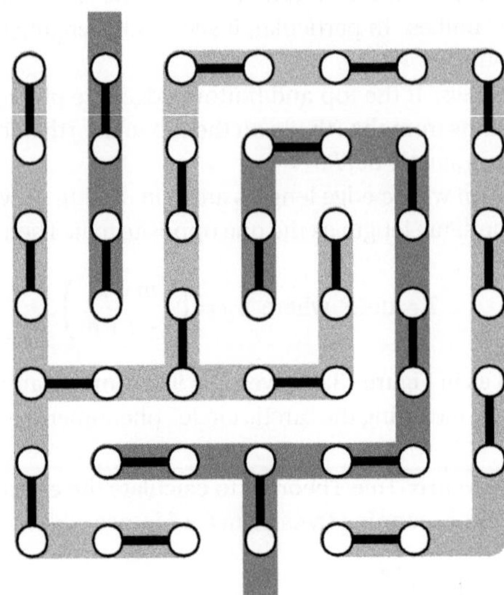

FIGURE 13.33: A set of one-to-one mappings between spanning trees on an $n \times n$ lattice, spanning forests on an $(n-1)\times(n-1)$ lattice where every tree touches the boundary, and perfect matchings on a $(2n-1)\times(2n-1)$ lattice with one corner removed.

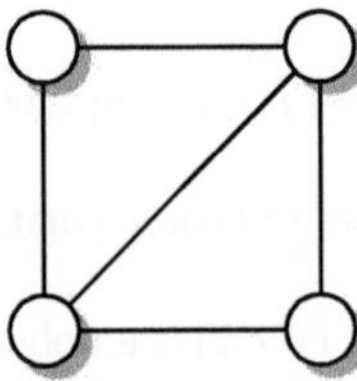

FIGURE 13.34: A graph with $k^4 - 5k^3 + 8k^2 - 4k$ proper k-colorings.

13.32 The Tutte polynomial. Consider the following recursive function of a graph G. Recall that an edge e is a *bridge* if cutting it would break G into two pieces.

$$T(G;x,y)=\begin{cases} x\,T(G\cdot e;x,y) & \text{if } e \text{ is a bridge} \\ y\,T(G-e;x,y) & \text{if } e \text{ is a self-loop} \\ T(G-e;x,y)+T(G\cdot e;x,y) & \text{otherwise} \\ 1 & \text{if } G \text{ has no edges}. \end{cases} \tag{13.51}$$

This is called the *Tutte polynomial*. For each graph G with n vertices, it is a polynomial in x and y of degree at most n. As an exercise, show that if G is a cycle of length n,

$$T(G;x,y)=x^{n-1}+x^{n-2}+\cdots+x+y=\frac{x^n-x}{x-1}+y\,.$$

It is not obvious that (13.51) gives the same result no matter which edge e we choose to remove or contract at each step. Show that this is the case, by proving that $T(G;x,y)$ can also be defined as

$$T(G;x,y)=\sum_S (x-1)^{c(S)-c(G)}\,(y-1)^{c(S)+|S|-n}\,. \tag{13.52}$$

Here S ranges over all spanning subgraphs of G: that is, subgraphs which include all of G's vertices, and an arbitrary subset of G's edges. Then $c(G)$ and $c(S)$ denote the number of connected components in G and S, including isolated vertices, and $|S|$ is the number of edges in S. Note that $c(S)+|S|-n$ is the total *excess* of S's connected components, where the excess of a connected graph with n vertices and m edges is $m-n+1$. For instance, a tree has excess 0 and a cycle has excess 1.

13.33 The many faces of Tutte. By setting x and y to various values, the Tutte polynomial includes many important quantities as special cases. Using either definition from the previous problem, show the following:

1. If G is connected, $T(G;1,1)$ is the number of spanning trees. More generally, $T(G;1,1)$ is the number of maximal forests, with one tree spanning each connected component of G.
2. $T(G;1,2)$ is the number of spanning subgraphs, with or without cycles.
3. $T(G;2,1)$ is the number of forests, i.e., acyclic subgraphs.
4. $T(G;2,0)$ is the number of acyclic orientations, i.e., the number of ways to orient the edges of G so that the resulting directed graph is acyclic. (In this case, proving that that $T(G;x,y)=T(G-e;x,y)+T(G\cdot e;x,y)$ if e is not a bridge or a self-loop is a little tricky.)
5. $T(G;0,2)$ is the number of *totally cyclic* orientations, i.e., those such that every edge is part of a directed cycle. If G is connected, show that these are the orientations that make G strongly connected, i.e., for any pair of vertices u,v there is a directed path from u to v and from v to u.
6. If G has m edges, $T(G;2,2)$ is 2^m. More generally, if $(x-1)(y-1)=1$, then

$$T(G;x,y)=x^m(x-1)^{n-m-1}\,.$$

7. If G is connected, its chromatic polynomial (see Problem 13.31) is given by

$$P(G,k)=(-1)^{n-1}k\,T(G;-(k-1),0)\,.$$

This last relation shows that calculating the Tutte polynomial for arbitrary x and y is #P-hard. In fact, this is true for all but a few values of x and y. 13.10

13.34 Tutte and percolation. In physics, *bond percolation* refers to a process in which each edge of a lattice is included or removed with some probability. Continuing from the previous two problems, let G be a connected graph with n vertices and m edges. For each edge e, we keep e with probability p and remove it with probability $1-p$. Show that the probability $Q_p(G)$ that the resulting graph is connected can be written in terms of the Tutte polynomial, as

$$Q_p(G) = p^{n-1}(1-p)^{m-n+1}\, T(G;1,1/(1-p)).$$

13.35 Tutte, Ising, and clusters of spins. Let $Z(G)$ denote the partition function of the ferromagnetic Ising model on a connected graph $G=(V,E)$, i.e.,

$$Z(G) = \sum_{\{s_i\}} \mathrm{e}^{\beta \sum_{(i,j)\in E} s_i s_j} = \sum_{\{s_i\}} \prod_{(i,j)\in E} \mathrm{e}^{\beta s_i s_j},$$

where the sum is over all $2^{|V|}$ states where $s_i = \pm 1$ for each $i \in V$. Show that $Z(G)$ is yet another case of the Tutte polynomial,

$$Z(G) = 2\mathrm{e}^{\beta(n-m-1)}\left(\mathrm{e}^{\beta}-\mathrm{e}^{-\beta}\right)^{n-1} T(G;x,y) \text{ where } x = \frac{\mathrm{e}^{\beta}+\mathrm{e}^{-\beta}}{\mathrm{e}^{\beta}-\mathrm{e}^{-\beta}},\ y = \mathrm{e}^{2\beta},$$

and that it can be written as a sum over all spanning subgraphs,

$$Z(G) = \sum_{S} 2^{c(S)}\left(\mathrm{e}^{\beta}-\mathrm{e}^{-\beta}\right)^{|S|} \mathrm{e}^{-\beta(m-|S|)}.$$

This expression for $Z(G)$ is called the *Fortuin–Kasteleyn representation* or the *random cluster model.*

13.36 High and low. There is a delightful symmetry, or duality, between the behavior of the two-dimensional Ising model at high and low temperatures. For each temperature below the phase transition, there is a corresponding temperature above it, at which the partition function has similar behavior.

Suppose we have an Ising model on an arbitrary graph (not necessarily planar). For each pair of neighboring sites i,j, define a variable $x_{ij} = s_i s_j$, and think of it as living on the edge between them. Then use the fact that if $x = \pm 1$,

$$\mathrm{e}^{\beta x} = (1 + x \tanh\beta)\cosh\beta$$

to write the partition function Z as a polynomial in the variables x_{ij}.

Now imagine multiplying this polynomial out. Each term is the product of some set of x_{ij}, which we can think of as some subset of the edges in the lattice. Show that if we sum over all the spins s_i, each such term is 2^n if it includes an even number of x_{ij} meeting at each vertex i, and zero otherwise. In other words, if we associate each term with a subgraph consisting of the edges corresponding to the x_{ij} appearing in that term, the subgraphs that contribute to Z are those in which every vertex has degree 2.

On the square lattice in particular, these subgraphs are exactly the legal polygon configurations that we defined in Section 13.7.3, but on the original graph rather than its dual. Conclude that

$$Z = (2\cosh^2\beta)^n\, g(\tanh\beta),$$

where $g(z)$ is the weighted sum defined in (13.47). Comparing this to (13.48) and matching the argument of $g(z)$ suggests that, for each β, we should define its dual β^* so that

$$\mathrm{e}^{-2\beta^*} = \tanh\beta.$$

Show that $\beta < \beta_c$ if and only if $\beta^* > \beta_c$ and vice versa. In particular, β_c is the unique solution to the requirement that the system be self-dual, $\beta_c = \beta_c^*$. This argument was used to derive β_c before the Ising model was fully solved.

13.7

13.37 Hard spheres on trees. In Problem 12.19, we met the hard sphere model. Given a graph $G=(V,E)$, each state is an independent set $S \subseteq V$, and has probability proportional to $\lambda^{|S|}$ for some fugacity λ. The partition function of this model is the sum of $\lambda^{|S|}$ over all independent sets,

$$Z = \sum_{S \subseteq V, \text{independent}} \lambda^{|S|} .$$

For instance, if G is a chain of three vertices then $Z = 1 + 3\lambda + \lambda^2$, since it has one independent set of size zero (the empty set), three of size one (single vertices), and one of size two (the two endpoints). If $\lambda = 1$, then Z is simply the total number of independent sets; if $\lambda > 1$, then Z is a weighted sum, with larger sets given more weight.

Show that, given G and λ, we can compute Z in polynomial time in the special case where G is a tree. Hint: use a dynamic programming approach like the one in Problem 3.25. Start by writing $Z = Z_0 + Z_1$ where Z_0 and Z_1 sum over the independent sets that exclude or include, respectively, the root of the tree.

13.38 More thermodynamics. The *Gibbs–Shannon entropy* of a probability distribution $p(x)$ (see Problem 12.3) is defined as

$$S = -\sum_x p(x) \ln p(x) .$$

Show that the entropy of the Boltzmann distribution can be written as

$$S = \beta\, \mathbb{E}[E] + \ln Z = -\frac{\partial F}{\partial T} ,$$

where Z is the partition function, $\mathbb{E}[E]$ is the average energy, and $F = -(1/\beta)\ln Z$ is the free energy.

13.39 Measuring magnetization with a little field. As in Section 12.1, we can define the magnetization of the Ising model as the average spin,

$$m = \frac{1}{n} \sum_i s_i .$$

How can we write the expected magnetization in terms of the partition function? One way to do this is to add an *external field* to the system, which gives each spin a positive or negative energy depending on whether it is down or up respectively. If the strength of this field is h, the total energy is then

$$E(S) = -\sum_{ij} s_i s_j - h \sum_i s_i . \tag{13.53}$$

Show that the expected magnetization as a function of h is then

$$\mathbb{E}[m] = \frac{1}{n\beta} \frac{\partial}{\partial h} \ln Z = -\frac{1}{n} \frac{\partial F}{\partial h} . \tag{13.54}$$

Unfortunately, no one knows how to solve the two-dimensional Ising model exactly for a nonzero external field. However, by considering an *infinitesimal* field, it is possible to calculate the spontaneous magnetization $\lim_{h \to 0} \mathbb{E}[|m|]$ that occurs below T_c. 13.7

13.40 An external field in one dimension. Modify the discussion of the one-dimensional Ising model in Section 13.7.2 to include an external field as in (13.53). As before, when n is large we can write

$$Z \sim \lambda_1^n$$

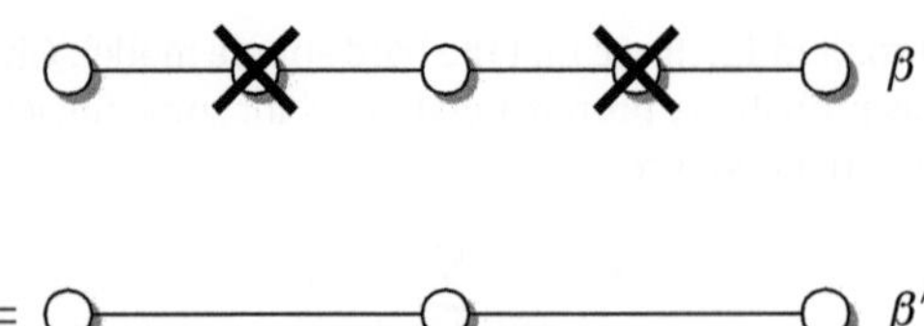

FIGURE 13.35: Renormalization in the one-dimensional Ising model. If we sum over half the sites, the remaining sites form an Ising model at a new value of β.

where λ_1 is the largest eigenvalue of the transfer matrix M. Show that

$$\lambda_1 = \mathrm{e}^{\beta}\left(\cosh\beta h + \sqrt{\mathrm{e}^{-4\beta} + \sinh^2\beta h}\right).$$

Apply (13.54) and conclude that the expected magnetization is

$$\mathbb{E}[m] = \frac{1}{\beta}\frac{\partial}{\partial h}\ln\lambda_1 = \frac{\sinh\beta h}{\sqrt{\mathrm{e}^{-4\beta} + \sinh^2\beta h}}.$$

For any constant β, this tends to $+1$ when $h \to \infty$, and to -1 when $h \to -\infty$. So, when the field is very strong, it forces all the spins to line up with it.

13.41 Correlations and eigenvalues. In Chapter 12 we defined the mixing time τ of a Markov chain as the number of steps it takes for its state to become roughly independent of its initial condition, and derived bounds on τ in terms of the eigenvalues of the transition matrix. In the same way, we can define the correlation length ξ as the distance apart that two sites have to be in order to be roughly independent, and calculate ξ from the transfer matrix M.

Specifically, suppose we have the one-dimensional Ising model. Analogous to Section 12.10, we write

$$\mathbb{E}[s_i s_j] \sim \mathrm{e}^{-\ell/\xi},$$

where $|i-j| = \ell$. Show that in fact this correlation is exactly

$$\mathbb{E}[s_i s_j] = \tanh^{\ell}\beta,$$

so the correlation length is

$$\xi = -\frac{1}{\ln\tanh\beta}.$$

How does ξ behave in the limits $\beta \to 0$ and $\beta \to \infty$?

13.42 Renormalization and scaling. An important idea in physics is *renormalization*, where we change the scale of a system, and ask how this change affects its parameters. For instance, consider the one-dimensional Ising model on a chain of n sites. Suppose that we ignore every other site as shown in Figure 13.35. The remaining sites again form an Ising model, but with a weaker interaction, or equivalently at a higher temperature.

How can we calculate the new value of β from the old one? First, consider the ratio between the probabilities that two adjacent sites are both up, or that one is up and the other is down. According to the Boltzmann distribution, this is

$$\frac{P(++)}{P(+-)} = \mathrm{e}^{2\beta}.$$

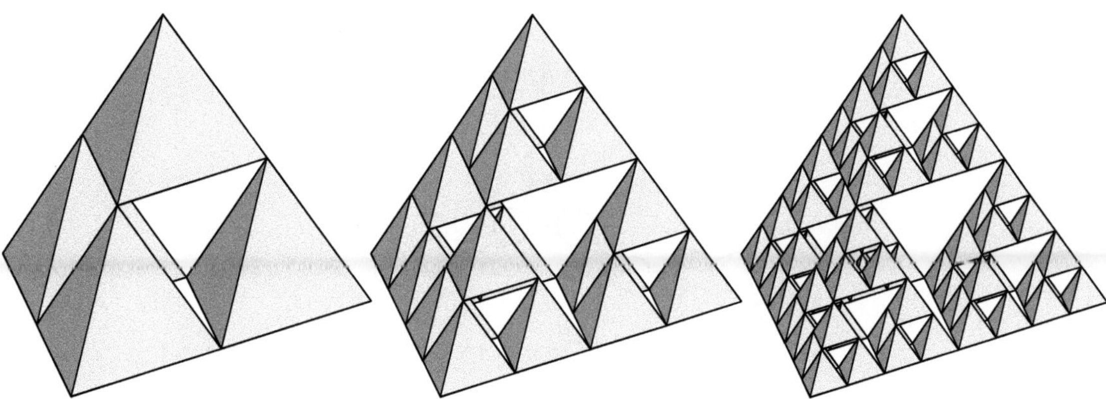

FIGURE 13.36: A three-dimensional fractal analogous to the Sierpiński gasket of page 83, on which the Ising model can be solved exactly using renormalization.

Now consider a pair of sites two steps apart. The total probability that they are both up, say, is the sum over the possible spins of the intervening site,

$$P(+*+) = P(+++) + P(+-+).$$

The effective β of the new model is then given by

$$\frac{P(+*+)}{P(+*-)} = \mathrm{e}^{2\beta'}.$$

Show that this gives

$$\beta' = \frac{1}{2}\ln\cosh 2\beta.$$

Since this process reduces the length scale by 2, the new correlation length should be half the old one. Use the results of the previous problem to show that indeed $\xi' = \xi/2$.

This same approach can be used to obtain exact results on some fractal lattices as well. For instance, consider an Ising model on the Sierpiński gasket depicted in Figure 3.28 on page 83. Show that in this case, dividing the length scale of the lattice by 2 (which divides the number of sites by 3) gives

$$\beta' = \frac{1}{4}\ln\left(\frac{\mathrm{e}^{9\beta} + 3\mathrm{e}^{\beta} + 4\mathrm{e}^{-3\beta}}{e^{5\beta} + 4\mathrm{e}^{\beta} + 3\mathrm{e}^{-3\beta}}\right).$$

A similar expression, with a larger number of terms, can be written for the three-dimensional version of the Sierpiński gasket shown in Figure 13.36.

Why can't we use renormalization to solve the two-dimensional Ising model exactly? And for lattices such as the Sierpiński gasket, what is the relationship between the fact that renormalization can be analyzed exactly, and that problems such as MAX-WEIGHT INDEPENDENT SET can be solved in polynomial time as in Problem 3.28?

13.43 Planar spin glasses. Recall that a *spin glass* is a generalized Ising model where, for each pair i, j of adjacent sites, we have an interaction strength J_{ij} which may be positive (ferromagnetic) or negative (antiferromagnetic). The energy is then

$$E = -\sum_{ij} J_{ij}\, s_i s_j.$$

Now consider the following problem:

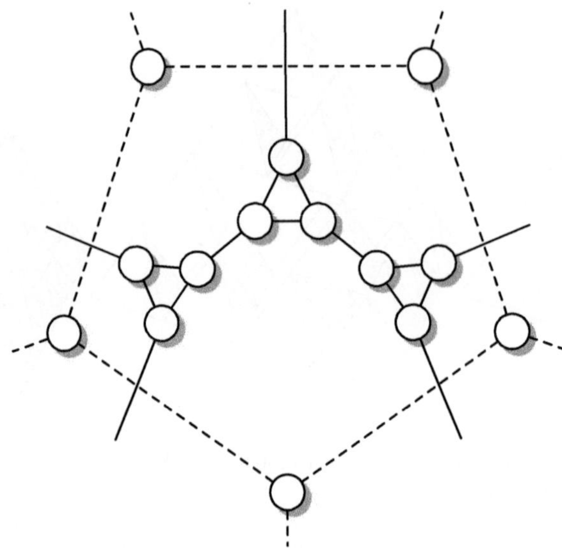

FIGURE 13.37: With the right gadgets and the right edge weights, we can convert any planar spin glass to counting weighted perfect matchings on a decorated dual graph, and then use the Pfaffian method to calculate its partition function exactly. Here we show the gadget for a pentagonal face.

ISING
Input: A graph G with interactions J_{ij} for each edge (i,j), and an inverse temperature β
Output: The partition function $Z(\beta)$

Show that the special case of ISING where G is planar is in P. To do this, generalize the approach of Section 13.7.3 to write the partition function $Z(\beta)$ of such a spin glass as the square root of a determinant. Hint: consider Figure 13.37.

Having shown that the planar case is in P, now assume that the J_{ij} are integers, and show that by calculating $Z(\beta)$ at a sufficiently low temperature $1/\beta$, we can find the ground state energy, i.e., the minimum of E over all states. As a bonus, show that MAX CUT is in P for planar graphs. How low does the temperature have to be for the ground state to dominate, no matter how many states are at the next higher temperature? (Note the similarity to Problem 13.5.)

13.44 Non-planar Ising models are hard. As a converse to the previous problem, show that the general case of ISING is #P-hard under Turing reductions, even in the ferromagnetic case where $J_{ij} = 1$ on every edge. Hint: if G has m edges, the energy E ranges from $-m$ to m. Show that $e^{\beta m} Z(\beta)$ is a polynomial function of e^{β} of degree at most $2m$. Then use the polynomial interpolation approach of Problem 13.13 to reduce from a known #P-complete problem.

Notes

13.1 The Matrix-Tree Theorem. The Matrix-Tree Theorem goes back to the work of Kirchhoff on electric circuits. The proof we give here is slightly modified from the book of Bollobás [119]. The all-minors version of the Matrix-Tree Theorem given in Problem 13.3 can be proved with a recurrence analogous to (13.3). Another nice proof can be found in [150].

13.2 Toda's Theorem. The fact that $\mathsf{P}^{\#\mathsf{P}}$ includes the polynomial hierarchy was proved by Toda [791], for which he received the Gödel Prize in 1998. The main step of the proof is to use the Valiant–Vazirani randomized reduction of Section 10.6.2 from existence to unique existence. This lets us convert any nonempty set of solutions, with high probability, to a set of odd size (since 1 is odd). In terms of complexity classes, this means that NP is contained in a rather exotic class called BP(⊕P), where ⊕ and BP are generalized quantifiers that demand, respectively, that the number of solutions is odd or forms a clear majority. Thus BP(⊕P) is the class of problems with polynomial-time checkable properties $B(x,y,z)$ where x is a yes-instance if, for at least 2/3 of the possible zs, there are an odd number of ys such that $B(x,y,z)$ holds, and where this is true of at most 1/3 of the possible zs if x is a no-instance.

By iterating this reduction and using several other ideas, Toda showed that the class BP(⊕P) can absorb any constant number of ∃s or ∀s, and that $\mathsf{BP}(\oplus\mathsf{P}) \subseteq \mathsf{P}^{\#\mathsf{P}}$. In fact, he showed the stronger statement that $\mathsf{BP}(\oplus\mathsf{P}) \subseteq \mathsf{P}^{\mathsf{PP}}$, where PP is the class of problems that ask whether the majority of possible witnesses work: in other words, whether $B(x,w)$ holds for a majority of the possible w. So if we can tell whether the majority of a haystack consists of needles, we can solve any problem in the polynomial hierarchy.

13.3 #P-completeness and the permanent. The class #P and the original proof that PERMANENT is #P-complete are due to Leslie Valiant [802, 803], who won the 2010 Turing Award for this and other work. The proof of PERMANENT's #P-completeness that we give here is by Ben-Dor and Halevi [98]. Another simple proof, which reduces from #VERTEX COVERS, appears in Kozen [502]. Parsimonious reductions between NP-complete problems were also considered by Simon [754].

In Valiant's original work, he defined #P-completeness with respect to Turing reductions as in Problem 13.13. Zankó [844] showed that PERMANENT is still #P-complete under *many–one* reductions, which call B only once. Indeed, she used the more conservative definition of counting reduction given in Problem 13.10. However, this would prevent us from using the proof of [98], since the permanent is 12^m times the number of solutions to the 3-SAT formula where m is the number of clauses.

13.4 Sampling and counting. The definitions of approximate counting algorithm and FPRAS were first given by Karp and Luby [464]. The equivalence between random sampling and approximate counting for self-reducible problems, and the trick of taking the median of multiple trials as in Problem 13.15, is from Jerrum, Valiant, and Vazirani [436]; see also Broder [138]. Earlier, Knuth [492] gave an algorithm for estimating the size of a search tree by walking down a random branch and taking the reciprocal of its probability.

The improved sampling algorithm of Problem 13.16, which works even when the counter is off by a polynomial factor, was given by Sinclair and Jerrum [756]. It was one of the first uses of conductance to prove that a Markov chain mixes rapidly.

Problem 13.1 is from Jerrum, Valiant, and Vazirani [436].

13.5 Random perfect matchings and approximating the permanent. The Markov chain on the set of perfect and near-perfect matchings was first described by Broder [138]. In 1989, Jerrum and Sinclair [432] showed that it mixes in polynomial time for graphs where the number of near-perfect matchings is only polynomially larger than the number of perfect ones. Kenyon, Randall, and Sinclair [476] showed that this holds, in particular, for lattices in any dimension, or more generally vertex-transitive graphs, as shown in Problem 13.17.

The proof of polynomial mixing in [432] uses canonical paths, in which we choose a single shortest path between each pair of states. Our proof here, where probability is spread across all shortest paths, is closer to that of Dagum, Luby, Mihail, and Vazirani [209].

In 2001, Jerrum, Sinclair, and Vigoda [435] devised the weighted version of this Markov chain that we sketch in the text, and thus found an FPRAS for #PERFECT MATCHINGS and 0-1 PERMANENT. They shared the Fulkerson Prize for this achievement in 2006. Earlier, Dyer, Frieze, and Kannan [258] shared this prize for a Markov chain algorithm for approximating another #P-complete problem, the volume of an n-dimensional convex polytope described by a set of constraints.

The best known algorithm for calculating the permanent *exactly* is still Ryser's algorithm [720], described in Problem 13.6, which takes $2^n \text{poly}(n)$ time.

13.6 Permanents and determinants of planar graphs. The technique of transforming the permanent of a lattice to a determinant by placing weights or orientations of its edges was found independently by the mathematical physicist Piet Kasteleyn [468] and by H. N. V. Temperley and Michael Fisher [786, 282]. They also derived the entropy of perfect matchings on the lattice. The expression (13.50) in Problem 13.26 for the entropy of rhombus tilings, and Problem 13.27 for the triangular antiferromagnet, are from Wannier [818, 819].

The discussion in Section 13.6.3 follows a manuscript by Propp [672]. A full treatment of the so-called permanent-determinant method, and the Pfaffian method, can be found in Kasteleyn [469]. A nice review can also be found in Kuperberg [508].

The permanent-determinant method is not the only source of exact asymptotics in statistical physics. For instance, Lieb [533] calculated the entropy density of the six-vertex ice model discussed in Section 12.6.4, or equivalently of 3-colorings of the square lattice, and found that the number of them grows as z^n where $z = (4/3)^{3/2}$. Astonishingly, the best way to understand this result leads one into the theory of quantum groups.

The fact that the number of perfect matchings of a graph with genus g can be written as a linear combination of 4^g Pfaffians as illustrated by Problem 13.25 was suggested by Regge and Zecchina [690] as a possible approach to the three-dimensional Ising model.

13.7 The Ising model. The Ising model was first proposed in 1920 by Wilhelm Lenz [525]. In 1925, his student Ernst Ising solved the one-dimensional version exactly [421] and mistakenly concluded that it could not explain the phase transition observed in three-dimensional materials. In 1936, Peierls [655] argued that in two or more dimensions, the Ising model has a phase transition at a nonzero temperature due to the tradeoff between the entropy of the patches of up and down spins and the energy of the boundaries between them.

The first exact solution of the two-dimensional Ising model was given by Onsager [641], who used sophisticated techniques to diagonalize the transfer matrix. Earlier, Kramers and Wannier [504] identified the critical temperature using the duality between high and low temperature that we discuss (in a different form) in Problem 13.36. The solution we give in the text, which maps each state to a perfect matching of a weighted planar graph, is from Fisher [283].

Using the exact solution of the Ising model, it is possible to derive the average magnetization,

$$\mathbb{E}\Big[\big|\sum_i s_i\big|/n\Big] = \left(1 - \frac{1}{\sinh^2 2\beta}\right)^{1/8} .$$

This hits zero at the phase transition where $\sinh 2\beta_c = 1$. While this formula was first announced by Onsager in 1944, it was not proved until eight years later when Yang [837] succeeded in analyzing the Ising model with a small external field as discussed in Problem 13.39. Montroll, Potts, and Ward [600] reproved this result using the Pfaffian method to analyze long-range correlations in the lattice. As they said of Onsager's announcement,

> In the days of Kepler and Galileo it was fashionable to announce a new scientific result through the circulation of a cryptogram which gave the author priority and his colleagues headaches. Onsager is one of the few moderns who operates in this tradition.

The random cluster model, which we describe in Problem 13.35 as a case of the Tutte polynomial, was described by Fortuin and Kasteleyn [287]. The dual expression for the partition function given in Problem 13.36 is also known as the *high-temperature expansion.* In the presence of an external field, vertices with odd degree are allowed if we give them an appropriate weight.

Jerrum and Sinclair [433] used the high-temperature expansion to give an approximation algorithm for the partition function for the ferromagnetic Ising model on an arbitrary graph, by devising a rapidly-mixing Markov chain on the set of all subgraphs. They also showed, as in Problem 13.44, that calculating the partition function exactly is #P-hard under Turing reductions. So except in the planar case as described in Problem 13.43, an approximation algorithm is almost certainly all we can hope for.

Building on [433], Randall and Wilson [683] showed that the random cluster model is self-reducible in a way that lets us sample states according to the Boltzmann distribution at any temperature. No such algorithm is known in the antiferromagnetic case, or for spin glasses with ferromagnetic and antiferromagnetic bonds. This is for good reason—if there were an FPRAS for the partition function in either of these cases, there would be a randomized polynomial-time algorithm for MAX CUT, and NP would be contained in RP.

The renormalization operation we discuss in Problem 13.42, which decimates the lattice and defines effective interactions between the remaining sites, appears in Fisher [281] along with several other local transformations. The fact that it lets us solve the Ising model exactly on some fractal lattices such as the Sierpiński gasket was first pointed out in [628] and [322]. The idea of renormalization plays a major role in quantum field theory, where it is used to understand how the strength of subatomic interactions diverges at the smallest scales.

13.8 Using random determinants to estimate the permanent. Problem 13.19, which shows that we can estimate the permanent by taking the determinant of a matrix with random signs, is from Godsil and Gutman [334]. Karmarkar, Karp, Lipton, Lovász, and Luby [461] showed that the variance of this estimator, in the worst case, is exponentially larger than the square of its expectation. Thus we would need an exponential number of trials to estimate the permanent within a constant factor.

This led to a series of results showing that if we replace each entry with a random complex number, or a random element of a nonabelian group, this ratio decreases to a milder exponential: see e.g. Barvinok [83], Chien, Rasmussen, and Sinclair [160], and Moore and Russell [611]. In the nonabelian case there are several ways to define the determinant. However, so far this line of research has not led to an alternative polynomial-time approximation algorithm for PERMANENT.

13.9 Determinants and nonintersecting paths. Problem 13.28 is from Gessel and Viennot [326], although the general theorem that the number of nonintersecting families of paths can be written as a determinant goes back to Lindström [535]. Earlier, MacMahon [550] showed that the number of rhombus tilings in a hexagon of sides ℓ, m, and n is

$$T = \prod_{i=0}^{\ell-1} \prod_{j=0}^{m-1} \prod_{k=0}^{n-1} \frac{i+j+k+2}{i+j+k+1} .$$

13.10 The Tutte polynomial. The Tutte polynomial was discovered by William Thomas Tutte, who called it the dichromatic polynomial, as a generalization of the chromatic polynomial. Jaeger, Vertigan, and Welsh [425] showed that for almost all pairs (x,y), calculating $T(G;x,y)$ is #P-hard under Turing reductions. Specifically, it is hard except along the hyperbola $(x-1)(y-1)=1$ and at the points $(x,y)=(1,1),(0,-1),(-1,0),(-1,-1),(\imath,-\imath),(-\imath,\imath),(\omega,\omega^*)$, and (ω^*,ω) where $\omega = e^{2\imath\pi/3}$. Their proof uses a polynomial interpolation scheme similar in spirit to that in Problem 13.13.

Chapter 14

When Formulas Freeze: Phase Transitions in Computation

God is subtle, but he is not malicious.
Albert Einstein

Up to now, we have focused on the worst-case behavior of problems. We have assumed that our instances are chosen by the adversary, and are cleverly designed to thwart our ability to solve them. After all, we can't say that an algorithm works in every case unless it works in the worst case. Moreover, in some settings—such as cryptography, the milieu in which modern computer science was born—there really is an adversary in the form of a human opponent.

But what if instances are chosen randomly, by a Nature who is indifferent rather than malevolent? For instance, suppose we construct a 3-SAT formula by choosing each clause randomly from all possible triplets of variables, and flipping a coin to decide whether or not to negate each one. Are such formulas typically satisfiable? If they are, are satisfying assignments easy to find?

Intuitively, the probability that these formulas are satisfiable decreases as the number of clauses increases. However, it is not obvious whether this happens continuously, or all at once. If experimental evidence and deep insights from physics are to be believed, these formulas undergo a *phase transition* from almost certain satisfiability to almost certain unsatisfiability when the number of constraints per variable crosses a critical threshold. This transition is similar in some respects to the freezing of water, or the Ising model's transition from magnetized to unmagnetized states at a critical temperature. Moreover, the difficulty of solving these instances—measured as the amount of backtracking we need to do, and the number of blind alleys we need to pursue—appears to be maximized at this transition.

Since they first appeared in the 1990s, these phenomena have produced a lively collaboration between computer scientists, mathematicians, and statistical physicists. Similar phase transitions exist for many other NP-complete problems. For GRAPH COLORING, random graphs appear to go from colorable to uncolorable at a certain value of the average degree. Similarly, INTEGER PARTITIONING goes from solvable to unsolvable when the number of integers is too small compared to the number of digits. How can we compute the thresholds at which these transitions take place? And what happens to the set of solutions as we approach it?

We begin this chapter by looking at some experimental results on random 3-SAT, and formulating the conjecture that a phase transition exists. To build up our skills, we then explore some simple phase transitions in random graphs. When the average degree of a vertex exceeds one, a *giant component* emerges, much like the transition when isolated outbreaks of a disease join together to become an epidemic. Further transitions, such as the emergence of a well-connected structure called the *k-core*, take place at larger values of the average degree. We show how to compute the size of these cores, and the degrees at which they first appear.

We then turn our attention to random k-SAT formulas. We show how to prove upper and lower bounds on the critical density of clauses—determining the threshold almost exactly in the limit of large k—by computing the average and variance of the number of satisfying assignments. We describe simple search algorithms as flows through state space, and track their progress with differential equations.

Finally, we sketch recent advances inspired by techniques in statistical physics. We describe algorithms that estimate the number of solutions by passing messages between variables and clauses until these messages reach a fixed point. We also describe additional phase transitions in the geometry of the set of solutions—densities at which the solutions fragment into clusters, and higher densities at which these clusters condense, freeze, and finally disappear.

The reader should be aware of two caveats. First, we do not claim that these random formulas are a good model of real-world instances, any more than random graphs are a good model of real social networks. Real SAT formulas arising from hardware verification, for instance, possess an enormous amount of structure, which simple probability distributions fail to capture, and which clever algorithms can exploit. Nevertheless, random graphs and formulas possess a great deal of interest in their own right, and they may give a good qualitative picture of how complexity works in the real world.

Second, there are places in this chapter where we do not attempt, and do not desire, to be completely rigorous. Our goal is to give the reader a taste of many different ideas and techniques, some of which can be made rigorous with additional work, and others for which proving them remains an area of active research. Happily, we can make a great deal of progress using ideas from probability theory which are quite elementary. If you have not worked with discrete probability before, we urge you to first consume

14.1 and enjoy Appendix A.3.

14.1 Experiments and Conjectures

If mathematics describes an objective world just like physics, there is no reason why inductive methods should not be applied in mathematics just the same as in physics.

Kurt Gödel (1951)

Let's begin by exploring random 3-SAT formulas experimentally. Using a simple search algorithm, we measure both the formulas' satisfiability and the algorithm's running time. Based on these results, we formulate a conjecture that there is a sharp phase transition from satisfiability to unsatisfiability, and address to what extent this conjecture can be proved.

14.2

14.1.1 *First Light*

We need to be precise about what we mean by a random formula. Our probability distribution—or as physicists say, our *ensemble*—consists of formulas with n variables and m clauses. For each clause, we choose uniformly from the possible $\binom{n}{3}$ triplets of variables, and flip a fair coin to decide whether or not to negate each one. Thus we choose each clause uniformly from the $2^3\binom{n}{3}$ possible 3-SAT clauses.

We do this *with replacement.* That is, we draw m times from a bag containing all possible clauses, putting each one back in the bag before drawing the next one. Thus we do not ensure that all the clauses are different. We call this model $F_3(n,m)$, or more generally $F_k(n,m)$ when we choose from the $2^k\binom{n}{k}$ k-SAT clauses.

We will be especially interested in the regime where these random formulas are *sparse*—where the ratio $\alpha = m/n$ of clauses to variables stays constant as n grows. As a first exercise, you can show that while there is no prohibition on the same clause appearing twice in the formula, this hardly ever happens.

Exercise 14.1 *Show that if $m = \alpha n$ where α is constant, the probability that any clause appears twice in $F_3(n,m)$ is $O(1/n)$. Hint: compute the expected number of pairs of identical clauses appearing in the formula, and use Markov's inequality from Appendix A.3.2.*

To explore the satisfiability and complexity of these formulas, we use a classic backtracking search algorithm: `DPLL`, named for Davis, Putnam, Logemann, and Loveland. As shown in Figure 14.1, this
algorithm works by choosing a variable x and assigning the two possible truth values to it. Each setting 14.3
results in a new formula on the remaining variables, and `DPLL` recursively checks whether either one is satisfiable. We can visualize the resulting sequence of trials and errors as a depth-first traversal of a search tree. At each node of the tree, some of the variables have already been set, and we have a formula ϕ on the unset variables. The two children of this node correspond to the formulas $\phi[x = \texttt{true}]$ and $\phi[x = \texttt{false}]$ resulting from setting the next variable.

Setting x `true` satisfies the clauses that contain x, so these clauses disappear from the formula. On the other hand, it shortens the clauses that contain $\overline{x}$, requiring them to be satisfied by one of their other variables. Similarly, setting x `false` satisfies the clauses containing $\overline{x}$ and shortens those containing x. Thus, at a given node in the search tree, ϕ contains a mix of 3-, 2-, and 1-variable clauses.

If a clause becomes empty, i.e., if all of its variables are given values that disagree with it, then we have a contradiction and `DPLL` is forced to backtrack and explore another branch of the tree. If every leaf of the tree has a contradiction, the original formula is unsatisfiable. On the other hand, if we find a leaf where every clause has been satisfied and the formula is empty, we have found a satisfying assignment. Thus `DPLL` either produces a satisfying assignment, or a proof—by exhaustively exploring the search tree—that none exists.

We are now in a position to measure, as a function of m and n, how often a random formula $F_3(n,m)$ is satisfiable. Figure 14.2 shows some experimental results. We can see that the key variable is $\alpha = m/n$, the density of clauses per variable. When α crosses a threshold $\alpha_c \approx 4.267$, the probability of satisfiability drops from 1 to 0. For finite n this drop takes place continuously, but it becomes steeper as n increases, suggesting that in the limit $n \to \infty$ it becomes a discontinuous jump.

```
DPLL
input: a SAT formula φ
output: is φ satisfiable?
begin
    if φ is empty then return "yes";
    if φ contains an empty clause then return "no";
    select an unset variable x ;
    if DPLL(φ[x = true]) = "yes" then return "yes";
    else if DPLL(φ[x = false]) = "yes" then return "yes";
    else return "no";
end
```

FIGURE 14.1: The DPLL backtracking search algorithm.

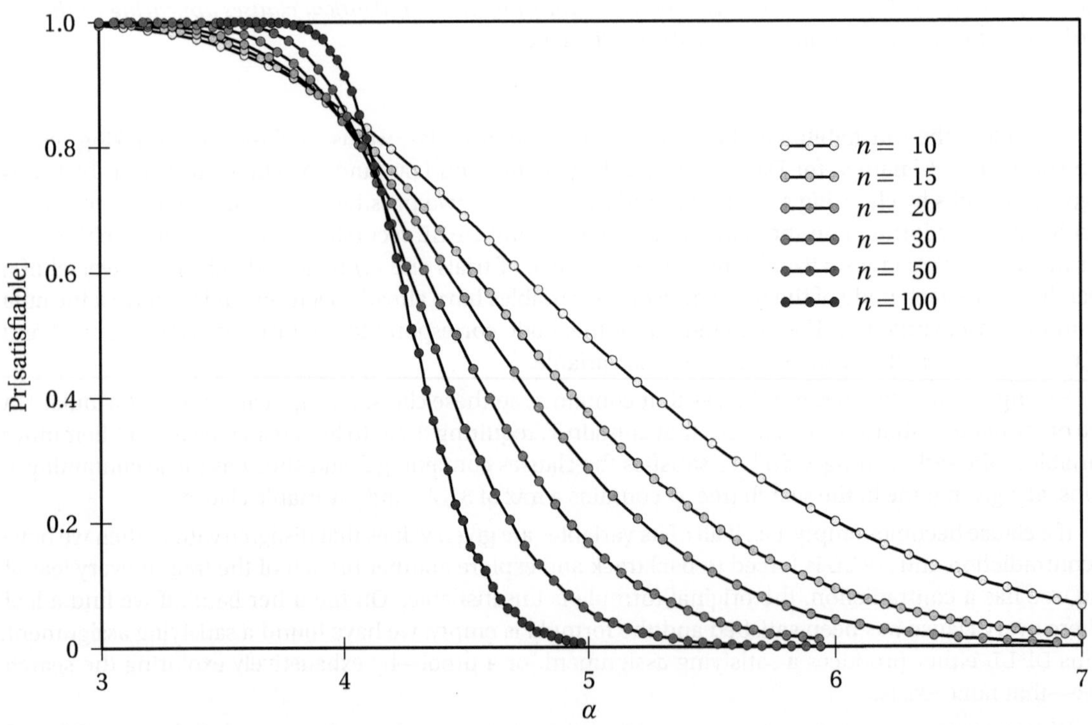

FIGURE 14.2: The probability that a random 3-SAT formula $F_3(n, m)$ is satisfiable as a function of the clause density $\alpha = m/n$, for various values of n. The sample size varies from 10^6 for $n = 10$ to 10^4 for $n = 100$.

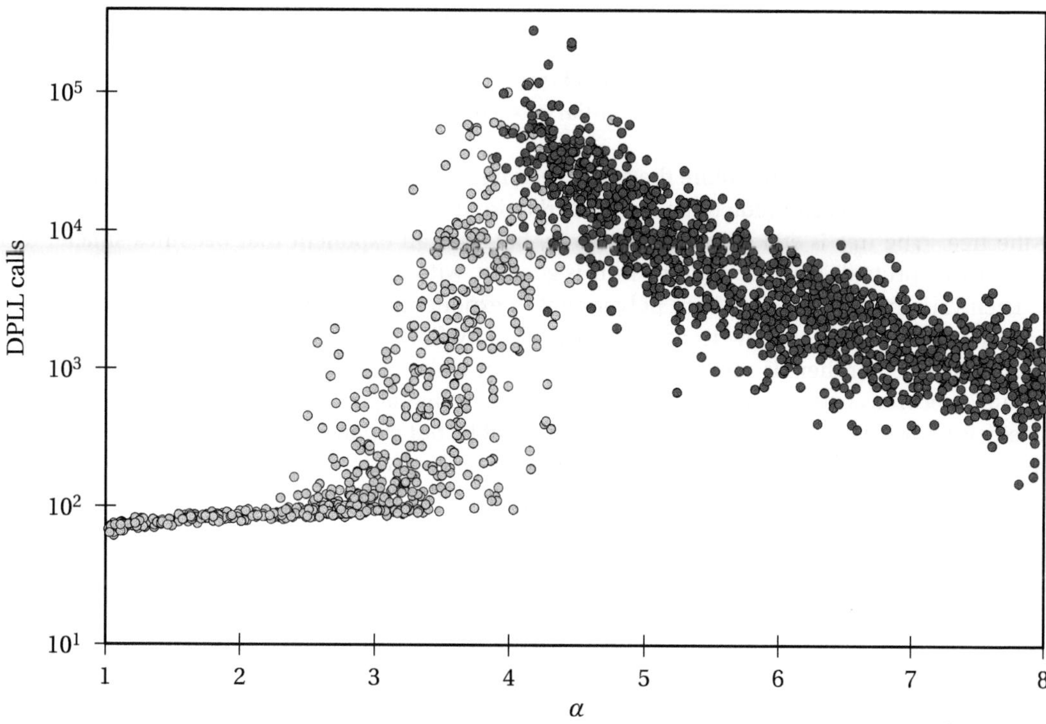

FIGURE 14.3: Number of recursive calls of DPLL on random 3-SAT formulas as a function of $\alpha = m/n$. Here $n = 100$. Light and dark dots represent satisfiable and unsatisfiable instances. Note that the y-axis is logarithmic.

14.1.2 Backtracking and Search Times

We can also ask about the running time of DPLL on these random formulas. How long does it take to either find a satisfying assignment, or confirm that none exists?

DPLL is actually a family of algorithms. Each has a different *branching rule* that determines which variable to set next and which truth value to give it first. For instance, if ϕ contains a *unit clause*—a clause with a single variable, (x) or $(\overline{x})$—then we should immediately set x to whichever value this clause demands. As we saw in Section 4.2.2, this *unit clause propagation* rule is powerful enough to solve 2-SAT in polynomial time. In our experiments we use a slight generalization called the *short clause rule*: choose a clause c from among the shortest ones, and choose x randomly from c's variables.

DPLL's running time is essentially the number of times it calls itself recursively, or the number of nodes of the search tree it explores. How does this depend on α? As Figure 14.3 shows, when α is small enough, DPLL finds a satisfying assignment with little or no backtracking. Intuitively, this is because each clause shares variables with just a few others, so there are few conflicts where satisfying one clause dissatisfies another one. In this regime, DPLL sets each variable once and only rarely reconsiders its decisions, and its running time is linear in n.

As α increases, so does the number of interconnections and conflicts between the clauses. At some value of α, which depends on the branching rule, many of our choices now lead to contradictions. This forces us to do an exponential amount of backtracking, exploring a larger fraction of the search tree as satisfying leaves become rarer. This exponential gets worse until we reach α_c and ϕ becomes unsatisfiable with high probability.

Beyond α_c, `DPLL` has to explore the entire search tree to prove that ϕ is unsatisfiable. However, as α continues to increase, contradictions appear earlier in the search process, pruning away many branches of the tree. The tree is still exponentially large, but with an exponent that becomes smaller as more branches are pruned. Thus the running time is greatest, with the most severe exponential, at $\alpha = \alpha_c$.

Each variant of `DPLL` goes from polynomial to exponential time at some density below α_c, and in Section 14.3 we will show how to calculate this density for several simple branching rules. A natural question is whether there is some density $\alpha_{exp} < \alpha_c$ at which the *intrinsic* complexity of random 3-SAT formulas becomes exponential, so that no polynomial-time algorithm can succeed. Formulas whose density is between α_{exp} and α_c would then be satisfiable, but exponentially hard to solve. Of course, we have no hope of proving such a thing, since we don't even know that $\mathsf{P} \neq \mathsf{NP}$. But there are hints from the geometry of the set of satisfying assignments, which we will discuss at the end of this chapter, that such a phase exists.

14.1.3 *The Threshold Conjecture*

Inspired by our experimental results, let's conjecture that there is a sudden transition from satisfiability to unsatisfiability at a critical density.

Conjecture 14.1 *For each $k \geq 2$, there is a constant α_c such that*

$$\lim_{n\to\infty} \Pr\left[F_k(n, m = \alpha n) \text{ is satisfiable}\right] = \begin{cases} 1 & \text{if } \alpha < \alpha_c \\ 0 & \text{if } \alpha > \alpha_c \,. \end{cases}$$

As in previous chapters, we say that something occurs *with high probability* if its probability is $1 - o(1)$, i.e., tending to 1 as n grows. Then this conjecture states that there is an α_c such that $F_k(n, m = \alpha n)$ is satisfiable with high probability if $\alpha < \alpha_c$, and unsatisfiable with high probability if $\alpha > \alpha_c$.

This conjecture is known to hold for $k = 2$, in which case $\alpha_c = 1$ (see Problem 14.17). As of 2015, it also holds when k is sufficiently large: that is, there is some constant k_0 such that it holds for all $k > k_0$. For
14.4 values such as $k = 3$, however, the closest anyone has come to proving it is the following.

Theorem 14.2 (Friedgut) *For each $k \geq 3$, there is a function $\alpha_c(n)$ such that, for any $\varepsilon > 0$,*

$$\lim_{n\to\infty} \Pr\left[F_k(n, m = \alpha n) \text{ is satisfiable}\right] = \begin{cases} 1 & \text{if } \alpha < (1-\varepsilon)\alpha_c(n) \\ 0 & \text{if } \alpha > (1+\varepsilon)\alpha_c(n) \,. \end{cases}$$

This theorem is tantalizingly close to Conjecture 14.1. However, we don't know that $\alpha_c(n)$ converges to a constant as $n \to \infty$. In principle, $\alpha_c(n)$ could oscillate somehow as a function of n.

14.5 Physically, this possibility is unthinkable. All our experience in statistical physics tells us that systems approach a *thermodynamic limit* as their size tends to infinity. Microscopic details disappear, and only bulk properties—in this case the density of clauses—affect their large-scale behavior. The freezing temperature of water doesn't depend on how much water we have.

Armed with the faith that the same is true for random k-SAT formulas, we will assume throughout this chapter that Conjecture 14.1 is true. We will then prove upper and lower bounds on α_c by showing that $F_k(n, m = \alpha n)$ is probably satisfiable, or probably unsatisfiable, if α is small enough or large enough.

There is a significant gap between what we believe and what we can prove. For 3-SAT, the best known rigorous bounds are currently

$$3.52 < \alpha_c < 4.49,$$

leaving a distressing amount of room around the presumably true value $\alpha_c \approx 4.267$. On the other hand, when k is large the gap between our upper and lower bounds becomes very narrow, and we know the k-SAT threshold quite precisely:

$$\alpha_c = (1 - o(1))2^k \ln 2, \tag{14.1}$$

where here $o(1)$ means something that tends to 0 in the limit $k \to \infty$.

We will prove these bounds using a variety of tools. When α is large enough, we will show that $F_k(n, m = \alpha n)$ is unsatisfiable by showing that the expected number of satisfying assignments, and therefore the probability that there are any, is exponentially small. Any such α is an upper bound on α_c.

For lower bounds, we will use two very different methods. When α is small enough, we can give a constructive proof that a satisfying assignment probably exists by showing that a simple search algorithm—such as a single branch of DPLL with a particular branching rule—probably finds one. However, these algorithmic techniques fail at the densities where backtracking becomes necessary, and where DPLL's running time becomes exponential.

To prove satisfiability above the densities where these algorithms work, we resort to *nonconstructive* methods. Like the techniques discussed in Section 6.7, these prove that satisfying assignments exist without telling us how to find them. In particular, we will prove (14.1) using the *second moment method*, which works by showing that the number of satisfying assignments has a large expectation but a small variance.

Our efforts to prove bounds on α_c will be aided by the following corollary of Theorem 14.2:

Corollary 14.3 *If for some constants α^* and $C > 0$*

$$\lim_{n\to\infty} \Pr[F_k(n, m = \alpha^* n) \text{ is satisfiable}] \geq C,$$

then, for any constant $\alpha < \alpha^*$,

$$\lim_{n\to\infty} \Pr[F_k(n, m = \alpha n) \text{ is satisfiable}] = 1.$$

Therefore, assuming Conjecture 14.1 holds, $\alpha_c \geq \alpha^*$.

In other words, if random formulas with a given density are satisfiable with positive probability, i.e., a probability bounded above zero as $n \to \infty$, then formulas with any smaller density are satisfiable with high probability. Similarly, if the probability of satisfiability at a given density is bounded below one, it is zero at any higher density. Such a result is called a *zero-one law.*

Exercise 14.2 *Derive Corollary 14.3 from Theorem 14.2.*

We can gain the skills we need to analyze these random formulas by practicing on one of the most basic random structures in mathematics—random graphs. These, and their phase transitions, are the subject of the next section.

14.2 Random Graphs, Giant Components, and Cores

The first example of a phase transition in computer science—or, more broadly, in combinatorics—came in the study of random graphs. While it is far simpler than the transition in SAT or GRAPH COLORING, it is a good way to learn some basic intuitions and techniques that will help us analyze random formulas.

14.2.1 Erdős–Rényi Graphs

Consider the following very simple model of a random graph. There are n vertices. For each of the $\binom{n}{2}$ pairs of vertices, we add an edge between them randomly and independently with probability p.

This model is denoted $G(n,p)$, and is usually called the Erdős–Rényi model after the mathematicians Paul Erdős and Alfréd Rényi. Unlike a social network where two people are much more likely to be con-
14.6 nected if they have a friend in common, it assumes that every pair of vertices is equally likely to be connected and that these events are independent of each other. Indeed, this model possesses none of the structure that real social, technological, or biological networks possess.

Nevertheless, these random graphs capture some of the qualitative aspects of these networks. In particular, they undergo a phase transition analogous to an epidemic in a social network, where small outbreaks join and spread across much of the population, or percolation in physics, where a path through a lattice suddenly appears when a critical fraction of vertices become occupied.

The average degree of a given vertex v is $p(n-1) \approx pn$, since v is connected to each of the other $n-1$ vertices with probability p. More generally, the probability that v has degree exactly j, or equivalently the expected fraction of vertices with degree j, is the binomial distribution

$$a_j = \binom{n-1}{j} p^j (1-p)^{n-1-j} .$$

This is called the *degree distribution*. As with our random formulas, we will be especially interested in *sparse* random graphs, where $p = c/n$ for some constant c. In the limit of large n, the degree distribution then becomes a Poisson distribution with mean c (see Appendix A.4.2):

$$a_j = \frac{\mathrm{e}^{-c}\, c^j}{j!} . \tag{14.2}$$

This distribution is peaked at c, and drops off rapidly for larger j.

Exercise 14.3 *Show that with high probability, $G(n, p = c/n)$ has no vertices of degree* $\ln n$ *or greater.*

Exercise 14.4 *Show that the expected number of triangles in $G(n, p = c/n)$ approaches $c^3/6$ as $n \to \infty$.*

A useful variant of the Erdős–Rényi model is $G(n,m)$. Here we fix the number of edges to be exactly m, choosing them uniformly from among the $\binom{n}{2}$ possible edges. In other words, $G(n,m)$ is chosen uniformly from all $\binom{\binom{n}{2}}{m}$ possible graphs with n vertices and m edges. Since the number of edges in $G(n,p)$ is tightly concentrated around its expectation $\mathbb{E}[m] = p\binom{n}{2} \approx pn^2/2 = cn/2$, for many purposes $G(n,m)$ and $G(n,p)$ are equivalent if we set $m = \mathbb{E}[m]$. In particular, Problem 14.1 shows that these two models have the same thresholds for many natural properties.

We can now describe the most basic phase transition that $G(n, p = c/n)$ undergoes. When c is very small, G consists of small components isolated from each other, almost all of which are trees. As c grows and we add more edges, these trees grow and connect with each other. Suddenly, at $c = 1$, many of them come together to form a *giant component*, which grows to engulf a larger and larger fraction of G as c increases further. In the next few sections, we look at the basic mechanics of this transition, and introduce ideas that will help us understand the phase transition in SAT later on.

14.2.2 *The Branching Process*

We start at some vertex v and grow a spanning tree outward from it—its neighbors, its neighbors' neighbors, and so on. If we think of these as v's children and grandchildren, then v's connected component consists of all its descendants, and the jth generation consists of the vertices whose distance from v is j.

The average number of children v has is c. One might think that the average number of grandchildren that each child u produces is $c-1$, since u's average degree is c and we already know that it is connected to its parent v. However, there are $n-2$ vertices besides u and v, and u is connected to each one independently of its edge with v. Thus the average number of neighbors u has in addition to v is $p(n-2)$, and when n is large this is essentially $pn = c$.

To put this differently, if we follow an edge from v, we reach a vertex u with probability proportional to u's degree i. If we then ask how many other neighbors u has, we find that $i-1$ is distributed according the same Poisson distribution (14.2) that we had for i in the first place. Algebraically,

$$\frac{i a_i}{\sum_{j=1}^{\infty} j a_j} = a_{i-1} . \tag{14.3}$$

Exercise 14.5 *Prove* (14.3) *explicitly from* (14.2). *Is it still true if the degree distribution is not Poisson?*

This lets us model v's connected component as a *branching process*—an abstract process where v is the progenitor, each node has a certain number of children, and v and its descendants form a family tree. This process ignores the possibility of edges other than those between parent and child, and it ignores the fact that number of potential new neighbors decreases as the tree grows. Thus it assumes that v's component is a tree, and slightly overestimates its size. Nevertheless, in the limit of large n it is a good model of most components as long as they are not too large, and a good model of v's neighborhood as long as we don't go too far.

The expected number of descendants in the jth generation is c^j, so the expected size of the entire tree, and thus of v's component, is the geometric sum

$$\mathbb{E}[s] = 1 + c + c^2 + \cdots .$$

If the branching process is *subcritical*, i.e., if $c < 1$, then this sum converges to a constant,

$$\mathbb{E}[s] = \frac{1}{1-c} .$$

As c increases, $\mathbb{E}[s]$ increases as well, until at $c = 1$ the branching process becomes critical and $\mathbb{E}[s]$ diverges. For $c > 1$ it is *supercritical*, and with positive probability v has an infinite number of descendants.

At this point the branching process is no longer an accurate model of v's connected component. Instead, v and its progeny form a giant component, containing $\Theta(n)$ vertices instead of $O(1)$.

Note that $\mathbb{E}[s]$ is the expected size of the connected component of a uniformly random vertex v. Equivalently, it is the expected size of a connected component where components of size s are chosen with probability proportional to s. If there are t components with sizes $s_1, \ldots, s_t$ then $\mathbb{E}[s]$ is proportional to the sum of their squares,

$$\mathbb{E}[s] = \frac{1}{n} \sum_{i=1}^{t} \mathbb{E}[s_i^2] \,. \tag{14.4}$$

Exercise 14.6 *Show that if v's connected component has size $o(\sqrt{n})$ then with high probability it is a tree. Hint: first build a breadth-first spanning tree rooted at v, and then ask whether any additional edges exist.*

Then use (14.4) *to show that if $c < 1$ the expected number of components in the graph that are not trees is $O(1)$. You may find it helpful to draw an analogy with the Birthday Problem discussed in Appendix A.3.3.*

We develop our picture of $G(n, p = c/n)$ further in the Problems, where we show that the following properties hold with high probability:

- When $c < 1$, the number of components of size s drops off exponentially for large s, as $n\lambda^s$ for some $\lambda < 1$, and the largest component has size $O(\log n)$. The expected number of cycles in the entire graph is $O(1)$, so almost all components are trees, and no components contain more than one cycle.

- When $c > 1$, there is a unique giant component of size $\gamma n + o(n)$ where $\gamma > 0$ is a function of c. The remaining components look like a random graph with $c < 1$, consisting almost entirely of trees of maximum size $O(\log n)$.

- Finally, if we are exactly at the critical point $c = 1$, the number of components of size s scales as a power law $ns^{-5/2}$, and the largest component is of size $O(n^{2/3})$.

Next, let's calculate the fraction γ of vertices that lie in the giant component. We will do this using two different arguments. Neither one is quite rigorous, but both can be made so. Moreover, they have rather different flavors, and they will teach us techniques that will help with random SAT formulas later on.

14.2.3 The Giant Component

Our first argument uses the branching process model. As c approaches 1, the expected size of v's family tree diverges, and if $c > 1$ then v has an infinite number of descendants with positive probability. As we suggested above, this divergence corresponds to v being in the giant component. In contrast, if v's descendants die out after a finite number of generations—which also happens with positive probability whenever c is finite—then v is in one of the small components disconnected from the giant, almost all of which are trees.

Let's call a node v in the branching process *abundant* if it has an infinite number of descendants. Recursively, v is abundant if it has at least one abundant child. Each child is abundant with the same

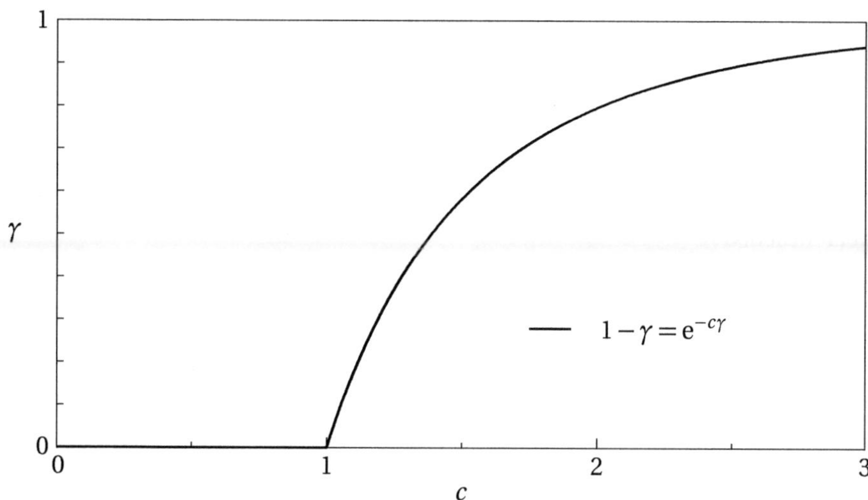

FIGURE 14.4: The fraction γ of the vertices contained in the giant component, as a function of the average degree c.

probability γ and these events are independent, so the number j of abundant children that v has is Poisson-distributed with mean $c\gamma$:

$$\Pr[v \text{ has } j \text{ abundant children}] = \frac{e^{-c\gamma}(c\gamma)^j}{j!}.$$

The probability that v is abundant is then the probability that $j > 0$. This gives

$$\gamma = 1 - e^{-c\gamma}. \tag{14.5}$$

This derivation may seem rather circular, but we can state it as follows. Suppose P is some property of nodes in the branching process, such that P is true of v if and only if it is true of at least one of v's children—say, the property of someday having a descendant in the New York Philharmonic. The probability γ that P is true of v must be a root of (14.5).

Thinking of the branching process as exploring v's neighborhood is a little confusing in this case. After all, if v has at least one neighbor in the giant component, then v and all its other neighbors are in the giant component as well. But we can imagine for a moment that v is missing from the graph, and that a fraction γ of the other vertices are in the giant component in v's absence. If we add v back in and connect it to some set of neighbors, v will be in the giant component if at least one of those neighbors is. If the presence or absence of v makes only an infinitesimal difference in γ, then γ must again be a root of (14.5). This style of thinking, where we imagine what properties v's neighborhood would have if v weren't there, is called the *cavity method* in physics. We will meet it again in Section 14.6.

You can check that for $c < 1$ the only nonnegative root of (14.5) is $\gamma = 0$, while for $c > 1$ there is a positive root as well. We show this root as a function of c in Figure 14.4. Since (14.5) is transcendental, we

```
Explore
begin
    label v Boundary ;
    label all other vertices Untouched ;
    while there are Boundary vertices do
        choose a Boundary vertex v ;
        label each of v's Untouched neighbors Boundary ;
        label v Reached ;
    end
end
```

FIGURE 14.5: An algorithm that explores v's connected component.

can't write γ in terms of familiar functions. However, it is easy enough to approximate γ at two extremes—when c is just above 1, and when c is very large. 14.

Exercise 14.7 *Show that if $c = 1+\varepsilon$, the size of the giant component is $\gamma = 2\varepsilon + O(\varepsilon^2)$. At the other extreme, show that when c is large, $\gamma \approx 1 - \mathrm{e}^{-c}$ and the giant component encompasses almost all of G.*

Thus γ starts at zero at the critical point $c = 1$ and increases continuously as c increases. In the language of statistical physics, the emergence of the giant component is a *second-order* phase transition.

14.2.4 The Giant Component Revisited

Let's compute γ again, but this time by exploring it exhaustively from within. This will let us show off an important tool: modeling the behavior of a simple algorithm with a system of differential equations.

Starting with a vertex v, we explore its connected component using an algorithm similar to that at the beginning of Section 3.4.1. At each point in time, each vertex is labeled Reached, Boundary, or Untouched. Reached means it is known to be in v's component and its neighborhood has been explored; Boundary means that it is in v's component but its neighbors are yet to be explored; and Untouched means that it is not yet known to be in v's component. The algorithm shown in Figure 14.5 explores G one vertex at a time, until there are no Boundary vertices left and v's entire component has been Reached.

Let R, B, and U denote the number of vertices of each type at a given point in time. In each step of the algorithm, R increases by 1, and the expected change in B and U is

$$\begin{aligned} \mathbb{E}[\Delta U] &= -pU = -cU/n \\ \mathbb{E}[\Delta B] &= pU - 1 = cU/n - 1\,. \end{aligned} \tag{14.6}$$

Let's explain each of these terms. There is an expected "flow" from U to B of pU vertices, since each Untouched vertex is connected to v with probability p. The -1 term in ΔB comes from changing the chosen vertex v from Boundary to Reached.

Of course, B and U are stochastic quantities, and fluctuate from one step of the algorithm to the next. However, if we scale everything down by n, we expect them to become concentrated around some continuous trajectory. In other words, let $r = R/n$ denote the fraction of G we have explored so far. Then we hope that there are real-valued functions $b(r)$ and $u(r)$ such that, with high probability, the values of B and U after the Rth step obey

$$\begin{aligned} B(R) &= b(R/n)\cdot n + o(n) \\ U(R) &= u(R/n)\cdot n + o(n). \end{aligned} \tag{14.7}$$

With this rescaling, at each step r increases by $1/n$ and the expected change in $b(r)$ and $u(r)$ is $O(1/n)$. If we assume that $\mathrm{d}b/\mathrm{d}r$ and $\mathrm{d}u/\mathrm{d}r$ are given exactly by these expectations, the stochastic difference equations (14.6) become a system of differential equations:

$$\begin{aligned} \frac{\mathrm{d}u}{\mathrm{d}r} &= \frac{\mathbb{E}[\Delta U]}{\Delta R} = -cu \\ \frac{\mathrm{d}b}{\mathrm{d}r} &= \frac{\mathbb{E}[\Delta B]}{\Delta R} = cu - 1. \end{aligned} \tag{14.8}$$

If we take these differential equations seriously, solving them with the initial conditions $b(0)=0$ and $u(0)=1$ gives

$$u(r) = \mathrm{e}^{-cr} \text{ and } b(r) = 1 - r - \mathrm{e}^{-cr}.$$

The fraction γ of vertices in the giant component is the value of r at which no boundary vertices remain and the algorithm stops. That is,

$$b(\gamma) = 1 - \gamma - \mathrm{e}^{-c\gamma} = 0,$$

and rearranging gives (14.5) as before.

Are these differential equations a valid approximation for Explore's behavior? For one thing, they only hold if v is in the giant component. Otherwise, the exploration process quickly dies out, B never becomes $\Theta(n)$, and the differential equations never get off the ground. One way to avoid this problem is to "seed" the process by initial labeling εn vertices Boundary for some small ε. We solve the differential equations with the initial conditions $b(0)=\varepsilon$ and $u(0)=1-\varepsilon$, and then take the limit $\varepsilon \to 0$.

But more generally, when can we model a stochastic process as a set of differential equations? When can a discrete process be described as a flow, with real-valued variables in continuous time? As we discuss further below, there are two key reasons why we can do this.

First, the expected changes in B and U depend only on their current values—not on the previous history of the algorithm, nor any details of the graph that we have learned in the past. One conceptual tool that helps clarify this is the *principle of deferred decisions*. Rather than thinking of the graph as created in advance, we think of its creator as making it up "on the fly" as we ask questions about it. In this case, each time we ask whether v is connected to a given Untouched vertex u, the creator can flip a biased coin and answer "yes" with probability p. We have never seen this part of the graph before, and we have no knowledge about it—so these are fresh coin flips, independent of everything that has gone before.

Second, the expected changes in B and U are functions of the rescaled variables $b = B/n$ and $u = U/n$. This allows us to write the expected derivative of b and u in terms of the same variables, giving a sensible system of differential equations. Moreover, because these functions are smooth, standard results from the theory of first-order differential equations tell us that this system has a unique solution. With some conditions on the tail of the distribution of ΔB and ΔU, it can then be shown that B/n and U/n stay concentrated around this solution with high probability. 14.8

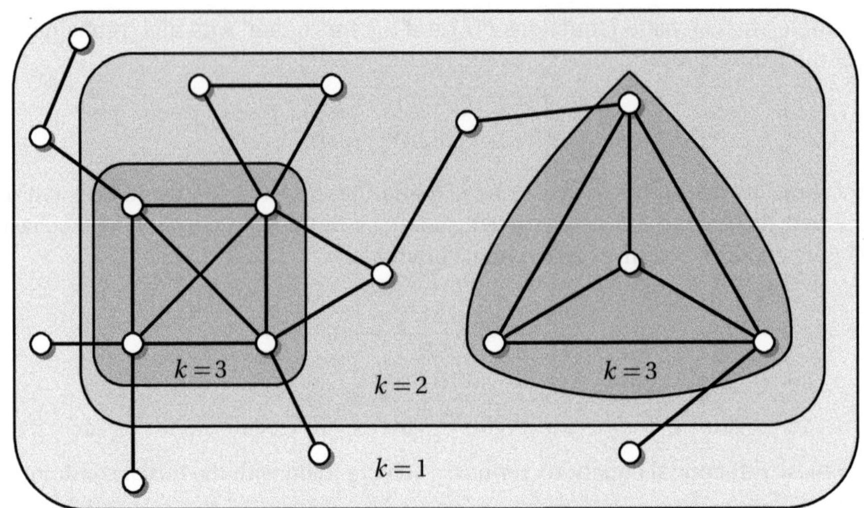

FIGURE 14.6: k-cores of a graph.

```
Prune(G)
input: a graph G
output: the k-core of G
begin
    while there are vertices of degree less than k do
        choose a vertex v of degree less than k ;
        remove v and all edges incident to v ;
    end
    return G ;
end
```

FIGURE 14.7: This algorithm prunes away vertices of degree less than k, until only the k-core is left.

14.2.5 The k-Core

Even if a random graph has a giant connected component, its internal connections might be relatively loose. If we ask whether $G(n, p = c/n)$ has a subgraph which is more densely connected in a certain sense, we get a series of phase transitions at larger values of c.

The *k-core* of a graph is the largest subgraph where every vertex has minimum degree k. In other words, it is the largest set S of vertices where each vertex in S is connected to at least k vertices in S. As Figure 14.6 shows, the 1-core is the set of all vertices of degree at least 1. Pruning away all the trees attached to the graph gives the 2-core, pruning away the vertices of degree 2 or less gives the 3-core, and so on. In general, if we run the algorithm in Figure 14.7, repeatedly removing vertices of degree less than k, the k-core is what remains. Of course, it could be the case that this algorithm prunes away the entire graph, in which case the k-core is empty.

For what values of c does $G(n, p = c/n)$ probably have a k-core? The case $k = 2$ is special, since any

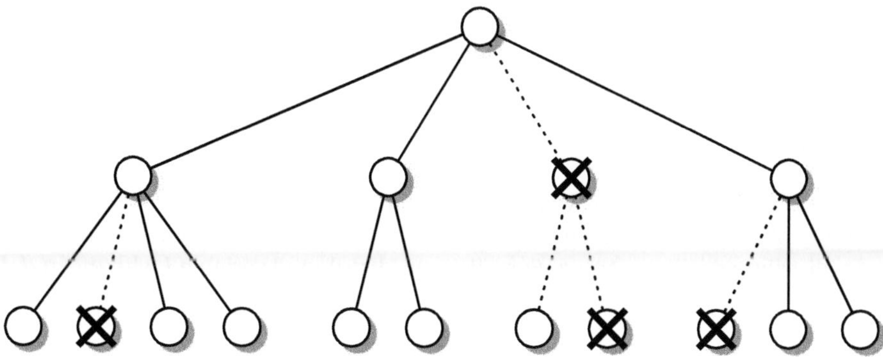

FIGURE 14.8: The branching process model of the 3-core. A vertex in the tree, other than the root, is well-connected if it has at least 2 well-connected children. To be in the 3-core, the root must have at least 3 well-connected children.

cycle is contained in the 2-core. Thus even for $c < 1$ the 2-core is nonempty, albeit of size $o(n)$, with positive probability. When $c > 1$ and there is a giant component, a constant fraction of vertices lie on a cycle, and the 2-core is of size $\Theta(n)$.

For $k \geq 3$, on the other hand, Problem 14.9 shows that, with high probability, the k-core is of size $\Theta(n)$ if it exists at all. Moreover, there is a threshold c_k^{core} such that $G(n, p = c/n)$ has a k-core of size $\Theta(n)$ if $c > c_k^{\text{core}}$, and no k-core at all if $c < c_k^{\text{core}}$.

One of our motivations for studying the k-core is the following.

Exercise 14.8 *Show that if a graph G does not have a k-core, it is k-colorable. Hint: run the algorithm of Figure 14.7 backwards.*

Analogous to Conjecture 14.1, we believe that for each $k \geq 3$ there is a phase transition GRAPH k-COLORING in random graphs, at some critical value c_k of the average degree. That is,

$$\lim_{n\to\infty} \Pr\left[G(n, p = c/n) \text{ is } k\text{-colorable}\right] = \begin{cases} 1 & \text{if } c < c_k \\ 0 & \text{if } c > c_k \,. \end{cases}$$

In that case Exercise 14.8 implies that $c_k \geq c_k^{\text{core}}$.

In the rest of this section, we will calculate the threshold c_k^{core} and the likely fraction γ_k of vertices in the k-core when it exists. As for the size of the giant component, we will do this in two different ways. First, we will use a branching process model to write an equation whose roots describe c_k^{core} and γ_k. Second, we will analyze the algorithm of Figure 14.7 using differential equations.

For the branching process model, consider Figure 14.8. As before, we place a vertex v at the root of a tree, with its neighbors and their neighbors drawn as its children and grandchildren. Now imagine the process of pruning vertices of degree less than k working its way up the tree. Like every parent, v hopes that its children are well-connected enough to survive this relentless social pruning—especially since v itself only survives if it has at least k well-connected children.

k	2	3	4	5	6	7	8
c_k^{core}	1	3.35	5.15	6.80	8.37	9.88	11.34
γ_k	0	0.27	0.44	0.54	0.60	0.65	0.68

TABLE 14.1: Thresholds for the emergence of the k-core, and the size with which it first appears.

Suppose that each of v's children is well-connected independently with probability q. Then analogous to the number of abundant children in Section 14.2.3, the number of well-connected children that v has is Poisson-distributed with mean cq. For a Poisson-distributed variable j with mean d, let

$$Q_{<k}(d) = \sum_{j=0}^{k-1} \frac{e^{-d} d^j}{j!} \tag{14.9}$$

denote the probability that j is less than k. Then the probability that v has at least k well-connected children, or equivalently the fraction of vertices in the k-core, is

$$\gamma_k = 1 - Q_{<k}(cq). \tag{14.10}$$

Now consider a vertex w somewhere down in the tree. Since it is already connected to its parent, it is well-connected if it has at least $k-1$ well-connected children. If each of its children is well-connected with the same probability q that w is, q must be a root of the equation

$$q = 1 - Q_{<k-1}(cq). \tag{14.11}$$

Note that setting $k=2$ gives (14.5), the equation for the size of the giant component, since in that case a vertex is well-connected—or as we called it there, abundant—if at least one of its children is.

Like (14.5), the equation (14.11) always has a root at $q=0$. The threshold c_k^{core} at which the k-core first appears is the smallest value of c for which it has a nonzero root as well. Above c_k^{core} there are two nonzero roots, and we will see below that the size of the core is the largest one.

For $k=2$ we have $c_2^{\text{core}}=1$, since except for $O(1)$ cycles the 2-core is part of the giant component. Like the giant component, its size increases continuously as a function of c (see Problem 14.15).

For $k \geq 3$, however, the k-core occupies a constant fraction of the graph when it first appears. For instance, $c_3^{\text{core}} \approx 3.351$, and as Figure 14.9 shows, the size of the 3-core jumps discontinuously from 0 to $\gamma_3 \approx 0.268$. The same happens for larger values of k, as shown in Table 14.1. Thus, unlike the emergence of the giant component, the emergence of the k-core for $k \geq 3$ is a *first-order* phase transition.

Exercise 14.9 *Plot the left and right sides of* (14.11) *for* $k=3$. *Find the threshold* c_3^{core} *where they first touch at a nonzero* q, *and use* (14.10) *to compute* γ_3.

Let's calculate c_k^{core} and γ_k again, using the technique of differential equations. First we define some terms. We call a vertex *light* if its current degree is less than k, and *heavy* if it is k or greater. Each step of the pruning algorithm in Figure 14.7 removes a light vertex and its edges, decreasing the degrees of its neighbors, and possibly causing some heavy vertices to become light.

To analyze the progress of this algorithm, it helps to think of each vertex of degree j as a collection of j half-edges or "spokes." The graph is a perfect matching of these spokes, where two spokes are partners if

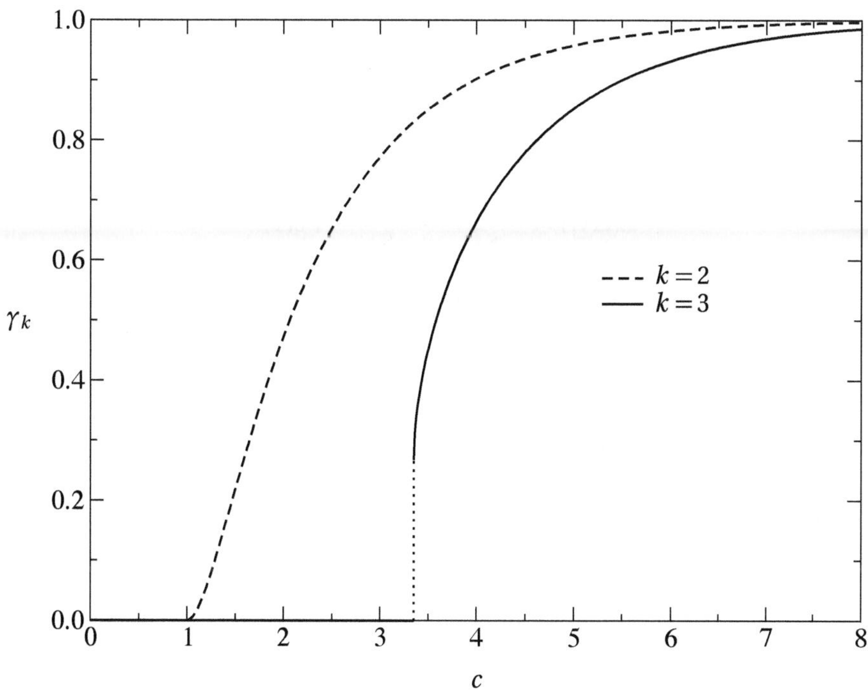

FIGURE 14.9: The fraction γ_k of vertices in the k-core as a function of the average degree c. The 2-core emerges continuously at $c=1$ as part of the giant component. In contrast, the 3-core emerges discontinuously at $c \approx 3.351$, where its size jumps from 0 to $\gamma_3 \approx 0.268$.

they form an edge. We discuss this type of random graph, where the degree of each vertex is fixed but the connections between them are random, in Problem 14.10. There is a technical hitch because matching the spokes randomly could create a multigraph with self-loops or multiple edges. However, the expected number of such edges is $O(1)$, and we ignore them here.

Our analysis will be simpler if we slow the pruning algorithm down in the following way. Rather than removing a light vertex and all its spokes at once, we remove one light spoke at a time. In that case, we can "smash" the light vertices, and simply think of them as a bag of light spokes—we don't need to keep track of how these spokes are gathered together into vertices. At each step we reach into the bag, choose a light spoke, and find out who its partner is. If its partner is another light spoke, we remove both from the bag. If instead its partner is a spoke of a heavy vertex v, we remove that spoke, thus decreasing v's degree. If v had degree k before, v has gone from heavy to light, so we add $k-1$ light spokes to the bag.

It's time to define our variables. Let S be the total number of spokes in the world, both heavy and light—in other words, the total degree of all the vertices. Let L be the number of light spokes. For each $j \geq k$, let V_j be the number of heavy vertices of degree j. Note that

$$S = L + \sum_{j=k}^{\infty} j V_j \,.$$

Then the expected change in these variables when we remove a light spoke is given by the following system of difference equations:

$$\Delta S = -2 \tag{14.12}$$

$$\mathbb{E}(\Delta L) = -1 - \frac{L}{S} + \frac{k(k-1)V_k}{S} + o(1) \tag{14.13}$$

$$\mathbb{E}(\Delta V_j) = -\frac{j V_j}{S} + \frac{(j+1)V_{j+1}}{S} + o(1) \,. \tag{14.14}$$

We explain these terms as follows. Removing the light spoke and its partner gives $\Delta S = -2$. Its partner is chosen uniformly from all the other spokes, so it is also light with probability L/S, and it belongs to a heavy vertex of degree j with probability $j V_j/S$. If $j = k$, this creates $k-1$ new light spokes, giving (14.13). In addition, if the partner is heavy then it causes a flow of heavy vertices from degree j to degree $j-1$, and from $j+1$ to j, giving (14.14). The facts that there are $S-1$ potential partners instead of S, and $L-1$ light ones instead of L, are absorbed by the $o(1)$ error term as long as S and L are $\Theta(n)$.

Analogous to Section 14.2.4, we now define rescaled variables $s = S/n$, $\ell = L/n$, $v_j = V_j/n$, and $t = T/n$, where T is the number of steps of this slow-motion algorithm we have performed so far. Setting the derivatives of these variables to their expectations gives

$$\frac{\mathrm{d}s}{\mathrm{d}t} = -2 \tag{14.15}$$

$$\frac{\mathrm{d}\ell}{\mathrm{d}t} = -1 - \frac{\ell}{s} + \frac{(k-1)k\, v_k}{s} \tag{14.16}$$

$$\frac{\mathrm{d}v_j}{\mathrm{d}t} = -\frac{j v_j}{s} + \frac{(j+1)v_{j+1}}{s} \quad \text{for all } j \geq k \,. \tag{14.17}$$

This is a system of differential equations on an infinite number of variables. However, we can reduce it to a finite system by thinking about this process from the point of view of the heavy vertices. As far as they are concerned, at a given point in time each of their spokes has disappeared with a certain probability. Thus their degrees are distributed just as in a random graph, but with an average degree λ that varies with time:

$$v_j(t) = \frac{\mathrm{e}^{-\lambda(t)}\, \lambda(t)^j}{j!} \,. \tag{14.18}$$

Initially $\lambda(0) = c$, but λ decreases as heavy spokes are removed. Substituting (14.18) into (14.17) yields (exercise!)

$$\frac{\mathrm{d}\lambda}{\mathrm{d}t} = -\frac{\lambda}{s} \,.$$

It's convenient to change our variable of integration from t to λ. Then (14.15) and (14.16) become

$$\frac{\mathrm{d}s}{\mathrm{d}\lambda} = \frac{2s}{\lambda} \tag{14.19}$$

$$\frac{\mathrm{d}\ell}{\mathrm{d}\lambda} = \frac{s}{\lambda} + \frac{\ell}{\lambda} - \frac{\mathrm{e}^{-\lambda}\lambda^{k-1}}{(k-2)!} \,. \tag{14.20}$$

Along with the initial condition $s(0)=c$, (14.19) implies that

$$s=\frac{\lambda^2}{c},$$

and substituting this into (14.20) gives a differential equation in a single variable,

$$\frac{\mathrm{d}\ell}{\mathrm{d}\lambda}=\frac{\lambda}{c}+\frac{\ell}{\lambda}-\frac{\mathrm{e}^{-\lambda}\lambda^{k-1}}{(k-2)!}. \tag{14.21}$$

If the initial degree distribution a_j is the Poisson distribution (14.2), the initial density of light spokes is

$$\ell(c)=\sum_{j=0}^{k-1} j a_j=\sum_{j=0}^{k-1} j\frac{\mathrm{e}^{-c}c^j}{j!}=c\sum_{j=0}^{k-2}\frac{\mathrm{e}^{-c}c^j}{j!}=cQ_{<k-2}(c),$$

where, as in (14.9), $Q_{<k-1}(c)$ is the probability that a Poisson-distributed random variable with mean c is less than $k-1$. Using the identity (another exercise!)

$$\frac{\mathrm{d}}{\mathrm{d}\lambda}Q_{<m}(\lambda)=Q_{<m-1}(\lambda)-Q_{<m}(\lambda)=-\frac{\mathrm{e}^{-\lambda}\lambda^{m-1}}{(m-1)!},$$

we finally obtain the solution to (14.21),

$$\ell(\lambda)=\lambda\left(\frac{\lambda}{c}+Q_{<k-1}(\lambda)-1\right). \tag{14.22}$$

Figure 14.10 shows the trajectory of the pruning algorithm as a function of λ/c for various values of c. At first, many light spokes are partnered with other light spokes and the number of light spokes drops rapidly. As light spokes become rarer, most of them are partnered with heavy ones, and we start adding bundles of $k-1$ light spokes to the bag. Indeed, when c is just below c_3^{core} the number of light spokes almost reaches zero, and then increases again as these bundles are added.

The pruning process ends when $\ell(\lambda)=0$ and there are no light edges left. According to (14.22), this happens at the root λ of the equation

$$\frac{\lambda}{c}=1-Q_{<k-1}(\lambda). \tag{14.23}$$

Specifically, λ is the largest root, since this is the first one we encounter as λ decreases. The size of the k-core is the number of heavy vertices remaining at that point,

$$\gamma_k=\sum_{j=k}^{\infty} v_j=\sum_{j=k}^{\infty}\frac{\mathrm{e}^{-\lambda}\lambda^j}{j!}=1-Q_{<k}(\lambda). \tag{14.24}$$

Looking back at our branching process analysis, these are just equations (14.11) and (14.10) again, where $\lambda=cq$ is the expected number of well-connected children. As before, c_k^{core} is the smallest value of c such that (14.11), or equivalently (14.23), has a nonzero root.

Now that we have gained some facility with branching processes and differential equations, let's turn 14.9
our attention to random k-SAT formulas, and algorithms that try to satisfy them.

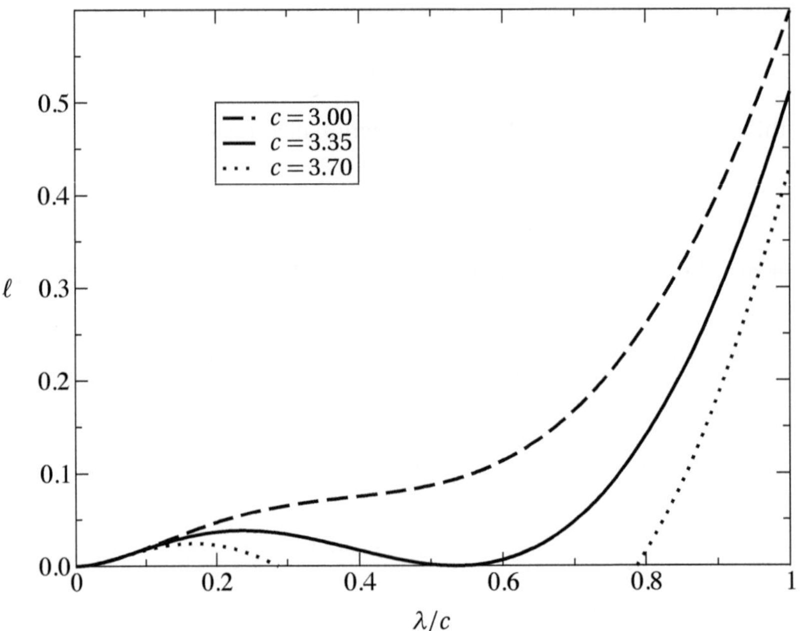

FIGURE 14.10: Trajectories of the pruning algorithm for $k = 3$, where c is above, below, or at the threshold c_3^{core}. At first $\lambda = c$, and then λ decreases, so time goes to the left. The algorithm stops when $\ell = 0$ and there are no light vertices left, and c_3^{core} is the smallest value of c such that this happens at some $\lambda \neq 0$.

14.3 Equations of Motion: Algorithmic Lower Bounds

In this section, we prove our first lower bounds on the threshold density α_c for 3-SAT. We do this by using differential equations to model simple search algorithms, and showing that they probably succeed up to a certain value of α. We start with one of the simplest possible algorithms, and then discuss to what extent we can make it smarter without losing our ability to analyze it.

14.3.1 Following a Single Branch

Consider the Unit Clause algorithm, or UC for short, shown in Figure 14.11. It goes through the variables one at a time, setting each variable permanently, and simplifying the formula as it goes along. If it creates a contradiction, it immediately gives up and returns "don't know" instead of backtracking. Thus it acts like a DPLL algorithm, but it only explores one branch of the search tree.

The branching rule that UC uses is extremely simple. Whenever it can, it chooses a variable x randomly from among the variables that have not yet been set, and gives it a random truth value. Its only concession to intelligence is to perform *unit clause propagation* as described in Section 4.2.2, immediately satisfying unit clauses—hence the name—whenever they exist.

It is hard to imagine a more simple-minded algorithm. Nevertheless, UC is good enough to succeed with positive probability on random 3-SAT formulas $F_3(n, m = \alpha n)$ as long as $\alpha < 8/3$. Thus it provides

```
UC
input: a 3-SAT formula φ
output: "satisfiable" or "don't know"
begin
    while φ contains unsatisfied clauses do
        if φ contains any unit clauses then
            choose c = (x) or (x̄) uniformly from the unit clauses ;
            satisfy c by setting x ;
        else
            choose x uniformly from the unset variables ;
            set x = true or x = false with equal probability ;
        remove or shorten the clauses containing x ;
        if φ contains an empty clause then return "don't know" ;
    end
    return "satisfiable";
end
```

FIGURE 14.11: The Unit Clause (UC) algorithm. If there are any unit clauses, it satisfies them immediately. Otherwise, it chooses a random variable and sets its truth value by flipping a coin.

the following lower bound on the satisfiability threshold,

$$\alpha_c \geq \frac{8}{3} = 2.666...$$

This is far below the conjectured value $\alpha_c \approx 4.267$, but it's a good start. Moreover, it shows that for densities up to 8/3 random formulas are not only satisfiable, but are easy to satisfy.

We will model UC's progress using differential equations, just as we did for the giant component and the k-core in Section 14.2. Let's define our variables. At a given point in time, some clauses have been satisfied and others have been shortened, so the remaining formula has clauses of various lengths. Let T be the number of variables set so far, so that $n-T$ variables are unset, and let S_3, S_2, and S_1 be the number of 3-clauses, 2-clauses, and unit clauses respectively.

Our analysis relies on the fact that, at all times throughout the algorithm, the remaining formula ϕ is uniformly random in the following sense. If I tell you S_3, S_2, S_1, and T, then ϕ is distributed just as if I constructed a random formula with $n-T$ variables and S_ℓ clauses of each length $1 \leq \ell \leq 3$. In other words, while we know how many variables there are left, and how many clauses of each length there are, we know nothing at all about these clauses.

Let's dwell on this a little longer. What do you need to know in order to keep track of S_ℓ for each ℓ, and compute its new value each time we set a variable x? How much of the formula do I need to reveal to you? I only need to show you the clauses that x appears in. For that matter, I don't even need to show you the entire clause—I just need to tell you its length, and whether it agrees or disagrees with x's value. Invoking the principle of deferred decisions, we can pretend that the other clauses, and the other literals appearing in x's clauses, haven't been chosen yet. If we were to reveal them all at once, we would find that they have been chosen with replacement from the $2^\ell \binom{n-T}{\ell}$ possible ℓ-SAT clauses on the unset variables, just as we chose $F_3(n,m)$ in the first place.

This kind of conditional randomness, where the remaining formula is uniformly random conditioned on a finite number of variables, is exactly what makes it possible to use the differential equation approach. We discuss for what kind of algorithms it holds in Section 14.3.3.

Now let's analyze the effect of each step on the variables. We call a step of the algorithm *forced* if it satisfies a unit clause, and *free* if it chooses a random variable. From the point of view of the rest of the formula, these two types of steps have exactly the same effect—namely, they set a random unset variable to a random value. For forced steps, this is because the formula itself is random, so each unit clause is chosen randomly from among the unset variables and is negated half the time. For free steps, it follows from the algorithm's own randomness.

The probability that the chosen variable x appears in a given 3-clause c is $3/(n-T)$, since each of c's three variables is chosen uniformly from the $n-T$ unset variables. (Actually, since these three variables are required to be distinct, this probability is slightly larger, but this makes no difference when n is large.) Moreover, when we give x a random value, half the time c is satisfied and removed from the formula, and the other half of the time it becomes a 2-clause.

Thus 3-clauses disappear at a rate $3S_3/(n-T)$, and $(3/2)S_3/(n-T)$ of these become 2-clauses. Similarly, the probability that x appears in a given 2-clause is $2/(n-T)$, so 2-clauses disappear at a rate $2S_2/(n-T)$. Putting this together, the expected change in S_3 and S_2 in a single step is given by

$$\begin{aligned} \mathbb{E}[\Delta S_3] &= -\frac{3S_3}{n-T} \\ \mathbb{E}[\Delta S_2] &= \frac{(3/2)S_3}{n-T} - \frac{2S_2}{n-T}. \end{aligned} \tag{14.25}$$

Rescaling our variables to $s_3 = S_3/n$, $s_2 = S_2/n$, and $t = T/n$ gives a system of differential equations,

$$\begin{aligned} \frac{\mathrm{d}s_3}{\mathrm{d}t} &= -\frac{3s_3}{1-t} \\ \frac{\mathrm{d}s_2}{\mathrm{d}t} &= \frac{(3/2)\,s_3 - 2s_2}{1-t}, \end{aligned} \tag{14.26}$$

and solving with the initial conditions $s_3(0) = \alpha$, $s_2(0) = 0$ gives

$$s_3(t) = \alpha(1-t)^3, \quad s_2(t) = \frac{3}{2}\alpha t(1-t)^2.$$

The reader will notice that S_1, the number of unit clauses, doesn't appear in these differential equations. The reason for this is that S_3 and S_2 are *extensive* variables, which change slowly relative to their size—they scale as $\Theta(n)$, and their expected changes are $O(1)$. Thus their densities s_3 and s_2 are nonzero, and change continuously with time. In contrast, S_1 is typically $O(1)$ and fluctuates by $O(1)$ from step to step as unit clauses are satisfied and new ones are created. This makes a continuous approximation for S_1 impossible, and obligates us to model it with a discrete random process—namely, a branching process.

On each free step, setting x shortens any 2-clause in which x appears with the opposite value, and creates a unit clause. In expectation, x appears in $2S_2/(n-T)$ 2-clauses, half of which disagree with its value. Therefore, the expected number of new unit clauses we create is

$$\lambda = \frac{1}{2}\frac{2S_2}{n-T} = \frac{s_2}{1-t} = \frac{3}{2}\alpha t(1-t). \tag{14.27}$$

But the same thing happens on the forced steps—satisfying one unit clause creates an expected number λ of new ones. Thus each free step sets off a cascade of forced steps, which continues until all the unit clauses are satisfied.

The expected number of steps in a cascade, including the free step that set it off, is a geometric series

$$1+\lambda+\lambda^2+\cdots.$$

If $\lambda < 1$ and the branching process is subcritical, this sum converges to $1/(1-\lambda)$. The cascade dies out after satisfying all the unit clauses, `UC` breathes a sigh of relief, and we can take another free step.

On the other hand, if $\lambda > 1$ then this sum diverges, and with positive probability the cascade explodes. The unit clauses proliferate like the heads of a hydra, growing faster than we can satisfy them. As soon as two of them demand opposite truth values from the same variable, we have a contradiction and `UC` fails. Specifically, the Birthday Problem of Appendix A.3.3 shows that this occurs with constant probability as soon as the number of unit clauses reaches $\Theta(\sqrt{n})$.

Thus the probability `UC` succeeds is positive if and only if $\lambda < 1$ throughout the operation of the algorithm. Maximizing (14.27) over t gives

$$\lambda_{\max} = \frac{3}{8}\alpha,$$

and this is less than 1 if $\alpha < 8/3$.

We note that even if $\lambda < 1$ for all t, the probability that `UC` succeeds tends to a constant smaller than 1 in the limit $n \to \infty$. Even if there are only a constant number of unit clauses in each cascade, the probability that two of them contradict each other is $\Theta(1/n)$. Integrating this over the $\Theta(n)$ steps of the algorithm gives a positive probability of failure. However, the probability of success is also positive, and Corollary 14.3 then implies that $F(n, m = \alpha n)$ is satisfiable with high probability whenever $\alpha < 8/3$.

This analysis also tells us something about when a `DPLL` algorithm needs to backtrack. Since `UC` is exactly like the first branch of a simple `DPLL` algorithm, we have shown that for $\alpha < 8/3$ we can find a satisfying assignment with no backtracking at all with positive probability. Indeed, it can be shown that, with high probability, only $o(n)$ backtracking steps are necessary, so `DPLL`'s running time is still linear in this regime.

In contrast, if $\alpha > 8/3$ the probability that a single branch succeeds is exponentially small. If the branches were independent, this would imply that it takes exponential time to find a satisfying assignment. The branches of a search tree are, in fact, highly correlated, but one can still compute the expected running time and show that it is exponential. As long as its branching rule is sufficiently simple, the differential equation method tells us how to locate the critical density where a given `DPLL` algorithm goes from linear to exponential time.

```
SC
input: a 3-SAT formula φ
output: "satisfiable" or "don't know"
begin
   while φ contains unsatisfied clauses do
      if φ contains any unit clauses then
         choose c = (x) or (x̄) uniformly from the unit clauses ;
         satisfy c by setting x ;
      else if there are any 2-clauses then
         choose c uniformly from among the 2-clauses ;
         choose x uniformly between c's two variables ;
         set x to the value that satisfies c ;
      else
         choose x uniformly from among the unset variables ;
         set x to a random value ;
      remove or shorten the clauses containing x ;
      if φ contains an empty clause then return "don't know" ;
   end
   return "satisfiable";
end
```

FIGURE 14.12: The SC (Short Clause) algorithm. In addition to satisfying unit clauses immediately, it prioritizes 2-clauses over 3-clauses.

14.3.2 *Smarter Algorithms*

Let's use differential equations to analyze an algorithm that is a little smarter than UC. Consider the algorithm shown in Figure 14.12, called SC for Short Clause. Unlike UC, where free steps simply set a random variable, SC gives priority to the 2-clauses, choosing and satisfying a random one if any exist. The goal is to satisfy as many 2-clauses as possible before they turn into unit clauses, and thus keep the ratio λ of the branching process as small as possible.

As for UC, we claim that at any point in the algorithm the remaining formula is uniformly random conditioned on S_3, S_2, S_1, and T. The only question is how these variables change in each step. At very low densities, the 2-clauses follow a subcritical branching process and $S_2 = O(1)$, but at the densities we care about we have $S_2 = \Theta(n)$. Thus we again model S_3 and S_2 with differential equations, and S_1 with a branching process.

Since every free step of SC is guaranteed to satisfy a 2-clause, the differential equations describing SC are the same as those in (14.26) for UC except that $\mathrm{d}s_2/\mathrm{d}t$ has an additional term $-p_{\text{free}}$, where p_{free} is the probability that the current step is free rather than forced. As in our discussion of UC, each free step of SC sets off a cascade of forced steps, creating $\lambda = s_2/(1-t)$ new unit clauses in expectation.

If we look at any interval in which λ stays relatively constant, the expected fraction of steps which are free is 1 divided by the total expected number of steps in this cascade, including the free step that started it. Thus

$$p_{\text{free}} = 1\Big/\Big(1+\lambda+\lambda^2+\cdots\Big) = 1-\lambda = 1-\frac{s_2}{1-t},$$

and the new system of differential equations is

$$\begin{aligned}\frac{ds_3}{dt} &= -\frac{3s_3}{1-t}\\ \frac{ds_2}{dt} &= \frac{(3/2)s_3-2s_2}{1-t} - p_{\text{free}} = \frac{(3/2)s_3-s_2}{1-t}-1.\end{aligned} \tag{14.28}$$

Solving with the initial conditions $s_3(0)=\alpha$ and $s_2(0)=0$ gives

$$\begin{aligned}s_3(t) &= \alpha(1-t)^3\\ s_2(t) &= (1-t)\left(\frac{3}{4}\alpha t(2-t)+\ln(1-t)\right),\end{aligned}$$

and so

$$\lambda = \frac{s_2}{1-t} = \frac{3}{4}\alpha t(2-t)+\ln(1-t).$$

Maximizing λ over t gives (exercise!)

$$\lambda_{\max} = \frac{3\alpha}{4} - \frac{1}{2}\left(1+\ln\frac{3\alpha}{2}\right),$$

which crosses 1 when $\alpha = 3.003$...

Thus SC gives an improved lower bound on the 3-SAT threshold of

$$\alpha_c > 3.003. \tag{14.29}$$

We can also consider a DPLL algorithm whose branching rule is based on SC rather than UC. Namely, after performing unit propagation, it chooses a variable x from a random 2-clause, and first tries setting x to the truth value that satisfies that clause. The running time of this algorithm goes from linear to exponential time at $\alpha = 3.003$ rather than 8/3, and this is borne out by experiment. Indeed, this is essentially the branching rule we used in Figure 14.3.

14.3.3 *When Can We Use Differential Equations? Card Games and Randomness*

We claimed that throughout the operation of algorithms like UC and SC, the remaining formula is uniformly random conditioned on just a few parameters, such as the number of variables and the number of clauses of each length. How can we justify this claim, and for what kinds of algorithm does it hold? Let's explore an analogy that we learned from Dimitris Achlioptas.

You and I are playing a card game. I have a deck of cards, each of which has a literal x or $\overline{x}$ on its face. I deal m hands of 3 cards each, and place these hands face-down on a table. On each turn, you may perform one of two kinds of moves. You may choose a card and ask me to turn it over, revealing what variable x is on it and whether or not it is negated. Or, you may simply call out a variable x. In both cases, I then turn over every card with x or $\overline{x}$ on it.

Next, you choose a truth value for x using any technique you like. I then remove every card with x or $\overline{x}$ on it from the table. If your chosen value agrees with the literal on a given card, I remove its entire hand from the table, because the corresponding clause has been satisfied. If your value disagrees with that card, I leave the rest of its hand on the table, because that clause has been shortened.

Any card which is turned face-up is removed before the next turn begins. Therefore, at the end of the turn, *all the remaining cards are face down.* As a result, while you know the number of hands of each size, you know nothing at all about the literals printed on them, and the remaining clauses are chosen from the unset variables just as randomly as the original clauses were. Using the principle of deferred decisions, rather than turning cards over, I might as well be printing random literals on them on the fly.

Clearly this game allows for a wide variety of algorithms. For instance, unit clause propagation consists of always turning over the last remaining card in a hand. If there are no hands with only one card left, `UC` calls out the name of a random variable, while `SC` chooses a hand with two remaining cards and turns over one of them. In addition to choosing your variable, you could choose its value in a variety of ways. For example, after I turn over all the cards with x or $\overline{x}$ on them, you could set x's value according to the majority of these cards.

We can analyze algorithms that take more information into account by adopting a more sophisticated model of random formulas. For instance, we can consider random formulas where we know the degree of each literal, i.e., how many clauses it appears in. This is like showing you the deck at the beginning of the game, and then shuffling it before I deal. You know how many cards with each literal are on the table, but not how these cards are permuted among the hands.

This picture allows us to analyze algorithms that choose variables and their values based on their degree. For instance, if x has many positive appearances and just a few negative ones, setting $x = \texttt{true}$ will satisfy many clauses and shorten just a few. Similarly, if you want to 3-color a graph, it makes sense to color the high-degree vertices first. Typically, these algorithms correspond to large systems of coupled differential equations which have to be integrated numerically, but this approach does indeed yield better lower bounds on α_c.

However, even these more elaborate models restrict us to algorithms that perform no backtracking. Backtracking is like peeking at the cards, or turning face-up cards face-down again. Doing this builds up a complicated body of information about the cards, creating a probability distribution that cannot be parametrized with just a few numbers. This is precisely why the branches of a `DPLL` search tree are correlated with each other, since following one branch gives us a great deal of information about how the next branch will fare. Thus as far as we know, the technique of differential equations is limited to linear-time algorithms that set each variable once.

14.10

14.4 Magic Moments

We have shown how to prove lower bounds on the satisfiability threshold with algorithms and differential equations. In this section, we will show how to prove upper and lower bounds using two simple yet powerful methods from discrete probability: the first and second moment methods. While they do not determine α_c for 3-SAT exactly, they work extremely well for k-SAT when k is large, confining α_c to a small interval near $2^k \ln 2$.

In addition, the second moment method gives an early clue about the structure of the set of satisfying assignments—namely, that at a certain density it falls apart into an exponential number of isolated clusters. This fits with a deep, albeit nonrigorous, treatment of this problem using the methods of statistical physics, which we explore in Sections 14.6 and 14.7. As an introduction to the first and second moment methods, we offer Appendix A.3.

14.4.1 *Great Expectations: First Moment Upper Bounds*

We begin by proving an upper bound on α_c for 3-SAT. Given a random formula $F_k(n, m = \alpha n)$, let Z be the number of satisfying assignments. Understanding the entire probability distribution of Z seems very difficult—after all, we don't know how to rigorously find the threshold α_c at which $\Pr[Z > 0]$ jumps from 1 to 0, or even prove that it exists.

On the other hand, as we discuss in Appendix A.3.2, the probability that Z is positive is at most its expectation:

$$\Pr[Z > 0] \leq \mathbb{E}[Z] . \tag{14.30}$$

And, as for many random variables with complicated distributions, $\mathbb{E}[Z]$ is very easy to calculate. There are 2^n possible truth assignments σ, and by symmetry they are all equally likely to satisfy the formula. The clauses are chosen independently, so the probability that a given σ satisfies all m of them is the mth power of the probability that it satisfies a particular one. Since a clause is violated by σ only if it disagrees with σ on all 3 of its variables, σ satisfies a random clause with probability 7/8. Thus the expected number of satisfying assignments is $2^n (7/8)^m$.

To see this more formally, we define an *indicator random variable* Z_σ for each truth assignment σ,

$$Z_\sigma = \begin{cases} 1 & \text{if } \sigma \text{ is a satisfying assignment} \\ 0 & \text{if it isn't.} \end{cases}$$

Then $\mathbb{E}[Z_\sigma]$ is the probability that σ is satisfying, and $Z = \sum_\sigma Z_\sigma$. Using linearity of expectation and the independence of the clauses then gives

$$\begin{aligned} \mathbb{E}[Z] &= \mathbb{E}\left[\sum_\sigma Z_\sigma\right] = \sum_\sigma \mathbb{E}[Z_\sigma] \\ &= \sum_\sigma \prod_c \Pr[\sigma \text{ satisfies } c] \\ &= 2^n (7/8)^m . \end{aligned}$$

Now setting $m = \alpha n$ gives

$$\mathbb{E}[Z] = \left(2(7/8)^\alpha\right)^n . \tag{14.31}$$

If $2(7/8)^\alpha < 1$ then $\mathbb{E}[Z]$ is exponentially small, and by (14.30) so is the probability that $F_k(n, m = \alpha n)$ is satisfiable. This gives us our first upper bound on the 3-SAT threshold,

$$\alpha_c \leq \frac{\ln 2}{\ln 8/7} \approx 5.191 .$$

This upper bound is far above the conjectured value $\alpha_c \approx 4.267$. And yet, we made no approximations—our computation (14.31) of the expected number of solutions is exact, and for $\alpha < 5.19$ it really is exponentially large. What's going on?

The answer is that in the range $\alpha_c < \alpha < 5.19$, random formulas usually have no solutions—but with exponentially small probability, they have exponentially many. This skews the probability distribution to such an extent that the expectation of Z is exponentially large, even though $Z = 0$ almost all the time.

There are various ways to suppress this exponential skewness. For instance, we can compute the expectation of the number of *locally maximal* satisfying assignments—that is, those where flipping any variable from `false` to `true` would violate some clause. We can also condition on the event that the formula is typical in certain ways, such as including roughly the expected number of variables of each degree. As Problems 14.25–14.27 shows, approaches like these yield better upper bounds on α_c.

14.11 But in order to prove a *lower* bound on α_c, it is not enough to show that Z has a large expectation. We also need to bound its variance, and that is exactly what we will do next.

14.4.2 Closing the Asymptotic Gap: the Second Moment

Let's move from 3-SAT to k-SAT, and focus on random formulas $F_k(n, m = \alpha n)$. Since the probability that a given truth assignment σ satisfies a random clause is $1 - 2^{-k}$, our calculation of the expected number of solutions easily generalizes to

$$\mathbb{E}[Z] = 2^n (1 - 2^{-k})^m = \left(2(1 - 2^{-k})^{\alpha}\right)^n . \tag{14.32}$$

This gives the following upper bound on the k-SAT threshold,

$$\alpha_c \le \frac{\ln 2}{-\ln(1 - 2^{-k})} . \tag{14.33}$$

As the following exercise shows, this bound grows exponentially with k:

Exercise 14.10 *Show that when k is large,* (14.33) *becomes*

$$\alpha_c \le 2^k \ln 2 - \frac{\ln 2}{2} - O(2^{-k}) . \tag{14.34}$$

Hint: use the Taylor series of $\ln(1 - \varepsilon)$.

What about a lower bound? As Problem 14.18 shows, we can generalize our analysis of the UC algorithm in Section 14.3.1 to show that

$$\alpha_c \ge \frac{2^k}{k} . \tag{14.35}$$

However, as far as we know, smarter algorithms only improve this by a constant. Thus there is a factor of k between our upper and lower bounds. How can we close this gap? Experimentally, α_c seems to be much closer to (14.34) than to (14.35). Can we prove a matching lower bound?

In this section, we describe how to prove that $\alpha_c = \Theta(2^k)$ using the *second moment method.* As shown in Appendix A.3.5, if we have a nonnegative random variable Z, the probability that it is positive is bounded below by

$$\Pr[Z > 0] \ge \frac{\mathbb{E}[Z]^2}{\mathbb{E}[Z^2]} . \tag{14.36}$$

Our goal is to show that, below a certain α, this ratio is bounded above zero. Equivalently, we want to show that Z has a large expectation, and that its variance $\mathbb{E}[Z^2]-\mathbb{E}[Z]^2$ is not too large. If it is nonzero with constant probability at some density α^*, then $F_k(n, m=\alpha n)$ is satisfiable with high probability for all $\alpha<\alpha^*$ by Corollary 14.3, and $\alpha_c \geq \alpha^*$.

We can calculate the second moment $\mathbb{E}[Z^2]$ using our indicator random variables Z_σ. For any pair (σ, τ) of truth assignments, $\mathbb{E}[Z_\sigma Z_\tau]$ is the probability that they are both satisfying, and we have

$$\begin{aligned}\mathbb{E}[Z^2] &= \mathbb{E}\left[\left(\sum_\sigma Z_\sigma\right)\left(\sum_\tau Z_\tau\right)\right] = \mathbb{E}\left[\sum_{\sigma,\tau} Z_\sigma Z_\tau\right] = \sum_{\sigma,\tau} \mathbb{E}[Z_\sigma Z_\tau] \\ &= \sum_{\sigma,\tau} \Pr[\sigma, \tau \text{ both satisfy } F_k(n,m)]\,.\end{aligned}$$

Thus $\mathbb{E}[Z^2]$ is the expected number of ordered pairs (σ, τ) of truth assignments that both satisfy the formula. Since the clauses are independent, we can write

$$\mathbb{E}[Z^2] = \sum_{\sigma,\tau} \left(\Pr[\sigma, \tau \text{ both satisfy a random clause } c]\right)^m\,. \tag{14.37}$$

However, the probability that σ and τ both satisfy c is not just the product of these probabilities. These events are correlated, and the correlations depends on how similar σ and τ are. To see this, note that

$$\Pr[\sigma, \tau \text{ both satisfy } c] = \Pr[\sigma \text{ satisfies } c]\, \Pr[\tau \text{ satisfies } c \mid \sigma \text{ satisfies } c]\,.$$

While the probability that σ satisfies c is the same for all σ, the conditional probability that τ does too is larger if σ and τ have a lot in common—that is, if they agree on most variables. Using the inclusion–exclusion principle (see Appendix A.3.1),

$$\Pr[\sigma, \tau \text{ both satisfy } c] = 1 - \Pr[\sigma \text{ violates } c] - \Pr[\tau \text{ violates } c] + \Pr[\sigma, \tau \text{ both violate } c]\,.$$

To compute these probabilities, let the *overlap* z be the number of variables on which σ and τ agree. Alternately, the Hamming distance between them is $n-z$. Then σ and τ both violate c if and only if (1) σ and τ agree on all k of c's variables, and (2) c disagrees with them on all k. If we ignore the requirement that the variables in each clause are distinct—which makes no difference when n is large—then the probability that both of these events happen is $(z/n)^k\, 2^{-k}$. Thus we have

$$\Pr[\sigma, \tau \text{ satisfy } c] = 1 - 2^{1-k} + (z/n)^k\, 2^{-k}\,.$$

In what follows, we will write $\zeta = z/n$ for the fraction of variables on which σ and τ overlap. Then we denote this probability

$$q(\zeta) = \Pr[\sigma, \tau \text{ satisfy } c] = 1 - 2^{1-k} + \zeta^k\, 2^{-k}\,. \tag{14.38}$$

We can now write (14.37) as a sum over all 2^n choices of σ, and over all $\binom{n}{z}$ choices of τ whose overlap with σ is z. Then

$$\mathbb{E}[Z^2] = 2^n \sum_{z=0}^{n} \binom{n}{z} q(z/n)^m\,. \tag{14.39}$$

Now that we have an expression for $\mathbb{E}[Z^2]$, when does all this work?

What matters is whether the sum in (14.39) is dominated by terms with overlap $z \approx n/2$. To see this, first note that if

$$p = 1 - 2^{-k}$$

denotes the probability that a single assignment satisfies a random clause, then

$$q(1/2) = p^2 .$$

In other words, if σ and τ agree on $n/2$ variables, they satisfy c with the same probability as if they were chosen independently. This stands to reason, since if they were independent their typical overlap would be $z = n/2 + O(\sqrt{n})$. Since $\binom{n}{n/2} \sim 2^n/\sqrt{n}$, if (14.39) is dominated by $\sqrt{n}$ terms near $z = n/2$ we then have

$$\mathbb{E}[Z^2] \sim \sqrt{n}\, 2^n \binom{n}{n/2} q(1/2)^m \sim (2^n p^m)^2 = \mathbb{E}[Z]^2 .$$

In that case $\mathbb{E}[Z]^2/\mathbb{E}[Z^2]$ is a constant, and (14.36) shows that the formula is satisfiable with constant probability.

To evaluate the sum (14.39), we start by applying Stirling's approximation $n! \approx \sqrt{2\pi n}\, n^n \mathrm{e}^{-n}$ to the binomial. As discussed in Appendix A.4.3, this gives

$$\binom{n}{z} \approx \frac{1}{\sqrt{2\pi n \zeta(1-\zeta)}} \mathrm{e}^{n h(\zeta)} ,$$

where $h(\zeta)$ is the entropy function,

$$h(\zeta) = -\zeta \ln \zeta - (1-\zeta) \ln(1-\zeta) . \qquad (14.40)$$

Thus if we set $m = \alpha n$ and define the functions

$$f(\zeta) = \frac{1}{\sqrt{\zeta(1-\zeta)}}$$

and

$$\phi(\zeta) = \ln 2 + h(\zeta) + \alpha \ln q(\zeta) ,$$

then (14.39) becomes

$$\mathbb{E}[Z^2] \approx \frac{1}{\sqrt{2\pi n}} \sum_{z=0}^{n} f(z/n) \mathrm{e}^{n\phi(z/n)} .$$

Finally, replacing the sum over z by an integral over ζ gives

$$\mathbb{E}[Z^2] \approx \sqrt{\frac{n}{2\pi}} \int_0^1 f(\zeta) \mathrm{e}^{n\phi(\zeta)} \,\mathrm{d}\zeta .$$

As described in Appendix A.6.1, asymptotic integrals of this kind can be approximated using Laplace's method. In the limit $n \to \infty$, the integral is dominated by an interval of width $1/\sqrt{n}$ around the $\zeta_{\max}$ that maximizes $\phi(\zeta)$. In this region, we can treat $e^{n\phi(\zeta)}$ as a Gaussian and $f(\zeta)$ as a constant. Applying (A.32) then gives

$$\int_0^1 f(\zeta)\,e^{n\phi(\zeta)}\,d\zeta \approx \sqrt{\frac{2\pi}{n\left|\phi''(\zeta_{\max})\right|}}\, f(\zeta_{\max})\,e^{n\phi(\zeta_{\max})}.$$

Happily, the factors of $\sqrt{n}$ cancel, and we have

$$\mathbb{E}[Z^2] \approx \frac{f(\zeta_{\max})}{\sqrt{\left|\phi''(\zeta_{\max})\right|}}\, e^{n\phi(\zeta_{\max})} = C e^{n\phi(\zeta_{\max})}, \tag{14.41}$$

for some constant C.

Mirroring our discussion above, the question is then whether ϕ is maximized at $\zeta = 1/2$. Since we have (exercise!)

$$e^{n\phi(1/2)} = (2p^{\alpha})^{2n} = \mathbb{E}[Z]^2,$$

if $\zeta_{\max} = 1/2$ then (14.36) yields

$$\Pr[Z > 0] \geq 1/C.$$

Having followed this beautiful exposition up to now, the reader will be crushed to see in Figure 14.13 that this approach simply fails. For any nonzero density α, the function $\phi(\zeta)$ has a positive derivative at $1/2$, and so $\phi(\zeta_{\max}) > \phi(1/2)$. Because of this, $\mathbb{E}[Z^2]$ is exponentially larger than $\mathbb{E}[Z]^2$, and the ratio $\mathbb{E}[Z]^2/\mathbb{E}[Z^2]$ is exponentially small.

The problem is that $q(\zeta)$ is proportional to the correlation between the events that σ and τ are satisfying assignments, and this correlation increases monotonically with the overlap ζ. In essence, there is an "attractive force" between satisfying assignments which encourages them to overlap as much as possible.

To see why this is, imagine choosing a truth assignment σ by flipping n coins, one for each variable x. If we want to maximize the chances that σ is satisfying, we should bias x's coin towards the majority of x's appearances in the formula so that it comes up `true` with some probability $p_i \neq 1/2$. But if we choose another truth assignment τ by flipping the same coins, σ and τ will agree on x_i with probability $p_i^2 + (1-p_i)^2 > 1/2$. Thus if we choose σ and τ uniformly from the set of satisfying assignments, their typical overlap is greater than $1/2$. In the next section, we will see how we can repair the second moment method by canceling this attraction out.

14.4.3 Symmetry Regained

The problem with SAT is that the more true literals there are in a clause, the happier it is. This breaks the symmetry between true and false, and creates an attraction between satisfying assignments.

We can restore this symmetry, and cancel out this attraction, by focusing on a slightly different problem: our old friend NAESAT from Chapter 5. Rather than demanding that at least one literal in each clause is true, NAESAT demands that at least one is true and at least one is false.

Just as we did for k-SAT, we can form a random NAE-k-SAT formula $F_k(n,m)$ by choosing with replacement from the $2^k\binom{n}{k}$ possible clauses. Since σ violates a NAE-k-SAT clause c if it disagrees *or agrees* with all of c's literals, the probability that it satisfies a random clause is

$$p = 1 - 2^{1-k}.$$

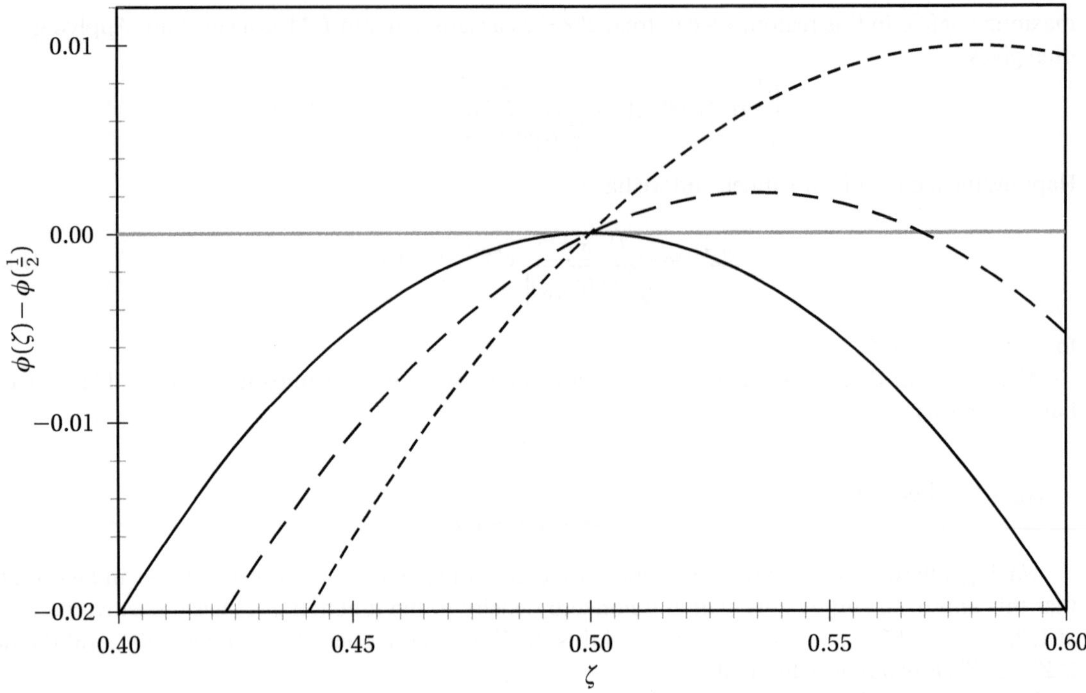

FIGURE 14.13: The function $\phi(\zeta)$ for 3-SAT at densities $\alpha=0$, 1, and 2, with the value $\phi(1/2)$ corresponding to $\mathbb{E}[Z]^2$ subtracted off. For $\alpha>0$, $\phi'(1/2)>0$ and $\phi(\zeta_{\max})>\phi(1/2)$. Thus $\mathbb{E}[Z^2]$ is exponentially larger than $\mathbb{E}[Z]^2$, and the second moment method fails.

As before, the expected number of satisfying assignments is $\mathbb{E}[X]=(2p^{\alpha})^n$. This gives the following upper bound for the NAE-k-SAT threshold,

$$\alpha_c^{\rm NAE} \le \frac{\ln 2}{-\ln(1-2^{1-k})} = 2^{k-1}\ln 2 - \frac{\ln 2}{2} - O(2^{-k}) . \qquad (14.42)$$

Note that this differs from the corresponding bound (14.34) for k-SAT by a factor of 2. This is essentially because NAE-k-SAT clauses forbid twice as many truth assignments as k-SAT clauses do.

Next we turn to the second moment. If τ is a satisfying assignment then its complement $\overline{\tau}$ is also satisfying. Since this flips the overlap of σ and τ from ζ to $1-\zeta$, the probability that two assignments with overlap ζ both satisfy a random clause is now symmetric around $\zeta=1/2$:

$$q(\zeta)=q(1-\zeta)=1-2^{2-k}+(\zeta^k+(1-\zeta)^k)2^{1-k} .$$

This is analogous to (14.38), except the third term is now the probability that σ and τ agree or disagree on all k variables.

As before, we define

$$\phi(\zeta)=\ln 2+h(\zeta)+\alpha\ln q(\zeta) .$$

k	3	4	5	6	7	8	9	10
upper	2.450	5.191	10.741	21.833	44.014	88.376	177.099	354.545
lower	3/2	49/12	9.973	21.190	43.432	87.827	176.570	354.027

TABLE 14.2: Upper and lower bounds, from the first and second moment methods respectively, on the threshold α_c^{NAE} for NAE-k-SAT. The gap between them approaches 1/2 as k increases.

This is a sum of an entropic term $h(\zeta)$, which is concave, and a *correlation* term $\alpha \ln q(\zeta)$, which is convex. When α is small enough, the entropy dominates, $\phi(\zeta)$ is maximized at $\zeta = 1/2$, and the second moment method succeeds. When α is too large, on the other hand, a pair of maxima appear on either side of $\zeta = 1/2$, and when these exceed $\phi(1/2)$ the second moment fails.

As Figure 14.14 shows, for $k = 3$ these side maxima are born continuously out of $\zeta = 1/2$. When $\phi''(1/2)$ changes from negative to positive, $\phi(1/2)$ changes from the maximum to a local minimum. The same is true for $k = 4$. For $k \geq 5$, on the other hand, there is a range of α with three local maxima, one at $\zeta = 1/2$ and one on either side. The second moment succeeds as long as $\phi(1/2)$ is still the global maximum.

By finding the value of α at which $\phi(1/2)$ ceases to be the global maximum, we can obtain a lower bound on the NAE-k-SAT threshold for each k. We show these bounds in Table 14.2. As k grows, the gap between the first moment upper bound and the second moment lower bound converges to 1/2, allowing us to pinpoint the NAE-k-SAT threshold almost exactly. Specifically, by approximating the locations of the side maxima—a calculus problem which we omit here—one can show that

$$2^{k-1}\ln 2 - \frac{1+\ln 2}{2} - O(2^{-k}) \leq \alpha_c^{\text{NAE}} \leq 2^{k-1}\ln 2 - \frac{\ln 2}{2}. \tag{14.43}$$

Thus NAESAT's symmetry allows us to determine its threshold to great precision. But what have we learned about our original quarry, the threshold α_c for k-SAT?

In fact, $\alpha_c^{\text{NAE}} \leq \alpha_c$ so our lower bounds on α_c^{NAE} are also lower bounds on α_c. To see this, note that if a NAESAT formula is satisfied, then so is the corresponding SAT formula—since, among other things, NAESAT requires that at least one literal in each clause is true. To put this differently, instead of defining Z in this section as the number of satisfying assignments of a random NAESAT formula, we could have defined it as the number of satisfying assignments of a random SAT formula whose complements are also satisfying. If such an assignment exists, the formula is satisfiable.

So combining (14.34) and (14.43), we have

$$2^{k-1}\ln 2 - O(1) \leq \alpha_c \leq 2^k \ln 2 .$$

This pins down α_c to within a factor of 2, and confirms that $\alpha_c = \Theta(2^k)$. In the next section, we use yet another technique to close this factor of 2, and determine α_c within a factor of $1 + o(1)$.

14.4.4 *Weighted Assignments: Symmetry Through Balance*

In the previous sections, we saw that the second moment method fails for k-SAT because of a lack of symmetry. We restored this symmetry by focusing on NAESAT, or equivalently by requiring that every satisfying assignment remain satisfying it we flip all its variables. But this global symmetry came at a

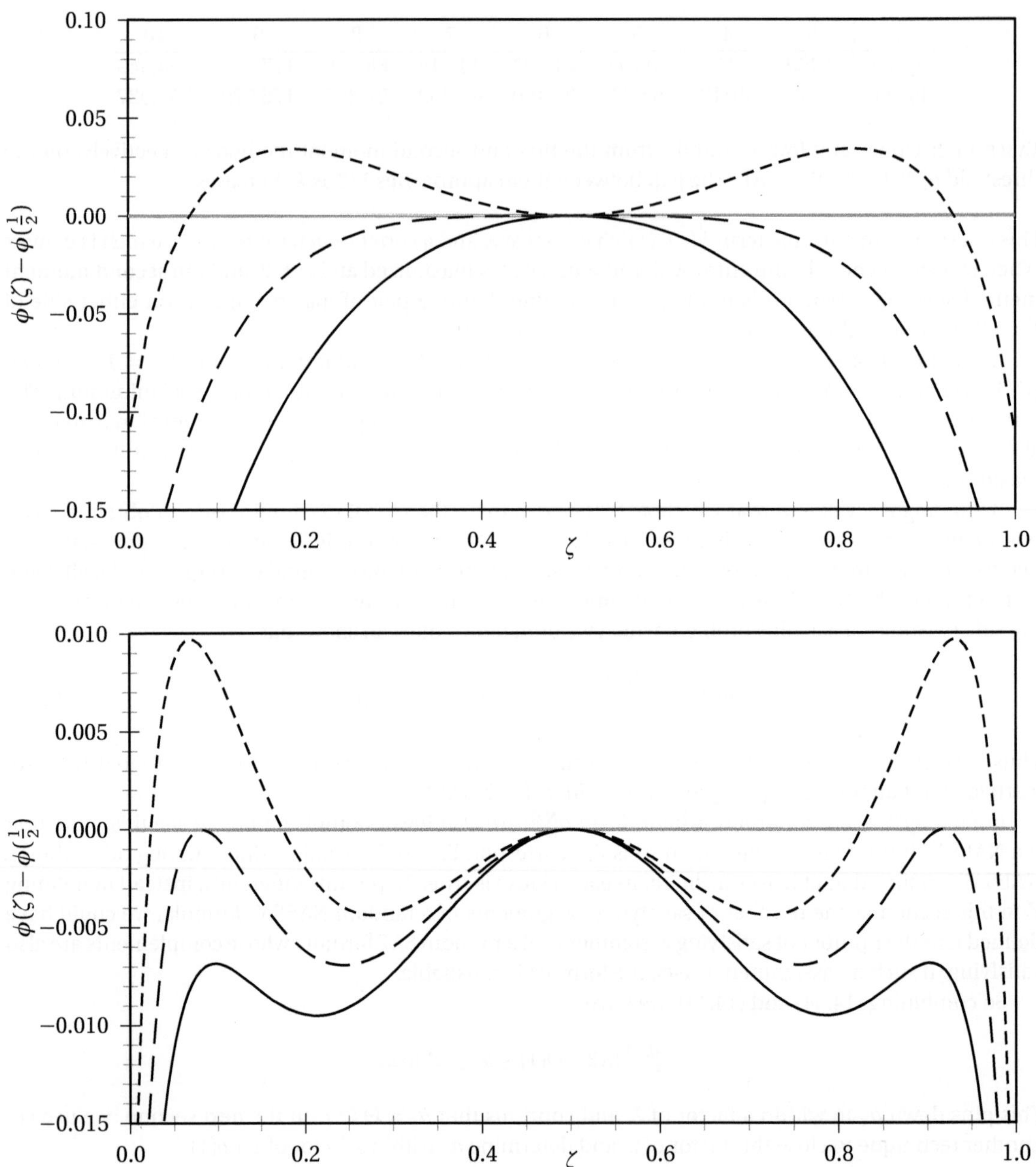

FIGURE 14.14: The function $\phi(\zeta)$ for NAE-k-SAT. Unlike $\phi(\zeta)$ for k-SAT, this function is symmetric around $\zeta = 1/2$, so when α is small enough $\phi(1/2)$ is the global maximum and the second moment method succeeds. Above, $k=3$ and $\alpha = 1$, 1.5, and 2. Below, $k=5$ and $\alpha = 9.8$, 9.973, and 10.2.

cost—it made the clauses twice as restrictive, and decreased the threshold by a factor of 2. In this section, we will restore symmetry in a more delicate way, and determine the k-SAT threshold more precisely.

Suppose we give each satisfying assignment σ a positive weight $w(\sigma)$, and let Z be the sum of all these weights. Then the formula is satisfiable whenever $Z > 0$. Our goal is then to cancel out the attraction between satisfying assignments by defining this weight correctly. Our move from SAT to NAESAT corresponds to setting $w(\sigma)=1$ if both σ and $\overline{\sigma}$ are satisfying and 0 otherwise. However, there is no reason to limit ourselves to integer-valued weights, since the second moment method works just as well for real-valued random variables.

The following choice turns out be especially good, and even optimal in a certain sense. Given a truth assignment σ, we define its weight as a product over all clauses c,

$$w(\sigma)=\prod_c w_c(\sigma), \tag{14.44}$$

where

$$w_c(\sigma)=\begin{cases} 0 & \text{if } \sigma \text{ does not satisfy } c \\ \eta^t & \text{if } \sigma \text{ makes } t \text{ of } c\text{'s literals true}. \end{cases}$$

We will compute the first and second moments of

$$Z_\eta=\sum_\sigma w(\sigma).$$

If $\eta=1$, then $w(\sigma)$ is simply 1 if σ is satisfying and 0 otherwise, and Z_η is just the number of satisfying assignments. If $\eta<1$, on the other hand, Z_η discounts assignments that satisfy too many literals. As we will see, the right value of η causes a balance between true literals and false ones, achieving the same symmetry that NAESAT achieves by fiat.

Let's calculate the first and second moments of Z_η. In analogy with our previous calculations, define $p_\eta=\mathbb{E}[w_c(\sigma)]$. Then summing over the number t of true literals, each of which is satisfied by σ with probability $1/2$, we have

$$p_\eta=\mathbb{E}[w_c(\sigma)]=2^{-k}\sum_{t=1}^{k}\binom{k}{t}\eta^t=\frac{(1+\eta)^k-1}{2^k}.$$

As before, linearity of expectation and the fact that clauses are chosen independently gives

$$\mathbb{E}[Z_\eta]=\sum_\sigma\prod_c\mathbb{E}[w_c(\sigma)]=2^n\,\mathbb{E}[w_c(\sigma)]^m=(2p_\eta^\alpha)^n.$$

For the second moment, analogous to (14.39) we have

$$\mathbb{E}[Z_\eta^2]=\sum_{\sigma,\tau}\mathbb{E}[w_c(\sigma)w_c(\tau)]=2^n\sum_{z=0}^{n}\binom{n}{z}q_\eta(z/n),$$

where, if σ and τ agree on a fraction ζ of variables,

$$q_\eta(\zeta)=\mathbb{E}[w_c(\sigma)w_c(t)]=\left(\zeta\left(\frac{1+\eta^2}{2}\right)+(1-\zeta)\eta\right)^k-2^{1-k}\left(\zeta+(1-\zeta)\eta\right)^k+2^{-k}\zeta^k. \tag{14.45}$$

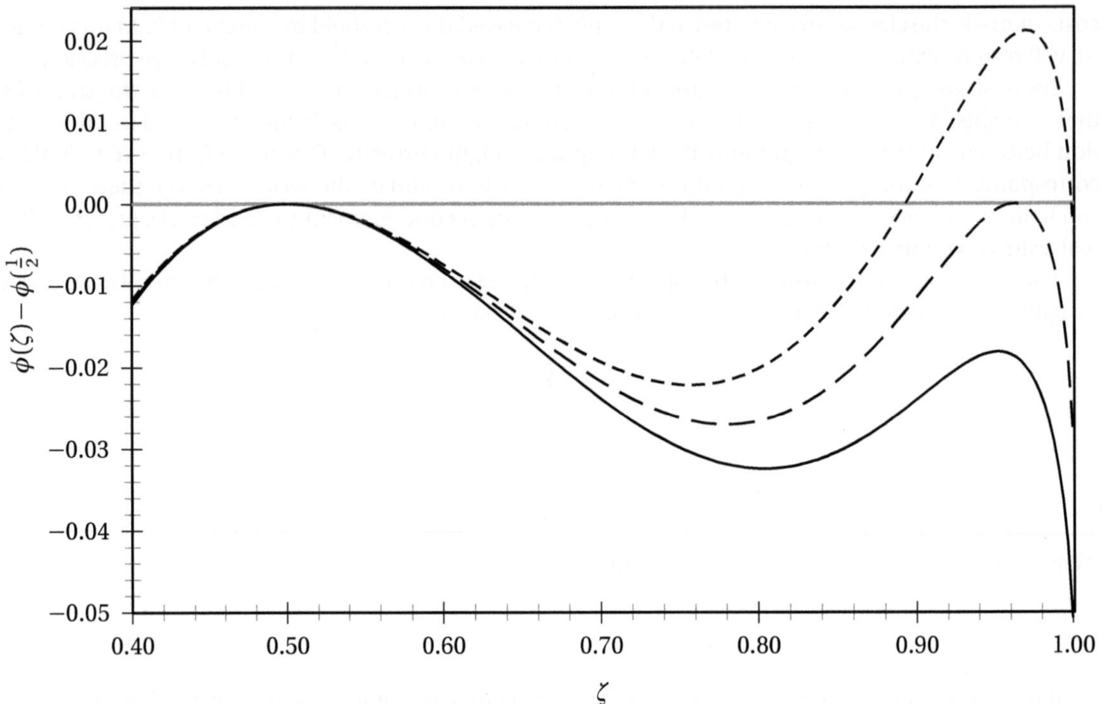

FIGURE 14.15: The function $\phi_\eta(\zeta)$ for k-SAT, with $k=5$ and $\alpha=17$, 17.617, and 18.3. Here η is tuned so that $\phi'(1/2)=0$, so $\zeta=1/2$ is a local maximum. Until the side maximum surpasses it, it is the global maximum and the second moment method succeeds.

Exercise 14.11 *Derive* (14.45) *using the inclusion–exclusion principle.*

As before, $q_\eta(1/2)=p_\eta^2$. Moreover, for the right value of η we have $q_\eta'(1/2)=0$, so that $q(\zeta)$ is locally symmetric—that is, symmetric up to second order—around $\zeta=1/2$. Taking derivatives and simplifying, this occurs when η is the unique positive real root of

$$(1-\eta)(1+\eta)^{k-1}=1\,. \tag{14.46}$$

For $k=3$, for instance, this gives $\eta=(\sqrt{5}-1)/2=0.618...$, the reciprocal of the golden ratio. Then if we define

$$\phi_\eta(\zeta)=\ln 2+h(\zeta)+\alpha\ln q_\eta(\zeta)\,,$$

then since $h(\zeta)$ is symmetric, for this value of η we have $\phi_\eta'(1/2)=0$ as well. As Figure 14.15 shows, because of this local symmetry, $\phi_\eta(1/2)$ is a local maximum over a wide range of α. Another local maximum appears near $\zeta=1$, but until it exceeds $\phi_\eta(1/2)$ the second moment succeeds and the formula is satisfiable with positive probability.

This approach gives the lower bounds on α_c shown in Table 14.3. Comparing them with the first moment upper bound (14.33), we find that the gap between them grows as $O(k)$—a tiny amount of un-

k	3	4	5	6	7	8	9	10
upper	5.191	10.741	21.833	44.014	88.376	177.099	354.545	709.436
lower	2.547	7.313	17.617	39.026	82.637	170.627	347.352	701.533

TABLE 14.3: Upper and lower bounds on the threshold α_c for k-SAT from (14.33) and the weighted second moment method. The gap between them narrows to $O(k)$ as k increases.

certainty compared to their asymptotic growth. These bounds confine α_c to the window

$$2^k \ln 2 - O(k) \le \alpha_c \le 2^k \ln 2 - O(1), \tag{14.47}$$

and prove that

$$\alpha_c = (1 + o(1)) 2^k \ln 2 .$$

Thus, in the limit of large k, we can determine α_c almost exactly.

We can think of this weighted second moment method in another way. In Section 14.4.3, we restored 14.12
symmetry by counting satisfying assignments whose complements are also satisfying. As Problem 14.29 shows, we can think of Z_η as counting satisfying assignments that are *balanced* in the sense that they satisfy close to half of the literals in the formula. This prevents us from setting each variable equal to its majority value—and by removing this temptation, we also eliminate the extent to which σ and τ are attracted to each other.

It is important to note that the second moment method is *nonconstructive.* It proves that a satisfying assignment probably exists, but it gives us no hints as to how we might find one. The method of differential equations, on the other hand, is constructive in the sense that it establishes that a certain algorithm actually has a good chance of finding a satisfying assignment.

The second moment method also gives us our first hint about the geometry of the set of solutions and how it changes as α increases. Ignoring the weighting factor η, the side peak in Figure 14.15 shows that at a certain density many pairs of solutions have a large overlap, or equivalently a small Hamming distance. The appearance of this peak corresponds to an additional phase transition below the satisfiability transition, where the set of solutions breaks apart into clusters. A pair of solutions typically has an overlap of $\zeta_1 n$ if they are in the same cluster and $\zeta_2 n$ if they are in different clusters, for some pair of constants $\zeta_1 > \zeta_2$. Thus clusters consist of sets of similar solutions, and are widely separated from each other.

The clustering phenomenon first emerged from statistical physics using methods that we will explore in Sections 14.6 and 14.7. As we will discuss there, we believe that clustering prepares the ground for the hardness of random SAT. But first, let's discuss another NP-complete problem, whose phase transition we can understand completely using the first and second moment methods.

14.5 The Easiest Hard Problem

The phase transition in random k-SAT is too complex to be completely characterized by the techniques we have seen so far. In the sections after this one, we will get a glimpse of its true structure, and what methods might establish it rigorously. But before we embark on that mission, let's discuss a phase transition whose critical point can be determined exactly with the first and second moment methods: INTEGER PARTITIONING.

14.5.1 *The Phase Transition*

In case you don't remember INTEGER PARTITIONING, we repeat its definition here:

INTEGER PARTITIONING

Input: A list $S = \{a_1, \dots, a_n\}$ of positive integers

Question: Is there a *balanced partition*, i.e., a subset $A \subseteq \{1, \dots, n\}$ such that $\left|\sum_{j \in A} a_j - \sum_{j \notin A} a_j\right| \leq 1$?

If you compare this definition with the one in Section 4.2.3, you'll notice a slight difference. There we called a partition *balanced* if the sum of the left and right sides are exactly equal. Here we allow the absolute value of their difference,

$$D = \left|\sum_{j \in A} a_j - \sum_{j \notin A} a_j\right|,$$

to be 0 or 1. We do this because if the a_j are random integers then the total weight $\sum_j a_j$ is odd half the time, in which case $D = 1$ is the best balance we can hope for. We have also changed our notation slightly—in this section we use n to denote the number of weights $|S|$ rather than the total number of bits describing the instance.

We showed that INTEGER PARTITIONING is NP-complete in Section 5.3.5. However, in Section 4.2.3 we learned that the only hard instances are those where the weights a_j are large—where each one is a b-bit integer where b grows as some function of n. So let's construct our random instances by choosing each a_j uniformly and independently from $\{0, 1, \dots, 2^{b-1}\}$. Intuitively, the larger b is the less likely it is that a balanced partition exists, since the possible values of D are spread over a larger range. We will see that there is a phase transition at a critical value $b_c \approx n$—that is, along the diagonal in Figure 4.7 on page 107.

We start with an intuitive argument. We have $D < n\,2^b$, so D has at most $b + \log_2 n$ bits. Each of these bits represents a constraint on the partition: if $D \leq 1$ then each of D's bits must be zero, except its least significant bit which can be 0 or 1.

Let's pretend that these $b + \log_2 n$ constraints are independent, and that each one is satisfied by a random choice of A with probability $1/2$. This is true of the b least significant bits, and it would be true of the remaining $\log_2 n$ bits if we could ignore correlations generated by carry bits when we add the a_j together. Since there are 2^n possible partitions, the expected number of balanced partitions is then

$$\mathbb{E}[Z] = 2^{n-(b+\log_2 n)}.$$

Setting $\mathbb{E}[Z] = 1$ suggests that the critical value of b is

$$b_c \approx n - \log_2 n.$$

However, this argument slightly overestimates the number of bits in D. If we flip a coin for each a_j, including or excluding it from the subset A with equal probability, then

$$D = \pm a_1 \pm a_2 \pm \cdots \pm a_n = \sum_{j=1}^{n} \sigma_j a_j,$$

where $\sigma_j = \pm 1$ for each j. The value of D is like the position after n steps of a random walk, where on the jth step we move a distance a_j to the left or right. If each of these steps were ± 1 instead of $\pm a_j$, D would scale as $\sqrt{n}$ (see Appendix A.4.4), so we have $D \sim \sqrt{n}\, 2^b$. In that case, D has $b + (1/2)\log_2 n$ bits, and again assuming that each one is zero with probability 1/2 gives

$$\mathbb{E}[Z] = 2^{n-(b+(1/2)\log_2 n)} = \Theta\left(\frac{2^{n-b}}{\sqrt{n}}\right). \tag{14.48}$$

Our prediction for b_c then becomes

$$b_c = n - \frac{1}{2}\log_2 n. \tag{14.49}$$

Just as we defined the ratio $\alpha = m/n$ of clauses to variables for k-SAT, let's define the ratio

$$\kappa = \frac{b}{n}. \tag{14.50}$$

Our nonrigorous argument above suggests that the critical value of κ is

$$\kappa_c = 1 - \frac{\log_2 n}{2n}.$$

The term $(\log_2 n)/(2n)$ is a *finite-size effect*. It doesn't affect the limiting value $\lim_{n\to\infty} \kappa_c = 1$, but it makes a significant difference when n is finite, and we need to take it into account when we compare our results with numerical experiments.

Figure 14.16 shows an experiment in which we fix b and vary n. We plot the probability that a set of n random 20-bit integers has a balanced partition. Setting $b = 20$, taking the finite-size effect into account, and solving for n, our argument predicts that this probability will jump from zero to one at $n \approx 22.24$. This agrees very well with our results. We also plot the running times of two algorithms, and find that both of them are maximized at this value of n. Just as for k-SAT, it seems that the hardest instances occur 14.13
at the transition.

So our expressions for b_c and κ_c appear to be correct. Let's prove them.

14.5.2 The First Moment

We can locate the phase transition in INTEGER PARTITIONING using the first and second moment methods. Let Z denote the number of perfectly balanced partitions, i.e., those with $D = 0$. We will deal with the case $D = 1$ later.

We use the same transformation described in Section 5.4.5, writing Z in terms of the integral (5.13):

$$Z = \frac{2^n}{2\pi}\int_{-\pi}^{\pi}\left(\prod_{j=1}^{n}\cos a_j\theta\right)\mathrm{d}\theta. \tag{14.51}$$

This expression makes it possible to average over the a_j. By linearity of expectation, the expectation of an integral is the integral of its expectation,

$$\mathbb{E}[Z] = \frac{2^n}{2\pi}\int_{-\pi}^{\pi}\mathbb{E}\left[\prod_{j=1}^{n}\cos a_j\theta\right]\mathrm{d}\theta.$$

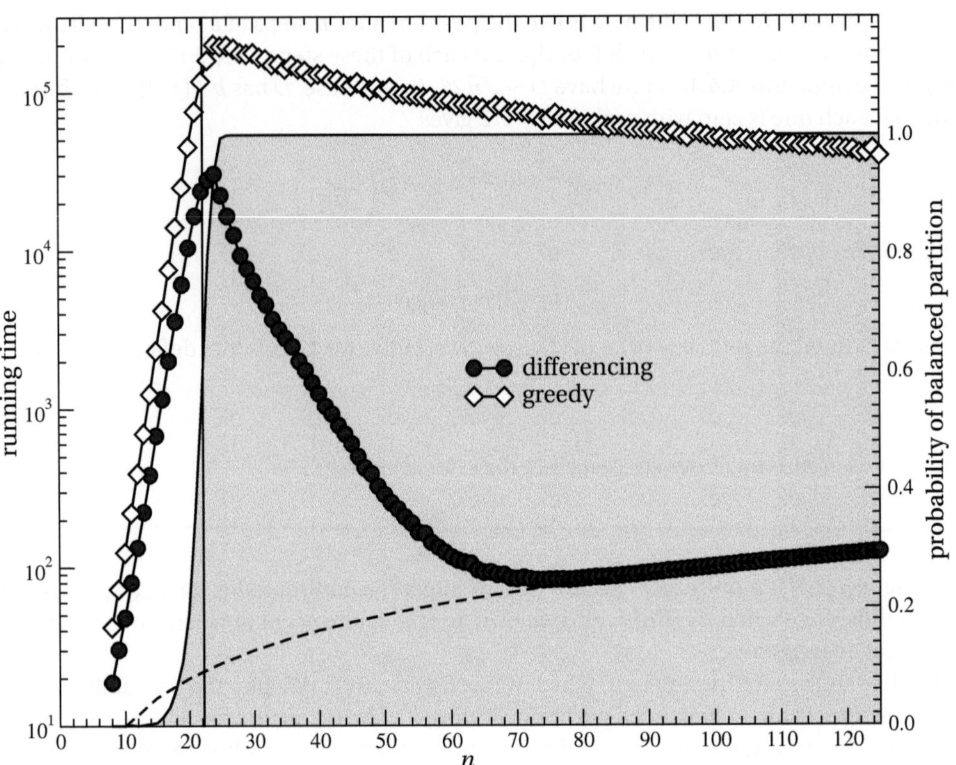

FIGURE 14.16: The phase transition in INTEGER PARTITIONING. We fix $b = 20$, so that the a_j are random 20-bit integers, and vary n. The curve with the gray area indicates the probability that a balanced partition exists, and the vertical line marks the transition predicted by solving (14.49) for n. We also plot the average running time of two algorithms, showing that they are maximized at the transition. The dashed line is n, the minimum number of steps needed to construct a single partition.

Since the a_j are independent, the expectation of the product is the product of the expectations. This gives

$$\mathbb{E}[Z] = \frac{2^n}{2\pi}\int_{-\pi}^{\pi} \mathbb{E}[\cos a\theta]^n \,\mathrm{d}\theta = \frac{2^n}{2\pi}\int_{-\pi}^{\pi} g^n(\theta)\mathrm{d}\theta\,, \tag{14.52}$$

where

$$g(\theta) = \mathop{\mathbb{E}}_{a}[\cos a\theta]\,.$$

Now let B denote 2^b. Since each a_j is chosen uniformly from $\{0, 1, \ldots, B-1\}$,

$$g(\theta) = \frac{1}{B}\sum_{a=0}^{B-1}\cos a\theta = \frac{1}{2B}\left(1 + \frac{\sin((B-1/2)\theta)}{\sin(\theta/2)}\right)\,. \tag{14.53}$$

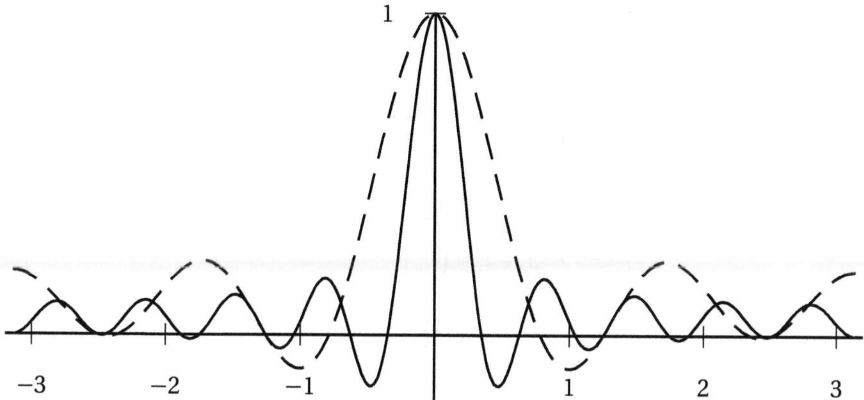

FIGURE 14.17: The function $g(\theta)$ for $B=5$ (dashed) and $B=10$ (solid).

The second equality (exercise!) follows from the identities $\cos\theta=(e^{\imath\theta}+e^{-\imath\theta})/2$ and $\sin\theta=(e^{\imath\theta}-e^{-\imath\theta})/2$ and the geometric series.

The integral (14.52) is of the same form as the ones we approximated in Section 14.4.2 using Laplace's method. As Figure 14.17 shows, $g(\theta)$ is maximized at $\theta=0$. Here we have (exercise!)

$$g(0)=1 \quad \text{and} \quad g''(0)=-\frac{1}{3}\left(1+O(1/B)\right)B^2, \tag{14.54}$$

and applying (A.33) from Appendix A.6.1 gives

$$\mathbb{E}[Z]\approx\frac{2^n}{2\pi}\sqrt{\frac{2\pi}{n\left|g''(0)\right|}}\approx\frac{2^n}{B}\frac{1}{\sqrt{(2/3)\pi n}}. \tag{14.55}$$

This approximation has two sources of multiplicative error. There is a factor of $1+O(1/n)$ from Laplace's method, and a much smaller error of $1+O(1/B)$ from (14.54). In addition, we only took g's global maximum at $\theta=0$ into account, and ignored the other maxima visible in Figure 14.17. As Problem 14.38 shows, the contribution of these other maxima is vanishingly small.

Setting $B=2^b$ gives $\mathbb{E}[Z]=\Theta(2^{n-b}/\sqrt{n})$ just as our nonrigorous argument predicted in (14.48). If we set $b=\kappa n$, the critical value of κ is bounded above by

$$\kappa_c\le 1-\frac{\log_2 n}{2n},$$

in the following sense:

Exercise 14.12 *Using* (14.55), *show that* $\Pr[Z>0]$ *is exponentially small if* $\kappa>1$. *In addition, show that if* $\kappa=1-a(\log_2 n)/n$, *then* $\Pr[Z>0]$ *is polynomially small if* $a<1/2$, *i.e.,* $O(n^{-c})$ *for some* c.

Thus at least as far as an upper bound is concerned, our nonrigorous argument—including the finite-size effect—is correct.

14.5.3 The Second Moment

To get a matching lower bound on κ_c, we compute the second moment $\mathbb{E}[Z^2]$, again employing the integral representation for the number Z of perfectly balanced partitions. Linearity of expectation and the independence of the a_j gives

$$\begin{aligned}\mathbb{E}[Z^2] &= \mathbb{E}\left[\left(\frac{2^n}{2\pi}\right)^2 \int_{-\pi}^{\pi}\int_{-\pi}^{\pi}\left(\prod_{j=1}^{n}\cos a_j\theta\,\cos a_j\varphi\right)\mathrm{d}\theta\,\mathrm{d}\varphi\right] \\ &= \left(\frac{2^n}{2\pi}\right)^2 \int_{-\pi}^{\pi}\int_{-\pi}^{\pi}\mathbb{E}\left[\prod_{j=1}^{n}\cos a_j\theta\,\cos a_j\varphi\right]\mathrm{d}\theta\,\mathrm{d}\varphi \\ &= \left(\frac{2^n}{2\pi}\right)^2 \int_{-\pi}^{\pi}\int_{-\pi}^{\pi}\mathbb{E}[\cos a\theta\,\cos a\varphi]^n\,\mathrm{d}\theta\,\mathrm{d}\varphi\,. \end{aligned} \tag{14.56}$$

Using the identity

$$\cos a\theta\,\cos a\varphi = \frac{\cos a(\theta-\varphi)+\cos a(\theta+\varphi)}{2}\,,$$

we can write (14.56) as

$$\mathbb{E}[Z^2] = \left(\frac{2^n}{2\pi}\right)^2 \int_{-\pi}^{\pi}\int_{-\pi}^{\pi}\left(\frac{g(\theta+\varphi)+g(\theta-\varphi)}{2}\right)^n \mathrm{d}\theta\,\mathrm{d}\varphi\,, \tag{14.57}$$

with $g(\theta)$ again defined as in (14.53).

We simplify this integral by changing variables to

$$\xi = \theta+\varphi \quad\text{and}\quad \eta = \theta-\varphi\,.$$

As Figure 14.18 shows, letting ξ and η range from $-\pi$ to π gives a diamond which occupies half the domain of integration of (14.57). Using the periodicity of g, i.e., the fact that $g(\theta+2\pi)=g(\theta)$, we can match up the triangles on the right of the figure and show that the entire integral is 1/2 the integral over this diamond. On the other hand, the Jacobian of our change of variables, i.e., the determinant of the matrix of first derivatives, is $\begin{vmatrix}1 & 1\\ 1 & -1\end{vmatrix} = 2$ and this cancels the factor of 1/2. Thus

$$\mathbb{E}[Z^2] = \left(\frac{2^n}{2\pi}\right)^2 \int_{-\pi}^{\pi}\int_{-\pi}^{\pi}\left(\frac{g(\xi)+g(\eta)}{2}\right)^n \mathrm{d}\xi\,\mathrm{d}\eta\,. \tag{14.58}$$

Figure 14.19 shows the integrand $h(\xi,\eta)=(g(\xi)+g(\eta))/2$. It has a global maximum of 1 at $(\xi,\eta)=(0,0)$, and two ridges of height 1/2 along the axes $\xi=0$ and $\eta=0$. The integral (14.58) is dominated by contributions from these sources, which we will now estimate.

For the peak, we use the two-dimensional version of Laplace's method (see Appendix A.6.1). The Hessian, or matrix of second derivatives, of h is

$$h'' = \begin{pmatrix}\partial^2 h/\partial\xi^2 & \partial^2 h/\partial\xi\,\partial\eta \\ \partial^2 h/\partial\eta\,\partial\xi & \partial^2 h/\partial\eta^2\end{pmatrix} = \begin{pmatrix} g''/2 & 0 \\ 0 & g''/2\end{pmatrix},$$

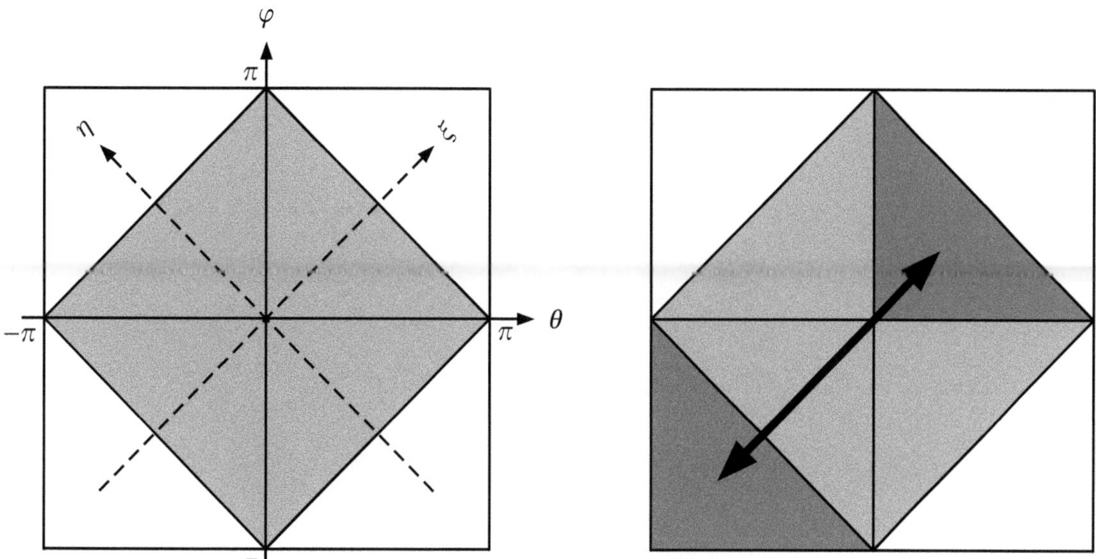

FIGURE 14.18: The domain of integration in (14.57). Corresponding triangles (right) give the same contribution to the integral.

so its determinant at $(0,0)$ is

$$\left|\det h''(0,0)\right| = \frac{1}{4}\left|g''(0)\right|^2 .$$

Applying (A.35) and comparing with the first moment (14.55) then gives, for the contribution to the integral (14.58) from the peak,

$$\left(\frac{2^n}{2\pi}\right)^2 \frac{2\pi}{n\,|\det h''(0,0)|} = 2\left(\frac{2^n}{2\pi}\sqrt{\frac{2\pi}{n\left|g''(0)\right|}}\right)^2 = 2\,\mathbb{E}[Z]^2 , \tag{14.59}$$

Thus the peak's contribution to the second moment is proportional to the first moment squared. This suggests that for most pairs of assignments A, A', the events that they are balanced are only weakly correlated. Essentially the only correlation comes from the fact that if A is perfectly balanced then $\sum_j a_j$ must be even. This doubles the probability that A' is balanced, and gives the factor of 2.

Next we estimate the contribution from the ridge along the η-axis, where ξ is close to zero. If we assume that $|\eta|$ is large compared to $1/B$, then (14.53) tells us that $g(\eta) = O(1/B)$. So ignoring the part of the ridge close to the peak at $(0,0)$, we take $g(\eta)$ to be zero. The integral over η then becomes trivial, giving a factor of 2π, and leaving us with

$$2\pi\left(\frac{2^n}{2\pi}\right)^2 \int_{-\pi}^{\pi} \left(\frac{g(\xi)}{2}\right)^n \mathrm{d}\xi = \frac{2^n}{2\pi}\int_{-\pi}^{\pi} g(\xi)^n \,\mathrm{d}\xi .$$

This integral is identical to that in (14.52), and it yields $\mathbb{E}[Z]$. This is no accident. This ridge corresponds to pairs of balanced assignments A, A' where $A = A'$, and the expected number of such pairs is $\mathbb{E}[Z]$.

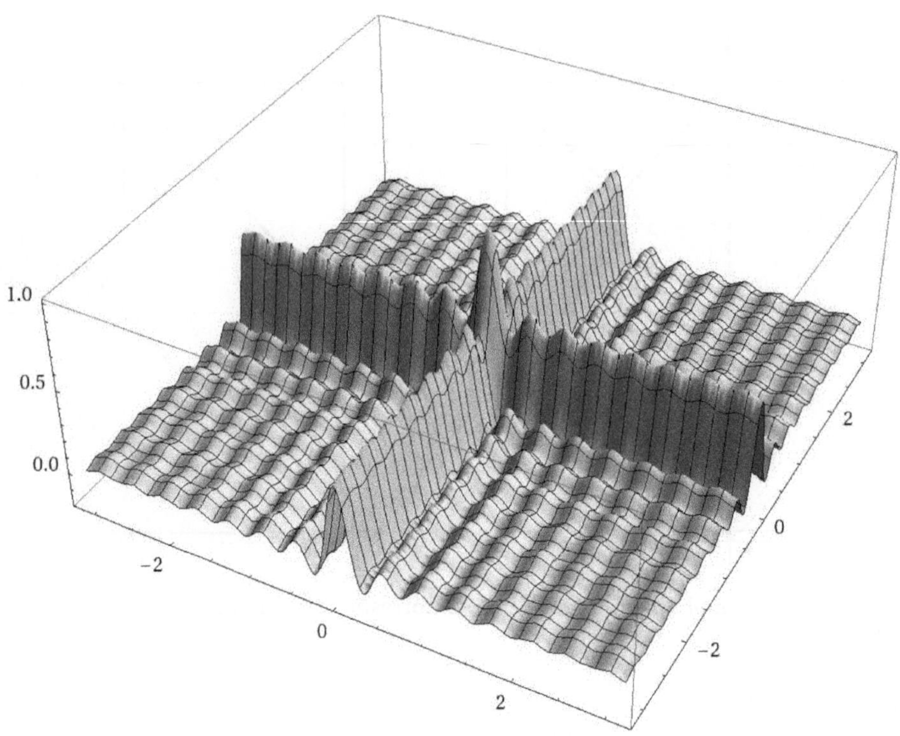

FIGURE 14.19: The function $h(\xi,\eta) = (g(\xi)+g(\eta))/2$ for $B = 16$.

The ridge along the ξ-axis corresponds to pairs of assignments A, A' where $A' = \overline{A}$, and gives the same contribution.

Combining the ridge's contributions with that of the peak (14.59) gives

$$\mathbb{E}[Z^2] \approx 2\,\mathbb{E}[Z]^2 + 2\,\mathbb{E}[Z]\,. \tag{14.60}$$

The $\approx$ here hides multiplicative errors of size $1+O(1/B)$, which we will ignore. The second moment method then gives the following lower bound on the probability that a perfectly balanced partition exists:

$$\Pr[Z>0] \geq \frac{\mathbb{E}[Z]^2}{\mathbb{E}[Z^2]} \approx \frac{1}{2}\,\frac{\mathbb{E}[Z]}{\mathbb{E}[Z]+1}\,.$$

This factor of $1/2$ is simply the probability that $\sum_j a_j$ is even. As Problem 14.39 shows, the probability that a partition with $D = 1$ exists is nearly identical to this. Combining these, the total probability that a balanced partition exists is bounded below by

$$\Pr[\text{a balanced partition exists}] \geq \frac{\mathbb{E}[Z]}{\mathbb{E}[Z]+1} \geq 1 - \frac{1}{\mathbb{E}[Z]}\,, \tag{14.61}$$

where in the second inequality we used $1/(1+z) \geq 1-z$. Thus the probability that a balanced partition exists is close to 1 whenever the expected number of such partitions is large.

Again setting $B = 2^b$ where $b = \kappa n$ and recalling that $\mathbb{E}[Z] = \Theta(2^{n-b}/\sqrt{n})$, you can now complete the following exercise:

Exercise 14.13 *Show that* Pr[a balanced partition exists] *is exponentially close to* 1 *if* $\kappa < 1$. *Then show that if* $\kappa = 1 - a(\log_2 n)/n$ *this probability is polynomially close to* 1 *if* $a > 1/2$, *i.e.,* $1 - O(n^{-c})$ *for some* c.

This proves our conjecture that

$$\kappa_c = 1 - \frac{\log_2 n}{2n}.$$

Thus we have rigorously located the phase transition—not just the critical ratio $\kappa = b/n$, but its finite-size corrections as well.

Why are these results so much sharper than those we obtained in Section 14.4 for k-SAT? The events that two truth assignments satisfy a random k-SAT formula are highly correlated unless their Hamming distance is $n/2$, but other pairs of assignments make a significant contribution to the second moment. As a result, the number of solutions can have such a large variance that it can be zero with high probability, even when its expectation is large—creating a gap between the first-moment upper bound and the second-moment lower bound.

In contrast, given any two partitions A and A', the events that they are balanced are nearly independent unless $A' = A$ or $A' = \overline{A}$. Even if the Hamming distance between A and A' is small, their totals differ by some random weights a_j, and this gives A' an essentially independent chance of being balanced whether or not A is. Indeed, Problem 14.40 shows that this independence becomes exact if we define a partition as balanced if $D \equiv_B 0$.

As Problem A.10 shows, the second moment method gives the bound (14.61) whenever Z is the sum of indicator random variables that are pairwise independent. In that case, $\Pr[Z > 0]$ is close to 1 or 0 if $\mathbb{E}[Z]$ is large or small respectively, so the first-moment upper bound and the second-moment lower bound coincide. Thus we can determine the threshold exactly whenever the correlations between solutions are sufficiently small.

We can be even more precise about how the phase transition in INTEGER PARTITIONING takes place. Let

$$\kappa = 1 - \frac{\log_2 n}{2n} + \frac{\lambda}{n},$$

where λ is a real number, and let $P(\lambda)$ denote the probability that a balanced partition exists.

Exercise 14.14 *Using our first- and second-moment calculations, place upper and lower bounds on* $P(\lambda)$. *In particular, show that* $0 < P(\lambda) < 1$ *for any finite* λ, *and that*

$$\lim_{\lambda \to +\infty} P(\lambda) = 0 \quad \textit{and} \quad \lim_{\lambda \to -\infty} P(\lambda) = 1$$

Thus we can determine the width of the window over which the transition occurs. For any pair of probabilities $0 < p_1 < p_2 < 1$, the probability that a balanced partition exists decreases from p_2 to p_1 as κ varies over an interval of size $\Theta(1/n)$. By computing higher moments, $\mathbb{E}[Z^r]$ for $r > 2$, it's possible to compute $P(\lambda)$ exactly.

14.14

14.6 Message Passing

> I will pursue the implications of this conjecture as far as possible, and not examine until later whether it is in fact true. Otherwise, the untimely discovery of a mistake could keep me from my undertaking...
>
> Johannes Kepler, *The Six-Cornered Snowflake*

For INTEGER PARTITIONING, the first and second moment bounds are sharp enough to pin down the phase transition exactly. But for random k-SAT, there is still a gap between them. Once the variance in the number of satisfying assignments becomes too large, we can no longer prove that one exists with high probability. If we want to get closer to the transition, we have to think harder about how variables interact and communicate through the constraints that connect them.

In this section and the next we will meet *message-passing algorithms*, including *belief propagation* and *survey propagation*. The idea is for the clauses and variables to send each other messages such as
14.15 "my other variables have failed me, I need you to be true," or "my other clauses need me, I can't satisfy you." By exchanging probability distributions of these messages, we can reach a fixed point which, in theory, encodes a great deal of information about the set of solutions.

Algorithms of this sort first appeared in machine learning and artificial intelligence, and were independently discovered by statistical physicists in their study of spin glasses and other disordered systems. While they are exact on trees, they can be deceived by loops. However, we believe that they are asymptotically exact on sparse random graphs and formulas, at least up to a certain density, since these structures are locally treelike. This belief is related, in turn, to the belief that correlations decay with distance, so that two variables far apart in the formula—like distant spins in the Ising model above its transition temperature—are essentially independent.

Moreover, there is a density above which these algorithms fail precisely because the correlations between variables become global and can no longer be ignored. The story of how this failure occurs, and how we can generalize these algorithms to move beyond it, gives us even more insight into the problem. It turns out that k-SAT has not just one phase transition, but several. While still in the satisfiable regime, the set of solutions breaks apart into exponentially many clusters. These clusters undergo further phase transitions called *condensation* and *freezing*, until they finally wink out of existence when the formula becomes unsatisfiable.

Making this picture rigorous has been a major mathematical challenge, and much progress has been made in the past few years. We will present it here with a physics attitude, taking for granted that it is correct, and embracing what it has to tell us about the structure of these problems and the geometry of their solutions.

14.6.1 Entropy and its Fluctuations

Let's review why the second moment method fails somewhere below the phase transition in k-SAT. There is a clause density α at which, while the formula is still satisfiable, the variance in the number of solutions Z is too large. Once the ratio $\mathbb{E}[Z]^2/\mathbb{E}[Z^2]$ becomes exponentially small, we can no longer prove that $\Pr[Z > 0]$ is bounded above zero.

In fact, for random k-SAT this is the case for any $\alpha > 0$. We saw in Sections 14.4.3 and 14.4.4 that we can "correct" the variance by focusing on NAE-k-SAT or weighting the satisfying assignments, but there seems to be a density beyond which no trick of this kind works. The fluctuations in Z are simply too large.

However, even if Z has huge fluctuations in absolute terms, that doesn't mean that the formula isn't satisfiable. Let's think about *multiplicative* fluctuations instead. Each time we add a random clause c to a 3-SAT formula, the number of satisfying assignments goes down, on average, by a factor of $p = 7/8$. But this factor is itself a random variable. For instance, if there are already many clauses that agree with c on its variables, then most satisfying assignments already satisfy c, and $p > 7/8$. Conversely, if many clauses disagree with c, then $p < 7/8$.

Suppose as before that we have n variables and $m = \alpha n$ clauses, and that the ith clause decreases Z by a factor p_i. Then Z is just 2^n times their product,

$$Z = 2^n \prod_{i=1}^{m} p_i \,.$$

If we imagine for a moment that the p_i are independent of each other—which is certainly not true, but is a useful cartoon—then the logarithm of Z is the sum of independent random variables,

$$\ln Z = n \ln 2 + \sum_{i=1}^{m} \ln p_i \,.$$

The Central Limit Theorem would then imply that $\ln Z$ obeys a normal distribution and is tightly concentrated around its expectation $\mathbb{E}[\ln Z]$.

Borrowing terminology from physics, we call $\mathbb{E}[\ln Z]$ the *entropy* and denote it S. Since Z is exponential in n, S is linear in n, and as in Section 13.6 we call the limit

$$s = \lim_{n \to \infty} \frac{S}{n} \,,$$

the *entropy density*. For instance, in our cartoon we have

$$s = \ln 2 + \alpha \mathbb{E}[\ln p_i] \,.$$

If the formula is unsatisfiable—which happens with nonzero probability even below the critical density—then $Z = 0$ and $\ln Z = -\infty$. To avoid this divergence, we will actually define the entropy as

$$S = \mathbb{E}[\ln \min(Z,1)] = \mathbb{E}[\ln Z \mid Z > 0] \Pr[Z > 0] \,,$$

and $s = \lim_{n \to \infty} S/n$.

If $\ln Z$ obeys a normal distribution around S, then Z obeys a log-normal distribution. The tails of a log-normal distribution are quite long, so Z is not very concentrated. Specifically, if with high probability we have

$$\ln Z = s n + o(n) \,,$$

then all we can say about Z is that it is within a *subexponential* factor of its typical value,

$$Z = \mathrm{e}^{sn} \, \mathrm{e}^{o(n)} \,.$$

This factor $e^{o(n)}$ could be very large—for a log-normal distribution it would be exponential in $\sqrt{n}$. In contrast, the second moment method demands that Z be within a *constant* factor of its expectation with constant probability (see Problem 14.41).

For that matter, the typical value of Z can be exponentially far from its expectation. Jensen's inequality (see Appendix A.3.6) shows that

$$e^{sn} = e^{\mathbb{E}[\ln Z]} \leq \mathbb{E}[e^{\ln Z}] = \mathbb{E}[Z] .$$

Equivalently, $(1/n)\ln \mathbb{E}[Z]$ is an upper bound on the entropy density,

$$s = \frac{1}{n}\mathbb{E}[\ln Z] \leq \frac{1}{n}\ln \mathbb{E}[Z] .$$

Physicists call $(1/n)\ln \mathbb{E}[Z]$ the *annealed* density, s_{annealed}. For k-SAT, our first-moment computation (14.32) gives

$$s_{\text{annealed}} = \ln 2 + \alpha \ln(1-2^{-k}) .$$

But usually $s < s_{\text{annealed}}$, in which case Z's typical value e^{sn} is exponentially smaller than $\mathbb{E}[Z]$. We have already seen this in Section 14.4.1, where we noticed that $\mathbb{E}[Z]$ can be exponentially large even at densities where $Z = 0$ with high probability.

Thus it is $\ln Z$, not Z, which is concentrated around its expectation. Moreover, if we could compute the entropy density as a function of α then we could determine the critical density α_c exactly. To see why, consider the following exercise:

Exercise 14.15 *Assuming that the threshold α_c exists, show that the entropy density s is positive for any $\alpha < \alpha_c$ and zero for any $\alpha > \alpha_c$. Hint: below α_c, show that with high probability the number of satisfying assignments is exponentially large, simply because there are $\Theta(n)$ variables that do not appear in any clauses. Above α_c, use the fact that even for satisfiable formulas we have $Z \leq 2^n$.*

14.16 Unfortunately, computing the entropy $\mathbb{E}[\ln Z]$ seems to be far more difficult than computing $\mathbb{E}[Z]$ or $\mathbb{E}[Z^2]$. But as we will see next, there is an ingenious technique that gives the right answer up to a certain value of α. In Section 14.7 we will see how and why it fails at that point, and how we can generalize it to get all the way—or so we think—to the transition at α_c.

14.6.2 The Factor Graph

> If what I say resonates with you, it is merely because we are both branches on the same tree.
>
> William Butler Yeats

In order to define our message-passing algorithm, we will think of a k-SAT formula as a bipartite graph. There are clause vertices and variable vertices, and each clause is connected to the variables it contains. Thus the degree of each clause vertex is k, and the degree of each variable vertex is the number of clauses it appears in.

We call this the *factor graph*, and show an example in Figure 14.20. We draw each edge as a solid or dashed line, depending on whether the variable appears positively or negatively—that is, unnegated or negated—in the clause.

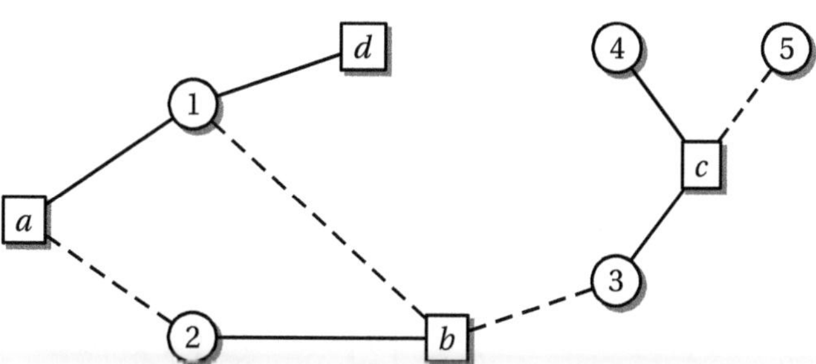

FIGURE 14.20: The factor graph of $(x_1 \vee \overline{x}_2)_a \wedge (\overline{x}_1 \vee x_2 \vee \overline{x}_3)_b \wedge (x_3 \vee x_4 \vee \overline{x}_5)_c \wedge (x_1)_d$. Positive and negative appearances of a variable correspond to solid and dashed lines respectively.

For a random k-SAT formula with $m = \alpha n$ clauses, the average degree of a variable is $km/n = k\alpha$ since there are k variables per clause. Like a sparse random graph $G(n, p = c/n)$, the variables' degrees are Poisson-distributed:

Exercise 14.16 *Show in analogy to* (14.2) *that in the limit of large n the degree distribution of the variables in $F_k(n, m = \alpha n)$ becomes Poisson with mean $k\alpha$.*

Also like a sparse random graph, the factor graph of a sparse random formula is locally treelike. Analogous to Problem 14.2, for almost all vertices, the smallest loop they belong to is of length $\Omega(\log n)$, so their neighborhoods are trees until we get $\Omega(\log n)$ steps away.

Emboldened by this locally tree-like structure, we will proceed in the following way. First we start with an algorithm that calculates Z exactly on formulas whose factor graphs are trees. We then use this same algorithm to estimate $\ln Z$ even when the factor graph has loops, hoping that it gives the right answer. Finally, we look at the behavior of this algorithm on random formulas, using the fact that their degree distributions are Poisson. We will be rewarded with a result for the entropy density s that is correct up to a certain density α, but breaks down at higher densities.

If we know that our tree-based algorithm is wrong above a certain α, why are we going through all this? The analysis of *why* it fails will reveal the full, and surprisingly rich, structure of random k-SAT and how the set of solutions changes as we approach the transition. Once we've understood this scenario, we can modify our algorithm and use it to locate the transition from satisfiability to unsatisfiability, as well as the various other phase transitions that we are about to discover. So let's get to work.

14.6.3 Messages that Count

No matter how high a tree grows, the falling leaves return to the root.

Malay proverb

Let's start by counting the number Z of satisfying assignments of a SAT formula whose factor graph is a tree. Our approach will look very similar at first to the dynamic programming algorithms we used in Problems 3.25 and 3.26 to find the MAX INDEPENDENT SET and MAX MATCHING, or the one in Problem 13.37

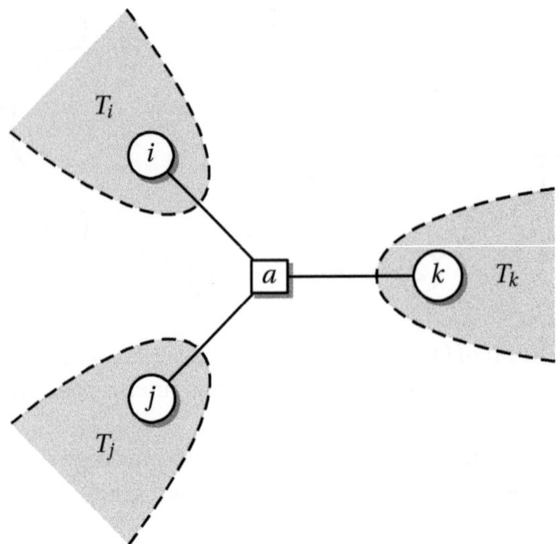

FIGURE 14.21: Removing a vertex from a tree breaks it into independent subtrees. Here we remove a clause vertex a, creating subtrees rooted at each of its variables $i, j, k \in \partial a$.

for counting independent sets. Namely, we use the fact that removing a vertex or an edge from a tree causes it to break apart into subtrees, and we can treat each subtree separately.

Exercise 14.17 *Suppose that the factor graph of a k-SAT formula is a tree. Show that it is satisfiable.*

We will use $a, b, \ldots$ to denote clauses and $i, j, \ldots$ to denote variables. Let's choose an arbitrary clause a and declare it to be the root of the tree. Let ∂a denote the set of a's neighbors in the factor graph, i.e., the set of variables appearing in a. Then if we remove a, the factor graph disintegrates into k subtrees T_i, each rooted at a variable $i \in \partial a$ as in Figure 14.21.

Let $Z_{i\to a}(x_i)$ denote the number of satisfying assignments of the subtree T_i assuming that the value of the variable i is x_i. Then the total number of satisfying assignments is the product of $Z_{i\to a}(x_i)$, summed over all values x_i such that the clause a is satisfied. We write this as

$$Z = \sum_{\mathbf{x}_{\partial a}} C_a(\mathbf{x}_{\partial a}) \prod_{i \in \partial a} Z_{i\to a}(x_i), \tag{14.62}$$

where $\mathbf{x}_{\partial a} = \{x_i, x_j, \ldots\}$ denotes the truth values of the k variables in ∂a and

$$C_a(\mathbf{x}_{\partial a}) = \begin{cases} 1 & \text{if clause } a \text{ is satisfied} \\ 0 & \text{otherwise.} \end{cases}$$

We will interpret $Z_{i\to a}(x_i)$ as a *message* that is sent by the variable i to the clause a. In terms of dynamic programming, we can think of it as the solution to the subproblem corresponding to the subtree T_i being passed toward the root.

The message $Z_{i\to a}(x_i)$ can be computed, in turn, from messages farther up the tree. Let ∂i denote the set of i's neighbors in the factor graph, i.e., the set of clauses in which i appears. For each $b \in \partial i$, let

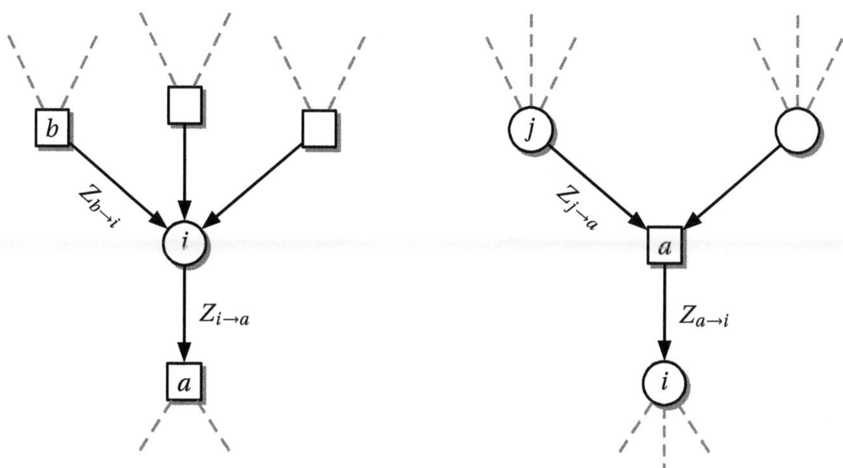

FIGURE 14.22: The flow of messages from the leaves towards the root of the factor graph. Left, each variable sends a message to its parent clause based on the messages it receives from its other clauses. Right, each clause sends a message to its parent variable based on the messages it receives from its other variables. Square vertices a, b correspond to clauses and round vertices i, j correspond to variables.

$Z_{b\to i}(x_i)$ denote the number of satisfying assignments of the subtree rooted at b assuming that i has value x_i. Then

$$Z_{i\to a}(x_i) = \prod_{b\in\partial i-a} Z_{b\to i}(x_i). \tag{14.63}$$

The messages on the right-hand side of (14.63) are computed from messages even farther up the tree. Analogous to (14.62) we have

$$Z_{a\to i}(x_i) = \sum_{\mathbf{x}_{\partial a-i}} C_a(\mathbf{x}_{\partial a}) \prod_{j\in\partial a-i} Z_{j\to a}(x_j). \tag{14.64}$$

Thus, as shown in Figure 14.22, variables send messages to clauses based on the messages they receive from the other clauses in which they appear, and clauses send messages to variables based on the messages they receive from the other variables they contain. (Whew!)

We evaluate (14.63) and (14.64) recursively until we reach a leaf, the base case of the recursion. Here the sets $\partial i - a$ and $\partial a - i$ are empty, but empty products evaluate to 1 by convention. Therefore the messages from the leaves are

$$Z_{i\to a}(x_i) = 1 \quad \text{and} \quad Z_{a\to i}(x_i) = C_a(x_i).$$

Note that a leaf a is a unit clause, in which case $C_a(x_i)$ is 1 or 0 depending on whether x_i satisfies a.

If we start at the leaves and work our way to the root, we know Z as soon as the root has received messages from all its neighbors. The total number of messages that we need to compute and send equals the number of edges in the factor graph—that is, the total number of literals in the formula—so our algorithm runs in essentially linear time.

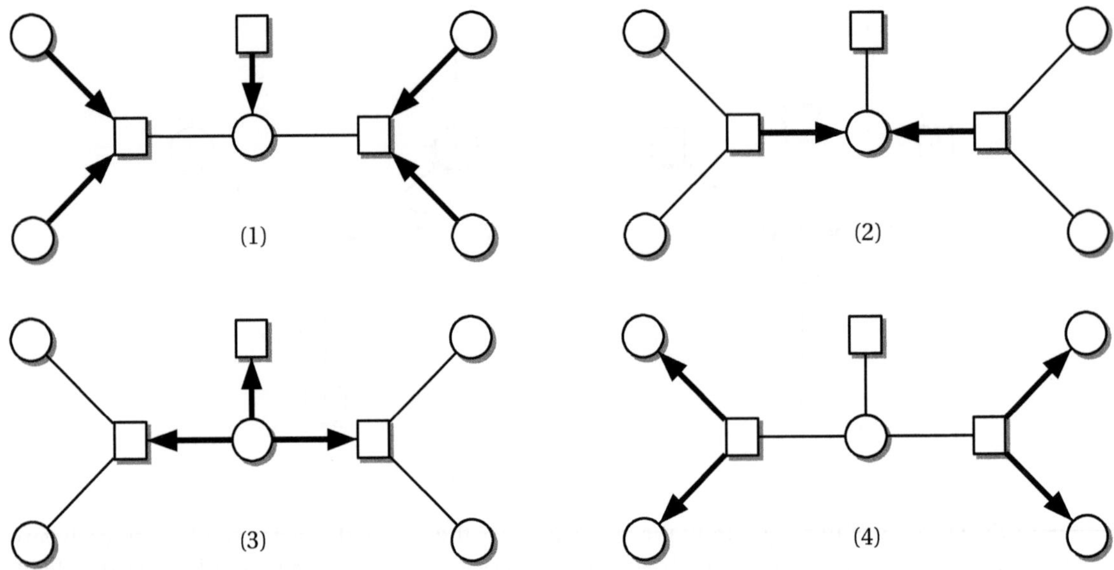

FIGURE 14.23: The global flow of messages from neighbor to neighbor. At step 2, the center vertex knows Z since it has received messages from all its neighbors. At step 4, every vertex knows Z.

But we don't actually need to choose a root at all. Imagine a parallel version of this algorithm, where each vertex sends a message to its neighbor as soon as it has received incoming messages from all of its other neighbors. As soon as *any* vertex has received messages from all its neighbors, we can compute Z. If this vertex is a clause we use (14.62). Similarly, if it is a variable i we can write

$$Z = \sum_{x_i} Z_i(x_i) = \sum_{x_i} \prod_{a \in \partial i} Z_{a \to i}(x_i), \tag{14.65}$$

where $Z_i(x_i)$ is the number of satisfying assignments where i has the value x_i. For that matter, once every vertex has received messages from all its neighbors, then every vertex "knows" the total number of assignments Z.

Figure 14.23 shows the resulting flow of messages. The leaves are the only vertices that can send a message without receiving one, so the flow starts at the leaves and moves inward. The center vertex is the first to receive messages from all its neighbors. In the next iteration it sends messages to all its neighbors, reversing the flow of messages from the center towards the leaves. At the end of the process, every vertex has received messages from all its neighbors.

This algorithm sends two messages along each edge, so the total number of messages is twice the number of edges. Moreover, as we will see next, this set of messages gives us much more than just the number of satisfying assignments Z.

14.6.4 Marginal Distributions

> The warning message we sent the Russians was a calculated ambiguity that would be clearly understood.
>
> Alexander Haig, US Secretary of State 1981–82

If we know the number Z of satisfying assignments of a SAT formula ϕ, we can compute the uniform probability distribution μ over all satisfying assignments,

$$\mu(\mathbf{x}) = \frac{1}{Z}\prod_a C_a(X_{\partial a}) = \begin{cases} Z^{-1} & \text{if } \mathbf{x} \text{ satisfies } \phi \\ 0 & \text{if it doesn't.} \end{cases} \tag{14.66}$$

The fraction of satisfying assignments in which i takes the value x_i is called the *marginal distribution* of the variable i, and we denote it $\mu_i(x_i)$. It tells us how biased a variable is in the ensemble of satisfying assignments. We can obtain $\mu_i(x_i)$ directly by summing μ over all the other variables' values,

$$\mu_i(x_i) = \sum_{\mathbf{x}-x_i} \mu(\mathbf{x}).$$

This sum has 2^{n-1} terms, one for each truth assignment in which i has the value x_i, so evaluating it directly would take exponential time. But using the messages we defined above, we can write

$$\mu_i(x_i) = \frac{1}{Z} Z_i(x_i) = \frac{1}{Z}\prod_{a\in\partial i} Z_{a\to i}(x_i),$$

and thus compute $\mu_i(x_i)$ in linear time.

These marginal distributions have many applications. For instance, suppose we want to sample a uniformly random satisfying assignment. In general, sampling from a high-dimensional distribution like μ is hard—but if we can compute the marginals then we can use the self-reducibility of SAT to set variables one at a time. We choose a variable x_i, give it a truth value drawn from μ_i, and simplify the formula by shortening or removing the clauses containing x_i. We then compute the marginals of the remaining variables in the resulting formula, set another variable, and so on until all the variables have been set.

We met this technique, often called *importance sampling*, in Section 13.4. As we said there, if we can tell what fraction of the leaves lie on each branch of a search tree, we can move down the branches with this probability and sample uniformly from the leaves.

The messages also allow us to calculate marginal distributions of subsets of variables, such as the set ∂a of variables appearing in the clause a,

$$\mu_a(\mathbf{x}_{\partial a}) = \sum_{\mathbf{x}-\mathbf{x}_{\partial a}} \mu(\mathbf{x}) = \frac{C_a(\mathbf{x}_{\partial a})}{Z}\prod_{i\in\partial a} Z_{i\to a}(x_i).$$

Exercise 14.18 *Show that the two-variable marginal $\mu_{ij}(x_i, x_j)$ can be computed in linear time if the factor graph is a tree. Hint: $\mu_{ij}(x_i, x_j) = \mu(x_i \,|\, x_j)\mu_j(x_j)$, where $\mu(x_i \,|\, x_j)$ is the conditional probability that i has value x_i given that j has value x_j.*

Since our message-passing algorithm can be thought of as a method of computing marginal distributions, perhaps it makes sense to use probability distributions for the messages themselves. Let's replace the integers $Z_{a\to i}(x_i)$ and $Z_{i\to a}(x_i)$ by the fraction of satisfying assignments of each subtree where i has a given value,

$$\mu_{a\to i}(x_i) = \frac{Z_{a\to i}(x_i)}{Z_{a\to i}(1)+Z_{a\to i}(0)}, \qquad \mu_{i\to a}(x_i) = \frac{Z_{i\to a}(x_i)}{Z_{i\to a}(1)+Z_{i\to a}(0)}.$$

These μ-messages are the marginals of formulas corresponding to the subtrees. Specifically, $\mu_{a\to i}(x_i)$ is the marginal distribution of the variable i in a formula where a is the only clause in which i appears. Similarly, $\mu_{i\to a}(x_i)$ is the marginal distribution of i in a modified formula, in which i appears in its other clauses but is disconnected from a.

As in (14.63) and (14.64), each vertex sends its neighbor a μ-message based on the messages it receives from its other neighbors. The only difference is that we now have to normalize our messages. We denote the normalization constants by $z_{a\to i}$ and $z_{i\to a}$, giving

$$\begin{aligned}\mu_{a\to i}(x_i) &= \frac{1}{z_{a\to i}} \sum_{\mathbf{x}_{\partial a-i}} C_a(\mathbf{x}_{\partial a}) \prod_{j\in\partial a-i} \mu_{j\to a}(x_j)\\ \mu_{i\to a}(x_i) &= \frac{1}{z_{i\to a}} \prod_{b\in\partial i-a} \mu_{b\to i}(x_i).\end{aligned} \tag{14.67}$$

We compute $z_{a\to i}$ and $z_{i\to a}$ by summing over both truth values x_i,

$$\begin{aligned}z_{a\to i} &= \sum_{\mathbf{x}_{\partial a}} C_a(\mathbf{x}_{\partial a}) \prod_{j\in\partial a-i} \mu_{j\to a}(x_j),\\ z_{i\to a} &= \sum_{x_i} \prod_{b\in\partial i-a} \mu_{b\to i}(x_i).\end{aligned} \tag{14.68}$$

The advantage of μ-messages over Z-messages is that the former are independent of n. That is, $\mu\in[0,1]$ while Z typically grows exponentially with n. This will prove very useful when we discuss distributions of messages later in this section.

The marginal distributions of a variable i, or the neighborhood of a clause a, in the original factor graph are given by

$$\mu_a(\mathbf{x}_{\partial a}) = \frac{C_a(\mathbf{x}_{\partial a})}{z_a} \prod_{i\in\partial a} \mu_{i\to a}(x_i) \tag{14.69a}$$

$$\mu_i(x_i) = \frac{1}{z_i} \prod_{a\in\partial i} \mu_{a\to i}(x_i), \tag{14.69b}$$

where z_i and z_a are normalization constants given by

$$\begin{aligned}z_a &= \sum_{\mathbf{x}_{\partial a}} C_a(\mathbf{x}_{\partial a}) \prod_{i\in\partial a} \mu_{i\to a}(x_i),\\ z_i &= \sum_{x_i} \prod_{a\in\partial i} \mu_{a\to i}(x_i).\end{aligned} \tag{14.70}$$

If we consider the set of satisfying assignments for the formula where a clause a has been removed, z_a is the fraction of these assignments that remain satisfying when we add a back in. That is, z_a is the total probability that a is satisfied, assuming that each of its variables i is chosen independently from $\mu_{i\to a}$.

Similarly, if we consider a formula where i is "split" into independent variables, one for each of its clauses, z_i is the fraction of satisfying assignments that remain when we merge these variables and force them to have the same value. Equivalently, z_i is the probability that we get the same value of x_i from each clause a if we draw from each $\mu_{a\to i}$ independently.

This assumption of independence is a crucial point. In (14.67) and (14.69) we assume that the messages received from each of a's variables, or each of i's clauses, are independent. More generally, for every vertex v in the factor graph we assume that *the only correlations between v's neighbors go through v*. Thus if we condition on the value of v—its truth value if v is a variable, or the fact that it is satisfied if v is a clause—its neighbors become independent. This is perfectly true if the factor graph is a tree, since conditioning on v's value effectively removes it and breaks the factor graph into independent subtrees. But as we will discuss in the next section, for factor graphs with loops this is a major assumption.

By substituting μ-messages for Z-messages we seem to have lost the ability to compute the total number of satisfying assignments. However, this is not the case. The following lemma shows that we can compute Z by gathering information from all the μ-messages:

Lemma 14.4 *If the factor graph of a* SAT *formula is a tree, the entropy $S=\ln Z$ can be computed from the messages $\mu_{i\to a}$ and $\mu_{a\to i}$ as*

$$S=\sum_a \ln z_a+\sum_i \ln z_i-\sum_{(i,a)} \ln z_{i,a}\,, \tag{14.71}$$

where z_i and z_a are given by (14.70), *$z_{i,a}$ is given by*

$$z_{i,a}=\sum_{x_i}\mu_{a\to i}(x_i)\,\mu_{i\to a}(x_i)\,,$$

and the third sum runs over all edges (i,a) of the factor graph.

Proof We will start by proving a somewhat different formula,

$$S=-\sum_a\sum_{\mathbf{x}_{\partial a}}\mu_a(\mathbf{x}_{\partial a})\ln\mu_a(\mathbf{x}_{\partial a})+\sum_i(|\partial i|-1)\sum_{x_i}\mu_i(x_i)\ln\mu_i(x_i)\,. \tag{14.72}$$

The entropy of a probability distribution μ is defined as $S=-\sum_{\mathbf{x}}\mu(\mathbf{x})\ln\mu(\mathbf{x})$, and this coincides with $S=\ln Z$ when $\mu=1/Z$ is the uniform distribution over Z objects. Thus (14.72) is equivalent to the claim that

$$\mu(\mathbf{x})=\frac{\prod_a\mu_a(\mathbf{x}_{\partial a})}{\prod_i\mu_i(x_i)^{|\partial i|-1}}\,. \tag{14.73}$$

We will prove (14.73) by induction. The reader can check that it holds for factor graphs with just one clause. So let's consider a tree with m clauses and assume that it holds for all trees with fewer than m clauses. For any clause a, we can write

$$\mu(\mathbf{x})=\mu(\mathbf{x}\,|\,\mathbf{x}_{\partial a})\,\mu_a(\mathbf{x}_{\partial a})\,.$$

Fixing the variables $\mathbf{x}_{\partial a}$ such that clause a is satisfied allows us to remove a from the factor graph. This leaves us with disconnected subtrees T_i, so the conditional probability factorizes as follows:

$$\mu(\mathbf{x}) = \mu_a(\mathbf{x}_{\partial a}) \prod_{i \in \partial a} \mu_{T_i}(\mathbf{x}_{T_i} \,|\, x_i) = \mu_a(\mathbf{x}_{\partial a}) \prod_{i \in \partial a} \frac{\mu_{T_i}(\mathbf{x}_{T_i}, x_i)}{\mu_i(x_i)}, \tag{14.74}$$

where $\mathbf{x}_{T_i}$ denotes the set of variables in T_i except x_i. Since each subtree has fewer than m clauses, we can apply our inductive assumption and write

$$\mu_{T_i}(\mathbf{x}_{T_i}, x_i) = \frac{\prod_{a \in T_i} \mu_a(\mathbf{x}_{\partial a})}{\mu_i(x_i)^{|\partial i|-2} \prod_{j \in T_i} \mu_j(x_j)^{|\partial j|-1}}.$$

Plugging this into (14.74) gives (14.73), and thus proves by induction that (14.73) holds for any factor graph which is a tree. Finally, we get from (14.72) to (14.71) by substituting (14.69) for the marginals and using (14.67). □

The expression (14.72) has a nice interpretation in terms of incremental contributions to the entropy. The sum over clauses adds the entropy of the variables in each clause, and the sum over variables corrects for the fact that the first sum counts each variable $|\partial i|$ times. Similarly, in (14.71) each term $\ln z_a$ or $\ln z_i$ represents the change in the entropy when we glue together the subtrees around a clause a or a variable i, and the term $\ln z_{i,a}$ corrects for the fact that each edge (i, a) has been counted twice in this gluing process.

Lemma 14.4 is a key ingredient for our computation of $\mathbb{E}[\ln Z]$ because it tells us how to represent the entropy in terms of the messages. But can we apply it to random k-SAT formulas, given that their factor graphs are not trees?

14.6.5 Loopy Beliefs

If we try to apply our message-passing algorithm to general factor graphs, i.e, those with loops, we encounter two problems. First, the neighbors of a vertex v can be connected to each other by a roundabout path, so they might be correlated in ways that don't go through v. In that case, we can't treat the messages from v's neighbors as independent—we can't assume that their joint distribution is simply the product of their marginals when conditioned on the value of v.

Second, the initial messages from the leaves aren't enough to fill the graph with messages as in Figure 14.23. If we wait for each vertex to receive trustworthy messages from all but one neighbor before it sends a message, many vertices never send any messages at all.

We respond to both these problems by throwing rigor to the winds. First, we will assume that v's neighbors are independent, even though they aren't. This is a reasonable assumption if most of the roundabout paths between v's neighbors are long—equivalently, if most vertices in the factor graph do not lie on a short cycle—and if the correlations between variables decay quickly with distance in the graph so that distant variables are effectively independent.

Second, rather than relying on messages from the leaves to get off the ground, we start the process with *random* initial messages $\mu_{i \to a}$ and $\mu_{a \to i}$. We then iterate according to the message-passing equations (14.67), repeatedly updating the messages each vertex sends out based on the messages it received from its neighbors on the previous step. Our hope is that this iteration reaches a fixed point, where the messages $\mu_{a \to i}$ and $\mu_{i \to a}$ are solutions of (14.67).

Exercise 14.19 *Show that on a tree of diameter d, iterating* (14.67) *converges to the correct marginals after d iterations, no matter what initial messages we start with.*

Even if we reach a fixed point, there is no guarantee that the μs are the correct marginals. They are locally self-consistent, but there might not even exist a joint distribution of the x_i consistent with them. However, they represent *beliefs* about the true marginals, and iterating them according to (14.67) is called *belief propagation*, or BP for short.

In the next section we embrace these beliefs, and use them to estimate the entropy density of random formulas. As we will see, they give the right answer precisely up to the density where our assumptions fail—where long-range correlations develop between the variables, so that the neighbors of a vertex in the factor graph are no longer independent.

14.6.6 Beliefs in Random k-SAT

As we discussed in Section 14.6.2, the factor graphs of random k-SAT formulas are locally treelike. Most vertices do not lie on any short cycles, and the roundabout paths connecting their neighbors to each other typically have length $\Omega(\log n)$. If we assume that correlations decay with distance, then the neighbors of each vertex v are effectively independent once we condition on the value of v, and the BP fixed point gives a good approximation for their marginals.

In this section we make exactly this assumption. We then show how to compute the fixed-point distribution of the messages, which are themselves probability distributions. A distribution of distributions sounds like a complicated mathematical object, but in our case it's just a distribution of real numbers.

First we strain the reader's patience with one more shift in notation. Given a variable i and a clause a in which it appears, we define $\eta_{i\to a}$ and $\widehat{\eta}_{a\to i}$ as the probability that i disagrees with a if its value is chosen according to $\mu_{i\to a}$ or $\mu_{a\to i}$ respectively. If i appears positively in a, this is the probability that i is false,

$$\eta_{i\to a} = \mu_{i\to a}(0) \quad \text{and} \quad \widehat{\eta}_{a\to i} = \mu_{a\to i}(0),$$

and if i is negated then it is the probability that i is true,

$$\eta_{i\to a} = \mu_{i\to a}(1) \quad \text{and} \quad \widehat{\eta}_{a\to i} = \mu_{a\to i}(1).$$

In order to write the BP equations in terms of these ηs, for each i and a we divide i's other clauses $\partial i - a$ into two groups: $S_{i,a}$ consists of the clauses in which i has the same sign (positive or negative) as it does in a, and $U_{i,a}$ consists of the clauses in which i appears with the opposite sign. See Figure 14.24 for an example.

We can now rewrite the BP equations (14.67) as (exercise!)

$$\widehat{\eta}_{a\to i} = \frac{1-\prod_{j\in\partial a-i}\eta_{j\to a}}{2-\prod_{j\in\partial a-i}\eta_{j\to a}} \tag{14.75a}$$

$$\eta_{i\to a} = \frac{\prod\limits_{b\in S_{i,a}}\widehat{\eta}_{b\to i}\prod\limits_{b\in U_{i,a}}(1-\widehat{\eta}_{b\to i})}{\prod\limits_{b\in S_{i,a}}\widehat{\eta}_{b\to i}\prod\limits_{b\in U_{i,a}}(1-\widehat{\eta}_{b\to i}) + \prod\limits_{b\in S_{i,a}}(1-\widehat{\eta}_{b\to i})\prod\limits_{b\in U_{i,a}}\widehat{\eta}_{b\to i}}\,. \tag{14.75b}$$

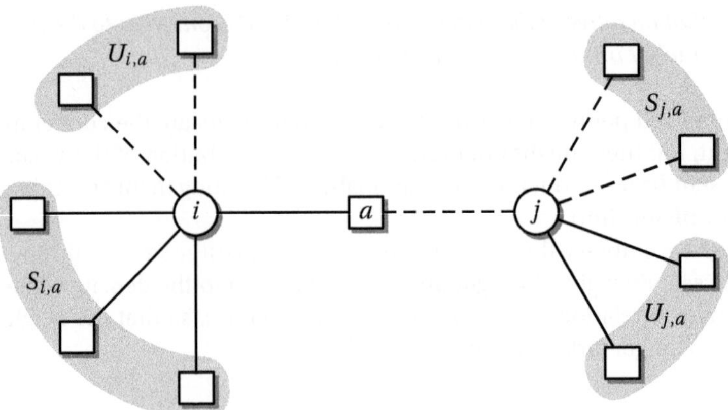

FIGURE 14.24: Given a variable i and a clause a in which it appears, we group i's other clauses into two sets: $S_{i,a}$, in which i appears with the same sign as in a, and $U_{i,a}$, in which it appears with the opposite sign. As before, dashed lines indicate negations.

For instance, $\widehat{\eta}_{a\to i}$ is the probability that i disagrees with a, assuming that a is the only clause in which i appears. The probability that a is not satisfied by any of its other variables j is the product $\prod_{j\in\partial a-i}\eta_{j\to a}$. If that occurs then $\widehat{\eta}_{a\to i}$ must be 0, forcing i to agree with a and satisfy it. On the other hand, if a is already satisfied by some other j then a doesn't care what value i takes, and $\widehat{\eta}_{a\to i}=1/2$. This is exactly what (14.75a) does. Similarly, (14.75b) causes i to send a "warning" to a if it is forced to disagree with a by its other clauses.

Let S_i and U_i be the set of clauses in ∂i in which i appears positively or negatively respectively. Then we can write the entropy in terms of the ηs and $\widehat{\eta}s$ by substituting the following expressions in (14.71):

$$\begin{aligned} z_a &= 1-\prod_{i\in\partial a}\eta_{i\to a} \\ z_i &= \prod_{a\in S_i}\widehat{\eta}_{a\to i}\prod_{a\in U_i}(1-\widehat{\eta}_{a\to i})+\prod_{a\in S_i}(1-\widehat{\eta}_{a\to i})\prod_{a\in U_i}\widehat{\eta}_{a\to i} \\ z_{i,a} &= \widehat{\eta}_{a\to i}\eta_{i\to a}+(1-\widehat{\eta}_{a\to i})(1-\eta_{i\to a}). \end{aligned} \tag{14.76}$$

Our next step is to understand the behavior of the BP equations on the factor graph of a random k-SAT formula. In a random formula, the messages are random variables themselves, and we need to determine how they are distributed. The key fact is that every variable i is equally likely to appear in each clause a. As a consequence, we can assume that the distributions of the messages $\eta_{i\to a}$ and $\widehat{\eta}_{a\to i}$ are independent of i and a.

In other words, we will assume that each message $\eta_{i\to a}$ a clause receives from its variables is chosen independently from some distribution of messages η. Similarly, each message $\widehat{\eta}_{a\to i}$ that a variable receives from its clauses is chosen independently from some distribution of messages $\widehat{\eta}$. All we have to do is find two distributions, one for η and one for $\widehat{\eta}$, that are consistent with the BP equations (14.75).

According to (14.75a), each $\widehat{\eta}$ depends on $k-1$ incoming ηs. So if $\eta_1, \ldots, \eta_{k-1}$ are drawn independently from the distribution of ηs, we can write

$$\widehat{\eta} \stackrel{d}{=} \frac{1-\eta_1 \cdots \eta_{k-1}}{2-\eta_1 \cdots \eta_{k-1}}. \tag{14.77a}$$

The sign $\stackrel{d}{=}$ denotes *equality in distribution*—that is, each side of the equation is a random variable and they have the same distribution.

The corresponding equation for the ηs involves a little more randomness. First we choose the numbers of clauses besides a that i appears in with each sign, setting $|S_{i,a}| = p$ and $|U_{i,a}| = q$ where p and q are independently Poisson-distributed with mean $k\alpha/2$. We then apply (14.75b), giving

$$\eta \stackrel{d}{=} \frac{\prod_{i=1}^{p} \widehat{\eta}_i \prod_{i=1}^{q} (1-\widehat{\eta}_{p+i})}{\prod_{i=1}^{p} \widehat{\eta}_i \prod_{i=1}^{q} (1-\widehat{\eta}_{p+i}) + \prod_{i=1}^{p} (1-\widehat{\eta}_i) \prod_{i=1}^{q} \widehat{\eta}_{p+i}}, \tag{14.77b}$$

where $\widehat{\eta}_i$ is drawn independently from the distribution of $\widehat{\eta}$s for each $1 \le i \le p+q$.

Before, we started with some initial set of messages and iterated (14.67) until we reached a fixed point. Now we start with a pair of initial distributions of messages, each of which is a probability distribution on the unit interval, and iterate (14.77a) and (14.77b) until we reach a fixed point—a pair of distributions that remain fixed if we use (14.77a) and (14.77b) to generate new $\widehat{\eta}$s and ηs.

In practice, we do this with a numerical method called *population dynamics*. We represent the distributions for η and $\widehat{\eta}$ by two lists or "populations" of numbers, $\eta_1, \ldots, \eta_N$ and $\widehat{\eta}_1, \ldots, \widehat{\eta}_N$ for some large N. In each step, we choose $k-1$ random entries from the η population and use (14.77a) to update an element in the $\widehat{\eta}$ population. Similarly, after choosing p and q from the Poisson distribution with mean $k\alpha/2$, we choose $p+q$ random entries from the $\widehat{\eta}$ population and use (14.77b) to update an element in the η population. After we have done enough of these steps, the two populations are good representatives of the fixed point distributions for η and $\widehat{\eta}$.

Once we have the fixed point distributions of η and $\widehat{\eta}$, we can use (14.76) to derive the distribution of the quantities appearing in our expression (14.71) for the entropy. The entropy density—or at least what BP thinks it is—is then

$$\begin{aligned} s(\alpha) &= \alpha \mathbb{E}[\ln z_a] + \mathbb{E}[\ln z_i] - k\alpha \mathbb{E}[\ln z_{i,a}] \\ &= \alpha \mathbb{E}\left[\ln\left(1-\prod_{i=1}^{k} \eta_i\right)\right] \\ &\quad + \mathbb{E}\left[\ln\left(\prod_{i=1}^{p} \widehat{\eta}_i \prod_{i=1}^{q} (1-\widehat{\eta}_{p+i}) + \prod_{i=1}^{p} (1-\widehat{\eta}_i) \prod_{i=1}^{q} \widehat{\eta}_{p+i}\right)\right] \\ &\quad - k\alpha\, \mathbb{E}\left[\ln\left(\widehat{\eta}\eta + (1-\widehat{\eta})(1-\eta)\right)\right], \end{aligned} \tag{14.78}$$

where each η or $\widehat{\eta}$ is chosen independently and p and q are again Poisson-distributed with mean $k\alpha/2$.

Figure 14.25 shows this value of the entropy density for random 3-SAT. The result is fairly convincing—at $\alpha = 0$ we have $s = \ln 2 = 0.6931\ldots$ as it should be, and $s(\alpha)$ decreases monotonically as α increases. It looks like $s(\alpha) = 0$ at $\alpha \approx 4.6$. Is this where the phase transition takes place?

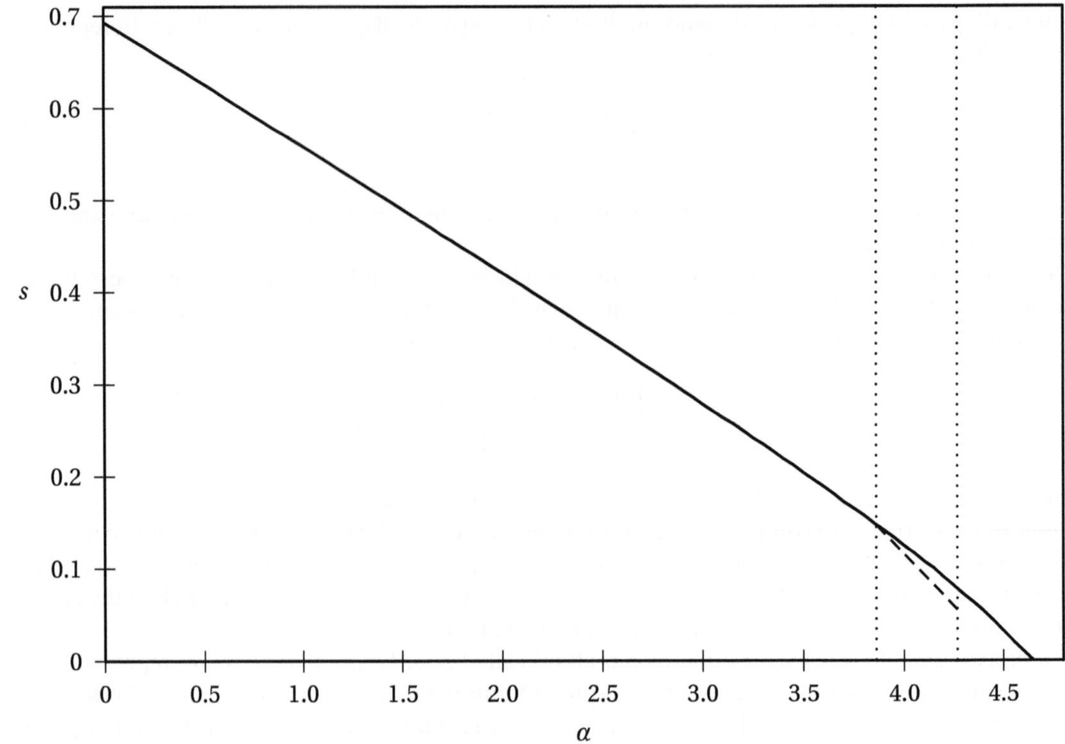

FIGURE 14.25: The entropy density for random 3-SAT according to belief propagation (solid line) and the correct value (dashed line). The first dotted line marks the density $\alpha_{\text{cond}} = 3.86$ beyond which belief propagation fails, and the second dotted line marks the satisfiability threshold $\alpha_c = 4.267$. Note that the true entropy drops to zero discontinuously at α_c.

Let α_{BP} denote the value of α at which $s(\alpha)$ vanishes. Problem 14.43 shows that α_{BP} can be written as the solution of a simple transcendental equation. For $k = 2$ this equation can be solved analytically, and the result

$$\alpha_{\text{BP}} = 1 \tag{14.79}$$

agrees with the rigorously known value $\alpha_c = 1$. For $k > 2$, the transcendental equation for $\alpha^\star$ can be solved numerically, and the result for $k = 3$ is $\alpha^\star = 4.6672805\ldots$. But this can't be α_c. It violates the upper bound $\alpha_c \leq 4.643$ derived in Problem 14.27, not to mention even sharper upper bounds. Thus belief propagation fails to locate the phase transition correctly. Something is wrong with our assumptions.

As we will see in the next section, we believe that BP yields the correct entropy up to some density $\alpha_{\text{cond}} < \alpha_c$, where $\alpha_{\text{cond}} \approx 3.86$ for $k = 3$. Beyond that point, the key assumption behind BP—namely, that the neighbors of each vertex are independent of each other—must break down. In other words, the variables become correlated across large distances in the factor graph. In the next section, we will see how these correlations are related to phase transitions in the geometry of the set of satisfying assignments.

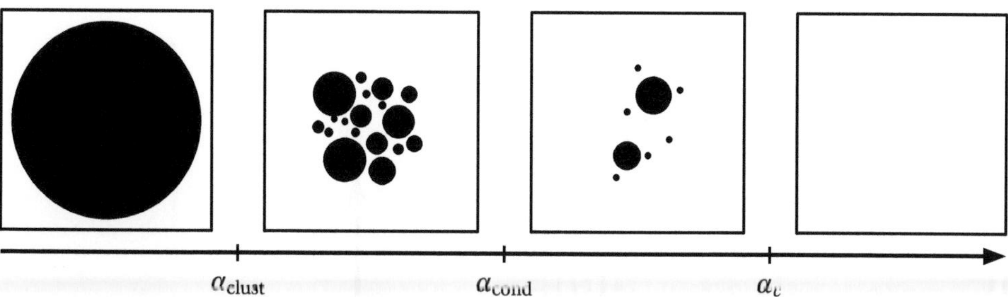

FIGURE 14.26: Phase transitions in random k-SAT. Below the clustering transition at α_{clust}, the set of satisfying assignments is essentially a single connected set. At α_{clust} it breaks apart into exponentially many clusters, each of exponential size, which are widely separated from each other. At the condensation transition α_{cond}, a subexponential number of clusters dominates the set. Finally, at α_c all clusters disappear and there are no solutions at all.

14.7 Survey Propagation and the Geometry of Solutions

The human heart likes a little disorder in its geometry.
Louis de Bernières

Belief propagation fails at a certain density because long-range correlations appear between the variables of a random formula. These correlations no longer decay with distance, so the messages that a clause or variable receives from its neighbors in the factor graph can no longer be considered independent—even though the roundabout paths that connect these neighbors to each other have length $\Omega(\log n)$.

As we will see in this section, these correlations are caused by a surprising geometrical organization of the set of solutions. Well below the satisfiability transition, there is another transition at which the set of solutions breaks apart into exponentially many *clusters*. Each cluster contains solutions that are similar to each other, and the clusters have a large Hamming distance between them.

As the density increases further, there are additional transitions—*condensation*, where a few clusters come to dominate the set of solutions, and *freezing*, where many variables in each cluster are forced to take particular values. We believe that condensation is precisely the point at which BP fails to give the right entropy density, and that freezing is the density at which random k-SAT problems become exponentially hard.

In order to locate these transitions, we need to fix belief propagation so that it can deal with long-range correlations. The resulting method is known as *survey propagation*, and it is one of the main contributions of statistical physics to the interdisciplinary field of random constraint satisfaction problems.

But first we need to address a few basic questions. What is a cluster? What exactly are those blobs in Figure 14.26? And what do they have to do with correlations? To answer these questions we return to a classic physical system: the Ising model.

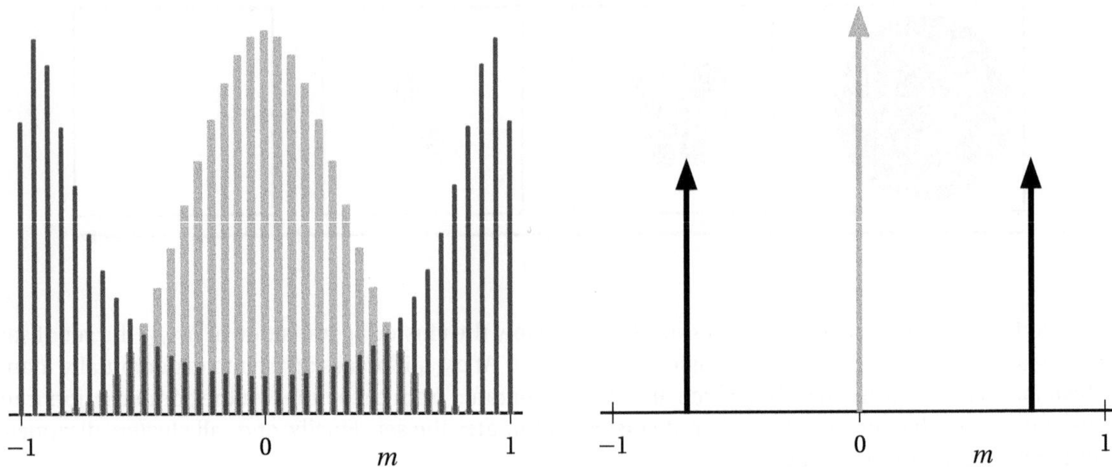

FIGURE 14.27: Left, the probability distribution of the magnetization m in a 6×6 Ising model (left). For $T > T_c$ the states are concentrated around $m = 0$ (gray), whereas for $T < T_c$ they split into two clusters (black), one with $m > 0$ and the other with $m < 0$. On the right, in the limit $n \to \infty$ the corresponding distributions become δ-functions, and the two clusters become disconnected.

14.7.1 Macrostates, Clusters, and Correlations

> The Milky Way is nothing else but a mass of innumerable stars planted together in clusters.
>
> Galileo Galilei

We have already seen a kind of clustering transition in Section 12.1 where we discussed the Ising model of magnetism. Recall that each site i has a spin $s_i = \pm 1$. Above the critical temperature T_c, typical states of the model are homogeneous, with roughly equal numbers of spins pointing up and down. In a large system, the distribution of the magnetization $m = (1/n)\sum_i s_i$ is concentrated around zero. Neighboring spins are likely to be equal, but if i and j are a distance r apart then their correlation $\mathbb{E}[s_i s_j]$ decays exponentially as a function of r.

Below T_c, on the other hand, the set of states splits into two groups or "macrostates," each one with a clear majority of up or down spins. The distribution of m splits into two peaks as shown in Figure 14.27, and the typical value of m is nonzero. By symmetry, if we average over the entire set of states then each spin is equally likely to be up or down, so $\mathbb{E}[s_i] = 0$. But any pair of spins i, j is probably both up or both down, so $\mathbb{E}[s_i s_j] > 0$ no matter how far apart they are.

In many ways, these macrostates act like the clusters in random k-SAT. The main difference is that in the Ising model there are just two of them, while in k-SAT there are exponentially many. Let's look at them from several points of view—with the caveat that while some parts of this picture have made rigorous, others have not.

Intra- and inter-cluster distances. There are constants $0 < \delta_1 < \delta_2 < 1$, such that the typical Hamming distance between two states is $\delta_1 n$ if they are in the same cluster and $\delta_2 n$ if they are in different clusters. In the Ising model, if the magnetization of the two clusters is $+m$ and $-m$, then (exercise!) $\delta_1 = (1-m^2)/2$ and $\delta_2 = (1+m^2)/2$.

In most of the rigorous work on the clustering transition, we define clusters in the strict sense that the Hamming distance between any two states in the same cluster is at most $\delta_1 n$ and that the distance between two states in different clusters is at least $\delta_2 n$. We already saw a hint of this in our second moment calculations in Section 14.4.

Energy barriers and slow mixing. Within each cluster of the Ising model, we can get from one state to another by flipping single spins. But to cross from one cluster to another we have to pass through high-energy states, with boundaries that stretch across the entire lattice. These states have exponentially small probability in the Boltzmann distribution, so it takes a natural Markov chain exponential time to cross them. In the limit of infinite system size, the probability of crossing from one cluster to the other becomes zero. At that point, the Markov chain is no longer ergodic—the state space has split apart into two separate components.

Similarly, to get from one cluster to another in k-SAT we typically have to go through truth assignments that violate $\Theta(n)$ clauses. If we think of the number of unsatisfied clauses as the energy, the clusters become valleys separated by high mountain crags—or in terms of Section 12.8, regions of state space with exponentially small conductance between them. A Markov chain based on local moves will take exponential time to explore this space, and sampling a uniformly random satisfying assignment becomes exponentially hard. For this reason, physicists call the appearance of clusters the *dynamical transition.*

Of course, it might be easy to find a solution even if it is hard to sample a uniformly random one—more about this later.

Boundary conditions and reconstruction. Imagine constructing an equilibrium state σ in the Ising model in some large but finite square. We erase the interior of the square leaving only the boundary of σ, and then sample a new equilibrium state σ' with that same boundary. To what extent does this let us reconstruct our original state σ? Clearly σ and σ' will be similar close to the boundary, but are they also similar deep in the interior?

As we discussed in Section 12.10, above T_c we have "spatial mixing," and the interior of the lattice is essentially independent of its boundary. But below T_c, the boundary conditions affect the lattice all the way to its heart. If our original state σ was mostly up then so is its boundary, and with high probability σ' will be mostly up as well. Even at the center of the square, the probability that σ and σ' agree at a given site will be greater than $1/2$.

We can ask the same question about random k-SAT. We start with a random satisfying assignment $\mathbf{x}$ and erase all its variables within a ball of radius r in the factor graph, leaving its values at the variables on the surface of this ball as its boundary condition. Below the clustering transition, if we select a new satisfying assignment $\mathbf{x}'$ with the same boundary, then x_i' and x_i are essentially independent for variables i deep inside this ball. But above the clustering transition, the reconstructed assignment $\mathbf{x}'$ will be in the same cluster as $\mathbf{x}$, and they will be highly correlated even at the center.

Within each cluster, correlations decay. The correlation, or rather covariance, of two spins in the Ising model is the expectation of their product minus the product of their expectations,

$$\mathbb{E}[s_i s_j] - \mathbb{E}[s_i]\,\mathbb{E}[s_j]\,.$$

Below T_c this is positive—we are in one cluster or the other, and in either case we probably have $s_i = s_j$ regardless of their distance. But this is only true if we average over the entire set of states, since then $\mathbb{E}[s_i] = \mathbb{E}[s_j] = 0$. If we condition on being in the cluster with magnetization m, then $\mathbb{E}[s_i] = \mathbb{E}[s_j] = m$. Once we subtract this bias away, the correlation becomes

$$\mathbb{E}[s_i s_j] - m^2 = \mathbb{E}\left[(s_i - m)(s_j - m)\right].$$

Now the correlation decays exponentially with distance, just as it does above T_c.

Similarly, in k-SAT there are long-range correlations if we average over the entire set of satisfying assignments, generated by the fact that we are in some cluster but we don't know which one. But if we condition on being in any particular cluster and take its biases into account, the correlations decay with distance. That suggests that within each cluster, the assumptions behind the BP equations are right after all, which brings us to...

Clusters are fixed points of the BP equations. If the reader solved Problem 12.5, she saw how the two clusters in the Ising model correspond to fixed points of the mean-field equations—values of the magnetization that are self-consistent if we derive the magnetization of each site from that of its neighbors. In the same way, each cluster in k-SAT is a fixed point of the BP equations—a set of biases or marginals for the variables, which is self-consistent if we derive the marginal of each variable from those of its neighbors in the factor graph.

14.17 Since there are no long-range correlations within each cluster, we can fix BP by applying it to individual clusters and summing up the results. Since each fixed point describes the marginal distributions of the variables, we will have to consider distributions of fixed points, and therefore distributions of distributions. But first we need to discuss how many clusters there are of each size, and how to tell where the clustering and condensation transitions take place.

14.7.2 *Entropies, Condensation, and the Legendre Transform*

In this section and the next we set up some machinery for computing the distribution of cluster sizes. We remind the reader that while this theory is very well-developed from the physics point of view, and while we are confident that it is correct, it has not yet been made rigorous.

In the clustered phase there are exponentially many clusters, and each cluster contains an exponential number of solutions. As with the total number of solutions, we expect the logarithms of these exponentially large numbers to be concentrated around their expectations. Therefore it makes sense to define the *internal entropy* of a cluster as

$$s = \frac{1}{n}\ln(\#\text{ of solutions in a cluster}),$$

and the *configurational entropy* as

$$\Sigma(s) = \frac{1}{n}\ln(\#\text{ of clusters of internal entropy } s).$$

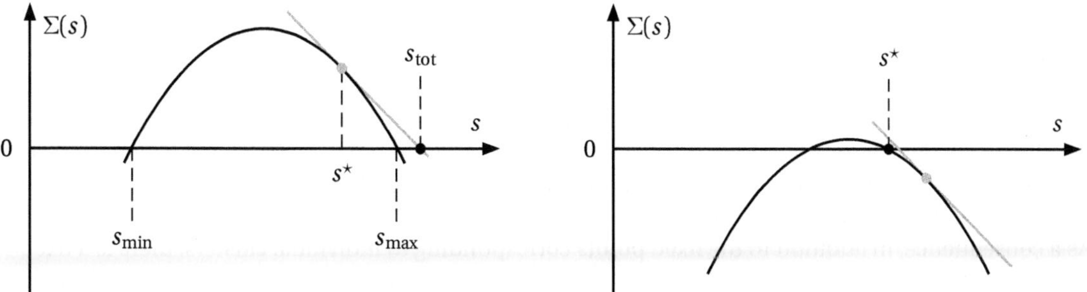

FIGURE 14.28: The total entropy s_{tot} depends on the $s^\star$ that maximizes $\Sigma(s)+s$ as long as $\Sigma(s^\star)\geq 0$. In the clustered phase (left), we have $\Sigma'(s^\star)=-1$ and $\Sigma(s^\star)>0$, so there are exponentially many clusters of size roughly $e^{s^\star n}$. In the condensed phase (right), the point s with $\Sigma'(s)=-1$ has $\Sigma(s)<0$. Then $\Sigma(s^\star)=0$ and the set of solutions is dominated by a subexponential number of clusters.

Both quantities are actually entropy densities, but for simplicity we call them entropies. Thus for each s there are $e^{\Sigma(s)n}$ clusters of size e^{sn}, where we ignore subexponential terms. The total number of solutions is

$$Z=e^{ns_{\text{tot}}}=\int_{s:\Sigma(s)\geq 0} e^{n(\Sigma(s)+s)}\,ds\,, \tag{14.80}$$

where s_{tot} is the total entropy density and we ignore subexponential terms.

Just as in our second moment calculations in Section 14.4, when n is large the integral in (14.80) is dominated by the value of s that maximizes the integrand. The total entropy density is then

$$s_{\text{tot}}=\Sigma(s^\star)+s^\star\,,$$

where

$$s^\star=\operatorname{argmax}\left(\Sigma(s)+s:\Sigma(s)\geq 0\right). \tag{14.81}$$

Note that if $\Sigma(s)$ is concave then this maximum is unique.

There are two possibilities. If $s^\star$ is in the interval where $\Sigma(s)>0$, then

$$\Sigma'(s^\star)=-1\,.$$

This is the *clustered phase.* The dominant contribution to Z comes from $e^{n\Sigma(s^\star)}$ clusters, each of which contains $e^{ns^\star}$ satisfying assignments, an exponentially small fraction of the total number. We show this scenario on the left of Figure 14.28. We based this figure on a toy model of clustering for which $\Sigma(s)$ can be computed explicitly, analyzed in Problem 14.44, but the curves for random k-SAT look very similar.

On the other hand, if there is no s with $\Sigma'(s)=-1$ in the interval where $\Sigma(s)\geq 0$, then $s^\star$ lies at the upper end of this interval as shown on the right of Figure 14.28. In this case, $\Sigma(s^\star)=0$, $s_{\text{tot}}=s^\star$, and Z is dominated by a subexponential number of clusters. This is called the *condensed phase,* since a few clusters contain most of the solutions. In fact, for any constant $\varepsilon>0$ there is a constant number of clusters that contain $1-\varepsilon$ of the satisfying assignments.

As we said in the previous section, within each cluster correlations decay with distance. Thus if we 14.18
condition on being in a particular cluster, the marginal distributions are fixed points of the BP equations.

If we start at precisely that point and iterate the BP equations then we will stay there, and BP will give us the internal entropy s of that cluster.

However, in the clustered phase, even the *total* correlations decay with distance. This follows from the fact that there are exponentially many clusters, all at roughly the same distance from each other and all of roughly the same size. Therefore, the overlap between a random pair of satisfying assignments is tightly concentrated. As Problem 14.45 shows, if the variance in the overlap is small, each variable can be noticeably correlated with just a few others.

As a consequence, in addition to the fixed points corresponding to individual clusters, there is a fixed point corresponding to the sum over all the clusters. If we start BP with *random* messages, we converge to this fixed point with high probability. It gives us the correct marginals, and the correct entropy density s_{tot}, for the entire formula. This is the curve in Figure 14.25 below the condensation transition at α_{cond}.

In the condensed phase, on the other hand, there are only a few clusters, so two random satisfying assignments are in the same cluster with constant probability. In that case the variance in their overlap is large, and Problem 14.45 shows that correlations must stretch across the factor graph.

Above the condensation transition, the BP equations may still converge to a fixed point, but it no longer represents the correct marginals of the variables. It corresponds to clusters of size s such that $\Sigma'(s) = -1$, but these clusters probably don't exist since $\Sigma(s) < 0$. This causes the true entropy $s_{\text{tot}} = s^\star$ to diverge from the BP entropy $\Sigma(s) + s$, shown as the dashed and solid lines in Figure 14.25.

Thus it is condensation, not clustering, which causes BP to fail. Happily, we can modify BP to make it work in even the condensed phase, and to determine the transition points α_{cond} and α_c. The idea is to ramp it up to a kind of "meta-belief propagation," whose messages are distributions of the messages we had before. This is what we will do next.

14.7.3 *Belief Propagation Reloaded*

A fool sees not the same tree that a wise man sees.
William Blake, *Proverbs of Hell*

It would be wonderful if we could compute the function $\Sigma(s)$ directly using some sort of message-passing algorithm. Instead, we will compute the following function $\Phi(m)$,

$$e^{n\Phi(m)} = \int_{s:\Sigma\geq 0} e^{n(\Sigma(s)+ms)}\,\mathrm{d}s\,. \tag{14.82}$$

Since e^{ms} is the mth power of the size of a cluster with internal entropy s, this is essentially the mth moment of the cluster size distribution.

In the limit $n \to \infty$ this integral is again dominated by the value of s that maximizes the integrand. Thus $\Phi(m)$ and $\Sigma(s)$ are related by

$$\Phi(m) = \Sigma(s) + ms\,,$$

where s is given implicitly by

$$\Sigma'(s) = -m\,.$$

This transformation from $\Sigma(s)$ to $\Phi(m)$ is known as the *Legendre transform*. As we discussed in the previous section, below the condensation transition the entropy s_{tot} is dominated by the s such that

$\Sigma'(s) = -1$, so setting $m = 1$ gives $s_{\text{tot}} = \Phi(1) = \Sigma(s) + s$. But other values of m let us probe the function $\Sigma(s)$ at other places, and this will let us determine the condensation and satisfiability thresholds. In particular, if we can compute $\Phi(m)$ for varying m, we can produce a series of pairs $(s, \Sigma(s))$ via the inverse Legendre transform

$$s = \Phi'(m) \quad \text{and} \quad \Sigma(s) = \Phi(m) - ms\,, \tag{14.83}$$

and thus draw graphs like those in Figure 14.28.

Exercise 14.20 *Derive* (14.83). *Keep in mind that s varies with m.*

Let's see how we can compute $\Phi(m)$ with a message-passing approach. We start by rewriting the BP equations (14.67) more compactly as

$$\begin{aligned}\mu_{a\to i} &= f_{a\to i}(\{\mu_{j\to a}\}_{j\in\partial a - i})\\ \mu_{i\to a} &= f_{i\to a}(\{\mu_{b\to i}\}_{b\in\partial i - a}).\end{aligned}$$

Clustering means that these equations have many solutions. Each solution $\mu = \{\mu_{i\to a}, \mu_{a\to i}\}$ represents a cluster with internal entropy s given by (14.71). Consider a probability distribution P_m over all these solutions in which each one is weighted with the factor e^{nms}, the mth power of the size of the corresponding cluster. We can write this as

$$P_m(\mu) = \frac{1}{Z(m)} \prod_a z_a^m \prod_i z_i^m \prod_{(i,a)} z_{i,a}^{-m}\, \delta\big(\mu_{a\to i} - f_{a\to i}(\{\mu_{j\to a}\})\big)\, \delta\big(\mu_{i\to a} - f_{i\to a}(\{\mu_{b\to i}\})\big)\,. \tag{14.84}$$

The δ-functions in (14.84) guarantee that μ is a solution of the BP equations, and the factors z_i^m, z_a^m and $z_{i,a}^{-m}$ give the weight e^{nms} according to (14.71). The normalization constant $Z(m)$ is

$$Z(m) = \sum_\mu \mathrm{e}^{nms} = \int_{s:\Sigma\geq 0} \mathrm{e}^{n(\Sigma(s)+ms)}\, \mathrm{d}s\,,$$

so $\Phi(m)$ is given by

$$\Phi(m) = \frac{1}{n} \ln Z(m).$$

We will use belief propagation to compute $Z(m)$ in much the same way that we computed $S = \ln Z$ for the uniform distribution on satisfying assignments. Bear in mind that the messages in this case are marginals of $P_m(\mu)$, i.e., probability distributions of BP messages $\mu = \{\mu_{i\to a}, \mu_{a\to i}\}$, which are probability distributions themselves. This distribution of distributions is called a *survey*, and the corresponding version of belief propagation is called *survey propagation* or SP for short. But at its heart, SP is just BP on distributions of distributions—that is, distributions of clusters, where each cluster is a distribution of satisfying assignments.

Besides this difference between SP and BP, there is a little technical issue. If $m \neq 0$ then $P_m(\mu)$ is a non-uniform distribution. Thus we need to generalize the message-passing approach to compute marginals for non-uniform distributions.

In Section 14.6.4 we used belief propagation to compute the normalization constant Z in the uniform distribution over all satisfying assignments,

$$\mu(\mathbf{x}) = \frac{1}{Z} \prod_a C_a(\mathbf{x}_{\partial a}).$$

But BP is more versatile than that. It can also deal with non-uniform distributions of the following form,

$$\mu(\mathbf{x}) = \frac{1}{Z} \prod_a \Psi_a(\mathbf{x}_{\partial a}), \tag{14.85}$$

where each weight $\Psi_a(\mathbf{x}_{\partial a})$ is an arbitrary nonnegative function of a subset ∂a of variables. For example, consider the weights

$$\Psi_a(\mathbf{x}_{\partial a}) = \mathrm{e}^{-\beta(1-C_a(\mathbf{x}_{\partial a}))} = \begin{cases} 1 & \text{if } \mathbf{x}_{\partial a} \text{ satisfies clause } a \\ \mathrm{e}^{-\beta} & \text{otherwise.} \end{cases}$$

Rather than prohibiting unsatisfying assignments completely, this choice of Ψ_a reduces their probability by a factor $\mathrm{e}^{-\beta}$ for each unsatisfied clause. Thus μ is the Boltzmann distribution on truth assignments at the temperature $T = 1/\beta$, where the "energy" of an assignment $\mathbf{x}$ is the number of unsatisfied clauses. At $T = 0$ or $\beta = \infty$, we get back the uniform distribution over satisfying assignments we had before.

In this more general setting, the clause vertices of the factor graph are called *function vertices*. It is easy to modify belief propagation to deal with these vertices—we just have to replace $C_a(\mathbf{x}_{\partial a})$ with $\Psi_a(\mathbf{x}_{\partial a})$ in (14.67), (14.68) and (14.70). In particular, the expression (14.71) for the entropy still holds, where the entropy is now $S = -\sum_{\mathbf{x}} \mu(\mathbf{x}) \ln \mu(\mathbf{x})$ rather than simply $\ln Z$.

We will use this generalized BP to compute $Z(m)$. First we rearrange (14.84) a little bit to get

$$P_m(\mu) = \frac{1}{Z(m)} \prod_a \Psi_a(\{\mu_{j\to a}, \mu_{a\to j}\}) \prod_i \Psi_i(\{\mu_{i\to b}, \mu_{b\to i}\}) \prod_{(i,a)} \Psi_{i,a}(\mu_{i\to a}\mu_{a\to i}), \tag{14.86}$$

where

$$\begin{aligned} \Psi_a(\{\mu_{j\to a}, \mu_{a\to j}\}) &= \prod_{j\in\partial a} \delta\left(\mu_{a\to j} - f_{a\to j}(\{\mu_{i\to a}\})\right) z_a^m \\ \Psi_i(\{\mu_{i\to b}, \mu_{b\to i}\}) &= \prod_{b\in\partial i} \delta\left(\mu_{i\to b} - f_{i\to b}(\{\mu_{a\to i}\})\right) z_i^m \\ \Psi_{i,a}(\mu_{i\to a}, \mu_{a\to i}) &= z_{i,a}^{-m}. \end{aligned} \tag{14.87}$$

Notice that we have three kinds of weights: one for each clause a, one for each variable i, and one for each pair (i,a). Thus we can represent (14.86) with a new factor graph defined as in Figure 14.29. Its variable vertices correspond to edges (i,a) of the old factor graph, and each one carries the pair of messages $(\mu_{i\to a}, \mu_{a\to i})$. Each clause a or variable i becomes a function vertex with weight Ψ_a or Ψ_i. Finally, we connect each variable vertex (i,a) to the function vertices i and a, and one more function vertex with weight $\Psi_{i,a}$.

For random formulas this new factor graph is locally treelike since the old one is, so we can compute $P_m(\mu)$ and $\Phi(m)$ much as we computed μ and s before. We send messages ν back and forth between the

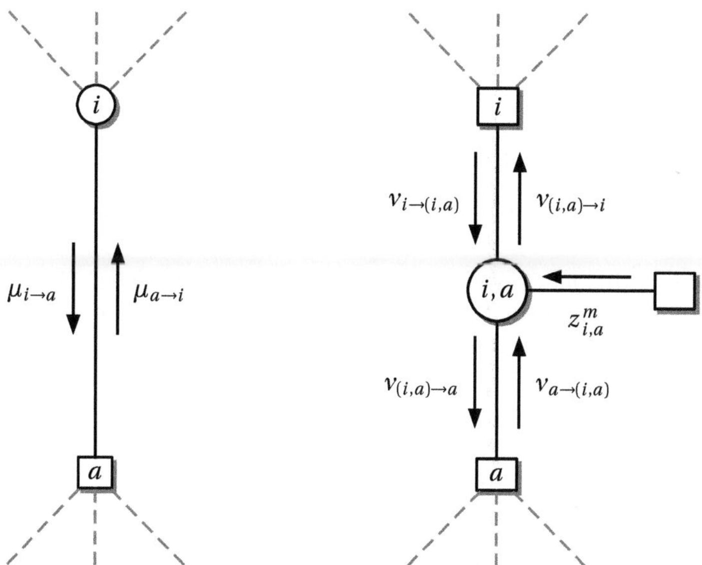

FIGURE 14.29: Messages along an edge in the original factor graph (left) and corresponding messages in the augmented factor graph for SP (right). The messages, or surveys, in the augmented graph are distributions of BP messages.

vertices of the new factor graph as in Figure 14.29. These messages are marginal distributions of pairs $(\mu_{i\to a}, \mu_{a\to i})$, which are themselves marginal distributions on binary variables. Thus the νs are surveys, and survey propagation is simply belief propagation on the new graph.

One can write update equations for the surveys analogous to (14.67), but they are lengthy and jam-packed with new notation. As such they don't comply with the aesthetic standards of this book—and once you have understood the basic idea behind BP, they're not that illuminating either. Therefore we will skip them, and the analog of (14.71) for $\Phi(m)$.

Just as we solved the BP equations with population dynamics, we solve the SP equations with a population of populations. The resulting numerics are challenging, but manageable. Let's discuss the results they yield.

14.7.4 *Results and Reconciliation with Rigor*

If we run survey propagation for many values of m, we get a curve $\Phi(m)$ whose shape depends on α. Below the clustering transition, $\Phi(m)$ is simply

$$\Phi(m) = m\, s_{\text{tot}},$$

which implies that $\Sigma(s) = 0$ and $s = s_{\text{tot}}$. That is, the set of satisfying assignments is dominated by a single cluster with entropy s_{tot}. Moreover, the value of s_{tot} is exactly the entropy predicted by belief propagation as shown in Figure 14.25.

For $\alpha > \alpha_{\text{clust}}$, we get a nonlinear function $\Phi(m)$ whose Legendre transform $\Sigma(s)$ is a concave function very similar to the function shown in Figure 14.28. We can then determine the condensation threshold

k	α_{clust}	α_{cond}	α_c
3		3.86	4.26675
4	9.38	9.547	9.93
5	19.16	20.80	21.12
6	36.53	43.08	43.4

TABLE 14.4: Transition points in random k-SAT.

α_{cond} by checking to see whether $\Sigma(s) > 0$ when $\Sigma'(s) < -1$, and the satisfiability threshold α_c where s_{tot} drops to zero. If we are only interested in locating these transitions, the following exercise shows that it suffices to compute Φ and Φ' at $m = 0$ and $m = 1$:

Exercise 14.21 *Show that α_{clust} and α_{cond} are the smallest and largest α such that $\Phi(1) - \Phi'(1) \geq 0$, and that α_c is the largest α such that $\Phi(0) \geq 0$.*

Table 14.4 lists the numerical values of the thresholds for some small values of k, found by solving the SP equations with population dynamics. The case $k = 3$ is special because it has no phase with an exponential number of dominating clusters. For large k, we can get series expansions for these thresholds in the small parameter 2^{-k}. The results are as follows:

14.19

$$\begin{aligned}
\alpha_{\text{clust}} &= \frac{2^k}{k}\left[\ln k + \ln\ln k + 1 + O\left(\frac{\ln\ln k}{\ln k}\right)\right] \\
\alpha_{\text{cond}} &= 2^k \ln 2 - \frac{3}{2}\ln 2 + O(2^{-k}) \\
\alpha_c &= 2^k \ln 2 - \frac{1}{2}(1 + \ln 2) + O(2^{-k}).
\end{aligned} \tag{14.88}$$

The numerical values for α_c from Table 14.4 agree very well with experiments, and they are consistent with known rigorous bounds. For large k, the series expansion (14.88) for α_c falls right into the asymptotic window (14.47) we derived using first and second moment bounds. Yet the question remains: how reliable are these results?

Our central assumptions were that long-range correlations between variables are only caused by clustering, and that within a single cluster correlations decay exponentially with distance. Neither assumption has been established rigorously, but there are some rigorous results that support the clustering picture. The following theorem tells us that at certain densities consistent with (14.88) there are exponentially many clusters, that they are separated by large Hamming distances, and that we can't get from one to another without violating many clauses:

14.20

Theorem 14.5 *There exist constants β, γ, ϑ and δ and a sequence $\varepsilon_k \to 0$ such that for*

$$(1+\varepsilon_k)\frac{2^k}{k}\ln k \leq \alpha \leq (1-\varepsilon_k)2^k \ln 2, \tag{14.89}$$

the set of satisfying assignments of $F_k(n, m = \alpha n)$ can be partitioned into clusters such that, with high probability,

- *the number of clusters is at least* $e^{\beta n}$,
- *each cluster contains at most an* $e^{-\gamma n}$ *fraction of all solutions,*
- *the Hamming distance between any two clusters is at least* δn, *and*
- *any path between solutions in distinct clusters violates at least* ϑn *clauses somewhere along the way.*

Note that (14.89) perfectly agrees with the leading terms of the large k series (14.88) for α_{clust} and α_c.

Theorem 14.5 is strong evidence that the clustering picture is true. Additional evidence comes from the fact that survey propagation can be turned into an algorithm that solves random instances of 3-SAT in essentially linear time for densities α quite close to α_c. Indeed, for many computer scientists it was the performance of this algorithm that first convinced them that the physicists' picture of random k-SAT had some truth to it.

However, we believe that neither survey propagation, nor any other polynomial-time algorithm, can 14.21
solve random k-SAT all the way up to the satisfiability threshold—that there is a range of densities where random instances are exponentially hard. To explain this and our intuition behind it, we have to discuss yet another phase transition in the structure of the set of solutions.

14.8 Frozen Variables and Hardness

> Despair is the price one pays for setting oneself an impossible aim. It is, one is told, the unforgivable sin, but it is a sin the corrupt or evil man never practices. He always has hope. He never reaches the freezing-point of knowing absolute failure.
>
> Graham Greene

We have seen that, below the satisfiability transition, there are several transitions in the structure of the satisfying assignments of a random formula. How do these transitions affect the hardness of these problems? Is there a range of densities α where we believe random SAT problems are exponentially hard?

Linear-time algorithms like those we studied in Section 14.3 all fail well below the satisfiability threshold. As Problem 14.18 shows, they succeed for k-SAT only up to some $\alpha = \Theta(2^k/k)$, and above this density the corresponding DPLL algorithms take exponential time as shown in Figure 14.3. Except for the factor of $\ln k$, this is roughly where the clustering transition takes place according to (14.88). Is it clustering that makes these problems hard?

As we remarked in Section 14.7.1, we believe that clustering does make it hard to sample uniformly random solutions, since the mixing time of reasonable Markov chains becomes exponential. However, this doesn't mean that it's hard to find a single solution. For instance, there are linear-time algorithms that solve GRAPH 3-COLORING on random graphs at average degrees above the clustering transition. So if not clustering, what is it exactly that keeps algorithms from working? 14.22

All the algorithms for SAT we have met in this book are "local" in some sense. DPLL algorithms (Section 14.1.1) set one variable at a time, and algorithms like WalkSAT (Section 10.3) flip one variable at a time. Both of them choose which variable to set or flip using fairly simple criteria—the number of clauses in which a variable appears, or how many clauses flipping it would satisfy.

Now imagine an instance of SAT for which many of the variables are *frozen*. That is, each of these variables has a "correct" value which it takes in all solutions. Imagine further that these correct values are the result of global interactions across the entire formula, and not simply consequences of these variables' immediate neighborhood. A DPLL algorithm is sure to set some of these variables wrong early on in its search tree, creating a subtree with no solutions. If there is no simple way to see what it did wrong, it has to explore this subtree before it realizes its mistake—forcing it to take exponential time.

As the following exercise shows, we can't literally have a large number of variables that are globally frozen in this sense:

Exercise 14.22 *Consider an instance of random k-SAT at some density $\alpha < \alpha_c$. Assume that $\Theta(n)$ variables are frozen, i.e., they take fixed values in all satisfying assignments. Show that adding a single random clause renders this instance unsatisfiable with finite probability. Therefore we have $\alpha_c < \alpha + \varepsilon$ for any $\varepsilon > 0$, a contradiction.*

But even if there are no variables that are frozen in *all* solutions, we could certainly have variables frozen within a cluster. Let's say that a variable x_i is frozen in a cluster C if x_i takes the same value in every solution in C, and call a cluster frozen if it has κn frozen variables for some constant $\kappa > 0$.

Intuitively, these frozen clusters spell doom for local algorithms. Image a DPLL algorithm descending into its search tree. With every variable it sets, it contradicts any cluster in which this variable is frozen with the opposite value. If every cluster is frozen then it contradicts a constant fraction of them at each step, until it has excluded every cluster from the branch ahead. This forces it to backtrack, so it takes exponential time.

It's also worth noting that if the clusters are a Hamming distance δn apart, the DPLL algorithm is limited to a single cluster as soon as it reaches a certain depth in the search tree. Once it has set $(1-\delta)n$ variables, all the assignments on the resulting subtree are within a Hamming distance δn of each other, so they can overlap with at most one cluster. If any of the variables it has already set are frozen in this cluster, and if it has set any of them wrong, it is already doomed.

Recent rigorous results strongly suggest that this is exactly what's going on. In addition to the other properties of clusters established by Theorem 14.5 at densities above $(2^k/k)\ln k$, one can show that almost all clusters have κn frozen variables for a constant $\kappa > 0$. Specifically, if we choose a cluster with probability proportional to its size, it has κn frozen variables with high probability. Equivalently, if we choose a uniformly random satisfying assignment, with high probability there are κn variables on which it agrees with every other solution in its cluster.

Conversely, it can be shown that algorithms based on setting one variable at a time using BP messages fail in this frozen region. But in a recent breakthrough, an algorithm was discovered that works at densities up to $(1-\varepsilon_k)(2^k/k)\ln k$, where $\varepsilon_k \to 0$ as $k \to \infty$. Thus for large k, it seems that algorithms end precisely where the frozen phase begins.

For large k, clustering and freezing take place at roughly the same density. In contrast, for small k they are widely separated, which explains why some algorithms can probe deep into the clustered phase. Figure 14.30 shows a refined picture of random k-SAT that includes frozen clusters. The freezing transition is defined by the density α_{rigid} at which the number of unfrozen clusters drops to zero.

14.23

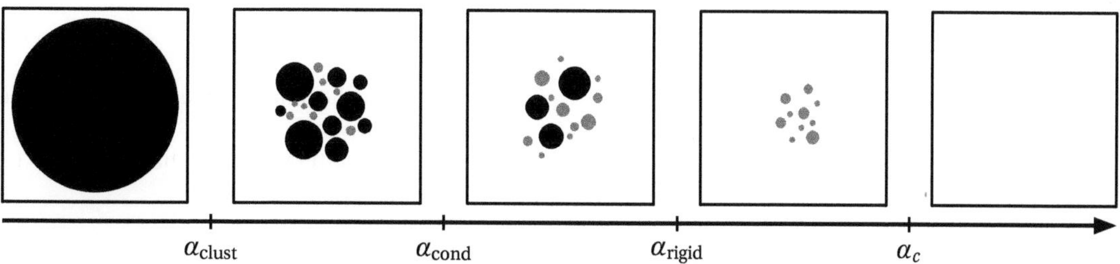

FIGURE 14.30: A refined phase diagram of random k-SAT. Gray blobs represent frozen clusters, i.e., those where $\Theta(n)$ variables take fixed values. Above α_{rigid}, almost all clusters are frozen, and we believe that this is responsible for the average-case hardness of random k-SAT.

For finite k we can determine the freezing point α_{rigid} numerically using survey propagation. A frozen variable corresponds to BP messages $\mu_{i\to a}(x) = \delta(x)$ or $\mu_{i\to a}(x) = \delta(1-x)$, demanding that i take a particular value with probability 1. If the fixed point surveys place a high weight on these "freeze!" messages, then almost every cluster is frozen. Numerical results for small k then give additional evidence for the link between freezing and hardness. For instance, random 3-SAT freezes at $\alpha_{\text{rigid}} = 4.254 \pm 0.009$, and the best known algorithms seem to stop working a hair's breadth below that point.

The clusters do not freeze all at once. Intuitively we expect frozen clusters to be smaller than unfrozen ones, and this is confirmed by survey propagation. At a given density between α_{clust} and α_{rigid} there is an s_{frozen} such that almost all clusters with internal entropy $s \leq s_{\text{frozen}}$ are frozen, while almost all with $s > s_{\text{frozen}}$ are still "liquids" where every variable can take either value. We can still solve problems in this middle range, but the solutions our algorithms find are always in the unfrozen clusters—yet more evidence that freezing is what makes problems hard. 14.24

We end, however, with a cautionary note. Consider random XORSAT, where each clause demands that its variables have a certain parity. Its clustering transition can be determined rigorously, and it freezes at exactly the same point. Yet we can solve it in polynomial time by treating an XORSAT formula as a system of linear equations mod 2, and solving this system using Gaussian elimination.

Of course, Gaussian elimination is not a local algorithm. It applies a series of global transformations that redefine the variables and simplify the problem until it becomes trivial. Indeed, if we run `DPLL` or `WalkSAT` on random XORSAT instances, we find that they take exponential time just as they do on random SAT in the frozen phase. 14.25

This proves that statistical properties alone, such as clustering or freezing in the set of solutions, are not enough to prove that a problem is outside P. On the other hand, it gives us another way to express our belief that P $\neq$ NP—that for NP-complete problems like k-SAT, there is no global algorithm that can magically rearrange the problem to make it easy. We are still waiting for a modern Euler to show us a way to cut through these haystacks, or prove that local search algorithms are essentially the best we can do.

Problems

It is easier to judge the mind of a man by his questions rather than his answers.

Pierre-Marc-Gaston, duc de Lévis (1808)

14.1 Equivalent models. We have two models of random graphs with average degree c: $G(n, p)$ where $p = c/n$, and $G(n, m)$ where $m = cn/2$. In this problem we will show that they have the same threshold behavior for the kinds of properties we care about.

A property of graphs is *monotone decreasing* if adding additional edges can make it false but never make it true. Examples of monotone properties include planarity and 3-colorability. Suppose that a monotone decreasing property P has a sharp threshold c^* in one of the two models. That is, P holds with high probability for any $c < c^*$, and with low probability for any $c > c^*$. Show that P has exactly the same threshold c^* in the other model.

Hint: use the fact that if we condition on the even that $G(n, p)$ has m edges then it is random according to $G(n, m)$. You may find the Chernoff bound of Appendix A.5.1 useful.

14.2 Locally treelike. For each $\ell \geq 3$, calculate the expected number of cycles of length ℓ in $G(n, p = c/n)$. Then sum over ℓ and show that if $c < 1$ the expected total number of vertices in all these cycles is $O(1)$. Hint: a set of 4 vertices can be connected to form a cycle in 3 different ways.

Using the same calculation, show that for any constant c, there is a constant B such that the total number of vertices in cycles of length $B \log n$ or less is $o(n)$. Therefore, for almost all vertices, the smallest cycle in which they participate is of size $A \log n$ or greater. In particular, almost all vertices have neighborhoods that are trees until we get $\Omega(\log n)$ steps away.

14.3 Small components. Here we compute the distribution of component sizes in $G(n, p = c/n)$ with $c < 1$. Consider the number of trees that span connected components of size k, which we denote T_k. This is an upper bound on the number of connected components of size k, and is a good estimate as long as most of these components are trees. Using Cayley's formula k^{k-2} for the number of ways to connect k vertices together in a tree (see Problem 3.36), show that

$$\mathbb{E}[T_k] \leq \binom{n}{k} k^{k-2} p^{k-1} (1-p)^{k(n-k)}. \tag{14.90}$$

Show that if $c < 1$, there is a $\lambda < 1$ such that

$$\mathbb{E}[T_k] = O(n\lambda^k),$$

assuming that $k = o(\sqrt{n})$. Using the first moment method, conclude that, with high probability, there are no components of size greater than $A \log n$ where A is a constant that depends on c (and diverges at $c = 1$). Combine this with Problem 14.2 to show that most components are trees. There is a gap in this proof, however—namely, our assumption that $k = o(\sqrt{n})$. Do you see how to fix this?

14.4 No bicycles. Use Problem 14.3 to prove that in a sparse random graph with $c < 1$, with high probability there are no *bicycles*—that is, there are no components with more than one cycle. Specifically, let the *excess* E of a connected graph be the number of edges minus the number of vertices. A component with a single cycle has $E = 0$, a bicycle has $E = 1$, and so on. Show that the expected number of components with excess E, and therefore the probability that any exist, is $\Theta(n^{-E})$. Hint: compare Exercise 14.6.

14.5 A power law at criticality. Use the same approach as in the previous problem to show that, at the critical point $c = 1$,

$$\mathbb{E}[T_k] \approx \frac{1}{\sqrt{2\pi}} \frac{n}{k^{5/2}} \left(1 + O(k^2/n)\right). \tag{14.91}$$

A priori, this expression is only accurate for $k = o(\sqrt{n})$ because of the $O(k^2/n)$ error terms. More importantly, and conversely to Exercise 14.6, components of size $\Omega(\sqrt{n})$ often contain loops. In that case they have multiple spanning trees, and T_k becomes an overestimate for the number of connected components. However, the power law (14.91) turns out to be correct even for larger values of k. Assuming that is true, show that with high probability there is no component of size n^β or greater for any $\beta > 2/3$.

14.6 Duality above and below the transition. Argue that if $c > 1$, the small components outside the giant component of $G(n, p = c/n)$ look just like a random graph with average degree $d < 1$, where c and d obey a beautiful duality relation:

$$c\mathrm{e}^{-c} = d\mathrm{e}^{-d}.$$

This equation has two roots. One is $d = c$, and the other gets small as c gets large. As a consequence, the small components get very small as the giant component takes up more and more of the graph, and almost all of them are trees of size $O(1)$.

Hint: reverse the branching process argument for the size of the giant component by conditioning on the event that v's branching process dies out after a finite number of descendants, i.e., that v is not abundant. Show that this gives a branching process where the average number of children is d.

14.7 High expectations and low probabilities. Let T and H denote the number of spanning trees and Hamiltonian cycles in $G(n, p)$. Show that if $p = c/n$ then $\mathbb{E}[T]$ is exponentially large whenever $c > 1$, and $\mathbb{E}[H]$ is exponentially large whenever $c > \mathrm{e}$. On the other hand, show that for any constant c the graph is disconnected with high probability, so with high probability both T and H are zero. This is another example of how expectations can be misleading—$\mathbb{E}[T]$ and $\mathbb{E}[H]$ are dominated by exponentially rare events where T and H are exponentially large.

14.8 A threshold for connectivity. Consider a random graph $G(n, p)$ where $p = (1+\varepsilon)\ln n/n$ for some $\varepsilon > 0$. Show that with high probability, this graph is connected. Hint: bound the expected number of subsets of $k \geq 1$ vertices which have no edges connecting any of them to the other $n - k$ vertices. You might find it useful to separate the resulting sum into those terms with $k < \delta n$ and those with $\delta n \leq k \leq n/2$ for some small $\delta > 0$.

Conversely, suppose that $p = (1-\varepsilon)\ln n/n$ for some $\varepsilon > 0$. Show that now with high probability $G(n, p)$ is not connected, for the simple reason that it probably has at least one isolated vertex. Thus the property of connectivity has a sharp transition at $p = \ln n/n$.

14.9 No small cores, and no small culprits for colorability. Show that in $G(n, p = c/n)$ for any constant c, there is a constant $\varepsilon > 0$ such that with high probability the smallest subgraph with minimum degree 3 has size at least εn. Therefore, the 3-core is either of size $\Theta(n)$ or is empty. As a consequence, even when we are above the 3-colorability threshold, the smallest non-3-colorable subgraph has size $\Theta(n)$. Thus non-colorability is a global phenomenon—it cannot be blamed on any small subgraphs.

Hint: consider the simpler property that a subgraph of size s has at least $3s/2$ edges. Then use the inequality $\binom{a}{b} \leq (\mathrm{e}a/b)^b$ to show that the expected number of such subgraphs is at most $(as/n)^{s/2}$ for some constant a.

14.10 The configuration model. In the Erdős–Rényi model, the degree of each vertex is a random variable. If we want to define a random graph with a particular degree distribution $\{a_j\}$, where there are exactly $a_j n$ vertices of each degree j, we need a different model. The *configuration model* works by making j copies of each vertex of degree j, corresponding to a "spoke" as in Section 14.2.5. We then form a graph by choosing a uniformly random perfect matching of all these spokes.

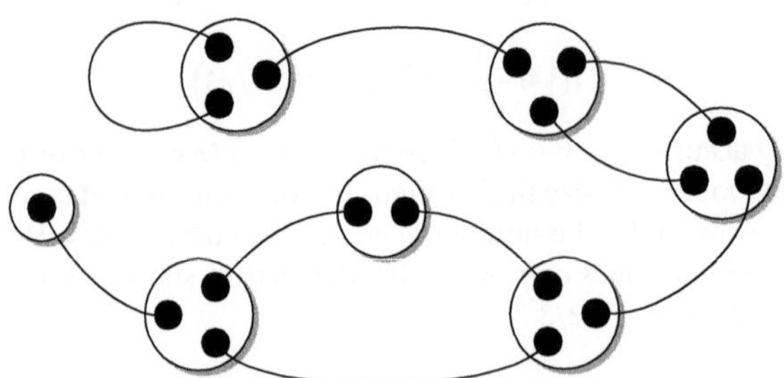

FIGURE 14.31: The configuration model. To create a random multigraph with a specific degree distribution, we create j copies or "spokes" for each vertex of degree j, and then attach them according to a random perfect matching.

Of course, this may lead some self-loops or multiple edges as shown in Figure 14.31. Thus this model produces a random multigraph. However, show that if we condition on the graph being simple, i.e., without self-loops or multiple edges, then every simple graph appears with the same probability. Moreover, show that if the degree distribution has a finite mean and variance—that is, if $\sum_j j a_j$ and $\sum_j j^2 a_j$ are both $O(1)$—then the expected number of self-loops and multiple edges in the entire graph is $O(1)$.

14.11 The emergence of the giant component in the configuration model. Like $G(n,p)$, the configuration model has a threshold at which a giant component appears. We can derive, at least heuristically, this threshold using a branching process argument. It's helpful to think of spokes, rather than vertices, as the individuals in this process. Each spoke s looks for a partner t, which is chosen uniformly from all possible spokes. If t belongs to a vertex of degree j, the other $j-1$ spokes of that vertex become the children of s. If the degree distribution is $\{a_j\}$, show that the average number of children is

$$\lambda = \frac{\sum_j j(j-1)a_j}{\sum_j j a_j} = \frac{\mathbb{E}[j^2]}{\mathbb{E}[j]} - 1 .$$

Then there is a giant component, with high probability, if and only if $\lambda > 1$. Note that this depends, not just on the average degree, but on its second moment. Finally, show that if the degree distribution is Poisson, $a_j = e^{-c} c^j / j!$, then $\lambda = c$ as before.

14.12 Random regular graphs are connected. Consider a random 3-regular multigraph G formed according to the configuration model as described in the previous problem. In other words, given n vertices where n is even, we choose a uniformly random perfect matching of $m = 3n$ copies. We will show that G is connected with high probability.

First show that the number of perfect matchings of m copies is

$$(m-1)(m-3)(m-5)\cdots 5\cdot 3\cdot 1 = \frac{m!}{2^{m/2}(m/2)!} ,$$

which is sometimes written $(m-1)!!$. Then show that the probability that a given set of t copies is matched only with itself, so that the remaining $m-t$ copies are matched with themselves, is

$$\binom{m/2}{t/2} \Big/ \binom{m}{t} .$$

Conclude that the expected number of sets of s vertices that are disconnected from the rest of the graph is

$$\binom{n}{s}\binom{3n/2}{3s/2}\bigg/\binom{3n}{3s} \le 1\bigg/\binom{n/2}{s/2},$$

where we use the inequality $\binom{a}{b}\binom{c}{d} \le \binom{a+c}{b+d}$ (which you should also prove). Finally, show that the probability that G is disconnected is at most

$$\sum_{s=2,4,6}^{n-2} 1\bigg/\binom{n/2}{s/2} = O(1/n).$$

Thus G is connected with probability $1-O(1/n)$. How does this generalize to random d-regular graphs for larger d?

14.13 Random regular graphs are expanders. Recall from Section 12.9 that a d-regular graph $G=(V,E)$ with n vertices is an *expander* if there is a constant $\mu>0$ such that, for any $S\subseteq V$ where $|S|\le n/2$, we have $|E(S,\overline{S})|/|S|\ge \mu d$ where $E(S,\overline{S})$ denotes the set of edges crossing from S to $\overline{S}$. In this problem, we will show that random 3-regular graphs are expanders with high probability. In particular, they almost always have diameter $O(\log n)$.

First, similar to the previous problem, show that in a uniformly random perfect matching of m things, the probability that a given set T of size t has exactly ℓ partnerships with things outside the set, so that the remaining $t-\ell$ things inside T are matched with each other, and similarly for the $m-t-\ell$ things outside T, is

$$2^{\ell}\binom{m/2}{(t-\ell)/2}\binom{(m-t+\ell)/2}{\ell}\bigg/\binom{m}{t}.$$

Use this to show that in a random 3-regular multigraph, the expected number of subsets $S\subseteq V$ of size $|S|=s$, such that exactly ℓ edges connect S to the rest of the graph, is

$$2^{\ell}\binom{n}{s}\binom{3n/2}{(3s-\ell)/2}\binom{(3n-3s+\ell)/2}{\ell}\bigg/\binom{3n}{3s}. \tag{14.92}$$

By setting $\ell\le s/10$ and summing over $1\le s\le n/2$, show that the expected number of subsets where $|S|\le n/2$ and fewer than $|S|/10$ edges lead out of S, is $o(1)$. Thus, with high probability, G is an expander with $\mu\ge 1/30$.

Hint: as in Problem 14.8, divide the sum over s into those terms where $s\le \varepsilon n$ and those where $s>\varepsilon n$, for some small ε. For the former, show that when s is small enough, (14.92) decreases geometrically as s increases. For the latter, use Stirling's approximation in the binomial—ignoring the polynomial terms—and show that (14.92) is exponentially small.

14.14 Ramsey doats and dozy doats. A classic puzzle states that at any party of six people, there are either three people who know each other, or three people who don't know each other (assuming these relations are symmetric). In graph-theoretic terms, if we color the edges of the complete graph K_6 red and green, there is always either a red triangle or a green triangle. For K_5, on the other hand, there is a coloring such that no monochromatic triangle exists.

More generally, the *Ramsey number* $R(k,\ell)$ is the smallest n such that, for any coloring of the edges of K_n, there is either a red clique K_k or a green clique K_ℓ. Thus $R(3,3)=6$. In this problem, we prove upper and lower bounds on $R(k,k)$. For an upper bound, prove that

$$R(k,\ell)\le R(k,\ell-1)+R(k-1,\ell),$$

and therefore that

$$R(k,\ell)\le\binom{k+\ell-2}{k-1}.$$

For $k=\ell$ this gives

$$R(k,k)\le\binom{2k-2}{k-1}=O\left(\frac{2^{2k}}{\sqrt{k}}\right). \tag{14.93}$$

Hint: let $n = R(k, \ell - 1) + R(k-1, \ell)$. Focus on a single vertex v in K_n, and consider its red and green edges.

For a lower bound, show that if the edges of K_n are colored randomly, the expected number of monochromatic cliques K_k is $o(1)$ whenever $n < 2^{k/2}\, k/(\sqrt{2}\mathrm{e})$. Thus

$$R(k,k) \geq \frac{2^{k/2}\, k}{\sqrt{2}\mathrm{e}} = \Omega(2^{k/2}\, k). \tag{14.94}$$

Thus, ignoring polynomial terms, $R(k,k)$ is somewhere in between $2^{k/2}$ and 2^{2k}. Surprisingly, these are essentially the best bounds known. At the time we write this, even $R(5,5)$ isn't known—it is at least 43 and at most 49.

14.15 The 2-core emerges slowly. Use (14.11) and (14.10), or equivalently (14.23) and (14.24), to show that the 2-core emerges quadratically when we cross the threshold $c_2^{\mathrm{core}} = 1$. Specifically, show that when $c = 1 + \varepsilon$, the size of the 2-core is

$$\gamma_2 = 2\varepsilon^2 - O(\varepsilon^3).$$

On the other hand, show that for $k \geq 3$, if $c = c_k^{\mathrm{core}} + \varepsilon$ then γ_k behaves as

$$\gamma_k = \gamma_k^* + A_k \sqrt{\varepsilon}$$

for some constant A_k, where γ_k^* is the size of the k-core at the threshold $c = c_k^{\mathrm{core}}$.

14.16 Pure literals and cores. Consider a SAT formula ϕ, and suppose that for some variable x, its appearances in ϕ are all positive or all negative. In that case, it does no harm to set x `true` or `false`, satisfying all the clauses it appears in. Show that this *pure literal rule* satisfies the entire formula unless it has a *core*, i.e., a subformula where every variable appears both positively and negatively. Then use a branching process as in Section 14.2.5 to argue that the threshold at which such a core appears in a random 3-SAT formula $F_3(n, m = \alpha n)$ is $\alpha \approx 1.637$, this being the smallest α such that

$$q = 1 - \mathrm{e}^{-3\alpha q^2/2}$$

has a positive root. Show also that the fraction of variables in the core when it first appears is $q^2 \approx 0.512$.

14.17 The 2-SAT threshold. Recall from Section 4.2.2 that a 2-SAT formula is satisfiable if and only if there is a contradictory loop of implications from some literal x to its negation $\overline{x}$ and back. We can view a random 2-SAT formula $F_2(n, m = \alpha n)$ as a kind of random directed graph with $2n$ vertices, and model all the literals implied by a given x or $\overline{x}$ as the set of vertices reachable from the corresponding vertex along directed paths.

We can model these paths as a branching process, corresponding exactly to the branching process of unit clauses in our analysis of UC. Calculate the ratio λ of this process as a function of α, and show that it becomes critical at $\alpha_c = 1$. Then argue that this is in fact where the satisfiability transition for 2-SAT occurs.

14.18 Algorithmic lower bounds for k-SAT. Consider applying the UC algorithm to random k-SAT formulas. Write down a system of difference equations for the number S_ℓ of ℓ-clauses for $2 \leq \ell \leq k$, transform this into a system of differential equations, and solve it. Then calculate for what α the branching process of unit clauses is subcritical at all times, and show that the threshold for k-SAT is bounded below by $\alpha_c \geq a_k 2^k/k$ where a_k converges to $\mathrm{e}/2$ as k goes to infinity.

14.19 An exact threshold for 1-IN-k SAT. In Problem 5.2 we showed that 1-IN-k SAT is NP-complete. Consider random formulas, where each clause consists of one of the $\binom{n}{k}$ possible sets of variables, each one is negated with probability 1/2, and exactly one of the resulting literals must be true.

Show that if we run UC on these formulas, we get exactly the same differential equations for the density s_ℓ of ℓ-clauses as in Problem 14.18. However, the rate λ at which we produce unit clauses is now much greater, since satisfying any literal in a clause turns its other $\ell - 1$ literals into unit clauses. Show that λ is maximized at $t = 0$, the beginning of the algorithm, and that UC succeeds as long as $\alpha < 1/\binom{k}{2}$, giving the lower bound $\alpha_c \geq 1/\binom{k}{2}$. Conversely, argue that if $\alpha > 1/\binom{k}{2}$ the branching process causes a large number of contradictions. Thus $\alpha_c \leq 1/\binom{k}{2}$, and we can determine α_c exactly.

14.20 Branching on clauses. Consider the Monien–Speckenmeyer algorithm from Problem 6.6, whose branching rule asks for the first literal that satisfies a random 3-clause. Show that the differential equations describing its first branch are

$$\begin{aligned}\frac{\mathrm{d}s_3}{\mathrm{d}t} &= -\frac{3s_3}{1-t} - p_{\text{free}} = \frac{-3s_3+s_2}{1-t} - 1 \\ \frac{\mathrm{d}s_2}{\mathrm{d}t} &= \frac{(3/2)s_3 - 2s_2}{1-t} .\end{aligned} \tag{14.95}$$

By integrating these equations and demanding that the maximum value of λ is less than 1, show that the first branch succeeds with positive probability if

$$\alpha < \alpha^* \approx 2.841 ,$$

suggesting that this is where the Monien–Speckenmeyer algorithm's running time goes from linear to exponential.

14.21 Greedy colorings. We can prove a lower bound on the 3-colorability transition by analyzing the following algorithm. At each step, each uncolored vertex v has a list of allowed colors. Whenever we color one of v's neighbors, we remove that color from its list. Analogous to unit clause propagation, whenever there is a 1-color vertex we take a *forced step* and immediately give it the color it wants. If there are no 1-color vertices, we take a *free step*, which consists of choosing randomly from among the 2-color vertices, flipping a coin to decide between its two allowed colors, and giving it that color. Finally, if there are no 2-color vertices, we choose a vertex randomly from among the 3-color vertices and give it a random color.

Suppose we run this algorithm on $G(n, p = c/n)$. First, argue that if this algorithm succeeds in coloring the giant component, then with high probability it will have no problem coloring the rest of the graph. Hint: show that it has no difficulty coloring trees.

Then, note that during the part of this algorithm that colors the giant component, we always have a 1- or 2-color vertex. Let $S_{1,2}$ be the number of 1- or 2-color vertices, and let S_3 be the number of 2- and 3-color vertices respectively. Assume that each 2-color vertex is equally likely to be missing each of the three colors. Then show that if $s_2 = S_{1,2}/n$ and $s_3 = S_3/n$, these obey the differential equations

$$\begin{aligned}\frac{\mathrm{d}s_3}{\mathrm{d}t} &= -c s_3 \\ \frac{\mathrm{d}s_2}{\mathrm{d}t} &= c s_3 - 1\end{aligned}$$

(note the similarity to the differential equations (14.8) for exploring the giant component). Under the same assumption, show that the rate of the branching process of forced steps is

$$\lambda = \frac{2}{3} c s_2$$

and conclude that $G(n, p = c/n)$ is 3-colorable with positive probability for $c < c^*$, where c^* is the root of

$$c - \ln c = \frac{5}{2} .$$

This gives $c^* \approx 3.847$, and shows that the critical degree for the 3-colorability transition is higher than that for the existence of a 3-core.

14.22 Counting colorings. Use the first moment method to give the following upper bound on the k-colorability threshold in $G(n, p=c/n)$:

$$c_k \leq \frac{2\ln k}{\ln k - \ln(k-1)} = 2k\ln k - \ln k - O\left(\frac{\ln k}{k}\right).$$

In particular, this gives an upper bound of 5.419 on the phase transition in GRAPH 3-COLORING.

Hint: consider a coloring of the vertices, and calculate the probability that there are no edges connecting vertices of the same color. Show that this probability is maximized when there are an equal number of vertices of each color. You may find it convenient to slightly change your model of $G(n,m)$ to one where the m edges are chosen independently with replacement from the $\binom{n}{2}$ possibilities.

14.23 Random negations. Each clause in our random formulas $F_k(n,m)$ consists of a random k-tuple of variables, each of which is negated with probability 1/2. Suppose instead that the set of variables in each clause is completely deterministic—for instance, we could have a lattice in which each clause involves a set of neighboring variables. Show that as long as we still negate each variable independently with probability 1/2, the expected number of satisfying assignments is $\mathbb{E}[X] = 2^n(1-2^{-k})^m$, just as in the original model. Note that this is true even if all m clauses are on the same triplet of variables!

14.24 When the model makes a difference. Our random formulas $F_k(n,m)$ always have exactly m clauses, just as $G(n,m)$ always has exactly m edges. An alternate model, analogous to $G(n,p)$, includes each of the $2^k\binom{n}{k}$ possible clauses independently with probability p. Find the value of p such that the expected number of clauses is αn. Then, calculate the expected number of solutions $\mathbb{E}[X]$ in this model, and show that the resulting first-moment upper bound on α_c is larger than that derived from $F_k(n,m)$. For instance, for $k=3$ we obtain $\alpha_c \leq 8\ln 2 \approx 5.545$. Given that the two models are equivalent in the limit $n \to \infty$, why is this upper bound so much weaker?

14.25 Locally maximal assignments. Given a random formula $F_3(n, m = \alpha n)$, let Y be the number of satisfying assignments σ that are *locally maximal.* That is, if we flip any variable from `false` to `true`, σ is no longer satisfying. Show that if the formula is satisfiable then $Y > 0$, so $\mathbb{E}[Y]$ is an upper bound on the probability it is satisfiable.

Now suppose σ is one of these assignments. For every `false` variable x in σ, there must be a *blocking clause c* which will be dissatisfied if we make x `true`. In other words, c agrees with σ on x, but disagrees with σ on its other two variables. Show that if we choose m clauses independently from those satisfied by σ, the probability that a given x is blocked by at least one of them is

$$1-\left(1-\frac{3}{7n}\right)^m \approx 1-e^{-3\alpha/7}.$$

Now, clearly the events that the false variables are blocked are negatively correlated—each clause can block at most one variable, and there are only a fixed number of clauses to go around. So, we can get an upper bound on $\mathbb{E}[Y]$ by assuming that these events are independent. By summing over the number f of false variables, show that this gives

$$\mathbb{E}[Y] = \left(\frac{7}{8}\right)^m \sum_{f=0}^{n} \binom{n}{f} \left(1-e^{-3\alpha/7}\right)^f.$$

Set $f = \varphi n$, approximate $\binom{n}{f}$ using the entropy function, maximize the resulting function of φ, and show that

$$\alpha_c \leq 4.667,$$

since $\mathbb{E}[Y]$ is exponentially small for large α. In the next two problems, we will improve this bound further by computing the probability that all false variables are blocked, without assuming that these events are independent.

14.26 Better bounds on bins. In preparation for improving our upper bound on the 3-SAT threshold, first we compute the probability of an exponentially unlikely event. Suppose we throw b balls into f bins. What is the probability $P_{\text{occ}}(b,f)$ that every bin is occupied?

We solve this problem using a trick called "Poissonization." First, we imagine that the number of balls that fall in each bin is independently chosen from a Poisson distribution with mean λ. Then for any λ, we can write

$$\begin{aligned}P_{\text{occ}}(b,f) &= \Pr[\text{all bins occupied} \mid b \text{ balls total}] \\ &= \Pr[b \text{ balls total} \mid \text{all bins occupied}]\,\frac{\Pr[\text{all bins occupied}]}{\Pr[b \text{ balls total}]}\,,\end{aligned}$$

where we use Bayes' rule $\Pr[A \mid B] = \Pr[A \wedge B]/\Pr[B]$. Now we tune λ so that the expected total number of balls, conditioned on the event that all the bins are occupied, is b. Show that if $b = \eta f$, this occurs when λ is the unique root of

$$\frac{\lambda}{1-\mathrm{e}^{-\lambda}} = \eta\,.$$

For this value of λ, $\Pr[b \text{ balls total} \mid \text{all bins occupied}] = \Theta(1/\sqrt{f})$.

Compute $\Pr[\text{all bins occupied}]$ using the fact that the bins are independent. To compute $\Pr[b \text{ balls total}]$, first prove that the *total* number of balls is Poisson-distributed with mean λf. Finally, use Stirling's approximation to show that

$$P_{\text{occ}}(b,f) \sim \left[(\mathrm{e}^{\lambda}-1)\left(\frac{\eta}{\mathrm{e}\lambda}\right)^{\eta}\right]^{f}, \tag{14.96}$$

where $\sim$ hides polynomial factors. For instance, show that in the limit $\eta \to 1$ we have $\lambda \to 0$, and if $b = s$ the probability that each ball lands in a different bin so that every bin is occupied is $\Theta(\mathrm{e}^{-f} f^{1/2})$. Show, moreover, that for any constant $\eta > 1$ this polynomial factor disappears, and (14.96) holds up to a multiplicative constant.

14.27 Counting locally maximal assignments exactly. With the help of the previous problem, we can now compute the expected number $\mathbb{E}[Y]$ of locally maximal satisfying assignments to within a polynomial factor—even, if we wish, within a constant. Let f be the number of false variables, and let b be the number of clauses that block some variable. Then show that

$$\mathbb{E}[Y] = \left(\frac{7}{8}\right)^{m} \sum_{f=0}^{n}\sum_{b=0}^{m} \binom{n}{f}\binom{m}{b}\left(\frac{3f}{7n}\right)^{b}\left(1-\frac{3f}{7n}\right)^{m-b} P_{\text{occ}}(b,f)\,.$$

Set $f = \varphi n$ and $b = \beta m$, and maximize the resulting function of φ and β to show that

$$\alpha_c \le 4.643\,.$$

14.28 Weighty assignments. Derive (14.46), and show that when k is large its unique positive real root is given by

$$\eta = 1 - 2^{1-k} - (k-1)2^{1-2k} + O(k^2 2^{-3k})\,.$$

14.29 Balanced assignments. Suppose that we wish to construct a random k-SAT formula that is guaranteed to be satisfied by the all-`true` assignment. To do this, we choose m random clauses, where each one is chosen from among the $2^k - 1$ clauses containing at least one positive literal. If we choose from these clauses uniformly, the majority of literals will be positive. Instead, suppose that we choose a clause with t positive literals with probability proportional to η^t. Show that if η is the root of (14.46) then the expected fraction of positive literals is exactly 1/2, so that the all-`true` assignment satisfies the formula in a balanced way. Conversely, argue that X_η essentially counts the number of satisfying assignments that satisfy $1/2 + o(n)$ of the literals in the formula.

14.30 Half and half, randomly. Recall HALF-AND-HALF k-SAT from Problem 5.33. As before, suppose we create a random formula with n variables and αn clauses, where each variable is negated with probability 1/2. Use the first moment method to prove an upper bound on the threshold. For instance, show that if $k=4$ then $\alpha_c \leq 0.707$.

Then use the second moment method to prove a lower bound on the threshold for $k=4$. Given two truth assignments with overlap ζ, show that the probability that they both satisfy a random clause is

$$q(\zeta)=\frac{3}{8}\left(\zeta^4+4\zeta^2(1-\zeta)^2+(1-\zeta)^4\right).$$

Defining the function $\phi(\zeta)$ as we did for NAESAT in Section 14.4.3, and graphing it for various α, show that these formulas are satisfiable with positive probability if $\alpha \leq 0.601$.

How does this generalize to larger k? Explore the second moment method numerically and make a guess about how the threshold behaves as a function of k. Does it match the first-moment upper bound when k is large?

14.31 INDEPENDENT SET in random graphs. This and the next few problems look at independent sets and matchings in random graphs. First, consider a random graph $G(n,p)$ with $p=1/2$. Show that, with high probability, its largest independent set—and, by symmetry, its largest clique—is of size at most $2\log_2 n$. Conversely, show that with high probability it has an independent set of size $\beta \log_2 n$ for any constant $\beta<2$.

Hint: use the first moment method to show that $G(n,1/2)$ probably doesn't have an independent set of size $2\log_2 n$. You may find it helpful to first derive the following from Stirling's approximation: if $k=o(\sqrt{n})$,

$$\binom{n}{k}=(1-o(1))\frac{1}{\sqrt{2\pi k}}\left(\frac{en}{k}\right)^k.$$

Then use the second moment method to show that an independent set of size $\beta \log_2 n$ probably exists for any $\beta<2$. Use the fact that most pairs of sets of size $o(\sqrt{n})$ are disjoint, so the events that they are independent sets in $G(n,p)$ are independent. Therefore, if the expected number of such sets is large then one exists with high probability. Generalize this to show that for $G(n,p)$ where p is constant, with high probability the size of the largest independent set is

$$(1-o(1))\frac{2\ln n}{-\ln(1-p)}.$$

This result can be tightened enormously. For each n and each constant p, there is an integer k such that the size of the largest independent set in $G(n,p)$ is, with high probability, either k or $k+1$. In other words, the likely size of the independent set is concentrated on just two integer values

14.32 Sparseness and independence. Generalize the first moment argument from the previous problem to the sparse case. Specifically, show that with high probability, the largest independent set in $G(n,p=c/n)$ is of size at most $\alpha n+o(n)$ where α is the root of

$$\frac{\alpha c}{2}+\ln\alpha=1.$$

In particular, show that there is a constant c_0 such that if $c>c_0$, there are no independent sets of size αn where

$$\alpha=\frac{2\ln c}{c}.$$

14.33 Greedy independent sets. Now consider a simple greedy algorithm for constructing an independent set in $G(n, p = c/n)$. At each step, we choose a random vertex, add it to the set, and remove it and its neighbors from the graph. Let $U(T)$ be the number of vertices remaining after T steps. Rescaling to $t = T/n$ and $u(t) = U/n$, derive the differential equation

$$\frac{\mathrm{d}u}{\mathrm{d}t} = -1 - cu .$$

Solve this equation, set $u(t) = 0$, and conclude that this algorithm typically finds an independent set of size αn where

$$\alpha = \frac{\ln(c+1)}{c} .$$

Observe that when c is large there is a factor of 2 between this and the upper bound given by Problem 14.32.

14.34 Greedy matchings. Here's a very simple-minded algorithm for finding a matching. At each step, choose a random edge, add it to the matching, and remove its endpoints and their edges from the graph. In a random graph, each step eliminates 2 vertices and 2γ edges where γ is the current average degree. Derive the following differential equation for γ, where t is the number of steps divided by n:

$$\frac{\mathrm{d}\gamma}{\mathrm{d}t} = -\frac{2(\gamma+1)}{1-2t} .$$

Solve this differential equation with the initial condition $\gamma(0) = c$, set $\gamma(t) = 0$, and conclude that this algorithm finds a matching in $G(n, p = c/n)$ with high probability with $tn - o(n)$ edges where

$$t = \frac{1}{2} - \frac{1}{2(c+1)} . \tag{14.97}$$

This approaches 1/2 as $c \to \infty$, indicating that $G(n, p = c/n)$ has a nearly-perfect matching when c is large.

14.35 Smarter matchings. Here's another algorithm for finding a matching in a random graph. At each step, choose a random vertex v. If v has degree 0, simply remove it. Otherwise, choose one of v's neighbors w randomly, add the edge between v and w to the matching, and remove both v and w from the graph.

Write the current number of vertices and edges as νn and μn respectively. Then the current average degree is $\gamma = 2\mu/\nu$, and the probability a random vertex has degree 0 is $\mathrm{e}^{-\gamma}$. Use this to write the following system of differential equations, where t is the number of steps and s is the number of edges in the matching, both divided by n:

$$\frac{\mathrm{d}s}{\mathrm{d}t} = 1 - \mathrm{e}^{-\gamma} , \qquad \frac{\mathrm{d}\nu}{\mathrm{d}t} = -(2 - \mathrm{e}^{-\gamma}), \qquad \frac{\mathrm{d}\mu}{\mathrm{d}t} = -\gamma(2 - \mathrm{e}^{-\gamma}).$$

Solve these differential equations (it might be helpful to change the variable of integration), set $\mu = 0$, and conclude that this algorithm finds a matching in $G(n, p = c/n)$ with high probability of size $sn - o(n)$ where

$$s = \frac{1}{2} - \frac{\ln(2 - \mathrm{e}^{-c})}{2c} . \tag{14.98}$$

Comparing with (14.97), we see that this algorithm performs slightly better than the previous one. Why?

14.36 Karp and Sipser find independent sets. Here's a smarter algorithm for finding independent sets in $G(n, p = c/n)$, which does much better than that of Problem 14.33 as long as c isn't too large. It is due to Karp and Sipser, and was one of the first uses of differential equations in computer science.

We use the fact that in any graph G, if a vertex v has degree 1 then there is some maximal independent set that includes v (prove this!) So, at each step, we choose a random vertex v of degree 1 (as long as there is one), remove it and its neighbor w, and decrement the degrees of w's neighbors as shown in Figure 14.32.

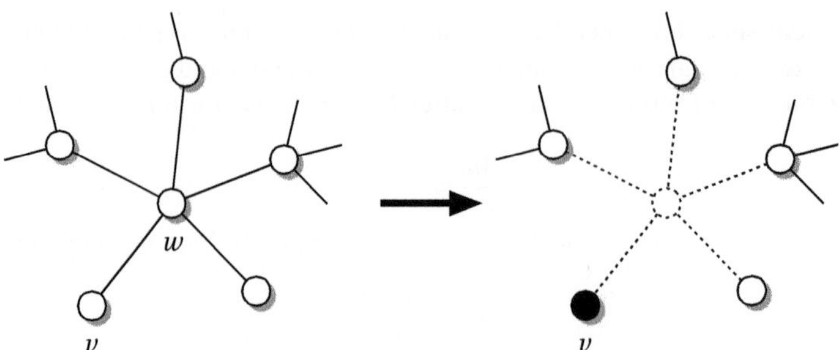

FIGURE 14.32: A step of the Karp–Sipser greedy algorithm. We choose a random vertex v of degree 1, add it to the independent set, remove its neighbor w, and decrement the degrees of w's neighbors.

As we did in our discussion of the k-core in Section 14.2.5, let's divide the vertices into "heavy" ones of degree 2 or more, and "light" ones of degree 1 or 0. The degree distribution of the heavy vertices remains Poissonian, so if a_j is the number of vertices of degree j divided by n we write

$$a_j = \frac{\alpha\beta^j}{j!} \text{ for } j \geq 2.$$

We keep track of a_0 and a_1 separately. We set the degree of the chosen vertex v to zero on each step and put all the degree-0 vertices in the independent set, including those whose initial degree is zero. Thus we find an independent set of size $a_0 n$.

Note that w is chosen randomly with probability proportional to its degree, since a vertex with degree j has j chances of being v's neighbor. So, let's add a little more notation: let μ be the total degree of all the vertices (i.e., the total number of "spokes") divided by n, and let δ be the average of w's degree, minus 1. Write expressions for μ and δ in terms of a_1, α, and β, and derive the identity

$$\delta = \frac{\alpha\beta^2 \mathrm{e}^\beta}{\mu}.$$

Now, given that a vertex of degree j vertex has δj chances on average of being one of w's neighbors and to have its degree decremented, and j chances of being chosen as w and simply removed, derive the following (initially daunting) system of differential equations, where t is the number of steps divided by n:

$$\frac{\mathrm{d}\alpha}{\mathrm{d}t} = \frac{\delta\beta\alpha}{\mu}, \quad \frac{\mathrm{d}\beta}{\mathrm{d}t} = -\frac{(\delta+1)\beta}{\mu}, \quad \frac{\mathrm{d}\mu}{\mathrm{d}t} = -2(\delta+1),$$
$$\frac{\mathrm{d}\delta}{\mathrm{d}t} = -\frac{\delta\beta}{\mu}, \quad \frac{\mathrm{d}a_0}{\mathrm{d}t} = 1 + \frac{\delta}{\mu}a_1, \quad \frac{\mathrm{d}a_1}{\mathrm{d}t} = -1 - \frac{\delta+1}{\mu}a_1 + \frac{\delta\alpha\beta^2}{\mu}.$$

These have the initial conditions $\alpha = \mathrm{e}^{-c}$, $\mu = \delta = \beta = c$, $a_0 = \mathrm{e}^{-c}$, and $a_1 = c\mathrm{e}^{-c}$.

In fact, we can focus on α, β, μ, and δ, and then solve for a_0 and a_1 separately. First use the facts that $(\ln \alpha)' = -\delta'$ and $(\ln \mu)' = 2(\ln \beta)'$ to derive these two invariants:

$$\alpha = \mathrm{e}^{-\delta} \text{ and } \frac{\mu}{c} = \left(\frac{\beta}{c}\right)^2.$$

Now change the variable of integration from t to s, where

$$\frac{\mathrm{d}s}{\mathrm{d}t} = \frac{\beta}{\mu},$$

to obtain a new system of differential equations,

$$\frac{d\beta}{ds} = -(\delta+1) \quad \text{and} \quad \frac{d\delta}{ds} = -\delta\,,$$

and conclude that

$$\beta(s) = c\mathrm{e}^{-s} - s \quad \text{and} \quad \delta(s) = c\mathrm{e}^{-s}\,.$$

Finally, solve for a_1 and a_0, obtaining

$$a_0(s) = \mathrm{e}^{-c\mathrm{e}^{-s}} + \mathrm{e}^{-s} s - \frac{s^2}{2c} \quad \text{and} \quad a_1(s) = \frac{\mathrm{e}^{-s}}{c}\left(c\mathrm{e}^{-c\mathrm{e}^{-s}} - s\right)(c - s\mathrm{e}^{s})\,.$$

Now, the algorithm proceeds until $a_1(s) = 0$ and we run out of degree-1 vertices. Examining (14.36), we see that one of the roots of $a_1(s)$ is the root of $s\mathrm{e}^s = c$. Using Lambert's function, we denote this $s = W(c)$. Show that when $c \le \mathrm{e}$ this is the only root, and at that point the fraction of heavy vertices is zero. Therefore, if $c \le \mathrm{e}$ then with 14.7
high probability the algorithm eats the entire graph except for $o(n)$ vertices, and finds an independent set occupying $a_0 n - o(n)$ vertices where

$$a_0 = a_0(W(c)) = \frac{1}{c}\left(W(c) + \frac{W(c)^2}{2}\right). \tag{14.99}$$

Moreover, since this algorithm is optimal as long as degree-1 vertices exist, this is an exact calculation of the likely size of the largest independent set in $G(n, p = c/n)$ when $c \le \mathrm{e}$.

If $c > \mathrm{e}$, on the other hand, show that $a_1(s)$ has another root $s < W(c)$ due to the factor $c\mathrm{e}^{-c\mathrm{e}^{-s}} - s$. At this point, there are no degree-1 vertices left but the graph still has a "core" of heavy vertices, and the algorithm would have to switch to another strategy. Show that if $c = \mathrm{e} + \varepsilon$, the fraction of vertices in this core is

$$\frac{12}{\mathrm{e}^2}\,\varepsilon + O(\varepsilon^{3/2})\,.$$

So, like the giant component but unlike the 3-core, it appears continuously at this phase transition.

14.37 Matching the leaves. Prove that if a graph G has a vertex v of degree 1, there is a MAX MATCHING which includes the edge connecting v to its neighbor. Consider an algorithm that chooses a random degree-1 vertex v, adds this edge to the matching, and removes both v and its neighbor from the graph. Use the analysis of the previous problem to show that, if $c \le \mathrm{e}$, with high probability this algorithm finds a matching with $tn - o(n)$ edges in $G(n, p = c/n)$ where

$$t = 1 - \frac{1}{c}\left(W(c) + \frac{W(c)^2}{2}\right),$$

and that this is exactly the likely size of the MAX MATCHING when $c \le \mathrm{e}$. Why is this 1 minus the size (14.99) of the independent set found by the Karp–Sipser algorithm of the previous problem?

14.38 Local extrema. Reconsider the integral (14.52) in Section 14.5.2. We used Laplace's method to evaluate this integral for large values of n, which relies on the concentration of $g^n(\theta)$ on its global maximum. In the case of (14.52), however, the function g depends very strongly on n via $B = 2^{\kappa n}$. This means that g has an exponential number of local extrema in the interval of integration, and these extrema are exponentially close to each other and to the global maximum at $\theta = 0$.

Prove, nevertheless, that our application of Laplace's method in Section 14.5.2 is correct. Hint: write g as a function of $y = B\theta$ and consider a Taylor series for $g(y)$ in $\varepsilon = 1/B$.

14.39 Balance and perfect balance. In a random instance of INTEGER PARTITIONING, where each a_j is chosen uniformly from $\{0,1,\ldots,B-1\}$, let X_0 and X_1 denote the number of partitions with $D=0$ and $D=1$ respectively. Show that

$$\Pr[X_1>0]=(1-O(1/B))\Pr[X_0>0]\,.$$

Hint: given a partition A, show that for most sets of weights $\{a_j\}$ such that A is perfectly balanced, there is a "partner" set $\{a'_j\}$ that make it nearly balanced.

14.40 Balance mod B. Consider the following variant of random INTEGER PARTITIONING. As before, each a_j is chosen uniformly and independently from $\{0,1,\ldots,B-1\}$. But now, we say that A is balanced if the difference $D=\sum_{j\in A}a_j-\sum_{j\notin A}a_j$ is zero mod B, or, equivalently, if

$$\sum_{j\in A}a_j\equiv_B\sum_{j\notin A}a_j\,.$$

Let X denote the number of such partitions. Show that the first moment is exactly $\mathbb{E}[X]=2^n/B$, and the second moment is exactly

$$\mathbb{E}[X^2]=\frac{2^n(2^n-2)}{B^2}+\frac{2^{n+1}}{B}\leq\mathbb{E}[X]^2+2\,\mathbb{E}[X]\,.$$

if B is odd, and

$$\mathbb{E}[X^2]=\frac{2^{n+1}(2^n-2)}{B^2}+\frac{2^{n+1}}{B}\leq 2\,\mathbb{E}[X]^2+2\,\mathbb{E}[X]\,.$$

is B is even. Hint: if B is odd, show that the events that two partitions A,A' are balanced are pairwise independent, unless $A'=A$ or $A'=\overline{A}$. If B is even, show that this is the case once we condition on the value of $\sum_j a_j \bmod 2$.

14.41 Concentration within a constant factor. If $\mathbb{E}[Z^2]$ is a constant times $\mathbb{E}[Z]^2$, the second moment method lets us prove that $Z>0$ with constant probability. In fact, we can do more: we can show that, with constant probability, Z is with a constant factor of its expectation. Prove that, for any $\theta\in[0,1]$ we have

$$\Pr[Z>\theta\,\mathbb{E}[Z]]\geq(1-\theta)^2\frac{\mathbb{E}[Z]^2}{\mathbb{E}[Z^2]}\,.$$

This is called the *Paley–Zygmund inequality*. Hint: modify the proof of Section A.3.5 by letting Y be the indicator random variable for the event that $Z>\theta\,\mathbb{E}[Z]$.

14.42 Messages that optimize. We can also use message-passing algorithms to solve optimization problems. Let

$$E(\mathbf{x})=\sum_a E_a(X_{\partial a})$$

denote the objective function, where each E_a depends on a subset ∂a of variables. For instance, $E(\mathbf{x})$ could be the number of unsatisfied clauses in a SAT formula. The variables $\mathbf{x}=(x_1,\ldots,x_n)$ are from a finite alphabet like $x_i\in\{0,1\}$. The corresponding factor graph consists of variable vertices plus constraint vertices for each E_a. Devise a message-passing algorithm to compute

$$E^*=\min_{\mathbf{x}}E(\mathbf{x})\quad\text{and}\quad\mathbf{x}^\star=\operatorname{argmin}E(X)$$

in linear time for factor graphs that are trees with bounded degree.

Hint: in (14.67) we defined the BP equations in terms of the "sum-product" rule. Try defining them in terms of a "min-sum" rule instead, analogous to the algorithm for SHORTEST PATH in Section 3.4.3.

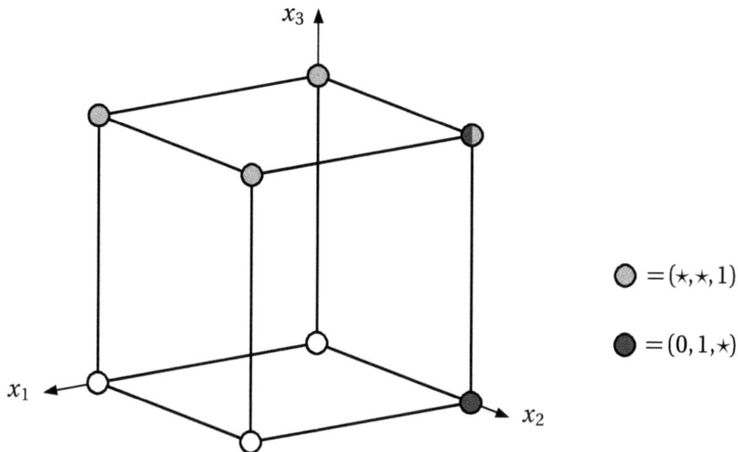

FIGURE 14.33: Binary cube for $n = 3$ and two of its subcubes.

14.43 The transition according to belief propagation. Let's calculate α_{BP}, the critical density of k-SAT according to belief propagation. The idea is that above this value, any solution of the distributional equations (14.77) has $\widehat{\eta} = 0$ with positive probability.

Let x denote the probability that $\widehat{\eta} = O(\varepsilon)$ and let y denote the probability that $\eta = 1 - O(\varepsilon)$, where the constants in these Os are chosen suitably, and consider the limit $\varepsilon \to 0$.

1. Show that (14.77a) implies $x = y^{k-1}$.
2. Let $p' \le p$ and $q' \le q$ denote the number of messages $\widehat{\eta} = O(\varepsilon)$ in (14.77b). Show that

$$\eta = \begin{cases} O(\varepsilon) & \text{if } p' > q' \\ O(1) & \text{if } p' = q' \\ \Omega(1-\varepsilon) & \text{if } p' < q' . \end{cases}$$

3. Show that at the fixed point $1 - 2y = e^{-k\alpha x} I_0(k\alpha x)$ where I_0 is the modified Bessel function of the first kind,

$$I_0(z) = \sum_{t=0}^{\infty} \frac{(z/2)^{2t}}{(t!)^2} .$$

4. Combine this with $x = y^{k-1}$ and write a transcendental equation for the density α_{BP} such that $y = 0$ is the only fixed point. Solve this equation numerically for $k = 2, 3, 4$, and show analytically that $\alpha_{BP}(2) = 1$.

14.44 The random subcube model. One way to understand the clustering transition is to consider a toy model of constraint satisfaction problems, where the set of solutions is clustered by construction. Such a problem is given by the *random subcube model*.

Recall from Section 12.3.1 that the vectors $\mathbf{x} \in \{0, 1\}^n$ form the corners of an n-dimensional hypercube. A subcube is specified by fixing some elements of $\mathbf{x}$ (the frozen variables) while letting the remaining variables take on both values 0 and 1 (the free variables). Figure 14.33 shows an example.

In the random subcube model, a subcube $A=(a_1,\ldots,a_n)$ is given by random entries a_i,

$$a_i = \begin{cases} 1 & \text{with probability } p/2, \\ 0 & \text{with probability } p/2, \\ \star & \text{with probability } 1-p. \end{cases}$$

The "wild card" symbol $\star$ denotes a free variable, and p is the probability that a variable is frozen.

The solution space of this model is *defined* as the union of $2^{(1-\alpha)n}$ random subcubes. If α is small, we have so many subcubes that their union essentially covers the whole cube, i.e., every vertex of the cube is a solution. As α gets larger the solution space gets smaller. More importantly, it breaks into clusters as the subcubes cease to overlap. The parts of the solution space that are still connected form the clusters. For $\alpha > \alpha_c = 1$, there are no solutions at all. Thus α plays a role similar to that of the density.

Show that the configurational entropy $\Sigma(s)$ as defined in Section 14.7.2 of the random subcube model is

$$\Sigma(s) = 1-\alpha - s\log_2 \frac{s}{1-p} - (1-s)\log_2 \frac{1-s}{p}.$$

This is the function that we used in Figure 14.28. Then show that in the limit $n \to \infty$ the random subcube model undergoes clustering and condensation transitions at

$$\begin{aligned} \alpha_{\text{clust}} &= \log_2(2-p) \\ \alpha_{\text{cond}} &= \frac{p}{p-2} + \log_2(2-p). \end{aligned}$$

Hints: for α_{clust} consider the average number of clusters that a random vertex of the hypercube belongs to. What is the internal entropy s of a cluster? Show that the number $N(s)$ of clusters with internal entropy s follows a binomial distribution. Use the first and second moment method to show that $N(s)$ is concentrated around its average value. This average value gives you $\Sigma(s)$, and from $\Sigma(s)$ you get α_{cond} by considering the value of s where $\Sigma'(s) = -1$.

14.45 Clusters and correlations. Consider two satisfying assignments $\mathbf{x}$ and $\mathbf{x}'$ of a random k-SAT formula. Show that the variance in their fractional overlap $\zeta = z/n$ can be expressed through the one- and two-variable marginals μ,

$$\operatorname{Var}\zeta = \frac{1}{n^2}\sum_{i,j}\sum_{x_i,x_j}\left(\mu^2(x_i,x_j) - \mu^2(x_i)\mu^2(x_j)\right). \tag{14.100}$$

Now suppose that $z = \zeta_1 n + O(\sqrt{n})$ if they are in the same cluster and $\zeta_2 n + O(\sqrt{n})$ if they are in different clusters. Argue that if there are exponentially many clusters of roughly the same size, $\operatorname{Var}\zeta = O(1/n)$. Show in that case that a typical variable x_i can have significant correlations with only $O(1)$ other variables.

In contrast, argue that in the condensed phase the expected number of variables that a typical x_i is strongly correlated with is $\Theta(n)$, so correlations stretch across the entire factor graph. Assuming for simplicity that the factor graph is r-regular for some constant r, show that x_i must be correlated with variables $\Omega(\log n)$ steps away.

Notes

14.1 Further reading. The study of phase transitions in computational problems is a young and highly active field, and many of the relevant ideas and results can only be found in the original research papers. Notable exceptions are the book by Mézard and Montanari [576] and the Ph.D. thesis of Lenka Zdeborová [845]. One collection of pedagogical contributions is [656]; another book [373] focuses on the phase transition in VERTEX COVER. For the probabilistic tools that are indispensable in this field we recommend *The Probabilistic Method* by Alon and Spencer [38].

14.2 Early experiments. In 1991, Cheeseman, Kanefsky, and Taylor published their paper "Where the *really* hard problems are" [156], giving experimental evidence of phase transitions in random NP-hard problems. Shortly after, Mitchell, Selman, and Levesque [586] presented the first numerical study of the phase transition in random k-SAT using the same DPLL algorithm that we used to produce Figure 14.3. In 1994, Kirkpatrick and Selman [482] used finite size scaling, a numerical technique from statistical physics, to show that the numerical experiments support a sharp transition in random k-SAT.

14.3 DPLL algorithms. The Davis–Putnam–Logemann–Loveland algorithm appeared in [223, 222]. As we discuss in the text, while any DPLL algorithm performs an exhaustive search, a great deal of cleverness can go into the branching rule that organizes this search. Recall Problem 6.6, for instance, where we do exponentially better than the naive running time of 2^n by branching on clauses rather than variables. We can also use heuristics from artificial intelligence to improve the running time in practice, such as the Brelaz heuristic we discuss in Note 14.10.

14.4 Threshold proofs. The threshold for 2-SAT was proved in [336, 176, 279]. As Problem 14.17 shows, this is essentially the same transition as the emergence of the giant component in a random graph.

The 2015 proof by Ding, Sly, and Sun [240] of the threshold conjecture for large k is enormously sophisticated, and is far more than just an existence proof. It amounts to proving that the survey propagation picture of Section 14.7 is largely correct, and connects survey propagation fixed points to a version of the second moment method conditioned on the distribution of neighborhoods in the factor graph. At a number of places, the arguments only go through if k is sufficiently large. Making the proof work for all $k \geq 3$ seems technically possible, but quite challenging.

14.5 Friedgut's Theorem. Theorem 14.2 is due to Ehud Friedgut [296]; he gives an accessible introduction in [297]. His results are much more general than k-SAT, and show that a wide variety of properties of random graphs and hypergraphs have sharp thresholds in this sense.

Specifically, let W be a property of random graphs, and let $\mu(p)$ be the probability that W is true of $G(n,p)$. We assume that W is *monotone increasing*, i.e., that adding an edge to G can only help make W true. For instance, W could be the property of non-3-colorability. We say that W has a *coarse threshold* if there is a function $p(n)$ and a constant $\varepsilon > 0$ such that $\mu((1-\varepsilon)p) > \varepsilon$ but $\mu((1+\varepsilon)p) < 1-\varepsilon$. In other words, there is some range of p in which $\mu(p)$ hovers between 0 and 1.

Friedgut proves that any property with a coarse threshold can be approximated to arbitrary precision by a *local* property, i.e., the existence of some small subgraph. Specifically, he shows that for any constant $\delta > 0$ there is a list of finite subgraphs $H_1, H_2, \ldots, H_m$ such that, if L is the property of containing one of these subgraphs, the total probability in $G(n,p)$ of the symmetric difference of W and L is at most δ. In particular, if $\nu(p)$ is the probability that $G(n,p)$ has property L, then $\left|\mu(p)-\nu(p)\right| \leq \varepsilon$ for all p. Moreover, the size of the subgraphs H_i can be bounded in terms of ε (the coarseness of the threshold) and δ (the quality of the approximation).

For instance, the property of non-2-colorability is well-approximated by the property of containing a cycle of length 3, or 5, or 7, and so on up to some odd ℓ. By increasing ℓ, we can approximate non-2-colorability within any constant δ. And indeed the probability that $G(n, p = c/n)$ is non-2-colorable does not have a sharp threshold, but increases continuously from 0 to 1 as c goes from 0 to 1.

Friedgut offers another version of his result, which is easier to apply in practice. If W has a coarse threshold, there is a finite subgraph H and a constant $\varepsilon > 0$ such that, for some p, adding a random copy of H to G increases $\mu(p)$ more than increasing p to $(1+\varepsilon)p$ does. That is, adding a single copy of H, located randomly in the vertices of G, makes more difference than adding $\varepsilon p\binom{n}{2}$ more random edges. Moreover, H is a subgraph that appears with reasonable probability in $G(n,p)$. Thus in the sparse regime where $p = O(1/n)$, H can have at most one cycle.

If no finite subgraph with this property exists, W must have a sharp threshold. For instance, Achlioptas and Friedgut [11] proved that k-colorability has a sharp threshold for $k \geq 3$ in the sense of Theorem 14.2 by showing that no finite subgraph with a single cycle has a large effect on the probability that $G(n,p)$ is k-colorable.

Friedgut's proof is deep and beautiful, and relies on the Fourier analysis of Boolean functions. Let $z_1,\ldots,z_m$ be a set of Boolean variables, such as whether a given edge in $G(n,p)$ is present or absent. Let $W(z_1,\ldots,z_m)$ be a monotone Boolean function, and suppose that each z_i is set independently to 1 or 0 with probability p or $1-p$. Roughly speaking, if the probability that $W=1$ increases smoothly as a function of p, there must be a small subset of variables $\{z_i\}$ with unusually large influence on W. These variables form the edges of the subgraph H.

14.6 Random graph models and the giant component. The $G(n,m)$ version of the Erdős–Rényi model appeared in [269]. However, the $G(n,p)$ model was in fact invented, at the same time, by Edgar Gilbert [328].

Mathematicians have achieved an incredibly precise picture of how the giant component emerges as the average degree c passes through a "scaling window" of width $\Theta(n^{-1/3})$ around the critical point $c=1$. For a detailed and rigorous picture, we refer the reader to the books by Bollobás [120] and Janson,Łuczak and Ruci ński [426]. The emergence and size of the giant component in the configuration model (Problem 14.11) was studied by Molloy and Reed [597], and independently in the physics literature by Newman, Strogatz, and Watts [632].

Although it is somewhat out of date at this point, there is an excellent review by Frieze and McDiarmid [300] regarding algorithmic questions about random graphs. They discuss the various heuristics for INDEPENDENT SET and MAX MATCHING on random graphs covered in Problems 14.33, 14.34, and 14.35, as well as phase transitions in connectivity, the existence of a HAMILTONIAN PATH, and other properties.

14.7 Lambert's W function. The size of the giant component can be written in closed form if we indulge ourselves in the use of Lambert's function $W(x)$, defined as the root of $w\mathrm{e}^w=x$. In that case, the size of the giant component can be written

$$\gamma=1+\frac{W(-c\mathrm{e}^{-c})}{c}\ .$$

Similarly, the lower bound on the 3-SAT threshold (14.29) derived from SC can be written as $\alpha^*=-(2/3)W_{-1}(-\mathrm{e}^{-3})$, the upper bound on the size of the independent set in $G(n,c/n)$ from Problem 14.32 can be written $\alpha=2W(\mathrm{e}c/2)/c$, and the lower bound on the GRAPH 3-COLORING threshold from Problem 14.21 can be written $c^*=-W_{-1}(-\mathrm{e}^{-5/2})$. Here W_{-1} refers to the -1st branch of W in the complex plane.

Given that W allows us to write the solution of many transcendental equations in closed form, a number of people advocate adopting it as a standard function, analogous to sin, cos, and log [385].

14.8 Differential equations for algorithms. Differential equations are a familiar tool in many fields, but showing rigorously that they can be used to approximate discrete random processes takes some work. The most general such result is from Wormald [834]. We give one version of his theorem here.

Theorem 14.6 (Wormald) *Let $Y_1(T),\ldots,Y_\ell(T)$ be a finite set of integer-valued random functions from $\mathbb{N}$ to $\mathbb{N}$. Suppose that for all i and all $T\geq 0$ we have*

$$\mathbb{E}[\Delta Y_i]=f_i(Y_1/n,\ldots,Y_\ell/n)+o(1)$$

where each function $f_i(y_1,\ldots,y_\ell)$ is Lipschitz continuous *on some open set $D\subseteq\mathbb{R}^\ell$. That is, there is a constant c such that for any $\mathbf{y},\mathbf{y}'\in D$ and for all i,*

$$\left|f_i(\mathbf{y})-f_i(\mathbf{y}')\right|\leq c\left\|\mathbf{y}-\mathbf{y}'\right\|.$$

where $\|\cdot\|$ denotes the Euclidean length of a vector. Finally, assume that large changes are very unlikely, $\Pr\left[|\Delta Y_i|>n^\beta\right]=o(n^{-3})$ *for some $\beta<1/3$. Then with high probability, for all i and all T,*

$$Y_i(T)=y_i(T/n)\cdot n+o(n),$$

where $y_i(t)$ is the unique solution to the corresponding system of differential equations,

$$\frac{\mathrm{d}y_i}{\mathrm{d}t}=f_i(y_1,\ldots,y_\ell),$$

with the initial conditions $y_i(0)=Y_i(0)/n$, as long as $\mathbf{y}(t)\in D$ for all $0\leq t\leq T/n$.

Armed with this result, Wormald has used differential equations to model many processes on random graphs.

The proof of this theorem works by dividing the steps of the process into, say, $n^{1/3}$ blocks of $n^{2/3}$ steps each. In each block, we subtract the differential equation's trajectory from the $Y_i(T)$ to obtain a *martingale*. This is a random process whose expected change at each step is zero—or, in this case, $o(1)$. We then use Azuma's inequality (see Appendix A.5.2) to bound this martingale, or equivalently the extent to which the $Y_i(T)$ deviate from the differential equation's trajectory. We then use the union bound over the $n^{1/3}$ blocks to show that none of them deviates from the trajectory significantly.

In computer science, the first analysis of an algorithm using differential equations is due to Karp and Sipser [466], who studied the heuristics for INDEPENDENT SET and MAX MATCHING on random graphs discussed in Problems 14.36 and 14.37. These heuristics were rediscovered by Weigt and Hartmann [822], who analyzed them using generating functions. Other important applications, besides the ones given for SAT and GRAPH COLORING given here, include the work of Mitzenmacher [588, 589] on load balancing, and Luby et al. [543] on error-correcting codes.

We note that in some cases, such as the exploration algorithm for the giant component and the behavior of UC on random 3-SAT, the full generality of Wormald's theorem is not necessary, and simpler arguments suffice.

14.9 The k-core. The concept of a k-core was introduced in graph theory by Bollobás [118]. The emergence of the giant k-core in $G(n, p = c/n)$ was proved by Pittel, Spencer, and Wormald [661]. Our presentation is similar in spirit to theirs, but it should be emphasized that turning these differential equations into a proof takes a great deal of additional work—especially because we have to follow the pruning process all the way down to $L = 0$, past the point where $L = \Theta(n)$ and the concentration arguments of Wormald's Theorem apply.

The pure literal rule of Problem 14.16 was analyzed by Broder, Frieze, and Upfal [139] and Luby, Mitzenmacher, and Shokrollahi [544]. The threshold for the appearance of the core of a SAT formula, as well as in various kinds of hypergraphs, was established rigorously by Molloy [596] using techniques similar to [661].

14.10 Algorithmic lower bounds. The UC and SC algorithms were first studied by Chao and Franco [153, 154] and Chvátal and Reed [176], along with an algorithm GUC that satisfies a clause chosen randomly from among the shortest ones. Frieze and Suen studied these algorithms on k-SAT as in Problem 14.18. Using a variant of SC, they showed that $\alpha_c \geq a_k 2^k / k$ where $\lim_{k \to \infty} a_k \approx 1.817$. The greedy algorithm for GRAPH 3-COLORING discussed in Problem 14.21 was analyzed by Achlioptas and Molloy [12], who called it 3-GL.

Frieze and Suen [301] also showed that at the densities where UC succeeds, with high probability the corresponding DPLL algorithm only backtracks $o(n)$ times and therefore runs in linear time. The precise probability that algorithms like UC and 3-GL succeed can be calculated analytically, and displays an interesting phase transition of its own at the density where the algorithm fails [301, 183, 438].

The running time of DPLL at higher densities was studied by Achlioptas, Beame, and Molloy [8] who found a regime below α_c where the running time is exponential with high probability. The expected running time for DPLL with various branching rules was computed by Cocco and Monasson [183], who showed that it becomes exponential at the density where the corresponding linear-time algorithm fails.

Problem 14.19, determining the 1-IN-k SAT threshold, is from Achlioptas, Chtcherba, Istrate, and Moore [9]. This makes it one of the few SAT-like problems for which the threshold can be computed exactly. On the other hand, these formulas are easy on both sides of the transition. They get harder if we change the probability that literals are negated; for a physics analysis, see [685].

The idea of prioritizing variables with the largest degree is called the *Brelaz heuristic*. Such algorithms were first analyzed with differential equations by Achlioptas and Moore [15]. By analyzing an algorithm that first colors vertices of high degree, they showed that $G(n, p = c/n)$ is 3-colorable with high probability for $c \leq 4.03$. Their analysis relies on the fact that, at all times throughout the algorithm, the uncolored part of the graph is uniformly random conditioned on its degree distribution, i.e., it is random in the configuration model of Problem 14.10.

For 3-SAT, similar algorithms were studied by Kaporis, Kirousis, and Lalas [453] and Hajiaghayi and Sorkin [360], who independently showed that $\alpha_c > 3.52$. At the time we write this, these are the best known lower bounds for the GRAPH 3-COLORING and 3-SAT thresholds.

We owe the "card game" metaphor to Achlioptas, whose review [7] of the differential equation method describes various improved algorithmic lower bounds for 3-SAT. Also recommended is [454].

14.11 First moment upper bounds. The upper bound $\alpha_c < 2^k \ln 2$ for k-SAT was first given by Franco and Paull [292]. The improved bounds given in Problems 14.25–14.27 where we count locally maximal satisfying assignments are from Dubois and Boufkhad [250].

By conditioning on the event that the formula is "typical" in the sense that the degree distribution of the variables—that is, the fraction of variables with a given number of positive and negative appearances—is within $o(1)$ of its expectation, Dubois, Boufkhad, and Mandler [251] improved the first moment upper bound for 3-SAT to 4.506. By applying this approach to the 2-core, and conditioning on the number of clauses with a given number of negations as well, Díaz, Kirousis, Mitsche, and Pérez-Giménez [235] improved this further to $\alpha_c < 4.4898$, which is currently the best known upper bound. See Kirousis, Stamatiou, and Zito [483] for a review of first moment bounds in general.

A rather different kind of result was given by Maneva and Sinclair [554]. They showed that, if the satisfying assignments are indeed clustered as we believe them to be, then $\alpha_c \leq 4.453$. Their technique involves counting satisfying *partial* assignments corresponding to clusters and weighting these in a certain way.

14.12 The second moment method. The first application of the second moment method to k-SAT was by Frieze and Wormald [302], who showed that $\alpha_c \sim 2^k \ln 2$ if k is allowed to grow as a function of n. Essentially, the asymmetry in $q(\zeta)$ at $\zeta = 1/2$ disappears when $k = \log_2 n + \omega(1)$.

The idea of restoring symmetry to the second moment method for finite k by focusing on NAE-k-SAT comes from Achlioptas and Moore [14]. They provided the first lower bound on α_c beyond the $O(2^k/k)$ algorithmic results, and proved the long-standing conjecture that $\alpha_c = \Theta(2^k)$. The refined argument using the weighted second moment method, showing that $\alpha_c = 2^k \ln k - O(k)$, is due to Achlioptas and Peres [17]. At the time we write this, their lower bounds are the best known for $k \geq 4$.

The second moment method was applied to GRAPH k-COLORING by Achlioptas and Naor [16]. Here the overlap ζ becomes a $k \times k$ matrix, requiring us to maximize a function of k^2 variables.

14.13 Algorithms and heuristics for INTEGER PARTITIONING. In Section 4.2.3 we saw a greedy approach to INTEGER PARTITIONING that places the largest remaining weight in the subset with the smaller total weight until all the weights are placed. Thus it tries to keep the difference between the subsets small at each step.

Another polynomial-time heuristic is the differencing method of Karmarkar and Karp [460], where we reduce the weights by repeatedly replacing the two largest numbers with the absolute value of their difference. Equivalently, we commit to placing these numbers in different subsets without deciding yet which subset each one will go in. With each differencing operation the number of weights decreases by one, and the last number is the final difference D.

Both heuristics can be used as a basis for a DPLL-style algorithm. At each branch point, we can try the greedy heuristic first, and place the weight in the other subset if this doesn't work. Similarly, we can replace the two largest weights either with their difference or their sum, giving the search tree shown in Figure 14.34. Korf [500] calls these algorithms the *complete greedy* and *complete differencing* algorithms respectively, and these are the algorithms whose average running times are depicted in Figure 14.16.

Both complete algorithms can be sped up by pruning branches of the search tree. We don't have to explore the offspring of a node if the largest weight is larger than the sum of all other weights, and a similar pruning rule can be applied to the complete greedy tree. The most effective rule, however, is simply to stop if you find a partition with $D \leq 1$. This rule leads to the decrease of the running time in Figure 14.16 when n exceeds its critical value.

The dashed line in Figure 14.16 shows that when n is large, the complete differencing algorithm finds a perfectly balanced partition in linear time, i.e., with a single descent through the search tree with little or no backtracking.

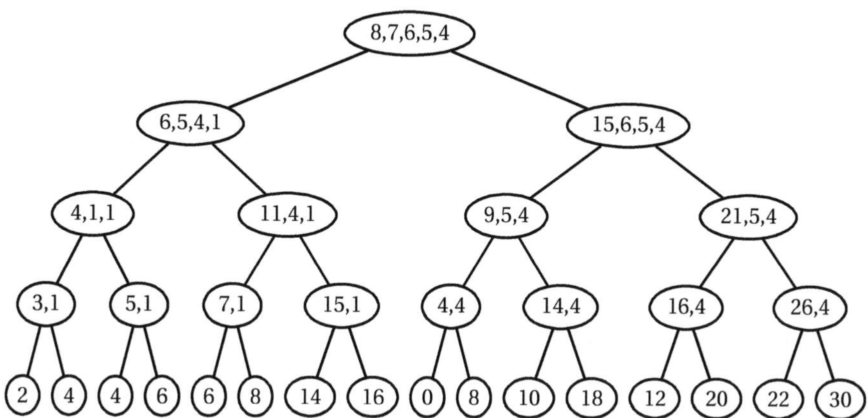

FIGURE 14.34: The differencing tree for INTEGER PARTITIONING. This example has a balanced partition, $\{8,7\}$ vs $\{6,5,4\}$.

Thus the Karmarkar–Karp heuristic finds balanced partitions with high probability in this regime. Boettcher and Mertens [117] showed that the typical difference achieved by this heuristic scales as

$$D_{KK} = B\, n^{-\frac{1}{2}\log_2 n}\,.$$

For $B = 2^{20}$ this gives $D_{KK} \leq 1$ for $n > 80$, agreeing very well with Figure 14.16. The greedy strategy can also find balanced partitions, but only for much larger values of n. See Mertens [571] for a quantitative analysis of this scenario.

14.14 The easiest hard problem. The investigation of the phase transition in random INTEGER PARTITIONING is an excellent example of a successful interdisciplinary endeavor. It was found numerically by computer scientists [323], solved analytically by a physicist [569], and established rigorously by mathematicians [126]. The whole story is told by Brian Hayes [384], from whom we borrowed the title of Section 14.5.

The rigorous mathematical analysis in [126] goes far beyond the proof of the location and the width of the phase transition that we saw in Sections 14.5.2 and 14.5.3. Using similar techniques one can calculate higher moments of the number of partitions with any given difference $D > 1$, and these moments provide a wealth of information about the typical properties of random instances of INTEGER PARTITIONING.

For example, suppose that $\kappa > \kappa_c$. In the absence of perfectly balanced partitions, we can ask for the smallest difference we can achieve. Let's call this minimum D_1, the second smallest D_2 and so on. Calculating higher moments reveals that the sequence $D_1, D_2, \ldots, D_\ell$ converges to a Poisson process, i.e., the distances between adjacent values D_j and D_{j+1} are statistically independent. Furthermore, the corresponding partitions are uncorrelated. Thus the closer you get to the optimum, the more erratic the landscape of differences appears, until any partition in your current neighborhood has an independently random value. This explains why algorithms based on local search perform very poorly on instances with $\kappa > \kappa_c$. Like the phase transition, this scenario had been established by nonrigorous arguments from physics [84, 85, 570] before it was proven rigorously [123, 124, 126].

14.15 Beliefs, cavities and replicas. In statistical physics, belief propagation is known as the *cavity method*. The idea is to imagine removing an atom from a material, and then consider what field its neighbors would generate in the resulting cavity. Here we create cavities in the factor graph by removing a clause or a variable, and the BP equations give us the "field" generated by its neighbors.

Belief propagation, assuming that correlations decay exponentially, is referred to as the *replica symmetric* approach. Taking clustering into account but assuming that each cluster is an ergodic component as in Section 14.7

is the *one-step replica symmetry breaking* or 1RSB approach. There can be more than just one step of symmetry breaking—the clusters can consist of subclusters and so on—and we can even have an infinite hierarchy of clusters and subclusters. This scenario of *full RSB* was invented by Parisi to solve the statistical mechanics of Ising spin glasses, i.e., the Sherrington–Kirkpatrick model where each edge of a complete graph has a random ferromagnetic or antiferromagnetic bond on it. In sparse or "dilute" models, where each spin interacts only with $O(1)$ other spins, one-step replica symmetry breaking is usually sufficient.

The full story, including many applications of belief propagation (RS, 1RSB and full RSB) to error-correcting codes, combinatorial optimization, spin glasses, and random k-SAT, can be found in Mézard and Montanari [576].

14.16 The zeroth moment. How can we compute $\mathbb{E}[\ln Z]$? We computed the second moment $\mathbb{E}[Z^2]$ by analyzing correlations between pairs of assignments, and in theory we could compute the rth moment $\mathbb{E}[Z^r]$ by analyzing sets of r assignments. One possible approach is then to write

$$\mathbb{E}[\ln Z] = \lim_{r \to 0} \frac{\mathbb{E}[Z^r] - 1}{r}.$$

There is indeed a technique in statistical physics that works by computing $\mathbb{E}[Z^r]$ for every integer r and then taking the rather mysterious limit $r \to 0$. This method is known as the replica trick (see previous note), because each factor in Z^r can be considered as the partition sum of a copy, or replica, of the system. As the reader can appreciate, there are many potential pitfalls here, not least the fact that $\mathbb{E}[Z^r]$ might not be an analytic function of r.

14.17 Uniqueness and fixed points. Our sketch of the phase transitions for random k-SAT in Figure 14.26 is far from complete. There is also a *unicity* transition at a density $\alpha_{\text{unique}} < \alpha_{\text{clust}}$, beyond which the fixed point of the BP equations is no longer unique. Between α_{unique} and α_{clust} there is a dominant fixed point to which the BP equations converge from almost all initial conditions, and almost all solutions lie in the corresponding cluster—but an exponentially small fraction of solutions lie in small clusters corresponding to the other fixed points.

14.18 Condensation. In the condensed phase, the internal entropy of the largest clusters is the largest $s^\star$ such that $\Sigma(s^\star) = 0$. The statistics of these clusters are determined by the derivative $\Sigma'(s^\star) = -m^\star$, describing how the number of clusters increases as s decreases. Note that $m^\star$ goes from 1 to 0 as α goes from α_{cond} to α_c.

Let $\mathrm{e}^{ns^\star - \delta_\ell}$ denote the size of the ℓth largest cluster, where $\delta_\ell = O(1)$. The probability that there is a cluster of size between $\mathrm{e}^{ns^\star - \delta}$ and $\mathrm{e}^{ns^\star - \delta - \mathrm{d}\delta}$ is $\mathrm{e}^{m^\star \delta} \mathrm{d}\delta$, so the δ_ℓ are generated by a Poisson process with rate $\mathrm{e}^{m^\star \delta}$. The relative size of the ℓth largest cluster is

$$w_\ell = \frac{\mathrm{e}^{\delta_\ell}}{\sum_{i=1}^N \mathrm{e}^{\delta_i}}.$$

One can show that if δ_ℓ is a Poisson process then w_ℓ is a Poisson–Dirichlet process [659], and we can compute all the moments of w_ℓ analytically. The second moment, for example, can be used to express the probability Y that two random solutions belong to the same cluster,

$$Y = \mathbb{E}\left[\sum_{\ell=1}^N w_\ell^2\right] = 1 - m^\star.$$

Below the condensation transition this probability is zero, and in the condensed phase it increases from zero at $\alpha = \alpha_{\text{cond}}$ to one at $\alpha = \alpha_c$.

The random subcube model from Problem 14.44, for which we can compute $\Sigma(s)$ analytically and demonstrate clustering and condensation explicitly, was introduced by Mora and Zdeborová [615].

14.19 Series expansions for thresholds. As Exercise 14.21 shows, the satisfiability threshold α_c is the density at which $\Phi(0)$ vanishes, so if we are just interested in α_{cond} we can restrict ourselves to the case $m = 0$. This simplifies the SP equations considerably, since the weights z_i^m, z_a^m and $z_{(i,a)}^m$ all disappear.

This approach was used in [572] to compute the values of α_c shown in Table 14.4 and the series expansion for α_c in (14.88), and we can compute additional terms in these series if we like. The pth-order asymptotic expansion for α_c in the parameter $\varepsilon = 2^{-k}$ reads

$$2^{-k}\alpha_c = \ln 2 + \sum_{i=1}^{p} \hat{\alpha}_i \varepsilon^i + o(\varepsilon^p), \tag{14.101}$$

The coefficients $\hat{\alpha}_i$ are polynomials in k of degree $2i-2$, and in [572] this series has been computed up to $p=7$. The values for α_{clust} and α_{cond} in Table 14.4 and their large k expansions are from [599].

14.20 Rigorous results on clustering. Let X_ζ be the number of pairs of satisfying assignments with an overlap ζ. One can use the first and second moment methods—which involves computing the *fourth* moment of the number of satisfying assignments Z—to prove that X_ζ is either zero or nonzero with high probability. This idea was introduced by Mézard, Mora, and Zecchina [577], who proved that for $k \geq 8$ there are densities at which pairs of solutions are either far apart or very close.

Further rigorous results on clustering that combine the second moment method with the Paley–Zygmund inequality (Problem 14.41) were given by Achlioptas and Ricci-Tersenghi [18]. Theorem 14.5, on the scaling of the clustering transition and additional properties of the clusters, was proved by Achlioptas and Coja-Oghlan [10].

For NAE-k-SAT, the physics picture of the typical number and size distribution of the clusters, as well as the condensation and satisfiability thresholds, in terms of the fixed points of the update equations for the distribution of surveys has been made rigorous by Sly, Sun, and Zhang [760].

14.21 Inspired decimations. If belief propagation yields the exact marginals, we can use it to sample uniformly random satisfying assignments as we discussed in Section 14.6.4. If we don't care about sampling but we want to find a satisfying assignment, BP can help even if the beliefs are only approximations of the marginals. The idea is to select the variable that is maximally biased, i.e., for which the belief is closest to true or false, set it to its most probable value, and simplify the formula. We then iterate until all variables are set.

This technique of *belief-inspired decimation* works up to some density which for $k=3$ equals α_{cond}, but for $k \geq 4$ is smaller than α_{clust}. For $k=4$, we have $\tilde{\alpha} \approx 9.25$, for example. Even at densities where BP gives the correct marginals for the initial formula, as we set variables and shorten clauses, the remaining formula moves into the condensed phase. At this point, BP incorrectly thinks that the marginals are close to uniform, while the true marginals for many variables are highly biased. Once frozen variables appear, we set some of them wrong with high probability, and the remaining formula becomes unsatisfiable. As a result, above a critical density $\Theta(2^k/k)$, belief-guided decimation fails to find a satisfying assignment. This process was described in the physics literature by Ricci-Tersenghi and Semerjian [702], and established rigorously by Coja-Oghlan [192] and Coja-Oghlan and Pachon-Pinzon [193].

We can use the same idea with survey propagation. *Survey-inspired decimation* selects the variable that is maximally biased according to the surveys and sets it accordingly. For $k=3$ this algorithm works up to densities very close to α_c. When it was published in 2002 by Mézard, Parisi, and Zecchina [578, 579] it was an enormous breakthrough, solving large random instances close to the threshold in a few seconds.

In the wake of this success, other algorithms have been investigated that, empirically, work in linear time almost up to α_c. One is a local search algorithm, somewhat like `WalkSAT` (Section 10.3) that flips variables in an effort to satisfy unsatisfied clauses [738]. Another is a version of survey-inspired decimation that backtracks in order to find clusters with as few frozen variables as possible [556]. None of these algorithms appear to work in the frozen phase, however, which for larger values of k is far below the satisfiability threshold.

14.22 Clustered colorings. As we mentioned in Note 14.10, an algorithm of Achlioptas and Moore colors random graphs $G(n, p = c/n)$ up to $c = 4.03$. However, it follows from results of Gerschenfeld and Montanari [325] that the clustered phase in GRAPH 3-COLORING starts at $c = 4$.

14.23 Freezing by degrees. The process of freezing actually takes place in three stages with sharp transitions between them. The first transition, called *rigidity*, occurs when the dominant clusters freeze. This is followed by *total* rigidity, where clusters of all sizes are frozen—but even in this phase there are exponentially many unfrozen clusters, where solutions can be easy to find. Finally we reach the freezing transition beyond which there are, with high probability, no unfrozen clusters at all. See Montanari, Ricci-Tersenghi, and Semerjian [599] for the detailed phase diagram for k-SAT, Zdeborová and Krząkala [846] for GRAPH k-COLORING, or their joint paper [506] on both problems.

The scaling $(2^k/k)\ln k$ for the rigidity threshold was predicted by Semerjian [740] using a message-passing approach, and established rigorously by Achlioptas and Coja-Oghlan [10] as part of Theorem 14.5.

As we comment in the text, DPLL algorithms with simple branching rules like those analyzed in Section 14.3 only work in polynomial time up to densities $\Theta(2^k/k)$. But in 2009, Amin Coja-Oghlan [191] presented an algorithm that works in linear time right up to the rigidity transition $(2^k/k)\ln k$. The preciseness of this result—without even a multiplicative gap between this density and the one established by Theorem 14.5—is compelling evidence that freezing marks the transition between easy and hard.

However, both [740] and Theorem 14.5 leave open the possibility that there are a small number of unfrozen clusters which algorithms could conceivably exploit, as in Note 14.24. So far, the lowest density at which we know that *every* cluster has frozen variables is $\Theta(2^k)$; see Achlioptas and Ricci-Tersenghi [18].

14.24 Whitening. Algorithms like survey-inspired decimation (Note 14.21) can work in polynomial time even in the rigid phase. The reason is that even if almost all clusters are frozen, there can be exponentially many clusters which are not. If algorithms can somehow zoom in on these clusters, our arguments for exponential time no longer apply.

There is a cute way to check that a solution belongs to a non-frozen cluster. We assign a "wild card" symbol $\star$ to variables that can take either value—that is, which belong only to clauses that are already satisfied by other variables, including possibly a variable already labeled $\star$. We iterate this until we reach a fixed point, where no new variable can be labeled $\star$. The variables that are not in the "wild card" state, i.e., that still have a definite truth value, are the frozen variables of the cluster containing the solution we started with. If *every* variable ends up in the wild card state, the solution belongs to an un-frozen cluster.

If we apply this "whitening" procedure to solutions found by algorithms deep in the clustered phase, we find that these solutions always belong to unfrozen clusters [136, 553, 738]. This can happen even when solutions from unfrozen clusters are exponentially rare [210].

14.25 Random XORSAT. In random 3-XORSAT the clustering and freezing transitions take place at the same density, and the size and number of clusters can be computed exactly. We can define the "core" of the formula by repeatedly removing variables that appear in a single clause, since we can always use such a variable to fix the parity of that clause. The clustering transition is precisely the density α_{clust} at which a nonzero core exists, and this is the root of a transcendental equation similar to those we wrote in Section 14.2.5 for the core of $G(n, p)$.

The core corresponds to a system of linear equations $Mv = w$, where M is a random $m_{\text{core}} \times n_{\text{core}}$ matrix with three ones in each row and at least two ones in each column. Each solution v corresponds to a cluster, so the number of clusters is $2^{n_{\text{core}}-r}$ where $r = \operatorname{rank} M$. With high probability these vectors v form an error-correcting code, so they have a Hamming distance $\Theta(n)$ between them.

Finally, if we start with one of these v and rebuild the entire formula from the core, we get a choice of two solutions each time we add a clause with two new variables. Thus the size of each cluster is 2^t where $t = (n - n_{\text{core}}) - (m - m_{\text{core}})$. See Dubois and Mandler [252] and Cocco, Dubois, Mandler, and Monasson [182].

Chapter 15

Quantum Computation

> It used to be supposed in Science that if everything were known about the Universe at any particular moment then we could predict what it will be through all the future... More modern science, however, has come to the conclusion that when we are dealing with atoms and electrons we are quite unable to know the exact state of them; our instruments being made of atoms and electrons themselves.
>
> Alan M. Turing

We live in a computationally powerful universe. The laws of physics fill the world with a rich palette of objects and interactions: particles with forces between them, electromagnetic waves that can be generated and felt, energies that can be held and released. These interactions let us store, process, and transmit information, using everything from the neurons in our heads to the books in our libraries and the computers in our pockets. Indeed, if the physics of our universe could not support computation, it's doubtful that it could support life.

But the laws of physics are stranger and more wonderful than we can see in our everyday lives. In the first quarter of the 20th century, physicists learned that the fabric of our material world—its atoms, electrons and photons—behave rather differently from the objects of classical physics, such as billiard balls and planets. This new physics is counterintuitive yet strangely beautiful. It teaches us that electrons can be in two places at once, particles can spread and diffract like waves, and objects light-years apart can be inextricably entangled with each other. Can quantum physics help us compute in fundamentally new ways?

In Chapter 7 we discussed the Church–Turing Thesis, the claim that any reasonable computing device can be simulated by a Turing machine, and therefore by the programming languages and computers we know. Let's consider a sharper claim, about the devices we can actually build in the physical world:

> *Physical Church–Turing Thesis*: Any function that can be computed in finite time by a physical device can be computed by a Turing machine.

This is equivalent to the claim that, given enough time, a classical computer can simulate any physical process with any precision we desire.

Unlike the Church–Turing Thesis, which is a philosophical position about what counts as an algorithm, the Physical Church–Turing Thesis is a scientific claim about the physical universe. It may or may not be true. It is even, in principle, amenable to experiment. If you came across a physical device that solved every case of the Halting Problem you asked it about, this would be strong evidence that some physical processes are uncomputable.

Personally, we believe that the Physical Church–Turing Thesis is correct. It is conceivable that there are exotic physical systems that can somehow cram an infinite amount of computation into a finite amount of space and time, but this seems unlikely. If spacetime is continuous even at the smallest scales,
15.1 it is certainly true that describing, or simulating, a physical system exactly takes an infinite amount of information. But that doesn't mean we can access these infinitely microscopic scales and use them to build a computer.

Now let's consider a stronger claim—not just that Turing machines can compute anything a physical device can, but that they can do so in roughly the same amount of time, where "roughly the same" means within a polynomial factor:

> *Strong Physical Church–Turing Thesis*: Any function that can be computed in polynomial time by a physical device can also be calculated in polynomial time by a Turing machine.

We believe that this is false. Indeed, we believe that there are physical devices that are exponentially more efficient than a Turing machine, and therefore exponentially hard for a Turing machine to simulate.

One of the first to consider this question was the Nobel prize-winning physicist Richard Feynman. In 1981, he asked whether quantum mechanical processes can be simulated by classical computers in time proportional to the volume of space and time we wish to simulate. He argued that this is impossible, and that simulating a quantum system takes an exponential amount of resources on a classical computer. To simulate quantum systems, he said, we need quantum computers instead. But if quantum computers can simulate quantum physics exponentially faster than classical computers can, maybe they can solve other problems exponentially faster as well.

Our understanding of quantum computation is at a very early stage, but what little we know is extremely exciting. In addition to being more powerful than their classical counterparts, many quantum algorithms possess an entrancing beauty, and force us to think about computation in ways we never dreamed of before. In this chapter, we will meet algorithms that break cryptosystems by listening to the harmonics of the integers, find needles in haystacks by rotating and reflecting around them, and tell who can win a game by sending a wave through a tree.
15.2

15.1 Particles, Waves, and Amplitudes

Much of the strangeness of the quantum world can be illustrated with a classic experiment. Suppose we fire a beam of electrons at a metal plate with two slits in it, which we call A and B. On the other side of the plate is a detecting screen, which makes a dot of light at each place an electron hits it. If we cover A and let the electrons pass through B, we see a bright stripe on the screen. The same thing happens, of course, if we cover B and let the electrons go through A.

What happens if we uncover both screens? We expect to see two bright stripes, one behind each slit. After all, the brightness at each point on the screen is proportional to the fraction of electrons that arrive

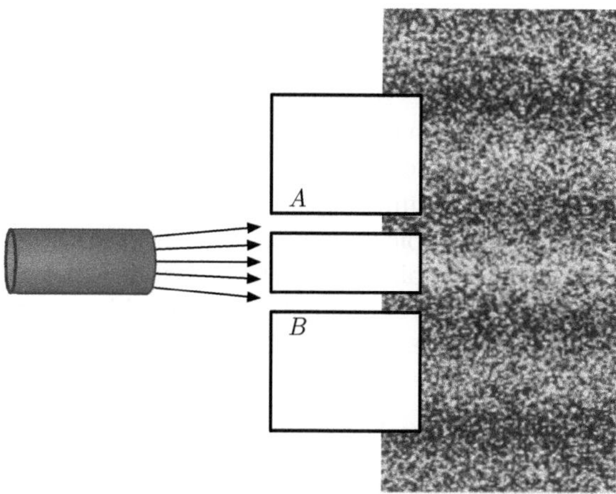

FIGURE 15.1: The two-slit experiment. We fire a beam of electrons through a pair of slits and onto a detecting screen. Instead of two bright stripes, we see an interference pattern of light and dark stripes.

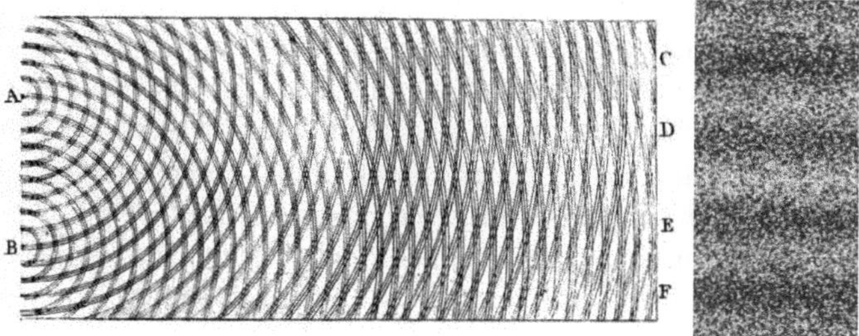

FIGURE 15.2: A sketch by Thomas Young showing waves of light or water emanating from two slits, A and B. At C, D, E, and F, these waves arrive with opposite phases, canceling out and creating dark stripes. At the points in between, they arrive with the same phase, creating bright stripes. For illustration, we align this sketch with the image from Figure 15.1.

there. For each point x, there are two ways that an electron can get there: one through A, and the other through B. Isn't the total probability that the electron arrives at x just the sum of these two probabilities?

As intuitive as this might sound, it is simply not true. Instead, we see a pattern of alternating light and dark stripes, as shown in Figure 15.1. This kind of *interference pattern* occurs in optics as well; Figure 15.2 shows a sketch by the 19th century physicist Thomas Young, who performed similar experiments with light. Waves emanate from both slits and spread out in concentric circles. If these waves are in phase with each other when they hit the screen, they form a single wave of greater intensity, like the photons in a laser beam. If they are out of phase, on the other hand, they cancel each other out. The same thing happens with ripples in a tank of water. Perhaps the electrons in the beam interact with each other, just as water molecules interact to form ripples in their surface?

The situation is stranger than that. In fact, the same pattern appears even when we do the experiment with one electron at a time. In that case, we can't explain the interference pattern in terms of electrons interacting with each other. In some sense, *each electron interacts with itself.* In the 19th century, we thought of electrons as little steel balls, with definite positions and velocities in space. But even a single electron acts to some extent like a wave, spreading out in space and interfering with itself at every point.

Moreover, this wave can't be described simply as a probability distribution. Rather than a real-valued probability, its value at each point is a complex number, called the *amplitude.* Each amplitude has both a length and an angle, or *phase,* in the complex plane. The probability associated with an amplitude a is $|a|^2$, its absolute value squared. But when we combine two paths with amplitudes a and b, we add their amplitudes before taking the absolute value, so the total probability is

$$P = |a+b|^2 .$$

If a and b have the same phase, such as at the point x in Figure 15.3, the paths interfere *constructively* and P is large. If they have opposite phases, such as at x', they interfere *destructively* and cancel. So when we add the amplitudes for the two ways the electron could get to each point on the screen, there are some where it shows up with a probability greater than their sum—and others where it never shows up at all.

15.3 This phenomenon is at the heart of quantum algorithms. Like an electron whose wave function spreads out in space, quantum computers have a kind of parallelism. They can be in a superposition of many states at once, corresponding to many possible solutions to a problem, or many possible paths through the computer's state space. By carefully designing how these states interfere, we can arrange for the wrong answers to cancel out, while creating a peak of probability at the right one.

There is one more version of the two-slit experiment that we should mention. Suppose we add a pair of detectors at the slits, which measure which slit the electron passes through. Now the interference pattern disappears: we see a pair of bright stripes, one behind each slit, just as we expected in the first place. The law of classical probability, in which the total probability of an event is the sum of the probabilities of all the ways it can happen, has been restored.

This process is called *decoherence.* As the following exercise shows, we can think of it as randomizing the phase of a quantum amplitude, and retaining information only about its absolute value.

Exercise 15.1 *Suppose that a and b are amplitudes corresponding to paths in a quantum process, such as the two paths to a given point in the two-slit experiment. Depending on their relative phase, and on whether they interfere constructively or destructively, show that their combined probability $P = |a+b|^2$ can lie anywhere in the range*

$$(|a|-|b|)^2 \le P \le (|a|+|b|)^2 .$$

Now suppose that their relative phase is random. For instance, suppose we multiply b by $e^{\imath\theta}$ where θ is a random angle between 0 and 2π. Show that averaging over θ gives

$$\mathop{\mathbb{E}}_{\theta} P = \mathop{\mathbb{E}}_{\theta} \left| a + e^{\imath\theta} b \right|^2 = |a|^2 + |b|^2 .$$

In other words, when the phase is random, the combined probability of the two paths is the sum of the two probabilities, just as in the classical case.

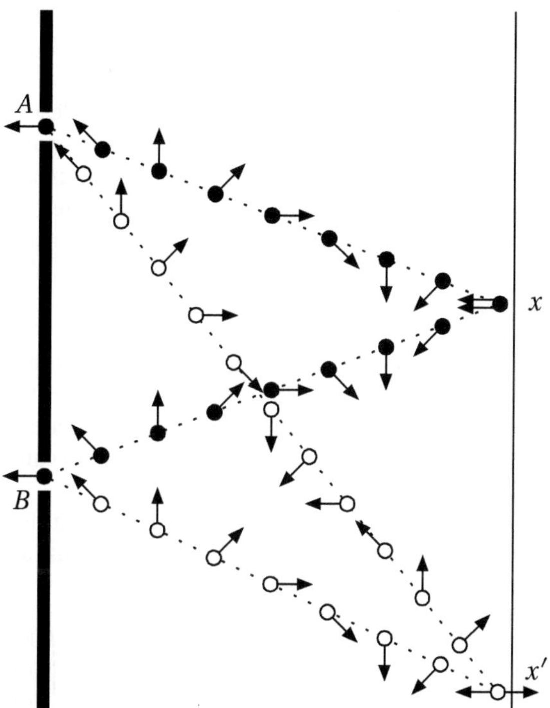

FIGURE 15.3: Constructive and destructive interference in the two-slit experiment. The phase of the electron rotates with time, so the phase of each path depends on its length. The two paths leading to x arrive with the same phase, so the electron has a large probability of hitting the screen there. The two paths leading to x' arrive with opposite phases, so they cancel and the total probability the electron hits the screen at that point is zero.

Decoherence helps explain why the world of our everyday experience looks classical rather than quantum. Quantum states are fragile. When they are allowed to interact with large, blundering objects like humans, detectors, or even stray electromagnetic waves, their phases quickly randomize, and they collapse into classical probability distributions. To build a quantum computer, we have to create a physical system where every interaction is carefully controlled—where the states within the computer interact strongly with each other, without becoming decohered by stray interactions with their environment. Clever people are performing amazing feats of physics and engineering in pursuit of this goal. This chapter is about what problems we can solve if they succeed.

15.2 States and Operators

In the classical world, we analyze our computers not in terms of physical quantities like voltages and currents, but in terms of the logical quantities—the bits—they represent. In the same way, to get from quantum physics to quantum computers, we need to consider quantum systems that can store digital information. The simplest such system is a quantum bit, or *qubit* for short. Like a classical bit, it can be true or false—but since it is quantum, it can be in a superposition of these.

In this section, we will write down the mathematical machinery for describing systems of qubits, and how their states can be transformed and measured. We will see that, mathematically speaking, quantum algorithms are not so different from classical randomized algorithms. The main difference is that they have complex-valued amplitudes instead of real-valued probabilities, allowing these amplitudes to interfere constructively or destructively, rather than simply add.

15.2.1 Back to the State Space

Let's start by viewing good-old-fashioned classical computation in a somewhat bizarre way. As we discussed in Chapter 8, a computer with m bits of memory has 2^m possible states. Let's view each of these states as a basis vector in a 2^m-dimensional vector space. Each elementary step transforms the computer's state by multiplying it by a $2^m \times 2^m$ matrix, and the entire program can be thought of as the $2^m \times 2^m$ matrix that is the product of all of its steps.

For instance, suppose we have a computer with 2 bits, x_1 and x_2. It has four possible states, which we write in the form $|x_1 x_2\rangle$. This notation $|\cdot\rangle$ is called a "ket," but it is really just a column vector. These states form a basis for a four-dimensional state space:

$$|00\rangle = \begin{pmatrix} 1 \\ 0 \\ 0 \\ 0 \end{pmatrix}, \; |01\rangle = \begin{pmatrix} 0 \\ 1 \\ 0 \\ 0 \end{pmatrix}, \; |10\rangle = \begin{pmatrix} 0 \\ 0 \\ 1 \\ 0 \end{pmatrix}, \; |11\rangle = \begin{pmatrix} 0 \\ 0 \\ 0 \\ 1 \end{pmatrix}.$$

We have ordered these vectors so that x_1 is the most significant bit and x_2 is the least significant, but this is an arbitrary choice.

Now imagine a step of the program that carries out the instruction

$$x_2 := x_1 .$$

This maps old states to new states in the following way:

$$|00\rangle \to |00\rangle , \; |01\rangle \to |00\rangle , \; |10\rangle \to |11\rangle , \; |11\rangle \to |11\rangle .$$

We can think of this as multiplying the state by a 4×4 matrix,

$$U = \begin{pmatrix} 1 & 1 & & \\ 0 & 0 & & \\ & & 0 & 0 \\ & & 1 & 1 \end{pmatrix},$$

where matrix entries not shown are zero. Each column of U has a single 1, since a deterministic computation gives exactly one new state for each old state.

Now suppose we have a *randomized* algorithm, with instructions like this:

With probability 1/2, set $x_2 := x_1$. Otherwise do nothing.

This corresponds to a stochastic matrix, i.e., one whose columns sum to 1:

$$U = \begin{pmatrix} 1 & 1/2 & & \\ 0 & 1/2 & & \\ & & 1/2 & 0 \\ & & 1/2 & 1 \end{pmatrix}.$$

Vectors in the state space now correspond to probability distributions. For instance, applying U to the state $|10\rangle$ gives the state

$$U|10\rangle = \begin{pmatrix} 0 \\ 0 \\ 1/2 \\ 1/2 \end{pmatrix} = \frac{1}{2}\left(|10\rangle + |11\rangle\right),$$

in which the computer has an equal probability of being in the state $|10\rangle$ or the state $|11\rangle$.

Of course, if our computer is classical, it is in one of these states or the other. But for many purposes, it makes sense to think of the "state" of the computer as a probability distribution, so that it is, in some sense, in both states at once. If the computer has m bits of memory, this probability distribution is a 2^m-dimensional vector, with one component for each of the 2^m possible truth assignments.

This picture of computation may seem needlessly complicated. Does it really make sense to say that if your computer has one gigabyte of memory, then its state is a $2^{2^{33}}$-dimensional vector, and each step of your program multiplies the state by a $2^{2^{33}}$-dimensional matrix? We agree that if our goal is to understand classical computation, this point of view is rather cumbersome. But the benefit of this approach is that quantum computers—which really *can* be in many states at once—are now just a small step away.

15.2.2 Bras and Kets

She was nae get o' moorland tips,
Wi' tawted ket, an' hairy hips...

Robert Burns, *Poor Mailie's Elegy*

Rather than a real-valued vector whose components are probabilities, the state of a quantum computer is a complex-valued vector whose components are called *amplitudes*. For instance, the state of a single qubit can be written

$$|v\rangle = \begin{pmatrix} v_0 \\ v_1 \end{pmatrix} = v_0|0\rangle + v_1|1\rangle,$$

where v_0 and v_1 are complex, and where

$$|0\rangle = \begin{pmatrix} 1 \\ 0 \end{pmatrix} \text{ and } |1\rangle = \begin{pmatrix} 0 \\ 1 \end{pmatrix}$$

are the basis vectors corresponding to the qubit being 0 (false) and 1 (true) respectively. We say that $|v\rangle$ is a *superposition* of the basis states $|0\rangle$ and $|1\rangle$ with amplitudes v_0 and v_1 respectively. Note that $|0\rangle$ is not the zero vector—it is a basis vector of length 1 in the state space.

Writing column vectors as "kets" is one half of a handy notational device invented by Paul Dirac. Here's the other half. We can also write a state as a row vector, or "bra" $\langle v|$. For a single qubit, this looks like

$$\langle v| = (v_0^*, v_1^*) = v_0^* \langle 0| + v_1^* \langle 1|.$$

The inner product of two vectors is a "bra-ket" (yes, this is what passes for humor among physicists):

$$\langle v | w \rangle = \sum_i v_i^* w_i .$$

Note that when changing a ket to a bra, we automatically take its complex conjugate, since this is essential when defining the inner product of complex-valued vectors. In particular, the inner product of a vector with itself is the square of its 2-norm, or its Euclidean length:

$$\langle v | v \rangle = \sum_i v_i^* v_i = \sum_i |v_i|^2 = \|v\|_2^2 .$$

15.2.3 *Measurements, Projections, and Collapse*

Physically, the states $|0\rangle$ and $|1\rangle$ might correspond to an electron having its magnetic field pointing north or south, or a photon being horizontally or vertically polarized, or any other pair of states that a quantum system can be in. By performing a measurement on this system, we can observe the qubit's truth value. If its state is $|v\rangle = v_0|0\rangle + v_1|1\rangle$, the probability that we observe $|0\rangle$ or $|1\rangle$ is the square of the absolute value of the corresponding amplitude,

$$P(0) = |v_0|^2 \quad \text{and} \quad P(1) = |v_1|^2 .$$

The total probability is $|v_0|^2 + |v_1|^2 = \langle v | v \rangle = 1$.

We can express these probabilities in terms of projection operators. Let Π_0 be the operator that projects onto $|0\rangle$,

$$\Pi_0 = \begin{pmatrix} 1 & 0 \\ 0 & 0 \end{pmatrix} .$$

The probability that we observe the truth value 0 is

$$P(0) = \|\Pi_0|v\rangle\|_2^2 = \langle v|\Pi_0|v\rangle , \tag{15.1}$$

where in the second equality we used the fact that $\Pi^2 = \Pi$ for any projection operator Π. If we define Π_1 similarly as the projection operator onto $|1\rangle$, these projection operators sum to the identity,

$$\Pi_0 + \Pi_1 = \mathbb{1} ,$$

so the total probability is

$$P(0) + P(1) = \langle v|\Pi_0|v\rangle + \langle v|\Pi_1|v\rangle = \langle v|(\Pi_0 + \Pi_1)|v\rangle = \langle v | v \rangle = 1 .$$

Let's say this one more way. We can also use bra-ket notation to write *outer* products. Given two states $|v\rangle = \sum_i v_i |i\rangle$ and$|w\rangle = \sum_i w_i |i\rangle$, their outer product $|v\rangle\langle w|$ is the matrix where

$$(|v\rangle\langle w|)_{ij} = \langle i | v \rangle \langle w | j \rangle = v_i w_j^* .$$

The projection operator corresponding to observing a basis state is its outer product with itself,

$$\Pi_0 = |0\rangle\langle 0| \quad \text{and} \quad \Pi_1 = |1\rangle\langle 1|\,,$$

so for instance

$$P(0) = \langle v|\Pi_0|v\rangle = \langle v\,|0\rangle\langle 0\,|\,v\rangle = |\langle v\,|0\rangle|^2\,.$$

In the two-slit experiment, the electron is in a superposition of two states, one for each slit it can go through. When we place detectors at the slits and measure which one of the states it is in, this superposition is destroyed and the electron behaves like a tiny billiard ball that passed through one slit or the other. This is true for qubits as well. After we have observed the truth value 0, the qubit has gone from the superposition $v_0|0\rangle + v_1|1\rangle$ to the observed state $|0\rangle$. We can think of this as applying the projection operator Π_0 and then normalizing the state so that the total probability is 1, giving

$$|0\rangle = \frac{\Pi_0|v\rangle}{\sqrt{\langle v|\Pi_0|v\rangle}}\,.$$

This process is sometimes called the *collapse of the wave function.* Opinions differ as to whether it is a real physical process or simply a convenient shorthand for calculation. Without weighing in too heavily on this debate, we point out that it is not so different mathematically from the classical probabilistic case. If we learn something about a probability distribution, from our point of view it "collapses" to the 15.4
distribution conditioned on that knowledge. This conditional distribution is its image under a projection operator, normalized so that the total probability is 1.

What happens if we perform a partial measurement, and only measure part of the system? Suppose we have a two-qubit state, i.e., a four-dimensional vector

$$|v\rangle = \begin{pmatrix} v_{00} \\ v_{01} \\ v_{10} \\ v_{11} \end{pmatrix} = v_{00}|00\rangle + v_{01}|01\rangle + v_{10}|10\rangle + v_{11}|11\rangle\,.$$

If we measure just the first qubit x_1, the probability that we will observe $|1\rangle$ is the total probability of the two states in which x_1 is true, i.e., $|v_{10}|^2 + |v_{11}|^2$. As in the one-qubit case above, this observation can be described by a projection operator Π that projects onto the states that will give us this outcome. In this example, we have

$$\Pi = \begin{pmatrix} 0 & & & \\ & 0 & & \\ & & 1 & \\ & & & 1 \end{pmatrix},$$

and we can write the probability that the first qubit is true as

$$P(x_1 = 1) = \||\Pi|v\rangle\|_2^2 = \langle v|\Pi|v\rangle\,.$$

After this observation, the state has collapsed to its image under Π and been renormalized, giving

$$|v'\rangle = \frac{\Pi|v\rangle}{\sqrt{\langle v|\Pi|v\rangle}} = \frac{v_{10}|10\rangle + v_{11}|11\rangle}{\sqrt{|v_{10}|^2 + |v_{11}|^2}}\,,$$

which is a superposition of $|10\rangle$ and $|11\rangle$. Similarly, if we observe the outcome $x_1 = 0$, the state collapses to a superposition of $|00\rangle$ and $|01\rangle$.

How much difference do complex numbers make? Since the probability that qubit is 0 or 1 depends only on the amplitudes' absolute value, we might think that their *phases*—their angles in the complex plane—don't matter. Indeed, if two states differ by an overall phase, so that $|v'\rangle = e^{\imath\theta}|v\rangle$ for some θ, they are physically identical.

However, the *relative* phases between v's components play a crucial role. For instance, the states

$$\frac{1}{\sqrt{2}}(|0\rangle + |1\rangle) \text{ and } \frac{1}{\sqrt{2}}(|0\rangle - |1\rangle)$$

are very different. If we measure the truth value of the qubit, we get $P(0) = P(1) = 1/2$ in either case—but as we will see below, this is not the only measurement we can perform.

15.2.4 Unitary Matrices

Each step of a classical randomized algorithm transforms the probability distribution by multiplying it by a matrix U. This matrix has to preserve the total probability, so like the transition matrices of Markov chains we discussed in Chapter 12, it must be *stochastic*. That is, it must have nonnegative entries, and each column must sum to 1.

What kind of transition matrix makes sense for a quantum computer? It again has to preserve the total probability, but now this is defined as $\|v\|_2^2 = \langle v \,|\, v\rangle$. What does this tell us about U? Given a matrix U, its *Hermitian conjugate* $U^\dagger = (U^T)^*$ is the complex conjugate of its transpose. If we transform the ket $|v\rangle$ to $|v'\rangle = U|v\rangle$, the bra $\langle v|$ becomes $\langle v'| = \langle v|U^\dagger$. If the total probability is preserved, we have

$$\langle v' \,|\, v'\rangle = \langle v|U^\dagger U|v\rangle = \langle v \,|\, v\rangle. \tag{15.2}$$

More generally, as Problem 15.1 shows, U preserves inner products. For all v and w,

$$\langle v' \,|\, w'\rangle = \langle v|U^\dagger U|w\rangle = \langle v \,|\, w\rangle.$$

It follows that

$$U^\dagger U = \mathbb{1},$$

and therefore

$$U^\dagger = U^{-1}.$$

Equivalently, the columns of U form an orthonormal basis. If $|u_i\rangle$ denotes the ith column,

$$\langle u_i \,|\, u_j\rangle = \begin{cases} 1 & \text{if } i = j \\ 0 & \text{if } i \neq j, \end{cases}$$

and similarly for the rows.

Matrices with this property are called *unitary*. In particular, they are invertible—they correspond to reversible physical processes, and can neither create nor destroy information. Geometrically, unitary matrices are a complex-valued generalization of the *orthogonal* matrices, which are real-valued rotations and reflections for which $U^T U = \mathbb{1}$. Every step of a quantum computer, no matter how complicated, is just a rotation or reflection in some high-dimensional space.

Exercise 15.2 *Show that if U is a unitary matrix, all its eigenvalues are on the unit circle in the complex plane. That is, they are of the form $e^{\imath\theta} = \cos\theta + \imath \sin\theta$ for some angle θ.*

Exercise 15.3 *Show that the unitary matrices U whose entries are 0s and 1s are exactly the permutation matrices—namely, those with a single 1 in each row and each column, so that multiplying a vector by U permutes its components. Such matrices correspond to reversible classical computation.*

Let's summarize what we have seen so far. On a purely mathematical level, we can go from classical randomized computation to quantum computation just by replacing real probabilities with complex amplitudes, 1-norms with 2-norms, and stochastic matrices with unitary ones. Of course, these changes make an enormous difference in the structure of quantum states and operations—and they lead to some very strange phenomena, which we will explore in the next section. But first, let's look at some simple unitary operators, and see how we can string them together to make quantum circuits and algorithms.

15.2.5 The Pauli and Hadamard Matrices, and Changing Basis

Three of the most important 2×2 unitary operators are the *Pauli matrices*:

$$X = \begin{pmatrix} 0 & 1 \\ 1 & 0 \end{pmatrix}, \quad Y = \begin{pmatrix} 0 & -\imath \\ \imath & 0 \end{pmatrix}, \quad Z = \begin{pmatrix} 1 & 0 \\ 0 & -1 \end{pmatrix}.$$

How do these act on our basis states $|0\rangle$ and $|1\rangle$? The effect of X is easy to describe—it acts like a classical NOT gate and flips the qubit's truth value.

On the other hand, Z's effect is fundamentally quantum. In a sense it leaves the truth value unchanged, but it induces a phase change of -1 if the qubit is true. That is, $|0\rangle$ and $|1\rangle$ are eigenvectors of Z with eigenvalues $+1$ and -1 respectively. Finally, $Y = \imath XZ$ switches the two truth values, and adds a phase change of $\pm\imath$ as well.

Another important one-qubit operation is the *Hadamard matrix*,

$$H = \frac{1}{\sqrt{2}} \begin{pmatrix} 1 & 1 \\ 1 & -1 \end{pmatrix}.$$

For instance, H transforms the states $|0\rangle$ and $|1\rangle$ to superpositions of $|0\rangle$ and $|1\rangle$, which we call $|+\rangle$ and $|-\rangle$:

$$H|0\rangle = |+\rangle = \frac{1}{\sqrt{2}}(|0\rangle + |1\rangle) \text{ and } H|1\rangle = |-\rangle = \frac{1}{\sqrt{2}}(|0\rangle - |1\rangle).$$

These are the eigenvectors of X, with eigenvalues $+1$ and -1 respectively. Since H is its own inverse, we can write

$$X = HZH \quad \text{and} \quad Z = HXH.$$

Thus H acts as a basis change, transforming Z to X and vice versa. In this basis, Z acts like a NOT gate, switching $|+\rangle$ and $|-\rangle$ just as X switches $|0\rangle$ and $|1\rangle$.

As we commented above, if we take $|+\rangle$ or $|-\rangle$ and measure the qubit's truth value, we have $P(0) = P(1) = 1/2$ in both cases. But applying H to $|+\rangle$ or $|-\rangle$ changes them back to $|0\rangle$ and $|1\rangle$. Thus these two states are as different as they possibly could be—we just lose this difference if we measure them without transforming them first.

We will refer to the states $|0\rangle, |1\rangle$ as the *computational basis*, or the Z-basis since they are the eigenvectors of Z. But this is just one of many bases we could use to describe a qubit's state space. If we want to measure a qubit $|v\rangle$ in the X-basis, for instance, along the basis vectors $|+\rangle$ and $|-\rangle$, we can do this by applying the basis change H and then measuring in the computational basis. The probability that we observe $|+\rangle$ or $|-\rangle$ is then $|\langle +|v\rangle|^2 = |\langle 0|Hv\rangle|^2$ and $|\langle -|v\rangle|^2 = |\langle 1|Hv\rangle|^2$ respectively. We can measure in any basis we like, as long as we can efficiently carry out the basis change between that basis and the computational one.

Exercise 15.4 *Show that the Pauli matrices have the following properties. First, they obey the cyclically symmetric relations $XY = \imath Z$, $YZ = \imath X$, and $ZX = \imath Y$. Secondly, they* anticommute *with each other: in other words, $XY = -YX$, $YZ = -ZY$, and $ZX = -XZ$. Finally, $X^2 = Y^2 = Z^2 = \mathbb{1}$.*

Exercise 15.5 *Write the projection operators Π_+ and Π_- that project onto X's eigenvectors.*

15.2.6 Multi-Qubit States and Tensor Products

How do we define states with multiple qubits? If $|u\rangle, |v\rangle$ are vectors of dimension n and m, their *tensor product* $|u\rangle \otimes |v\rangle$ is an (nm)-dimensional vector whose components are the products of those of $|u\rangle$ and $|v\rangle$. For instance, the tensor product of two one-qubit states looks like

$$\begin{pmatrix} u_0 \\ u_1 \end{pmatrix} \otimes \begin{pmatrix} v_0 \\ v_1 \end{pmatrix} = \begin{pmatrix} u_0 v_0 \\ u_0 v_1 \\ u_1 v_0 \\ u_1 v_1 \end{pmatrix}.$$

In particular, our two-qubit basis vectors are tensor products of one-qubit ones. For instance,

$$|01\rangle = \begin{pmatrix} 0 \\ 1 \\ 0 \\ 0 \end{pmatrix} = \begin{pmatrix} 1 \\ 0 \end{pmatrix} \otimes \begin{pmatrix} 0 \\ 1 \end{pmatrix} = |0\rangle \otimes |1\rangle.$$

If $|u\rangle$ and $|v\rangle$ are independent probability distributions then $|u\rangle \otimes |v\rangle$ is the joint distribution, where the probability of each combination of values is the product of their probabilities. Similarly, if $|u\rangle$ and $|v\rangle$ are two qubits, the amplitudes in their joint state $|u\rangle \otimes |v\rangle$ are the products of their amplitudes.

We will see, however, that most states with two or more qubits can't be written as tensor products. Quantum states can be *entangled*, so that their qubits are not independent of each other. This is analogous to the fact that a joint probability distribution can have correlations, so that it is not a product of independent distributions. However, quantum entanglement turns out to be much weirder than classical correlations can be.

We can also define the tensor product of two operators. If we have two qubits x_1, x_2 and we apply two operators A and B to x_1 and x_2 respectively, the resulting operator is their tensor product $A \otimes B$. We can write this as a 4×4 matrix, in which each 2×2 block consists of a copy of B multiplied by an entry of A:

$$A \otimes B = \left(\begin{array}{c|c} A_{11} B & A_{12} B \\ \hline A_{21} B & A_{22} B \end{array} \right) = \begin{pmatrix} A_{11} B_{11} & A_{11} B_{12} & A_{12} B_{11} & A_{12} B_{12} \\ A_{11} B_{21} & A_{11} B_{22} & A_{12} B_{21} & A_{12} B_{22} \\ A_{21} B_{11} & A_{21} B_{12} & A_{22} B_{11} & A_{22} B_{12} \\ A_{21} B_{21} & A_{21} B_{22} & A_{22} B_{21} & A_{22} B_{22} \end{pmatrix},$$

where we have arbitrarily ordered our basis so that x_1 is the most significant qubit and x_2 is the least significant. For instance, the tensor product of the Hadamard matrix with itself is

$$H \otimes H = \frac{1}{\sqrt{2}} \left(\begin{array}{c|c} H & H \\ \hline H & -H \end{array} \right) = \frac{1}{2} \begin{pmatrix} 1 & 1 & 1 & 1 \\ 1 & -1 & 1 & -1 \\ 1 & 1 & -1 & -1 \\ 1 & -1 & -1 & 1 \end{pmatrix}.$$

Exercise 15.6 *Show that*

$$(A \otimes B)(|u\rangle \otimes |v\rangle) = A|u\rangle \otimes B|v\rangle$$

and

$$(\langle u| \otimes \langle v|)(|u'\rangle \otimes |v'\rangle) = \langle u|u'\rangle\langle v|v'\rangle.$$

Exercise 15.7 *Show that the two-qubit state $\frac{1}{\sqrt{2}}(|00\rangle + |11\rangle)$ is entangled. That is, it is not the tensor product of any pair of one-qubit states.*

15.2.7 *Quantum Circuits*

When we discuss classical computation, we don't use elementary logical gates. Thanks to the hard work of our 20th-century forebears, we can describe classical algorithms at the software level—using high-level pseudocode—rather than at the hardware level. For quantum computing, however, we are at a very early stage of understanding what higher-level programming constructs we can use. So, while doubtless our descendants will pity us—just as we pity our grandparents who programmed with punch cards and wires—we still often describe quantum algorithms in terms of circuits, where each gate consists of an elementary quantum operation.

Let's introduce a notation for quantum circuits. We draw a one-qubit operator U like this:

$$-\boxed{U}-.$$

This gate takes a one-qubit input $|v\rangle$, entering through the "wire" on the left, and transforms it to $U|v\rangle$. We compose two such gates by stringing them together,

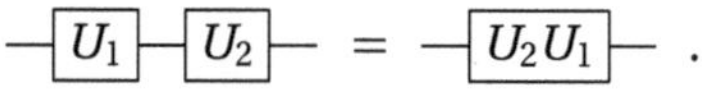

$$-\boxed{U_1}-\boxed{U_2}- \;=\; -\boxed{U_2 U_1}- \;.$$

If we have two qubits, applying two operators in parallel gives their tensor product,

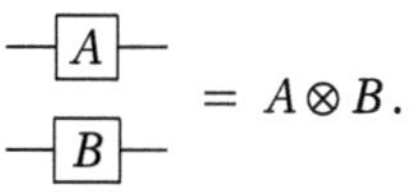

$$\begin{array}{c} -\boxed{A}- \\ -\boxed{B}- \end{array} = A \otimes B.$$

A major benefit of these circuit diagrams is that certain algebraic identities become obvious, as the following exercise shows:

Exercise 15.8 *If A, B, C, D are matrices of the same size, show that*

$$(A \otimes B)(C \otimes D) = (AC) \otimes (BD).$$

Hint: consider the diagram

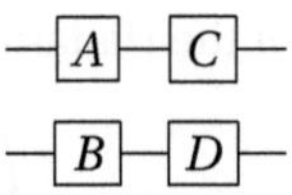

15.5

To get any computation off the ground, our qubits need to interact with each other. One of the simplest and most useful quantum gates is the "controlled-NOT," which originated in reversible computation. It takes two qubits $|a\rangle$ and $|b\rangle$ as its inputs, and flips the truth value of $|b\rangle$ if $|a\rangle$ is true:

$$|a,b\rangle \to |a, b \oplus a\rangle$$

where $\oplus$ denotes exclusive OR, or equivalently addition mod 2. As a matrix, this is

$$\begin{pmatrix} 1 & 0 & & \\ 0 & 1 & & \\ & & 0 & 1 \\ & & 1 & 0 \end{pmatrix} = \left(\begin{array}{c|c} \mathbb{1} & \\ \hline & X \end{array} \right).$$

Diagrammatically, we represent a controlled-NOT like this:

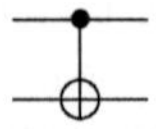

The top qubit is the "control" a, and the bottom qubit is the "target" b. But the next exercise shows that these names are naive. There is no such thing as a one-way flow of information in quantum mechanics.

Exercise 15.9 *Show that if we transform to the X-basis, a controlled-*NOT *gate goes the other way:*

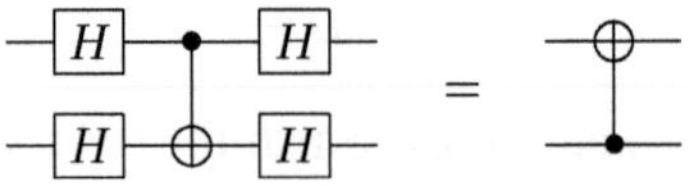

Exercise 15.10 *Show that the circuit*

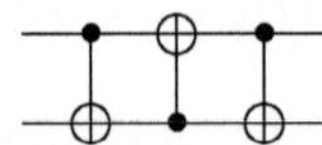

switches the two qubits.

More generally, given any one-qubit operator U, we can define a *controlled-U* gate that applies U to the target qubit if the control qubit is true. We draw this as

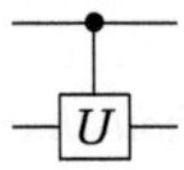

$$= \left(\begin{array}{c|c} \mathbb{1} & \\ \hline & U \end{array} \right).$$

Thus we could also call the controlled-NOT gate the controlled-X gate.

Another useful gate is the *controlled phase shift*, which induces a phase shift if both qubits are true and otherwise leaves the state unchanged. This gate is symmetric with respect to the two qubits, and we write

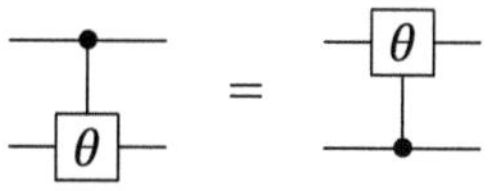

$$= \begin{pmatrix} 1 & & & \\ & 1 & & \\ & & 1 & \\ & & & e^{\iota\theta} \end{pmatrix}.$$

We can write the diagonal entries of this matrix as $e^{\iota x_1 x_2 \theta}$ where $x_1, x_2 \in \{0,1\}$. In particular, setting $\theta = \pi$ gives a controlled-Z gate.

Classically, a handful of basic logical operations suffice to carry out any computation we desire. In particular, if we have AND, OR, and NOT gates—or, for that matter, just NANDs or NORs—we can build a circuit for any Boolean function. What unitary operators do we need to carry out universal quantum computation, and implement any unitary operator?

Since quantum operators live in a continuous space, we say that a set S of quantum gates is *universal* if, for any 2^n-dimensional unitary matrix U and any ε, there is a circuit on n qubits made of these gates that approximates U to within an error ε. One such set consists of the controlled-NOT, the Hadamard, and the one-qubit gate 15.6

$$Z^{1/4} = \begin{pmatrix} 1 & 0 \\ 0 & e^{\iota\pi/4} \end{pmatrix}.$$

In general, the number of gates in our circuit has to increase as ε decreases, just as we need larger denominators to find better rational approximations to a real number. But even among families of matrices U that can be carried out exactly by a finite circuit, some require more gates than others. That's where computational complexity comes in.

15.3 Spooky Action at a Distance

> (secret, secret, close the doors!) we always have had a great deal of difficulty in understanding the world view that quantum mechanics represents. At least I do, because I'm an old enough man that I haven't got to the point that this stuff is obvious to me. Okay, I still get nervous with it.
>
> Richard Feynman

One of the strangest aspects of quantum mechanics is *entanglement*, in which the states of two distant quantum systems are inextricably linked. While parts of a classical system can be correlated with each other, quantum entanglement is a much more curious phenomenon, and has led to some of the deepest debates about the nature of physical reality.

In this section we describe how quantum gates like the controlled-NOT can create entangled pairs of qubits, how measuring these qubits can produce paradoxical results, and how entanglement can—or can't—be used to communicate across vast distances. These phenomena have more to do with quantum information or communication than computation per se, but they help reveal the counterintuitive behavior of quantum systems.

15.3.1 *Cloning and Entanglement*

Suppose that we apply a controlled-NOT to a state in which the target qubit is $|0\rangle$. Since we flip the target qubit if the control qubit is $|1\rangle$, it seems that the effect is to produce two copies of the control qubit:

$$\begin{array}{ccc} |v\rangle & \text{---}\bullet\text{---} & |v\rangle \\ |0\rangle & \text{---}\oplus\text{---} & |v\rangle \end{array}$$

This is certainly true if $|v\rangle = |0\rangle$ or $|1\rangle$. But what if $|v\rangle$ is a superposition of these two? If each qubit is independently in the state $|v\rangle = v_0|0\rangle + v_1|1\rangle$, their joint state would be the tensor product

$$|v\rangle \otimes |v\rangle = v_0^2|00\rangle + v_0 v_1|01\rangle + v_1 v_0|10\rangle + v_1^2|11\rangle\,.$$

But since the amplitudes of $|v\rangle \otimes |v\rangle$ are quadratic functions of those of $|v\rangle$, no linear operator—let alone a unitary one—can transform $|v\rangle$ to $|v\rangle \otimes |v\rangle$. That is, there is no matrix M such that, for all $|v\rangle$,

$$M\big(|v\rangle \otimes |0\rangle\big) = M\begin{pmatrix} v_0 \\ 0 \\ v_1 \\ 0 \end{pmatrix} = \begin{pmatrix} v_0^2 \\ v_0 v_1 \\ v_1 v_0 \\ v_1^2 \end{pmatrix}.$$

This result is called the *No-Cloning Theorem.*

15.7 This inability to copy quantum information seems rather strange at first. In classical computation, we take it for granted that we can copy one bit to another. For instance, we routinely feed the output of one logical gate into the inputs of multiple gates, simply by splitting a wire into several branches.

However, a kind of No-Cloning Theorem also holds for classical randomized algorithms. If x is a bit in a randomized algorithm, we can't create another bit y which has the same distribution as x *but is independent of* x, since their joint probabilities $P(x,y) = P(x)P(y)$ would be quadratic functions of x's probabilities. We can clone truth values, but not probability distributions of truth values.

If it doesn't copy a qubit, what *does* the controlled-NOT do in this case? Rather than $|v\rangle \otimes |v\rangle$, the state it generates is

$$v_0|00\rangle + v_1|11\rangle = \begin{pmatrix} v_0 \\ 0 \\ 0 \\ v_1 \end{pmatrix}.$$

These two qubits always have the same truth value, as opposed to $|v\rangle \otimes |v\rangle$ where all four combinations are possible. Therefore, if we measure the truth value of one qubit, we also learn the truth value of the other one. To focus on a specific example, consider the state

$$|\psi\rangle = \frac{1}{\sqrt{2}}\big(|00\rangle + |11\rangle\big)\,. \tag{15.3}$$

We could generate $|\psi\rangle$ from an initial state of $|00\rangle$ using the following circuit:

$$\left.\begin{array}{cc} |0\rangle & \text{---}\boxed{H}\text{---}\bullet\text{---} \\ |0\rangle & \text{-----------}\oplus\text{---} \end{array}\right\}|\psi\rangle$$

As Exercise 15.7 showed, this two-qubit state is *entangled.* That is, there is no way to write it as the tensor product of two one-qubit states. This entanglement is the source of some very surprising phenomena. To explore them, let's welcome two famous characters onto the stage: Alice and Bob.

15.3.2 Alice, Bob, and Bell's Inequalities

> How wonderful that we have met with a paradox.
> Now we have some hope of making progress!
>
> Niels Bohr

Suppose two qubits are in the entangled state $|\psi\rangle = \frac{1}{\sqrt{2}}(|00\rangle+|11\rangle)$. We give one of these qubits to Alice and the other to Bob, and they fly off in opposite directions until they are a light-year apart. If Bob measures his qubit, he immediately knows the value of Alice's qubit as well. To put it differently, his measurement collapses not just his qubit, but also Alice's, no matter how far apart they are. Has some sort of physical effect flown instantaneously across the distance between them?

At first glance, there's nothing paradoxical going on here. We create long-range correlations in the classical world all the time. For instance, we could write two copies of a letter, seal them in envelopes, and give one each to Alice and Bob. If Bob opens his envelope, he immediately knows what's written on Alice's letter as well, even if it is a light-year away. Bob hasn't affected Alice's letter physically—he has simply learned something about it.

A priori, we might expect quantum states to be like unopened letters, which appear to be in a superposition just because we don't know their contents yet. If that were true, the only thing that collapses when we measure quantum states would be our uncertainty about them. The entanglement between Alice's and Bob's qubits would simply be a form of correlation between their underlying states.

In this section, we will see a clever proof that this is not the case. There are experiments that Alice and Bob can perform that yield correlations stronger than any pair of classical states, *no matter how they are correlated*, could produce. There is simply no way to think of each qubit as having an underlying state that determines the outcome of the measurements that Alice or Bob could perform. They are inextricably entangled in a way that classical systems cannot be.

These results call into question some of our most cherished assumptions about the physical world. To understand them, we need to delve a little deeper into the physics and mathematics of measurement.

In quantum mechanics, each physical quantity or *observable*—such as energy, momentum, or angular momentum—is associated with a linear operator M. This operator is *Hermitian*, i.e., $M^\dagger = M$. As the following exercise shows, this means that its eigenvalues are real and that its eigenvectors form a basis.

Exercise 15.11 *Suppose that $M^\dagger = M$. Show that M's eigenvalues are real, and that any pair of eigenvectors $|v\rangle, |w\rangle$ with different eigenvalues are orthogonal, i.e., that $\langle v\,|\,w\rangle = 0$. Hint: consider the product $\langle v|M|u\rangle$.*

If we measure a state v in this basis, we observe one of M's eigenvalues λ according to the probability distribution

$$P(\lambda) = \langle v|\Pi_\lambda|v\rangle\,,$$

where Π_λ projects onto the subspace spanned by the eigenvector(s) with eigenvalue λ. In particular, since

$$M = \sum_\lambda \lambda \Pi_\lambda\,,$$

the expectation of λ is

$$\mathbb{E}[\lambda] = \sum_\lambda \lambda P(\lambda) = \langle v|M|v\rangle\,. \tag{15.4}$$

Since this is the expected value of the observable corresponding to M, we will call it $\mathbb{E}[M]$ below.

Now imagine that Alice and Bob's qubits each consist of a particle such as a photon. We can think of the Z operator as measuring the photon's spin, or angular momentum, around the z axis. Its eigenvalue is $+1$ or -1 depending on whether this spin is counterclockwise or clockwise. Similarly, the X operator measures the spin around the x-axis.

Exercise 15.12 *Check that the Pauli matrices X, Y, and Z are Hermitian as well as unitary. What does this imply about their eigenvalues?*

Classically, angular momentum is a vector, and we can measure its components along both the z-axis and the x-axis. But two operators have the same eigenvectors if and only if they commute. Since $XZ = -ZX$ and their eigenvalues are nonzero, X and Z have no eigenvectors in common—there is no state for which they both have well-defined values. More generally, we can't measure noncommuting observables simultaneously. This is known as *Heisenberg's uncertainty principle*, which states, for instance, that we
15.8 can't measure both a particle's position and its momentum.

On the other hand, if Alice measures her qubit in one basis, and Bob measures his qubit in another, the two corresponding operators commute. For instance, Alice can measure her qubit in the Z-basis or the X-basis by applying one of these operators:

$$Z_A = Z \otimes \mathbb{1} \text{ or } X_A = X \otimes \mathbb{1}\,.$$

Each of these applies Z or X to Alice's qubit—the one on the left of the tensor product—while leaving Bob's unchanged. Similarly, Bob can measure his qubit in either the H-basis or the H'-basis, applying

$$H_B = \mathbb{1} \otimes H \text{ or } H'_B = \mathbb{1} \otimes H'\,,$$

where

$$H' = XHX = \frac{1}{\sqrt{2}}\begin{pmatrix} -1 & 1 \\ 1 & 1 \end{pmatrix}.$$

Geometrically, H and H' measure the spin around axes that are an angle $\pi/4$ away from the z-axis and x-axis as shown in Figure 15.4.

While we can't measure Z_A and X_A simultaneously, we can certainly measure, say, Z_A and H_B simultaneously, since they commute: $Z_A H_B = H_B Z_A = Z \otimes H$. The same is true of X_A and H'_B, and so on. If Alice and Bob send us their results, we can check to see how often they agree, and how correlated their measurements are.

As Problems 15.7 and 15.8 show, if the bases that Alice and Bob use are an angle θ apart, the correlation between their measurements is $\cos\theta$. Since $\theta = \pi/4$ for each pair of bases except Z_A and H'_B, for which $\theta = 3\pi/4$, we have

$$\mathbb{E}[Z_A H_B] = \mathbb{E}[X_A H_B] = \mathbb{E}[X_A H'_B] = -\mathbb{E}[Z_A H'_B] = \frac{1}{\sqrt{2}}\,.$$

But something funny is going on. Consider the following quantity:

$$W = Z_A H_B + X_A H_B + X_A H'_B - Z_A H'_B\,. \tag{15.5}$$

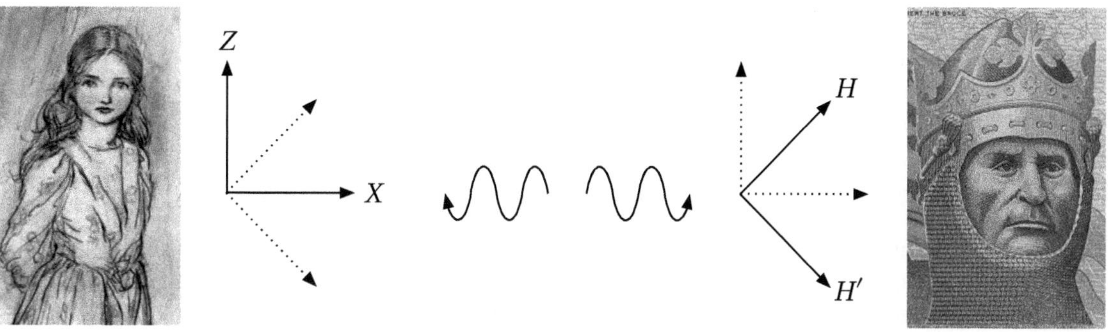

FIGURE 15.4: Alice and Bob. Alice can measure her qubit in either the Z-basis or the X-basis, and Bob can measure his in the H-basis or the H'-basis. The resulting measurements are more correlated than any classical probability distribution, correlated or not, can account for.

Since the expectation of the sum is the sum of the expectations, we have

$$\mathbb{E}[W] = \frac{4}{\sqrt{2}} = 2\sqrt{2}\,.$$

On the other hand, we can factor W as follows,

$$W = Z_A\left(H_B - H'_B\right) + X_A\left(H_B + H'_B\right). \tag{15.6}$$

Since each of the four operators appearing in W yields $+1$ or -1, there are two possibilities. If $H'_B = H_B$ then $W = 2X_AH_B$, and if $H'_B = -H_B$ then $W = 2Z_AH_B$. In either case we have

$$W \in \{\pm 2\}\,.$$

Certainly the expectation of W has to lie between -2 and $+2$, and so

$$|\mathbb{E}[W]| \leq 2\,. \tag{15.7}$$

This is one form of *Bell's inequality*. It holds for any set of four random variables Z_A, X_A, H_B, H'_B that take values in $\{\pm 1\}$, and which are chosen according to any joint probability distribution, with arbitrary correlations between them. Yet it is flatly contradicted by (15.5). How can this be?

When we factored W as in (15.6), we assumed that the Z_A appearing in Z_AH_B, for instance, is the same as the Z_A appearing in $Z_AH'_B$. That is, we assumed that whether or not Alice chooses to measure her photon along the z-axis, there is some value Z_A that *she would observe if she did*, and that this value is not affected by Bob's choice of basis a light-year away.

Physically, we are assuming that the universe is *realistic*, i.e., that the outcome of Alice's experiments has a probability distribution that depends on the state of her photon, and that it is *local*, i.e., that influences cannot travel faster than light. The fact that Bell's inequality is violated—and this violation has been shown very convincingly in the laboratory—tells us that one or both of these assumptions is false.

Einstein called this phenomenon "spukhafte Fernwirkung," or as it is often translated, "spooky action at a distance." He found it profoundly disturbing, and he's not the only one. It seems that we must either accept that physics is nonlocal, which violates special relativity, or that the outcome of a measurement is not a consequence of a system's state. Opinions differ on the best way to resolve this paradox. For now, let's take for granted that Bob really can affect Alice's qubit by collapsing it from afar. Can he use this effect to send Alice a message?

15.9

15.3.3 *Faster-Than-Light Communication, Mixed States, and Density Matrices*

If Bob's choices have an instantaneous effect on Alice's qubit, no matter how far away she is, it stands to reason that they could use this effect to communicate faster than light. Of course, given special relativity, this means that they can also communicate backwards in time, which would be even more useful. Let's see if we can make this work.

Suppose, once again, that Alice and Bob share an entangled pair of qubits, $|\psi\rangle = \frac{1}{\sqrt{2}}(|00\rangle + |11\rangle)$. If Bob measures his qubit in the Z-basis, observing $|0\rangle$ or $|1\rangle$, then Alice's qubit will be in the same state. The following exercise shows that $|\psi\rangle$ is identical to the entangled state we would get if we used $|+\rangle$ and $|-\rangle$ instead of $|0\rangle$ and $|1\rangle$:

Exercise 15.13 *Show that*

$$|\psi\rangle = \frac{1}{\sqrt{2}}(|++\rangle + |--\rangle) = \frac{1}{\sqrt{2}}(|00\rangle + |11\rangle).$$

Show this both by direct calculation and by showing that

$$(H \otimes H)|\psi\rangle = |\psi\rangle.$$

Hint: start with a circuit that generates $|\psi\rangle$, and use Exercise 15.9.

Therefore, if Bob measures in the X-basis, observing $|+\rangle$ or $|-\rangle$ with equal probability, Alice's qubit will again be in the same state as Bob's.

Bob can't control the outcome of his measurement, but he can control which basis he measures in. So he attempts to communicate a bit of information to Alice in the following way. If he wants to send a 0, he measures his qubit in the Z-basis, and if he wants to send a 1 he measures in the X-basis. In the first case, the state of Alice's qubit is $|v\rangle = |0\rangle$ or $|1\rangle$, each with probability 1/2. In the second case, it is $|v\rangle = |+\rangle$ or $|-\rangle$, again with equal probability. The question is whether Alice can distinguish between these two scenarios. Is there is any measurement she can perform where the outcomes have different probability distributions depending on which basis Bob used?

As we discussed in Section 15.2.6, if Alice measures her qubit, the probability that she observes a given outcome is $P = \langle v|\Pi|v\rangle$ for some projection operator Π. If Bob measured in the Z-basis, the probability she observes this outcome is the average of P over $|v\rangle = |0\rangle$ and $|1\rangle$,

$$P = \frac{1}{2}\left(\langle 0|\Pi|0\rangle + \langle 1|\Pi|1\rangle\right),$$

while if he uses the X-basis the average probability is

$$P = \frac{1}{2}\left(\langle +|\Pi|+\rangle + \langle -|\Pi|-\rangle\right).$$

But these two expressions are exactly the same. The first sums over the diagonal entries of Π in the Z-basis, and the second sums over the diagonal entries in the X-basis. The trace of a matrix is basis-independent, so in both cases we have

$$P = \frac{1}{2}\,\mathrm{tr}\,\Pi\,.$$

In other words, no matter what measurement Alice performs, she sees the same probability distribution of outcomes regardless of which basis Bob measured in. His choice of basis has precisely zero effect on anything Alice can observe—just as if physics were local after all.

It's important to note that the "state" of Alice's qubit in, say, the first case is not a quantum superposition of $|0\rangle$ and $|1\rangle$. It is a *probabilistic mixture* of these two quantum states, where each state has a real probability rather than a complex amplitude. How can we represent such a mixture mathematically?

Let's start with a *pure state*, i.e., a quantum state represented by a complex vector $|v\rangle$. If we define its *density matrix* ρ as the outer product of $|v\rangle$ with itself,

$$\rho = |v\rangle\langle v|\,,$$

then we can write the probability of an observation associated with a projection operator Π as

$$P = \langle v|\Pi|v\rangle = \mathrm{tr}\,\Pi\rho\,.$$

Now suppose we have a *mixed state*, i.e., a probability distribution over pure states $|v\rangle$ where each one appears with probability $P(v)$. If we define its density matrix as

$$\rho = \sum_v P(v)\,|v\rangle\langle v|\,,$$

then the average probability of this observation can again be written as $\mathrm{tr}\,\Pi\rho$ since

$$P = \sum_v P(v)\,\langle v|\Pi|v\rangle = \mathrm{tr}\,\Pi\rho\,.$$

Exercise 15.14 *Show that the density matrix ρ of any mixed state obeys $\mathrm{tr}\,\rho = 1$. Show also that ρ is a projection operator, i.e., that $\rho^2 = \rho$, if and only if it is a pure state, i.e., if $\rho = |v\rangle\langle v|$ for some $|v\rangle$.*

Now let's calculate the density matrix of Alice's qubit in the two scenarios. It is the same regardless of which basis Bob measures in, since

$$\rho = \frac{1}{2}\big(|0\rangle\langle 0| + |1\rangle\langle 1|\big) = \frac{1}{2}\big(|+\rangle\langle +| + |-\rangle\langle -|\big) = \frac{1}{2}\mathbb{1}\,.$$

This is called the *completely mixed state.* Since the identity matrix $\mathbb{1}$ is the same in every basis, ρ is a uniform probability distribution over all basis states, no matter which basis we measure it in. Again, Bob's choice of basis has no effect on the probabilities of Alice's measurements.

We have shown that Alice and Bob can't use "spooky action at a distance" to communicate. This suggests that it might not be so spooky after all—or rather, that it might not really be an action. To an empiricist, an effect is not real unless it can be measured, so there is really no sense in which Bob can affect Alice's qubit. However, the fact that Bell's inequality is violated forces us to admit that something more than classical correlation is going on. We leave the reader to ponder this, and move on to some practical uses for entanglement.

15.3.4 Using Entanglement to Communicate

Even though entanglement doesn't let us communicate faster than light, it is still a useful resource for quantum communication. Consider the following technique, called *superdense coding*.

Before they leave for their respective journeys, Alice and Bob buy a supply of entangled pairs of qubits at their local space station, each of which is in the state $|\psi\rangle = \frac{1}{\sqrt{2}}(|00\rangle + |11\rangle)$. They each take one qubit from each pair, shake hands, and then fly light-years apart.

Later on, Alice wants to send Bob two classical bits about her travels. Let's call these bits a and b. She takes her half of an entangled pair, applies Z if a is true, and then applies X if b is true. Let's call the resulting two-qubit state $|\psi_{a,b}\rangle$. This yields one of four possible states, where Alice's qubit is on the left:

$$\begin{aligned}
|\psi_{0,0}\rangle &= |\psi\rangle = \frac{1}{\sqrt{2}}(|00\rangle + |11\rangle) \\
|\psi_{0,1}\rangle &= X_A|\psi\rangle = \frac{1}{\sqrt{2}}(|10\rangle + |01\rangle) \\
|\psi_{1,0}\rangle &= Z_A|\psi\rangle = \frac{1}{\sqrt{2}}(|00\rangle - |11\rangle) \\
|\psi_{1,1}\rangle &= X_A Z_A|\psi\rangle = \frac{1}{\sqrt{2}}(|10\rangle - |01\rangle) .
\end{aligned}$$

Alice then sends her qubit to Bob, so that he possesses the entire two-qubit state.

The four states $|\psi_{a,b}\rangle$ are called the ***Bell basis.*** They are orthogonal, so Bob can learn a and b by performing an appropriate measurement. Specifically, by applying a unitary operator U, Bob can change the Bell basis $|\psi_{a,b}\rangle$ to the computational basis $|a,b\rangle$, and read a and b from the qubit's truth values. We ask you to design a quantum circuit for U in Problem 15.13. Thus this protocol lets Alice send Bob two bits at a time, or vice versa.

How much does this cost? Each time Alice and Bob do this, they use up one of their entangled pairs. Let's say that they use one bit of entanglement, or one "ebit." In addition, they actually have to send a qubit across the space between them. If we think of ebits, qubits, and classical bits as resources that we can use for communication, this gives us the following inequality:

$$1 \text{ qubit} + 1 \text{ ebit} \geq 2 \text{ bits} .$$

If an entangled pair allows us to encode two classical bits into one qubit, can it do the reverse? Indeed it can. Suppose that in addition to her half of the entangled pair, Alice has a qubit in the state $|v\rangle$. This gives us a three-qubit state

$$|\Psi\rangle = |v\rangle \otimes |\psi\rangle = |v\rangle \otimes \frac{1}{\sqrt{2}}(|00\rangle + |11\rangle) ,$$

where Alice holds the first and second qubit and Bob holds the third. A little calculation shows that this state can be rewritten in the following form:

$$\begin{aligned}
|\Psi\rangle = \frac{1}{2}\big(&|\psi_{0,0}\rangle \otimes |v\rangle \\
&+ |\psi_{0,1}\rangle \otimes X|v\rangle \\
&+ |\psi_{1,0}\rangle \otimes Z|v\rangle \\
&+ |\psi_{1,1}\rangle \otimes ZX|v\rangle\big) ,
\end{aligned}$$

Here Alice holds $|\psi_{a,b}\rangle$ and Bob holds $|v\rangle$, $X|v\rangle$, $Z|v\rangle$, or $ZX|v\rangle$.

Exercise 15.15 *Confirm that* $|\Psi\rangle$ *can be written this way.*

Alice measures her two qubits in the Bell basis $|\psi_{a,b}\rangle$, thus collapsing $|\Psi\rangle$ to one of these four states. Having learned a and b, she transmits those two classical bits to Bob. He then applies X to his qubit if b is true and Z to his qubit if a is true, in that order. This *disentangles* his qubit from Alice's, and puts it in the state $|v\rangle$.

Note that the qubit $|v\rangle$ has not been cloned. Bob ends up with $|v\rangle$, but Alice's copy of $|v\rangle$ has been measured and hence destroyed. We can say that $|v\rangle$ has been "teleported" from Alice to Bob. In theory, we could do this even with quantum states composed of many particles. But before you jump into the quantum teleporter, keep in mind that it destroys the original.

Quantum teleportation is especially astonishing because Alice manages to send a quantum state to Bob *without knowing it herself.* If Alice tried to measure her qubit in order to learn its amplitudes, her first measurement would collapse it to one basis state or the other—and since qubits can't be cloned, she can only measure it once. In any case, no finite amount of classical information can completely convey a qubit, since the real and imaginary parts of its amplitudes have an infinite number of digits. Instead, Alice sends her qubit "through" the entangled pair, giving Bob two classical bits as directions for how to extract it. The fact that we can do this gives us another inequality,

$$1 \text{ ebit} + 2 \text{ bits} \geq 1 \text{ qubit}.$$

Since superdense coding and quantum teleportation were discovered in the early 1990s, a great deal more has been learned about the capacity of different types of quantum communication. For one especially charming result, we refer the reader to Problems 15.14 and 15.15.

But enough about communication. It's time to delve into computation, and show some examples of what quantum algorithms can do. 15.10

15.4 Algorithmic Interference

Quantum systems possess a fascinating kind of parallelism—the ability of an electron to pass through two slits at once, or of a computer's qubits to be in a superposition of many states. Since unitary operators are linear, they act on every component of the state simultaneously. In some sense, this lets a quantum algorithm test many possible solutions to a problem at once.

However, as we will see, parallelism alone doesn't account for the power of quantum computation. Even classical randomized algorithms can explore many solutions at once in a probabilistic sense. What makes quantum computation unique is the combination of parallelism with *interference*: the fact that complex amplitudes can combine in phase or out of phase. By applying the right unitary transformation, we can use constructive interference to make the right answers more likely, while using destructive interference to cancel out the wrong ones.

In this section, we will show how interference lets a quantum computer learn some things about a function by asking fewer questions than a classical computer would need. This will culminate in Simon's problem, which quantum computers can solve exponentially faster than classical ones can. Then, in Section 15.5, we will show how some of the same ideas lead to Shor's FACTORING algorithm, and how to break public-key cryptography.

15.4.1 *Two for One: Deutsch's Problem*

Suppose we have a function f. Classically, if we want to learn something about its values $f(x)$, we have to evaluate it, or *query* it, once for each value of x we care about. We can think of each query as asking an oracle to give us $f(x)$, or calling a subroutine that calculates $f(x)$. Either way, if we want to learn $f(x)$ for N different values of x, this costs us N queries. In quantum mechanics, on the other hand, we can apply f to a superposition of many inputs at once. This allows us, in some sense, to obtain $f(x)$ for multiple values of x in parallel.

Let's start with the simplest possible case, where f takes a single bit x of input and returns a single bit of output. Consider the following problem, which was first proposed in 1985 by David Deutsch: are f's two values $f(0)$ and $f(1)$ the same or different? Equivalently, what is their parity $f(0) \oplus f(1)$? Classically, in order to solve this problem we have to query $f(0)$ and $f(1)$ separately. But quantum mechanically, a single query suffices.

First let's fix what we mean by a query in a quantum algorithm. If we have a qubit y, we can't simply set $y = f(x)$ as we would in the classical case, since this would destroy whatever bit of information was stored in y before. Instead, we query $f(x)$ reversibly by flipping y if $f(x)$ is true. This gives the following unitary operator, where $|x,y\rangle$ is shorthand for $|x\rangle \otimes |y\rangle$:

$$U_f |x,y\rangle = |x, y \oplus f(x)\rangle \, .$$

We represent U_f in a quantum circuit as follows:

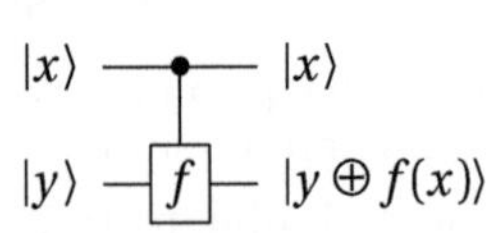

$|x\rangle$ —•— $|x\rangle$

$|y\rangle$ —[f]— $|y \oplus f(x)\rangle$

For instance, if $f(0) = 0$ and $f(1) = 1$ then U_f is an ordinary controlled-NOT, while if $f(0) = 1$ and $f(1) = 0$ then U_f flips $|y\rangle$ if and only if $|x\rangle$ is false.

Now suppose we prepare $|x\rangle$ in a uniform superposition of its possible values, i.e., in the state $|+\rangle = \frac{1}{\sqrt{2}}(|0\rangle + |1\rangle)$, and that we prepare $|y\rangle$ in the state $|0\rangle$. Then

$$U_f (|+\rangle \otimes |0\rangle) = \frac{1}{\sqrt{2}} \left(|0, f(0)\rangle + |1, f(1)\rangle \right) . \tag{15.8}$$

With a single query, we have created a state that contains information about both $f(0)$ and $f(1)$. We seem to have gotten two queries for the price of one.

Unfortunately, this state is less useful than one might hope. If we measure x, we see either $|0, f(0)\rangle$ or $|1, f(1)\rangle$, each with probability 1/2. Thus we learn either $f(0)$ or $f(1)$, but we have no control over which. We could do just as well classically by choosing x randomly and querying $f(x)$.

If we want to do better than classical randomized computation, we need to use interference. Let's create a state in which $f(x)$ is encoded in the phase of $|x\rangle$'s amplitude, instead of in the value of $|y\rangle$. To do this, first note that we can write the query operator U_f as

$$U_f |x,y\rangle = |x\rangle \otimes X^{f(x)} |y\rangle \, .$$

Now prepare $|y\rangle$ in the state $|-\rangle = \frac{1}{\sqrt{2}}(|0\rangle - |1\rangle)$ instead of $|0\rangle$. Since $|-\rangle$ is an eigenvector of X with eigenvalue -1, we have

$$U_f |x,-\rangle = (-1)^{f(x)} |x,-\rangle \, .$$

Again preparing $|x\rangle$ in a uniform superposition $|+\rangle$ then gives

$$U_f(|+\rangle \otimes |-\rangle) = \frac{1}{\sqrt{2}}\left((-1)^{f(0)}|0\rangle + (-1)^{f(1)}|1\rangle\right) \otimes |-\rangle.$$

Unlike the state in (15.8), the two qubits are now unentangled. Since we can ignore an overall phase, we can factor out $(-1)^{f(0)}$ and write the state of the first qubit as

$$|\psi\rangle = \frac{1}{\sqrt{2}}\left((-1)^{f(0)}|0\rangle + (-1)^{f(1)}|1\rangle\right) \equiv \frac{1}{\sqrt{2}}\left(|0\rangle + (-1)^{f(0)\oplus f(1)}|1\rangle\right). \tag{15.9}$$

This is $|+\rangle$ if $f(0) = f(1)$, and $|-\rangle$ if $f(0) \neq f(1)$. Thus if we measure $|\psi\rangle$ in the X-basis, we can learn the parity $f(0) \oplus f(1)$ while only calling the query operator once.

We can express this algorithm as the following circuit,

$|+\rangle$ —●— $|\pm\rangle$

$|-\rangle$ —[f]— $|-\rangle$

If we wish to start and end in the computational basis, we can add Hadamard gates on either side, giving

$|0\rangle$ —[H]—●—[H]— $|f(0) \oplus f(1)\rangle$ (15.10)

$|1\rangle$ —[H]—[f]—[H]— $|1\rangle$

Note the similarity to Exercise 15.9, in which a controlled-NOT goes the other way when transformed to the X-basis.

This algorithm is the first example of a useful trick in quantum computing, called *phase kickback*. We prepare the "output" in an eigenvector, and its eigenvalue affects the phase of the "input." Instead of measuring the output, we throw it away, and learn about the function by measuring the input. This may seem strange. But as Exercise 15.9 showed, information in quantum mechanics always flows both ways.

Before we proceed, let's talk a little more about these quantum queries. How do we actually implement U_f? And if we can implement it, why can't we just look at its matrix entries, and learn whatever we want to know about f?

Suppose we have a device, such as a classical computer, which calculates f. We feed x into the input, activate the device, and then read $f(x)$ from the output. Like any other physical object, this device is governed by the laws of quantum mechanics. For that matter, our classical computers are just quantum computers that spend most of their time in classical states. So we are free to feed it a quantum input rather than a classical one, by preparing the input in a superposition of multiple values of x. We can then implement U_f by arranging for the device to flip a bit y if $f(x) = 1$.

We are assuming here that, after each query, the device's internal state is unentangled from the input. Otherwise, it will partially measure the input state and destroy the quantum superposition over its values. To avoid entanglement, and to be re-usable, the device must return to exactly the same state it had before the query.

This is not the case for the computers on our desks. In addition to the "official" bits of their memory, their state also includes the thermal fluctuations of their particles. If we try to return the memory to its initial state by erasing it, this act of erasure transfers information, and therefore entanglement, to these thermal bits in the form of heat. In principle this obstacle can be overcome by performing the computation reversibly, and returning the memory to its original state without erasing any bits.

However, even if we have an idealized, reversible device in our hands that computes f, this doesn't
15.5 give us instant knowledge of all of f's values. It takes time to feed each value of x into the input of the device, and observe the output $f(x)$. Similarly, being able to implement U_f doesn't mean we can look inside it and see its matrix entries. We have to probe it by applying it to a quantum state and then performing some measurement on the result.

Many of the quantum algorithms we discuss in this chapter treat a function f as a "black box," and try to discover something about f by querying it. When dealing with such problems, we will think of their complexity as the number of queries we make to f, or equivalently, the number of times U_f appears in our circuit. This model makes sense whenever the values $f(x)$ are difficult to compute or obtain, or simply if we want to assume as little as possible about f's internal structure. With this in mind, let's move on to functions of many bits, rather than one.

15.4.2 Hidden Parities and the Quantum Fourier Transform

Suppose that f's input is an n-bit string $\mathbf{x} = x_1 x_2 \cdots x_n$, or equivalently, an n-dimensional vector whose components are integers mod 2. For now, we will assume that f still gives a single bit of output. We define the query operator U_f as

$$U_f|\mathbf{x}, y\rangle = |\mathbf{x}, y \oplus f(\mathbf{x})\rangle\,,$$

where $\mathbf{x}$ is stored in the first n qubits and y is the $(n+1)$st.

Just as we did in the case $n = 1$, let's prepare the input in a uniform superposition over all possible $\mathbf{x}$. This is easy, since

$$\frac{1}{\sqrt{2^n}}\sum_{\mathbf{x}}|\mathbf{x}\rangle = \underbrace{|+\rangle \otimes \cdots \otimes |+\rangle}_{n \text{ times}}\,.$$

If you don't have a supply of $|+\rangle$s, you can start with n qubits in the state $|0\rangle$ and apply a Hadamard gate to each one.

Now let's use the phase kickback trick again and see what we can learn from it. If we prepare y in the eigenvector $|-\rangle$ and apply U_f, we get

$$U_f \frac{1}{\sqrt{2^n}}\sum_{\mathbf{x}}|\mathbf{x}, -\rangle = \frac{1}{\sqrt{2^n}}\sum_{\mathbf{x}} U_f|\mathbf{x}, -\rangle = \frac{1}{\sqrt{2^n}}\sum_{\mathbf{x}}(-1)^{f(\mathbf{x})}|\mathbf{x}, -\rangle = |\psi\rangle \otimes |-\rangle,$$

where

$$|\psi\rangle = \frac{1}{\sqrt{2^n}}\sum_{\mathbf{x}}(-1)^{f(\mathbf{x})}|\mathbf{x}\rangle\,. \tag{15.11}$$

Like the state $|\psi\rangle$ in (15.9) that we used to solve Deutsch's problem, this state contains the values of $f(\mathbf{x})$ in the phases of $\mathbf{x}$'s amplitudes. What information about f can we extract from it? And in what basis should we measure it?

Stepping back a little, suppose we have a state of the form

$$|\psi\rangle = \sum_{\mathbf{x}} a_{\mathbf{x}}|\mathbf{x}\rangle\,.$$

We can think of its 2^n amplitudes $a_{\mathbf{x}}$ as a function a that assigns a complex number to each vector $\mathbf{x}$. Just as for a function defined on the integers or the reals, we can consider a's Fourier transform. That is, we can write a as a linear combination of basis functions, each of which oscillates at a particular frequency.

In this case, each "frequency" is, like $\mathbf{x}$, an n-dimensional vector mod 2. The basis function with frequency $\mathbf{k}$ is

$$(-1)^{\mathbf{k}\cdot\mathbf{x}} = (-1)^{k_1 x_1 + \cdots + k_n x_n} ,$$

and the corresponding basis state is

$$|\mathbf{k}\rangle = \frac{1}{\sqrt{2^n}} \sum_{\mathbf{x}} (-1)^{\mathbf{k}\cdot\mathbf{x}} |\mathbf{x}\rangle .$$

Note that we have normalized $|k\rangle$ so that $\|\mathbf{k}\|^2 = 1$.

As we discussed in Section 3.2.3, the Fourier transform is just a unitary change of basis, where we write $|\psi\rangle$ as a linear combination of the $|\mathbf{k}\rangle$s instead of the $|\mathbf{x}\rangle$s:

$$|\psi\rangle = \sum_{\mathbf{k}} \tilde{a}_{\mathbf{k}} |\mathbf{k}\rangle .$$

The amplitudes $\tilde{a}_{\mathbf{k}}$ are the Fourier coefficients of $a_{\mathbf{x}}$,

$$\tilde{a}_{\mathbf{k}} = \langle \mathbf{k} | \psi \rangle = \frac{1}{\sqrt{2^n}} \sum_{\mathbf{x}} (-1)^{\mathbf{k}\cdot\mathbf{x}} a_{\mathbf{x}} . \tag{15.12}$$

This Fourier transform is precisely the kind of thing quantum computers are good at. Let Q denote the unitary matrix that transforms the $|\mathbf{x}\rangle$ basis to the $|\mathbf{k}\rangle$ basis. We can write its entries as a product over the n bits of $\mathbf{k}$ and $\mathbf{x}$,

$$Q_{\mathbf{k},\mathbf{x}} = \langle \mathbf{k} | \mathbf{x} \rangle = \frac{1}{\sqrt{2^n}} (-1)^{\mathbf{k}\cdot\mathbf{x}} = \prod_{i=1}^{n} \frac{1}{\sqrt{2}} (-1)^{k_i x_i} .$$

Since the entries of the Hadamard gate are $H_{k,x} = \frac{1}{\sqrt{2}}(-1)^{kx}$, we can write Q as the tensor product

$$Q = \underbrace{H \otimes \cdots \otimes H}_{n \text{ times}} .$$

Thus we can implement Q simply by applying H to each qubit.

We call Q the *quantum Fourier transform*, or QFT for short. To be precise, it is the QFT on the group $\mathbb{Z}_2^n$, the set of n-dimensional vectors mod 2. We will consider QFTs over other groups later. We stress, however, that Q is not a quantum algorithm for the Fourier transform, since the Fourier coefficients $\tilde{a}_{\mathbf{k}}$ are not given to us as its output. Rather, they are encoded in $|\psi\rangle$'s amplitudes, and must be revealed by measurement in the Fourier basis.

Specifically, let's return to the state of $|\psi\rangle$ of (15.11). The following circuit prepares ψ and then applies Q, where for illustration $n = 3$:

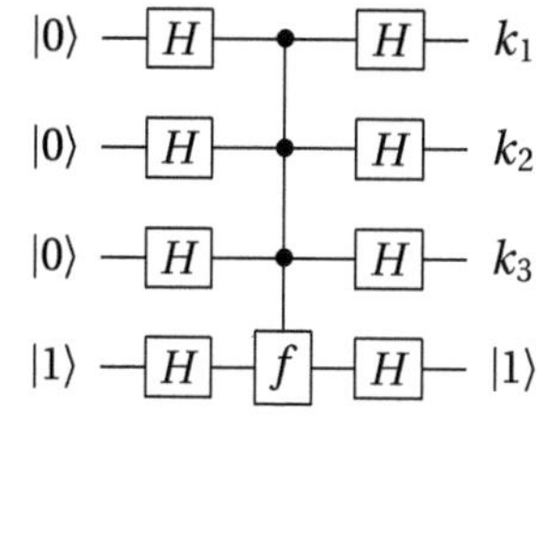

$|0\rangle$ — H — • — H — k_1

$|0\rangle$ — H — • — H — k_2

$|0\rangle$ — H — • — H — k_3 (15.13)

$|1\rangle$ — H — f — H — $|1\rangle$

This leaves us with the transformed state $Q|\psi\rangle \otimes |1\rangle$, in which the bits $k_1 k_2 \cdots k_n$ of the frequency are written in the computational basis. If we measure these bits, we observe a given $\mathbf{k}$ with probability

$$P(\mathbf{k}) = \left|\langle \mathbf{k} | \psi \rangle\right|^2 = |\tilde{a}_{\mathbf{k}}|^2 ,$$

where

$$a_{\mathbf{x}} = \frac{1}{\sqrt{2^n}}(-1)^{f(\mathbf{x})} \quad \text{and} \quad \tilde{a}_{\mathbf{k}} = \frac{1}{2^n}\sum_{\mathbf{x}}(-1)^{\mathbf{k}\cdot\mathbf{x}+f(\mathbf{x})} .$$

Compare this with the case $n=1$, namely the circuit (15.10) for Deutsch's problem. There the two possible frequencies were $k=0$ and $k=1$. If $f(0)=f(1)$ we had $|\tilde{a}_0|^2=1$ and $|\tilde{a}_1|^2=0$, and vice versa if $f(0)\neq f(1)$.

We now have everything we need to solve two more problems proposed in the early days of quantum computing. One, the *Deutsch–Jozsa problem*, is admittedly artificial, but it illustrates what the quantum Fourier transform can do. Suppose I promise you that one of the following two statements is true:

1. f is constant, i.e., $f(\mathbf{x})$ is the same for all $\mathbf{x}$.
2. f is balanced, i.e., $f(\mathbf{x})=0$ for half of all $\mathbf{x}$, and $f(\mathbf{x})=1$ for the other half.

The question is, which one? The trick is to consider the probability $P(\mathbf{0})$ of observing the frequency $\mathbf{0} = (0,\ldots,0)$. The corresponding Fourier coefficient is

$$\tilde{a}_{\mathbf{0}} = \frac{1}{\sqrt{2^n}}\sum_{\mathbf{x}} a_{\mathbf{x}} = \frac{1}{2^n}\sum_{\mathbf{x}}(-1)^{f(\mathbf{x})} .$$

This is just the average of $(-1)^{f(\mathbf{x})}$ over all $\mathbf{x}$, which is ± 1 if f is constant and 0 if f is balanced. Therefore,

$$P(\mathbf{0}) = |\tilde{a}_{\mathbf{0}}|^2 = \begin{cases} 1 & \text{if } f \text{ is constant} \\ 0 & \text{if } f \text{ is balanced.} \end{cases}$$

Assuming my promise is true, one query of f and one observation of $\mathbf{k}$ is all it takes. Just return "constant" if $k_i=0$ for all i, and return "balanced" otherwise.

We can use the same setup to solve another problem, posed by Bernstein and Vazirani. Again I make a promise about f: this time, that $f(\mathbf{x}) = \mathbf{k}\cdot\mathbf{x} \bmod 2$ for some $\mathbf{k}$. Equivalently, I promise that $f(\mathbf{x})$ is the parity of some fixed subset of $\mathbf{x}$'s components, namely those x_i where $k_i=1$. In this case, $|\psi\rangle$ is precisely the basis state $|\mathbf{k}\rangle$, so measuring $|\psi\rangle$ in the Fourier basis reveals $\mathbf{k}$ with probability 1. As for the Deutsch and Deutsch–Jozsa problems, one quantum query is enough to discover $\mathbf{k}$.

In contrast, Problem 15.19 shows that classical computers need n queries to solve the Bernstein–Vazirani problem. This gives us our first problem where quantum computing reduces the number of queries by more than a constant. Our first *exponential* speedup is right around the corner.

15.11

Exercise 15.16 *Give a randomized classical algorithm for the Deutsch–Jozsa problem where the average number of queries is* 3 *or less.*

Exercise 15.17 *Several early papers generated the state* $|\psi\rangle$ *of* (15.11) *in a more elaborate way. The idea is to compute* $f(\mathbf{x})$*, apply the* Z *operator, and then disentangle the target qubit from the input qubits by "uncomputing"* $f(\mathbf{x})$*:*

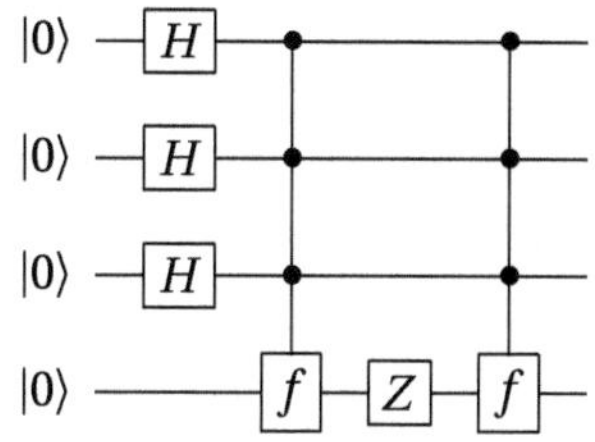

```
|0⟩ —[H]——●————————●——
|0⟩ —[H]——●————————●——
|0⟩ —[H]——●————————●——
|0⟩ ——————[f]—[Z]—[f]——
```

Show that this works, although it uses two queries instead of one.

15.4.3 Hidden Symmetries and Simon's Problem

We close this section with Simon's problem. This was the first case in which quantum computers were proved to be exponentially faster than classical ones, and was also an important precursor to Shor's algorithm for FACTORING. It illustrates one of quantum computation's most important strengths—its ability to detect a function's symmetries.

Once again, let f be a function of n-dimensional vectors $\mathbf{x}$ whose coefficients are integers mod 2. But now, rather than giving a single bit of output, $f(\mathbf{x})$ returns a symbol in some set S. We don't care much about the structure of S, or the specific values f takes. We just care about f's symmetries—that is, which values of $\mathbf{x}$ have the same $f(\mathbf{x})$.

Specifically, I promise that each input $\mathbf{x}$ has exactly one partner $\mathbf{x}'$ such that $f(\mathbf{x}) = f(\mathbf{x}')$. Moreover, I promise that there is a fixed vector $\mathbf{y}$ such that

$$f(\mathbf{x}) = f(\mathbf{x}') \text{ if and only if } \mathbf{x}' = \mathbf{x} \oplus \mathbf{y}. \tag{15.14}$$

For instance, if $n = 3$ and $\mathbf{y} = 101$, then

$$f(000) = f(101),\ f(001) = f(100),\ f(010) = f(111),\ f(011) = f(110),$$

and these four values of f are distinct. How many times do we need to query f to learn $\mathbf{y}$?

As before, we define a query operator U_f which writes $f(\mathbf{x})$ reversibly by XORing it with another set of qubits. If we prepare $\mathbf{x}$ in a uniform superposition and prepare the "output" qubits in the state $|\mathbf{0}\rangle = |0,\ldots,0\rangle$, this gives

$$U_f \frac{1}{\sqrt{2^n}} \sum_{\mathbf{x}} |\mathbf{x}, \mathbf{0}\rangle = \frac{1}{\sqrt{2^n}} \sum_{\mathbf{x}} |\mathbf{x}, f(\mathbf{x})\rangle .$$

Now suppose we measure the value of $f(\mathbf{x})$. We will observe one of the 2^{n-1} possible outputs f_0, all of which are equally likely. After we have done so, the state has collapsed to a superposition of those $\mathbf{x}$ with $f(\mathbf{x}) = f_0$. According to my promise, the resulting state is $|\psi\rangle \otimes |f_0\rangle$ where

$$|\psi\rangle = \frac{1}{\sqrt{2}} \left(|\mathbf{x}_0\rangle + |\mathbf{x}_0 \oplus \mathbf{y}\rangle \right)$$

for some $\mathbf{x}_0$.

This state $|\psi\rangle$ is a superposition over two states $\mathbf{x_0}$ and $\mathbf{x_0}' = \mathbf{x}_0 \oplus \mathbf{y}$. For those unschooled in quantum mechanics, it is very tempting to think that we can just add these states together, and get $\mathbf{y} = \mathbf{x_0} \oplus \mathbf{x_0}'$. However, this kind of interaction between two terms in a superposition just isn't allowed. Instead, we have to perform some measurement on $|\psi\rangle$, and learn something about $|\mathbf{y}\rangle$ from the result.

One thing we shouldn't do is measure $|\psi\rangle$ in the $\mathbf{x}$-basis, since this would just yield a uniformly random $\mathbf{x}$. We should measure $|\psi\rangle$ is some other basis instead. The right choice is—you guessed it—the Fourier basis. According to (15.12), applying the QFT gives the amplitudes

$$\tilde{a}_{\mathbf{k}} = \frac{1}{\sqrt{2^n}} \frac{1}{\sqrt{2}} \left((-1)^{\mathbf{k}\cdot\mathbf{x}_0} + (-1)^{\mathbf{k}\cdot(\mathbf{x}_0 \oplus \mathbf{y})} \right) = \frac{(-1)^{\mathbf{k}\cdot\mathbf{x}_0}}{\sqrt{2^{n-1}}} \left(\frac{1+(-1)^{\mathbf{k}\cdot\mathbf{y}}}{2} \right). \tag{15.15}$$

Note that $\mathbf{x}_0$, and therefore the value f_0 we happened to observe, just affects the overall phase. This phase disappears when we calculate the probability $P(\mathbf{k})$ that we observe a given frequency vector $\mathbf{k}$,

$$P(\mathbf{k}) = |\tilde{a}_{\mathbf{k}}|^2 = \frac{1}{2^{n-1}} \left(\frac{1+(-1)^{\mathbf{k}\cdot\mathbf{y}}}{2} \right)^2 .$$

This probability depends on whether $\mathbf{k}\cdot\mathbf{y}$ is 0 or 1, or geometrically, whether $\mathbf{k}$ is perpendicular to $\mathbf{y}$ or not. We can write it as follows,

$$P(\mathbf{k}) = \begin{cases} 1/2^{n-1} & \text{if } \mathbf{k} \perp \mathbf{y} \\ 0 & \text{otherwise}. \end{cases}$$

Thus we observe a frequency vector $\mathbf{k}$ chosen randomly from the 2^{n-1} vectors perpendicular to $\mathbf{y}$. The set of such vectors forms an $(n-1)$-dimensional subspace, and we can determine $\mathbf{y}$ as soon as we have observed a set of $\mathbf{k}$s that span this subspace. As Problem 15.20 shows, this happens with constant probability after n observations, and with high probability after $n + o(n)$ of them. So, we can find $\mathbf{y}$ by querying f about n times.

Classically, on the other hand, it takes an exponential number of queries to determine $\mathbf{y}$. Using the Birthday Problem from Appendix A.3.3, Problem 15.21 shows that it takes about $\sqrt{2^n} = 2^{n/2}$ queries to find a pair $\mathbf{x}, \mathbf{x}'$ such that $f(\mathbf{x}) = f(\mathbf{x}')$. Until then, we have essentially no information about $\mathbf{y}$.

Simon's algorithm shows that quantum computers can solve certain problems—at least in the black-box model, where the number of queries is what matters—exponentially faster than classical computers can. It is also our first example of a *Hidden Subgroup Problem*, where a function f is defined on a group G and our goal is to find the subgroup H that describes f's symmetries. In this case, $G = \mathbb{Z}_2^n$ and $H = \{\mathbf{0}, \mathbf{y}\}$. In the next section, we will see how Shor's algorithm solves FACTORING by finding the symmetry—in particular, the periodicity—of a function defined on the integers.

15.5 Cryptography and Shor's Algorithm

It is fair to say that much of the interest in quantum computation stems from Peter Shor's discovery, in 1994, that quantum computers can solve FACTORING in polynomial time. In addition to being one of the most fundamental problems in mathematics, FACTORING is intimately linked with modern cryptography.

In this section, we describe the RSA public-key cryptosystem and why an efficient algorithm for FACTORING would break it. We then show how Shor's algorithm uses the quantum Fourier transform, and a little number theory, to solve FACTORING in polynomial time. We go on to show that quantum computers can also solve the DISCRETE LOG problem, and break another important cryptographic protocol.

15.5.1 *The RSA Cryptosystem*

> It may be roundly asserted that human ingenuity cannot concoct a cipher which human ingenuity cannot resolve.
>
> Edgar Allen Poe

Several times in this book, we have discussed *one-way functions*—functions f such that we can compute $f(x)$ from x in polynomial time, but we can't invert f and compute x from $f(x)$. For instance, we believe that modular exponentiation, $f(x)=a^x \bmod p$, is one such function.

Now suppose that f has a *trapdoor*: a secret key d which, if you know it, lets you invert f easily. Then I can arrange a *public-key cryptosystem*, in which anyone can send me messages that only I can read. I publicly announce how to compute f, and anyone who wants to send me a message m can encrypt it in the form $f(m)$. But only I know the secret key d, so only I can invert $f(m)$ and read the original message.

This is the idea behind the *RSA cryptosystem*, named after Rivest, Shamir, and Adleman who published it in 1977. Today, it is the foundation for much of electronic commerce. Here's how it works.

I publish an n-bit integer N, which is the product of two large primes p,q that I keep to myself, and an 15.12
n-bit integer e. Thus my public key consists of the pair (N,e). If you want to send me a message written as an n-bit integer m, you encrypt it using the following function:

$$f(m)=m^e \bmod N\,.$$

I have a private key d known only to me, such that

$$f^{-1}(m)=m^d \bmod N\,.$$

In other words, for all m we have

$$(m^e)^d = m^{ed} \equiv_N m\,, \tag{15.16}$$

so I can decrypt your message m^e simply by raising it to the dth power. Both encryption and decryption can be done in polynomial time using the repeated-squaring algorithm for MODULAR EXPONENTIATION in Section 3.2.2.

For what pairs e,d does (15.16) hold? Recall Fermat's Little Theorem, which we saw in Sections 4.4.1 and 10.8. It states that if p is prime, then for all a with $1\le a<p$,

$$a^{p-1}\equiv_p 1\,.$$

What happens if we replace the prime p with an arbitrary number N? Let $\mathbb{Z}_N^*$ denote the set of numbers between 1 and $N-1$ that are mutually prime to N. This set forms a group under multiplication. As we discuss in Appendix A.7, for any element a of a group G we have

$$a^{|G|}=1\,.$$

If p is prime then $|\mathbb{Z}_p^*|=p-1$, recovering the Little Theorem.

More generally, $|\mathbb{Z}_N^*|$ is given by Euler's *totient function* $\varphi(N)$. For instance, for $N=15$ we have

$$\mathbb{Z}_{15}^*=\{1,2,4,7,8,11,13,14\}\,,$$

so $\varphi(15)=8$. Thus for any N and any $a \in \mathbb{Z}_N^*$,

$$a^{\varphi(N)} \equiv_N 1. \tag{15.17}$$

It follows that (15.16) holds whenever

$$ed \equiv_{\varphi(N)} 1.$$

This implies that e and d are mutually prime to $\varphi(N)$, and that $d = e^{-1}$ is the multiplicative inverse of e in the group $\mathbb{Z}_{\varphi(N)}^*$. Note that for (15.16) to hold, we also need the message m to be in $\mathbb{Z}_N^*$. However, if $N = pq$ where p and q are large primes, the probability that m isn't mutually prime to N is at most $1/p + 1/q$, which is vanishingly small.

How hard is it to break this system, and decrypt messages that are intended for someone else? Cracking my RSA key would consist of deriving my private key d from my public key (N, e). Let's give this problem a name:

> RSA KEYBREAKING
>
> Input: Integers N, e with $\gcd(e, \varphi(N)) = 1$
>
> Output: $d = e^{-1}$ in $\mathbb{Z}_{\varphi(N)}^*$

If you know φ, you can find inverses in $\mathbb{Z}_\varphi^*$ in polynomial time using a version of Euclid's algorithm (see Problem 2.2). Thus RSA KEYBREAKING can be reduced to the problem TOTIENT of finding $\varphi(N)$ given N.

TOTIENT, in turn, can be reduced to FACTORING for the following reason. If $N = pq$ where p and q are distinct primes,

$$\varphi(N) = (p-1)(q-1). \tag{15.18}$$

Exercise 15.18 *Prove* (15.18). *Better yet, work out its generalization in Problem 15.25.*

Thus we have a chain of reductions

$$\text{RSA KEYBREAKING} \le \text{TOTIENT} \le \text{FACTORING}.$$

Do these reductions go the other way? The following exercise shows that TOTIENT is just as hard as FACTORING, at least for the case $N = pq$:

Exercise 15.19 *Suppose $N = pq$ where p and q are prime. Show that, given N and $\varphi(N) = (p-1)(q-1)$, we can derive p and q.*

In addition, Problem 15.28 shows that if we can solve RSA KEYBREAKING—or, for that matter, if we can obtain even a single pair (e, d) where $ed \equiv_{\varphi(N)} 1$—then there is a randomized polynomial-time algorithm which factors N. This gives

$$\text{FACTORING} \le_{\text{RP}} \text{RSA KEYBREAKING},$$

where $\le_{\text{RP}}$ denotes a randomized polynomial-time reduction as in Section 10.9.

Thus if we allow randomized algorithms, FACTORING and RSA KEYBREAKING are equally hard. However, we could conceivably decrypt individual messages, extracting the eth root of m^e, without possessing the private key d. Thus decrypting one message at a time might be easier than RSA KEYBREAKING, which
15.13 breaks the public key once and for all.

The best known classical algorithm for factoring n-bit integers has running time $\exp\left(O(n^{1/3} \log^{2/3} n)\right)$. This is much better than simple approaches like trial division or the sieve of Eratosthenes, but it is still exponential in n. It is generally believed that FACTORING is not in P, and therefore that RSA KEYBREAKING 15.14
is hard for classical computers to break. What about quantum ones?

15.5.2 *From Factoring to Order-Finding*

Shor's algorithm solves FACTORING by finding the periodicity of a certain number-theoretic function. We have already seen a hint of this idea in Section 10.8.2 where we discussed the Miller–Rabin test for PRIMALITY, but we present it here from scratch.

We start by asking about the square roots of 1. For any N, there are at least two elements $y \in \mathbb{Z}_N^*$ such that $y^2 \equiv_N 1$, namely 1 and $-1 \equiv_N N-1$. But if N is divisible by two distinct odd primes, there are at least two more square roots of 1 in $\mathbb{Z}_N^*$, which we call the *nontrivial* ones. If we can find one of them, we can obtain a proper divisor of N. To see this, suppose that $y^2 \equiv_N 1$. Then for some k,

$$y^2 - 1 = (y+1)(y-1) = kN.$$

If $y \not\equiv_N \pm 1$, neither $y+1$ nor $y-1$ is a multiple of N. In that case, $\gcd(y+1, N)$ and $\gcd(y-1, N)$ must each be proper divisors of N.

How can we find a nontrivial square root of 1? Given an integer $c \in \mathbb{Z}_N^*$, its *order* is the smallest $r > 0$ such that

$$c^r \equiv_N 1.$$

In other words, r is the periodicity of the function

$$f(x) = c^x \bmod N.$$

This gives us a potential algorithm for FACTORING. Suppose that we can find the order r of a given c. If we're lucky, r is even and $c^{r/2}$ is a square root of 1. We know that $c^{r/2} \not\equiv 1$ since by definition r is the smallest such power. If we're even luckier, $c^{r/2} \not\equiv_N -1$ and $c^{r/2}$ is a nontrivial root.

For example, let $N = 21$ and let $c = 2$. The powers of 2 mod 21 are

x	0	1	2	3	4	5	6	7	8	9	10	11	12	$\cdots$
$c^x \bmod N$	1	2	4	8	16	11	1	2	4	8	16	11	1	$\cdots$

The order is $r = 6$, since $2^6 = 64 \equiv_{21} 1$ is the first time this sequence repeats. Then $c^{r/2} = 2^3 = 8$ is one of the nontrivial square roots of 1 in $\mathbb{Z}_{21}^*$, and $\gcd(9, 21)$ and $\gcd(7, 21)$ are N's factors 3 and 7.

This doesn't always work. If $c = 5$, its order is $r = 6$, but $c^{r/2} = 20 \equiv_N -1$. As another example, if $c = 16$ then its order $r = 3$ is odd. Let's call $c \in \mathbb{Z}_N^*$ *good* if its order r is even and $c^{r/2} \not\equiv_N -1$. How can we find a good c?

Simple—just choose c randomly. To be more precise, first choose c randomly from $\{0, \ldots, N-1\}$, and then check that c is mutually prime to N. If it isn't, $\gcd(c, N)$ is a proper divisor of N and we can declare victory already—so we might as well assume that c is a uniformly random element of $\mathbb{Z}_N^*$. If a large fraction of the elements of $\mathbb{Z}_N^*$ are good, we only need to try this a few times before we find a good one. The following theorem, which we prove in Problem 15.29, shows that this is true.

Theorem 15.1 *If N is divisible by ℓ distinct odd primes, then at least a fraction $1-1/2^{\ell-1}$ of the elements of $\mathbb{Z}_N^*$ are good.*

Thus if N is divisible by two or more odd primes, a random element of $\mathbb{Z}_N^*$ is good with probability at least $1/2$. Then we can find a good c, and a proper divisor of N, with $O(1)$ trials on average.

This covers every case except where N is even, prime, or a prime power. As Problem 10.41 shows, we can tell if N is a prime power in polynomial time, and we can certainly tell if it is even, so we can find a divisor in those cases too.

If we can find a proper divisor d of N, we can apply the same algorithm recursively to d and N/d and break them down further. If N has n bits, it has only $O(n)$ prime factors (see Exercise 4.21) so we only need to do this $O(n)$ times. Finally, we can use the polynomial-time algorithm for PRIMALITY to tell when we have broken N all the way down to the primes.

We can summarize all this as a chain of reductions,

$$\text{FACTORING} \leq \text{NONTRIVIAL SQUARE ROOT OF } 1 \leq_{\mathsf{RP}} \text{ORDER FINDING}.$$

So how can we find the order of a given c? Equivalently, how can we find the periodicity of the function $f(x) = c^x \bmod N$?

Let's leave the number theory behind and treat $f(x)$ as a black box. Given a function that repeats with periodicity r, how can we find r? If we query it classically, computing $f(1)$, $f(2)$, and so on, it will take r queries before we see the sequence repeat. This is bad news, since in our application r could be exponentially large.

But periodicity is a kind of symmetry, and Simon's algorithm showed us that quantum computers can detect some kinds of symmetry exponentially faster than a classical computer can. In the next section, we will see exactly how Shor's algorithm does this.

15.5.3 *Finding Periodicity with the Quantum Fourier Transform*

Suppose that f is a function from the integers to some arbitrary set S, and that f is periodic in the following sense. There is an r such that, for all x, x',

$$f(x) = f(x') \text{ if and only if } x \equiv_r x'. \tag{15.19}$$

Intuitively, we can learn about f's oscillations by measuring its Fourier transform, and this is exactly what we will do. Just as the QFT on the group $\mathbb{Z}_2^n$ helped us solve Simon's problem, we can find r using the QFT on the integers, or rather on $\mathbb{Z}_M$, the group of integers mod M for some M.

What should M be? As we will see below, if r divides M then the observed frequency is always a multiple of a fundamental frequency M/r, allowing us to determine r from a few observations. Since r divides $\varphi(N)$, we would set $M = \varphi(N)$ if we could—but if we knew $\varphi(N)$, we would already know how to factor N. For now, we will do the analysis as if r divides M. Happily, in the next section we will see that everything still works as long as M is sufficiently large.

We start by creating a uniform superposition over all x in $\mathbb{Z}_M$, and then using the unitary operator U_f to write $f(x)$ on another set of qubits. This gives the state

$$\frac{1}{\sqrt{M}} \sum_{x=0}^{M-1} |x, f(x)\rangle.$$

We then measure $f(x)$, and observe some random value f_0. This collapses the state to $|\psi\rangle\otimes|f_0\rangle$ where $|\psi\rangle$ is the uniform superposition over the x such that $f(x)=f_0$. According to (15.19), this is precisely the set of x that are equivalent mod r to some x_0. There are M/r such values of x, so we have

$$|\psi\rangle = \frac{1}{\sqrt{M/r}} \sum_{x\equiv_r x_0} |x\rangle = \sqrt{\frac{r}{M}} \sum_{t=0}^{M/r-1} |tr+x_0\rangle . \tag{15.20}$$

The quantum Fourier transform over $\mathbb{Z}_M$ is exactly the unitary operator we described in Section 3.2.3,

$$Q_{kx} = \frac{1}{\sqrt{M}}\,\omega^{kx},$$

where

$$\omega = \mathrm{e}^{2\imath\pi/M}$$

is the Mth root of 1 in the complex plane. Analogous to (15.12), applying Q to a state $|\psi\rangle=\sum_x a_x|x\rangle$ gives amplitudes $\tilde{a}_k$ that are Fourier coefficients,

$$\tilde{a}_k = \frac{1}{\sqrt{M}} \sum_{x=0}^{M-1} \omega^{kx} a_x .$$

Below we will discuss how to implement Q efficiently, i.e., with a polynomial number of quantum gates.

The Fourier coefficients of the state (15.20) are

$$\tilde{a}_k = \frac{\sqrt{r}}{M} \sum_{t=0}^{M/r-1} \omega^{k(tr+x_0)} = \frac{\sqrt{r}}{M}\,\omega^{kx_0} \sum_{t=0}^{M/r-1} \omega^{krt} . \tag{15.21}$$

As in (15.15), the value $f_0=f(x_0)$ that we observed only affects the overall phase. When we measure $|\psi\rangle$ in the Fourier basis, the probability $P(k)$ that we observe a given frequency k is then

$$P(k) = |\tilde{a}_k|^2 = \frac{r}{M^2} \left| \sum_{t=0}^{M/r-1} \omega^{krt} \right|^2 .$$

There are now two cases. If k is a multiple of M/r, which we have assumed here is an integer, then $\omega^{kr}=1$. In that case each term in the sum over t is 1, so these amplitudes interfere constructively and $P(k)=1/r$. On the other hand, if k is not a multiple of M/r then they interfere destructively, rotating in the complex plane and canceling out, and $P(k)=0$. Thus

$$P(k) = \begin{cases} 1/r & \text{if } k \text{ is a multiple of } M/r \\ 0 & \text{otherwise.} \end{cases}$$

Note that all r multiples of M/r between 0 and $M-1$ are equally likely, so we can write $k=bM/r$ where b is chosen randomly from $0,\ldots,r-1$. In acoustic terms, M/r is the fundamental frequency, and k is chosen randomly from its harmonics.

If each observation gives us a random harmonic, how long does it take to learn the fundamental frequency? Suppose we observe the frequencies $k_1 = b_1 M/r$, $k_2 = b_2 M/r$, and so on. As soon as the b_i are relatively prime, i.e., $\gcd(b_1, b_2, \ldots) = 1$, then $\gcd(k_1, k_2, \ldots) = M/r$. With constant probability even the first two observations suffice, since as Problem 10.26 shows, b_1 and b_2 are mutually prime with probability $6/\pi^2$ when r is large. Thus we can determine M/r from a constant number of observations on average, and we simply divide M by M/r to obtain r.

In this analysis, the interference was either completely constructive or completely destructive, because of our assumption that r divides M. We relax this assumption in the next section. While the analysis is a little more difficult, it involves some machinery that is quite beautiful in its own right.

15.5.4 Tight Peaks and Continued Fractions

What happens when the "fundamental frequency" M/r is not an integer? We will show that when M is large enough, the probability distribution $P(k)$ is still tightly peaked around multiples of M/r. Then with a little extra effort, we can again find the periodicity r.

First let ℓ be the number of x between 0 and $M-1$ such that $x \equiv_r x_0$, and note that ℓ is either $\lfloor M/r \rfloor$ or $\lceil M/r \rceil$. Preparing a uniform superposition over $\mathbb{Z}_M$ and measuring $f(x)$ then gives the state

$$|\psi\rangle = \frac{1}{\sqrt{\ell}} \sum_{t=0}^{\ell-1} |rt + x_0\rangle .$$

Applying the QFT as before, (15.21) becomes

$$\tilde{a}_k = \frac{1}{\sqrt{M\ell}} \sum_{t=0}^{\ell-1} \omega^{k(x_0+rt)} = \frac{\omega^{kx_0}}{\sqrt{M\ell}} \sum_{t=0}^{\ell-1} \omega^{krt} . \tag{15.22}$$

This is a geometric series, and summing it gives

$$\tilde{a}_k = \frac{\omega^{kx_0}}{\sqrt{M\ell}} \left(\frac{1-\omega^{kr\ell}}{1-\omega^{kr}} \right) .$$

The probability $P(k)$ that we observe the frequency k is then

$$P(k) = |\tilde{a}_k|^2 = \frac{1}{M\ell} \frac{\left|1-\omega^{kr\ell}\right|^2}{\left|1-\omega^{kr}\right|^2} . \tag{15.23}$$

Let's rewrite this in terms of the phase angle θ between successive terms in the series (15.22). That is,

$$\omega^{kr} = \mathrm{e}^{\imath\theta} \text{ where } \theta = \frac{2\pi kr}{M} . \tag{15.24}$$

Then using the identity

$$\left|1-\mathrm{e}^{\imath\theta}\right|^2 = 4\sin^2\frac{\theta}{2} ,$$

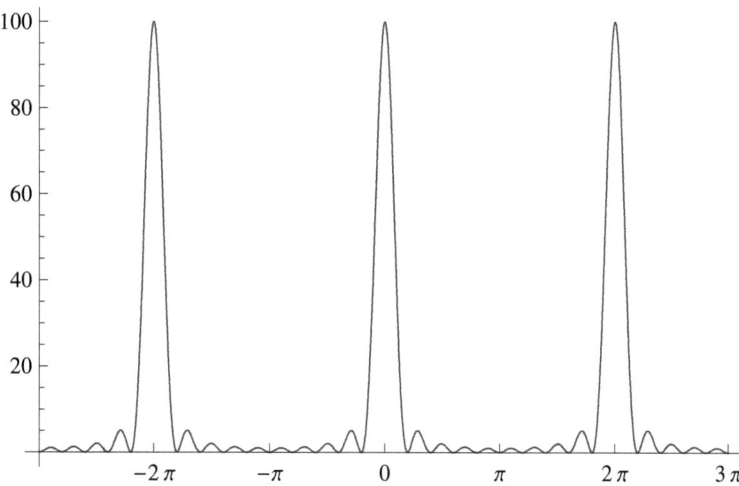

FIGURE 15.5: A plot of the function $\sin^2(\ell\theta/2)/\sin^2(\theta/2)$ appearing in the probability $P(k)$. It peaks with a value of ℓ^2 at the multiples of 2π, and falls off rapidly for other values of θ. Here $\ell = 10$.

we can rewrite (15.23) as

$$P(k) = \frac{1}{M\ell}\left(\frac{4\sin^2\ell\theta/2}{4\sin^2\theta/2}\right) = \frac{1}{M\ell}\left(\frac{\sin^2\ell\theta/2}{\sin^2\theta/2}\right). \tag{15.25}$$

Intuitively, if θ is small enough, i.e., close enough to an integer multiple of 2π, the terms in the sum (15.22) interfere constructively and $P(k)$ is large. Otherwise, their phases rotate rapidly in the complex plane and interfere destructively, making $P(k)$ very small. As Figure 15.5 shows, this intuition is correct. The function $\sin^2(\ell\theta/2)/\sin^2(\theta/2)$ appearing in (15.25) is tightly peaked at the multiples of 2π, so $P(k)$ is as well. How tight are these peaks, and how close to them are our observations likely to be?

In the simpler case where r divides M, θ is exactly a multiple of 2π when k is a multiple of M/r. So let's assume that k is the closest integer to a multiple of M/r. That is, assume that for some integer b we have

$$\left|k - \frac{bM}{r}\right| \le 1/2, \tag{15.26}$$

so that k is either $\lfloor bM/r \rfloor$ or $\lceil bM/r \rceil$. Plugging this in to (15.24) shows that

$$|\theta - 2\pi b| \le \frac{\pi r}{M}. \tag{15.27}$$

So if M is much larger than r, θ is indeed very close to a multiple of 2π.

Now write $\ell = M/r$. The fact that it is, instead, $\lfloor M/r \rfloor$ or $\lceil M/r \rceil$ makes only a minute difference in what follows. If we write

$$\phi = \frac{\ell}{2}(\theta - 2\pi b) = \frac{M}{2r}(\theta - 2\pi b),$$

then combining (15.25) and (15.27) gives

$$P(k) = \frac{1}{M\ell}\left(\frac{\sin^2 \phi}{\sin^2 \phi/\ell}\right) \quad \text{where } |\phi| \le \pi/2. \tag{15.28}$$

For ϕ in this range, a little calculus gives the inequality

$$\frac{\sin^2 \phi}{\sin^2 \phi/\ell} \ge \frac{4}{\pi^2}\ell^2, \tag{15.29}$$

so that (15.28) gives

$$P(k) \ge \frac{4}{\pi^2}\frac{\ell}{M} = \frac{4}{\pi^2}\frac{1}{r}. \tag{15.30}$$

Recall that in the simpler case where r divides M, the observed frequency k equals each of the r multiples of M/r with probability $1/r$. What (15.30) says is that k is the integer closest to each of these multiples with probability just a constant smaller than $1/r$. To put this differently, if we sum over all r multiples, the probability that k is the integer closest to one of them is at least $4/\pi^2$.

Let's assume from now on that k is indeed one of these closest integers. We can rewrite (15.26) as

$$\left|\frac{k}{M} - \frac{b}{r}\right| \le \frac{1}{2M}, \tag{15.31}$$

To put this differently, b/r is a good approximation to k/M, but with a smaller denominator $r < M$. Thus we can phrase the problem of finding r as follows: given a real number $y = k/M$, how can we find a rational approximation $b/r \approx y$ with high accuracy but a small denominator?

There is a beautiful theory of such approximations, namely the theory of *continued fractions*. For example, consider the following expression:

$$\pi = 3 + \cfrac{1}{7 + \cfrac{1}{15 + \cfrac{1}{1 + \cfrac{1}{292 + \cdots}}}}.$$

The coefficients $3, 7, 15, 1, 292, \ldots$ can be obtained by iterating the Gauss map $g(x) = 1/x \bmod 1$, which we encountered in Problem 2.7, and writing down the integer part $\lfloor 1/x \rfloor$ at each step.

If we cut this fraction off at 3, 7, 15, and so on, we get a series of rational approximations for π which oscillate around the correct value:

15.15

$$3, \frac{22}{7}, \frac{333}{106}, \frac{355}{113}, \frac{103993}{33102}, \ldots$$

Rationals obtained from y's continued fraction expansion in this way are called *convergents*. It turns out that the convergents include all the best rational approximations to y, in the following sense. We say that a rational number b/r is an *unusually good* approximation of y if

$$\left|y - \frac{b}{r}\right| < \frac{1}{2r^2}.$$

We call this unusually good because, for a typical denominator r, the closest that b/r can get to y is about $1/r$. The following classic theorem of number theory was proved by Lagrange:

Theorem 15.2 *If b/r is an unusually good approximation for y, it is one of y's convergents.*

See Problem 15.24 for a proof.

Looking back at (15.31), we see that b/r is an unusually good approximation to k/M whenever $M > r^2$. Since $r < N$ where N is the number we're trying to factor, it suffices to take $M > N^2$. So we measure the frequency k, use the continued fraction expansion of k/M to generate all the convergents with denominators smaller than N, and check the denominator of each one. As soon as we find a denominator r such that $f(r) = f(0)$, we're done.

If b and r are not mutually prime, the convergent b/r reduces to a fraction whose denominator is some proper divisor of r. However, if b is chosen uniformly from $0, \ldots, r-1$, then b is mutually prime to r with probability $\varphi(r)/r$ where φ is Euler's totient function. Problem 15.26 guides you through an elementary proof of the following fact: there is a constant $C > 0$ such that

$$\frac{\varphi(r)}{r} \geq \frac{C}{\log r}. \tag{15.32}$$

Since r has $O(n)$ bits, the probability that b is mutually prime to r is then

$$\frac{\varphi(r)}{r} = \Omega(1/\log r) = \Omega(1/n),$$

so the number of attempts we need to make is just $O(n)$. In fact, we really only need $O(\log n)$ attempts because of the stronger result that, for some $C' > 0$, 15.16

$$\frac{\varphi(r)}{r} \geq \frac{C'}{\log\log r}. \tag{15.33}$$

This can be proved using the Prime Number Theorem, which bounds the density of the primes.

There's only one thing missing from this algorithm. How can we carry out the quantum Fourier transform on $\mathbb{Z}_M$? That is, how can we implement the unitary operator Q with an efficient quantum circuit? The good news is that, having gone through all this analysis, we are free to take any value of M greater than N^2, and can set M to make the QFT as simple as possible. As we will see next, if M is a power of 2 we can apply a divide-and-conquer approach analogous to the Fast Fourier Transform of Section 3.2.3, and carry out the QFT with a polynomial number of elementary gates.

15.5.5 From the FFT to the QFT

In Section 3.2.3, we showed how to carry out the Fourier transform recursively, by dividing a list of length M into two sublists of size $M/2$ and Fourier transforming each one. We can use the same idea in the quantum setting. That is, if Q_M denotes the QFT on $\mathbb{Z}_M$, we can write a quantum circuit for Q_M recursively in terms of that for $Q_{M/2}$.

Our goal is to transform from the $|x\rangle$ basis to the $|k\rangle$ basis:

$$Q_M|x\rangle = \frac{1}{\sqrt{M}} \sum_{k=0}^{M-1} \omega_M^{kx} |k\rangle,$$

where once again $\omega_M = e^{2\imath\pi/M}$ denotes the Mth root of 1. Let's write x and k in the following way:

$$x = 2x' + x_0 \quad \text{and} \quad k = \frac{M}{2}\,k_0 + k',$$

where $x_0, k_0 \in \{0,1\}$ and $0 \le x', k' < M/2$. In particular, suppose that $M = 2^m$ and we write x and k in binary. Then x_0 is the least significant bit of x, k_0 is the most significant bit of k, and x' and k' are the remaining $m-1$ bits of x and k respectively.

In order to use the same m qubits for both x and k, it will be convenient to store the bits of x in reverse order, starting with the least significant. Thus we will write $|x\rangle = |x_0, x'\rangle$ and $|k\rangle = |k_0, k'\rangle$, so that the first qubit contains x_0 at the beginning of the process and ends up containing k_0. The remaining $m-1$ qubits start out containing x' and end up containing k'.

We can write the phase factor ω_M^{kx} as a product of three terms,

$$\omega_M^{kx} = \omega_{M/2}^{k'x'}\,\omega_M^{k'x_0}\,(-1)^{k_0 x_0},$$

where we used the facts $\omega_M^{M/2} = -1$ and $\omega_M^2 = \omega_{M/2}$. These three terms will correspond to three steps of our algorithm. First, we apply $Q_{M/2}$ to $|x'\rangle$, transforming it to a superposition of $|k'\rangle$ while leaving x_0 unchanged:

$$(\mathbb{1} \otimes Q_{M/2})\,|x_0, x'\rangle = \frac{1}{\sqrt{M/2}} \sum_{k'=0}^{M/2-1} \omega_{M/2}^{k'x'}\,|x_0, k'\rangle.$$

We then apply a "twiddle operator" W, which applies a phase shift that depends on x_0 and k':

$$W\,|x_0, k'\rangle = \omega_M^{k'x_0}\,|x_0, k'\rangle.$$

We will discuss how to implement W shortly. Finally, we apply a Hadamard gate to the first qubit, transforming x_0 to k_0:

$$(H \otimes \mathbb{1})\,|x_0, k'\rangle = \frac{1}{\sqrt{2}} \sum_{k_0=0,1} (-1)^{k_0 x_0}\,|k_0, k'\rangle.$$

Putting all this together, we have

$$\begin{aligned}
|x\rangle = |x_0, x'\rangle &\xrightarrow{\mathbb{1}\otimes Q_{M/2}} \frac{1}{\sqrt{M/2}} \sum_{k'=0}^{M/2-1} \omega_{M/2}^{k'x'}\,|x_0, k'\rangle \\
&\xrightarrow{W} \frac{1}{\sqrt{M/2}} \sum_{k'=0}^{M/2-1} \omega_{M/2}^{k'x'}\,\omega_M^{k'x_0}\,|x_0, k'\rangle \\
&\xrightarrow{H\otimes\mathbb{1}} \frac{1}{\sqrt{M/2}}\frac{1}{\sqrt{2}} \sum_{x_0=0,1} \sum_{k'=0}^{M/2-1} \omega_{M/2}^{k'x'}\,\omega_M^{k'x_0}\,(-1)^{k_0 x_0}\,|k_0, k'\rangle \\
&= \frac{1}{\sqrt{M}} \sum_{k=0}^{M-1} \omega_M^{kx}\,|k\rangle.
\end{aligned}$$

Thus we can write

$$Q_M = (H \otimes \mathbb{1})\,W\,(\mathbb{1} \otimes Q_{M/2}). \tag{15.34}$$

If we look back at Section 3.2.3, we see that each of these steps corresponds exactly to a step in the classical FFT. Applying $Q_{M/2}$ corresponds to Fourier transforming the sublists where x is odd or even—that is, where x_0 is 0 or 1. Here, using the power of quantum parallelism, we handle both these sublists at once. The operator W applies a phase shift $\omega_M^{k'}$ when x_0 is true, and this is exactly the "twiddle factor" appearing in (3.7) when x is odd. The same is true for the factor $(-1)^{k_0}$ induced by the Hadamard when x_0 is true.

Let's look more closely at the twiddle operator W. Since it has no effect unless $x_0 = 1$, we should think of it as a controlled phase-shift gate, or a series of such gates, with x_0 as their control bit. If we write k' in binary,

$$k' = 2^{m-2}k_1 + 2^{m-3}k_2 + \cdots + k_m = (M/4)k_1 + (M/8)k_2 + \cdots + k_{m-1},$$

we can write the phase shift as

$$\omega_M^{k'x_0} = e^{\imath\pi k_1 x_0/2}\, e^{\imath\pi k_2 x_0/4} \cdots e^{2\imath\pi k_{m-1} x_0/M}.$$

Each one of these is a controlled phase shift gate as described in Section 15.2.7, which applies a phase shift of $e^{\imath\pi/2^j}$ if both x_0 and k_j are true.

Now that we know how to implement W, we can turn (15.34) into a recursive description of a circuit. For example, the QFT for $M = 2^m$ where $m = 4$ is given by

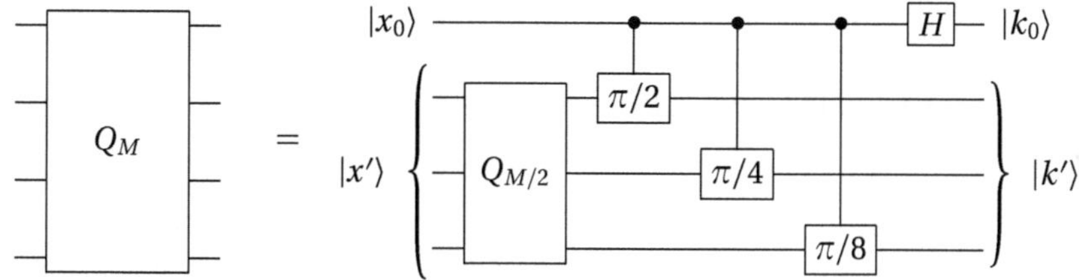

We unfold this recursively until we get to the base case $m = 1$ and $M = 2$, for which Q_2 is simply the Hadamard gate H. This gives

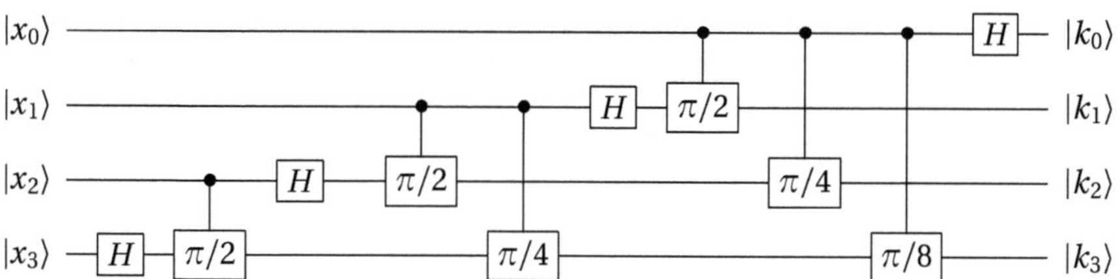

How many gates do these circuits have? Let $T(m)$ be the number of elementary gates in the circuit for Q_M where $M = 2^m$. We can see from this picture that

$$T(m) = T(m-1) + m.$$

Given the base case $T(1) = 1$, we have

$$T(m) = 1 + 2 + 3 + \cdots + m = O(m^2),$$

and the total number of gates is polynomial in the number of bits of M. To ensure that M is large enough for the continued fraction technique of Section 15.5.4 to work, we take M to be the smallest power of 2 greater than N^2. If N has n bits then $m \le 2n$, and we can implement the QFT for $\mathbb{Z}_M$ with $O(n^2)$ gates.

Wrapping all this up, there is a randomized quantum algorithm for ORDER FINDING, and therefore, by the reductions in Section 15.5.2, for FACTORING. This algorithm succeeds after a polynomial number of trials, and each one takes a polynomial number of elementary quantum operators.

Let's define a quantum complexity class, analogous to the class BPP we defined for randomized computation in Section 10.9:

> BQP, for "bounded-error quantum polynomial time," is the class of problems that can be solved by polynomial-time quantum algorithms that give the right answer at least 2/3 of the time.

Then we have shown that

FACTORING is in BQP.

On a practical level—where "practical" is defined rather broadly—this means that quantum computers can break the leading modern cryptosystem. More importantly, it shows that quantum physics helps us solve one of the oldest and most fundamental problems in mathematics—a problem that has challenged us for thousands of years.

15.17

15.5.6 *Key Exchange and* DISCRETE LOG

> Don't let him know she liked them best,
> For this must ever be
> A secret, kept from all the rest,
> Between yourself and me.
>
> Lewis Carroll, *Alice in Wonderland*

Shor discovered a quantum algorithm for another important number-theoretic problem, DISCRETE LOG. Like his algorithm for FACTORING, it uses the quantum Fourier transform as its main algorithmic tool, and we present it here. But first, let's describe a clever cryptographic protocol for having a private conversation even when anyone can overhear, and how solving DISCRETE LOG would let an eavesdropper break it.

One day, Alice calls Bob on the phone. "Hello!" she says. "I have something important to tell you, and I'm worried someone might be listening in. Shall we speak Navajo?"

"Good idea!" Bob replies. "But wait... whoever is listening just heard us say that. What if they speak Navajo too?" It seems impossible for Alice and Bob to agree on a secret key that will allow them to communicate privately, if the only communication channel they have is public. But in fact there is a way they can do this, if we assume that certain functions are hard to invert.

Suppose that Alice and Bob have publicly agreed on an n-bit prime p and a primitive root $a \in \mathbb{Z}_p^*$, i.e., an a such that every element of $\mathbb{Z}_p^*$ can be written $a^x \bmod p$ for some x. Alice picks x and Bob picks y, where x and y are uniformly random n-bit integers that they keep to themselves. Alice sends $a^x \bmod p$ to Bob, and Bob sends $a^y \bmod p$ to Alice. They can compute these in polynomial time using the algorithm of Section 3.2.2 for MODULAR EXPONENTIATION.

Now that she knows a^y, Alice can raise it to the xth power, and get $a^{xy} \bmod p$. Similarly, Bob can raise a^x to the yth power, and get the same thing. Now that both of them have $a^{xy} \bmod p$, they can use it as a secret key to encrypt the rest of their conversation—say, by XORing its bits with each n-bit packet they send each other. This trick is called *Diffie–Hellman key exchange.*

Now suppose that Eve has been listening in all this time. Eve knows a^x and a^y, as well as a and p. She 15.18
is faced with the following problem:

BREAKING KEY EXCHANGE
Input: A n-bit prime p, a primitive root a, and $a^x, a^y \in \mathbb{Z}_p^*$
Output: a^{xy}

Eve could solve this problem if she could determine x and y from a^x and a^y—that is, if she could determine their logarithms base a in $\mathbb{Z}_p^*$. This is the DISCRETE LOG problem, which we first encountered in Section 3.2.2. This gives the reduction

$$\text{BREAKING KEY EXCHANGE} \leq \text{DISCRETE LOG}.$$

Note that since x and y are chosen uniformly, in order for our key exchange scheme to be secure we need DISCRETE LOG to be hard almost all the time, rather than just in the worst case. We made a similar assumption in Sections 11.1.4 and 11.4.3, where we used DISCRETE LOG in protocols for bit commitment and pseudorandom number generation.

Given everything we've seen so far, the quantum algorithm for DISCRETE LOG is easy to describe. Given g and a, we want to find the x such that $g = a^x$. We define a function of two variables,

$$f(s,t) = g^s a^{-t} = a^{xs-t}.$$

The exponents s, t belong to the group $\mathbb{Z}_{p-1}$. We create a uniform superposition over all pairs s, t,

$$\frac{1}{p-1} \sum_{s,t=0}^{p-2} |s,t\rangle$$

Calculating and measuring $f(s,t)$ collapses this state to the uniform superposition over the pairs s, t such that $f(s,t) = a^{t_0}$ for some t_0. Since a is a primitive root, a^x is a one-to-one function from $\mathbb{Z}_{p-1}$ to $\mathbb{Z}_p^*$. So these are the $p-1$ pairs such that $xs - t = t_0$, and writing $t = xs - t_0$ we have

$$|\psi\rangle = \frac{1}{\sqrt{p-1}} \sum_{s=0}^{p-2} |s, xs - t_0\rangle.$$

These pairs form a line in the s, t plane with slope x and intercept $-t_0$. Our goal is to obtain the slope x.

Once again, we apply the quantum Fourier transform, but now a two-dimensional one. Let Q_{p-1} denote the QFT on $\mathbb{Z}_{p-1}$. If we apply $Q_{p-1} \otimes Q_{p-1}$, the amplitude of each frequency vector $|k,\ell\rangle$ is

$$\tilde{a}_{k,\ell} = \frac{1}{p-1} \sum_{s,t=0}^{p-2} \omega^{(k,\ell)\cdot(s,t)} a_{s,t} = \frac{\omega^{-\ell t_0}}{(p-1)^{3/2}} \sum_{s=0}^{p-2} \left(\omega^{(k,\ell)\cdot(1,x)}\right)^s,$$

where $\omega = e^{2\imath\pi/(p-1)}$ and $(k,\ell)\cdot(s,t) = ks + \ell t$. The probability $P(k,\ell)$ that we observe a given frequency vector is then

$$P(k,\ell) = \left|\tilde{a}_{k,\ell}\right|^2 = \frac{1}{(p-1)^3}\left|\sum_{s=0}^{p-2}\left(\omega^{(k,\ell)\cdot(1,x)}\right)^s\right|^2 = \begin{cases} \frac{1}{p-1} & \text{if } (k,\ell)\cdot(1,x) \equiv_{p-1} 0 \\ 0 & \text{otherwise}. \end{cases}$$

In Simon's algorithm, the observed frequency vector $\mathbf{k}$ was chosen uniformly from the subspace perpendicular to $\mathbf{y}$. Now (k,ℓ) is chosen randomly from the $p-1$ vectors that are perpendicular, mod $p-1$, to $(1,x)$. Since

$$(k,\ell)\cdot(1,x) = k + \ell x \equiv_{p-1} 0\,,$$

if ℓ is invertible in $\mathbb{Z}_{p-1}$, we can write x as

$$x = -k\ell^{-1} \bmod p-1\,.$$

We can find ℓ^{-1} in polynomial time using Euclid's algorithm as in Problem 2.2, so we can determine x as soon as we observe a frequency vector (k,ℓ) where ℓ is invertible.

Since ℓ is uniformly random (exercise!) the probability that it is invertible is $\varphi(p-1)/(p-1)$. By (15.32) this probability is $\Omega(1/n)$, and in fact it is $\Omega(1/\log n)$ by (15.33). Thus it takes just $O(\log n)$ trials to find such an invertible ℓ. We skip the details of implementing Q_{p-1}, but we can do so to high accuracy with poly(n) elementary gates. Thus DISCRETE LOG, and therefore BREAKING KEY EXCHANGE, are in BQP.

Like RSA encryption, Diffie–Hellman key exchange will cease to be secure whenever quantum computers are built. In addition, the pseudorandom number generator described in Section 11.4.1 may no longer be secure, in the sense that a quantum computer might be able to distinguish its output from a string of truly random numbers.

15.6 Graph Isomorphism and the Hidden Subgroup Problem

We have seen two problems, FACTORING and DISCRETE LOG, that quantum computers can solve in polynomial time but that, as far as we know, classical computers cannot. How much more powerful are quantum computers than classical ones? What other problems are in BQP that we believe are outside P?

Our understanding of quantum algorithms is at a very early stage. Nevertheless, as we will see in Section 15.7, it seems unlikely that they can efficiently solve NP-complete problems. Therefore our current guess is that BQP does not contain NP. If this is correct, then the best targets for quantum algorithms are problems that are "just outside" P—those which don't seem to be in P, but which don't seem to be NP-complete either.

As we discussed in Section 6.6, we know that, unless P = NP, there are an infinite number of problems that lie in this middle ground. However, there are relatively few "naturally occurring" candidates. Besides FACTORING and DISCRETE LOG, one of the few others is GRAPH ISOMORPHISM, the problem of telling whether two graphs are topologically identical. We repeat its definition here:

GRAPH ISOMORPHISM

Input: Two graphs $G_1 = (V_1, E_1)$, $G_2 = (V_2, E_2)$

Question: Is there a permutation $\pi : V_1 \to V_2$ such that $\pi(G_1) = G_2$: that is, such that $(u,v) \in E_1$ if and only if $(\pi(u), \pi(v)) \in E_2$?

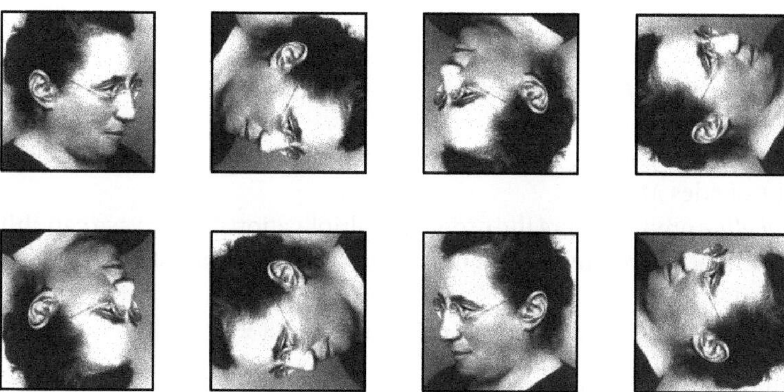

FIGURE 15.6: The rotations and reflections of a square form a group with 8 elements. Pictured is Emmy Noether, who established a deep relationship between symmetries and conservation laws in physics.

If such a permutation exists, we write $G_1 \cong G_2$.

We showed in Section 11.1 that GRAPH ISOMORPHISM is in the class $\mathsf{NP} \cap \mathsf{coAM}$, since GRAPH NONISOMORPHISM has a simple interactive proof. Thus we believe that GRAPH ISOMORPHISM is not NP-complete. However, while many special cases of GRAPH ISOMORPHISM are in P, such as planar graphs and graphs of constant degree, there is no known polynomial-time algorithm that works for general graphs.

15.19

At first blush, GRAPH ISOMORPHISM seems very different from number-theoretic problems like FACTORING and DISCRETE LOG. However, Simon's and Shor's algorithms work by detecting the periodicities or symmetries of a function. We will show in this section that GRAPH ISOMORPHISM can, in principle, be solved the same way, by treating it as an instance of a more general problem called HIDDEN SUBGROUP. While this has not yet led to an efficient quantum algorithm for GRAPH ISOMORPHISM, it has produced some very interesting results along the way. We will use a little bit of group theory—if this is new to you, we recommend Appendix A.7 for a brief introduction.

15.6.1 Groups of Symmetries

> *H*s are never upside down, except when they're sideways.
>
> Rosemary Moore

When we say that an object is symmetric, we mean that there are transformations, or *automorphisms*, that leave it unchanged. For instance, if we rotate a square by $\pi/2$ or reflect it around one of its axes, it looks just the same as it did before. The inverse of an automorphism or the composition of two automorphisms is an automorphism, so the set of automorphisms forms a group. In this example, the automorphism group of the square has a total of 8 elements as shown in Figure 15.6, and is called the *dihedral group* (see Appendix A.7).

Similarly, when we say that a function $f(x)$ defined on a group G is symmetric, we mean that $f(x)$ stays the same if we transform x in some way. If f is periodic, shifting x by some h leaves $f(x)$ unchanged.

The set of all such h forms a subgroup,

$$H = \{h \in G : f(x) = f(x+h) \text{ for all } x \in G\}.$$

In Simon's algorithm, $G = \mathbb{Z}_2^n$ and $H = \{\mathbf{0}, \mathbf{y}\}$. In Shor's algorithm, $G = \mathbb{Z}_M$ and H consists of the multiples of r, assuming that r divides M.

To include *nonabelian* groups where the order of multiplication matters, we can shift x by multiplying on the right by an element h. We then define the *automorphism group* of f as the set of h such that this leaves $f(x)$ unchanged:

$$H = \{h \in G : f(x) = f(xh) \text{ for all } x \in G\}. \tag{15.35}$$

As in Simon's and Shor's algorithms, we assume that $f(x) = f(x')$ if and only if x and x' differ by an element of h. Then our goal is to determine H from a small number of queries to f. We can phrase this as the following problem:

> HIDDEN SUBGROUP
>
> Input: A function $f : G \to S$ with the property that there is a subgroup $H \subseteq G$ such that $f(x) = f(x')$ if and only if $x' = xh$ for some $h \in H$
>
> Output: The subgroup H

All the exponential speedups in quantum algorithms that we have seen so far work by solving cases of HIDDEN SUBGROUP. Moreover, they do this with only a polynomial number of queries, even though groups like $\mathbb{Z}_M$ or $\mathbb{Z}_2^n$ are exponentially large.

Exercise 15.20 *State Shor's algorithm for* DISCRETE LOG *as a case of* HIDDEN SUBGROUP. *What is G, what is f, and what is H?*

How can we state GRAPH ISOMORPHISM in terms of symmetry? Let S_n denote the group of all $n!$ permutations of n objects. If Γ is a graph with n vertices, then as in Section 11.1.2 its automorphism group $\mathrm{Aut}(\Gamma)$ is the subgroup of S_n consisting of permutations of its vertices that leave it unchanged,

$$\mathrm{Aut}(\Gamma) = \{\sigma \in S_n : \sigma(\Gamma) = \Gamma\}.$$

Now suppose we place G_1 and G_2 side-by-side as in Figure 15.7. What is the automorphism group of the combined graph $G_1 \cup G_2$? If $G_1 \ncong G_2$, its only symmetries are those which permute G_1 and G_2 separately, so

$$\mathrm{Aut}(G_1 \cup G_2) = \mathrm{Aut}(G_1) \times \mathrm{Aut}(G_2).$$

On the other hand, if $G_1 \cong G_2$ then half the symmetries of $G_1 \cup G_2$ switch G_1 and G_2, so $\mathrm{Aut}(G_1 \cup G_2)$ is twice as large as the subgroup $\mathrm{Aut}(G_1) \times \mathrm{Aut}(G_2)$. This gives

$$\frac{|\mathrm{Aut}(G_1 \cup G_2)|}{|\mathrm{Aut}(G_1)|\,|\mathrm{Aut}(G_2)|} = \begin{cases} 1 & \text{if } G_1 \ncong G_2 \\ 2 & \text{if } G_1 \cong G_2 . \end{cases}$$

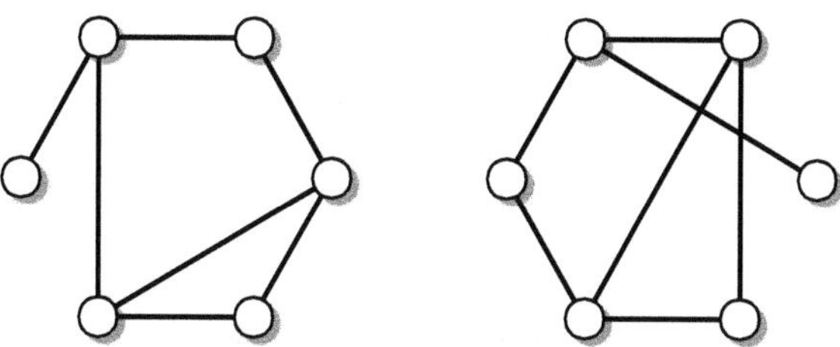

FIGURE 15.7: A pair of graphs G_1 and G_2. The automorphism group of the combined graph $G_1 \cup G_2$ depends on whether or not G_1 and G_2 are isomorphic.

Thus if we can determine the automorphism group of a graph Γ, or even estimate its size, we can solve GRAPH ISOMORPHISM. We can reduce this to HIDDEN SUBGROUP by defining a function f from S_n to the set of all graphs with n vertices, where $f(\sigma)$ is the result of permuting Γ with σ:

$$f(\sigma) = \sigma(\Gamma).$$

Then $f(\sigma\tau) = f(\sigma)$ if and only if $\tau(\Gamma) = \Gamma$, and f's automorphism group as defined in (15.35) is Aut(Γ).

This shows that we can solve GRAPH ISOMORPHISM if we can solve HIDDEN SUBGROUP in the case where G is the permutation group S_n. Can we follow in Simon's and Shor's footsteps and find a quantum algorithm that does this?

15.6.2 Coset States and the Nonabelian Fourier Transform

In Simon's and Shor's algorithms, we start by creating a uniform superposition over the group G and querying f on this superposition of inputs. This gives the state

$$\frac{1}{\sqrt{|G|}} \sum_{x \in G} |x, f(x)\rangle .$$

We then measure $f(x)$ and observe some value f_0. This collapses the state to $|\psi\rangle \otimes |f_0\rangle$, where $|\psi\rangle$ is the uniform superposition over the set of x such that $f(x) = f_0$. According to the premise of HIDDEN SUBGROUP, the set of such x forms a *coset* of H, i.e., a set of the form $x_0 H = \{x_0 h : h \in H\}$ where $f(x_0) = f_0$. So in general we have

$$|\psi\rangle = \frac{1}{\sqrt{|H|}} \sum_{h \in H} |x_0 h\rangle . \tag{15.36}$$

Since all values of f_0 are observed with equal probability, we can think of x_0 as being chosen uniformly. A state of the form (15.36) is called a *coset state*. In the case of GRAPH ISOMORPHISM, the hidden subgroup H is the automorphism group of $G_1 \cup G_2$, and all we need to do is determine H's size.

For simplicity, let's focus on the case where G_1 and G_2 are *rigid*, with no nontrivial symmetries of their own. That is, $\text{Aut}(G_1) = \text{Aut}(G_2) = \{1\}$ where 1 is the identity permutation. Then $H = \text{Aut}(G_1 \cup G_2)$ is either $\{1\}$ or $\{1, \tau\}$ where τ is a permutation that swaps the two graphs, and the coset state is

$$|\psi\rangle = \begin{cases} \frac{1}{\sqrt{2}}\left(|\sigma\rangle + |\sigma\tau\rangle\right) & \text{if } G_1 \cong G_2 \\ |\sigma\rangle & \text{if } G_1 \ncong G_2 , \end{cases} \tag{15.37}$$

where $\sigma \in S_n$ is uniformly random. To solve GRAPH ISOMORPHISM, at least for rigid graphs, all we ask is the ability to distinguish between these two cases. What basis should we use to measure $|\psi\rangle$, and thus learn about H?

The Fourier basis, you say. Well, of course we agree. But what basis is that? On the integers or the reals, the Fourier transform measures the frequencies at which a function oscillates. But what do "frequency" and "oscillation" mean for functions defined on the set of all permutations?

Let's review the various Fourier transforms we have seen so far. On $\mathbb{Z}_2^n$, the Fourier basis functions are of the form

$$\phi_{\mathbf{k}}(\mathbf{x}) = (-1)^{\mathbf{k}\cdot\mathbf{x}}.$$

On $\mathbb{Z}_M$, they are

$$\phi_k(x) = \omega^{kx} \quad \text{where} \quad \omega = e^{2\imath\pi/M}.$$

What these have in common is that, for each frequency $\mathbf{k}$ or k, the function $\phi_k(x)$ is a *homomorphism* from the group into the complex numbers $\mathbb{C}$. That is,

$$\phi_k(x+y) = \phi_k(x)\,\phi_k(y).$$

Both $\mathbb{Z}_2^n$ and $\mathbb{Z}_M$ are *abelian* groups, in which the binary operation is commutative: $x+y=y+x$. For any abelian group G, there are exactly $|G|$ such homomorphisms from G to $\mathbb{C}$, and they form an orthogonal basis for the vector space of all superpositions over G. It is the homomorphic nature of these basis functions that gives the Fourier transform all the properties that we know and love, such as the fact that the Fourier transform of the convolution of two functions is the product of their Fourier transforms.

The permutation group S_n, on the other hand, is nonabelian—the order of multiplication matters. If we swap $1 \leftrightarrow 2$ and then $2 \leftrightarrow 3$, we get the rotation $1 \to 3 \to 2$, but if do these things in the opposite order we get $1 \to 2 \to 3$. Since multiplication in $\mathbb{C}$ is commutative, any homomorphism from S_n to $\mathbb{C}$ must "forget" this kind of information.

Moreover, the vector space in which $|\psi\rangle$ lives, of superpositions of elements of S_n, has dimension $|S_n| = n!$. So in order to define a Fourier transform, we need $n!$ different basis functions. But there are only two homomorphisms from S_n to $\mathbb{C}$: the trivial one which sends every element to 1, and the parity that sends even and odd permutations to $+1$ and -1 respectively.

To define a Fourier basis for a nonabelian group G, we need to go beyond homomorphisms from G to $\mathbb{C}$, and instead look at homomorphisms from G to the group U_d of d-dimensional unitary matrices. Such homomorphisms are called *representations* of G, and the Fourier basis functions we are used to in the abelian case are just the special case where $d=1$.

For example, suppose that $G = S_3$. There are two one-dimensional representations, namely the trivial one and the parity. But there is also a two-dimensional representation, which permutes the three corners of a triangle by rotating and reflecting the plane as shown in Figure 15.8.

Representations like these give a geometric picture of the group in terms of rotations and reflections in some—possibly high-dimensional—space. For another example, consider Figure 15.9. For every even permutation σ of the 5 colors, there is a rotation that maps the 5 tetrahedra onto each other according to σ. This gives a three-dimensional representation of A_5, the subgroup of S_5 consisting of permutations with even parity.

While it would take us too far afield, there is a beautiful theory of Fourier transforms for nonabelian groups, in which representations play the role of frequencies. Specifically, each basis vector $|\rho,i,j\rangle$ corresponds to the i,j entry of some representation ρ. For instance, for S_3 we have $3! = 6$ basis vectors,

1 3 2

$$\rho(1) = \begin{pmatrix} 1 & \\ & 1 \end{pmatrix}$$

$$\rho(1 \leftrightarrow 2) = \begin{pmatrix} 1 & \\ & -1 \end{pmatrix}$$

$$\rho(1 \to 2 \to 3 \to 1) = \begin{pmatrix} -1/2 & -\sqrt{3}/2 \\ \sqrt{3}/2 & -1/2 \end{pmatrix}$$

FIGURE 15.8: The two-dimensional representation of S_3. For the identity permutation, it gives the identity matrix; for the permutation that swaps 1 and 2, it reflects around the x-axis; and for the cyclic permutation $1 \to 2 \to 3 \to 1$, it rotates the plane counterclockwise by $2\pi/3$.

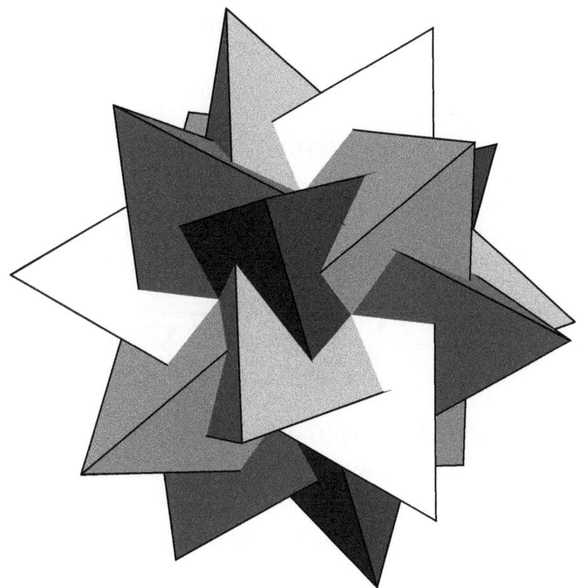

FIGURE 15.9: A three-dimensional representation of A_5, the group of even-parity permutations of 5 objects. See if you can tell how to switch two pairs of colors simultaneously.

corresponding to the trivial representation, the parity, and the four matrix entries of the two-dimensional representation. For many groups, including S_n, it is even possible to carry out the QFT efficiently, using nonabelian versions of the FFT and "quantizing" them in ways analogous to Section 15.5.5.

Sadly, all this machinery doesn't give us what we want. While the general approach of generating a coset state and measuring it in the Fourier basis does solve the HIDDEN SUBGROUP problem for some families of nonabelian groups, it fails for S_n. The situation is even worse than that. It can be shown that there is no measurement at all that we can perform on the coset state (15.37) that will distinguish an isomorphic pair of graphs from a non-isomorphic pair. Specifically, no matter what measurement we perform, the probability distribution of the outcomes we get in these two cases are exponentially close, so it would take an exponential number of experiments to distinguish them.

Some hope yet remains. It is known that if we have many copies of the coset state—that is, the tensor product of many $|\psi\rangle$s, where each one is a superposition over a randomly chosen coset—then there is a measurement that tells us whether the two graphs are isomorphic or not. However, this measurement must be highly entangled. In other words, rather than an independent series of measurements on the $|\psi\rangle$s, it is a complicated joint measurement along a basis where each basis vector corresponds to an entangled state.

However, while we can write this measurement down mathematically, we do not know how, or if, it can be carried out efficiently, i.e., by a quantum circuit with a polynomial number of gates. At the time we write this, nearly every proposed family of algorithms has been proved to fail. Our current intuition is that, if there is a quantum algorithm for GRAPH ISOMORPHISM, it doesn't work by reducing to HIDDEN SUBGROUP first.

15.20

15.6.3 *Quantum Sampling and the Swap Test*

There is another possible approach to GRAPH ISOMORPHISM. Suppose that, given a graph G with n vertices, we can generate the uniform superposition over all graphs G' such that $G' \cong G$, or equivalently over all the graphs we can get by permuting the vertices of G. If there are M such graphs, this is

$$|\psi(G)\rangle = \frac{1}{\sqrt{M}} \sum_{G' \cong G} |G'\rangle = \frac{\sqrt{M}}{n!} \sum_{\sigma \in S_n} |\sigma(G)\rangle . \tag{15.38}$$

For two graphs G_1, G_2, the states $|\psi(G_1)\rangle$ and $|\psi(G_2)\rangle$ are identical if $G_1 \cong G_2$, and orthogonal if $G_1 \ncong G_2$.

To distinguish between these two cases, we use a variant of phase kickback called the *swap test*. Given two quantum states $|\psi_1\rangle$ and $|\psi_2\rangle$, we want to tell whether $\langle\psi_1|\psi_2\rangle$ is 1 or 0, or more generally to estimate $\left|\langle\psi_1|\psi_2\rangle\right|^2$. We start with their tensor product, along with a qubit in the state $|+\rangle = \frac{1}{\sqrt{2}}(|0\rangle + |1\rangle)$:

$$|+\rangle \otimes |\psi_1\rangle \otimes |\psi_2\rangle .$$

We now apply a *controlled swap gate*, which swaps $|\psi_1\rangle$ and $|\psi_2\rangle$ if the first qubit is true, and otherwise leaves them alone. This gives

$$\frac{1}{\sqrt{2}} \big(|0\rangle \otimes |\psi_1\rangle \otimes |\psi_2\rangle + |1\rangle \otimes |\psi_2\rangle \otimes |\psi_1\rangle \big) .$$

Finally, we measure the qubit in the X-basis. Problem 15.33 shows that we observe $|+\rangle$ or $|-\rangle$ with probabilities

$$P(+) = \frac{1+\left|\langle\psi_1|\psi_2\rangle\right|^2}{2} \quad \text{and} \quad P(-) = \frac{1-\left|\langle\psi_1|\psi_2\rangle\right|^2}{2}. \tag{15.39}$$

Thus if $|\psi_1\rangle$ and $|\psi_2\rangle$ are identical, we always observe $|+\rangle$, but if they are orthogonal we observe $|+\rangle$ or $|-\rangle$ with equal probability. If we can generate the states $|\psi_1\rangle = |\psi(G_1)\rangle$ and $|\psi_2\rangle = |\psi(G_2)\rangle$, we can use this swap test to tell whether G_1 and G_2 are isomorphic or not.

Given a graph G, can we generate $|\psi(G)\rangle$? In the classical case, it's easy to generate a uniform distribution over all the graphs isomorphic to G—just generate a random permutation and apply it to G's vertices. Similarly, in the quantum setting, we can create a uniform superposition over S_n, and then apply a controlled-permutation operator to G. However, this generates the entangled state

$$\frac{1}{n!}\sum_{\sigma\in S_n} |\sigma\rangle \otimes |\sigma(G)\rangle\,.$$

To obtain $|\psi(G)\rangle$ we have to *disentangle* $\sigma(G)$ from σ. To do this we have to "uncompute" σ from $\sigma(G)$, telling what permutation takes us from G to a given $\sigma(G)$. But this seems to be just as hard as solving GRAPH ISOMORPHISM in the first place.

There is an interesting generalization of this problem, called *quantum sampling*. Given a probability distribution $P(x)$, such as the equilibrium distribution of a rapidly mixing Markov chain, can we generate the quantum state $\sum_x \sqrt{P(x)}\,|x\rangle$? In the classical case, if P_1 and P_2 are probability distributions over an exponentially large state space, we have no idea how to tell whether they are identical or disjoint—but if we can generate the analogous quantum states, the swap test lets us do this. 15.21

15.7 Quantum Haystacks: Grover's Algorithm

Shor's algorithm is a spectacular example of the power of quantum computing. But FACTORING and DISCRETE LOG seem rather special, in that they are reducible to finding the periodicity of a number-theoretic function. Is there a generic way in which quantum computers can solve search problems more quickly than classical ones? For instance, suppose there is a needle hidden in a haystack of size N. Classically, we have to look at all N blades of hay. Can a quantum computer do better?

In 1996, Lov Grover discovered a quantum algorithm that finds the needle with just $O(\sqrt{N})$ queries. While this improvement is merely quadratic, it makes an enormous difference if N is exponentially large. Imagine that we have a SAT formula with 80 variables and a unique satisfying assignment. If we compare Grover's algorithm with a brute-force search, the number of truth assignments we have to try decreases from 2^{80} to 2^{40}. At a rate of 10^9 assignments per second, this reduces our running time from 38 million years to 18 minutes.

On the other hand, changing N to $\sqrt{N}$ can't change an exponential to a polynomial—it can only improve the exponent. Moreover, as we will see in this section, Grover's algorithm is optimal. This means that if we want to solve an NP-complete problem in polynomial time, we have to find something about the specific structure of that problem that a quantum algorithm can exploit. Just as with classical computation, our current belief is that NP-complete problems are just as hard as haystacks, and that even quantum computers cannot solve them in polynomial time. In terms of complexity classes, we believe that $\mathsf{NP} \not\subseteq \mathsf{BQP}$.

15.7.1 *Rotating towards the Needle*

Suppose there is a haystack with N locations—in quantum terms, a state space with N basis vectors $|j\rangle$ where $1 \le j \le N$. There is a single needle hidden at some location i, and we want to find it. We can query the haystack by looking at a location j and seeing if the needle is there. As in Section 15.4, each such query corresponds to applying a unitary operator that flips a qubit y if $j = i$:

$$|j\rangle \otimes |y\rangle \rightarrow \begin{cases} |j\rangle \otimes X|y\rangle & \text{if } j = i \\ |j\rangle \otimes |y\rangle & \text{if } j \neq i. \end{cases}$$

By preparing y in the eigenvector $|-\rangle$ we can use the phase kickback trick, and treat each query as this N-dimensional operator instead:

$$U|j\rangle = \begin{cases} -|j\rangle & \text{if } j = i \\ |j\rangle & \text{if } j \neq i. \end{cases}$$

This operator is almost identical to the identity, differing only at $U_{ii} = -1$.

How many queries do we need to find the needle? In other words, how many times do we have to apply U in order to obtain a state for which some measurement will tell us what i is? Grover's clever idea is to adapt, after each query, the superposition of locations at which we query the haystack. We do this by applying a "diffusion operator" D which has $(2/N) - 1$ on the diagonal and $2/N$ everywhere else:

$$D = \frac{2}{N} \begin{pmatrix} 1 & 1 & \cdots \\ 1 & 1 & \\ \vdots & & \ddots \end{pmatrix} - \mathbb{1}. \tag{15.40}$$

It is not obvious at first that D is unitary, so we recommend the following exercise.

Exercise 15.21 *Show that Grover's operator D is unitary, and that $D^2 = \mathbb{1}$.*

The algorithm starts in the uniform superposition over all N states, $|u\rangle = \frac{1}{\sqrt{N}} \sum_{j=1}^{N} |j\rangle$. We then alternately apply U and D. After t iterations, we have the state

$$|\psi\rangle = \underbrace{DU\,DU \cdots DU}_{t \text{ times}} |u\rangle = (DU)^t |u\rangle.$$

Marvelously, alternating between U and D causes the amplitude at the needle's location $|i\rangle$ to interfere constructively, while the amplitude at all other locations cancels out. After just $t = O(\sqrt{N})$ queries, if we measure $|\psi\rangle$ we observe $|i\rangle$ with high probability.

How does this work? Consider Figure 15.10, where we reflect a state $|\psi\rangle$ first around a vector $|u\rangle$ and then around another vector $|v\rangle$. The composition of these two reflections is a rotation. Specifically, if the angle between $|u\rangle$ and $|v\rangle$ is θ, this rotates $|\psi\rangle$ by 2θ.

For any vector $|u\rangle$ of unit length, the operator R_u that reflects around $|u\rangle$ can be written

$$R_u = 2|u\rangle\langle u| - \mathbb{1}. \tag{15.41}$$

That is, R_u negates the component perpendicular to $|u\rangle$ and keeps the component parallel to $|u\rangle$ the same. To see this, try the following exercise.

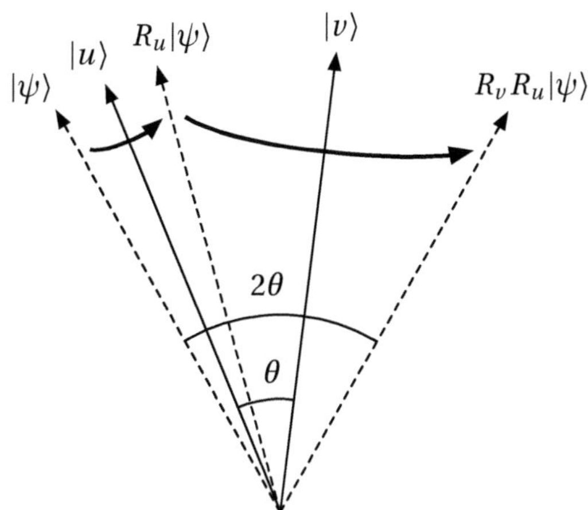

FIGURE 15.10: If we reflect $|\psi\rangle$ around two vectors, $|u\rangle$ and then $|v\rangle$, we rotate it by twice the angle θ between them.

Exercise 15.22 *Show that if* $|\psi\rangle = a|u\rangle + b|w\rangle$ *where* $\langle u\,|\,w\rangle = 0$, *then*

$$R_u|\psi\rangle = a|u\rangle - b|w\rangle\,,$$

and that R_u *is unitary.*

Looking again at (15.40), we see that D is just the reflection operator around the uniform superposition $|u\rangle$,

$$D = 2|u\rangle\langle u| - \mathbb{1} = R_u\,,$$

since $|u\rangle\langle u|$ is the N-dimensional matrix where every entry is $1/N$. As Problem 15.35 shows, this description also helps us implement D efficiently using elementary quantum gates. Similarly, up to a sign, the query operator reflects around the basis vector $|i\rangle$ corresponding to the needle's location:

$$U = \mathbb{1} - 2|i\rangle\langle i| = -R_i\,. \tag{15.42}$$

What happens when we iteratively apply U and D? While Grover's algorithm takes place in an N-dimensional space, the state $|\psi\rangle$ always lies in a two-dimensional subspace—namely, the space of vectors in which all the $|j\rangle$s other than $|i\rangle$ have the same amplitude. Let's call this subspace V. It is spanned by $|i\rangle$ and $|u\rangle$, or equivalently by $|i\rangle$ and the uniform superposition over all the states other than $|i\rangle$, which we denote $|v\rangle$:

$$|v\rangle = \frac{1}{\sqrt{N-1}}\sum_{j\neq i}|j\rangle = \frac{1}{\sqrt{N-1}}\left(\sqrt{N}|u\rangle - |i\rangle\right).$$

Since $|i\rangle$ and $|v\rangle$ are perpendicular, we can write the projection operator onto V as $\mathbb{1}_V = |i\rangle\langle i| + |v\rangle\langle v|$. As far as vectors in V are concerned, we can replace $\mathbb{1}$ with $\mathbb{1}_V$ in (15.42), giving

$$U = |v\rangle\langle v| - |i\rangle\langle i| = 2|v\rangle\langle v| - \mathbb{1} = R_v\,.$$

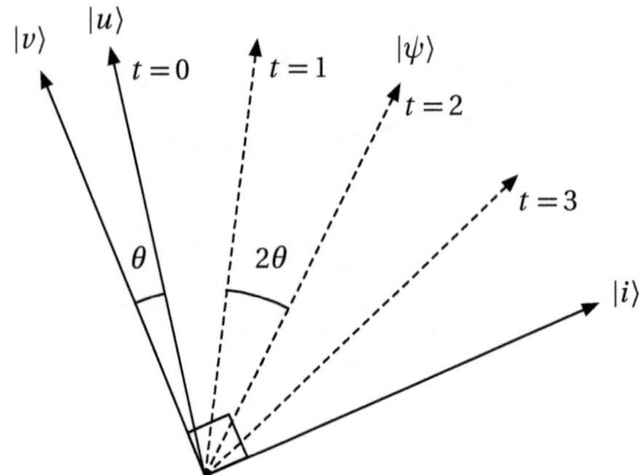

FIGURE 15.11: Rotating $|\psi\rangle$ from the initial uniform state $|u\rangle$ towards the state $|i\rangle$ marking the needle's location.

Thus each iteration of DU rotates $|\psi\rangle$ by 2θ, where θ is the angle between $|u\rangle$ and $|v\rangle$. As shown in Figure 15.11, our goal is to rotate $|\psi\rangle$ so that it is close to $|i\rangle$. Initially $|\psi\rangle = |u\rangle$, which is close to $|v\rangle$ when N is large. The angle between $|v\rangle$ and $|i\rangle$ is $\pi/2$, so we need $t = (\pi/2)/(2\theta) = \pi/(4\theta)$ steps to bring $|\psi\rangle$ close to $|i\rangle$. The fact that $|\psi\rangle$ starts at $|u\rangle$ instead of $|v\rangle$ means that the initial angle is actually $\pi/2-\theta$, but this only saves us half a step.

All that remains is to calculate the angle θ between $|u\rangle$ and $|v\rangle$. We can obtain θ from their inner product,

$$\langle u\,|\,v\rangle = \cos\theta = \sqrt{1-\frac{1}{N}} = 1-\frac{1}{2N} - O(N^{-2}).$$

Matching this with the Taylor series $\cos\theta = 1-\theta^2/2+O(\theta^4)$ then gives

$$\theta = \frac{1}{\sqrt{N}} + O(N^{-3/2}),$$

and so

$$t = \frac{\pi}{4\theta} = \frac{\pi}{4}\sqrt{N} - O(N^{-1/2}).$$

If we round t to the nearest integer, $|\psi\rangle$ will be within an angle θ of $|i\rangle$, so the probability of observing i will be

$$P(i) = \left|\langle\psi\,|\,i\rangle\right|^2 \geq \cos^2\theta = 1-O(\theta^2) = 1-O(1/N).$$

So if we perform $t = (\pi/4)\sqrt{N}$ iterations of DU and then measure $|\psi\rangle$, we observe the basis vector $|i\rangle$ with high probability. Note that if we run the algorithm for too many steps, $|\psi\rangle$ rotates back down to a uniform state, and then round and round again.

Exercise 15.23 *We assumed here that there is a unique solution to our search problem—or that there is exactly one needle. Show that if there are x needles, θ is given by*

$$\cos\theta = \sqrt{1 - \frac{x}{N}},$$

and therefore that we can find a solution, where all solutions are equally likely, after $O\left(\sqrt{N/x}\right)$ queries.

Exercise 15.24 *Show that if $N = 4$ and there is a unique solution, or more generally if $x/N = 1/4$, Grover's algorithm finds a solution with probability 1 after just one iteration.*

Exercise 15.23 shows that the number of times we should iterate Grover's algorithm before measuring $|\psi\rangle$ depends on the number of needles in the haystack. But what if we don't know how many there are? In Problem 15.37, we modify Grover's algorithm so that it can deal with an unknown number of solutions.

Grover's algorithm is wonderful. But is it the best possible? Can we find the needle with even fewer queries, say N^α for some $\alpha < 1/2$? In the next section we will see that this is impossible—that where haystacks are concerned, a quadratic speedup is the best that quantum computers can achieve.

15.7.2 Entangling with the Haystack

There are many proofs that Grover's algorithm is optimal, i.e., that any algorithm needs to query the haystack $\Omega(\sqrt{N})$ times to find the needle. We focus here on a proof that is especially enlightening. In addition to giving us the result we want, it will let us explore the nature of entanglement more deeply.

If the haystack consists of a physical system with which our quantum queries can interact, it must be a quantum system. This means that it could be in a superposition of many different states, corresponding to different locations of the needle. To find the needle, our computer must measure the haystack—but this means that the computer and the haystack must become entangled.

In Section 10.5, we proved lower bounds on classical randomized algorithms by letting the adversary choose the worst possible probability distribution of instances. In the same way, we can prove lower bounds on quantum algorithms by letting him choose the worst possible *superposition* of instances. If we can show that some superposition requires t queries to solve, then t is a lower bound on the worst case as well.

To make this idea work, we endow the haystack with an N-dimensional state space. If the haystack is in state $|i\rangle$, the needle is at location i. Intuitively, the needle is hardest to find when every location has the same amplitude, so we will use the uniform superposition $|u\rangle = \frac{1}{\sqrt{N}}\sum_i |i\rangle$ to prove our lower bound.

For simplicity we assume that the computer's state space is also N-dimensional, with one state for each possible location. We also assume that the algorithm works perfectly, so that after running the algorithm the computer is in precisely the state $|i\rangle$. Both of these assumptions can be relaxed without changing the proof very much.

At the beginning of the algorithm, the computer is in some initial state $|\psi_0\rangle$. The computer and the haystack are unentangled, so their joint state is a tensor product,

$$|\Psi_{\text{init}}\rangle = |u\rangle \otimes |\psi_0\rangle .$$

As the algorithm proceeds, the state of the computer comes to depend on the state of the haystack, so their joint state becomes entangled. If $|\psi_i\rangle$ is the computer's state in the case where the haystack's state is $|i\rangle$, their joint state is a sum of tensor products,

$$\frac{1}{\sqrt{N}}\sum_i |i\rangle \otimes |\psi_i\rangle . \tag{15.43}$$

At the end of the algorithm we have $|\psi_i\rangle = |i\rangle$, so their final joint state is

$$|\Psi_{\text{final}}\rangle = \frac{1}{\sqrt{N}}\sum_i |i\rangle \otimes |i\rangle .$$

Like the entangled state $\frac{1}{\sqrt{2}}(|00\rangle + |11\rangle)$ shared by Alice and Bob in Section 15.3, the state $|\Psi_{\text{final}}\rangle$ is *maximally entangled*—the two systems are always in the same state. Our goal is to prove an upper bound on the amount by which each query can increase the entanglement, and therefore a lower bound on the number of queries we need to get from the unentangled state $|\Psi_{\text{init}}\rangle$ to $|\Psi_{\text{final}}\rangle$.

How can we quantify the amount of entanglement between two systems? The idea is that measuring one of them also measures the other to some extent. We can track this process using the density matrix formalism of Section 15.3.3, by calculating the mixed state ρ that one system will be in if we measure the other. If they are only slightly entangled, then ρ is close to a pure state. However, if they are maximally entangled like $|\Psi_{\text{final}}\rangle$ or Alice and Bob's pair of qubits, then measuring one measures the other completely. If all states are equally likely then ρ is the completely mixed state, i.e., the classical uniform distribution.

Recall that for a pure state $|v\rangle$, the density matrix is $|v\rangle\langle v|$. Its diagonal entries $\rho_{ii} = v_i v_i^* = |v_i|^2$ are the probabilities associated with $|v\rangle$'s components, while its off-diagonal entries $\rho_{ij} = v_i v_j^*$ contain information about their relative phase. As entanglement increases, the off-diagonal entries of ρ disappear, leaving only the diagonal ones. This process, called *decoherence*, removes all the quantum phase information, until only classical probabilities remain.

Problem 15.39 shows how to calculate the density matrix whenever the joint state is of the form (15.43). In our case, the density matrix of the haystack is

$$\rho_{ij} = \frac{1}{N}\langle \psi_j | \psi_i \rangle . \tag{15.44}$$

Note that ρ is unchanged by any step of the algorithm that doesn't make a query, since unitary operations on the computer's state space preserve the inner products $\langle \psi_j | \psi_i \rangle$. At the beginning of the algorithm we have $|\psi_i\rangle = |\psi_0\rangle$ for all i, and the haystack is in the pure state

$$\rho = |u\rangle\langle u| = \frac{1}{N}\begin{pmatrix} 1 & 1 & \cdots \\ 1 & 1 & \\ \vdots & & \ddots \end{pmatrix} . \tag{15.45}$$

As the computer learns more about the haystack, ψ_i and ψ_j become different for $i \neq j$, and the off-diagonal entries ρ_{ij} decrease. By the end of the algorithm, $|\psi_i\rangle$ and $|\psi_j\rangle$ are orthogonal for $i \neq j$, and the haystack is in the completely mixed state

$$\rho = \frac{1}{N}\begin{pmatrix} 1 & 0 & \cdots \\ 0 & 1 & \\ \vdots & & \ddots \end{pmatrix} = \frac{1}{N}\mathbb{1} . \tag{15.46}$$

How does each query change the computer's state $|\psi_i\rangle$, and therefore the haystack's density matrix ρ? The query operator U is now a unitary operator acting on the joint state. Specifically,

$$U\left(|i\rangle \otimes |\psi_i\rangle\right) = |i\rangle \otimes U_i|\psi_i\rangle\,,$$

where U_i is the query operator from (15.42),

$$U_i = \mathbb{1} - 2|i\rangle\langle i|\,.$$

The reader might ask whether some other query operator would allow us to find the needle more quickly. However, since this operator imposes the largest possible phase shift when the computer looks at the right location—namely, -1, or an angle π in the complex plane—it can be shown to be optimal.

Applying (15.44), we see that U changes the haystack's density matrix from ρ to ρ' where

$$\begin{aligned}
\rho'_{ij} &= \frac{1}{N}\langle U_j\psi_j | U_i\psi_i\rangle \\
&= \frac{1}{N}\left(\langle\psi_j| - 2\langle\psi_j|j\rangle\langle j|\right)\left(|\psi_i\rangle - 2|i\rangle\langle i|\psi_i\rangle\right) \\
&= \frac{1}{N}\left(\langle\psi_j|\psi_i\rangle - 2\langle\psi_j|i\rangle\langle i|\psi_i\rangle - 2\langle\psi_j|j\rangle\langle j|\psi_i\rangle + 4\langle\psi_j|j\rangle\langle j|i\rangle\langle i|\psi_i\rangle\right).
\end{aligned}$$

For $i = j$, there is no change at all, since ρ_{ii} is always $(1/N)\left|\psi_i\right|^2 = 1/N$. For $i \neq j$, we have $\langle j\,|\,i\rangle = 0$, eliminating the fourth term. Thus the off-diagonal terms decrease by

$$\rho'_{ij} = \rho_{ij} - \Delta_{ij} \quad \text{where} \quad \Delta_{ij} = \frac{2}{N}\left(\langle\psi_j|i\rangle\langle i|\psi_i\rangle + \langle\psi_j|j\rangle\langle j|\psi_i\rangle\right). \tag{15.47}$$

To measure our progress, we will adopt a very simple measure of entanglement—namely, the sum of all of ρ's entries,

$$S = \sum_{ij} \rho_{ij}\,.$$

From (15.45) and (15.46), we see that $S = N$ at the beginning of the algorithm and $S = 1$ at the end. According to (15.47), each query changes S by

$$S' = S - \Delta \quad \text{where} \quad \Delta = \sum_{i,j:i\neq j} \Delta_{ij}\,.$$

We will show that $|\Delta| = O(\sqrt{N})$. Since S must decrease from N to 1, we will then need $\Omega(\sqrt{N})$ queries.

To bound $|\Delta|$, first note that when we sum over all i and j, the two terms in Δ_{ij} become complex conjugates of each other. Using the triangle inequality, we can then write

$$|\Delta| \leq \frac{4}{N}\left|\sum_{i,j:i\neq j}\langle\psi_j|i\rangle\langle i|\psi_i\rangle\right| \leq \frac{4}{N}\sum_j\left|\sum_{i\neq j}\langle\psi_j|i\rangle\langle i|\psi_i\rangle\right|.$$

Bounding the inner sum over i using the Cauchy–Schwarz inequality (see Appendix A.2.3) gives

$$\begin{aligned}|\Delta| &\le \frac{4}{N}\sqrt{\sum_i \left|\langle\psi_i|i\rangle\right|^2} \times \sum_j \sqrt{\sum_{i\neq j} \left|\langle\psi_j|i\rangle\right|^2} \\ &= \frac{4}{N}\sqrt{\sum_i p_i} \times \sum_j \sqrt{1-p_j}\,, \end{aligned} \tag{15.48}$$

where $p_i = \left|\langle\psi_i|i\rangle\right|^2$ denotes the probability that observing the computer gives the right answer if the haystack is in state $|i\rangle$. Applying Cauchy–Schwarz again gives

$$\begin{aligned}|\Delta| &\le \frac{4}{N}\sqrt{\sum_i p_i} \times \sqrt{N}\sqrt{\sum_j (1-p_j)} \\ &= 4\sqrt{Np(1-p)}\,,\end{aligned}$$

where $p = (1/N)\sum_i p_i$ is the average probability that the computer finds the needle if its location is uniformly random.

Since $\sqrt{p(1-p)} \le 1/2$ for all $0 \le p \le 1$, we have

$$|\Delta| \le 2\sqrt{N}\,. \tag{15.49}$$

Since S has to decrease from N to 1, this shows that the minimum number of queries is at least $\sqrt{N}/2$ when N is large, and therefore that any algorithm must make $\Omega(\sqrt{N})$ queries.

15.22 What about the constant in front of $\sqrt{N}$? In (15.49), we pessimistically assumed that $\sqrt{p(1-p)} = 1/2$. However, this is only true when $p = 1/2$, so for most of the algorithm the rate of decoherence is slower than (15.49) suggests. Using another type of argument, one can show that the constant $\pi/4$ in Grover's algorithm is in fact optimal if we want to find the solution with high probability. However, if we want to minimize the *expected* number of queries and we are willing to take the risk that the algorithm might take longer, Problem 15.46 shows that we can do slightly better.

This *quantum adversary* method has been used to prove lower bounds on the number of queries for a wide variety of problems. We explore another technique for proving lower bounds, the *polynomial method*, in Problems 15.40 through 15.45.

15.23

15.8 Quantum Walks and Scattering

We end this chapter with another algorithmic idea. In Chapters 12 and 13, we showed how random walks can explore the space of possible solutions to a problem, generating a random solution in polynomial time. Here we consider the quantum version of these walks, where waves with complex amplitudes, rather than real probabilities, propagate through space.

As always, the key difference is interference. By canceling out in some places and adding up in others, these quantum walks can spread through space faster than their classical cousins. By propagating down a tree and interfering with themselves in complex ways, they can even tell us who has a winning strategy in a two-player game.

The easiest way to think of this kind of algorithm is as a physical process taking place in continuous time. As we will see, each one works by designing a matrix called the Hamiltonian, which governs the interactions in the system and acts as the transition matrix of the walk. We then simply run Schrödinger's equation, the basic equation of quantum physics, for a certain amount of time.

We start by looking at the classical version of this process. If you are not already familiar with classical random walks and with the Poisson and Gaussian distributions, you may find Appendix A.4 useful.

15.8.1 *Walking the Line*

As a warm-up, we consider the classical random walk on the line, but in continuous time. Suppose we have an infinite line, with one vertex for each integer. Its adjacency matrix is

$$H = \frac{1}{2}\begin{pmatrix} \ddots & & & & & \\ & 0 & 1 & & & \\ & 1 & 0 & 1 & & \\ & & 1 & 0 & 1 & \\ & & & 1 & 0 & \\ & & & & & \ddots \end{pmatrix}. \tag{15.50}$$

We have normalized this to make it stochastic. In particular, it is the transition matrix of a random walk where we step to the left or the right with equal probability. If we write the current probability distribution as a column vector P, the distribution after t steps is $H^t P$.

Now suppose that we perform a random walk in continuous time. In each infinitesimal time interval of width $\mathrm{d}t$, there is a probability $\mathrm{d}t$ of changing the current probability distribution P_t to HP_t. Thus, to first order in $\mathrm{d}t$, we have

$$P_{t+\mathrm{d}t} = \mathrm{d}t\, HP_t + (1-\mathrm{d}t)P_t\,. \tag{15.51}$$

This gives us a linear differential equation,

$$\frac{\mathrm{d}}{\mathrm{d}t}P_t = (H - \mathbb{1})\,P_t\,.$$

Its solution is just what you expect,

$$P_t = \mathrm{e}^{(H-\mathbb{1})t}\,P_0\,. \tag{15.52}$$

However, $\mathrm{e}^{(H-\mathbb{1})t}$ is a *matrix* exponential, defined using the Taylor series:

$$\begin{aligned} \mathrm{e}^{(H-\mathbb{1})t} &= \mathrm{e}^{-t}\mathrm{e}^{Ht} = \mathrm{e}^{-t}\left(\mathbb{1} + Ht + \frac{(Ht)^2}{2} + \frac{(Ht)^3}{6} + \cdots\right) \\ &= \sum_{j=0}^{\infty} \frac{\mathrm{e}^{-t}t^j}{j!}H^j\,. \end{aligned} \tag{15.53}$$

We can interpret (15.53) as follows. First, choose j from a Poisson distribution with mean t, i.e., with probability $P(j) = \mathrm{e}^{-t}t^j/j!$. Then take j steps of the discrete-time random walk by applying H^j.

Given an initial distribution $P(0)$, how can we use (15.52) to calculate the distribution P_t at time t? We can do this using Fourier analysis, just as we did for the discrete-time walk on the cycle in Section 12.7.2. The Fourier basis vectors are $v_k = \mathrm{e}^{\imath kx}$ where k ranges from $-\pi$ to π, and decomposing P_t in this basis gives

$$P_t(x) = \frac{1}{2\pi} \int_{-\pi}^{\pi} \widetilde{P}_t(k)\,\mathrm{e}^{\imath kx}\,\mathrm{d}k\,.$$

If our initial distribution is concentrated at the origin, with $P_0(0)=1$ and $P_0(x)=0$ for all $x \neq 0$, its Fourier distribution is uniform and $\widetilde{P}_0(k)=1$ for all k.

The Fourier basis vectors are eigenvectors of H. Since $(Hv)_x = (v_{x-1} + v_{x+1})/2$ for any vector v, we have

$$H v_k = \lambda_k v_k\,,$$

where the eigenvalue is

$$\lambda_k = \frac{\mathrm{e}^{-\imath k} + \mathrm{e}^{\imath k}}{2} = \cos k\,.$$

It follows that v_k is also an eigenvector of $\mathrm{e}^{(H-\mathbb{1})t}$,

$$\mathrm{e}^{(H-\mathbb{1})t}\, v_k = \mathrm{e}^{(\lambda_k-1)t}\, v_k\,.$$

Applying $\mathrm{e}^{(H-\mathbb{1})t}$ to P_0 then gives

$$\begin{aligned} P_t(x) &= \mathrm{e}^{(H-\mathbb{1})t}\,\frac{1}{2\pi}\int_{-\pi}^{\pi} \mathrm{e}^{\imath kx}\,\mathrm{d}k \\ &= \frac{1}{2\pi}\int_{-\pi}^{\pi} \mathrm{e}^{(H-\mathbb{1})t}\,\mathrm{e}^{\imath kx}\,\mathrm{d}k \\ &= \frac{1}{2\pi}\int_{-\pi}^{\pi} \mathrm{e}^{(\cos k-1)t}\,\mathrm{e}^{\imath kx}\,\mathrm{d}k\,. \end{aligned} \tag{15.54}$$

The integral (15.54) looks difficult to evaluate, but it is easy to approximate when t is large. As the probability distribution spreads out, low frequencies dominate, and we can use the second-order Taylor series $\cos k \approx 1 - k^2/2$ for small k. At the cost of a small error, we can also extend the integral over k to the entire real line, giving

$$P_t(x) \approx \frac{1}{2\pi}\int_{-\infty}^{\infty} \mathrm{e}^{-tk^2/2}\,\mathrm{e}^{\imath kx}\,\mathrm{d}k = \frac{1}{\sqrt{2\pi t}}\,\mathrm{e}^{-x^2/(2t)}\,.$$

Thus $P_t(x)$ approaches a Gaussian distribution with variance t and width $O(\sqrt{t})$. If we carry out this process on a line or a cycle of length n rather than on the infinite line, the mixing time—the time it takes to reach a nearly uniform distribution—is $\Theta(n^2)$. This is the same result we obtained in Section 12.7 and Problem 12.36 for the discrete-time walk.

15.8.2 *Schrödinger's Equation and Quantum Diffusion*

Our next task is to define quantum walks analogous to the classical walk we just analyzed. First we need to review a little more physics. Throughout this chapter, we have described the steps of a quantum computer as unitary operators. But where do these unitary operators come from?

Since the eigenvalues of a unitary operator U are of the form $\mathrm{e}^{\imath\theta}$, applying it to a state rotates the amplitude ψ of each eigenvector by some angle θ in the complex plane. But it takes time to do this. In an infinitesimal time interval $\mathrm{d}t$, we can only rotate the state by an infinitesimal angle $\theta = \lambda\,\mathrm{d}t$. To first order in $\mathrm{d}t$ we have

$$\mathrm{e}^{\imath\lambda\,\mathrm{d}t} = 1 + \imath\lambda\,\mathrm{d}t\,,$$

so, analogous to (15.51), the amplitude evolves as

$$\psi_{t+\mathrm{d}t} = \psi_t + \imath\lambda\psi_t\,\mathrm{d}t\,.$$

We can write this as a differential equation,

$$\frac{\mathrm{d}}{\mathrm{d}t}\psi_t = \imath\lambda\psi_t\,. \tag{15.55}$$

Now suppose that we have a matrix H with the same eigenvectors as U, but with real eigenvalues λ. Then we can express the differential equation (15.55) for all eigenvectors at once in the following way:

$$\frac{\mathrm{d}}{\mathrm{d}t}|\Psi_t\rangle = \imath H|\Psi_t\rangle\,. \tag{15.56}$$

This is known as *Schrödinger's equation,* although physicists usually write it with $-\imath$ instead of $\imath$. Assuming that H does not change with time, its solution is 15.24

$$|\Psi_t\rangle = \mathrm{e}^{\imath Ht}\,|\Psi_0\rangle\,.$$

Again using the matrix exponential, we see that running Schrödinger's equation for a time interval t applies the following unitary operator to the state Ψ,

$$\begin{aligned}\mathrm{e}^{\imath Ht} &= \mathbb{1} + \imath Ht - \frac{(Ht)^2}{2} - \imath\frac{(Ht)^3}{6} + \cdots \\ &= \sum_{j=0}^{\infty}\frac{(\imath t)^j}{j!}H^j\,.\end{aligned} \tag{15.57}$$

This operator is unitary as long as H is Hermitian, i.e., $H^\dagger = H$:

Exercise 15.25 *Show that* $\mathrm{e}^{\imath Ht}$ *is unitary for all real* t *if and only if* H *is Hermitian.*

In physics, H is called the *Hamiltonian.* It describes all the interactions in the system, and all the transitions it can make from one state to another. In our setting, we can take H to be a weighted adjacency matrix of an undirected graph, normalized as in (15.50) so that its maximum eigenvalue is 1.

For the quantum walk on the line, we can solve for the amplitude $\Psi_t(x)$ just as we solved for $P_t(x)$ in the classical case. The eigenvectors of H are again the Fourier basis vectors $e^{\imath kx}$, with eigenvalues $\lambda = \cos k$. Writing Ψ_t in this basis gives

$$\Psi_t(x) = \frac{1}{2\pi}\int_{-\pi}^{\pi} \widetilde{\Psi}_t(k)\,e^{\imath kx}\,dk\,.$$

As before, we assume that the initial state is concentrated at the origin, $\Psi_0(0) = 1$ and $\Psi_0(x) = 0$ for all $x \neq 0$. Then $\widetilde{\Psi}_0(k) = 1$ for all k, and applying $e^{\imath\lambda t}$ to each eigenvector gives

$$\Psi_t(x) = \frac{1}{2\pi}\int_{-\pi}^{\pi} e^{\imath\lambda t}\,e^{\imath kx}\,dk = \frac{1}{2\pi}\int_{-\pi}^{\pi} e^{\imath t\cos k}\,e^{\imath kx}\,dk\,. \tag{15.58}$$

Except for the factor of $\imath$, and the replacement of $\lambda - 1$ with λ which gives an overall phase change, this looks just like the integral (15.54) for the classical random walk. Does it result in a similar probability distribution?

As Figure 15.12 shows, the probability distribution of the quantum walk is vastly different from the classical one. Rather than being confined to a Gaussian of width $\sqrt{t}$, the probability stretches across over the interval $[-t, t]$, and is modulated by rapid oscillations. Problem 15.50 shows that if we ignore these oscillations, the distribution scales asymptotically as

$$P_t(x) = |\Psi_t(x)|^2 \sim \frac{1}{\sqrt{t^2 - x^2}}\,. \tag{15.59}$$

Since it spreads out quadratically faster than in the classical case, the mixing time on a line or cycle of length n is now just $\Theta(n)$ instead of $\Theta(n^2)$.

What's going on? As in the classical case, the continuous-time walk includes paths of many different lengths, corresponding to different powers of H in the matrix exponential. But now, looking at (15.57), we see that paths of different lengths j contribute with different phases $\imath^j$. Combinatorially speaking, the vast majority of paths still return to within $O(\sqrt{t})$ of the origin—but now they interfere destructively, suppressing their total probability. On the other hand, far from the origin we have constructive interference, making the probability much greater than it would be in the tail of a Gaussian.

As another example, Figure 15.13 shows a quantum walk on the square lattice. Instead of a two-dimensional Gaussian with width $\sqrt{t}$, its probability distribution is roughly uniform over a circle of radius $\Theta(t)$, so its mixing time on an $\ell \times \ell$ lattice is $O(\ell)$ instead of $O(\ell^2)$. Since quantum walks mix faster than classical ones, they can achieve certain search and sampling tasks faster than the classical random walks of Chapter 12.

15.25 Yet another kind of algorithm probes the properties of a graph by running a quantum walk on it. As we are about to see, we can tell who has a winning strategy in a game by telling whether the tree of possible moves is transparent or reflective.

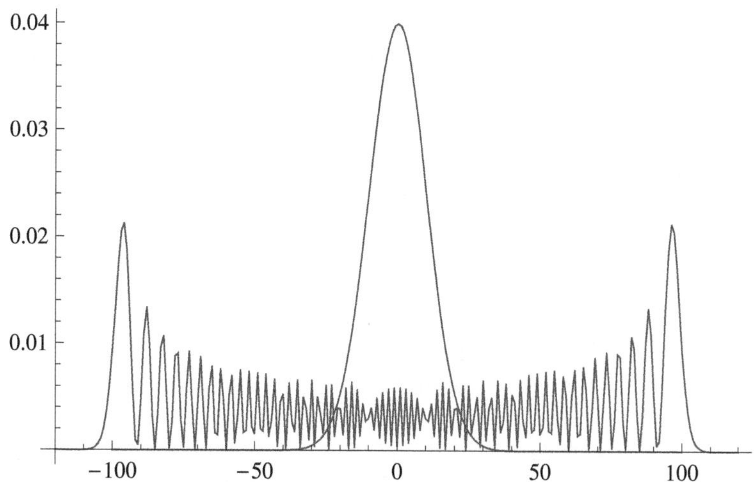

FIGURE 15.12: The quantum and classical walks on the line. The classical walk gives a Gaussian distribution of width $O(\sqrt{t})$, while the probability distribution $|\Psi|^2$ of the quantum walk is roughly uniform over the interval $[-t,t]$. Here $t=100$.

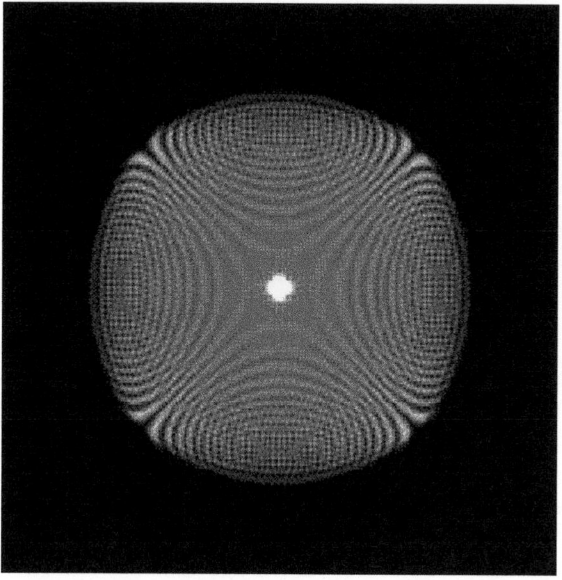

FIGURE 15.13: The probability distribution of a discrete-time quantum walk on the square lattice. Except for a bright spot in the center, the probability is roughly uniform over a circle of radius $t/\sqrt{2}$. Here $t=100$.

15.8.3 *Scattering off a Game Tree*

In Section 8.6.3, we discussed the problem of telling who has a winning strategy in a two-player game by recursively exploring the tree of possible paths that the game can take. Following the approach of Section 10.5, let's say that a position as `true` or `false` if it is a win or a loss for the current player, assuming that both players play optimally. Since a winning position is one where at least one move creates a loss for the other player, we can think of each position in the tree as a NAND gate, which returns `true` if at least one of its inputs is `false`. Our goal is then to evaluate the entire tree and see if the root is `true`.

In 2007, Edward Farhi, Jeffrey Goldstone, and Sam Gutmann found an astonishing quantum algorithm for this problem. We start with the quantum walk on the line as described in the previous section. We then attach the root of the game tree to one vertex on the line, and try to send a wave through it. Incredibly, whether this wave is reflected or transmitted depends on the logical value of the tree. If the root is `true` and the current player has a winning strategy, the wave bounces off—and if the root is `false`, it passes through.

To see how this works, let's start with the Hamiltonian (15.50) for the walk on the line, and attach an object to the origin. In the simplest case, this object consists of a self-loop as shown in Figure 15.14, giving an amplitude α with which the quantum walk stays at the origin instead of moving left or right. This gives

$$H = \begin{pmatrix} \ddots & & & & \\ & 0 & 1/2 & & \\ & 1/2 & \alpha & 1/2 & \\ & & 1/2 & 0 & \\ & & & & \ddots \end{pmatrix}. \tag{15.60}$$

Before this object came along, there were two eigenvectors with the same eigenvalue $\lambda = \cos k$, namely $e^{\imath kx}$ and $e^{-\imath kx}$, corresponding to right-moving and left-moving waves respectively. But now if we send a right-moving wave through the origin, part of it will be transmitted, and part of it will be reflected as a left-moving wave. We can write the combination of all three waves as a single state,

$$v(x) = \begin{cases} e^{\imath kx} + Re^{-\imath kx} & x \le 0 \\ Te^{\imath kx} & x \ge 0, \end{cases} \tag{15.61}$$

where R and T are the reflection and transmission amplitudes respectively. We then demand that this combined state be an eigenvector of the new Hamiltonian, so that it constitutes a steady flow of reflection and transmission consistent with Schrödinger's equation. There is one such eigenvector for each frequency k.

Requiring that v is an eigenvector lets us solve for R and T as a function of the self-loop amplitude α. First, taking $x = 0$ in (15.61) gives

$$R = T - 1.$$

Then applying the eigenvalue equation $Hv = \lambda v$ at the origin gives

$$\alpha v_0 + \frac{v_1 + v_{-1}}{2} = \lambda v_0. \tag{15.62}$$

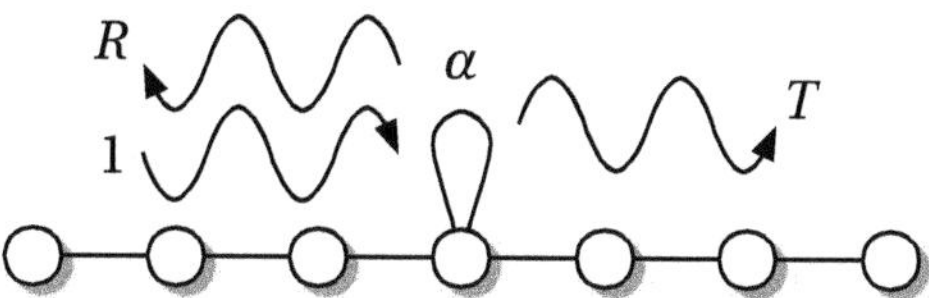

FIGURE 15.14: Adding a self-loop to the Hamiltonian at the origin. A wave entering from the left is transmitted with amplitude T and reflected with amplitude R, giving the eigenvector (15.61).

Substituting (15.61) gives

$$\alpha T + T e^{\imath k} - \imath \sin k = T \cos k \,,$$

and solving for T gives

$$T = \frac{1}{1 - \imath \alpha / \sin k} \,.$$

The probability with which the wave is transmitted is then

$$|T|^2 = \frac{1}{1 + \alpha^2 / \sin^2 k} \,.$$

When $\alpha = 0$ and the self-loop is absent, $|T|^2 = 1$ and the origin is transparent. As α increases, $|T|^2$ decreases and the self-loop becomes more and more reflective, until at $\alpha = \infty$ we have $|T|^2 = 0$ and the wave is reflected completely.

What happens when we scatter off a more complicated object? Suppose that the origin is attached to a "tadpole," an edge with amplitude β leading to a vertex that has a self-loop with amplitude α. Call this additional vertex u. Applying the eigenvalue equation $Hv = \lambda v$ at u and at the origin gives

$$\alpha v_u + \beta v_0 = \lambda v_u$$
$$\beta v_u + \frac{v_1 + v_{-1}}{2} = \lambda v_0 \,.$$

Eliminating v_u, we get

$$\frac{\beta^2}{\lambda - \alpha} v_0 + \frac{v_1 + v_{-1}}{2} = \lambda v_0 \,.$$

But this is exactly the equation (15.62) for a self-loop of amplitude α', where

$$\alpha' = \frac{\beta^2}{\lambda - \alpha} \,. \tag{15.63}$$

In general, for any object consisting of a tree with self-loops on its leaves, this "tadpole rule" lets us contract a branch of the tree and replace it with an equivalent self-loop as in Figure 15.15. If there are two tadpoles with $\beta = 1$ and self-loops α_1 and α_2, we can contract them both and add the resulting amplitudes together. This gives a single self-loop with amplitude

$$\alpha' = \frac{1}{\lambda - \alpha_1} + \frac{1}{\lambda - \alpha_2} \,. \tag{15.64}$$

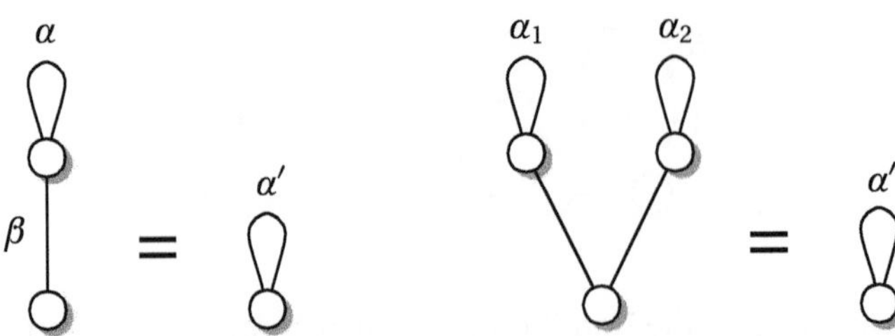

FIGURE 15.15: The tadpole rule (15.63) contracts a leaf of the tree into a self-loop. With (15.64), we can contract two leaves at once.

Note that α' depends on λ, so the entire object might be much more reflective for one eigenvector than for another.

What does all this have to do with NAND trees? Let's set $\lambda = 0$. Since $\lambda = \cos k$, this means that $k = \pi/2$. Then (15.64) gives

$$\alpha' = -\frac{1}{\alpha_1} - \frac{1}{\alpha_2}.$$

Thus α' is infinite if either α_1 or α_2 is zero, and $\alpha' = 0$ if $\alpha_1 = \alpha_2 = \infty$. But if ∞ and 0 represent `true` and `false` respectively, this means that α' is `true` if and only if at least one of α_1 and α_2 is `false`. Thus for the eigenvector with $\lambda = 0$, (15.64) acts exactly like a NAND gate with inputs α_1, α_2 and output α'.

This lets us treat the NAND tree as shown in Figure 15.16. All edges have weight $\beta = 1$. To get things started, a `false` leaf is a vertex without a self-loop, i.e., with $\alpha = 0$. A `true` leaf is connected to an additional node without a self-loop, which according to (15.63) is equivalent to a self-loop with $\alpha = \infty$. By repeatedly applying (15.64), we can contract the entire tree down to a self-loop with amplitude α at the origin. If the tree evaluates to `false`, then $\alpha = 0$ and it transmits the wave. If it is `true`, then $\alpha = \infty$ and it reflects.

How can we turn all this into an algorithm? We can't literally bounce the eigenvector with eigenvalue $\lambda = 0$ off the tree, since this would require a state distributed over the infinite line. Given finite resources, we can only create a *wave packet* of some finite length L:

$$\Psi(x) = \begin{cases} \frac{1}{\sqrt{L}}\, e^{\imath \pi x/2} & -L < x \le 0 \\ 0 & \text{otherwise}. \end{cases}$$

This packet moves to the right at unit speed, so after running Schrödinger's equation for time $t = L$ we can measure its position and tell whether it has been reflected or transmitted. How large does L, and therefore t, have to be?

According to Heisenberg's uncertainty principle, the width of the wave packet is inversely proportional to the spread of its frequencies in the Fourier basis. Specifically, Problem 15.51 shows that in the Fourier basis, a constant fraction of the probability lies on eigenvectors such that $|\lambda| \le \pi/L$. Thus both the width of the packet and the running time of the algorithm are inversely proportional to the largest $|\lambda|$ for which the NAND tree still works—that is, the largest $|\lambda|$ such that contracting the tree according to (15.64) leads to a clear difference in α, depending on whether it is `true` or `false`.

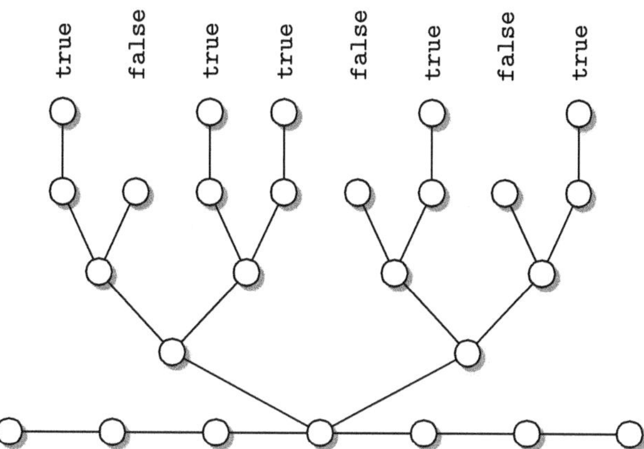

FIGURE 15.16: We can treat the NAND tree as a graph, in which true leaves have an additional vertex attached to them.

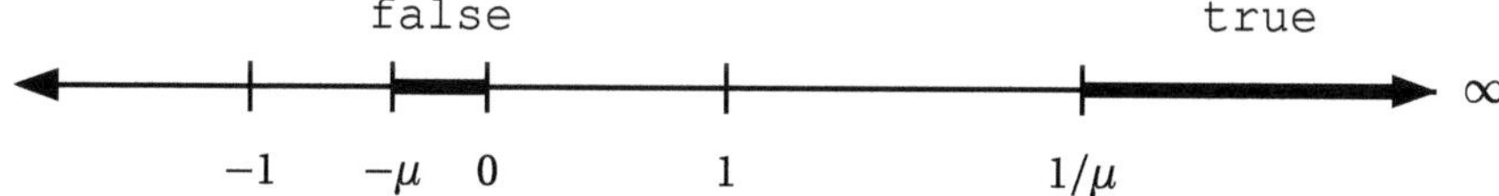

FIGURE 15.17: When $\lambda > 0$, the intervals $[-\mu, 0]$ and $[1/\mu, +\infty]$ correspond to `false` and `true` values respectively, for some $\mu > 0$ which changes from one level of the tree to the next.

How small does $|\lambda|$ have to be? Let's take $\lambda > 0$, since the case $\lambda < 0$ is similar. Equation (15.64) still acts like a NAND gate, but the logical values `true` and `false` no longer correspond exactly to α being infinite or zero. Instead, α will be in one of the two intervals shown in Figure 15.17,

$$\begin{aligned} -\mu \leq \alpha \leq 0 \quad & \text{for a } \texttt{false} \text{ node} \\ \alpha \geq 1/\mu \quad & \text{for a } \texttt{true} \text{ one,} \end{aligned}$$

for some small $\mu > 0$. For instance, contracting a `true` leaf with the tadpole rule gives a self-loop with $\alpha = 1/\lambda$, so initially we have $\mu = \lambda$. Each time we use (15.64), μ gets slightly larger, making these intervals wider. Our goal is to bound how quickly μ grows, so we can still distinguish `true` from `false` even after contracting the entire tree.

Let's assume that this correspondence holds with some value of μ for the inputs to a NAND gate, and see with what value μ' it holds for the output. For convenience, we assume that there is some $\varepsilon > 0$ such that at every level of the tree we have

$$\lambda, \mu \leq \varepsilon .$$

There are three cases to check. First, suppose that both inputs are `true`, so that $\alpha_1, \alpha_2 \geq 1/\mu$. Then the output should be `false`, so $\alpha' \geq -\mu'$.

Looking at (15.64), we see that α' is a decreasing function of both α_1 and α_2. So we can get a lower bound on α', and thus an upper bound on μ', by setting $\alpha_1 = \alpha_2 = 1/\mu$. Plugging these values into (15.64) gives

$$\alpha' \geq \frac{2}{\lambda - 1/\mu} = -\mu' \quad \text{where} \quad \mu' = \frac{2\mu}{1-\lambda\mu} \leq \frac{2\mu}{1-\varepsilon^2}.$$

Similarly, if both inputs are `false`, we set $\alpha_1 = \alpha_2 = -\mu$. The output should be `true`, and so

$$\alpha' \geq \frac{2}{\lambda+\mu} = -\frac{1}{\mu'} \quad \text{where} \quad \mu' = \frac{\mu+\lambda}{2}.$$

Finally, if one leaf is `true` and the other is `false`, we set $\alpha_1 = 1/\mu$ and $\alpha_2 = -\mu$. The output should again be `true`, giving

$$\alpha' \geq \frac{1}{\lambda - 1/\mu} + \frac{1}{\lambda+\mu} = -\frac{1}{\mu'} \quad \text{where} \quad \mu' = \frac{(\mu+\lambda)(1-\lambda\mu)}{1-2\lambda\mu-\mu^2} \leq \frac{\mu+\lambda}{1-3\varepsilon^2}.$$

Looking over these equations, we see that the first case is the most dangerous—if both leaves are `true`, then μ roughly doubles. Happily, in this case the output of the NAND gate is `false`, so this can only occur at half of the nodes along any path from a leaf to the root. So, we expect that μ doubles once for every two levels of the tree. In other words, if μ_j denotes the value of μ after we have contracted j levels of the tree, we expect that μ_{j+2} is roughly twice μ_j.

To confirm this, let's take the worst case of the above equations, in which every `true` node has one `true` input and one `false` one. Note that this is also the worst case for the classical randomized algorithm, as we discussed in Section 10.5. Applying this to a tree of depth two and simplifying gives

$$\mu_{j+2} \leq \frac{2}{1-6\varepsilon^2}(\mu_k + \lambda) \leq (2+\varepsilon)(\mu_j + \lambda). \tag{15.65}$$

We assumed in the second inequality that $\varepsilon \leq 0.08$, but this is fine since we can take ε to be any small constant. Taking the initial value $\mu_0 = \lambda$ and iterating (15.65) implies the following upper bound,

$$\mu_j \leq 3(2+\varepsilon)^{j/2}\lambda. \tag{15.66}$$

Exercise 15.26 *Confirm* (15.65). *Then solve for* μ_j *exactly and confirm* (15.66).

Now suppose that the tree has N nodes and depth $k = \log_2 N$. Setting $j = k$, we find that at the root of the tree we have

$$\mu_k \leq 3(2+\varepsilon)^{k/2}\lambda = 3N^{\frac{1}{2}\log_2(2+\varepsilon)}\lambda \leq 3N^{\frac{1}{2}+\varepsilon}\lambda,$$

where we used the generous upper bound $\log_2(2+\varepsilon) \leq 1+2\varepsilon$. If we want to ensure that $\mu_k \leq \varepsilon$, we have to take

$$\lambda \leq \frac{\varepsilon}{3}N^{-(\frac{1}{2}+\varepsilon)} = \Theta\left(N^{-(\frac{1}{2}+\varepsilon)}\right).$$

Therefore, the length of the wave packet, and the running time of our algorithm, obey

$$t \sim L = \Theta\left(N^{\frac{1}{2}+\varepsilon}\right).$$

By making ε arbitrarily small, we can make this exponent as close to $1/2$ as we like, with a slight cost in the leading constant. In fact, our analysis here was deliberately crude—with a more careful calculation, one can get rid of ε completely, and obtain an algorithm that runs in time $\Theta(N^{1/2})$.

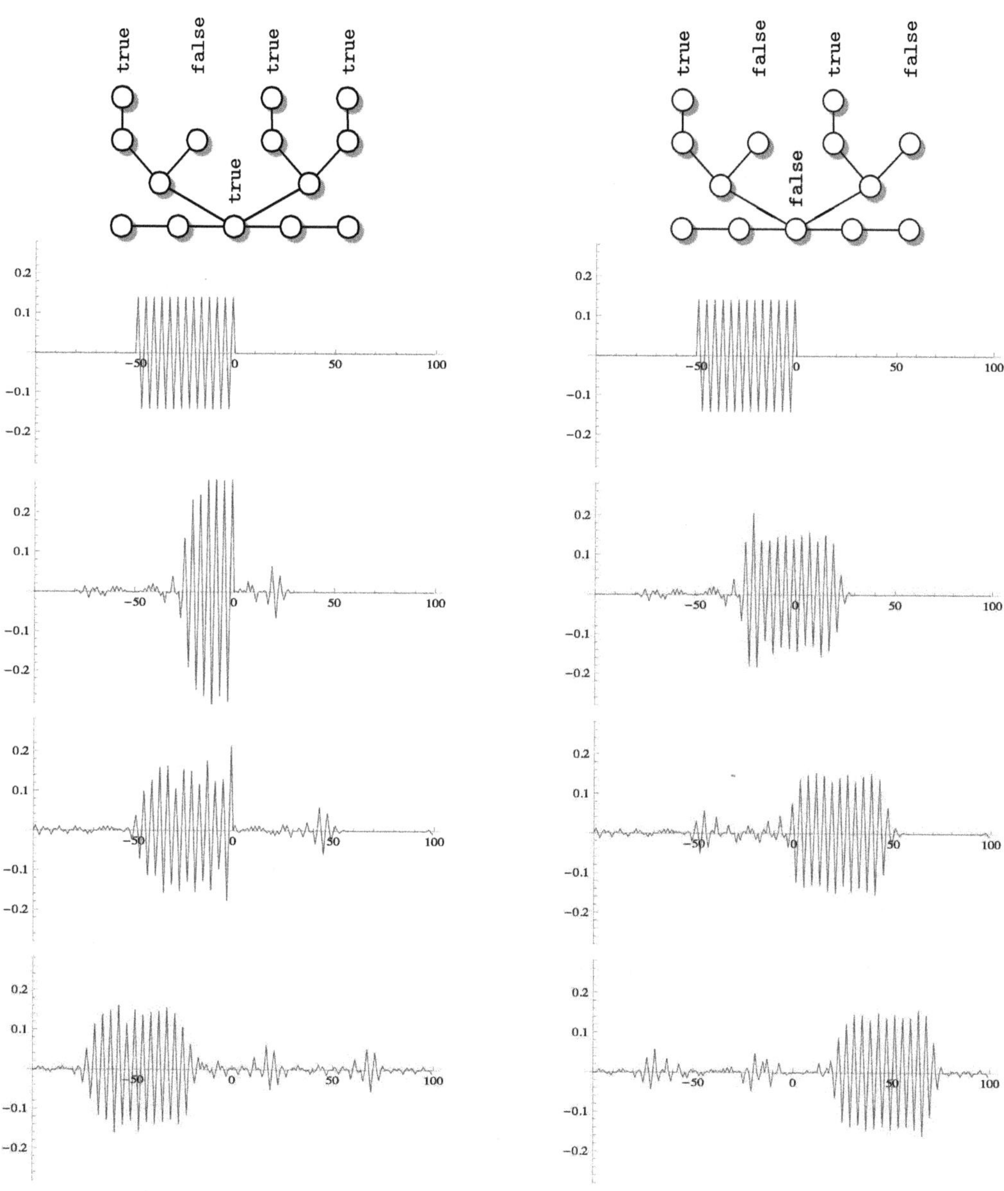

FIGURE 15.18: The Farhi–Goldstone–Gutmann algorithm in action. On the left, the NAND tree is `true`, and the wave packet is reflected. On the right, it is `false`, and the packet is transmitted. The width of the packet is $L = 50$. Top to bottom, we take snapshots at $t = 0, 25, 50$, and 75. We plot the real part of the amplitude in order to show the phase oscillations.

We show an example of this algorithm at work in Figure 15.18. On the left, the tree is `true`, and the wave packet is reflected. On the right, we change the tree to `false` by altering the truth value of one leaf, and the packet is transmitted instead.

How does this algorithm compare with classical algorithms for the NAND tree? We showed in Section 10.5 that the best possible randomized algorithm queries $\Theta(N^{0.753})$ leaves. Comparing the number of queries with the running time of a continuous-time algorithm may seem like comparing apples and oranges. However, we can think of this quantum algorithm as querying a *Hamiltonian oracle*, in which case the number of queries is proportional to t. Moreover, it is possible to transform this continuous-time algorithm into a discrete-time one carried out by a quantum circuit, in which the number of queries is again $\Theta(N^{1/2})$. While we focused here on balanced binary trees, this algorithm can also be generalized to evaluate any tree or Boolean formula, with essentially the same running time.

15.26 The Farhi–Goldstone–Gutmann algorithm shows that, in the right setting, even the simplest physical processes have surprising computational power. We have been thinking about quantum computing for less than twenty years, and there is every reason to believe that new algorithmic ideas will continue to emerge. As they do, we will learn more about the nature of complexity, and about the computational power of the universe in which we live.

Problems

> You have nothing to do but mention the quantum theory, and people will take your voice for the voice of science, and believe anything.
>
> George Bernard Shaw

15.1 Inner products. Show that if $\langle v|U^\dagger U|v\rangle = \langle v \mid v\rangle$ for all vectors $|v\rangle$, then $\langle v|U^\dagger U|w\rangle = \langle v \mid w\rangle$ for all $|v\rangle$ and $|w\rangle$, and therefore that $U^\dagger U = \mathbb{1}$.

15.2 Simulating complex numbers. Consider the following mapping from complex numbers to real 2×2 matrices:

$$c = re^{\imath\theta} \to r\begin{pmatrix} \cos\theta & -\sin\theta \\ \sin\theta & \cos\theta \end{pmatrix} = \begin{pmatrix} \operatorname{Re} c & -\operatorname{Im} c \\ \operatorname{Im} c & \operatorname{Re} c \end{pmatrix}.$$

Show that this mapping respects addition, multiplication, and inverses (except for zero). Formally, it is an isomorphism from the field $\mathbb{C}$ to the real-valued, invertible 2×2 matrices. Conclude that n-dimensional unitary matrices can be represented by real-valued $2n$-dimensional ones. Thus complex numbers, strictly speaking, are not necessary to quantum mechanics—but negative numbers are.

15.3 The square root of NOT. Construct a unitary matrix V such that $V^2 = X$. Such a V can be called the "square root of NOT," since it switches a qubit's truth value if we apply it twice. From a classical point of view, this is quite strange!

15.4 Building three-qubit gates. We can define controlled-U gates acting on three qubits $|a,b,c\rangle$, which apply U to the target qubit c only if both a and b are true. Describe these as 8×8 matrices. Then show that we can simulate these with two-qubit gates as follows. If $V^2 = U$,

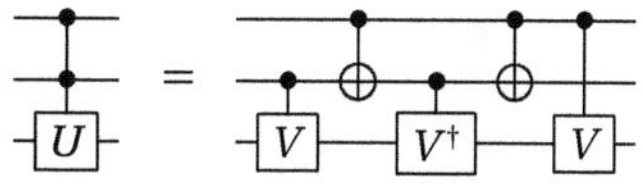

In particular, if V is a square root of X as in Problem 15.3, this gives us a *Toffoli gate*, which flips c if both a and b are true. Interestingly, in classical reversible computation we need three-bit gates like these in order to achieve universality.

15.6

15.5 No cloning. Suppose that we have a unitary operator which makes copies of two different quantum states $|v\rangle$ and $|w\rangle$. In other words, for some "blank" state $|b\rangle$, we have

$$U(|v\rangle \otimes |b\rangle) = |v\rangle \otimes |v\rangle$$
$$U(|w\rangle \otimes |b\rangle) = |w\rangle \otimes |w\rangle .$$

Show that $\langle v|w\rangle$ is either 1 or 0, so $|v\rangle$ and $|w\rangle$ are either identical or orthogonal.

15.6 Commuting gates in parallel. Show that if D_1 and D_2 are diagonal matrices, we can apply them to a qubit simultaneously by first "copying" it with a controlled-NOT, and then "uncopying" it:

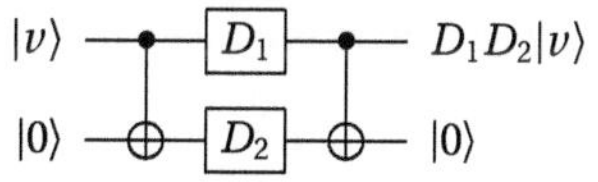

More generally, show that if U_1 and U_2 are unitary operators which commute with each other, we can perform both simultaneously by first applying a basis change that diagonalizes them.

15.7 Correlations between Alice and Bob. Suppose that A and B are Hermitian operators, and that they commute with each other so that they can be measured simultaneously. Show that their product AB is also Hermitian. Conclude that, given a state $|\psi\rangle$, the covariance of the observed eigenvalues of A and B is given by

$$\mathbb{E}[AB] = \langle\psi|AB|\psi\rangle .$$

Now, suppose that $|\psi\rangle$ is the entangled pair $\frac{1}{\sqrt{2}}(|00\rangle + |11\rangle)$. Show in this case that if $A_1 = A \otimes \mathbb{1}$ and $B_2 = \mathbb{1} \otimes B$, then

$$\mathbb{E}[A_1 B_2] = \frac{1}{2}\,\mathrm{tr}(AB^T) .$$

15.8 Alice and Bob at an angle. Consider the operator

$$S^\theta = (\cos\theta)Z + (\sin\theta)X = \begin{pmatrix} \cos\theta & \sin\theta \\ \sin\theta & -\cos\theta \end{pmatrix} .$$

Geometrically, this operator measures the spin of a particle around an axis that lies in the x-z plane and is an angle θ away from the z axis. For instance, $S^{\pi/4} = H$, $S^{\pi/2} = X$, and $S^{3\pi/4} = H' = XHX$. Show that the eigenvectors of S^θ are

$$|v_+\rangle = \left(\cos\frac{\theta}{2}\right)|0\rangle + \left(\sin\frac{\theta}{2}\right)|1\rangle \text{ and } |v_-\rangle = \left(\sin\frac{\theta}{2}\right)|0\rangle - \left(\cos\frac{\theta}{2}\right)|1\rangle .$$

Now suppose that Alice and Bob share the entangled state $|\psi\rangle = \frac{1}{\sqrt{2}}(|00\rangle + |11\rangle)$ and that they measure their qubits in bases that are an angle θ apart. Using Problem 15.7, show that the correlation between their measurements is

$$\mathbb{E}[Z_A S_B^\theta] = \cos\theta ,$$

and thus verify the claim in Section 15.3.2 that

$$\mathbb{E}[Z_A H_B] = \mathbb{E}[X_A H_B] = \mathbb{E}[X_A H'_B] = -\mathbb{E}[Z_A B'_2] = \frac{1}{\sqrt{2}}\,.$$

15.9 Maximally entangled in any basis. Let v_+ and v_- be the eigenvectors of S^θ defined in Problem 15.8, and let $\psi = \frac{1}{\sqrt{2}}(|00\rangle + |11\rangle)$. Generalize Exercise 15.13 by showing that, for all θ,

$$\frac{1}{\sqrt{2}}(|v_+ v_+\rangle + |v_- v_-\rangle) = |\psi\rangle\,.$$

More generally, show that if U is any real-valued one-qubit unitary operator, applying U to both qubits leaves $|\psi\rangle$ unchanged up to an overall phase:

$$(U \otimes U)|\psi\rangle = e^{\imath\phi}|\psi\rangle\,,$$

for some phase angle ϕ.

15.10 Singlet symmetries. Many discussions of the EPR paradox involve the "singlet" state

$$|\psi\rangle = \frac{1}{\sqrt{2}}(|01\rangle - |10\rangle)\,,$$

in which the two qubits have opposite spin and there is a relative phase of -1 between the two terms. This state has even more symmetry than the entangled pair we used in the text. Show that for any one-qubit unitary operator U, we have

$$(U \otimes U)|\psi\rangle = e^{\imath\phi}|\psi\rangle\,,$$

for some ϕ. It follows that $|\psi\rangle$ is completely basis-independent—in physics parlance, its total angular momentum is zero. What is ϕ?

15.11 Tsirelson's inequality. In Section 15.3.2, Alice measures her qubit with Z_A or X_A, and Bob measures his with H_B or H'_B. The quantity $W = Z_A H_B + X_A H_B + X_A H'_B - Z_A H'_B$ then has an expectation $\mathbb{E}[W] = 2\sqrt{2}$, greater than classical random variables could attain. By choosing different pairs of bases, can we design an experiment in which $\mathbb{E}[W]$ is even larger?

In fact, this degree of correlation is as large as possible. Given two operators A and B, define their *commutator* as $[A, B] = AB - BA$. Now let A, A', B, B' be Hermitian operators such that $A^2 = (A')^2 = B^2 = (B')^2 = \mathbb{1}$, and such that $[A, B] = [A', B] = [A', B'] = [A, B'] = 0$. If we define

$$W = AB + A'B + A'B' - AB'\,,$$

show that W is Hermitian, and that

$$W^2 = 4 + [A, A'][B, B']\,. \tag{15.67}$$

Now define the norm of an operator M as the maximum, over all vectors v, of how much M increases v's length:

$$\|M\| = \max_v \frac{\langle v|M|v\rangle}{\langle v|v\rangle}\,.$$

Prove that the norm obeys the following inequalities,

$$\|M_1 M_2\| \le \|M_1\|\,\|M_2\| \quad \text{and} \quad \|M_1 + M_2\| \le \|M_1\| + \|M_2\|\,.$$

Then show that if M is is Hermitian, $\|M\|$ is just the maximum of $|\lambda|$ over its eigenvalues λ (we also used this fact in Problem 12.39) and that $\left\|M^2\right\| = \|M\|^2$.

Combine these facts with (15.67) to prove *Tsirelson's inequality*,

$$\mathbb{E}[W] \leq 2\sqrt{2}.$$

Thus Alice and Bob's experiment violates Bell's inequality as much as possible.

15.12 Eigenbells. We know that X and Z do not commute. However, show that the two-qubit operators $X \otimes X$ and $Z \otimes Z$ *do* commute. If two operators commute, they have the same eigenvalues (prove this if you don't already know it!) So, find four eigenvectors $|v\rangle$ such that $(X \otimes X)|v\rangle = \pm|v\rangle$ and $(Z \otimes Z)|v\rangle = \pm|v\rangle$, with all four combinations of eigenvalues.

15.13 Circuits for the Bell basis. Design a two-qubit circuit for the unitary operator U that converts the Bell basis states $|\psi_{a,b}\rangle$ defined in Section 15.3.4 to the computational basis states $|a,b\rangle$. You may use H, the Pauli matrices, and the controlled-NOT.

15.14 Communicating with cobits. If we apply a controlled-NOT from a control qubit $|v\rangle$ to a target qubit in the state $|0\rangle$, we create the entangled state $v_0|00\rangle + v_1|11\rangle$. Let us say that the second qubit is a "coherent copy" of the first.

Now suppose that Alice and Bob have the ability to carry out this operation, where Alice has the control qubit and Bob has the target qubit (and where Bob's qubit is required to start in the state $|0\rangle$). Let us say that Alice sends Bob a "coherent bit" or "cobit" each time they do this.

Clearly, a cobit is at least as useful for communication as an ebit, since Bob's half of the entangled pair $|\psi\rangle$ is simply a coherent copy of Alice's qubit $|+\rangle$. Similarly, a qubit is at least as useful as a cobit, since Alice could make a coherent copy of her own and send it to Bob. Thus

$$1 \text{ ebit} \leq 1 \text{ cobit} \leq 1 \text{ qubit}.$$

In this problem and the next, we will show that the power of a cobit is exactly half-way in between these two! In other words,

$$1 \text{ cobit} = \frac{1 \text{ ebit} + 1 \text{ qubit}}{2}.$$

First, recall that in superdense coding, Alice applies X or Z to her qubit depending on whether two classical bits, a and b, are true. Suppose instead that, in addition to her half of the entangled pair, Alice has two qubits $|a\rangle$ and $|b\rangle$, and that she uses them to apply a controlled-X gate and a controlled-Z gate to her entangled qubit before sending it to Bob. Show that if Bob now applies the same basis change U as in Problem 15.13, mapping the Bell basis to the computational basis, these two qubits become coherent copies of $|a\rangle$ and $|b\rangle$. Thus Alice has sent Bob two cobits.

In terms of circuits, if Alice and Bob start with an entangled pair $|\psi\rangle = \frac{1}{\sqrt{2}}(|00\rangle + |11\rangle)$, these two circuits generate the same state:

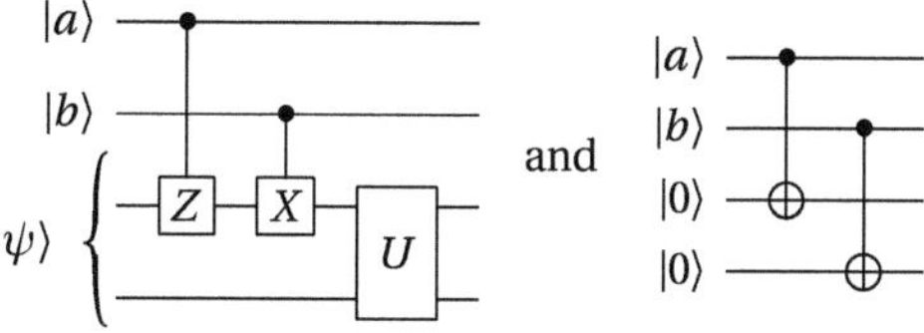

From this we can conclude that

$$1 \text{ qubit} + 1 \text{ ebit} \geq 2 \text{ cobits}. \tag{15.68}$$

15.15 Cobits, conversely. As a converse to the previous problem, consider the coherent version of quantum teleportation. Again, Alice and Bob have an entangled pair $|\psi\rangle = \frac{1}{\sqrt{2}}(|00\rangle + |11\rangle)$ and Alice has an additional qubit $|v\rangle$. However, rather than measuring her two qubits in the Bell basis and sending Bob the resulting classical bits, she transforms the Bell basis $|\Psi_{a,b}\rangle$ to the computational basis $|a,b\rangle$ by applying the operator U from Problem 15.13.

She then sends two cobits, applying controlled-NOTs from a to b to two additional qubits that Bob has prepared in the state $|0\rangle$. Now that these qubits contain coherent copies of a and b, Bob applies controlled-X and controlled-Z gates from them to his half of the entangled pair, disentangling it from the other qubits and putting it in the state $|v\rangle$. As a circuit, this looks like

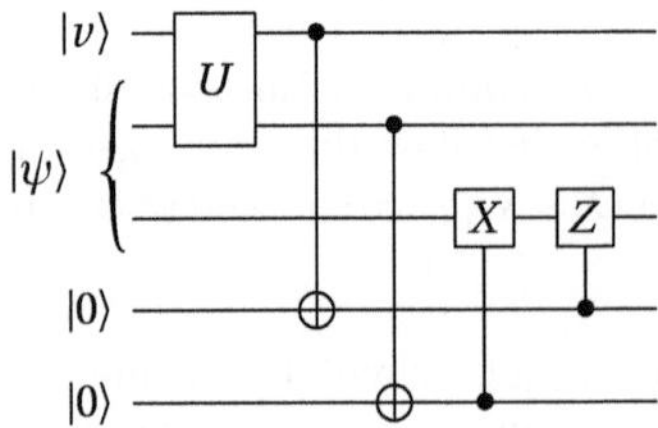

Show that when we're done, $|v\rangle$ has been teleported to Bob, but each of the qubits that started in the state $|0\rangle$ now forms an entangled pair $|\psi\rangle$ with one of Alice's qubits. So, we have sent a qubit using an ebit and two cobits, but we have two ebits left over! This gives

$$1 \text{ ebit} + 2 \text{ cobits} \geq 1 \text{ qubit} + 2 \text{ ebits}.$$

We can recycle one of these ebits, using it over and over again as a "catalyst" to convert 2 cobits to an qubit and an ebit. Asymptotically, when we repeat this protocol many times, this gives

$$2 \text{ cobits} \geq 1 \text{ qubit} + 1 \text{ ebit}. \tag{15.69}$$

Along with (15.68), this proves that

$$1 \text{ cobit} = \frac{1 \text{ ebit} + 1 \text{ qubit}}{2}.$$

15.16 Eve listens in. Suppose that Alice is trying to communicate with Bob using superdense coding, but that Eve has intercepted Alice's qubit on its way to Bob. Can Eve tell what bits Alice is trying to send—that is, whether Alice applied $\mathbb{1}$, Z, X, or ZX to her qubit before sending it? Note that this qubit is still entangled with Bob's.

15.17 Cloning violates relativity. Argue that if Alice could clone her half of an entangled pair of qubits, then Alice and Bob could communicate faster than light.

15.18 Deutsch's algorithm. In his 1985 paper, Deutsch gave a different algorithm for his problem than the one we describe in Section 15.4.1. We start with $|x\rangle$ in the state $|+\rangle$ and $|y\rangle$ in the state $|0\rangle$, and generate the state on the right-hand side of (15.8). Now suppose that we measure $|y\rangle$ in the X-basis. Show that if we observe $|-\rangle$, we can then measure $|x\rangle$ in the X-basis and learn whether $f(0) = f(1)$ or not.

On the other hand, show that if we observe $|+\rangle$ when measuring $|y\rangle$, nothing further can be learned. Thus this algorithm succeeds with probability 1/2, and the expected number of queries is 2, no better than the classical case.
15.11

15.19 Solving Bernstein–Vazirani classically. Show that in order to solve the Bernstein–Vazirani problem classically, determining $\mathbf{k}$ from $f(\mathbf{x}) = \mathbf{k} \cdot \mathbf{x}$, n queries are both necessary and sufficient. Hint: how many functions $f(\mathbf{x})$ of this form are there?

15.20 Spanning Simon. Let V be a d-dimensional subspace of the set of n-dimensional vectors whose components are integers mod 2. I select vectors $\mathbf{k}_1, \mathbf{k}_2, \ldots, \mathbf{k}_t$ independently at random from V, until they span all of V. Show that the number of vectors I need is roughly d in the following three senses. First, show that the probability that t such vectors span V is at least $1-2^{d-t}$, which is $1-o(1)$ when, say, $t = d + \log d$. Second, show that there is a constant

$$c = \sum_{i=1}^{\infty} \frac{1}{2^i - 1} < 1.607\,,$$

such that the expected number of vectors I need is $t = d + c$. Third, show that even when t is exactly d, these vectors span the space with probability at least

$$\prod_{i=1}^{\infty} \left(1 - \frac{1}{2^i}\right) > 0.288\,.$$

Hint: for the first part, calculate the expected number of vectors in V that are perpendicular to all the **k**s, and use the first moment method from Appendix A.3.2. For the second and third parts, calculate the probability p that the next **k** we choose is not in the subspace spanned by the previous ones, if the previous ones span a subspace of dimension $d - i$.

15.21 Simon's birthday. Consider a classical algorithm for Simon's problem, where we seek to learn **y** by querying $f(\mathbf{x})$ for a series of t inputs **x**. Show that if **y** is chosen randomly, with high probability it will take $t = \Theta(2^{n/2})$ queries before we find a pair of inputs $\mathbf{x}, \mathbf{x}'$ with the same value of f, no matter what method we use to choose the inputs **x**. Moreover, if $t = 2^{an}$ for some $a < 1/2$, there are exponentially many possible values of **y** that are consistent with what we have seen so far.

Hint: use the first moment method of the Birthday Problem from Appendix A.3.3. Why is it enough to assume that **y** is random, even if the inputs **x** aren't?

15.22 Continued fractions. Consider the continued fraction

$$x = c_0 + \cfrac{1}{c_1 + \cfrac{1}{c_2 + \cdots}}\,.$$

Let p_n/q_n denote its nth convergent. Prove that for $n \geq 2$, these numerators and denominators are given by the recurrences

$$\begin{aligned} p_n &= c_n p_{n-1} + p_{n-2} \\ q_n &= c_n q_{n-1} + q_{n-2}\,. \end{aligned} \tag{15.70}$$

with appropriate initial values. Use this to show that

$$C_n := \begin{pmatrix} p_n & p_{n-1} \\ q_n & q_{n-1} \end{pmatrix} = \begin{pmatrix} c_0 & 1 \\ 1 & 0 \end{pmatrix} \begin{pmatrix} c_1 & 1 \\ 1 & 0 \end{pmatrix} \cdots \begin{pmatrix} c_n & 1 \\ 1 & 0 \end{pmatrix}. \tag{15.71}$$

Finally, prove the following properties of the convergents:

1. C_n is unimodular, i.e., $\det C_n = \pm 1$,
2. the sequence of denominators q_n is monotonically increasing, and
3. $a_n = q_n x - p_n$ is an alternating sequence and $|a_n|$ is monotonically decreasing.

15.23 Rational approximations. The convergents of continued fractions are the best rational approximations to a real number x in the following sense. A rational p/q with $\gcd(p,q)=1$ is called a *locally best approximation* to x if

$$|qx-p| < |q'x-p'|$$

for all rationals $p'/q' \neq p/q$ with $q' \leq q$. Prove that p/q is a locally best approximation to x if and only if it is a convergent of x.

Hint: you will need the properties of the convergents we showed in the previous problem. In particular, unimodularity implies that any integer vector (p,q) can be written as an integer linear combination of (p_n,q_n) and (p_{n-1},q_{n-1}). Start by proving that a convergent is a locally best approximation. The converse is easily proven by contradiction.

15.24 Unusually good approximations. Prove Theorem 15.2, i.e., show that if p and q are integers such that

$$\left|x-\frac{p}{q}\right| < \frac{1}{2q^2}, \tag{15.72}$$

then p/q is a convergent of the continued fraction expansion of x. Hint: show that (15.72) implies that p/q is a locally best approximation to x as defined in Problem 15.23.

15.25 Euler's totient. Suppose the prime factorization of N is

$$N = p_1^{t_1} p_2^{t_2} \cdots p_\ell^{t_\ell}.$$

Show that $\varphi(N)$, the number of integers between 1 and $N-1$ which are mutually prime to N, is given by the following generalization of (15.18):

$$\varphi(N) = N\left(1-\frac{1}{p_1}\right)\cdots\left(1-\frac{1}{p_\ell}\right) = (p_1-1)p_1^{t_1-1}\cdots(p_\ell-1)p_\ell^{t_\ell-1}. \tag{15.73}$$

15.26 More about the totient. Prove (15.32), i.e., show that there is a constant $C>0$ such that, for any N,

$$\frac{\varphi(N)}{N} \geq \frac{C}{\log N}.$$

Then use Problem 10.24 to provide a "back of the envelope" argument for the stronger bound (15.33),

$$\frac{\varphi(N)}{N} \geq \frac{C'}{\log\log N}.$$

Hint: for the first part, start from (15.73). Then divide N's prime divisors into two groups, the small primes $p \leq \log_2 N$ and the large primes $p > \log_2 N$. Use the fact (Exercise 4.21) that the number of distinct prime factors is at most $\log_2 N$ to show that the product of $1-1/p$ over the large divisors is bounded from below by a constant. For the small primes, pessimistically assume that *every* integer up to $\log N$ is a divisor.

15.27 Another way to break RSA with the totient. Suppose that we can compute Euler's totient function φ. Show that another way we can obtain Alice's decryption key d from e and N, and thus solve RSA KEYBREAKING, is given by

$$d = e^{\varphi(\varphi(N))-1} \bmod \varphi(N).$$

15.28 RSA KEYBREAKING is as hard as FACTORING. Suppose that we can solve RSA KEYBREAKING. Show that given the encryption and decryption keys e and d, we can find a nontrivial square root of 1 in randomized polynomial time, and therefore a proper divisor of N, as long as N is divisible by two distinct odd primes. Therefore, FACTORING $\leq_{\mathrm{RP}}$ RSA KEYBREAKING.

Hint: since $ed \equiv_{\varphi(N)} 1$, we know that $ed - 1 = k\varphi(N)$ for some k. By the generalized Little Theorem (15.17), this means that $c^{ed-1} \equiv_N 1$ for all $c \in \mathbb{Z}_N^*$. As in the Miller–Rabin test for PRIMALITY, we can consider the sequence $c^{(ed-1)/2}, c^{(ed-1)/4}, \ldots$, up to $c^{(ed-1)/2^s}$, where 2^s is the largest power of 2 that divides $\varphi(N)$. The first element of this sequence that differs from 1 is a square root of 1. Using an argument analogous to Problem 10.40, show that it is a nontrivial square root with probability at least $1/2$.

15.29 Shor's algorithm works at least half the time. Prove Theorem 15.1. The following hints will guide you through a proof. First, let us assume that N is odd and can be factored as follows:

$$N = q_1 q_2 \cdots q_\ell ,$$

where the q_i are distinct odd primes or powers thereof. The Chinese Remainder Theorem (see Appendix A.7.3) states that c is uniquely determined by the set $\{c_i = c \bmod q_i\}$. Moreover, if c is uniformly random in $\mathbb{Z}_N^*$ then each c_i is uniformly random in $\mathbb{Z}_{q_i}^*$.

The order r of c is the smallest $r > 0$ such that, for all i,

$$c_i^r \equiv_{q_i} 1 .$$

For each i, let r_i be the order of c_i in $\mathbb{Z}_{q_i}^*$. Show that r is the lowest common multiple of the r_i. In particular, r is even if any of the r_i are.

Since q_i is a prime or a prime power, $\mathbb{Z}_{q_i}^*$ is cyclic (see Appendix A.7) and $+1$ and -1 are the only square roots of 1. Assuming r is even, it follows that

$$c_i^{r/2} \equiv_{q_i} \pm 1 .$$

Show that c is good as long as $c_i^{r/2} \equiv_{q_i} +1$ for at least one i.

For each i there is a t_i such that 2^{t_i} is the largest power of 2 that divides r_i. Show that c is bad if and only if all these t_i are the same. Finally, show that if c_i is chosen randomly from $\mathbb{Z}_{q_i}$, the probability that t_i takes any particular value is at most $1/2$. Therefore, the probability that all t_i are the same is at most $1/2^{\ell-1}$.

Hint: let G be a cyclic group of order n, and let 2^s be the largest power of 2 that divides N. In our case, $n = \varphi(q_i)$, which is even since q_i is odd. Show that $1/2$ of G's elements have order n, $1/4$ of them have order $n/2$, and so on, until the remainder have odd order $n/2^s$.

15.30 Square roots are hard. Consider the problem of extracting square roots in $\mathbb{Z}_N^*$:

SQUARE ROOT
Input: An integer N and a $y \in \mathbb{Z}_N^*$
Question: An integer x such that $x^2 \equiv_N y$, if there is one

Show that this problem is just as hard as FACTORING, in the sense that

$$\text{FACTORING} \leq_{\mathrm{RP}} \text{SQUARE ROOT} .$$

Show that this is true even if an adversary can choose which square root x to report for each input y. Hint: what happens if you choose $x \in \mathbb{Z}_N^*$ randomly and ask the adversary for the square root of x^2?

15.31 Kickback with qudits. Let $|x\rangle$ represent a "qudit," i.e., a basis vector in a d-dimensional space where $x \in \{0,\ldots,d-1\}$ is an integer mod d. Let $|k\rangle$ represent the Fourier basis vector

$$|k\rangle = \frac{1}{\sqrt{d}} \sum_{x=0}^{d-1} \omega^{kx} |x\rangle\,,$$

where $\omega = e^{2\pi\imath/d}$. Now consider a unitary operator U on two qudits described by the following action on the basis vectors:

$$U|x,y\rangle = |x,y+x\rangle\,.$$

Show that in the Fourier basis, this has the following effect:

$$U|k,\ell\rangle = |k-\ell,\ell\rangle\,.$$

This is a generalization of Exercise 15.9, since setting $d=2$ tells us that a controlled-NOT gate goes the other way in the X-basis.

15.32 Love in Kleptopia. Here is a variation on Diffie–Hellman key exchange. While it is less efficient, it has its own charms.

Alice and Bob publicly agree on a prime p, and they choose random numbers x and y respectively which they keep secret. Alice has a message m that she wants to transmit securely to Bob. Alice sends m^x to Bob; Bob raises this to the yth power, and sends m^{xy} back to Alice; Alice raises this to the x^{-1}th power, and sends m^y back to Bob; and finally Bob raises this to the y^{-1}th power and obtains m. All this exponentiation, of course, is done mod p.

To break this scheme, Eve has to solve the following problem: given m^x, m^{xy}, and m^y for an unknown x and y, determine m. Show that this problem can be reduced to DISCRETE LOG, so Eve can break it if she has a quantum computer.

Caroline Calderbank turned this scheme into a wonderful puzzle, called "Love in Kleptopia":

> Jan and Maria have fallen in love (via the Internet) and Jan wishes to mail her a ring. Unfortunately, they live in the country of Kleptopia where anything sent through the mail will be stolen unless it is enclosed in a padlocked box. Jan and Maria each have plenty of padlocks, but none to which the other has a key. How can Jan get the ring safely into Maria's hands?

At first this seems impossible. Just as Eve can overhear everything Alice says to Bob, the Kleptocrats can steal anything Jan sends to Maria, unless it is locked inside a box that she can't open either. Try this puzzle out on a friend unfamiliar with cryptography!

15.33 The swap test. Derive the probabilities given by (15.39) for observing $|+\rangle$ or $|-\rangle$ in the swap test. Hint: let S be the swap operator, defined by

$$S(|\psi_1\rangle \otimes |\psi_2\rangle) = |\psi_2\rangle \otimes |\psi_1\rangle\,.$$

Now write $|\psi_1\rangle \otimes |\psi_2\rangle$ as the sum of a symmetric and an asymmetric part,

$$|\psi_1\rangle \otimes |\psi_2\rangle = |\psi_{\text{sym}}\rangle + |\psi_{\text{asym}}\rangle\,.$$

where $|\psi_{\text{sym}}\rangle$ and $|\psi_{\text{asym}}\rangle$ are eigenvectors of S with eigenvalues $+1$ and -1 respectively. Show that if we apply the controlled-swap operator, which applies S if the input qubit is true, then $|\psi_1\rangle \otimes |\psi_2\rangle$ goes to

$$\big(|+\rangle \otimes |\psi_{\text{sym}}\rangle\big) + \big(|-\rangle \otimes |\psi_{\text{asym}}\rangle\big)\,.$$

Conclude that if we measure the control qubit, $P(+) = \left|\psi_{\text{sym}}\right|^2$ and $P(-) = \left|\psi_{\text{asym}}\right|^2$. Show that this implies (15.39).

15.34 Trace estimation. Consider the following circuit, where U is a d-dimensional unitary operator and $|\psi\rangle$ is a d-dimensional state:

$|0\rangle$ —[H]—●—[H]—

$|\psi\rangle$ ————[U]————

Show that if we measure the control qubit after applying this circuit, we observe $|0\rangle$ and $|1\rangle$ with probability

$$P(0)=\frac{1}{2}\left(1+\operatorname{Re}\langle\psi|U|\psi\rangle\right) \quad\text{and}\quad P(1)=\frac{1}{2}\left(1-\operatorname{Re}\langle\psi|U|\psi\rangle\right).$$

Now suppose that instead of a pure state $|\psi\rangle$, we give this circuit the completely mixed state, with density matrix $\rho=(1/d)\mathbb{1}$. Show that in this case we have

$$P(0)=\frac{1}{2}\left(1+\frac{1}{d}\operatorname{Re}\operatorname{tr}U\right) \quad\text{and}\quad P(1)=\frac{1}{2}\left(1-\frac{1}{d}\operatorname{Re}\operatorname{tr}U\right).$$

Then modify this scheme slightly so that these probabilities depend on $\operatorname{Im}\operatorname{tr}U$. This gives us a quantum algorithm for estimating the trace of a unitary matrix, where the only quantum resource consists of a single qubit.

Hint: the completely mixed state is equivalent to a random pure state, chosen uniformly from the d-dimensional 15.27
complex sphere, or simply from some arbitrary basis.

15.35 Grover's diffusion. Show that if $N=2^n$, Grover's diffusion operator D can be implemented with poly(n) elementary quantum gates. Feel free to use additional work qubits, as long as you return them to their original state and disentangle them from the other qubits before you finish. Hint: what does D look like in the Fourier basis?

15.36 Flip around the average. In his original papers, Grover described D as an "inversion around the average." Show that if $|\psi\rangle=\sum_j a_j|j\rangle$ and $D|\psi\rangle=\sum_j a'_j|j\rangle$, then

$$a'_j=\overline{a}-(a_j-\overline{a}) \text{ where } \overline{a}=\frac{1}{N}\sum_j a_j.$$

In other words, a'_j is as far above the average amplitude $\overline{a}$ as a_j was below it, or vice versa.

Now suppose there is a unique solution i. Show that as long as $a_i\le 1/\sqrt{2}$, each iteration of DU increases a_i by at least $\sqrt{2/N}$ in the limit of large N. Conclude that measuring after t iterations, for some $t=O(\sqrt{N})$, will produce the solution $|i\rangle$ with probability at least 1/2.

15.37 Finding solutions when we don't know how many there are. Suppose that we are running Grover's algorithm, but we don't know how many solutions there are. Show that there is a probabilistic algorithm that uses $O(\sqrt{N})$ queries that finds a solution with constant probability. Hint: what happens if we rotate $|\psi\rangle$ by a random angle?

15.38 The collision problem. Suppose we have a two-to-one function $f:\{1,\ldots,N\}\to\{1,\ldots,N/2\}$. In other words, for every $y\in\{1,\ldots,N/2\}$ there are exactly two elements $x,x'\in\{1,\ldots,N\}$ such that $f(x)=f(x')=y$. Such a pair (x,x') is called a *collision.*

We wish to find a collision with as few queries to f as possible. First show that there is a classical randomized algorithm that uses $O(\sqrt{N})$ queries. Hint: use the Birthday Problem from Appendix A.3.3.

Then show that there is a quantum algorithm that uses $O(N^{1/3})$ queries. Hint: choose a random subset $K\subset\{1,\ldots,N\}$ of a certain size, and check classically whether any pair of elements $x,x'\in K$ collide. If not, use Grover's algorithm to find one of the $|K|$ elements of $\{1,\ldots,N\}-K$ which collide with some element of K. Finally, find the optimal size $|K|$.

15.39 Partial traces. Suppose that two systems are in an entangled pure state, which we can write

$$|\Psi\rangle = \sum_{i,k} a_{ik} |i\rangle \otimes |k\rangle .$$

where $|i\rangle$ and $|k\rangle$ range over orthogonal basis vectors for the two systems' state spaces. Their joint density matrix is $R = |\Psi\rangle\langle\Psi|$, or

$$R_{ik,j\ell} = a_{ik} a^*_{j\ell} .$$

Show that if we measure the second system, the density matrix of the first becomes

$$\rho_{ij} = \sum_k R_{ik,jk} = \sum_k a_{ik} a^*_{jk} .$$

This operation, where we sum over the diagonal on one pair of indices while another pair remains unsummed, is called a *partial trace.*

Hint: if we measure the second system and observe $|k\rangle$, the first system collapses to a particular pure state $|v_k\rangle$. Sum over the resulting density matrices $|v_k\rangle\langle v_k|$, weighting each one with the probability of observing k. Now write $|\Psi\rangle$ in the following form,

$$|\Psi\rangle = \sum_i a_i |i\rangle \otimes |\psi_i\rangle ,$$

where $\left|\psi_i\right|^2 = 1$, so that $a_{ik} = a_i \langle k | \psi_i \rangle$. Then show that the density matrix of the first system is given by

$$\rho_{ij} = a_i a^*_j \langle \psi_j | \psi_i \rangle .$$

Note that this holds regardless of how many dimensions the second system is in, or what basis we use to measure it.

15.40 The polynomial method. Suppose we have a function $f : \{1,\ldots,n\} \to \{0,1\}$. If we use the phase kickback trick, querying f corresponds to applying the unitary operator

$$\begin{pmatrix} (-1)^{f(1)} & & & \\ & (-1)^{f(2)} & & \\ & & \ddots & \\ & & & (-1)^{f(n)} \end{pmatrix} .$$

We can think of the values of f as n variables $x_i = f(i)$. Show that, no matter what other unitary operators it uses, any algorithm that makes t queries to f returns "yes" with probability $P(x_1,\ldots,x_n)$, where P is a polynomial of degree at most $2t$. Hint: if $x \in \{0,1\}$, then $(-1)^x = 1 - 2x$.

15.41 Symmetric polynomials. Continuing the previous problem, suppose that the property of f we are trying to discover is symmetric with respect to permutations of the variables. That is, it depends only the number of solutions i such that $f(i) = 1$, namely $x = x_1 + \cdots + x_n$. In that case, we're free to permute the variables randomly before we run the algorithm. The average probability that it returns "yes" is then

$$P_{\text{sym}}(x_1,\ldots,x_n) = \frac{1}{n!} \sum_{\sigma \in S_n} P(x_{\sigma(1)},\ldots,x_{\sigma(n)}) .$$

Show that there is a polynomial $Q(x)$ of degree at most $2t$ such that, whenever $x_i \in \{0,1\}$ for all i,

$$P_{\text{sym}}(x_1,\ldots,x_n) = Q(x_1 + \cdots + x_n) .$$

Hint: use the fact that if $x_i \in \{0,1\}$ then $x_i^k = x_i$ for any $k \ge 1$.

15.42 Grover and Chebyshev. Show that if there are x solutions, the probability Grover's algorithm finds one of them after t iterations is exactly

$$P_t(x) = \sin^2(2t+1)\theta \text{ where } \cos\theta = \sqrt{1-\frac{x}{N}}.$$

Using the identity $\cos 2\alpha = \cos^2 2\alpha - \sin^2 2\alpha$, rewrite this as

$$P_t(x) = \frac{1}{2}\left[1-\cos\left((2t+1)\cos^{-1}\left(1-\frac{2x}{N}\right)\right)\right].$$

Now use de Moivre's formula $\cos k\alpha + \imath \sin k\alpha = (\cos\alpha + \imath \sin\alpha)^n$ to show that

$$\cos(k\cos^{-1} y) = \frac{\left(y+\imath\sqrt{1-y^2}\right)^k + \left(y-\imath\sqrt{1-y^2}\right)^k}{2}.$$

Show that this is, in fact, a polynomial of y of degree k. It is called the *Chebyshev polynomial* $T_k(y)$. Conclude that $P_t(x)$ is a polynomial of degree $2t+1$. There is no contradiction with Problem 15.40, since we perform one more query at the end of the algorithm to confirm that we have found a solution.

15.43 Markov and Chebyshev prove that Grover is optimal. Consider the following theorem of Markov:

Theorem 15.3 *Let $Q(x)$ be a polynomial of degree d. Suppose that $Q(x) \in [y_0, y_1]$ for all $x \in [x_0, x_1]$. Then*

$$\max_{x\in[x_0,x_1]} \left|Q'(x)\right| \le d^2 \frac{y_1 - y_0}{x_1 - x_0}.$$

This bound is tight for the Chebyshev polynomials $T_d(x)$, since $T_d(x) \in [-1,+1]$ for all $x \in [-1,+1]$ and $\left|T_d'\right| = d^2$ at $x = \pm 1$.

Grover's algorithm clearly lets us tell whether the number x of solutions is zero or one, since in the latter case it finds the unique solution with high probability. Let's prove that any such algorithm requires $\Omega(\sqrt{n})$ queries. As in Problem 15.41, the probability the algorithm returns "yes" is a polynomial $Q(x)$ where x is the total number of solutions. Moreover, we know that $Q(x) \in [0,1]$ at all the integer values of x between 0 and n. Using Theorem 15.3, show that

$$\max_{x\in[0,n]} \left|Q'(x)\right| \ge \frac{d^2/n}{1-d^2/n}.$$

Hint: if $Q(x) \in [0,1]$ when x is an integer, how far outside this interval can it be for noninteger values of x?

Now suppose that the algorithm distinguishes the case $x=1$ from the case $x=0$ correctly with probability 2/3, say. (We make no assumptions about what it does for other values of x.) Then we have $Q(0) \le 1/3$ and $Q(1) \ge 2/3$, so $Q' \ge 1/3$ for some x between 0 and 1. Conclude that $d \ge \sqrt{n}/2$, so the algorithm needs at least $\sqrt{n}/4$ queries. Furthermore, use this argument to show that for any symmetric black-box problem, quantum computing can offer at most a quadratic speedup. 15.23

15.44 Block sensitivity. Suppose that we have a Boolean function f of n variables $z_1, \ldots, z_n$. For each $\mathbf{z} = (z_1, \ldots, z_n)$ and each subset $S \subseteq \{1, \ldots, n\}$, let $\mathbf{z} \oplus S$ denote the assignment $\mathbf{z}'$ where $z_i' = \overline{z}_i$ if $i \in S$ and $z_i' = z_i$ if $i \notin S$. Now let us say that *S flips f at* $\mathbf{z}$ if $f(\mathbf{z} \oplus S) \ne f(\mathbf{z})$. For each $\mathbf{z}$, define $b(\mathbf{z})$ as the size of the largest family of disjoint sets $S_1, \ldots, S_b \subseteq \{1, \ldots, n\}$ such that every S_i flips f at $\mathbf{z}$. Then the *block sensitivity* $b(f)$ is the maximum of $b(\mathbf{z})$ over all $\mathbf{z}$.

Show that if a quantum algorithm calculates f correctly with probability 2/3, it must perform $\Omega(\sqrt{b(f)})$ queries. Hint: show that if there is a polynomial P that approximates $f(\mathbf{z})$ within 1/3 for all $\mathbf{z} \in \{0,1\}^n$, there is a polynomial Q on $b(f)$ variables $y_1, \ldots, y_{b(f)}$ of the same degree that approximates $\textsc{or}(y_1, \ldots, y_{b(f)})$.

15.45 Querying the parity. Suppose we have a function $f:\{1,\ldots,n\}\to\{0,1\}$. We wish to determine whether the number of inputs i such that $f(i)=1$ is even or odd. Use Problem 15.41 to show that any quantum algorithm that answers this question correctly with probability at least 2/3 requires at least $n/2$ queries. This shows that there are black-box problems for which quantum computing can only provide a constant speedup. Then show that there is, in fact, a quantum algorithm that uses exactly $n/2$ queries.

15.46 Grover, a little faster. Suppose that we run Grover's algorithm for $a\sqrt{N}$ steps. Show that the probability $\left|\langle\psi\,|\,i\rangle\right|^2$ that we observe the unique solution is essentially $\sin^2 2a$. Rather than taking $a=\pi/4$ so that $P(a)$ is very close to 1, we could take a risk and stop the algorithm early. However, if this fails, we have to start over and try again. Show that if we pursue this strategy and optimize a, we can reduce the expected number of steps to $b\sqrt{N}$ where $b\approx 0.69$, roughly 12% faster than Grover's algorithm.

15.47 Analog Grover. Here's a continuous-time version of Grover's algorithm. Suppose there is a Hamiltonian which has an unknown eigenvector $|v\rangle$ with eigenvalue 1, and that all other eigenvectors have eigenvalue 0. In other words, $H=|v\rangle\langle v|$. Our goal is to find $|v\rangle$, starting with some known vector $|u\rangle$. Let r denote the inner product $\langle u\,|\,v\rangle=r$. Now consider the combined Hamiltonian

$$H'=|u\rangle\langle u|+|v\rangle\langle v|\,.$$

Show that the eigenvectors of H' are $|u\rangle\pm|v\rangle$, and that Schrödinger's equation gives

$$\mathrm{e}^{-\imath H't}|u\rangle=\mathrm{e}^{-\imath t}\left((\cos rt)|u\rangle-(\imath\sin rt)|v\rangle\right).$$

Thus at time $t=\pi/(2r)$, the state will be equal $|v\rangle$ up to a phase. In particular, if $|v\rangle$ is chosen from a set of N basis vectors and $|u\rangle$ is the uniform superposition, then $r=1/\sqrt{N}$ and $t=(\pi/2)\sqrt{N}$.

15.25 **15.48 Grover as a quantum walk.** Let H be the normalized adjacency matrix of the complete graph with N vertices,

$$H=\frac{1}{N}\begin{pmatrix}1 & 1 & \cdots \\ 1 & 1 & \\ \vdots & & \ddots\end{pmatrix}.$$

Show that $\mathrm{e}^{\imath Ht}$ is exactly Grover's operator D for some t, perhaps up to a phase factor. Thus calling D a "diffusion operator" makes sense after all. More generally, show that if $H=|v\rangle\langle v|$ for some v, there is some t for which $\mathrm{e}^{\imath Ht}$ is, up to a phase, the reflection operator R_v around $|v\rangle$.

15.49 Walking on the line, exactly. Let H be the normalized adjacency matrix of the line given in (15.50). By evaluating the matrix exponentials e^{Ht} and $\mathrm{e}^{\imath Ht}$, show that the probability distribution of the continuous-time classical walk on the line, and the amplitude of the quantum version, are given exactly by

$$P_t(x)=\sum_{j=0}^{\infty}\frac{\mathrm{e}^{-t}t^j}{j!}\frac{1}{2^j}\binom{j}{(j+x)/2}\quad\text{and}\quad\psi_t(x)=\sum_{j=0}^{\infty}\frac{(\imath t)^j}{j!}\frac{1}{2^j}\binom{j}{(j+x)/2}.$$

These are identical to the integrals (15.54) and (15.58) respectively.

15.50 Walking on the line, asymptotically. Apply the method of stationary phase (see Appendix A.6.2) to the integral (15.58) for the continuous-time quantum walk on the line, and derive the asymptotic form (15.59) for the probability distribution. To put this differently, show that $P_t(x)$ can be approximated as

$$|\Psi_t(x)|^2\sim\frac{f(x/t)}{t}\quad\text{where}\quad f(a)=\frac{1}{\sqrt{1-a^2}},$$

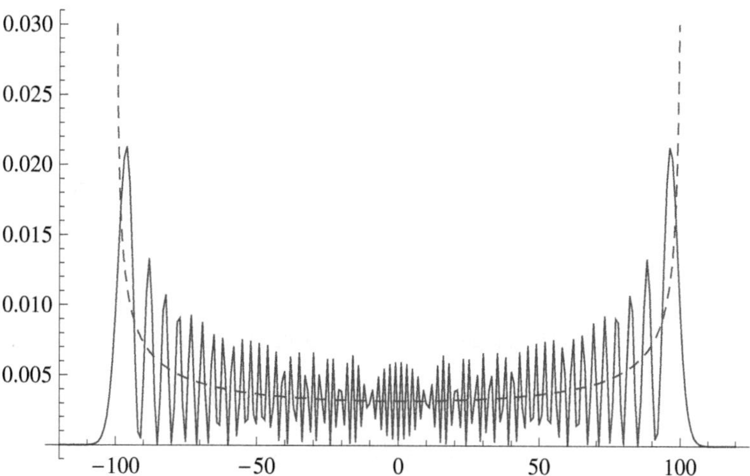

FIGURE 15.19: The quantum walk on the line and the approximation (dashed) given by the method of stationary phase. Here $t = 100$.

if $|x| \le t$, and that $P_t(x)$ is exponentially small if $|x| > t$. This approximation, appropriately normalized, is shown in Figure 15.19. Furthermore, show that the peaks at $x = \pm t$ have width $t^{1/3}$ and height $t^{-2/3}$. In other words, show that if $|x| = t - O(t^{1/3})$, then $P_t(x) \sim t^{-2/3}$. Thus the total probability residing in these peaks scales as $t^{-1/3}$.

15.51 Wave packets. Consider a wave packet of length L to the left of the origin with frequency ω,

$$\Psi(x) = \begin{cases} \frac{1}{\sqrt{L}}\, \mathrm{e}^{\imath \omega x} & -L < x \le 0 \\ 0 & \text{otherwise}. \end{cases}$$

The total probability among the frequencies k such that $|\omega - k| \le \pi/L$ is

$$P = \frac{1}{2\pi} \int_{k-\pi/L}^{k+\pi/L} \left|\widetilde{\Psi}(k)\right|^2 \mathrm{d}k\,,$$

where

$$\widetilde{\Psi}(k) = \sum_{x=-\infty}^{\infty} \Psi(x)\,\mathrm{e}^{\imath k x}\,.$$

Show that P is bounded below by a constant. Then take $\omega = \pi/2$ and $\lambda = \cos k$, and conclude that with constant probability the observed eigenvalue obeys $|\lambda| \le \pi/L$. Hint: you may find the inequality (15.29) useful.

15.52 Majority trees. In the spirit of the scattering algorithm for the NAND tree, we could ask for a similar algorithm that evaluates a majority tree. As in Problem 10.21, this is a trinary tree in which each node has a MAJ gate, whose truth value is equal to the majority of its three inputs.

Consider the gadget in Figure 15.20. The three input nodes are connected to the output node with edges of weight $1/\sqrt{3}$. They are also connected to an additional vertex v with edges whose amplitudes are 1, $\omega = \mathrm{e}^{2\pi\imath/3}$, and $\omega^2 = \mathrm{e}^{-2\pi\imath/3}$ respectively. Since the Hamiltonian has to be Hermitian, the weights of these edges are conjugated when we traverse them in the opposite order.

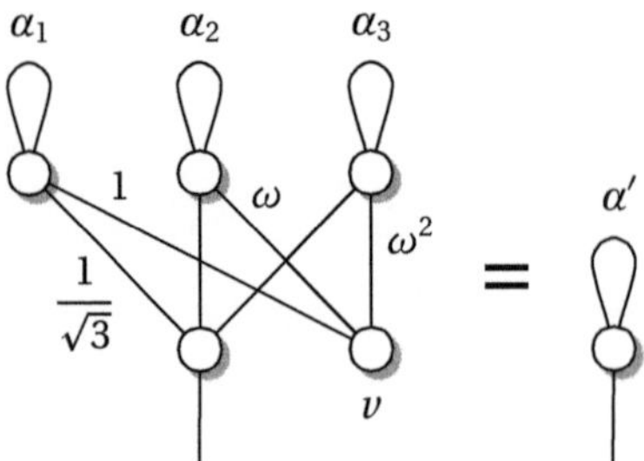

FIGURE 15.20: The gadget for a scattering algorithm for the MAJ tree.

Show that at $\lambda = 0$, we can contract this gadget to a self-loop with weight

$$\alpha' = -\frac{\alpha_1 + \alpha_2 + \alpha_3}{\alpha_1\alpha_2 + \alpha_2\alpha_3 + \alpha_1\alpha_3}.$$

Associating $\alpha = \infty$ and $\alpha = 0$ with `true` and `false` as before, show that this acts like a MAJ gate, perhaps with an extra negation thrown in. With some additional work, this gives a $\Theta(N^{1/2})$ quantum algorithm for a majority tree with N leaves. Classically, the optimal algorithm is not known, but it is known to require at least $\Omega(N^\beta)$ queries where $\beta \geq \log_3 7/3 = 0.771...$

15.53 Optimal measurements. The most general version of a quantum measurement is a *positive operator-valued measurement*, or POVM. This is a list of matrices $\{M_1, \ldots, M_t\}$ such that $\sum_{i=1}^{t} M_i = \mathbb{1}$, where each M_i is positive semidefinite (i.e., real, symmetric, and with nonnegative eigenvalues). Given a mixed state ρ, the POVM has outcome i with probability $\mathrm{tr}(M_i \rho)$. Show that these probabilities sum to 1.

Now suppose that we have t mixed states $\rho_1, \ldots, \rho_t$, where each ρ_i occurs with probability p_i. We want to maximize the probability that the POVM correctly identifies the state—that is, the probability that it gives outcome i when given ρ_i, averaged over i. Show that finding the POVM that does this is an instance of SEMIDEFINITE PROGRAMMING (see Section 9.7.3), or its generalization to multiple matrices.

15.28

Notes

15.1 The physical Church–Turing Thesis. There are some fanciful circumstances in which the physical Church–Turing Thesis might not hold. For instance, one can contemplate a "Zeno computer" which performs the first step of computation in 1 second, the next in 1/2 second, the next in 1/4 second, and so on until at 2 seconds it has done an infinite number of steps. Similarly, in certain exotic spacetimes you can experience an infinite redshift, so that from your point of view your loyal students carry out an infinite amount of computation—say, searching for counterexamples to Goldbach's Conjecture—before you pass an event horizon. In the words of Pitowsky [660], when they grow old, or become professors, they transmit the holy task to their own disciples. From your point of view, they search through all possibilities in a finite amount of time. If their signal hasn't arrived by then, you disintegrate with a smile, knowing that the conjecture is true.

Another possibility is a physical system whose parameters, or initial conditions, are uncomputable real numbers, i.e., numbers whose binary expansions encode the truth table of an uncomputable function. Such a system could "compute" this function, or access it as an oracle, by extracting these bits over time. But it would need infinite precision, and a total lack of noise, to do so.

Barring this sort of silliness, the physical Church–Turing Thesis seems convincing to us. Moreover, while quantum computers seem more powerful than classical ones, it seems reasonable to posit some upper bound on what

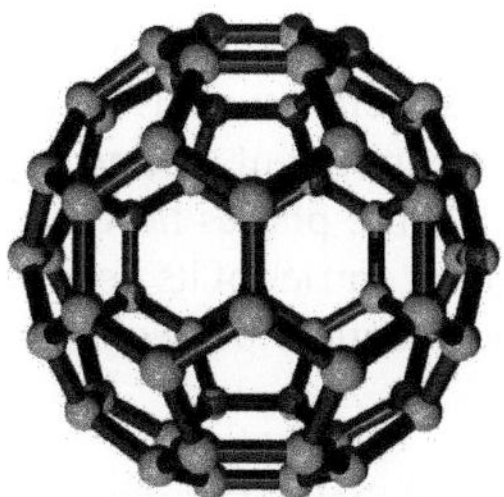

FIGURE 15.21: Interference in the two-slit experiment has been observed even for large molecules, such as these soccer-ball-shaped fullerenes.

a physical system can compute in polynomial time. Aaronson [2] suggests that we treat the postulate that physical systems cannot solve NP-complete problems in polynomial time as an axiom about the physical world. We would then have serious doubts about any model of physics that violates this postulate, just as we would if it allowed faster-than-light communication. For instance, Abrams and Lloyd [6] showed that variants of quantum mechanics that allow nonlinear operations can solve #P-complete problems in polynomial time, and we could take this as evidence that quantum mechanics is indeed linear.

15.2 Further reading. Choosing one chapter's worth of topics from this rapidly-growing field is not easy. We have completely neglected several major areas, including quantum cryptography, quantum error-correcting codes, adiabatic quantum computation, and topological quantum computation. The reader can fill these gaps with the help of the following excellent texts: Nielsen and Chuang [633], Mermin [568], Kaye, Laflamme, and Mosca [471], and the lecture notes of Preskill [668]. We have, of course, stolen from all these sources shamelessly. Our quantum circuits are drawn with the LaTeX package `Qcircuit` by Steve Flammia and Bryan Eastin.

15.3 The two-slit experiment. The two-slit experiment was performed for light and for water waves in the first few years of the 19th century by Thomas Young [842]. It was finally performed for electrons in 1961 by Jönsson [447]. The screen image in Figure 15.1 is from a 1989 experiment by Tonomura et al. [793], who used an electron beam weak enough so that individual electrons can only interfere with themselves. More recently, it has been shown even for large molecules, such as the fullerenes or "buckyballs" composed of 60 Carbon atoms shown in Figure 15.21, and the even larger flourofullerenes $C_{60}F_{48}$ [51, 359].

Young gives a fine description of destructive interference in his 1803 paper:

> From the experiments and calculations which have been premised, we may be allowed to infer, that homogeneous light, at certain equal distances in the direction of its motion, is possessed of opposite qualities, capable of neutralising or destroying each other, and of extinguishing the light, where they happen to be united...

Isaac Newton and Christiaan Huygens argued over whether light was "corpuscular" or "undulatory"; that is, whether it consists of particles or waves. Hopefully they would have enjoyed learning that light, and everything else, is both.

15.4 The collapse of the wave function. Early formulations of quantum mechanics, such as the Copenhagen interpretation, treated the collapse of the wave function as a physical event. However, this leads to the rather bizarre point of view that there are two kinds of events in physics: "measurements", where the wave function collapses, and everything else, where it evolves according to a unitary operator. Since we think of measurements as things done by a human experimenter, this has led to a lot of confusion about the role of consciousness in physical law.

A more modern, though not unanimous, point of view is that measurements are physical processes like any other. When a quantum system interacts with a large, complicated system—such as a macroscopic measuring apparatus, or a human brain, or simply the atoms and photons of the environment around it—information about the relative phases of the components of its state is quickly lost. When these phases become effectively random, the system *decoheres*: probabilities add as in Exercise 15.1, the off-diagonal entries of its density matrix disappear as described in Section 15.7.2, and the classical state emerges. For an introduction to decoherence, along with a clear historical commentary, we recommend the review of Zurek [851].

15.5 Reversible computation. In 1961, Landauer [517] showed that any computer, no matter how it is built, must release heat whenever it performs an irreversible operation and destroys a bit of information. However, Bennett [100] showed that these thermal losses can in principle be avoided, since any computation can be carried out reversibly. His idea was to keep track of the history of the computation, copy the result by XOR-ing it onto a register whose initial value is zero, and then undo the computation, erasing the history as we go.

In quantum computation, reversible gates become essential since quantum dynamics is unitary, and hence reversible, at least in a closed system. Bennett's ideas reappeared in the quantum context, where we disentangle an output qubit from the inputs by "uncomputing" as in Exercise 15.17.

Fredkin and Toffoli [293, 792] showed that arbitrary reversible computations can be carried out if we can implement a few simple reversible gates. Two-bit gates such as the controlled-NOT are not sufficient, however. To be computationally universal, we need reversible gates that act on three bits, such as the Toffoli gate of Problem 15.4. Another universal gate is the Fredkin gate, a controlled swap gate that switches the values of b and c if a is true.

15.6 Universal quantum gates and Turing machines. DiVincenzo [245] showed that two-qubit gates are enough for universal quantum computation. Lloyd [536] and, independently, Deutsch, Barenco, and Ekert [229] showed that almost any two-qubit gate suffices. Barenco et al. [78] showed that the controlled-NOT and the set of all one-qubit gates suffice, and Problem 15.4 is taken from that paper. The universal gate set that we mention here, consisting of the controlled-NOT, H, and $Z^{1/4}$, is from Boykin et al. [132].

While the circuit model is convenient for describing quantum computation, one can define quantum versions of other models as well, such as the Turing machine. Yao [841] showed that these models are equivalent as long as the circuits belong to a uniform family. The trick of Problem 15.6 was used by Moore and Nilsson [604] to show that certain quantum circuits can be efficiently parallelized.

15.7 No cloning. The No-Cloning Theorem was pointed out by Wootters and Zurek [833]. Problem 15.17 is from [398], which gave a flawed proposal for faster-than-light communication by assuming that cloning is possible.

15.8 Heisenberg's uncertainty principle. To a non-physicist, the statement that position and moment do not commute is quite mysterious at first. Here's the idea. The momentum of a traveling wave whose amplitude is proportional to $e^{\imath\omega x}$ turns out to be its frequency ω times Planck's constant $\hbar$. Since taking the derivative with respect to x pulls down a factor of $\imath\omega$, the momentum is the eigenvalue of the following operator:

$$p = -\imath\hbar \frac{\mathrm{d}}{\mathrm{d}x}.$$

Then for any function f we have

$$xpf = -\imath\hbar x \frac{\mathrm{d}f}{\mathrm{d}x},$$

while

$$pxf = -\imath\hbar \frac{\mathrm{d}}{\mathrm{d}x}(xf) = -\imath\hbar\left(x\frac{\mathrm{d}f}{\mathrm{d}x} + f\right).$$

The commutator $[p,x] = px - xp$ measures the extent to which p and x do not commute. We have $(px - xp)f = -\imath\hbar f$, or $[p,x] = -\imath\hbar$. In the limit $\hbar \to 0$, p and x commute, and physics becomes classical.

15.9 The EPR paradox and Bell's inequalities. The particular version of Bell's inequality we give in (15.7) is called the CHSH inequality, after Clauser, Horne, Shimony, and Holt, who stated it in 1969 [179]. Tsirelson's inequality, given in Problem 15.11, is proved in [177]; our presentation is taken from [668].

In their 1935 paper [264], Einstein, Podolsky and Rosen argued that if two systems are entangled, we can learn the values of two physical quantities P and Q even if they don't commute, by measuring one system in one basis and its entangled partner in the other. They give the example of two particles traveling with opposite momentum: if we measure the position of one of them and the momentum of the other then we arguably know the position and momenta of both, even though the position and momentum operators do not commute. They write

> Indeed, one would not arrive at our conclusion if one insisted that two or more physical quantities can be regarded as simultaneous elements of reality *only when they can be simultaneously measured or predicted.* On this point of view, since either one or the other, but not both simultaneously, of the quantities P and Q can be predicted, they are not simultaneously real. This makes the reality of P and Q depend upon the process of measurement carried out on the first system, which does not disturb the second system in any way. No reasonable definition of reality could be expected to permit this.

The fact that Bell's inequality is violated shows that this philosophical position, called *local realism,* is incompatible with quantum mechanics. One option is to give up realism, and abandon the belief that the outcome of a measurement is a function of the state of the world (even a nondeterministic one). Equivalently, we can give up *counterfactual definiteness*: namely, the belief that questions like "what would Alice have observed if she had measured X instead of Z?" have an answer. Alternately, we can give up *locality,* the belief that physical influences can travel no faster than light. However, given what we know about relativity, this is arguably an even higher price to pay.

The fact that Bell's inequality is violated, and that the world really cannot be both local and realistic, has been shown through an increasingly convincing series of experiments. Aspect, Dalibard, and Roger [59] carried out an experiment where Alice and Bob's bases are chosen randomly while the photons are in transit, so that any correlations would have to travel faster than light, and Tittel, Brendel, Zbinden, and Gisin [790] created entangled pairs of photons separated by more than 10 kilometers of fiber-optic cable.

For an even starker violation of local realism, consider the following scenario of Mermin [567], who called it an "all-or-nothing demolition of the elements of reality." It is a simplified version of an argument by Greenberger, Horne, and Zeilinger [352]. We start with the entangled three-qubit state

$$|\psi\rangle = \frac{1}{2}\left(|000\rangle - |111\rangle\right).$$

Suppose that these three qubits are possessed by three physicists, named Alice, Bob, and Claire, who are light-years apart from each other. Since $|\psi\rangle$ is an eigenvector of $X\otimes X\otimes X$ with eigenvalue -1, if all three measure their qubits in the X-basis and multiply their observations together, they always get -1.

Now suppose that two of the three physicists measure their qubits in the Y-basis, and the third one uses the X-basis. In this case, the product of their observations is $+1$. In other words, $|\psi\rangle$ is an eigenvector of $Y\otimes Y\otimes X$ with eigenvalue $+1$. By symmetry, this is true for $Y\otimes X\otimes Y$ and $X\otimes Y\otimes Y$ as well.

Now let us suppose, along with Einstein, that Alice's qubit has definite values of X and Y (realism), and that these values cannot be affected by Bob and Claire's choices of basis (locality). Since $Y^2 = 1$, this would lead us to believe that

$$X\otimes X\otimes X = (Y\otimes Y\otimes X)\cdot(Y\otimes X\otimes Y)\cdot(X\otimes Y\otimes Y).$$

But exactly the opposite is true. In other words,

$$X\otimes X\otimes X = -(Y\otimes Y\otimes X)\cdot(Y\otimes X\otimes Y)\cdot(X\otimes Y\otimes Y).$$

In particular, for the state $|\psi\rangle$ the product of the three Xs is -1, but the product of any two Ys and an X is $+1$.

15.10 Superdense coding and quantum teleportation. Superdense coding was discovered by Bennett and Wiesner in 1992 [104], and quantum teleportation was discovered the following year by Bennett et al. [102]. The idea of sending a "cobit" and Problems 15.14 and 15.15 are from Harrow [369].

The precise rate at which we can send information over a given kind of quantum channel is still being worked out. Unlike the classical setting, if we run two quantum channels in parallel we can sometimes achieve a rate greater than the sum of the two; see Hastings [380].

15.11 Deutsch, Jozsa, and Bernstein–Vazirani. The first few papers on quantum computation were characterized by a search for basic algorithmic ideas, including parallelism, phase kickback, and the Fourier transform. Deutsch [228] introduced the phrase "quantum parallelism" in 1985 [228], along with the problem of finding the parity $f(0) \oplus f(1)$ of a one-bit function. However, in that paper he gave the algorithm of Problem 15.18, which only succeeds with probability 1/2. The version we give here, which uses phase kickback to solve this problem with a single query, first appeared in print in Cleve, Ekert, Machhiavello, and Mosca [180], and was found independently by Tapp.

The idea of measuring the average of $(-1)^f$ by applying a Hadamard gate to each qubit first appeared in Deutsch and Jozsa [230] in 1992. In 1993, Bernstein and Vazirani [107] recognized this operation as the Fourier transform over $\mathbb{Z}_2^n$. In their original versions, both these algorithms queried $f(x)$ twice as in Exercise 15.17. The one-query versions of these algorithms, which use phase kickback, appeared in [180].

15.12 RSA cryptography. The idea of a public-key cryptosystem based on a one-way function with a trapdoor was put forward publicly in 1976 by Diffie and Hellman [238], but without a practical implementation. The RSA system was published in 1977 by Ron Rivest, Adi Shamir, and Len Adleman [704]. It was first presented to the public by an article in *Scientific American* by Martin Gardner [317], who began his column with the prescient statement that "the transfer of information will probably be much faster and cheaper by electronic mail than by conventional postal systems." While RSA is still believed to be secure overall, it is known to be insecure in certain cases, such as when e or d is too small, or if p and q are too close to $\sqrt{N}$. A review of these results can be found in [470].

There is an interesting parallel history in which these techniques were developed in secret a few years earlier by mathematicians in the British intelligence agency GCHQ. In 1970, James Ellis suggested the idea of "non-secret" (public-key) encryption as a theoretical possibility. In 1973, Clifford Cocks proposed the special case of RSA where the encryption key is $e = N$. However, these schemes were dismissed as impractical at the time because of the amount of computation they require.

Diffie and Hellman [238] also pointed out that a public-key cryptosystem lets you prove your identity, and sign digital documents securely, by using it in reverse. Suppose you wish to sign a document D, and authenticate it as having come from you. You encrypt D using your secret key d and append your signature $D^d \bmod N$ to the document. Anyone can decrypt your signature using your public key e, and check that $D^{de} \bmod N = D$. However, without knowing d, it seems hard for anyone else to produce your signature. In practice, rather than encrypting the entire document, you could apply some publicly-known hash function to map D down to a shorter string.

15.13 RSA vs. Factoring. Problem 15.28, showing that RSA Keybreaking is just as hard as Factoring, is from Crandall and Pomerance [205, Ex. 5.27]. We can define the problem of decrypting individual RSA messages as follows:

RSA Decryption
Input: Integers N, e, and m'
Output: An integer m such that $m' \equiv_N m^e$

This is equivalent to extracting eth roots in $\mathbb{Z}_N^*$. At the time we write this, the relationship between Factoring and RSA Decryption is not entirely clear. However, Aggarwal and Maurer [21] showed in 2009 that if there is a "generic ring algorithm" for RSA Decryption—that is, an efficient algorithm that uses the algebraic operations of addition, subtraction, multiplication, and division, and can branch based on whether two elements of $\mathbb{Z}_N$ are equal or not—then there is also an efficient algorithm for FACTORING. It follows that, if FACTORING is hard, any efficient algorithm for RSA DECRYPTION must manipulate the bits of its inputs in a non-algebraic way.

15.14 Classical algorithms for FACTORING and DISCRETE LOG. The best known classical algorithm for FACTORING is the so-called Number Field Sieve; see Crandall and Pomerance [205] for a review. An analogous algorithm for DISCRETE LOG was given by Gordon [347]. Subject to certain number-theoretic assumptions, both run in time $\exp\left(n^{1/3}(\log n)^{2/3}\right)$. Interestingly, while number-theoretic ideas that help us solve one of these problems often help solve the other, no polynomial-time reduction between them is known.

15.15 Rational approximations. Several of these approximations for π appeared in antiquity, and were obtained by bounding the area of a circle with that of an inscribed or circumscribed polygon. Archimedes proved that $\pi < 22/7$, and in the 16th century Adriaan Anthonisz proved that $\pi > 333/106$. However, more than a thousand years earlier, the Chinese astronomer Zu Chongzhi gave the approximation $\pi \approx 355/113$, which is accurate to 6 decimal places. The error in this approximation is so small because the next denominator in the continued fraction, 292, is so large.

15.16 Mertens' Theorem. The sharpest result on the probability that a randomly chosen number $0 < b < r$ is mutually prime to r is

$$\liminf_{r\to\infty} \frac{\varphi(r)}{r} = \frac{e^{-\gamma}}{\ln \ln r},$$

where $\gamma = 0.577\ldots$ is the Euler–Mascheroni constant. This is a corollary to Mertens' Theorem (no relation). A proof of Mertens' Theorem can be found in most textbooks on number theory, such as Hardy and Wright [367]. For an elementary proof, we refer the reader to Yaglom and Yaglom [836].

15.17 Simon's and Shor's algorithms. Simon's algorithm [753] and Shor's algorithms for FACTORING and DISCRETE LOG [750] both appeared in 1994. Shor received the Nevanlinna Prize in 1998 and the Gödel Prize in 1999.

The efficient quantum Fourier transform for $\mathbb{Z}_M$ where M is a power of 2 was found by Coppersmith [201]. Another approach to order-finding, due to Kitaev [484], uses an approach called *phase estimation*, in which we combine phase kickback and the QFT to estimate the eigenvalue of an eigenvector.

Shor's algorithm has been implemented in the laboratory: using a nuclear magnetic resonance quantum computer with 7 qubits, Vandersypen et al. [808] succeeded in factoring the number 15.

Another important algorithm in this family is Hallgren's algorithm [363] for solving a Diophantine equation known as Pell's equation: given an integer d, find integers x and y such that $x^2 - dy^2 = 1$, so that x/y is a good approximation to $\sqrt{d}$. (Like many things in science, this is named after the wrong person: it was studied by Brahmagupta in 628 A.D., and the cases $d = 2$ and $d = 3$ much earlier than that.) Hallgren's algorithm is similar to Shor's in that it uses Fourier sampling to find the periodicity of a function defined on the reals. However, constructing this function is highly nontrivial, and uses machinery from the theory of algebraic number fields and principal ideals.

15.18 Key exchange. The key exchange scheme described in Section 15.5.6 was published in 1976 by Diffie and Hellman [238]. While we have not done so here, Hellman urges us to call it Diffie–Hellman–Merkle key exchange, saying that it was only a "quirk of fate" that Merkle is not commonly given a share of the credit for it [395]. While he was not a coauthor of [238], Merkle [565] considered analogous schemes in which Alice sends a menu of puzzles to Bob, each of which contains a key, and Bob solves a randomly chosen puzzle.

As with RSA cryptography, this approach to key exchange was discovered in parallel by a GCHQ mathematician, namely Malcolm Williamson, who described it in a classified memo in 1976. Earlier, in 1974, Williamson described the variant given in Problem 15.32. The "Love in Kleptopia" puzzle is quoted from Peter Winkler's collection *Mathematical Mind-Benders* [828].

Our example of using the Navajo language for secure communication is based on history. The United States used more than 400 Navajo "code talkers" in World War II to communicate securely in the Pacific. The code talkers' efforts remained classified until 1968, and they were awarded the Congressional Gold Medal in 2001.

15.19 Graph Isomorphism. The best known classical algorithm for GRAPH ISOMORPHISM runs in $\exp\left(\sqrt{n \log n}\right)$ time; see Babai [62] and Babai and Luks [67], as well as Spielman [769]. However, many special cases of GRAPH ISOMORPHISM are in P, including the case where the maximum degree of the graphs is a constant [548, 67]. The algorithm for that case uses sophisticated ideas in computational group theory to build an isomorphism between the two graphs, starting at a pair of corresponding edges and building outward from there.

15.20 The nonabelian Hidden Subgroup Problem and post-quantum cryptography. Efficient QFTs have been found for a wide variety of nonabelian groups, in the sense that the number of steps is $\text{polylog}(|G|)$ for a group G, and therefore polynomial in n for a group of exponential size. See Beals [89] and Moore, Rockmore, and Russell [607] for the most general results. For an introduction to representation theory and nonabelian Fourier analysis, we recommend the text by Fulton and Harris [304].

Efficient algorithms for HIDDEN SUBGROUP are known for some nonabelian groups, e.g. [72, 71, 227, 298, 423, 608]. However, all of these families are only "slightly nonabelian" in the sense that they can be built from a few abelian pieces. For the permutation group S_n, which is highly nonabelian, Hallgren, Russell, and Ta-Shma [365] showed that "weak Fourier sampling," where we measure just the representation ρ but not the row and column i, j, cannot solve the case relevant to GRAPH ISOMORPHISM (although it can solve HIDDEN SUBGROUP in the case where H is normal). Grigni, Schulman, Vazirani, and Vazirani [353] then showed that measuring i and j in a random basis doesn't work. The power of "strong Fourier sampling," in which we measure i and j in a basis of our choosing, remained open for several years, until Moore, Russell, and Schulman [612] showed that no basis, and more generally no measurement on single coset states, can distinguish whether two graphs are isomorphic or not.

Earlier, Ettinger, Høyer, and Knill [271] showed that for any group G there exists a measurement on the tensor product of $O(\log|G|)$ coset states that solves HIDDEN SUBGROUP. However, Hallgren et al. [364] showed for S_n, only a globally-entangled measurement on $\Omega(\log n!) = \Omega(n \log n)$ coset states can succeed. One of the few known families of quantum algorithms that perform such a measurement is the "quantum sieve" of Kuperberg [509]. However, Moore, Russell, and Śniady [613] showed that no such sieve algorithm can work for GRAPH ISOMORPHISM.

Another problem believed to be outside P but not NP-complete, UNIQUE SHORTEST LATTICE VECTOR, reduces to HIDDEN SUBGROUP on the dihedral group; see Regev [688]. While Kuperberg's sieve solves this case of HIDDEN SUBGROUP in subexponential time, no polynomial-time algorithm is known. Problems like these might form the foundation for cryptosystems which we can carry out today, but which will remain secure even in the presence of quantum computers [689]. Another family of candidate for such "post-quantum cryptosystems" are the McEliece and Niederreiter cryptosystems, which seem resistant to HIDDEN SUBGROUP-type attacks [241].

15.21 The swap test and quantum sampling. The swap test appeared in Buhrman, Cleve, Watrous, and de Wolf [144] and Gottesman and Chuang [348] in quantum schemes for fingerprinting and digital signatures, and in Watrous [820] in quantum proof schemes for properties of groups.

As we alluded to in the text, no similar way to estimate the overlap between two classical probability distributions exists. For instance, imagine two distributions on $\{0,1\}^n$, each of which is uniform on a subset of size 2^{n-1} where these subsets are either identical or disjoint. If we sample from these distributions classically, then according to the Birthday Paradox (see Appendix A.3.3) it will take about $2^{n/2}$ samples before we gain any evidence about which of these is the case.

Quantum sampling algorithms for statistical physics and Markov chains, that produce the pure state $\sum_x \sqrt{P(x)}\,|x\rangle$ where $P(x)$ is the Boltzman distribution of the Ising model, say, or the equilibrium distribution of a rapidly mixing Markov chain, were proposed by Lidar and Biham [532]. For recent results, see Somma, Boixo, Barnum, and Knill [767], or Wocjan and Abeyesinghe [829].

15.22 Grover's algorithm. Grover's algorithm appeared in [356], where he gave the simple argument of Problem 15.36. Its running time was analyzed precisely by Boyer, Brassard, Høyer, and Tapp [131], who also discussed the case where the number of solutions is not known. Using the fact that θ depends on the number of solutions as in Exercise 15.23, Brassard, Høyer, and Tapp gave a quantum algorithm for approximately counting the solutions [134].

The first proof that Grover's algorithm is optimal actually predated Grover's algorithm. Using a geometrical argument, Bennett, Bernstein, Brassard, and Vazirani [101] showed that $\Omega(\sqrt{N})$ queries are necessary. Zalka [843] refined this argument to show that the constant $\pi/4$ is optimal, although he also pointed out that we can improve the expected number of queries as in Problem 15.46.

The "quantum adversary" argument we give here, using entanglement between the computer and the oracle, is from Ambainis [40]. His paper handles the more general case where the computer has a larger state space, and where the algorithm succeeds with probability $1-\varepsilon$ instead of 1. Bose, Rallan, and Vedral [129] gave a similar argument using the mutual information between the two systems, also known as the von Neumann entropy $\text{tr}(\rho \ln \rho)$, as their measure of entanglement.

Many other adversary techniques have been developed since then. Barnum, Saks, and Szegedy [82] used spectral methods and semidefinite programming, Laplante and Magniez [519] used Kolmogorov complexity, andŠpalek and Szegedy [768] showed that these are equivalent. The most powerful known adversary techniques are those of Høyer, Lee, andŠpalek [413].

There is a vast literature on quantum algorithms for black-box problems that are polynomially faster than their classical counterparts. We limit ourselves here to the $O(n^{1/3})$ algorithm of Problem 15.38 for finding collisions in a two-to-one function. It is from Brassard, Høyer, and Tapp [135], and Aaronson and Shi [4] showed that it is optimal.

15.23 The polynomial method. In the classical setting, the idea of approximating a Boolean function f with a polynomial goes back to Minsky and Papert's book *Perceptrons* [584]. In particular, they proved the fact shown in Problem 15.41 that symmetrizing a multivariate polynomial $P(x_1,\ldots,x_n)$ gives a univariate polynomial $Q(x_1+\cdots+x_n)$ of at most the same degree. Bounds relating the number of steps in a classical decision tree, either deterministic or randomized, to the degree of this polynomial were given by Nisan [634], who also introduced the notion of block sensitivity discussed in Problem 15.44.

In the quantum setting, the fact that the degree of this polynomial is at most twice the number of queries was shown by Beals et al. [90], who used it to show that quantum computing can provide at most a polynomial speedup for black-box problems. The argument of Problem 15.43, though not its quantum application, is due to Nisan and Szegedy [638]; Markov's theorem can be found in [159, Sec. 3.7]. Improved bounds for symmetric functions were given by Paturi [650] and Nayak and Wu [624]. The fact that a quantum computer needs $N/2$ queries to find the parity (Problem 15.45) was proved independently by Beals et al. [90] and Farhi, Goldstone, Gutmann, and Sipser [275].

15.24 Schrödinger's equation. For deep reasons connecting classical and quantum mechanics, the Hamiltonian H turns out to be the energy operator. For instance, if Ψ describes a plane wave with frequency ω moving at the speed of light, we have

$$\Psi(x,t) = e^{\imath\omega(x/c-t)}.$$

and since $\frac{\partial}{\partial t}\Psi = -\imath\omega\Psi$, the energy H is simply the frequency ω. This fits with Einstein's explanation of the photoelectric effect, in which the energy of a photon is proportional to its frequency. Note that we are using units in which Planck's constant $\hbar$ is 1.

15.25 Quantum walks. Continuous-time quantum walks were studied by Farhi and Gutmann [277] and Childs, Farhi, and Gutmann [162]. The continuous-time version of Grover's algorithm appearing in Problem 15.47 was given by Farhi and Gutmann [276].

Discrete-time quantum walks first appeared in computer science in Watrous [821], who studied them as part of a quantum algorithm for UNDIRECTED REACHABILITY. Note that in order to define a discrete-time quantum walk, we have to keep track of the particle's direction as well as its position. We then alternately apply a unitary "coin" operator that changes the direction while leaving the position fixed, and a shift operator that changes the position. For the walk on the plane shown in Figure 15.13, we used the $N = 4$ case of Grover's diffusion operator to change between the four directions.

The fact that the quantum walk on the cycle of size n mixes in time $\Theta(n)$ rather than $\Theta(n^2)$ as in the classical case was shown by Ambainis et al. [42] and Aharonov, Ambainis, Kempe, and Vazirani [24] using different notions of the mixing time. The precise asymptotic behavior of the probability distribution of the quantum walk on the line, using the method of stationary phase as in Problem 15.50, appeared in [42]. The integrals (15.54) and (15.58) can also be expressed exactly using Bessel functions; see e.g. [499].

Continuous and discrete quantum walks were studied by Moore and Russell [609] and Kempe [472] on the hypercube, and by Gerhardt and Watrous [324] on the permutation group. It's tempting to think that the latter might help us solve the quantum sampling problem of Section 15.6.3 and thus solve GRAPH ISOMORPHISM, but this doesn't appear to be the case.

Childs et al. [161] defined a black-box problem for which quantum walks give an exponential speedup; it is somewhat artificial, but the fact that such a problem exists is very interesting. Ambainis [41] gave a walk-based algorithm for telling whether all n elements of a list are distinct with only $O(n^{2/3})$ queries, and Aaronson and Shi [4] showed that this is optimal. Quantum walks that search a d-dimensional lattice were given by Ambainis, Kempe, and Rivosh [44] and by Childs and Goldstone [163]. Magniez, Santha, and Szegedy [551] used a quantum walk whether a graph with n vertices has a triangle with $O(n^{13/10})$ queries; the best known lower bound for this problem is $\Omega(n)$.

15.26 The NAND tree and generalizations. The Farhi–Goldstone–Gutmann algorithm for the NAND tree appeared in [274], using the Hamiltonian oracle model proposed by Mochon [591]. In our presentation we identify the origin with the root of the tree, while they connect it with an edge. As a result, in their paper the wave is transmitted or reflected if the tree is `true` or `false` respectively.

We glossed over the fact that scattering states of the form (15.61) are not the only eigenvectors of the Hamiltonian. There are also "bound states" where the amplitude falls off exponentially on either side of the origin, $v(x) \propto e^{-k|x|}$. However, it can be shown that these bound states are observed only a small fraction of the time.

A discrete-time version of the NAND tree algorithm, along with a generalization to any Boolean formula made of AND and OR gates, was given by Ambainis et al. [43]. Reichardt and Š palek[691] designed the algorithm for majority trees described in Problem 15.52. They also gave systematic methods for designing gadgets for other types of trees, including any three-input gate.

15.27 Trace estimation and the power of one qubit. The trace estimation algorithm of Problem 15.34 is due to Knill and Laflamme [491]. They define a complexity class DQC1 consisting of the problems we can solve with bounded probability given access to a single "clean qubit," namely those problems that can be reduced to estimating the trace of a unitary operator. This class contains a number of interesting problems that seem difficult classically, including estimating the Jones polynomial of the *trace closure* of a braid, i.e., the knot or link formed by attaching the bottom of each strand of a braid to its top; see Aharonov, Jones, and Landau [25] and Shor and Jordan [751].

15.28 Optimal measurements. The fact that the optimal measurement to identify a given set of mixed states is an instance of SEMIDEFINITE PROGRAMMING was recognized by Eldar, Megretski, and Verghese [265]. For some cases of HIDDEN SUBGROUP, this optimal measurement can be computed exactly, and coincides with the so-called "pretty good measurement" from quantum optics; see Bacon, Childs, and van Dam [72], and Moore and Russell [610].

Appendix A

Mathematical Tools

> Can one learn mathematics by reading it? I am inclined to say no. Reading has an edge over listening because reading is more active—but not much. Reading with pencil and paper on the side is very much better—it is a big step in the right direction. The very best way to read a book, however, with, to be sure, pencil and paper on the side, is to keep the pencil busy on the paper and throw the book away.
>
> Paul Halmos

In this appendix, we provide a mathematical toolbox that we use throughout the book—asymptotic notation, approximation methods and inequalities, the theory of discrete probability, the binomial, Gaussian, and Poisson distributions, methods for approximating asymptotic integrals, and the theory of finite groups. Where possible, we look "under the hood," so you can see how the relevant theorems are proved.

A.1 The Story of O

Like physics, the theory of computational complexity is concerned with scaling—how the resources we need to solve a problem increase with its size. In order to make qualitative distinctions between how different functions grow, we use O and its companion symbols, the trademarks of asymptotic analysis.

Suppose that a careful analysis of an algorithm reveals that its running time on instances of size n is

$$T(n) = an^2 + bn + c$$

for some constants a, b, c where $a > 0$. These constants depend on the details of the implementation, the hardware we're running the algorithm on, and the definition of elementary operations. What we really care about is that when n is large, $T(n)$ is dominated by its quadratic term. In particular, there is a constant d such that, for all $n > 0$,

$$T(n) \leq dn^2 .$$

We write this as $T(n) = O(n^2)$, and read "T is *big-oh* of n^2."

The general definition of O is the following. Let f and g be two functions defined on the natural numbers. We say that $f(n) = O(g(n))$, or $f = O(g)$ for short, if there are constants C and n_0 such that

$$f(n) \leq Cg(n) \quad \text{for all } n > n_0 . \tag{A.1}$$

We can also express this in terms of the limiting ratio between f and g: there is a constant C such that

$$\limsup_{n\to\infty} \frac{f(n)}{g(n)} \le C\,.$$

Using the same definition, we can say that $f(n)=O(1)$ if $f(n)$ is bounded by a constant. We can also use O to state that a real-valued function $f(n)$ decays to zero at a certain rate as $n \to \infty$. For instance, $f(n)=O(1/n)$ means that $f(n)\le C/n$ for some constant C.

Here are some examples:

$$\begin{aligned}
an^2+bn+c &= O(n^k) \text{ for any } k\ge 2 \\
\sqrt{3n+10} &= O(\sqrt{n}) \\
\log(n^7) &= O(\log n) \\
n^k &= O(2^n) \text{ for any constant } k \\
n! &= O(n^n) \\
\mathrm{e}^{\sin n} &= O(1) \\
\mathrm{e}^{-n} &= O(n^{-c}) \text{ for any } c>0
\end{aligned} \tag{A.2}$$

We can also use O inside arithmetic expressions. For instance, the precise version of Stirling's approximation for the factorial is

$$n! = \left(1+O(n^{-1})\right)\sqrt{2\pi n}\, n^n\, \mathrm{e}^{-n}\,. \tag{A.3}$$

This means that, in the limit $n\to\infty$, the multiplicative error in this approximation is at most proportional to n^{-1}. In other words, there is a constant C such that

$$n! = (1+\varepsilon)\sqrt{2\pi n}\, n^n\, \mathrm{e}^{-n} \quad\text{where}\quad |\varepsilon| < Cn^{-1}\,.$$

Note that here $O(n^{-1})$ denotes an error that could be positive or negative.

Exercise A.1 *Show that if $f_1=O(g)$ and $f_2=O(g)$ then $f_1+f_2=O(g)$.*

Exercise A.2 *Show that the relation O is transitive. That is, if $f=O(g)$ and $g=O(h)$ then $f=O(h)$.*

Exercise A.3 *When we say $f(n)=O(\log n)$, why don't we need to state the base of the logarithm?*

Exercise A.4 *What is wrong with the following argument? For any k, we have $kn=O(n)$. Hence*

$$\sum_{k=1}^{n} kn = \sum_{k=1}^{n} O(n) = nO(n) = O(n^2)\,.$$

Exercise A.5 *Is $2^{O(n)}$ the same as $O(2^n)$? Why or why not?*

Other sciences, such as physics, often use $f = O(g)$ to indicate that f is proportional to g when n is large. However, in computer science it denotes an upper bound, and most of the bounds shown in (A.2) are rather generous. You can think of $f = O(g)$ as the statement "f grows at most as fast as g does," or "asymptotically, ignoring multiplicative constants, $f \leq g$."

The analogous symbols for $\geq$ and $=$ are Ω and Θ respectively. We say that $f = \Omega(g)$ if and only if $g = O(f)$. In other words, there exist constants $C > 0$ and n_0 such that

$$f(n) \geq C g(n) \quad \text{for all } n > n_0 . \tag{A.4}$$

Alternately, $f = \Omega(g)$ if there is a constant $C > 0$ such that

$$\liminf_{n\to\infty} \frac{f(n)}{g(n)} \geq C .$$

Thus $f(n)$ grows at least as fast, ignoring constants, as $g(n)$ does. We write $f = \Theta(g)$ if $f = O(g)$ and $g = O(f)$. Typically, this means that there is a constant $C > 0$ such that

$$\lim_{n\to\infty} \frac{f(n)}{g(n)} = C ,$$

so that $f(n)$ is proportional to $g(n)$ when n is large. It is possible, though rare in practice, for this ratio to oscillate between two constants without ever settling down.

We also have asymptotic versions of $<$ and $>$. We say $f = o(g)$ if $f = O(g)$ but $f \neq \Theta(g)$, i.e., f is dwarfed by g in the limit $n \to \infty$:

$$\limsup_{n\to\infty} \frac{f(n)}{g(n)} = 0 ,$$

On the other side, we write $f = \omega(g)$ if $g = o(f)$, or equivalently

$$\liminf_{n\to\infty} \frac{f(n)}{g(n)} = \infty ,$$

Exercise A.6 *Which of the examples in* (A.2) *remain correct if we replace O by Θ? In which cases can we replace O with o?*

Exercise A.7 *Give functions f and g such that $f = \Theta(g)$ but $2^f = o(2^g)$, or conversely, functions such that $f = o(g)$ but $\log f = \Theta(\log g)$.*

Exercise A.8 *For each pair of functions, state whether their relationship is $f = o(g)$, $f = \Theta(g)$, or $f = \omega(g)$.*

1. $f(n) = \log \sqrt{n}$, $g(n) = \log(n^2)$
2. $f(n) = 3^{n/2}$, $g(n) = 2^n$
3. $f(n) = 2^n$, $g(n) = n^{\log n}$
4. $f(n) = 2^{n+\log n}$, $g(n) = 2^n$.

We will often give coarser classifications of functions by grouping various classes of functions together as in Table A.1. For instance, the class poly(n) of polynomial functions is the union, over all constant c, of the class $O(n^c)$, and we call a function *exponential* if it is $2^{\text{poly}(n)}$.

Exercise A.9 *Give several examples of functions $f(n)$ that are superpolynomial but subexponential.*

class	definition
polylogarithmic	$f = O(\log^c n)$ for some constant c
polynomial	$f = O(n^c)$ for some constant c, or $n^{O(1)}$
superpolynomial	$f = \omega(n^c)$ for every constant c, or $n^{\omega(1)}$
subexponential	$f = o(2^{n^\varepsilon})$ for every $\varepsilon > 0$
exponential	$f = \Omega(2^{n^\varepsilon})$ for some $\varepsilon > 0$, or $f = 2^{\mathrm{poly}(n)}$

TABLE A.1: Coarse-grained classes of functions in asymptotic analysis.

A.2 Approximations and Inequalities

A.2.1 *Norms*

Given a real number $r > 0$, the *r-norm* of a vector $\mathbf{v}$ is

$$\|\mathbf{v}\|_r = \Big(\sum_i |v_i|^r\Big)^{1/r}.$$

When $r = 2$, this is the standard Euclidean length. When $r = 1$, it becomes the "Manhattan metric" $\sum_i |v_i|$, so called because it's the total number of blocks we have to travel north, south, east, or west to reach our destination. In the limit $r \to \infty$, the r-norm becomes the max norm,

$$\|v\|_{\max} = \max_i |v_i|.$$

For any $r \geq 1$ the r-norm is *convex* in the sense that interpolating between two points doesn't take you farther away from the origin. That is, for any $0 \leq \lambda \leq 1$,

$$\left\|\lambda x + (1-\lambda) y\right\|_r \leq \max\left(\|x\|_r, \|y\|_r\right).$$

The "unit disk" with respect to the r-norm, i.e., the set of vectors $\mathbf{v}$ such that $\|\mathbf{v}\|_r \leq 1$, is a diamond for $r = 1$, a circle for $r = 2$, and a square for $r = \infty$. The Danish mathematician and poet Piet Hein was particularly fond of the case $r = 5/2$. He felt that a "superellipse," the set of points (x, y) such that $ax^r + by^r = 1$, was a good shape for tables and town squares [315]. See Figure A.1.

A.2.2 *The Triangle Inequality*

Suppose I have two vectors x, y as shown in Figure A.2. The *triangle inequality* is the fact that the length of their sum is at most the sum of their lengths,

$$\left\|x + y\right\|_2 \leq \|x\|_2 + \left\|y\right\|_2,$$

where this holds with equality if and only if x and y are parallel. More generally, for any set of vectors v_i,

$$\left\|\sum v\right\|_2 \leq \sum \|v\|_2.$$

The triangle inequality holds for the r-norm for any $r \geq 1$. If these quantities are simply numbers rather than vectors, it holds if $|v|$ denotes the absolute value.

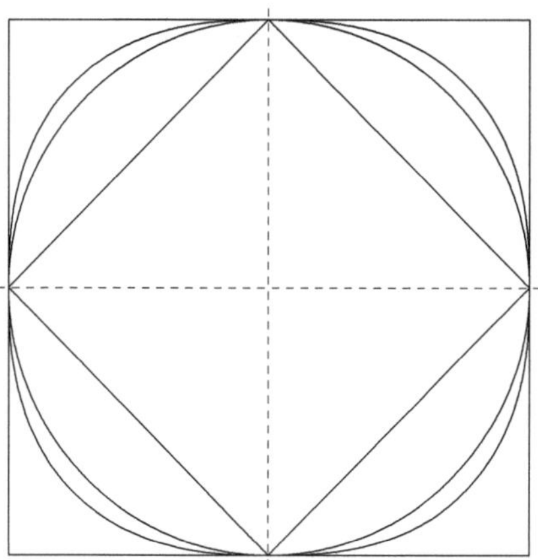

FIGURE A.1: The unit disk with respect to the r-norm, for $r = 1$ (diamond), $r = 2$ (circle), $r = 5/2$ (Piet Hein's superellipse), and $r = \infty$ (square).

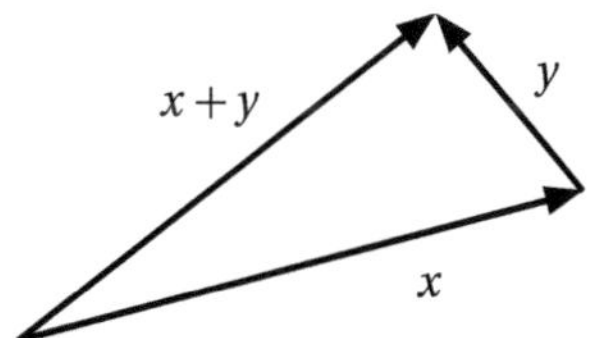

FIGURE A.2: The triangle inequality.

A.2.3 *The Cauchy–Schwarz Inequality*

Another geometrical fact is that the inner product, or dot product, of two vectors is at most the product of their lengths:

$$x^T y = \sum_i x_i y_i \le \|x\|_2 \|y\|_2 .$$

Using Pythagoras' theorem and squaring both sides gives the *Cauchy–Schwarz inequality*,

$$\left(\sum_i x_i y_i\right)^2 \le \left(\sum_i x_i^2\right)\left(\sum_i y_i^2\right), \tag{A.5}$$

where equality holds only if x and y are parallel.

The Cauchy–Schwarz inequality holds for generalized inner products as well. If we have a set of non-negative real coefficients a_i, then

$$\left(\sum_i a_i x_i y_i\right)^2 \le \left(\sum_i a_i x_i^2\right)\left(\sum_i a_i y_i^2\right). \tag{A.6}$$

To see this, just write $x_i' = \sqrt{a_i}\, x_i$ and $y_i' = \sqrt{a_i}\, y_i$ and apply (A.5) to x' and y'.

The Cauchy–Schwarz inequality is often useful even for sums that don't look like inner products. For instance,

$$\left(\sum_{i=1}^{N} v_i\right)^2 \leq N \sum_{i=1}^{N} v_i^2 ,$$

since we can think of the sum on the left-hand side as the inner product of v with an N-dimensional vector $(1,\ldots,1)$. This lets us bound the 1-norm of an N-dimensional vector in terms of its 2-norm,

$$\|v\|_1 = \sum_{i=1}^{N} |v_i| \leq \sqrt{N \sum_{i=1}^{N} v_i^2} = \sqrt{N}\,\|v\|_2 . \tag{A.7}$$

On the other hand, we also have

$$\|v\|_2 \leq \|v\|_1 . \tag{A.8}$$

Exercise A.10 *Show that the extreme cases of the inequalities* (A.7) *and* (A.8) *occur when v is a basis vector or when all its components are equal. Hint: consider the inscribed and circumscribed spheres of an octahedron.*

A.2.4 Taylor Series

Given an analytic function $f(x)$ we can approximate it by a Taylor series,

$$f(x) = f(0) + f'(0)x + \frac{1}{2}f''(0)x^2 + \frac{1}{6}f'''(0)x^3 + \cdots = \sum_{k=0}^{\infty} \frac{1}{k!} f^{(k)}(0)\,x^k ,$$

where $f^{(k)} = \mathrm{d}^k f/\mathrm{d}x^k$ is the kth derivative.

Since e^x is its own derivative this gives the following series for the exponential,

$$\mathrm{e}^x = 1 + x + \frac{x^2}{2} + \frac{x^3}{6} + \cdots = \sum_{k=0}^{\infty} \frac{x^k}{k!} .$$

Another helpful series is

$$-\ln(1-x) = x + \frac{x^2}{2} + \frac{x^3}{3} + \cdots = \sum_{k=0}^{\infty} \frac{x^k}{k} ,$$

which gives us, for instance,

$$(1+x)^y = \mathrm{e}^{y\ln(1+x)} = \mathrm{e}^{y(x - O(x^2))} = \mathrm{e}^{xy}\left(1 - O(x^2 y)\right) .$$

In some cases the Taylor series gives a firm inequality. For instance, for all x we have

$$1 + x \leq \mathrm{e}^x .$$

For $x > 0$, this follows since all the higher-order terms in e^x are positive. For $-1 < x < 0$ their signs alternate, but they decrease in size geometrically.

A.2.5 *Stirling's Approximation*

Stirling gave a very useful approximation for the factorial,

$$n! = \left(1 + O(1/n)\right)\sqrt{2\pi n}\, n^n \mathrm{e}^{-n}. \tag{A.9}$$

We prove this in Problem A.28. Some simpler approximations can be useful as well. For instance,

$$n! > n^n \mathrm{e}^{-n},$$

and this gives us a useful upper bound on binomial coefficients $\binom{n}{k}$,

$$\binom{n}{k} \le \frac{n^k}{k!} \le \left(\frac{\mathrm{e}n}{k}\right)^k. \tag{A.10}$$

As Problem A.13 shows, this approximation is tight as long as $k = o(\sqrt{n})$.

A.3 Chance and Necessity

> The most important questions in life are, for the most part, really only problems of probability.
>
> Pierre-Simon Laplace

Here we collect some simple techniques in discrete probability, which we use in Chapter 10 and beyond. Most of the reasoning we present here is quite elementary, but even these techniques have some surprisingly powerful applications.

A.3.1 *ANDs and ORs*

The *union bound* is the following extremely simple observation. If A and B are events, the probability that one or the other of them occurs is bounded by

$$\Pr[A \vee B] \le \Pr[A] + \Pr[B]. \tag{A.11}$$

More generally, if we have a set of events A_i, the probability that at least one occurs is bounded by

$$\Pr\left[\bigvee_i A_i\right] \le \sum_i \Pr[A_i].$$

The union bound (A.11) holds with equality only when when $\Pr[A \wedge B] = 0$, i.e., when A and B are disjoint so that they never occur simultaneously. We can correct the union bound by taking this possibility into account, and this gives the *inclusion–exclusion principle*:

$$\Pr[A \vee B] = \Pr[A] + \Pr[B] - \Pr[A \wedge B]. \tag{A.12}$$

Think of this as calculating the area $A \cup B$ in a Venn diagram. Adding the areas of A and B counts the area of $A \cap B$ twice, and we correct this by subtracting it once.

In general, if we have a set S of events $E_1, \ldots, E_n$, the probability that none of them occurs is an alternating sum over all subsets $T \subseteq S$ of the probability that every event in T holds:

$$\Pr\left[\overline{\bigvee_{i=1}^{n} E_i}\right] = \sum_{T \subseteq \{1,\ldots,n\}} (-1)^{|T|} \Pr\left[\bigwedge_{i \in T} E_i\right]. \tag{A.13}$$

There are many ways to prove this—see Problem A.1 for one. Note that if $T = \emptyset$ then $\Pr\left[\bigwedge_{i \in T} E_i\right] = 1$.

A.3.2 Expectations and Markov's and Chebyshev's Inequalities

Even when a random variable X is complicated, subtle, and full of correlations, it is often quite easy to calculate its average, or *expectation*, which we denote $\mathbb{E}[X]$. The reason for this is the so-called *linearity of expectation*, the fact that the expectation of a sum is the sum of the expectations:

$$\mathbb{E}[X+Y] = \mathbb{E}[X] + \mathbb{E}[Y].$$

This is true even if X and Y are correlated. For instance, if h is the average height of a child, the average of the sum of two children's heights is $2h$, even if they are siblings.

The expectation of the product, on the other hand, is not generally the product of the expectations: $\mathbb{E}[XY]$ is larger or smaller than $\mathbb{E}[X]\,\mathbb{E}[Y]$ if their correlation is positive or negative respectively. For instance, the average of the product of two siblings' heights is greater than h^2.

Once we know the expectation of a random variable, a *concentration inequality* seeks to show that it is probably not too far from its expectation. The simplest and weakest of these is *Markov's inequality*. Let X be a nonnegative random variable. Then for any λ, the probability that X is at least λ times its expectation is bounded by

$$\Pr[X \geq \lambda\,\mathbb{E}[X]] \leq 1/\lambda. \tag{A.14}$$

The proof is easy: if $X \geq t$ with probability at least p and X is never negative, then $\mathbb{E}[X] \geq pt$.

The *variance* of a random variable X is the expected squared difference between X and its expectation, $\operatorname{Var} X = \mathbb{E}[(X - \mathbb{E}[X])^2]$. Some simple things you should know about the variance are given in the following two exercises.

Exercise A.11 *Show that* $\operatorname{Var} X = \mathbb{E}[X^2] - \mathbb{E}[X]^2$.

Exercise A.12 *Two random variables X, Y are* independent *if the joint probability distribution $P[X,Y]$ is the product $P[X]P[Y]$. Show that if X and Y are independent, then* $\mathbb{E}[XY] = \mathbb{E}[X]\,\mathbb{E}[Y]$ *and* $\operatorname{Var}(X+Y) = \operatorname{Var} X + \operatorname{Var} Y$.

Markov's inequality can be used to prove another simple inequality, showing that if $\operatorname{Var} X$ is small then X is probably close to its expectation:

Exercise A.13 *Derive Chebyshev's inequality,*

$$\Pr\big[|X - \mathbb{E}[X]| > \delta\big] \leq \frac{\operatorname{Var} X}{\delta^2}. \tag{A.15}$$

Hint: apply Markov's inequality to the random variable $(X - \mathbb{E}[X])^2$.

Chebyshev's inequality is a good example of a fact we will see many times—that half the battle is choosing which random variable to analyze.

Now suppose that X is a nonnegative integer which counts the number of objects of some kind. Then Markov's inequality implies

$$\Pr[X>0]=\Pr[X\geq 1]\leq \mathbb{E}[X]. \tag{A.16}$$

An even easier way to see this is

$$\Pr[X>0]=\sum_{x=1}^{\infty}\Pr[X=x]\leq \sum_{x=0}^{\infty}x\Pr[X=x]=\mathbb{E}[X].$$

We can use this to prove that a given type of object probably doesn't exist. If the expected number of such objects is $o(1)$, then (A.16) implies that $\Pr[X>0]=o(1)$. Then with probability $1-o(1)$ we have $X=0$ and there is no such object. This is often called the *first moment method*, since the kth moment of X is $\mathbb{E}[X^k]$.

A.3.3 The Birthday Problem

As an application of some of these techniques, consider the so-called Birthday Problem. How many people do we need to have in a room before it becomes likely that two of them have the same birthday? Assume that each person's birthday is uniformly random, i.e., that it is chosen from the 365 possibilities (ignoring leap years) with equal probability. Most people guess that the answer is roughly half of 365, but this is very far off the mark.

If there are n people and y days in the year, there are $\binom{n}{2}$ possible pairs of people. For each pair, the probability that they have the same birthday is $1/y$. Thus by the union bound, the probability that at least one pair of people have the same birthday is at most

$$\binom{n}{2}\frac{1}{y}\approx\frac{n^2}{2y}.$$

This is small until $n\approx\sqrt{2y}$, or roughly 27 in the case $y=365$. This is just an upper bound on this probability, but it turns out that $\sqrt{2y}$ is essentially the right answer.

Let's repeat this calculation using the first moment method. Let B denote the number of pairs of people with the same birthday. By linearity of expectation, $\mathbb{E}[B]$ is the sum over all pairs i,j of the probability p_{ij} that i and j have the same birthday. Once again, there are $\binom{n}{2}$ pairs and $p_{ij}=1/y$ for all of them, so

$$\mathbb{E}[B]=\binom{n}{2}\frac{1}{y}\approx\frac{n^2}{2y}. \tag{A.17}$$

This calculation may seem exactly the same as our union bound above, but there is an important difference. That was an upper bound on the probability that any pair share a birthday, while this is an *exact* calculation of the expected number of pairs that do.

Exercise A.14 *Suppose there are three people, Albus, Bartholemew, and Cornelius, and consider the possibility that one or more pairs were born on the same day of the week. These events are positively correlated: for instance, if this is true of two pairs then it is true of all three, so we never have $B=2$. Nevertheless, confirm that* (A.17) *with $n=3$ and $y=7$ gives the expectation of B exactly.*

A slightly more formal way to write this is to define an *indicator random variable* for each pair of people, associated with the event that they have the same birthday:

$$X_{ij} = \begin{cases} 1 & \text{if } i \text{ and } j \text{ have the same birthday} \\ 0 & \text{if they don't.} \end{cases}$$

Then

$$B = \sum_{i<j} X_{ij} .$$

Since $\mathbb{E}[X_{ij}] = p_{ij}$, linearity of expectation gives

$$\mathbb{E}[B] = \mathbb{E}\left[\sum_{i<j} X_{ij}\right] = \sum_{i<j} \mathbb{E}[X_{ij}] = \sum_{i<j} p_{ij} = \binom{n}{2}\frac{1}{y} .$$

Now that we have computed $\mathbb{E}[B]$, we can use the first moment method. If $n = o(\sqrt{y})$ then $\mathbb{E}[B] = o(1)$, so the probability that there is a pair of people with the same birthday is $o(1)$. Thus we can say that *with high probability*, i.e., with probability $1 - o(1)$, all the birthdays are different.

It's important to realize that even if the expected number of pairs is large, that doesn't prove that there probably is one. In other words, a lower bound on $\mathbb{E}[B]$ does not, in and of itself, give a lower bound on $\Pr[B > 0]$. This the because the distribution of B could have a very *heavy tail*, in which it is zero almost all the time, but is occasionally enormous. Such a variable has a very high variance, so we can eliminate this possibility by placing an upper bound on $\mathbb{E}[B^2]$. This is the point of the *second moment method* described in Section A.3.5. The Birthday Problem is simple enough, however, that we can prove that a pair with the same birthday really does appear with high probability when $n \approx \sqrt{2y}$; see Problem A.6.

A.3.4 Coupon Collecting

Another classic problem in discrete probability is the *Coupon Collector's Problem.* To induce children to demand that their parents buy a certain kind of cereal, the cereal company includes a randomly chosen toy at the bottom of each box. If there are n different toys and they are equally likely, how many boxes of cereal do I have to buy for my daughter before she almost certainly has one of each? Clearly I need to buy more than n boxes—but how many more?

Suppose I have bought b boxes of cereal so far, and let T be the number of toys I still don't have. The probability I am still missing a particular toy is $(1 - 1/n)^b$, since I get that toy with probability $1/n$ each time I buy a box. By linearity of expectation, the expected number of toys I am missing is

$$\mathbb{E}[T] = n(1 - 1/n)^b < n\mathrm{e}^{-b/n} .$$

This is 1 when $b = n \ln n$. Moreover, if $b = (1 + \varepsilon) n \ln n$ for some $\varepsilon > 0$, then $\mathbb{E}[T] < n^{-\alpha} = o(1)$, and by Markov's inequality, with high probability my daughter has a complete collection.

Another type of result we can obtain is the *expected* number of boxes it takes to obtain a complete collection. First consider the following exercise:

Exercise A.15 *Suppose I have a biased coin which comes up heads with probability p. Let t be the number of times I need to flip it to get the first head. Show that $\mathbb{E}[t] = 1/p$. First do this the hard way, by showing that the probability the first head comes on the t th flip is $P(t) = (1-p)^{t-1}p$ and summing the series $\sum_t P(t)t$. Then do it the easy way, by showing that $\mathbb{E}[t] = p + (1-p)(\mathbb{E}[t]+1)$.*

Now suppose that I am missing i of the toys. The probability that the next box of cereal adds a new toy to my collection is i/n, so the average number of boxes I need to buy to get a new toy is n/i. For instance, if I am a new collector I gain a new toy in the very first box, while if my collection is complete except for one last toy, it will take me an average of n boxes to get it.

By linearity of expectation, the expected number of boxes it takes to get a complete collection is then the sum of n/i over all i,

$$\mathbb{E}[b] = \frac{n}{1} + \frac{n}{2} + \frac{n}{3} + \cdots + \frac{n}{n} = n\left(1 + \frac{1}{2} + \frac{1}{3} + \cdots + \frac{1}{n}\right) = nH_n \approx n \ln n\,.$$

Here $H_n = \sum_{i=1}^{n} 1/i$ is the n harmonic number, which diverges like $\ln n$ as n grows.

Note the difference between this result, which gives the expected number of boxes I need, and the previous one, which gave a number of boxes after which I have all the toys with high probability.

A.3.5 Bounding the Variance: the Second Moment Method

The first moment method is an excellent way to prove that certain things probably don't exist. If X is the number of some kind of thing then $\Pr[X > 0] \le \mathbb{E}[X]$, so if $\mathbb{E}[X]$ is small then probably $X = 0$ and none of these things exist.

How can we derive a *lower* bound on $\Pr[X > 0]$, and thus prove that one of these things probably does exist? As mentioned above, it is not enough to show that $\mathbb{E}[X]$ is large, since this could be due to X occasionally being enormous. To exclude this possibility, we can bound X's variance, or equivalently its second moment $\mathbb{E}[X^2]$.

One way to do this is to apply Chebyshev's inequality (A.15), as the following exercise suggests:

Exercise A.16 *Suppose $\mathbb{E}[X] > 0$. Show that*

$$\Pr[X > 0] \ge 1 - \frac{\operatorname{Var} X}{\mathbb{E}[X]^2} = 2 - \frac{\mathbb{E}[X^2]}{\mathbb{E}[X]^2}\,.$$

However, we will derive the following inequality, which is often much stronger:

$$\Pr[X > 0] \ge \frac{\mathbb{E}[X]^2}{\mathbb{E}[X^2]}\,. \tag{A.18}$$

Thus we can show that $X > 0$ is likely whenever the second moment is not much larger than the square of the expectation. This is called the *second moment method.*

We prove (A.18) using the Cauchy–Schwarz inequality from Appendix A.2.3. If X and Y depend on some random variable i with a probability distribution p_i, we can think of $\mathbb{E}[XY]$ as an inner product,

$$\mathbb{E}[XY] = \sum_i p_i X_i Y_i$$

and using (A.6) gives

$$\mathbb{E}[XY]^2 \le \left(\sum_i p_i X_i^2\right)\left(\sum_i p_i Y_i^2\right) = \mathbb{E}[X^2]\,\mathbb{E}[Y^2]. \tag{A.19}$$

Now let Y be the indicator random variable for the event $X > 0$, i.e., $Y = 1$ if $X > 0$ and $Y = 0$ if $X = 0$. Using the facts that $X = XY$, $Y^2 = Y$, and $\mathbb{E}[Y] = \Pr[X > 0]$, the Cauchy–Schwarz inequality (A.19) gives

$$\mathbb{E}[X]^2 = \mathbb{E}[XY]^2 \le \mathbb{E}[X^2]\,\mathbb{E}[Y^2] = \mathbb{E}[X^2]\,\mathbb{E}[Y] = \mathbb{E}[X^2]\Pr[X > 0].$$

Dividing both sides by $\mathbb{E}[X^2]$ completes the proof of (A.18).

How do we calculate the second moment? Suppose that X is a sum of indicator random variables X_i, each of which is associated with an event i. Then

$$\mathbb{E}[X^2] = \mathbb{E}\left[\sum_{i,j} X_i X_j\right] = \sum_{i,j}\mathbb{E}[X_i X_j] = \sum_{i,j}\Pr[i \wedge j].$$

In other words, $\mathbb{E}[X^2]$ is the sum over all ordered pairs of events i, j of the probability that they both occur. We can write this as the probability of i times the conditional probability of j given i,

$$\Pr[i \wedge j] = \Pr[i]\Pr[j \mid i],$$

which gives

$$\mathbb{E}[X^2] = \sum_i \Pr[i] \sum_j \Pr[j \mid i]. \tag{A.20}$$

To calculate the conditional probability $\Pr[j \mid i]$, we need to know how these events are correlated. In particular, if knowing that i occurs makes it more likely that j occurs, then $\Pr[j \mid i] > \Pr[j]$. Pairs of events for which this effect is strong make a larger contribution to the second moment. In many applications, these events are only weakly correlated. Then $\mathbb{E}[X^2] = (1 + o(1))\mathbb{E}[X]^2$, and X is concentrated around its expectation.

However, in Chapter 14 we consider events that are strongly correlated, such as the events that two truth assignments both satisfy the same random formula. In such cases, we can often prove that $\mathbb{E}[X^2]$ is at most $C\,\mathbb{E}[X]^2$ for some constant C, and therefore that $\Pr[X > 0] \ge 1/C$.

A.3.6 *Jensen's Inequality*

Let x be a random variable, and let f be a function. In general, $\mathbb{E}[f(x)] = f(\mathbb{E}[x])$ only if x's probability distribution is concentrated at a single value or if f is a straight line. If x has some variance around its most likely value, and if f has some curvature, then $\mathbb{E}[f(x)]$ and $f(\mathbb{E}[x])$ can be quite different.

We say that $f(x)$ is *convex* if, for any x_1, x_2 and any $\lambda \in [0,1]$ we have

$$f(\lambda x_1 + (1-\lambda)x_2) \le \lambda f(x_1) + (1-\lambda) f(x_2).$$

In other words, if we draw a line segment between any two points on the graph of f as in Figure A.3, then f lies at or below this line segment. For a continuous function, we can say equivalently that the second

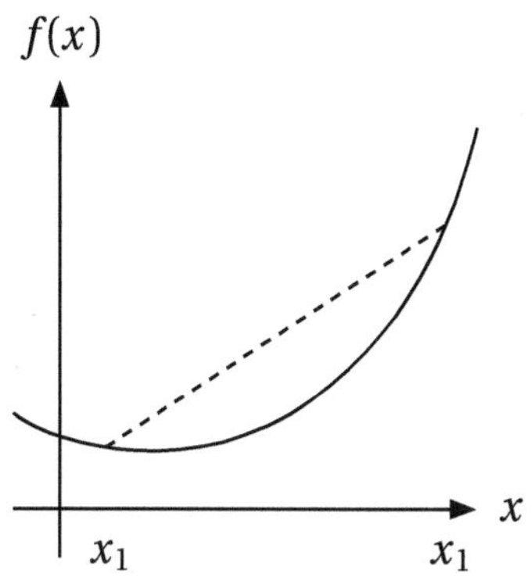

FIGURE A.3: A convex function.

derivative f'' is nonnegative. *Jensen's inequality* states that for any convex function f and any probability distribution on x,

$$\mathbb{E}[f(x)] \geq f(\mathbb{E}[x]) .$$

For instance, $\mathbb{E}[x^2] \geq \mathbb{E}[x]^2$ and $\mathbb{E}[e^x] \geq e^{\mathbb{E}[x]}$.

Exercise A.17 *Prove Jensen's inequality for the case where x takes two different values, x_1 and x_2 with probability p and $1-p$ respectively.*

Similarly, $\mathbb{E}[f(x)] \leq f(\mathbb{E}[x])$ for any *concave* function f, i.e., a function whose second derivative is less than or equal to zero. For instance, $\mathbb{E}[\ln x] \leq \ln \mathbb{E}[x]$ and $\mathbb{E}[\sqrt{x}] \leq \sqrt{\mathbb{E}[x]}$.

A.4 Dice and Drunkards

He uses statistics as a drunken man uses lamp-posts... for support rather than illumination.

Andrew Lang

They reel to and fro, and stagger like a drunken man, and are at their wit's end.

King James Bible, Psalm 107:27

Many random processes generate probability distributions of numbers and positions—for instance, the number of heads in a sequence of coin flips, or the position we end up in when we take a random walk. In this section we discuss these distributions, how concentrated they are around their expectations, and how to estimate the time that it takes a random walk to travel a certain distance from its starting point.

A.4.1 *The Binomial Distribution*

Suppose I flip a series of n coins, each one of which comes up heads with probability p and tails with probability $1-p$. The probability that exactly x of them come up heads is the number of subsets of size x, times the probability that these all come up heads, times the probability that all the others come up tails:

$$P(x) = \binom{n}{x} p^x (1-p)^{n-x} .$$

This is called the *binomial distribution*, and is often written $\mathrm{Bin}(n,p)$. We can also think of x as the sum of n independent random variables,

$$x = \sum_{i=1}^{n} y_i \quad \text{where} \quad y_i = \begin{cases} 1 & \text{with probability } p \\ 0 & \text{with probability } 1-p . \end{cases}$$

By linearity of expectation (and common sense!) the expectation of the binomial distribution is $\mathbb{E}[x] = pn$. The next exercise asks you to calculate its variance.

Exercise A.18 *Show that the variance of the binomial distribution* $\mathrm{Bin}(n,p)$ *is* $p(1-p)n$.

A.4.2 *The Poisson Distribution*

Suppose that I toss m balls randomly into n bins. What is the probability that the first bin has exactly x balls in it? This is a binomial distribution where $p = 1/n$, so we have

$$P(x) = \binom{m}{x} \left(\frac{1}{n}\right)^x \left(1 - \frac{1}{n}\right)^{m-x} . \tag{A.21}$$

Now suppose that $m = cn$ for some constant c. The average number of balls in any particular bin is c, independent of n, so $P(x)$ should become independent of n in the limit $n \to \infty$. What distribution does $P(x)$ converge to?

First we claim that for any constant x, in the limit $m \to \infty$ we have

$$\binom{m}{x} \approx \frac{m^x}{x!} .$$

Then (A.21) becomes

$$P(x) = \lim_{n\to\infty} \frac{c^x}{x!} \left(1 - \frac{1}{n}\right)^{cn-x} = \frac{\mathrm{e}^{-c} c^x}{x!} . \tag{A.22}$$

As Problem A.13 shows, this limit holds whenever $x = o(\sqrt{m})$, and in particular for any constant x.

This distribution (A.22) is called the *Poisson distribution*, and you should check that it has mean c. It occurs, for instance, as the degree distribution of a sparse random graph in Section 14.2. In physics, if I am using a Geiger counter to measure a radioactive material, and if in each infinitesimal time interval $\mathrm{d}t$ there is an probability $p\,\mathrm{d}t$ of hearing a click, the total number of clicks I hear in T seconds is Poisson-distributed with mean pT. Just as in the balls-in-bins example, this is an extreme case of the binomial distribution—there are a large number of events, each one of which occurs with very small probability.

A.4.3 Entropy and the Gaussian Approximation

Let's rephrase the binomial distribution a bit, and ask for the probability that the fraction x/n of heads that come up is some a between 0 and 1. Applying Stirling's approximation (A.9) to the binomial and simplifying gives

$$\binom{n}{an} \approx \frac{1}{\sqrt{2\pi a(1-a)n}} \left(\frac{1}{a^a (1-a)^{1-a}} \right)^n = \frac{1}{\sqrt{2\pi a(1-a)n}} \, \mathrm{e}^{n\,h(a)} . \tag{A.23}$$

Here $h(a)$ is the *Gibbs–Shannon entropy*,

$$h(a) = -a \ln a - (1-a)\ln(1-a) .$$

We can think of $h(a)$ as the average amount of information generated by a random coin, which comes up heads with probability a and tails with probability $1-a$. Its maximum value is $h(1/2) = \ln 2$, since a fair coin generates one bit of information per toss.

Applying (A.23) to the binomial distribution gives

$$P(an) = \frac{1}{\sqrt{2\pi a(1-a)n}} \, \mathrm{e}^{n\,h(a\|p)} , \tag{A.24}$$

where $h(a\|p)$ is a weighted version of the Shannon entropy,

$$h(a\|p) = -a \ln \frac{a}{p} - (1-a) \ln \frac{1-a}{1-p} .$$

We can think of $-h(a\|p)$ as a kind of distance between the probability distribution $(p, 1-p)$ and the observed distribution $(a, 1-a)$. It is called the *Kullback–Leibler divergence.* It is minimized at $a = p$, where $h(a\|p) = 0$. The second-order Taylor series around p then gives

$$h(p+\varepsilon\|p) = -\frac{1}{2} \frac{\varepsilon^2}{p(1-p)} + O(\varepsilon^3) ,$$

and the second-order term is an upper bound.

Let's plug this back in to (A.24), assume that the slowly-varying part $1/\sqrt{2\pi a(1-a)n}$ is constant for $a \approx p$. Then setting $\delta = \varepsilon n$ gives

$$P(pn+\delta) \approx \frac{1}{\sqrt{2\pi p(1-p)n}} \exp\left(-\frac{1}{2} \frac{\delta^2}{p(1-p)n} \right) .$$

The *Gaussian* or *normal* distribution with mean zero and variance v is

$$p(y) = \frac{1}{\sqrt{2\pi v}} \, \mathrm{e}^{-(1/2) y^2 / v} . \tag{A.25}$$

Thus when n is large, the deviation δ of a binomial random variable from its mean pn is distributed as a Gaussian with mean zero and variance $p(1-p)n$. This is one manifestation of the Central Limit Theorem, which states that the sum of any set of independent random variables with bounded variance converges to a Gaussian. In particular, it is tightly concentrated around its mean in a sense we will see in Appendix A.5.1.

A.4.4 *Random Walks*

One example of a binomial distribution is a random walk on the line. I have had too much to drink. At each step, I stumble one unit to the left or right with equal probability. If I start at the origin and take t steps, the number r of steps in which I move to the right is binomially distributed with $p=1/2$, and I end at a position x if $r=(x+t)/2$. The probability that this happens is

$$P(x)=2^{-t}\binom{t}{(x+t)/2}\approx\frac{2}{\sqrt{2\pi t}}\,\mathrm{e}^{-(1/2)x^2/t}\,.$$

Except for the factor of 2, which appears since only the values of x with the proper parity, odd or even, can occur, this is a Gaussian with mean zero and variance t. Since the width of this distribution is $O(\sqrt{t})$, intuitively it takes about $t\sim n^2$ steps for me to get n steps away from the origin—or to reach the origin from an initial position n steps away.

Let's prove a precise result to this effect. Suppose we have a random walk on a finite line, ranging from 0 to n. We move left or right with equal probability unless we are at the right end, where we have no choice but to move to the left. If we ever reach the origin, we stay there.

Let $t(x)$ be the expected time it takes to reach the origin, given that we start at position x where $0\le x\le n$. Since $t(x)=1+\mathbb{E}[t(x')]$, where x' is wherever we'll be on the next step, for any $0<x<n$ we have the following recurrence:

$$t(x)=1+\frac{t(x-1)+t(x+1)}{2}\,.$$

Combined with the boundary conditions $t(0)=0$ (since we're already there) and $t(n)=f(n-1)+1$ (since we bounce off the right end), the unique solution is

$$t(x)=x(2n-x)\le n^2\,. \tag{A.26}$$

Exercise A.19 *Prove* (A.26).

Of course, knowing the expected time to reach the origin isn't the same as knowing a time by which we will reach it with high probability. To get a result of that form, we can cleverly focus on a different random variable. While the expectation of x stays the same at each step, the expectation of $\sqrt{x}$ decreases. Specifically, a little algebra gives

$$\mathbb{E}\left[\sqrt{x}\right]=\frac{1}{2}\left(\sqrt{x+1}+\sqrt{x-1}\right)\le\sqrt{x}\left(1-\frac{1}{8x^2}\right)\le\sqrt{x}\,\mathrm{e}^{-1/8x^2}\,.$$

Now let's consider a walk on the half-infinite line, where if we ever touch the origin we stay there. If our initial position is n, then after t steps we have

$$\mathbb{E}\left[\sqrt{x}\right]\le\sqrt{n}\,\mathrm{e}^{-t/8n^2}\,.$$

The probability that we still haven't reached the origin is the probability that $\sqrt{x}\ge 1$, and by Markov's inequality this is at most $\mathbb{E}\left[\sqrt{x}\right]$. Setting $t=8n^2\ln n$, say, gives $\mathbb{E}\left[\sqrt{x}\right]\le 1/\sqrt{n}$, so we have reached the origin with high probability.

A.5 Concentration Inequalities

The Markov and Chebyshev inequalities discussed in Section A.3.2 state, in a very weak fashion, that a random variable is probably not too far from its expectation. For distributions such as the binomial or Poisson distributions we can get much tighter results. In this section, we prove several inequalities showing that a random variable is close to its expectation with high probability.

A.5.1 Chernoff Bounds

One classic type of concentration inequality is the *Chernoff bound.* It asserts that if x is chosen from the binomial distribution $\text{Bin}(n,p)$, the probability that x differs from its expectation by a factor $1 \pm \delta$ decreases exponentially as a function of δ and n.

To bound the probability that $x \geq t$ for a given t, we start with a clever choice of random variable. We apply Markov's inequality to $\mathrm{e}^{\lambda x}$, rather than to x itself. This gives

$$\Pr[x \geq t] = \Pr[\mathrm{e}^{\lambda x} \geq \mathrm{e}^{\lambda t}] \leq \frac{\mathbb{E}[\mathrm{e}^{\lambda x}]}{\mathrm{e}^{\lambda t}}. \tag{A.27}$$

This inequality is true for any value of λ, so we are free to choose λ however we like. Below, we will select λ in order to make our bounds as tight as possible.

The expectation $\mathbb{E}[\mathrm{e}^{\lambda x}]$ is called the *moment generating function* of a distribution. It is especially easy to calculate for the binomial distribution $\text{Bin}(n,p)$. Recall that x is the sum of n independent random variables y_i, each of which is 1 with probability p, and 0 with probability $1-p$. Thus we can write

$$\mathrm{e}^{\lambda x} = \mathrm{e}^{\lambda \sum_{i=1}^n y_i} = \prod_{i=1}^n \mathrm{e}^{\lambda y_i}.$$

The y_i are independent, so the expectation of this product is the product of these expectations. Since

$$\mathbb{E}[\mathrm{e}^{\lambda y_i}] = p\mathrm{e}^{\lambda} + (1-p),$$

this gives

$$\mathbb{E}[\mathrm{e}^{\lambda x}] = \prod_{i=1}^n \mathbb{E}[\mathrm{e}^{\lambda y_i}] = \left(p\mathrm{e}^{\lambda} + (1-p)\right)^n = \left(1 + (\mathrm{e}^{\lambda} - 1)p\right)^n.$$

Since $1 + z \leq \mathrm{e}^z$, we can bound this as

$$\mathbb{E}[\mathrm{e}^{\lambda x}] \leq \mathrm{e}^{(\mathrm{e}^{\lambda}-1)pn} = \mathrm{e}^{(\mathrm{e}^{\lambda}-1)\mathbb{E}[x]}. \tag{A.28}$$

Exercise A.20 *Show that if x is Poisson-distributed then* (A.28) *is exact. That is, its moment generating function is*

$$\mathbb{E}[\mathrm{e}^{\lambda x}] = \mathrm{e}^{(\mathrm{e}^{\lambda}-1)\mathbb{E}[x]},$$

We are interested in bounding the probability that x is a factor $1+\delta$ larger than its expectation. So we set $t = (1+\delta)\mathbb{E}[x]$ and plug (A.28) into (A.27), giving

$$\Pr\left[x \geq (1+\delta)\mathbb{E}[x]\right] \leq \frac{\mathrm{e}^{(\mathrm{e}^{\lambda}-1)\mathbb{E}[x]}}{\mathrm{e}^{\lambda(1+\delta)\mathbb{E}[x]}} = \left(\mathrm{e}^{\mathrm{e}^{\lambda} - \lambda(1+\delta) - 1}\right)^{\mathbb{E}[x]}. \tag{A.29}$$

As stated above, we can now tune λ to turn (A.29) into the tightest bound possible. At this point, we ask the reader to take over.

Exercise A.21 *Find the value of λ that minimizes the right-hand side of* (A.29) *and show that, for any $\delta \ge 0$,*

$$\Pr[x \ge (1+\delta)\mathbb{E}[x]] \le \left(\frac{e^{\delta}}{(1+\delta)^{1+\delta}}\right)^{\mathbb{E}[x]}.$$

Then use the inequalities

$$\delta - (1+\delta)\ln(1+\delta) \le \begin{cases} -\delta^2/3 & \text{if } 0 \le \delta \le 1 \\ -\delta/3 & \text{if } \delta > 1 \end{cases}$$

to prove the following theorem:

Theorem A.1 *For any binomial random variable x,*

$$\Pr[x \ge (1+\delta)\mathbb{E}[x]] \le \begin{cases} e^{-\delta^2 \mathbb{E}[x]/3} & \text{if } 0 \le \delta \le 1 \\ e^{-\delta \mathbb{E}[x]/3} & \text{if } \delta > 1. \end{cases}$$

Note that when δ is small this bound decays exponentially with δ^2, just like the Gaussian approximation for the binomial in Section A.4.3. Indeed, it is identical to (A.25) except that we replace the constant $1/2$ in the exponent with the weaker $1/3$ to obtain a valid inequality in the interval $\delta \in [0,1]$.

Similarly, we can bound the probability that x is a factor $1-\delta$ less than its expectation. Applying Markov's inequality to the random variable $e^{-\lambda x}$ gives

$$\Pr[x \le t] \le e^{\lambda t}\, \mathbb{E}[e^{-\lambda x}].$$

Exercise A.22 *Show that*

$$\Pr[x \le (1-\delta)\mathbb{E}[x]] \le \left(e^{e^{-\lambda}+\lambda(1-\delta)-1}\right)^{\mathbb{E}[x]}.$$

Minimize the right-hand side as a function of λ and use the inequality

$$-\delta - (1-\delta)\ln(1-\delta) \le \delta^2/2 \quad \text{if} \quad 0 \le \delta \le 1,$$

to prove the following theorem:

Theorem A.2 *For any binomial random variable x, if $\delta \ge 0$ then*

$$\Pr[x \le (1-\delta)\mathbb{E}[x]] \le e^{-\mathbb{E}[x]\delta^2/2}.$$

A common use of the Chernoff bound is to show that, if a randomized algorithm gives the correct yes-or-no answer with probability greater than $1/2$, we can get the correct answer with high probability by running it multiple times and taking the majority of its answers. Consider the following exercise:

Exercise A.23 *Suppose a coin comes up heads with probability $1/2+\varepsilon$ for some $\varepsilon > 0$. Show that if we flip it n times, the majority of these flips will be heads with probability $1 - e^{\Omega(\varepsilon^2 n)}$.*

A.5.2 Martingales and Azuma's inequality

The Chernoff bounds of the previous section show that if x is the sum of n independent random variables y_i then x is probably close to its expectation. In this section, we show that under certain circumstances, this is true even if the y_i are correlated.

First, let's prove another type of Chernoff bound. Here the y_i are still independent, but each one has its own probability distribution. All we ask is that the y_i are bounded, and that each one has zero expectation.

Theorem A.3 *Let $x = \sum_{i=1}^n y_i$ where each y_i is chosen independently from some probability distribution p_i such that $|y_i| \le 1$ and $\mathbb{E}[y_i] = 0$. Then for any $a \ge 0$,*

$$\Pr[|x| > a] \le 2\mathrm{e}^{-a^2/(2n)}.$$

Proof We again use the moment generating function. For each i, we have

$$\mathbb{E}[\mathrm{e}^{\lambda y_i}] = \int_{-1}^{1} p_i(y_i)\,\mathrm{e}^{\lambda y_i}\,\mathrm{d}y_i. \tag{A.30}$$

The function $\mathrm{e}^{\lambda y}$ is convex, as in Section A.3.6. Thus for any $y \in [-1,1]$ we have

$$\mathrm{e}^{\lambda y} \le \frac{1+y}{2}\,\mathrm{e}^{\lambda} + \frac{1-y}{2}\,\mathrm{e}^{-\lambda} = \cosh\lambda + y\sinh\lambda,$$

since this is the value we would get by drawing a straight line between $y = +1$ and $y = -1$ and interpolating. This lets us bound the integral in (A.30) as follows,

$$\begin{aligned}\mathbb{E}[\mathrm{e}^{\lambda y_i}] &\le \int_{-1}^{1} p_i(y_i)\left(\cosh\lambda + y_i\sinh\lambda\right)\,\mathrm{d}y_i \\ &= \cosh\lambda + \mathbb{E}[y_i]\sinh\lambda = \cosh\lambda.\end{aligned} \tag{A.31}$$

Since the y_i are independent, the moment generating function of x is bounded by

$$\mathbb{E}[\mathrm{e}^{\lambda x}] = \prod_{i=1}^{n}\mathbb{E}[\mathrm{e}^{\lambda y_i}] \le (\cosh\lambda)^n.$$

Applying Markov's inequality to the random variable $\mathrm{e}^{\lambda x}$ then gives, for any $\lambda > 0$,

$$\Pr[x > a] = \Pr[\mathrm{e}^{\lambda x} > \mathrm{e}^{\lambda a}] \le \frac{(\cosh\lambda)^n}{\mathrm{e}^{\lambda a}}.$$

Using the inequality

$$\cosh\lambda \le \mathrm{e}^{\lambda^2/2},$$

we can bound this further as

$$\Pr[x > an] \le \mathrm{e}^{\lambda^2 n/2 - \lambda a}.$$

This is minimized when $\lambda = a/n$, in which case

$$\Pr[x > a] \le e^{-a^2/(2n)}.$$

By symmetry, the same bound applies to $\Pr[x < -a]$, and by the union bound we have $\Pr[|x| > a] \le 2e^{-a^2/(2n)}$ as stated. □

Now suppose that the y_i are correlated rather than independent. If we think of choosing them one at a time, then each y_i has a probability distribution $p(y_i|y_1,\ldots,y_{i-1})$ conditioned on all the ones before it. Each of these conditional distributions has an expectation $\mathbb{E}[y_i|y_1,\ldots,y_{i-1}]$. But if these conditional expectations are all zero, and if we have $|y_i| \le 1$ as before, then exactly the same proof goes through. In other words,

Theorem A.4 *Let $x = \sum_{i=1}^n y_i$, where the y_i are chosen from some joint probability distribution such that $|y_i| \le 1$ and $\mathbb{E}[y_i|y_1,\ldots,y_{i-1}] = 0$. Then for any $a \ge 0$,*

$$\Pr[|x| > a] \le 2e^{-a^2/(2n)}.$$

Proof Let $x_j = \sum_{i=1}^j y_i$. We will show by induction on j that the moment generating function of x_j is at most $(\cosh\lambda)^j$, just as in Theorem A.3. Then if we take $j = n$, the analytic part of the proof will work just as before.

Assume by induction that

$$\mathbb{E}[e^{\lambda x_{j-1}}] \le (\cosh\lambda)^{j-1}.$$

Now, for any $y_1,\ldots,y_{j-1}$, the conditional expectation of $e^{\lambda y_j}$ is bounded by

$$\mathbb{E}[e^{\lambda y_j} \,|\, y_1,\ldots,y_{j-1}] \le \cosh\lambda,$$

just as in (A.31). Therefore, increasing the number of variables from $j-1$ to j multiplies the moment generating function by at most a factor of $\cosh\lambda$:

$$\begin{aligned}\mathbb{E}[e^{\lambda x_j}] &= \mathbb{E}[e^{\lambda x_{j-1}}\, e^{\lambda y_j}] \\ &= \mathop{\mathbb{E}}_{y_1,\ldots,y_{j-1}}\left[e^{\lambda x_{j-1}}\, \mathbb{E}[e^{\lambda y_j} \,|\, y_1,\ldots,y_{j-1}]\right] \\ &\le (\cosh\lambda)\, \mathbb{E}[e^{\lambda x_{j-1}}] \\ &\le (\cosh\lambda)^j.\end{aligned}$$

We leave the base case $j = 0$ to the reader. □

The requirement that each y_i have zero conditional expectation may seem artificial. However, it comes up in a large family of probabilistic processes. A *martingale* is a sequence of random variables $x_0, x_1, x_2, \ldots$ such that, while each x_t can depend on all the previous ones, its expectation is equal to the previous one:

$$\mathbb{E}[x_t \,|\, x_0,\ldots,x_{t-1}] = x_{t-1}.$$

If we define $y_i = x_i - x_{i-1}$ then $x_n = x_0 + \sum_{i=1}^n y_i$, and each y_i has zero conditional expectation. Thus we can restate Theorem A.4 as follows:

Theorem A.5 (Azuma's inequality) *Let the sequence $x_0, x_1, \ldots$ be a martingale with $|x_i - x_{i-1}| \le 1$ for all i. Then for any $a \ge 0$,*

$$\Pr[|x_n - x_0| > a] \le 2\mathrm{e}^{-a^2/(2n)}.$$

For example, the position x_t of a random walk after t steps, where $x_t - x_{t-1} = +1$ or -1 with equal probability, is a martingale. If the random walk is biased, so that $x_t = x_{t-1} + 1$ or $x_{t-1} - 1$ with probability p or $1-p$ respectively, we can turn it into a martingale by subtracting away the expected change. Thus z_t is a martingale where

$$z_t = x_t - (2p-1)t.$$

Of course, x_t and z_t are sums of independent random variables, so we could show that they are concentrated around their expectations using Chernoff bounds instead. The power of Azuma's inequality is that it proves concentration even when these variables are based on correlated events, such as the steps of the algorithms for 3-SAT we analyze in Section 14.3.

A.6 Asymptotic Integrals

There are two chapters in this book where we need to understand the asymptotic behavior of certain integrals. In Chapter 14, these integrals give the second moment for the number of satisfying assignments of a random formula. In Chapter 15, they give the probability distribution of a quantum walk. Even though the first kind of integral is real and the second is complex, their analysis is similar.

A.6.1 Laplace's Method

In Chapter 14, our calculations of the second moment of the number of solutions of a random SAT formula involve integrals of the form

$$I = \int_a^b f(x)\,\mathrm{e}^{n\phi(x)}\,\mathrm{d}x,$$

where $f(x)$ and $\phi(x)$ are smooth functions. In the limit $n \to \infty$, the integrand is sharply peaked. That is, the integral is dominated by values of x close to the $x_{\max}$ that maximizes ϕ, and the contributions from all other x are exponentially smaller. We can estimate I by computing the height and width of this peak.

Assume for simplicity that ϕ has a unique maximum $\phi_{\max} = \phi(x_{\max})$ in the interval $[a,b]$ and that $a < x_{\max} < b$. Assume also that this maximum is quadratic, i.e., that $\phi''(x_{\max}) < 0$. Using the second-order Taylor series for $\phi(x)$ near $x_{\max}$ and writing ϕ'' for $\phi''(x_{\max})$ to save ink, we have

$$\phi(x_{\max} + y) = \phi(x_{\max}) - \frac{1}{2}\left|\phi''\right| y^2 + O(y^3).$$

Comparing with (A.25), we see that $\mathrm{e}^{n\phi}$ is proportional to a Gaussian whose mean is $x_{\max}$ and whose variance is $1/(n\left|\phi''\right|)$. Its width scales as $1/\sqrt{n\left|\phi''\right|} \sim 1/\sqrt{n}$, and its height is $\mathrm{e}^{n\phi_{\max}}$.

Since f is smooth, as n increases and the width of the peak decreases, $f = f(x_{\max})$ becomes essentially constant on this interval. Then

$$I \approx f(x_{\max})\,\mathrm{e}^{n\phi_{\max}} \int_{-\infty}^{\infty} \mathrm{e}^{-(1/2)n\left|\phi''\right| y^2}\,\mathrm{d}y = \sqrt{\frac{2\pi}{n\left|\phi''\right|}}\, f(x_{\max})\,\mathrm{e}^{n\phi_{\max}}. \qquad \text{(A.32)}$$

This approximation is called *Laplace's method.* Its multiplicative error is $1+O(1/n)$. If there are multiple maxima where ϕ is equally large, I receives a contribution from each one.

In Section 14.5 we use another convenient form of this approximation. Suppose $\phi(x)=\ln g(x)$. Since $g'=0$ at a maximum, we have

$$\phi'' = \frac{g''}{g_{\max}} .$$

So we can also write

$$\int_a^b f(x)\, g(x)^n \,\mathrm{d}x \approx \sqrt{\frac{2\pi}{n\left|g''/g_{\max}\right|}}\, f(x_{\max})\, {g_{\max}}^n . \tag{A.33}$$

Laplace's method generalizes easily to higher-dimensional integrals. Suppose $\mathbf{x}$ is a d-dimensional vector and ϕ has a unique maximum $\phi_{\max}=\phi(\mathbf{x}_{\max})$. If the *Hessian*, i.e., the matrix of second derivatives

$$(\phi'')_{ij} = \frac{\partial^2 \phi}{\partial x_i\, \partial x_j} ,$$

is nonsingular, then

$$\int f(\mathbf{x})\, \mathrm{e}^{n\phi(\mathbf{x})} \,\mathrm{d}\mathbf{x} \approx \sqrt{\frac{(2\pi)^d}{n^d \left|\det \phi''\right|}}\, f(\mathbf{x}_{\max})\, \mathrm{e}^{n\phi_{\max}} , \tag{A.34}$$

and so

$$\int f(\mathbf{x})\, g(\mathbf{x})^n \,\mathrm{d}\mathbf{x} \approx \sqrt{\frac{(2\pi)^d}{n^d \left|\det(g''/g_{\max})\right|}}\, f(\mathbf{x}_{\max})\, {g_{\max}}^n . \tag{A.35}$$

We can also consider the case where $\phi''(x_{\max})=0$ but some higher-order derivative is nonzero. We say that $x_{\max}$ is a *pth-order stationary point* if the pth derivative $\phi^{(p)}(x_{\max})$ is nonzero but all lower derivatives are zero. In that case, the integral is dominated by an interval of width $n^{-1/p}$ around $x_{\max}$ instead of $n^{-1/2}$, so

$$I \sim \frac{1}{(n\phi^{(p)}(x_{\max}))^{1/p}} .$$

Since a maximum on the real line often becomes a saddle point in the complex plane, physicists call Laplace's method the *saddle-point method.*

A.6.2 The Method of Stationary Phase

In our study of quantum walks in Chapter 15, we use a generalization of Laplace's method which applies to complex-valued integrals where the integrand oscillates rapidly in the complex plane. Consider an integral of the form

$$I = \int_a^b \mathrm{e}^{\imath n\phi(x)} \,\mathrm{d}x .$$

In the limit $n\to\infty$, the contributions from x with $\phi(x)\neq 0$ are exponentially small, since the phase oscillations cause destructive interference and the integral cancels out. Since we get constructive interference from any interval where $\phi(x)$ is roughly constant, I is dominated by values of x near *stationary points* x_0 where $\phi'(x_0)=0$. The width of this interval is roughly $1/\sqrt{n|\phi''|}$, just as in Laplace's method.

Suppose there is a unique stationary point x_0 in the interval $[a,b]$, and write $\phi_0 = \phi(x_0)$. Using the second-order Taylor series for ϕ just as we did in the Laplace method gives

$$I \approx \mathrm{e}^{\imath n\phi_0} \int_{-\infty}^{\infty} \mathrm{e}^{\imath(1/2)n\phi'' x^2}\,\mathrm{d}x = \sqrt{\frac{2\imath\pi}{n\phi''}}\,\mathrm{e}^{\imath n\phi_0}. \tag{A.36}$$

If there are multiple stationary points in the interval $[a,b]$ then I receives a contribution from each one. Since each one has a phase $\mathrm{e}^{\imath n\phi_0}$, these contributions can interfere constructively or destructively. Indeed, as the parameters of the integral vary, there are often rapid oscillations, or *moiré patterns*, in its overall value. You can see these patterns in Figures 15.12 and 15.13.

As for the real-valued Laplace approximation, if x_0 is a pth-order stationary point then $I \sim n^{-1/p}$. This happens, for instance, at the extreme points of the one-dimensional quantum walk in Problem 15.50.

A.7 Groups, Rings, and Fields

> The different branches of Arithmetic—Ambition, Distraction, Uglification, and Derision.
>
> Lewis Carroll, *Alice in Wonderland*

The familiar operations of arithmetic, addition and multiplication, have certain properties in common. First, they are *associative*: for any a,b,c we have $(a+b)+c = a+(b+c)$ and $(ab)c = a(bc)$. Second, they have an *identity*, something which leaves other things unchanged: $a+0 = 0+a = a$ and $1\cdot a = a\cdot 1 = a$. Finally, each a has an *inverse*, which combined with a gives the identity: $a+(-a)=0$ and $aa^{-1}=1$.

A general structure of this kind is called a *group*. A group G is a set with a *binary operation*, a function from $G \times G$ to G, which we write $a \cdot b$. This binary operation obeys the following three axioms:

1. (associativity) For all $a,b,c \in G$, $(a\cdot b)\cdot c = a\cdot(b\cdot c)$.

2. (identity) There exists an element $1 \in G$ such that, for all $a \in G$, $a\cdot 1 = 1\cdot a = a$.

3. (inverses) For each $a \in G$, there is an $a^{-1} \in G$ such that $a\cdot a^{-1} = a^{-1}\cdot a = 1$.

Addition and multiplication possess another property, namely *commutativity*: $a+b = b+a$ and $ab = ba$. Groups of this kind are called *abelian*, after the mathematician Niels Henrik Abel. (Having something named after you in mathematics is a great honor, but the greatest honor is having it in lower case.) However, many of the most interesting groups are *nonabelian*, meaning that the order of multiplication matters.

Here are some common groups. Which ones are abelian?

1. $(\mathbb{Z},+)$, the integers with addition
2. $(\mathbb{Z}_n,+)$, the integers mod n with addition
3. $(\mathbb{Z}_n^*,\times)$, the integers mod n that are mutually prime to n, with multiplication
4. $(\mathbb{C},+)$, the complex numbers with addition
5. $(\mathbb{C},\times)$, the nonzero complex numbers with multiplication
6. U_d, the unitary $d \times d$ matrices: that is, those for which $U^{-1} = U^\dagger$, where as described in Chapter 15, $U^\dagger$ is the complex conjugate of U's transpose
7. D_4, the dihedral group shown in Figure 15.6, consisting of the 8 rotations and reflections of a square
8. S_n, the group of $n!$ permutations of n objects, where we multiply two permutations by composing them
9. A_p, the set of functions $f(x) = ax + b$ where $a \in \mathbb{Z}_p^*$ and $b \in \mathbb{Z}_p$, where we multiply two functions by composing them.

A.7.1 *Subgroups, Lagrange, and Fermat*

A *subgroup* of a group G is a subset $H \subseteq G$ that is closed under multiplication. That is, if $a, b \in H$ then $ab \in H$. For infinite groups, we also need to require that if $a \in H$ then $a^{-1} \in H$. For example, for any integer r, the multiples of r form a subgroup $r\mathbb{Z} = \{\ldots,-2r,-r,0,r,2r,\ldots\}$ of the integers $\mathbb{Z}$.

A *coset* of a subgroup is a copy of it that has been shifted by multiplying it by some element—in other words, a set of the form $aH = \{ah : h \in H\}$ for some $a \in G$. For instance, if $G = \mathbb{Z}$ and $H = r\mathbb{Z}$, the coset aH is the set of integers equivalent to a mod r,

$$a + r\mathbb{Z} = \{a + tr : t \in \mathbb{Z}\} = \{b \in \mathbb{Z} : b \equiv_r a\}.$$

We call aH a *left* coset. In an abelian group, $aH = Ha$, but in a nonabelian group the right cosets Ha are typically not left cosets. For another example, in the group $\mathbb{R}^2$ of two-dimensional vectors with addition, each straight line passing through the origin is a subgroup, and its cosets are lines parallel to it.

Now let's prove something interesting. First, we recommend the following two exercises, which you should prove using nothing but the axioms of group theory:

Exercise A.24 *Show that all the cosets of a subgroup have the same size, i.e., that $|aH| = |H|$ for all $a \in G$.*

Exercise A.25 *Show that two cosets are either identical or disjoint, i.e., for any a,b we have $aH = bH$ or $aH \cap bH = \emptyset$. Hint: consider whether the "difference" $a^{-1}b$ between a and b is in H.*

These two results imply that there are $|G|/|H|$ cosets of H. This implies *Lagrange's Theorem*:

Theorem A.6 *If G is a finite group and $H \subseteq G$ is a subgroup, then $|H|$ divides $|G|$.*

Given an element a, the *cyclic subgroup* generated by a is $\langle a \rangle = \{1, a, a^2, \ldots\}$, the set of all powers of a. Define the *order* of a as the smallest r such that $a^r = 1$. If G is finite, then r exists:

Exercise A.26 *Let G be a finite group and let $a \in G$. Show that there is an r such that $a^r = 1$.*

Thus in a finite group, $\langle a \rangle = \{1, a, a^2, \ldots, a^{r-1}\}$. Since $|\langle a \rangle| = r$, we get the following corollary to Theorem A.6:

Corollary A.7 *If G is a finite group, the order of any element a divides $|G|$.*

If a has order r then $a^{kr} = 1^k = 1$ for any k, so this implies that $a^{|G|} = 1$. In particular, Euler's totient function $\varphi(n) = |\mathbb{Z}_n^*|$ is the number of integers mod n that are mutually prime to n. Then for any a we have $a^{\varphi(n)} \equiv_n 1$, giving us the generalization of Fermat's Little Theorem which we use in Section 15.5.1. If n is prime, we have $\varphi(n) = n - 1$, giving the original version of Fermat's Little Theorem we use in Section 4.4.1.

If G is cyclic, an element $a \in G$ such that $G = \langle a \rangle$ is called a *generator.* For historical reasons, a generator of $\mathbb{Z}_n^*$ is also called a *primitive root* mod n.

A.7.2 Homomorphisms and Isomorphisms

A *homomorphism* is a map from one group to another that preserves the structure of multiplication. In other words, $\phi : G \to H$ is a homomorphism if and only if $\phi(a \cdot b) = \phi(a) \cdot \phi(b)$ for all $a, b \in G$. For instance, the function $\phi(x) = e^x$ is a homomorphism from $(\mathbb{C}, +)$ to $(\mathbb{C}, \times)$ since $e^{x+y} = e^x \times e^y$.

For another example, recall that the *parity* of a permutation is even or odd depending on how many times we need to swap a pair of objects to perform it. For instance, a permutation which rotates three objects $1 \to 2 \to 3 \to 1$ has even parity, since we can perform it by first swapping objects 2 and 3 and then swapping objects 1 and 2. Now define $\pi(\sigma)$ as $+1$ if σ is even, and -1 if σ is odd. Then π is a homomorphism from S_n to $(\mathbb{C}, \times)$.

Another important example, mentioned in Section 15.6, is the Fourier basis function $\phi_k(x) = \omega^{kx}$ where $\omega = e^{2\imath\pi/n}$ is the nth root of 1 in the complex plane. For any integer k, this is a homomorphism from $\mathbb{Z}_n$ to $(\mathbb{C}, \times)$.

A subgroup H is *normal* if for all $a \in G$ and all $h \in H$, we have $a^{-1}ha \in H$, or more compactly, if $a^{-1}Ha = H$ for all a. Although we don't use it in the book, the following is a nice exercise:

Exercise A.27 *The* kernel *of a homomorphism ϕ is the set $K = \{a \in G : \phi(a) = 1\}$. Show that K is a normal subgroup.*

An *isomorphism* is a homomorphism that is one-to-one. For example, if a is an element of order n, then the cyclic group $G = \langle a \rangle$ is isomorphic to $\mathbb{Z}_n$, since the map $\phi(x) = a^x$ is an isomorphism from $\mathbb{Z}_n$ to G. If two groups G, H are isomorphic, we write $G \cong H$.

A.7.3 The Chinese Remainder Theorem

> Suppose we have an unknown number of objects. When counted in threes, two are left over; when counted in fives, three are left over; and when counted in sevens, two are left over. How many objects are there?
>
> Sun Zi, ca. 400 A.D.

In this section, we discuss one of the oldest theorems in mathematics. In honor of its origins in China, it is known as the Chinese Remainder Theorem. We will state it here using modern language.

Given two groups G_1, G_2, their Cartesian product $G_1 \times G_2$ is a group whose elements are ordered pairs (a_1, a_2) with $a_1 \in G_1$ and $a_2 \in G_2$. We define multiplication componentwise, $(a_1, a_2) \cdot (b_1, b_2) = (a_1 b_1, a_2 b_2)$. Consider the following theorem:

Theorem A.8 *Suppose p and q are mutually prime. Then*

$$\mathbb{Z}_{pq} \cong \mathbb{Z}_p \times \mathbb{Z}_q .$$

By applying Theorem A.8 inductively, we get the full version:

Theorem A.9 (Chinese Remainder Theorem) *Let $q_1, q_2, \ldots, q_\ell$ be mutually prime and let $n = \prod_{i=1}^{\ell} q_i$. Then*

$$\mathbb{Z}_n \cong \mathbb{Z}_{q_1} \times \mathbb{Z}_{q_2} \times \cdots \times \mathbb{Z}_{q_\ell} .$$

In particular, let $n = p_1^{t_1} p_2^{t_2} \cdots p_\ell^{t_\ell}$ be the prime factorization of n. Then

$$\mathbb{Z}_n \cong \mathbb{Z}_{p_1^{t_1}} \times \mathbb{Z}_{p_2^{t_2}} \times \cdots \times \mathbb{Z}_{p_\ell^{t_\ell}} .$$

Here is a more traditional restatement of Theorem A.9. It tells us that Sun Zi's riddle has a solution x, where $0 \le x < 3 \times 5 \times 7$:

Theorem A.10 (Chinese Remainder Theorem, traditional version) *Suppose that $q_1, q_2, \ldots, q_\ell$ are mutually prime, let $n = \prod_{i=1}^{\ell} q_i$, and let $r_1, r_2, \ldots, r_\ell$ be integers such that $0 \le r_i < q_i$ for each i. Then there exists a unique integer x in the range $0 \le x < n$ such that $x_i \equiv r_i \bmod q_i$ for each i.*

It follows that for any integer x, the value of $x \bmod n$ is determined uniquely by $x \bmod p^t$ for each of the prime power factors p^t of n. For instance, the following table shows how each element $x \in \mathbb{Z}_{12}$ is determined by $x \bmod 3$ and $x \bmod 4$:

	0	1	2	3
0	0	9	6	3
1	4	1	10	7
2	8	5	2	11

Notice how we move diagonally through the table as we increment x, winding around when we hit the bottom or right edge. Since 3 and 4 are mutually prime, these windings cover the entire table before we return to the origin $x = 0$.

To prove Theorem A.8, we start by noting that the function $f : \mathbb{Z}_{pq} \to \mathbb{Z}_p \times \mathbb{Z}_q$ defined by

$$f(x) = (x \bmod p, x \bmod q),$$

is a homomorphism. To prove that it is an isomorphism, we need to show that there is an x such that $f(x) = (1,0)$, and another such that $f(x) = (0,1)$—in this example, 4 and 9—since these two elements generate $\mathbb{Z}_p \times \mathbb{Z}_q$. In other words, we need to complete the following exercise, which is the two-factor case of Problem 5.26:

Exercise A.28 *Suppose that $n = pq$ where p and q are mutually prime. Show that there is an x such that $x \equiv_p 1$ and $x \equiv_q 0$, and another x such that $x \equiv_p 0$ and $x \equiv_q 1$. Hint: recall that p and q are mutually prime if and only if there are integers a, b such that*

$$ap + bq = 1.$$

Once Theorem A.8 is proved, Theorem A.9 follows by induction on the number of factors ℓ.

A.7.4 Rings and Fields

Addition and multiplication have another important property, the *distributive law*: $a(b+c) = ab + ac$. A *ring* R is a set with two binary operations, $+$ and $\cdot$, such that

1. both $+$ and $\cdot$ are associative,
2. $(R, +)$ is an Abelian group, whose identity we denote 0, and
3. the distributive law $a \cdot (b+c) = (a \cdot b) + (a \cdot c)$ holds.

Examples of rings include $\mathbb{R}$ or $\mathbb{C}$ with the usual addition and multiplication, $\mathbb{Z}_n$ with addition and multiplication mod n, and the set of polynomials over a variable x with integer or rational coefficients.

The following exercise gives us the familiar fact that, if adding zero leaves things alone, then multiplying by zero annihilates them:

Exercise A.29 *Show that, in any ring, we have $0 \cdot a = a \cdot 0 = 0$ for all a.*

Therefore, if R has more than one element, 0 cannot have a multiplicative inverse.

A *field* is a ring with the additional property that all nonzero elements have inverses. Equivalently, the nonzero elements form a group under multiplication. Infinite examples include $\mathbb{R}$, $\mathbb{C}$, the rationals $\mathbb{Q}$ with the usual operations, and the set of rational functions—that is, functions of the form $f(x) = g(x)/h(x)$ where g and h are polynomials.

The main type of *finite* field we meet in this book is $\mathbb{F}_p$, the set of integers mod p where p is a prime, with addition and multiplication mod p. It is a field since every $a \in \mathbb{F}_p^* = \{1, 2, \ldots, p-1\}$ is mutually prime to p, and so has a multiplicative inverse b such that $ab \equiv_p 1$.

There is also a finite field $\mathbb{F}_q$ of size q for each prime power $q = p^t$. Addition in $\mathbb{F}_q$ is isomorphic to $\mathbb{Z}_p^t$, or to addition of t-dimensional vectors mod p. But if $t > 1$, defining multiplication is a little more complicated—it is not simply multiplication mod q. Instead, it is defined in terms of polynomials of degree t with coefficients in $\mathbb{F}_p$. For instance, we can define the finite field $\mathbb{F}_8 = \mathbb{F}_{2^3}$ as the set of polynomials $f(x) = a_2x^2 + a_1x + a_0$ where $a_0, a_1, a_2 \in \mathbb{F}_2$. When we add two polynomials, their coefficients add, so $(\mathbb{F}_8, +) \cong \mathbb{Z}_2^3$. But when we multiply them, we take the result modulo some polynomial Q, such as $Q = x^3 + x + 1$. Then $x^{-1} = x^2 + 1$, since

$$x(x^2+1) = x^3 + x \equiv_Q 1 .$$

Exercise A.30 *Compute the powers of x in this presentation of $\mathbb{F}_8$, and show that $x^7 = 1$.*

Since Q is *irreducible* mod 2—that is, it cannot be written as a product of polynomials of lower degree with coefficients in $\mathbb{F}_2$—the resulting ring is a field, just as $\mathbb{Z}_p$ is a field if and only if p is prime. Surprisingly, all finite fields of size q are isomorphic, where of course we demand that an isomorphism between fields preserves both the additive and the multiplicative structure.

A.7.5 Polynomials and Roots

For any ring R we can define $R[x]$, the ring of polynomials over a variable x with coefficients in R. Any polynomial $f(x) \in R[x]$ defines a function from R to R in the obvious way. We say $f(x)$ has degree d if

$$f(x) = a_d x^d + a_{d-1}x^{d-1} + \cdots + a_1 x + a_0 . \tag{A.37}$$

for some coefficients $a_0, \ldots, a_d \in R$.

Now let's prove the following classic fact:

Theorem A.11 *If R is a field, a polynomial $f \in R[x]$ of degree d has at most d roots.*

Proof First note that for any $r \in R$, there is set of coefficients $b_0, \ldots, b_d$ such that

$$f(x) = b_d(x-r)^d + b_{d-1}(x-r)^{d-1} + \cdots + b_1(x-r) + b_0 .$$

To see this, start with (A.37) and set b_d equal to the leading coefficient a_d. Then $f(x) - b_d(x-r)^d$ is a polynomial of degree $d-1$. We set b_{d-1} equal to its leading coefficient, and so on. We continue decreasing the degree until we are left with a constant function b_0.

Now suppose that r is a root of $f(x)$, i.e., that $f(r) = 0$. In that case $b_0 = 0$. If R is a field, we can then divide $f(x)$ by $x - r$, factoring it as

$$f(x) = (x-r)\left(b_d(x-r)^{d-1} + b_{d-1}(x-r)^{d-2} + \cdots + b_1\right) = (x-r)f'(x).$$

Since $f'(x)$ is a polynomial of degree $d-1$, it follows by induction that $f(x)$ has at most d roots. □

Note that if R is a ring but not a field, Theorem A.11 does not necessarily hold. For instance, in $\mathbb{Z}_6$ the polynomial $f(x) = 2x$ has two roots, $x = 0$ and $x = 3$.

Problems

A mathematical problem should be difficult in order to entice us, yet not completely inaccessible, lest it mock at our efforts. It should be to us a guidepost on the mazy paths to hidden truths, and ultimately a reminder of our pleasure in the successful solution.

David Hilbert

A.1 Inclusion and exclusion. Here is a combinatorial proof of the inclusion–exclusion principle (A.13). First, let's separate it into one term for each of the 2^n possible cases of which E_i are true and which E_i are false—that is, one term for each piece of the Venn diagram. For a subset $V \subseteq \{1,\ldots,n\}$, let

$$P_V = \Pr\left[\left(\bigwedge_{i\in V} E_i\right) \wedge \left(\bigwedge_{i\notin V} \overline{E}_i\right)\right].$$

Then show that (A.13) can be written

$$\Pr\left[\overline{\bigvee_{i=1}^{n} E_i}\right] = \sum_{V\subseteq\{1,\ldots,n\}} P_V \sum_{T\subseteq V} (-1)^{|T|} = \sum_{V\subseteq\{1,\ldots,n\}} P_V \sum_{j=0}^{|V|} (-1)^j \binom{|V|}{j}. \tag{A.38}$$

Finally, show that

$$\sum_{j=0}^{m} (-1)^j \binom{m}{j} = \begin{cases} 1 & \text{if } m=0 \\ 0 & \text{if } m>0, \end{cases}$$

so that we can rewrite (A.38) as

$$\Pr\left[\overline{\bigvee_{i=1}^{n} E_i}\right] = P_\emptyset .$$

A.2 A night at the opera. A classic puzzle describes n opera-goers, whose manteaux become randomly permuted by an incompetent coat checker, and asks for the probability that not a single one of them receives the correct manteau. Equivalently, if we choose randomly from among the $n!$ permutations, what is the probability that we obtain a *derangement*, i.e., a permutation with no fixed points? Use the inclusion–exclusion principle (A.13) to show that this probability is exactly

$$P = 1 - 1 + \frac{1}{2} - \frac{1}{3!} + \cdots + \frac{(-1)^n}{n!} = \sum_{i=0}^{n} \frac{(-1)^i}{i!},$$

which converges quickly to $1/e$. Hint: given a set of i opera-goers, what is the probability that those i (and possibly some others) all get back their manteaux?

A.3 Fixed points are Poisson. Using the previous problem, show that for any constant x the probability that there are exactly x fixed points in a random permutation of n objects approaches $P(x) = 1/(ex!)$, the Poisson distribution with mean 1, as n tends to infinity. Therefore, the number of people who get their manteaux back is distributed almost as if each of these events occurred independently with probability $1/n$.

A.4 The first cycle. Given a permutation σ, a k-*cycle* is a set of k objects each of which is mapped to the next by σ: that is, a set $t_1, t_2, \ldots, t_k$ such that $\sigma(t_i) = t_{(i+1) \bmod k}$. Given a random permutation of n objects, show that the length of the cycle containing the first object is uniformly distributed from 1 to n.

A.5 Harmonic cycles. Show that the expected number of cycles in a random permutation of n objects is exactly the nth harmonic number,

$$H_n = \sum_{k=1}^{n} \frac{1}{k} \approx \ln n \,.$$

For instance, of the 6 permutations of 3 objects, two of them consist of a single cycle of all 3, three of them consist of a 2-cycle and a 1-cycle (i.e., a fixed point), and the identity consists of three 1-cycles. Thus the average is $(2/6)\cdot 1 + (3/6)\cdot 2 + (1/6)\cdot 3 = 11/6 = H_3$. Hint: use linearity of expectation. What is the expected number of k-cycles for each k?

A.6 Different birthdays. If there are n people in a room and there are y days in the year, show that the probability that everyone has a different birthday is at most $e^{-n(n-1)/2y}$. Combined with the discussion in Sections A.3.1 and A.3.2, we then have

$$1 - \frac{n(n-1)}{2y} \le \Pr[\text{everyone has a different birthday}] \le e^{-n(n-1)/2y} \,.$$

Hint: imagine assigning a birthday to one person at a time. Write the probability that each one avoids all the previous birthdays as a product, and use the inequality $1 - x \le e^{-x}$. If $y = 365$, for what n is this upper bound roughly 1/2?

A.7 A birthday triple. If there are n people in a room and y days in the year, how large can n be before there is a good chance that some group of *three* people all have the same birthday? Answer within Θ.

A.8 Rules of engagement. There are n suitors waiting outside your door. Each one, when you let them in, will make a proposal of marriage. Your goal is to accept the best one. You adopt the following strategy: you invite them in in random order. After interviewing and rejecting the first r suitors for some $r \le n$, you accept the first one that is better than any of those r.

Show that the probability you nab the best one is exactly

$$P = \frac{1}{n} \sum_{i=r+1}^{n} \frac{r}{i-1} \,.$$

Hint: if the best suitor is the ith one you invite in, you will accept him or her if the best among the first $i-1$ was among the first r. Then replace this sum with an integral to get

$$P \approx \int_{\alpha}^{1} \frac{\alpha}{x}\, dx = \alpha \ln(1/\alpha) .$$

where $\alpha = r/n$. By maximizing P as a function of α, conclude that the best strategy is to reject the first n/e suitors and that this strategy succeeds with probability $1/e$.

A.9 A lark ascending. Consider the definition of an increasing subsequence of a permutation given in Problem 3.22. Now consider a random permutation of n objects. Show that there is some constant A such that, with high probability, the longest increasing subsequence is of size $k < A\sqrt{n}$.

Hint: one way to generate a random permutation is to choose n real numbers independently from the unit interval $[0,1]$ and rank them in order. Given a set of k such numbers, what is the probability that they are in increasing order? You might find the bound (A.10) on the binomial useful.

It turns out that the likely length of the longest increasing subsequence in a random permutation converges to $2\sqrt{n}$. This is known as Ulam's problem, and its exact solution involves some very beautiful mathematics [32].

A.10 Pairwise independence. Suppose that X is a sum of indicator random variables X_i which are *pairwise independent*: that is, $\mathbb{E}[X_i X_j] = \mathbb{E}[X_i]\,\mathbb{E}[X_j]$ for all $i \neq j$. Show that in this case the second moment method gives

$$\Pr[X > 0] \geq \frac{\mathbb{E}[X]}{1 + \mathbb{E}[X]} \geq 1 - \frac{1}{\mathbb{E}[X]},$$

so that $\Pr[X > 0]$ is close to 1 whenever $\mathbb{E}[X]$ is large.

Show that this is the case for the Birthday Problem, where i and j represent different pairs of people which may or may not overlap. In other words, show that the positive correlations described in the text don't appear until we consider sets of three or more pairs. Conclude that the probability that some pair of people has the same birthday is at least 1/2 when $n = \sqrt{2y} + 1$.

A.11 The case of the missing coupon. Repeat the previous problem in the case where the X_i are negatively correlated, so that $\mathbb{E}[X_i X_j] \leq \mathbb{E}[X_i]\,\mathbb{E}[X_j]$. Use this to show that in the Coupon Collector's problem, with high probability we are still missing at least one toy when $b = (1 - \varepsilon) n \ln n$ for any $\varepsilon > 0$. Combined with the discussion in Section A.3.4, this shows that the probability of a complete collection has a sharp phase transition from 0 to 1 when b passes $n \ln n$.

A.12 The black pearls. Suppose I have a necklace with n beads on it, where each bead is randomly chosen to be black or white with probability 1/2. I want to know how likely it is that somewhere on the necklace there is a string of k consecutive black beads. Let X_i be the indicator random variable for the event that there is a string of k black beads starting at bead i. Then the number of strings is $X = \sum_i X_i$, where we count overlapping strings separately. For instance, a string of 5 black beads contains 3 strings of length 3.

By calculating $\mathbb{E}[X_i]$, show that with high probability there are no strings of length $(1+\varepsilon)\log_2 n$ for any $\varepsilon > 0$. Then calculate the second moment $\mathbb{E}[X_i^2]$ by considering the correlations between overlapping strings. Show that strings of length $k = \log_2 n$ exist with probability at least 1/4, and strings of length $k = (1-\varepsilon)\log_2 n$ exist with high probability for any $\varepsilon > 0$.

A.13 The replacements. If we have a barrel of n objects, choosing k of them *without replacement* means to take an object out of the barrel, then another, and so on until we have chosen k objects. Thus we end up with one of the $\binom{n}{k}$ possible subsets of k objects. Choosing *with replacement*, in contrast, means tossing each object back in the barrel before we choose the next one. Show that if $k = o(\sqrt{n})$,

$$\binom{n}{k} = (1 - o(1)) \frac{n^k}{k!},$$

and therefore there is little or no distinction between choosing with or without replacement. Hint: use the Birthday Problem.

A.14 Heavy bins. Suppose that I toss $m = cn$ balls randomly into n bins, where c is a constant. Prove that with high probability the largest number of balls in any bin is $o(\log n)$.

A.15 Convoluted sums. If x and y are independent random variables whose probability distributions are $P(x)$ and $Q(y)$ respectively, the probability distribution of their sum $z = x + y$ is the convolution of P and Q:

$$R(z) = \sum_x P(x) Q(z - x).$$

We write $R = P \star Q$ as in Problem 3.15. Show that if P and Q are Poisson distributions with means c and d respectively, then $P \star Q$ is a Poisson distribution with mean $c + d$. Show this analytically by calculating the sum. Then come up with a "physical" explanation that makes it obvious.

A.16 Poisson and Gauss. Consider the Poisson distribution $P(x)$ with mean c where c is large. Show that it can be approximated by a Gaussian for $x \approx c$. What is its variance?

A.17 Chernoff for biased coins. Show that Theorems A.1 and A.2 hold in the more general case where x is the sum of any number of independent random variables x_i, each of which is 1 or 0 with probability p_i and $1-p_i$ respectively. Hint: show that the moment generating function of x is

$$\mathbb{E}[\mathrm{e}^{\lambda x}] = \prod_i \left(1 + p_i(\mathrm{e}^{\lambda} - 1)\right) \le \mathrm{e}^{(\mathrm{e}^{\lambda}-1)\mathbb{E}[x]} .$$

A.18 Chernoff's tail. Consider the Chernoff bound of Theorem A.1. When δ is large, the probability that x is $1+\delta$ times its expectation is not exponentially small in δ^2. However, it falls off faster than the simple exponential of Theorem A.1. Show that for $\delta > 1$,

$$\Pr[x > (1+\delta)\mathbb{E}[x]] \le \mathrm{e}^{-(\delta \ln \delta)\mathbb{E}[x]/2} ,$$

and that asymptotically the constant $1/2$ can be replaced with 1 when δ is large.

A.19 Azuma with big steps and little ones. Generalize Theorems A.3 and A.4 as follows. If each y_i is bounded by $|y_i| \le c_i$ for some c_i, then for any $a \ge 0$,

$$\Pr[|x| > a] \le 2\exp\left(-a^2 \Big/ \left(2\sum_i c_i^2\right)\right) .$$

A.20 Concentrated cycles. Once again, consider a random permutation of n objects. Now that we know the expected number of cycles, let's prove that the number of cycles is $O(\log n)$ with high probability. Specifically, show that for every constant a, there is a b such that the probability there are more than $b \log n$ cycles is less than n^{-a}.

Hint: imagine constructing the cycles one by one, starting with the cycle containing the first object. Use Problem A.4 to argue that with constant probability, each cycle "eats" at least 1/2 of the remaining objects. Then use Chernoff bounds to argue that after $b \log n$ cycles, with high probability there are no objects left.

A.21 Visits to the origin. Suppose I take a random walk on the infinite line where I start at the origin and move left or right with equal probability at each step. Let x_n be my position after n steps. First, show that $\mathbb{E}[|x_n|]$ is the expected number of times I visited the origin in the previous $n-1$ steps. Then, show that $\mathbb{E}[|x_n|] \le \sqrt{\mathrm{Var}\, x_n} = \sqrt{n}$. Conclude that the expected number of times we visit the origin in n steps is $O(\sqrt{n})$. What does this say about the average number of steps we take in between visits to the origin?

A.22 Sticky ends. Suppose I perform a random walk on a finite line ranging from 0 to n. At each step I move left or right with equal probability, but if I reach either end I stick there forever. Show that my initial position is x, the probability that I will end up stuck at n instead of at 0 is exactly x/n.

Prove this in two ways. First, write a recurrence similar to (A.26) for this probability $p(x)$ as a function of x, and solve it with the boundary conditions $p(0)=0$ and $p(n)=1$. Second, use the fact that while my position changes, its expectation does not. In other words, it is a martingale as discussed in Section A.5.2.

A.23 Distance squared minus time. Let x_t be the position of a random walk where $x_t - x_{t-1} = +1$ or -1 with equal probability. Show that the sequence

$$y_t = x_t^2 - t$$

is a martingale.

A.24 Scaling away the bias. Let x_t be the position of a biased random walk where $x_t - x_{t-1} = +1$ or -1 with probability p or $1-p$ respectively. We saw in Section A.5.2 one way to turn x_t into a martingale. Here is another: show that the sequence

$$y_t = \left(\frac{1-p}{p}\right)^{x_t}$$

is a martingale.

A.25 Pólya's urn. There are some balls in an urn. Some are red, and some are blue. At each step, I pick a ball uniformly at random from the urn. I then place it back in the urn, along with an new ball the same color as the one I picked. Show that the fraction of red balls in the urn is a martingale. Therefore, if I continue until all the balls are the same color, the probability that I end up with all red balls is equal to the initial fraction of balls that are red.

A.26 Resisting infinity. Roughly speaking, a random walk on a d-dimensional lattice consists of d independent one-dimensional random walks. Argue that the probability we return to the origin after n steps is $\Theta(n^{-d/2})$. Then, justify the following claim: if $d=1$ or $d=2$, a random walk returns to the origin an infinite number of times with probability 1, but if $d \geq 3$, with probability 1 it escapes to infinity after only a finite number of visits to the origin. Doyle and Snell [248] have written a beautiful monograph on the following analogy: if I have a d-dimensional lattice with a 1 Ohm resistor in each edge, the total resistance between the origin and infinity is infinite if $d<3$ and finite if $d \geq 3$. If you know how to calculate resistances in serial and parallel, show this in a model where we replace the lattice with a series of concentric spheres.

A.27 Fibonacci, Pascal, and Laplace. Show that the nth Fibonacci number can be written

$$F_n = \sum_{j=0}^{\lfloor n/2 \rfloor} \binom{n-j}{j}.$$

Then use Stirling's approximation and the Laplace method to recover the result

$$F_n = \Theta(\varphi^n),$$

where $\varphi = (1+\sqrt{5})/2$ is the golden ratio. This is probably the most roundabout possible way to prove this, but it's a good exercise.

A.28 Laplace and Stirling. One nice application of Laplace's method is to derive the precise version of Stirling's approximation for the factorial (A.9). We can do this using the Gamma function,

$$n! = \Gamma(n+1) = \int_0^\infty x^n \, \mathrm{e}^{-x} \, \mathrm{d}x$$

and noticing that the integrand $x^n \, \mathrm{e}^{-x}$ is tightly peaked for large n. Write the integrand in the form

$$x^n \, \mathrm{e}^{-x} = \mathrm{e}^{n\phi(x)},$$

Taking derivatives, show that $\phi(x)$ is maximized at $x_{\max} = n$, where $\mathrm{e}^{\phi} = n/\mathrm{e}$ and $\phi'' = -n^2$. In this case ϕ depends weakly on n, but the method of approximating the integrand with a Gaussian is otherwise the same. Then apply (A.32) and show that

$$n! = (1+O(1/n)) \sqrt{\frac{2\pi}{n\left|\phi''(x_{\max})\right|}} \, \mathrm{e}^{n\phi(x_{\max})} = (1+O(1/n)) \sqrt{2\pi n} \, n^n \, \mathrm{e}^{-n}.$$

A.29 Let X be a random variable that takes nonnegative integer values. Show that

$$\mathbb{E}[X] = \sum_{x=0}^{\infty} \Pr[X > x].$$

References

[1] Scott Aaronson. Is P versus NP formally independent? *Bulletin of the European Association for Theoretical Computer Science*, **81**:109–136, 2003. [218, 221, 293]

[2] Scott Aaronson. NP-complete problems and physical reality. *SIGACT News*, **36**(1):30–52, 2005. [903]

[3] Scott Aaronson. Lower bounds for local search by quantum arguments. *SIAM Journal on Computing*, **35**(4):804–824, 2006. [222]

[4] Scott Aaronson and Yaoyun Shi. Quantum lower bounds for the collision and the element distinctness problems. *Journal of the ACM*, **51**(4):595–605, 2004. [909, 910]

[5] Scott Aaronson and Avi Wigderson. Algebrization: a new barrier in complexity theory. In *Proc. 40th Symposium on Theory of Computing*, pages 731–740. 2008. [220]

[6] Daniel S. Abrams and Seth Lloyd. Nonlinear quantum mechanics implies polynomial-time solution for NP-complete and #P problems. *Physical Review Letters*, **81**(18):3992–3995, 1998. [903]

[7] Dimitris Achlioptas. Lower bounds for random 3-SAT via differential equations. *Theoretical Computer Science*, **265**(1–2):159–185, 2001. [814]

[8] Dimitris Achlioptas, Paul Beame, and Michael Molloy. A sharp threshold in proof complexity yields lower bounds for satisfiability search. *Journal of Computer and System Sciences*, **68**(2):238–268, 2004. [813]

[9] Dimitris Achlioptas, Arthur Chtcherba, Gabriel Istrate, and Cristopher Moore. The phase transition in 1-in-k SAT and NAE 3-SAT. In *Proc. 12th Symposium on Discrete Algorithms*, pages 721–722. 2001. [813]

[10] Dimitris Achlioptas and Amin Coja-Oghlan. Algorithmic barriers from phase transitions. In *Proc. 49th FOCS*, pages 793–802. 2008. [817, 818]

[11] Dimitris Achlioptas and Ehud Friedgut. A sharp threshold for k-colorability. *Random Structures & Algorithms*, **14**(1):63–70, 1999. [811]

[12] Dimitris Achlioptas and Michael Molloy. The analysis of a list-coloring algorithm on a random graph. In *Proc. 38th Symposium on Foundations of Computer Science*, pages 204–212. 1997. [813]

[13] Dimitris Achlioptas, Michael Molloy, Cristopher Moore, and Frank Van Bussel. Sampling grid colorings with fewer colors. In *Proc. 6th Latin American Symposium on Theoretical Informatics*, pages 80–89. 2004. [649]

[14] Dimitris Achlioptas and Cristopher Moore. The asymptotic order of the random k-SAT threshold. In *Proc. 43rd Symposium on Foundations of Computer Science*, pages 779–788. 2002. [814]

[15] Dimitris Achlioptas and Cristopher Moore. Almost all graphs with average degree 4 are 3-colorable. *Journal of Computer and System Sciences*, **67**(2):441–471, 2003. [813]

[16] Dimitris Achlioptas and Assaf Naor. The two possible values of the chromatic number of a random graph. In *Proc. 36th Symposium on Theory of Computing*, pages 587–593. 2004. [814]

[17] Dimitris Achlioptas and Yuval Peres. The threshold for random k-SAT is $2^k(\ln 2 - o(k))$. In *Proc. 35th Symposium on Theory of Computing*, pages 223–231. 2003. [814]

[18] Dimitris Achlioptas and Federico Ricci-Tersenghi. On the solution-space geometry of random constraint satisfaction problems. In *Proc. 38th Symposium on Theory of Computing*, pages 130–139. 2006. [817, 818]

[19] Wilhelm Ackermann. Zum Hilbertschen Aufbau der reellen Zahlen. *Mathematische Annalen*, **99**:118–133, 1928. [294]

[20] Andrew Adamatzky, editor. *Collision-Based Computing*. Springer-Verlag, 2002. [298]

[21] Divesh Aggarwal and Ueli M. Maurer. Breaking RSA generically is equivalent to factoring. In *Proc. EUROCRYPT*, pages 36–53. 2009. [906]

[22] Manindra Agrawal and Somenath Biswas. Primality and identity testing via Chinese remaindering. In *Proc. 40th Symposium on Foundations of Computer Science*, pages 202–209. 1999. [506]

[23] Manindra Agrawal, Neeraj Kayal, and Nitin Saxena. PRIMES is in P. *Annals of Mathematics*, **160**:781–793, 2004. [506]

[24] Dorit Aharonov, Andris Ambainis, Julia Kempe, and Umesh V. Vazirani. Quantum walks on graphs. In *Proc. 33rd Symposium on Theory of Computing*, pages 50–59. 2001. [910]

[25] Dorit Aharonov, Vaughan Jones, and Zeph Landau. A polynomial quantum algorithm for approximating the Jones polynomial. *Algorithmica*, **55**(3):395–421, 2009. [910]

[26] Martin Aigner and Günther M. Ziegler. *Proofs from the Book*. Springer-Verlag, 3rd edition, 2003. [13]

[27] Miklós Ajtai. Σ_1^1-formula on finite structures. *Annals of Pure and Applied Logic*, **24**:1–48, 1983. [220]

[28] David Aldous. Minimization algorithms and random walk on the d-cube. *Annals of Probability*, **11**(2):403–413, 1983. [222]

[29] David Aldous. Random walks on finite groups and rapidly mixing Markov chains. *Séminaire de Probabilités*, **XVII**:243–297, 1983. [643]

[30] David Aldous. The random walk construction of uniform spanning trees and uniform labelled trees. *SIAM Journal on Discrete Mathematics*, **3**(4):450–465, 1990. [645]

[31] David Aldous and Persi Diaconis. Shuffling cards and stopping times. *American Mathematical Monthly*, **93**(5):333–348, 1986. [643, 644]

[32] David Aldous and Persi Diaconis. Longest increasing subsequences: From patience sorting to the Baik–Deift–Johansson theorem. *Bulletin (New Series) of the American Mathematical Society*, **36**(4):413–432, 1999. [940]

[33] Romas Aleliunas, Richard M. Karp, Richard J. Lipton, László Lovász, and Charles Rackoff. Random walks, universal traversal sequences, and the complexity of maze problems. In *Proc. 20th Symposium on Foundations of Computer Science*, pages 218–223. 1979. [645]

[34] Werner Alexi, Benny Chor, Oded Goldreich, and Claus-Peter Schnorr. RSA and Rabin functions: Certain parts are as hard as the whole. *SIAM Journal on Computing*, **17**(2):194–209, 1988. [561]

[35] W. R. Alford, Andrew Granville, and Carl Pomerance. There are infinitely many Carmichael numbers. *Annals of Mathematics*, **140**:703–722, 1994. [505]

[36] Victor Allis. *Searching for Solutions in Games and Artificial Intelligence*. Ph.D. thesis, University of Limburg, Maastricht, The Netherlands, 1994. [36]

[37] Noga Alon. Eigenvalues and expanders. *Combinatorica*, **6**(2):83–96, 1986. [648, 649]

[38] Noga Alon and Joel H. Spencer. *The Probabilistic Method*. John Wiley & Sons, 2000. [810]

[39] Carme Àlvarez and Raymond Greenlaw. A compendium of problems complete for symmetric logarithmic space. *Computational Complexity*, **9**(2):123–145, 2000. [349]

[40] Andris Ambainis. Quantum lower bounds by quantum arguments. *Journal of Computer and System Sciences*, **64**(4):750–767, 2002. [909]

[41] Andris Ambainis. Quantum walk algorithm for element distinctness. *SIAM Journal on Computing*, **37**(1):210–239, 2007. [910]

[42] Andris Ambainis, Eric Bach, Ashwin Nayak, Ashvin Vishwanath, and John Watrous. One-dimensional quantum walks. In *Proc. 33rd Symposium on Theory of Computing*, pages 37–49. 2001. [910]

[43] Andris Ambainis, Andrew M. Childs, Ben Reichardt, RobertŠpa lek, and Shengyu Zhang. Any AND–OR formula of size n can be evaluated in time $n^{1/2+o(1)}$ on a quantum computer. In *Proc. 48th Symposium on Foundations of Computer Science*, pages 363–372. 2007. [910]

[44] Andris Ambainis, Julia Kempe, and Alexander Rivosh. Coins make quantum walks faster. In *Proc. 16th Symposium on Discrete Algorithms*, pages 1099–1108. 2005. [910]

[45] Nina Amenta and Günter M. Ziegler. Deformed products and maximal shadows of polytopes. In *Advances in Discrete and Computational Geometry*, vol. 223 of *Contemporary Mathematics*, pages 57–90. Amer. Math. Soc., 1999. [446]

[46] Nesmith C. Ankeny. The least quadratic non-residue. *Annals of Mathematics*, **55**(1):65–72, 1952. [505]

[47] Kurt M. Anstreicher. Linear programming and the Newton barrier flow. *Mathematical Programming*, **41**:367–373, 1988. [448]

[48] Kenneth Appel and Wolfgang Haken. Every planar map is four colorable. part i. discharging. *Illinois Journal of Mathematics*, **21**:429–490, 1977. [125]

[49] Kenneth Appel, Wolfgang Haken, and John Koch. Every planar map is four colorable. part ii. reducibility. *Illinois Journal of Mathematics*, **21**:491–567, 1977. [125]

[50] David L. Applegate, Robert E. Bixby, Vašek Chvátal, and William J. Cook. *The Traveling Salesman Problem*. Princeton University Press, 2007. [442]

[51] Markus Arndt, Olaf Nairz, Julian Vos-Andreae, Claudia Keller, Gerbrand van der Zouw, and Anton Zeilinger. Wave–particle duality of C_{60} molecules. *Nature*, **401**:680–682, 1999. [903]

[52] Sanjeev Arora. Polynomial time approximation scheme for Euclidean TSP and other geometric problems. *Journal of the ACM*, **45**(5):753–782, 1998. [443]

[53] Sanjeev Arora and Boaz Barak, editors. *Computational Complexity: A Modern Approach.* Cambridge University Press, 2009. [504, 561, 562]

[54] Sanjeev Arora, Boaz Barak, and David Steurer. Subexponential algorithms for UNIQUE GAMES and related problems. In *Proc. 51st Annual IEEE Symposium on Foundations of Computer Science*, page 563. 2010. [503]

[55] Sanjeev Arora, Michelangelo Grigni, David Karger, Philip Klein, and Andrzej Woloszyn. A polynomial-time approximation scheme for weighted planar graph TSP. In *Proc. 9th Annual ACM-SIAM Symposium on Discrete Algorithms*, pages 33–41. 1998. [444]

[56] Sanjeev Arora, Carsten Lund, Rajeev Motwani, Madhu Sudan, and Mario Szegedy. Proof verification and the hardness of approximation problems. *Journal of the ACM*, **45**(3):501–555, 1998. [561]

[57] Sanjeev Arora and Shmuel Safra. Probabilistic checking of proofs: A new characterization of NP. *Journal of the ACM*, **45**(1):70–122, 1998. [561]

[58] C. W. Ashley. *The Ashley Book of Knots.* Doubleday, 1944. [126]

[59] Alain Aspect, Jean Dalibard, and Gérard Roger. Experimental test of Bell's inequalities using time-varying analyzers. *Physical Review Letters*, **49**(25):1804–1807, 1982. [905]

[60] Giorgio Ausiello, Pierluigi Crescenzi, Giorgio Gambosi, Viggo Kann, Alberto Marchetti-Spaccamela, and Marco Protasi. *Complexity and Approximation. Combinatorial Optimization Problems and Their Approximability Properties.* Springer-Verlag, 1999. [442]

[61] László Babai. Monte-Carlo Algorithms in Graph Isomorphism Testing. Université de Montréal technical report, D.M.S. No. 79-10. [506]

[62] László Babai. On the complexity of canonical labeling of strongly regular graphs. *SIAM Journal on Computing*, **9**(1):212–216, 1980. [908]

[63] László Babai. Trading group theory for randomness. In *Proc. 17th Symposium on Theory of Computing*, pages 421–429. 1985. [560]

[64] László Babai and Lance Fortnow. A characterization of $\#P$ arithmetic straight line programs. In *Proc. 31st Symposium on Foundations of Computer Science*, pages 26–34. 1990. [561]

[65] László Babai, Lance Fortnow, Leonid A. Levin, and Mario Szegedy. Checking computations in polylogarithmic time. In *Proc. 23rd Symposium on Theory of Computing*, pages 21–31. 1991. [561]

[66] László Babai, Lance Fortnow, and Carsten Lund. Non-deterministic exponential time has two-prover interactive protocols. *Computational Complexity*, **1**(1):3–40, 1991. [561]

[67] László Babai and Eugene M. Luks. Canonical labeling of graphs. In *Proc. 15th Symposium on Theory of Computing*, pages 171–183. 1983. [908]

[68] László Babai and Shlomo Moran. Arthur–Merlin games: A randomized proof system, and a hierarchy of complexity classes. *Journal of Computer and System Sciences*, **36**(2):254–276, 1988. [560]

[69] Charles Babbage. *Passages from the Life of a Philosopher.* Longman, Green, Longman, Roberts, and Green, 1864. Reprinted by Rutgers University Press, Martin Campbell-Kelly, Editor, 1994. [37]

[70] Eric Bach. Explicit bounds for primality testing and related problems. *Mathematics of Computation*, **55**(191):355–380, 1990. [505]

[71] Dave Bacon, Andrew M. Childs, and Wim van Dam. From optimal measurement to efficient quantum algorithms for the hidden subgroup problem over semidirect product groups. In *Proc. 46th Symposium on Foundations of Computer Science*, pages 469–478. 2005. [908]

[72] Dave Bacon, Andrew M. Childs, and Wim van Dam. Optimal measurements for the dihedral hidden subgroup problem. *Chicago Journal of Theoretical Computer Science*, 2006. Article 2. [908, 910]

[73] Brenda S. Baker. Approximation algorithms for NP-complete problems on planar graphs. *Journal of the ACM*, **41**(1):153–180, 1994. [444]

[74] Theodore Baker, John Gill, and Robert Solovay. Relativizations of the $P = ?NP$ problem. *SIAM Journal on Computing*, **4**:431–442, 1975. [219]

[75] José L. Balcázar, Josep Díaz, and Joaquim Gabarró. *Structural Complexity I.* Springer-Verlag, 2nd edition, 1995. [220, 502]

[76] W. W. Rouse Ball. *Mathematical Recreations and Essays.* Macmillan, 1960. Revised by H. S. M. Coxeter. [89]

[77] Francisco Barahona. On the computational complexity of Ising spin glass models. *Journal of Physics A: Mathematical and General*, **15**:3241–3253, 1982. [171]

[78] Adriano Barenco, Charles H. Bennett, Richard Cleve, David P. DiVincenzo, Norman H. Margolus, Peter W. Shor, Tycho Sleator, John A. Smolin, and Harald Weinfurter. Elementary gates for quantum computation. *Physical Review A*, **52**(5):3457–3467, 1995. [904]

[79] Hans Pieter Barendregt. *The Lambda Calculus: Its Syntax and Semantics.* North-Holland, 1984. [295]

[80] Henk Barendregt and Erik Barendsen. Introduction to Lambda Calculus. Technical report, Department of Computer Science, Catholic University of Nijmegen, 1994. [295]

[81] Gill Barequet, Micha Moffie, Ares Ribó, and Günter Rote. Counting polyominoes on twisted cylinders. In *Proc. 3rd European Conference on Combinatorics, Graph Theory, and Applications,* pages 369–374. 2005. [169]

[82] Howard Barnum, Michael E. Saks, and Mario Szegedy. Quantum query complexity and semi-definite programming. In *Proc. 18th IEEE Conference on Computational Complexity,* pages 179–193. 2003. [909]

[83] Alexander I. Barvinok. Polynomial time algorithms to approximate permanents and mixed discriminants within a simply exponential factor. *Random Structures & Algorithms,* **14**(1):29–61, 1999. [721]

[84] Heiko Bauke, Silvio Franz, and Stephan Mertens. Number partitioning as random energy model. *Journal of Statistical Mechanics: Theory and Experiment,* **2004**(4):P04003, 2004. [815]

[85] Heiko Bauke and Stephan Mertens. Universality in the level statistics of disordered systems. *Physical Review E,* **70**:025102(R), 2004. [815]

[86] R. J. Baxter. Colorings of a hexagonal lattice. *Journal of Mathematical Physics,* **11**:784–789, 1970. [646]

[87] D.A. Bayer and J.C. Lagarias. Karmarkar's linear programming algorithm and Newton's method. *Mathematical Programming,* **50**:291–330, 1991. [448]

[88] Dave Bayer and Persi Diaconis. Tracking the dovetail shuffle to its lair. *Annals of Applied Probability,* **2**(2):294–313, 1992. [644]

[89] Robert Beals. Quantum computation of Fourier transforms over symmetric groups. In *Proc. 29th Symposium on Theory of Computing,* pages 48–53. 1997. [908]

[90] Robert Beals, Harry Buhrman, Richard Cleve, Michele Mosca, and Ronald de Wolf. Quantum lower bounds by polynomials. *Journal of the ACM,* **48**(4):778–797, 2001. [909]

[91] Paul Beame, Stephen A. Cook, Jeff Edmonds, Russell Impagliazzo, and Toniann Pitassi. The relative complexity of NP search problems. *Journal of Computer and System Sciences,* **57**(1):3–19, 1998. [221]

[92] Danièle Beauquier, Maurice Nivat, Eric Rémila, and John Michael Robson. Tiling figures of the plane with two bars. *Computational Geometry,* **5**:1–25, 1995. [169]

[93] Richard Beigel and David Eppstein. 3-coloring in time $O(1.3289^n)$. *Journal of Algorithms,* **54**(2):168–204, 2005. [503]

[94] Mihir Bellare, Don Coppersmith, Johan Håstad, Marcos A. Kiwi, and Madhu Sudan. Linearity testing in characteristic two. *IEEE Transactions on Information Theory,* **42**(6):1781–1795, 1996. [561]

[95] Richard Bellman. *Dynamic Programming.* Princeton University Press, 1957. [90]

[96] Amir Ben-Amram. Tighter constant-factor time hierarchies. *Information Processing Letters,* **87**(1):37–44, 2003. [219]

[97] Shai Ben-David and Shai Halevi. On the provability of P vs. NP. Technical Report 699, Department of Computer Science, Technion, Haifa, 1991. [221]

[98] Amir Ben-Dor and Shai Halevi. Zero-one permanent is #p-complete, a simpler proof. In *Proc. 2nd Israel Symposium on Theory of Computing Systems,* pages 108–117. 1993. [719]

[99] Michael Ben-Or, Shafi Goldwasser, Joe Kilian, and Avi Wigderson. Multi-prover interactive proofs: How to remove intractability assumptions. In *Proc. 20th Symposium on Theory of Computing,* pages 113–131. 1988. [561]

[100] Charles H. Bennett. Logical reversibility of computation. *IBM Journal of Research and Development,* **17**(6):525–532, 1973. [904]

[101] Charles H. Bennett, Ethan Bernstein, Gilles Brassard, and Umesh V. Vazirani. Strengths and weaknesses of quantum computing. *SIAM Journal on Computing,* **26**(5):1510–1523, 1997. [909]

[102] Charles H. Bennett, Gilles Brassard, Claude Crépeau, Richard Jozsa, Asher Peres, and William K. Wootters. Teleporting an unknown quantum state via dual classical and Einstein–Podolsky–Rosen channels. *Physical Review Letters,* **70**(13):1895–1899, 1993. [906]

[103] Charles H. Bennett and John Gill. Relative to a random oracle A, $P^A \neq NP^A \neq co-NP^A$ with probability 1. *SIAM Journal on Computing,* **10**(1):96–113, 1981. [219]

[104] Charles H. Bennett and Stephen J. Wiesner. Communication via one- and two-particle operators on Einstein–Podolsky–Rosen states. *Physical Review Letters,* **69**(20):2881–2884, 1992. [906]

[105] Robert Berger. The undecidability of the domino problem. *Memoirs of the American Mathematical Society,* **66**, 1966. [299]

[106] Elwyn R. Berlekamp, John Horton Conway, and Richard K. Guy. *Winning Ways for Your Mathematical Plays.* A. K. Peters, 2001. [348]

[107] Ethan Bernstein and Umesh V. Vazirani. Quantum complexity theory. In *Proc. 25th Symposium on Theory of Computing,* pages 11–20. 1993. [906]

[108] A. Bertoni, G. Mauri, and N. Sabadini. Simulations among classes of random access machines and equivalence among numbers succinctly represented. *Annals of Discrete Mathematics*, **25**:65–90, 1985. [38]

[109] Dimitris Bertsimas and John N. Tsitsiklis. *Introduction to Linear Optimization*. Athena Scientific, 1997. [445]

[110] Dimitris Bertsimas and Robert Weismantel. *Optimization Over Integers*. Dynamic Ideas, 2005. [448]

[111] Norman L. Biggs, E. Keith Lloyd, and Robin J. Wilson. *Graph Theory, 1736–1936*. Oxford University Press, 1976. [13]

[112] Henk W. J. Blöte and Henk J. Hilhorst. Roughening transitions and the zero-temperature triangular Ising antiferromagnet. *Journal of Physics A: Mathematical and General*, **15**:L631–L637, 1982. [646]

[113] Lenore Blum, Manuel Blum, and Mike Shub. A simple unpredictable pseudo-random number generator. *SIAM Journal on Computing*, **15**(2):364–383, 1986. [562]

[114] Manuel Blum. Coin flipping by telephone. In *Proc. CRYPTO*, pages 11–15. 1981. [561]

[115] Manuel Blum, Michael Luby, and Ronitt Rubinfeld. Self-testing/correcting with applications to numerical problems. *Journal of Computer and System Sciences*, **47**(3):549–595, 1993. [561]

[116] Manuel Blum and Silvio Micali. How to generate cryptographically strong sequences of pseudo-random bits. *SIAM Journal on Computing*, **13**(4):850–864, 1984. [561, 562]

[117] Stefan Boettcher and Stephan Mertens. Analysis of the Karmarkar–Karp differencing algorithm. *European Physics Journal B*, **65**:131–140, 2008. [815]

[118] Béla Bollobás. The evolution of sparse graphs. In B. Bollobás, editor, *Graph Theory and Combinatorics: Proceedings of the Cambridge Combinatorial Conference in Honour of Paul Erdös*, pages 35–57. Academic Press, 1984. [813]

[119] Béla Bollobás. *Modern Graph Theory*. Springer-Verlag, 1998. [13, 718]

[120] Béla Bollobás. *Random Graphs*. Cambridge University Press, 2nd edition, 2001. [812]

[121] John Adrian Bondy and U. S. R. Murty. *Graph Theory with Applications*. North-Holland, 1976. [348]

[122] Maria Elisa Sarraf Borelli and Louis H. Kauffman. The Brazilian knot trick. In David Wolfe and Tom Rodgers, editors, *Puzzler's Tribute: A Feast for the Mind*, pages 91–96. 2002. [126]

[123] C. Borgs, J. Chayes, S. Mertens, and C. Nair. Proof of the local REM conjecture for number partitioning I: Constant energy scales. *Random Structures & Algorithms*, **34**:217–240, 2009. [815]

[124] C. Borgs, J. Chayes, S. Mertens, and C. Nair. Proof of the local REM conjecture for number partitioning II: Growing energy scales. *Random Structures & Algorithms*, **34**:241–284, 2009. [815]

[125] Christian Borgs, Jennifer Chayes, Alan M. Frieze, Jeong Han Kim, Prasad Tetali, Eric Vigoda, and Van H. Vu. Torpid mixing of some Monte Carlo Markov Chain algorithms in statistical physics. In *Proc. 40th Symposium on Foundations of Computer Science*, pages 218–229. 1999. [649]

[126] Christian Borgs, Jennifer Chayes, and Boris Pittel. Phase transition and finite-size scaling for the integer partitioning problem. *Random Structures & Algorithms*, **19**(3–4):247–288, 2001. [815]

[127] Karl-Heinz Borgwardt. The average number of pivot steps required by the simplex method is polynomial. *Zeitschrift für Operations Research*, **26**:157–177, 1982. [446]

[128] Otakar Borůvka. On a minimal problem. *Práce Moravské Přídovědecké Společnosti*, **3**:37–58, 1926. [92]

[129] Sougato Bose, Luke Rallan, and Vlatko Vedral. Communication capacity of quantum computation. *Physical Review Letters*, **85**:5448–5451, 2000. [909]

[130] J. N. Boyd and P. N. Raychowdhury. Linear programming applied to a simple circuit. *American Journal of Physics*, **48**(5):376–378, 1980. [445]

[131] Michel Boyer, Gilles Brassard, Peter Høyer, and Alain Tapp. Tight bounds on quantum searching. *Fortschritte Der Physik*, **46**(4–5):493–505, 1998. [909]

[132] P. Oscar Boykin, Tal Mor, Matthew Pulver, Vwani P. Roychowdhury, and Farrokh Vatan. A new universal and fault-tolerant quantum basis. *Information Processing Letters*, **75**(3):101–107, 2000. [904]

[133] C. V. Boys. A new analytical engine. *Nature*, **81**(2070):14–15, 1909. [291]

[134] Gilles Brassard, Peter Høyer, and Alain Tapp. Quantum counting. In *Proc. 25th International Colloquium on Automata, Languages and Programming*, pages 820–831. 1998. [909]

[135] Gilles Brassard, Peter Høyer, and Alain Tapp. Quantum cryptanalysis of hash and claw-free functions. In *Proc. 3rd Latin American Symposium on Theoretical Informatics*, pages 163–169. 1998. [909]

[136] Alfredo Braunstein and Riccardo Zecchina. Survey propagation as local equilibrium equations. *Journal of Statistical Mechanics: Theory and Experiment*, **2004**:P06007, 2004. [818]

[137] Andrei Broder. Generating random spanning trees. In *Proc. 30th Symposium on Foundations of Computer Science*, pages 442–447. 1989. [645]

[138] Andrei Z. Broder. How hard is it to marry at random? (on the approximation of the permanent). In *Proc. 18th Symposium on Theory of Computing*, pages 50–58. 1986. Erratum in Proc. 20th Symposium on Theory of Computing, page 551, 1988. [719]

[139] Andrei Z. Broder, Alan M. Frieze, and Eli Upfal. On the satisfiability and maximum satisfiability of random 3-CNF formulas. In *Proc. 4th Symposium on Discrete Algorithms*, pages 322–330. 1993. [813]

[140] Rowland Leonard Brooks. On colouring the nodes of a network. *Proc. Cambridge Philosophical Society*, **37**:194–197, 1941. [121]

[141] Stephen G. Brush. Foundations of statistical mechanics 1845–1915. *Archive for History of Exact Sciences*, **34**(4):145–183, 1967. [643]

[142] Russ Bubley and Martin E. Dyer. Path coupling: a technique for proving rapid mixing in Markov chains. In *Proc. 38th Symposium on Foundations of Computer Science*, pages 223–231. 1997. [644]

[143] Russ Bubley, Martin E. Dyer, and Catherine Greenhill. Beating the 2δ bound for approximately counting colourings: a computer-assisted proof of rapid mixing. In *Proc. 9th Symposium on Discrete Algorithms*, pages 355–363. 1998. [649]

[144] Harry Buhrman, Richard Cleve, John Watrous, and Ronald de Wolf. Quantum fingerprinting. *Physical Review Letters*, **87**(16):167902, 2001. [908]

[145] Andrei A. Bulatov. A dichotomy theorem for constraint satisfaction problems on a 3-element set. *Journal of the ACM*, **53**(1):66–120, 2006. [170]

[146] Cristian Calude and Solomon Marcus. The first example of a recursive function which is not primitive recursive. *Historia Mathematica*, **6**:380–384, 1979. [294]

[147] Larry Carter and Mark N. Wegman. Universal classes of hash functions. *Journal of Computer and System Sciences*, **18**(2):143–154, 1979. [504]

[148] Filippo Cesi, G. Guadagni, Fabio Martinelli, and R. H. Schonmann. On the 2D stochastic Ising model in the phase coexistence region close to the critical point. *Journal of Statistical Physics*, **85**:55–102, 1996. [649]

[149] Jean-Luc Chabert, editor. *A History of Algorithms: From the Pebble to the Microchip.* Springer-Verlag, 1999. [36]

[150] Seth Chaiken. A combinatorial proof of the all minors matrix tree theorem. *SIAM Journal on Algebraic and Discrete Methods*, **3**(3):319–329, 1982. [718]

[151] Ashok K. Chandra, Dexter C. Kozen, and Larry J. Stockmeyer. Alternation. *Journal of the ACM*, **28**(1):114–133, 1981. [348]

[152] Richard Chang, Benny Chor, Oded Goldreich, Juris Hartmanis, Johan Håstad, Desh Ranjan, and Pankaj Rohatgi. The random oracle hypothesis is false. *Journal of Computer and System Sciences*, **49**(1):24–39, 1994. [219]

[153] Ming-Te Chao and John V. Franco. Probabilistic analysis of two heuristics for the 3-satisfiability problem. *SIAM Journal on Computing*, **15**(4):1106–1118, 1986. [813]

[154] Ming-Te Chao and John V. Franco. Probabilistic analysis of a generalization of the unit clause literal selection heuristic for the k-satisfiability problem. *Information Science*, **51**:289–314, 1990. [813]

[155] Jeff Cheeger. A lower bound for the lowest eigenvalue of the Laplacian. In R. C. Gunning, editor, *Problems in Analysis: A Symposium in Honor of S. Bochner*, pages 195–199. Princeton University Press, 1970. [648]

[156] Peter Cheeseman, Bob Kanefsky, and William M. Taylor. Where the *really* hard problems are. In J. Mylopoulos and R. Rediter, editors, *Proc. of IJCAI-91*, pages 331–337. Morgan Kaufmann, 1991. [811]

[157] Fang Chen, László Lovász, and Igor Pak. Lifting Markov chains to speed up mixing. In *Proc. 31st Annual Symposium on Theory of Computing*, pages 275–281. 1999. [650]

[158] Xi Chen and Xiaotie Deng. Settling the complexity of two-player Nash equilibrium. In *Proc. 47th Symposium on Foundations of Computer Science*, pages 261–272. 2006. [221]

[159] E. W. Cheney. *Introduction to Approximation Theory.* McGraw-Hill, 1966. [909]

[160] Steve Chien, Lars Eilstrup Rasmussen, and Alistair Sinclair. Clifford algebras and approximating the permanent. *Journal of Computer and System Sciences*, **67**(2):263–290, 2003. [721]

[161] Andrew M. Childs, Richard Cleve, Enrico Deotto, Edward Farhi, Sam Gutmann, and Daniel A. Spielman. Exponential algorithmic speedup by a quantum walk. In *Proc. 39th Symposium on Theory of Computing*, pages 59–68. 2003. [910]

[162] Andrew M. Childs, Edward Farhi, and Sam Gutmann. An example of the difference between quantum and classical random walks. *Quantum Information Processing*, **1**(1–2):35–43, 2002. [909]

[163] Andrew M. Childs and Jeffrey Goldstone. Spatial search and the Dirac equation. *Physical Review A*, **70**(4):042312, 2004. [910]

[164] Noam Chomsky. Three models for the description of language. *IRE Transactions on Information Theory*, **2**:113–124, 1956. [93]

[165] Timothy Y. Chow. Almost-natural proofs. In *Proc. 49th Symposium on Foundations of Computer Science*, pages 86–91. 2008. [222]

[166] Nicos Christofides. Worst-case analysis of a new heuristic for the travelling salesman problem. Technical Report 388, CMU, 1976. [443]

[167] Maria Chudnovsky, Gérard Cornuéjols, Xinming Liu, Paul Seymour, and Kristina Vušković. Recognizing Berge graphs. *Combinatorica*, **25**(2):143–186, 2005. [449]

[168] Maria Chudnovsky, Neil Robertson, Paul D. Seymour, and Robin Thomas. The strong perfect graph theorem. *Annals of Mathematics*, **164**:51–229, 2006. [449]

[169] Fan R. K. Chung. *Spectral Graph Theory*. American Mathematical Society, 1997. [649]

[170] Kai-Min Chung, Omer Reingold, and Salil P. Vadhan. $s-t$ connectivity on digraphs with a known stationary distribution. In *Proc. 22nd IEEE Conf. on Computational Complexity*, pages 236–249. 2007. [649]

[171] Alonzo Church. A set of postulates for the foundation of logic. *Annals of Mathematics*, **33**(2):346–366, 1932. [295]

[172] Alonzo Church. A set of postulates for the foundation of logic (second paper). *Annals of Mathematics*, **34**(4):839–864, 1933. [295]

[173] Alonzo Church. An unsolvable problem of elementary number theory. *American Journal of Mathematics*, **58**(2):345–363, 1936. Presented to the American Mathematical Society, April 19, 1935. [293, 295]

[174] Alonzo Church and J. Barkley Rosser. Some properties of conversion. *Transactions of the American Mathematical Society*, **39**(3):472–482, 1936. [295]

[175] Vasek Chvátal. On certain polytopes associated with graphs. *Journal of Combinatorial Theory, Series B*, **18**:138–154, 1975. [449]

[176] Vasek Chvátal and Bruce Reed. Mick gets some (the odds are on his side). In *Proc. 33rd Symposium on Foundations of Computer Science*, pages 620–627. 1992. [811, 813]

[177] Boris S. Cirel'son. Quantum generalizations of Bell's inequality. *Letters in Mathematical Physics*, **4**:93–100, 1980. [905]

[178] Jens Clausen. Teaching duality in linear programming—the multiplier approach. In *Proc. 4th Meeting of the Nordic MPS Section*, pages 137–148. Publication 97/1, Department of OR, Århus University, 1996. [445]

[179] John F. Clauser, Michael A. Horne, Abner Shimony, and Richard A. Holt. Proposed experiment to test local hidden-variable theories. *Physical Review Letters*, **23**(15):880–884, 1969. [905]

[180] Richard Cleve, Artur Ekert, Chiara Macchiavello, and Michele Mosca. Quantum algorithms revisited. *Proceedings of the Royal Society of London A*, **454**:339–354, 1998. [906]

[181] Alan Cobham. The intrinsic computational difficulty of functions. In *Proc. 1964 Congress for Logic, Mathematics, and Philosophy of Science*, pages 24–30. North-Holland, 1964. [39]

[182] Simone Cocco, Olivier Dubois, Jacque Mandler, and Remi Monasson. Rigorous decimation-based construction of ground pure states for spin-glass models on random lattices. *Physical Review Letters*, **90**(4):047205, 2003. [818]

[183] Simone Cocco and Remi Monasson. Analysis of the computational complexity of solving random satisfiability problems using branch and bound search algorithms. *European Physical Journal B*, **22**:505–531, 2001. [813]

[184] John Cocke and Marvin Minsky. Universality of tag systems with $P=2$. *Journal of the ACM*, **11**(1):15–20, 1964. [298]

[185] Aviad Cohen and Avi Wigderson. Dispersers, deterministic amplification, and weak random sources. *Proc. 30th Symposium on Foundations of Computer Science*, pages 14–19, 1989. [648]

[186] E. G. D. Cohen. Boltzmann and statistical mechanics. In *Proceedings of the International Meeting "Boltzmann's Legacy: 150 Years After His Birth"*, pages 9–23. Atti della Accademia Nazionale dei Lincei, 1997. [643]

[187] Henry Cohn, Noam Elkies, and James Propp. Local statistics for random domino tilings of the Aztec diamond. *Duke Mathematics Journal*, **85**:117–116, 1996. [645]

[188] Henry Cohn, Robert D. Kleinberg, Balázs Szegedy, and Christopher Umans. Group-theoretic algorithms for matrix multiplication. In *Proc. 46th Symposium on Foundations of Computer Science*, pages 379–388. 2005. [38]

[189] Henry Cohn, Michael Larsen, and James Propp. The shape of a typical boxed plane partition. *New York Journal of Mathematics*, **4**:137–165, 1998. [645]

[190] Henry Cohn and Christopher Umans. A group-theoretic approach to fast matrix multiplication. In *Proc. 44th Symposium on Foundations of Computer Science*, pages 438–449. 2003. [38]

[191] Amin Coja-Oghlan. A better algorithm for random k-SAT. *SIAM Journal on Computing*, **39**:2823–2864, 2010. [818]

[192] Amin Coja-Oghlan. On belief propagation guided decimation for random k-SAT. In *Proc. 22nd Symposium on Discrete Algorithms*, pages 957–966. 2011. [817]

[193] Amin Coja-Oghlan and Angelica Y. Pachon-Pinzon. The decimation process in random k-SAT. *SIAM Journal on Discrete Mathematics*, **26**(4):1471–1509, 2012. [817]

[194] John H. Conway. Unpredictable iterations. In *Proceedings of the Number Theory Conference, Boulder, Colorado*, pages 49–52. 1972. [298]

[195] John H. Conway. Fractran: a simple universal programming language for arithmetic. In T. M. Cover and B. Gopinath, editors, *Open Problems in Communication and Computation*, pages 4–26. Springer-Verlag, 1987. [297]

[196] Matthew Cook. Universality in elementary cellular automata. *Complex Systems*, **15**:1–40, 2004. [299]

[197] Stephen A. Cook. A hierarchy for nondeterministic time complexity. In *Proc. 4th Symposium on Theory of computing*, pages 187–192. 1972. [219]

[198] Steven Cook. The complexity of theorem-proving procedures. In *Proc. 3rd Symposium on Theory of Computing*, pages 151–158. 1971. [168]

[199] James W. Cooley, Peter A. W. Lewis, and Peter D. Welch. Historical notes on the Fast Fourier Transform. *IEEE Transactions on Audio and Electroacoustics*, **AU-15**(2):76–79, 1967. [90]

[200] James W. Cooley and John W. Tukey. An algorithm for the machine calculation of complex Fourier series. *Mathematics of Computing*, **19**(2):297–301, 1965. [90]

[201] Don Coppersmith. An approximate Fourier transform useful in quantum factoring, 1994. IBM Research Report RC 19642. [907]

[202] Don Coppersmith and Shmuel Winograd. Matrix multiplication via arithmetic progressions. *Journal of Symbolic Computation*, **9**(3):251–280, 1990. [38]

[203] Thomas H. Cormen, Charles E. Leiserson, Ronald L. Rivest, and Clifford Stein. *Introduction to Algorithms*. The MIT Press, 2nd edition, 2001. [89]

[204] Michel Cosnard, Max H. Garzon, and Pascal Koiran. Computability properties of low-dimensional dynamical systems. In *Proc. 10th Annual Symposium on Theoretical Aspects of Computer Science*, pages 365–373. 1993. [299]

[205] Richard Crandall and Carl Pomerance. *Prime Numbers: A Computational Perspective*. Springer-Verlag, 2nd edition, 2005. [502, 504, 506, 906, 907]

[206] Marcel Crâşmaru and John Tromp. Ladders are PSPACE-complete. In *2nd International Conference on Computers and Games*, pages 241–249. 2000. [348]

[207] Joseph C. Culberson. Sokoban is PSPACE-complete. In *Proc. International Conference on Fun with Algorithms*, pages 65–76. 1998. [349]

[208] Haskell B. Curry. An analysis of logical substitution. *American Journal of Mathematics*, **51**(3):363–384, 1929. [294]

[209] Paul Dagum, Michael Luby, Milena Mihail, and Umesh V. Vazirani. Polytopes, permanents and graphs with large factors. In *Proc. 29th Symposium on Foundations of Computer Science*, pages 412–421. 1988. [719]

[210] Luca Dall'Asta, Abolfazl Ramezanpour, and Riccardo Zecchina. Entropy landscape and non-Gibbs solutions in constraint satisfactiob problems. *Physical Review E*, **77**(3):031118, 2008. [818]

[211] G. C. Danielson and C. Lanczos. Some improvements in practical Fourier analysis and their application to X-ray scattering from liquids. *Journal of the Franklin Institute*, **233**:365–380, 1942. [90]

[212] Evgeny Dantsin, Andreas Goerdt, Edward A. Hirsch, Ravi Kannan, Jon M. Kleinberg, Christos H. Papadimitriou, Prabhakar Raghavan, and Uwe Schöning. A deterministic $(2-2/(k+1))^n$ algorithm for k-SAT based on local search. *Theoretical Computer Science*, **289**(1):69–83, 2002. [503]

[213] G. Dantzig, R. Fulkerson, and S. Johnson. Solution of a large-scale travelling salesman problem. *Operations Research*, **2**(4):393–410, 1954. [448]

[214] George Bernard Dantzig. Maximization of a linear function of variables subject to linear inequalities. In Tjalling C. Koopmans, editor, *Activity Analysis of Production and Allocation*, pages 339–347. John Wiley & Sons, 1951. [445]

[215] Sanjoy Dasgupta, Christos H. Papadimitriou, and Umesh Vazirani. *Algorithms*. McGraw-Hill, 2006. [13, 37, 89, 92, 445, 447, 503]

[216] Constantinos Daskalakis, Paul W. Goldberg, and Christos H. Papadimitriou. The complexity of computing a nash equilibrium. In *Proc. 38th Symposium on Theory of Computing*, pages 71–78. 2006. [221]

[217] J. C. Angles d'Auriac, M. Preissmann, and R. Rammal. The random field Ising model: algorithmic complexity and phase transition. *Journal de Physique Lettres*, **46**:L–173–L–180, 1985. [93]

[218] Guy David and Carlos Tomei. The problem of the calissons. *American Mathematical Monthly*, **96**(5):429–431, 1989. [646]

[219] A. M. Davie and A. J. Stothers. Improved bound for the complexity of matrix multiplication. *Proc. Roy. Soc. Edinburgh Sect. A*, **143**:351–369, 2013. [38]

[220] Brian Davies. Whither Mathematics. *Notices of the AMS*, **52**:1350–1356, 2005. [13]

[221] Martin Davis, editor. *The Undecidable: Basic Papers on Undecidable Propositions, Unsolvable Problems and Computable Functions*. Raven Press, 1965. [296]

[222] Martin Davis, George Logemann, and Donald W. Loveland. A machine program for theorem proving. *Communications of the ACM*, **5**(7):394–397, 1962. [811]

[223] Martin Davis and Hilary Putnam. A computing procedure for quantification theory. *Journal of the ACM*, **7**(1):201–215, 1962. [811]

[224] Martin Davis, Hilary Putnam, and Julia Robinson. The decision problem for exponential Diophantine equations. *Annals of Mathematics*, **74**(3):425–436, 1961. [292]

[225] Paolo del Santo and Giorgio Strano. Observational evidence and the evolution of Ptolemy's lunar model. *Nuncius Annali di Storia della Scienza*, **11**(1):93–122, 1996. [90]

[226] R. A. DeMillo and R. J. Lipton. A probabilistic remark on algebraic program testing. *Information Processing Letters*, **7**(4):193–195, 1978. [504]

[227] Aaron Denney, Cristopher Moore, and Alexander Russell. Finding conjugate stabilizer subgroups in PSL$(2,q)$ and related groups. *Quantum Information and Computation*, **10**:282–291, 2010. [908]

[228] David Deutsch. Quantum theory, the Church-Turing principle and the universal quantum computer. *Proceedings of the Royal Society of London A*, **400**:97–117, 1985. [906]

[229] David Deutsch, Adriano Barenco, and Artur Ekert. Universality in quantum computation. *Proceedings of the Royal Society A*, **449**:669–677, 1995. [904]

[230] David Deutsch and Richard Jozsa. Rapid solution of problems by quantum computation. *Proceedings of the Royal Society of London A*, **439**:553–558, 1992. [906]

[231] Persi Diaconis. *Group Representations in Probability and Statistics*. Institute of Mathematical Statistics, 1988. [644, 646]

[232] Persi Diaconis. The cutoff phenomenon in finite Markov chains. *Proceedings of the National Academy of Science*, **93**:1659–1664, 1996. [650]

[233] Persi Diaconis, Susan Holmes, and Radford M. Neal. Analysis of a nonreversible Markov chain sampler. *Annals of Applied Probability*, **10**(3):726–752, 2000. [650]

[234] Persi Diaconis and Daniel Stroock. Geometric bounds for eigenvalues of Markov chains. *Annals of Applied Probability*, **1**:36–61, 1991. [648]

[235] Josep Díaz, Lefteris M. Kirousis, Dieter Mitsche, and Xavier Pérez-Giménez. A new upper bound for 3-SAT. In *Proc. Foundations of Software Technology and Theoretical Computer Science*, pages 163–174. 2008. [814]

[236] Josep Díaz and Jacobo Torán. Classes of bounded nondeterminism. *Mathematical Systems Theory*, **23**(1):21–32, 1990. [221]

[237] Martin Dietzfelbinger. *Primality Testing in Polynomial Time: From Randomized Algorithms to "PRIMES is in P"*. Springer-Verlag, 2004. [502]

[238] Whitfield Diffie and Martin E. Hellman. New directions in cryptography. *IEEE Transactions on Information Theory*, **IT-22**(6):644–654, 1976. [906, 907]

[239] Edsger W. Dijkstra. A note on two problems in connexion with graphs. *Numerische Mathematik*, **1**:269–271, 1959. [92]

[240] Jian Ding, Allan Sly, and Nike Sun. Proof of the satisfiability conjecture for large k. In *Proc. 47th Symposium on Theory of Computing*, pages 59–68. 2015. [811]

[241] Hang Dinh, Cristopher Moore, and Alexander Russell. McEliece and Niederreiter cryptosystems that resist quantum attack. In *Proc. CRYPTO*. 2011. [908]

[242] Hang Dinh and Alexander Russell. Quantum and randomized lower bounds for local search on vertex-transitive graphs. In *Proc. 12th RANDOM*, pages 385–401. 2008. [222]

[243] E. A. Dinic. Algorithm for solution of a problem of maximum flow in a network with power estimation. *Soviet Mathematics Doklady*, **11**:1277–1280, 1970. [93]

[244] Irit Dinur. The PCP theorem by gap amplification. In *Proc. 38th Symposium on Theory of Computation*, pages 241–250. 2006. [562]

[245] David P. DiVincenzo. Two-bit gates are universal for quantum computation. *Physical Review A*, **51**(2):1015–1022, 1995. [904]

[246] Jozef Dodziuk. Difference equations, isoperimetric inequality and transience of certain random walks. *Transactions of the American Mathematical Society*, **284**(2):787–794, 1984. [648]

[247] Apostolos Doxiadis and Christos H. Papadimitriou. *Logicomix: An Epic Search For Truth*. Bloomsbury, 2009. [292]

[248] Peter G. Doyle and J. Laurie Snell. *Random Walks and Electrical Networks*. Mathematical Association of America, 1984. Available online at arxiv.org/abs/math/0001057. [943]

[249] Stuart E. Dreyfus. *Dynamic Programming and the Calculus of Variations*. Academic Press, 1965. [91]

[250] Olivier Dubois and Yacine Boufkhad. A general upper bound for the satisfiability threshold of random r-SAT formulae. *Journal of Algorithms*, **24**(2):395–420, 1997. [814]

[251] Olivier Dubois, Yacine Boufkhad, and Jacques Mandler. Typical random 3-SAT formulae and the satisfiability threshold. In *Proc. 11th Symposium on Discrete Algorithms*, pages 126–127. 2000. [814]

[252] Olivier Dubois and Jacques Mandler. The 3-XORSAT threshold. In *Proc. 43rd Symposium on Foundations of Computer Science*, pages 769–778. 2002. [818]

[253] Henry Dudeney. *The Canterbury Puzzles (And Other Curious Problems)*. Thomas Nelson and Sons, 1907. [89]

[254] Martin Dyer and Catherine Greenhill. A more rapidly mixing Markov chain for graph colorings. *Random Structures & Algorithms*, **13**:285–317, 1998. [649]

[255] Martin Dyer and Catherine Greenhill. Random walks on combinatorial objects. In J. D. Lamb and D. A. Preece, editors, *Surveys in Combinatorics*, pages 101–136. Cambridge University Press, 1999. [643]

[256] Martin E. Dyer and Alan M. Frieze. Randomly colouring graphs with lower bounds on girth and maximum degree. *Random Structures & Algorithms*, **23**:167–179, 2003. [644]

[257] Martin E. Dyer, Alan M. Frieze, Thomas P. Hayes, and Eric Vigoda. Randomly coloring constant degree graphs. In *Proc. 45th Symposium on Foundations of Computer Science*, pages 582–589. 2004. [644]

[258] Martin E. Dyer, Alan M. Frieze, and Ravi Kannan. A random polynomial time algorithm for approximating the volume of convex bodies. *Journal of the ACM*, **38**(1):1–17, 1991. [719]

[259] Martin E. Dyer, Alistair Sinclair, Eric Vigoda, and Dror Weitz. Mixing in time and space for lattice spin systems: A combinatorial view. *Random Structures & Algorithms*, **24**(4):461–479, 2004. [649]

[260] Jack Edmonds. Paths, trees and flowers. *Canadian Journal of Mathematics*, **17**:449–467, 1965. [39, 170]

[261] Jack Edmonds and Ellis L. Johnson. Matching, Euler tours and the Chinese postman. *Mathematical Programming*, **5**(1):88–124, 1973. [170]

[262] John Edmonds. Matroids and the greedy algorithm. *Mathematical Programming*, **1**:127–136, 1971. [93]

[263] John Edmonds and Richard M. Karp. Theoretical improvements in algorithmic efficiency for network flow problems. *Journal of the ACM*, **19**(2):248–264, 1972. [93]

[264] Albert Einstein, Boris Podolsky, and Nathan Rosen. Can quantum-mechanical description of reality be considered complete? *Physical Review*, **47**:777–780, 1935. [905]

[265] Y. C. Eldar, A. Megretski, and G. C. Verghese. Optimal detection of symmetric mixed quantum states. *IEEE Transactions on Information Theory*, **50**(6):1198–1207, 2004. [910]

[266] P. Elias, A. Feinstein, and Claude E. Shannon. Note on maximum flow through a network. *IRE Transactions on Information Theory*, **IT-2**:117–119, 1956. [93]

[267] David Eppstein. Improved algorithms for 3-coloring, 3-edge-coloring, and constraint satisfaction. In *Proc. 12th Symposium on Discrete Algorithms*, pages 329–337. 2001. [219]

[268] Paul Erdős and Mark Kac. The Gaussian law of errors in the theory of additive number theoretic functions. *American Journal of Mathematics*, **62**:738–742, 1940. [497]

[269] Paul Erdős and Alfréd Rényi. On the evolution of random graphs. *Matematikai Kutató Intézetének Közleményei. Magyar Tudományos Akadémia*, **5**:17–61, 1960. [812]

[270] Paul Erdős and George Szekeres. A combinatorial problem in geometry. *Compositio Mathematica*, **2**:463–470, 1935. [93]

[271] Mark Ettinger, Peter Høyer, and Emanuel Knill. The quantum query complexity of the hidden subgroup problem is polynomial. *Information Processing Letters*, **91**(1):43–48, 2004. [908]

[272] L. Euler. Solutio problematis ad geometrian situs pertinentis. *Commetarii Academiae Scientiarum Imperialis Petropolitanae*, **8**:128–140, 1736. [13]

[273] Shimon Even and Robert E. Tarjan. A combinatorial problem which is complete in polynomial space. *Journal of the ACM*, **23**(4):710–719, 1976. [349]

[274] Edward Farhi, Jeffrey Goldstone, and Sam Gutmann. A quantum algorithm for the Hamiltonian NAND tree. *Theory of Computing*, **4**(1):169–190, 2008. [910]

[275] Edward Farhi, Jeffrey Goldstone, Sam Gutmann, and Michael Sipser. A limit on the speed of quantum computation in determining parity. *Physical Review Letters*, **81**:5442–5444, 1998. [909]

[276] Edward Farhi and Sam Gutmann. Analog analogue of a digital quantum computation. *Physical Review A*, **57**(4):2403–2406, 1998. [909]

[277] Edward Farhi and Sam Gutmann. Quantum computation and decision trees. *Physical Review A*, **58**(2):915–928, 1998. [909]

[278] Uriel Feige, Shafi Goldwasser, László Lovász, Shmuel Safra, and Mario Szegedy. Interactive proofs and the hardness of approximating cliques. *Journal of the ACM*, **43**(2):268–292, 1996. [561]

[279] Wenceslas Fernandez de la Vega. On random 2-SAT, 1992. Manuscript. [811]

[280] Sabino José Ferreira and Alan Sokal. Antiferromagnetic Potts models on the square lattice: A high-precision Monte Carlo study. *Journal of Statistical Physics*, **96**:461–530, 1999. [649]

[281] Michael E. Fisher. Transformations of Ising models. *Physical Review*, **113**(4):969–981, 1959. [721]

[282] Michael E. Fisher. Statistical mechanics of dimers on a plane lattice. *Physical Review*, **124**(6):1664–1672, 1961. [720]

[283] Michael E. Fisher. On the dimer solution of planar Ising models. *Journal of Mathematical Physics*, **7**(10):1776–1781, 1966. [720]

[284] Michael E. Fisher and John Stephenson. Statistical mechanics of dimers on a plane lattice II: Dimer correlations and monomers. *Physical Review*, **132**:1411–1431, 1963. [645]

[285] Gary William Flake and Eric B. Baum. Rush Hour is PSPACE-complete, or "why you should generously tip parking lot attendants". *Theoretical Computer Science*, **270**(1–2):895–911, 2002. [349]

[286] L. R. Ford and D. R. Fulkerson. Maximal flow through a network. *Canadian Journal of Mathematics*, **8**:399–404, 1956. [93]

[287] Kees Fortuin and Pieter W. Kasteleyn. On the random cluster model I. Introduction and relation to other models. *Physica*, **57**(4):536–564, 1972. [645, 720]

[288] Jean Baptiste Joseph Fourier. Solution d'une question particulière du calcul des inégalités. *Nouveau Bulletin des Sciences par la Societé Philomatique de Paris*, pages 99–100, 1826. [445]

[289] Aviezri S. Fraenkel, M. R. Garey, David S. Johnson, T. Schaefer, and Yaacov Yesha. The complexity of checkers on an $N \times N$ board. In *Proc. 19th Symposium on Foundations of Computer Science*, pages 55–64. 1978. [349]

[290] Aviezri S. Fraenkel and David Lichtenstein. Computing a perfect strategy for $n \times n$ chess requires time exponential in n. *Journal of Combinatorial Theory, Series A*, **31**(2):199–214, 1981. [349]

[291] Aviezri S. Fraenkel, Edward R. Scheinerman, and Daniel Ullman. Undirected edge geography. *Theoretical Computer Science*, **112**:371–381, 1993. [348]

[292] John Franco and Marvin Paull. Probabilistic analysis of the Davis–Putnam procedure for solving the satisfiability problem. *Discrete Applied Mathematics*, **5**(1):77–87, 1983. [814]

[293] Edward Fredkin and Tommaso Toffoli. Conservative logic. *International Journal of Theoretical Physics*, **21**:219–253, 1982. [904]

[294] Rusins Freivalds. Fast probabilistic algorithms. In *Proc. 8th Symposium on Mathematical Foundations of Computer Science*, pages 57–69. 1979. [504]

[295] Robert M. Freund. *Introduction to Semidefinite Programming (SDP)*. MIT Open Courseware, 2004. [448]

[296] Ehud Friedgut. Sharp thresholds of graph properties, and the k-SAT problem. *Journal of the American Mathematical Society*, **12**(4):1017–1054, 1999. Appendix by Jean Bourgain. [811]

[297] Ehud Friedgut. Hunting for sharp thresholds. *Random Structures & Algorithms*, **26**(1–2):37–51, 2005. [811]

[298] Katalin Friedl, Gábor Ivanyos, Frédéric Magniez, Miklos Santha, and Pranab Sen. Hidden translation and orbit coset in quantum computing. In *Proc. 35th Symposium on Theory of Computing*, pages 1–9. 2003. [908]

[299] Oliver Friedmann, Thomas Dueholm Hansen, and Uri Zwick. Subexponential lower bounds for randomized pivoting rules for solving linear programs. To appear at STOC'11, 2010. [446]

[300] Alan M. Frieze and Colin McDiarmid. Algorithmic theory of random graphs. *Random Structures & Algorithms*, **10**(1–2):5–42, 1997. [812]

[301] Alan M. Frieze and Stephen Suen. Analysis of two simple heuristics on a random instance of k-SAT. *Journal of Algorithms*, **20**(2):312–355, 1996. [813]

[302] Alan M. Frieze and Nicholas C. Wormald. Random SAT: A tight threshold for moderately growing k. *Combinatorica*, **25**(3):297–305, 2005. [814]

[303] D. R. Fulkerson. Blocking and anti-blocking pairs of polyhedra. *Mathematical Programming*, **1**:168–194, 1971. [449]

[304] William Fulton and Joe Harris. *Representation Theory: A First Course*. Springer-Verlag, 1991. [908]

[305] Martin Fürer. The tight deterministic time hierarchy. In *Proc. 14th Symposium on Theory of computing*, pages 8–16. 1982. [219]

[306] Martin Fürer. Faster integer multiplication. In *Proc. 39th Symposium on Theory of Computing*, pages 57–66. 2007. [37]

[307] Merrick L. Furst, James B. Saxe, and Michael Sipser. Parity, circuits, and the polynomial-time hierarchy. *Mathematical Systems Theory*, **17**(1):13–27, 1984. [220]

[308] Timothy Furtak, Masashi Kiyomi, Takeaki Uno, and Michael Buro. Generalized Amazons is PSPACE-complete. In *Proc. 19th International Joint Conference on Artificial Intelligence*, pages 132–137. 2005. [350]

[309] Ofer Gabber and Zvi Galil. Explicit constructions of linear-sized superconcentrators. *Journal of Computer and System Sciences*, **22**:407–420, 1981. [649]

[310] Harold N. Gabow and Oded Kariv. Algorithms for edge coloring bipartite graphs and multigraphs. *SIAM Journal on Computing*, **11**(1):117–129, 1982. [93]

[311] David Gale, Harold W. Kuhn, and Albert W. Tucker. Linear programming and the theory of games. In Tjalling C. Koopmans, editor, *Activity Analysis of Production and Allocation*, pages 317–329. John Wiley & Sons, 1951. [447]

[312] Zvi Galil and Kunsoo Park. Dynamic programming with convexity, concavity, and sparsity. *Theoretical Computer Science*, **92**(1):49–76, 1992. [90]

[313] David Galvin and Dana Randall. Torpid mixing of local Markov chains on 3-colorings of the discrete torus. In *Proc. 40th Symposium on Foundations of Computer Science*, pages 218–229. 2007. [650]

[314] Robin Gandy. The confluence of ideas in 1936. In Rolf Herken, editor, *The Universal Turing Machine. A Half-Century Survey*. Springer-Verlag, 1995. [296]

[315] Martin Gardner. Piet Hein's superellipse. In *Mathematical Carnival: A New Round-Up of Tantalizers and Puzzles from Scientific American*, pages 240–254. Vintage, 1977. [914]

[316] Martin Gardner. The binary Gray code. In *Knotted Doughnuts and Other Mathematical Entertainments*, pages 11–27. W. H. Freeman, 1986. [89]

[317] Martin Gardner. A new kind of cipher that would take millions of years to break. In *Scientific American*, pages 120–124. August 1977. [906]

[318] M. R. Garey, R. L. Graham, and D. S. Johnson. Some NP-complete geometric problems. In *STOC '76: Proc. 8th Annual ACM Symposium on Theory of Computing*, pages 10–22. 1976. [443]

[319] M. R. Garey, R. L. Graham, and D. S. Johnson. The complexity of computing Steiner minimal trees. *SIAM Journal on Applied Mathematics*, **32**(4):835–859, 1977. [171]

[320] Carl Friedrich Gauss. Nachlass, theoria interpolationis methodo nova tractata. In *Carl Friedrich Gauss Werke, Band 3*, pages 265–330. Königliche Gesellschaft der Wissenschaften, 1866. [90]

[321] John Geanakoplos. Nash and Walras equilibrium via Brouwer. *Economic Theory*, **21**(2–3):585–603, 2003. [221]

[322] Yuval Gefen, Benoit B. Mandelbrot, and Amnon Aharony. Critical phenomena on fractal lattices. *Physical Review Letters*, **45**(11):855–858, 1980. [721]

[323] Ian P. Gent and Toby Walsh. Phase transitions and annealed theories: number partitioning as a case study. In W. Wahlster, editor, *Proc. of ECAI-96*, pages 170–174. John Wiley & Sons, 1996. [815]

[324] Heath Gerhardt and John Watrous. Continuous-time quantum walks on the symmetric group. In *Proc. 7th International Workshop on Randomization and Approximation Techniques in Computer Science*, pages 290–301. 2003. [910]

[325] Antoine Gerschenfeld and Andrea Montanari. Reconstruction for models on random graphs. In *Proc. 48th Symposium on Foundations of Computer Science*, pages 194–204. 2007. [818]

[326] Ira M. Gessel and Xavier Gérard Viennot. Determinants, paths, and plane partitions, 1989. people.brandeis.edu/~gessel/homepage/papers/pp.pdf. [721]

[327] Shayan Oveis Gharan, Amin Saberi, and Mohit Singh. A randomized rounding approach to the Traveling Salesman Problem. In *Proc. 52nd Symposium on Foundations of Computer Science*, pages 550–559. 2011. [443]

[328] E. N. Gilbert. Random graphs. *The Annals of Mathematical Statistics*, **30**(4):1141–1144, 12 1959. [812]

[329] E. W. Gilbert. Theory of shuffling. *Bell Laboratories Technical Memorandum*, 1955. [644]

[330] John Gill. Computational complexity of probabilistic Turing machines. *SIAM Journal on Computing*, **6**(4):675–695, 1977. [506]

[331] Philip E. Gill, Walter Murray, Michael A. Saunders, J. A. Tomlin, and Margaret H. Wright. On projected Newton barrier methods for linear programming and an equivalence to Karmarkar's projective method. *Mathematical Programming*, **36**:183–209, 1986. [448]

[332] Roy J. Glauber. Time-dependent statistics of the Ising model. *Journal of Mathematical Physics*, **4**(2):294–307, 1963. [644]

[333] F. Glover and M. Laguna. Tabu search. In C. R. Reeves, editor, *Modern Heuristic Techniques for Combinatorial Problems*, pages 70–150. John Wiley & Sons, 1996. [503]

[334] C. D. Godsil and Ivan Gutman. On the matching polynomial of a graph. In *Algebraic Methods in Graph Theory*, pages 241–249. North-Holland, 1981. [721]

[335] Michael X. Goemans and David P. Williamson. Improved approximation algorithms for maximum cut and satisfiability problems using semidefinite programming. *Journal of the ACM*, **42**:1115–1145, 1995. [503]

[336] Andreas Goerdt. A threshold for unsatisfiability. *Journal of Computer and System Sciences*, **53**(3):469–486, 1996. [811]

[337] Leslie Ann Goldberg, Russell A. Martin, and Mike Paterson. Strong spatial mixing with fewer colors for lattice graphs. *SIAM Journal on Computing*, **35**(2):486–517, 2005. [649]

[338] Oded Goldreich. *Modern Cryptography, Probabilistic Proofs and Pseudorandomness*. Springer-Verlag, 1999. [560]

[339] Oded Goldreich, Shafi Goldwasser, and Silvio Micali. How to construct random functions. *Journal of the ACM*, **33**(4):792–807, 1986. [562]

[340] Oded Goldreich and Leonid A. Levin. A hard-core predicate for all one-way functions. In *Proc. 21st Symposium on Theory of Computing*, pages 25–32. 1989. [561]

[341] Oded Goldreich, Silvio Micali, and Avi Wigderson. Proofs that yield nothing but their validity and a methodology of cryptographic protocol design. In *Proc. 27th Symposium on Foundations of Computer Science*, pages 174–187. 1986. [560, 561]

[342] Shafi Goldwasser, Silvio Micali, and Charles Rackoff. The knowledge complexity of interactive proof systems. In *Proc. 17th Symposium on Theory of Computing*, pages 291–304. 1985. [560, 561]

[343] Shafi Goldwasser and Michael Sipser. Private coins versus public coins in interactive proof systems. In *Proc. 18th Symposium on Theory of Computing*, pages 59–68. 1986. [560]

[344] Solomon W. Golomb. *Polyominoes: Puzzles, Patterns, Problems, and Packings*. Princeton University Press, 1994. [168]

[345] Clóvis Caesar Gonzaga. A simple presentation of Karmarkar's algorithm. Technical report, Dept. of Mathematics, Federal University of Santa Catarina, Brazil, 2002. [448]

[346] I. J. Good. The interaction algorithm and practical Fourier analysis. *Journal of the Royal Statistical Society, Series B*, **20**(2):361–372, 1958. [90]

[347] Daniel M. Gordon. Discrete logarithms in gf(p) using the number field sieve. *SIAM Journal on Discrete Mathematics*, **6**(1):124–138, 1993. [907]

[348] Daniel Gottesman and Isaac Chuang. Quantum digital signatures, 2001. arxiv.org/abs/quant-ph/0105032. [908]

[349] R. L. Graham. Bounds on multiprocessing timing anomalies. *SIAM Journal on Applied Mathematics*, **17**(2):416–429, 1969. [442]

[350] Andrew Granville. It is easy to determine whether a given integer is prime. *Bulletin (New Series) of the American Mathematical Society*, **42**(1):3–38, 2004. [506]

[351] C. G. Gray and Edwin F. Taylor. When action is not least. *American Journal of Physics*, **75**(5):434–458, 2007. [92]

[352] Daniel M. Greenberger, Michael A. Horne, and Anton Zeilinger. Going beyond Bell's theorem. In *Bell's Theorem, Quantum Theory, and Conceptions of the Universe*, pages 73–76. Kluwer Academic, 1989. [905]

[353] Michelangelo Grigni, Leonard J. Schulman, Monica Vazirani, and Umesh V. Vazirani. Quantum mechanical algorithms for the nonabelian hidden subgroup problem. *Combinatorica*, **24**(1):137–154, 2004. [908]

[354] Jonathan L. Gross and Jay Yellen, editors. *Handbook of Graph Theory*. CRC Press, 2004. [13]

[355] Jerrold W. Grossman and R. Suzanne Zeitman. An inherently iterative computation of Ackermann's function. *Theoretical Computer Science*, **57**:327–330, 1988. [294]

[356] Lov K. Grover. A fast quantum mechanical algorithm for database search. In *Proc. 28th Symposium on Theory of Computing*, pages 212–219. 1996. [909]

[357] Yu. Sh. Gurevich and I. O. Koriakov. Remarks on Berger's paper on the domino problem. *Siberian Mathematical Journal*, **13**:459–463, 1972. [299]

[358] Richard K. Guy. Conway's prime producing machine. *Mathematical Magazine*, **56**(1):26–33, 1983. [297]

[359] Lucia Hackermüller, Stefan Uttenthaler, Klaus Hornberger, Elisabeth Reiger, Björn Brezger, Anton Zeilinger, and Markus Arndt. Wave nature of biomolecules and fluorofullerenes. *Physical Review Letters*, **91**(9):090408, 2003. [903]

[360] Mohammad Taghi Hajiaghayi and Gregory B. Sorkin. The satisfiability threshold for random 3-SAT is at least 3.52. Technical report, 2003. arxiv.org/abs/math/0310193. [814]

[361] Wolfgang Haken. Theorie der Normalflächen, ein Isotopiekriterium für den Kreisknoten. *Acta Mathematica*, **105**:245–375, 1961. [126]

[362] Thomas C. Hales. A proof of the Kepler conjecture. *Annals of Mathematics*, **162**(3):1065–1185, 2005. [13]

[363] Sean Hallgren. Polynomial-time quantum algorithms for Pell's equation and the principal ideal problem. In *Proc. 34th Symposium on Theory of Computing*, pages 653–658. 2002. [907]

[364] Sean Hallgren, Cristopher Moore, Martin Rötteler, Alexander Russell, and Pranab Sen. Limitations of quantum coset states for graph isomorphism. In *Proc. 38th Symposium on Theory of Computing*, pages 604–617. 2006. [908]

[365] Sean Hallgren, Alexander Russell, and Amnon Ta-Shma. The hidden subgroup problem and quantum computation using group representations. *SIAM Journal on Computing*, **32**(4):916–934, 2003. [908]

[366] James E. Hanson and James P. Crutchfield. Computational mechanics of cellular automata: an example. *Physica D*, **103**:169–189, 1997. [299]

[367] G. H. Hardy and E. M. Wright. *An Introduction to the Theory of Numbers*. Clarendon Press, 1954. [907]

[368] T. E. Harris and F. S. Ross. Fundamentals of a method for evaluating rail net capacities, 1955. Research Memorandum RM–1573, The RAND Corporation. [xiii, 93]

[369] Aram Harrow. Coherent communication of classical messages. *Physical Review Letters*, **92**(9):097902, 2004. [906]

[370] Juris Hartmanis. Gödel, von Neumann and the $P =?NP$ problem. *Bulletin of the European Association for Theoretical Computer Science*, **38**:101–107, 1989. [218]

[371] Juris Hartmanis and Janos Simon. On the power of multiplication in random-access machines. In *Proc. 15th Annu. IEEE Sympos. Switching Automata Theory*, pages 13–23. 1974. [38]

[372] Juris Hartmanis and Richard E. Stearns. On the computational complexity of algorithms. *Transactions of the American Mathematical Society*, **117**:285–306, 1965. [38, 219]

[373] Alexander K. Hartmann and Martin Weigt. *Phase Transitions in Combinatorial Optimization Problems.* John Wiley & Sons, 2005. [810]

[374] David Harvey and Joris Van Der Hoeven. Integer multiplication in time $O(n \log n)$. Preprint <hal-02070778>, 2019. [37]

[375] Joel Hass and Jeffrey C. Lagarias. The number of Reidemeister moves needed for unknotting. *Journal of the American Mathematical Society*, **14**:399–428, 2001. [126]

[376] Joel Hass, Jeffrey C. Lagarias, and Nicholas Pippenger. The computational complexity of knot and link problems. *Journal of the ACM*, **46**:185–211, 1999. [126]

[377] Johan Håstad. Almost optimal lower bounds for small depth circuits. In *Proc. 18th Symposium on Theory of Computing*, pages 6–20. 1986. [220]

[378] Johan Håstad. Some optimal inapproximability results. *Journal of the ACM*, **48**(4):798–859, 2001. [444, 561]

[379] Johan Håstad, Russell Impagliazzo, Leonid A. Levin, and Michael Luby. A pseudorandom generator from any one-way function. *SIAM Journal on Computing*, **28**(4):1364–1396, 1999. [562]

[380] Matt B. Hastings. Superadditivity of communication capacity using entangled inputs. *Nature Physics*, **5**:255, 2009. [906]

[381] Julian Havil. *Gamma: Exploring Euler's Constant.* Cambridge University Press, 2003. [504]

[382] Julian Havil. *Nonplussed!* Princeton University Press, 2007. [297]

[383] Brian Hayes. Can't get no satisfaction. *American Scientist*, **85**(2):108–112, March–April 1997. [125]

[384] Brian Hayes. The easiest hard problem. *American Scientist*, **90**(2):113–117, March–April 2002. [815]

[385] Brian Hayes. Why W? *American Scientist*, **93**(2):104–108, March–April 2005. [812]

[386] Thomas P. Hayes. Randomly coloring graphs of girth at least five. In *Proc. 44th Symposium on Foundations of Computer Science*, pages 618–627. 2003. [644]

[387] Thomas P. Hayes and Alistair Sinclair. A general lower bound for mixing of single-site dynamics on graphs. In *Proc. 46th Symposium on Foundations of Computer Science*, pages 511–520. 2005. [644]

[388] Thomas P. Hayes and Eric Vigoda. A non-Markovian coupling for randomly sampling colorings. In *Proc. 44th Symposium on Foundations of Computer Science*, pages 618–627. 2003. [645]

[389] Thomas P. Hayes and Eric Vigoda. Coupling with the stationary distribution and improved sampling for colorings and independent sets. *Annals of Applied Probability*, **16**(3):1297–1318, 2006. [644]

[390] Robert A. Hearn. Amazons, Konane, and Cross Purposes are PSPACE-complete. In Michael H. Albert and Richard J. Nowakowski, editors, *Games of No Chance 3.* Cambridge University Press, 2007. [350]

[391] Robert A. Hearn and Erik D. Demaine. The nondeterministic constraint logic model of computation: reductions and applications. In *Proc. 29th International Colloquium on Automata, Languages and Programming*, pages 401–413. 2002. [349]

[392] Robert A. Hearn and Erik D. Demaine. PSPACE-completeness of sliding-block puzzles and other problems through the nondeterministic constraint logic model of computation. *Theoretical Computer Science*, **343**(1–2):72–96, 2005. [349]

[393] Michael T. Heideman, Don H. Johnson, and C. Sidney Burrus. Gauss and the history of the fast Fourier transform. *Archive for History of Exact Sciences*, **34**(3):265–277, 1985. [90]

[394] Pavol Hell, David Kirkpatrick, Jan Kratochvíl, and Igor Kříž. On restricted two-factors. *SIAM Journal on Discrete Mathematics*, **1**(4):472–484, 1988. [449]

[395] Martin E. Hellman. An overview of public key cryptography. *IEEE Communications Magazine*, **40**(5):42–49, 2002. [907]

[396] Paul Helman, Bernard M.E. Moret, and Henry D. Shapiro. An exact characterization of greedy structures. *SIAM Journal on Discrete Mathematics*, **6**(2):274–283, May 1993. [93]

[397] F. C. Hennie and Richard Edwin Stearns. Two-tape simulation of multitape Turing machines. *Journal of the ACM*, **13**(4):533–546, 1966. [219]

[398] Nick Herbert. FLASH–a superluminal communicator based upon a new kind of quantum measurement. *Foundations of Physics*, **12**(12):1171–1179, 1982. [904]

[399] Timon Hertli. 3-SAT faster and simpler - unique-SAT bounds for PPSZ hold in general. *SIAM Journal on Computing*, **43**(2):718–729, 2014. [503]

[400] C. Hierholzer. Über die Möglichkeit, einen Linienzug ohne Wiederholung und ohne Unterbrechung zu umfahren. *Mathematische Annalen*, **6**:30–32, 1873. [13]

[401] D. Hilbert and W. Ackermann. *Grundzüge der theoretischen Logik.* Springer-Verlag, 1928. [291]

[402] Andreas M. Hinz. An iterative algorithm for the Tower of Hanoi with four pegs. *Computing*, **42**:133–140, 1989. [89]

[403] C. A. R. Hoare and D. C. S. Allison. Incomputability. *Computing Surveys*, **4**(3):169–178, 1972. [293]

[404] Charles Anthony Richard Hoare. Algorithm 63: Partition, and Algorithm 64: Quicksort. *Communications of the ACM*, **4**:321–322, 1961. [89]

[405] Dorit S. Hochbaum and David B. Shmoys. Using dual approximations for scheduling problems: theoretical and practical results. *Journal of the ACM*, **34**(1):144–162, 1987. [442]

[406] Alan J. Hoffman and Joseph B. Kruskal. Integral boundary points of convex polyhedra. In H. W. Kuhn and A. W. Tucker, editors, *Linear Inequalities and Related Systems*, pages 223–246. Princeton University Press, 1956. [448]

[407] Douglas R. Hofstadter. *Gödel, Escher, Bach: An Eternal Golden Braid.* Basic Books, 1979. [292]

[408] Shlomo Hoory, Nathan Linial, and Avi Wigderson. Expander graphs and their applications. *Bulletin of the American Mathematical Society*, **43**(4):439–561, 2006. [648]

[409] Holger Hoos. An adaptive noise mechanism for WalkSAT. In *Proc. 17th National Conference on Artificial Intelligence*, pages 655–660. 2002. [503]

[410] John E. Hopcroft, Jacob T. Schwartz, and Micha Sharir. On the complexity of motion planning for multiple independent objects: PSPACE-hardness of the Warehouseman's Problem. *International Journal of Robotics Research*, **3**(4):76–88, 1984. [349]

[411] Edward Hordern. *Sliding Piece Puzzles.* Oxford University Press, 1986. [349]

[412] Ellis Horowitz and Sartaj Sahni. Exact and approximate algorithms for scheduling nonidentical processors. *Journal of the ACM*, **33**(2):317–327, 1976. [126]

[413] Peter Høyer, Troy Lee, and RobertŠpalek. Negative weights ma ke adversaries stronger. In *Proc. 39th Symposium on Theory of Computing*, pages 526–535. 2007. [909]

[414] Cor A. J. Hurkens and Gerhard J. Woeginger. On the nearest neighbor rule for the traveling salesman problem. *Operations Research Letters*, **32**:1–4, 2004. [443]

[415] Oscar H. Ibarra. A note concerning nondeterministic tape complexities. *Journal of the ACM*, **19**(4):608–612, 1972. [219]

[416] Neil Immerman. Nondeterministic space is closed under complementation. *SIAM Journal on Computing*, **17**(5):935–938, 1988. [347]

[417] Russell Impagliazzo. A personal view of average-case complexity. In *Structure in Complexity Theory Conference*, pages 134–147. 1995. [562]

[418] Russell Impagliazzo, Leonid A. Levin, and Michael Luby. Pseudo-random generation from one-way functions. In *Proc. 21st Annual Symposium on Theory of Computing*, pages 12–24. 1989. [562]

[419] Russell Impagliazzo and Avi Wigderson. P = BPP if E requires exponential circuits: Derandomizing the XOR lemma. In *Proc. 29th Symposium on Theory of Computing*, pages 220–229. 1997. [562]

[420] Russell Impagliazzo and David Zuckerman. How to recycle random bits. *Proc. 30th Symposium on Foundations of Computer Science*, pages 248–253, 1989. [562, 649]

[421] Ernst Ising. Beitrag zur Theorie des Ferromagnetismus. *Zeitschrift für Physik*, **31**:253–258, 1925. [720]

[422] Alon Itai, Christos H. Papadimitriou, and Jayme Luiz Szwarcfiter. Hamilton paths in grid graphs. *SIAM Journal on Computing*, **11**(4):676–686, 1982. [443]

[423] Gábor Ivanyos, Luc Sanselme, and Miklos Santha. An efficient quantum algorithm for the hidden subgroup problem in nil-2 groups. In *Proc. 8th Latin American Symposium on Theoretical Informatics*, pages 759–771. 2008. [908]

[424] Shigeki Iwata and Takumi Kasai. The Othello game on an $n \times n$ board is PSPACE-complete. *Theoretical Computer Science*, **123**(2):329–340, 1994. [350]

[425] François Jaeger, Dirk Vertigan, and Dominic J. A. Welsh. On the computational complexity of the Jones and Tutte polynomials. *Mathematical Proceedings of the Cambridge Philosophical Society*, **108**(1):35–53, 1990. [721]

[426] Svante Janson, Tomasz Luczak, and Andrzej Ruciński. *Random Graphs.* John Wiley & Sons, 2000. [812]

[427] Vojtěch Jarník. On a certain minimal problem. *Práce Moravské Přídovědecké Společnosti*, **6**:57–63, 1930. [93]

[428] T. S. Jayram, Ravi Kumar, and D. Sivakumar. Two applications of information complexity. In *Proc. 35th Symposium on Theory of Computing*, pages 673–682. 2003. [504]

[429] Iwan Jensen and Anthony J. Guttmann. Statistics of lattice animals (polyominoes) and polygons. *Journal of Physics A: Mathematical and General*, **33**:L257–L263, 2000. [169]

[430] Robert G. Jeroslow. The simplex algorithm with the pivot rule of maximizing criterion improvement. *Discrete Mathematics*, **4**(4):367–377, 1973. [446]

[431] Mark Jerrum. A very simple algorithm for estimating the number of k-colourings of a low-degree graph. *Random Structures & Algorithms*, **7**:157–165, 1995. [644]

[432] Mark Jerrum and Alistair Sinclair. Approximating the permanent. *SIAM Journal on Computing,* **18**:1149–1178, 1989. [648, 719]

[433] Mark Jerrum and Alistair Sinclair. Polynomial-time approximation algorithms for the Ising model. *SIAM Journal on Computing,* **22**(5):1087–1116, 1993. [720, 721]

[434] Mark Jerrum and Alistair Sinclair. The Markov chain Monte Carlo method: an approach to approximate counting and integration. In *Approximations for NP-hard Problems,* pages 482–520. PWS Publishing, 1996. [643]

[435] Mark Jerrum, Alistair Sinclair, and Eric Vigoda. A polynomial-time approximation algorithm for the permanent of a matrix with nonnegative entries. *Journal of the ACM,* **51**(4):671–697, 2004. [648, 719]

[436] Mark Jerrum, Leslie Valiant, and Vijay Vazirani. Random generation of combinatorial structures from a uniform distribution. *Theoretical Computer Science,* **43**:169–188, 1986. [719]

[437] D. Jeulin. Dead leaves models: from space tesselation to random functions. In *Proc. Symposium on Advances in Theory and Applications of Random Sets,* pages 137–156. World Scientific, 1997. [645]

[438] Haixia Jia and Cristopher Moore. How much backtracking does it take to color random graphs? Rigorous results on heavy tails. In *Proc. 10th International Conference on Principles and Practice of Constraint Programming,* pages 742–746. 2004. [813]

[439] S. Jimbo and A. Maruoka. Expanders obtained from affine transformations. *Combinatorica,* **7**(4):343–355, 1987. [649]

[440] William Jockusch, James Propp, and Peter Shor. Random domino tilings and the arctic circle theorem, 1995. arxiv.org/abs/math/9801068. [645]

[441] David S. Johnson. NP-Completeness column: An ongoing guide. *Journal of Algorithms,* **8**:285–303, 1987. [40]

[442] David S. Johnson, Christos H. Papadimitriou, and Mihalis Yannakakis. How easy is local search? *Journal of Computer and System Sciences,* **37**(1):79–100, 1988. [221]

[443] J. P. Jones and Yuri V. Matijasevič. Register machine proof of the theorem on exponential Diophantine representation of enumerable sets. *The Journal of Symbolic Logic,* **49**(3):818–829, 1984. [292]

[444] J. P. Jones and Yuri V. Matijasevič. Proof of recursive unsolvability of Hilbert's tenth problem. *American Mathematical Monthly,* **98**(8):689–709, 1991. [292]

[445] Neil D. Jones. Space-bounded reducibility among combinatorial problems. *Journal of Computer and System Sciences,* **11**:68–85, 1975. [347]

[446] Neil D. Jones. Constant time factors do matter. In *Proc. 25th Symposium on Theory of Computing,* pages 602–611. 1993. [219]

[447] Claus Jönsson. Electron diffraction at multiple slits. *Zeitschrift für Physik,* **161**:454–474, 1961. [903]

[448] James Joyce. *Ulysses.* Sylvia Beach, 1922. [125]

[449] Valentine Kabanets and Russell Impagliazzo. Derandomizing polynomial identity tests means proving circuit lower bounds. *Computational Complexity,* **13**(1–2):1–46, 2004. [562]

[450] Adam Kalai. Generating random factored numbers, easily. *Journal of Cryptology,* **16**(4):287–289, 2003. [506]

[451] Gil Kalai and Daniel J. Kleitman. A quasi-polynomial bound for the diameter of graphs of polyhedra. *Bulletin of the American Mathematical Society,* **26**(2):315–316, 1992. [445]

[452] Leonid Vitaliyevich Kantorovich. Mathematical methods of organizing and planning production. *Management Science,* **6**:366–422, 1960. [445]

[453] Alexis Kaporis, Lefteris Kirousis, and Efthimios Lalas. Selecting complementary pairs of literals. In *Proc. LICS 03 Workshop on Typical Case Complexity and Phase Transitions.* 2003. [814]

[454] Alexis C. Kaporis, Lefteris M. Kirousis, and Yannis C. Stamatiou. Proving conditional randomness using the principle of deferred decision. In Percus et al. [656], pages 179–183. [814]

[455] Anatolii Karatsuba and Yu. Ofman. Multiplication of multidigit numbers by automata. *Doklady Akademii Nauk SSSR,* **145**:293–294, 1962. [37]

[456] David R. Karger. Global min-cuts in RNC, and other ramifications of a simple min-cut algorithm. In *Proc. 4th Symposium on Discrete Algorithms,* pages 21–30. 1993. [502]

[457] David R. Karger and Clifford Stein. An $\tilde{O}(n^2)$ algorithm for minimum cuts. In *Proc. 25th Symposium on Theory of Computing,* pages 757–765. 1993. [502]

[458] Jarkko Kari and Cristopher Moore. Rectangles and squares recognized by two-dimensional automata. In *Theory Is Forever: Essays Dedicated to Arto Salomaa on the Occasion of His 70th Birthday,* pages 134–144. Springer-Verlag, 2004. [298]

[459] Narendra Karmarkar. A new polynomial-time algorithm for linear programming. *Combinatorica,* **4**(4):373–395, 1984. [447]

[460] Narendra Karmarkar and Richard M. Karp. The differencing method of set partitioning. Technical Report UCB/CSD 81/113, Computer Science Division, University of California, Berkeley, 1982. [814]

[461] Narendra Karmarkar, Richard M. Karp, Richard J. Lipton, László Lovász, and Michael Luby. A Monte-Carlo algorithm for estimating the permanent. *SIAM Journal on Computing,* **22**(2):284–293, 1993. [721]

[462] Richard M. Karp. Reducibility among combinatorial problems. In *Complexity of Computer Computations*, pages 85–104. Plenum Press, 1972. [168]

[463] Richard M. Karp and Richard J. Lipton. Some connections between nonuniform and uniform complexity classes. In *Proc. 12th Symposium on Theory of Computing*, pages 302–309. 1980. [219]

[464] Richard M. Karp and Michael Luby. Monte Carlo algorithms for enumeration and reliability problems. In *Proc. 24th Symposium on Foundations of Computer Science*, pages 56–64. 1983. [719]

[465] Richard M. Karp and Michael O. Rabin. Efficient randomized pattern-matching algorithms. *IBM Journal of Research and Development*, **31**(2):249–260, 1987. [504]

[466] Richard M. Karp and Michael Sipser. Maximum matchings in sparse random graphs. In *Proc. 22nd Symposium on Foundations of Computer Science*, pages 364–375. 1981. [813]

[467] Marek Karpinski, Michael Lampis, and Richard Schmied. New inapproximability bounds for TSP. *Journal of Computer and System Sciences*, **81**(8):1665–1677, 2015. [445]

[468] Pieter W. Kasteleyn. The statistics of dimers on a lattice: I. The number of dimer arrangements on a quadratic lattice. *Physica*, **27**:1209–1225, 1961. [720]

[469] Pieter W. Kasteleyn. Graph theory and crystal physics. In Frank Harary, editor, *Graph Theory and Theoretical Physics*, pages 43–110. Academic Press, 1967. [720]

[470] Stefan Katzenbeisser. *Recent Advances in RSA Cryptography*. Kluwer Academic, 2001. [906]

[471] Phillipe R. Kaye, Raymond Laflamme, and Michele Mosca. *An Introduction to Quantum Computing*. Oxford University Press, 2007. [903]

[472] Julia Kempe. Discrete quantum walks hit exponentially faster. In *Proc. 7th International Workshop on Randomization and Approximation Techniques in Computer Science*, pages 354–369. 2003. [910]

[473] Wilfrid S. Kendall and Elke Thönnes. Perfect simulation in stochastic geometry. *Pattern Recognition*, **32**(9):1569–1586, 1999. [645]

[474] E. S. Kennedy and Victor Roberts. The planetary theory of Ibn al-Shāṭir. *Isis*, **50**(3):227–235, 1959. [90]

[475] Claire Kenyon and Richard Kenyon. Tiling a polygon with rectangles. In *Proc. 33rd Symposium on Foundations of Computer Science*, pages 610–619. 1992. [169]

[476] Claire Kenyon, Dana Randall, and Alistair Sinclair. Approximating the number of monomer-dimer coverings of a lattice. *Journal of Statistical Physics*, **83**:637–659, 1996. [719]

[477] Leonid G. Khachiyan. A polynomial algorithm in linear programming. *Doklady Akademii Nauk SSSR*, **224**:1093–1096, 1979. (English Translation:*Soviet Mathematics Doklady* **20** 191–194). [448]

[478] Subhash Khot. On the power of unique 2-prover 1-round games. In *Proc. 34th Symposium on Theory of Computing*, pages 767–775. 2002. [503]

[479] Subhash Khot, Guy Kindler, Elchanan Mossel, and Ryan O'Donnell. Optimal inapproximability results for MAX-CUT and other 2-variable CSPs? In *Proc. 45th Symposium on Foundations of Computer Science*, pages 146–154. 2004. [503]

[480] Subhash Khot and Oded Regev. Vertex cover might be hard to approximate to within $2-\varepsilon$. *Journal of Computer and System Sciences*, **74**(3):335–349, 2008. [503]

[481] Scott Kirkpatrick, C. D. Gelatt, and M. P. Vecchi. Optimization by simulated annealing. *Science*, **220**:671–680, 1983. [503]

[482] Scott Kirkpatrick and Bart Selman. Critical behavior in the satisfiability of random Boolean expressions. *Science*, **264**:1297–1301, 1994. [811]

[483] Lefteris M. Kirousis, Yannis C. Stamatiou, and Michele Zito. The unsatisfiability threshold conjecture: the techniques behind upper bound improvements. In *Computational Complexity and Statistical Physics*, pages 159–178. Oxford University Press, 2006. [814]

[484] Alexei Yu. Kitaev, A. H. Shen, and Mikhail N. Vyalyi. *Classical and Quantum Computation*. American Mathematical Society, 2002. [907]

[485] D. A. Klarner and R. L. Rivest. A procedure for improving the upper bound for the number of n-ominoes. *Canadian Journal of Mathematics*, **25**:585–602, 1973. [169]

[486] Victor L. Klee and George James Minty. How good is the Simplex algorithm? In Oved Shisha, editor, *Inequalities III: Proceedings of the 3rd Symposium on Inequalities*, pages 159–175. Academic Press, 1972. [446]

[487] Stephen C. Kleene. A theory of positive integers in formal logic. *American Journal of Mathematics*, **57**(1–2):153–173, 219–244, 1935. [295]

[488] Stephen C. Kleene. General recursive functions of natural numbers. *Mathematische Annalen*, **112**(5):727–742, 1936. [294]

[489] Stephen C. Kleene. The theory of recursive functions, approaching its centennial. *Bulletin of the American Mathematical Society*, **5**(1):43–61, 1981. [294]

[490] Jon Kleinberg and Eva Tardos. *Algorithm Design.* Addison-Wesley, 2005. [89]

[491] E. Knill and R. Laflamme. Power of one bit of quantum information. *Physical Review Letters,* **81**(25):5672–5675, 1998. [910]

[492] Donald E. Knuth. Estimating the efficiency of backtrack programs. *Mathematics of Computation,* **29**(129):121–136, 1975. [719]

[493] Donald E. Knuth. Algorithms in modern mathematics and computer science. In Andrei P. Ershov and Donald E. Knuth, editors, *Proceedings on Algorithms in Modern Mathematics and Computer Science, Urgench, Uzbek SSR, September 16–22, 1979,* pages 82–99. Springer, 1981. ISBN 3-540-11157-3. [36]

[494] Donald E. Knuth. The sandwich theorem. *Electronic Journal of Combinatorics,* **1**:1–48, 1994. [449]

[495] Donald E. Knuth. *The Art of Computer Programming, Volume 2: Seminumerical Algorithms.* Addison-Wesley, 1997. [36]

[496] Donald E. Knuth. *The Art of Computer Programming, Volume 3: Sorting and Searching.* Addison-Wesley, 1998. [89]

[497] Donald E. Knuth and Michael F. Plass. Breaking paragraphs into lines. *Software—Practice and Experience,* **11**:1119–1184, 1981. Reprinted as Chapter 3 of Knuth, *Digital Typography,* Center for the Study of Language and Information, 1999, pp. 67–155. [91]

[498] Jiří Kolafa. Monte Carlo study of the three-state square Potts antiferromagnet. *Journal of Physics A: Mathematical and General,* **17**:L777–L781, 1984. [645]

[499] Norio Konno. Limit theorem for continuous-time quantum walk on the line. *Physical Review E,* **72**:026113, 2005. [910]

[500] Richard E. Korf. A complete anytime algorithm for number partitioning. *Artificial Intelligence,* **106**:181–203, 1998. [814]

[501] Bernhard Korte, László Lovász, and Rainer Schrader. *Greedoids.* Springer-Verlag, 1991. [93]

[502] Dexter Kozen. *The Design and Analysis of Algorithms.* Springer-Verlag, 1991. [719]

[503] Dexter Kozen and Shmuel Zaks. Optimal bounds for the change-making problem. *Theoretical Computer Science,* **123**:377–388, 1994. [93]

[504] Hendrik A. Kramers and Gregory H. Wannier. Statistics of the two-dimensional ferromagnet. Part I. *Physical Review,* **60**(3):252–262, 1941. [720]

[505] Joseph B. Kruskal, Jr. On the shortest spanning subtree of a graph and the traveling salesman problem. *Proceedings of the American Mathematical Society,* **7**:48–50, 1956. [92]

[506] Florent Krząkala, Andrea Montanari, Federico Ricci-Tersenghi, Guilhem Semerjian, and Lenka Zdeborová. Gibbs states and the set of solutions of random constraint satisfaction problems. *Proceedings of the National Academy of Sciences,* **104**(25):10318–10323, 2007. [818]

[507] Manfred Kudlek and Yurii Rogozhin. A universal Turing machine with 3 states and 9 symbols. In *Proc. 5th International Conference on Developments in Language Theory,* pages 311–318. 2001. [295]

[508] Greg Kuperberg. An exploration of the permanent-determinant method. *Electronic Journal of Combinatorics,* **5**(1):R46, 1988. [720]

[509] Greg Kuperberg. A subexponential-time quantum algorithm for the dihedral hidden subgroup problem. *SIAM Journal on Computing,* **35**(1):170–188, 2005. [908]

[510] Greg Kuperberg. Knottedness is in NP, modulo GRH. *Advances in Mathematics,* **256**(Supplement C):493–506, 2014. [126]

[511] Kazimierz Kuratowski. Sur le problème des courbes gauches en topologie. *Fundamenta Mathematicae,* **15**:271–283, 1930. [39]

[512] Mei-Ko Kwan. Graphic programming using odd or even points. *Chinese Mathematics,* **1**:273–277, 1962. [13]

[513] M. Lackenby. The efficient certification of knottedness and Thurston norm, April 2016. arxiv.org/abs/1604.00290. [126]

[514] Marc Lackenby. A polynomial upper bound on Reidemeister moves. *Annals of Mathematics,* **182**:491–564, 2015. [126]

[515] Richard E. Ladner. On the structure of polynomial time reducibility. *Journal of the ACM,* **22**(1):155–171, 1975. [170, 220]

[516] J. C. Lagarias. The $3x+1$ problem and its generalization. *American Mathematical Monthly,* **92**(1):3–25, 1985. Updated versions: arxiv.org/abs/math/0309224 and arxiv.org/abs/math/0608208. [298]

[517] Rolf Landauer. Irreversibility and heat generation in the computing process. *IBM Journal of Research and Development,* **5**:183–191, 1961. [904]

[518] Klaus-Jörn Lange, Birgit Jenner, and Bernd Kirsig. The logarithmic alternation hierarchy collapses. In *Proc. 14th International Colloquium on Automata, Languages and Programming,* pages 531–541. 1987. [348]

[519] Sophie Laplante and Frédéric Magniez. Lower bounds for randomized and quantum query complexity using Kolmogorov arguments. *SIAM Journal on Computing,* **38**(1):46–62, 2008. [909]

[520] E. L. Lawler, J. K. Lenstra, A. H. G. Rinnoy Kan, and D. B. Shmoys, editors. *The Traveling Salesman Problem.* Wiley-Interscience Series in Discrete Mathematics and Optimization. John Wiley & Sons, 1985. [442]

[521] Gregory F. Lawler and Alan D. Sokal. Bounds on the l^2 spectrum for Markov chains and Markov processes: A generalization of Cheeger's inequality. *Transactions of the American Mathematical Society,* **309**(2):557–580, 1988. [648]

[522] François Le Gall. Powers of tensors and fast matrix multiplication. In *Proceedings of the 39th International Symposium on Symbolic and Algebraic Computation*, ISSAC '14, pages 296–303. 2014. [38]

[523] Joel L. Lebowitz. Statistical mechanics: a selective review of two central issues. *Reviews of Modern Physics*, **71**:S346–S357, 1999. [643]

[524] Hendrik W. Lenstra, Jr. and Carl Pomerance. Primality testing with Gaussian periods, 2005. www.math.dartmouth.edu/~carlp/PDF/complexity12.pdf. [506]

[525] W. Lenz. Beitrag zu Verständnis der magnetischen Erscheinungen in festen Körpern. *Physikalische Zeitschrift*, **21**:613–615, 1920. [720]

[526] David A. Levin, Yuval Peres, and Elizabeth L. Wilmer. *Markov Chains and Mixing Times*. American Mathematical Society, 2009. [643]

[527] Leonid A. Levin. Universal search problems. *Problems of Information Transmission (translation of Problemy Peredachi Informatsii)*, **9**(3):265–266, 1973. [168]

[528] L. S. Levitov. Equivalence of the dimer resonating-valence-bond problem to the quantum roughening problem. *Physical Review Letters*, **64**:92–94, 1990. [646]

[529] Harry R. Lewis and Christos H. Papadimitriou. Symmetric space-bounded computation. *Theoretical Computer Science*, **19**(2):161–187, 1982. [349]

[530] Philip M. Lewis II, Richard Edwin Stearns, and Juris Hartmanis. Memory bounds for the recognition of context-free and context-sensitive languages. In *IEEE Conference Record on Switching Circuit Theory and Logical Design*, pages 191–202. 1965. [347]

[531] David Lichtenstein and Michael Sipser. Go is polynomial-space hard. *Journal of the ACM*, **27**(2):393–401, 1980. [348]

[532] Daniel A. Lidar and Ofer Biham. Simulating Ising spin glasses on a quantum computer. *Physical Review E*, **56**(3):3661–3681, 1997. [908]

[533] Elliott H. Lieb. The residual entropy of square ice. *Physical Review*, **162**(1):162–172, 1967. [646, 720]

[534] Kristian Lindgren, Cristopher Moore, and Mats G. Nordahl. Complexity of two-dimensional patterns. *Journal of Statistical Physics*, **91**:909–951, 1998. [298]

[535] Bernt Lindström. On the vector representation of induced matroids. *Bulletin of the London Mathematical Society*, **5**:85–90, 1973. [721]

[536] Seth Lloyd. Almost any quantum logic gate is universal. *Physical Review Letters*, **75**(2):346–349, 1995. [904]

[537] Douglas L. Long and Avi Wigderson. The discrete logarithm hides $o(\log n)$ bits. *SIAM Journal on Computing*, **17**(2):363–372, 1988. [561]

[538] László Lovász. Normal hypergraphs and the perfect graph conjecture. *Discrete Mathematics*, **2**(3):253–267, 1972. [449]

[539] László Lovász. On determinants, matchings and random algorithms. In *Proc. Conference on Algebraic, Arithmetic, and Categorial Methods in Computation Theory*. 1979. [505]

[540] László Lovász. On the Shannon capacity of a graph. *IEEE Transactions on Information Theory*, **25**:1–7, 1979. [449]

[541] László Lovász. *An Algorithmic Theory of Numbers, Graphs, and Convexity*. Society for Industrial and Applied Mathematics, 1986. [449]

[542] Alexander Lubotzky, Ralph Phillips, and Peter Sarnak. Ramanujan graphs. *Combinatorica*, **8**(3):261–277, 1988. [649]

[543] Michael Luby, Michael Mitzenmacher, M. Amin Shokrollahi, Daniel A. Spielman, and Volker Stemann. Practical loss-resilient codes. In *Proc. 29th Symposium on Theory of Computing*, pages 150–159. 1997. [813]

[544] Michael Luby, Michael Mitzenmacher, and Mohammad Amin Shokrollahi. Analysis of random processes via And–Or tree evaluation. In *Proc. 9th Symposium on Discrete Algorithms*, pages 364–373. 1998. [813]

[545] Michael Luby, Dana Randall, and Alistair Sinclair. Markov Chain algorithms for planar lattice structures. *SIAM Journal on Computing*, **31**(1):167–192, 2001. [646]

[546] Édouard Lucas. *Recreations Mathematiques*, vol. III. Gauthier-Villars et fils, 1893. [89]

[547] R. Duncan Luce and Howard Raiffa. *Games and Decisions: Introduction and Critical Survey*. Dover, 1957. [503]

[548] Eugene M. Luks. Isomorphism of graphs of bounded valence can be tested in polynomial time. *Journal of Computer and System Sciences*, **25**(1):42–65, 1982. [908]

[549] Carsten Lund, Lance Fortnow, Howard J. Karloff, and Noam Nisan. Algebraic methods for interactive proof systems. *Journal of the ACM*, **39**(4):859–868, 1992. [561]

[550] Percy A. MacMahon. *Combinatory Analysis*. Cambridge University Press, 1915–16. [721]

[551] Frédéric Magniez, Miklos Santha, and Mario Szegedy. Quantum algorithms for the triangle problem. *SIAM Journal on Computing*, **37**(2):413–424, 2007. [910]

[552] Kenneth L. Manders and Leonard Adleman. NP-complete decision problems for binary quadratics. *Journal of Computer and System Sciences*, **16**:168–184, 1978. [169]

[553] Elitza N. Maneva, Elchanan Mossel, and Martin J. Wainwright. A new look at survey propagation and its generalizations. *Journal of the ACM*, **54**(4):2–41, 2007. [818]

[554] Elitza N. Maneva and Alistair Sinclair. On the satisfiability threshold and clustering of solutions of random 3-SAT formulas. *Theoretical Computer Science*, **407**(1–3):359–369, 2008. [814]

[555] G. A. Margulis. Explicit constructions of concentrators. *Problems of Information Transmission*, **9**(4):71–80, 1973. [649]

[556] Raffaele Marino, Giorgio Parisi, and Federico Ricci-Tersenghi. The backtracking survey propagation algorithm for solving random k-SAT problems. *Nature Communications*, **7**:12996, 2016. [817]

[557] Russell A. Martin and Dana Randall. Sampling adsorbing staircase walks using a new Markov chain decomposition method. In *Proc. 41st Symposium on Foundations of Computer Science*, pages 492–502. 2000. [649]

[558] Fabio Martinelli. Lectures on Glauber dynamics for discrete spin models. *Lecture Notes on Mathematics*, **1717**:93–191, 1998. [649]

[559] Yuri Matiyasevich. Enumerable sets are Diophantine. *Doklady Akademii Nauk SSSR*, **191**(2):279–282, 1970. English translation in *Soviet Mathematics Doklady* **11**(2):354–358, 1970. [292]

[560] Yuri Matiyasevich. *Hilbert's Tenth Problem*. The MIT Press, 1993. [292]

[561] Jiří Matoušek and Bernd Gärtner. *Understanding and Using Linear Programming*. Springer-Verlag, 2007. [392, 445]

[562] David McAllester, Bart Selman, and Henry Kautz. Evidence for invariants in local search. In *Proc. 14th National Conference on Artificial Intelligence*, pages 321–326. 1997. [503]

[563] John McCarthy. Recursive functions of symbolic expressions and their computation by machine, part I. *Communications of the ACM*, **3**(4):184–195, 1960. [295]

[564] Nimrod Megiddo and Christos H. Papadimitriou. On total functions, existence theorems and computational complexity. *Theoretical Computer Science*, **81**(2):317–324, 1991. [221]

[565] Ralph C. Merkle. Secure communications over insecure channels. *Communications of the ACM*, **21**(4):294–299, 1978. [907]

[566] Ralph C. Merkle and Martin E. Hellman. Hiding information and signatures in trapdoor knapsacks. *IEEE Transactions on Information Theory*, **24**(5):525–530, 1978. [125]

[567] N. David Mermin. What's wrong with these elements of reality? *Physics Today*, **43**:9–11, 1990. [905]

[568] N. David Mermin. *Quantum Computer Science: An Introduction*. Cambridge University Press, 2007. [903]

[569] Stephan Mertens. Phase transition in the number partitioning problem. *Physical Review Letters*, **81**(20):4281–4284, November 1998. [815]

[570] Stephan Mertens. Random costs in combinatorial optimization. *Physical Review Letters*, **84**(6):1347–1350, February 2000. [815]

[571] Stephan Mertens. The easiest hard problem: number partitioning. In Percus et al. [656], pages 125–139. [815]

[572] Stephan Mertens, Marc Mézard, and Riccardo Zecchina. Threshold values of random K-SAT from the cavity method. *Random Structures & Algorithms*, **28**:340–373, 2006. [817]

[573] Nicholas Metropolis, Arianna W. Rosenbluth, Marshall N. Rosenbluth, Augusta H. Teller, and Edward Teller. Equation of state calculation by fast computing machines. *Journal of Chemical Physics*, **21**(6):1087–1092, 1953. [644]

[574] Albert R. Meyer and Dennis M. Ritchie. The complexity of loop programs. In *Proc. 22nd National ACM Conference*, pages 465–469. 1967. [294]

[575] Albert R. Meyer and Larry J. Stockmeyer. The equivalence problem for regular expressions with squaring requires exponential space. In *Proc. 13th Symposium on Switching and Automata Theory*, pages 125–129. 1972. [218]

[576] Marc Mézard and Andrea Montanari. *Information, Physics, and Computation*. Oxford University Press, 2009. [810, 816]

[577] Marc Mézard, Thierry Mora, and Riccardo Zecchina. Clustering of solutions in random satisfiability problem. *Physical Review Letters*, **94**:197205, 2005. [817]

[578] Marc Mézard, Giorgio Parisi, and Riccardo Zecchina. Analytic and algorithmic solution of random satisfiability problems. *Science*, **297**:812–815, 2002. [817]

[579] Marc Mézard and Riccardo Zecchina. Random k-satisfiability problem: From an analytic solution to an efficient algorithm. *Physical Review E*, **66**(5):056126, 2002. [817]

[580] E. Milková, J. Nešetřil, and H. Nešetřilová. Otakar Borůvka on minimum spanning tree problem: Translation of both the 1926 papers, comments, history. *Discrete Mathematics*, **233**:3–36, 2001. [92]

[581] Gary L. Miller. Riemann's Hypothesis and tests for primality. *Journal of Computer and System Sciences*, **13**(3):300–317, 1976. [506]

[582] Marvin Minsky. Recursive unsolvability of Post's problem of "Tag" and other topics in theory of Turing machines. *Annals of Mathematics*, **74**(3):437–455, 1961. [298]

[583] Marvin Minsky. *Computation: Finite and Infinite Machines.* Prentice-Hall, Inc., 1967. [297, 298]

[584] Marvin L. Minsky and Seymour A. Papert. *Perceptrons.* The MIT Press, 1969. [909]

[585] George J. Minty. A comment on the shortest-route problem. *Operations Research*, **5**:724, 1957. [447]

[586] David Mitchell, Bart Selman, and Hector Levesque. Hard and easy distributions of SAT problems. In *Proc. 10th AAAI, San Jose, CA*, pages 249–465. AAAI Press, July 1992. [811]

[587] Joseph S. B. Mitchell. A PTAS for TSP with neighborhoods among fat regions in the plane. In *Proc. 18th Symposium on Discrete Algorithms*, pages 11–18. 2007. [443]

[588] Michael Mitzenmacher. Studying balanced allocations with differential equations. *Combinatorics, Probability and Computing*, **8**(5):473–482, 1999. [813]

[589] Michael Mitzenmacher. Analyses of load stealing models based on families of differential equations. *Theory of Computing Systems*, **34**(1):77–98, 2001. ISSN 1432-4350. [813]

[590] Michael Mitzenmacher and Eli Upfal. *Probability and Computing: Randomized Algorithms and Probabilistic Analysis.* Cambridge University Press, 2005. [502, 643]

[591] Carlos Mochon. Hamiltonian oracles. *Physical Review A*, **75**:042313, 2007. [910]

[592] Liesbeth De Mol. Closing the circle: an analysis of Emil Post's early work. *Bulletin of Symbolic Logic*, **12**(2):267–289, 2006. [298]

[593] Liesbeth De Mol. Tag systems and Collatz-like functions. *Theoretical Computer Science*, **390**(1):92–101, 2008. [298]

[594] Michael Molloy. Very rapidly mixing Markov chains for 2δ-colourings and for independent sets in a 4-regular graph. *Random Structures & Algorithms*, **18**:101–115, 2001. [644]

[595] Michael Molloy. The Glauber dynamics on colorings of a graph with high girth and maximum degree. *SIAM Journal on Computing*, **33**(3):721–737, 2004. [644]

[596] Michael Molloy. Cores in random hypergraphs and Boolean formulas. *Random Structures & Algorithms*, **27**(1):124–135, 2005. [813]

[597] Michael Molloy and Bruce A. Reed. A critical point for random graphs with a given degree sequence. *Random Structures & Algorithms*, **6**(2/3):161–180, 1995. [812]

[598] Burkhard Monien and Ewald Speckenmeyer. Natural proofs. *Discrete Applied Mathematics*, **10**:287–295, 1985. [219]

[599] Andrea Montanari, Federico Ricci-Tersenghi, and Guilhem Semerjian. Clusters of solutions and replica symmetry breaking in random k-satisfiability. *Journal of Statistical Mechanics: Theory and Experiment*, **2008**(04):P04004, 2008. [817, 818]

[600] Elliott W. Montroll, Renfrey B. Potts, and John C. Ward. Correlations and spontaneous magnetization of the two-dimensional Ising model. *Journal of Mathematical Physics*, **4**(2):308–322, 1963. [720]

[601] Cristopher Moore. Undecidability and unpredictability in dynamical systems. *Physical Review Letters*, **64**:2354–2357, 1990. [299]

[602] Cristopher Moore. Generalized shifts: undecidability and unpredictability in dynamical systems. *Nonlinearity*, **4**:199–230, 1991. [299]

[603] Cristopher Moore and Mark E. J. Newman. Height representation, critical exponents, and ergodicity in the four-state triangular Potts antiferromagnet. *Journal of Statistical Physics*, **99**:629–660, 2000. [646]

[604] Cristopher Moore and Martin Nilsson. Parallel quantum computation and quantum codes. *SIAM Journal on Computing*, **31**(3):799–815, 2001. [904]

[605] Cristopher Moore, Mats G. Nordahl, Nelson Minar, and Cosma R. Shalizi. Vortex dynamics and entropic forces in antiferromagnets and antiferromagnetic Potts models. *Physical Review E*, **60**:5344–5351, 1999. [645]

[606] Cristopher Moore and John Michael Robson. Hard tiling problems with simple tiles. *Discrete and Computational Geometry*, **26**(4):573–590, 2001. [169]

[607] Cristopher Moore, Daniel N. Rockmore, and Alexander Russell. Generic quantum Fourier transforms. *ACM Transactions on Algorithms*, **2**(4):707–723, 2006. [908]

[608] Cristopher Moore, Daniel N. Rockmore, Alexander Russell, and Leonard J. Schulman. The power of strong Fourier sampling: quantum algorithms for affine groups and hidden shifts. *SIAM Journal on Computing*, **37**(3):938–958, 2007. [908]

[609] Cristopher Moore and Alexander Russell. Quantum walks on the hypercube. In *Proc. 6th International Workshop on Randomization and Approximation Techniques in Computer Science*, pages 164–178. 2002. [910]

[610] Cristopher Moore and Alexander Russell. For distinguishing conjugate hidden subgroups, the Pretty Good Measurement is as good as it gets. *Quantum Information and Computation*, **7**:752–765, 2007. [910]

[611] Cristopher Moore and Alexander Russell. Approximating the permanent via nonabelian determinants. *SIAM Journal on Computing*, **41**(2):332–355, 2012. [721]

[612] Cristopher Moore, Alexander Russell, and Leonard J. Schulman. The symmetric group defies strong Fourier sampling. In *Proc. 46th Symposium on Foundations of Computer Science*, pages 479–490. 2005. [908]

[613] Cristopher Moore, Alexander Russell, and Piotr Sniady. On the impossibility of a quantum sieve algorithm for graph isomorphism. In *Proc. 39th Symposium on Theory of Computing*, pages 536–545. 2007. [908]

[614] Jonathan Michael Moore. *The Kaliningrad Region: Keystone of the Soviet Baltic.* Master's thesis, George Washington University, 1990. [13]

[615] Thierry Mora and Lenka Zdeborová. Random subcubes as a toy model for constraint satisfaction problems. *Journal of Statistical Physics*, **131**:1121–1138, 2008. [816]

[616] Bernard M. E. Moret. *The Theory of Computation.* Addison-Wesley Publishing Company, 1998. [168]

[617] Rajeev Motwani and Prabhakar Raghavan. *Randomized Algorithms.* Cambridge University Press, 1995. [502]

[618] Ketan Mulmuley and Milind A. Sohoni. Geometric complexity theory I: an approach to the P vs. NP and related problems. *SIAM Journal on Computing*, **31**(2):496–526, 2001. [220]

[619] Ketan Mulmuley, Umesh V. Vazirani, and Vijay V. Vazirani. Matching is as easy as matrix inversion. In *Proc. 19th Symposium on Theory of Computing*, pages 345–354. 1987. [504]

[620] Denis Naddef. The Hirsch conjecture is true for (0,1)-polytopes. *Mathematical Programming*, **45**(1):109–110, 1989. ISSN 0025-5610. [446]

[621] Moni Naor. Bit commitment using pseudorandomness. *Journal of Cryptology*, **4**(2):151–158, 1991. [561]

[622] John C. Nash. The (Dantzig) simplex method for linear programming. *Computing in Science and Engineering*, **2**(1):29–31, 2000. [445]

[623] John F. Nash. Equilibrium points in n-person games. *Proceedings of the National Academy of Sciences*, **36**:48–49, 1950. [503]

[624] Ashwin Nayak and Felix Wu. The quantum query complexity of approximating the median and related statistics. In *Proc. 31st Symposium on Theory of Computing*, pages 384–393. 1999. [909]

[625] Turlough Neary and Damien Woods. P-completeness of cellular automaton rule 110. In *Proc. 33rd International Colloquium on Automata, Languages and Programming*, pages 132–143. 2006. [299]

[626] Turlough Neary and Damien Woods. Four small universal Turing machines. *Fundamenta Informaticae*, **91**(1):123–144, 2009. [295]

[627] Turlough Neary and Damien Woods. Small weakly universal Turing machines. In *Proc. 17th Intl. Symp. on Fundamentals of Computation Theory*, pages 262–273. 2009. [295]

[628] David R. Nelson and Michael E. Fisher. Soluble renormalization groups and scaling fields for low-dimensional lsing systems. *Annals of Physics*, **91**:226–274, 1975. [721]

[629] George L. Nemhauser and Leslie E. Trotter, Jr. Properties of vertex packing and independence system polyhedra. *Mathematical Programming*, **6**:48–61, 1974. [448]

[630] Arkady S. Nemirovsky and David B. Yudin. *Problem Complexity and Method Efficiency in Optimization.* John Wiley & Sons, 1983. [448]

[631] Mark E. J. Newman and Gerard T. Barkema. *Monte Carlo Methods in Statistical Physics.* Oxford University Press, 1999. [643]

[632] Mark E. J. Newman, Steven H. Strogatz, and Duncan J. Watts. Random graphs with arbitrary degree distributions and their applications. *Physical Review E*, **64**, 2001. [812]

[633] Michael A. Nielsen and Isaac L. Chuang. *Quantum Computation and Quantum Information.* Cambridge University Press, 2000. [903]

[634] Noam Nisan. CREW PRAMs and decision trees. In *Proc. 21st Symposium on Theory of Computing*, pages 327–335. 1989. [909]

[635] Noam Nisan. Pseudorandom generators for space-bounded computation. *Combinatorica*, **12**(4):449–461, 1992. [562]

[636] Noam Nisan. RL $\subseteq$ SC. In *Proc. 24th Symposium on Theory of Computing*, pages 619–623. 1992. [562]

[637] Noam Nisan, Tim Roughgarden, Eva Tardos, and Vijay V. Vazirani, editors. *Algorithmic Game Theory.* Cambridge University Press, 2007. [221]

[638] Noam Nisan and Mario Szegedy. On the degree of Boolean functions as real polynomials. In *Proc. 24th Symposium on Theory of Computing*, pages 462–467. 1992. [909]

[639] Noam Nisan and Amnon Ta-Shma. Symmetric logspace is closed under complement. In *Proc. 27th Symposium on Theory of Computing*, pages 140–146. 1995. [349]

[640] Noam Nisan and Avi Wigderson. Hardness vs. randomness. *Journal of Computer and System Sciences*, **49**(2):149–167, 1994. [562]

[641] Lars Onsager. Crystal statistics. I. A two-dimensional model with an order-disorder transition. *Physical Review*, **65**(3-4):117–149, 1944. [720]

[642] James G. Oxley. *Matroid Theory*. Oxford University Press, 1992. [93]

[643] Christos H. Papadimitriou. The Euclidean traveling salesman problem is NP-complete. *Theoretical Computer Science*, **4**:237–244, 1977. [443]

[644] Christos H. Papadimitriou. Games against nature. *Journal of Computer and System Sciences*, **31**(2):288–301, 1985. [560]

[645] Christos H. Papadimitriou. On selecting a satisfying truth assignment. In *Proc. 32nd Symposium on Foundations of Computer Science*, pages 163–169. 1991. [503]

[646] Christos H. Papadimitriou. *Computational Complexity*. Addison-Wesley, 1994. [168]

[647] Christos H. Papadimitriou. On the complexity of the parity argument and other inefficient proofs of existence. *Journal of Computer and System Sciences*, **48**(3):498–532, 1994. [221]

[648] Christos H. Papadimitriou, Alejandro A. Schäffer, and Mihalis Yannakakis. On the complexity of local search. In *Proc. 22nd Symposium on Theory of Computing*, pages 438–445. 1990. [221]

[649] Christos H. Papadimitriou and Santosh Vempala. On the approximability of the traveling salesman problem. In *Proceedings of the 32nd Annual ACM Symposium on Theory of Computing*, pages 126–133. 2000. [445]

[650] Ramamohan Paturi. On the degree of polynomials that approximate symmetric Boolean functions. In *Proc. 24th Symposium on Theory of Computing*, pages 468–474. 1992. [909]

[651] Ramamohan Paturi, Pavel Pudlak, Michael E. Saks, and Francis Zane. An improved exponential-time algorithm for k-SAT. In *Proc. 39th Symposium on Foundations of Computer Science*, pages 628–637. 1998. [503]

[652] Judea Pearl. Asymptotic properties of minimax trees and game-searching procedures. *Artificial Intelligence*, **14**:113–138, 1980. [504]

[653] Judea Pearl. The solution for the branching factor of the alpha beta pruning algorithm and its optimality. *Communications of the ACM*, **25**:559–564, 1982. [504]

[654] David Pearson. A polynomial-time algorithm for the change-making problem. *Operations Research Letters*, **33**:231–234, 2004. [93]

[655] Rudolf Peierls. On Ising's model of ferromagnetism. *Proceedings of the Cambridge Philosophical Society, Mathematical and Physical Sciences*, **32**:477–481, 1936. [720]

[656] Allon Percus, Gabriel Istrate, and Cristopher Moore, editors. *Computational Complexity and Statistical Physics*. Santa Fe Institute Studies in the Sciences of Complexity. Oxford University Press, 2006. [810, 960, 964]

[657] Rózsa Péter. Konstruktion nichtrekursiver Funktionen. *Mathematische Annalen*, **111**:42–60, 1935. [294]

[658] Charles Petzold. *The Annotated Turing*. Wiley Publishing, Inc., 2008. [296]

[659] Jim Pitman and Marc Yor. The two–parameter Poisson–Dirichlet distribution derived from a stable subordinator. *Annals of Probability*, **25**(2):855–900, 1997. [816]

[660] Itamar Pitowsky. The physical Church thesis and physical computational complexity. *Iyyun*, **39**:81–99, 1990. [902]

[661] Boris Pittel, Joel Spencer, and Nicholas C. Wormald. Sudden emergence of a giant k-core in a random graph. *Journal of Combinatorial Theory, Series B*, **67**(1):111–151, 1996. [813]

[662] David A. Plaisted. Some polynomial and integer divisibility problems are NP-hard. *SIAM Journal on Computing*, **7**(4):458–464, 1978. [169]

[663] Emil L. Post. Finite combinatory processes—formulation 1. *Journal of Symbolic Logic*, **1**:103–105, 1936. [298]

[664] Emil L. Post. Formal reductions of the general combinatorial decision problem. *American Journal of Mathematics*, **65**(2):197–215, 1943. [298]

[665] Emil L. Post. Recursively enumerable sets of positive integers and their decision problems. *Bulletin of the American Mathematical Society*, **50**:284–316, 1944. [298]

[666] Emil L. Post. A variant of a recursively unsolvable problem. *Bulletin of the American Mathematical Society*, **52**:264–268, 1946. [298]

[667] Vaughan R. Pratt. Every prime has a succinct certificate. *SIAM Journal on Computing*, **4**:214–220, 1975. [126]

[668] John Preskill. Lecture Notes for Physics 229: Quantum Information and Computation. www.theory.caltech.edu/people/preskill/ph229/notes/book.ps. [903, 905]

[669] William H. Press, Saul A. Teukolsky, William T. Vetterling, and Brian P. Flannery. *Numerical Recipes. The Art of Scientific Computing*. Cambridge University Press, 3rd edition, 2007. [445]

[670] Robert C. Prim. Shortest connection networks and some generalizations. *Bell System Technical Journal*, **36**:1389–1401, 1957. [93]

[671] Hans Jürgen Prömel and Angelika Steger. *The Steiner Tree Problem.* Vieweg Advanced Lectures in Mathematics, 2002. [171]

[672] James Propp. Dimers and Dominoes, 1992. jamespropp.org/domino.ps.gz. [720]

[673] James Propp and David Wilson. Exact sampling with coupled Markov chains and applications to statistical mechanics. *Random Structures & Algorithms,* **9**:223–252, 1996. [645]

[674] Heinz Prüfer. Neuer Beweis eines Satzes über Permutationen. *Archiv für Mathematik und Physik,* **27**:742–744, 1918. [93]

[675] Michael O. Rabin. Digital signatures and public-key functions as intractable as factorization, 1979. Technical memo, MIT/LCS/TR-212. [561]

[676] Michael O. Rabin. Probabilistic algorithm for testing primality. *Journal of Number Theory,* **12**(1):128–138, 1980. [506]

[677] Charles M. Rader. Discrete Fourier transforms when the number of data samples is prime. *Proceedings of the IEEE,* **56**:1107–1108, 1968. [90]

[678] Jaikumar Radhakrishnan and Madhu Sudan. On Dinur's proof of the PCP theorem. *Bulletin of the American Mathematical Society,* **44**(1):19–61, 2007. [562]

[679] Richard Rado. A note on independence functions. *Proceedings of the London Mathematical Society,* **7**:300–320, 1957. [93]

[680] Tibor Rado. On non-computable functions. *Bell System Technical Journal,* **41**(3):877–884, 1962. [297]

[681] Dana Randall. Rapidly mixing Markov chains with applications in computer science and physics. *Computing in Science and Engineering,* **8**(2):30–41, 2006. [643]

[682] Dana Randall and Prasad Tetali. Analyzing Glauber dynamics by comparison of Markov chains. *Journal of Mathematical Physics,* **41**:1598–1615, 2000. [646]

[683] Dana Randall and David Wilson. Sampling spin configurations of an Ising system. In *Proc. 10th Symposium on Discrete Algorithms,* pages 959–960. 1999. [721]

[684] Brian Randell. From analytical engine to electronic digital computer: the contributions of Ludgate, Torres, and Bush. *Annals of the History of Computing,* **4**(4):327–341, 1982. [291]

[685] Jack Raymond, Andrea Sportiello, and Lenka Zdeborová. Phase diagram of the 1-in-3 satisfiability problem. *Physical Review E,* **76**(1):011101, 2007. [813]

[686] Alexander A. Razborov and Steven Rudich. Natural proofs. *Journal of Computer and System Sciences,* **55**(1):24–35, 1997. [220, 562]

[687] Kenneth W. Regan. Understanding the Mulmuley–Sohoni approach to P vs. NP. *Bulletin of the EATCS,* **78**:86–99, 2002. [220]

[688] Oded Regev. Quantum computation and lattice problems. *SIAM Journal on Computing,* **33**(3):738–760, 2004. [908]

[689] Oded Regev. On lattices, learning with errors, random linear codes, and cryptography. In *Proc. 37th Symposium on Theory of Computing,* pages 84–93. 2005. [908]

[690] Tullio Regge and Riccardo Zecchina. Combinatorial and topological approach to the 3D Ising model. *Journal of Physics A: Math. Gen,* **33**:741–761, 2000. [720]

[691] Ben Reichardt and RobertŠpalek. Span-program-based quantu m algorithm for evaluating formulas. In *Proc. 40th Symposium on Theory of Computing,* pages 103–112. 2008. [910]

[692] John H. Reif. Complexity of the mover's problem and generalizations. In *Proc. 20th Symposium on Foundations of Computer Science,* pages 421–427. 1979. [349]

[693] John H. Reif. Symmetric complementation. *Journal of the ACM,* **31**(2):401–421, 1984. [349]

[694] John H. Reif, J. D. Tygar, and Akitoshi Yoshida. The computability and complexity of optical beam tracing. In *Proc. 31st Annual Symposium on Foundations of Computer Science,* pages 106–114. 1990. [299]

[695] Gerhard Reinelt. *The Travelling Salesman. Computational Solutions for TSP Applications.* Springer-Verlag, 1994. [442]

[696] Omer Reingold. Undirected ST-connectivity in log-space. In *Proc. 37th Symposium on Theory of Computing,* pages 376–385. 2005. [349]

[697] Omer Reingold, Luca Trevisan, and Salil P. Vadhan. Pseudorandom walks on regular digraphs and the RL vs. L problem. In *Proc. 38th Symposium on Theory of Computing,* pages 457–466. 2006. [649]

[698] Omer Reingold, Salil Vadhan, and Avi Wigderson. Entropy waves, the zig-zag graph product, and new constant-degree expanders. *Annals of Mathematics,* **155**(1):157–187, 2002. [349, 649]

[699] Stefan Reisch. Gobang ist PSPACE-vollständig. *Acta Informatica,* **13**:59–66, 1980. [350]

[700] Stefan Reisch. Hex ist PSPACE-vollständig. *Acta Informatica,* **15**:167–191, 1981. [350]

[701] James Renegar. A polynomial-time algorithm, based on Newton's method, for linear programming. *Mathematical Programming,* **40**:59–93, 1988. [448]

[702] Federico Ricci-Tersenghi and Guilhem Semerjian. On the cavity method for decimated random constraint satisfaction problems and the analysis of belief propagation guided decimation algorithms. *Journal of Statistical Mechanics: Theory and Experiment,* **2009**(09):P09001, 2009. [817]

[703] Henry Gordon Rice. Classes of recursively enumerable sets and their decision problems. *Transactions of the American Mathematical Society*, **74**(2):358–366, March 1953. [293]

[704] Ronald L. Rivest, Adi Shamir, and Leonard Adleman. A method for obtaining digital signatures and public-key cryptosystems. *Communications of the ACM*, **21**(2):120–126, 1978. [906]

[705] Neil Robertson and Paul D. Seymour. Graph minors. XX. Wagners conjecture. *Journal of Combinatorial Theory, Series B*, **92**(2):325–357, 2004. [40]

[706] Julia Robinson. Existential definability in arithmetic. *Transactions of the American Mathematical Society*, **72**(3):437–449, 1952. [292]

[707] Raphael M. Robinson. Recursion and double recursion. *Bulletin of the American Mathematical Society*, **54**:987–993, 1948. [294]

[708] Raphael M. Robinson. Undecidability and nonperiodicity for tilings of the plane. *Inventiones Mathematicae*, **12**:177–209, 1971. [299]

[709] John Michael Robson. The complexity of Go. In R. E. A. Mason, editor, *Information Processing*, pages 413–417. 1983. [348]

[710] John Michael Robson. Combinatorial games with exponential space complete decision problems. In *Proc. Mathematica Foundations of Computer Science*, pages 498–506. 1984. [348]

[711] John Michael Robson. *N* by *N* Checkers is Exptime Complete. *SIAM Journal on Computing*, **13**(2):252–267, 1984. [349]

[712] Daniel N. Rockmore. The FFT—an algorithm the whole family can use. *Computing in Science and Engineering*, **2**(1):60–64, 2000. [90]

[713] Daniel N. Rockmore. *Stalking the Riemann Hypothesis: The Quest to Find the Hidden Law of Prime Numbers*. Pantheon Books, 2005. [505]

[714] Yurii Rogozhin. Small universal Turing machines. *Theoretical Computer Science*, **168**(2):215–240, 1996. [295]

[715] Daniel J. Rosenkrantz, Richard E. Stearns, and Philip M. Lewis II. An analysis of several heuristics for the traveling salesman problem. *SIAM Journal on Computing*, **6**(3):563–581, 1977. [443]

[716] Eyal Rozenman and Salil P. Vadhan. Derandomized squaring of graphs. In *Proc. 9th RANDOM*, pages 436–447. 2005. [649]

[717] Steven Rudich and Avi Wigderson, editors. *Computational Complexity Theory*. American Mathematical Society, 2003. [220]

[718] Carl Runge. über die Zerlegung empirisch gegebener periodischer Funktionen in Sinuswellen. *Zeitschrift für Mathematische Physik*, **48**:443–456, 1903. [90]

[719] Bertrand Russell. *The Philosophy of Logical Atomism*. Open Court Publishing Company, 1985. First published in 1918 and 1924. [292]

[720] Herbert J. Ryser. *Combinatorial Mathematics*. Mathematical Association of America, 1963. [720]

[721] Sartaj Sahni and Teofilo Gonzales. P-complete approximation problems. *Journal of the ACM*, **23**(3):555–565, 1976. [444]

[722] Sartaj K. Sahni. Algorithms for scheduling independent tasks. *Journal of the ACM*, **23**(1):116–127, 1976. [442]

[723] Michael Saks and Avi Wigderson. Probabilistic Boolean trees and the complexity of evaluating game trees. In *Proc. 27th Symposium on Foundations of Computer Science*, pages 29–38. 1986. [504]

[724] Jesús D. Salas and Alan D. Sokal. Absence of phase transition for antiferromagnetic Potts models via the Dobrushin uniqueness theorem. *Journal of Statistical Physics*, **86**:551–579, 1997. [644]

[725] Francisco Santos. A counterexample to the Hirsch conjecture. *Annals of Mathematics*, **176**:383–412, 2012. [445]

[726] Walter J. Savitch. Relationships between nondeterministic and deterministic tape complexities. *Journal of Computer and System Sciences*, **4**:177–192, 1970. [347]

[727] Thomas J. Schaefer. The complexity of satisfiability problems. In *Proc. 10th Symposium on Theory of Computing*, pages 216–226. 1978. [170]

[728] Thomas J. Schaefer. On the complexity of some two-person perfect-information games. *Journal of Computer and System Sciences*, **16**:185–225, 1978. [348]

[729] Jonathan Schaeffer, Neil Burch, Yngvi Björnsson, Akihiro Kishimoto, Martin Müller, Robert Lake, Paul Lu, and Steve Sutphen. Checkers is solved. *Science*, page 1144079, 2007. [14]

[730] Jonathan Schaeffer, Joseph Culberson, Norman Treloar, Brent Knight, Paul Lu, and Duane Szafron. Reviving the game of checkers. In D. Levy and D. Beal, editors, *Heuristic Programing in Artifical Intelligence; The Second Computer Olimpiad*, pages 119–136. Ellis Horwood, 1991. [36]

[731] Moses Schönfinkel. Über die Bausteine der mathematischen Logik. *Mathematische Annalen*, **92**:305–316, 1924. English translation in Jean van Heijenoort, editor, *From Frege to Gödel: A Source Book in Mathematical Logic, 1879–1931*. Harvard University Press, 1967. [294]

[732] Arnold Schönhage. On the power of random access machines. In *Proc. 6th International Colloquium on Automata, Languages and Programming*, pages 520–529. Springer-Verlag, 1979. [38]

[733] Arnold Schönhage and Volker Strassen. Schnelle Multiplikation großer Zahlen. *Computing*, **7**:281–292, 1971. [37]

[734] Uwe Schöning. A probabilistic algorithm for k-SAT and constraint satisfaction problems. In *Proc. 40th Symposium on Foundations of Computer Science*, pages 410–414. 1999. [503]

[735] Alexander Schrijver. On the history of combinatorial optimization (till 1960). In K. Aardal, G.L. Nemhauser, and R. Weismantel, editors, *Discrete Optimization*, pages 1–68. Elsevier, 2005. [92, 93]

[736] Jacob T. Schwartz. Fast probabilistic algorithms for verification of polynomial identities. *Journal of the ACM*, **27**(4):701–717, 1980. [504]

[737] Abraham Seidenberg. A simple proof of the Theorem of Erdős and Szekeres. *Journal of the London Mathematical Society*, **34**:352, 1959. [93]

[738] Sakari Seitz, Mikko Alava, and Pekka Orponen. Focused local search for random 3-satisfiability. *Journal of Statistical Mechanics: Theory and Experiment*, **2005**(06):P06006, 2005. [817, 818]

[739] Bart Selman, Henry Kautz, and Bram Cohen. Noise strategies for local search. In *Proc. 11th National Conference on Artificial Intelligence*, pages 337–343. 1994. [503]

[740] Guilhem Semerjian. On the freezing of variables in random constraint satisfaction problems. *Journal of Statistical Physics*, **130**:251–293, 2008. [818]

[741] Jeffrey Shallit. Origins of the analysis of the Euclidean algorithm. *Historia Mathematica*, **21**:401–419, 1994. [36]

[742] Jeffrey Shallit. What this country needs is an 18¢ piece. *Mathematical Intelligencer*, **25**(2):20–23, 2003. [93]

[743] Adi Shamir. How to share a secret. *Communications of the ACM*, **22**(11):612–613, 1979. [505]

[744] Adi Shamir. A polynomial time algorithm for breaking the basic Merkle–Hellman cryptosystem. *IEEE Transactions on Information Theory*, **30**(5):699–704, 1984. [126]

[745] Claude E. Shannon. A mathematical theory of communication. *Bell System Technical Journal*, **27**:379–423 and 623–656, 1948. [643]

[746] Claude E. Shannon. Programming a computer for playing chess. *Philosophical Magazine*, **41**:256–275, 1950. [36]

[747] Alexander Shen. IP = PSPACE: simplified proof. *Journal of the ACM*, **39**(4):878–880, 1992. [561]

[748] David Sherrington and Scott Kirkpatrick. Solvable model of a spin-glass. *Physical Review Letters*, **35**(26):1792–1796, 1975. [171]

[749] Naum Zuselevich Shor. Cut-off method with space extension in convex programming problems. *Kibernetika*, **6**:94–95, 1977. [448]

[750] Peter W. Shor. Algorithms for quantum computation: discrete logarithms and factoring. In *Proc. 35th Symposium on Foundations of Computer Science*, pages 124–134. 1994. [907]

[751] Peter W. Shor and Stephen P. Jordan. Estimating Jones polynomials is a complete problem for one clean qubit. *Quantum Information and Computation*, **8**(8/9):681–714, 2008. [910]

[752] Hava T. Siegelmann and Eduardo D. Sontag. On the computational power of neural nets. In *Proc. Computational Learning Theory*, pages 440–449. 1992. [299]

[753] Daniel R. Simon. On the power of quantum computation. In *Proc. 35th Symposium on Foundations of Computer Science*, pages 116–123. 1994. [907]

[754] Janos Simon. On the difference between one and many (preliminary version). In *Proc. 4th International Colloquium on Automata, Languages and Programming*, pages 480–491. 1977. [719]

[755] Alistair Sinclair. Improved bounds for mixing rates of Markov chains and multicommodity flow. *Combinatorics, Probability and Computing*, **1**:351–370, 1992. [648]

[756] Alistair Sinclair and Mark Jerrum. Approximate counting, uniform generation and rapidly mixing Markov chains. *Information and Computation*, **82**:93–133, 1989. [648, 719]

[757] Michael Sipser. A complexity theoretic approach to randomness. In *Proc. 15th Symposium on Theory of Computing*, pages 330–335. 1983. [560]

[758] Michael Sipser. The history and status of the P versus NP question. In *Proc. 24th Symposium on Theory of Computing*, pages 603–618. 1992. [218]

[759] Michael Sipser. *Introduction to the Theory of Computation*. Course Technology, 2005. [168, 347]

[760] Allan Sly, Nike Sun, and Yumeng Zhang. The number of solutions for random regular NAE-SAT. In *Proc. 57th Symposium on Foundations of Computer Science*, pages 724–731. 2016. [817]

[761] Steven Smale. On the average number of steps of the simplex method of linear programming. *Mathematical Programming*, **27**:241–262, 1983. [446]

[762] Steven Smale. Mathematical problems for the next century. *The Mathematical Intelligencer*, **20**(2):7–15, 1998. [218]

[763] Raymond M. Smullyan. *To Mock a Mockingbird.* Oxford University Press, 2000. [295]

[764] Marc Snir. Lower bounds on probabilistic decision trees. *Theoretical Computer Science,* **38**:69–82, 1985. [504]

[765] Alan Sokal. A personal list of unsolved problems concerning lattice gases and anti-ferromagnetic Potts models. *Markov Processes and Related Fields,* **7**:21–38, 2001. [649]

[766] Robert M. Solovay and Volker Strassen. A fast Monte-Carlo test for primality. *SIAM Journal on Computing,* **6**(1):84–86, 1977. [506]

[767] R. D. Somma, S. Boixo, H. Barnum, and E. Knill. Quantum simulations of classical annealing processes. *Physical Review Letters,* **101**(13):130504, 2008. [908]

[768] RobertŠpalek and Mario Szegedy. All quantum adversary metho ds are equivalent. *Theory of Computing,* **2**(1):1–18, 2006. [909]

[769] Daniel A. Spielman. Faster isomorphism testing of strongly regular graphs. In *Proc. 28th Symposium on Theory of Computing,* pages 576–584. 1996. [908]

[770] Daniel A. Spielman and Shang-Hua Teng. Smoothed analysis of algorithms: why the simplex algorithm usually takes polynomial time. *Journal of the ACM,* **51**(3):385–463, 2004. [446]

[771] Daniel A. Spielman and Shang-Hua Teng. Smoothed analysis of algorithms and heuristics: Progress and open questions. In Luis M. Pardo, Allan Pinkus, Endre Süli, and Michael J. Todd, editors, *Foundations of Computational Mathematics,* pages 274–342. Cambridge University Press, 2005. [446]

[772] Paul G. Spirakis and Chee-Keng Yap. Strong NP-hardness of moving many discs. *Information Processing Letters,* **19**(1):55–59, 1984. [349]

[773] Richard Edwin Stearns, Juris Hartmanis, and Philip M. Lewis II. Hierarchies of memory limited computations. In *Proc. 6th Symposium on Switching Circuit Theory and Logical Design,* pages 179–190. 1965. [219]

[774] Lewis Benjamin Stiller. *Exploiting Symmetry on Parallel Architectures.* Ph. D. Thesis, Johns Hopkins University, 1995. [14]

[775] John Stillwell. Emil Post and his anticipation of Gödel and Turing. *Mathematics Magazine,* **77**(1):3–14, 2004. [298]

[776] Larry J. Stockmeyer. The polynomial-time hierarchy. *Theoretical Computer Science,* **3**(1):1–22, 1977. [218]

[777] Larry J. Stockmeyer and Albert R. Meyer. Word problems requiring exponential time. In *Proc. 5th Symposium on Theory of Computing,* pages 1–9. 1973. [348]

[778] Paul K. Stockmeyer. Variations on the four-post Tower of Hanoi puzzle. *Congressus Numerantium,* **102**:3–12, 1994. [89]

[779] James A. Storer. On the complexity of chess. *Journal of Computer and System Sciences,* **27**(1):77–100, 1983. [349]

[780] Gabriel Sudan. Sur le nombre transfini ω^ω. *Bulletin Mathématique de la Société Roumaine des Sciences,* **30**:11–30, 1927. [294]

[781] Ivan Hal Sudborough and A. Zalcberg. On families of languages defined by time-bounded random access machines. *SIAM Journal on Computing,* **5**(2):217–230, 1976. [219]

[782] Mario Szegedy. In how many steps the k peg version of the Towers of Hanoi game can be solved? In *Proc. 16th Symposium on Theoretical Aspects of Computer Science,* pages 356–361. 1999. [89]

[783] Róbert Szelepcsényi. Nondeterministic space is closed under complementation. *Acta Informatica,* **26**:279–284, 1988. [347]

[784] Peter Guthrie Tait. *Collected Scientific Papers,* vol. 2. Cambridge University Press, 1890. [121]

[785] Michael Tarsi. Optimal search on some game trees. *Journal of the ACM,* **3**:389–396, 1983. [504]

[786] H. N. V. Temperley and Michael E. Fisher. Dimer problem in statistical mechanics—an exact result. *Philosophical Magazine,* **6**:1061–1063, 1961. [720]

[787] Robin Thomas. An update on the Four-Color Theorem. *Notices of the American Mathematical Society,* **45**(7):848–859, 1998. www.ams.org/notices/199807/thomas.pdf. [125]

[788] A. G. Thomason. Hamiltonian cycles and uniquely edge colourable graphs. *Annals of Discrete Mathematics,* **3**:259–268, 1978. [221]

[789] Ken Thompson. Retrograde analysis of certain endgames. *Journal of the International Computer Chess Association,* **9**(3):131–139, 1986. [14]

[790] Wolfgang Tittel, J. Brendel, Hugo Zbinden, and Nicolas Gisin. Violation of Bell inequalities by photons more than 10 km apart. *Physical Review Letters,* **81**:3563–3566, 1998. [905]

[791] Seinosuke Toda. PP is as hard as the polynomial-time hierarchy. *SIAM Journal on Computing,* **20**(5):865–877, 1991. [719]

[792] Tommaso Toffoli. Reversible computing. In J. W. de Bakker and Jan van Leeuwen, editors, *Proc. 7th. Colloquium on Automata, Languages and Programming,* pages 632–644. Springer-Verlag, 1980. [904]

[793] Akira Tonomura, Junji Endo, Tsuyoshi Matsuda, Takeshi Kawasaki, and Hiroshi Ezawa. Demonstration of single-electron build-up of an interference pattern. *American Journal of Physics,* **57**:117–120, 1989. [xiv, 903]

[794] Andre L. Toom. The complexity of a scheme of functional elements simulating the multiplication of integers. *Doklady Akademii Nauk SSSR,* **150**:496–498, 1963. English translation in *Soviet Mathematics* **3**:714–716, 1963. [37]

[795] John Tromp and Gunnar Farnebäck. Combinatorics of Go. In *5th International Conference on Computers and Games*, pages 84–99. 2006. [348]

[796] Richard J. Trudeau. *Introduction to Graph Theory*. Dover Publications Inc., 1994. [13]

[797] Alan M. Turing. On computable numbers, with an application to the Entscheidungsproblem. *Proceedings of the London Mathematical Society*, **42**(236):230–265, 1936. [293, 296]

[798] Alan M. Turing. Computing machinery and intelligence. *Mind*, **59**(236):433–460, 1950. [296, 297]

[799] Alan M. Turing. The chemical basis of morphogenesis. *Philosophical Transactions of the Royal Society of London, Series B*, **237**(641):37–72, 1952. [297]

[800] Alan M. Turing. Intelligent machinery. In B. Jack Copeland, editor, *The Essential Turing*, pages 395–432. Oxford University Press, 2004. [297]

[801] William Thomas Tutte. The factorization of linear graphs. *Journal of the London Mathematical Society*, **22**:107–111, 1947. [505]

[802] Leslie G. Valiant. The complexity of computing the permanent. *Theoretical Computer Science*, **8**:189–201, 1979. [719]

[803] Leslie G. Valiant. The complexity of enumeration and reliability problems. *SIAM Journal on Computing*, **8**(3):410–421, 1979. [719]

[804] Leslie G. Valiant and Vijay V. Vazirani. NP is as easy as detecting unique solutions. *Theoretical Computer Science*, **47**(3):85–93, 1986. [504]

[805] Henk van Beijeren. Exactly solvable model for the roughening transition of a crystal surface. *Physical Review Letters*, **38**:993–996, 1977. [646]

[806] Peter van Emde Boas. Machine models and simulation. In Jan van Leeuwen, editor, *Handbook of Theoretical Computer Science*, vol. A, pages 1–66. Elsevier, 1990. [38]

[807] Lieven Vandenberghe and Stephen Boyd. Semidefinite programming. *SIAM Review*, **38**(1):49–95, 1996. [448]

[808] Lieven M. K. Vandersypen, Matthias Steffen, Gregory Breyta, Costantino S. Yannoni, Mark H. Sherwood, and Isaac L. Chuang. Experimental realization of Shor's quantum factoring algorithm using nuclear magnetic resonance. *Nature*, **414**(6866):883–887, 2001. [907]

[809] Umesh Vazirani. Rapidly mixing Markov chains. In Béla Bollobás, editor, *Probabilistic Combinatorics and Its Applications*, pages 99–121. American Mathematical Society, 1991. [643]

[810] Vijay V. Vazirani. *Approximation Algorithms*. Springer-Verlag, 2003. [442]

[811] Arthur F. Veinott and George B. Dantzig. Integral extreme points. *SIAM Review*, **10**(3):371–372, 1968. [448]

[812] Santosh S. Vempala. *The Random Projection Method*. American Mathematical Society, 2004. [503]

[813] Eric Vigoda. Improved bounds for sampling colorings. *Journal of Mathematical Physics*, **41**:1555–1569, 2000. [644]

[814] John von Neumann. Zur Theorie der Gesellschaftsspiele. *Mathematische Annalen*, **100**:295–320, 1928. [447, 503]

[815] John von Neumann. *Theory of Self-Reproducing Automata*. University of Illinois Press, 1966. Edited by Arthur W. Burks. [296]

[816] Klaus Wagner. Über eine Eigenschaft der ebenen Komplexe. *Mathematische Annalen*, **114**:570–590, 1937. [39]

[817] Hao Wang. Proving theorems by pattern recognition II. *Bell System Technical Journal*, **40**:1–42, 1961. [299]

[818] Gregory H. Wannier. Antiferromagnetism, the triangular Ising net. *Physical Review*, **79**:357–364, 1950. [720]

[819] Gregory H. Wannier. Errata. Antiferromagnetism. The triangular Ising net. *Physical Review B*, **7**(11):5017, 1973. [720]

[820] John Watrous. Succinct quantum proofs for properties of finite groups. In *Proc. 41st Symposium on Foundations of Computer Science*, pages 537–546. 2000. [908]

[821] John Watrous. Quantum simulations of classical random walks and undirected graph connectivity. *Journal of Computer and System Sciences*, **62**(2):376–391, 2001. [909]

[822] Martin Weigt and Alexander K. Hartmann. Number of guards needed by a museum: a phase transition in vertex covering of random graphs. *Physical Review Letters*, **84**(26):6118–6121, 2000. [813]

[823] Norbert Wiener. *The Human Use of Human Beings: Cybernetics and Society*. Houghton Mifflin, 1950. [36]

[824] Keith B. Wiley and Lance R. Williams. Representation of interwoven surfaces in $2\frac{1}{2}$-D drawing. In *Proc. Conference on Human Factors in Computing Systems*. 2006. [126]

[825] Herbert Wilf. *generatingfunctionology*. Academic Press, 1994. [641]

[826] Virginia Vassilevska Williams. Multiplying matrices faster than Coppersmith–Winograd. In *Proc. 44th STOC*, pages 887–898. 2012. [38]

[827] David B. Wilson. Mixing times of lozenge tiling and card shuffling Markov chains. *Annals of Applied Probability*, **14**(1):274–325, 2004. [646]

[828] Peter Winkler. *Mathematical Mind-Benders*. A. K. Peters, 2007. [907]

[829] Pawel Wocjan and Anura Abeyesinghe. Speedup via quantum sampling. *Physical Review A*, **78**(4):042336, 2008. [908]

[830] David Wolfe. Go endgames are PSPACE-hard. In *More Games of No Chance*, pages 125–136. Cambridge University Press, 2002. [348]

[831] Stephen Wolfram. Statistical mechanics of cellular automata. *Reviews of Modern Physics*, **55**:601–644, 1983. [170]

[832] Stephen Wolfram. *A New Kind of Science*. Wolfram Media, 2002. [299]

[833] William K. Wootters and Wojciech H. Zurek. A single quantum cannot be cloned. *Nature*, **299**:802–803, 1982. [904]

[834] Nicholas C. Wormald. Differential equations for random processes and random graphs. *Annals of Applied Probability*, **5**(4):1217–1235, 1995. [812]

[835] Celia Wrathall. Complete sets and the polynomial-time hierarchy. *Theoretical Computer Science*, **3**(1):23–33, 1977. [218]

[836] A. M. Yaglom and I. M. Yaglom. *Challenging Mathematical Problems With Elementary Solutions*, vol. 2. Dover Publications Inc., 1987. [907]

[837] C. N. Yang. The spontaneous magnetization of a two-dimensional Ising model. *Physical Review*, **85**(5):808–816, 1952. [720]

[838] Andrew Chi-Chih Yao. Probabilistic computations: toward a unified measure of complexity. In *Proc. 18th Symposium on Foundations of Computer Science*, pages 222–227. 1977. [503]

[839] Andrew Chi-Chih Yao. Some complexity questions related to distributive computing. In *Proc. 11th Symposium on Theory of Computing*, pages 209–213. 1979. [504]

[840] Andrew Chi-Chih Yao. Theory and applications of trapdoor functions. In *Proc. 23rd Symposium on Foundations of Computer Science*, pages 80–91. 1982. [562]

[841] Andrew Chi-Chih Yao. Quantum circuit complexity. In *Proc. 34th Symposium on Foundations of Computer Science*, pages 352–361. 1993. [904]

[842] Thomas Young. The Bakerian lecture: experiments and calculations relative to physical optics. *Philosophical Transactions of the Royal Society*, **94**:1–16, 1804. [903]

[843] Christof Zalka. Grover's quantum searching algorithm is optimal. *Physical Review A*, **60**:2746–2751, 1999. [909]

[844] Viktória Zankó. #P-completeness via many-one reductions. *International Journal of Foundations of Computer Science*, **2**(1):77–82, 1991. [719]

[845] Lenka Zdeborová. Statistical physics of hard optimization problems. *Acta Physica Slovaka*, **59**(3):169–303, 2009. [810]

[846] Lenka Zdeborová and Florent Krząkala. Phase transitions in the coloring of random graphs. *Physical Review E*, **76**:031131, 2007. [818]

[847] W. Zheng and S. Sachdev. Sine–Gordon theory of the non-Néel phase of two-dimensional quantum antiferromagnetcs. *Physical Review B*, **40**:2704–2707, 1989. [646]

[848] Günther M. Ziegler. *Lectures on Polytopes*. Springer-Verlag, 2006. [446]

[849] Richard E. Zippel. Probabilistic algorithms for sparse polynomials. *Lecture Notes in Computer Science*, **72**:216–226, 1979. [504]

[850] Richard E. Zippel. *Effective Polynomial Computation*. Kluwer Academic, 1993. [504]

[851] Wojciech H. Zurek. Decoherence, einselection, and the quantum origins of the classical. *Reviews of Modern Physics*, **75**(3):715–775, 2003. [904]

Index